# The Geometry of Physics

This book is intended to provide a working knowledge of those parts of exterior differential forms, differential geometry, algebraic and differential topology, Lie groups, vector bundles, and Chern forms that are essential for a deeper understanding of both classical and modern physics and engineering. Included are discussions of analytical and fluid dynamics, electromagnetism (in flat and curved space), thermodynamics, elasticity theory, the geometry and topology of Kirchhoff's electric circuit laws, soap films, special and general relativity, the Dirac operator and spinors, and gauge fields, including Yang–Mills, the Aharonov–Bohm effect, Berry phase, and instanton winding numbers, quarks, and the quark model for mesons. Before a discussion of abstract notions of differential geometry, geometric intuition is developed through a rather extensive introduction to the study of surfaces in ordinary space; consequently, the book should be of interest also to mathematics students.

This book will be useful to graduate and advance undergraduate students of physics, engineering, and mathematics. It can be used as a course text or for self-study.

This Third Edition includes a new overview of Cartan's exterior differential forms. It previews many of the geometric concepts developed in the text and illustrates their applications to a single extended problem in engineering; namely, the Cauchy stresses created by a small twist of an elastic cylindrical rod about its axis.

THEODORE FRANKEL received his Ph.D. from the University of California, Berkeley. He is currently Emeritus Professor of Mathematics at the University of California, San Diego.

# The Geometry of Physics

## An Introduction

*Third Edition*

**Theodore Frankel**

*University of California, San Diego*

CAMBRIDGE
UNIVERSITY PRESS

**CAMBRIDGE**
**UNIVERSITY PRESS**

University Printing House, Cambridge CB2 8BS, United Kingdom

One Liberty Plaza, 20th Floor, New York, NY 10006, USA

477 Williamstown Road, Port Melbourne, VIC 3207, Australia

314-321, 3rd Floor, Plot 3, Splendor Forum, Jasola District Centre, New Delhi - 110025, India

79 Anson Road, #06-04/06, Singapore 079906

Cambridge University Press is part of the University of Cambridge.

It furthers the University's mission by disseminating knowledge in the pursuit of education, learning and research at the highest international levels of excellence.

www.cambridge.org
Information on this title: www.cambridge.org/9781107602601



First published 1997
Revised paperback edition 1999
Second edition 2004
Reprinted 2006, 2007 (twice), 2009
Third edition 2012
Reprinted 2017

*A catalogue record for this publication is available from the British Library*

*Library of Congress Cataloging in Publication data*
Frankel, Theodore, 1929–
The geometry of physics : an introduction / Theodore Frankel. – 3rd ed.
p. cm.
Includes bibliographical referencs and index.
ISBN 978-1-107-60260-1 (pbk.)
1. Geometry, Differential. 2. Mathematical physics. I. Title.
QC20.7.D52F73 2011
530.15'636 – dc23 2011027890

ISBN 978-1-107-60260-1 Paperback

*For*
*Thom-kat, Mont, Dave*
*and*
*Jonnie*

*and*

*In fond memory of*
*Raoul Bott*
*1923–2005*

*Photograph of Raoul by Montgomery Frankel*

# Contents

# I Manifolds, Tensors, and Exterior Forms

# Preface to the Third Edition

A main addition introduced in this third edition is the inclusion of an Overview

**An Informal Overview of Cartan's Exterior Differential Forms, Illustrated with an Application to Cauchy's Stress Tensor**

which can be read before starting the text. This appears at the beginning of the text, before Chapter 1. The only prerequisites for reading this overview are sophomore courses in calculus and basic linear algebra. Many of the geometric concepts developed in the text are previewed here and these are illustrated by their applications to a single extended problem in engineering, namely the study of the Cauchy stresses created by a small twist of an elastic cylindrical rod about its axis.

The new shortened version of Appendix A, dealing with elasticity, requires the discussion of Cauchy stresses dealt with in the Overview. The author believes that the use of Cartan's vector valued exterior forms in elasticity is more suitable (both in principle and in computations) than the classical tensor analysis usually employed in engineering (which is also developed in the text.)

The new version of Appendix A also contains contributions by my engineering colleague Professor Hidenori Murakami, including his treatment of the Truesdell stress rate. I am also very grateful to Professor Murakami for many very helpful conversations.

# Preface to the Third Edition

# Preface to the Second Edition

This second edition differs mainly in the addition of three new appendices: C, D, and E. Appendices C and D are applications of the elements of representation theory of compact Lie groups.

Appendix C deals with applications to the flavored quark model that revolutionized particle physics. We illustrate how certain observed mesons (pions, kaons, and etas) are described in terms of quarks and how one can "derive" the mass formula of Gell-Mann/Okubo of 1962. This can be read after Section 20.3b.

Appendix D is concerned with isotropic hyperelastic bodies. Here the main result has been used by engineers since the 1850s. My purpose for presenting proofs is that the hypotheses of the Frobenius–Schur theorems of group representations are exactly met here, and so this affords a compelling excuse for developing representation theory, which had not been addressed in the earlier edition. An added bonus is that the group theoretical material is applied to the three-dimensional rotation group $SO(3)$, where these generalities can be pictured explicitly. This material can essentially be read after Appendix A, but some brief excursion into Appendix C might be helpful.

Appendix E delves deeper into the geometry and topology of compact Lie groups. Bott's extension of the presentation of Morse theory that was given in Section 14.3c is sketched and the example of the topology of the Lie group $U(3)$ is worked out in some detail.

# Preface to the Second Edition

# Preface to the Revised Printing

In this reprinting I have introduced a new appendix, Appendix B, Harmonic Chains and Kirchhoff's Circuit Laws. This appendix deals with a finite-dimensional version of Hodge's theory, the subject of Chapter 14, and can be read at any time after Chapter 13. It includes a more geometrical view of cohomology, dealt with entirely by matrices and elementary linear algebra. A bonus of this viewpoint is a systematic "geometrical" description of the Kirchhoff laws and their applications to direct current circuits, first considered from roughly this viewpoint by Hermann Weyl in 1923.

I have corrected a number of errors and misprints, many of which were kindly brought to my attention by Professor Friedrich Heyl.

Finally, I would like to take this opportunity to express my great appreciation to my editor, Dr. Alan Harvey of Cambridge University Press.

# Preface to the Revised Printing

# Preface to the First Edition

The basic ideas at the foundations of point and continuum mechanics, electromagnetism, thermodynamics, special and general relativity, and gauge theories are geometrical, and, I believe, should be approached, by both mathematics and physics students, from this point of view.

This is a textbook that develops some of the geometrical concepts and tools that are helpful in understanding classical and modern physics and engineering. The mathematical subject material is essentially that found in a first-year graduate course in differential geometry. This is not coincidental, for the founders of this part of geometry, among them Euler, Gauss, Jacobi, Riemann and Poincaré, were also profoundly interested in "natural philosophy."

Electromagnetism and fluid flow involve line, surface, and volume integrals. Analytical dynamics brings in multidimensional versions of these objects. In this book these topics are discussed in terms of **exterior differential forms**. One also needs to differentiate such integrals with respect to time, especially when the domains of integration are changing (circulation, vorticity, helicity, Faraday's law, etc.), and this is accomplished most naturally with aid of the **Lie derivative**. Analytical dynamics, thermodynamics, and robotics in engineering deal with **constraints**, including the puzzling nonholonomic ones, and these are dealt with here via the so-called Frobenius theorem on differential forms. All these matters, and more, are considered in Part One of this book.

Einstein created the astonishing principle **field strength = curvature** to explain the gravitational field, but if one is not familiar with the classical meaning of surface curvature in ordinary 3-space this is merely a tautology. Consequently I introduce **differential geometry** before discussing general relativity. **Cartan's** version, in terms of exterior differential forms, plays a central role. Differential geometry has applications to more down-to-earth subjects, such as soap bubbles and periodic motions of dynamical systems. Differential geometry occupies the bulk of Part Two.

Einstein's principle has been extended by physicists, and now all the field strengths occurring in elementary particle physics (which are required in order to construct a

**Lagrangian**) are discussed in terms of curvature and **connections**, but it is the curvature of a **vector bundle**, that is, the *field* space, that arises, not the curvature of space–time. The symmetries of the quantum field play an essential role in these **gauge theories**, as was first emphasized by Hermann Weyl, and these are understood today in terms of **Lie groups**, which are an essential ingredient of the vector bundle. Since many quantum situations (charged particles in an electromagnetic field, Aharonov–Bohm effect, Dirac monopoles, Berry phase, Yang–Mills fields, instantons, etc.) have analogues in elementary differential geometry, we can use the geometric methods and pictures of Part Two as a guide; a picture *is* worth a thousand words! These topics are discussed in Part Three.

**Topology** is playing an increasing role in physics. A physical problem is "well posed" if there *exists* a solution and it is *unique*, and the topology of the configuration (spherical, toroidal, etc.), in particular the singular **homology groups**, has an essential influence. The **Brouwer degree**, the **Hurewicz homotopy groups**, and **Morse theory** play roles not only in modern gauge theories but also, for example, in the theory of "defects" in materials.

Topological methods are playing an important role in field theory; versions of the **Atiyah–Singer index theorem** are frequently invoked. Although I do not develop this theorem in general, I *do* discuss at length the most famous and elementary example, the **Gauss–Bonnet–Poincaré** theorem, in two dimensions and also the meaning of the **Chern characteristic classes**. These matters are discussed in Parts Two and Three.

The Appendix to this book presents a nontraditional treatment of the **stress tensors** appearing in continuum mechanics, utilizing exterior forms. In this endeavor I am greatly indebted to my engineering colleague Hidenori Murakami. In particular Murakami has supplied, in Section **g** of the Appendix, some typical computations involving stresses and strains, but carried out with the machinery developed in this book. We believe that these computations indicate the efficiency of the use of forms and Lie derivatives in elasticity. The material of this Appendix could be read, except for some minor points, after Section **9.5**.

Mathematical applications to physics occur in at least two aspects. Mathematics is of course the principal tool for solving technical analytical problems, but increasingly it is also a principal guide in our understanding of the basic structure and concepts involved. Analytical computations with elliptic functions *are* important for certain technical problems in rigid body dynamics, but one could not have begun to understand the dynamics before Euler's introducing the moment of inertia tensor. I am very much concerned with the basic concepts in physics. A glance at the Contents will show in detail what mathematical and physical tools are being developed, but frequently physical applications appear also in Exercises. My main philosophy has been to attack physical topics as soon as possible, but only after effective mathematical tools have been introduced. By analogy, one *can* deal with problems of velocity and acceleration after having learned the definition of the derivative as the limit of a quotient (or even before, as in the case of Newton), but we all know how important the *machinery* of calculus (e.g., the power, product, quotient, and chain rules) is for handling specific problems. In the same way, it is a mistake to talk seriously about thermodynamics before understanding that a total differential equation in more than two dimensions need not possess an integrating factor.

In a sense this book is a "final" revision of sets of notes for a year course that I have given in La Jolla over many years. My goal has been to give the reader a *working* knowledge of the tools that are of great value in geometry and physics and (increasingly) engineering. For this it is *absolutely essential* that the reader work (or at least attempt) the Exercises. *Most of the problems are simple and require simple calculations. If you find calculations becoming unmanageable, then in all probability you are not taking advantage of the machinery developed in this book.*

This book is intended primarily for two audiences, first, the physics or engineering student, and second, the mathematics student. My classes in the past have been populated mostly by first-, second-, and third-year graduate students in physics, but there have also been mathematics students and undergraduates. The only real *mathematical* prerequisites are *basic* linear algebra and some familiarity with calculus of several variables. Most students (in the United States) have these by the beginning of the third undergraduate year.

All of the physical subjects, with two exceptions to be noted, are preceded by a brief introduction. The two exceptions are analytical dynamics and the quantum aspects of gauge theories.

Analytical (Hamiltonian) dynamics appears as a problem set in Part One, with very little motivation, for the following reason: the problems form an ideal application of exterior forms and Lie derivatives and involve no knowledge of physics. Only in Part Two, after geodesics have been discussed, do we return for a discussion of analytical dynamics from first principles. (Of course most physics and engineering students will already have seen *some* introduction to analytical mechanics in their course work anyway.) The significance of the Lagrangian (based on special relativity) is discussed in Section **16.4** of Part Three when changes in dynamics are required for discussing the effects of electromagnetism.

An introduction to quantum mechanics would have taken us too far afield. Fortunately (for me) only the simplest quantum ideas are needed for most of our discussions. I would refer the reader to Rabin's article [R] and Sudbery's book [Su] for excellent introductions to the quantum aspects involved.

Physics and engineering readers would profit *greatly* if they would form the habit of translating the vectorial and tensorial statements found in their customary reading of physics articles and books into the language developed in this book, and using the newer methods developed here in their own thinking. (By "newer" I mean methods developed over the last one hundred years!)

As for the mathematics student, I feel that this book gives an overview of a large portion of differential geometry and topology that should be helpful to the mathematics graduate student in this age of very specialized texts and absolute rigor. The student preparing to specialize, say, in differential geometry *will* need to augment this reading with a more rigorous treatment of some of the subjects than that given here (e.g., in Warner's book [Wa] or the five-volume series by Spivak [Sp]). The mathematics student should also have exercises devoted to showing what can go wrong if hypotheses are weakened. I make no pretense of worrying, for example, about the differentiability classes of mappings needed in proofs. (Such matters are studied more carefully in the book [A, M, R] and in the encyclopedia article [T, T]. This latter article (and the accompanying one by Eriksen) are also excellent for questions of historical priorities.) I hope that mathematics students will enjoy the discussions of the physical subjects even if they know very little physics; after all, physics is *the* source of interesting vector fields. Many of the "physical" applications are useful even if they are thought of as simply giving explicit examples of rather abstract concepts. For example, Dirac's equation in *curved space* can be considered as a nontrivial application of the method of connections in associated bundles!

This *is* an introduction and there is much important mathematics that is not developed here. Analytical questions involving existence theorems in partial differential equations, Sobolev spaces, and so on, are missing. Although complex manifolds are defined, there is no discussion of Kaehler manifolds nor the algebraic–geometric notions used in string theory. Infinite dimensional manifolds are not considered. On the physical side, topics are introduced usually only if I felt that geometrical ideas would be a great help in their understanding or in computations.

I have included a small list of references. Most of the articles and books listed have been referred to in this book for specific details. The reader will find that there are many good books on the subject of "geometrical physics" that are not referred to here, primarily because I felt that the development, or sophistication, or notation used was sufficiently different to lead to, perhaps, more confusion than help in the first stages of their struggle. A book that I feel is in very much the same spirit as my own is that by Nash and Sen [N, S]. The standard reference for differential geometry is the two-volume work [K, N] of Kobayashi and Nomizu.

Almost every section of this book begins with a question or a quotation which may concern anything from the main thrust of the section to some small remark that should not be overlooked.

A term being defined will usually appear in **bold type**.

I wish to express my gratitude to Harley Flanders, who introduced me long ago to exterior forms and de Rham's theorem, whose superb book [Fl] was perhaps the first to awaken scientists to the use of exterior forms in their work. I am indebted to my chemical colleague John Wheeler for conversations on thermodynamics and to Donald Fredkin for helpful criticisms of earlier versions of my lecture notes. I have already expressed my deep gratitude to Hidenori Murakami. Joel Broida made many comments on earlier versions, and also prevented my Macintosh from taking me over. I've had many helpful conversations with Bruce Driver, Jay Fillmore, and Michael Freedman. Poul Hjorth made many helpful comments on various drafts and also served as "beater," herding physics students into my course. Above all, my colleague Jeff Rabin used my notes as the text in a one-year graduate course and made many suggestions and corrections. I have also included corrections to the 1997 printing, following helpful remarks from Professor Meinhard Mayer.

Finally I am grateful to the many students in my classes on geometrical physics for their encouragement and enthusiasm in my endeavor. Of course none of the above is responsible for whatever inaccuracies undoubtedly remain.

OVERVIEW

# An Informal Overview of Cartan's Exterior Differential Forms, Illustrated with an Application to Cauchy's Stress Tensor

## Introduction

### 0.a. Introduction

My goal in this overview is to introduce **exterior calculus** in a *brief* and *informal* way that leads directly to their use in engineering and physics, both in basic physical concepts and in specific engineering calculations. The presentation will be very informal. Many times a proof will be omitted so that we can get quickly to a calculation. In some "proofs" we shall look only at a typical term.

The chief mathematical prerequisites for this overview are sophomore courses dealing with basic linear algebra, partial derivatives, multiple integrals, and tangent vectors to parameterized curves, but not necessarily "vector calculus," i.e., curls, divergences, line and surface integrals, Stokes' theorem, .... These last topics will be sketched here using Cartan's "exterior calculus."

We shall take advantage of the fact that most engineers live in euclidean 3-space $\mathbb{R}^3$ with its everyday metric structure, but we shall try to use methods that make sense in much more general situations. Instead of including exercises we shall consider, in the section **Elasticity and Stresses**, one main example and illustrate *everything* in terms of this example but hopefully the general principles will be clear. This engineering example will be the following. Take an elastic circular cylindrical rod of radius $a$ and length $L$, described in cylindrical coordinates $r, \theta, z$, with the ends of the cylinder at $z = 0$ and $z = L$. Look at this same cylinder except that it has been axially twisted through an angle $kz$ proportional to the distance $z$ from the fixed end $z = 0$.

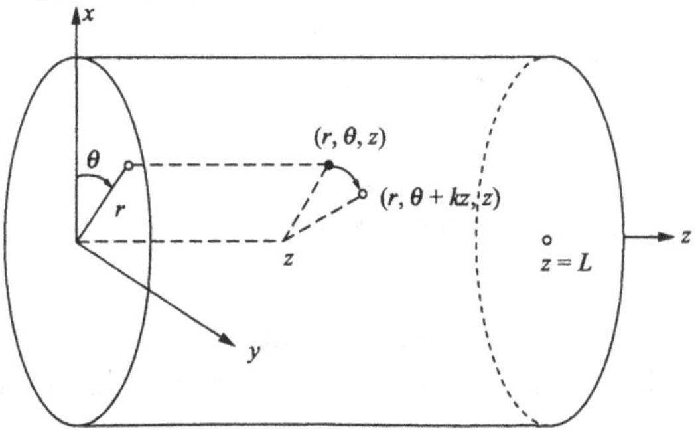

We shall *neglect gravity* and investigate the **stresses** in the cylinder in its final twisted state, in the first approximation, i.e., where we put $k^2 = 0$. Since "stress" and "strain" are "tensors" (as Cauchy and I will show) this is classically treated via "tensor analysis." The final equilibrium state involves surface integrals and the tensor divergence of the Cauchy stress tensor. Our main tool will *not* be the usual *classical* tensor analysis (Christoffel symbols $\Gamma^i_{jk} \ldots$, etc.) but rather **exterior differential forms** (first used in the nineteenth century by Grassmann, Poincaré, Volterra, ..., and developed especially by **Elie Cartan**), which, I believe, is a far more appropriate tool.

We are very much at home with cartesian coordinates but curvilinear coordinates play a very important role in physical applications, and the fact that there are *two distinct types of vectors* that arise in curvilinear coordinates (and, even more so, in *curved spaces*) that appear identical in cartesian coordinates *must* be understood, not only when making calculations but also in our understanding of the basic ingredients of the physical world. We shall let $x^i$, and $u^i$, $i = 1, 2, 3$, be **general** (curvilinear) coordinates, in euclidean 3 dimensional space $\mathbb{R}^3$. *If cartesian coordinates are wanted, I will say so explicitly.*

## Vectors, 1-Forms, and Tensors

### 0.b. Two Kinds of Vectors

There are two kinds of vectors that appear in physical applications and it is important that we distinguish between them. First there is the familiar "arrow" version.

Consider $n$ dimensional euclidean space $\mathbb{R}^n$ with cartesian coordinates $x^1, \ldots, x^n$ and local (perhaps curvilinear) coordinates $u^1, \ldots, u^n$.

**Example:** $\mathbb{R}^2$ with cartesian coordinates $x^1 = x$, $x^2 = y$, and with polar coordinates $u^1 = r$, $u^2 = \theta$.

**Example:** $\mathbb{R}^3$ with cartesian coordinates $x, y, z$ and with cylindrical coordinates $R, \Theta, Z$.

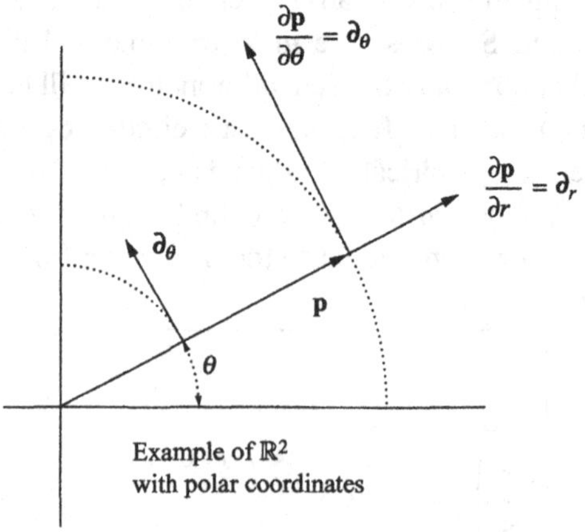


Example of $\mathbb{R}^2$
with polar coordinates

Let $\mathbf{p}$ be the position vector from the origin of $\mathbb{R}^n$ to the point $p$. In the curvilinear coordinate system $u$, the coordinate curve $C_i$ through the point $p$ is the curve where all

$u^j$, $j \neq i$, are constants, and where $u^i$ is used as parameter. Then the tangent vector to this curve in $\mathbb{R}^n$ is

$$\partial \mathbf{p}/\partial u^i \quad \text{which we shall abbreviate to} \quad \boldsymbol{\partial}_i \quad \text{or} \quad \boldsymbol{\partial}/\partial u^i$$

At the point $p$ these $n$ vectors $\boldsymbol{\partial}_1, \ldots, \boldsymbol{\partial}_n$ form a **basis** for all vectors in $\mathbb{R}^n$ based at $p$. Any vector $\mathbf{v}$ at $p$ has a unique expansion with curvilinear coordinate *components* $(v^1, \ldots, v^n)$

$$\mathbf{v} = \Sigma_i v^i \boldsymbol{\partial}_i = \Sigma_i \boldsymbol{\partial}_i v^i$$

We prefer the last expression with the components to the *right* of the basis vectors since it is traditional to put the vectorial *components* in a *column* matrix, and we can then form the matrices

$$\boldsymbol{\partial} = (\boldsymbol{\partial}_1, \ldots, \boldsymbol{\partial}_n) \quad \text{and} \quad v = \begin{pmatrix} v^1 \\ \cdot \\ \cdot \\ \cdot \\ v^n \end{pmatrix} = (v^1 \ldots v^n)^T$$

($T$ denotes transpose) and then we can write the matrix expression (with $\mathbf{v}$ a $1 \times 1$ matrix)

$$\mathbf{v} = \boldsymbol{\partial} v \tag{0.1}$$

Please *beware* though that in $\boldsymbol{\partial}_i v^i$ or $(\boldsymbol{\partial}/\partial u^i) v^i$ or $\mathbf{v} = \boldsymbol{\partial} v$, the bold $\boldsymbol{\partial}$ does not differentiate the component term to the right; it is merely the symbol for a basis vector. Of course we can still differentiate a function $f$ along a vector $\mathbf{v}$ by *defining*

$$\mathbf{v}(f) := (\Sigma_i \boldsymbol{\partial}_i v^i)(f) = \Sigma_i \partial/\partial u^i (f) v^i := \Sigma_i (\partial f/\partial u^i) v^i$$

replacing the basis vector $\boldsymbol{\partial}/\partial u^i$ with bold $\boldsymbol{\partial}$ by the partial differential operator $\partial/\partial u^i$ and then applying to the function $f$. A vector *is* a first order differential operator on functions!

In cylindrical coordinates $R$, $\Theta$, $Z$ in $\mathbb{R}^3$ we have the basis vectors $\boldsymbol{\partial}_R = \boldsymbol{\partial}/\partial R$, $\boldsymbol{\partial}_\Theta = \boldsymbol{\partial}/\partial \Theta$, and $\boldsymbol{\partial}_Z = \boldsymbol{\partial}/\partial Z$.

Let $\mathbf{v}$ be a vector at a point $p$. We can always find a curve $u^i = u^i(t)$ through $p$ whose velocity vector there is $\mathbf{v}$, $v^i = du^i/dt$. Then if $u'$ is a second coordinate system about $p$, we then have $v'^j = du'^j/dt = (\partial u'^j/\partial u^i) du^i/dt = (\partial u'^j/\partial u^i) v^i$. Thus the **components** of a vector transform under a change of coordinates by the rule

$$v'^j = \Sigma_i (\partial u'^j/\partial u^i) v^i \quad \text{or as matrices} \quad v' = (\partial u'/\partial u) v \tag{0.2}$$

where $(\partial u'/\partial u)$ is the Jacobian matrix. This is the **transformation law** for the components of a **contravariant** vector, or **tangent** vector, or simply **vector**.

There is a second, different, type of vector. In linear algebra we learn that to each vector space $V$ (in our case the space of all vectors at a point $p$) we can associate its

**dual** vector space $V^*$ of all real **linear** functionals $\alpha : V \to \mathbb{R}$ . In coordinates, $\alpha(\mathbf{v})$ is a number

$$\alpha(\mathbf{v}) = \Sigma_i a_i v^i$$

for unique numbers $(a_i)$. We shall explain why $i$ is a *sub*script in $a_i$ shortly.

The most familiar linear functional is the **differential** of a function $df$. As a function on vectors it is defined by the derivative of $f$ along $v$

$$df(\mathbf{v}) := \mathbf{v}(f) = \Sigma_i(\partial f/\partial u^i)v^i \quad \text{and so} \quad (df)_i = \partial f/\partial u^i$$

Let us write $df$ in a much more familiar form. In elementary calculus there is mumbo-jumbo to the effect that $du^i$ is a function of pairs of points: it gives you the difference in the $u^i$ coordinates between the points, and the points do not need to be close together. What is *really* meant is

> $du^i$ is the **linear functional** that reads off the $i$th component of any vector $\mathbf{v}$ with respect to the basis vectors of the coordinate system $u$
>
> $$du^i(\mathbf{v}) = du^i(\Sigma_j \boldsymbol{\partial}_j v^j) := v^i$$

Note that this agrees with $du^i(\mathbf{v}) = \mathbf{v}(u^i)$ since $\mathbf{v}(u^i) = (\Sigma_j \boldsymbol{\partial}_j v^j)(u^i) = \Sigma_j(\partial u^i/\partial u^j)v^j = \Sigma_j \delta^i_j v^j = v^i$.

Then we can write

$$df(\mathbf{v}) = \Sigma_i(\partial f/\partial u^i)v^i = \Sigma_i(\partial f/\partial u^i)du^i(\mathbf{v})$$

i.e.,

$$df = \Sigma_i(\partial f/\partial u^i)du^i$$

as usual, except that now both sides have meaning as linear functionals on vectors.

**Warning:** We shall see that this is *not* the gradient vector of $f$!

It is very easy to see that $du^1, \ldots, du^n$ form a basis for the space of linear functionals at each point of the coordinate system $u$, since they are linearly independent. In fact, this basis of $V^*$ is the **dual basis** to the basis $\boldsymbol{\partial}_1, \ldots, \boldsymbol{\partial}_n$, meaning

$$du^i(\boldsymbol{\partial}_j) = \delta^i_j$$

Thus in the coordinate system $u$, every linear functional $\alpha$ is of the form

$$\alpha = \Sigma_i a_i(u)du^i \quad \text{where} \quad \alpha(\boldsymbol{\partial}_j) = \Sigma_i a_i(u)du^i(\boldsymbol{\partial}_j) = \Sigma_i a_i(u)\delta^i_j = a_j$$

is the $j$th **component** of $\alpha$.

We shall see in Section ©.i that it is *not* true that every $\alpha$ is equal to $df$ for some $f$!

Corresponding to (©.1) we can write the matrix expansion for a linear functional as

$$\alpha = (a_1, \ldots, a_n)(du^1, \ldots, du^n)^T = a\, du \tag{©.3}$$

i.e., $a$ is a **row** matrix and $du$ is a column matrix!

If $V$ is the space of contravariant vectors at $p$, then $V^*$ is called the space of **covariant** vectors, or covectors, or **1-forms** at $p$. Under a change of coordinates, using the chain rule, $\alpha = a'\,du' = a\,du = (a)(\partial u/\partial u')(du')$, and so

$$a' = a(\partial u/\partial u') = a(\partial u'/\partial u)^{-1} \qquad \text{i.e.,} \quad a'_j = \Sigma_i a_i(\partial u^i/\partial u'^j) \tag{0.4}$$

which should be compared with (0.2). This is the law of transformation of components of a covector.

Note that by definition, if $\alpha$ is a covector and $\mathbf{v}$ is a vector, then the value

$$\alpha(\mathbf{v}) = av = \Sigma_i a_i v^i$$

is **invariant**, i.e., independent of the coordinates used. This also follows, from (0.2) and (0.4)

$$\alpha(\mathbf{v}) = a'v' = a(\partial u/\partial u')(\partial u'/\partial u)v = a(\partial u'/\partial u)^{-1}(\partial u'/\partial u)v = av$$

Note that a vector can be considered as a linear functional on covectors,

$$\mathbf{v}(\alpha) := \alpha(\mathbf{v}) = \Sigma_i a_i v^i$$

## 0.c. Superscripts, Subscripts, Summation Convention

First the **summation convention**. Whenever we have a single term of an expression with any number of indices up and down, e.g., $T^{abc}{}_{de}$, if we rename one of the **lower** indices, say $d$ so that it becomes the same as one of the **upper** indices, say $b$, and if we then sum over this index, the result, call it $S$,

$$\Sigma_b T^{abc}{}_{be} = S^{ac}{}_e$$

is called a **contraction** of $T$. The index $b$ has disappeared (it was a summation or "dummy" index on the left expression; you could have called it anything). This process of summing over a repeated index *that occurs as both a subscript and a superscript* occurs so often that we shall omit the summation sign and merely write, for example, $T^{abc}{}_{be} = S^{ac}{}_e$. This "Einstein convention" does *not* apply to two upper or two lower indices. Here is why.

We have seen that if $\alpha$ is a covector, and if $\mathbf{v}$ is a vector then $\alpha(\mathbf{v}) = a_i v^i$ is an invariant, independent of coordinates. But if we have another vector, say $\mathbf{w} = \boldsymbol{\partial} w$ then $\Sigma_i v^i w^i$ *will not be invariant*

$$\Sigma_i v'^i w'^i = v'^T w' = [(\partial u'/\partial u)v]^T(\partial u'/\partial u)w = v^T(\partial u'/\partial u)^T(\partial u'/\partial u)w$$

will not be equal to $v^T w$, for all $\mathbf{v}, \mathbf{w}$ unless $(\partial u'/\partial u)^T = (\partial u'/\partial u)^{-1}$, i.e., unless the coordinate change matrix is an **orthogonal** matrix, as it is when $u$ and $u'$ are cartesian coordinate systems.

Our **conventions** regarding the **components** of vectors and covectors

$$(\text{contravariant} \Rightarrow \text{index up}) \text{ and } (\text{covariant} \Rightarrow \text{index down}) \tag{**}$$

*help us avoid errors*! For example, in calculus, the differential equations for curves of **steepest ascent** for a function $f$ are written in cartesian coordinates as

$$dx^i/dt = \partial f/\partial x^i$$

but these equations cannot be correct, say, in spherical coordinates, since we cannot equate the *contravariant* components $v^i$ of the velocity vector with the *covariant* components of the differential $df$; they transform in different ways under a (nonorthogonal) change of coordinates. We shall see the correct equations for this situation in Section 0.d.

**Warning:** Our convention (*∗) applies only to the **components** of vectors and covectors. In $\alpha = a_i dx^i$, the $a_i$ are the components of a single covector $\alpha$, while each individual $dx^i$ is itself a basis covector, *not* a component. The summation convention, however, always holds.

I cringe when I see expressions like $\Sigma_i v^i w^i$ in noncartesian coordinates, for the notation is informing me that I have misunderstood the "variance" of one of the vectors.

## 0.d. Riemannian Metrics

One *can* identify vectors and covectors by introducing an *additional* structure, but the identification will depend on the structure chosen. The metric structure of ordinary euclidean space $\mathbb{R}^3$ is based on the fact that we can measure angles and lengths of vectors and scalar products $\langle , \rangle$. The arc length of a curve $C$ is

$$\int_C ds$$

where $ds^2 = dx^2 + dy^2 + dz^2$ in *cartesian* coordinates. In curvilinear coordinates $u$ we have, putting $dx^k = (\partial x^k/\partial u^i)du^i$, and then

$$ds^2 = \Sigma_k (dx^k)^2 = \Sigma_{i,j} g_{ij} du^i du^j = g_{ij} du^i du^j \tag{0.5}$$

where

$$\begin{aligned} g_{ij} &= \Sigma_k (\partial x^k/\partial u^i)(\partial x^k/\partial u^j) \\ &= \langle \partial \mathbf{p}/\partial u^i, \partial \mathbf{p}/\partial u^j \rangle \text{ (since the } x \text{ coordinates are cartesian)} \end{aligned}$$

$$g_{ij} = \langle \boldsymbol{\partial}_i, \boldsymbol{\partial}_j \rangle = g_{ji}$$

and generally

$$\langle \mathbf{v}, \mathbf{w} \rangle = g_{ij} v^i w^j \tag{0.6}$$

For example, consider the plane $\mathbb{R}^2$ with cartesian coordinates $x^1 = x, x^2 = y$, and polar coordinates $u^1 = r, u^2 = \theta$. Then

$$\begin{bmatrix} g_{xx} = 1 & g_{xy} = 0 \\ g_{yx} = 0 & g_{yy} = 1 \end{bmatrix} \quad \text{i.e.,} \quad \begin{bmatrix} g_{xx} & g_{xy} \\ g_{yx} & g_{yy} \end{bmatrix} = \begin{bmatrix} 1 & 0 \\ 0 & 1 \end{bmatrix}$$

Then, from $x = r\cos\theta$, $dx = dr\cos\theta - r\sin\theta\, d\theta$, etc., we get $ds^2 = dr^2 + r^2\, d\theta^2$,

$$\begin{bmatrix} g_{rr} = 1 & g_{r\theta} = 0 \\ g_{\theta r} = 0 & g_{\theta\theta} = r^2 \end{bmatrix} \quad \text{i.e.,} \quad \begin{bmatrix} g_{rr} & g_{r\theta} \\ g_{\theta r} & g_{\theta\theta} \end{bmatrix} = \begin{bmatrix} 1 & 0 \\ 0 & r^2 \end{bmatrix} \tag{0.7}$$

which is "evident" from the picture

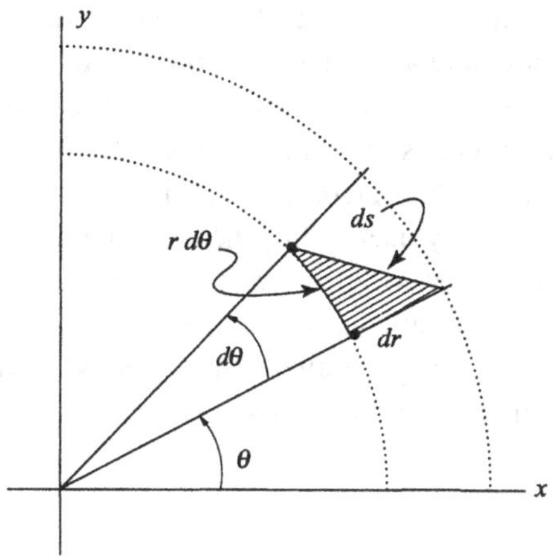


In **spherical** coordinates a picture shows $ds^2 = dr^2 + r^2\, d\theta^2 + r^2\sin^2\theta\, d\varphi^2$, where $\theta$ is co-latitude and $\varphi$ is co-longitude, so $(g_{ij}) = \operatorname{diag}(1, r^2, r^2\sin^2\theta)$. In **cylindrical** coordinates, $ds^2 = dR^2 + R^2\, d\Theta^2 + dZ^2$, with $(g_{ij}) = \operatorname{diag}(1, R^2, 1)$.

Let us look again at the expression (0.5). If $\alpha$ and $\beta$ are 1-forms, i.e., linear functionals, *define* their **tensor product** $\alpha \otimes \beta$ to be the function of (ordered) **pairs** of vectors defined by

$$\alpha \otimes \beta(\mathbf{v}, \mathbf{w}) := \alpha(\mathbf{v})\beta(\mathbf{w}) \tag{0.8}$$

In particular

$$(du^i \otimes du^k)(\mathbf{v}, \mathbf{w}) := v^i w^k$$

Likewise $(\boldsymbol{\partial}_i \otimes \boldsymbol{\partial}_j)(\alpha, \beta) = a_i b_j$ (why?).

$\alpha \otimes \beta$ is a *bilinear* function of **v** and **w**, i.e., it is linear in each vector when the other is unchanged. A **second rank covariant tensor** is just such a bilinear function and in the coordinate system $u$ it can be expressed as

$$\Sigma_{i,j} a_{ij}\, du^i \otimes du^j$$

where the coefficient matrix $(a_{ij})$ is written with indices down. Usually the tensor product sign $\otimes$ is omitted (in $du^i \otimes du^j$ but *not* in $\alpha \otimes \beta$). For example, the metric

$$ds^2 = g_{ij}\, du^i \otimes du^j = g_{ij}\, du^i du^j \tag{0.5$'$}$$

is a second rank covariant tensor that is **symmetric**, i.e., $g_{ji} = g_{ij}$. We may write

$$ds^2(\mathbf{v}, \mathbf{w}) = \langle \mathbf{v}, \mathbf{w} \rangle$$

It is easy to see that under a change of coordinates $u' = u'(u)$, demanding that $ds^2$ be independent of coordinates, $g'_{ab}du'^a du'^b = g_{ij}du^i du^j$, yields the transformation rule

$$g'_{ab} = (\partial u^i/\partial u'^a)g_{ij}(\partial u^j/\partial u'^b) \tag{©.9}$$

for the components of a second rank *covariant* tensor.

**Remark:** We have been using the euclidean metric structure to construct $(g_{ij})$ in any coordinate system, but there are times when other structures are more appropriate. For example, when considering some delicate astronomical questions, a metric from Einstein's general relativity yields more accurate results. When dealing with complex analytic functions in the upper half plane $y > 0$, Poincaré found that the planar metric $ds^2 = (dx^2 + dy^2)/y^2$ was very useful. In general, when some second rank covariant tensor $(g_{ij})$ is used in a metric $ds^2 = g_{ij}dx^i dx^j$ (in which case it must be symmetric and positive definite), this metric is called a **Riemannian metric**, after Bernhard Riemann, who was the first to consider this generalization of **Gauss**' thoughts.

Given a Riemannian metric, one can associate to each (contravariant) vector $\mathbf{v}$ a covector $\upsilon$ by

$$\upsilon(\mathbf{w}) = \langle \mathbf{v}, \mathbf{w} \rangle$$

for *all* vectors $\mathbf{w}$, i.e.,

$$\upsilon_j w^j = \upsilon^k g_{kj} w^j \quad \text{and so} \quad \upsilon_j = \upsilon^k g_{kj} = g_{jk}\upsilon^k$$

In *components*, it is traditional to use the same letter for the covector as for the vector

$$\upsilon_j = g_{jk}\upsilon^k$$

there being no confusion since the covector has the subscript. We say that "we lower the contravariant index" by means of the covariant metric tensor $(g_{jk})$.

Similarly, since $(g_{jk})$ is the matrix of a positive definite quadratic form $ds^2$, it has an inverse matrix, written $(g^{jk})$, which can be shown to be a **contravariant** second rank symmetric tensor (a bilinear function of pairs of covectors given by $g^{jk}a_j b_k$). Then for each covector $\alpha$ we can associate a vector $\mathbf{a}$ by $a^i = g^{ij}a_j$, i.e., we *raise the covariant index* by means of the contravariant metric tensor $(g^{jk})$.

The **gradient vector** of a function $f$ is defined to be the vector **grad** $f = \nabla f$ associated to the covector $df$, i.e., $df(\mathbf{w}) = \langle \nabla \mathbf{f}, \mathbf{w} \rangle$

$$(\nabla f)^i := g^{ij}\partial f/\partial u^j$$

Then the correct version of the equation of steepest ascent considered at the end of section ©.c is

$$du^i/dt = (\nabla f)^i = g^{ij}\partial f/\partial u^j$$

in *any* coordinates. For example, in polar coordinates, from (©.7), we see $g^{rr} = 1$, $g^{\theta\theta} = 1/r^2$, $g^{r\theta} = 0 = g^{\theta r}$.

### $\mathbb{O}$.e. Tensors

We shall consider examples rather than generalities.

(i) A tensor of the third rank, twice contravariant, once covariant, is locally of the form

$$A = \partial_i \otimes \partial_j A^{ij}{}_k \otimes du^k$$

It is a trilinear function of pairs of covectors $\alpha = a_i du^i$, $\beta = b_j du^j$, and a single vector $\mathbf{v} = \partial_k v^k$

$$A(\alpha, \beta, \mathbf{v}) = a_i b_j A^{ij}{}_k v^k$$

summed, of course, on all indices. Its components transform as

$$A'^{ef}{}_g = (\partial u'^e/\partial u^i)(\partial u'^f/\partial u^j) A^{ij}{}_k (\partial u^k/\partial u'^g)$$

(When I was a lad I learned the mnemonic "*co low, primes below.*")

If we **contract** on $i$ and $k$, the result $B^j := A^{ij}{}_i$ are the components of a contravariant **vector**

$$\begin{aligned} B'^f = A'^{ef}{}_e &= A^{ij}{}_k(\partial u'^f/\partial u^j)(\partial u^k/\partial u'^e)(\partial u'^e/\partial u^i) \\ &= A^{ij}{}_k(\partial u'^f/\partial u^j)\delta^k{}_i = A^{ij}{}_i(\partial u'^f/\partial u^j) = (\partial u'^f/\partial u^j)B^j \end{aligned}$$

(ii) A **linear transformation** is a second rank ("mixed") tensor $P = \partial_i P^i{}_j \otimes du^j$. Rather than thinking of this as a real valued bilinear function of a covector and a vector, we usually consider it as a *linear function taking vectors into vectors* (called a vector valued 1-form in Section $\mathbb{O}$.n)

$$P(\mathbf{v}) = [\partial_i P^i{}_j \otimes du^j](\mathbf{v}) := \partial_i P^i{}_j\{du^j(\mathbf{v})\} = \partial_i P^i{}_j v^j$$

i.e., the usual

$$[P(\mathbf{v})]^i = P^i_j v^j$$

Under a coordinate change, $(P^i{}_j)$ transforms as $P' = (\partial u'/\partial u)P(\partial u'/\partial u)^{-1}$, as usual. If we contract we obtain a scalar (invariant), $\operatorname{tr} P := P^i{}_i$ , the **trace** of $P$. $\operatorname{tr} P' = \operatorname{tr} P(\partial u'/\partial u)^{-1}(\partial u'/\partial u) = \operatorname{tr} P$.

**Beware:** If we have a twice covariant tensor $G$ (a "bilinear form"), for example, a metric $(g_{ij})$, then $\Sigma_k g_{kk}$ is *not a scalar*, although it is the trace of the matrix; see for example, equation ($\mathbb{O}$.7). This is because the transformation law for the matrix $G$ is, from ($\mathbb{O}$.9), $G' = (\partial u/\partial u')^T G(\partial u/\partial u')$ and $\operatorname{tr} G' \neq \operatorname{tr} G$ generically.

## Integrals and Exterior Forms

### $\mathbb{O}$.f. Line Integrals

We illustrate in $\mathbb{R}^3$ with any coordinates $x$. For simplicity, let $C$ be a smooth "oriented" or "directed" curve, the image under $F : [a,b] \subset \mathbb{R}^1 \to C \subset \mathbb{R}^3$ (which is read

"$F$ maps the interval $[a,b]$ on $\mathbb{R}^1$ into the curve $C$ in $\mathbb{R}^3$") with $F(a)$ for some $p$ and $F(b)$ for some $q$.

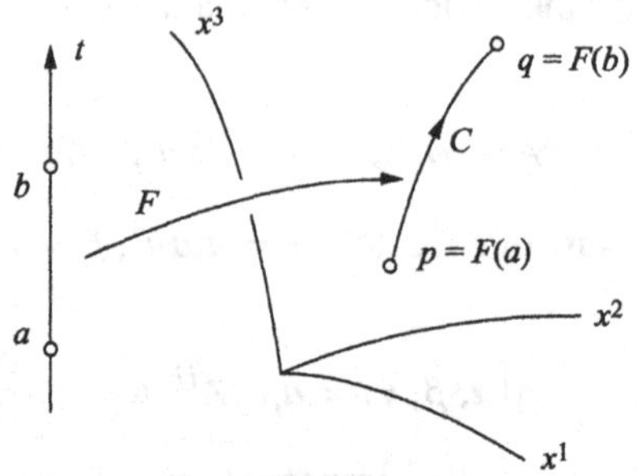


If $\alpha = \alpha^1 = a_i(x)dx^i$ is a 1-form, a covector, in $\mathbb{R}^3$, we define the line integral $\int_C \alpha$ as follows.

Using the parameterization $x^i = F^i(t)$ of $C$, we define

$$\int_C \alpha^1 = \int_C a_i(x)dx^i := \int_a^b a_i(x(t))(dx^i/dt)dt = \int_a^b \alpha(d\mathbf{x}/dt)dt \qquad (0.10)$$

We say that we *pull back* the form $\alpha^1$ (that lives in $\mathbb{R}^3$ ) to a 1-form on the parameter space $\mathbb{R}^1$, called the **pull-back** of $\alpha$, denoted by $F^*(\alpha)$

$$F^*(\alpha) = \alpha(d\mathbf{x}/dt)dt = a_i(x(t))(dx^i/dt)dt$$

and then take the *ordinary* integral $\int_a^b \alpha(d\mathbf{x}/dt)dt$. It is a classical theorem that the result is *independent of the parameterization* of $C$ chosen, so long as the resulting curve has the same orientation. This will become "apparent" from the usual geometric interpretation that we now present.

In the definition there has been no mention of *arc length* or *scalar product*. Suppose now that a Riemannian metric (e.g., the usual metric in $\mathbb{R}^3$) is available. Then to $\alpha$ we may associate its contravariant vector $\mathbf{A}$. Then $\alpha(d\mathbf{x}/dt) = \langle \mathbf{A}, d\mathbf{x}/dt\rangle = \langle \mathbf{A}, d\mathbf{x}/ds\rangle(ds/dt)$ where $s = s(t)$ is the arclength parameter along $C$. Then $F^*(\alpha) = \alpha(d\mathbf{x}/dt)dt = \langle \mathbf{A}, d\mathbf{x}/ds\rangle ds$. But $\mathbf{T} := d\mathbf{x}/ds$ is the **unit** tangent vector to $C$ since $g_{ij}(dx^i/ds)(dx^j/ds) = (g_{ij}dx^i dx^j)/(ds^2) = 1$. Thus

$$F^*(\alpha) = \langle \mathbf{A}, \mathbf{T}\rangle ds = \|\mathbf{A}\|\|\mathbf{T}\| \cos\angle(\mathbf{A}, \mathbf{T})ds$$

and so

$$\int_C \alpha = \int_C A_{\tan} ds \qquad (0.11)$$

is geometrically the integral of the tangential component of $\mathbf{A}$ with respect to the arc length parameter along $C$. This "shows" independence of the parameter $t$ chosen, but to evaluate the integral one would *usually* just use (0.10) which involves no metric at all!

**Moral:** The integrand in a line integral is naturally a **1-form,** *not* a vector.

For example, in *any* coordinates, force is often a 1-form $f^1$ since a basic measure of force is given by a line integral $W = \int_C f^1 = \int_C f_k dx^k$ which measures the **work** done by the force along the curve $C$, and this does not require a metric. Frequently there is a force **potential** $V$ such that $f^1 = dV$, exhibiting $f$ *explicitly* as a covector. (In this case, from (0.10), $W = \int_C f^1 = \int_C dV = \int_a^b dV(d\mathbf{x}/dt)dt = \int_a^b (\partial V/\partial x^i)(dx^i/dt)dt =$

$\int_a{}^b\{dV(\mathbf{x}(t)/dt\}dt = V[\mathbf{x}(b)] - V[\mathbf{x}(a)] = V(q) - V(p)$.) Of course metrics do play a large role in mechanics. In Hamiltonian mechanics, a particle of mass $m$ has a kinetic energy $T = mv^2/2 = mg_{ij}\dot{x}^i\dot{x}^j/2$ (where $\dot{x}^i$ is $dx^i/dt$) and its **momentum** is defined by $p_k = \partial(T - V)/\partial\dot{x}^k$. When the potential energy is independent of $\dot{\mathbf{x}} = d\mathbf{x}/dt$, we have $p_k = \partial T/\partial\dot{x}^k = (1/2)mg_{ij}(\delta^i{}_k\dot{x}^j + \dot{x}^i\delta^j{}_k) = (m/2)(g_{kj}\dot{x}^j + g_{ik}\dot{x}^i) = mg_{kj}\dot{x}^j$. Thus in this case $p$ is $m$ times the *covariant version* of the velocity vector $d\mathbf{x}/dt$.

The momentum 1-form "$p_i dx^i$" on the "phase space" with coordinates $(x, p)$ plays a *central role* in all of Hamiltonian mechanics.

## 0.g. Exterior 2-Forms

We have already defined the **tensor** product $\alpha^1 \otimes \beta^1$ of two 1-forms to be the bilinear form $\alpha^1 \otimes \beta^1(\mathbf{v}, \mathbf{w}) = \alpha^1(\mathbf{v})\beta^1(\mathbf{w})$. We now define a more *geometrically* significant **wedge** or **exterior product** $\alpha \wedge \beta$ to be the *skew symmetric* bilinear form

$$\alpha^1 \wedge \beta^1 := \alpha^1 \otimes \beta^1 - \beta^1 \otimes \alpha^1$$

and thus

$$du^j \wedge du^k(\mathbf{v}, \mathbf{w}) = v^j w^k - v^k w^j = \begin{vmatrix} du^j(\mathbf{v}) & du^j(\mathbf{w}) \\ du^k(\mathbf{v}) & du^k(\mathbf{w}) \end{vmatrix} \tag{0.12}$$

In **cartesian** coordinates $x, y, z$ in $\mathbb{R}^3$, see the figure below, $dx \wedge dy(\mathbf{v}, \mathbf{w})$ is $\pm$ the area of the parallelogram spanned by the projections of $\mathbf{v}$ and $\mathbf{w}$ into the $x,y$ plane, the plus sign used only if proj($\mathbf{v}$) and proj($\mathbf{w}$) describe the same orientation of the plane as the basis vectors $\boldsymbol{\partial}_x$ and $\boldsymbol{\partial}_y$.

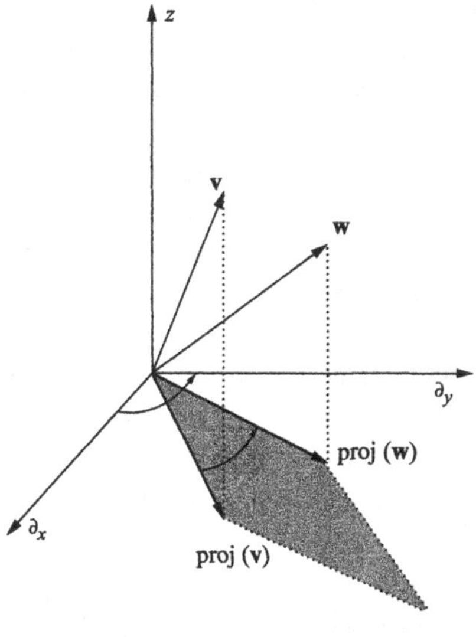


Let now $x^i, i = 1, 2, 3$ be *any* coordinates in $\mathbb{R}^3$. Note that

$$dx^j \wedge dx^k = -dx^k \wedge dx^j \quad \text{and} \quad dx^k \wedge dx^k = 0 \quad \text{(no sum!)} \tag{0.13}$$

The most general **exterior 2-form** is of the form $\beta^2 = \Sigma_{i<j} b_{ij}\, dx^i \wedge dx^j$ where $b_{ji} = -b_{ij}$. In $\mathbb{R}^3$, $\beta^2 = b_{12}\, dx^1 \wedge dx^2 + b_{23} dx^2 \wedge dx^3 + b_{13}\, dx^1 \wedge dx^3$, or, as we prefer, for

reasons soon to be evident,

$$\beta^2 = b_{23}dx^2 \wedge dx^3 + b_{31}dx^3 \wedge dx^1 + b_{12}\, dx^1 \wedge dx^2 \tag{0.14}$$

An exterior 2-form is a skew symmetric covariant tensor of the second rank in the sense of Section 0.d. We frequently will omit the *term* "exterior," but *never* the wedge $\wedge$.

## 0.h. Exterior $p$-Forms and Algebra in $\mathbb{R}^n$

The **exterior algebra** has the following properties. We have already discussed 1-forms and 2-forms. An (exterior) $p$-form $\alpha^p$ in $\mathbb{R}^n$ is a completely skew symmetric multilinear function of $p$-tuples of vectors $\alpha(\mathbf{v}_1, \ldots, \mathbf{v}_p)$ that changes sign whenever two vectors are interchanged. In any coordinates $x$, for example, the 3-form $dx^i \wedge dx^j \wedge dx^k$ in $\mathbb{R}^n$ is defined by

$$dx^i \wedge dx^j \wedge dx^k(\mathbf{A}, \mathbf{B}, \mathbf{C}) := \begin{vmatrix} dx^i(\mathbf{A}) & dx^i(\mathbf{B}) & dx^i(\mathbf{C}) \\ dx^j(\mathbf{A}) & dx^j(\mathbf{B}) & dx^j(\mathbf{C}) \\ dx^k(\mathbf{A}) & dx^k(\mathbf{B}) & dx^k(\mathbf{C}) \end{vmatrix} = \begin{vmatrix} A^i & B^i & C^i \\ A^j & B^j & C^j \\ A^k & B^k & C^k \end{vmatrix} \tag{0.15}$$

When the coordinates are cartesian the interpretation of this is similar to that in (0.12). Take the three vectors at a given point $x$ in $\mathbb{R}^n$, project them down into the 3 dimensional affine subspace of $\mathbb{R}^n$ spanned by $\boldsymbol{\partial}_i$, $\boldsymbol{\partial}_j$, and $\boldsymbol{\partial}_k$ at $x$, and read off $\pm$ the 3-volume of the parallelopiped spanned by the projections, the $+$ used only if the projections define the same orientation as $\boldsymbol{\partial}_i$, $\boldsymbol{\partial}_j$, and $\boldsymbol{\partial}_k$.

Clearly any interchange of a single pair of $dx$ will yield the negative, and thus *if the same $dx^i$ appears twice the form will vanish*, just as in (0.12), similarly for a $p$-form. The most general 3-form is of the form $\alpha^3 = \Sigma_{i<j<k} a_{ijk} dx^i \wedge dx^j \wedge dx^k$. In $\mathbb{R}^3$ there is only one nonvanishing 3-form, $dx^1 \wedge dx^2 \wedge dx^3$ and its multiples. In *cartesian* coordinates this is the **volume form** $\mathrm{vol}^3$, but in spherical coordinates we know that $dr \wedge d\theta \wedge d\phi$ does *not* yield the euclidean volume element, which is $r^2 \sin\theta\, dr \wedge d\theta \wedge d\phi$. We will discuss this soon. Note further that all $p > n$ forms in $\mathbb{R}^n$ vanish since there are always repeated $dx$ in each term.

We take the **exterior product** of a $p$-form $\alpha$ and a $q$-form $\beta$, yielding a $p+q$ form $\alpha \wedge \beta$ by expressing them in terms of the $dx$, using the usual algebra (including the associative law), except that the product of $dx$ is anticommutative, $dx \wedge dy = -dy \wedge dx$. For examples in $\mathbb{R}^3$ with any coordinates

$$\begin{aligned} \alpha^1 \wedge \gamma^1 &= (a_1\, dx^1 + a_2\, dx^2 + a_3\, dx^3) \wedge (c_1\, dx^1 + c_2\, dx^2 + c_3\, dx^3) \\ &= \cdots (a_2\, dx^2) \wedge (c_1\, dx^1) + \cdots + (a_1\, dx^1) \wedge (c_2\, dx^2) + \cdots \\ &= (a_2c_3 - a_3c_2)\, dx^2 \wedge dx^3 + (a_3c_1 - a_1c_3)\, dx^3 \wedge dx^1 \\ &\quad + (a_1c_2 - a_2c_1)\, dx^1 \wedge dx^2 \end{aligned}$$

which in *cartesian* coordinates has the components of the vector product $\mathbf{a} \times \mathbf{c}$. Also we have

$$\begin{aligned}\alpha^1 \wedge \beta^2 &= (a_1\, dx^1 + a_2\, dx^2 + a_3\, dx^3) \wedge (b_{23}\, dx^2 \wedge dx^3 \\ &\quad + b_{31}\, dx^3 \wedge dx^1 + b_{12}\, dx^1 \wedge dx^2) \\ &= (a_1 b_1 + a_2 b_2 + a_3 b_3) dx^1 \wedge dx^2 \wedge dx^3\end{aligned}$$

(where we use the notation $b_1 := b_{23}, b_2 = b_{31}, b_3 = b_{12}$, but *only* in cartesian coordinates) with component $\mathbf{a} \cdot \mathbf{b}$ in cartesian coordinates. The $\wedge$ product in cartesian $\mathbb{R}^3$ yields both the dot $\cdot$ and the cross $\times$ products of vector analysis!! The $\cdot$ and $\times$ products of vector analysis have strange expressions when curvilinear coordinates are used in $\mathbb{R}^3$, but the form expressions $\alpha^1 \wedge \beta^2$ and $\alpha^1 \wedge \gamma^1$ are always the same. Furthermore, the $\times$ product is nasty since it is not associative, $\mathbf{i} \times (\mathbf{i} \times \mathbf{j}) \neq (\mathbf{i} \times \mathbf{i}) \times \mathbf{j}$.

By counting the number of interchanges of pairs of $dx$ one can see the **commutation rule**

$$\alpha^p \wedge \beta^q = (-1)^{pq} \beta^q \wedge \alpha^p \tag{0.16}$$

## 0.i. The Exterior Differential $d$

First a remark. If $\mathbf{v} = \boldsymbol{\partial}_a v^a$ is a contravariant vector field, then generically $(\partial v^a / \partial x^b) = Q^a{}_b$ *do not yield the components of a tensor* in curvilinear coordinates, as is easily seen from looking at the transformation of $Q$ under a change of coordinates and using (0.2). It is, however, always possible, in $\mathbb{R}^n$ and in any coordinates, to take a very important **exterior** derivative $d$ of $p$-forms. We define $d\alpha^p$ to be a $p + 1$ form, as follows; $\alpha$ is a sum of forms of the type $a(x) dx^i \wedge dx^j \wedge \cdots \wedge dx^k$. Define

$$\begin{aligned}d[a(x)dx^i \wedge dx^j \wedge \ldots \wedge dx^k] &= da \wedge dx^i \wedge dx^j \wedge \ldots \wedge dx^k \\ &= \Sigma_r (\partial a / \partial x^r) dx^r \wedge dx^i \wedge dx^j \wedge \ldots \wedge dx^k\end{aligned} \tag{0.17}$$

(in particular $d[dx^i \wedge dx^j \wedge \ldots \wedge dx^k] = 0$), and then sum over all the terms in $\alpha^p$. In particular, in $\mathbb{R}^3$ in *any* coordinates

$$\begin{aligned}df^0 &= df = (\partial f/\partial x^1) dx^1 + (\partial f/\partial x^2) dx^2 + (\partial f/\partial x^3) dx^3 \\ d\alpha^1 &= d(a_1\, dx^1 + a_2\, dx^2 + a_3\, dx^3) = (\partial a_1/\partial x^2) dx^2 \wedge dx^1 + (\partial a_1/\partial x^3) dx^3 \wedge dx^1 + \cdots \\ &= [(\partial a_3/\partial x^2) - (\partial a_2/\partial x^3)] dx^2 \wedge dx^3 + [(\partial a_1/\partial x^3) - (\partial a_3/\partial x^1)] dx^3 \wedge dx^1 \\ &\quad + [(\partial a_2/\partial x^1) - (\partial a_1/\partial x^2)] dx^1 \wedge dx^2 \qquad\qquad (0.18) \\ d\beta^2 &= d(b_{23}\, dx^2 \wedge dx^3 + b_{31}\, dx^3 \wedge dx^1 + b_{12}\, dx^1 \wedge dx^2) \\ &= [(\partial b_{23}/\partial x^1) + (\partial b_{31}/\partial x^2) + (\partial b_{12}/\partial x^3)] dx^1 \wedge dx^2 \wedge dx^3\end{aligned}$$

In *cartesian* coordinates we then have correspondences with vector analysis, using again $b_1 := b_{23}$ etc.,

$$df^0 \Leftrightarrow \nabla \mathbf{f} \cdot \mathbf{dx} \quad d\alpha^1 \Leftrightarrow (\text{curl } \mathbf{a}) \cdot \text{“}\mathbf{dA}\text{”} \quad d\beta^2 \Leftrightarrow \text{div } \mathbf{B} \text{ “dvol”} \tag{0.19}$$

the quotes, for example, “**dA**” being used since this is not really the differential of a 1-form. We shall make this correspondence precise, in any coordinates, later. Exterior differentiation of **exterior forms** does essentially grad, curl and divergence with a *single general formula* (©.17)!! Also, this machinery works in $\mathbb{R}^n$ as well. Furthermore, *d does not require a metric*. On the other hand, without a metric (and hence without cartesian coordinates), one *cannot* take the curl of a contravariant vector field. Also to take the *divergence* of a vector field requires at least a specified "volume form." These will be discussed in more detail later in section ©.n.

There are two fairly easy but very important properties of the differential $d$:

$$\begin{aligned} d^2\alpha^p &:= d\, d\alpha^p = 0 \text{ (which says curl grad} = 0 \text{ and div curl} = 0 \text{ in } \mathbb{R}^3) \\ d(\alpha^p \wedge \beta^q) &= d\alpha \wedge \beta + (-1)^p \alpha \wedge d\beta \end{aligned} \tag{©.20}$$

For example, in $\mathbb{R}^3$ with function (0-form) $f$, $df = (\partial f/\partial x)dx + (\partial f/\partial y)dy + (\partial f/\partial z)dz$, and then $d^2 f = (\partial^2 f/\partial x\ \partial y)dy \wedge dx + \cdots + (\partial^2 f/\partial y\ \partial x)dx \wedge dy + \cdots = 0$, since $(\partial^2 f/\partial y\ \partial x) = (\partial^2 f/\partial x\ \partial y)$.

Note then that a *necessary* condition for a $p$-form $\beta^p$ to be the differential of some $(p-1)$-form, $\beta^p = d\alpha^{p-1}$, is that $d\beta = d\, d\alpha = 0$. (What does this say in vector analysis in $\mathbb{R}^3$ ?)

Also, we know that in cartesian $\mathbb{R}^3$, $\alpha^1 \wedge \beta^1 \Leftrightarrow \mathbf{a} \times \mathbf{b}$ is a 2-form, $d(\alpha \wedge \beta) \Leftrightarrow$ div $\mathbf{a} \times \mathbf{b}$ (from (©.19)), and $d\alpha \Leftrightarrow$ curl $\mathbf{a}$, and we know $\alpha^1 \wedge \gamma^2 = \gamma^2 \wedge \alpha^1 \Leftrightarrow \mathbf{a} \cdot \mathbf{c}$. Then (©.20), in cartesian coordinates, says *immediately* that $d(\alpha \wedge \beta) = d\alpha \wedge \beta - \alpha \wedge d\beta$, i.e.,

$$\text{div}\ \mathbf{a} \times \mathbf{b} = (\text{curl}\ \mathbf{a}) \cdot \mathbf{b} - \mathbf{a} \cdot (\text{curl}\ \mathbf{b}) \tag{©.21}$$

## ©.j. The Push-Forward of a Vector and the Pull-Back of a Form

Let $F$: $\mathbb{R}^k \to \mathbb{R}^n$ be any differentiable map of $k$-space into $n$-space, where any values of $k$ and $n$ are permissible. Let $(u^1, \ldots, u^k)$ be any coordinates in $\mathbb{R}^k$, let $(x^1, \ldots, x^n)$ be any coordinates in $\mathbb{R}^n$. Then $F$ is described by $n$ functions $x^i = F^i(u) = F^i(u^1, \ldots, u^r, \ldots, u^k)$ or briefly $x^i = x^i(u)$.

The "pull-back" of a **function** (0-form) $\phi = \phi(x)$ on $\mathbb{R}^n$ is the function $F^*\phi = \phi(x(u))$ on $\mathbb{R}^k$, i.e., the function on $\mathbb{R}^k$ whose value at $u$ is simply the value of $\phi$ at $x = F(u)$.

Given a vector $\mathbf{v}_0$ at the point $u_0 \in \mathbb{R}^k$ we can "push forward" the vector to the point $x_0 = F(u_0) \in \mathbb{R}^n$ by means of the so-called "differential of $F$," written $F_*$, as follows. Let $u = u(t)$ be any curve in $\mathbb{R}^k$ with $u(0) = u_0$ and velocity at $u_0 = [du/dt]_0$ equal to the given $\mathbf{v}_0$. (For example, in terms of the coordinates $u$, you may use the curve defined by $u^r(t) = u_0{}^r + v_0{}^r t$ .) Then the image curve $x(t) = x(u(t))$ will have velocity vector at $t = 0$ called $F_*[\mathbf{v}_0]$ given by the chain rule,

$$[F_*(\mathbf{v}_0)]^i := dx^i(u(t))/dt]_0 = [\partial x^i/\partial u^r]_{u(0)}[du^r/dt]_0 = [\partial x^i/\partial u^r]_{u(0)} \mathbf{v}_0{}^r$$

Briefly

$$[F_*(\mathbf{v})]^i = (\partial x^i/\partial u^r) v^r$$

Then

$$F_*[v^r \boldsymbol{\partial}/\partial u^r] = v^r\ \boldsymbol{\partial}/\partial x^i (\partial x^i/\partial u^r), \tag{©.22}_*$$

and so

$$F_*\boldsymbol{\partial}_r = F_*[\partial/\partial u^r] = [\boldsymbol{\partial}/\partial x^i](\partial x^i/\partial u^r) = \boldsymbol{\partial}_i(\partial x^i/\partial u^r)$$

*is again simply the chain rule.*

Given any $p$-form $\alpha$ at $x \in \mathbb{R}^n$, we define the **pull-back** $F^*(\alpha)$ to be the $p$-form at each pre-image point $u \in F^{-1}(x)$ of $\mathbb{R}^k$ by

$$(F^*\alpha)(\mathbf{v}, \ldots, \mathbf{w}) := \alpha(F_*\mathbf{v}, \ldots, F_*\mathbf{w}) \tag{©.23}$$

For the 1-form $dx^i$, $F^*dx^i$ must be of the form $a_s du^s$; using $dx^i(\boldsymbol{\partial}_j) = \delta^i{}_j$ we get

$$(F^*dx^i)(\boldsymbol{\partial}_r) = dx^i[\boldsymbol{\partial}_j(\partial x^j/\partial u^r)] = \partial x^i/\partial u^r = (\partial x^i/\partial u^s)du^s(\boldsymbol{\partial}_r)$$

and so

$$F^*dx^i = (\partial x^i/\partial u^s)du^s \tag{©.22$^*$}$$

is again simply the chain rule.

It can be shown in general that $F^*$ operating on forms satisfies

$$F^*(\alpha^p \wedge \beta^q) = (F^*\alpha) \wedge (F^*\beta)$$

and

$$F^*d\alpha = dF^*\alpha \tag{©.24}$$

For example, $F^*dx^i = dF^*(x^i) = dx^i(u) = (\partial x^i/\partial u^s)du^s$, as we have just seen.

For $p$-forms we shall use the same procedure but also use the fact that $F^*$ commutes with exterior product, $F^*(\alpha \wedge \beta) = (F^*\alpha) \wedge (F^*\beta)$. For simplicity we shall just illustrate the idea for the case when $\beta^2$ is a 2-form in $\mathbb{R}^n$ and $F: \mathbb{R}^3 \to \mathbb{R}^n$. For more simplicity we just consider a typical term $b_{23}(x)dx^2 \wedge dx^3$ of $\beta$.

$$\begin{aligned} F^*[b_{23}(x)dx^2 \wedge dx^3] &:= [F^*b_{23}(x)][F^*dx^2] \wedge [F^*dx^3] \\ &:= b_{23}(x(u))[(\partial x^2/\partial u^a)du^a] \\ &\quad \wedge [(\partial x^3/\partial u^c)du^c] \quad \text{(summed on } a \text{ and } c) \end{aligned}$$

Now $(\partial x^2/\partial u^a)du^a = (\partial x^2/\partial u^1)du^1 + (\partial x^2/\partial u^2)du^2 + (\partial x^2/\partial u^3)du^3$ with a similar expression for $(\partial x^3/\partial u^c)du^c$. Taking their $\wedge$ product and using (©.13)

$$\begin{aligned} &[(\partial x^2/\partial u^1)du^1 + (\partial x^2/\partial u^2)du^2 + (\partial x^2/\partial u^3)du^3] \wedge [(\partial x^3/\partial u^1)du^1 \\ &\quad + (\partial x^3/\partial u^2)du^2 + (\partial x^3/\partial u^3)du^3] \\ &= (\partial x^2/\partial u^1)du^1 \wedge (\partial x^3/\partial u^2)du^2 + (\partial x^2/\partial u^1)du^1 \wedge (\partial x^3/\partial u^3)du^3 \\ &\quad + (\partial x^2/\partial u^2)du^2 \wedge (\partial x^3/\partial u^1)du^1 + (\partial x^2/\partial u^2)du^2 \wedge (\partial x^3/\partial u^3)du^3 \\ &\quad + (\partial x^2/\partial u^3)du^3 \wedge (\partial x^3/\partial u^1)du^1 + (\partial x^2/\partial u^3)du^3 \wedge (\partial x^3/\partial u^2)du^2 \\ &= [(\partial x^2/\partial u^2)(\partial x^3/\partial u^3) - (\partial x^2/\partial u^3)(\partial x^3/\partial u^2)]du^2 \wedge du^3 \\ &\quad + [(\partial x^2/\partial u^1)(\partial x^3/\partial u^3) - (\partial x^2/\partial u^3)(\partial x^3/\partial u^1)]du^1 \wedge du^3 \\ &\quad + [(\partial x^2/\partial u^1)(\partial x^3/\partial u^2) - (\partial x^2/\partial u^2)(\partial x^3/\partial u^1)]du^1 \wedge du^2 \end{aligned}$$

and so

$$F^*[b_{23}(x)dx^2 \wedge dx^3] = b_{23}(x(u))\Sigma_{a<c}[\partial(x^2, x^3)/\partial(u^a, u^c)]du^a \wedge du^c$$

where

$$\partial(x, y)/\partial(u, v) = \begin{vmatrix} \partial x/\partial u & \partial x/\partial v \\ \partial y/\partial u & \partial y/\partial v \end{vmatrix}$$

is the usual **Jacobian determinant**. In general, for pulling back a $p$-form on $\mathbb{R}^n$ to $\mathbb{R}^k$ via $F$: $\mathbb{R}^k \to \mathbb{R}^n$ we use

$$F^*(dx^i \wedge \ldots \wedge dx^j) = \Sigma_{a<\ldots<r}[\partial(x^i, \ldots \partial x^j)/\partial(u^a, \ldots, u^r)]du^a \wedge \ldots \wedge du^r \tag{$\mathbb{0}$.22}$$ **

This procedure will play a key role in our discussion of surface integrals, see ($\mathbb{0}$.25).

($\mathbb{0}$.20) and ($\mathbb{0}$.24) contribute to what makes forms so powerful and useful, compared to vector fields. The push-forward $F_*$ associated to a map $F$: $\mathbb{R}^k \to \mathbb{R}^n$ will map a vector $\mathbf{v}$ at $u \in \mathbb{R}^k$ to a vector $F_*\mathbf{v}$ at $x = F(u)$. But let $\mathbf{v}$ be a vector **field**, say on all of $\mathbb{R}^k$ and suppose $F$ is not 1:1. Let $u' \neq u$ and $F(u') = x = F(u)$. Then generically $F_*\mathbf{v}(u')$ will not agree with $F_*\mathbf{v}(u)$, and so $F_*\mathbf{v}$ will *not* be a well defined vector **field** on $\mathbb{R}^n$. On the other hand, if $\alpha$ is a $p$-form at $x$, then $F^*\alpha$ will define a unique form at $u$ and another form at $u'$. If $\alpha^p$ is a well defined $p$-form field on $\mathbb{R}^n$ then $F^*\alpha$ is a well defined $p$-form field on $\mathbb{R}^k$. For fields the tools ($\mathbb{0}$.24) are then available.

Note that when $F : \mathbb{R}^n \to \mathbb{R}^n$ is the **identity** map, using two sets of coordinates, for example, $(r, \theta)$ and $(x, y)$ in the plane, and where the identity map $F = I$ is $x = r\cos\theta$, $y = r\sin\theta$ in $\mathbb{R}^2$, then the pull-back $F^*\alpha$ is simply expressing the form $\alpha$, given in coordinates $x$ in terms of the new coordinates $u$.

Finally note that ($\mathbb{0}$.23) makes sense when $\alpha$ is a **covariant** $p$-tensor even if it is not an exterior form, i.e., even if $\alpha$ is not completely skew symmetric. The pull-back of the Riemannian metric tensor $g$, $g(\mathbf{v}, \mathbf{w}) = g_{ij}v^i w^j$ plays a central role in elasticity, as will be seen in Section $\mathbb{0}$.p. The pull-back of the quadratic form $g_{ij}\, dx^i dx^j$ is again just the application of the chain rule. Of course ($\mathbb{0}$.24) does not make sense if $\alpha$ is not an exterior form.

## $\mathbb{0}$.k. Surface Integrals and "Stokes' theorem"

We illustrate with a surface $V^2$ in $\mathbb{R}^3$. Assume, for example, that $\mathbb{R}^3$ has the "right handed orientation." Assume that $V^2$ is also "oriented" meaning that at each point $p$ of $V$ there is a preferred sense of rotation of the tangent plane at $p$ (indicated in the figure below by a circular arrow), and this sense varies continuously on $V$. For example, if $V$ has a continuous choice of normal vector everywhere (unlike a Möbius band) then the right hand rule for $\mathbb{R}^3$ will yield an orientation for $V$.

We are going to define $\int_V \beta^2$ for any 2-form $\beta$ on $\mathbb{R}^3$. If $V$ is sufficiently small we may choose a parameterization of all of $V$ that yields the same orientation as $V$, i.e., we ask for a smooth 1:1 map

$$F : \text{region } S^2 \subset \text{some } \mathbb{R}^2 \to \text{onto } V^2 \subset \mathbb{R}^3 \qquad x^i = x^i(t^1, t^2)$$

(If $V$ is too large for such a parameterization, break it up into smaller pieces and add up the individual resulting integrals.) We picture the resulting $t^1$, $t^2$ coordinate curves on $V$ as engraved on $V$ just as latitude and longitude curves are engraved on globes of the Earth. We demand that the sense of rotation from the engraved $t^1$ curve to the $t^2$ curve on $V$ (i.e., from $F_*\boldsymbol{\partial}_1$ to $F_*\boldsymbol{\partial}_2$) is the same as the given orientation arrow on $V$. We say $V = F(S)$.

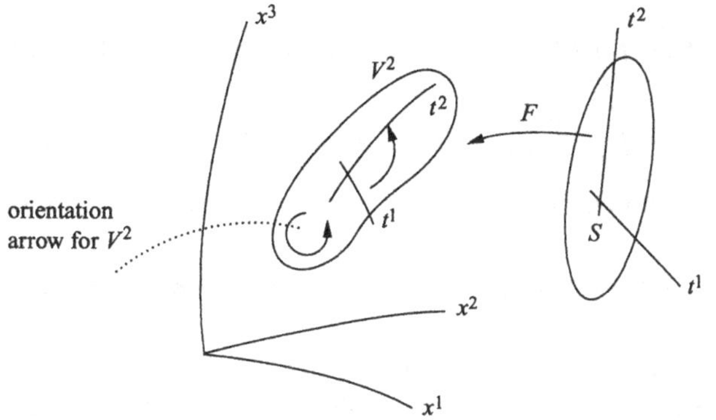


We now define

$$\int_V b_{23}\, dx^2 \wedge dx^3 + b_{31}\, dx^3 \wedge dx^1 + b_{12}\, dx^1 \wedge dx^2 = \int_V \beta^2 = \int_{F(S)} \beta^2 := \int_S F^*\beta$$

reducing the problem to defining the integral of the pull-back of $\beta$ over $S$. First write this out, but for simplicity we just look at the term $b_{31}(x)dx^3 \wedge dx^1$. As in (0.22)**

$$\begin{aligned}\int_S F^*(b_{31}(x)dx^3 \wedge dx^1) &:= \int_S b_{31}(x(t))[(\partial x^3/\partial t^a)dt^a \wedge (\partial x^1/\partial t^b)dt^b] \\ &= \int_S b_{31}(x(t))[\partial(x^3, x^1)/\partial(t^1, t^2)]dt^1 \wedge dt^2 \\ &:= \int_S b_{31}(x(t))[\partial(x^3, x^1)/\partial(t^1, t^2)]dt^1 dt^2\end{aligned}$$

and where the very last integral, with no $\wedge$, is the **usual double integral** over a region $S$ in the $t^1, t^2$ plane. Thus

$$\begin{aligned}&\int_V \beta^2 = \int_{F(S)} \beta^2 = \int_S F^*\beta^2 \\ &\quad := \int_S \{b_{23}(x(t))[\partial(x^2, x^3)/\partial(t^1, t^2)] + b_{31}(x(t))[\partial(x^3, x^1)/\partial(t^1, t^2)] \\ &\qquad + b_{12}(x(t))[\partial(x^1, x^2)/\partial(t^1, t^2)]\}dt^1 dt^2 \end{aligned} \tag{0.25}$$

Note that one does not need to commit this to memory. One merely uses the chain rule in calculus and $dt^1 \wedge dt^2 = -dt^2 \wedge dt^1$ to get an integral over a region in the $t^1, t^2$ plane, then omit the $\wedge$ and evaluate the resulting double integral.

**Interpretation:** In *cartesian* coordinates with the usual metric in $\mathbb{R}^3$, associate to $\beta^2$ the vector

$$\mathbf{B} = (B^1 = b_{23},\ B^2 = b_{31},\ B^3 = b_{12})^T$$

$\mathbf{n} = [\partial\mathbf{x}/\partial t^1] \times [\partial\mathbf{x}/\partial t^2]$ is a normal to the surface with components

$$([\partial(x^2, x^3)/\partial(t^1, t^2)], [\partial(x^3, x^1)/\partial(t^1, t^2)], [\partial(x^1, x^2)/\partial(t^1, t^2)])^T$$

Just as in the case of a curve, where $\|d\mathbf{x}/dt\|dt$ is the element of arc length $ds$, so in the case of a surface, where $\partial\mathbf{x}/\partial t^1$ and $\partial\mathbf{x}/\partial t^2$ span a parallelogram of area $\|(\partial\mathbf{x}/\partial t^1) \times (\partial\mathbf{x}/\partial t^2)\| = \|\mathbf{n}\|$, we have the area element "$dA$" $= \|\mathbf{n}\|dt^1dt^2$. Our integral (©.25) then becomes

$$\int_V \beta^2 = \iint_S \langle \mathbf{B}, \mathbf{n}\rangle dt^1dt^2 = \iint_S \|\mathbf{B}\|\,\|\mathbf{n}\| \cos\angle(\mathbf{B}, \mathbf{n})dt^1dt^2$$
$$= \iint_V B_{\text{normal}} \text{"}dA\text{"} \quad \text{(classically)}$$

and this shows further that the integral $\int_V \beta$ is in fact independent of the parameterization $F$ used.

Note again that our form version (©. 25) requires no metric or area element.

**Moral:** The integrand in a surface integral is naturally a 2-form, not a vector.

One integrates exterior $p$-forms over oriented $p$ dimensional "surfaces" $V^p$. If $V^p$ is not a "closed" surface it will generically have a $(p-1)$ dimensional oriented boundary, written $\partial V$. For example, if $V^2$ is oriented, then the circular orientation arrow near the boundary curve of $V$ will yield a "direction" for $\partial V$ ( see the surface integral picture above)

$$\textbf{Stokes' Theorem} \qquad \int_V d\beta^{p-1} = \oint_{\partial V} \beta^{p-1} \qquad (©.26)$$

is perhaps the World's Most Beautiful Formula. The vector analysis versions, using (©.19), include not only Stokes' theorem (really due to **William Thomson, Lord Kelvin**) when $p = 2$ and $V^2$ is an oriented surface and $\partial V$ is its closed curve boundary, but also **Gauss'** divergence theorem when $p = 3$, $V^3$ is a bounded region in space and $\partial V$ is its closed surface boundary. For a proof see Chapter 3.

### ©.l. Electromagnetism, or, Is it a Vector or a Form?

For simplicity we consider electric and magnetic fields caused by charges, currents, and magnets in a vacuum (without polarizations,...)

**Electric field intensity E:** The work done in moving a particle with charge $q$ along a curve $C$ is classically $W = \int_C q\, \mathbf{E} \cdot d\mathbf{r}$ but really $w = q\int_C \mathcal{E}^1 = q\int_C E_1\, dx^1 + E_2\, dx^2 + E_3\, dx^3$. The electric field intensity is a 1-form $\mathcal{E}^1 = E_1\, dx^1 + E_2\, dx^2 + E_3\, dx^3$.

**Electric field D:** The charge $Q$ contained in a region $V^3$ with boundary $\partial V$ is classically given by $4\pi Q(V^3) = \iint_{\partial V} \mathbf{D} \cdot d\mathbf{A} = \iiint_V \text{div } \mathbf{D} \text{ vol}$, but really

$$\iint_{\partial V} \mathfrak{D}^2 = \iiint_V d\mathfrak{D} = 4\pi Q(V^3) = 4\pi \iiint_V \rho \text{vol}^3$$

where $\rho$ is the charge density. Stokes' theorem thus yields **Gauss' law**

$$d\mathfrak{D}^2 = 4\pi\ \rho \text{vol}^3$$

$\mathfrak{D}^2$ is a 2-form version of $\mathcal{E}^1$ . In *cartesian* coordinates $\mathfrak{D}^2 = E_1\, dx^2 \wedge dx^3 + E_2\, dx^3 \wedge dx^1 + E_3\, dx^1 \wedge dx^2$.

**Magnetic field intensity B:** Faraday's law says classically, for a *fixed* surface $V^2$, $\oint_{\partial V} \mathbf{E} \cdot d\mathbf{r} = -d/dt \iint_V \mathbf{B} \cdot d\,\mathbf{A}$. Really $\oint_{\partial V} \mathcal{E}^1 = -d/dt \iint_V \mathcal{B}^2$. The magnetic field intensity is a 2-form $\mathcal{B}^2$ and **Faraday's law** says

$$d\mathcal{E}^1 = -\partial\mathcal{B}^2/\partial t$$

where $\partial\mathcal{B}^2/\partial t$ means take the time derivative of the components of $\mathcal{B}^2$. Another axiom states that

$$\text{div } \mathbf{B} = 0 = d\mathcal{B}^2$$

**Magnetic field H: Ampère–Maxwell** says classically $\oint_{C=\partial V} \mathbf{H} \cdot d\mathbf{r} = 4\pi \iint_V \mathbf{j} \cdot d\mathbf{A} + d/dt \iint_V \mathbf{D} \cdot d\mathbf{A}$ where $V^2$ is **fixed** and **j** is the current vector. Really

$$\oint_{C=\partial V} \mathcal{H}^1 = 4\pi \iint_V \mathfrak{j}^2 + d/dt \iint_V \mathcal{D}^2$$

and thus

$$d\mathcal{H}^1 = 4\pi \mathfrak{j}^2 + \partial\mathcal{D}^2/\partial t$$

where $\mathfrak{j}^2$ is the current 2-form whose integral over $V^2$ (with a preferred normal direction) measures the time rate of charge passing through $V^2$ in that direction. $\mathcal{H}^1$ is a 1-form version of $\mathcal{B}^2$. In *cartesian* coordinates

$$\mathcal{H}^1 = B_{23}dx^1 + B_{31}dx^2 + B_{12}dx^3$$

**Heaviside–Lorentz force:** Classically the electromagnetic force acting on a particle of charge $q$ moving with velocity **v** is given by $\mathbf{f} = q(\mathbf{E} + \mathbf{v} \times \mathbf{B})$. We have seen that force and the electric field should be 1-forms, $\mathfrak{f}^1 = q(\mathcal{E}^1 + ??)$. **v** is definitely a vector, and $\mathcal{B}$ is a 2-form! We now discuss this dilemma raised by the vector product $\times$ and its resolution will play a large role in our discussion of elasticity also.

## 0.m. Interior Products

We are at home with the fact $\alpha^1 \wedge \beta^1$ is a 2-form replacement for a $\times$ product of vectors in $\mathbb{R}^3$, but if we had started out with two vectors **A** and **B** it would require a metric to change them to 1-forms. It turns out there is also a 1-form replacement that is frequently more useful, and will resolve the Lorentz force problem.

In $\mathbb{R}^n$, if **v** is a vector and $\beta^p$ is a $p$-form, $p > 0$, we define the **interior product** of **v** and $\beta$ to be the $(p-1)$-form $i_\mathbf{v}\beta$ (sometimes we write $i(\mathbf{v})\beta$) with values

$$i_\mathbf{v}\beta^p(\mathbf{A}_2, \ldots, \mathbf{A}_p) := \beta^p(\mathbf{v}, \mathbf{A}_2, \ldots, \mathbf{A}_p) \tag{0.27}$$

(It can be shown that this is a contraction, $(i_\mathbf{v}\beta)_{bc\ldots} = v^i \beta_{ibc\ldots}$). This is a form since it clearly is multilinear in $\mathbf{A}_2, \ldots, \mathbf{A}_p$, since $\beta$ is, and changes sign under each interchange of the $A$, and is defined independent of any coordinates. In the case of a 1-form $\beta$, $i_\mathbf{v}\beta$ is the 0-form (function)

$$i_\mathbf{v}\beta^1 = \beta^1(\mathbf{v}) = b_i v^i$$

which is equal to $\langle \mathbf{v}, \mathbf{b} \rangle$ in any Riemannian metric. Look at $i_{\mathbf{v}}(\alpha^1 \wedge \beta^1)$:

$$i_{\mathbf{v}}(\alpha^1 \wedge \beta^1)(\mathbf{C}) = (\alpha^1 \wedge \beta^1)(\mathbf{v}, \mathbf{C}) = \alpha(\mathbf{v})\beta(\mathbf{C}) - \alpha(\mathbf{C})\beta(\mathbf{v})$$
$$= (i_{\mathbf{v}}\alpha)\beta(\mathbf{C}) - (i_{\mathbf{v}}\beta)\alpha(\mathbf{C}) = [(i_{\mathbf{v}}\alpha)\beta - (i_{\mathbf{v}}\beta)\alpha](\mathbf{C})$$

A more tedious calculation shows the general product rule

$$i_{\mathbf{v}}(\alpha^p \wedge \beta^q) = [i_{\mathbf{v}}(\alpha^p)] \wedge \beta^q + (-1)^p \alpha^p \wedge [i_{\mathbf{v}}\beta^q] \qquad (\mathbb{0}.28)$$

just as for the differential $d$ (see ($\mathbb{0}$.20)).

## $\mathbb{0}$.n. Volume Forms and Cartan's Vector Valued Exterior Forms

Let $x$, $y$ be positively oriented cartesian coordinates in $\mathbb{R}^2$. The area 2-form in the cartesian plane is $\text{vol}^2 = dx \wedge dy$, but in polar coordinates we have $\text{vol}^2 = r dr \wedge d\theta$. Looking at ($\mathbb{0}$.7) we note that $r = \sqrt{g}$, where

$$g := \det(g_{ij}) \qquad (\mathbb{0}.29)$$

In any Riemannian metric, in any oriented $\mathbb{R}^n$, we define the volume $n$-form to be

$$\text{vol}^n := \sqrt{g} dx^1 \wedge \ldots \wedge dx^n \qquad (\mathbb{0}.30)$$

in any positively oriented curvilinear coordinates. It can be shown that this is indeed an $n$-form (modulo some question of orientation that I do not wish to consider here). In spherical coordinates in $\mathbb{R}^3$ we get, since $(g_{ij}) = \text{diag}(1, r^2, r^2\sin^2\theta)$, the familiar $\text{vol}^3 = r^2 \sin\theta dr \wedge d\theta \wedge d\phi$.

Note now the following in $\mathbb{R}^3$ in any coordinates. For any vector $\mathbf{v}$

$$i_{\mathbf{v}}\text{vol}^3 = i_{\mathbf{v}}\sqrt{g} dx^1 \wedge dx^2 \wedge dx^3 = \sqrt{g} i_{\mathbf{v}}(dx^1 \wedge dx^2 \wedge dx^3)$$

Now apply the product rule ($\mathbb{0}$.28) repeatedly

$$i_{\mathbf{v}}(dx^1 \wedge dx^2 \wedge dx^3) = v^1 dx^2 \wedge dx^3 - dx^1 \wedge i_{\mathbf{v}}(dx^2 \wedge dx^3)$$
$$= v^1 dx^2 \wedge dx^3 - dx^1 \wedge [v^2 dx^3 - v^3 dx^2]$$
$$= v^1 dx^2 \wedge dx^3 - v^2 dx^1 \wedge dx^3 + v^3 dx^1 \wedge dx^2$$

and so

$$i_{\mathbf{v}}\text{vol}^3 = \sqrt{g}[v^1 dx^2 \wedge dx^3 + v^2 dx^3 \wedge dx^1 + v^3 dx^1 \wedge dx^2] \qquad (\mathbb{0}.31)$$

is the **2-form version** of a vector $\mathbf{v}$ in $\mathbb{R}^3$ with a volume form $\text{vol}^3$.

**Remark:** For a surface $V^2$ in Riemannian $\mathbb{R}^3$, with **unit** normal vector field $\mathbf{n}$, it is easy to see that $i_{\mathbf{n}}\text{vol}^3$ is the **area 2-form** for $V^2$. Simply look at its value on a pair of vectors $(\mathbf{A}, \mathbf{B})$ **tangent** to $V$; $i_{\mathbf{n}}\text{vol}^3(\mathbf{A}, \mathbf{B}) = \text{vol}^3(\mathbf{n}, \mathbf{A}, \mathbf{B})$ is the area spanned by $\mathbf{A}$ and $\mathbf{B}$.

Comparing ($\mathbb{0}$.31) with ($\mathbb{0}$.14) we see that the most general 2-form $\beta^2$ in $\mathbb{R}^3$ (with $\text{vol}^3$), in any coordinates, is of the form

$$\beta^2 = i_{\mathbf{b}}\text{vol}^3 \quad \text{where } b^1 = b_{23}/\sqrt{g}, \text{ etc.} \qquad (\mathbb{0}.14)'$$

In electromagnetism,

$$\mathcal{D}^2 = i_{\mathbf{E}}\text{vol}^3$$

The same procedure works for an $(n-1)$ form in $\mathbb{R}^n$. Note that this does not require an entire metric tensor, we use only the *volume element*. If we have a *distinguished volume form* (i.e., if we have a coordinate independent notion of the volume spanned by a "positively oriented" $n$-tuple of vectors in $\mathbb{R}^n$), even if it is not derived from a metric, we shall use the same notation in positively oriented coordinates, as given in (©.30)

$$\text{vol}^n = \sqrt{g}dx^1 \wedge \ldots \wedge dx^n$$

where $\sqrt{g} > 0$ is now merely some coefficient function dependent on the choice of volume form and the coordinates used. (Warning: this notation is my own and is not standard.)

If we have a volume form, we can define the **divergence of a vector field v** as follows

$$\begin{aligned}(\text{div } \mathbf{v})\text{vol}^n :&= d(i_{\mathbf{v}}\text{vol}^n) = d\{\sqrt{g}[v^1\, dx^2 \wedge dx^3 \wedge \ldots \wedge dx^n \\ &\quad - v^2\, dx^1 \wedge dx^3 \wedge \ldots \wedge dx^n + \cdots]\} \\ &= [\partial(v^1\sqrt{g})/\partial x^1 + \partial(v^2\sqrt{g})/\partial x^2 + \cdots]\, dx^1 \wedge \ldots \wedge dx^n\end{aligned}$$

i.e.,

$$\text{div } \mathbf{v} = (1/\sqrt{g})\partial/\partial x^i(\sqrt{g}v^i) \qquad (©.32)$$

If, furthermore, the volume form comes from a Riemannian metric we can define the **Laplacian** of a **function** $f$ by

$$\nabla^2 f := \Delta f := \text{div } \nabla f = (1/\sqrt{g})\partial/\partial x^i(\sqrt{g}g^{ij}\partial f/\partial x^j)) \qquad (©.33)$$

We now wish to consider the notion of vector or $\times$ product in more detail. We have seen in Section ©.h that in $\mathbb{R}^3$ in *any* coordinates the 2-form

$$\begin{aligned}\alpha^1 \wedge \gamma^1 &= (a_1\, dx^1 + a_2\, dx^2 + a_3\, dx^3) \wedge (c_1\, dx^1 + c_2\, dx^2 + c_3\, dx^3) \\ &= (a_2c_3 - a_3c_2)dx^2 \wedge dx^3 + (a_3c_1 - a_1c_3)dx^3 \wedge dx^1 + (a_1c_2 - a_2c_1)dx^2 \wedge dx^3\end{aligned}$$

corresponds to the cross product $\mathbf{a} \times \mathbf{c}$ in *cartesian* coordinates, and this 2-form version is ideal when considering surface integrals in any coordinates.

We shall now give a 1-form version of $\mathbf{a} \times \mathbf{b}$, we write $(\mathbf{a} \times \mathbf{b})_*$, which will be very useful in line integrals and in our later sections considering electromagnetism and elasticity.

In $\mathbb{R}^3$ with a $\text{vol}^3$, and in any coordinates, we define

$(\mathbf{a} \times \mathbf{b})_*$ is the unique **1-form** defined by $(\mathbf{a} \times \mathbf{b})_*(\mathbf{c}) := \text{vol}^3(\mathbf{a}, \mathbf{b}, \mathbf{c})$

for every vector $\mathbf{c}$. If we have a metric, then $(\mathbf{a} \times \mathbf{b})_*\,(\mathbf{c}) = (\mathbf{a} \times \mathbf{b}) \cdot (\mathbf{c}) = \text{vol}^3(\mathbf{a}, \mathbf{b}, \mathbf{c})$ gives the usual definition of the **vector a** $\times$ **b**, but clearly the 1-form version is more basic since it does not require a metric. (Question: how would you define a $\times$-product of $n-1$ vectors in an $\mathbb{R}^n$ with a $\text{vol}^n$?)

Note

$$\mathrm{vol}^3(\mathbf{a},\mathbf{b},\mathbf{c}) = -\mathrm{vol}^3(\mathbf{b},\mathbf{a},\mathbf{c}) = (-i_{\mathbf{b}}\,\mathrm{vol}^3)(\mathbf{a},\mathbf{c}) = -\beta^2(\mathbf{a},\mathbf{c}) = (-i_{\mathbf{a}}\beta^2)(\mathbf{c})$$

where $\beta^2 = i_{\mathbf{b}}$ vol is the 2-form version of **b**. Thus in *any* coordinates with a $\mathrm{vol}^3$

$$(\mathbf{a}\times\mathbf{b})_* = -i_{\mathbf{a}}\beta^2 = -i_{\mathbf{a}}[i_{\mathbf{b}}\,\mathrm{vol}^3] \qquad (\mathbb{O}.34)$$

which, from ($\mathbb{O}$.31)

$$\begin{aligned}(\mathbf{a}\times\mathbf{b})_* &= -i(a^1\partial_1 + a^2\partial_2 + a^3\partial_3)\sqrt{g}[b^1\,dx^2\wedge dx^3 + b^2\,dx^3\wedge dx^1 + b^3\,dx^1\wedge dx^2]\\ &= \sqrt{g}[(a^2b^3 - a^3b^2)dx^1 + (a^3b^1 - a^1b^3)dx^2 + (a^1b^2 - a^2b^1)dx^3]\end{aligned}$$

*Now* we can write the Lorentz force law of Section $\mathbb{O}$.1

$$f^1 = q(\mathcal{E}^1 - i_{\mathbf{v}}\mathcal{B}^2)$$

Finally, an important restatement of the cross product in $\mathbb{R}^3$. We are going to follow **Elie Cartan** and use 2-forms whose values on pairs of vectors are not numbers but rather vectors or covectors. Let $\chi_*^{(2)} = \chi_*$ be the covector-valued 2-form with value the covector $\chi_*$ $(\mathbf{a},\mathbf{b})$: $= (\mathbf{a}\times\mathbf{b})_*$ . The $j^{\text{th}}$ component of this covector is

$$\chi_*(\mathbf{a},\mathbf{b})_j = (\mathbf{a}\times\mathbf{b})_j = (\mathbf{a}\times\mathbf{b})_*(\partial_j) = \mathrm{vol}^3(\partial_j,\mathbf{a},\mathbf{b}) = [i(\partial_j)\mathrm{vol}^3](\mathbf{a},\mathbf{b})$$

Thus

$$\chi_* = dx^j \otimes \chi_j = dx^j \otimes [i(\partial_j)\mathrm{vol}^3] \qquad (\mathbb{O}.35)_*$$

Note the $\otimes$ *not* $\wedge$. By definition, the value of the 2-form $\chi_*$ on the pair of vectors **a, b** is not a number, but rather the 1-form

$$\chi_*(\mathbf{a},\mathbf{b}) = [\mathrm{vol}^3(\partial_j,\mathbf{a},\mathbf{b})]dx^j$$

With a Riemannian metric, the **contravariant** version is the vector valued 2-form

$$\chi^* = \partial_i \otimes g^{ij} i(\partial_j)\mathrm{vol}^3 \qquad (\mathbb{O}.35)^*$$

This is the 2-form that, when applied to the pair of vectors, yields $\mathbf{a}\times\mathbf{b}$. In cartesian coordinates we can write it symbolically as the column of 2-forms

$$[dy\wedge dz \;\; dz\wedge dx \;\; dx\wedge dy]^T$$

whose value on a pair of vectors $(\mathbf{a},\mathbf{b})$ is the column of components of $\mathbf{a}\times\mathbf{b}$.

## $\mathbb{O}$.o. Magnetic Field for Current in a Straight Wire

This simple example illustrates much of what we have done. Consider a steady current **j** in a thin straight wire of infinite length.

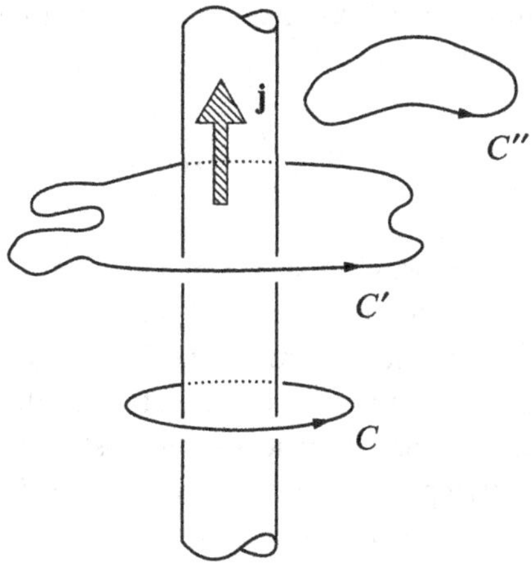


Since the current is steady we have Ampère's law $\oint_{C=\partial V} \mathcal{H}^1 = 4\pi \iint_V j^2$. Looking at three surfaces bounded respectively by $C$, $C'$, and $C''$ and the flux of current through them, we have

$$\oint_C \mathcal{H}^1 = 4\pi j = \oint_{C'} \mathcal{H}^1$$

while $\oint_{C''} \mathcal{H}^1 = 0$. Introducing cylindrical coordinates, we can guess immediately that $\mathcal{H}^1 = 2j\, d\theta$ *in the region outside the wire*, for it has the correct integrals. We *require*, however, that div $\mathbf{B} = 0 = d\mathcal{B}^2$. Now $\mathcal{B}^2 = i_{\mathbf{H}}\, \text{vol}^3$ where $\mathbf{H}$ is the contravariant version of the 1-form $\mathcal{H}$. The metric for cylindrical coordinates is diag$(1, r^2, 1)$ and $H_\theta = 2j$ is the only nonzero component of our guess $\mathcal{H}^1$, hence $H^\theta = g^{\theta\theta} H_\theta$ (no sum) $= (1/r^2)2j$. Then $\mathcal{B}^2 = i_{\mathbf{H}}\text{vol}^3$ becomes

$$\mathcal{B}^2 = (2j/r^2) i(\partial_\theta) r\, dr \wedge d\theta \wedge dz = -(2j/r) dr \wedge dz = d[-2j(\ln r) dz]$$

Clearly $d\mathcal{B} = 0$, as required and, in fact, $[-2j(\ln r)dz]$ is a "magnetic **potential**" 1-form $\alpha^1$ outside the wire, $\mathcal{B}^2 = d\alpha^1$. Another choice is $\alpha^1 = 2jz/r\, dr$.

## Elasticity and Stresses

### $\mathfrak{O}$.p. Cauchy Stress, Floating Bodies, Twisted Cylinders, and Strain Energy

In learning the sciences examples are of more use than precepts.

Isaac Newton, *Arithmetica Universalis* (1707)

We look at our cylinder $B$ and its twisted version $F(B)$ in Section $\mathfrak{O}$.a, but *first* we shall use *cartesian* coordinates $x^i$. Consider any small surface $V$ in $F(B)$ passing through a point $p$ and let $\mathbf{n}$ be a normal to $V$ at $p$. Then because of the twisting, the material on the side of $V$ towards which $\mathbf{n}$ is pointing, exerts a force $\mathbf{f}$ on the material on the other side of $V$. Cauchy's "first theorem" states that this force is reversed if we replace $\mathbf{n}$ by

–**n**, and further this (contravariant) force is given by integrating a vector valued 2-form $\mathfrak{t}$ over $V$ (*not* Cauchy's language)

$$\mathbf{f} \text{ on } V = \partial_a \left[ \int_V t^{ab} i(\partial_b) \text{vol}^3 \right]$$

where **t** is the "Cauchy stress tensor." A sketch of a proof of Cauchy's theorem will be given in Section ©.q. Cauchy's "second theorem" says $t^{ab} = t^{ba}$ and a proof sketch is given in Section ©.r. (The fact that the stress force is reversed if **n** is replaced by $-\mathbf{n}$ informs us (see Section 2.8f) that the stress form is technically a "pseudo-form." )

As a warm-up check of our machinery, let us look first at an example of the simplest type of stress from elementary physics. In the case of a **nonviscous fluid**, given a very small parallelogram spanned by **v** and **w** and with normal $\mathbf{n} = \mathbf{v} \times \mathbf{w}$, the fluid on the side to which **n** is pointing exerts a force on the other side approximated by $-p\mathbf{v} \times \mathbf{w}$, where $p$ is the hydrostatic **pressure**. From (©.35) the stress vector valued 2-form is given by $\mathfrak{t} = -\partial_i \otimes p g^{ij} i(\partial_j) \text{vol}^3$. In a pool with cartesian coordinates $x, y, z$, with the origin at the surface and $z$ pointing down, look at a floating body $B$, with portion $B'$ below the water surface, with surface normal pointing out of $B$. While Archimedes knew the result, *we* need to practice with our new tools.

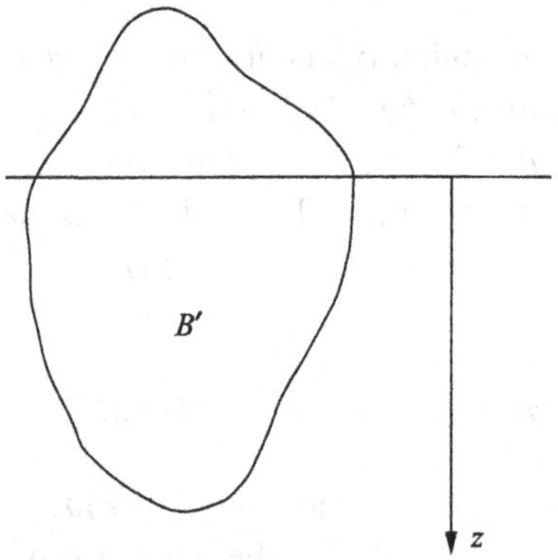


Then the total stress force exerted on $\partial B$ from water of *constant* density $\rho$ outside $B$ is, with $g^{ij} = \delta^{ij}$ and $p = \rho g z$

$$\begin{aligned} \mathbf{f} = \partial_i \int_{\partial B'} t^{ij} i(\partial_j) \text{vol}^3 &= -\partial_i \int_{\partial B'} p \delta^{ij}\, i(\partial_j) \text{vol}^3 \\ &= -\partial_x \int_{\partial B'} \rho g z\, dy \wedge dz - \partial_y \int_{\partial B'} \rho g z\, dz \wedge dx - \partial_z \int_{\partial B'} \rho g z\, dx \wedge dy \end{aligned}$$

where we have included the part of $\partial B'$ at water level $z = 0$, even though there is no water there, since $\rho g z = 0$ there and we get a 0 contribution from it. We shall evaluate the surface integrals by applying Stokes' theorem (©.26) to $B'$. The three 2-forms $\rho g z\, dy \wedge dz$, etc, apply only to the outside of $B'$ since there is no water inside $B'$. To apply Stokes' theorem to $B'$, we must *extend* these 2-forms from the boundary of $B'$ mathematically to the inside of $B'$, *in any smooth way that we wish*, and we choose the same forms as are given outside $B'$, with $\rho = \rho_{\text{water}}$ again! Then by Stokes

$$\mathbf{f} = -\boldsymbol{\partial}_x \int_{B'} d[\rho g z\, dy \wedge dz] \quad -\boldsymbol{\partial}_y \int_{B'} d[\rho g z\, dz \wedge dx] \quad -\boldsymbol{\partial}_z \int_{B'} d[\rho g z\, dx \wedge dy]$$
$$= -\boldsymbol{\partial}_z \int_{B'} \rho g\, dx \wedge dy \wedge dz = -\boldsymbol{\partial}_z W'$$

where $W'$ is the weight of the water displaced by $B'$. Equilibrium demands this must equal the weight of the whole body $B$. Thus a floating body displaces its own weight in water. EUREKA!

**Back to our twisted cylinder:** Introduce cylindrical coordinates $(X^A) = (R, \Theta, Z)$ for the untwisted cylinder $B$. Next, introduce an *identical* set of coordinates $(x^a) = (r, \theta, z)$ and use the capitalized coordinates for a point in the untwisted body and $r, \theta, z$ for the coordinates of the image point under the twist $F$. Thus $F$ is described by $r = R, \theta = \Theta + kZ$, and $z = Z$, where $k$ is a constant. We need to determine the **Cauchy** vector valued stress 2-form $\mathfrak{t} = \boldsymbol{\partial}_a \otimes \mathfrak{t}_a = \boldsymbol{\partial}_a \otimes t^{ab} i(\boldsymbol{\partial}_b)\mathrm{vol}^3$ on $F(B)$ in terms of the twisting forces and the material from which $B$ is made. We shall do this by first pulling this 2-form back to the untwisted body $B$ by the following procedure; we **pull** back the 2-forms $\mathfrak{t}^a$ by $F^*$ and we **push** the vectors $\boldsymbol{\partial}_a$ back to $B$ by the inverse $(F^{-1})_*$, which exists since $F$ is a 1:1 deformation. The resulting vector valued 2-form on $B$ is

$$\mathfrak{S} = [(F^{-1})_*(\boldsymbol{\partial}_a)] \otimes F^*\mathfrak{t}^a = (F^{-1})_*(\boldsymbol{\partial}_a) \otimes F^*[t^{ab} i(\boldsymbol{\partial}_b)\mathrm{vol}^3]$$

which is of the form

$$\mathfrak{S} = \boldsymbol{\partial}_A \otimes \mathfrak{S}^A = \boldsymbol{\partial}_A \otimes \mathrm{S}^{AB} i(\boldsymbol{\partial}_B)\mathrm{vol}^3 \tag{0.36}$$

called the **second Piola–Kirchhoff** vector valued stress 2-form. We shall relate *this* form to the twist $F$ by a generalization of Hooke's law.

We need to know how this twist $F$ has stretched lengths and changed angles in the body, and this is described as follows. The euclidean metric is $dS^2 = (dR^2 + R^2\, d\Theta^2 + dZ^2) = ds^2 = (dr^2 + r^2\, d\theta^2 + dz^2)$. The pull-back (last paragraph of Section 0.j) of $ds^2$ under the twist $F$ is given by the chain rule

$$F^*ds^2 = F^*(dr^2 + r^2\, d\theta^2 + dz^2) = dR^2 + R^2[(\partial\theta/\partial\Theta)d\Theta + (\partial\theta/\partial Z)dZ]^2 + dZ^2$$
$$= dR^2 + R^2[d\Theta + k\, dZ]^2 + dZ^2$$
$$= dR^2 + R^2[d\Theta^2 + 2k\, d\Theta dZ + k^2\, dZ^2] + dZ^2$$

Recall what this is saying. At a point $R, \Theta, Z$ of the untwisted body, given two vectors $\mathbf{A}, \mathbf{B}$, we have not only the scalar product $\langle \mathbf{A}, \mathbf{B}\rangle = dS^2(\mathbf{A}, \mathbf{B})$ but also the scalar product of the images after the twist, i.e., from (0.23), $ds^2(F_*\mathbf{A}, F_*\mathbf{B}) =: (F^*ds^2)(\mathbf{A}, \mathbf{B})$. Then *one* measure of how much the twist $F$ is distorting distances and angles is defined by the **Lagrange deformation tensor**

$$E := \tfrac{1}{2}[(F^*ds^2) - dS^2] \tag{0.37}$$

The quadratic form (covariant second rank tensor) $E$ is determined by its square matrix.

How do the stresses depend on the deformations? In our twisting case we have $E = kR^2\, d\Theta\, dZ + \frac{1}{2}k^2R^2\, dZ^2$. We will work only to the *first approximation for small*

$k$, i.e., we shall put $k^2 = 0$, so $E = kR^2\, d\Theta\, dZ = \frac{1}{2}kR^2(d\Theta\, dZ + dZ\, d\Theta)$. We write the components as a symmetric matrix

$$(E_{IJ}) = \begin{bmatrix} 0 & 0 & 0 \\ 0 & 0 & kR^2/2 \\ 0 & kR^2/2 & 0 \end{bmatrix}$$

The mixed version, using $E^A{}_B = G^{AI}E_{IB}$ and $(G^{KL}) = \mathrm{diag}(1, 1/R^2, 1)$, is the (nonsymmetric)

$$(E^A{}_B) = \begin{bmatrix} 0 & 0 & 0 \\ 0 & 0 & k/2 \\ 0 & kR^2/2 & 0 \end{bmatrix}$$

and thus tr $E = E^A{}_A = 0$ "mod $k^2$," i.e., putting $k^2 = 0$. Finally, putting $E^{AB} = E^A{}_I G^{IB}$

$$(E^{AB}) = \begin{bmatrix} 0 & 0 & 0 \\ 0 & 0 & k/2 \\ 0 & k/2 & 0 \end{bmatrix}$$

**Linear** elasticity assumes a linear, vastly generalized "Hooke's law" relating the stress $S$ to the deformation $E$. Assuming the body is **isotropic** (i.e., the material has no special internal directional structure such as grains in wood), it can then be shown (e.g., equation (D.9)), that there are then only two "elastic constants" $\mu$ and $\lambda$ relating $S$ to $E$

$$S^{AB} = 2\mu E^{AB} + \lambda(\mathrm{tr}\, E)G^{AB} \tag{$\mathfrak{o}$.38}$$

and so

$$(S^{AB}) = \begin{bmatrix} 0 & 0 & 0 \\ 0 & 0 & \mu k \\ 0 & \mu k & 0 \end{bmatrix}$$

This gives rise to the second Piola–Kirchhoff vector valued stress 2-form on the *undeformed* body

$$\begin{aligned} \mathfrak{S} := \partial_I \otimes S^{IJ} i(\partial_J)\mathrm{VOL}^3 &= \partial_I \otimes S^{IJ} i(\partial_J) R\, dR \wedge d\Theta \wedge dZ \\ &= [\partial_\Theta \otimes S^{\theta Z} i(\partial_Z) + \partial_Z \otimes S^{Z\theta} i(\partial_\theta)] R\, dR \wedge d\Theta \wedge dZ \\ \mathfrak{S} &= \mu k R[\partial_\Theta \otimes dR \wedge d\Theta + \partial_Z \otimes dZ \wedge dR] \end{aligned} \tag{$\mathfrak{o}$.38$'$}$$

Finally, the **Cauchy stress vector valued 2-form** $\mathfrak{t}$ on the "current" deformed body from ($\mathfrak{o}$.36), is $\mathfrak{t} = F_* \partial_A \otimes (F^{-1})^* S^A$. Using $F^{-1}$ defined by $R = r$, $\Theta = \theta - kz$, $Z = z$, we get

$$\begin{aligned} \mathfrak{t} &= \mu k r[\partial_\theta \otimes (F^{-1})^*(dR \wedge d\Theta) + \partial_z \otimes (F^{-1})^*(dZ \wedge dR)] \\ &= \mu k r[\partial_\theta \otimes dr \wedge (d\theta - k\, dz) + \partial_z \otimes dz \wedge dr] \quad \text{and discarding } k^2 \\ \mathfrak{t} &= \mu k r[\partial_\theta \otimes dr \wedge d\theta + \partial_z \otimes dz \wedge dr] \end{aligned} \tag{$\mathfrak{o}$.39}$$

To get correct "dimensions" for force we use the "physical" components of force, i.e., we normalize the (already orthogonal) basis vectors. Since $g_{rr} = 1 = g_{zz}$, $\partial_r$ and

$\boldsymbol{\partial}_z$ are unit vectors, call them $\mathbf{e}_r$ and $\mathbf{e}_z$. But $g_{\theta\theta} = r^2$, and so $\boldsymbol{\partial}_\theta$, by (©.6), has length $r$, and so we put $\mathbf{e}_\theta = r^{-1}\boldsymbol{\partial}_\theta$. We make no changes to the form parts $dr$, $d\theta$, and $dz$

$$\mathfrak{t} = \mu k r^2 \mathbf{e}_\theta \otimes dr \wedge d\theta + \mu k r \mathbf{e}_z \otimes dz \wedge dr \tag{©.40}$$

We shall now see the consequences of this Cauchy stress. Look first at the lateral surface $r = a$. Then $dr = 0$ there and so $\mathfrak{t} = 0$ on this surface. *This means that no external "traction" on this part of the boundary is needed for this twisting.*

Now look at the end boundary at $z = L$. From (©.40) we have stress from outside

$$\mu k r^2 \mathbf{e}_\theta \otimes dr \wedge d\theta$$

acting in the $\mathbf{e}_\theta$ direction. This has to be supplied by **external tractions** since there is no part of the body past its ends. What is the **moment** of the traction? We have a disk, radius $a$, a force of magnitude $\mu k r^2\, dr\, d\theta$ acting in the $\mathbf{e}_\theta$ direction on an infinitesimal "rectangle" of "sides" $dr$ and $d\theta$. The **moment** about the $z$ axis is $r(\mu k r^2) dr\, d\theta$, and so the total moment is $\mu k \iint r^3 dr\, d\theta = \mu k (a^4/4) 2\pi = \pi \mu k a^4/2$. If the total twist at $z = L$ is an angle of twist $\alpha = kL$, then the total moment required is $\pi \mu a^4 \alpha / 2L$. An opposite moment is required at $z = 0$. An experiment could yield the value of $\mu$.

In the case of the floating body, treated near the beginning of our Section ©.p, our argument *really* showed the following. Take any blob of fluid $B''$ surrounded by fluid at rest under the surface $z = 0$. Then the hydrostatic stress (pressure) on $\partial B''$ due to the water surrounding $B''$ produced a "body force" that supported the weight of the water in $B''$. We now show that in the case of our twisted cylinder, to order $k$,

the Cauchy stresses produce no internal **body** forces inside the cylinder.

Look at an internal portion $B$ of the cylinder, with boundary $\partial B$. The Cauchy stress acting on $B$ from outside $B$ derives from the vector valued 2-form in (©.40) at points of $\partial B$. For *total* stress force on $\partial B$, we cannot just integrate this because it makes no sense to add vectors like $\mathbf{e}_\theta$ at different points. There is no problem with the $\mathbf{e}_z$ components because $\mathbf{e}_z$ is a constant vector field in $\mathbb{R}^3$. So let us express the unit vector $\mathbf{e}_\theta$ in terms of the constant basis $\mathbf{e}_x$ and $\mathbf{e}_y$. Again we leave the cylindrical coordinate 2-forms alone. Now

$$\boldsymbol{\partial}/\boldsymbol{\partial}\theta = (\partial x/\partial\theta)\boldsymbol{\partial}/\boldsymbol{\partial}x + (\partial y/\partial\theta)\boldsymbol{\partial}/\boldsymbol{\partial}y = (-r\sin\theta)\mathbf{e}_x + (r\cos\theta)\mathbf{e}_y$$

and $\mathbf{e}_\theta = r^{-1}(\boldsymbol{\partial}/\boldsymbol{\partial}\theta) = -\mathbf{e}_x \sin\theta + \mathbf{e}_y \cos\theta$, and so (©.40) becomes

$$\mathfrak{t} = \mu k r^2(-\mathbf{e}_x \sin\theta + \mathbf{e}_y \cos\theta) \otimes dr \wedge d\theta + \mu k r \mathbf{e}_z \otimes dz \wedge dr$$

Then, with constant basis, $\iint_{\partial B} \mathbf{e}_x \mu k r^2 \sin\theta\, dr \wedge d\theta = \mathbf{e}_x \iint_{\partial B} \mu k r^2 \sin\theta\, dr \wedge d\theta$, etc., and so

$$\iint_{\partial B} \mathfrak{t} = -\mathbf{e}_x \iint_{\partial B} \mu k r^2 \sin\theta\, dr \wedge d\theta + \mathbf{e}_y \iint_{\partial B} \mu k r^2 \cos\theta\, dr \wedge d\theta$$
$$+ \mathbf{e}_z \iint_{\partial B} \mu k r\, dz \wedge dr$$

But, even though $d\theta$ is not defined at the axis $r = 0$, each integral $= 0$; e.g., $\iint_{\partial B} (r\sin\theta) dr \wedge r d\theta = \iint_{\partial B} y\, dx \wedge dy = \iiint_B dy \wedge dx \wedge dy = 0$.

It is a fact, alas, that this simple approach will not work to higher order, keeping terms of order $k^2$. One cannot realize such a simple twist; other deformations are required (see [Mu]).

I would like to emphasize one point brought out in the calculation above. When *integrating* vector valued exterior forms, such as Cauchy's $\boldsymbol{\partial}_i \otimes t^{ij} i(\boldsymbol{\partial}_j)\mathrm{vol}^3$, we were forced to make a change to a constant basis for the vector part, $\boldsymbol{\partial}_i = \mathbf{e}_a A^a{}_i$, but kept the cylindrical exterior forms, yielding

$$\iint_{\partial B} \mathbf{e}_a \otimes A^a_i t^{ij} i(\boldsymbol{\partial}_j)\mathrm{vol}^3 = \mathbf{e}_a \iint_{\partial B} A^a{}_i t^{ij} i(\boldsymbol{\partial}_j)\mathrm{vol}^3 = \mathbf{e}_a \iiint_B d[A^a{}_i t^{ij} i(\boldsymbol{\partial}_j)\mathrm{vol}^3]$$

and our exterior differential completely avoids Christoffel symbols and tensor divergence of $(t^{ij})$ in curvilinear coordinates, that appear in tensor treatments.

Finally, let us compute the work done by the traction acting on the face $Z = L$, moving each point $(R, \Theta)$ to the point $(R, \Theta + \alpha)$. Let $0 \le \beta \le \alpha$. The traction force on the small "rectangle" of sides $dR$, $d\Theta$ at $(R, \Theta + \beta)$ has, from ($\mathbb{O}$.38′), covariant component approximately $f_\Theta\, dR\, d\Theta = g_{\Theta\Theta}\mu k_\beta R\, dR\, d\Theta = \mu k_\beta R^3\, dR\, d\Theta$, where $k_\beta = \beta/L$. The work done in moving this rectangle from $\beta = 0$ to $\beta = \alpha$ is approximately $(dR\, d\Theta)\int_0^\alpha (\mu R^3 \beta/L)\, d\beta = (dR\, d\Theta)\mu R^3\alpha^2/2L$. Thus the total work done in the twist of the face is $W = (\mu\alpha^2/2L)\iint R^3\, dR\, d\Theta = \pi\mu\, a^4\alpha^2/4L$. In most common materials (**hyperelastic**), in particular for our isotropic body, this work yields a **strain energy** of the same amount $W$, that is stored in the twisted body. Furthermore, for hyperelastic bodies, this can be computed from an integral over the undeformed body (see Sections A.d and D.a),

$$W = \tfrac{1}{2}\iiint S^{AB} E_{AB} \mathrm{VOL}^3$$

and the reader can verify this in our example using $E$ and $S$ given before and after ($\mathbb{O}$.38).

This is one reason for our choice, at the beginning of this section, of considering stress force as being contravariant, rather than covariant. Note that a **metric** $ds^2 = g_{ij}\, dx^i\, dx^j$ can be thought of as the **covector valued 1-form** $dx^i \otimes g_{ij}\, dx^j$ whose value on any vector $\mathbf{v}$ is the covariant version of $\mathbf{v}$, $dx^i \otimes g_{ij}\, dx^j(\mathbf{v}) = dx^i g_{ij} v^j = v_i\, dx^i$. Likewise, the Lagrange deformation tensor can be thought of as a covector valued 1-form

$$\mathcal{E} = dX^I \otimes E_{IJ}\, dX^J = dX^I \otimes \mathcal{E}^{(1)}{}_I$$

The stress tensor is a vector valued 2-form $\mathcal{S} = \boldsymbol{\partial}_A \otimes S^{AB} i(\boldsymbol{\partial}_B)\mathrm{VOL}^3 = \boldsymbol{\partial}_A \otimes \mathcal{S}^{(2)A}$. It is natural then to construct a **scalar** valued 3-form by introducing a **new product** $\mathcal{S}(\wedge)\mathcal{E}$ by taking the wedge product of the forms in both and evaluating the 1-form $dX^I$ of $\mathcal{E}$ on the vector $\boldsymbol{\partial}_A$ of $\mathcal{S}$

$$\mathcal{S}(\wedge)\mathcal{E} := dX^I(\boldsymbol{\partial}_A)[S^{(2)A} \wedge \mathcal{E}^{(1)}{}_I] = \mathcal{S}^{(2)A} \wedge \mathcal{E}^{(1)}{}_A$$

which is easily seen, since the two forms are of complementary dimension, to be the integrand of the strain energy $W$

$$\mathcal{S}(\wedge)\mathcal{E} = [S^{AB} i(\boldsymbol{\partial}_B)\mathrm{VOL}^3] \wedge E_{AJ}\, dX^J = S^{AB} E_{AB} \mathrm{VOL}^3$$

$$W = \tfrac{1}{2}\iiint \mathcal{S}(\wedge)\,\mathcal{E}$$

While work in particle mechanics pairs a force covector $(f_i)$ with a contravariant tangent vector $(dx^i/dt)$ to a curve, work done by traction in elasticity pairs the contravariant stress force 2-form $\mathcal{S}$ with the covector valued deformation 1-form $\mathfrak{S}$, to yield a scalar valued 3-form. (Warning: the notation -($\wedge$)- does not appear in the literature.)

## $\mathfrak{O}$.q. Sketch of Cauchy's "First Theorem"

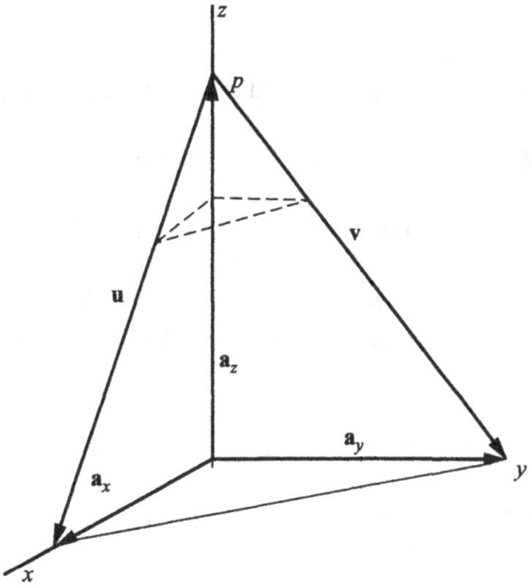


Consider a plane through a point $p$ on the $z$ axis of a cartesian coordinate system. This plane generically cuts the $x$ and $y$ axes at two points, yielding two vectors $\mathbf{u}$ and $\mathbf{v}$ that span the "roof" of a solid tetrahedron $T$, as in the figure above. The coordinate vectors $\mathbf{a}_x, \mathbf{a}_y, \mathbf{a}_z$ are *not* necessarily of the same length. The material outside $T$ exerts a stress force, call it $\frac{1}{2}\mathfrak{t}(\mathbf{u}, \mathbf{v})$ across the roof ($\frac{1}{2}$ because the roof is not a parallelogram). $(\mathbf{u}, \mathbf{v})$ tells us not only the roof, but also $\mathbf{u}, \mathbf{v}$, in that order is describing the normal pointing out of $T$. Likewise $\frac{1}{2}\mathfrak{t}(\mathbf{v}, \mathbf{u})$ describes a force that the material in $T$ exerts on material outside $T$. $\mathfrak{t}(\mathbf{v}, \mathbf{u}) = -\mathfrak{t}(\mathbf{u}, \mathbf{v})$ can be seen by considering the equilibrium of a small thin disk with faces parallel to the plane spanned by $\mathbf{u}$ and $\mathbf{v}$. This is the first part of Cauchy's first theorem.

Stress forces act also on the coordinate faces. We now let the tetrahedron $T$ shrink to the point $p$ by moving the $x, y$ plane up to the point $p$, the dashed triangle showing an intermediate position for the bottom face. At each stage the proportions of $T$ are preserved. As the vertical edge $\|\mathbf{a}_z\|$ shrinks to 0, the stress **forces** on the faces vanish as their areas, i.e., as $\|\mathbf{a}_z\|^2$ while the body forces, for example, gravity, if present, vanish as the volume, i.e., as $\|\mathbf{a}_z\|^3$. We will *neglect the body forces* for vanishingly small $T$.

For our small $T$ to be in equilibrium we must have, neglecting body forces

$$\begin{aligned} \mathfrak{t}(\mathbf{u}, \mathbf{v}) + \mathfrak{t}(\mathbf{a}_z, \mathbf{a}_y + \mathfrak{t}(\mathbf{a}_x, \mathbf{a}_z) + \mathfrak{t}(\mathbf{a}_y, \mathbf{a}_x) &\approx 0 \\ \mathfrak{t}(\mathbf{u}, \mathbf{v}) &\approx -\mathfrak{t}(\mathbf{a}_z, \mathbf{a}_y) - \mathfrak{t}(\mathbf{a}_x, \mathbf{a}_z) - \mathfrak{t}(\mathbf{a}_y, \mathbf{a}_x) \\ \mathfrak{t}(\mathbf{u}, \mathbf{v}) &\approx \mathfrak{t}(\mathbf{a}_y, \mathbf{a}_z) + \mathfrak{t}(\mathbf{a}_z, \mathbf{a}_x) + \mathfrak{t}(\mathbf{a}_x, \mathbf{a}_y) \end{aligned} \tag{$\mathfrak{O}$.41}$$

Look at the first term $\mathfrak{t}(\mathbf{a}_y, \mathbf{a}_z)$. The normal to the pair $\mathbf{a}_y, \mathbf{a}_z$ is in the positive $x$ direction and so the area form for the $y, z$ face is $dy \wedge dz$. Let $\langle \mathfrak{t}_{yz} \rangle$ be the **area vector average** of the vector $\mathfrak{t}(\mathbf{a}_y, \mathbf{a}_z)$, so

$$\mathfrak{t}(\mathbf{a}_y, \mathbf{a}_z) = \langle \mathfrak{t}_{yz} \rangle dy \wedge dz(\mathbf{a}_y, \mathbf{a}_z)$$

Now note that for projected areas, $dy \wedge dz(\mathbf{u}, \mathbf{v}) = dy \wedge dz(\mathbf{a}_x - \mathbf{a}_z, -\mathbf{a}_z + \mathbf{a}_y) = dy \wedge dz(-\mathbf{a}_z, -\mathbf{a}_z) + dy \wedge dz(-\mathbf{a}_z, \mathbf{a}_y) = dy \wedge dz(-\mathbf{a}_z, \mathbf{a}_y) = -dy \wedge dz(\mathbf{a}_z, \mathbf{a}_y) = dy \wedge dz(\mathbf{a}_y, \mathbf{a}_z)$. Thus

$$dy \wedge dz\,(\mathbf{a}_y, \mathbf{a}_z) = dy \wedge dz(\mathbf{u}, \mathbf{v}) \quad \text{and so} \quad \mathfrak{t}(\mathbf{a}_y, \mathbf{a}_z) = \langle \mathfrak{t}_{yz} \rangle dy \wedge dz(\mathbf{u}, \mathbf{v})$$

and similarly for the other faces in (0.41). We then have

$$\mathfrak{t}(\mathbf{u}, \mathbf{v}) \approx \langle \mathfrak{t}_{yz} \rangle dy \wedge dz(\mathbf{u}, \mathbf{v}) + \langle \mathfrak{t}_{zx} \rangle dz \wedge dx(\mathbf{u}, \mathbf{v}) + \langle \mathfrak{t}_{xy} \rangle dx \wedge dy(\mathbf{u}, \mathbf{v}) \tag{0.42}$$

Now as $T$ shrinks to the point $p$ the average $\langle \mathfrak{t}_{yz} \rangle$ tends to a **vector** $\mathfrak{t}^x(p) = \mathfrak{t}^1(p)$ at $p$, etc. We can then approximate the stress in (0.42), for a very small parallelogram at $p$ spanned by $\mathbf{u}$ and $\mathbf{v}$

$$\mathfrak{t}(\mathbf{u}, \mathbf{v}) \approx [\mathfrak{t}^x(p) \otimes dy \wedge dz + \mathfrak{t}^y(p) \otimes dz \wedge dx + \mathfrak{t}^z(p) \otimes dx \wedge dy](\mathbf{u}, \mathbf{v})$$

which suggests Cauchy's theorem, that for any surface $V^2$ with normal direction prescribed, the stress across $V$ is given by a vector valued integral of the form

$$\int_V \mathfrak{t}^x(x, y, z) \otimes dy \wedge dz + \mathfrak{t}^y(x, y, z) \otimes dz \wedge dx + \mathfrak{t}^z(x, y, z) \otimes dx \wedge dy$$

with Cauchy vector valued stress 2-form

$$\mathfrak{t} = \partial_i \otimes t^{ij} i(\partial_j)\mathrm{vol}^3 \tag{0.42$_{\mathrm{Cauchy}}$}$$

but this is not the way it is written in engineering texts. Consider first just the surface integral of a 2-form $\beta^2 = i(\mathbf{b})\mathrm{vol}^3$ over a surface $V^2 \subset \mathbb{R}^3$ (using any coordinates $x^i$), with unit normal vector field $\mathbf{n}$ and covector version the 1-form $n_* = n_i\, dx^i$. Then, when applied to two vectors $\mathbf{v}$ and $\mathbf{w}$ tangent to $V$, "$dA$" $(\mathbf{v}, \mathbf{w}) := \mathrm{vol}(\mathbf{n}, \mathbf{v}, \mathbf{w}) = [i(\mathbf{n})\mathrm{vol}]$ $(\mathbf{v}, \mathbf{w})$ is the area spanned by $\mathbf{v}$ and $\mathbf{w}$. Then we can write, with $\mathbf{b}_{\tan}$ the tangential part of $\mathbf{b}$

$$\int_V \beta = \int_V i(\mathbf{b})\mathrm{vol}^3 = \int_V i[(\mathbf{b} \cdot \mathbf{n})\mathbf{n} + \mathbf{b}_{\tan}]\mathrm{vol} = \int_V (\mathbf{b} \cdot \mathbf{n})[i(\mathbf{n})\mathrm{vol}]$$

since $\mathrm{vol}(\mathbf{b}_{\tan}, \mathbf{v}, \mathbf{w}) = 0$ for three tangent vectors to $V^2$. Then

$$\int_V \beta = \int_V i(\mathbf{b})\mathrm{vol}^3 = \int_V (\mathbf{b} \cdot \mathbf{n})[i(\mathbf{n})\mathrm{vol}] = \int_V (\mathbf{b} \cdot \mathbf{n}) dA = \int_V b^j n_j\, dA$$

Likewise, on a surface $V^2$, engineering texts write the stress

$$t^{ij} n_j\, dA \quad \text{instead of} \quad t^{ij} i(\partial_j)\mathrm{vol}^3$$

## O.r. Sketch of Cauchy's "Second Theorem," Moments as Generators of Rotations

For Cauchy's second theorem, the symmetry of the stress tensor $t^{ij} = t^{ji}$, we shall consider only the simplest case of a deformed body, at **rest** and in **equilibrium** with its external tractions on its boundary, and with no external body forces (like gravity) considered. We employ *cartesian* coordinates throughout. Then, since $g_{ij} = \delta_{ij}$, tensorial indices may be raised and lowered indiscriminately and we can use the summation convention for *all* repeated indices.

Let $B$ be any sub-body in the *interior* of the body, with boundary $\partial B$. Then the (assumed vanishing) total stress force *covector* on $B$ yields

$$0 = \int_{\partial B} \{dx^c\} \otimes t_c{}^b i(\partial b)\mathrm{vol}^3 = \{dx^c\} \int_{\partial B} \mathfrak{t}_c = \{dx^c\} \int_B d\mathfrak{t}_c$$

where we use the braces { } just to remind us that the basis form to the left of $\otimes$ is a constant covector that plays no role in the integral. Since this holds for every interior $B$ we must have

$$d\mathfrak{t}_c = dt_c{}^b i(\partial_b)\mathrm{vol}^3 = 0 \quad \text{for each } c \tag{O.43}$$

which classically is written as a divergence $\partial t_c{}^b/\partial x^b = 0$.

For equilibrium we must also have that the total **moment** of stress forces on $\partial B$ must vanish. Now the moment about the origin, of a force $\mathbf{f}$ at position vector $\mathbf{r}$ is, in elementary point mechanics, $\mathbf{r} \times \mathbf{f}(\mathbf{r})$, but this expression makes no sense in more than 3 dimensions. But moments and torques surely make sense in any euclidean $\mathbb{R}^n$, indicating that we have not understood *mathematically* the notion of moment. Now in cartesian coordinates in $\mathbb{R}^n$, if we replace $\mathbf{r}$ and $\mathbf{f}(\mathbf{r})$ by 1-forms $\mathfrak{r} = x^a\, dx^a$ and $\mathfrak{f} = f_c(\mathbf{r})dx^c$, then $\mathfrak{r} \wedge \mathfrak{f}$ does make sense as a 2-form *at the origin* of $\mathbb{R}^n$ and its components, in the case of $\mathbb{R}^3$, coincide with those of $\mathbf{r} \times \mathbf{f}(\mathbf{r})$. There is a more important point. A moment about the origin 0 of $\mathbb{R}^n$ is *physically* a "generator" of a rotation about 0. Let us see why a 2-form at the origin of $\mathbb{R}^n$, with components forming a skew symmetric matrix, also is associated to a rotation there.

Let $g(t)$ be a 1-parameter group (i.e., $g(t)\, g(s) = g(t+s)$, and $g(0) = \mathrm{I}$) of **rotations** of $\mathbb{R}^n$ about the origin. Since each $g(t)$ is an "orthogonal" matrix, $g(t)\, g(t)^T = \mathrm{I}$, where $T$ is transpose. Differentiate with respect to $t$ (indicated by an overdot) and put $t = 0$. Then

$$0 = \dot{g}(0)g(0)^T + g(0)\dot{g}(0)^T = \dot{g}(0) + \dot{g}(0)^T$$

says that $A := \dot{g}(0)$ (the so-called "infinitesimal **generator**" of the 1-parameter group $g(t)$), is a skew symmetric $n \times n$ matrix, and so defines a 2-form $\mathfrak{A} = \Sigma_{j<k} A_{jk} dx^j \wedge dx^k$ at the origin. For example, a 1-parameter group of rotations about the $z$ axis of $\mathbb{R}^3$ is, with $\omega$ a constant,

$$g(t) = \begin{bmatrix} \cos(\omega t) & -\sin(\omega t) & 0 \\ \sin(\omega t) & \cos(\omega t) & 0 \\ 0 & 0 & 1 \end{bmatrix} \quad \text{and has generator} \quad A = \dot{g}(0) = \begin{bmatrix} 0 & -\omega & 0 \\ \omega & 0 & 0 \\ 0 & 0 & 0 \end{bmatrix}$$

with associated 2-form $\mathfrak{a} = -\omega\, dx \wedge dy$ at the origin. If $\mathbf{v}$ is a vector at the origin, then $A\mathbf{v}$ is the vector $(A\mathbf{v})_j = A_{jk}v^k = -v^k A_{kj}$, i.e., the covector version of $A\mathbf{v}$ is $-i(\mathbf{v})\mathfrak{a}$.

Conversely, if $A$ is a skew symmetric $n \times n$ matrix at the origin (a 2-form at the origin), then $A$ generates a 1-parameter group of rotations $g(t)$ by means of the exponential matrix

$$g(t) = e^{tA} = \exp tA := \Sigma_k t^k A^k / k!$$

(it is an orthogonal matrix since $g(t)^T = \exp tA^T = \exp(-tA) = g(-t) = g^{-1}(t)$). **A 2-form at the origin of $\mathbb{R}^n$ generates a 1 parameter group of rotations about the origin of $\mathbb{R}^n$**. (Linear algebra also shows that the generator of $e^{tA}$ is $[d/dt e^{tA}]_{t=0} = Ae^0 = A$.)

Thus to each moment of a force $\mathbf{f}$ about the origin of $\mathbb{R}^n$ we may attach the generator of its rotations, i.e., a 2-form at the origin, which is simply a skew symmetric $n \times n$ matrix.

Then with our sub-body $B$ of an elastic body in $\mathbb{R}^3$, the Cauchy stress **covector** valued 2-form yields an "area covector force density" with "components" the 2-forms $\mathfrak{t}_c = t_c{}^b\, i(\partial_b)\mathrm{vol}^3$ at points of the boundary $\partial B$. The "moment about an origin (chosen inside $B$)" density, on $\partial B$, has *cartesian* "components" the matrix of 2-forms

$$\mathfrak{m}_{ac} = [x^a t_c{}^b - x^c t_a{}^b] i(\partial_b)\mathrm{vol}^3 = x^a \mathfrak{t}_c - x^c \mathfrak{t}_a$$

Thus the **total moment** about the **origin** due to these stress forces on $\partial B$ is the 2-form at the origin $\Sigma_{a<c} M_{ac}\, dx^a \wedge dx^c$ with components the matrix of numbers

$$M_{ac} = \int_{\partial B} [x^a \mathfrak{t}_c - x^c \mathfrak{t}_a] = \int_B d[x^a \mathfrak{t}_c - x^c \mathfrak{t}_a]$$

which, from (©.43) (i.e., assuming no external body forces), is

$$M_{ac} = \int_B dx^a \wedge \mathfrak{t}_c - dx^c \wedge \mathfrak{t}_a$$

In most common elastic materials, this must vanish if there are to be no "couple stresses" without applied internal torque sources. Since this holds for any portion $B$ we must have

$$dx^a \wedge \mathfrak{t}_c = dx^c \wedge \mathfrak{t}_a \tag{©.44}$$

Since these are 3-forms in $\mathbb{R}^3$,

$$dx^a \wedge \mathfrak{t}_c = dx^a \wedge t_c{}^b\, i(\partial_b)\mathrm{vol}^3 = t_c{}^a \mathrm{vol}^3 \tag{©.44'}$$

For example, in $\mathbb{R}^3$ with $a = 2$ and $c = 1$,

$$\begin{aligned} dx^2 \wedge t_1{}^b[i(\partial_b)dx^1 \wedge dx^2 \wedge dx^3] &= dx^2 \wedge t_1{}^2[i(\partial_2)dx^1 \wedge dx^2 \wedge dx^3] \\ &= -dx^2 \wedge t_1{}^2[i(\partial_2)dx^2 \wedge dx^1 \wedge dx^3] \\ &= -dx^2 \wedge t_1{}^2\, dx^1 \wedge dx^3 \\ &= -t_1{}^2\, dx^2 \wedge dx^1 \wedge dx^3 = t_1{}^2\, dx^1 \wedge dx^2 \wedge dx^3 \\ &= t_1{}^2 \mathrm{vol}^3 \end{aligned}$$

(©.44) then yields $t_c{}^a\text{vol}^3 = t_a{}^c\text{vol}^3$, and since the coordinates are cartesian we have

$$t^{ca} = t^{ac} \tag{©.45}$$

Since the Cauchy stress $t$ is a tensor, this symmetry holds in *any* coordinate system. This is Cauchy's second theorem.

**Warning:** In Section ©.p we allowed and encouraged the use of different coordinates for the 2-form part and the value part of the stress vector valued 2-form

$$\boldsymbol{\partial}_i \otimes t^{ij} i(\boldsymbol{\partial}_j)\text{vol}^3 = \mathbf{e}_a \otimes A^a_i t^{ij} i(\boldsymbol{\partial}_j)\text{vol}^3 =: \mathbf{e}_a \otimes \tau^{aj} i(\boldsymbol{\partial}_j)\text{vol}^3$$

The left index "$a$" on $\tau$ is associated with the **e** basis and the right index "$j$" is associated with the $\boldsymbol{\partial}$ basis. (Think, for example, of **e** as cartesian and $\boldsymbol{\partial}$ as cylindrical.) Does the fact that $t$ is symmetric, $t^T = t$, insure that $\tau = At$ is also ? No!

$$\tau^T = (At)^T = t^T A^T = tA^T = A^{-1}\tau A^T \neq \tau \quad \text{generically}$$

## ©.s. A Remarkable Formula for Differentiating Line, Surface, and . . ., Integrals

Let **v** be a **time independent** vector field in a coordinate patch $U$ of $\mathbb{R}^n$ with any coordinates $x^i$. Roughly speaking, i.e., omitting some technicalities, by integrating the differential equations $dx^i/dt = v^i(x)$ we can move along the integral curves of **v** for $t$ seconds yielding a "flow" $\phi_t : U \to \mathbb{R}^n$. Since **v** is time independent, the $\phi_t$ form a 1 parameter commutative group of mappings, $\phi_t\phi_h = \phi_{t+h}$ and $\phi_0$ is the identity map. Let $V^r$ be an oriented $r$ dimensional "submanifold" of $U$. For examples, $V^1$ is an oriented curve , $V^2$ is an oriented 2 dimensional surface, . . . . $V^r$ is the kind of object over which one integrates an exterior $r$-form $\alpha = \alpha^r$ (a **scalar** valued, not vector valued form), yielding the number $\int_V \alpha^r$. As time changes, the flow moves $V$ from $V(0) = V$ to $V(t) = \phi_t(V)$. We consider only the simplest case where the r-form $\alpha$ is time independent. How does the integral change in time? The answer can be shown (see Section 4.3a) to be

$$d/dt|_{t=0} \int_{V(t)} \alpha^r = \int_V \mathcal{L}_{\mathbf{v}}\alpha^r \tag{©.46}$$

where the $r$-form $\mathcal{L}_{\mathbf{v}}\alpha^r$, the **Lie derivative** of the form $\alpha$, is defined via the pull-backs

$$\begin{aligned}[\mathcal{L}_{\mathbf{v}}\alpha^r](\text{at } x) :&= [d/dt]_{t=0}\phi_t{}^*[\alpha^r(\text{at } \phi_t x)] \\ &= \lim_{t\to 0}\{\phi_t{}^*[\alpha^r(\text{at } \phi_t x)] - \alpha^r(\text{at } x)\}/t \end{aligned} \tag{©.47}$$

Furthermore, there is a remarkable expression for computing the Lie derivative of any form, given by the **Henri Cartan** (son of Elie Cartan) **formula**

$$\mathcal{L}_{\mathbf{v}}\alpha^r = i_{\mathbf{v}}(d\alpha^r) + d(i_{\mathbf{v}}\alpha^r) \tag{©.48}$$

Thus (©.46) and Stokes say

$$d/dt|_{t=0} \int_{V(t)} \alpha^r = \int_V \mathcal{L}_{\mathbf{v}}\alpha^r = \int_V i_{\mathbf{v}}\, d\alpha + \int_{\partial V} i_{\mathbf{v}}\alpha \tag{©.49}$$

Consider for example the case of a line integral in $\mathbb{R}^3$, which we also write in classical form in cartesian coordinates. $V^1$ is then a curve $C$ starting at point $P$ and ending at point $Q$. Symbolically $\partial C = Q - P$. Classically $\alpha = \mathbf{a} \cdot d\mathbf{x}$. Then $i_{\mathbf{v}}\alpha$ is the 0-form, i.e., function $v \cdot \mathbf{a}$, and $\int_{\partial C} \mathbf{v} \cdot \mathbf{a}$ is *by definition* simply $(\mathbf{v} \cdot \mathbf{a})(Q) - (\mathbf{v} \cdot \mathbf{a})(P)$. This is the second "integral" in (©.49). Also, $d\alpha^1$ is the 2-form version of the vector curl $\mathbf{a}$, and so $i_{\mathbf{v}}\, d\alpha$, from (©.34), is the 1-form version of $-\mathbf{v} \times \text{curl}\, \mathbf{a}$. We then have, in the classical version

$$d/dt]_{t=0} \int_{C(t)} \mathbf{a} \cdot d\mathbf{x} = -\int_C [\mathbf{v} \times \text{ curl } \mathbf{a}] \cdot d\mathbf{x} + (\mathbf{v} \cdot \mathbf{a})(Q) - (\mathbf{v} \cdot \mathbf{a})(P)$$

The reader might enjoy computing the rates of change of surface and volume integrals

$$\int_S \mathbf{b} \cdot \mathbf{n}\, dA \quad \text{and} \quad \int_M \text{vol}^3$$

**A final remark about time dependent flows and forms.** In the real world, vector fields and forms are frequently time dependent. Consider, for example, $\mathbb{R}^n$ with local coordinates $\mathbf{x} = (x^i)$, and let $\alpha^r$ be an $r$-form (with components that may be time $t$ dependent) and $\mathbf{v} = \boldsymbol{\partial}_i v^i(t, \mathbf{x})$. We may again solve the differential equations $d\mathbf{x}/dt = \mathbf{v}(t, \mathbf{x})$ to get maps $\phi_t$ but (as discussed in Section 4.3b) generically they will not satisfy the crucial $\phi_a \circ \phi_b = \phi_{a+b}$. To circumvent this we introduce the space $\mathbb{R} \times \mathbb{R}^n$ with $n+1$ local coordinates $(x^0 = t, x^i)$, $1 \le i \le n$, that is, we enlarge the space $\mathbb{R}^n$ to $\mathbb{R}^{n+1}$ by introducing time as another dimension. We then augment the original vector field $\mathbf{v}$ on $\mathbb{R}^n$ to the new field $\nu\,(t, \mathbf{x}) = \boldsymbol{\partial}_t + \mathbf{v}(t, \mathbf{x})$ on $\mathbb{R}^1 \times \mathbb{R}^n$. Then it is shown in Theorem (4.42) that we get new maps $\phi_t : \mathbb{R}^1 \times \mathbb{R}^n \to \mathbb{R}^1 \times \mathbb{R}^n$ that do form a flow, and if $V = V_0$ is an $r$ dimensional submanifold of the $\mathbb{R}^n$ slice $t = 0$, then $V(a) = \phi_a V$ is in slice $t = a$, and (©.49) is replaced by

$$\begin{aligned} d/dt|_{t=0} \int_{V(t)} \alpha &= \int_V \mathcal{L}_\nu \alpha = \int_V i(\nu) d\alpha + \int_V d[i(\nu)\alpha] \\ &= \int_V (\partial\alpha/\partial t) + i_{\mathbf{v}}\mathbf{d}\alpha + \mathbf{d}i_{\mathbf{v}}\alpha \end{aligned} \tag{©.50}$$

(note $i_{\mathbf{v}} = i(\mathbf{v})$ uses the original vector field $\mathbf{v}$, not the augmented $\nu = \mathbf{v} + \boldsymbol{\partial}_t$). The bold $\mathbf{d}$ is the "spatial" exterior differential of $\mathbb{R}^n$ (keeping $t$ constant) and $\partial\alpha/\partial t$ is the $r$-form (with no $dt$ term) where each term of $\alpha$

$$a_{i\ldots j}(x, t) dx^i \wedge \ldots \wedge dx^j$$

is replaced by

$$[\partial a_{i\ldots j}(x, t)/\partial t]_{t=0}\, dx^i \wedge \ldots \wedge dx^j$$

For example, (©.50) tells us that Faraday's law of section ©.1 says that for a *moving* surface $V^2(t)$

$$d/dt \iint_{V(t)} \mathcal{B}^2 = -\oint_{\partial V} (\mathcal{E} - i_{\mathbf{v}}\mathcal{B}) = -\oint_{\partial V} (\mathbf{E} + \mathbf{v} \times \mathbf{B}) \cdot d\mathbf{x}$$

is the line integral of the *electromotive force* along the boundary curve.

Applications to fluid flows, vorticity, and magnetohydrodynamics can be seen in Section 4.3c.

PART ONE

# Manifolds, Tensors, and Exterior Forms

CHAPTER 1

# Manifolds and Vector Fields

Better is the end of a thing than the beginning thereof.

Ecclesiastes 7:8

As students we learn differential and integral calculus in the context of euclidean space $\mathbb{R}^n$, but it is necessary to apply calculus to problems involving "curved" spaces. Geodesy and cartography, for example, are devoted to the study of the most familiar curved surface of all, the surface of planet Earth. In discussing maps of the Earth, latitude and longitude serve as "coordinates," allowing us to use calculus by considering functions on the Earth's surface (temperature, height above sea level, etc.) as being functions of latitude and longitude. The familiar Mercator's projection, with its stretching of the polar regions, vividly informs us that these coordinates are badly behaved at the poles: that is, that they are not defined everywhere; they are not "global." (We shall refer to such coordinates as being "local," even though they might cover a huge portion of the surface. Precise definitions will be given in Section 1.2.) Of course we may use two sets of "polar" projections to study the Arctic and Antarctic regions. With these three maps we can study the entire surface, provided we know how to relate the Mercator to the polar maps.

We shall soon define a "manifold" to be a space that, like the surface of the Earth, can be covered by a family of local coordinate systems. *A manifold will turn out to be the most general space in which one can use differential and integral calculus with roughly the same facility as in euclidean space.* It should be recalled, though, that calculus in $\mathbb{R}^3$ demands special care when curvilinear coordinates are required.

The most familiar manifold is $N$-dimensional euclidean space $\mathbb{R}^N$, that is, the space of ordered $N$ tuples $(x^1, \ldots, x^N)$ of real numbers. Before discussing manifolds in general we shall talk about the more familiar (and less abstract) concept of a submanifold of $\mathbb{R}^N$, generalizing the notions of curve and surface in $\mathbb{R}^3$.

## 1.1. Submanifolds of Euclidean Space

What is the configuration space of a rigid body fixed at one point of $\mathbb{R}^n$?

### 1.1a. Submanifolds of $\mathbb{R}^N$

Euclidean space, $\mathbb{R}^N$, is endowed with a global coordinate system $(x^1, \dots, x^N)$ and is the most important example of a manifold.

In our familiar $\mathbb{R}^3$, with coordinates $(x, y, z)$, a locus $z = F(x, y)$ describes a (2-dimensional) surface, whereas a locus of the form $y = G(x)$, $z = H(x)$, describes a (1-dimensional) curve. We shall need to consider higher-dimensional versions of these important notions.

A subset $M = M^n \subset \mathbb{R}^{n+r}$ is said to be an $n$-dimensional **submanifold** of $\mathbb{R}^{n+r}$, if *locally* $M$ can be described by giving $r$ of the coordinates differentiably in terms of the $n$ remaining ones. This means that given $p \in M$, a neighborhood of $p$ on $M$ can be described in *some* coordinate system $(x, y) = (x^1, \dots, x^n, y^1, \dots, y^r)$ of $\mathbb{R}^{n+r}$ by $r$ differentiable functions

$$y^\alpha = f^\alpha(x^1, \dots, x^n), \qquad \alpha = 1, \dots r$$

We abbreviate this by $y = f(x)$, or even $y = y(x)$. We say that $x^1, \dots, x^n$ are **local (curvilinear) coordinates** for $M$ near $p$.

**Examples:**

(i) $y^1 = f(x^1, \dots, x^n)$ describes an $n$-dimensional submanifold of $\mathbb{R}^{n+1}$.

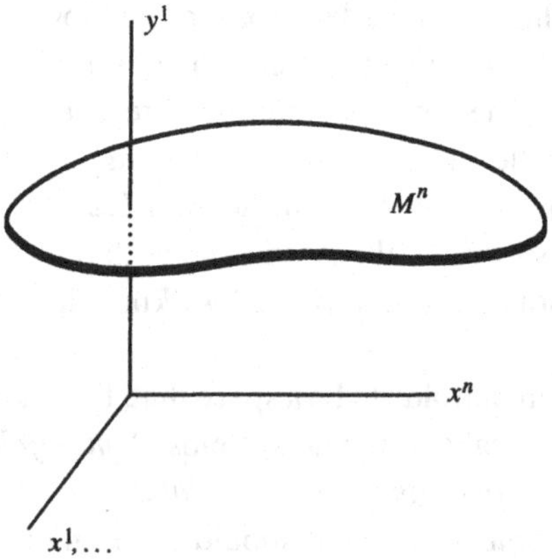


Figure 1.1

In Figure 1.1 we have drawn a portion of the submanifold $M$. This $M$ is the **graph** of a function $f : \mathbb{R}^n \to \mathbb{R}$, that is, $M = \{(\mathbf{x}, y) \in \mathbb{R}^{n+1} \mid y = f(\mathbf{x})\}$. When $n = 1$, $M$ is a curve; while if $n = 2$, it is a surface.

(ii) The *unit sphere* $x^2 + y^2 + z^2 = 1$ in $\mathbb{R}^3$. Points in the northern hemisphere can be described by $z = F(x, y) = (1 - x^2 - y^2)^{1/2}$ and this function is differentiable everywhere except at the equator $x^2 + y^2 = 1$. Thus $x$ and $y$ are local coordinates for the northern hemisphere except at the equator. For points on the equator one can solve for $x$ or $y$ in terms of the others. If we have solved for $x$ then $y$ and $z$ are the two local coordinates. For points in the southern hemisphere one can use the negative square root for $z$. The unit sphere in $\mathbb{R}^3$ is a 2-dimensional submanifold of $\mathbb{R}^3$. We note that we have *not* been able to describe the *entire* sphere by expressing one of the coordinates, say $z$, in terms of the two remaining ones, $z = F(x, y)$. We settle for local coordinates.

More generally, given $r$ functions $F^\alpha(x_1, \dots, x_n, y_1, \dots, y_r)$ of $n + r$ variables, we may consider the locus $M^n \subset \mathbb{R}^{n+r}$ defined by the equations

$$F^\alpha(x, y) = c^\alpha, \qquad (c^1, \dots, c^r) \text{ constants}$$

If the **Jacobian** determinant

$$\left[\frac{\partial(F^1, \dots, F^r)}{\partial(y^1, \dots, y^r)}\right](x_0, y_0)$$

at $(x_0, y_0) \in M$ of the locus is not 0, the **implicit function theorem** assures us that locally, near $(x_0, y_0)$, we may solve $F^\alpha(x, y) = c^\alpha$, $\alpha = 1, \dots, r$, for the $y$'s in terms of the $x$'s

$$y^\alpha = f^\alpha(x^1, \dots, x^n)$$

We may say that "a portion of $M^n$ near $(x_0, y_0)$ is a submanifold of $\mathbb{R}^{n+r}$." If the Jacobian $\neq 0$ *at all points of the locus*, then the entire $M^n$ is a submanifold.

Recall that the Jacobian condition arises as follows. If $F^\alpha(x, y) = c^\alpha$ can be solved for the $y$'s differentiably in terms of the $x$'s, $y^\beta = y^\beta(x)$, then if, for fixed $i$, we differentiate the identity $F^\alpha(x, y(x)) = c^\alpha$ with respect to $x^i$, we get

$$\frac{\partial F^\alpha}{\partial x^i} + \sum_\beta \left[\frac{\partial F^\alpha}{\partial y^\beta}\right] \frac{\partial y^\beta}{\partial x^i} = 0$$

and

$$\frac{\partial y^\beta}{\partial x^i} = -\sum_\alpha \left(\left[\frac{\partial F}{\partial y}\right]^{-1}\right)^\beta{}_\alpha \left[\frac{\partial F^\alpha}{\partial x^i}\right]$$

provided the subdeterminant $\partial(F^1, \dots, F^r)/\partial(y^1, \dots, y^r)$ is not zero. (Here $([\partial F/\partial y]^{-1})^\beta{}_\alpha$ is the $\beta\alpha$ entry of the inverse to the matrix $\partial F/\partial y$; we shall use the convention that for matrix indices, the index to the *left* always is the *row* index, whether it is up or down.) This *suggests* that if the indicated Jacobian is nonzero then we might indeed be able to solve for the $y$'s in terms of the $x$'s, and the implicit function theorem confirms this. The (nontrivial) *proof* of the implicit function theorem can be found in most books on real analysis.

Still more generally, suppose that we have $r$ functions of $n+r$ variables, $F^\alpha(x^1, \dots, x^{n+r})$. Consider the locus $F^\alpha(x) = c^\alpha$. Suppose that at each point $x_0$ of the locus the Jacobian *matrix*

$$\left(\frac{\partial F^\alpha}{\partial x^i}\right) \qquad \alpha = 1, \dots, r \qquad i = 1, \dots, n + r$$

has rank $r$. Then the equations $F^\alpha = c^\alpha$ define an $n$-dimensional submanifold of $\mathbb{R}^{n+r}$, since we may locally solve for $r$ of the coordinates in terms of the remaining $n$.

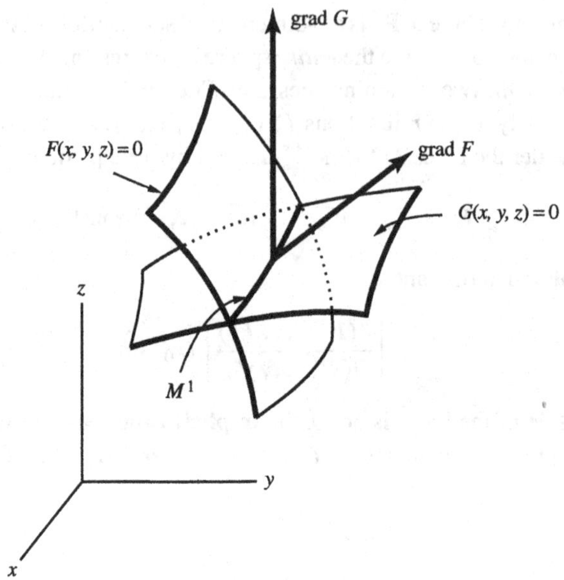


**Figure 1.2**

In Figure 1.2, two surfaces $F = 0$ and $G = 0$ in $\mathbb{R}^3$ intersect to yield a curve $M$.

The simplest case is *one* function $F$ of $N$ variables $(x^1, \ldots, x^N)$. If *at each point of the locus* $F = c$ there is always *at least one partial derivative that does not vanish*, then the Jacobian (row) matrix $[\partial F/\partial x^1, \partial F/\partial x^2, \ldots, \partial F/\partial x^N]$ has rank 1 and we may conclude that *this locus is indeed an* $(N-1)$*-dimensional submanifold of* $\mathbb{R}^N$. This criterion is easily verified, for example, in the case of the 2-sphere $F(x, y, z) = x^2 + y^2 + z^2 = 1$ of Example (ii). The column version of this row matrix is called in calculus the gradient vector of $F$. In $\mathbb{R}^3$ this vector

$$\begin{bmatrix} \frac{\partial F}{\partial x} \\ \frac{\partial F}{\partial y} \\ \frac{\partial F}{\partial z} \end{bmatrix}$$

is orthogonal to the locus $F = 0$, and we may conclude, for example, that if this gradient vector has a nontrivial component in the $z$ direction at a point of $F = 0$, then locally we can solve for $z = z(x, y)$.

A submanifold of dimension $(N-1)$ in $\mathbb{R}^N$, that is, of "**codimension**" 1, is called a **hypersurface**.

**(iii)** The $x$ axis of the $xy$ plane $\mathbb{R}^2$ can be described (perversely) as the locus of the quadratic $F(x, y) := y^2 = 0$. Both partial derivatives vanish on the locus, the $x$ axis, and our criteria would not allow us to say that the $x$ axis is a 1-dimensional submanifold of $\mathbb{R}^2$. Of course the $x$ axis *is* a submanifold; we should have used the usual description $G(x, y) := y = 0$. Our Jacobian criteria are *sufficient* conditions, not necessary ones.

**(iv)** The locus $F(x, y) := xy = 0$ in $\mathbb{R}^2$, consisting of the union of the $x$ and $y$ axes, is not a 1-dimensional submanifold of $\mathbb{R}^2$. It seems "clear" (and can be proved) that in a neighborhood of the intersection of the two lines we are not going to be able to describe the locus in the form of $y = f(x)$ or $x = g(y)$, where $f$, $g$, are differentiable functions. The best we can say is that this locus *with the origin removed* is a 1-dimensional submanifold.

### 1.1b. The Geometry of Jacobian Matrices: The "Differential"

The **tangent space** to $\mathbb{R}^n$ at the point $x$, written here as $\mathbb{R}^n_x$, is by definition the vector space of all vectors in $\mathbb{R}^n$ based at $x$ (i.e., it is a copy of $\mathbb{R}^n$ with origin shifted to $x$).

Let $x^1, \ldots, x^n$ and $y^1, \ldots, y^r$ be coordinates for $\mathbb{R}^n$ and $\mathbb{R}^r$ respectively. Let $F : \mathbb{R}^n \to \mathbb{R}^r$ be a **smooth** map. ("Smooth" ordinarily means infinitely differentiable. For our purposes, however, it will mean differentiable at least as many times as is necessary in the present context. For example, if $F$ is once continuously differentiable, we may use the chain rule in the argument to follow.) In coordinates, $F$ is described by giving $r$ functions of $n$ variables

$$y^\alpha = F^\alpha(x) \qquad \alpha = 1, \ldots, r$$

or simply $y = F(x)$. We will frequently use the more dangerous notation $y = y(x)$.

Let $y_0 = F(x_0)$; the Jacobian *matrix* $(\partial y^\alpha/\partial x^i)(x_0)$ has the following significance.

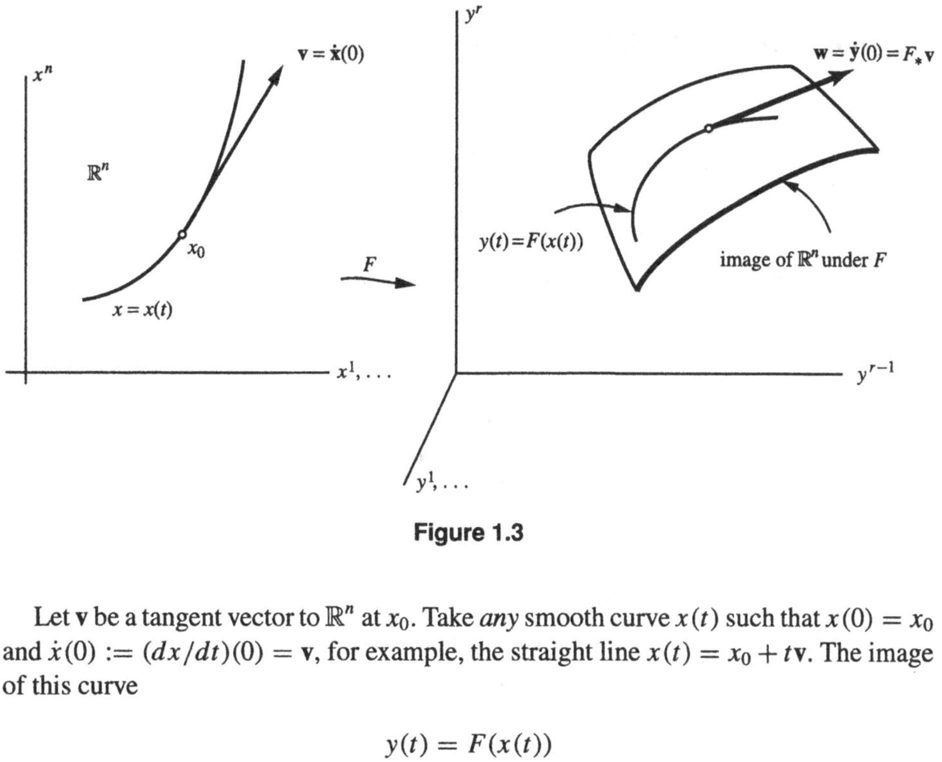


**Figure 1.3**

Let $\mathbf{v}$ be a tangent vector to $\mathbb{R}^n$ at $x_0$. Take *any* smooth curve $x(t)$ such that $x(0) = x_0$ and $\dot{x}(0) := (dx/dt)(0) = \mathbf{v}$, for example, the straight line $x(t) = x_0 + t\mathbf{v}$. The image of this curve

$$y(t) = F(x(t))$$

has a tangent vector $\mathbf{w}$ at $y_0$ given by the chain rule

$$w^\alpha = \dot{y}^\alpha(0) = \sum_{i=1}^n \left(\frac{\partial y^\alpha}{\partial x^i}\right)(x_0)\dot{x}^i(0) = \sum_{i=1}^n \left(\frac{\partial y^\alpha}{\partial x^i}\right)(x_0)v^i$$

The assignment $\mathbf{v} \mapsto \mathbf{w}$ is, from this expression, independent of the curve $x(t)$ chosen, and defines a *linear transformation*, the **differential** of $F$ at $x_0$

$$F_* : \mathbb{R}^n_{x_0} \to \mathbb{R}^r_{y_0} \qquad F_*(\mathbf{v}) = \mathbf{w} \tag{1.1}$$

whose matrix is simply the Jacobian matrix $(\partial y^\alpha/\partial x^i)(x_0)$. *This interpretation of the Jacobian matrix, as a linear transformation sending tangents to curves into tangents to the image curves under $F$, can sometimes be used to replace the direct computation of matrices.* This philosophy will be illustrated in Section 1.1d.

### 1.1c. The Main Theorem on Submanifolds of $\mathbb{R}^N$

The main theorem is a geometric interpretation of what we have discussed. Note that the statement "$F$ has rank $r$ at $x_0$," that is, $[\partial y^\alpha/\partial x^i](x_0)$ has rank $r$, is geometrically the statement that the differential

$$F_* : \mathbb{R}^n_{x_0} \to \mathbb{R}^r_{y_0=F(x_0)}$$

is **onto** or "surjective"; that is, given any vector $\mathbf{w}$ at $y_0$ there is at least one vector $\mathbf{v}$ at $x_0$ such that $F_*(\mathbf{v}) = \mathbf{w}$. We then have

**Theorem (1.2):** *Let $F : \mathbb{R}^{r+n} \to \mathbb{R}^r$ and suppose that the locus*

$$F^{-1}(y_0) := \{x \in \mathbb{R}^{r+n} \mid F(x) = y_0\}$$

*is not empty. Suppose further that for all $x_0 \in F^{-1}(y_0)$*

$$F_* : \mathbb{R}^{n+r}_{x_0} \to \mathbb{R}^r_{y_0}$$

*is onto. Then $F^{-1}(y_0)$ is an $n$-dimensional submanifold of $\mathbb{R}^{n+r}$.*

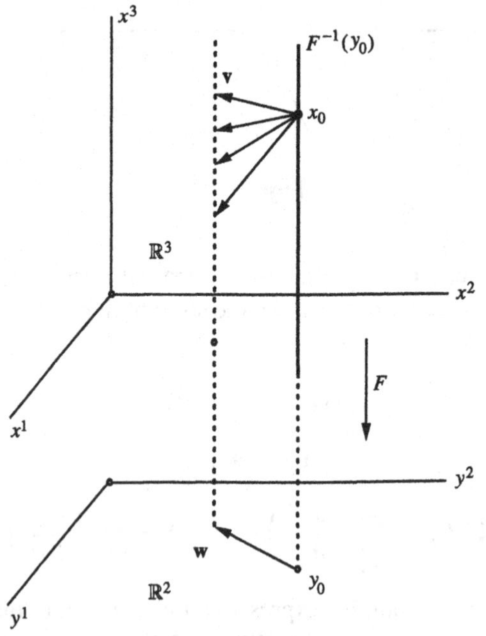


Figure 1.4

The best example to keep in mind is the linear "projection" $F : \mathbb{R}^3 \to \mathbb{R}^2$, $F(x^1, x^2, x^3) = (x^1, x^2)$, that is, $y^1 = x^1$ and $y^2 = x^2$. In this case, $x^3$ serves as global coordinate for the submanifold $x^1 = y_0^1$, $x^2 = y_0^2$, that is, the vertical line.

### 1.1d. A Nontrivial Example: The Configuration Space of a Rigid Body

Assume a rigid body has one point, the origin of $\mathbb{R}^3$, fixed. By comparing a cartesian right-handed system fixed in the body with that of $\mathbb{R}^3$ we see that the configuration of the body at any time is described by the rotation matrix taking us from the basis of $\mathbb{R}^3$ to the basis fixed in the body. The configuration space of the body is then the **rotation group** SO(3), that is, the $3 \times 3$ real matrices $x = (x_{ij})$ such that

$$x^T = x^{-1} \quad \text{and} \quad \det x = 1$$

where $T$ denotes transpose. (If we omit the determinant condition, the group is the full **orthogonal** group, O(3).) By assigning (in some fixed order) the nine coordinates $x_{11}, x_{12}, \ldots, x_{33}$ to *any* matrix $x$, we see that the space of all $3 \times 3$ real matrices, $M(3 \times 3)$, is the euclidean space $\mathbb{R}^9$. The group O(3) is then the locus in this $\mathbb{R}^9$ defined by the equations $x^T x = I$, that is, by the system of nine quadratic equations $(i, k)$

$$(i, k) \qquad \sum_{j=1}^{3} x_{ji} x_{jk} = \delta_{ik}$$

We then have the following situation. The configuration of the body at time $t$ can be represented by a point $x(t)$ in $\mathbb{R}^9$, but in fact the point $x(t)$ lies on the locus O(3) in $\mathbb{R}^9$. We shall see shortly that *this locus is in fact a 3-dimensional submanifold* of $\mathbb{R}^9$. As time $t$ evolves, the point $x(t)$ traces out a curve on this 3-dimensional locus. Since O(3) *is* a submanifold, we shall see, in Section 10.2c from the principle of least action, that this path is a very special one, a "geodesic" on the submanifold O(3), and this in turn will yield important information on the existence of periodic motions of the body even when the body is subject to an unusual potential field. All this depends on the fact that O(3) is a submanifold, and we turn now to the proof of this crucial result.

Note first that since $x^T x$ is a symmetric matrix, equation $(i, k)$ is the same as equation $(k, i)$; there are, then, only 6 independent equations. This suggests the following. Let

$$\mathrm{Sym}^6 := \{x \in M(3 \times 3) \mid x^T = x\}$$

be the space of all *symmetric* $3 \times 3$ matrices. Since this is defined by the three *linear* equations $x_{ik} - x_{ki} = 0$, $i \neq k$, we see that $\mathrm{Sym}^6$ is a 6-dimensional linear subspace of $\mathbb{R}^9$; that is, it can be considered as a copy of $\mathbb{R}^6$. To exhibit O(3) as a locus in $\mathbb{R}^9$, we consider the map

$$F : \mathbb{R}^9 \to \mathbb{R}^6 = \mathrm{Sym}^6 \quad \text{defined by } F(x) = x^T x - I$$

O(3) is then the locus $F^{-1}(0)$. Let $x_0 \in F^{-1}(0) = \mathrm{O}(3)$. We shall show that $F_* : \mathbb{R}^9_{x_0} \to \mathrm{Sym}^6$ is onto.

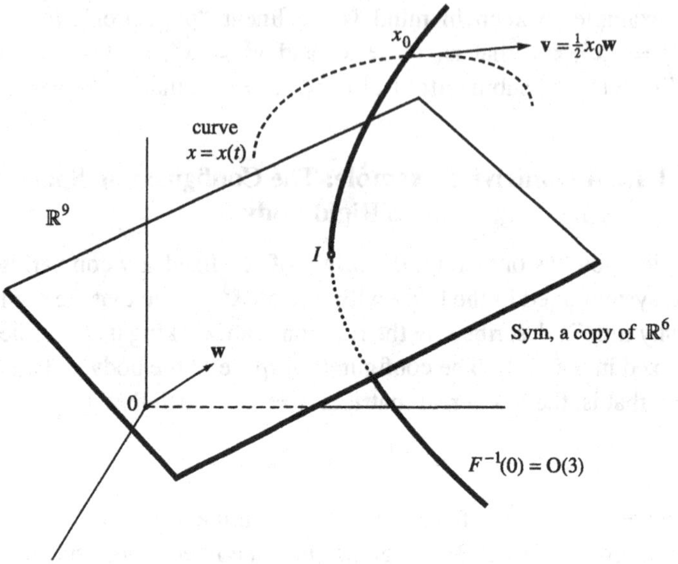


**Figure 1.5**

Let $\mathbf{w}$ be tangent to $\text{Sym}^6$ at the zero matrix. As usual, we identify a vector at the origin of $\mathbb{R}^n$ with its endpoint. *Then* $\mathbf{w}$ *is itself a symmetric matrix.* We must find $\mathbf{v}$, a tangent vector to $\mathbb{R}^9$ at $x_0$, such that $F_*\mathbf{v} = \mathbf{w}$. Consider a general curve $x = x(t)$ of matrices such that $x(0) = x_0$; its tangent vector at $x_0$ is $\dot{x}(0)$. The image curve

$$F(x(t)) = x(t)^T x(t) - I$$

has tangent at $t = 0$ given by

$$\frac{d}{dt}[F(x(t))]_{t=0} = \dot{x}(0)^T x_0 + x_0^T \dot{x}(0)$$

We wish this quantity to be $\mathbf{w}$. You should verify that it is sufficient to satisfy the matrix equation $x_0^T \dot{x}(0) = \mathbf{w}/2$. Since $x_0 \in O(3)$, $x_0^T = x_0^{-1}$ and we have as solution the matrix product $\mathbf{v} = \dot{x} = x_0\mathbf{w}/2$. Thus $F_*$ *is* onto at $x_0$ and by our main theorem O(3)$= F^{-1}(0)$ is a $(9-6) = 3$-dimensional submanifold of $\mathbb{R}^9$.

What about the subset SO(3) of O(3)? Recall that each orthogonal matrix has determinant $\pm 1$, whereas SO(3) consists of those orthogonal matrices with determinant $+1$. The mapping

$$\det : \mathbb{R}^9 \to \mathbb{R}$$

that sends each matrix $x$ into its determinant is continuous (it is a cubic polynomial function of the coordinates $x_{ik}$) and consequently the two subsets of O(3) where det is $+1$ and where det is $-1$ must be separated. This means that SO(3) itself must have the property that it is locally described by giving 6 of the coordinates in terms of the remaining 3, that is, SO(3) is a 3-dimensional submanifold of $\mathbb{R}^9$.

*Thus the configuration space of a rigid body with one point fixed is the group* SO(3). *This is a 3-dimensional submanifold of* $\mathbb{R}^9$. *Each point of this configuration space lies in some local curvilinear coordinate system.*

In physics books the coordinates in an $n$-dimensional configuration space are usually labeled $q^1, \ldots, q^n$. For SO(3) physicists usually use the three "Euler angles" as coordinates. These coordinates do not cover all of SO(3) in the sense that they become singular at certain points, just as polar coordinates in the plane are singular at the origin.

## Problems

**1.1(1)** Investigate the locus $x^2 + y^2 - z^2 = c$ in $\mathbb{R}^3$, for $c > 0$, $c = 0$, and $c < 0$. Are they submanifolds? What if the origin is omitted? Draw all three loci, for $c = 1, 0, -1$, in one picture.

**1.1(2)** SO($n$) is defined to be the set of all *orthogonal* $n \times n$ matrices $x$ with $\det x = 1$. The preceding discussion of SO(3) extends immediately to SO($n$). What is the dimension of SO($n$) and in what euclidean space is it a submanifold?

**1.1(3)** Is the **special linear group**

$$\mathrm{Sl}\,(n) := \{n \times n \text{ real matrices } x \mid \det x = 1\}$$

a submanifold of some $\mathbb{R}^N$? Hint: You will need to know something about $\partial/\partial x_{ij}$ (det $x$); expand the determinant by the $j^{\text{th}}$ column.This is an example where it might be easier to deal directly with the Jacobian matrix rather than the differential.

**1.1(4)** Show, in $\mathbb{R}^3$, that if the cross product of the gradients of $F$ and $G$ has a nontrivial component in the $x$ direction at a point of the intersection of $F = 0$ and $G = 0$, then $x$ can be used as local coordinate for this curve.

## 1.2. Manifolds

In learning the sciences examples are of more use than precepts.

Newton, *Arithmetica Universalis* (1707)

The notion of a "topology" will allow us to talk about "continuous" functions and points "neighboring" a given point, in spaces where the notion of distance and metric might be lacking.

The cultivation of an intuitive "feeling" for manifolds is of more importance, at this stage, than concern for topological details, but some basic notions from point set topology are helpful. The reader for whom these notions are new should approach them as one approaches a new language, with some measure of fluency, it is hoped, coming later.

In Section 1.2c we shall give a technical (i.e., complete) definition of a manifold.

### 1.2a. Some Notions from Point Set Topology

The **open ball** in $\mathbb{R}^n$, of radius $\epsilon$, centered at $\mathbf{a} \in \mathbb{R}^n$ is

$$B_{\mathbf{a}}(\epsilon) = \{\mathbf{x} \in \mathbb{R}^n \mid \| \mathbf{x} - \mathbf{a} \| < \epsilon\}$$

The **closed ball** is defined by

$$\overline{B}_{\mathbf{a}}(\epsilon) = \{\mathbf{x} \in \mathbb{R}^n \mid \| \mathbf{x} - \mathbf{a} \| \leq \epsilon\}$$

that is, the closed ball is the open ball with its edge or boundary included.

A set $U$ *in* $\mathbb{R}^n$ is declared **open** if given any $\mathbf{a} \in U$ there is an open ball of some radius $r > 0$, centered at $\mathbf{a}$, that lies entirely in $U$. Clearly each $B_{\mathbf{b}}(\epsilon)$ is open if $\epsilon > 0$ (take $r = (\epsilon - \| \mathbf{b} - \mathbf{a} \|)/2$), whereas $\overline{B}_{\mathbf{b}}(\epsilon)$ is not open because of its boundary points. $\mathbb{R}^n$ itself is trivially open. The empty set is technically open since there are no points $\mathbf{a}$ in it.

A set $F$ *in* $\mathbb{R}^n$ is declared **closed** if its complement $\mathbb{R}^n - F$ is open. It is easy to check that each $\overline{B}_{\mathbf{a}}(\epsilon)$ is a closed set, whereas the open ball is not. Note that the entire space $\mathbb{R}^n$ is both open and closed, since its complement is empty.

It is immediate that the *union* of *any* collection of open sets in $\mathbb{R}^n$ is an open set, and it is not difficult to see that the *intersection* of any *finite* number of open sets in $\mathbb{R}^n$ is open.

We have described explicitly the "usual" open sets in euclidean space $\mathbb{R}^n$. What do we mean by an open set in a more general space? We shall define the notion of open set axiomatically.

A **topological space** is a set $M$ with a distinguished collection of subsets, to be called the **open** sets. These open sets must satisfy the following.

1. Both $M$ and the empty set are open.
2. If $U$ and $V$ are open sets, then so is their intersection $U \cap V$.
3. The union of any collection of open sets is open.

These open subsets "define" **the topology** of $M$. □

(A different collection might define a different topology.) Any such collection of subsets that satisfies 1, 2, and 3 is eligible for defining a topology in $M$. In our introductory discussion of open balls in $\mathbb{R}^n$ we also defined the collection of open subsets of $\mathbb{R}^n$. These define the topology of $\mathbb{R}^n$, the "usual" topology. An example of a "perverse" topology on $\mathbb{R}^n$ is the **discrete** topology, in which *every* subset of $\mathbb{R}^n$ is declared open! In discussing $\mathbb{R}^n$ in this book we shall always use the usual topology.

A subset of $M$ is **closed** if its complement is open.

Let $A$ be any subset of a topological space $M$. Define a topology for the space $A$ (the **induced** or **subspace** topology) by declaring $V \subset A$ to be an open subset of $A$ provided $V$ is the intersection of $A$ with some open subset $U$ of $M$, $V = A \cap U$. These sets *do* define a topology for $A$. For example, let $A$ be a line in the plane $\mathbb{R}^2$. An open ball in $\mathbb{R}^2$ is simply a disc without its edge. This disc either will not intersect $A$ or will intersect $A$ in an "interval" that does not contain its endpoints. This interval will be an open set in the induced topology on the line $A$. It can be shown that any open set in $A$ will be a union of such intervals.

Any *open* set in $M$ that contains a point $x \in M$ will be called a **neighborhood** of $x$.

If $F{:}M \to N$ is a map of a topological space $M$ into a topological space $N$, we say that $F$ is **continuous** if for every open set $V \subset N$, the **inverse image** $F^{-1}V := \{x \in M \mid F(x) \in V\}$ is open in $M$. (This reduces to the usual $\epsilon$, $\delta$ definition in the case where $M$ and $N$ are euclidean spaces.) The map sending all of $\mathbb{R}^n$ into a single point of $\mathbb{R}^m$ is an example showing that a continuous map need not send open sets into open sets.

If $F:M \to N$ is one to one (1 : 1) and onto, then the inverse *map* $F^{-1}:N \to M$ exists. If further both $F$ and $F^{-1}$ are continuous, we say that $F$ is a **homeomorphism** and that $M$ and $N$ are homeomorphic. A homeomorphism takes open (closed) sets into open (closed) sets. Homeomorphic spaces are to be considered to be "the same" as topological spaces; we say that they are "topologically the same." It can be proved that $\mathbb{R}^n$ and $\mathbb{R}^m$ are homeomorphic if and only if $m = n$.

The technical definition of a manifold requires two more concepts, namely "Hausdorff" and "countable base." We shall not discuss these here since they will not arise *explicitly* in the remainder of the book. The reader is referred to [S] for questions concerning point set topology.

There is one more concept that plays a very important role, though not needed for the definition of a manifold; the reader may prefer to come back to this later on when needed.

A topological space $X$ is called **compact** if from *every* covering of $X$ by open sets one can pick out a *finite* number of the sets that still covers $X$. For example, the open interval (0,1), considered as a subspace of $\mathbb{R}$, is *not* compact; we cannot extract a finite subcovering from the open covering given by the sets $U_n = \{x \mid 1/n < x < 1\} n = 1, 2, \ldots$.

On the other hand, the closed interval [0,1] is a compact space. In fact, it is shown in every topology book that *any subset $X$ of $\mathbb{R}^n$* (with the induced topology) *is compact if and only if*

1. $X$ is a *closed* subset of $\mathbb{R}^n$,
2. $X$ is a *bounded* subset, that is, $\| \mathbf{x} \| <$ some number $c$, for all $\mathbf{x} \in X$.

Finally we shall need two properties of continuous maps. First

*The continuous image of a compact space is itself compact.*

**PROOF:** If $f : G \to M$ is continuous and if $\{U_i\}$ is an open cover of $f(G) \subset M$, then $\{f^{-1}(U_i)\}$ is an *open* cover of $G$. Since $G$ is compact we can extract a finite open subcover $\{f^{-1}(U_\alpha)\}$ of $G$, and then $\{U_\alpha\}$ is a finite subcover of $f(G)$. □

Furthermore

*A continuous real-valued function $f : G \to \mathbb{R}$ on a compact space $G$ is bounded.*

**PROOF:** $f(G)$ is a compact subspace of $\mathbb{R}$, and thus is closed and bounded. □

### 1.2b. The Idea of a Manifold

An $n$-dimensional (differentiable) manifold $M^n$ (briefly, an $n$-manifold) is a topological space that is locally $\mathbb{R}^n$ in the following sense. It is covered by a family of local (curvilinear) coordinate systems $\{U; x_U^1, \ldots, x_U^n\}$, consisting of open sets or "patches" $U$ and coordinates $x_U$ in $U$, such that a point $p \in U \cap V$ that lies in two coordinate patches will have its two sets of coordinates related differentiably

$$x_V^i(p) = f_{VU}^i(x_U^1, \ldots, x_U^n) \qquad i = 1, 2, \ldots, n. \tag{1.3}$$

(If the functions $f_{VU}$ are $C^\infty$, that is, infinitely differentiable, or real analytic, ..., we say that $M$ is $C^\infty$, or real analytic, ....) There are more requirements; for example, we shall demand that each coordinate patch is homeomorphic to some open subset of $\mathbb{R}^n$. Some of these requirements will be mentioned in the following examples, but details will be spelled out in Section 1.2c.

**Examples:**

**(i)** $M^n = \mathbb{R}^n$, covered by a single coordinate system. The condition (1.3) is vacuous.

**(ii)** $M^n$ is an open ball in $\mathbb{R}^n$, again covered by one patch.

**(iii)** The *closed* ball in $\mathbb{R}^n$ is *not* a manifold. It can be shown that a point on the edge of the ball can never have a neighborhood that is homeomorphic to an *open* subset of $\mathbb{R}^n$. For example, with $n = 1$, a half open interval $0 \le x < 1$ in $\mathbb{R}^1$ can never be homeomorphic to an open interval $0 < x < 1$ in $\mathbb{R}^1$.

**(iv)** $M^n = S^n$, the unit *sphere* in $\mathbb{R}^{n+1}$. We shall illustrate this with the familiar case $n = 2$. We are dealing with the locus $x^2 + y^2 + z^2 = 1$.

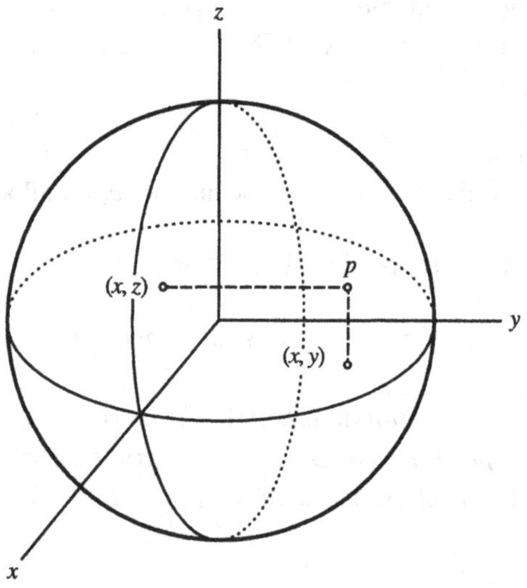


**Figure 1.6**

Cover $S^2$ with six "open" subsets (patches)

$$U_x+ = \{p \in S^2 \mid x(p) > 0\} \qquad U_x- = \{p \in S^2 \mid x(p) < 0\}$$
$$U_y+ = \{p \in S^2 \mid y(p) > 0\} \qquad U_y- = \{p \in S^2 \mid y(p) < 0\}$$
$$U_z+ = \{p \in S^2 \mid z(p) > 0\} \qquad U_z- = \{p \in S^2 \mid z(p) < 0\}$$

The point $p$ illustrated sits in $[U_x+] \cap [U_y+] \cap [U_z+]$. Project $U_z+$into the $xy$ plane; this introduces $x$ and $y$ as curvilinear coordinates in $U_z+$.

Do similarly for the other patches. For $p \in [U_y+] \cap [U_z+]$, $p$ is assigned the two sets of coordinates $\{(u_1, u_2) = (x, z)\}$ and $\{(v_1, v_2) = (x, y)\}$ arising from the two projections

$$\pi_{xz} : U_y \to xz \text{ plane} \quad \text{and} \quad \pi_{xy} : U_z \to xy \text{ plane}$$

These are related by $v_1 = u_1$ and $v_2 = +[1 - u_1^2 - u_2^2]^{1/2}$; these are differentiable functions provided $u_1^2 + u_2^2 < 1$, and this is satisfied since $p \in U_y+$.

$S^2$ is "locally $\mathbb{R}^2$." The indicated point $p$ has a neighborhood (in the topology of $S^2$ induced as a subset of $\mathbb{R}^3$) that is homeomorphic, via the projection $\pi_{xy}$, say, to an open subset of $\mathbb{R}^2$ (in this case an open subset of the $xy$ plane). We say that a manifold is **locally euclidean.**

If two sets of coordinates are related differentiably in an overlap we shall say that they are **compatible**. On $S^2$ we could introduce, in addition to the preceding coordinates, the usual spherical coordinates $\theta$ and $\phi$, representing colatitude and longitude. They do not work for the entire sphere (e.g., at the poles) but where they do work they are compatible with the original coordinates.

We could also introduce (see Section 1.2d) coordinates on $S^2$ via stereographic projection onto the planes $z = 1$ and $z = -1$, again failing at the south and north pole, respectively, but otherwise being compatible with the previous coordinates. On a manifold we should allow the use of *all* coordinate systems that are compatible with those that originally were used to define the manifold. Such a collection of compatible coordinate systems is called a **maximal atlas.**

**(v)** If $M^n$ is a manifold with local coordinates $\{U; x^1, \dots, x^n\}$ and $W^r$ is a manifold with local coordinates $\{V; y^1, \dots, y^r\}$, we can form the **product manifold**

$$L^{n+r} = M^n \times W^r = \{(p, q) \mid p \in M^n \text{ and } q \in W^r\}$$

by using $x^1, \dots, x^n, y^1, \dots, y^r$ as local coordinates in $U \times V$.

$S^1$ is simply the unit circle in the plane $\mathbb{R}^2$; it has a local coordinate $\theta = \tan^{-1}(y/x)$, using any branch of the multiple-valued function $\theta$. One must use at least two such coordinates (branches) to cover $S^1$. "Topologically" $S^1$ is conveniently represented by an interval on the real line $\mathbb{R}$ with endpoints identified; by this we mean that there is a homeomorphism between these two models. In order to talk about a homeomorphism

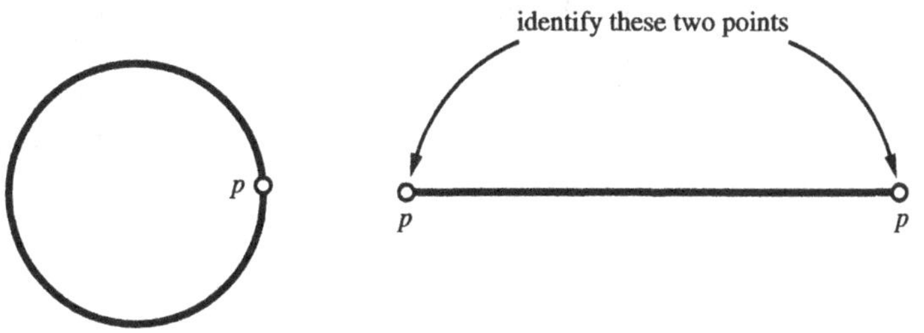


**Figure 1.7**

we would first have to define the topology in the space consisting of the interval with endpoints identified; it clearly is not the same space as the interval without the identification. To define a topology, we may simply consider the map $F : [0 \le \theta \le 2\pi] \to \mathbb{R}^2 = \mathbb{C}$ defined by $F(\theta) = e^{i\theta}$. It sends the endpoints $\theta = 0$ and $\theta = 2\pi$ to the point $p = 1$ on the unit circle in the complex plane. This map is $1:1$ and onto if we identify the endpoints. The unit circle has a topology induced from that of the plane, built up from little curved intervals. We can construct open subsets of the interval by taking the inverse images under $F$ of such sets. (What then is a neighborhood of the endpoint $p$?) By using this topology we force $F$ to be a homeomorphism.

$S^1$ is the configuration space for a rigid *pendulum* constrained to oscillate in the plane

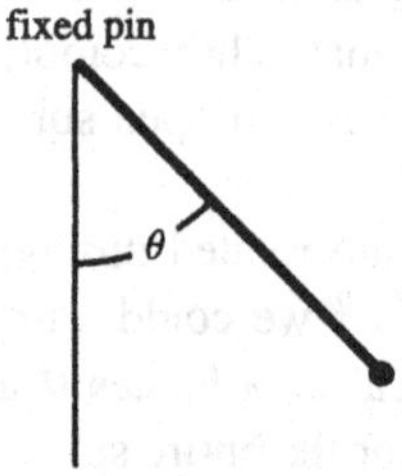


Figure 1.8

The $n$-dimensional **torus** $T^n := S^1 \times S^1 \times \cdots \times S^1$ has local coordinates given by the $n$-angular parameters $\theta^1, \ldots, \theta^n$. Topologically it is the $n$ cube (the product of $n$ intervals) with identifications. For $n = 2$

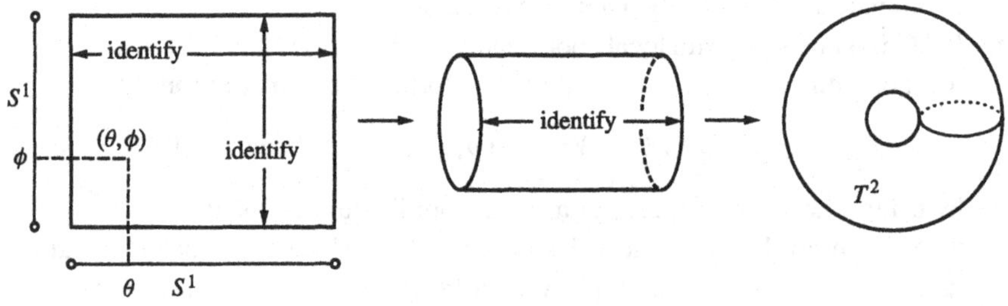


Figure 1.9

$T^2$ is the configuration space of a planar *double pendulum*. It might be thought that it is simpler to picture the double pendulum itself rather than the seemingly abstract version of a 2-dimensional torus. We shall see in Section 10.2d that this abstract picture allows us to conclude, for example, that *a double pendulum, in an arbitrary potential field, always has periodic motions in which the upper pendulum makes p revolutions while the lower makes q revolutions.*

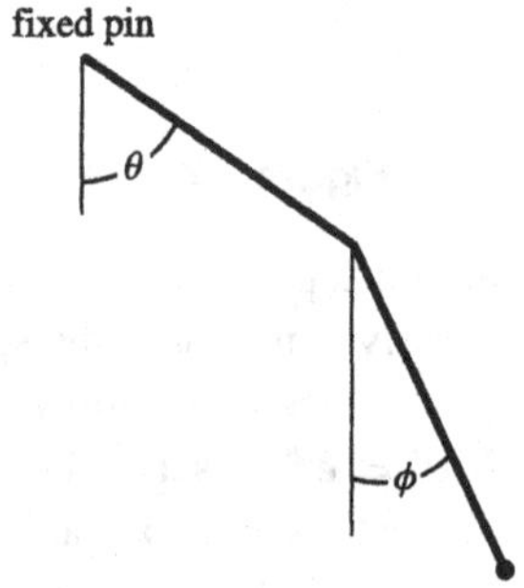


Figure 1.10

(vi) The **real projective** $n$ **space** $\mathbb{R}P^n$ is the space of all *unoriented* lines $L$ through the origin of $\mathbb{R}^{n+1}$. We illustrate with the *projective plane* of lines through the origin of $\mathbb{R}^3$.

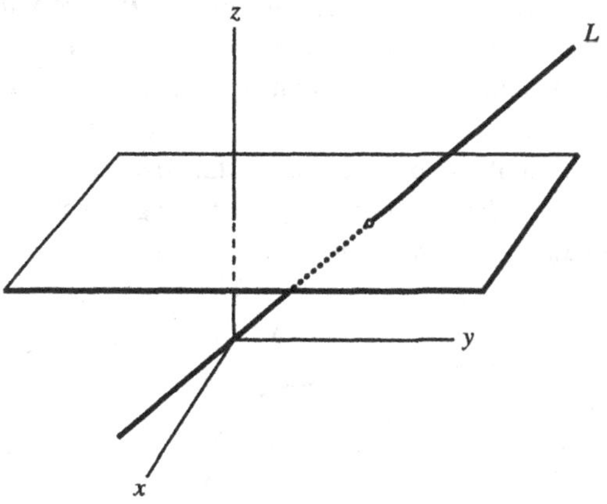


**Figure 1.11**

Such a line $L$ is completely determined by any point $(x, y, z)$ on the line, other than the origin, but note that $(ax, ay, az)$ represents the same line if $a \neq 0$. We should really use the ratios of coordinates to describe a line. We proceed as follows.

We cover $\mathbb{R}P^2$ by three sets:

$$U_x := \text{those lines not lying in the } yz \text{ plane}$$
$$U_y := \text{those lines not lying in the } xz \text{ plane}$$
$$U_z := \text{those lines not lying in the } xy \text{ plane}$$

Introduce coordinates in the $U_z$ patch; if $L \in U_z$, choose any point $(x, y, z)$ on $L$ other than the origin and define (since $z \neq 0$)

$$u_1 = \frac{x}{z}, \qquad u_2 = \frac{y}{z}$$

Do likewise for the other two patches. In Problem 1.2(1) you are asked to show that these patches make $\mathbb{R}P^2$ into a 2-dimensional manifold.

These coordinates are the most convenient for analytical work. Geometrically, the coordinates $u_1$ and $u_2$ are simply the $xy$ coordinates of the point where $L$ intersects the plane $z = 1$.

Consider a point in $\mathbb{R}P^2$; it represents a line through the origin 0. Let $(x, y, z)$ be a point other than the origin that lies on this line. We may represent this line by the triple $[x, y, z]$, called the **homogeneous coordinates** of the point in $\mathbb{R}P^2$ where we must identify $[x, y, z]$ with $[\lambda x, \lambda y, \lambda z]$ for all $\lambda \neq 0$. They are not true coordinates in our sense.

We have suceeded in "parameterizing" the *set* of undirected lines through the origin by means of a manifold, $M^2 = \mathbb{R}P^2$ . *A manifold is a generalized parameterization of some set of objects.* $\mathbb{R}P^2$ *is* the set of undirected lines through the origin; each point of $\mathbb{R}P^2$ is an entire line in $\mathbb{R}^3$ and $\mathbb{R}P^2$ is a *global* object. If, however, one insists on describing a particular line $L$ by coordinates, that is, pairs of *numbers* $(u, v)$, then this can, in general, only be done *locally*, by means of the manifold's local coordinates.

Note that if we had been considering *directed* lines, then the manifold in question would have been the sphere $S^2$, since each directed line $\vec{L}$ could be uniquely defined by the "forward" point where $\vec{L}$ intersects the unit sphere. An *un*directed line meets $S^2$ in a pair of antipodal points; $\mathbb{R}P^2$ *is topologically* $S^2$ *with antipodal points identified.*

We can now construct a topological model of $\mathbb{R}P^2$ that will allow us to identify certain spaces we shall meet as projective spaces. Our model will respect the topology; that is, "nearby points" in $\mathbb{R}P^2$ (that is, nearby lines in $\mathbb{R}^3$) will be represented by nearby points in the model, but we won't be concerned with the differentiability of our procedure. Also it will be clear that certain natural "distances" will not be preserved; *in the rigorous definition of manifold,* to be given shortly, *there is no mention of metric notions such as distance or area or angle.*

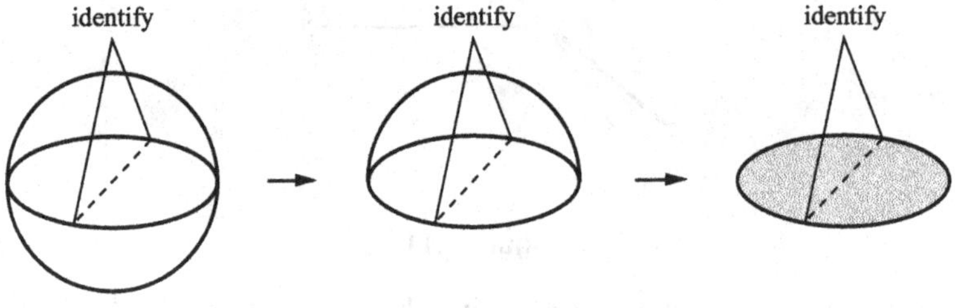


**Figure 1.12**

In the sphere with antipodal points identified, we may discard the entire southern hemisphere (exclusive of the equator) of redundant points, leaving us with the northern hemisphere, the equator, and with antipodal points only on the equator identified. We may then project this onto the disc in the plane. *Topologically* $\mathbb{R}P^2$ *is the unit disc in the plane with antipodal points on the unit circle identified.*

Similarly, $\mathbb{R}P^n$ is topologically the unit $n$ sphere $S^n$ in $\mathbb{R}^{n+1}$ with antipodal points identified, and this in turn is the solid $n$-dimensional unit ball in $\mathbb{R}^n$ with antipodal points on the boundary unit $(n-1)$ sphere identified.

**(vii)** It is a fact that *every submanifold of* $\mathbb{R}^n$ *is a manifold.* We verified this in the case of $S^2 \subset \mathbb{R}^3$ in Example (ii). In 1.1d we showed that the rotation group SO(3) is a 3-dimensional submanifold of $\mathbb{R}^9$. A convenient topological model is constructed as follows. Use the "right-hand rule" to associate the endpoint of the vector $\theta\mathbf{r}$ to the rotation through an angle $\theta$ (in radians) about an axis descibed by the unit vector $\mathbf{r}$. Note, however, that the rotation $\pi\mathbf{r}$ is exactly the same as the rotation $-\pi\mathbf{r}$ and $(\pi+\alpha)\mathbf{r}$ is the same as $-(\pi-\alpha)\mathbf{r}$. The collection of all rotations then can be represented by the points in the solid ball of radius $\pi$ in $\mathbb{R}^3$ with antipodal points on the sphere of radius $\pi$ identified; $SO(3)$ *can be identified with the real projective space* $\mathbb{R}P^3$.

**(viii)** The *Möbius band Mö* is the space obtained by identifying the left and right hand edges of a sheet of paper after giving it a "half twist"

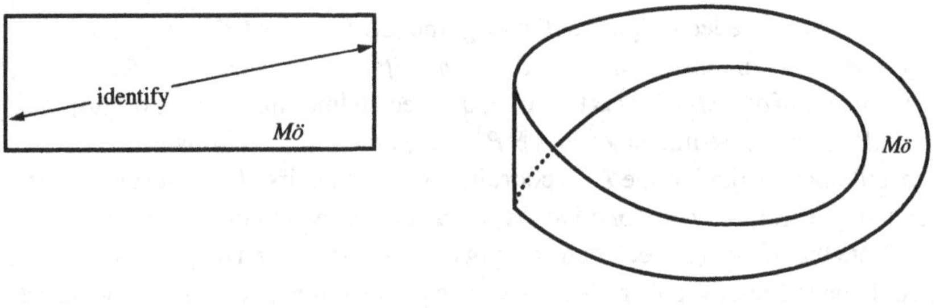


**Figure 1.13**

If one omits the edge one can see that *Mö* is a 2-dimensional submanifold of $\mathbb{R}^3$ and is therefore a 2-manifold. You should verify (i) that the Möbius band sits naturally as the shaded "half band" in the model of $\mathbb{R}P^2$ consisting of $S^2$ with antipodal points identified, and (ii) that this half band is the same as the full band. The *edge* of the

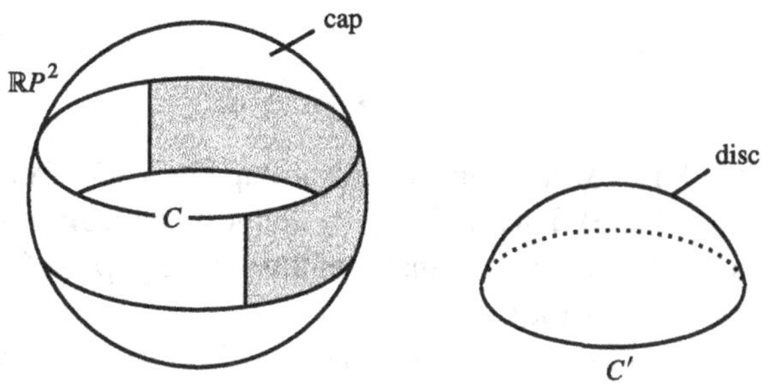


**Figure 1.14**

Möbius band consists of a *single* closed curve $C$ that can be pictured as the "upper" edge of this full band in $\mathbb{R}P^2$. Note that the indicated "cap" is topologically a 2-dimensional disc with a circular edge $C'$. If we observe that the lower cap is the same as the upper, we conclude that *if we take a 2-disc and sew its edge to the single edge of a Möbius band, then the resulting space is topologically the projective plane*! We may say that $\mathbb{R}P^2$ is Mö with a 2-disc attached along its boundary. Although the actual sewing, say with cloth, cannot be done in ordinary space $\mathbb{R}^3$ (the cap would have to slice through itself), this sewing *can* be done in $\mathbb{R}^4$, where there is "more room."

### 1.2c. A Rigorous Definition of a Manifold

Let $M$ be any *set* (without a topology) that has a covering by subsets $M = U \cup V \cup \ldots$, where each subset $U$ is in $1:1$ correspondence $\phi_U : U \to \mathbb{R}^n$ with an *open* subset $\phi_U(U)$ of $\mathbb{R}^n$.

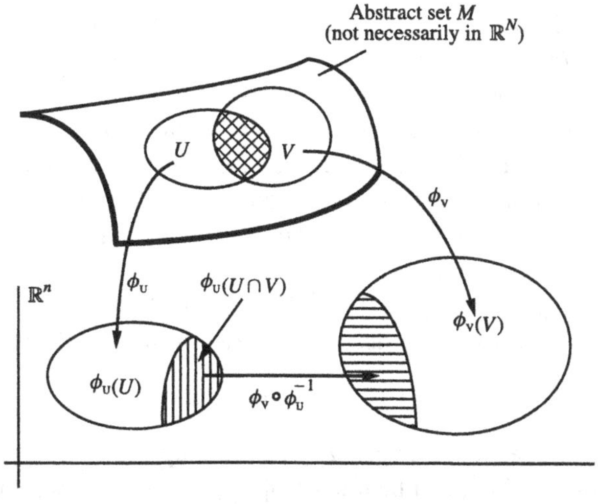


**Figure 1.15**

We require that each $\phi_U(U \cap V)$ be an open subset of $\mathbb{R}^n$. We require that the overlap maps

$$f_{VU} = \phi_V \circ \phi_U^{-1} : \phi_U(U \cap V) \to \mathbb{R}^n \tag{1.4}$$

that is,

$$\phi_U(U \cap V) \stackrel{\phi_U^{-1}}{\to} M \stackrel{\phi_V}{\to} \mathbb{R}^n$$

be differentiable (we know what it means for a map $\phi_V \circ \phi_U^{-1}$ from an open set of $\mathbb{R}^n$ to $\mathbb{R}^n$ to be differentiable). Each pair $U, \phi_U$ defines a **coordinate patch** on $M$; to $p \in U \subset M$ we may assign the $n$ coordinates of the point $\phi_U(p)$ in $\mathbb{R}^n$. For this reason we shall call $\phi_U$ a **coordinate map**.

Take now a maximal atlas of such coordinate patches; see Example (iv). Define a **topology** in the set $M$ by declaring a subset $W$ of $M$ to be open provided that given any $p \in W$ there is a coordinate chart $U, \phi_U$ such that $p \in U \subset W$. If the resulting topology for $M$ is Hausdorff and has a countable base (see [S] for these technical conditions) we say that $M$ is an $n$-dimensional differentiable manifold. We say that a *map* $F : \mathbb{R}^p \to \mathbb{R}^q$ is of class $C^k$ if all $k^{\text{th}}$ partial derivatives are continuous. It is of class $C^\infty$ if it is of class $C^k$ for all $k$. We say that a *manifold* $M^n$ is of class $C^k$ if its overlap maps $f_{VU}$ are of class $C^k$. Likewise we have the notion of a $C^\infty$ manifold. An analytic manifold is one whose overlap functions are analytic, that is, expandable in power series.

Let $F : M^n \to \mathbb{R}$ be a real-valued function on the manifold $M$. Since $M$ is a topological space we know from 1.2a what it means to say that $F$ is continuous. We say that $F$ is **differentiable** if, when we express $F$ in terms of a local coordinate system $(U, x)$, $F = F_U(x^1, \ldots, x^n)$ is a differentiable function of the coordinates $x$. Technically this means that that when we compose $F$ with the inverse of the coordinate map $\phi_U$

$$F_U := F \circ \phi_U^{-1}$$

(recall that $\phi_U$ is assumed $1 : 1$) we obtain a real-valued function $F_U$ defined on a portion $\phi_U(U)$ of $\mathbb{R}^n$, and we are asking that this function be differentiable. Briefly speaking, *we envision the coordinates $x$ as being engraved on the manifold $M$*, just as we see lines of latitude and longitude engraved on our globes. A function on the Earth's surface is continuous or differentiable if it is continuous or differentiable when expressed in terms of latitude and longitude, at least if we are away from the poles. Similarly with a manifold.With this understood, *we shall usually omit the process of replacing $F$ by its composition $F \circ \phi_U^{-1}$, thinking of $F$ as directly expressible as a function $F(x)$ of any local coordinates.*

Consider the real projective plane $\mathbb{R}P^2$, Example (vi) of Section 1.2b. In terms of homogeneous coordinates we may define a map $(\mathbb{R}^3 - 0) \to \mathbb{R}P^2$ by

$$(x, y, z) \to [x, y, z]$$

At a point of $\mathbb{R}^3$ where, for example, $z \neq 0$ we may use $u = x/z$ and $v = y/z$ as local coordinates in $\mathbb{R}P^2$, and then our map is given by the two smooth functions $u = f(x, y, z) = x/z$ and $v = g(x, y, z) = y/z$.

## 1.2d. Complex Manifolds: The Riemann Sphere

A **complex manifold** is a set $M$ together with a covering $M = U \cup V \cup \ldots$, where each subset $U$ is in 1 : 1 correspondence $\phi_U : U \to \mathbb{C}^n$ with an open subset $\phi_U(U)$ of complex $n$-space $\mathbb{C}^n$. We then require that the overlap maps $f_{VU}$ mapping sets in $\mathbb{C}^n$ into sets in $\mathbb{C}^n$ be *complex analytic*; thus if we write $f_{VU}$ in the form $w^k = w^k(z^1, \ldots, z^n)$ where $z^k = x^k + iy^k$ and $w^k = u^k + iv^k$, then $u^k$ and $v^k$ satisfy the Cauchy–Riemann equations with respect to each pair $(x^r, y^r)$. Briefly speaking, each $w^k$ can be expressed entirely in terms of $z^1, \ldots, z^n$, with no complex conjugates $\overline{z}^r$ appearing. We then proceed as in the real case in 1.2c. The resulting manifold is called an $n$-dimensional complex manifold, although its topological dimension is $2n$.

Of course the simplest example is $\mathbb{C}^n$ itself. Let us consider the most famous non-trivial example, the **Riemann sphere** $M^1$.

The complex plane $\mathbb{C}$ (topologically $\mathbb{R}^2$) comes equipped with a global complex coordinate $z = x+iy$. It is a complex 1-dimensional manifold $\mathbb{C}^1$. To study the behavior of functions at "$\infty$" we introduce a point at $\infty$, to form a new manifold that is topologically the 2-sphere $S^2$. We do this by means of stereographic projection, as follows.

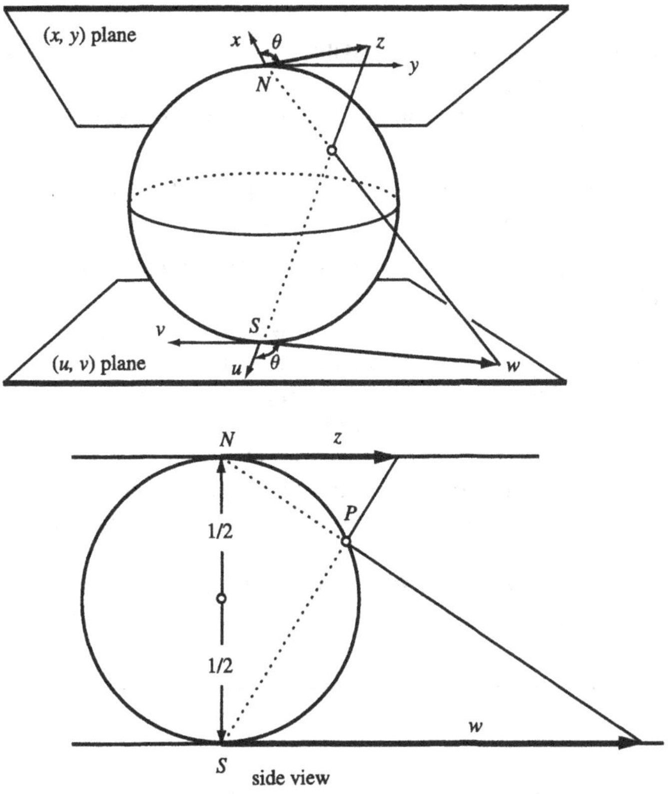


**Figure 1.16**

In the top part of the figure we have a sphere of radius $1/2$, resting on a $w = u + iv$ plane, with a tangent $z = x+iy$ plane at the north pole. Note that we have oriented these two tangent planes to agree with the usual orientation of $S^2$ (questions of orientation will be discussed in Section 2.8).

Let $U$ be the subset of $S^2$ consisting of all points except for the south pole, let $V$ be the points other than the north pole, let $\phi_U$ and $\phi_V$ be stereographic projections of $U$ and $V$ from the south and north poles, respectively, onto the $z$ and the $w$ planes. In this way we assign to any point $p$ other than the poles two complex coordinates, $z = |z|e^{i\theta}$ and $w = |w|e^{-i\theta}$. From the bottom of the figure, which depicts the planar section in the plane holding the two poles and the point $p$, one reads off from elementary geometry that $|w| = 1/|z|$, and consequently

$$w = f_{VU}(z) = \frac{1}{z} \tag{1.5}$$

gives the relation between the two sets of coordinates. Since this is complex analytic in the overlap $U \cap V$, we may consider $S^2$ as a 1-dimensional complex manifold, the Riemann sphere. The point $w = 0$ (the south pole) represents the point $z = \infty$ that was missing from the original complex plane $\mathbb{C}$.

Note that the two sets of real coordinates $(x, y)$ and $(u, v)$ make $S^2$ into a real analytic manifold.

## Problems

**1.2(1)** Show that $\mathbb{R}P^2$ is a differentiable 2-manifold by looking at the transition functions.

**1.2(2)** Give a coordinate covering for $\mathbb{R}P^3$, pick a pair of patches, and show that the overlap map is differentiable.

**1.2(3)** **Complex projective *n*-space** $\mathbb{C}P^n$ is defined to be the space of *complex* lines through the origin of $\mathbb{C}^{n+1}$. To a point $(z_0, z_1, \ldots, z_n)$ in $(\mathbb{C}^{n+1} - 0)$ we associate the line consisting of all *complex* multiples $\lambda\, (z_0, z_1, \ldots, z_n)$ of this point, $\lambda \in \mathbb{C}$. We call $[z_0, z_1, \ldots, z_n]$ the homogeneous coordinates of this line, that is, of this point in $\mathbb{C}P^n$; thus $[z_0, z_1, \ldots, z_n] = [\mu z_0, \mu z_1, \ldots, \mu z_n]$ for all $\mu \in (\mathbb{C} - 0)$. If $z_p \neq 0$ on this line, we may associate to this point $[z_0, z_1, \ldots, z_n]$ its $n$ complex $U_p$ coordinates $z_0/z_p, z_1/z_p, \ldots, z_n/z_p$, with $z_p/z_p$ omitted.

Show that $\mathbb{C}P^2$ is a complex manifold of complex dimension 2.

Note that $\mathbb{C}P^1$ has complex dimension 1, that is, real dimension 2. For $z_1 \neq 0$ the $U_1$ coordinate of the point $[z_0, z_1]$ is $z = z_0/z_1$, whereas if $z_0 \neq 0$ the $U_0$ coordinate is $w = z_1/z_0$. These two patches cover $\mathbb{C}P^1$ and in the intersection of these two patches we have $w = 1/z$. *Thus $\mathbb{C}P^1$ is nothing other than the Riemann sphere*!

## 1.3. Tangent Vectors and Mappings

What do we mean by a "critical point" of a map $F : M^n \to V^r$?

We are all acquainted with vectors in $\mathbb{R}^N$. A tangent vector to a *submanifold* $M^n$ of $\mathbb{R}^N$, at a given point $p \in M^n$, *is simply the usual velocity vector* $\dot{x}$ *to some parameterized curve* $x = x(t)$ *of* $\mathbb{R}^N$ *that lies on* $M^n$. On the other hand, a *manifold* $M^n$, as defined in the previous section, is a rather abstract object that need not be given as a subset of $\mathbb{R}^N$. For example, the projective plane $\mathbb{R}P^2$ was defined to be the space of lines through the origin of $\mathbb{R}^3$, that is, a point in $\mathbb{R}P^2$ is an entire *line* in $\mathbb{R}^3$; if $\mathbb{R}P^2$ were a submanifold of $\mathbb{R}^3$ we would associate a *point* of $\mathbb{R}^3$ to each point of $\mathbb{R}P^2$. We will be forced to define what we mean by a tangent vector to an abstract manifold. This definition will coincide with the previous notion in the case that $M^n$ is a submanifold of $\mathbb{R}^N$. The fact that we understand tangent vectors to submanifolds is a powerful psychological tool, for it can be shown (though it is not elementary) that *every manifold can be realized as a submanifold of some* $\mathbb{R}^N$. In fact, Hassler Whitney, one of the most important contributors to manifold theory in the twentieth century, has shown that every $M^n$ can be realized as a submanifold of $\mathbb{R}^{2n}$. Thus although we cannot "embed" $\mathbb{R}P^2$ in $\mathbb{R}^3$ (recall that we had a difficulty with sewing in 1.2b, Example (vii) ), it can be embedded in $\mathbb{R}^4$. It is surprising, however, that for many purposes it is of little help to use the fact that $M^n$ can be embedded in $\mathbb{R}^N$, and we shall try to give definitions that are "intrinsic," that is, independent of the use of an embedding. Nevertheless, we shall not hesitate to use an embedding for purposes of visualization, and in fact most of our examples will be concerned with submanifolds rather than manifolds.

A good reference for manifolds is [G, P]. The reader should be aware, however, that these authors deal only with manifolds that are *given* as subsets of some euclidean space.

## 1.3a. Tangent or "Contravariant" Vectors

We motivate the definition of vector as follows. Let $p = p(t)$ be a curve lying on the manifold $M^n$; thus $p$ is a map of some interval on $\mathbb{R}$ into $M^n$. In a coordinate system $(U, x_U)$ about the point $p_0 = p(0)$ the curve will be described by $n$ functions $x_U^i = x_U^i(t)$, which will be assumed differentiable. The "velocity vector" $\dot{p}(0)$ was classically described by the $n$-tuple of real numbers $dx_U^1/dt]_0, \ldots, dx_U^n/dt]_0$. If $p_0$ also lies in the coordinate patch $(V, x_V)$, then this same velocity vector is described by another $n$-tuple $dx_V^1/dt]_0, \ldots, dx_V^N/dt]_0$, related to the first set by the chain rule applied to the overlap functions (1.3), $x_V = x_V(x_U)$,

$$\left.\frac{dx_V^i}{dt}\right]_0 = \sum_{j=1}^{n} \left(\frac{\partial x_V^i}{\partial x_U^j}\right)(p_0)\left(\frac{dx_U^j}{dt}\right)_0$$

This suggests the following.

**Definition:** A **tangent vector,** or **contravariant vector,** or simply a **vector** at $p_0 \in M^n$, call it $\mathbf{X}$, assigns to each coordinate patch $(U, x)$ holding $p_0$, an $n$-tuple of real numbers

$$(X_U^i) = (X_U^1, \ldots, X_U^n)$$

such that if $p_0 \in U \cap V$, then

$$X_V^i = \sum_j \left[\frac{\partial x_V^i}{\partial x_U^j}(p_0)\right] X_U^j \tag{1.6}$$

If we let $X_U = (X_U^1, \ldots, X_U^n)^T$ be the column of vector "components" of $\mathbf{X}$, we can write this as a matrix equation

$$X_V = c_{VU} X_U \tag{1.7}$$

where the **transition function** $c_{VU}$ is the $n \times n$ Jacobian matrix evaluated at the point in question.

The term contravariant is traditional and is used throughout physics, and we shall use it even though it conflicts with the modern mathematical terminology of "categories and functors."

### 1.3b. Vectors as Differential Operators

In euclidean space an important role is played by the notion of differentiating a function $f$ with respect to a vector at the point $p$

$$D_{\mathbf{v}}(f) = \frac{d}{dt}[f(p + t\mathbf{v})]_{t=0} \tag{1.8}$$

and if $(x)$ is any cartesian coordinate system we have

$$D_{\mathbf{v}}(f) = \sum_j \left[\frac{\partial f}{\partial x^j}\right](p) v^j$$

This is the motivation for a similar operation on functions on any manifold $M$. A real-valued function $f$ defined on $M^n$ near $p$ can be described in a local coordinate system $x$ in the form $f = f(x^1, \ldots, x^n)$. (Recall, from Section 1.2c, that we are really dealing with the function $f \circ \phi_U^{-1}$ where $\phi_U$ is a coordinate map.) If $\mathbf{X}$ is a vector at $p$ we define the derivative of $f$ with respect to the vector $\mathbf{X}$ by

$$\mathbf{X}_p(f) := D_{\mathbf{X}}(f) := \sum_j \left[\frac{\partial f}{\partial x^j}\right](p) X^j \tag{1.9}$$

This seems to depend on the coordinates used, although it should be apparent from (1.8) that this is not the case in $\mathbb{R}^n$. We must show that (1.9) defines an operation that is independent of the local coordinates used. Let $(U, x_U)$ and $(V, x_V)$ be two coordinate systems. From the chain rule we see

$$D_{\mathbf{X}}^V(f) = \sum_j \left(\frac{\partial f}{\partial x_V^j}\right) X_V^j = \sum_j \left(\frac{\partial f}{\partial x_V^j}\right) \sum_i \left(\frac{\partial x_V^j}{\partial x_U^i}\right) X_U^i$$
$$= \sum_i \left(\frac{\partial f}{\partial x_U^i}\right) X_U^i = D_{\mathbf{X}}^U(f)$$

This illustrates a basic point. *Whenever we define something by use of local coordinates, if we wish the definition to have intrinsic significance we must check that it has the same meaning in all coordinate systems.*

Note then that there is a 1 : 1 correspondence between tangent vectors $\mathbf{X}$ to $M^n$ at $p$ and first-order differential operators (on differentiable functions defined near $p$) that take the special form

$$\mathbf{X}_p = \sum_j X^j \frac{\partial}{\partial x^j}\bigg]_p \tag{1.10}$$

in a local coordinate system $(x)$. From now on, *we shall make no distinction between a vector and its associated differential operator*. Each one of the $n$ operators $\partial/\partial x^i$ then defines a vector, written $\boldsymbol{\partial}/\boldsymbol{\partial} x^i$, at each $p$ in the coordinate patch.

The $i^{\text{th}}$ component of $\boldsymbol{\partial}/\boldsymbol{\partial} x^\alpha$ is, from (1.9), given by $\delta^i_\alpha$ (where the Kronecker $\delta^i_\alpha$ is 1 if $i = \alpha$ and 0 if $i \neq \alpha$). On the other hand, consider the $\alpha^{\text{th}}$ **coordinate curve** through a point, the curve being parameterized by $x^\alpha$. This curve is described by $x^i(t) = \text{constant}$ for $i \neq \alpha$ and $x^\alpha(t) = t$. The velocity vector for this curve at parameter value $t$ has components $dx^i/dt = \delta^i_\alpha$. *The $j^{\text{th}}$ coordinate vector $\boldsymbol{\partial}/\boldsymbol{\partial} x^j$ is the velocity vector to the $j^{\text{th}}$ coordinate curve parameterized by $x^j$!* If $M^n \subset \mathbb{R}^N$, and if $\mathbf{r} = (y^1, \ldots, y^N)^T$ is the usual position vector from the origin, then $\boldsymbol{\partial}/\boldsymbol{\partial} x^j$ would be written *classically* as $\partial\mathbf{r}/\partial x^j$,

$$\frac{\boldsymbol{\partial}}{\boldsymbol{\partial} x^j} = \frac{\partial \mathbf{r}}{\partial x^j} = \left(\frac{\partial y^1}{\partial x^j}, \ldots, \frac{\partial y^N}{\partial x^j}\right)^T \tag{1.11}$$

A familiar example will be given in the next section.

## 1.3c. The Tangent Space to $M^n$ at a Point

It is evident from (1.6) that the sum of two vectors at a point, defined in terms of their $n$-tuples, is again a vector at that point, and that the product of a vector by a scalar, that is, a real number, is again a vector.

**Definition:** The **tangent space** to $M^n$ at the point $p \in M^n$, written $M^n_p$, is the real vector space consisting of all tangent vectors to $M^n$ at $p$. If $(x)$ is a coordinate system holding $p$, then the $n$ vectors

$$\frac{\boldsymbol{\partial}}{\boldsymbol{\partial} x^1}\bigg]_p, \ldots, \frac{\boldsymbol{\partial}}{\boldsymbol{\partial} x^n}\bigg]_p$$

form a basis of this $n$-dimensional vector space (as is evident from (1.10)) and this basis is called a **coordinate basis** or **coordinate frame**.

*If $M^n$ is a submanifold of $\mathbb{R}^N$, then $M^n_p$ is the usual n-dimensional affine subspace of $\mathbb{R}^N$ that is "tangent" to $M^n$ at p, and this is the picture to keep in mind.*

A vector **field** on an open set $U$ will be the differentiable assignment of a vector $\mathbf{X}$ to each point of $U$; in terms of local coordinates

$$\mathbf{X} = \sum_j X^j(x) \frac{\boldsymbol{\partial}}{\boldsymbol{\partial} x^j}$$

where the components $X^j$ are differentiable functions of $(x)$. In particular, each $\boldsymbol{\partial}/\boldsymbol{\partial} x^j$ is a vector field in the coordinate patch.

**Example:**

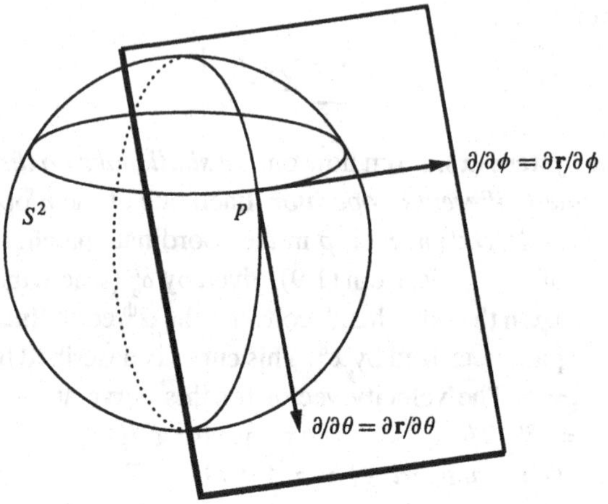


tangent space to $S^2$ at $p = S_p^2$

**Figure 1.17**

We have drawn the unit 2-sphere $M^2 = S^2$ in $\mathbb{R}^3$ with the usual spherical coordinates $\theta$ and $\phi$ ($\theta$ is colatitude and $-\phi$ is longitude). The equations defining $S^2$ are $x = \sin\theta\cos\phi$, $y = \sin\theta\sin\phi$, and $z = \cos\theta$. The coordinate vector $\partial/\partial\theta = \partial\mathbf{r}/\partial\theta$ is the velocity vector to a line of longitude, that is, keep $\phi$ constant and parameterize the meridian by "time" $t = \theta$. $\partial/\partial\phi$ has a similar description. Note that *these two vectors at p do not live in $S^2$, but rather in the linear space $S_p^2$ attached to $S^2$ at p*. Vectors at $q \neq p$ live in a different vector space $S_q^2$.

**Warning**: Because $S^2$ is a submanifold of $\mathbb{R}^3$ and because $\mathbb{R}^3$ carries a familiar metric, it makes sense to talk about the length of tangent vectors to this particular $S^2$; for example, we would say that $\| \partial/\partial\theta \| = 1$ and $\| \partial/\partial\phi \| = \sin\theta$. However, the definition of a manifold given in 1.2c does not require that $M^n$ be given as some specific subset of some $\mathbb{R}^N$; *we do not have the notion of length of a tangent vector to a general manifold*. For example, the configuration space of a thermodynamical system might have coordinates given by pressure $p$, volume $v$, and temperature $T$, and the notions of the *lengths* of $\partial/\partial p$, and so on, seem to have no physical significance. If we wish to talk about the "length" of a vector on a manifold we shall be forced to introduce an *additional structure* on the manifold in question. The most common structure so used is called a *Riemannian structure*, or *metric*, which will be introduced in Chapter 2. See Problem 1.3 (1) at this time.

### 1.3d. Mappings and Submanifolds of Manifolds

Let $F : M^n \to V^r$ be a map from one manifold to another. In terms of local coordinates $x$ near $p \in M^n$ and $y$ near $F(p)$ on $V^r$ $F$ is described by $r$ functions of $n$ variables $y^\alpha = F^\alpha(x^1, \dots, x^n)$, which can be abbreviated to $y = F(x)$ or $y = y(x)$. If, as we shall assume, the functions $F^\alpha$ are differentiable functions of the $x$'s, we say that $F$ is differentiable. As usual, such functions are, in particular, continuous.

When $n = r$, we say that $F$ is a **diffeomorphism** provided $F$ is $1:1$, onto, and if, in addition, $F^{-1}$ is also differentiable. Thus such an $F$ is a differentiable homeomorphism (see 1.2a) with a differentiable inverse. (If $F^{-1}$ does exist and the Jacobian determinant does not vanish, $\partial(y^1, \dots, y^n)/\partial(x^1, \dots, x^n) \neq 0$, then the inverse function theorem of advanced calculus (see 1.3e) would assure us that the inverse *is* differentiable.)

The map $F : \mathbb{R} \to \mathbb{R}$ given by $y = x^3$ is a differentiable homeomorphism, but it is not a diffeomorphism since the inverse $x = y^{1/3}$ is not differentiable at $x = 0$.

We have already discussed submanifolds of $\mathbb{R}^n$ but now we shall need to discuss submanifolds of a manifold. A good example is the equator $S^1$ of $S^2$.

**Definition:** $W^r \subset M^n$ is an **(embedded) submanifold** of the manifold $M^n$ provided $W$ is *locally* described as the common locus

$$F^1(x^1, \dots, x^n) = 0, \dots, F^{n-r}(x^1, \dots, x^n) = 0$$

of $(n - r)$ differentiable functions that are independent in the sense that the Jacobian matrix $[\partial F^\alpha / \partial x^i]$ has rank $(n - r)$ at each point of the locus.

The implicit function theorem assures us that $W^r$ can be locally described (after perhaps permuting some of the $x$ coordinates ) as a locus

$$x^{r+1} = f^{r+1}(x^1, \dots x^r), \ \dots, \ x^n = f^n(x^1, \dots, x^r)$$

It is not difficult to see from this (as we saw in the case $S^2 \subset \mathbb{R}^3$) that *every embedded submanifold of $M^n$ is itself a manifold*!

Later on we shall have occasion to discuss submanifolds that are not "embedded," but for the present we shall assume "embedded" without explicit mention.

**Definition:** The **differential** $F_*$ of the map $F : M^n \to V^r$ has the same meaning as in the case $\mathbb{R}^n \to \mathbb{R}^r$ discussed in 1.1b. $F_* : M^n_p \to V^r_{F(p)}$ is the linear transformation defined as follows. For $\mathbf{X} \in M^n_p$, let $p = p(t)$ be a curve on $M$ with $p(0) = p$ and with velocity vector $\dot{p}(0) = \mathbf{X}$. Then $F_*\mathbf{X}$ is the velocity vector $d/dt\{F(p(t))\}_{t=0}$ of the image curve at $F(p)$ on $V$. This vector is independent of the curve $p = p(t)$ chosen (as long as $\dot{p}(0) = \mathbf{X}$). The matrix of this linear transformation, in terms of the bases $\partial/\partial x$ at $p$ and $\partial/\partial y$ at $F(p)$, is the Jacobian matrix

$$(F_*)^\alpha{}_i = \frac{\partial F^\alpha}{\partial x^i}(p) = \frac{\partial y^\alpha}{\partial x^i}(p)$$

The main theorem on submanifolds is exactly as in euclidean space (Section 1.1c).

**Theorem (1.12):** *Let $F : M^n \to V^r$ and suppose that for some $q \in V^r$ the locus $F^{-1}(q) \subset M^n$ is not empty. Suppose further that $F_*$ is onto, that is, $F_*$ is of rank $r$, at each point of $F^{-1}(q)$. Then $F^{-1}(q)$ is an $(n-r)$-dimensional submanifold of $M^n$.*

**Example:** Consider a 2-dimensional torus $T^2$ (the surface of a doughnut), embedded in $\mathbb{R}^3$.

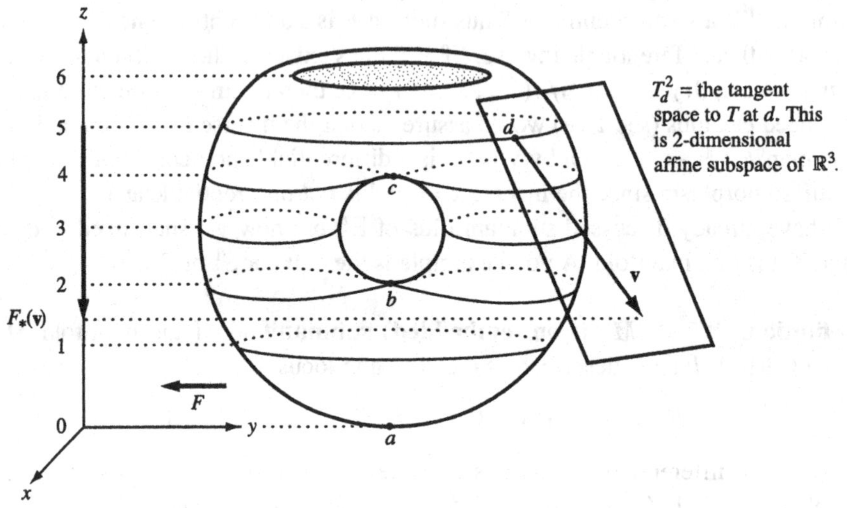


**Figure 1.18**

We have drawn it smooth with a flat top (which is supposed to join *smoothly* with the rest of the torus). Define a differentiable map (function) $F : T^2 \to \mathbb{R}$ by $F(p) = z$, the height of the point $p \in T^2$ above the $z$ plane ($\mathbb{R}$ is being identified with the $z$ axis). Consider a point $d \in T$ and a tangent vector $\mathbf{v}$ to $T$ at $d$. Let $p = p(t)$ be a curve on $T$ such that $p(0) = d$ and $\dot{p}(0) = \mathbf{v}$. The image curve in $\mathbb{R}$ is described in the coordinate $z$ for $\mathbb{R}$ by $z(t) = z(p(t))$, and it is clear from the geometry of $T^2 \subset \mathbb{R}^3$ that $\dot{z}(0)$ is simply the $z$ component of the spatial vector $\mathbf{v}$. In other words $F_*(\mathbf{v})$ *is the projection of* $\mathbf{v}$ *onto the $z$ axis*. Note then that $F_*$ will be onto at each point $p \in T^2$ for which the tangent plane $T^2(p)$ is *not* horizontal, that is, at all points of $T^2$ except $a \in F^{-1}(0)$, $b \in F^{-1}(2)$, $c \in F^{-1}(4)$, and the entire flat top $F^{-1}(6)$.

From the main theorem, we may conclude that $F^{-1}(z)$ is a 1-dimensional submanifold of the torus for $0 \le z \le 6$ except for $z = 0, 2, 4$, and 6, and this is indeed "verified" in our picture. (We have drawn the inverse images of $z = 0, 1, \ldots, 6$.) Notice that $F^{-1}(2)$, which looks like a figure 8, is *not* a submanifold; a neighborhood of the point $b$ on $F^{-1}(2)$ is topologically a cross $+$ and thus no neighborhood of $b$ is topologically an open interval on $\mathbb{R}$.

**Definition:** If $F : M^n \to V^r$ is a differentiable map between manifolds, we say that

(i) $x \in M$ is a **regular point** if $F_*$ maps $M_x^n$ onto $V_{F(x)}^r$; otherwise we say that $x$ is a **critical point.**

(ii) $y \in V^r$ is a **regular value** provided *either* $F^{-1}(y)$ is empty, or $F^{-1}(y)$ consists entirely of regular points. Otherwise $y$ is a **critical value.**

Our main theorem on submanifolds can then be stated as follows.

**Theorem (1.13):** *If $y \in V^r$ is a regular value, then $F^{-1}(y)$ either is empty or is a submanifold of $M^n$ of dimension $(n - r)$.*

Of course, if $x$ is a critical *point* then $F(x)$ is a critical *value*. In our toroidal example, Figure 1.18, all values of $z$ other than 0, 2, 4, and 6 are regular. The critical points on $T^2$ consist of $a$, $b$, $c$, and the entire flat top of $T^2$. These latter critical points thus fill up a positive area (in the sense of elementary calculus) on $T^2$. Note however, that the image of this 2-dimensional set of critical points consists of the single critical value $z = 6$. The following theorem assures us that the critical *values* of a map form a "small" subset of $V^r$; the critical values cannot fill up any open set in $V^r$ and they will have "measure" 0. We will not be precise in defining "almost all"; roughly speaking we mean, in some sense, "with probability 1."

**Sard's Theorem (1.14):** *If $F : M^n \to V^r$ is sufficiently differentiable, then almost all values of $F$ are regular values, and thus for almost all points $y \in V^r$, $F^{-1}(y)$ either is empty or is a submanifold of $M^n$ of dimension $(n - r)$.*

By sufficiently differentiable, we mean the following. If $n \leq r$, we demand that $F$ be of differentiability class $C^1$, whereas if $n - r = k > 0$, we demand that $F$ be of class $C^{k+1}$. The proof of Sard's theorem is delicate, especially if $n > r$; see, for example, [$A, M, R$].

## 1.3e. Change of Coordinates

The inverse function theorem is perhaps the most important theoretical result in all of differential calculus.

**The Inverse Function Theorem (1.15):** *If $F : M^n \to V^n$ is a differentiable map between manifolds of the same dimension, and if at $x_0 \in M$ the differential $F_*$ is an isomorphism, that is, it is* 1 : 1 *and onto, then $F$ is a local diffeomorphism near $x_0$.*

This means that there is a neighborhood $U$ of $x$ such that $F(U)$ is open in $V$ and $F : U \to F(U)$ is a diffeomorphism. This theorem is a powerful tool for introducing new coordinates in a neighborhood of a point, for it has the following consequence.

**Corollary (1.16):** *Let $x^1, \ldots, x^n$ be local coordinates in a neighborhood $U$ of the point $p \in M^n$. Let $y^1, \ldots, y^n$ be any differentiable functions of the $x$'s ( thus yielding a map:$U \to \mathbb{R}^n$) such that*

$$\frac{\partial(y^1, \ldots, y^n)}{\partial(x^1, \ldots, x^n)}(p) \neq 0$$

*Then the $y$'s form a coordinate system in some (perhaps smaller) neighborhood of $p$.*

For example, when we put $x = r\cos\theta$, $y = r\sin\theta$, we have $\partial(x, y)/\partial(r, \theta) = r$, and so $\partial(r, \theta)/\partial(x, y) = 1/r$. This shows that polar coordinates are good coordinates in a neighborhood of any point of the plane other than the origin.

It is important to realize that *this theorem is only local*. Consider the map $F : \mathbb{R}^2 \to \mathbb{R}^2$ given by $u = e^x \cos y$, $v = e^x \sin y$. This is of course the complex analytic map $w = e^z$. The real Jacobian $\partial(u, v)/\partial(x, y)$ never vanishes (this is reflected in the complex Jacobian $dw/dz = e^z$ never vanishing). Thus $F$ is locally $1 : 1$. It is not globally so since $e^{z+2\pi ni} = e^z$ for all integers $n$. $u$, $v$ form a coordinate system not in the whole plane but rather in any strip $a \le y < a + 2\pi$.

The inverse function theorem and the implicit function theorem are essentially equivalent, the proof of one following rather easily from that of the other. The proofs are fairly delicate; see for example, $[A, M, R]$.

## Problems

**1.3(1)** What would be wrong in defining $\| \mathbf{X} \|$ in an $M^n$ by

$$\| \mathbf{X} \|^2 = \sum_j (X_U^j)^2 \ ?$$

**1.3(2)** Lay a 2-dimensional torus flat on a table (the $xy$ plane) rather than standing as in Figure 1.18. By inspection, what are the critical points of the map $T^2 \to \mathbb{R}^2$ projecting $T^2$ into the $xy$ plane?

**1.3(3)** Let $M^n$ be a submanifold of $\mathbb{R}^N$ that does not pass through the origin. Look at the critical points of the function $f : M \to \mathbb{R}$ that assigns to each point of $M$ the square of its distance from the origin. Show, using local coordinates $u^1, \ldots, u^n$, that a point is a critical point for this distance function iff the position vector to this point is normal to the submanifold.

## 1.4. Vector Fields and Flows

Can one solve $dx^i/dt = \partial f/\partial x^i$ to find the curves of steepest ascent?

### 1.4a. Vector Fields and Flows on $\mathbb{R}^n$

A vector field on $\mathbb{R}^n$ assigns in a differentiable manner a vector $\mathbf{v}_p$ to each $p$ in $\mathbb{R}^n$. In terms of cartesian coordinates $x^1, \ldots, x^n$

$$\mathbf{v} = \sum_j v^j(x) \frac{\partial}{\partial x^j}$$

where the components $v^j$ are differentiable functions. Classically this would be written simply in terms of the cartesian components $\mathbf{v} = (v^1(x), \ldots, v^n(x))^T$.

Given a "stationary" (i.e., time-independent) flow of water in $\mathbb{R}^3$, we can construct the 1-parameter family of maps

$$\phi_t : \mathbb{R}^3 \to \mathbb{R}^3$$

where $\phi_t$ takes the molecule located at $p$ when $t = 0$ to the position of the same molecule $t$ seconds later. Since the flow is time-independent

$$\phi_s(\phi_t(p)) = \phi_{s+t}(p) = \phi_t(\phi_s(p))$$

and (1.17)

$$\phi_{-t}(\phi_t(p)) = p, \quad \text{i.e., } \phi_{-t} = \phi_t^{-1}$$

We say that this defines a 1-**parameter group** of maps. Furthermore, if each $\phi_t$ is differentiable, then so is each $\phi_t^{-1}$, and so *each $\phi_t$ is a diffeomorphism*. We shall call such a family simply a **flow.** Associated with any such flow is a time-independent **velocity field**

$$\mathbf{v}_p := \left.\frac{d\phi_t(p)}{dt}\right]_{t=0}$$

In terms of coordinates we have

$$v^j(p) = \left.\frac{dx^j(\phi_t(p))}{dt}\right]_{t=0}$$

which will usually be written

$$v^j(x) = \frac{dx^j}{dt}$$

Thought of as a differential operator on functions $f$

$$\mathbf{v}_p(f) = \sum_j v^j(p)\frac{\partial f}{\partial x^j} = \sum_j \frac{dx^j}{dt}\frac{\partial f}{\partial x^j}$$
$$= \left.\frac{d}{dt}f(\phi_t(p))\right]_{t=0}$$

is the derivative of $f$ along the "streamline" through $p$.

We thus have the almost trivial observation that to each flow $\{\phi_t\}$ we can associate the velocity vector field. The converse result, perhaps the most important theorem relating calculus to science, states, roughly speaking, that to each vector field $\mathbf{v}$ in $\mathbb{R}^n$ one may associate a flow $\{\phi_t\}$ having $\mathbf{v}$ as its velocity field, and that $\phi_t(p)$ can be found by solving the system of ordinary differential equations

$$\frac{dx^j}{dt} = v^j(x^1(t), \ldots, x^n(t)) \tag{1.18}$$

with initial conditions

$$x(0) = p$$

Thus one finds the **integral curves** of the preceding system, and $\phi_t(p)$ says, "Move along the integral curve through $p$ (the 'orbit' of $p$) for time $t$." We shall now give a precise statement of this "fundamental theorem" on the existence of solutions of ordinary differential equations. For details one can consult [A, M, R; chap. 4], where this result is proved in the context of Banach spaces rather than $\mathbb{R}^n$. I recommend highly chapters 4 and 5 of Arnold's book [A2].

**The Fundamental Theorem on Vector Fields in $\mathbb{R}^n$ (1.19):** *Let* $\mathbf{v}$ *be a* $C^k$ *vector field,* $k \geq 1$ *(each component* $v^j(x)$ *is of differentiability class* $C^k$*) on an open subset* $U$ *of* $\mathbb{R}^n$*. This can be written* $\mathbf{v} : U \to \mathbb{R}^n$ *since* $\mathbf{v}$ *associates to each* $x \in U$ *a point* $\mathbf{v}(x) \in \mathbb{R}^n$*. Then for each* $p \in U$ *there is a curve* $\gamma$ *mapping an interval* $(-b, b)$ *of the real line into* $U$

$$\gamma : (-b, b) \to U$$

*such that*

$$\frac{d\gamma(t)}{dt} = v(\gamma(t)) \quad \textit{and} \quad \gamma(0) = p$$

*for all* $t \in (-b, b)$*. (This says that* $\gamma$ *is an integral curve of* $\mathbf{v}$ *starting at* $p$*.) Any two such curves are equal on the intersection of their* $t$*-domains ("uniqueness"). Moreover, there is a neighborhood* $U_p$ *of* $p$*, a real number* $\epsilon > 0$*, and a* $C^k$ *map*

$$\Phi : U_p \times (-\epsilon, \epsilon) \to \mathbb{R}^n$$

*such that the curve* $t \in (-\epsilon, \epsilon) \mapsto \phi_t(q) := \Phi(q, t)$ *satisfies the differential equation*

$$\frac{\partial}{\partial t}\phi_t(q) = \mathbf{v}(\phi_t(q))$$

*for all* $t \in (-\epsilon, \epsilon)$ *and* $q \in U_p$*. Moreover, if* $t$*,* $s$*, and* $t + s$ *are all in* $(-\epsilon, \epsilon)$*, then*

$$\phi_t \circ \phi_s = \phi_{t+s} = \phi_s \circ \phi_t$$

*for all* $q \in U_p$*, and thus* $\{\phi_t\}$ *defines a local 1-parameter "group" of diffeomorphisms, or* ***local flow****.*

The term *local* refers to the fact that $\phi_t$ is defined only on a subset $U_p \subset U \subset \mathbb{R}^n$. The word "group" has been put in quotes because this family of maps does not form a group in the usual sense. In general (see Problem 1.4 (1)), the maps $\phi_t$ are only defined for small $t$, $-\epsilon < t < \epsilon$; that is, **the integral curve through a point q need only exist for a small time**. Thus, for example, if $\epsilon = 1$, then although $\phi_{1/2}(q)$ exists neither $\phi_1(q)$ nor $\phi_{1/2} \circ \phi_{1/2}$ need exist; the point is that $\phi_{1/2}(q)$ need not be in the set $U_p$ on which $\phi_{1/2}$ is defined.

**Example:** $\mathbb{R}^n = \mathbb{R}$, the real line, and $v(x) = x d/dx$. Thus $v$ has a single component $x$ at the point with coordinate $x$. Let $U = \mathbb{R}$. To find $\phi_t$ we simply solve the differential equation

$$\frac{dx}{dt} = x \qquad \text{with initial condition } x(0) = p$$

to get $x(t) = e^t p$, that is, $\phi_t(p) = e^t p$. In this example the map $\phi_t$ is clearly defined on all of $M^1 = \mathbb{R}$ and for all time $t$. It can be shown that this is true for any *linear* vector field

$$\frac{dx^j}{dt} = \sum_k a^j_k x^k$$

defined on *all* of $\mathbb{R}^n$.

Note that if we solved the differential equation $dx/dt = 1$ on the real line with the origin deleted, that is, on the *manifold* $M^1 = \mathbb{R} - 0$, then the solution curve starting at $x = -1$ at $t = 0$ would exist for all times less than 1 second, but $\phi_1$ would not exist; the solution simply runs "off" the manifold because of the missing point. One might think that if we avoid dealing with pathologies such as digging out a point from $\mathbb{R}^1$, then our solutions would exist for all time, but as you shall verify in Problem 1.4(1) this is not the case. The growth of the vector field can cause a solution curve to "leave" $\mathbb{R}^1$ in a finite amount of time.

We have required that the vector field $\mathbf{v}$ be differentiable. Uniqueness can be lost if the field $\mathbf{v}$ is only continuous. For example, again on the real line, consider the differential equation $dx/dt = 3x^{2/3}$. The usual solutions are of the form $x(t) = (t-c)^3$, but there is also the "singular" solution $x(t) = 0$ identically. This is a reflection of the fact that $x^{2/3}$ is not differentiable when $x = 0$.

## 1.4b. Vector Fields on Manifolds

If $\mathbf{X}$ is a $C^k$ vector field on an open subset $W$ of a *manifold* $M^n$ then we can again recover a 1-parameter local group $\phi_t$ of diffeomorphisms for the following reasons. If $W$ is contained in a single coordinate patch $(U, x_U)$ we can proceed just as in the case $\mathbb{R}^n$ earlier since we can use the local coordinates $x_U$. Suppose that $W$ is not contained in a single patch. Let $p \in W$ be in a coordinate overlap, $p \in U \cap V$. In $U$ we can solve the differential equations

$$\frac{dx_U^j}{dt} = X_U^j(x_U^1, \ldots, x_U^n)$$

as before. In $V$ we solve the equations

$$\frac{dx_V^j}{dt} = X_V^j(x_V^1, \ldots, x_V^n)$$

Because of the transformation rule (1.6), the right-hand side of this last equation is $\sum_k [\partial x_V^j / \partial x_U^k] X_U^k$; the left-hand side is, by the chain rule, $\sum_k [\partial x_V^j / \partial x_U^k] dx_U^k / dt$. Thus, *because of the transformation rule for a contravariant vector, the two differential equations say exactly the same thing.* Using uniqueness, we may then patch together the $U$ and the $V$ solutions to give a local solution in $W$.

**Warning**: Let $f : M^n \to \mathbb{R}$ be a differentiable function on $M^n$. In elementary mathematics it is often said that the $n$-tuple

$$\left[\frac{\partial f}{\partial x^1}, \ldots, \frac{\partial f}{\partial x^n}\right]^T$$

form the components of a vector field "grad $f$." However, if we look at the transformation properties in $U \cap V$, by the chain rule

$$\frac{\partial f}{\partial x_V^j} = \sum_k \left[\frac{\partial x_U^k}{\partial x_V^j}\right] \frac{\partial f}{\partial x_U^k}$$

and this is *not* the rule for a contravariant vector. One sees then that a proposed differential equation for "steepest ascent," $dx/dt =$"grad $f$," that is,

$$\frac{dx_U^j}{dt} = \frac{\partial f}{\partial x_U^j} \quad \text{in } U \quad \text{and} \quad \frac{dx_V^j}{dt} = \frac{\partial f}{\partial x_V^j} \quad \text{in } V$$

*would not say the same thing in two overlapping patches*, and consequently *would not yield a flow* $\phi_t$! In the next chapter we shall see how to deal with $n$-tuples that transform as "grad $f$."

### 1.4c. Straightening Flows

Our version of the fundamental theorem on the existence of solutions of differential equations, as given in the previous section, is not the complete story; see [$A$, $M$, $R$, theorem 4.1.14] or [A2, chap. 4] for details of the following. The map $(p, t) \to \phi_t(p)$ depends smoothly on the initial condition $p$ and on the time of flow $t$. This has the following consequence. (Since our result will be local, it is no loss of generality to replace $M^n$ by $\mathbb{R}^n$.) *Suppose that the vector field* $\mathbf{v}$ *does not vanish at the point* $p$. Then of course it doesn't vanish in some neighborhood of $p$ in $M^n$. Let $W^{n-1}$ be a hypersurface, that is, a submanifold of codimension 1, that passes through $p$. Assume that $W$ is **transversal** to $\mathbf{v}$, that is, the vector field $\mathbf{v}$ is not tangent to $W$.

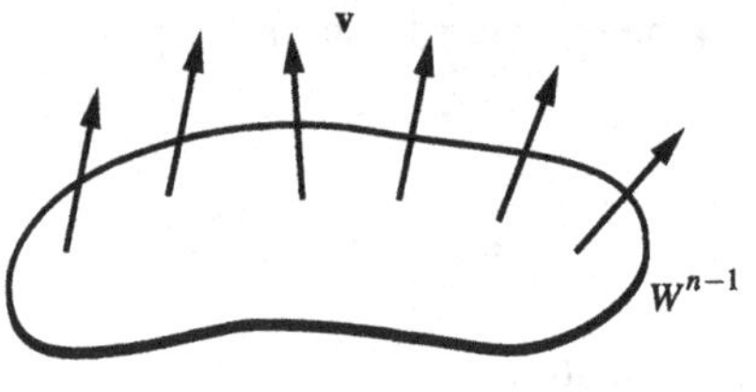


Figure 1.19

Let $u^1, \ldots, u^{n-1}$ be local coordinates for $W$, and let $p_u$ be the point on $W$ with local coordinates $u$. Then $\phi_t(p_u)$ is the point $t$ seconds along the orbit of $\mathbf{v}$ through $p_u$. This point can be described by the $n$-tuple $(u, t)$. The fundamental theorem states that if $W$ is sufficiently small and if $t$ is also sufficiently small, then $(u, t)$ *can be used as* (curvilinear) *coordinates for some n-dimensional neighborhood of* $p$ *in* $M^n$. To see this we shall apply the inverse function theorem. We thus consider the map $L : W^{n-1} \times (-\epsilon, \epsilon) \to M^n$ given by $L(u, t) = \phi_t(p_u)$. We compute the differential of this map *at the origin* $u = 0$ of the coordinates on $W^{n-1}$. Then by the geometric meaning of $L_*$, and since $\phi_0(p) = p$

$$L_*\left(\frac{\partial}{\partial u^1}\right) = \frac{\partial}{\partial u}[\phi_0(u, 0, \ldots, 0)]_0 = \left.\frac{\partial p_{(u,0,\ldots,0)}}{\partial u}\right|_{u=0} = \frac{\partial}{\partial u^1}$$

Likewise $L_*(\partial/\partial u^i) = \partial/\partial u^i$, for $i = 1, \ldots, n-1$. Finally

$$L_*(\mathbf{v}) = \frac{\partial}{\partial t}\phi_t(p_0) = \mathbf{v}$$

Thus $L_*$ is the identity linear transformation, and by Corollary (1.16) we may use $u^1, \ldots, u^{n-1}, t$ as local coordinates for $M^n$ near $p_0$.

It is then clear that *in these new local coordinates* near $p$, the flow defined by the vector field $\mathbf{v}$ is simply $\phi_S : (u, t) \to (u, s+t)$ and the vector field $\mathbf{v}$ in terms of $\partial/\partial u^1, \ldots, \partial/\partial u^{n-1}, \partial/\partial t$, is simply $\mathbf{v} = \partial/\partial t$. We have "straightened out" the flow!

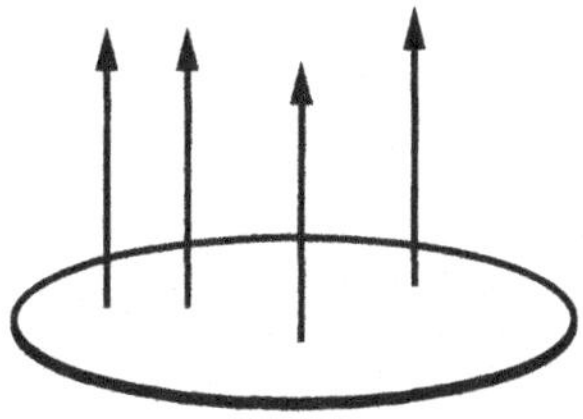

**Figure 1.20**

This says that near a **nonsingular** point of $\mathbf{v}$, that is, a point where $\mathbf{v} \neq 0$, coordinates $u^1, \ldots, u^n$ can be introduced such that the original system of differential equations $dx^1/dt = v^1(x), \ldots, dx^n/dt = v^n(x)$ becomes

$$\frac{du^1}{dt} = 0, \ldots, \frac{du^{n-1}}{dt} = 0, \qquad \frac{du^n}{dt} = 1 \tag{1.20}$$

**Thus all flows near a nonsingular point are qualitatively the same**! In a sense this result is of theoretical interest only, for in order to introduce the new coordinates $u$ *one must solve the original system of differential equations*. The theoretical interest is, however, considerable. For example, $u^1 = c_1, \ldots, u^{n-1} = c_{n-1}$, are $(n-1)$ "first integrals," that is, constants of the motion, for the system (1.20). We conclude that near any nonsingular point of any system there are $(n-1)$ first integrals, $u^1(x) = c^1, \ldots, u^{n-1}(x) = c_{n-1}$ (but of course, we might have to solve the original system to write down explicitly the functions $u^j$ in terms of the $x$'s).

## Problems

**1.4(1)** Consider the quadratic vector field problem on $\mathbb{R}$, $v(x) = x^2 d/dx$. You must solve the differential equation

$$\frac{dx}{dt} = x^2 \quad \text{and} \quad x(0) = p$$

Consider, as in the statement of the fundamental theorem, the case when $U_p$ is the set $1/2 < x < 3/2$. Find the largest $\epsilon$ so that $\Phi : U_p \times (-\epsilon, \epsilon) \to \mathbb{R}$ is defined; that is, find the largest $t$ for which the integral curve $\phi_t(q)$ will be defined for all $1/2 < q < 3/2$.

**1.4(2)** In the complex plane we can consider the differential equations $dz/dt = 1$, where $t$ is real. The integral curves are of course lines parallel to the real axis.This can also be considered a differential equation on the $z$ patch of the Riemann sphere of Section 1.2d. Extend this differential equation to the entire sphere by writing out the equivalent equation in the $w$ patch. Write out the general solution $w = w(t)$ in the neighborhood of $w = 0$, and draw in particular the solutions starting at $i$, $\pm 1$, and $-i$.

Figure 1.22

CHAPTER 2

# Tensors and Exterior Forms

IN Section 1.4b we considered the $n$-tuple of partial derivatives of a single function $\partial F/\partial x^j$ and we noticed that this $n$-tuple does not transform in the same way as the $n$-tuple of components of a vector. These components $\partial F/\partial x^j$ transform as a new type of "vector." In this chapter we shall talk of the general notion of "tensor" that will include both notions of vector and a whole class of objects characterized by a transformation law generalizing 1.6. We shall, however, strive to define these objects and operations on them "intrinsically," that is, in a basis-free fashion. We shall also be very careful in our use of sub- and superscripts when we express components in terms of bases; *the notation is designed to help us recognize intrinsic quantities when they are presented in component form and to help prevent us from making blatant errors.*

## 2.1. Covectors and Riemannian Metrics

How *do* we find the curves of steepest ascent?

### 2.1a. Linear Functionals and the Dual Space

Let $E$ be a real vector space. Although for some purposes $E$ may be infinite-dimensional, we are mainly concerned with the finite-dimensional case. Although $\mathbb{R}^n$, as the space of real $n$-tuples $(x^1, \ldots, x^n)$, comes equipped with a distinguished basis $(1, 0, 0, \ldots, 0)^T, \ldots$, the general $n$-dimensional vector space $E$ has no basis prescribed.

Choose a basis $\mathbf{e}_1, \ldots, \mathbf{e}_n$ for the $n$-dimensional space $E$. Then a vector $\mathbf{v} \in E$ has a unique expansion

$$\mathbf{v} = \sum_j \mathbf{e}_j v^j = \sum_j v^j \mathbf{e}_j$$

where the $n$ real numbers $v^j$ are the **components** of $\mathbf{v}$ with respect to the given basis. For algebraic purposes, *we prefer the first presentation*, where we have put the "scalars" $v^j$ to the right of the basis elements. We do this for several reasons, but mainly so that *we can use matrix notation*, as we shall see in the next paragraph. *When dealing* *with calculus, however, this notation is awkward.* For example, in $\mathbb{R}^n$ (thought of as a manifold), we can write the standard basis at the origin as $\mathbf{e}_j = \boldsymbol{\partial}/\boldsymbol{\partial} x^j$ (as in Section 1.3c); then our favored presentation would say $\mathbf{v} = \sum_j \boldsymbol{\partial}/\boldsymbol{\partial} x^j\ v^j$, making it appear, incorrectly, that we are differentiating the components $v^j$. *We shall employ the bold* $\boldsymbol{\partial}$ *to remind us that we are not differentiating the components in this expression.* Sometimes we will simply use the traditional $\sum_j v^j \mathbf{e}_j$.

We shall use the matrices

$$\mathbf{e} = (\mathbf{e}_1, \ldots, \mathbf{e}_n) \quad \text{and} \quad v = (v^1, \ldots, v^n)^T$$

The first is a symbolic *row* matrix since each entry is a vector rather than a scalar. Note that in the matrix $v$ we are preserving the traditional notation of representing the components of a vector by a *column* matrix. We can then write our preferred representation as a matrix product

$$\mathbf{v} = \mathbf{e}\, v \tag{2.1}$$

where $\mathbf{v}$ is a $1 \times 1$ matrix. As usual, we see that the $n$-dimensional vector space $E$, *with a choice of basis*, is isomorphic to $\mathbb{R}^n$ under the correspondence $\mathbf{v} \to (v^1, \ldots, v^n) \in \mathbb{R}^n$, but that this isomorphism is "unnatural," that is, dependent on the choice of basis.

**Definition:** A (real) **linear functional** $\alpha$ on $E$ is a real-valued linear function $\alpha$, that is, a linear transformation $\alpha : E \to \mathbb{R}$ from $E$ to the 1-dimensional vector space $\mathbb{R}$. Thus

$$\alpha(a\mathbf{v} + b\mathbf{w}) = a\alpha(\mathbf{v}) + b\alpha(\mathbf{w})$$

for real numbers a, b, and vectors $\mathbf{v}$, $\mathbf{w}$.

By induction, we have, for any basis $\mathbf{e}$

$$\alpha\left(\sum \mathbf{e}_j v^j\right) = \sum \alpha(\mathbf{e}_j) v^j \tag{2.2}$$

This is simply of the form $\sum a_j v^j$ (where $a_j := \alpha(\mathbf{e}_j)$), and this is a linear function of the components of $\mathbf{v}$. Clearly if $\{a_j\}$ are any real numbers, then $\mathbf{v} \mapsto \sum a_j v^j$ defines a linear functional on all of $E$. Thus, *after* one has picked a basis, *the most general linear functional on the finite-dimensional vector space E is of the form*

$$\alpha(\mathbf{v}) = \sum a_j v^j \quad \text{where } a_j := \alpha(\mathbf{e}_j) \tag{2.3}$$

**Warning:** A linear functional $\alpha$ on $E$ is not itself a member of $E$; that is, $\alpha$ is not to be thought of as a *vector* in $E$. This is especially obvious in infinite-dimensional cases. For example, let $E$ be the vector space of all continuous real-valued functions $f : \mathbb{R} \to \mathbb{R}$ of a real variable $t$. The **Dirac functional** $\delta_0$ is the linear functional on $E$ defined by

$$\delta_0(f) = f(0)$$

You should convince yourself that $E$ *is* a vector space and that $\delta_0$ *is* a linear functional on $E$. *No one would confuse* $\delta_0$, *the Dirac* $\delta$ *"function," with a continuous function, that is, with an element of E.* In fact $\delta_0$ is not a function on $\mathbb{R}$ at all. Where, then, do the linear functionals live?

**Definition:** The collection of all linear functionals $\alpha$ on a vector space $E$ form a new vector space $E^*$, the **dual space** to $E$, under the operations

$$(\alpha + \beta)(\mathbf{v}) := \alpha(\mathbf{v}) + \beta(\mathbf{v}), \qquad \alpha, \beta \in E^*, \qquad \mathbf{v} \in E$$
$$(c\alpha)(\mathbf{v}) := c\alpha(\mathbf{v}), \qquad c \in \mathbb{R}$$

We shall see in a moment that if $E$ is $n$-dimensional, then so is $E^*$.

If $\mathbf{e}_1, \ldots, \mathbf{e}_n$ is a basis of $E$, we define the **dual basis** $\sigma^1, \ldots, \sigma^n$ of $E^*$ by first putting

$$\sigma^i(\mathbf{e}_j) = \delta^i{}_j$$

and then "extending $\sigma$ by linearity," that is,

$$\sigma^i\left(\sum_j \mathbf{e}_j v^j\right) = \sum_j \sigma^i(\mathbf{e}_j) v^j = \sum_j \delta^i{}_j v^j = v^i$$

Thus $\sigma^i$ *is the linear functional that reads off the* $i^{th}$ *component* (with respect to the basis $\mathbf{e}$) *of each vector* $\mathbf{v}$.

Let us verify that the $\sigma$'s do form a basis. To show linear independence, assume that a linear combination $\sum a_j \sigma^j$ is the 0 functional. Then $0 = \sum_j a_j \sigma^j(\mathbf{e}_k) = \sum_j a_j \delta^j{}_k = a_k$ shows that all the coefficients $a_k$ vanish, as desired. To show that the $\sigma$'s span $E^*$, we note that if $\alpha \in E^*$ then

$$\alpha(\mathbf{v}) = \alpha\left(\sum \mathbf{e}_j v^j\right) = \sum \alpha(\mathbf{e}_j) v^j$$
$$= \sum \alpha(\mathbf{e}_j)\sigma^j(\mathbf{v}) = \left(\sum \alpha(\mathbf{e}_j)\sigma^j\right)(\mathbf{v})$$

Thus the two linear functionals $\alpha$ and $\sum \alpha(\mathbf{e}_j)\sigma^j$ must be the same!

$$\alpha = \sum_j \alpha(\mathbf{e}_j)\sigma^j \tag{2.4}$$

This very important equation shows that the $\sigma$'s do form a basis of $E^*$.

In (2.3) we introduced the $n$-tuple $a_j = \alpha(\mathbf{e}_j)$ for each $\alpha \in E^*$. From (2.4) we see $\alpha = \sum a_j \sigma^j$. $a_j$ defines the $j^{\text{th}}$ **component** of $\alpha$.

If we introduce the matrices

$$\sigma = (\sigma^1, \ldots, \sigma^n)^T \quad \text{and} \quad a = (a_1, \ldots, a_n)$$

then we can write

$$\alpha = \sum_j a_j \sigma^j = a\sigma \tag{2.5}$$

Note that the components of a linear functional are written as a *row* matrix $a$.

If $\beta = (\beta_{iR})$ is a matrix of linear functionals and if $\mathbf{f} = (\mathbf{f}_{Rs})$ is a matrix of vectors, then by $\beta\mathbf{f} = \beta\,(\mathbf{f})$ we shall mean the matrix of scalars

$$\beta(\mathbf{f})_{is} := \sum_R \beta_{iR}\,(\mathbf{f}_{Rs})$$

Note then that $\sigma\mathbf{e}$ is the identity $n \times n$ matrix, and then equation (2.3) says

$$\alpha(\mathbf{v}) = (a\sigma)(\mathbf{e}v) = a(\sigma\mathbf{e})v = av$$

## 2.1b. The Differential of a Function

**Definition:** The dual space $M_p^{n*}$ to the tangent space $M_p^n$ at the point $p$ of a manifold is called the **cotangent space**.

Recall from (1.10) that on a manifold $M^n$, a vector $\mathbf{v}$ at $p$ is a differential operator on functions defined near $p$.

**Definition:** Let $f : M^n \to \mathbb{R}$. The **differential** of $f$ at $p$, written $df$, is the linear functional $df : M_p^n \to \mathbb{R}$ defined by

$$df(\mathbf{v}) = \mathbf{v}_p(f) \tag{2.6}$$

Note that we have defined $df$ *independent of any basis*. In local coordinates, $\mathbf{e}_j = \partial/\partial x^j]_p$ defines a basis for $M_p^n$ and

$$df\left(\sum v^j \frac{\partial}{\partial x^j}\right) = \sum v^j(p)\frac{\partial f}{\partial x^j}(p)$$

is clearly a linear function of the components of $\mathbf{v}$. In particular, we may consider the differential of a coordinate function, say $x^i$

$$dx^i\left(\frac{\partial}{\partial x^j}\right) = \frac{\partial x^i}{\partial x^j} = \delta^i{}_j$$

and

$$dx^i\left(\sum_j v^j \frac{\partial}{\partial x^j}\right) = \sum_j v^j dx^i\left(\frac{\partial}{\partial x^j}\right) = v^i$$

Thus, *for each i, the linear functional $dx^i$ reads off the $i^{th}$ component of any vector* $\mathbf{v}$ (expressed in terms of the coordinate basis). In other words

$$\sigma^i = dx^i$$

yields, for $i = 1, \ldots, n$, *the dual basis to the coordinate basis.* $dx^1, \ldots, dx^n$ form a basis for the cotangent space $M_p^{n*}$.

The most general linear functional is then expressed in coordinates, from (2.5) as

$$\alpha = \sum_j \alpha\left(\frac{\partial}{\partial x^j}\right) dx^j = \sum_j a_j\, dx^j \tag{2.7}$$

**Warning:** We shall call an expression such as (2.7) a **differential form**. In elementary calculus it is called simply a "differential." We shall not use this terminology since, as we learned in calculus, not every differential form is the differential of a function; that is, it need not be "exact." We shall discuss this later on in great detail.

*The definition of the differential of a function reduces to the usual concept of differential as introduced in elementary calculus.* Consider for example $\mathbb{R}^3$ with its usual cartesian coordinates $x = x^1$, $y = x^2$, and $z = x^3$. The differential is *there* traditionally defined in two steps.

First, the differential of an "independent" variable, that is, a coordinate function, say $dx$, is a function of ordered pairs of points. If $P = (x, y, z)$ and $Q = (x', y', z')$ then $dx$ is defined to be $(x' - x)$. Note that this is the same as our expression $dx\ (Q - P)$, where $(Q - P)$ is now the *vector* from $P$ to $Q$. The elementary definition in $\mathbb{R}^3$ takes advantage of the fact that a vector in the manifold $\mathbb{R}^3$ is determined by its endpoints, which again are in the manifold $\mathbb{R}^3$. This makes no sense in a general manifold; you cannot subtract *points* on a manifold.

Second, the differential *df* of a "dependent" variable, that is, a function $f$, is defined to be the function on pairs of points given by

$$\left(\frac{\partial f}{\partial x}\right)dx + \left(\frac{\partial f}{\partial x}\right)dy + \left(\frac{\partial f}{\partial z}\right)dz$$

Note that this is exactly what we would get from (2.7)

$$df = \sum df\left(\frac{\partial}{\partial x^i}\right)dx^i = \sum\left(\frac{\partial f}{\partial x^i}\right)dx^i$$

Our definition makes no distinction between independent and dependent variables, and makes sense in any manifold.

Our coordinate expression for $df$ obtained previously holds in any manifold

$$df = \sum_j \left(\frac{\partial f}{\partial x^j}\right)dx^j \tag{2.8}$$

A linear functional $\alpha : M^n_p \to \mathbb{R}$ is called a **covariant vector**, or **covector**, or **1-form**. A differentiable assignment of a covector to each point of an open set in $M^n$ is locally of the form

$$\alpha = \sum_j a_j(x)\, dx^j$$

and would be called a covector **field**, and so on; $df = \sum_j (\partial f/\partial x^j)dx^j$ is an example. Thus the numbers $\partial f/\partial x^1, \ldots, \partial f/\partial x^n$ form the components *not* of a vector field but rather of a covector field, the differential of $f$. We remarked in our warning in paragraph 1.4c that these numbers are called the components of the "gradient vector" in elementary mathematics, but we shall *never* say this. It is important to realize that the local expression (2.8) holds in *any* coordinate system; for example, in spherical coordinates for $\mathbb{R}^3$, $f = f(r, \theta, \phi)$ and

$$df = \left(\frac{\partial f}{\partial r}\right)dr + \left(\frac{\partial f}{\partial \theta}\right)d\theta + \left(\frac{\partial f}{\partial \phi}\right)d\phi$$

and no one would call $\partial f/\partial r$, $\partial f/\partial \theta$, $\partial f/\partial \phi$ the components of the gradient vector in spherical coordinates! They are the components of the covector or 1-form $df$. The gradient *vector* grad $f$ will be defined in the next section after an additional structure is introduced.

Under a change of local coordinates the chain rule yields

$$dx_V{}^i = \sum_j \left( \frac{\partial x_V{}^i}{\partial x_U{}^j} \right) dx_U{}^j \tag{2.9}$$

and for a general covector $\sum_i a^V{}_i dx_V{}^i = \sum_{ij} a^V{}_i (\partial x_V{}^i / \partial x_U{}^j)\, dx_U{}^j$ must be the same as $\sum_j a^U{}_j dx_U{}^j$. We then must have

$$a^U{}_j = \sum_i a^V{}_i \left( \frac{\partial x_V{}^i}{\partial x_U{}^j} \right) \tag{2.10}$$

But $\sum_j (\partial x_V{}^i / \partial x_U{}^j)(\partial x_U{}^j / \partial x_V{}^k) = \partial x_V{}^i / \partial x_V{}^k = \delta^i{}_k$ shows that $\partial x_U / \partial x_V$ is the inverse matrix to $\partial x_V / \partial x_U$. Equation (2.10) is, in matrix form, $a^U = a^V \, (\partial x_V / \partial x_U)$, and this yields $a^V = a^U \, (\partial x_U / \partial x_V)$, or

$$a^V{}_i = \sum_j a^U{}_j \left( \frac{\partial x_U{}^j}{\partial x_V{}^i} \right) \tag{2.11}$$

This is the *transformation rule* for the components of a *covariant* vector, and should be compared with (1.6). In the notation of (1.7) we may write

$$a^V = a^U \, c_{UV} = a^U \, c^{-1}_{VU} \tag{2.12}$$

**Warning:** Equation (1.6) tells us how the components of a *single* contravariant vector transform under a change of coordinates. Equation (2.11), likewise, tells us how the components of a *single* 1-form $\alpha$ transform under a change of coordinates. This should be compared with (2.9). This latter tells us how the $n$-coordinate 1-forms $dx_V{}^1, \ldots, dx_V{}^n$ are related to the $n$-coordinate 1-forms $dx_U{}^1, \ldots, dx_U{}^n$. In a sense we could say that the $n$-tuple of *covariant vectors* $(dx^1, \ldots, dx^n)$ transforms as do the *components* of a single *contravariant* vector. *We shall never use this terminology.*

See Problem 2.1 (1) at this time.

## 2.1c. Scalar Products in Linear Algebra

Let $E$ be an $n$-dimensional vector space with a given **inner** (or **scalar**) **product** $\langle\, , \rangle$. Thus, for each pair of vectors $\mathbf{v}, \mathbf{w}$ of $E$, $\langle \mathbf{v}, \mathbf{w} \rangle$ is a real number, it is linear in each entry when the other is held fixed (i.e., it is *bi*linear), and it is symmetric $\langle \mathbf{v}, \mathbf{w} \rangle = \langle \mathbf{w}, \mathbf{v} \rangle$. Furthermore $\langle\, , \rangle$ is **nondegenerate** in the sense that if $\langle \mathbf{v}, \mathbf{w} \rangle = 0$ for all $\mathbf{w}$ then $\mathbf{v} = \mathbf{0}$; that is, the only vector "orthogonal" to every vector is the zero vector. If, further, $\| \mathbf{v} \|^2 := \langle \mathbf{v}, \mathbf{v} \rangle$ is positive when $\mathbf{v} \neq \mathbf{0}$, we say that the inner product is positive definite, but to accommodate relativity we shall *not* always demand this.

If $\mathbf{e}$ is a basis of $E$, then we may write $\mathbf{v} = \mathbf{e}v$ and $\mathbf{w} = \mathbf{e}w$. Then

$$\begin{aligned}\langle \mathbf{v}, \mathbf{w} \rangle &= \langle \sum_i \mathbf{e}_i v^i, \sum_j \mathbf{e}_j w^j \rangle \\ &= \sum_i v^i \langle \mathbf{e}_i, \sum_j \mathbf{e}_j w^j \rangle = \sum_i \sum_j v^i \langle \mathbf{e}_i, \mathbf{e}_j \rangle w^j\end{aligned}$$

If we define the matrix $G = (g_{ij})$ with entries

$$g_{ij} := \langle \mathbf{e}_i, \mathbf{e}_j \rangle$$

then

$$\langle \mathbf{v}, \mathbf{w} \rangle = \sum_{ij} v^i g_{ij} w^j \tag{2.13}$$

or

$$\langle \mathbf{v}, \mathbf{w} \rangle = vGw$$

The matrix $(g_{ij})$ is briefly called the **metric tensor**. This nomenclature will be explained in Section 2.3.

Note that when $\mathbf{e}$ is an **orthonormal basis**, that is, when $g_{ij} = \delta^i_j$ is the identity matrix (and this can happen only if the inner product is positive definite), then $\langle \mathbf{v}, \mathbf{w} \rangle = \sum_j v^j w^j$ takes the usual "euclidean" form. If one restricted oneself to the use of orthonormal bases, one would never have to introduce the matrix $(g_{ij})$, and this is what is done in elementary linear algebra.

By hypothesis, $\langle \mathbf{v}, \mathbf{w} \rangle$ is a linear function of $\mathbf{w}$ when $\mathbf{v}$ is held fixed. Thus if $\mathbf{v} \in E$, the function $\nu$ defined by

$$\nu(\mathbf{w}) = \langle \mathbf{v}, \mathbf{w} \rangle \tag{2.14}$$

is a linear functional, $\nu \in E^*$. Thus to each vector $\mathbf{v}$ in the inner product space $E$ we may associate a covector $\nu$; we shall call $\nu$ the **covariant version** of the vector $\mathbf{v}$. In terms of any basis $\mathbf{e}$ of $E$ and the dual basis $\sigma$ of $E^*$ we have from (2.4)

$$\begin{aligned}
\nu &= \sum_j \nu_j \sigma^j = \sum_j \nu(\mathbf{e}_j)\sigma^j \\
&= \sum_j \langle \mathbf{v}, \mathbf{e}_j \rangle \sigma^j \\
&= \sum_j \langle \sum_i \mathbf{e}_i v^i, \mathbf{e}_j \rangle \sigma^j \\
&= \sum_j (\sum_i v^i g_{ij}) \sigma^j
\end{aligned}$$

Thus the covariant version of the vector $\mathbf{v}$ has components $\nu_j = \sum_i v^i g_{ij}$ and *it is traditional in "tensor analysis" to use the same letter v rather than $\nu$*. Thus we write for the components of the covariant version

$$v_j = \sum_i v^i g_{ij} = \sum_i g_{ji} v^i \tag{2.15}$$

since $g_{ij} = g_{ji}$. The *sub*script $j$ in $v_j$ tells us that we are dealing with the covariant version; in tensor analysis one says that we have "lowered the upper index $i$, making it a $j$, by means of the metric tensor $g_{ij}$." We shall also call the $(v_j)$, with abuse of language, **the covariant components of the contravariant vector v**.

Note that if $\mathbf{e}$ is an orthonormal basis then $v_j = v^j$.

In our finite-dimensional inner product space $E$, every linear functional $\nu$ is the covariant version of some vector $\mathbf{v}$. Given $\nu = \sum_j v_j \sigma^j$ we shall find $\mathbf{v}$ such that $\nu(\mathbf{w}) = \langle \mathbf{v}, \mathbf{w} \rangle$ for all $\mathbf{w}$. For this we need only solve (2.15) for $v^i$ in terms of the given $v_j$. Since $G = (g_{ij})$ is assumed nondegenerate, the inverse matrix $G^{-1}$ must exist and is again symmetric. We shall denote the entries of this inverse matrix by the same letters $g$ but written with superscripts

$$G^{-1} = (g^{ij})$$

Then from (2.15) we have

$$v^i = \sum_j g^{ij} v_j \tag{2.16}$$

yields the contravariant version $\mathbf{v}$ of the covector $\nu = \sum_j v_j \sigma^j$. Again we call $(v^i)$ the contravariant components of the covector $\nu$.

Let us now compare the contravariant and covariant components of a vector $\mathbf{v}$ in a simple case. First of all, we have immediately

$$v_j = \nu(\mathbf{e}_j) = \langle \mathbf{v}, \mathbf{e}_j \rangle \tag{2.17}$$

and then $v^i = \sum_j g^{ij} v_j = \sum_j g^{ij} \langle \mathbf{v}, \mathbf{e}_j \rangle$. Thus although we always have $\mathbf{v} = \sum_i v^i \mathbf{e}_i$,

$$\mathbf{v} = \sum_i \Big( \sum_j g^{ij} \langle \mathbf{v}, \mathbf{e}_j \rangle \Big) \mathbf{e}_i$$

replaces the euclidean $\mathbf{v} = \sum_i \langle \mathbf{v}, \mathbf{e}_i \rangle \mathbf{e}_i$ that holds when the basis is orthonormal. Consider, for instance, the plane $\mathbb{R}^2$, where we use a basis $\mathbf{e}$ that consists of *unit* but not orthogonal vectors.

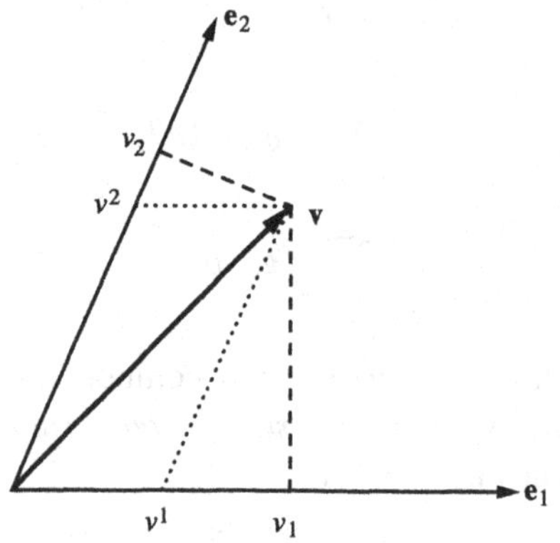


**Figure 2.1**

We must make some final remarks about linear functionals. It is important to realize that given an $n$-dimensional vector space $E$, whether or not it has an inner product, one can always construct the dual vector space $E^*$, and the construction has nothing to do with a basis in $E$. If a basis $\mathbf{e}$ *is* picked for $E$, *then* the dual basis $\sigma$ for $E^*$ is determined. There is then an **isomorphism**, that is, a 1:1 correspondence between $E^*$ and $E$ given by $\sum a_j\sigma^j \to \sum a_j\mathbf{e}_j$, but this isomorphism is said to be "unnatural" since if we change the basis in $E$ the correspondence will change. We shall *never* use this correspondence. Suppose now that an inner product has been introduced into $E$. As we have seen, there is another correspondence $E^* \to E$ that is independent of basis; namely to $\nu \in E^*$ we associate the unique vector $\mathbf{v}$ such that $\nu(\mathbf{w}) = \langle \mathbf{v}, \mathbf{w}\rangle$; we may write $\nu = \langle \mathbf{v}, \bullet\rangle$. In terms of a basis we are associating to $\nu = \sum v_i\sigma^i$ the vector $\sum v^i\mathbf{e}_i$. Then we know that each $\sigma^i$ can be represented as $\sigma^i = \langle \mathbf{f}_i, \bullet\rangle$; that is, there is a unique vector $\mathbf{f}_i$ such that $\sigma^i(\mathbf{w}) = \langle \mathbf{f}_i, \mathbf{w}\rangle$ for all $\mathbf{w} \in E$. Then $\mathbf{f} = \{\mathbf{f}_i\}$ is a new basis of the original vector space $E$, sometimes called the basis of $E$ dual to $\mathbf{e}$, and we have $\langle \mathbf{f}_i, \mathbf{e}_j\rangle = \delta^i_j$. Although this new basis *is* used in applied mathematics, *we shall not do so*, for there is a very powerful calculus that has been developed for *co*vectors, a calculus that cannot be applied to vectors!

### 2.1d. Riemannian Manifolds and the Gradient Vector

A **Riemannian metric** on a manifold $M^n$ assigns, in a differentiable fashion, a positive definite inner product $\langle , \rangle$ in each tangent space $M^n_p$. If $\langle , \rangle$ is only nondegenerate (i.e., $\langle \mathbf{u}, \mathbf{v}\rangle = 0$ for all $\mathbf{v}$ only if $\mathbf{u} = \mathbf{0}$) rather than positive definite, then we shall call the resulting structure on $M^n$ a **pseudo-Riemannian** metric. A manifold with a (pseudo-) Riemannian metric is called a (pseudo-) **Riemannian manifold**.

In terms of a coordinate basis $\mathbf{e}_i = \partial_i := \partial/\partial x^i$ we then have the differentiable matrices (the "metric tensor")

$$g_{ij}(x) = \left\langle \frac{\partial}{\partial x^i}, \frac{\partial}{\partial x^j} \right\rangle$$

as in (2.13). In an overlap $U \cap V$ we have

$$g^V_{ij} = \left\langle \frac{\partial}{\partial x_V{}^i}, \frac{\partial}{\partial x_V{}^j} \right\rangle \tag{2.18}$$

$$= \langle \sum_r \left(\frac{\partial x_U{}^r}{\partial x_V{}^i}\right)\partial^U_r, \sum_s \left(\frac{\partial x_U{}^s}{\partial x_V{}^j}\right)\partial^U_s \rangle$$

$$g^V_{ij} = \sum_{rs} \left(\frac{\partial x_U{}^r}{\partial x_V{}^i}\right)\left(\frac{\partial x_U{}^s}{\partial x_V{}^j}\right) g^U_{rs}$$

This is the transformation rule for the components of the metric tensor.

**Definition:** If $M^n$ is a (pseudo-) Riemannian manifold and $f$ is a differentiable function, the **gradient vector**

$$\text{grad } f = \nabla f$$

is the contravariant vector associated to the covector $df$

$$df(\mathbf{w}) = \langle \nabla f, \mathbf{w}\rangle \tag{2.19}$$

In coordinates

$$(\nabla f)^i = \sum_j g^{ij} \frac{\partial f}{\partial x^j}$$

Note then that $\| \nabla f \|^2 := \langle \nabla f, \nabla f \rangle = df(\nabla f) = \sum_{ij} (\partial f/\partial x^i) g^{ij} (\partial f/\partial x^j)$. We see that $df$ and $\nabla f$ will have the same components if the metric is "euclidean," that is, if the coordinates are such that $g^{ij} = \delta^i_j$.

**Example (special relativity): Minkowski space** is, as we shall see in Chapter 7, $\mathbb{R}^4$ but endowed with the pseudo-Riemannian metric given in the so-called inertial coordinates $t = x^0$, $x = x^1$, $y = x^2$, $z = x^3$, by

$$\begin{aligned} g_{ij} = \left\langle \frac{\partial}{\partial x^i}, \frac{\partial}{\partial x^j} \right\rangle &= 1 \qquad \text{if } i = j = 1, 2, \text{ or } 3 \\ &= -c^2 \qquad \text{if } i = j = 0, \qquad \text{where } c \text{ is the speed of light} \\ &= 0 \qquad \text{otherwise} \end{aligned}$$

that is, $(g_{ij})$ is the $4 \times 4$ diagonal matrix

$$(g_{ij}) = \mathrm{diag}(-c^2, 1, 1, 1)$$

Then

$$df = \left(\frac{\partial f}{\partial t}\right) dt + \sum_{j=1}^{3} \left(\frac{\partial f}{\partial x^j}\right) dx^j$$

is classically written in terms of components

$$df \sim \left[ \frac{\partial f}{\partial t}, \frac{\partial f}{\partial x}, \frac{\partial f}{\partial y}, \frac{\partial f}{\partial z} \right]$$

but

$$\nabla f = -\frac{1}{c^2} \left(\frac{\partial f}{\partial t}\right) \partial_t + \sum_{j=1}^{3} \left(\frac{\partial f}{\partial x^j}\right) \partial_j$$

$$\nabla f \sim \left[ -\frac{1}{c^2} \frac{\partial f}{\partial t}, \frac{\partial f}{\partial x}, \frac{\partial f}{\partial y}, \frac{\partial f}{\partial z} \right]^T$$

(It should be mentioned that the famous **Lorentz transformations** in general are simply *the* changes of coordinates in $\mathbb{R}^4$ that leave the origin fixed and *preserve the form* $-c^2t^2 + x^2 + y^2 + z^2$, just as orthogonal transformations in $\mathbb{R}^3$ are those transformations that preserve $x^2 + y^2 + z^2$!)

## 2.1e. Curves of Steepest Ascent

The gradient vector in a Riemannian manifold $M^n$ has much the same meaning as in euclidean space. If $\mathbf{v}$ is a unit vector at $p \in M$, then the derivative of $f$ with respect to $\mathbf{v}$ is $\mathbf{v}(f) = \sum (\partial f/\partial x^j) v^j = df(\mathbf{v}) = \langle \nabla f, \mathbf{v} \rangle$. Then Schwarz's inequality (which holds for a positive definite inner product), $|\mathbf{v}(f)| = |\langle \nabla f, \mathbf{v} \rangle| \leq \| \nabla f \| \| \mathbf{v} \| = \| \nabla f \|$, shows that $f$ has a maximum rate of change in the direction of $\nabla f$. If $f(p) = a$, then the **level set** of $f$ through $p$ is the subset defined by

$$M^{n-1}(a) := \{x \in M^n | \, f(x) = a\}$$

A good example to keep in mind is the torus of Figure 1.18. If $df$ does not vanish at $p$ then $M^{n-1}(a)$ is a submanifold in a neighborhood of $p$. If $x = x(t)$ is a curve in this level set through $p$ then its velocity vector there, $dx/dt$, is "annihilated" by $df$; $df(dx/dt) = 0$ since $f(x(t))$ is constant. We are tempted to say that $df$ is "orthogonal" to the tangent space to $M^{n-1}(a)$ at $p$, but this makes no sense since $df$ is not a vector. *Its contravariant version* $\nabla f$ *is, however, orthogonal to this tangent space* since $\langle \nabla f, dx/dt\rangle = df(dx/dt) = 0$ for all tangents to $M^{n-1}(a)$ at $p$. We say that $\nabla f$ is orthogonal to the level sets.

Finally recall that we showed in paragraph 1.4b that one does not get a well-defined flow by considering the local differential equations $dx^i/dt = \partial f/\partial x^i$; one simply cannot equate a contravariant vector $dx/dt$ with a covariant vector $df$. However it makes good sense to write $dx/dt = \nabla f$; that is, the "correct" differential equations are

$$\frac{dx^i}{dt} = \sum_j g^{ij}\left(\frac{\partial f}{\partial x^j}\right)$$

The integral curves are then tangent to $\nabla f$, and so are orthogonal to the level sets $f =$ constant. How does $f$ change along one of these "curves of steepest ascent"? Well, $df/dt = df(dx/dt) = \langle \nabla f,\ \nabla f\rangle$. Note then that if we solve *instead* the differential equations

$$\frac{dx}{dt} = \frac{\nabla f}{\|\nabla f\|^2}$$

(i.e., we move along the same curves of steepest ascent but at a different speed) then $df/dt = 1$. *The resulting flow has then the property that in time t it takes the level set* $f = a$ *into the level set* $f = a + t$. Of course this result need only be true locally and for small $t$ (see 1.4a). Such a motion of level sets into level sets is called a **Morse deformation**. For more on such matters see [M, chap. 1].

## Problems

**2.1(1)** If **v** is a vector and $\alpha$ is a covector, compute directly in coordinates that $\sum a_i^V v_V^i = \sum a_j^U v_U^j$. What happens if **w** is another vector and one considers $\sum v^i w^i$?

**2.1(2)** Let $x$, $y$, and $z$ be the usual cartesian coordinates in $\mathbb{R}^3$ and let $u^1 = r$, $u^2 = \theta$ (colatitude), and $u^3 = \phi$ be spherical coordinates.

**(i)** Compute the metric tensor components for the spherical coordinates

$$g_{r\theta} := g_{12} = \left\langle \frac{\partial}{\partial r}, \frac{\partial}{\partial \theta} \right\rangle \quad \text{etc.}$$

(Note: Don't fiddle with matrices; just use the chain rule $\partial/\partial r = (\partial x/\partial r)\partial/\partial x + \cdots$)

**(ii)** Compute the coefficients $(\nabla f)^j$ in

$$\nabla f = (\nabla f)^r \frac{\partial}{\partial r} + (\nabla F)^\theta \frac{\partial}{\partial \theta} + (\nabla f)^\phi \frac{\partial}{\partial \phi}$$

(iii) Verify that $\partial/\partial r$, $\partial/\partial\theta$, and $\partial/\partial\phi$ are orthogonal, but that not all are unit vectors. Define the unit vectors $\mathbf{e}'_j = (\partial/\partial u^j)/ \parallel \partial/\partial u^j \parallel$ and write $\nabla f$ in terms of this orthonormal set

$$\nabla f = (\nabla f)'^r \mathbf{e}'_r + (\nabla f)'^\theta \mathbf{e}'_\theta + (\nabla f)'^\phi \mathbf{e}'_\phi$$

These new components of grad $f$ are the usual ones found in all physics books (they are called the **physical** components); *but we shall have little use for such components*; $df$, as given by the simple expression $df = (\partial f/\partial r)\, dr + \cdots$, frequently has all the information one needs!

---

## 2.2. The Tangent Bundle

What is the space of velocity vectors to the configuration space of a dynamical system?

### 2.2a. The Tangent Bundle

The **tangent bundle**, $TM^n$, to a differentiable manifold $M^n$ is, by definition, the collection of all tangent vectors at all points of $M$.

Thus a "point" in this new space consists of a pair $(p, \mathbf{v})$, where $p$ is a point of $M$ and $\mathbf{v}$ is a tangent vector to $M$ at the point $p$, that is, $\mathbf{v} \in M^n_p$. Introduce local coordinates in $TM$ as follows. Let $(p, \mathbf{v}) \in TM^n$. $p$ lies in some local coordinate system $U$, $x^1, \ldots, x^n$. At $p$ we have the coordinate basis $(\partial_i = \partial/\partial x^i)$ for $M^n_x$. We may then write $\mathbf{v} = \sum_i v^i \partial_i$. Then $(p, \mathbf{v})$ is completely described by the $2n$-tuple of real numbers

$$x^1(p), \ldots, x^n(p), v^1, \ldots, v^n$$

*The 2n-tuple (x,v) represents the vector* $\sum_j v^j \partial_j$ *at p.* In this manner we associate $2n$ local coordinates to each tangent vector to $M^n$ that is based in the coordinate patch $(U, x)$. Note that the first $n$-coordinates, the $x$'s, take their values in a portion $U$ of $\mathbb{R}^n$, whereas the second set, the $v$'s, fill out an entire $\mathbb{R}^n$ since there are no restrictions on the components of a vector. This $2n$-dimensional coordinate patch is then of the form $(U \subset \mathbb{R}^n) \times \mathbb{R}^n \subset \mathbb{R}^{2n}$. Suppose now that the point $p$ also lies in the coordinate patch $(U', x')$. Then the same point $(p, \mathbf{v})$ would be described by the new $2n$-tuple

$$x'^1(p), \ldots, x'^n(p), v'^1, \ldots, v'^n$$

where

$$x'^i = x'^i(x^1, \ldots, x^n) \tag{2.20}$$

and

$$v'^i = \sum_j \left[\frac{\partial x'^i}{\partial x^j}\right](p) v^j$$

We see then that $TM^n$ *is a 2n-dimensional differentiable manifold*!

We have a mapping

$$\pi : TM \to M \qquad \pi(p, \mathbf{v}) = p$$

called **projection** that assigns to a vector tangent to $M$ the point in $M$ at which the vector sits. In local coordinates,

$$\pi(x^1, \ldots, x^n, v^1, \ldots, v^n) = (x^1, \ldots, x^n)$$

It is clearly differentiable.

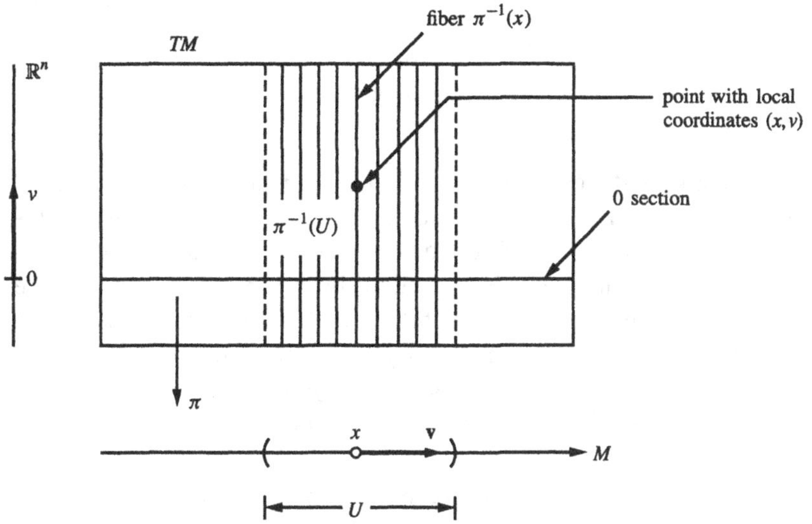


Figure 2.2

We have drawn a schematic diagram of the tangent bundle $TM$. $\pi^{-1}(x)$ represents all vectors tangent to $M$ at $x$, and so $\pi^{-1}(x) = M_x^n$ is a copy of the vector space $\mathbb{R}^n$. It is called "the **fiber** over $x$." Our picture makes it seem that $TM$ is the product space $M \times \mathbb{R}^n$, *but this is not so*! Although we do have a global projection $\pi : TM \to M$, *there is no projection map* $\pi' : TM \to \mathbb{R}^n$.

*A point in TM represents a tangent vector to M at a point p but there is no way to read off the components of this vector until a coordinate system (or basis for $M_p$) has been designated at the point at which the vector is based!*

Locally of course we may choose such a projection; if the point is in $\pi^{-1}(U)$ then by using the coordinates in $U$ we may read off the components of the vector. Since $\pi^{-1}(U)$ is topologically $U \times \mathbb{R}^n$ we say that the tangent bundle $TM$ is **locally a product**.

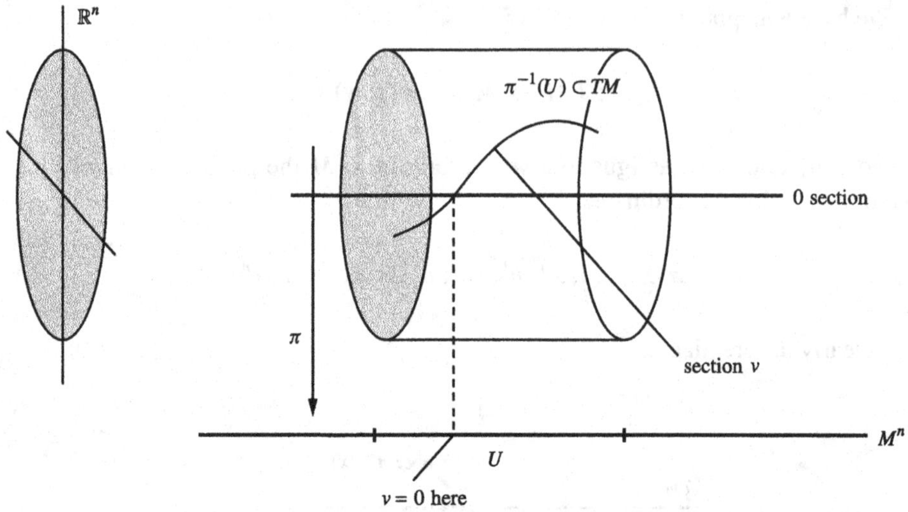


**Figure 2.3**

A vector field $\mathbf{v}$ on $M$ clearly assigns to each point $x$ in $M$ a point $\mathbf{v}(x)$ in $\pi^{-1}(x) \subset TM$ that "lies over $x$." Thus a vector field can be considered as a map $v : M \to TM$ such that $\pi \circ v$ is the identity map of $M$ into $M$. As such it is called a **(cross) section** of the tangent bundle. In a patch $\pi^{-1}(U)$ it is described by $v^i = v^i(x^1, \ldots, x^n)$ and the image $v(M)$ is then an $n$-dimensional submanifold of the $2n$-dimensional manifold $TM$. A special section, the 0 section (corresponding to the identically 0 vector field), always exists. Although different coordinate systems will yield perhaps different components for a given vector, they will all agree that the 0-vector will have all components 0.

**Example:** In mechanics, the configuration of a dynamical system with $n$ degrees of freedom is usually described as a point in an $n$-dimensional manifold, the **configuration space**. The coordinates $x$ are usually called $q^1, \ldots, q^n$, the "generalized coordinates." For example, if we are considering the motion of two mass points on the real line, $M^2 = \mathbb{R} \times \mathbb{R}$ with coordinates $q^1, q^2$ (one for each particle). The configuration space need not be euclidean space. For the planar double pendulum of paragraph 1.2b (v), the configuration space is $M^2 = S^1 \times S^1 = T^2$. For the *spatial* single pendulum $M^2$ is the 2-sphere $S^2$ (with center at the pin). A tangent vector to the configuration space $M^n$ is thought of, in mechanics, as a velocity vector; its components with respect to the coordinates $q$ are written $\dot{q}_1, \ldots, \dot{q}_n$ rather than $v^1, \ldots, v^n$. These are the **generalized velocities**. Thus $TM$ is the space of all generalized velocities, but there is no standard name for this space in mechanics (it is *not* the phase space, to be considered shortly).

## 2.2b. The Unit Tangent Bundle

If $M^n$ is a Riemannian manifold (see 2.1d) then we may consider, in addition to $TM$, the space of all *unit* tangent vectors to $M^n$. Thus in $TM$ we may restrict ourselves to the subset $T_0M$ of points $(x, \mathbf{v})$ such that $\| \mathbf{v} \|^2 = 1$. If we are in the coordinate patch

$(x^1, \dots, x^n, v^1, \dots, v^n)$ of $TM$, then this **unit tangent bundle** is locally defined by

$$T_0M^n : \sum_{ij} g_{ij}(x)v^i v^j = 1$$

In other words, we are looking at the locus in $TM$ defined locally by putting the single function $f(x, v) = \sum_{ij} g_{ij}(x)v^i v^j$ equal to a constant. The local coordinates in $TM$ are $(x, v)$. Note, using $g_{ij} = g_{ji}$, that

$$\frac{\partial f}{\partial v^k} = 2\sum_j g_{kj}(x)v^j$$

Since $\det(g_{ij}) \neq 0$, we conclude that not all $\partial f/\partial v^k$ can vanish on the subset $v \neq 0$, and thus *$T_0M^n$ is a $(2n-1)$-dimensional submanifold of $TM^n$*! In particular $T_0M$ is itself a manifold.

In the following figure, $\mathbf{v}_0 = \mathbf{v}/\parallel \mathbf{v} \parallel$.

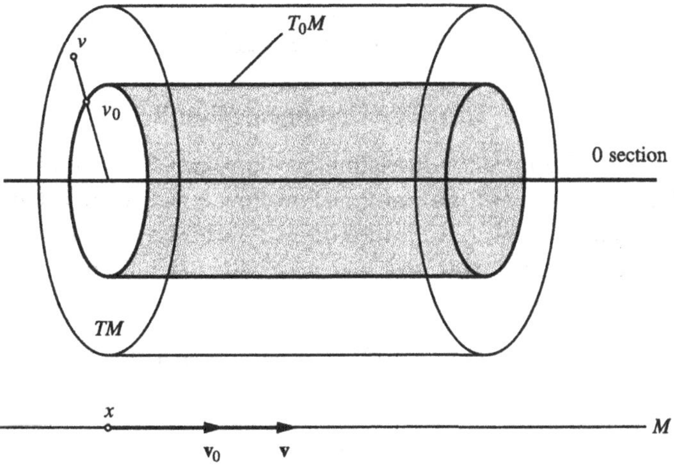


**Figure 2.4**

**Example:** $T_0S^2$ is the space of unit vectors tangent to the unit 2-sphere in $\mathbb{R}^3$.

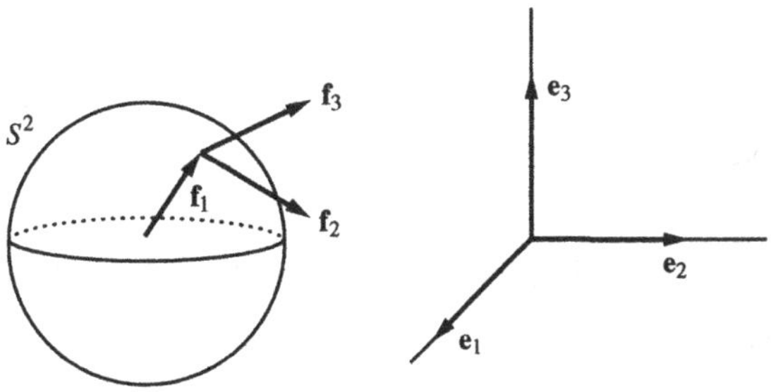


**Figure 2.5**

Let $\mathbf{v} = \mathbf{f}_2$ be a unit tangent vector to the unit sphere $S^2 \subset \mathbb{R}^3$. It is based at some point on $S^2$, described by a unit vector $\mathbf{f}_1$. Using the right-hand rule we may put $\mathbf{f}_3 = \mathbf{f}_1 \times \mathbf{f}_2$.

It is clear that by this association, there is a 1:1 correspondence between unit tangent vectors $\mathbf{v}$ to $S^2$ (i.e., to a point in $T_0S^2$) and such orthonormal triples $\mathbf{f}_1, \mathbf{f}_2, \mathbf{f}_3$. Translate these orthonormal vectors to the origin of $\mathbb{R}^3$ and compare them with a fixed right-handed orthonormal basis $\mathbf{e}$ of $\mathbb{R}^3$. Then $\mathbf{f}_i = \mathbf{e}_j R^j{}_i$ for a unique rotation matrix $R \in SO(3)$. In this way we have set up a 1:1 correspondence $T_0S^2 \to SO(3)$. It also seems evident that the topology of $T_0S^2$ is the same as that of $SO(3)$, meaning roughly that nearby unit vectors tangent to $S^2$ will correspond to nearby rotation matrices; precisely, we mean that $T_0S^2 \to SO(3)$ is a diffeomorphism. We have seen in 1.2b(vii) that $SO(3)$ is topologically projective space.

*The unit tangent bundle $T_0S^2$ to the 2-sphere is topologically the 3-dimensional real projective 3-space $T_0S^2 \sim \mathbb{R}P^3 \sim SO(3)$.*

## 2.3. The Cotangent Bundle and Phase Space

What *is* phase space?

### 2.3a. The Cotangent Bundle

The cotangent bundle to $M^n$ is by definition the space $T^*M^n$ of all *covectors* at all points of $M$. A point in $T^*M$ is a pair $(x, \alpha)$ where $\alpha$ is a covector at the point $x$. If $x$ is in a coordinate patch $U, x^1, \dots, x^n$, then $dx^1, \dots, dx^n$, gives a basis for the cotangent space $M_x^{n*}$, and $\alpha$ can be expressed as $\alpha = \sum a_i(x)dx^i$. Then $(x, \alpha)$ is completely described by the $2n$-tuple

$$x^1(x), \dots, x^n(x), a_1(x), \dots, a_n(x)$$

*The 2n-tuple (x, a) represents the covector $\sum a_i\, dx^i$ at the point x.* If the point $p$ also lies in the coordinate patch $U', x'^1, \dots, x'^n$, then

$$x'^i = x'^i(x^1, \dots, x^n)$$

and (2.21)

$$a'_i = \sum_j \left[\frac{\partial x^j}{\partial x'^i}\right](x)a_j$$

*$T^*M^n$ is again a 2n-dimensional manifold.* We shall see shortly that the phase space in mechanics is the cotangent bundle to the configuration space.

### 2.3b. The Pull-Back of a Covector

Recall that the differential $\phi_*$ of a smooth map $\phi : M^n \to V^r$ has as matrix the Jacobian matrix $\partial y/\partial x$ in terms of local coordinates $(x^1, \dots, x^n)$ near $x$ and $(y^1, \dots, y^r)$ near $y = \phi(x)$. Thus, in terms of the coordinate bases

$$\phi_*\left(\frac{\partial}{\partial x^j}\right) = \sum_R \left(\frac{\partial y^R}{\partial x^j}\right)\frac{\partial}{\partial y^R} \tag{2.22}$$

Note that if we think of vectors as differential operators, then for a function $f$ near $y$

$$\phi_*\left(\frac{\partial}{\partial x^j}\right)(f) = \sum_R \left(\frac{\partial y^R}{\partial x^j}\right)\left(\frac{\partial f}{\partial y^R}\right)$$

simply says, "*Apply the chain rule to the composite function* $f \circ \phi$, that is, $f(y(x))$."

**Definition:** Let $\phi : M^n \to V^r$ be a smooth map of manifolds and let $\phi(x) = y$. Let $\phi_* : M_x \to V_y$ be the differential of $\phi$. The **pull-back** $\phi^*$ is the linear transformation taking *covectors* at $y$ into *covectors* at $x$, $\phi^* : V(y)^* \to M(x)^*$, defined by

$$\phi^*(\beta)(\mathbf{v}) := \beta(\phi_*(\mathbf{v})) \tag{2.23}$$

for all covectors $\beta$ at $y$ and vectors $\mathbf{v}$ at $x$.

Let $(x^i)$ and $(y^R)$ be local coordinates near $x$ and $y$, respectively. The bases for the tangent vector spaces $M_x$ and $V_y$ are given by $(\partial/\partial x^j)$ and $(\partial/\partial y^R)$. Then

$$\begin{aligned}
\phi^*\beta &= \sum_j \phi^*(\beta)\left(\frac{\partial}{\partial x^j}\right)dx^j = \sum_j \beta\left(\phi_*\frac{\partial}{\partial x^j}\right)dx^j \\
&= \sum_j \beta\left(\sum_R \left(\frac{\partial y^R}{\partial x^j}\right)\frac{\partial}{\partial y^R}\right)dx^j \\
&= \sum_{jR} \left(\frac{\partial y^R}{\partial x^j}\right)\beta\left(\frac{\partial}{\partial y^R}\right)dx^j \\
&= \sum_{jR} b_R\left(\frac{\partial y^R}{\partial x^j}\right)dx^j, \qquad \text{where } \beta = \sum_R b_R dy^R
\end{aligned}$$

Thus

$$\phi^*(\beta) = \sum_{jR} b_R\left(\frac{\partial y^R}{\partial x^j}\right)dx^j \tag{2.24}$$

In terms of matrices, the *differential* $\phi_*$ is given by the Jacobian matrix $\partial y/\partial x$ acting on *columns* $v$ at $x$ from the *left*, whereas the *pull-back* $\phi^*$ is given by the same matrix acting on *rows* $b$ at $y$ from the *right*. (If we had insisted on writing covectors also as columns, then $\phi^*$ acting on such columns from the left would be given by the *transpose* of the Jacobian matrix.)

$\phi^*(dy^S)$ is given immediately from (2.24); since $dy^S = \sum_R \delta^S{}_R dy^R$

$$\phi^*(dy^S) = \sum_j \left(\frac{\partial y^S}{\partial x^j}\right)dx^j \tag{2.25}$$

This *is again simply the chain rule applied to the composition* $y^S \circ \phi$!

**Warning:** Let $\phi : M^n \to V^r$ and let $\mathbf{v}$ be a vector *field* on $M$. It may very well be that there are two distinct points $x$ and $x'$ that get mapped by $\phi$ to the same point $y = \phi(x) = \phi(x')$. Usually we shall have $\phi_*(\mathbf{v}(x)) \neq \phi_*(\mathbf{v}(x'))$ since the field $\mathbf{v}$ need have no relation to the map $\phi$. In other words, $\phi_*(\mathbf{v})$ does not yield a well defined vector field on $V$ (does one pick $\phi_*(\mathbf{v}(x))$ or $\phi_*(\mathbf{v}(x'))$ at $y$?). $\phi_*$ *does not take vector fields* *into vector fields.* (There is an exception if $n = r$ and $\phi$ is 1:1.) On the other hand, *if $\beta$ is a covector field on $V^r$, then $\phi^*\beta$ is always a well-defined covector field on $M^n$*; $\phi^*(\beta(y))$ yields a definite covector at each point $x$ such that $\phi(x) = y$. As we shall see, this fact makes covector fields easier to deal with than vector fields.

See Problem 2.3 (1) at this time.

### 2.3c. The Phase Space in Mechanics

In Chapter 4 we shall study Hamiltonian dynamics in a more systematic fashion. For the present we wish merely to draw attention to certain basic aspects that seem mysterious when treated in most physics texts, largely because they draw no distinction there between vectors and covectors.

Let $M^n$ be the configuration space of a dynamical system and let $q^1, \dots, q^n$ be local generalized coordinates. For simplicity, we shall restrict ourselves to time-independent Lagrangians. The Lagrangian $L$ is then a function of the generalized coordinates $q$ and the generalized velocities $\dot{q}$, $L = L(q, \dot{q})$. It is important to realize that $q$ and $\dot{q}$ are $2n$-independent coordinates. (Of course *if* we consider a specific path $q = q(t)$ in configuration space then the Lagrangian along this evolution of the system is computed by putting $\dot{q} = dq/dt$.) Thus the Lagrangian $L$ is to be considered as a function on the space of generalized velocities, that is, *$L$ is a real-valued function on the tangent bundle to $M$*,

$$L : TM^n \to \mathbb{R}$$

We shall be concerned here with the transition from the Lagrangian to the Hamiltonian formulation of dynamics. Hamilton was led to define the functions

$$p_i(q, \dot{q}) := \frac{\partial L}{\partial \dot{q}^i} \tag{2.26}$$

We shall only be interested in the case when $\det(\partial p_i / \partial \dot{q}^j) \neq 0$. In many books (2.26) is looked upon merely as a change of coordinates in $TM$; that is, one switches from coordinates $q, \dot{q}$, to $q, p$. Although this is technically acceptable, it has the disadvantage that the $p$'s do not have the direct geometrical significance that the coordinates $\dot{q}$ had. Under a change of coordinates, say from $q_U$ to $q_V$ in configuration space, there is an associated change in coordinates in $TM$

$$\begin{aligned} q_V &= q_V(q_U) \\ \dot{q}_U^j &= \sum_i \left(\frac{\partial q_U^j}{\partial q_V^i}\right)\dot{q}_V^i \end{aligned} \tag{2.27}$$

This is the meaning of the tangent bundle! Let us see now how the $p$'s transform.

$$p_i^V := \frac{\partial L}{\partial \dot{q}_V^i} = \sum_j \left\{ \left(\frac{\partial L}{\partial q_U^j}\right)\left(\frac{\partial q_U^j}{\partial \dot{q}_V^i}\right) + \left(\frac{\partial L}{\partial \dot{q}_U^j}\right)\left(\frac{\partial \dot{q}_U^j}{\partial \dot{q}_V^i}\right) \right\}$$

However, $q_V$ does not depend on $\dot{q}_U$; likewise $q_U$ does not depend on $\dot{q}_V$, and therefore the first term in this sum vanishes. Also, from (2.27),

$$\frac{\partial \dot{q}_U^j}{\partial \dot{q}_V^i} = \frac{\partial q_U^j}{\partial q_V^i} \tag{2.28}$$

Thus

$$p_i^V = \sum_j p_j^U \left( \frac{\partial q_U^j}{\partial q_V^i} \right) \tag{2.29}$$

and so the $p$'s represent then not the components of a vector on the configuration space $M^n$ but rather a *covector*. The $q$'s and $p$'s then are to be thought of not as local coordinates in the tangent bundle but as coordinates for the *cotangent* bundle. Equation (2.26) is then to be considered not as a change of coordinates in $TM$ but rather as the local description of a *map*

$$p : TM^n \to T^*M^n \tag{2.30}$$

*from the tangent bundle to the cotangent bundle.* We shall frequently call $(q^1, \ldots, q^n, p_1, \ldots, p_n)$ the local coordinates for $T^*M^n$ (even when we are not dealing with mechanics). This space $T^*M$ of covectors to the configuration space is called in mechanics the **phase space** of the dynamical system.

Recall that there is no *natural* way to identify vectors on a manifold $M^n$ with covectors on $M^n$. We have managed to make such an identification, $\sum_j \dot{q}^j \partial/\partial q^j \to \sum_j (\partial L/\partial \dot{q}^j) dq^j$, by introducing an extra structure, a Lagrangian function. $TM$ and $T^*M$ exist as soon as a manifold $M$ is given. We may (locally) identify these spaces by giving a Lagrangian function, but of course the identification changes with a change of $L$, that is, a change of "dynamics."

Whereas the $\dot{q}$'s of $TM$ are called generalized velocities, the $p$'s are called generalized **momenta.** This terminology is suggested by the following situation. The Lagrangian is frequently of the form

$$L(q, \dot{q}) = T(q, \dot{q}) - V(q)$$

where $T$ is the kinetic energy and $V$ the potential energy. $V$ is usually independent of $\dot{q}$ and $T$ is frequently a positive definite symmetric quadratic form in the velocities

$$T(q, \dot{q}) = \frac{1}{2} \sum_{jk} g_{jk}(q) \dot{q}^j \dot{q}^k \tag{2.31}$$

For example, in the case of two masses $m_1$ and $m_2$ moving in one dimension, $M = \mathbb{R}^2$, $TM = \mathbb{R}^4$, and

$$T = \frac{1}{2} m_1 (\dot{q}^1)^2 + \frac{1}{2} m_2 (\dot{q}^2)^2$$

and the "mass matrix" $(g_{ij})$ is the diagonal matrix diag$(m_1, m_2)$.

In (2.31) we have generalized this simple case, allowing the "mass" terms to depend on the positions. For example, for a single particle of mass $m$ moving in the plane, we have, using cartesian coordinates, $T = (1/2)m[\dot{x}^2 + \dot{y}^2]$, but if polar coordinates are used we have $T = (1/2)m[\dot{r}^2 + r^2\dot{\theta}^2]$ with the resulting mass matrix diag$(m, mr^2)$. In the general case,

$$p_i = \frac{\partial L}{\partial \dot{q}^i} = \frac{\partial T}{\partial \dot{q}^i} = \sum_j g_{ij}(q) \dot{q}^j \tag{2.32}$$

Thus, if we think of $2T$ as defining a *Riemannian metric* on the configuration space $M^n$

$$\langle \dot{q}, \dot{q} \rangle = \sum_{ij} g_{ij}(q)\dot{q}^i\dot{q}^j$$

then the kinetic energy represents half the length squared of the velocity vector, and the *momentum p* is by (2.32) simply *the covariant version of the velocity vector* $\dot{q}$. In the case of the two masses on $\mathbb{R}$ we have

$$p_1 = m_1\dot{q}^1 \quad \text{and} \quad p_2 = m_2\dot{q}^2$$

are indeed what everyone calls the momenta of the two particles.

The tangent and cotangent bundles, $TM$ and $T^*M$, exist for any manifold $M$, *independent of mechanics*. They are distinct geometric objects. If, however, $M$ is a Riemannian manifold, we may define a diffeomorphism $TM^n \to T^*M^n$ that sends the coordinate patch $(q, \dot{q})$ to the coordinate patch $(q, p)$ by

$$p_i = \sum_j g_{ij}\dot{q}^j$$

with inverse

$$\dot{q}^i = \sum_j g^{ij} p_j$$

We did just this in mechanics, where the metric tensor was chosen to be that defined by the kinetic energy quadratic form.

### 2.3d. The Poincaré 1-Form

Since $TM$ and $T^*M$ are diffeomorphic, it might seem that there is no particular reason for introducing the more abstract $T^*M$, but this is not so. *There are certain geometrical objects that live naturally on $T^*M$, not TM.* Of course these objects can be brought back to $TM$ by means of our identifications, but this is not only frequently awkward, it would also depend, say, on the specific Lagrangian or metric tensor employed.

Recall that "1-form" is simply another name for covector. We shall show, with Poincaré, that there is a well-defined 1-form field on every cotangent bundle $T^*M$. This will be a linear functional defined on each tangent vector to the $2n$-dimensional manifold $T^*M^n$, *not* $M$.

**Theorem (2.33):** *There is a globally defined 1-form on every cotangent bundle $T^*M^n$, the* ***Poincaré 1-form*** $\lambda$. *In local coordinates $(q, p)$ it is given by*

$$\lambda = \sum_i p_i dq^i$$

(Note that the most general 1-form on $T^*M$ is locally of the form $\sum_i a_i(q, p)dq^i + \sum_i b_i(q, p)dp_i$, and also note that the expression given for $\lambda$ cannot be considered a 1-form on the manifold $M$ since $p_i$ is not a function on $M$!)

**PROOF:** We need only show that $\lambda$ is well defined on an overlap of local coordinate patches of $T^*M$. Let $(q', p')$ be a second patch. We may restrict ourselves to coordinate changes of the form (2.21), for that is how the cotangent bundle was defined. Then

$$dq'^i = \sum_j \left\{ \left(\frac{\partial q'^i}{\partial q^j}\right) dq^j + \left(\frac{\partial q'^i}{\partial p_j}\right) dp_j \right\}$$

But from (2.21), $q'$ is independent of $p$, and the second sum vanishes. Thus

$$\sum_i p'_i dq'^i = \sum_i p_i' \sum_j \left(\frac{\partial q'^i}{\partial q^j}\right) dq^j = \sum_j p_j dq^j \quad \square$$

There is a simple *intrinsic* definition of the form $\lambda$, that is, a definition not using coordinates. Let $A$ be a point in $T^*M$; we shall define the 1-form $\lambda$ at $A$. $A$ represents a 1-form $\alpha$ at a point $x \in M$. Let $\pi : T^*M^n \to M^n$ be the *projection* that takes a point $A$ in $T^*M$, to the point $x$ at which the form $\alpha$ is located. Then the pull-back $\pi^*\alpha$ defines a 1-form at each point of $\pi^{-1}(x)$, in particular at $A$. $\lambda$ at $A$ is precisely this form $\pi^*\alpha$!

Let us check that these two definitions are indeed the same. In terms of local coordinates $(q)$ for $M$ and $(q, p)$ for $T^*M$ the map $\pi$ is simply $\pi(q, p) = (q)$. The point $A$ with local coordinates $(q, p)$ represents the form $\sum_j p_j dq^j$ at the point $q$ in $M$. Compute the pull-back (i.e., use the chain rule)

$$\begin{aligned}
\pi^*\left(\sum_i p_i dq^i\right) &= \sum_i p_i \pi^*(dq^i) \\
&= \sum_i p_i \sum_j \left\{ \left(\frac{\partial q^i}{\partial q^j}\right) dq^j + \left(\frac{\partial q^i}{\partial p_j}\right) dp_j \right\} \\
&= \sum_i p_i \sum_j \delta^i_j dq^j = \sum_i p_i dq^i = \lambda \quad \square
\end{aligned}$$

*As we shall see when we discuss mechanics, the presence of the Poincaré 1-form field on $T^*M$ and the capability of pulling back 1-form fields under mappings endow $T^*M$ with a powerful tool that is not available on $TM$.*

## Problems

**2.3(1)** Let $F : M^n \to W^r$ and $G : W^r \to V^s$ be smooth maps. Let $x$, $y$, and $z$ be local coordinates near $p \in M$, $F(p) \in W$, and $G(F(p)) \in V$, respectively. We may consider the composite map $G \circ F : M \to V$.

**(i)** Show, by using bases $\partial/\partial x, \partial/\partial y$, and $\partial/\partial z$, that

$$(G \circ F)_* = G_* \circ F_*$$

**(ii)** Show, by using bases $dx$, $dy$, and $dz$, that

$$(G \circ F)^* = F^* \circ G^*$$

**2.3(2)** Consider the tangent bundle to a manifold $M$.

(i) Show that under a change of coordinates in $M$, $\partial/\partial q$ depends on both $\partial/\partial q'$ and $\partial/\partial \dot{q}'$.
(ii) Is the locally defined vector field $\sum_j \dot{q}^j \partial/\partial q^j$ well defined on all of $TM$?
(iii) Is $\sum_j \dot{q}^j \partial/\partial \dot{q}^j$ well defined?
(iv) If any of the above in (ii), (iii) is well defined, can you produce an intrinsic definition?

---

## 2.4. Tensors

How does one construct a field strength from a vector potential?

### 2.4a. Covariant Tensors

In this paragraph we shall again be concerned with linear algebra of a vector space $E$. Almost all of our applications will involve the vector space $E = M_x^n$ of tangent vectors to a manifold at a point $x \in E$. Consequently we shall denote a basis $\mathbf{e}$ of $E$ by $\partial = (\partial_1, \dots, \partial_n)$, with dual basis $\sigma = dx = (dx^1, \dots, dx^n)^T$. It should be remembered, however, that most of our constructions are simply linear algebra.

**Definition:** A **covariant tensor of rank r** is a **multilinear** real-valued function

$$Q : E \times E \times \cdots \times E \to \mathbb{R}$$

of $r$-tuples of vectors, multilinear meaning that the function $Q(\mathbf{v}_1, \dots, \mathbf{v}_r)$ is linear in each entry provided that the remaining entries are held fixed.

We emphasize that *the values of this function must be independent of the basis in which the components of the vectors are expressed.*

A *covariant vector* is a covariant tensor of rank 1. When $r = 2$, a multilinear function is called bilinear, and so forth. Probably the most important covariant second-rank tensor is the *metric tensor* $G$, introduced in 2.1c:

$$G(\mathbf{v}, \mathbf{w}) = \langle \mathbf{v}, \mathbf{w} \rangle = \sum_{ij} g_{ij} v^i w^j$$

is clearly bilinear (and is assumed independent of basis).

We need a systematic notation for indices. Instead of writing $i, j, \dots, k$, we shall write $i_1, \dots, i_p$.

In components, we have, by multilinearity,

$$\begin{aligned} Q(\mathbf{v}_1, \dots, \mathbf{v}_r) &= Q\left( \sum_{i_1} v_1^{i_1} \partial_{i_1}, \dots, \sum_{i_r} v_r^{i_r} \partial_{i_r} \right) \\ &= \sum_{i_1} v_1^{i_1} Q\left( \partial_{i_1}, \dots, \sum_{i_r} v_r^{i_r} \partial_{i_r} \right) = \dots \\ &= \sum_{i_1, \dots, i_r} v_1^{i_1} \dots v_r^{i_r} Q(\partial_{i_1}, \dots, \partial_{i_r}) \end{aligned}$$

That is,

$$Q(\mathbf{v}_1, \ldots, \mathbf{v}_r) = \sum_{i_1,\ldots,i_r} Q_{i_1,\ldots,i_r} v_1^{i_1} \ldots v_r^{i_r}$$

where (2.34)

$$Q_{i_1,\ldots,i_r} := Q(\partial_{i_1}, \ldots, \partial_{i_r})$$

We now introduce a very useful notational device, the *Einstein summation convention.* In any single term involving indices, a summation is implied over any index that appears as both an upper (contravariant) and a lower (covariant) index. For example, in a matrix $A = (a^i{}_j)$, $a^i{}_i = \sum_i a^i{}_i$ is the trace of the matrix. With this convention we can write

$$Q(\mathbf{v}_1, \ldots, \mathbf{v}_r) = Q_{i_1,\ldots,i_r} v_1^{i_1} \ldots v_r^{i_r} \tag{2.35}$$

The collection of all covariant tensors of rank $r$ forms a vector space under the usual operations of addition of functions and multiplication of functions by real numbers. These simply correspond to addition of their *components* $Q_{i,\ldots,j}$ and multiplication of the components by real numbers. The number of components in such a tensor is clearly $n^r$. This vector space is the space of covariant $r^{\text{th}}$ rank tensors and will be denoted by

$$E^* \otimes E^* \otimes \cdots \otimes E^* = \otimes^r E^*$$

If $\alpha$ and $\beta$ are covectors, that is, elements of $E^*$, we can form the second-rank covariant tensor, the **tensor product** of $\alpha$ and $\beta$, as follows. We need only tell how $\alpha \otimes \beta : E \times E \to \mathbb{R}$.

$$\alpha \otimes \beta(\mathbf{v}, \mathbf{w}) := \alpha(\mathbf{v})\, \beta(\mathbf{w})$$

In components, $\alpha = a_i dx^i$ and $\beta = b_j dx^j$, and from (2.34)

$$(\alpha \otimes \beta)_{ij} = \alpha \otimes \beta(\partial_i, \partial_j) = \alpha(\partial_i)\beta(\partial_j) = a_i b_j$$

$(a_i b_j)$, where $i, j = 1, \ldots, n$, form the components of $\alpha \otimes \beta$. See Problem 2.4 (1) at this time.

### 2.4b. Contravariant Tensors

Note first that a contravariant vector, that is, an element of $E$, can be considered as a linear functional on covectors by defining

$$\mathbf{v}(\alpha) := \alpha(\mathbf{v})$$

In components $\mathbf{v}(\alpha) = a_i v^i$ is clearly linear in the components of $\alpha$.

**Definition:** A **contravariant tensor of rank** $s$ is a multilinear real valued function $T$ on $s$-tuples of *covectors*

$$T : E^* \times E^* \times \cdots \times E^* \to \mathbb{R}$$

As for covariant tensors, we can show immediately that for an $s$-tuple of 1-forms $\alpha_1, \ldots, \alpha_s$

$$T(\alpha_1, \ldots, \alpha_s) = a_{1\ i_1} \ldots a_{s\ i_s} T^{i_1 \ldots i_s}$$

where (2.36)

$$T^{i_1 \ldots i_s} := T(dx^{i_1}, \ldots, dx^{i_s})$$

We write for this space of contravariant tensors

$$E \otimes E \otimes \cdots \otimes E := \otimes^s E$$

Contravariant vectors are of course contravariant tensors of rank 1. An example of a second-rank contravariant tensor is the inverse to the metric tensor $G^{-1}$, with components $(g^{ij})$,

$$G^{-1}(\alpha, \beta) = g^{ij}\ a_i b_j$$

(see 2.1c). Does the matrix $g^{ij}$ really define a *tensor* $G^{-1}$? The local expression for $G^{-1}$ $(\alpha, \beta)$ given is certainly bilinear, but are the values really independent of the coordinate expressions of $\alpha$ and $\beta$? Note that the vector **b** associated to $\beta$ is coordinate-independent since $\beta(\mathbf{v}) = \langle \mathbf{v}, \mathbf{b} \rangle$, and the metric $\langle , \rangle$ is coordinate-independent. But then $G^{-1}(\alpha, \beta) = g^{ij} a_i b_j = a_i b^i = \alpha(\mathbf{b})$ is indeed independent of coordinates, and $G^{-1}$ is a tensor.

Given a pair $\mathbf{v}$, $\mathbf{w}$ of contravariant vectors, we can form their tensor product $\mathbf{v} \otimes \mathbf{w}$ in the same manner as we did for covariant vectors. It is the second-rank contravariant tensor with components $(\mathbf{v} \otimes \mathbf{w})^{ij} = v^i w^j$. As in Problem 2.4 (1) we may then write

$$G = g_{ij}\, dx^i \otimes dx^j \quad \text{and} \quad G^{-1} = g^{ij} \partial_i \otimes \partial_j \tag{2.37}$$

## 2.4c. Mixed Tensors

The following definition in fact includes that of covariant and contravariant tensors as special cases when $r$ or $s = 0$.

**Definition:** A **mixed tensor**, $r$ times covariant and $s$ times contravariant, is a real multilinear function $W$

$$W : E^* \times E^* \times \cdots \times E^* \times E \times E \times \cdots \times E \to \mathbb{R}$$

on $s$-tuples of covectors and $r$-tuples of vectors.

By multilinearity

$$W(\alpha_1, \ldots, \alpha_s, \mathbf{v}_1, \ldots, \mathbf{v}_r) = a_{1\ i_1} \ldots a_{s\ i_s}\ W^{i_1 \ldots i_s}{}_{j_1 \ldots j_r} v_1^{j_1} \ldots v_r^{j_r}$$

where (2.38)

$$W^{i_1 \ldots i_s}{}_{j_1 \ldots j_r} := W(dx^{i_1}, \ldots, \partial_{j_r})$$

A second-rank mixed tensor arises from a *linear transformation* $\mathbf{A} : E \to E$. Define $W_A : E^* \times E \to \mathbb{R}$ by $W_A(\alpha, \mathbf{v}) = \alpha(\mathbf{A}\mathbf{v})$. *Let* $A = (A^i_j)$ be the matrix of $\mathbf{A}$, that is, $\mathbf{A}(\boldsymbol{\partial}_j) = \boldsymbol{\partial}_i A^i_j$. The components of $W_A$ are given by

$$W_A{}^i{}_j = W_A(dx^i, \boldsymbol{\partial}_j) = dx^i(\mathbf{A}(\boldsymbol{\partial}_j)) = dx^i(\boldsymbol{\partial}_k A^k{}_j) = \delta^i_k A^k{}_j = A^i{}_j$$

*The matrix of the mixed tensor* $W_A$ is simply the matrix of $\mathbf{A}$! Conversely, given a mixed tensor $W$, once covariant and once contravariant, we can define a linear transformation $\mathbf{A}$ by saying $\mathbf{A}$ is that unique linear transformation such that $W(\alpha, \mathbf{v}) = \alpha(\mathbf{A}\mathbf{v})$. Such an $\mathbf{A}$ exists since $W(\alpha, \mathbf{v})$ *is* linear in $\mathbf{v}$. We shall not distinguish between a linear transformation $\mathbf{A}$ and its associated mixed tensor $W_A$; a linear transformation $\mathbf{A}$ *is* a mixed tensor with components $(A^i{}_j)$.

Note that in components the bilinear form has a pleasant matrix expression

$$W(\alpha, \mathbf{v}) = a_i A^i{}_j v^j = aA\, v$$

The tensor product $\mathbf{w} \otimes \beta$ of a vector and a covector is the mixed tensor defined by

$$(\mathbf{w} \otimes \beta)(\alpha, \mathbf{v}) = \alpha(\mathbf{w})\beta(\mathbf{v})$$

As in Problem 2.4 (1)

$$\mathbf{A} = A^i{}_j \boldsymbol{\partial}_i \otimes dx^j = \boldsymbol{\partial}_i \otimes A^i{}_j\, dx^j$$

In particular, the *identity* linear transformation is

$$I = \boldsymbol{\partial}_i \otimes dx^i \tag{2.38}$$

and its components are of course $\delta^i_j$.

Note that we have written matrices $A$ in three different ways, $A_{ij}$, $A^{ij}$, and $A^i{}_j$. The first two define bilinear forms (on $E$ and $E^*$, respectively)

$$A_{ij} v^i w^j \quad \text{and} \quad A^{ij} a_i b_j$$

and only the last is the matrix of a linear transformation $\mathbf{A} : E \to E$. A point of confusion in elementary linear algebra arises since the matrix of a linear transformation there is usually written $A_{ij}$ and they make no distinction between linear transformations and bilinear forms. We *must* make the distinction. In the case of an inner product space $E$, $\langle , \rangle$ we may relate these different tensors as follows. Given a linear transformation $\mathbf{A} : E \to E$, that is, a mixed tensor, we may associate a covariant bilinear form $A'$ by

$$A'(\mathbf{v}, \mathbf{w}) := \langle \mathbf{v}, \mathbf{A}\mathbf{w} \rangle = v^i\, g_{ij} A^j{}_k w^k$$

Thus $A'_{ik} = g_{ij} A^j{}_k$. Note that we have "lowered the index $j$, making it a $k$, by means of the metric tensor." *In tensor analysis one uses the same letter*; that is, instead of $A'$ one merely writes $A$,

$$A_{ik} := g_{ij} A^j{}_k \tag{2.39}$$

It is clear from the placement of the indices that we now have a covariant tensor. This is the matrix of the covariant bilinear form associated to the linear transformation $\mathbf{A}$. In general its components differ from those of the mixed tensor, *but they coincide when the basis is orthonormal*, $g_{ij} = \delta^i_j$. Since orthonormal bases are almost always used in elementary linear algebra, they may dispense with the distinction.

In a similar manner one may associate to the linear transformation **A** a contravariant bilinear form

$$\bar{A}(\alpha, \beta) = a_i A^i{}_j g^{jk} b_k$$

whose matrix of components would be written

$$A^{ik} = A^i{}_j g^{jk} \tag{2.40}$$

Recall that the components of a second-rank tensor always form a matrix such that the *left*-most index denotes the *row* and the *right*-most index the *column*, *independent of whether the index is up or down.*

A final remark. The metric tensor $\{g_{ij}\}$, being a covariant tensor, does *not* represent a linear transformation of $E$ into itself. However, it *does* represent a linear transformation from $E$ to $E^*$, sending the vector with components $v^j$ into the covector with components $g_{ij} v^j$.

## 2.4d. Transformation Properties of Tensors

As we have seen, a mixed tensor $W$ has components (with respect to a basis $\boldsymbol{\partial}$ of $E$ and the dual basis $dx$ of $E^*$) given by

$$W^{i\dots j}{}_{k\dots l} = W(dx^i, \dots, dx^j, \boldsymbol{\partial}_k, \dots, \boldsymbol{\partial}_l).$$

Under a change of bases, $\boldsymbol{\partial}'_l = \boldsymbol{\partial}_s(\partial x^s / \partial x'^l)$ and $dx'^i = (\partial x'^i / \partial x^c)\, dx^c$ we have, by multilinearity,

$$\begin{aligned} W'^{i\dots j}{}_{k\dots l} &= W(dx'^i, \dots, dx'^j, \boldsymbol{\partial}'_k, \dots, \boldsymbol{\partial}'_l) \qquad (2.41\text{a}) \\ &= \left(\frac{\partial x'^i}{\partial x^c}\right) \cdots \left(\frac{\partial x'^j}{\partial x^d}\right) \left(\frac{\partial x^r}{\partial x'^k}\right) \cdots \left(\frac{\partial x^s}{\partial x'^l}\right) W^{c\dots d}{}_{r\dots s} \end{aligned}$$

Similarly, for covariant $Q$ and contravariant $T$ we have

$$Q'_{i\dots j} = \left(\frac{\partial x^k}{\partial x'^i}\right) \cdots \left(\frac{\partial x^l}{\partial x'^j}\right) Q_{k\dots l} \tag{2.41b}$$

and

$$T'^{i\dots j} = \left(\frac{\partial x'^i}{\partial x^k}\right) \cdots \left(\frac{\partial x'^j}{\partial x^l}\right) T^{k\dots l} \tag{2.41c}$$

*Classical tensor analysts dealt not with multilinear functions, but rather with their components.* They would say that a mixed tensor assigns, to each basis of $E$, a collection of "components" $W^{i\dots j}{}_{k\dots l}$ such that under a change of basis the components transform by the law (2.41a). This is a convenient terminology generalizing (2.1).

**Warning:** A linear transformation (mixed tensor) $A$ has eigenvalues $\lambda$ determined by the equation $A\mathbf{v} = \lambda\mathbf{v}$, that is, $A^i_j v^j = \lambda v^i$, but a covariant second-rank tensor $Q$ does not. This is evident just from our notation; $Q_{ij}\, v^j = \lambda v^i$ makes no sense since $i$ is a covariant index on the left whereas it is a contravariant index on the right. Of course we *can* solve the linear equations $Q_{ij}\, v^j = \lambda v^i$ as in linear algebra; that is, we solve the secular equation $\det(Q - \lambda I) = 0$, but the point is that *the solutions* $\lambda$

*depend on the basis* used. Under a change of basis, the transformation rule (2.41b) says $Q'_{ij} = (\partial x^k/\partial x'^i) Q_{kl} (\partial x^l/\partial x'^j)$. Thus we have

$$Q' = \left(\frac{\partial x}{\partial x'}\right)^T Q \left(\frac{\partial x}{\partial x'}\right)$$

and the solutions of $\det[Q' - \lambda I] = 0$ in general differ from those of $\det[Q - \lambda I] = 0$. (In the case of a *mixed* tensor $W$, the transpose $T$ is replaced by the inverse, yielding an invariant equation $\det(W' - \lambda I) = \det(W - \lambda I)$.) *It thus makes no intrinsic sense to talk about the eigenvalues or eigenvectors of a quadratic form.* Of course *if we have a metric tensor* $g$ given, to a covariant matrix $Q$ we may form the mixed version $g^{ij} Q_{jk} = W^i{}_k$ and then find the eigenvalues of this $W$. This is equivalent to solving

$$Q_{ij} v^j = \lambda g_{ij} v^j$$

and this requires

$$\det(Q - \lambda g) = 0$$

It is easy to see that this equation is independent of basis, as is clear also from our notation. We may call these eigenvalues $\lambda$ **the eigenvalues of the quadratic form with respect to the given metric** $g$. This situation arises in the problems of *small oscillations of a mechanical system*; see Problem 2.4(4).

## 2.4e. Tensor Fields on Manifolds

A (differentiable) tensor field on a manifold has components that vary differentiably. A Riemannian metric $(g_{ij})$ is a very important second-rank covariant tensor field.

Tensors are important on manifolds because we are frequently required to construct expressions by using local coordinates, yet we wish our expressions to have an intrinsic *meaning* that all coordinate systems will agree upon.

Tensors in physics usually describe physical fields. For example, Einstein discovered that the metric tensor $(g_{ij})$ in 4-dimensional space–time describes the gravitational field, to be discussed in Chapter 11. (This is similar to describing the Newtonian gravitational field by the scalar Newtonian potential function $\phi$.) Different observers will usually use different local coordinates in 4-space. By making measurements with "rulers and clocks," each observer can in principle measure the components $g_{ij}$ for their coordinate system. Since the metric of space–time is assumed to have physical significance (Einstein's discovery), although two observers will find different components in their systems, the two sets of components $g_{ij}$ and $g'_{ij}$ will be related by the transformation law for a covariant tensor of the second rank. The observers will then want to describe and *agree* on the *strength* of the gravitational field, and this will involve derivatives of their metric components, just as the Newtonian strength is measured by grad $\phi$. By "agree," we mean, presumably, that the strengths will again be components of some tensor, perhaps of higher rank. In the Newtonian case the field is described by a scalar $\phi$ and the strength is a vector, grad$(\phi)$. We shall see that this is not at all a trivial task. We shall illustrate this point with a far simpler example; this example will be dealt with more extensively later on, after we have developed the appropriate tools.

Space–time is some manifold $M$, perhaps not $\mathbb{R}^4$. Electromagnetism is described locally by a "vector potential," that is, by some vector field. It is not usually clear in the texts whether the vector is contravariant or covariant; recall that even in Minkowski space there are differences in the components of the covariant and contravariant versions of a vector field (see 2.1d). As you will learn in Problem 2.4(3), there is good reason to assume that *the vector potential is a covector* $\alpha = A_j dx^j$.

In the following we shall use the popular notations $\partial_i \phi := \partial\phi/\partial x^i$, and $\partial'_i \phi = \partial\phi/\partial x'^i$.

The electromagnetic *field strength* will involve derivatives of the $A$'s, but it will be clear from the following calculation that the expressions

$$\partial_i A_j$$

do *not* form the components of a second-rank tensor!

**Theorem (2.42):** *If $A_j$ are the components of a covariant vector on any manifold, then*

$$F_{ij} := \partial_i A_j - \partial_j A_i$$

*form the components of a second-rank covariant tensor.*

PROOF: We need only verify the transformation law in (2.42). Since $\alpha = A_j dx^j$ *is* a covector, we have $A'_j = (\partial'_j x^l) A_l$ and so

$$\begin{aligned}
F'_{ij} &= \partial'_i A'_j - \partial'_j A'_i = \partial'_i \{(\partial'_j x^l) A_l\} - \partial'_j \{(\partial'_i x^l) A_l\} \\
&= (\partial'_j x^l)(\partial'_i A_l) + [(\partial'_i \partial'_j x^l) A_l] - (\partial'_i x^l)(\partial'_j A_l) - (\partial'_j \partial'_i x^l) A_l \\
&= (\partial'_j x^l)(\partial_r A_l)(\partial'_i x^r) - (\partial'_i x^l)(\partial_r A_l)(\partial'_j x^r)
\end{aligned}$$

(and since $r$ and $l$ are dummy summation indices)

$$\begin{aligned}
&= (\partial'_i x^l)(\partial'_j x^r)(\partial_l A_r - \partial_r A_l) \\
&= (\partial'_i x^l)(\partial'_j x^r) F_{lr} \quad \square
\end{aligned}$$

Note that the term in brackets [ ] is what prevents $\partial_i A_j$ itself from defining a tensor. Note also that if our manifold were $\mathbb{R}^n$ and if we restricted ourselves to *linear* changes of coordinates, $x'^i = L^i{}_j x^j$, then $\partial_i A_j$ *would* transform as a tensor. One *can* talk about objects that transform as tensors with respect to some restricted class of coordinate systems; a *cartesian* tensor is one based on cartesian coordinate systems, that is, on orthogonal changes of coordinates. For the present we shall allow *all* changes of coordinates. In our electromagnetic case, $(F_{ij})$ *is* the **field strength tensor**.

Our next immediate task will be the construction of a mathematical machine, the "exterior calculus," that will allow us systematically to generate "field strengths" generalizing (2.42).

## Problems

**2.4(1)** Show that the second-rank tensor given in components by $a_i b_j dx^i \otimes dx^j$ has the same values as $\alpha \otimes \beta$ on any pair of vectors, and so

$$\alpha \otimes \beta = a_i\, b_j dx^i \otimes dx^j$$

**2.4(2)** Let $\mathbf{A} : E \to E$ be a linear transformation.

**(i)** Show by the transformation properties of a mixed tensor that the trace $\mathrm{tr}(\mathbf{A}) = A^i{}_i$ is indeed a scalar, that is, is independent of basis.

**(ii)** Investigate $\sum_i A_{ii}$.

**2.4(3)** Let $\mathbf{v} = v^i \partial_i$ be a contravariant vector field on $M^n$.

**(i)** Show by the transformation properties that $v_j = g_{ji} v^i$ yields a covariant vector.

For the following you will need to use the chain rule

$$\frac{\partial}{\partial x'^i}\left(\frac{\partial x'^j}{\partial x^k}\right) = \sum_r \left(\frac{\partial^2 x'^j}{\partial x^r \partial x^k}\right)\left(\frac{\partial x^r}{\partial x'^i}\right)$$

**(ii)** Does $\partial_j\, v^i$ yield a tensor?

**(iii)** Does $(\partial_i\, v^j - \partial_j\, v^i)$ yield a tensor?

**2.4(4)** Let $(q = 0, \dot{q} = 0)$ be an equilibrium point for a dynamical system, that is, a solution of Lagrange's equations $d/dt(\partial L/\partial \dot{q}^k) = \partial L/\partial q^k$ for which $q$ and $\dot{q}$ are identically 0. Here $L = T - V$ where $V = V(q)$ and where $2T = g_{ij}(q)\dot{q}^i \dot{q}^j$ is assumed positive definite. Assume that $q = 0$ is a nondegenerate minimum for $V$; thus $\partial V/\partial q^k = 0$ and the Hessian matrix $Q_{jk} = (\partial^2 V/\partial q^j \partial q^k)(0)$ is positive definite. For an approximation of small motions near the equilibrium point one assumes $q$ and $\dot{q}$ are small and one discards all cubic and higher terms in these quantities.

**(i)** Using Taylor expansions, show that Lagrange' s equations in our quadratic approximation become

$$g_{kl}(0)\ddot{q}^l = -Q_{kl} q^l$$

One may then find the eigenvalues of $Q$ with respect to the kinetic energy metric $g$; that is, we may solve $\det(Q - \lambda g) = 0$. Let $y = (y^1, \ldots, y^n)$ be an (constant) eigenvector for eigenvalue $\lambda$, and put $q^i(t) := sin(t\sqrt{\lambda})y^i$.

**(ii)** Show that $q(t)$ satisfies Lagrange's equation in the quadratic approximation, and hence the eigendirection $y$ yields a small harmonic oscillation with frequency $\omega = \sqrt{\lambda}$. The direction $y$ yields a **normal mode** of vibration.

**(iii)** Consider the double planar pendulum of Figure 1.10, with coordinates $q^1 = \theta$ and $q^2 = \phi$, arm lengths $l_1 = l_2 = l$, and masses $m_1 = 3, m_2 = 1$. Write down $T$ and $V$ and show that in our quadratic approximation we have

$$g = l^2 \begin{bmatrix} 4 & 1 \\ 1 & 1 \end{bmatrix} \quad \text{and} \quad Q = gl \begin{bmatrix} 4 & 0 \\ 0 & 1 \end{bmatrix}$$

Show that the normal mode frequencies are $\omega_1 = (2g/3l)^{1/2}$ and $\omega_2 = (2g/l)^{1/2}$ with directions $(y^1, y^2) = (\theta, \phi) = (1, 2)$ and $(1, -2)$.

## 2.5. The Grassmann or Exterior Algebra

How can we define an *oriented* area spanned by two vectors in $\mathbb{R}^n$?

### 2.5a. The Tensor Product of Covariant Tensors

Before the middle of the nineteenth century, Grassmann introduced a new "algebra" whose product is a vast generalization of the scalar and vector products in use today in vector analysis. In particular it is applicable in space of any dimension. Before discussing this "Grassmann product" it is helpful to consider a simpler product, special cases of which we have used earlier. In 2.4 we defined the vector space $\otimes^p E^*$ of covariant $p$-tensors (i.e., tensors of rank $p$) over the vector space $E$; these covariant tensors were simply $p$-linear maps $\alpha : E \times \cdots \times E \to \mathbb{R}$. We now define the "tensor" product of a covariant $p$-tensor and a covariant $q$-tensor.

**Definition:** If $\alpha \in \otimes^p E^*$ and $\beta \in \otimes^q E^*$, then their **tensor product** $\alpha \otimes \beta$ is the covariant $(p+q)$-tensor defined by

$$\alpha \otimes \beta(\mathbf{v}_1, \ldots, \mathbf{v}_{p+q}) := \alpha(\mathbf{v}_1, \ldots, \mathbf{v}_p)\beta(\mathbf{v}_{p+1}, \ldots, \mathbf{v}_{p+q})$$

### 2.5b. The Grassmann or Exterior Algebra

**Definition:** An **(exterior) *p*-form** is a covariant $p$-tensor $\alpha \in \otimes^p E^*$ that is antisymmetric (= skew symmetric = alternating)

$$\alpha(\ldots \mathbf{v}_r, \ldots, \mathbf{v}_s, \ldots) = -\alpha(\ldots \mathbf{v}_s, \ldots, \mathbf{v}_r, \ldots)$$

in each pair of entries.

In particular, the value of $\alpha$ will be 0 if the same vector appears in two different entries. The collection of all $p$-forms is a vector space

$$\bigwedge^p E^* = E^* \wedge E^* \wedge \ldots \wedge E^* \subset \otimes^p E^*$$

By definition, $\bigwedge^1 E^* = E^*$ is simply the space of 1-forms. It is convenient to make the special definition $\bigwedge^0 E^* := \mathbb{R}$, that is, 0-forms are simply scalars. A 0-form field on a manifold is a differentiable *function.*

We need again to simplify the notation. We shall use the notion of a "multiindex," $I = (i_1, \ldots, i_p)$; the number $p$ of indices appearing will usually be clear from the context. Furthermore, we shall denote the $p$-tuple of vectors $(\mathbf{v}_{i_1}, \ldots, \mathbf{v}_{i_p})$ simply by $\mathbf{v}_I$.

Let $\alpha \in \bigwedge^p E^*$ be a $p$-form, and let $\boldsymbol{\partial}$ be a basis of $E$. Then by (2.34) (i.e., multilinearity) $\alpha$ is determined by its $n^p$ components

$$a_I = a_{i_1,\ldots,i_p} = \alpha(\boldsymbol{\partial}_{i_1}, \ldots, \boldsymbol{\partial}_{i_p}) = \alpha(\boldsymbol{\partial}_I)$$

By skew symmetry

$$a_{i_1,\ldots,i_r,\ldots,i_s,\ldots,i_p} = -a_{i_1,\ldots,i_s,\ldots,i_r,\ldots,i_p}$$

Thus $\alpha$ is completely determined by its values $\alpha(\partial_{i_1}, \ldots \partial_{i_p})$ where the indices are in strictly increasing order. When the indices in $I$ are in increasing order, $i_1 < i_2 < \ldots, < i_p$ we shall write $\underset{\to}{I}$

$$\underset{\to}{I} = (i_1 < \ldots < i_p)$$

The number of distinct $\underset{\to}{I} = (i_1 < \ldots < i_p)$ is the combinatorial symbol, that is,

$$\dim \bigwedge^p E^* = n!/p!(n-p)!$$

In particular, the dimension of the space of $n$-forms, where $n = \dim E$, is 1; any $n$-form is determined by its value on $(\partial_1, \ldots, \partial_n)$. Furthermore, since a repeated $\partial_i$ will give 0, $\bigwedge^p E^*$ is 0-dimensional if $p > n$. *There are no nontrivial p-forms on an n-manifold when $p > n$.*

We now wish to define a product of exterior forms. Clearly, if $\alpha$ is a $p$-form and $\beta$ is a $q$-form then $\alpha \otimes \beta$ is a $(p+q)$ *tensor* that is skew symmetric in the first $p$ and last $q$ entries, but need not be skew symmetric in all entities; that is, it need not be a $(p+q)$ form. Grassmann defined a new product $\alpha \wedge \beta$ that is indeed a form. To motivate the definition, consider the case of 1-forms $\alpha^1$ and $\beta^1$ (the superscripts are not tensor indices; they are merely to remind us that the forms are 1-forms). If we put

$$\alpha^1 \wedge \beta^1 := \alpha \otimes \beta - \beta \otimes \alpha$$

that is,

$$\alpha \wedge \beta(\mathbf{v}, \mathbf{w}) = \alpha(\mathbf{v})\beta(\mathbf{w}) - \beta(\mathbf{v})\alpha(\mathbf{w})$$

then $\alpha \wedge \beta$ is then not only a tensor, it is a 2-form. In a sense, we have taken the tensor product of $\alpha$ and $\beta$ and skew-symmetrized it. Define a "generalized Kronecker delta" symbol as follows

$$\begin{aligned} \delta^I_J &:= 1 && \text{if } J = (j_1, \ldots, j_r) \text{ is an } \textit{even} \text{ permutation of } I = (i_1, \ldots, i_r) \\ &= -1 && \text{if } J \text{ is an } \textit{odd} \text{ permutation of } I \\ &= 0 && \text{if } J \text{ is not a permutation of } I \end{aligned}$$

For examples, $\delta^{126}_{621} = -1$, $\delta^{126}_{623} = 0$, $\delta^{126}_{612} = 1$.

We can then define the usual permutation symbols

$$\epsilon_I = \epsilon_{i_1,\ldots,in} = \epsilon^I := \delta^I_{12,\ldots,n}$$

describing whether the $n$ indices $i_1, \ldots, i_n$ form an even or odd permutation of $1, \ldots, n$. This appears in the definition of the determinant of a matrix

$$\det A = \epsilon_I A^{i_1}{}_1 A^{i_2}{}_2 \ldots A^{i_n}{}_n$$

(From this one can see that the $\epsilon$ symbol does *not* define a tensor. For in $\mathbb{R}^2$, if $\epsilon_{ij}$ defined a covariant tensor, we would have $1 = \epsilon'_{12} = \epsilon_{rs}(\partial x^r/\partial x'^1)(\partial x^s/\partial x'^2) = \det(\partial x/\partial x')$, which is only equal to $\epsilon_{12} = 1$ if $\det(\partial x/\partial x') = 1$.)

We now define the **exterior** or **wedge** or **Grassmann** product

$$\wedge : \bigwedge^p E^* \times \bigwedge^q E^* \to \bigwedge^{p+q} E^*$$

of forms. Let $\alpha^p$ and $\beta^q$ be forms. We define $\alpha^p \wedge \beta^q$ to be the $(p+q)$-form with values on $(p+q)$-tuples of vectors $\mathbf{v}_I$, $I = (i_1, \ldots, i_{(p+q)})$ given as follows. Let $\underset{\to}{J} = (j_1 < \ldots < j_p)$ and $\underset{\to}{K} = (k_1 < \ldots < k_q)$ be subsets of $I$. Then

$$\alpha \wedge \beta(\mathbf{v}_I) := \sum_{\underset{\to}{K}} \sum_{\underset{\to}{J}} \delta_I^{JK} \alpha(\mathbf{v}_J) \beta(\mathbf{v}_K)$$

or (2.43)

$$(\alpha \wedge \beta)_I = \sum_{\underset{\to}{K}} \sum_{\underset{\to}{J}} \delta_I^{JK} \alpha_J \beta_K$$

For example, if $\dim E = 5$, and if $\mathbf{e}_1, \ldots, \mathbf{e}_5$ is a basis for $E$

$$\begin{aligned}(\alpha^2 \wedge \beta^1)_{523} = \alpha^2 \wedge \beta^1(\mathbf{e}_5, \mathbf{e}_2, \mathbf{e}_3) &= \sum_{r<s} \sum_{t} \delta_{523}^{rst} \alpha_{rs} \beta_t \\ &= \delta_{523}^{235} \alpha_{23} \beta_5 + \delta_{523}^{253} \alpha_{25} \beta_3 + \delta_{523}^{352} \alpha_{35} \beta_2 \\ &= \alpha_{23} \beta_5 - \alpha_{25} \beta_3 + \alpha_{35} \beta_2\end{aligned}$$

In general, one checks easily that $\alpha \wedge \beta$ is multilinear. Also, since $\delta^R_{i \ldots j \ldots k \ldots l} = -\delta^R_{i \ldots k \ldots j \ldots l}$ we see that $\alpha \wedge \beta$ is again skew symmetric. The wedge product, however, is not commutative in general.

$$\begin{aligned}(\beta^q \wedge \alpha^p)_I &= \sum_{\underset{\to}{J}} \sum_{\underset{\to}{K}} \delta_I^{KJ} \beta_K \alpha_J \\ &= (-1)^{pq} \sum_{\underset{\to}{J}} \sum_{\underset{\to}{K}} \delta_I^{JK} \alpha_J \beta_K\end{aligned}$$

since $KJ \to JK$ requires $pq$ transpositions. Thus,

$$\alpha^p \wedge \beta^q = (-1)^{pq} \beta^q \wedge \alpha^p \tag{2.44}$$

In particular, for forms of *odd* degree, $\alpha^{2p+1} \wedge \alpha^{2p+1} = 0$. Thus

$$dx \wedge dy = -dy \wedge dx \quad \text{and} \quad dx \wedge dx = 0 \tag{2.45}$$

We may consider the vector space of all forms of all degrees over $E^*$

$$\overset{*}{\bigwedge} E^* := \left(\overset{0}{\bigwedge} E^* = \mathbb{R}\right) \oplus \left(\overset{1}{\bigwedge} E^* = E^*\right) \oplus \ldots \oplus \left(\overset{n}{\bigwedge} E^*\right)$$

This is the **Grassmann** or **exterior algebra** over $E^*$, and

$$\dim \overset{*}{\bigwedge} E^* = \binom{n}{0} + \binom{n}{1} + \cdots + \binom{n}{n} = 2^n$$

It is crucial for computational purposes that the Grassmann algebra is distributive and associative. It is trivial to show distributivity; associativity will follow from the following very useful result.

**Lemma (2.46):**

$$\sum_{\underset{\to}{J}} \delta_M^{IJ} \delta_J^{KL} = \delta_M^{IKL}$$

**PROOF:** $I, K, L$, and $M$ are all fixed. Since $J$ is in increasing order, there is at most one term on the left-hand side, namely when $J$ is some permutation of $KL$. One then simply verifies that the preceding formula is correct in the cases when $J$ is an even and an odd permutation of $KL$. □

One can now verify that the exterior product is associative. Let $M$ be any $(p+q+r)$ multiindex. Look at the component $[\alpha^p \wedge (\beta^q \wedge \gamma^r)]_M$. Then

$$\begin{aligned}[\alpha^p \wedge (\beta^q \wedge \gamma^r)]_M &= \sum_{\underrightarrow{IJ}} \delta^{IJ}_M \alpha_I (\beta \wedge \gamma)_J \\ &= \sum_{\underrightarrow{IJ}} \delta^{IJ}_M \alpha_I \sum_{\underrightarrow{KL}} \delta^{KL}_J \beta_K \gamma_L \\ &= \sum_{\underrightarrow{IKL}} \delta^{IKL}_M \alpha_I \beta_K \gamma_L\end{aligned}$$

It is clear that one would get the same expression for $[(\alpha \wedge \beta) \wedge \gamma]$.

The same type of computation would show that if $\alpha_{(1)}, \ldots, \alpha_{(r)}$ are all 1-forms and if $\mathbf{v}_{(1)}, \ldots, \mathbf{v}_{(r)}$ is any $r$-tuple of vectors, then

$$\begin{aligned}\alpha_{(1)} \wedge \ldots \wedge \alpha_{(r)}(\mathbf{v}_{(1)}, \ldots, \mathbf{v}_{(r)}) &= \sum_I \delta^I_{12\ldots r} \alpha_{(1)}(\mathbf{v}_{i(1)}) \ldots \alpha_{(r)}(\mathbf{v}_{i(r)}) \\ &= \det[\alpha_{(j)}(\mathbf{v}_i)] \qquad (2.47)\end{aligned}$$

Let $\sigma^1, \ldots, \sigma^n$ be the basis of 1-forms dual to $\mathbf{e}_1, \ldots, \mathbf{e}_n$. If we write

$$\sigma^I \quad \text{for } \sigma^{i_1} \wedge \ldots \wedge \sigma^{i_r}$$

then we have

$$\sigma^I(\mathbf{e}_J) = \delta^I_J \qquad (2.48)$$

since this is certainly true, from (2.47), when $I$ and $J$ are increasing.

The reader should see Problem 2.5 (1) at this time. This problem says that

$$\alpha^p = \sum_{\underrightarrow{I}} a_I \sigma^I$$

where (2.49)

$$a_I = a_{i_1 \ldots i_p} := \alpha(\mathbf{e}_I)$$

is skew symmetric in $i_1, \ldots, i_p$. The $a_I$ are the "components of the covariant tensor $\alpha$ with respect to the basis $\sigma^1, \ldots, \sigma^n$ of $E^*$." Thus the most general 2-form in $\mathbb{R}^3$ is of the form

$$\begin{aligned}\beta^2 &= \sum_{i<j} b_{ij} dx^i \wedge dx^j = b_{12} dx^1 \wedge dx^2 + b_{13} dx^1 \wedge dx^3 + b_{23} dx^2 \wedge dx^3 \\ &= b_{23} dx^2 \wedge dx^3 + b_{31} dx^3 \wedge dx^1 + b_{12} dx^1 \wedge dx^2 \qquad (2.50)\end{aligned}$$

We shall see in a moment why we prefer this expression. The reader should see Problem 2.5 (2) at this point.

### 2.5c. The Geometric Meaning of Forms in $\mathbb{R}^n$

Let us look at the geometrical meaning of exterior forms in $E = \mathbb{R}^n$ *in the special case when the coordinates* $x^1, \ldots, x^n$ *are cartesian*; that is, we shall employ the euclidean metric of $\mathbb{R}^n$. The coordinate vectors $\{\boldsymbol{\partial}_i\}$ form an orthonormal basis of $E$, with dual basis $\{dx^i\}$ for $E^*$. We already know that for these 1-forms $dx^i(\mathbf{v}) = v^i$, that is, $dx^i$ *reads off the* $i^{\text{th}}$ *component of* $\mathbf{v}$. Next,

$$dx^i \wedge dx^j(\mathbf{v}, \mathbf{w}) = dx^i(\mathbf{v})dx^j(\mathbf{w}) - dx^j(\mathbf{v})dx^i(\mathbf{w})$$

$$= \begin{vmatrix} v^i & w^i \\ v^j & w^j \end{vmatrix}$$

$= \pm$ the area of the parallelogram spanned by the projections $\pi(\mathbf{v})$, $\pi(\mathbf{w})$ of the vectors $\mathbf{v}, \mathbf{w}$ into the $x^i x^j$ plane; the $+$ sign is used if these projections determine the same orientation of the plane as do $\boldsymbol{\partial}_i$ and $\boldsymbol{\partial}_j$. (We shall discuss the notion of orientation more thoroughly in Section 2.8.)

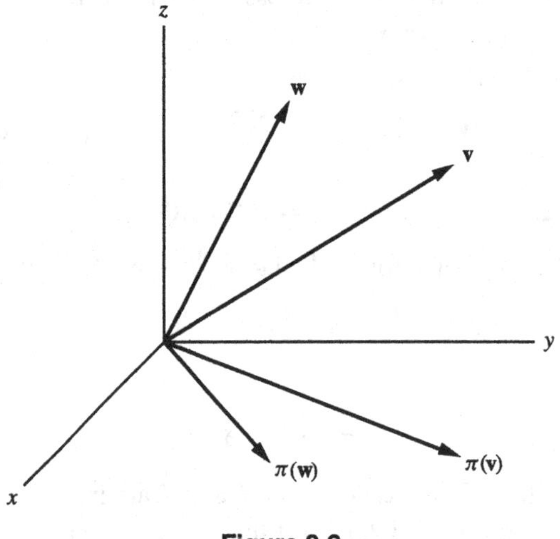


Figure 2.6

In the figure, $dx \wedge dy(\mathbf{v}, \mathbf{w})$ is the negative of the area of the parallelogram spanned by $\pi(\mathbf{v})$ and $\pi(\mathbf{w})$. Likewise, from (2.47),

$$dx^{i_1} \wedge \ldots \wedge dx^{i_p}(\mathbf{v}_1, \ldots, \mathbf{v}_p)$$

$= \pm$ the $p$-dimensional volume of the parallelopiped spanned by the projections of these vectors into the $x^{i_1} \ldots x^{i_p}$ coordinate plane; the $+$ sign is used only if these projected vectors define the same orientation as does $\boldsymbol{\partial}_{i_1}, \ldots, \boldsymbol{\partial}_{i_p}$.

### 2.5d. Special Cases of the Exterior Product

Let $\tau^1, \ldots, \tau^n$ be any $n$-tuple of 1-forms, and expand each in terms of a basis (we are not assuming any scalar product)

$$\tau^i = T^i{}_j \sigma^j$$

Then $\tau^1 \wedge \ldots \wedge \tau^n = \sum_J T^1{}_{j_1} \ldots T^n{}_{j_n} \sigma^{j_1} \wedge \ldots \wedge \sigma^{j_n}$

$$= \sum_J T^1{}_{j_1} \ldots T^n{}_{j_n} \delta^{j_1 \ldots j_n}_{12 \ldots n} \sigma^1 \wedge \ldots \wedge \sigma^n$$

that is,

$$\tau^1 \wedge \ldots \wedge \tau^n = (\det T) \sigma^1 \wedge \ldots \wedge \sigma^n \tag{2.51}$$

*Exterior products yield a coordinate-free expression for the determinant!* For this reason the wedge product is very convenient for discussing linear dependence.

**Theorem (2.52):** *The p 1-forms $\tau^1, \ldots, \tau^p$ are linearly dependent iff*

$$\tau^1 \wedge \ldots \wedge \tau^p = 0$$

PROOF: If $\tau^r = \sum_{i \neq r} a_i \tau^i$ then $\tau^1 \wedge \ldots \wedge \tau^r \wedge \ldots \wedge \tau^p$ will be a sum of terms, each having a repeated $\tau^i$, and so the product will vanish. On the other hand, if the $\tau$'s are linearly independent we may complete them to a basis $\tau^1, \ldots, \tau^n$. Let $\mathbf{f}_1, \ldots, \mathbf{f}_n$ be the dual basis. From (2.47) we have $\tau^1 \wedge \ldots \wedge \tau^p \wedge \ldots \wedge \tau^n(\mathbf{f}_1, \ldots, \mathbf{f}_n) = 1$, showing that $\tau^1 \wedge \ldots \wedge \tau^p \neq 0$. □

## 2.5e. Computations and Vector Analysis

For computations using forms we may use the usual rules of arithmetic except that the commutative law is replaced by (2.44). In particular $dx \wedge dy = -dy \wedge dx$ and $dx \wedge dx = 0$. Consider $\mathbb{R}^3$ as a 3-manifold with *any* (perhaps curvilinear) coordinate system $x^1, x^2, x^3$. Let $f$ be a 0-form, that is, a function of $x$, and let $a_i$, $b_i$, and $c_{ij}$ be functions. Then

$$\alpha^1 = a_1 dx^1 + a_2 dx^2 + a_3 dx^3 \quad \text{and} \quad \beta^1 = b_1 dx^1 + b_2 dx^2 + b_3 dx^3$$

are 1-forms

$$\begin{aligned} \gamma^2 &= c_{23} dx^2 \wedge dx^3 + c_{31} dx^3 \wedge dx^1 + c_{12} dx^1 \wedge dx^2 \\ &:= c_1 dx^2 \wedge dx^3 + c_2 dx^3 \wedge dx^1 + c_3 dx^1 \wedge dx^2 \end{aligned}$$

is a 2-form, and

$$\omega^3 = dx^1 \wedge dx^2 \wedge dx^3$$

is a 3-form.

(In cartesian coordinates $\omega^3$ is the "volume form," but note that, for example, in spherical coordinates $r^2 \sin\theta dr \wedge d\theta \wedge d\phi$ is the volume form; these matters will be discussed later.)

As we shall see, these are familiar expressions used in vector analysis *in the case when the coordinates are cartesian*, involving line, surface, and volume integrals, where they are usually written, for example, as $\alpha = \mathbf{a} \cdot d\mathbf{x}$ and $\gamma = \mathbf{c} \cdot d\mathbf{S}$, and $\omega = dV$. We then have

$$\begin{aligned}
\alpha^1 \wedge \beta^1 &= (a_1 dx^1 + a_2 dx^2 + a_3 dx^3) \wedge (b_1 dx^1 + b_2 dx^2 + b_3 dx^3) \\
&= a_1 b_1 dx^1 \wedge dx^1 + \cdots + a_2 b_3 dx^2 \wedge dx^3 + \cdots + a_3 b_2 dx^3 \wedge dx^2 \\
&= 0 + \cdots + (a_2 b_3 - a_3 b_2) dx^2 \wedge dx^3 \\
&= (a_2 b_3 - a_3 b_2) dx^2 \wedge dx^3 + (a_3 b_1 - a_1 b_3) dx^3 \wedge dx^1 \\
&\quad + (a_1 b_2 - a_2 b_1) dx^1 \wedge dx^2
\end{aligned}$$

In *cartesian* coordinates this says

$$(\mathbf{a} \cdot d\mathbf{x}) \wedge (\mathbf{b} \cdot d\mathbf{x}) = (\mathbf{a} \times \mathbf{b}) \cdot d\mathbf{S}$$

but note that the three components of $\alpha \wedge \beta$, which make sense in any coordinate system, are *not the components of the cross product in curvilinear coordinates*! The exterior product replaces the notion of $\times$ product (which is not associative; $\mathbf{i} \times (\mathbf{i} \times \mathbf{j}) \neq (\mathbf{i} \times \mathbf{i}) \times \mathbf{j}$). We shall see the exact correspondence between exterior forms and vector analysis in Section 2.9b.

## Problems

**2.5(1)** Show that if $\alpha^p$ is any $p$-form, we have the expansion

$$\begin{aligned}
\alpha^p &= \sum_{\underrightarrow{I}} \alpha^p(\mathbf{e}_I)\sigma^I \\
&= \sum_{\underrightarrow{I}} \alpha(\mathbf{e}_{i_1}, \ldots, \mathbf{e}_{i_p})\sigma^{i_1} \wedge \ldots \wedge \sigma^{i_p}
\end{aligned}$$

(Hint: Check values of both sides on $\mathbf{e}_{\underrightarrow{J}}$.)

**2.5(2)** Show that in $\mathbb{R}^n$, if $i < j < k$, then

$$(\alpha^1 \wedge \beta^2)_{ijk} = a_i b_{jk} + a_k b_{ij} + a_j b_{ki}$$

that is, one writes down $a_i b_{jk}$ and then one *cyclically permutes* the indices $i$, $j$, $k$. Investigate $\alpha^1 \wedge \beta^{n-1}$ in $\mathbb{R}^n$, paying special care to the parity of $n$.

**2.5(3)** In $\mathbb{R}^3$, compute $\alpha^1 \wedge \gamma^2$ and $\alpha^1 \wedge \beta^1 \wedge \rho^1$, where $\rho$ is a 1-form, and relate these results to vector analysis.

## 2.6. Exterior Differentiation

Does one *ever* need to write out curl **A** in curvilinear coordinates?

### 2.6a. The Exterior Differential

In Section 2.4e we saw that if $A = A_i(x)dx^i$ is a covariant vector field on a manifold, that is, a 1-form, then $F_{ij} = \partial_i A_j - \partial_j A_i$ are the components of a covariant second-order tensor that is clearly skew symmetric. Thus

$$F := \sum_{i<j} (\partial_i A_j - \partial_j A_i) dx^i \wedge dx^j$$

is an exterior 2-form. We then have a way of "differentiating" a 1-form, obtaining a 2-form. We also showed that the expressions $\{\partial_i A_j\}$ themselves do *not* form the components of a tensor. Problem 2.4 (3) indicated that it does not seem to be possible to differentiate a *contravariant* vector field and obtain a tensor field. In this chapter we shall define a differential operator $d$ that will always take exterior $p$-form fields into exterior $(p+1)$-form fields. In a sense then, *covariant skew symmetric tensors have a richer structure than tensors in general*, and this richer structure plays an essential role in physics.

Recall that if $f$ is a function, that is, a 0-form, then its differential $df = (\partial_i f)dx^i$ is a 1-form. Also, equation (2.44) says that $\alpha^0 \wedge \beta^p = \beta^p \wedge \alpha^0$. For this reason *one ordinarily does not put a wedge* $\wedge$ *in a product involving a 0-form.*

**Theorem (2.53):** *There is a unique operator,* ***exterior differentiation****,*

$$d : \bigwedge^p M^n \to \bigwedge^{p+1} M^n$$

*satisfying*

*(i)* $d$ *is additive,* $d(\alpha + \beta) = d\alpha + d\beta$.

*(ii)* $d\alpha^0$ *is the usual differential of the function* $\alpha^0$.

*(iii)* $d(\alpha^p \wedge \beta^q) = d\alpha^p \wedge \beta^q + (-1)^p \alpha^p \wedge d\beta^q$.

*(iv)* $d^2\alpha := d(d\alpha) = 0$, *for all forms* $\alpha$.

PROOF: We shall first define an operator $d_x$, using a local coordinate system $x$, and then show that this operator is in fact independent of the coordinate system.

*Step I.* If $f$ is a 0-form, define $d_x f = df = (\partial_i f)dx^i$. We know in fact that $df$ is independent of coordinates: Its coordinate-free definition is $df(\mathbf{v}) = \mathbf{v}(f)$; see (2.6). Condition (ii) has been satisfied.

*Step II.* If $a$ is a function, define, for $I = (i_1, \ldots, i_p)$

$$d_x[a(x)dx^I] = da \wedge dx^I = (\partial_j a)dx^j \wedge dx^I$$

We then define $d_x$ on any $p$-form in the coordinate patch $x$ by additivity

$$d_x \sum_{\underset{\rightarrow}{I}} a_I(x)dx^I = \sum_{\underset{\rightarrow}{I}} da_I \wedge dx^I$$

Condition (i) is automatically satisfied. Consider condition (iii). Let $J = (j_1, \dots, j_q)$. Then

$$
\begin{aligned}
d_x[\sum_{\underset{\to}{I}} a_I dx^I \wedge \sum_{\underset{\to}{J}} b_J dx^J] &= d_x \sum_{\underset{\to}{IJ}} a_I b_J dx^I \wedge dx^J \\
&= \sum_{\underset{\to}{IJ}} (da_I b_J + a_I db_J) \wedge dx^I \wedge dx^J \\
&= \sum_{\underset{\to}{I}} da_I \wedge dx^I \wedge \sum_{\underset{\to}{J}} b_J dx^J \\
&\quad + \sum_{\underset{\to}{I}} a_I dx^I \wedge \sum_{\underset{\to}{J}} (-1)^p db_J \wedge dx^J
\end{aligned}
$$

since $db_J \wedge dx^I = (-1)^p dx^I \wedge db_J$ involves $p$ interchanges. (iii) is satisfied.

To verify (iv), note that if $f$ is a function, then

$$
\begin{aligned}
d_x(d_x(f)) &= d_x \sum_i (\partial_i f) dx^i = \sum_i d_x(\partial_i f) \wedge dx^i = \sum_{ij} (\partial^2_{ij} f) dx^j \wedge dx^i \\
&= \cdots + \left( \frac{\partial^2 f}{\partial x^r \partial x^s} \right) dx^r \wedge dx^s \cdots + \left( \frac{\partial^2 f}{\partial x^s \partial x^r} \right) dx^s \wedge dx^r + \cdots = 0
\end{aligned}
$$

(It is a general and very useful fact that if $A^{\dots i \dots j \dots}_{\dots r \dots s \dots}$ is symmetric in $i$, $j$ and skew symmetric in $r$, $s$ then the contraction $A^{\dots i \dots j \dots}_{\dots i \dots j \dots} = 0$.)

Then from (iii), for *any* functions $f$, $g$, not simply for coordinate functions, we have

$$
d_x(df \wedge dg) = 0
$$

and by induction

$$
d_x(df \wedge dg \wedge \cdots \wedge dh) = 0 \tag{2.54}
$$

Then, for any $p$-form $\alpha$

$$
d_x^2 \alpha = d_x^2 \sum_{\underset{\to}{I}} a_I dx^I = d_x \sum_{\underset{\to}{I}} da_I \wedge dx^I = 0
$$

We have now defined an operator $d_x$ in each coordinate patch $x$ and it satisfies (i), (ii), (iii), and (iv). Let $y$ be another coordinate patch that overlaps $x$, and let $d_y$ be the corresponding differential. Then, since $d_y$ again coincides with $d_x$ on functions, in particular coordinate functions, we have, from (iii) and (2.54),

$$
\begin{aligned}
d_y \sum_{\underset{\to}{I}} a_I(x) dx^I &= \sum_{\underset{\to}{I}} d_y a_I[x(y)] \wedge dx^I \\
&= \sum_{\underset{\to}{I}} da_I \wedge dx^I \\
&= d_x \sum_{\underset{\to}{I}} a_I(x) dx^I
\end{aligned}
$$

Thus $d := d_y = d_x$ is well defined, independent of coordinates.

As to uniqueness, any operator $d'$ satisfying (i), (ii), (iii), and (iv) must satisfy

$$d' \sum_I a_I(x) dx^I = \sum_I da_I \wedge dx^I = d \sum_I a_I(x) dx^I \quad \square$$

### 2.6b. Examples in $\mathbb{R}^3$

Let $\mathbf{x} = x, y, z$ be *any* (perhaps curvilinear) coordinate system in $\mathbb{R}^3$. Then the differential of a function $f = f^0$ is

$$df^0 = \left(\frac{\partial f}{\partial x}\right) dx + \left(\frac{\partial f}{\partial y}\right) dy + \left(\frac{\partial f}{\partial z}\right) dz$$

If the coordinates are *cartesian*, then the components are the components of the gradient of $f$,

$$df = \nabla f \cdot d\mathbf{x}$$

If, in general coordinates

$$\alpha^1 = a_1(\mathbf{x}) dx + a_2(\mathbf{x}) dy + a_3(\mathbf{x}) dz$$

then

$$\begin{aligned} d\alpha^1 &= da_1 \wedge dx + da_2 \wedge dy + da_3 \wedge dz \\ &= \left[\left(\frac{\partial a_1}{\partial x}\right) dx + \left(\frac{\partial a_1}{\partial y}\right) dy + \left(\frac{\partial a_1}{\partial z}\right) dz\right] \wedge dx \\ &\quad + \left[\left(\frac{\partial a_2}{\partial x}\right) dx + \left(\frac{\partial a_2}{\partial y}\right) dy + \left(\frac{\partial a_2}{\partial z}\right) dz\right] \wedge dy \\ &\quad + \left[\left(\frac{\partial a_3}{\partial x}\right) dx + \left(\frac{\partial a_3}{\partial y}\right) dy + \left(\frac{\partial a_3}{\partial z}\right) dz\right] \wedge dz \\ &= (\partial_y a_3 - \partial_z a_2) dy \wedge dz + (\partial_z a_1 - \partial_x a_3) dz \wedge dx \\ &\quad + (\partial_x a_2 - \partial_y a_1) dx \wedge dy \end{aligned}$$

In cartesian coordinates the components are the components of the curl of the vector **A**,

$$d(\mathbf{A} \cdot d\mathbf{x}) = (\text{curl}\, \mathbf{A}) \cdot d\mathbf{S}$$

Finally, for a 2-form $\beta$ (writing $b_{23} = b_1, b_{31} = b_2, b_{12} = b_3$)

$$\begin{aligned} d\beta^2 &= d[b_1 dy \wedge dz + b_2 dz \wedge dx + b_3 dx \wedge dy] \\ &= db_1 \wedge dy \wedge dz + db_2 \wedge dz \wedge dx + db_3 \wedge dx \wedge dy \\ &= [\partial_x b_1 + \partial_y b_2 + \partial_z b_3] dx \wedge dy \wedge dz \end{aligned}$$

whose single component in cartesian coordinates is the divergence of the vector **B**,

$$d(\mathbf{B} \cdot d\mathbf{S}) = \text{div}\, \mathbf{B}\, dV$$

$d^2 = 0$ in any coordinate system; in cartesian coordinates this yields the famous curl grad $= 0$ and div curl $= 0$.

It is important to realize that *it is no more difficult to compute d in a curvilinear coordinate system than in a cartesian one*. For example, in spherical coordinates, for 1-form $\alpha = Pdr + Qd\theta + Rd\phi$

$$\begin{aligned} d[Pdr + Qd\theta + Rd\phi] &= dP \wedge dr + dQ \wedge d\theta + dR \wedge d\phi \\ &= (\partial_\theta R - \partial_\phi Q)d\theta \wedge d\phi + (\partial_\phi P - \partial_r R)d\phi \wedge dr \\ &\quad + (\partial_r Q - \partial_\theta P)dr \wedge d\theta \end{aligned}$$

Note that $(P, Q, R)$ form the components of a *covariant* vector, $\alpha$, and that the three components of $d\alpha^1$ do *not* form the components of the curl of a vector; they are the components of a second-rank covariant skew symmetric tensor. We shall see in Section 2.9 that it is possible to identify 2-forms in $\mathbb{R}^3$ (with a given metric) with contravariant vectors and then the vector identified with $d\alpha$ is the curl of the contravariant version of $\alpha$. This is not only an extremely awkward procedure, it serves no purpose, for we maintain that *there is never any reason to take the curl of a contravariant vector. In situations where the "curl" of a "vector" is required, the "vector" will most naturally appear in covariant form (i.e., it will be a 1-form $\alpha$), and then $d\alpha$ is all that is required.* For example, the electric field measures the force on a unit charge that is at rest. Force, being the time rate of change of momentum, appears naturally as a covector (see (2.29)) and so the electric field is a 1-form $\mathcal{E}^1$. Then Faraday's law really states that $d\mathcal{E}^1$ is the negative of the time rate of change of the magnetic field 2-form $\mathcal{B}^2$. These matters will be discussed in Section 3.5.

## 2.6c. A Coordinate Expression for *d*

Let $\alpha^p = \sum_{\underrightarrow{L}} a_L dx^L$ be a $p$-form; then $d\alpha^p = \sum_{\underrightarrow{L}} (da_L) \wedge dx^L$. Now $da_L$ is the 1-form whose $j^{\text{th}}$ component is $(da_L)_j = \partial_j a_L$. Also $dx^L$ is the $p$-form with components $(dx^L)_K = \delta^L_K$. Then from (2.43) we get

$$(d\alpha)_I = \sum_{\underrightarrow{L}} (da_L \wedge dx^L)_I = \sum_{\underrightarrow{L}} \sum_{j,\underrightarrow{K}} \delta_I^{jK} (\partial_j a_L) \delta^L_K$$

that is,

$$(d\alpha)_I = \sum_{j\underrightarrow{K}} \delta_I^{jK} (\partial_j a_K) \tag{2.55}$$

Thus for $I$ increasing

$$\begin{aligned} (d\alpha^p)_{\underrightarrow{I}} &= \sum_{j,\underrightarrow{K}} \delta^{jk_1\ldots k_p}_{i_1\ldots i_{(p+1)}} \partial_j a_{k_1\ldots k_p} \\ &= \partial_{i_1} a_{i_2\ldots i_{(p+1)}} - \partial_{i_2} a_{i_1 i_3 \ldots i_{(p+1)}} + \cdots \end{aligned}$$

Hence

$$(d\alpha^p)_I = \sum_r (-1)^{r+1} \partial_{i_r} a_{i_1 \ldots \widehat{i_r} \ldots i_{(p+1)}} \tag{2.56}$$

where the hat $\widehat{}$ over $i_r$ means omit $i_r$. We can also write

$$(d\alpha^p)_I = \sum_r (-1)^{r+1} \partial_{i_r}[\alpha^p(\partial_{i_1}, \ldots \widehat{\partial}_{i_r} \ldots \partial_{i_{(p+1)}})] \tag{2.57}$$

If, for example, $\alpha = \sum_i a_i dx^i$ is a 1-form on $M^n$, from (2.55)

$$(d\alpha^1)_{ij} = \partial_i a_j - \partial_j a_i \tag{2.58}$$

and this of course was the procedure used for defining the field strength in (2.42).

If $\beta^2 = \sum_{i<j} b_{ij} dx^i \wedge dx^j$ is a 2-form in an $M^n$, from (2.56)

$$(d\beta^2)_{i<j<k} = (\partial_i b_{jk} + \partial_k b_{ij} + \partial_j b_{ki}) \tag{2.59}$$

**Problem**

**2.6(1)** Relabel the components of a 3-form $\beta^3$ in $\mathbb{R}^4$ (as we did for a 2-form in $\mathbb{R}^3$, $b_{12} = b_3, \ldots$) to get a divergencelike expression for $d\beta^3$. Guess what should be done for $\beta^{n-1}$ in $\mathbb{R}^n$. Watch for the parity of $n$.

## 2.7. Pull-Backs

What are the deformation tensors that arise in elasticity theory?

### 2.7a. The Pull-Back of a Covariant Tensor

Let $F : M^n \to W^r$ be a differentiable map. Sometimes we shall write $M \xrightarrow{F} W$. In local coordinates $x$ for $M$ and $y$ for $W$ we have $y^j = F^j(x)$, or briefly $y = y(x)$.

If $f : W \to \mathbb{R}$ is a smooth function (0-form) on $W$ we define its pull-back to $M$, written $F^*f$, to be the composition $f \circ F : M \to \mathbb{R}$, that is, $M \xrightarrow{F} W \xrightarrow{f} \mathbb{R}$.

$$(F^*f)(x) = (f \circ F)(x) = f(y(x))$$

This is a real-valued function on $M$, $M \xrightarrow{f \circ F} \mathbb{R}$. *One can always pull back a function on $W$. If $F$ has an inverse $G = F^{-1}$* then one can "push forward" a function $h$ on $M$ to yield a function $h \circ F^{-1}$ on $W$, $W \xrightarrow{G} M \xrightarrow{g} \mathbb{R}$, but it should be clear that one cannot in general expect to push forward a function on $M$ to get a function on $W$, unless $F^{-1}$ exists.

For future needs, we exhibit here how a vector $\mathbf{v}$ at $x$ of $M$, as a differential operator, acts on the pull-back of a function.

$$\mathbf{v}(F^*f) = \mathbf{v}[f\{y(x)\}] = v^i \frac{\partial}{\partial x^i}[f\{y(x)\}]$$

$$= v^i \left(\frac{\partial y^j}{\partial x^i}\right)\left(\frac{\partial f}{\partial y^j}\right)$$

$$\mathbf{v}(F^*f) = (F_*\mathbf{v})(f) = df(F_*\mathbf{v}) \tag{2.60}$$

Now let $\alpha^p$ be a *covariant tensor* at $y$ in $W$. We have just defined the pull-back of $\alpha^p$ when $p = 0$. When $p = 1$, that is, when $\alpha$ is a 1-form, its pull-back was defined in (2.23). We now define in general the **pull-back** of a covariant tensor by

$$F^*\alpha^p(\mathbf{v}_1, \ldots, \mathbf{v}_p) := \alpha^p(F_*\mathbf{v}_1, \ldots, F_*\mathbf{v}_p) \tag{2.61}$$

It is clear that $F^*\alpha$ is alternating if $\alpha$ is; that is, the pull-back of a $p$-form on $W$ is a $p$-form on $M$

$$F^* : \bigwedge^p W \to \bigwedge^p M$$

*Unless otherwise indicated, by pull-back we shall mean the pull-back of an exterior form.*

In our warning following (2.25) we pointed out that one cannot push forward a *contravariant* vector field on $M$ to yield a vector field on $W$. The ability to pull back covariant tensors endows these tensors with a crucial operation that is not available to the contravariant ones. *It is difficult to overemphasize the importance of this advantage.*

It is clear from (2.61) that $F^*$ is additive; that is, $F^*$ of a sum is the sum of the $F^*$'s. This is further enhanced by the following two properties: The pull-back of a product of forms is the product of the pull-backs, and the pull-back of the exterior derivative of a form is the derivative of the pull-back. We proceed to these matters, for they are crucial to writing down coordinate expressions economically.

**Theorem (2.62):** *$F^*$ is an algebra homomorphism, that is,*

$$F^*(\alpha \wedge \beta) = (F^*\alpha) \wedge (F^*\beta)$$

For proof see Problem 2.7(1).

It is even simpler to prove that for any tensor product of covariant tensors

$$F^*(\alpha \otimes \beta) = (F^*\alpha) \otimes (F^*\beta) \tag{2.63}$$

**Theorem (2.64):** *$F^*$ commutes with exterior differentiation, $d \circ F^* = F^* \circ d$,*

$$F^*(d\alpha) = d(F^*\alpha)$$

PROOF: When $\alpha = \alpha^0$ is a function $f$ on $W$ near $F(x)$ and $\mathbf{v}$ is tangent vector to $M$ at $x$, we have from (2.60) and (2.23)

$$d(F^*f)(\mathbf{v}) = \mathbf{v}(F^*f) = df(F_*(\mathbf{v})) = (F^*(df))(\mathbf{v})$$

Thus (2.64) has been proved when $\alpha$ is a 0-form. When $\alpha$ is a $p$-form, we have

$$d \circ F^* \sum_{\underset{\rightarrow}{J}} a_J(y) dy^{j_1} \wedge \cdots \wedge dy^{j_p}, \text{ which from (2.62)}$$

$$= d \sum_{\underset{\rightarrow}{J}} (F^*a_J(y))(F^*dy^{j_1}) \wedge \cdots \wedge (F^*dy^{j_p}) =$$

(since (2.64) has been proved for 0-forms)

$$= d\sum_{\underline{J}}(F^*a_J(y))d(F^*y^{j_1})\wedge\ldots\wedge d(F^*y^{j_p})$$

$$= \sum_{\underline{J}}(dF^*a_J(y))\wedge d(F^*y^{j_1})\wedge\ldots\wedge d(F^*y^{j_p})$$

$$= \sum_{\underline{J}}(F^*da_J)\wedge(F^*dy^{j_1})\wedge\ldots\wedge(F^*dy^{j_p})$$

$$= F^*\sum_{\underline{J}}(da_J)\wedge dy^{j_1}\wedge\ldots\wedge dy^{j_p}$$

$$= F^*\circ d\sum_{\underline{J}} a_J(y)dy^{j_1}\wedge\ldots\wedge dy^{j_p}$$

as desired. □

Explicitly, with $I=(i_1,\ldots,i_p)$, $F^*d(y^J)=F^*(dy^{j_1}\wedge\ldots\wedge dy^{j_p})=\sum_I(\partial y^{j_1}/\partial x^{i_1})\ldots(\partial y^{j_p}/\partial x^{i_p})dx^I$. But $dx^I=\sum_{\underline{L}}\delta^I_L dx^L$ (we are merely putting the $dx$'s in increasing order; for each given $I$ there is only one nonzero term in the sum on the right). Then

$$F^*d(y^J)=\sum_{\underline{L}}\left\{\sum_I\left(\frac{\partial y^{j_1}}{\partial x^{i_1}}\right)\cdots\left(\frac{\partial y^{j_p}}{\partial x^{i_p}}\right)\delta^I_L\right\}dx^L$$

$$=\sum_{\underline{L}}\det\left\{\frac{\partial(y^J)}{\partial(x^L)}\right\}dx^L$$

Thus we have

$$F^*d(y^J)=\sum_{\underline{L}}\det\left\{\frac{\partial(y^J)}{\partial(x^L)}\right\}dx^L$$

and so

$$F^*\alpha^p=F^*\sum_{\underline{J}}a_Jdy^J=\sum_{\underline{L}}a^*{}_L(x)dx^L$$

where (2.65)

$$a^*{}_L(x):=\sum_{\underline{J}}a_J(y(x))\det\left\{\frac{\partial(y^J)}{\partial(x^L)}\right\}$$

Let, for example, $M^2$ be a surface in $\mathbb{R}^3$, that is, a 2-dimensional submanifold. We have the **inclusion map**, $i: M\to\mathbb{R}^3$, which is a fancy way of saying that any point of $M$ is also a point in $\mathbb{R}^3$. If $\mathbf{v}$ is a tangent vector to $M$, then $i_*\mathbf{v}$ is simply the same vector $\mathbf{v}$, considered as a vector in $\mathbb{R}^3$. If $\beta^2$ is a 2-form on $\mathbb{R}^3$, then the pull-back of $\beta$ to $M$ is the 2-form $i^*\beta$ whose value on the pair $\mathbf{v}$, $\mathbf{w}$ of tangent vectors to $M$ is given simply by $i^*\beta(\mathbf{v},\mathbf{w})=\beta(i_*\mathbf{v},i_*\mathbf{w})=\beta(\mathbf{v},\mathbf{w})$. In other words, *$i^*\beta$ in this case of inclusion is the same form $\beta$, but we restrict its domain to vectors that are tangent to $M$*. This same situation holds whenever $M^n$ is a submanifold of another manifold. If $\mathbf{u} = (u, v)$ are local coordinates in $M^2$ and $\mathbf{x} = (x, y, z)$ are coordinates for $\mathbb{R}^3$, then

$$\begin{aligned} i^*\beta &= i^*[b_1(\mathbf{x})dy \wedge dz + b_2(\mathbf{x})dz \wedge dx + b_3(\mathbf{x})dx \wedge dy] \\ &= \left[b_1(\mathbf{x}(\mathbf{u}))\frac{\partial(y, z)}{\partial(u, v)} + b_2(\mathbf{x}(\mathbf{u}))\frac{\partial(z, x)}{\partial(u, v)} + b_3(\mathbf{x}(\mathbf{u}))\frac{\partial(x, y)}{\partial(u, v)}\right] du \wedge dv \end{aligned}$$

See Problem 2.7(2) at this time.

Another way to get this coordinate expression for $i^*\beta$ is to compute directly, using the fact that $i^*$ commutes with exterior products and differentiation. For example, putting $\mathbf{x} = (x, y, z)$ and $\mathbf{u} = (u, v)$

$$\begin{aligned} i^*(b_1 dy \wedge dz) &= b_1(\mathbf{x}(\mathbf{u}))i^*(dy) \wedge i^*(dz) \\ &= b_1(\mathbf{x}(\mathbf{u}))\left[\left(\frac{\partial y}{\partial u}\right)du + \left(\frac{\partial y}{\partial v}\right)dv\right] \wedge \left[\left(\frac{\partial z}{\partial u}\right)du + \left(\frac{\partial z}{\partial v}\right)dv\right] \\ &= b_1(\mathbf{x}(\mathbf{u}))\left[\left(\frac{\partial y}{\partial u}\right)\left(\frac{\partial z}{\partial v}\right) - \left(\frac{\partial y}{\partial v}\right)\left(\frac{\partial z}{\partial u}\right)\right] du \wedge dv \end{aligned}$$

Two final remarks. First, if $F : M^n \to M^n$ is the *identity map* but expressed in different coordinates, that is, if $y = y(x)$ is simply a change of coordinates, then $\alpha = F^*\alpha$ is simply expressing the form $\alpha$ in the two coordinate systems. For example, if $u, v, w$ are curvilinear coordinates in $\mathbb{R}^3$ then from either (2.65) or from (2.51) we see

$$dx \wedge dy \wedge dz = \left[\frac{\partial(x, y, z)}{\partial(u, v, w)}\right] du \wedge dv \wedge dw$$

Finally, we have defined the Poincaré 1-form $\lambda = p_i dq^i$ in phase space $T^*M^n$ (see (2.33)). We then define the **Poincaré 2-form** by

$$\omega^2 = d\lambda = dp_i \wedge dq^i \tag{2.66}$$

This form, as we shall see, plays a most important role in Hamiltonian mechanics. If $F : \mathbb{R}^2 \to T^*M^n$ is a 2-dimensional surface in phase space, then the pull back of $\omega$ to $\mathbb{R}^2$ (whose coordinates are $u, v$) is the 2-form

$$F^*\omega = \{u, v\} du \wedge dv$$

where

$$\{u, v\} := \sum_i \frac{\partial(p_i, q^i)}{\partial(u, v)} \tag{2.67}$$

defines the **Lagrange bracket** of the functions $u$ and $v$.

## 2.7b. The Pull-Back in Elasticity

Consider an elastic body $\mathcal{B}$ in $\mathbb{R}^3$ and a deformation $\mathcal{B}' = F(\mathcal{B})$ of this body. To describe this we shall let $X_1, X_2, X_3$ be cartesian coordinates in $\mathbb{R}^3$ and the deformation will be described by functions $x^i = x^i(X)$. We may think of $X$ and $x$ as being two identical Cartesian coordinate systems in $\mathbb{R}^3$. A point with coordinates $X$ in $\mathcal{B}$ will be sent into the point with coordinates $x$ in $\mathcal{B}'$. We shall try to follow a common practice of denoting quantities associated with the undeformed body by capital letters, and those of the deformed body with lower case.

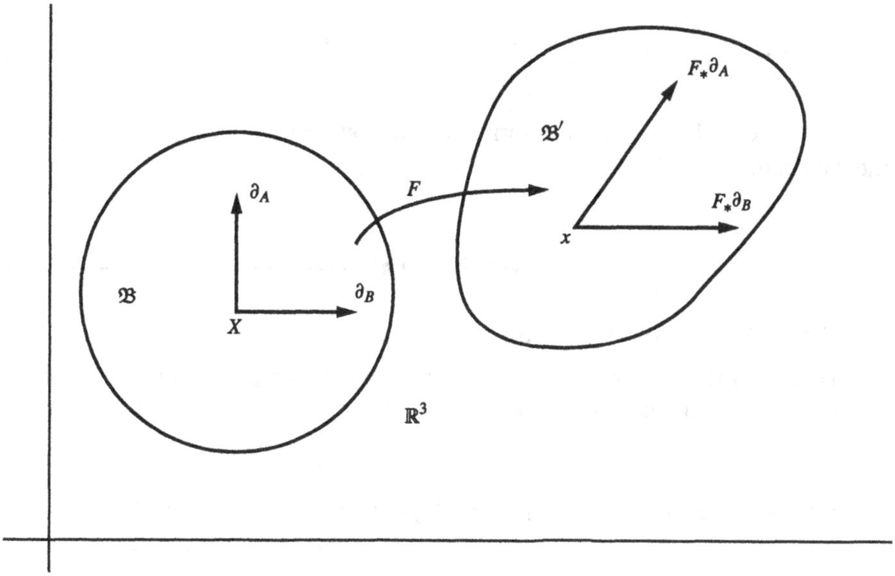


**Figure 2.7**

Under the deformation, the orthonormal pair $\boldsymbol{\partial}_A, \boldsymbol{\partial}_B$ at $X$ is sent, by the differential of $F$ at $X$, into a pair of vectors $F_*\boldsymbol{\partial}_A, F_*\boldsymbol{\partial}_B$ at $x$.

The metric tensor of $\mathbb{R}^3$ can be written $dS^2 = G_{AB}(X)dX^A \otimes dX^B$, meaning $dS^2(\mathbf{V}, \mathbf{W}) = G_{AB}V^AW^B$. *It is traditional to omit the tensor product sign* $\otimes$ *when dealing with symmetric tensors*. Thus at $X$, since the coordinates are cartesian,

$$dS^2 = G_{AB}dX^AdX^B = \delta_{AB}dX^AdX^B = \sum_A (dX^A)^2$$

and this is the usual expression for "arc length" in elementary calculus, $ds^2 = dx^2 + dy^2 + dz^2$. This will be discussed at great length in Part Two.

We may also write this *same tensor*, at the point $x$, as $ds^2 = \sum_a (dx^a)^2$. For the pull-back under $F$ we have, from (2.63),

$$F^*(ds^2) = \sum_a \left[\sum_A \left(\frac{\partial x^a}{\partial X^A}\right)dX^A\right] \otimes \left[\sum_B \left(\frac{\partial x^a}{\partial X^B}\right)dX^B\right]$$

$$= \sum_{aAB} \left(\frac{\partial x^a}{\partial X^A}\right)\left(\frac{\partial x^a}{\partial X^B}\right)dX^AdX^B$$

This tensor,

$$F^*(ds^2) = \sum_{AB} \left(\frac{\partial \mathbf{x}}{\partial X^A}\right) \cdot \left(\frac{\partial \mathbf{x}}{\partial X^B}\right)dX^AdX^B \tag{2.68}$$

when applied to the pair $\partial_C, \partial_D$, reads off the scalar product of the pair $F_*\partial_C, F_*\partial_D$, and is called the **right Cauchy–Green tensor** $C$

$$C := F^*(ds^2)$$

One measure of the deformation taking place is given by the **Lagrange deformation tensor**

$$\frac{1}{2}[F^*(ds^2) - dS^2] = \frac{1}{2}\left[\sum_{AB}\left(\frac{\partial \mathbf{x}}{\partial X^A}\right) \bullet \left(\frac{\partial \mathbf{x}}{\partial X^B}\right) - \delta_{AB}\right] dX^A dX^B \tag{2.69}$$

A more general discussion of deformations in continuum mechanics will be found in the Appendix to this book.

## Problems

**2.7(1)** Prove (2.62). [Hint: Use (2.43)].

**2.7(2)** Let $\mathbf{x}$ be cartesian coordinates for $\mathbb{R}^3$. Then the 2-form $\beta$ is of the form $\beta = \mathbf{b} \bullet d\mathbf{S}$. Show that in the coordinate patch $(u, v)$ of the surface $M^2 \subset \mathbb{R}^3$ we have

$$i^*\beta = \mathbf{b} \bullet \mathbf{n} du \wedge dv \tag{2.70}$$

where $\mathbf{n} := \mathbf{x}_u \times \mathbf{x}_v := (\partial\mathbf{x}/\partial u) \times (\partial\mathbf{x}/\partial v)$ is a (nonunit) normal to $M$.

## 2.8. Orientation and Pseudoforms

Leave your shoes, labeled $R$ and $L$, and take a long trip around the universe. Is it possible that your right foot will only fit into your left shoe when you return?

### 2.8a. Orientation of a Vector Space

Let $\mathbf{e} = (\mathbf{e}_1, \ldots, \mathbf{e}_n)$ and $\mathbf{f} = (\mathbf{f}_1, \ldots, \mathbf{f}_n)$ be two bases of a vector space $E$; we can then write $\mathbf{f} = \mathbf{e}P$, that is, $\mathbf{f}_i = \mathbf{e}_j P^j{}_i$, for a unique nonsingular matrix $P$. We say that $\mathbf{e}$ and $\mathbf{f}$ have the same (resp. opposite) **orientation** if det $P$ is positive (resp. negative). (It is easy to see, from the continuity of the function $P \to \det(P)$, that if a basis $\mathbf{e}$ is continuously deformed into a basis $\mathbf{f}$ *while remaining a basis at each stage*, then both bases have the same orientation.)

The collection of all bases of $E$ then falls naturally into two equivalence classes of bases. (For example, the tangent space to our physical 3-space at a given point is a 3-dimensional vector space, and we have the two classes of bases defined by using either the right- or the left-hand rule.) We **orient** a vector space by declaring one of the two classes of bases to be positive; the other class then consists of negatively oriented bases. In our 3-space it is usual to declare the right-handed bases to be positively oriented, but we could just as well have the left-handed bases as positive. It should be clear that except for our prejudices about right and left, neither choice is any more "natural" than the other. This is especially clear if we consider a 2-dimensional case instead. If we draw a "positive" basis for a sheet of paper by using an $xy$ coordinate system where, as is usual, we rotate through a right angle counterclockwise from $x$ to $y$, then if we view the sheet of paper from the reverse side we see that this basis requires us to rotate clockwise from $x$ to $y$.

To orient a 2-dimensional vector space is to declare one of the two possible senses of rotation about the origin to be positive. Given an oriented plane and a positively oriented basis $\mathbf{e}_1, \mathbf{e}_2$, the positive sense of rotation goes from the first basis vector to the second through the unique angle that is less than a straight angle.

$\mathbb{R}^n$, as a space of $n$-tuples, comes equipped with a natural basis $\mathbf{e}_1 = (1, 0, \dots, 0)^T$, and so on, but it is important to realize that most vector spaces we shall encounter do not have distinguished bases and consequently do not have a *natural* choice of orientation!

## 2.8b. Orientation of a Manifold

Now consider a manifold $M^n$. Of course we may orient each tangent space $M^n_x$ haphazardly, but for many purposes it would help if we could do this in a "continuous" or "coherent" fashion. For example, let $U_x$ be a coordinate patch with coordinates $x$. Then we may orient each tangent space at each point of $U_x$ by declaring the bases $\boldsymbol{\partial} = (\boldsymbol{\partial}_1, \dots, \boldsymbol{\partial}_n)$ to be positively oriented. We have then oriented all the tangent spaces at all points of the patch $U_x$. If a point lies in an overlap $U_x \cap U_y$ of two patches, the two bases are related by $\boldsymbol{\partial}_y = \boldsymbol{\partial}_x(\partial x/\partial y)$, and thus the two orientations agree if and only if the Jacobian determinant is positive.

We shall say that a manifold $M^n$ is **orientable** if we can cover $M$ by coordinate patches having positive Jacobians in each overlap. We can then declare the given coordinate bases to be positively oriented, and we then say that we have **oriented** the manifold. Briefly speaking, *if a manifold is orientable it is then possible to pick out, in a continuous fashion, an orientation for each tangent space $M^n_x$ to $M^n$. Conversely, if it is possible to pick out continuously an orientation in each tangent space, we can (by permuting $x_1$ and $x_2$ if necessary) assume that the coordinate frames in each coordinate patch have the chosen orientation and $M^n$ must be orientable.*

It should be clear that *if $M$* is connected and orientable, then there are exactly two different ways to orient it. Of course if $M$ can be covered by a single coordinate patch it is then orientable. Möbius discovered that there are manifolds that are not orientable and we shall consider this in a moment.

Let $p$ and $q$ be two points of a manifold $M^n$. Let $C$ be any curve joining these two points, $p = C(0)$ and $q = C(1)$. Given a frame $\mathbf{e}(0)$ at $C(0)$ we can extend this frame, in many ways, to yield a frame $\mathbf{e}(t)$ at $C(t)$ for all $0 \le t \le 1$ such that the assignment $t \mapsto \mathbf{e}_i(t)$ is continuous (we do *not* ask that $\mathbf{e}(t_1) = \mathbf{e}(t_2)$ whenever $C(t_1) = C(t_2)$). For example, if $C(t)$ lies in a coordinate patch $U_x$ for $0 \le t \le a$, we can insist that the components of the fields $\mathbf{e}_i(t)$ with respect to the coordinate basis $\boldsymbol{\partial}$ be constant. We can extend past $t = a$ by using perhaps a different patch that holds the next portion of the curve, and so forth. In this way we can, in a continuous fashion, transport a frame at $p$ to a frame at $q$. Although this process is in no sense unique, it is easy to see that the *orientation* of the frame $\mathbf{e}(1)$ at the end $q = C(1)$ of the curve is uniquely determined by the orientation $\mathbf{e}(0)$ at the begining $p = C(0)$, and the reader should verify this. In other words, we have *unique transport of orientation along a curve*. We do not claim that the resulting orientation at $q$ is independent of the *curve* $C$ joining it to $p$. If, however, $M$ is orientable, we may cover $M$ with coordinate patches having positive Jacobians in their overlaps; it is then clear that if $\mathbf{e}(0)$ is positively oriented then $\mathbf{e}(1)$ will also be positively oriented, independent of the curve $C$. It follows that if, in a manifold, transport of orientation can lead to opposing results when applied to two different curves joining $p$ and $q$, then $M$ cannot be orientable. Thus *if transport of orientation about some closed curve leads to a reversal of orientation on return to the starting point, then $M^n$ must be nonorientable*!

The Möbius band is thus clearly nonorientable.

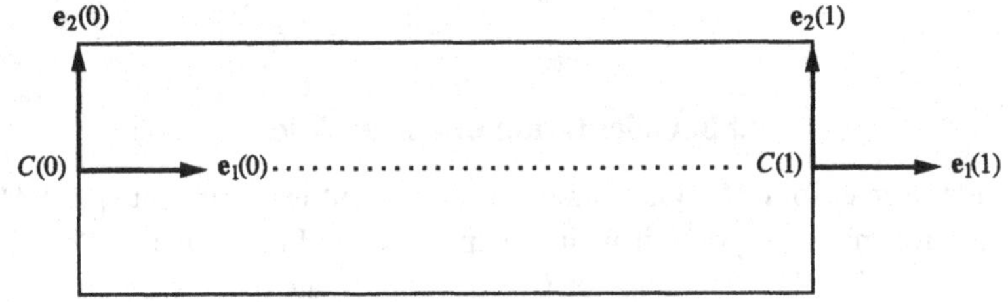


**Figure 2.8**

In this figure we have transported a frame along the midcircle of the Möbius band. By the identifications defining the Möbius band we see that $\mathbf{e}_1(1) = \mathbf{e}_1(0)$ and $\mathbf{e}_2(1) = -\mathbf{e}_2(0)$, and thus orientation is reversed on going around the midcircle.

This example of the Möbius band is but a special case of a very general situation involving "identifications." An accurate treatment of this subject would take us too long; we hope to convey the ideas by means of an example. Before this, we must discuss an important criterion for orientability of a hypersurface (i.e., a submanifold of codimension 1) of an orientable manifold.

### 2.8c. Orientability and 2-Sided Hypersurfaces

Let $M^n$ be a submanifold of $W^r$. A vector field *along* $M$ is a continuous tangent vector field *to* $W$ that is defined at all points of $M$ (it need not be defined at other points). A vector field $\mathbf{N}$ along $M$ is **transverse** to $M$ if it is never tangent to $M$; in particular it is never 0 on $M$.

> We say that a *hypersurface* $M^n$ in $W^{n+1}$ is **2-sided** in $W$ if there is a (continuous) transverse vector field $\mathbf{N}$ defined along $M$.

A surface $M^2$ in $\mathbb{R}^3$ has at each point a pair of oppositely pointing unit normals. Suppose that it is possible to make a continuous choice $\mathbf{N}$ for the entire surface. $\mathbf{N}$ is then a transversal field to $M^2$ and $M^2$ is 2-sided in $\mathbb{R}^3$. For example, the 2-sphere $S^2$ is the complete boundary of a solid ball, and consequently it makes sense to talk of the *outward pointing* unit normal. On the other hand, it is a famous fact that the Möbius band is "1-sided"; that is, there is no way to make a continuous selection of unit normal field. (If we choose a normal at a point of the midcircle of the band and transport it continuously once around the circle, we find on returning to the starting point that the normal has returned to its negative.) If one can define continuously a unit normal field to a surface in $\mathbb{R}^3$ then the surface must be orientable, for we could then make a continuous choice of orientation in each tangent space as follows. $\mathbb{R}^3$ is orientable and so we can choose an orientation of $\mathbb{R}^3$, say the right-handed one. We can then declare a basis $\mathbf{e}_1, \mathbf{e}_2$ of tangent vectors to $M^2$ to be positively oriented if $\mathbf{N}, \mathbf{e}_1, \mathbf{e}_2$ forms a positively oriented basis in $\mathbb{R}^3$.

More generally, *if $M^n$ is a 2-sided hypersurface of an orientable manifold $W^{n+1}$, then $M^n$ is itself orientable*!

We must emphasize the difference between orientability and 2-sidedness. Orientability is an *intrinsic* property of a manifold $M^n$; whether $M^n$ is 2-sided in $W^{n+1}$ *depends on W and on how M is embedded in W*. For example, if $M^n$ is any manifold, orientable or not, consider the product manifold $W^{n+1} = M^n \times \mathbb{R}$, with local coordinates $(x)$ from $M$ and a global coordinate $t$ from $\mathbb{R}$. Then $M^n$ considered as the submanifold defined by $t = 0$ is automatically a 2-sided hypersurface of $W^{n+1}$ with transverse vector field $\partial/\partial t$. Thus the Möbius band Mö is 1-sided in $\mathbb{R}^3$ but it is a 2-sided hypersurface of Mö $\times \mathbb{R}$.

## 2.8d. Projective Spaces

We have seen in Section 1.2b(vi) that the real projective plane $\mathbb{R}P^2$ is the 2-sphere $S^2$ with antipodal points identified. Since $S^2$ is 2-sided in $\mathbb{R}^3$ it is orientable; we declare a basis $\mathbf{e}_1, \mathbf{e}_2$ of tangent vectors to $S^2$ to be positively oriented provided $\mathbf{N}, \mathbf{e}_1, \mathbf{e}_2$, is a right-handed basis of $\mathbb{R}^3$, where $\mathbf{N}$ is the outward pointing normal to the sphere. Note that the antipodal map $a : S^2 \to S^2$ is simply the restriction to $S^2$ of the reversal map $r : \mathbb{R}^3 \to \mathbb{R}^3$, $r(\mathbf{x}) = -\mathbf{x}$, and *in 3 dimensions the reversal map reverses orientation of space*. Thus if $\mathbf{N}, \mathbf{e}_1, \mathbf{e}_2$, is right-handed at the north pole $n$ then $-\mathbf{e}_1, -\mathbf{e}_2, -\mathbf{N}$ is left-handed at the south pole $s$. But $-\mathbf{N}$ is the outward pointing normal at $s$, and so $-\mathbf{e}_1, -\mathbf{e}_2$ is *negatively* oriented at the south pole of $S^2$. This means, since $S^2$ is orientable, that if the basis $\mathbf{e}_1, \mathbf{e}_2$ at $n$ is transported along a curve $C$ on $S^2$ to $s$ (the pair remaining tangent to $S^2$ and independent) then the resulting basis $\mathbf{f}_1, \mathbf{f}_2$ has the opposite orientation as $-\mathbf{e}_1, -\mathbf{e}_2$ there. But the basis $-\mathbf{e}_1, -\mathbf{e}_2$ at $s$ represents, *on* $\mathbb{R}P^2$, exactly the same basis $\mathbf{e}_1, \mathbf{e}_2$ at $n$, and the arc $C$ on $S^2$ becomes a *closed* curve $C'$ on $\mathbb{R}P^2$ that starts and stops at $n$. This means that on transporting the basis $\mathbf{e}_1, \mathbf{e}_2$ at $n$ along $C'$ on $\mathbb{R}P^2$, one returns to an oppositely oriented basis. Thus $\mathbb{R}P^2$ *is not orientable*! Note that the crucial point in the preceding argument was that $\mathbb{R}P^2$ is obtained from the orientable $S^2$ by identifying points by means of the antipodal map, *and this map reverses orientation on $S^2$*.

In Problem 2.8(1) you are asked to show that $\mathbb{R}P^n$ *is not orientable if n is even*. We shall see later on that *odd-dimensional projective spaces are in fact orientable*.

## 2.8e. Pseudoforms and the Volume Form

The differential forms and vectors considered so far have not involved the notion of orientation of *space*. However, roughly half of the "forms," "vectors," and "scalars" that occur in physics are in fact "pseudo-objects" that make sense only when an orientation is prescribed. The magnetic field pseudovector **B** is perhaps the most famous example, and we shall discuss this later.

Consider ordinary 3-space $\mathbb{R}^3$ with its euclidean metric. We would like to define the "volume 3-form" $\text{vol}^3$ to be the form that assigns to any triple of vectors the volume of the parallelopiped spanned by the vectors; in particular $\text{vol}(X, Y, Z)$ should be 1 if $X$, $Y$, and $Z$ are orthonormal. But if vol is to be a form we must then have $\text{vol}(Y, X, Z) = -1$, and yet $Y$, $X$, and $Z$ are orthonormal. We have asked too much of vol. In some books they get around this by taking absolute value $|\text{vol}(Y, X, Z)|$, but this does great harm to the machinery of forms that we have labored to develop. What we could do is require that $\text{vol}(X, Y, Z) = 1$ if the triple is an orthonormal *right-handed* system. This makes the volume form orientation-dependent. There is a serious drawback to this definition; what if we are dealing with a space that is not orientable? The physical space in which we live is, according to general relativity, curved and perhaps not orientable. If you leave your shoes (labeled "right" and "left") at home and take a very long trip, it may very well be that on returning home your right foot will fit only into your shoe labeled "left." The term "right- handed" might not have an unambiguous meaning in the large, just as rotation in "the clockwise sense" has no meaning on the Möbius band.

We compromise by defining a new type of form (called "form of odd kind" by its inventor Georges de Rham) differing from our usual forms (of "even kind") in a way that will not seriously harm our machinery.

First note that the assignment of an *orientation* to a vector space $E$ is the same as the assignment of a *function* $o$ on *bases* of $E$ whose values are the two integers $\pm 1$; $o(\mathbf{e}) = +1$ iff the basis $\mathbf{e}$ has the given orientation. If $(x)$ is a coordinate system, we shall write $o(x)$ rather than $o(\boldsymbol{\partial}_x)$.

**Definition:** A **pseudo-$p$-form** $\alpha$ on a vector space $E$ assigns, *for each orientation* $o$ of $E$, an exterior $p$-form $\alpha_o$ such that if the orientation is reversed the exterior form is replaced by its negative

$$\alpha_{-o} = -\alpha_o$$

A pseudo-$p$-form on a manifold $M^n$ assigns a pseudo-$p$-form $\alpha$ to each tangent space $M^n_x$ in a smooth fashion; that is, if $(x)$ is a coordinate system in a patch then if we take the orientation $o$ in this patch defined by $o(\boldsymbol{\partial}_x) = +1$, we demand that the (ordinary) exterior form $\alpha_o$ be smooth.

For example, let us write down a **volume form** for $\mathbb{R}^3$ (we shall give a general definition later on). Let $x, y, z$, be a cartesian coordinate system in $\mathbb{R}^3$ (it may be right- or left-handed). Then the volume (pseudo) form is

$$\text{vol}^3 := o(\boldsymbol{\partial}_x, \boldsymbol{\partial}_y, \boldsymbol{\partial}_z) dx \wedge dy \wedge dz$$

Thus *if $o$ is the right-handed orientation* of $\mathbb{R}^3$, and if the coordinate system is right-handed then $\text{vol}_o = dx \wedge dy \wedge dz$, whereas if the coordinate system is left-handed $\text{vol}_o = -dx \wedge dy \wedge dz = dy \wedge dx \wedge dz$.

Similarly we can define pseudovectors, pseudoscalars, and so on, *pseudo* always referring to a *change of sign with a change of orientation*. For example, the magnetic field about a current carrying infinite straight wire circulates about the wire, *but the sense of circulation is undetermined*! If we employ the usual right-handed orientation of $\mathbb{R}^3$, then the field (by *definition*) circulates about the wire in the sense of a right-hand screw, whereas if we use the left-handed orientation the direction is in the sense of a left-hand screw. This indecisiveness *cannot be avoided*; it stems from the definition of the magnetic field, (see (3.36)), and the fact that a "sense" can be assigned to a $\times$ product of vectors $\mathbf{v} \times \mathbf{w}$ only after an orientation is chosen. Thus $\mathbf{B}$ *is not a true vector, but rather changes into its negative when the orientation of* $\mathbb{R}^3$ *is reversed*; $\mathbf{B}$ is a *pseudovector.*

**Warning:** We have defined vectors, forms, orientation and pseudoforms in a manner that is independent of coordinate systems. For example, in $\mathbb{R}^3$ we may assign the right-hand orientation and still employ a left-handed cartesian coordinate system. This is usually *not* done in physics books. In physics one usually does not talk about the orientation of $\mathbb{R}^3$ but rather the orientation of a *particular coordinate system* being employed. Where in this book we would say that a vector is unchanged under a change of orientation and a pseudovector $\mathbf{B}$ changes into $-\mathbf{B}$ if the orientation of $\mathbb{R}^3$ is reversed, a physicist would usually say, for example, that if $A^i$ and $B^i$ are the *components* of a vector $\mathbf{A}$ and a pseudovector $\mathbf{B}$ in a cartesian coordinate system $x, y, z$, then the components of $\mathbf{A}$ and $\mathbf{B}$ in the reversed system $-x, -y, -z$, are $-A^i$ and $B^i$. This *is* saying the same thing as in our definition.

## 2.8f. The Volume Form in a Riemannian Manifold

Let $p$ be a point in the Riemannian manifold $M^n$. The **volume (pseudo)-$n$-form** $\text{vol}^n$ is by definition the unique $n$-form that assigns to an orientation $o$ of the tangent space $M^n_p$ and a *positively* oriented orthonormal basis $\mathbf{e}$ the value $+1$. (Recall that an $n$-form is determined by its value on a single basis.) Let us find the coordinate expression for $\text{vol}^n$.

Clearly, if $(x)$ is a coordinate system that is orthonormal at $p$, that is, $(\partial_i)$ are orthonormal, then

$$\text{vol} = o(x) dx^1 \wedge \ldots \wedge dx^n$$

is the volume form at $p$, since this form, when applied to $(\partial_x)$, yields $o(x)$.

Let $(y)$ be *any* coordinate system holding $p$. Choose any coordinate system $(x)$ that is orthonormal at $p$. (This can be done as follows. Let $\mathbf{e}$ be an orthonormal basis at $p$ and let $(z)$ be any coordinate system near $p$. Then $\mathbf{e} = \partial_z P$ for a unique nonsingular $P$. Now define coordinates $x$ by $z^j = P^j{}_i x^i$. We then have

$$\frac{\partial}{\partial x^i} = \left(\frac{\partial z^j}{\partial x^i}\right) \frac{\partial}{\partial z^j} = \left(\frac{\partial}{\partial z^j}\right) P^j{}_i = \mathbf{e}_i$$

at $p$, as desired.) Then, at $p$

$$\begin{aligned} \text{vol}^n &= o(x) dx^1 \wedge \ldots \wedge dx^n = o(x) \frac{\partial(x)}{\partial(y)} dy^1 \wedge \ldots \wedge dy^n \\ &= o(y) \left| \frac{\partial(x)}{\partial(y)} \right| dy^1 \wedge \ldots \wedge dy^n \end{aligned}$$

Now at $p$ we have (in the notation of Section 2.7b) $ds^2 = \delta_{rs}dx^r dx^s = g_{ij}(y)dy^i dy^j$, where

$$g_{ij}(y) = \left(\frac{\partial x^r}{\partial y^i}\right)\delta_{rs}\left(\frac{\partial x^s}{\partial y^j}\right) = \sum_r \left(\frac{\partial x^r}{\partial y^i}\right)\left(\frac{\partial x^r}{\partial y^j}\right)$$

Thus if we define, for each Riemannian metric tensor $g_{ij}(y)$,

$$g(y) := \det[g_{ij}(y)] \tag{2.71}$$

we have

$$g(y) = \det\left[\sum_r \left(\frac{\partial x^r}{\partial y^i}\right)\left(\frac{\partial x^r}{\partial y^j}\right)\right]$$

$$= \det\left[\left(\frac{\partial x}{\partial y}\right)^T\left(\frac{\partial x}{\partial y}\right)\right] = \left[\det\left(\frac{\partial x}{\partial y}\right)\right]^2$$

and consequently $|\partial(x)/\partial(y)| = \sqrt{g}(y)$ and

$$\text{vol}^n = o(y)\sqrt{g}(y)dy^1 \wedge \ldots \wedge dy^n \tag{2.72}$$

*is the coordinate expression for the volume form.* Since the coordinates $(x)$ do not appear anywhere in this expression, (2.72) gives the volume form at *each* point of the $(y)$ coordinate patch. If we write, as we do for any form, $\text{vol}^n = \text{vol}^n_{12\ldots n}\, dy^1 \wedge \ldots \wedge dy^n$, we see that

$$\text{vol}^n_{i_1 i_2 \ldots i_n} = o(y)\sqrt{g}(y)\epsilon_{i_1 i_2 \ldots i_n} \tag{2.73}$$

*It is traditional to omit the orientation function* $o(y)$, and we shall do so when no confusion can arise.

Note that since $\text{vol}^n$ is a pseudo-$n$-form, we conclude that

$$\sqrt{g}(y)\epsilon_{i_1 i_2 \ldots i_n}$$

are the components of an $n^{\text{th}}$ rank covariant pseudotensor, but, as we noticed in Section 2.5 b, *the permutation symbol itself is not a tensor!*

## Problems

**2.8(1)** Show that even dimensional projective spaces are not orientable.

**2.8(2)** Show that a 1-sided hypersurface $M^n$ of an orientable manifold $W^{n+1}$ is not orientable. (Hint: Transport of a normal about *some* closed curve on $M$ must reverse this normal (why?). Now transport a basis of $W$ about this same curve.)

**2.8(3)** Use Problem **2.1(2)** to compute the volume 3-form of $\mathbb{R}^3$ in spherical coordinates.

## 2.9. Interior Products and Vector Analysis

What is the *precise* relationship between exterior forms and vector analysis in $\mathbb{R}^3$?

### 2.9a. Interior Products and Contractions

We know that if $\alpha$ is a covariant vector and $\mathbf{v}$ is a contravariant vector then $\alpha(\mathbf{v}) = a_i v^i$ is a scalar. Also, if $\mathbf{A}$ is a linear transformation, that is, a mixed tensor that is once covariant and once contravariant, then the trace $\mathrm{tr}(\mathbf{A}) = A^i{}_i$ is also a scalar. In fact we have a general remark, whose proof is requested in Problem 2.9(1).

**Theorem (2.74):** *If $T^{\cdots i\cdots}_{\cdots j\cdots}$ are the components of a mixed tensor, p times contravariant and q times covariant, then the* **contraction** *on a pair of indices i, j, defined by the components $\sum_i T^{\cdots i\cdots}_{\cdots i\cdots}$ defines a tensor $(p-1)$ times contravariant and $(q-1)$ times covariant.*

If $\mathbf{v}$ is a vector and $\alpha$ is a $p$-form, then their tensor product has components $v^j a_{i_1\dots i_p}$ and consequently the contraction $v^j a_{j i_2\dots i_p}$ defines a covariant tensor, and it is clearly a $(p-1)$-form. There is, however, a special machinery for contracting vectors and forms, and we turn now to this "interior product."

**Definition:** If $\mathbf{v}$ is a vector and $\alpha$ is a $p$-form, their **interior product** $(p-1)$-form $i_\mathbf{v}\alpha$ is defined by

$$
\begin{aligned}
&i_\mathbf{v}\alpha^0 = \mathbf{0} && \text{if } \alpha \text{ is a 0-form}\\
&i_\mathbf{v}\alpha^1 = \alpha(\mathbf{v}) && \text{if } \alpha \text{ is a 1-form}\\
&i_\mathbf{v}\alpha^\mathbf{p}(\mathbf{w_2}, \dots, \mathbf{w_p}) = \alpha^\mathbf{p}(\mathbf{v}, \mathbf{w_2}, \dots, \mathbf{w_p}) && \text{if } \alpha \text{ is a } p\text{-form}
\end{aligned}
$$

Clearly $i_{\mathbf{A}+\mathbf{B}} = i_\mathbf{A} + i_\mathbf{B}$ and $i_{a\mathbf{A}} = a i_\mathbf{A}$. Sometimes we shall write $i(\mathbf{v})$.

**Theorem (2.75):** $i_\mathbf{v} : \bigwedge^\mathbf{p} \to \bigwedge^{\mathbf{p}-1}$ *is an* ***antiderivation****, that is,*

$$i_\mathbf{v}(\alpha^p \wedge \beta^q) = [i_\mathbf{v}\alpha^p] \wedge \beta^q + (-1)^p \alpha^p \wedge [i_\mathbf{v}\beta^q]$$

(Note that exterior differentiation is also an antiderivation.)

PROOF: Let us write $\mathbf{v} = \mathbf{w}_1$. Then

$$
\begin{aligned}
i_\mathbf{v}(\alpha \wedge \beta)(\mathbf{w}_2, \dots, \mathbf{w}_{p+q}) &= \alpha \wedge \beta(\mathbf{w}_1, \mathbf{w}_2, \dots, \mathbf{w}_{p+q})\\
&= \sum_{\underset{\to}{I}\,\underset{\to}{J}} \delta^{IJ}_{1\dots(p+q)}\alpha(\mathbf{w}_I)\beta(\mathbf{w}_J) = \sum_{\underset{\to}{I}\,\underset{\to}{J}\,1\in I} + \sum_{\underset{\to}{I}\,\underset{\to}{J}\,1\in J}\\
&= \sum_{i_2<\dots<i_p}\sum_{\underset{\to}{J}} \delta^{1 i_2\dots i_p J}_{1\dots(p+q)}\alpha(\mathbf{w}_1, \mathbf{w}_{i_2}, \dots \mathbf{w}_{i_p})\beta(\mathbf{w}_J)\\
&\quad + \sum_{\underset{\to}{I}}\sum_{j_2<\dots<j_q} \delta^{I 1 j_2\dots j_q}_{1\dots(p+q)}\alpha(\mathbf{w}_I)\beta(\mathbf{w}_1, \mathbf{w}_{j_2}, \dots \mathbf{w}_{j_q})
\end{aligned}
$$

$$= \sum_{1\neq i_2<\ldots<i_p} \sum_{J-\{1\}} \delta^{i_2\ldots i_p J}_{2\ldots(p+q)} [i_{\mathbf{v}}\alpha](\mathbf{w}_{i_2},\ldots,\mathbf{w}_{i_p})\beta(\mathbf{w}_J)$$

$$+ \sum_{I-\{1\}} \sum_{1\neq j_2<\ldots<j_q} (-1)^p \delta^{I j_2\ldots j_q}_{2\ldots(p+q)} \alpha(\mathbf{w}_I)[i_{\mathbf{v}}\beta](\mathbf{w}_{j_2},\ldots,\mathbf{w}_{j_q})$$

$$= [(i_{\mathbf{v}}\alpha)\wedge\beta + (-1)^p \alpha\wedge(i_{\mathbf{v}}\beta)](\mathbf{w}_2,\ldots,\mathbf{w}_{p+q}) \quad \square$$

**Theorem (2.76):** *In components we have*

$$i_{\mathbf{v}}\alpha = \sum_{i_2<\ldots<i_p} \sum_j v^j a_{ji_2<\ldots<i_p} dx^{i_2} \wedge \ldots \wedge dx^{i_p}$$

*that is,*

$$(i_{\mathbf{v}}\alpha)_{i_2<\ldots<i_p} = \sum_j v^j a_{ji_2<\ldots<i_p}$$

*or*

$$[i_{\mathbf{v}}\alpha]_K = v^j \alpha_{jK}$$

Thus *the interior product of* $\mathbf{v}$ *and* $\alpha$ *is simply the contraction with the first index of* $\alpha$! For proof of (2.76) see Problem 2.10(2).

We also have the very easy $i_{\mathbf{v}} c\alpha = c i_{\mathbf{v}}\alpha = i_{c\mathbf{v}}\alpha$ for a real number $c$.

Before proceeding, we should mention that exterior algebra and calculus and interior products, and so on, all can be applied to pseudoforms as well. It should be clear, for example, if $\alpha$ is a pseudoform, then so is $d\alpha$. Also, if $\beta$ is also a pseudoform then $\alpha\wedge\beta$ is a (true) form, and if $\mathbf{v}$ is a vector then $i_{\mathbf{v}}\beta$ is a pseudoform, and so on.

## 2.9b. Interior Product in $\mathbb{R}^3$

In 2.5e we mentioned that in $\mathbb{R}^3$ with *cartesian coordinates* one can associate to a vector $\mathbf{v}$ a 1-form $\sum_i v^i dx^i$ and also a 2-form $v^1 dx^2\wedge dx^3 + v^2 dx^3\wedge dx^1 + v^3 dx^1\wedge dx^2$. *These correspondences do not make sense in general coordinates*; for instance, two different coordinate systems will yield different 1-forms associated to a given vector $\mathbf{v}$ (not just different coordinate expressions). We wish to give a correct correspondence that works in any coordinates. We have already done this for 1-forms in a Riemannian manifold; associated to the vector $\mathbf{v} = v^i \boldsymbol{\partial}_i$ is the covector $\nu = v_i dx^i$, where $v_i = g_{ij}v^j$. (We will write $\nu = \langle\ ,\mathbf{v}\rangle$ since $\nu(\mathbf{w}) = \langle\mathbf{w},\mathbf{v}\rangle$.) We shall indicate this correspondence simply by

$$\mathbf{v} \Leftrightarrow \nu^1 = v_1 dx^1 + v_2 dx^2 + v_3 dx^3$$

What is the 2-form corresponding to $\mathbf{v}$? We claim $\mathbf{v} \Leftrightarrow$ the pseudo-2-form $\nu^2 := i_{\mathbf{v}}\,\mathrm{vol}^3$. Let us look at the coordinate expression for this interior product. In curvilinear coordinates $u$ (with $\boldsymbol{\partial}_i = \partial/\partial u^i$, and omitting the orientation function $o$) we have the volume form (2.72) and

$$i_{\mathbf{v}}\sqrt{g}(u) du^1\wedge du^2\wedge du^3 = \sqrt{g}\sum_j v^j i_{\boldsymbol{\partial}_j}(du^1\wedge du^2\wedge du^3)$$

Repeated use of (2.75) then gives

$$\begin{aligned} &i(\partial_j)(du^1 \wedge du^2 \wedge du^3) \\ &\quad = i(\partial_j)(du^1)du^2 \wedge du^3 - du^1 \wedge i(\partial_j)(du^2) \wedge du^3 + du^1 \wedge du^2 i(\partial_j)(du^3) \\ &\quad = du^1(\partial_j)du^2 \wedge du^3 - du^2(\partial_j)du^1 \wedge du^3 + du^3(\partial_j)du^1 \wedge du^2 \\ &\quad = \delta^1{}_j du^2 \wedge du^3 - \delta^2{}_j du^1 \wedge du^3 + \delta^3{}_j du^1 \wedge du^2 \end{aligned}$$

Thus to the vector $\mathbf{v}$ we associate the *pseudo*-2-form

$$\mathbf{v} \Leftrightarrow \nu^2 := i_{\mathbf{v}} \operatorname{vol}^3$$

where (2.77)

$$i_{\mathbf{v}} \operatorname{vol}^3 = \sqrt{g}(v^1 du^2 \wedge du^3 + v^2 du^3 \wedge du^1 + v^3 du^1 \wedge du^2)$$

is the correct replacement for $v^1 dx^2 \wedge dx^3 + v^2 dx^3 \wedge dx^1 + v^3 dx^1 \wedge dx^2$. Note, conversely, that if

$$\beta^2 = b_{23} du^2 \wedge du^3 + b_{31} du^3 \wedge du^1 + b_{12} du^1 \wedge du^2$$

is a *pseudo*-2-form, then we may associate to it a vector $\mathbf{B}$ with components

$$B^1 = \frac{b_{23}}{\sqrt{g}}, \qquad B^2 = \frac{b_{31}}{\sqrt{g}}, \qquad B^3 = \frac{b_{12}}{\sqrt{g}} \tag{2.78}$$

Two things should be noted about (2.77). First, of course $i_{\mathbf{v}} \operatorname{vol}^3$ does not use the full Riemannian structure of $\mathbb{R}^3$; rather only the volume form is used. Second, the same procedure will work in any manifold $M^n$ having some distinguished volume form (not necessarily coming from a Riemannian metric)

$$\operatorname{vol}^n = \rho(u) du^1 \wedge \ldots \wedge du^n \tag{2.79}$$

where $\rho \neq 0$. To the vector $\mathbf{v}$ we may associate the *pseudo*-$(n-1)$-form

$$\mathbf{v} \Leftrightarrow \nu^{n-1} := i_{\mathbf{v}} \operatorname{vol}^n \tag{2.80}$$

One can easily work out the coordinate expression for this form, as in (2.77).

Back now to $\mathbb{R}^3$. Given a pair of vectors $\mathbf{v}$, $\mathbf{w}$, with associated covectors $\nu^1 = \langle\ ,\mathbf{v}\rangle$ and $\omega^1 = \langle\ ,\mathbf{w}\rangle$, we know that

$$\langle \mathbf{v}, \mathbf{w} \rangle = i_{\mathbf{v}} \omega^1 \tag{2.81}$$

We can also associate to our vectors their pseudo-2-forms $\nu^2$ and $\omega^2$. In cartesian coordinates we know that $\nu^1 \wedge \omega^2$ is a 3-form whose coefficient is again $\langle \mathbf{v}, \mathbf{w} \rangle$. We claim that in general we have

$$\nu^1 \wedge \omega^2 = \langle \mathbf{v}, \mathbf{w} \rangle \operatorname{vol}^3 \tag{2.82}$$

We give two proofs. For the first we simply notice that both sides are pseudo-3-forms. Since they are equal in cartesian coordinates they are always equal.

Our second proof illustrates the machinery of interior products.

$$\begin{aligned}\nu^1 \wedge \omega^2 &= \nu^1 \wedge i_{\mathbf{w}}\,\mathrm{vol}^3 = i_{\mathbf{w}}(\mathrm{vol}^3) \wedge \nu^1 \\ &= i_{\mathbf{w}}(\mathrm{vol}^3 \wedge \nu^1) + \mathrm{vol}^3 \wedge i_{\mathbf{w}}\nu^1 \\ &= i_{\mathbf{w}}(\nu^1)\,\mathrm{vol}^3 \qquad \text{(Why?)}\end{aligned}$$

What about the $\times$ product of the vectors? We know that in cartesian coordinates, the 2-form $\nu^1 \wedge \omega^1$ has as coefficients the three components of $\mathbf{v} \times \mathbf{w}$. We should like then to say that $\nu^1 \wedge \omega^1$ is the 2-form associated to the vector $\mathbf{v} \times \mathbf{w}$, but we only have a *pseudo*-2-form associated to a vector. Thus we should say that the *pseudo*vector $\mathbf{v} \times \mathbf{w}$ is associated to the 2-form $\nu^1 \wedge \omega^1$

$$i_{\mathbf{v}\times\mathbf{w}}\,\mathrm{vol}^3 = \nu^1 \wedge \omega^1 \tag{2.83}$$

This makes sense when we recall that the direction of $\mathbf{v} \times \mathbf{w}$ is given usually by the right-hand rule; that is, it uses the orientation of $\mathbb{R}^3$. Although not usually mentioned in elementary books, the vector product is **defined** in $\mathbb{R}^3$ as follows: $\mathbf{v} \times \mathbf{w}$ is the unique *pseudovector* such that

$$\langle(\mathbf{v} \times \mathbf{w}), \mathbf{c}\rangle = \mathrm{vol}^3(\mathbf{v}, \mathbf{w}, \mathbf{c}) \tag{2.84}$$

for each vector $\mathbf{c}$.

We may ask now for the 1-form version of $\mathbf{v} \times \mathbf{w}$, that is, the *pseudo*-1-form associated to the vector product. We claim

$$-i_{\mathbf{v}}\omega^2 \qquad \text{is the covariant version of } \mathbf{v} \times \mathbf{w} \tag{2.85}$$

This follows from (2.84)

$$\begin{aligned}\langle \mathbf{v} \times \mathbf{w}, \mathbf{c}\rangle &= \mathrm{vol}^3(\mathbf{v}, \mathbf{w}, \mathbf{c}) = -\,\mathrm{vol}^3(\mathbf{w}, \mathbf{v}, \mathbf{c}) \\ &= -[i_{\mathbf{w}}(\mathrm{vol}^3)](\mathbf{v}, \mathbf{c}) = -\omega^2(\mathbf{v}, \mathbf{c}) \\ &= [-i_{\mathbf{v}}\omega^2](\mathbf{c})\end{aligned}$$

## 2.9c. Vector Analysis in $\mathbb{R}^3$

Vector algebra in $\mathbb{R}^3$ is easily handled by use of interior and exterior products; the only question is, should one associate to a vector $\mathbf{B}$ its 1-form $\beta^1 = \langle\ ,\mathbf{B}\rangle$ or its 2-form $\beta^2 = i_{\mathbf{B}}\,\mathrm{vol}^3$? For example, consider an expansion of the vector triple product $\mathbf{A} \times (\mathbf{B} \times \mathbf{C})$. The following works. Let $\mathbf{B} \Leftrightarrow \beta^1$, $\mathbf{C} \Leftrightarrow \gamma^1$. Then

$$\begin{aligned}\mathbf{A} \times (\mathbf{B} \times \mathbf{C}) &\Leftrightarrow -i_{\mathbf{A}}(\beta^1 \wedge \gamma^1) = [-i_{\mathbf{A}}(\beta^1)]\gamma^1 + \beta^1[i_{\mathbf{A}}\gamma^1] \\ &\Leftrightarrow -\langle \mathbf{A}, \mathbf{B}\rangle \mathbf{C} + \langle \mathbf{A}, \mathbf{C}\rangle \mathbf{B}\end{aligned}$$

the familiar vector identity.

So much for vector algebra! Now for calculus. We already know that

$$df = \langle\ , \nabla f\rangle$$

We **define** curl **A** by using $\mathbf{A} \Leftrightarrow \alpha^1$ and then curl $\mathbf{A} \Leftrightarrow d\alpha^1$

$$d\alpha^1 = i_{\text{curl}\,\mathbf{A}}\,\text{vol}^3 \tag{2.86}$$

and **define** div **B** by using $\mathbf{B} \Leftrightarrow \beta^2$ and

$$d\beta^2 = (\text{div}\,\mathbf{B})\,\text{vol}^3 \tag{2.87}$$

for these are surely identities when expressed in cartesian coordinates. Note that in (2.87), since **B** is a vector, $\beta^2$ is a pseudoform. Since $\text{vol}^3$ is a pseudoform we conclude that div **B** is a (true) scalar. On the other hand, if **A** is a vector then curl **A** must be a *pseudo*vector!

**Warning:** Given a vector field **A**, one *can* write out the components of the vector curl **A** in a curvilinear coordinate system; one takes **A**, one converts it to a 1-form $\alpha^1$ using the metric tensor $g_{ij}$ (this is generally complicated), then takes $d\alpha^1$, and then uses (2.78). To my knowledge, however, there is no reason for *ever* writing out the components of the *vector* curl **A** in curvilinear coordinates; *if the expression curl A appears, it is a sure sign that the vector in question was not the contravariant* **A** *but rather the covariant vector* $\alpha^1 \Leftrightarrow \mathbf{A}$! But then $d\alpha^1$ is as simple to write down in curvilinear coordinates as in cartesian. A similar remark applies to the components of grad $f$ in curvilinear coordinates; $df$ is all that is needed.

It is a different story with div **B**. div **B** is the *scalar* coefficient of $\text{vol}^3$ in (2.87), and its expression in coordinates $u$ *is* needed. Since $\mathbf{B} \Leftrightarrow i_{\mathbf{B}}\,\text{vol}^3$ (and omitting the orientation function $o$)

$$\begin{aligned}
d[i_{\mathbf{B}}\,\text{vol}^3] &= d[\sqrt{g}b^1du^2 \wedge du^3 + \sqrt{g}b^2du^3 \wedge du^1 + \sqrt{g}b^3du^1 \wedge du^2] \\
&= [\frac{\partial}{\partial u^1}(\sqrt{g}b^1) + \frac{\partial}{\partial u^2}(\sqrt{g}b^2) + \frac{\partial}{\partial u^3}(\sqrt{g}b^3)]du^1 \wedge du^2 \wedge du^3 \\
&= \frac{1}{\sqrt{g}}\frac{\partial}{\partial u^i}[\sqrt{g}b^i]\sqrt{g}du^1 \wedge du^2 \wedge du^3
\end{aligned}$$

Thus

$$\text{div}\,\mathbf{B} = \frac{1}{\sqrt{g}}\frac{\partial}{\partial u^i}[\sqrt{g}b^i] \tag{2.88}$$

Note again that only the volume form appears, not the full metric tensor.

We define the **Laplacian** of a function $f$ by

$$\begin{aligned}
\nabla^2 f = \Delta f &:= \text{div}(\text{grad}\, f) \\
&= \frac{1}{\sqrt{g}}\frac{\partial}{\partial u^i}\left[\sqrt{g}g^{ij}\left(\frac{\partial f}{\partial u^j}\right)\right]
\end{aligned} \tag{2.89}$$

To continue with vector identities it is useful to associate a pseudo-3-form to each scalar $f$, namely

$$f \Leftrightarrow f\,\text{vol}^3$$

Then, for example, from (2.82)

$$\begin{aligned}\operatorname{div}(\mathbf{A}\times\mathbf{B}) \Leftrightarrow \operatorname{div}(\mathbf{A}\times\mathbf{B})\,\mathrm{vol}^3 &= d(\alpha^1\wedge\beta^1) = d\alpha^1\wedge\beta^1 - \alpha^1\wedge d\beta^1\\ &= \langle \operatorname{curl}\mathbf{A},\mathbf{B}\rangle\,\mathrm{vol}^3 - \langle \mathbf{A},\operatorname{curl}\mathbf{B}\rangle\,\mathrm{vol}^3\\ &\Leftrightarrow \langle \operatorname{curl}\mathbf{A},\mathbf{B}\rangle - \langle \mathbf{A},\operatorname{curl}\mathbf{B}\rangle\end{aligned}$$

## 2.10. Dictionary

Let

$$\begin{aligned}\mathrm{vol}^3 &= dx\wedge dy\wedge dz = \text{volume form}\\ \text{0-form } f &= \text{function } f\\ \text{1-form } \alpha^1 &= \text{covariant expression for a vector } \mathbf{A}\\ \text{1-form } \gamma^1 &= \text{covariant expression for a vector } \mathbf{C}\\ \text{2-form } \beta^2 & \text{ be associated to a vector } \mathbf{B} \text{ through}\\ \beta^2 &= i_{\mathbf{B}}\,\mathrm{vol}\end{aligned}$$

Then we may make the following rough, *symbolic* identifications

$$\begin{aligned}\alpha^1\wedge\gamma^1 &= i_{\mathbf{A}\times\mathbf{C}}\,\mathrm{vol}^3 \Leftrightarrow \mathbf{A}\times\mathbf{C}\\ \alpha^1\wedge\beta^2 &= \mathbf{A}\bullet\mathbf{B}\,\mathrm{vol}^3 \Leftrightarrow \mathbf{A}\bullet\mathbf{B}\\ i_{\mathbf{C}}\alpha^1 &= \mathbf{C}\bullet\mathbf{A}\\ i_{\mathbf{C}}\beta^2 &\Leftrightarrow -\mathbf{C}\times\mathbf{B}\\ df &\Leftrightarrow \operatorname{grad} f\\ d\alpha^1 &= i_{\operatorname{curl}\mathbf{A}}\,\mathrm{vol}^3 \Leftrightarrow \operatorname{curl}\mathbf{A}\\ d\beta^2 &= \operatorname{div}\mathbf{B}\,\mathrm{vol}^3 \Leftrightarrow \operatorname{div}\mathbf{B}\\ d i_{\operatorname{grad} f}\,\mathrm{vol}^3 &= (\nabla^2 f)\,\mathrm{vol}^3 \Leftrightarrow \nabla^2 f\end{aligned}$$

## Problems

**2.10(1)** Prove (2.74).

**2.10(2)** Prove (2.76).

**2.10(3)** Compute $\nabla^2 f$ in spherical coordinates.

**2.10(4)** Derive the following identities *using forms*

**(i)** $\operatorname{grad}(fg) = f\operatorname{grad} g + g\operatorname{grad} f$

**(ii)** $\operatorname{div}(f\mathbf{B}) = f\operatorname{div}\mathbf{B} + \langle \operatorname{grad} f, \mathbf{B}\rangle$

**(iii)** $\operatorname{curl}(f\mathbf{A}) = f\operatorname{curl}\mathbf{A} + \operatorname{grad} f\times\mathbf{A}$

**(iv)** $\langle \mathbf{A}\times\mathbf{B}, \mathbf{C}\times\mathbf{D}\rangle = \ldots ?$

**2.10(5)** Use (2.73) and invoke (2.76) twice to show

$$\mathbf{v}\times\mathbf{B} \Leftrightarrow \sqrt{g}\sum_k v^i B^j \epsilon_{ijk} dx^k$$

CHAPTER 3

# Integration of Differential Forms

EXTERIOR differential forms occur implicitly in all aspects of physics and engineering because *they are the natural objects appearing as integrands of line, surface, and volume integrals as well as the n-dimensional generalizations* required in, for example, Hamiltonian mechanics, relativity, and string theories. We shall see in this chapter that *one does not integrate vectors; one integrates forms.* If there is extra structure available, for example, a Riemannian metric, then it is possible to rephrase an integration, say of exterior 1-forms or 2-forms, in terms of a vector integrations involving "arc lengths" or "surface areas," but we shall see that even in this case we are *complicating* a basically simple situation. *If a line integral of a vector occurs in a problem, then usually a deeper look at the situation will show that the vector in question was in fact a covector, that is, a 1-form!* For example (and this will be discussed in more detail later), the strength of the electric field can be determined by the work done in moving a unit charge very slowly along a small path, that is, by a line integral. The electric field strength is a 1-form.

Integration of a *pseudo*form proceeds in a way that differs slightly from that for a (true) form. We shall consider pseudoforms later on.

## 3.1. Integration over a Parameterized Subset

How does one integrate the Poincaré 2-form $\omega$ over a surface in phase space?

### 3.1a. Integration of a $p$-Form in $\mathbb{R}^p$

We are familiar with the notion of a multiple integral of a *function* $f$ over a region in $\mathbb{R}^p$

$$\int_U f(u)du^1 \dots du^p$$

(Of course we shall assume that the integral makes sense; for example, this will be the case if $U$ is a closed ball and $f$ is continuous on $U$.) *This integral does not involve any notion of orientation*, and *it is immaterial in which order the $du^i$'s appear.*

We now define the integral of a *p-form* $\alpha^p = a(u)du^1 \wedge \ldots \wedge du^p$ over an **oriented** region $(U, o) \subset \mathbb{R}^p$.

$$\int_{(U,o)} \alpha = \int_{(U,o)} a(u)du^1 \wedge \ldots \wedge du^p \tag{3.1}$$

$$:= o(u) \int_U a(u)du^1 \ldots du^p$$

where the last integral is the ordinary multiple integral of the function $a$ over the region $U$, disregarding the orientation, and where $o(u) = \pm 1$, the $+$ sign being chosen if and only if the coordinate basis

$$\left( \frac{\partial}{\partial u^1}, \ldots, \frac{\partial}{\partial u^p} \right)$$

has the same orientation as given by $o$. Clearly the integral of a $p$-form changes into its negative if the orientation of $U$ is reversed

$$\int_{(U,-o)} \alpha = -\int_{(U,o)} \alpha \tag{3.2}$$

We shall see shortly that the definition (3.1), in spite of its appearance, is in fact independent of the coordinates $u$ used in $\mathbb{R}^p$.

### 3.1b. Integration over Parameterized Subsets

We define an **oriented parameterized *p*-subset** of a manifold $M^n$ to be a pair $(U, o; F)$ consisting of an oriented region $(U, o)$ in $\mathbb{R}^p$ and a differentiable map

$$F : U \to M^n$$

We shall also call the point set $F(U) \subset M^n$ a $p$-subset.

When $p = 1$ we simply have a *curve* on $M^n$ with a *specific* parameterization, expressed locally by $x^i = x^i(t)$, and when $p = 2$ we have a *surface* on $M^n$ again with a specific parameterization $x^i = x^i(u, v)$.

It should be noted that *we make no requirements on the rank of the differential* of the map $F$; for example, it may be that the curve has a vanishing tangent vector, $dx/dt = 0$, at some or perhaps all parameter values $t$. Consequently, the $p$-subset $F(U)$ need not have dimension $p$ everywhere (that is why we do not use the term $p$-dimensional subset, rather than $p$-subset). In the most important cases, $F_*$ *will* have rank $p$ "almost everywhere." For example, the map $\mathbb{R}^2 \to \mathbb{R}^3$ defined by $F(\theta, \phi) = (\sin\theta\cos\phi, \sin\theta\sin\phi, \cos\theta)$ defines a parameterized 2-subset of $\mathbb{R}^3$ that covers the unit sphere an infinity of times, and with $F_*$ of rank 2 everywhere except at the poles, that is, the lines $\theta = n\pi$ of $\mathbb{R}^2$.

If $\alpha^p$ is a $p$-form on $M^n$, defined at least in some neighborhood of the image $F(U)$ of $U$, we *define* the integral of $\alpha^p$ over the oriented parameterized $p$-subset by

$$\int_{(U,o;F)} \alpha^p := \int_{(U,o)} F^*\alpha^p \tag{3.3}$$

Thus we pull the form $\alpha^p$ back to the oriented region $(U, o)$ and integrate there by means of (3.1). In all detail

$$\begin{aligned}\int_{(U,o;F)} \alpha^p &:= \int_{(U,o)} F^*\alpha^p \\ &= \int_{(U,o)} (F^*\alpha^p)\left[\frac{\partial}{\partial u^1}, \ldots, \frac{\partial}{\partial u^p}\right] du^1 \wedge \ldots \wedge du^p \\ &= o(u) \int_U (F^*\alpha^p)\left[\frac{\partial}{\partial u^1}, \ldots, \frac{\partial}{\partial u^p}\right] du^1 \ldots du^p\end{aligned} \tag{3.4}$$

Note that we can also write this as

$$\int_{(U,o;F)} \alpha^p = o(u) \int_U \alpha^p \left[F_* \frac{\partial}{\partial u^1}, \ldots, F_* \frac{\partial}{\partial u^p}\right] du^1 \ldots du^p \tag{3.5}$$

### 3.1c. Line Integrals

Consider a curve $C : \mathbf{x} = \mathbf{F}(t)$, for $a \le t \le b$, in $\mathbb{R}^3$ (with $x$ any coordinates), oriented so that $d/dt$ defines the positive orientation in $U = \mathbb{R}^1$. If $\alpha^1 = a_1(x)dx^1 + a_2(x)dx^2 + a_3(x)dx^3$ is a 1-form on $\mathbb{R}^3$ then its integral or **line integral** over $C$ becomes

$$\begin{aligned}\int_C \alpha^1 &= \int_C \sum_i a_i(x)dx^i \\ &= \int_a^b F^*\left[\sum_i a_i(x)dx^i\right] \\ &= \int_a^b \left[\sum_i a_i(x(t))\frac{dx^i}{dt}\right] dt\end{aligned} \tag{3.6}$$

*Thus* (3.3) *is the usual rule for evaluating a line integral over an oriented parameterized curve*! We may write this as

$$\int_C \alpha^1 = \int_a^b \alpha^1\left(\frac{d\mathbf{x}}{dt}\right) dt \tag{3.7}$$

and so the integral of a 1-form over an oriented parameterized curve $C$ is simply the ordinary integral of the function that assigns to the parameter $t$ the value of the 1-form on the velocity vector at $\mathbf{x}(t)$. This of course is simply (3.5), since $F_*(d/dt) = d\mathbf{x}/dt$.

*Note that there is no mention of arc length nor dot product.* If we *wish* to use a Riemannian metric in $\mathbb{R}^3$, for example, if the $x$'s are cartesian coordinates, then to the 1-form $\alpha^1$ is associated the contravariant vector $\mathbf{A}$ and (3.6) or (3.7) says

$$\int_C \alpha^1 = \int_a^b \mathbf{A} \bullet \left(\frac{d\mathbf{x}}{dt}\right) dt \tag{3.8}$$

If the coordinates are not cartesian, then although (3.7) remains the same,

$\int_a^b a_i\,(dx^i/dt)dt$, (3.8) becomes the more complicated

$$\int_a^b [g_{ij}A^j]\left(\frac{dx^i}{dt}\right)dt$$

Thus if one insists on integrating a vector over a curve, rather than a 1-form, one is going to need a Riemannian metric to convert the contravariant vector first into a covariant one, that is, a 1-form. Line integrals of 1-forms do not involve a metric, whereas integrals of vectors *must* involve one!

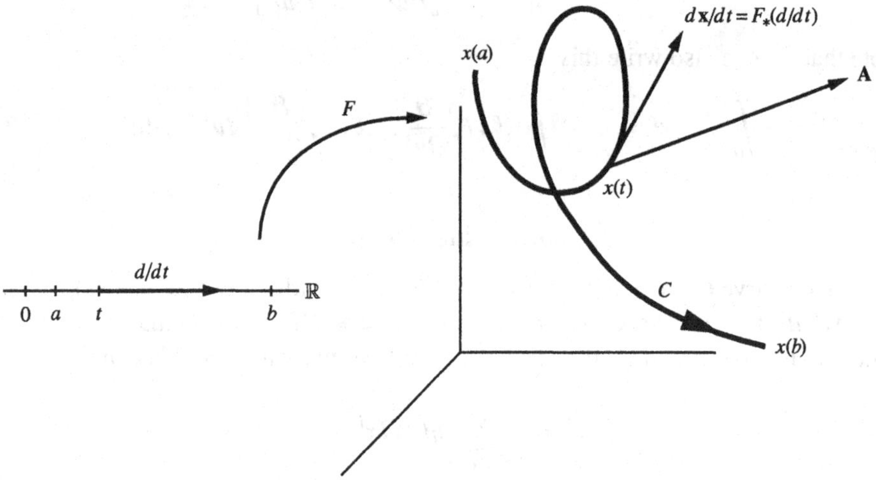


**Figure 3.1**

Use of a Riemannian metric allows us to write a line integral in the more *usual form*

$$\int_C \alpha^1 = \int_C \mathbf{A}\cdot d\mathbf{x} \tag{3.9}$$

$$= \int_a^b \mathbf{A}\cdot\left(\frac{d\mathbf{x}}{dt}\right)dt$$

$$= \int_a^b \|\mathbf{A}\|\,\left\|\frac{d\mathbf{x}}{dt}\right\|\,\cos\angle\left(\mathbf{A},\frac{d\mathbf{x}}{dt}\right)dt$$

$$= \int_0^L A_t ds$$

where $A_t$ is the tangential component of $\mathbf{A}$, $ds := \| d\mathbf{x}/dt \|\; dt$ is the element of arc length, and $L$ is the length of the curve. Although this *appears* simpler than (3.6), to compute using (3.9) one would have to introduce a parameterization, leading effectively back to (3.6)! There *are* times when one needs to compute the arc length of a curve, but, usually, *it is completely irrelevant to either the computation or the concept of a line integral! Line (and, as we shall see, surface) integrals are independent of any metric notions in space.* This is one case where the usual elementary treatment given in many calculus texts is harmful and misleading and should have been discarded long ago.

### 3.1d. Surface Integrals

Consider now an oriented parameterized surface in $\mathbb{R}^3$, with $x$ *any* coordinate system.

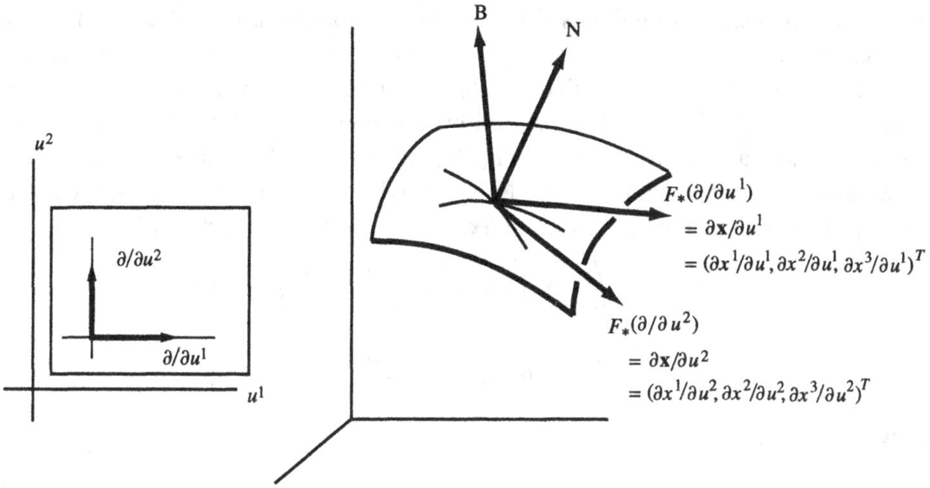


Figure 3.2

Suppose that $\partial/\partial u^1, \partial/\partial u^2$ has the given orientation $o$. Let $\beta^2$ be a 2-form on $\mathbb{R}^3$ and put $b_1 = b_{23},\ b_2 = b_{31},\ b_3 = b_{12}$. Then, as in (2.65)

$$\int_{F(U)} \beta^2 = \int_{F(U)} b_1 dx^2 \wedge dx^3 + b_2 dx^3 \wedge dx^1 + b_3 dx^1 \wedge dx^2$$
$$= \int_U \left[ \sum_{i<j} b_{ij}(x(u)) \frac{\partial(x^i, x^j)}{\partial(u^1, u^2)} \right] du^1 du^2 \tag{3.10}$$

or, as in (3.5),

$$\int_{F(U)} \beta^2 = \int_U \beta^2 \left( \frac{\partial \mathbf{x}}{\partial u^1}, \frac{\partial \mathbf{x}}{\partial u^2} \right) du^1 du^2 \tag{3.11}$$

Suppose that one insists on writing this in terms of the vector, or rather the *pseudo*vector **B**, associated to $\beta^2$

$$\int_{F(U)} \beta^2 = \int_U [i_{\mathbf{B}} \mathrm{vol}^3] \left( \frac{\partial \mathbf{x}}{\partial u^1}, \frac{\partial \mathbf{x}}{\partial u^2} \right) du^1 du^2$$
$$= \int_U \mathrm{vol}^3 \left( \mathbf{B}, \frac{\partial \mathbf{x}}{\partial u^1}, \frac{\partial \mathbf{x}}{\partial u^2} \right) du^1 du^2 \tag{3.12}$$

Recall that an orientation of $U \subset \mathbb{R}^2$ has already been given (it is inherent in the definition of the surface integral), but not one for $\mathbb{R}^3$. Since both $\mathrm{vol}^3$ and **B** change sign under a change of orientation of $\mathbb{R}^3$, it is clear that (3.12) is *independent of the choice of orientation of* $\mathbb{R}^3$.

We now proceed to the usual expression of (3.12). *Choose an orientation of* $\mathbb{R}^3$ and let $x$ be a positively oriented cartesian coordinate system for this chosen orientation. (In our Figure 3.2 we have perversely chosen a left-handed orientation.)

In the "classical" case discussed in elementary texts, the surface is **regular**; that is, the *map F has maximal rank* and thus the coordinate vectors $\partial\mathbf{x}/\partial u^1$, $\partial\mathbf{x}/\partial u^2$ are linearly independent. In this case we can transfer the orientation $o$ from the "parameter plane" $U \subset \mathbb{R}^3$ to the surface $F(U)$; since $\boldsymbol{\partial}/\boldsymbol{\partial} u^1$, $\boldsymbol{\partial}/\boldsymbol{\partial} u^2$ are positively oriented in $U$ we declare $\partial\mathbf{x}/\partial u^1$, $\partial\mathbf{x}/\partial u^2$ to define the positive orientation for $F(U)$. We then pick the unique unit normal $\mathbf{N}$ such that $\mathbf{N}$, $\partial\mathbf{x}/\partial u^1$, $\partial\mathbf{x}/\partial u^2$ is positively oriented in $\mathbb{R}^3$. We then have a unique decomposition $\mathbf{B} = (\mathbf{B} \bullet \mathbf{N})\mathbf{N} + \mathbf{T}$, where $\mathbf{T}$ is tangent to the surface (and consequently is a linear combination of $\partial\mathbf{x}/\partial u^1$ and $\partial\mathbf{x}/\partial u^2$). From (3.12)

$$\int_{F(U)} \beta^2 = \int_U \mathrm{vol}^3\left((\mathbf{B}\bullet\mathbf{N})\mathbf{N}, \frac{\partial\mathbf{x}}{\partial u^1}, \frac{\partial\mathbf{x}}{\partial u^2}\right) du^1 du^2$$

$$= \int_U (\mathbf{B}\bullet\mathbf{N})[i_{\mathbf{N}}\,\mathrm{vol}^3]\left(\frac{\partial\mathbf{x}}{\partial u^1}, \frac{\partial\mathbf{x}}{\partial u^2}\right) du^1 du^2$$

Now

$$i_{\mathbf{N}}\,\mathrm{vol}^3 \tag{3.13}$$

is simply the **area 2-form** for the surface, for its value on the (positively oriented) pair of tangent vectors $\partial\mathbf{x}/\partial u^1$, $\partial\mathbf{x}/\partial u^2$ is simply the area of the parallelogram spanned by them, $\| (\partial\mathbf{x}/\partial u^1) \times (\partial\mathbf{x}/\partial u^2) \|$. We shall write (with a classical abuse of notation since $dS$ is *not* the differential of a form)

$$dS := [i_{\mathbf{N}}\,\mathrm{vol}^3]\left(\frac{\partial\mathbf{x}}{\partial u^1}, \frac{\partial\mathbf{x}}{\partial u^2}\right) du^1 du^2 \tag{3.14}$$

$$= \| \mathbf{n} \| \, du^1 du^2$$

where $\mathbf{n} = (\partial\mathbf{x}/\partial u^1) \times (\partial\mathbf{x}/\partial u^2)$ is the (non-unit) normal to the surface. $B_n := \mathbf{B}\bullet\mathbf{N}$ is the normal component of $\mathbf{B}$. Thus we have the usual expression for the surface integral

$$\int_{F(U)} \beta^2 = \int_U B_n dS \tag{3.15}$$

This can all be said as follows. Given a *pseudo*vector $\mathbf{B}$ and an *oriented* parameterized surface in $\mathbb{R}^3$, choosing an orientation of $\mathbb{R}^3$ simultaneously picks out a specific vector field $\mathbf{B}$ and a definite unit normal $\mathbf{N}$. Then $\int_U B_n dS$ is the desired surface integral.

Surface integrals arise in higher dimensional manifolds. For example, in Hamiltonian mechanics, one sometimes needs to integrate the Poincaré 2-form $\omega$ over an arbitrary parameterized surface $q = q(u, v)$, $p = p(u, v)$ in phase space.

$$\iint \omega = \iint dp_j \wedge dq^j = \iint \sum_j \left[\frac{\partial(p_j, q^j)}{\partial(u, v)}\right] du \wedge dv$$

$$= \iint \{u, v\} du dv$$

becomes an integral of the Lagrange bracket of $u$ and $v$ (see (2.67)). Note that *there is no mention of a Riemannian metric, dot products, nor area elements*!

### 3.1e. Independence of Parameterization

We have defined our integral in terms of a *parameterized* subset of an $M^n$. What if we decide to consider the same subset (i.e., point set in $M^n$) but parameterized in a different fashion. We claim that if, in a sense to be prescribed later, the orientations are the same then the integrals will be the same; that is, the integral is independent of the parameterization. This is "clear" in the case of line or surface integrals in $\mathbb{R}^3$, for in $\mathbb{R}^3$ with the standard metric our integrals have been put in the geometric form $\int A_t ds$ or $\int B_n dS$. These involve length or area integrations, and so the original parameterizations have "disappeared." It is not easy to make this proof "honest" in the case of surface or higher dimensional integrals. We shall instead give a general proof relying directly on the famous Jacobi formula for change of variables in a multiple integral (whose proof is not trivial).

First, what do we mean by an orientation preserving reparameterization? Let $F : (U \subset \mathbb{R}^p) \to M^n$ be an oriented parameterized $p$-subset of a manifold $M^n$. We say that $G : (V \subset \mathbb{R}^p) \to M^n$ is a **reparameterization** of this subset if there is an *orientation preserving* diffeomorphism $H : U \to V$ such that $F = G \circ H$, that is, $F(u) = G[H(u)]$, or, in terms of local coordinates $x$ for $M^n$, $F(u) = x(v(u))$.

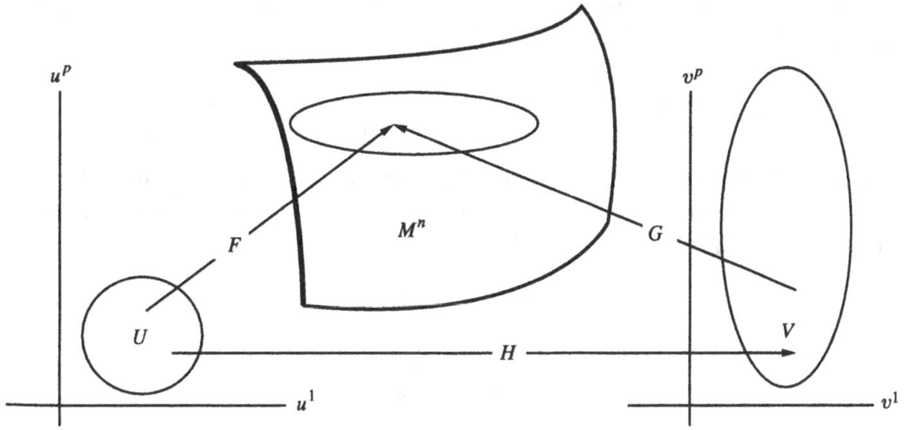


**Figure 3.3**

Since $H$ is orientation preserving, $H$ is of the form $v = H(u) = v(u)$ where

$$\frac{\partial(v)}{\partial(u)} = \frac{\partial(v^1, \ldots, v^p)}{\partial(u^1, \ldots, u^p)} > 0$$

provided $u$ and $v$ are positively oriented coordinates for $U$ and $V$, respectively.

Recall now *Jacobi's formula*. If $H : U \to V$ is a diffeomorphism of *unoriented* regions then

$$\int_{V=H(U)} f(v) dv^1 \ldots dv^p = \int_U f[H(u)] \left| \frac{\partial(v)}{\partial(u)} \right| du^1 \ldots du^p \tag{3.16}$$

(note the absolute value of the Jacobian determinant).

Now we can consider our integrals of forms. If $G$ is a reparameterization of $F$ (with positively oriented coordinates $u$ and $v$ in $U$ and $V$, respectively) and $x$ are local coordinates on $M^n$

$$\int_{(V,G)} \alpha^p = \int_V G^*\alpha^p = \int_V G^*[a_{\underline{I}}(x)dx^I]$$

$$= \int_V a_{\underline{I}}[x(v)]\left[\frac{\partial(x^I)}{\partial(v)}\right]dv^1\ldots dv^p$$

$$= \int_U a_{\underline{I}}[x(v(u))]\left[\frac{\partial(x^I)}{\partial(v)}\right]\left|\frac{\partial(v)}{\partial(u)}\right|du^1\ldots du^p$$

$$= \int_U a_{\underline{I}}[F(u)]\frac{\partial(x^I)}{\partial(u)}du^1\ldots du^p = \int_U F^*\alpha^p = \int_{(U,F)} \alpha^p$$

which shows that the integral *is* independent of the parameterization.

### 3.1f. Integrals and Pull-Backs

Let $\phi : M^n \to W^r$ be a smooth map of manifolds, and let $F : U \to M^n$ be an oriented parameterized $p$-subset of $M^n$. Then clearly $\psi = \phi \circ F : U \to W^r$ is an oriented parameterized $p$-subset of $W^r$. Then if $\alpha^p$ is a $p$-form on $W^r$, we have, from Problem 2.3(1)

$$\int_{(U,\psi)} \alpha^p = \int_U \psi^*\alpha^p = \int_U (\phi\circ F)^*\alpha^p = \int_U F^*\circ\phi^*\alpha^p = \int_{(U,F)} \phi^*\alpha^p$$

We shall write briefly $\sigma$ for the oriented subset $(U, F)$ of $M^n$ and then $(U, \psi) = (U, \phi\circ F)$ will be written simply as $\phi(\sigma)$, a subset of $W^r$. We then have the **general pull-back formula** (generalizing (3.3))

$$\phi : M^n \to W^r$$

$$\int_{\phi(\sigma)} \alpha^p = \int_\sigma \phi^*\alpha^p \tag{3.17}$$

In words, the integral of a form over the image $\phi(\sigma) \subset W^r$ of a subset $\sigma \subset M^n$ is the integral of the pull-back of the form over $\sigma$.

### 3.1g. Concluding Remarks

Again I must remark that (3.10) is ordinarily much simpler than (3.15). Of course there *are* very special situations when (3.15) is simpler. For example, let our surface be the unit sphere. Consider the vector $\mathbf{B} = \mathbf{x}$, the position vector. Then (3.15) gives immediately $\int \mathbf{x}\cdot\mathbf{N}dS = \int 1dS = 4\pi$. This is "simpler" because we already know the area of $S^2$.

Finally, note that we have only defined the integral of a form over an oriented parameterized subset of a manifold $M^n$, and these subsets are basically covered by a single coordinate system. We would ideally like to integrate $p$-forms over $p$-dimensional submanifolds of $M^n$. We shall discuss this in our next section.

## Problems

**3.1(1)** Let us say that a parameterized $p$-subset $(U, F)$ of $M^n$ is "irregular" at $u_0$ if rank $F < p$ at $u_0$. Show that if $\alpha^p$ is a form at such a $u_0$ then $F^*\alpha^p = 0$.

**3.1(2)** We know that $dS = \| \mathbf{n} \| \, du^1 du^2$. Show that in cartesian coordinates $x$ for $\mathbb{R}^3$

$$\mathbf{n} = \frac{\partial(x^2, x^3)}{\partial(u^1, u^2)} \frac{\partial}{\partial x^1} + \frac{\partial(x^3, x^1)}{\partial(u^1, u^2)} \frac{\partial}{\partial x^2} + \frac{\partial(x^1, x^2)}{\partial(u^1, u^2)} \frac{\partial}{\partial x^3}$$

and so $\| \mathbf{n} \|^2 = \sum_{i<j} [\partial(x^i, x^j)/\partial(u^1, u^2)]^2$

Show that when the surface is simply the graph of a function, that is,

$$x^1 = u^1, \qquad x^2 = u^2, \qquad x^3 = f(x^1, x^2)$$

we recover the classical expression for the area element. What do we get for the area element when the surface is given in the form $F(x, y, z) = 0$ and we assume that we can solve for $z$ in terms of $x, y$?

The following problem investigates the area element for a hypersurface and may be omitted.

**3.1(3)** The formula $dS = \| \mathbf{n} \| \, du^1 du^2$ followed from the fact that the area spanned by $\partial\mathbf{x}/\partial u^1$ and $\partial\mathbf{x}/\partial u^2$ is the length of the $\times$ product $(\partial\mathbf{x}/\partial u^1) \times (\partial\mathbf{x}/\partial u^2)$. Although we cannot define a vector $\mathbf{A}_1 \times \mathbf{A}_2$ for a pair of vectors in $\mathbb{R}^n$ we *can* define a generalized $\times$ product of $(n-1)$ vectors in $\mathbb{R}^n$ as follows (see (2.84)):

$\mathbf{A}_1 \times \ldots \times \mathbf{A}_{n-1}$ is the unique (pseudo) vector $\mathbf{B}$ such that

$\mathbf{C} \bullet \mathbf{B} = \mathrm{vol}^n(\mathbf{C}, \mathbf{A}_1, \ldots, \mathbf{A}_{n-1})$ for each vector $\mathbf{C}$

**(i)** Show that $\mathbf{B}$ is orthogonal to $\mathbf{A}_1, \ldots, \mathbf{A}_{n-1}$.

Suppose we consider a hypersurface of $\mathbb{R}^n$ parameterized by $u^1, \ldots, u^{n-1}$. Let $\mathbf{n} := (\partial\mathbf{x}/\partial u^1) \times \cdots \times (\partial\mathbf{x}/\partial u^{n-1})$ where the $x$'s are cartesian coordinates for $\mathbb{R}^n$, and let $\mathbf{N}$ be the unit vector in the direction of $\mathbf{n}$.

**(ii)** Show that we can then express the $(n-1)$-dimensional area element $dS^{n-1} := [i_{\mathbf{N}} \mathrm{vol}^n](\partial\mathbf{x}/\partial u^1, \ldots, \partial\mathbf{x}/\partial u^{n-1}) du^1 \ldots du^{n-1}$ as

$$dS^{n-1} = \| \mathbf{n} \| \, du^1 \ldots du^{n-1}$$

**(iii)** Let $i(\mathbf{v}) := i_{\mathbf{v}}$. Show that we can also say that *the covariant version* in $\mathbb{R}^n$ *of the vector* $\mathbf{n}$ is the 1-form

$$\langle \,, \mathbf{n} \rangle = i\left(\frac{\partial\mathbf{x}}{\partial u^{n-1}}\right) \circ \ldots \circ i\left(\frac{\partial\mathbf{x}}{\partial u^1}\right) \mathrm{vol}^n$$

(It is interesting that this 1-form *uses only the volume form*, not the metric of $\mathbb{R}^n$, and it vanishes on vectors tangent to the hypersurface.)

**(iv)** Now in *cartesian* coordinates, $\mathrm{vol}^n$ has components given by the permutation symbol (see 2.73). Use (2.73) repeatedly to show that

$$\langle \,, \mathbf{n} \rangle_j = \epsilon_{i_1 .. i_{(n-1)} j} \left(\frac{\partial x^{i_1}}{\partial u^1}\right) \cdots \left(\frac{\partial x^{i_{(n-1)}}}{\partial u^{n-1}}\right)$$

$$= \frac{\partial(x^1, x^2, \ldots \widehat{x^j}, \ldots x^n)}{\partial(u^1, \ldots, u^{n-1})} =: D_j$$

where $D_j$ is the determinant of the Jacobian matrix with the $j^{th}$ row omitted. We conclude

$$dS^{n-1} = [\sum_j D_j^2]^{1/2} du^1 \dots du^{n-1}$$

**(v)** Show that if the $x$ coordinates are not necessarily cartesian, with metric tensor $(g_{ij})$, then the correct formula for $\| \mathbf{n} \|$ is given by

$$\| \mathbf{n} \|^2 = g(x) g^{ij} D_i D_j$$

(this is also the correct expression in a Riemannian manifold).

## 3.2. Integration over Manifolds with Boundary

Does every manifold carry a Riemannian metric?

In 3.1 we defined how one integrates a (true) $p$-form over an oriented parameterized subset of a manifold. We would like to be able to integrate over objects that cannot be covered by a single parameterized subset, for example $p$-dimensional oriented submanifolds. A common way of doing this is indicated in the following figure.

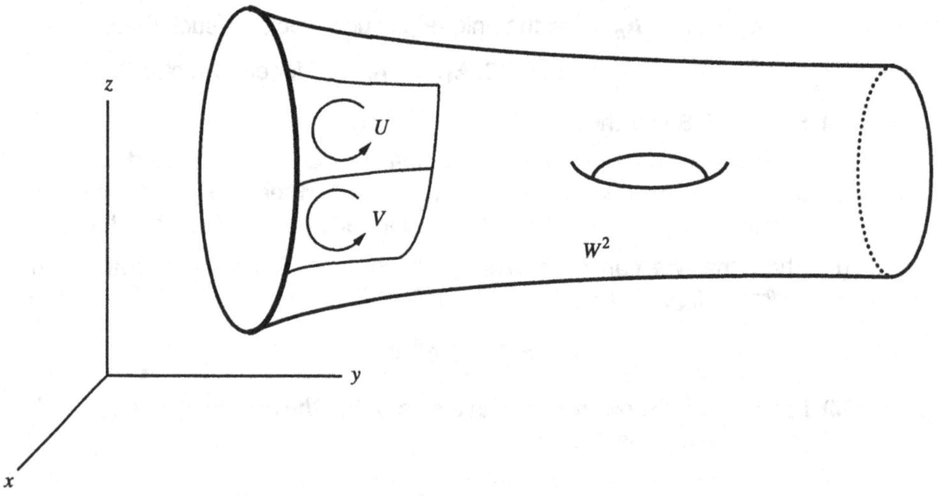


Figure 3.4

We have indicated a submanifold $W^2$ of $\mathbb{R}^3$ together with its boundary. It is oriented and we have indicated its orientation by giving the positive sense of rotation. We wish to integrate a 2-form $\beta^2$ of $\mathbb{R}^3$ over this object. We first *restrict* the form $\beta$ to the submanifold $W$: thus if $i : W \to \mathbb{R}^3$ is the inclusion map, we consider the pull-back $i^*\beta$ instead of $\beta$. This restricted form $i^*\beta$ has the same values on tangent vectors to $W$ as the original form $\beta$. We then break up $W^2$ into a finite union of coordinate patches that overlap only at edges or vertices. A theorem (whose proof is difficult) on "triangulations" shows that this can always be done. We have indicated two of the patches (as drawn, we can use $y$ and $z$ as local coordinates in each). We can assume that the coordinates $u$ in $U$, $v$ in $V$, and so forth, are such that the orientation of the patches agrees with the given orientation of $W^2$ (in our drawing, $y$, $z$, in that order yield the given orientation). We know how to integrate $i^*\beta^2$ over each of these patches, for if $\phi_U : U \to \mathbb{R}^2$ is the coordinate map for $U$, as in 1.2c, $\phi_U^{-1} : \phi_U(U) \to W^2$ is our parameterized map. We then compute these integrals and add the results. This is the integral of $\beta^2$ over $W^2$.

We emphasize that this is a perfectly acceptable way, and in fact the usual way to evaluate the integral. For theoretical purposes, however, we wish to define the integral in a different way. *Instead of breaking the object $W$ up into nonoverlapping coordinate regions, we shall rather write the form $i^*\beta$ as a sum $i^*\beta = \sum_U \beta_U$ of differential forms $\beta_U$, each of which vanishes outside its associated coordinate patch $U$* (this requires a "partition of unity"; see 3.2b). This is simpler than triangulating $W$ since we no longer demand that the patches fit together carefully. We know how to integrate $\beta_U$ over the oriented patch $U$. The integral of $\beta_U$ over $W$ should then be the same as the integral of $\beta_U$ over $U$, since $\beta_U$ is zero outside $U$. Then we shall define the integral of $\beta$ over $W$ to be the sum of the integrals of the $\beta_U$ over their patches $U$.

We now proceed with this program. Our first step is to generalize the notion of manifold so as to be able to include, as in Figure 3.4, the boundary of the object.

### 3.2a. Manifolds with Boundary

The *closed* 3-ball $\| \mathbf{x} \| \leq 1$ in $\mathbb{R}^3$ is not a 3-manifold, for although interior points, (i.e., points for which $\| \mathbf{x} \| < 1$) do have neighborhoods diffeomorphic to open balls in $\mathbb{R}^3$, $\| \mathbf{u} \| < 1$, points on the boundary 2-sphere have neighborhoods that resemble half open balls, $\| \mathbf{v} \| < 1$ and $v^3 \geq 0$.

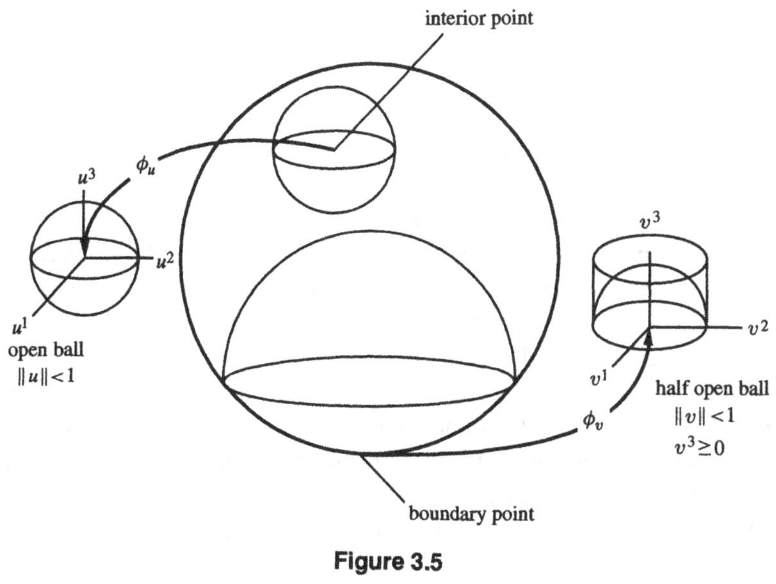


Figure 3.5

We shall check that boundary points *do* have such neighborhoods, as this illustrates a typical use of the inverse function theorem. For simplicity we consider the south pole on the boundary 2-sphere. This sphere, near the pole, can be described as $z + f(x, y) = 0$, where $f(x, y) = \sqrt{(1 - x^2 - y^2)}$. Thus a neighborhood of the south pole *in the closed unit ball* is given, say, by $x^2 + y^2 < \epsilon$ together with $0 \le z + f(x, y) < \delta$ where $\epsilon$ and $\delta$ are positive. The "bottom" boundary consists of a curved disc, a portion of the unit sphere. We would like to straighten this into a flat disc. Consider the three functions $v^1 = x$, $v^2 = y$, and $v^3 = z + f(x, y)$. From $dv^1 \wedge dv^2 \wedge dv^3 = dx \wedge dy \wedge dz$, that is, $\partial(v^1, v^2, v^3))/\partial(x, y, z) = 1 \neq 0$, we conclude (see Corollary (1.16)) that the $v$'s form a smooth coordinate system for $\mathbb{R}^3$ near the south pole. Thus the above neighborhood of the south pole described can be described by $(v^1)^2 + (v^2)^2 < \epsilon$ and $0 \le v^3 < \delta$, which is a cylindrical "can" (with sides and top removed) in a $v^1, v^2, v^3$ space (see the figure). By then removing the points in the can with $\| \mathbf{v} \| \ge \epsilon$ we have the desired half open ball.

Briefly speaking, **an $n$-manifold with boundary** $M^n$ has an **interior** that is a genuine $n$-manifold, and a **boundary** or **edge**, usually written

$$\partial M$$

Points on the boundary have neighborhoods diffeomorphic not to open sets in $\mathbb{R}^n$ but rather to half open sets, that is, sets of the form $\| \mathbf{v} \| < \epsilon$ and $0 \le v^n < \delta$. We still call such a neighborhood a coordinate patch. For more details the reader may consult [G, P, p. 57] or [A, M, R, p. 406]. It is an important fact that the boundary or edge $\partial M$ is itself always an $(n - 1)$-dimensional manifold *without boundary*, although it need not be connected; that is, it may consist of several disjoint manifolds, as in Figure 3.4. Local coordinates for $\partial M$ are given by the $v^1, \ldots, v^{n-1}$. In the example of the closed ball, $v^1 = x$ and $v^2 = y$ are local coordinates for $\partial M = S^2$ near the south pole.

Of course if the boundary is empty, $\partial M = \phi$, $M$ is a genuine manifold.

Concepts such as orientability and 1-sidedness apply to manifolds with boundary as well. An actual Möbius band constructed from a sheet of paper is a surface with boundary, the boundary in this case consisting of a single closed curve diffeomorphic to a circle $S^1$.

### 3.2b. Partitions of Unity

We discussed some elementary point set topology in Section 1.2a. Some further notions will, I hope, be helpful even if only lightly touched upon. If you find this discussion too brief to follow, you should consider the special familiar case of $\mathbb{R}^n$ rather than an abstract manifold. In $\mathbb{R}^n$ an open ball (i.e., a ball without its boundary sphere) centered at a point $x$ is the most important example of a neighborhood of $x$. Given a point $p$ in an $M^n$, let $\{U, x^i\}$ be a coordinate patch with origin at $p$. Then the set where $\sum (x^i)^2 < \epsilon^2$ is an open $\epsilon$-ball neighborhood of $p$ on $M^n$.

A point $x$ in $M^n$ is an **accumulation point** of a subset $A$ of $M^n$ provided *every* neighborhood of $x$ contains at least one point in $A$ other than $x$ itself. It is a fact that if one adjoins to $A$ all of its accumulation points, then the resulting set, called the **closure of A**, is a closed subset; its complement is open. (It is a fact that a subset of a topological space is closed if and only if it contains all of its accumulation points.)

Recall that a real-valued function $f : M \to \mathbb{R}$ is continuous if the inverse image of every open set in $\mathbb{R}$ is itself open in $M$. The nonzero real numbers clearly form an open subset of $\mathbb{R}$, and so the subset of $M$ where $f \neq 0$ is an open subset of $M$, being $f^{-1}(\mathbb{R} - 0)$. The *closure* of this set is called the **support** of $f$. Note that $f$ may be 0 at some points of the support of $f$. For example, for the function whose graph is given

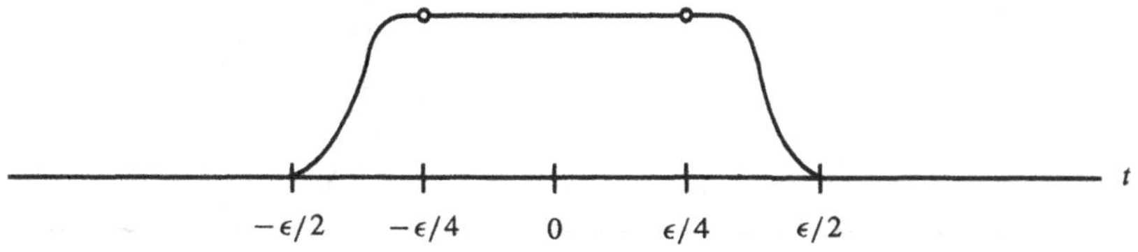


**Figure 3.6**

in Figure 3.6, the support is all $t$ with $| t | \leq \epsilon/2$. Similarly, we can define the support of any tensor field on $M$ as the closure of the set of points on $M$ where the tensor is different from 0.

Given a point $p \in M^n$, it is easy to construct an $n$-form on $M^n$ whose support is contained in an $\epsilon$-ball neighborhood of $p$. Let $p$ be the origin of local coordinates $x$, and let $f = f(t)$ be the function whose graph is depicted in Figure 3.6. This is an example of a **bump function.** We can then define an $n$-form $\omega^n$ on $M^n$, a **bump form,** by putting $\| x \|^2 = \sum (x^i)^2$ and

$$\omega^n := f(\| x \|) dx^1 \wedge \ldots \wedge dx^n, \qquad \text{for } x \text{ in the ball } \| x \| \leq \epsilon$$

and

$$\omega^n = 0 \quad \text{for } x \text{ outside the ball}$$

Now for the notion of a partition of unity. We shall restrict ourselves to manifolds (perhaps with boundary) that can be covered by a finite number of coordinate patches. In fact this restriction is not necessary, but we would have to be more careful (see [G, P, p. 52]).

Given a finite covering $\{U_\alpha\}, \alpha = 1, \ldots, N$, of $M^n$ by coordinate patches $U_\alpha$, a **partition of unity** subordinate to this covering will exhibit $N$ real-valued differentiable functions $f_\alpha : M^n \to \mathbb{R}$ having the following properties.

1. $f_\alpha(x) \geq 0$, all $\alpha$ and all $x$
2. the support of $f_\alpha$ is a (closed) subset of the patch $U_\alpha$ (in particular $f_\alpha$ vanishes outside $U_\alpha$).
3. $\sum_\alpha f_\alpha(x) = 1$ for all $x$ in $M^n$.

Such partitions always exist (it is clear that only the third condition is going to be difficult); they are constructed in the general case in [G, P]. We shall, instead, illustrate the construction in the simplest possible case. Let $M^1$ be the closed unit interval [0, 1] on $\mathbb{R}$. This is a 1-dimensional manifold with boundary consisting of the two endpoints.

Consider the covering given by the two patches $U_1 = \{x \mid 0 \le x < 3/4\}$ and $U_2 = \{x \mid 1/2 < x \le 1\}$.

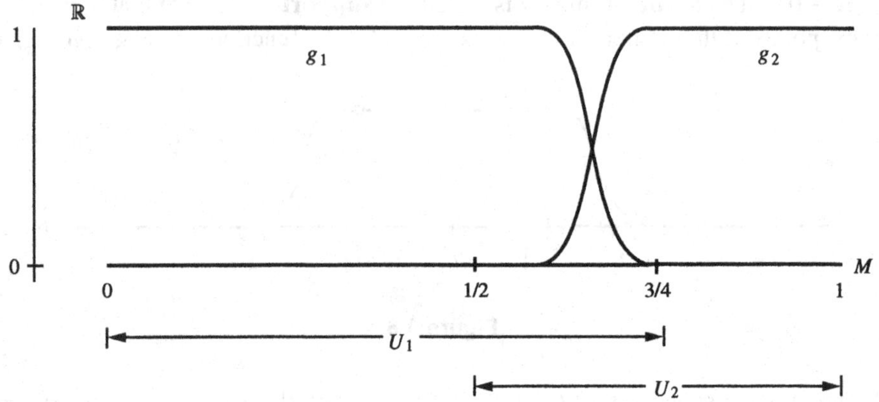


Figure 3.7

We first construct two bump functions $g_1$ and $g_2$ whose supports are in $U_1$ and $U_2$, respectively, and such that they do not vanish simultaneously. We have indicated their graphs in the figure. Since $g_1(x) + g_2(x) > 0$ everywhere on $M^1$ we may define

$$f_\alpha(x) = \frac{g_\alpha(x)}{[g_1(x) + g_2(x)]} \qquad \alpha = 1, 2$$

yielding the desired partition, $\sum_\alpha f_\alpha(x) = 1$. It is evident that keeping the $g$'s from all vanishing simultaneously might be difficult in a general covering of an $M^n$, but it can be done.

### 3.2c. Integration over a Compact Oriented Submanifold

Recall from Section 1.2a that a topological space is *compact* if from every open cover one may extract a finite subcover. This means in particular that *every compact manifold can be covered by a finite number of coordinate patches.* If it is a subset of $\mathbb{R}^n$, then it is compact iff it is *closed* (as a point set) and *bounded.* Thus $M^1 = \mathbb{R}$ is not compact since it is not bounded. $M^1 = (0, 1]$, the half open interval $\{x \mid 0 < x \le 1\}$, is not compact; see 1.2a. On the other hand, the closed interval $[0, 1]$ is a compact manifold with boundary, being a closed, bounded subset of $\mathbb{R}$.

The Möbius band in $\mathbb{R}^3$ including its edge is compact, but without its edge it is not a closed subset and is thus not compact. The 2-sphere $S^2$ is a compact manifold. The closed ball in $\mathbb{R}^3$ is a compact 3-manifold with boundary.

**Warning:** The Möbius band *without its edge*, when considered as a *subset* of $\mathbb{R}^3$, is not a closed subset of $\mathbb{R}^3$, and is thus not compact. The same set, but considered as a manifold or a *topological space* in its own right (with the induced topology), *is* closed, as are all topological spaces (this is because its complement is the empty set, which is open; see 1.2a). In this topology, however, the strip is not compact.

We first define the integral of a $p$-form $\beta^p$ over a compact $p$-dimensional oriented manifold (with or without boundary) $V^p$, that is, the integral of a form of maximal degree. Let $\{U(\alpha)\}, \alpha = 1, \ldots, N$, be a finite covering of $V^p$ by coordinate patches, each positively oriented. Let $\{f_\alpha\}$ be a partition of unity subordinate to this covering. Since each such chart is an oriented parameterized $p$-subset we then know how to evaluate $\int_{U(\alpha)} f_\alpha \beta^p$. We then define

$$\int_V \beta^p := \sum_\alpha \int_{U(\alpha)} f_\alpha \beta^p \tag{3.18}$$

It is easy to show then that the integral so defined is independent of the coordinate cover and partition of unity employed (see [B, T, p. 30]). Of course the crucial ingredient is $\sum_\alpha f_\alpha = 1$.

Finally, if $M^n$ is any manifold and if $\beta^p$ is a $p$-form on $M^n$, we define the integral of $\beta^p$ over any compact oriented $p$-dimensional submanifold $V^p \subset M^n$ (perhaps with boundary) by

$$\int_V \beta^p := \int_V i^* \beta^p \tag{3.19}$$

where $i : V^p \to M^n$ is the inclusion map (note that $i^*\beta^p$ is a $p$-form on the oriented manifold $V^p$).

We emphasize again that one does not really evaluate integrals by means of a partition of unity; it is merely a powerful theoretical tool, as we shall see.

### 3.2d. Partitions and Riemannian Metrics

If a manifold $M^n$ is a submanifold of some $\mathbb{R}^N$ we may let $i : M^n \to \mathbb{R}^N$ be the inclusion map. If we let $ds^2 = \sum_i (dy^i)^2$ be the usual Riemannian metric of $\mathbb{R}^N$, then the pull-back or "restriction" $i^*ds^2$ will be a Riemannian metric on $M^n$, the "induced" metric. For example, if a surface $M^2$ in $\mathbb{R}^3$ is given in the form $z = z(x, y)$, then we may use $x, y$ as coordinates for $M^2$ and then

$$\begin{aligned} i^*(dx^2 + dy^2 + dz^2) &= dx^2 + dy^2 + [z_x dx + z_y dy]^2 \\ &= [1 + z_x^2]dx^2 + 2z_x z_y dxdy + [1 + z_y^2]dy^2 \end{aligned} \tag{3.20}$$

How can we assign a Riemannian metric to a manifold that is not sitting in $\mathbb{R}^N$? Let $\{U_\alpha, x_\alpha^i\}$ be a coordinate cover for $M^n$ (again assumed finite for simplicity). In each patch $U_\alpha$ we may (artificially) introduce a metric $ds_\alpha^2 = \sum_\alpha (dx_\alpha^i)^2$, but of course $ds_\alpha^2$ need not be the same as $ds_\beta^2$ in $U_\alpha \cap U_\beta$. If, however, we introduce a partition of unity $\{f_\alpha\}$ subordinate to the cover we may define a Riemannian metric for $M^n$ by

$$ds^2 = \sum_\alpha f_\alpha ds_\alpha^2$$

(Note that $f_\alpha ds_\alpha^2$ makes sense on all of $M^n$ since $f_\alpha = 0$ outside $U_\alpha$.) Although this metric is again highly artificial, it does show that any manifold admits *some* Riemannian metric. This is a typical example of how a partition of unity is used to splice together local objects to form a global one.

## 3.3. Stokes's Theorem

$$\int_V d\omega^{p-1} = \int_{\partial V} \omega^{p-1}$$

### 3.3a. Orienting the Boundary

Let $M^n$ be an oriented manifold with nonempty boundary $\partial M$; we state again that $\partial M$ is an $(n-1)$-dimensional manifold without boundary. A triangle is not a 2-manifold with boundary since its boundary is only piecewise differentiable.

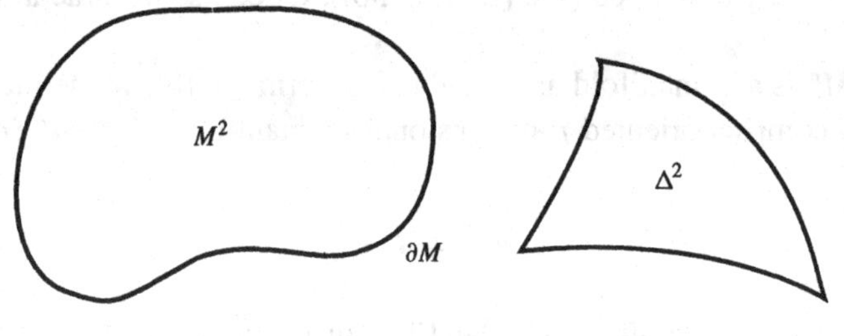


Figure 3.8

Given the orientation of $M^n$ we can orient the boundary $\partial M^n$ as follows. Let $\mathbf{e}_2, \ldots, \mathbf{e}_n$ span the *tangent space* to $\partial M^n$ at $x$. Let $\mathbf{N}$ be a tangent vector to $M^n$ at

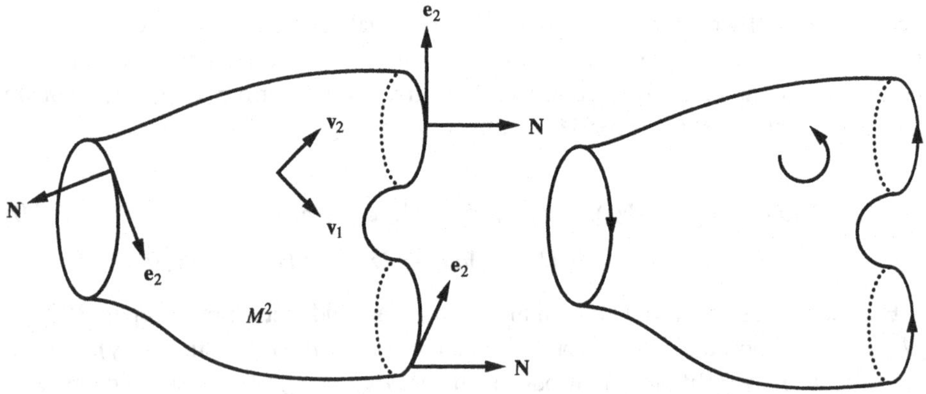


Figure 3.9

$x$ that is transverse to $\partial M^n$ and points *out* of $M^n$. We then declare that $\mathbf{e}_2, \ldots, \mathbf{e}_n$ is positively oriented for $\partial M^n$ provided $\mathbf{N}, \mathbf{e}_2, \ldots, \mathbf{e}_n$ is positively oriented with respect to the given orientation of $M^n$. In Figure 3.9, we have indicated the positive orientation for $M^2$ by the basis $\mathbf{v}_1, \mathbf{v}_2$; then the indicated $\mathbf{e}_2$ is positively oriented for the 1-dimensional manifold $\partial M$. In the right-hand figure we indicate the orientation of $M^2$ by describing the positive sense of rotation and the orientations of the boundary curves by simply giving arrows. Although this works only for 2-manifolds we shall use the same sort of symbolic picture even for $n$-manifolds.

### 3.3b. Stokes's Theorem

**Theorem (3.21):** *Let $V^p \subset M^n$ be a compact oriented submanifold with boundary $\partial V$ in a manifold $M^n$. Let $\omega^{p-1}$ be a continuously differentiable $(p-1)$-form on $M^n$. Then*

$$\int_V d\omega^{p-1} = \int_{\partial V} \omega^{p-1}$$

Versions of this for $p = 2$ and 3 in $\mathbb{R}^3$ were proved in the first half of the eighteenth century by Ampere, Lord Kelvin, Green, Gauss and others. (Unfortunately Kelvin's theorem is traditionally attributed to Stokes.) The general theorem stated previously is again called Stokes's theorem.

PROOF OF STOKES'S THEOREM: Let $i : V^p \to M^n$ be the inclusion map. Then from (3.19) and (2.64) we have

$$\int_V d\omega^{p-1} = \int_V i^* d\omega^{p-1} = \int_V di^*\omega^{p-1}$$

and also

$$\int_{\partial V} \omega^{p-1} = \int_{\partial V} i^*\omega^{p-1}$$

Thus to prove (3.21) we need only prove the same formula where $\omega$ is replaced by $i^*\omega$. In other words, it is sufficient to prove

$$\int_V d\beta^{p-1} = \int_{\partial V} \beta^{p-1}$$

for any continuously differentiable form $\beta^{p-1}$ on $V^p$, forgetting $M^n$ altogether!

Since $V^p$ is compact we may choose a *finite* cover of $V^p$ by coordinate patches $\{V(\alpha)\}$. Let $1 = \sum_\alpha f_\alpha$ be the associated partition of unity; we may then write $\beta = \sum_\alpha \beta_\alpha$, $\beta_\alpha = f_\alpha \beta$. Then

$$\int_V d\beta^{p-1} = \int_V d\sum_\alpha \beta_\alpha = \sum_\alpha \int_{V(\alpha)} d\beta_\alpha^{p-1}$$

and

$$\int_{\partial V} \beta^{p-1} = \sum_\alpha \int_{\partial V} \beta_\alpha^{p-1}$$

We see then that we need only prove

$$\int_{V(\alpha)} d\beta_\alpha^{p-1} = \int_{\partial V} \beta_\alpha^{p-1} \tag{3.22}$$

for the form $\beta_\alpha^{p-1}$ whose support lies in $V(\alpha)$. There are two cases.

*Case (i)*: $V(\alpha)$ is a full coordinate patch lying in the interior of $V$, that is, disjoint from the boundary of $V$.

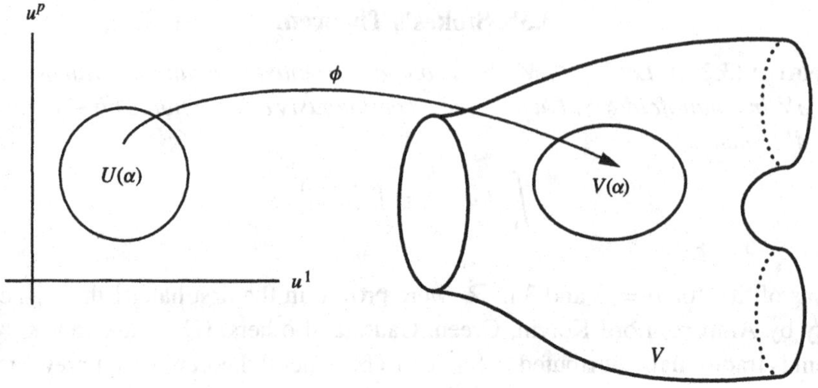


**Figure 3.10**

Then, when everything is expressed in terms of the parameterization $\phi : U(\alpha) \to V(\alpha)$

$$\int_{V(\alpha)=\phi U(\alpha)} d\beta_\alpha = \int_{U(\alpha)} \phi^* d\beta_\alpha = \int_{U(\alpha)} d(\phi^* \beta_\alpha)$$

Denote $\phi^* \beta_\alpha$ by $\gamma^{p-1}$.

$$\phi^* \beta_\alpha = \gamma^{p-1} = \sum_i (-1)^{i-1} \gamma_i du^1 \wedge \ldots \wedge \widehat{du^i} \wedge \ldots \wedge du^p$$

Then

$$\begin{aligned}\int_{U(\alpha)} d\gamma^{p-1} &= \sum_i (-1)^{i-1} \int_{U(\alpha)} d(\gamma_i du^1 \wedge \ldots \wedge \widehat{du^i} \wedge \ldots \wedge du^p) \\ &= \sum_i (-1)^{i-1} \int_{U(\alpha)} \left(\frac{\partial \gamma_i}{\partial u^i}\right) du^i \wedge du^1 \wedge \ldots \wedge \widehat{du^i} \wedge \ldots \wedge du^p \\ &= \sum_i \int_{U(\alpha)} \left(\frac{\partial \gamma_i}{\partial u^i}\right) du^1 \wedge \ldots \wedge du^p \qquad (3.23)\end{aligned}$$

We may assume that the coordinate patch $V(\alpha)$ carries the positive orientation of $V$. Then the last integral becomes an ordinary multiple integral and since the support of $d\phi^* \beta_\alpha$ lies entirely in $U(\alpha)$, we may replace $U(\alpha)$ in the right-hand integral by all of $\mathbb{R}^p$.

$$\begin{aligned}\int_{U(\alpha)} d\gamma^{p-1} &= \sum_i \int_{\mathbb{R}^p} \left(\frac{\partial \gamma_i}{\partial u^i}\right) du^1 \ldots du^p \\ &= \sum_i \int_{\mathbb{R}^{p-1}} du^1 \ldots \widehat{du^i} \ldots du^p \int_{-\infty}^{\infty} \left(\frac{\partial \gamma_i}{\partial u^i}\right) du^i = 0\end{aligned}$$

since $\gamma_i$ vanishes outside $U(\alpha)$. Thus the left-hand side of (3.22) vanishes. But the right-hand side of (3.22) vanishes since $\partial V$ does not meet the support of $\beta_\alpha$ in the case considered. This finishes Case (i).

*Case (ii)*: $V(\alpha)$ is a "half patch" that meets the boundary.

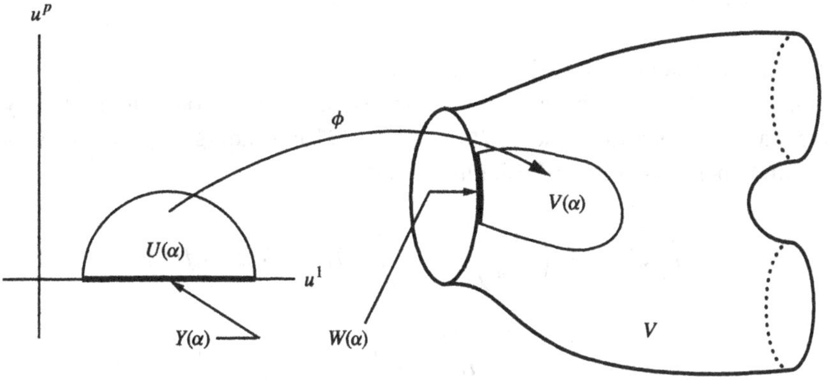


**Figure 3.11**

We proceed exactly as in case (i), reaching (3.23). The only nonvanishing term here is $i = p$ since the other terms will involve $\int_{-\infty}^{\infty}(\partial\gamma_i/\partial u^i)du^i$, which again vanishes if $i < p$. Thus

$$\begin{aligned}\int_{V(\alpha)} d\beta_\alpha &= \int_{U(\alpha)} \left(\frac{\partial\gamma_p}{\partial u^p}\right) du^1 \ldots du^p \\ &= \int_{\mathbb{R}^{p-1}} du^1 \ldots du^{p-1} \int_0^\infty \left(\frac{\partial\gamma_p}{\partial u^p}\right) du^p \\ &= \int_{\mathbb{R}^{p-1}} [\gamma_p(\infty) - \gamma_p(0)] du^1 \ldots du^{p-1} \\ &= -\int_{\mathbb{R}^{p-1}} \gamma_p(u^1, \ldots, u^{p-1}, 0) du^1 \ldots du^{p-1} \qquad (3.24)\end{aligned}$$

If we restrict $\phi : U(\alpha) \to V$ to the subset $Y$ of $U(\alpha)$ defined by $u^p = 0$ we get a $(p-1)$-dimensional coordinate patch $W(\alpha)$ for $\partial V$; $\phi(Y) = W$; see the preceding figure. Then the support of $\beta_\alpha$ meets $\partial V$ in $W$, and so

$$\begin{aligned}\int_{\partial V} \beta_\alpha &= \int_{W=\phi(Y)} \beta_\alpha = \int_Y \phi^*\beta_\alpha = \int_Y \gamma \\ &= \int_Y \sum_i (-1)^{i-1}\gamma_i(u^1, \ldots u^p) du^1 \wedge \ldots \wedge \widehat{du^i} \wedge \ldots \wedge du^p\end{aligned}$$

But $u^p = 0$ on $Y$ and so $du^p = 0$ and the only surviving term is

$$\int_{\partial V} \beta_\alpha = \int_Y (-1)^{p-1}\gamma_p(u^1, \ldots u^{p-1}, 0) du^1 \wedge \ldots \wedge du^{p-1}$$

Now since $\partial/\partial u^1, \ldots, \partial/\partial u^p$ is positively oriented on $V$ (by assumption), and $-\partial/\partial u^p$ is the outward pointing normal to $\partial V$ we conclude from Section 3.3a that $\partial/\partial u^1, \ldots, \partial/\partial u^{p-1}$ carries the orientation $(-1)^p$ on $\partial V$ (there is one minus sign for $-\partial/\partial u^p$ and $p-1$ minus signs to get $\partial/\partial u^p$ into the first position). Consequently

$$\int_{\partial V} \beta_\alpha = (-1)^p \int_Y (-1)^{p-1} \gamma_p(u^1, \ldots u^{p-1}, 0) du^1 \ldots du^{p-1}$$

Since this coincides with (3.24) we are finished.

Finally a note about the case $p = 1$. An oriented 1-manifold with boundary is simply a curve $C$ starting at some $P = x(a) \in M^n$ and ending at $Q = x(b) \in M^n$. The fundamental theorem of calculus says that

$$\int_C df = \int_C \left(\frac{\partial f}{\partial x^i}\right) dx^i = \int_a^b (\partial f/\partial x^i)(dx^i/dt)dt$$
$$= \int_a^b \left[\frac{d\{f[x(t)]\}}{dt}\right] dt = f(Q) - f(P)$$

If we define the oriented boundary of $C$ to be $\partial C = Q - P$ and define $f(\partial C) = f(Q) - f(P)$, then formally Stokes's theorem holds even when $p = 1$. It is then simply the fundamental theorem of calculus! □

## Problems

**3.3(1)** Write out in full in coordinates what (3.21) says in $\mathbb{R}^3$ for $p = 2$ and 3.

**3.3(2)** Write out in full in coordinates what (3.21) says in $\mathbb{R}^4$ for $p = 2, 3$, and 4.

## 3.4. Integration of Pseudoforms

How do we measure "flux"?

We would like to integrate pseudo-$p$-forms $\beta^p$ of $M^n$ over parameterized subsets $F: U \to M^n, U \subset \mathbb{R}^p$. If we orient $U$, we would like $F^*\beta$ to be a well-defined form on $U$, but $\beta$ is really a pair of forms $\pm\beta$ on $M^n$ and we would have to have a prescription for picking out one of the $\beta$'s to pull back. In general there is no way of accomplishing this; we would need, somehow, a way of picking out an orientation of $M^n$ near $F(u)$ whenever we pick an orientation of $U$, and if $M^n$ is nonorientable this might be impossible. If one can associate an orientation on $M^n$ near $F(u)$ whenever one assigns an orientation to $U$, the map is said to be **oriented** (de Rham). This is a restriction on the map $F$ and *in general one cannot pull back a pseudoform! We are not going to be able to integrate a pseudoform over an oriented submanifold*, as we can with a true form.

### 3.4a. Integrating Pseudo-$n$-Forms on an $n$-Manifold

We claim that any *pseudo-n*-form $\omega^n$ can be integrated over *any* compact $n$-dimensional manifold $M^n$, orientable or not! First note that if $U$ is a coordinate patch on such an $M^n$, then we can define $\int_U \omega^n$ as follows. Pick an orientation of $U$; this picks out a specific choice for $\omega^n$ and then the integral of the form $\omega^n$ over the oriented region $U$ is performed just as in the case of a true form. Note that if we had chosen the opposite orientation of $U$, then the integral would be unchanged since although the region of integration would have its orientation reversed we would also automatically have picked out the negative $-\omega^n$ of the previous form. One can then define the integral of $\omega^n$ over all of $M^n$ by use again of a partition of unity as in (3.18).

This should not be surprising. Certainly the Möbius band has an area and this can be computed using its area pseudo-2-form.

### 3.4b. Submanifolds with Transverse Orientation

Let $V^p$ be a $p$-dimensional submanifold of a manifold $M^n$. At each point $x$ of $V^p$ the tangent space to $M^n$ is of the form $M^n_x = V^p_x \oplus N^{n-p}$, where the vectors in $N$ are transverse to $V^p$. Let us say that $V^p$ is **transverse orientable** if each transversal $N^{n-p}$ can be oriented continuously as a function of the point $x$ in $V^p$. If $V^p$ is a **framed** submanifold, that is, if one can find $(n-p)$ continuous linearly independent vector fields on $V^p$ that are transverse to $V^p$, then clearly $V^p$ is transverse orientable.

Since every manifold carries a Riemannian metric (see 3.2d) one can always replace "transverse" by "normal" in some Riemannian metric.

Note that if $V^{n-1}$ is a *hypersurface*, then $V$ is framed if and only if $V$ is 2-sided (see 2.8c). It is also clear that in the case of a hypersurface, transverse orientability is equivalent to being framed by a normal vector field; in particular, the Möbius band in $\mathbb{R}^3$ is not transverse orientable. For $V^p \subset M^n$ for $p < n$, however, transverse orientability is a weaker condition than being framed.

Given a point $x$ on $V^p$ we may (since $V^p$ is an embedded submanifold, see 1.3d) introduce coordinates $x^1, \dots, x^n$ near this point $x = 0$ on $M^n$ (in a patch $W$) such that $V^p \cap W$ is defined by $x^\alpha = f^\alpha(x^1, \dots, x^p), \alpha = p+1, \dots, n$. Then the $n-p$ coordinate vectors $\mathbf{N}_\alpha = \partial/\partial x^\alpha$ are defined in $W$ and are transverse to $V^p$ at $V^p \cap W$. *A sufficiently small piece of a submanifold can always be framed* and is thus transverse orientable. $V^p \cap W$ is a coordinate patch for $V^p$; in fact $x^1, \dots, x^p$ could be used as local coordinates there. In particular, given an orientation for $V^p \cap W$, we can always find $p$ tangent vector fields $\mathbf{X}_1, \dots, \mathbf{X}_p$ that are positively oriented in this patch and these vector fields can be extended to all of $W$ by keeping their components constant as we move off $V$. We may then define an orientation of $W$ by insisting that $\mathbf{N}_{p+1}, \dots, \mathbf{N}_n, \mathbf{X}_1, \dots, \mathbf{X}_p$ define the positive orientation. Thus *to an orientation of $V^p \cap W$ on $V^p$ we may associate an orientation of $W$* on $M^n$, and thus if $\beta^p$ is a pseudo-$p$-form on $W$, we may pull it back to a pseudo-$p$-form $i^*\beta^p$ on $V^p \cap W$. To say that $V^p$ is transverse orientable is to say that we can patch these local constructions together in a coherent or continuous fashion. (We shall certainly fail in the case of a Möbius band in $\mathbb{R}^3$.) In summary, if $\beta^p$ is a pseudoform in $W$, we may pull back this form via the inclusion map $i : V^p \to M^n$ to yield a pseudo-$p$-form $i^*\beta^p$ on $V^p \cap W$ and *if $V^p$ is transverse orientable we may pull back a pseudo-p-form $\beta^p$ of $M^n$ to $i^*\beta^p$ on all of $V^p$.*

### 3.4c. Integration over a Submanifold with Transverse Orientation

Let $i: V^p \to M^n$ be a submanifold of the compact manifold $M^n$ (perhaps with boundary) with transverse orientation, and let $\beta^p$ be a pseudo-$p$-form on $M^n$. We have seen in the previous section that we may pull this pseudo-$p$-form back to $i^*\beta^p$ on $V^p$. Let $\{U(\alpha)\}$ be a finite coordinate cover of $V^p$ with associated partition of unity $\{f_\alpha\}$. Then we define (since $i^*\beta$ is a $p$-form on $V^p$)

$$\int_V \beta^p := \sum_\alpha \int_{U(\alpha)} (i^*\beta^p) f_\alpha \tag{3.25}$$

In summary, we have the following contrast.

> A true $p$-form on $M^n$ is always integrated over an oriented submanifold $V^p$, whereas a pseudo-$p$-form $\beta^p$ is always integrated over a submanifold $V^p$ with transverse orientation.

Consider, for example, the Möbius band $V^2$ sitting in $\mathbb{R}^3$ and one also in Mö$\times\mathbb{R}$. If $\beta^2$ is a true 2-form on $\mathbb{R}^3$ or Mö$\times\mathbb{R}$, then we cannot define the integral of $\beta^2$ over either Möbius band since the Möbius band is not orientable. If $\beta^2$ is a pseudo-2-form then we cannot integrate $\beta^2$ over the strip in $\mathbb{R}^3$ since this strip is 1-sided, and we cannot pull $\beta^2$ back to the strip. On the other hand Mö is 2-sided in Mö$\times\mathbb{R}$ (see 2.8c), and thus we can integrate $\beta^2$ over Mö $\subset$ Mö $\times\,\mathbb{R}$ once we have chosen one of the two possible normals $\partial/\partial t$ or $-\partial/\partial t$, where $t$ is the coordinate in $\mathbb{R}$.

In the case of a surface integral of a pseudo-2-form $\beta^2$ in $\mathbb{R}^3$ we have the following simple prescription. Let $F(U)$ be an *unoriented* parameterized surface in $\mathbb{R}^3$ *with a prescribed unit normal* **N**. We know that $\beta^2$ is of the form $\beta^2 = i_{\mathbf{B}} \mathrm{vol}^3$ for a unique (true) vector **B**. Then $\mathbf{B} \bullet \mathbf{N}$ is a true scalar and from (3.25) and (3.15)

$$\int_{F(U),\mathbf{N}} \beta^2 = \int_U \mathbf{B} \bullet \mathbf{N} dS = \int_U B_n dS \tag{3.26}$$

This is sometimes called the **flux** of **B** through the surface with given normal **N**. This result is independent of any choice of orientation of $\mathbb{R}^3$ or of orientation of the surface. *Only the normal was prescribed.*

Let $\alpha^1$ be a pseudo-1-form and $F(I)$ an *unoriented* curve with framing in $\mathbb{R}^3$; thus there are two mutually orthogonal unit normals $\mathbf{N}_1$ and $\mathbf{N}_2$ defined along the curve $F(I)$. (We shall see in Section 16.1d that such a framing exists for any curve in $\mathbb{R}^3$.) Let **A** be the contravariant pseudovector associated to the pseudoform $\alpha^1$. If we pick out arbitrarily an orientation, that is, a direction, for the curve $F(I)$, then a specific vector **A** is chosen through the orientation of $\mathbb{R}^3$ determined by the triple $\mathbf{N}_1, \mathbf{N}_2, \mathbf{T}$, where **T** is the unit tangent vector to the directed curve. We then have for a line integral

$$\int_{F(I),\mathbf{N}_1,\mathbf{N}_2} \alpha^1 = \int_I \mathbf{A} \bullet \mathbf{T} ds \tag{3.27}$$

and this is again independent of the orientation chosen for the curve.

### 3.4d. Stokes's Theorem for Pseudoforms

Let $\omega^{n-1}$ be a pseudo-$(n-1)$-form on a compact unoriented manifold $M^n$ with boundary. Then $d\omega^{n-1}$ is a pseudo-$n$-form on $M^n$ and we may compute the integral $\int_M d\omega$ as in 3.4a. Now $\partial M$ *has a natural transverse orientation* in $M^n$ since there is clearly an outward pointing transversal $\mathbf{N}$; *if* $M^n$ has a Riemannian metric we may even choose $\mathbf{N}$ to be a unit normal. In any case we may then form the integral $\int_{\partial M} \omega^{p-1}$ (we have omitted indicating the transversal since it will always be assumed to be the outward one). The proof of Stokes's theorem in the previous section carries over to yield again $\int_M d\omega = \int_{\partial M} \omega^{p-1}$, but we emphasize that *no orientation* has been assumed for $M$!

If you are used to proving Stokes's theorem by breaking up $M^n$ into nonoverlapping patches $U, V, \ldots$, you are familiar with the cancellations in $\int \omega$ over boundaries common to two adjacent patches. This still happens with pseudoforms in spite of the arbitrariness in picking orientations in the patches.

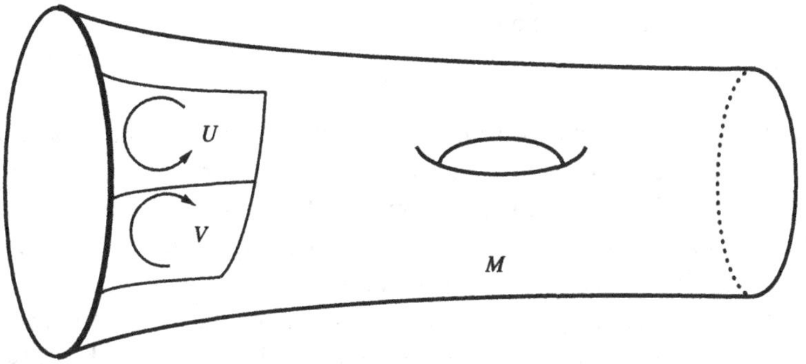


**Figure 3.12**

In Figure 3.12 we have given opposite orientations to the patches $U, V$ for the evaluations of $\int_U d\omega^{n-1}$ and $\int_V d\omega^{n-1}$. It *appears* as if the boundary integrals along the common part of their boundaries would not cancel, but this is not so since the $\omega$'s used in $U$ and $V$ would be negatives of each other!

Suppose now that $V^p$ is a compact submanifold with boundary of $M^n$, and suppose that $V$ is tranverse oriented in $M$: for simplicity we shall assume that $V$ has a normal framing $\mathbf{N}_1, \ldots, \mathbf{N}_{n-p}$. Let $\mathbf{n}$ be the unit vector that is tangent to $V$, normal to $\partial V$, and points out of $V$. Then we may frame $\partial V$ by using $\mathbf{N}_1, \ldots, \mathbf{N}_{n-p}, \mathbf{n}$. Thus a *transverse orientation of V leads in a natural way to a transverse orientation for its boundary* $\partial V$! With this understood we may state

**Stokes's Theorem (3.28):** *Let $\beta^{p-1}$ be a pseudo-$(p-1)$-form on any manifold $M^n$. Let $V^p$ be a compact transverse oriented submanifold (with boundary) of $M^n$. Then*

$$\int_V d\beta^{p-1} = \int_{\partial V} \beta^{p-1}$$

The proof is similar to that given for true forms. We emphasize that no orientation is required for $V^p$ or $M^n$.

## 3.5. Maxwell's Equations

Suppose that our space is really a 3-torus $T^3$. How does the electric field behave when a constant current is sent through a wire loop?

### 3.5a. Charge and Current in Classical Electromagnetism

We accept as a primitive notion the **charge** $Q$ on a particle and we assume that there is a 3-form $\sigma^3$ defined in $\mathbb{R}^3$ whose integral over any region $U$ will yield the charge contained in the region

$$Q(U) = \int_U \sigma^3 \tag{3.29}$$

We shall assume that $Q(U)$ is a scalar independent of the orientation of $\mathbb{R}^3$. This means that $\sigma^3$ is a *pseudoform*. Note that (3.29) does not require and is *independent of the use of any Riemannian metric* in space. If we do introduce a Riemannian metric, say the standard euclidean one, then we have

$$\sigma^3 = \rho(x)\,\mathrm{vol}^3 \tag{3.30}$$

where $\rho$ is the charge density 0-form (a scalar). Note that to define $\rho$ only a volume form is required, not a full metric. In the following, whenever $\mathrm{vol}^3$ or some object constructed from a Riemannian metric appears, it will be assumed that a choice of volume form or metric has been made, but it is intriguing to note which objects (such as $\sigma^3$) do not require these extraneous structures.

Let $W^2$ be a 2-sided surface. If we prescribe one of the two sides, that is, if $W$ is transverse oriented by, say, a transverse vector field $\mathbf{N}$, then we shall also *assume* that the *rate at which charge is crossing* $W$ (in the sense indicated by $\mathbf{N}$) is given by integrating a (necessarily pseudo-) 2-form $\mathcal{j}^2$, the **current 2-form**

$$\int_W \mathcal{j}^2 \tag{3.31}$$

We *assume that charge is conserved*; thus if $W^2 = \partial U^3$ is the boundary of a fixed compact region $U$ (with outward pointing transversal $\mathbf{N}$), then the rate at which charge is leaving $U$, $\int_{\partial U} \mathcal{j}^2$, must equal the rate of decrease of charge inside $U$,

$$-\frac{d}{dt}\int_U \sigma^3 = -\int_U \frac{\partial \sigma^3}{\partial t} = \int_{\partial U} \mathcal{j}^2$$

This must be true for *each* region $U$. If $\mathcal{j}^2$ is continuously differentiable we have $\int_{\partial U} \mathcal{j}^2 = \int_U \mathbf{d}\mathcal{j}^2$, and so

$$\frac{\partial \sigma^3}{\partial t} + \mathbf{d}\mathcal{j}^2 = 0 \tag{3.32}$$

We have introduced here two notational devices. First

We have used a bold $\mathbf{d}$ to emphasize that this exterior derivative is *spatial*, not using differentiation with respect to time; this distinction will be important when considering space–time later on.

Second

We have defined the **time derivative** of an exterior form by simply differentiating each component

$$\frac{\partial}{\partial t}[a_I(x,t)dx^I] := \left(\frac{\partial a_I}{\partial t}\right)dx^I \tag{3.33}$$

Since $\mathcal{j}^2$ is a pseudo-2-form we can associate a **current vector J** such that $\mathcal{j}^2 = i_{\mathbf{J}}\,\mathrm{vol}^3$. We can then write (3.32), using (2.87), as the "equation of continuity"

$$\frac{\partial \rho}{\partial t} + \operatorname{div}\mathbf{J} = 0 \tag{3.34}$$

In many cases the current is a **convective current**, meaning that **J** is of the form

$$\mathbf{J} = \rho\mathbf{v} \tag{3.35}$$

where **v** is the velocity of a charged fluid. In this case, in cartesian coordinates,

$$\mathcal{j}^2 = \rho[v^1 dy \wedge dz + v^2 dz \wedge dx + v^3 dx \wedge dy]$$

and by inserting a factor $\sqrt{g}$ we have the correct expression in any coordinates (see (2.77)).

### 3.5b. The Electric and Magnetic Fields

We isolate the effects of the electromagnetic field by assuming that no other external forces, such as gravity, are present. The electric and magnetic fields are defined operationally. In the following we shall use the euclidean metric and cartesian coordinates of $\mathbb{R}^3$ (where there is no blatant distinction between covariant and contravariant vectors) and then we shall put the results in a form independent of the metric.

We suppose units chosen so that the *velocity of light is unity*, $c = 1$. The electromagnetic force on a *point* mass of charge $q$ moving with velocity **v** is given by the (Heaviside–) **Lorentz force** law

$$\mathbf{F} = q[\mathbf{E} + \mathbf{v} \times \mathbf{B}] \tag{3.36}$$

Thus to determine the electric field **E** at a point $x$ and instant $t$, we measure the force on a unit charge *at rest* at the point $x$. To get **B**, we then measure immediately the forces on unit charges at $x$ that are moving with velocity vectors **i**, **j**, and **k**. This information will determine **B** since **E** has already been determined. Thus the Lorentz force law serves to **define** the fields **B** and **E**! It is interesting that the "correct" magnetic force $q\mathbf{v} \times \mathbf{B}$ was first written down by Heaviside only in 1889! (For a history of electromagnetism I recommend Whittaker's book [W].)

The force **F** has a direction that is independent of orientation of $\mathbb{R}^3$ and so must be a true vector. Since $q$ is a scalar both **E** and $\mathbf{v} \times \mathbf{B}$ must be vectors. But the velocity **v** is certainly a vector, and so **B** *must be a pseudovector* whose sense is orientation-dependent (agreeing with our discussion in 2.8e)!

We shall now redefine the electric and magnetic fields to free them from cartesian analysis and orientation. First note that force naturally enters in line integrals when computing work, and in fact force can be measured by looking at the work expended. We then prefer to consider force as a 1-form $f^1$. This is in agreement with our considering force as the time derivative of momentum and the fact that momentum is to be considered as covariant; see (2.32). From (3.35) we are then to consider the covariant versions of **E** and $\mathbf{v} \times \mathbf{B}$. We think then of the electric field as again a 1-form $\mathcal{E}^1$. To the pseudovector **B** in euclidean $\mathbb{R}^3$ we may associate the *true* 2-form $\mathcal{B}^2$ defined by

$$\mathcal{B}^2 = i_{\mathbf{B}} \operatorname{vol}^3$$

and then the magnetic force covector is $-qi_{\mathbf{v}}\mathcal{B}^2$; see (2.85). We consider the magnetic 2-form $\mathcal{B}^2$ as being more basic than the pseudovector **B**, since $\mathcal{B}$ is independent of the choice of volume form. We then have for the **Lorentz force covector**

$$f^1 = q(\mathcal{E}^1 - i_{\mathbf{v}}\mathcal{B}^2) \tag{3.37}$$

and *this* equation is independent of any metric or orientation.

Our view is then that *the electric field intensity is given by a 1-form* $\mathcal{E}^1$ *and the magnetic field intensity is given by a 2-form* $\mathcal{B}^2$. In *any* coordinates

$$\mathcal{E}^1 = E_1 dx^1 + E_2 dx^2 + E_3 dx^3$$

and (3.38)

$$\mathcal{B}^2 = B_{23} dx^2 \wedge dx^3 + B_{31} dx^3 \wedge dx^1 + B_{12} dx^1 \wedge dx^2$$

If we introduce a metric, then we may consider the associated vector field **E** and the pseudovector **B**. The pseudovector **B** has components $B^1 = B_{23}/\sqrt{g}$, and so on. See Problem 3.5(1) at this time.

### 3.5c. Maxwell's Equations

First some terminology.

A **closed manifold** is a compact manifold without boundary.

The 2-sphere and torus are familiar examples in $\mathbb{R}^3$. We have the 2:1 continuous map $S^2 \to \mathbb{R}P^3$ of the 2-sphere onto the projective plane, and so $\mathbb{R}P^2$ is compact. $\mathbb{R}P^2$ is a closed manifold that is not a submanifold of $\mathbb{R}^3$.

We accept the following empirical laws governing the electromagnetic field in $\mathbb{R}^3$. The name given to the first law is traditional and will be better understood after Gauss's law is given.

**The Absence of Magnetic Charges.** For each compact oriented region $U^3$ in $\mathbb{R}^3$ we have

$$\iint_{\partial U} \mathcal{B}^2 = 0 \tag{3.39}$$

Assume that the field $\mathcal{B}^2$ has continuous first partial derivatives. Then $\iiint_U \mathbf{d}\mathcal{B}^2 = \iint_{\partial U} \mathcal{B}^2 = 0$. Since this is true for arbitrarily small regions $U$ we conclude that

$$\mathbf{d}\mathcal{B}^2 = 0 \tag{3.39'}$$

which is simply the familiar vector analysis statement div $\mathbf{B} = 0$ (see (2.87)).

**Faraday's law.** Let $V^2$ be a compact oriented surface with boundary $\partial V^2$. Then

$$\oint_{\partial V} \mathcal{E}^1 = -\iint_V \frac{\partial \mathcal{B}^2}{\partial t} \tag{3.40}$$

If $\mathcal{E}^1$ has continuous first partial derivatives we may conclude that $\iint_V \mathbf{d}\mathcal{E}^1 + \partial\mathcal{B}^2/\partial t = 0$ for all such surfaces $V^2$. By applying this to small rectangles parallel to the $xy$, $xz$, and $yz$ planes we may conclude

$$\mathbf{d}\mathcal{E}^1 = -\frac{\partial \mathcal{B}^2}{\partial t} \tag{3.40'}$$

which is the vector statement curl $\mathbf{E} = -\partial \mathbf{B}/\partial t$.

**Warning:** Equation (3.40) holds for any surface, moving or not. However, the right-hand side can be written $-d/dt \iint_V \mathcal{B}^2$, that is, as a time rate of change of flux of $\mathcal{B}^2$, only if the surface is fixed in space. We shall see (Problem 4.3(4)) that in the case of a moving surface we may write $\oint_{\partial V}[\mathcal{E}^1 - i_{\mathbf{v}}\mathcal{B}^2] = -d/dt \iint_V \mathcal{B}^2$.(3.40′) of course holds under all circumstances.

For the remaining equations we must *assume a Riemannian metric* in $\mathbb{R}^3$. (We shall see later on that our 3-space does inherit a Riemannian metric, the one we use in daily life, from the space–time structure of general relativity.)

We may then introduce two *pseudoforms*

$$*\mathcal{E} := i_{\mathbf{E}} \,\text{vol}^3 = \sqrt{g}(E^1 dx^2 \wedge dx^3 + E^2 dx^3 \wedge dx^1 + E^3 dx^1 \wedge dx^2) \tag{3.41}$$

and

$$*\mathcal{B} := \langle\ ,\mathbf{B}\rangle = B_1 dx^1 + B_2 dx^2 + B_3 dx^3$$

Note that $*\mathcal{E}$ is a 2-form and $*\mathcal{B}$ is the 1-form version of $\mathbf{B}$.

**Gauss's law.** If $U^3$ is any compact region

$$\iint_{\partial U} *\mathcal{E} = 4\pi \iiint_U \sigma^3 = 4\pi Q(U) \tag{3.42}$$

measures the charge contained in $U$.

We again conclude, when $\mathcal{E}$ is continuously differentiable, that

$$\mathbf{d}{*\mathcal{E}} = 4\pi\sigma^3 \tag{3.42'}$$

or div $\mathbf{E} = 4\pi\rho$.

**Ampere–Maxwell law.** If $M^2$ is a compact 2-sided surface with prescribed normal, then

$$\oint_{\partial M} *\mathcal{B} = \iint_M 4\pi \mathcal{j}^2 + \frac{\partial *\mathcal{E}}{\partial t} \tag{3.43}$$

Thus

$$\mathbf{d}{*\mathcal{B}} = 4\pi \mathcal{j}^2 + \frac{\partial *\mathcal{E}}{\partial t} \tag{3.43'}$$

(assuming $\mathcal{B}$ continuously differentiable) with vector expression curl $\mathbf{B} = 4\pi\mathbf{J}+\partial\mathbf{E}/\partial t$.

Note that the integral versions of Maxwell's equations are more general than the partial differential equation versions since spatial derivatives do not appear in the equations. In particular, their continuity is of no concern!

### 3.5d. Forms and Pseudoforms

There is a general rule of thumb concerning forms versus pseudoforms; a *form* measures an *intensity* whereas a *pseudoform* measures a *quantity*. $\mathcal{E}$ and $\mathcal{B}$ measure the intensities of the electric and magnetic fields (they are "field strengths"). $\sigma^3$ measures the quantity of charge, as does $*\mathcal{E}$ through (3.42). $j^2$ measures essentially the quantity of charge passing through a (transverse oriented) surface in unit time. In Ampere's law, $\mathbf{d}*\mathcal{B} = 4\pi j^2$, $\mathbf{d}*\mathcal{B}$ measures again this flux of charge.

Our conclusions, however, about intensities and quantities must be reversed when dealing with a pseudo-quantity, i.e., a quantity whose sign reverses when the orientation of space is reversed. If this quantity is represented by integrating a 3-form over an oriented region, then the form must, by our definition of integration, be a *true* form. For example, in section 16.4e we shall discuss the hypothetical Dirac magnetic monopole. When such magnetic charge distributions are allowed, the Maxwell equation $\mathbf{d}\mathcal{B} = 0$ should be replaced by $\mathbf{d}\mathcal{B} = \varphi\, \text{vol}^3$, where $\varphi$ is the magnetic charge density, $\mathbf{d}\mathcal{B}$ is a true 3-form, $\varphi$ is a pseudo-scalar, and the total magnetic charge in a region, a pseudo-quantity, is given by the integral of this true 3-form over the oriented region. Furthermore, the classical "definition" of the magnetic field strength $\mathbf{B}(\mathbf{x})$, before the Heaviside–Lorentz force law was known, was the force acting on a "magnetic pole" of unit charge at the point $\mathbf{x}$. Thus the work done against the magnetic field in transporting a magnetic pole of charge $q$ along a curve is the true scalar given by the line integral $\int q*\mathcal{B}$. In terms of these hypothetical poles, the magnetic field strength is measured by the pseudo-form $*\mathcal{B}$ or contravariantly by the pseudo-vector $\mathbf{B}$. Thus magnetic field strength, when measured by a (true) electric charge, is given by the true 2-form $\mathcal{B}$, but when measured by a magnetic pseudo-charge it is given by the pseudo-1-form $*\mathcal{B}$.

## Problems

**3.5(1)** If the magnetic field is a 2-form, not a vector, how do you explain the *curves* generated by iron filings near a bar magnet (i.e., the **B** lines) when we have not informed the magnet of which metric we are using?

**3.5(2)** *Assume that Maxwell's equations* (3.39′), (3.40′), (3.42′), and (3.43′) for $\mathcal{B}$ and $\mathcal{E}$ *hold in every 3-manifold* $M^3$, not just $\mathbb{R}^3$. This will be discussed in more detail in Chapter 14.

The 3-dimensional torus $T^3$ is obtained from the solid unit cube in $\mathbb{R}^3$ by identifying opposite faces pairwise; for example, top and bottom faces are identified by identifying $(x, y, 0)$ with $(x, y, 1)$, and so on. Note then that each *face* has its opposite edges also identified; thus on the bottom face, $(x, 0, 0)$ is identified with

$(x, 1, 0)$. *In this way we see that each face of the cube becomes a 2-torus.* We have indicated the top (= bottom) $T^2$ = Top.

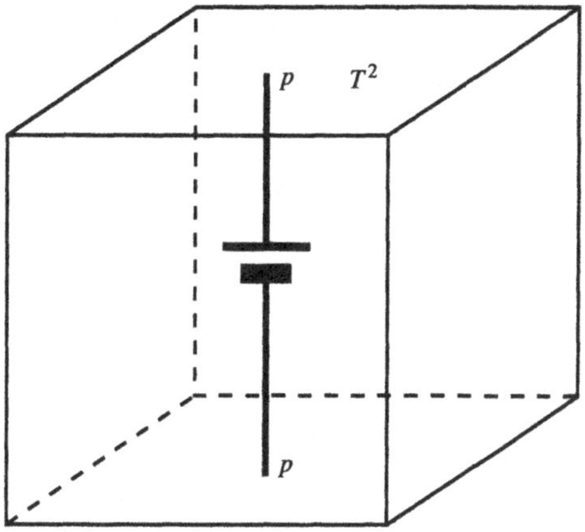


**Figure 3.13**

Consider a current flux of magnitude $j$ through the top torus for all times $t \geq 0$; $\iint_{\text{Top}} j^2 = j$. We can realize this by attaching a battery (delivering a current $j$) at time $t = 0$ to a *closed* wire loop that pierces the top face. Show that for $t \geq 0$

$$\left| \iint_{\text{Top}} *\mathcal{E} \right| = 4\pi jt$$

and thus, unlike the case of a wire loop carrying a constant current in $\mathbb{R}^3$, *the electric field must tend to infinity, with time, at some points of the torus*!

(**Warning:** The top torus $T^2$ is **not** the boundary of any 3-dimensional region!)

On the other hand, it can be shown, though it is more difficult, that if one has a loop that yields no net flux of current through the top, side, or back toroidal faces, for example, if the loop lies in the interior of the cube or if it can be "contracted to a point" in the torus, then a constant current will lead to an electric field that must remain bounded for all time. Thus *the behavior of the electric field is dependent on the "topological position" of the loop.* (It can be shown that the magnetic field remains bounded in all cases.) In a sense, given a closed 2-sided mathematical surface such as Top, and a closed wire loop that pierces it exactly once, the surface will increasingly resist a current through the wire by forcing an electric field to be generated, via Ampere-Maxwell, that will oppose the e.m.f. in the wire. On the other hand, an ordinary closed surface, one that bounds a 3-dimensional region $U$, can never be pierced exactly once by a wire loop; if the loop pierces the surface and enters the region $U$ then it must eventually leave the region, resulting in a zero net flow of current through the surface. For this and other strange behavior in spaces other than $\mathbb{R}^3$, see [D, F]. We shall have more to say about topology in Chapters 13 and 14.

CHAPTER 4

# The Lie Derivative

## 4.1. The Lie Derivative of a Vector Field

Walk one mile east, then north, then west, then south. Have you really returned?

### 4.1a. The Lie Bracket

Let **X** and **Y** be a pair of vector fields on a manifold $M^n$ and let $\phi(t) = \phi_t$ be the local flow generated by the field **X** (see 1.4a). Then $\phi_t x$ is the point $t$ seconds along the integral curve of **X**, the "orbit" of $x$, that starts at time 0 at the point $x$. We shall compare the vector $\mathbf{Y}_{\phi_t x}$ at that point with the result of pushing $\mathbf{Y}_x$ to the point $\phi_t x$ by means of the differential $\phi_{t*}$. The **Lie derivative of Y with respect to X** is defined to

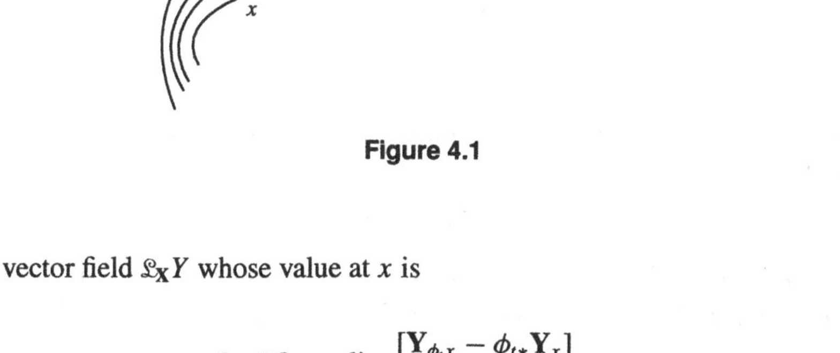

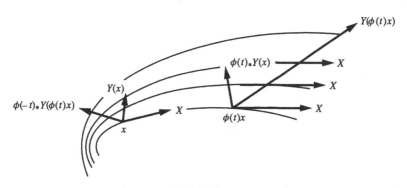

Figure 4.1

be the vector field $\mathcal{L}_{\mathbf{X}}Y$ whose value at $x$ is

$$[\mathcal{L}_{\mathbf{X}}Y]_x := \lim_{t\to 0} \frac{[\mathbf{Y}_{\phi_t x} - \phi_{t*}\mathbf{Y}_x]}{t} \tag{4.1}$$

$$= \lim_{t\to 0} \phi_{t*} \frac{[\phi_{-t*}\mathbf{Y}_{\phi_t x} - \mathbf{Y}_x]}{t}$$

$$= \lim_{t\to 0} \frac{[\phi_{-t*}\mathbf{Y}_{\phi_t x} - \mathbf{Y}_x]}{t} \tag{4.2}$$

since $\phi_{0*}$ is the identity. We must first show that the limit exists. In the process we shall discover an important alternative interpretation of the Lie derivative. First we shall need a very useful version of the mean value theorem in our context. In a sense this is a replacement for a Taylor expansion along the orbit of $x$.

**Hadamard's Lemma (4.3):** *Let $f$ be a continuously differentiable function defined in a neighborhood $U$ of $x$. Then for sufficiently small $t$, there is a function $g_t$, continuously differentiable in $t$ and points near $x$, such that*

$$g_0(x) = \mathbf{X}_x(f)$$

*and*

$$f(\phi_t x) = f(x) + t g_t(x)$$

*that is,*

$$f \circ \phi_t = f + t g_t$$

If we accept this for the moment we may proceed with the existence of the limit. At $x$

$$[\mathcal{L}_{\mathbf{X}}\mathbf{Y}](f) = \lim_{t\to 0} \frac{[\mathbf{Y}_{\phi_t x} - \phi_{t*}\mathbf{Y}_x]}{t}(f)$$

which from (2.60) is

$$= \lim_{t\to 0} \frac{[\mathbf{Y}_{\phi_t x}(f) - \mathbf{Y}_x(f \circ \phi_t)]}{t}$$

$$= \lim_{t\to 0} \frac{[\mathbf{Y}_{\phi_t x}(f) - \mathbf{Y}_x(f + t g_t)]}{t}$$

$$= \lim_{t\to 0} \frac{[\mathbf{Y}_{\phi_t x}(f) - \mathbf{Y}_x(f)]}{t} - \lim_{t\to 0} \mathbf{Y}_x(g_t)$$

$$= \mathbf{X}_x\{\mathbf{Y}(f)\} - \mathbf{Y}_x(\lim_{t\to 0} g_t)$$

$$= \mathbf{X}_x\{\mathbf{Y}(f)\} - \mathbf{Y}_x\{\mathbf{X}(f)\}$$

Thus not only have we shown that the limit exists, but also we have the alternative expression

$$\mathcal{L}_{\mathbf{X}}\mathbf{Y} = [\mathbf{X}, \mathbf{Y}] \tag{4.4}$$

where the **Lie bracket** $[\mathbf{X}, \mathbf{Y}] = -[\mathbf{Y}, \mathbf{X}]$ is the vector field whose differential operator is the **commutator** of the operators for $\mathbf{X}$ and $\mathbf{Y}$

$$[\mathbf{X}, \mathbf{Y}]_x f := \mathbf{X}_x\{\mathbf{Y}(f)\} - \mathbf{Y}_x\{\mathbf{X}(f)\} \tag{4.5}$$

In particular, for any two coordinates x,y we have

$$\mathcal{L}_{\partial/\partial x}\frac{\partial}{\partial y} = 0$$

In Problem 4.1(1) you are asked to show that by expressing the right-hand side of (4.5) in local coordinates one gets

$$[\mathbf{X}, \mathbf{Y}]^i = \sum_j \left\{ X^j\left(\frac{\partial Y^i}{\partial x^j}\right) - Y^j\left(\frac{\partial X^i}{\partial x^j}\right)\right\} \tag{4.6}$$

We remark that (4.2) can be written

$$\mathcal{L}_{\mathbf{X}}\mathbf{Y}_x = \left\{\frac{d}{dt}(\phi_{-t})_*\mathbf{Y}_{\phi_t x}\right\}_{t=0} \tag{4.7}$$

Note that $(\phi_{-t})_*\mathbf{Y}_{\phi_t x}$ is a vector that is always based at the point $x$.

**Proof of Hadamard's Lemma:** Define $F(t, x) = (f \circ \phi_t)(x)$. Fix $t$ and $x$ and put $\mathcal{F}(s) = F(st, x)$. Then

$$(f \circ \phi_t)(x) - f(x) = \mathcal{F}(1) - \mathcal{F}(0) = \int_0^1 \mathcal{F}'(s)ds$$

$$= \int_0^1 \frac{d}{ds}F(st, x)ds = \int_0^1 tF_1(st, x)ds$$

where $F_1$ denotes derivative with respect to the first variable. Thus if we define

$$g_t(x) := \int_0^1 F_1(st, x)ds$$

then

$$(f \circ \phi_t)(x) - f(x) = tg_t(x)$$

Furthermore

$$g_0(x) = \int_0^1 F_1(0, x)ds = F_1(0, x)$$

$$= \lim_{t\to 0}\frac{[F(t, x) - F(0, x)]}{t}$$

$$= \lim_{t\to 0}\frac{[(f \circ \phi_t)(x) - f(x)]}{t} = \mathbf{X}_x(f) \quad \square$$

### 4.1b. Jacobi's Variational Equation

If, in (4.6), we use the fact that $X^j = dx^j/dt$ along the orbit, we can write

$$[\mathcal{L}_X Y]^i = \frac{dY^i}{dt} - \sum_j \left(\frac{\partial X^i}{\partial x^j}\right)Y^j \tag{4.8}$$

We then notice that this makes sense even when **Y** is a vector field that is defined only along the orbit $\phi(t)x$ of the vector field **X**! (4.1) and (4.7) also make sense in this case.

The same derivation that yielded (4.5) will yield (4.8) and *we shall accept (4.8) in this extended sense.*

This equation thus even applies in the case when the vector field $\mathbf{X}$ vanishes at the point $x$. In this case the vector $\mathbf{Y}_{\phi_t x}$ is a time-dependent vector based forever at the point $x$; note then that $\mathcal{L}_{\mathbf{X}}Y$ need not vanish at $x$. For example, consider the vector field $\mathbf{X} = -y\partial/\partial x + x\partial/\partial y$ in $\mathbb{R}^2$, vanishing at the origin. The flow $\phi_t$ generated by $\mathbf{X}$ satisfies $dx/dt = -y$ and $dy/dt = x$

$$\begin{bmatrix} x(t) \\ y(t) \end{bmatrix} = \begin{bmatrix} \cos t & -\sin t \\ \sin t & \cos t \end{bmatrix} \begin{bmatrix} x \\ y \end{bmatrix} = \phi_t \begin{bmatrix} x \\ y \end{bmatrix}$$

Since $\phi$ is linear, $\phi_{t*} = \phi_t$. Let $\mathbf{Y} = \partial/\partial x$ sit at the origin; then $\mathcal{L}_{\mathbf{X}}Y$ is the vector at the origin given by $d/dt\{\phi_{-t*}\partial/\partial x\}_{t=0}$. In components

$$\begin{bmatrix} 0 & 1 \\ -1 & 0 \end{bmatrix} \begin{bmatrix} 1 \\ 0 \end{bmatrix} = \begin{bmatrix} 0 \\ -1 \end{bmatrix}$$

and so $\mathcal{L}_{\mathbf{X}}\partial/\partial x = -\partial/\partial y$.

In the case when $\mathbf{Y}$ is defined only along an orbit of $\mathbf{X}$, it makes no sense to consider $\mathcal{L}_{\mathbf{Y}}\mathbf{X}$, since $\mathbf{Y}$ has no integral curves. We shall reserve the notation $[\mathbf{X}, \mathbf{Y}] = -[\mathbf{Y}, \mathbf{X}]$ for the case in which both $\mathbf{X}$ and $\mathbf{Y}$ are vector fields defined in an open subset in $M^n$.

We shall say that a vector field $\mathbf{Y}$ defined along an orbit of $\mathbf{X}$ is **invariant** (under the flow generated by $\mathbf{X}$) provided

$$\mathbf{Y}_{\phi_t x} = \phi_{t*}\mathbf{Y}_x$$

From (4.1) we see that $\mathbf{Y}$ then satisfies the Jacobi **variational equations**

$$[\mathcal{L}_X Y]^i = \frac{dY^i}{dt} - \sum_j \left(\frac{\partial X^i}{\partial x^j}\right) Y^j = 0 \tag{4.9}$$

The reason for this classical terminology is the following. Classically one worked only in $\mathbb{R}^n$. Consider a solution curve $x = x(t)$ to the differential equation $dx/dt = \mathbf{X}_x$ that starts at the initial point $x(0)$. To discuss the stability of solutions, one would then, in classical language, consider a second integral curve $y = y(t)$ that starts at an "infinitesimally nearby" $y(0) = x(0) + \delta x(0)$. One would then write this solution in the form $y(t) = x(t) + \delta x(t)$. The solution curve $y$ is called a variation of the solution $x$, and $\delta x$ is called an infinitesimal **variation vector**. Now $dx/dt = \mathbf{X}(x)$ and $d(x + \delta x)/dt = \mathbf{X}(x + \delta x)$ are both satisfied.

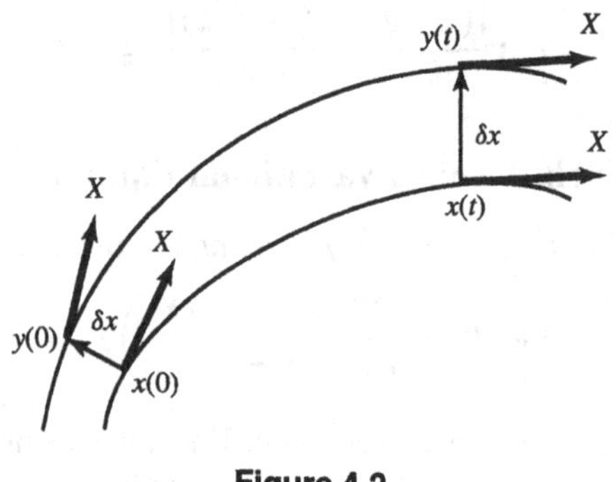


**Figure 4.2**

Subtracting, $\delta(dx/dt) := d(x+\delta x)/dt - dx/dt = d(\delta x)/dt$ becomes

$$\frac{d(\delta x^i)}{dt} = X^i_{x+\delta x} - X^i_x = \sum_j \left(\frac{\partial X^i}{\partial x^j}\right)_{x(t)} \delta x^j + \Delta^i$$

where $\Delta^i$ contains terms of higher order in $\delta \mathbf{x}$. This is a nonlinear system of ordinary differential equations for the infinitesimal variation vector $\delta x$; it is assumed that the base solution $x = x(t)$ is known. If we linearize this system, that is, throw away the high-order terms $\Delta$, we obtain the "infinitesimal" variational equations. Finally if we denote $\delta x$ by $\mathbf{Y}$ we return to the equations (4.9). In our development of (4.9) the *vector field* $\mathbf{Y}$ *replaces the obscure notion of infinitesimally near points*. Instead of seeing how two nearby points are pushed along by the flow, we observe how a vector $\mathbf{Y}$ at $x(0)$ is pushed by the differential $\phi_{t*}$. This differential, being the linear approximation to $\phi_{t*}$, leads to a linear equation for $\mathbf{Y}$ along the orbit $x(t)$.

If $x = x(t)$ is a given solution to the system $dx/dt = \mathbf{X}_x$, and if $\mathbf{Y}_0$ is a vector at the point $x(0)$, then there is a unique solution to the variational equations

$$\frac{dY^i}{dt} = \sum_j \left[\frac{\partial X^i}{\partial x^j}\right]_{x(t)} Y^j$$

with (4.10)

$$Y^i(0) = Y^i_0$$

and, since this system is linear, this solution exists for all $t$ for which the integral curve $x(t)$ is defined. $\mathbf{Y}$ is sometimes called a **Jacobi field** along the solution $x$.

We can also reinterpret (4.1) as follows. Let $\mathcal{Y}_{\phi_t x} := \phi_{t*}\mathbf{Y}_x$ be the Jacobi field along the orbit with initial value $\mathbf{Y}_x$. Then

$$\mathcal{L}_\mathbf{X}\mathbf{Y} = \frac{d}{dt}[\mathbf{Y}_{\phi_t x} - \mathcal{Y}_{\phi_t x}]_{t=0} \tag{4.11}$$

**Warning:** Neither side of (4.10) has intrinsic meaning, independent of coordinates; for instance, we know that $\partial X^i/\partial x^j$ do not form the components of a tensor. Nevertheless, (4.10) has intrinsic meaning since it expresses $\mathcal{L}_\mathbf{X}\mathbf{Y} = 0$, and $\mathcal{L}_\mathbf{X}\mathbf{Y}$ *is* a vector field (defined without the use of coordinates).

### 4.1c. The Flow Generated by [X, Y]

Let $\mathbf{X}$ and $\mathbf{Y}$ be vector fields on $M^n$. Let $\phi(t)$ and $\psi(t)$ be the flows generated by $\mathbf{X}$ and $\mathbf{Y}$. [$\mathbf{X}$, $\mathbf{Y}$] is also a vector field; what is its flow? We claim that the flow generated by [$\mathbf{X}$, $\mathbf{Y}$] is in the following sense the commutator of the two flows. Let $x \in M^n$.

**Theorem (4.12):** *Let* $\sigma$ *be the curve*

$$\sigma(t) := \psi_{-t} \circ \phi_{-t} \circ \psi_t \circ \phi_t x$$

*Then for any smooth function f*

$$[\mathbf{X}, \mathbf{Y}]_x f = \lim_{t\to 0} \frac{f[\sigma(\sqrt{t})] - f[\sigma(0)]}{t}$$

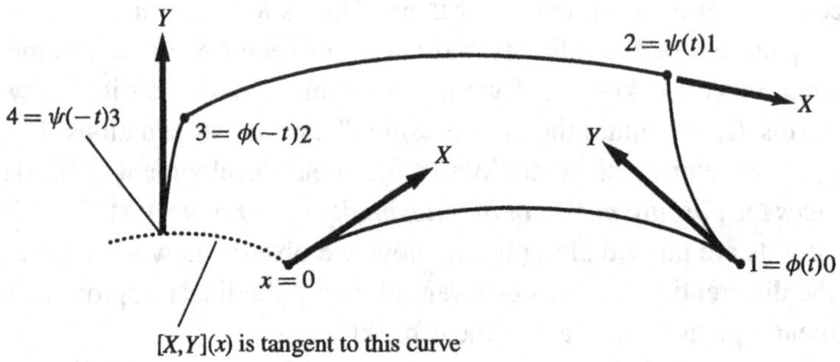


**Figure 4.3**

PROOF (Richard Faber): As in the preceding figure, let 0, 1, 2, 3, 4 be the vertices of the broken integral curves of **X** and **Y**. Let $f$ be a smooth function. Form

$$f(\sigma(t)) - f(0) = [f(4) - f(3)] + [f(3) - f(2)]$$
$$+ [f(2) - f(1)] + [f(1) - f(0)]$$

By Taylor's theorem, letting $\mathbf{X}_0$ denote $\mathbf{X}(0)$, and so on,

$$f(1) - f(0) = t\mathbf{X}_0(f) + \left(\frac{t^2}{2}\right)\mathbf{X}_0\{\mathbf{X}(f)\} + O(3) \tag{i}$$

where $O(3)(t)/t^2 \to 0$ as $t \to 0$. Also

$$f(2) - f(1) = t\mathbf{Y}_1(f) + \left(\frac{t^2}{2}\right)\mathbf{Y}_1\{\mathbf{Y}(f)\} + O(3)$$

Note $\mathbf{Y}_1\{\mathbf{Y}(f)\} = \mathbf{Y}_0\{\mathbf{Y}(f)\} + t\mathbf{X}_0[\mathbf{Y}_t\{\mathbf{Y}(f)\}] + O(2)$, where $\mathbf{Y}_t\{\mathbf{Y}(f)\}$ is the function $t \to \mathbf{Y}_{\phi_t 0}\{\mathbf{Y}(f)\}$. Thus

$$f(2) - f(1) = t\mathbf{Y}_1(f) + \left(\frac{t^2}{2}\right)\mathbf{Y}_0\{\mathbf{Y}(f)\} + O(3) \tag{ii}$$

Likewise

$$f(3) - f(2) = -t\mathbf{X}_2(f) + \left(\frac{t^2}{2}\right)\mathbf{X}_0\{\mathbf{X}(f)\} + O(3) \tag{iii}$$

and

$$f(4) - f(3) = -t\mathbf{Y}_3(f) + \left(\frac{t^2}{2}\right)\mathbf{Y}_0\{\mathbf{Y}(f)\} + O(3) \tag{iv}$$

Adding (i) through (iv) we get

$$f(4) - f(0) = t[\mathbf{X}_0(f) + \mathbf{Y}_1(f) - \mathbf{X}_2(f) - \mathbf{Y}_3(f)]$$
$$+ t^2[\mathbf{X}_0\{\mathbf{X}(f)\} + \mathbf{Y}_0\{\mathbf{Y}(f)\}] + O(3)$$

But

$$\begin{aligned}\mathbf{X}_2(f) - \mathbf{X}_0(f) &= \mathbf{X}_2(f) - \mathbf{X}_1(f) + \mathbf{X}_1(f) - \mathbf{X}_0(f)\\ &= t\mathbf{Y}_1\{\mathbf{X}(f)\} + O(2) + t\mathbf{X}_0\{\mathbf{X}(f)\} + O(2)\\ &= t\mathbf{Y}_0\{\mathbf{X}(f)\} + t\mathbf{X}_0\{\mathbf{X}(f)\} + O(2) \qquad \text{(v)}\end{aligned}$$

Also

$$\begin{aligned}\mathbf{Y}_3(f) - \mathbf{Y}_1(f) &= \mathbf{Y}_3(f) - \mathbf{Y}_2(f) + \mathbf{Y}_2(f) - \mathbf{Y}_1(f)\\ &= -t\mathbf{X}_2\{\mathbf{Y}(f)\} + O(2) + t\mathbf{Y}_1\{\mathbf{Y}(f)\} + O(2)\\ &= -t\mathbf{X}_0\{\mathbf{Y}(f)\} + t\mathbf{Y}_0\{\mathbf{Y}(f)\} + O(2) \text{ (from (v))}\end{aligned}$$

Thus

$$f(4) - f(0) = t^2[\mathbf{X}_0\{\mathbf{Y}(f)\} - \mathbf{Y}_0\{\mathbf{X}(f)\}] + O(3)$$

and then

$$\frac{f\{\sigma(t)\} - f\{\sigma(0)\}}{t^2} \to \mathbf{X}_0\{\mathbf{Y}(f)\} - \mathbf{Y}_0\{\mathbf{X}(f)\}$$

as $t \to 0$. This concludes the proof. □

We may write, in terms of a right-handed derivative,

$$\mathcal{L}_{\mathbf{X}}\mathbf{Y} = [\mathbf{X}, \mathbf{Y}] = \frac{d}{dt_+}\sigma(\sqrt{t})]_{t=0} \qquad (4.13)$$

**Corollary (4.14):** *Suppose that the vector fields* **X** *and* **Y** *on $M^n$ are tangent to a submanifold $V^p$ of $M^n$ at all points of $V^p$. Then since the orbits of* **X** *and* **Y** *that start at $x \in V^p$ will remain on $V^p$, we conclude that the curve $t \mapsto \sigma(t)$, starting at $x$, also lies on $V^p$ and therefore the vector* $[\mathbf{X}, \mathbf{Y}]$ *is also tangent to $V^p$.*

**Warning:** Many books use a sign convention opposite to ours for the bracket $[\mathbf{X}, \mathbf{Y}]$.

## Problems

**4.1(1)** Prove (4.6).

**4.1(2)** Prove Corollary (4.14) by introducing coordinates for $M^n$ such that $V^p$ is locally defined by $x^{p+1} = 0, \ldots, x^n = 0$, and then using (4.6).

**4.1(3)** Consider the unit 2-sphere with the usual coordinates and metric $ds^2 = d\theta^2 + \sin^2\theta\, d\phi^2$. The two coordinate vector fields $\partial_\theta$ and $\partial_\phi$ have, of course, a vanishing Lie bracket. Give a graphical verification of this by examining the "closure" of the "rectangle" of orbits used in the Theorem (4.12). Now consider the *unit* vector fields $\mathbf{e}_\theta$ and $\mathbf{e}_\phi$ associated to the coordinate vectors. Compute $[\mathbf{e}_\theta, \mathbf{e}_\phi]$ and illustrate this misclosure graphically. Verify Theorem (4.12) in this case.

---

## 4.2. The Lie Derivative of a Form

If a flow deforms some attribute, say volume, how does one measure the deformation?

### 4.2a. Lie Derivatives of Forms

If $\mathbf{X}$ is a vector field with local flow $\phi(t)$ and if $f$ is a function, we shall define the Lie derivative of $f$ with respect to $\mathbf{X}$ by $\mathcal{L}_\mathbf{X} f := \mathbf{X}(f) = \sum_i X^i \partial f/\partial x^i$. Thus at $x$, from 2.7a,

$$\mathcal{L}_\mathbf{X} f = \frac{d}{dt} f[\phi_t x]_{t=0} = d/dt[\phi_t^* f]_{t=0} \tag{4.15}$$

This simply describes how $f$ changes along the orbits of $\mathbf{X}$.

If $\alpha^p$ is a $p$-form we define, putting $\alpha_x = \alpha(x)$

$$\mathcal{L}_\mathbf{X}\alpha^p := \frac{d}{dt}[\phi_t^*\alpha^p]_{t=0} \tag{4.16}$$

$$= \lim_{t\to 0} \frac{\phi_t^*\alpha_{\phi_t x} - \alpha_x}{t}$$

By this we mean the following. Let $\mathbf{Y}_1, \ldots, \mathbf{Y}_p$ be vectors at $x$. Then

$$\left[\frac{d}{dt}\phi_t^*\alpha^p\right](\mathbf{Y}_1, \ldots, \mathbf{Y}_p) := \frac{d}{dt}[\phi_t^*\alpha^p(\mathbf{Y}_1, \ldots, \mathbf{Y}_p)] \tag{4.17}$$

$$= \frac{d}{dt}\{\alpha^p[\phi_{t*}\mathbf{Y}_1, \ldots, \phi_{t*}\mathbf{Y}_p]\}$$

In particular, if we extend the vectors $\mathbf{Y}_i$ to be *invariant* fields along the orbit through $x$, $\phi_{t*}\mathbf{Y}_x = \mathbf{Y}_{\phi_t x}$, then we can write

$$\mathcal{L}_\mathbf{X}\alpha^p(\mathbf{Y}_1, \ldots, \mathbf{Y}_p) = \frac{d}{dt}[\alpha^p_{\phi_t x}(\mathbf{Y}_1, \ldots, \mathbf{Y}_p)]_{t=0} \tag{4.18}$$

that is

> $\mathcal{L}_\mathbf{X}\alpha(\mathbf{Y}_1, \ldots, \mathbf{Y}_p)$ measures the derivative (as one moves along the orbit of $\mathbf{X}$) of the value of $\alpha$ evaluated on a $p$-tuple of vector fields $\mathbf{Y}$ that are invariant under the flow generated by $\mathbf{X}$.

The reader should note that although one cannot pull back a pseudoform by means of a general map, one can do so if the map is a diffeomorphism, or a 1-parameter group of such, that is, a flow. Thus it makes sense to talk about the Lie derivative of a pseudoform. For example, if

$$\alpha^n = \text{vol}^n = \sqrt{g}\,dx^1 \wedge dx^2 \wedge \ldots \wedge dx^n$$

is the volume form for a Riemannian $M^n$ and if $\mathbf{X}$ is a vector field on $M^n$, then $\mathcal{L}_\mathbf{X} \text{vol}^n$ is the $n$-form that reads off the rate of change of volume of a parallelopiped spanned by $n$ vectors that are pushed forward by the flow $\phi_t$. Schematically

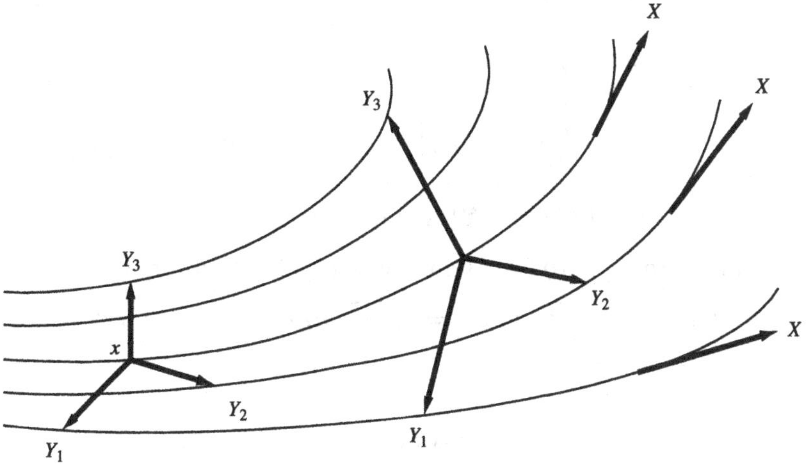


**Figure 4.4**

In other words, $\mathcal{L}_\mathbf{X} \text{vol}^n$ *measures how volumes are changing under the flow* $\phi_t$ *generated by* $\mathbf{X}$. One usually thinks of $\text{vol}^n$ as a given form; then $\mathcal{L}_\mathbf{X} \text{vol}^n$ is "really" describing a property of the vector field $\mathbf{X}$, namely, how the flow generated by $\mathbf{X}$ is distorting volumes!

We need convenient methods for computing Lie derivatives. First note that for a $(p+q)$-tuple $\mathbf{Y}_I$ and their "push-forwards" $\phi_{t*}\mathbf{Y}_I$

$$\begin{aligned}
\mathcal{L}_\mathbf{X}(\alpha^p \wedge \beta^q)(\mathbf{Y}_I) &= \frac{d}{dt}[\alpha^p \wedge \beta^q(\phi_{t*}\mathbf{Y}_I)]_{t=0} \\
&= \frac{d}{dt}\sum_{\underset{\to}{K}}\sum_{\underset{\to}{J}} \delta_I^{JK}\alpha(\phi_{t*}\mathbf{Y}_J)\beta(\phi_{t*}\mathbf{Y}_K)_{t=0} \\
&= \sum_{\underset{\to}{K}}\sum_{\underset{\to}{J}} \delta_I^{JK}\frac{d}{dt}[\alpha(\phi_{t*}\mathbf{Y}_J)]\beta(\mathbf{Y}_K) \\
&\quad + \sum_{\underset{\to}{K}}\sum_{\underset{\to}{J}} \delta_I^{JK}\alpha(\mathbf{Y}_J)\frac{d}{dt}[\beta(\phi_{t*}\mathbf{Y}_K)]_{t=0}
\end{aligned}$$

and so $\mathcal{L}_\mathbf{X}$ is a "derivation" (to be discussed shortly),

$$\mathcal{L}_\mathbf{X}(\alpha^p \wedge \beta^q) = (\mathcal{L}_\mathbf{X}\alpha^p) \wedge \beta^q + \alpha^p \wedge (\mathcal{L}_\mathbf{X}\beta^q) \tag{4.19}$$

**Theorem (4.20):** $\mathcal{L}_\mathbf{X}$ *commutes with exterior differentiation* $d$

$$\mathcal{L}_\mathbf{X} \circ d = d \circ \mathcal{L}_\mathbf{X}$$

PROOF: We first verify this for 0-forms, that is, functions $f$. In our computations we shall omit indications of location, such as, $x$ or $\phi_t x$. Also, all derivatives with respect to time will be evaluated at $t = 0$. Let $\mathbf{Y}$ be a fixed vector at $x \in M^n$. From (2,60)

$$\mathcal{L}_{\mathbf{X}}(df)(\mathbf{Y}) = \frac{d}{dt}\{[\phi_t^* df](\mathbf{Y})\} = \frac{d}{dt}\{df[\phi_{t*}\mathbf{Y}]\}$$

$$= \frac{d}{dt}\{\mathbf{Y}[\phi_t^* f]\}$$

$$= \mathbf{Y}\left\{\frac{d}{dt}[f \circ \phi(t)]\right\} \qquad \text{(since } \mathbf{Y} \text{ is time-independent)}$$

$$= \mathbf{Y}\{\mathbf{X}(f)\} = \mathbf{Y}\{\mathcal{L}_{\mathbf{X}}(f)\} = [d\mathcal{L}_{\mathbf{X}}(f)](\mathbf{Y})$$

and we have verified (4.20) for 0-forms. When applied to $p$-forms

$$\mathcal{L}_{\mathbf{X}} d\alpha^p = \mathcal{L}_{\mathbf{X}} d\sum a_I dx^I = \mathcal{L}_{\mathbf{X}} \sum da_I \wedge dx^{i_1} \wedge \ldots \wedge dx^{i_p}$$

$$= \sum (\mathcal{L}_{\mathbf{X}} da_I) \wedge dx^{i_1} \wedge \ldots \wedge dx^{i_p}$$

$$+ \sum da_I \wedge (\mathcal{L}_{\mathbf{X}} dx^{i_1}) \wedge \ldots \wedge dx^{i_p} + \cdots$$

$$= \sum d(\mathcal{L}_{\mathbf{X}} a_I) \wedge dx^{i_1} \wedge \ldots \wedge dx^{i_p}$$

$$+ \sum da_I \wedge d(\mathcal{L}_{\mathbf{X}} x^{i_1}) \wedge \ldots \wedge dx^{i_p} + \cdots$$

$$= d\sum (\mathcal{L}_{\mathbf{X}} a_I) dx^{i_1} \wedge \ldots \wedge dx^{i_p}$$

$$+ d\sum a_I d(\mathcal{L}_{\mathbf{X}} x^{i_1}) \wedge \ldots \wedge dx^{i_p} + \ldots$$

$$= d\sum (\mathcal{L}_{\mathbf{X}} a_I) dx^{i_1} \wedge \ldots \wedge dx^{i_p}$$

$$+ d\sum a_I (\mathcal{L}_{\mathbf{X}} dx^{i_1}) \wedge \ldots \wedge dx^{i_p} + \ldots$$

$$= d\mathcal{L}_{\mathbf{X}} \sum a_I dx^I = d\mathcal{L}_{\mathbf{X}} \alpha^p \quad \square$$

In particular, we have

$$\mathcal{L}_{\mathbf{X}} dx^i = d\mathcal{L}_{\mathbf{X}} x^i = d\{\mathbf{X}(x^i)\} = dX^i \tag{4.21}$$

Thus if $t$ is any one of the coordinate functions $x^j$ we have $\mathcal{L}_{\partial/\partial t} dx^i = 0$. Hence if $\alpha^p$ is any $p$-form and if $t$ is a coordinate function

$$\mathcal{L}_{\partial/\partial t} \alpha^p = \mathcal{L}_{\partial/\partial t} a_{\underrightarrow{I}} dx^I = \left(\frac{\partial a_{\underrightarrow{I}}}{\partial t}\right) dx^I = \frac{\partial \alpha^p}{\partial t} \tag{4.22}$$

simply differentiates the coefficients with respect to the coordinate!

See Problem 4.2(1) at this time.

### 4.2b. Formulas Involving the Lie Derivative

Let $\bigwedge^p M^n$ be the space of $p$-forms on $M^n$. This is an infinite dimensional vector space since the components are functions. A linear map A: $\bigwedge^p M^n \to \bigwedge^{p+r} M^n$ is said to be a **derivation** if $r$ is even and

$$A(\alpha^p \wedge \beta^q) = (A\alpha^p) \wedge \beta^q + \alpha^p \wedge (A\beta^q) \qquad \text{(e.g., } \mathcal{L}_{\mathbf{X}})$$

and is said to be an **antiderivation** if $r$ is odd and

$$A(\alpha^p \wedge \beta^q) = (A\alpha^p) \wedge \beta^q + (-1)^p \alpha^p \wedge (A\beta^q) \qquad \text{(e.g., } d \text{ and } i_{\mathbf{X}})$$

Suppose we know the value of a derivation or antiderivation on any function and on $d$ of any function. Since the general $p$-form is of the form $\alpha^p = \sum a_I(x) dx^{i_1} \wedge \ldots \wedge dx^{i_p}$, we then know the value of $A$ on any form:

> If $A$ and $B$ are both derivations or antiderivations, then to prove $A\alpha^p = B\alpha^p$ for all forms we need only prove this for $\alpha$ a function and for $\alpha = d$ (a function).

See Problem 4.2(2).

The following is perhaps the most often used formula involving Lie derivatives.

**H. Cartan's Formula (4.23):** When acting on exterior forms

$$\mathcal{L}_{\mathbf{X}} = i_{\mathbf{X}} \circ d + d \circ i_{\mathbf{X}}$$

PROOF: Both sides are derivations, by Problem 4.2(2). We need only verify (4.23) on functions and differentials of functions.

On functions, $i_{\mathbf{X}} f = 0$ and $i_{\mathbf{X}} df = \mathbf{X}(f) = \mathcal{L}_{\mathbf{X}}(f)$; we have verified the function case. On differentials of functions

$$[i_{\mathbf{X}} d + d i_{\mathbf{X}}] df = d i_{\mathbf{X}}(df) = d[i_{\mathbf{X}}(df)] = d[\mathbf{X}(f)]$$

$$= d\mathcal{L}_{\mathbf{X}}(f) = \mathcal{L}_{\mathbf{X}} df \quad \square$$

**Theorem (4.24):** *When applied to forms*

$$\mathcal{L}_{\mathbf{X}} \circ i_{\mathbf{Y}} - i_{\mathbf{Y}} \circ \mathcal{L}_{\mathbf{X}} = i_{[\mathbf{X},\mathbf{Y}]}$$

The reader is asked to supply the proof in Problem 4.2(3).

The following is an intrinsic (i.e., coordinate-free) expression for the exterior derivative of a 1-form. *It is extremely useful.*

**Theorem (4.25):** *Let $\alpha^1$ be a 1-form and let $\mathbf{X}_x$ and $\mathbf{Y}_x$ be vectors at $x$. Extend these vectors in any smooth way to be fields near $x$. Then*

$$d\alpha^1(\mathbf{X}_x, \mathbf{Y}_x) = \mathbf{X}_x\{\alpha^1(\mathbf{Y})\} - \mathbf{Y}_x\{\alpha^1(\mathbf{X})\} - \alpha^1([\mathbf{X}, \mathbf{Y}])$$

PROOF: We shall use (4.23) and (4.24)

$$d\alpha(\mathbf{X}, \mathbf{Y}) = \{i_{\mathbf{X}} d\alpha\}(\mathbf{Y}) = \{\mathcal{L}_{\mathbf{X}}\alpha - d i_{\mathbf{X}}\alpha\}(\mathbf{Y}) = i_{\mathbf{Y}}\mathcal{L}_{\mathbf{X}}\alpha - \mathbf{Y}\{\alpha(\mathbf{X})\}$$

$$= \mathcal{L}_{\mathbf{X}} i_{\mathbf{Y}}\alpha - i_{[\mathbf{X},\mathbf{Y}]}\alpha - \mathbf{Y}\{\alpha(\mathbf{X})\}$$

$$= \mathcal{L}_{\mathbf{X}}\alpha(\mathbf{Y}) - \alpha([\mathbf{X}, \mathbf{Y}]) - \mathbf{Y}\{\alpha(\mathbf{X})\}$$

$$= \mathbf{X}\{\alpha(\mathbf{Y})\} - \alpha([\mathbf{X}, \mathbf{Y}]) - \mathbf{Y}\{\alpha(\mathbf{X})\} \quad \square$$

See Problem 4.2(4) at this time.

The following proposition says that if Y's are vector fields, one can differentiate the function $\alpha^p(\mathbf{Y}_1, \dots, \mathbf{Y}_p) = a_I(x) Y_1^{i_1} \dots Y_p^{i_p}$ by using a "Leibniz" rule for Lie derivative.

**Theorem (4.26):** *For a form $\alpha^p$ and vector fields* $\mathbf{X}, \mathbf{Y}_1, \dots, \mathbf{Y}_p$ *we have*

$$\mathbf{X}\{\alpha^p(\mathbf{Y}_1, \dots, \mathbf{Y}_p)\} = \{\mathcal{L}_\mathbf{X}\alpha^p\}(\mathbf{Y}_1, \dots, \mathbf{Y}_p) + \sum_r \alpha^p(\mathbf{Y}_1, \dots, (\mathcal{L}_\mathbf{X}\mathbf{Y}_r), \dots, \mathbf{Y}_p)$$

PROOF: For 1-forms we have

$$\{\mathcal{L}_\mathbf{X}\alpha\}(\mathbf{Y}) = i_\mathbf{Y}\mathcal{L}_\mathbf{X}\alpha = \mathcal{L}_\mathbf{X} i_\mathbf{Y}\alpha - \alpha([\mathbf{X}, \mathbf{Y}]) = \mathbf{X}\{\alpha(\mathbf{Y})\} - \alpha(\mathcal{L}_\mathbf{X}\mathbf{Y})$$

as desired. By induction, assuming true for $(p-1)$-forms,

$$\{\mathcal{L}_\mathbf{X}\alpha\}(\mathbf{Y}_1, \dots, \mathbf{Y}_p) = i_{\mathbf{Y}_1}\{\mathcal{L}_\mathbf{X}\alpha\}(\mathbf{Y}_2, \dots, \mathbf{Y}_p) = \{\mathcal{L}_\mathbf{X} i_{\mathbf{Y}_1}\alpha - i_{[\mathbf{X}, \mathbf{Y}_1]}\alpha\}(\mathbf{Y}_2, \dots, \mathbf{Y}_p)$$

But $i_{\mathbf{Y}_1}\alpha$ is a $(p-1)$-form and so we may apply (4.26) to compute $\{\mathcal{L}_\mathbf{X} i_{\mathbf{Y}_1}\alpha\}(\mathbf{Y}_2, \dots, \mathbf{Y}_p)$. This will complete the proof. □

Finally, we have a formula that generalizes (4.25) to $p$-forms. For vector fields $\mathbf{Y}_0, \dots, \mathbf{Y}_p$

$$d\alpha^p(\mathbf{Y}_0, \dots, \mathbf{Y}_p) = \sum_r (-1)^r \mathbf{Y}_r\{\alpha^p(\mathbf{Y}_0, \dots, \hat{\mathbf{Y}}_r, \dots, \mathbf{Y}_p)\} + \sum_{r<s} (-1)^{r+s} \alpha^p([\mathbf{Y}_r, \mathbf{Y}_s], \dots \hat{\mathbf{Y}}_r, \dots, \hat{\mathbf{Y}}_s, \dots, \mathbf{Y}_p) \tag{4.27}$$

This can again be proved by induction. Note that from the left-hand side we see that this result depends only on the values of the Y's at the given point!

## 4.2c. Vector Analysis Again

Let $\mathrm{vol}^n$ be a **volume form** for an $M^n$, that is, a pseudo-$n$-form that never vanishes on any basis of tangent vectors. If **X** is a vector field on $M^n$, the divergence of $X$ is the scalar div **X** **defined** by the formula

$$\mathcal{L}_\mathbf{X}\,\mathrm{vol}^n = (\mathrm{div}\,\mathbf{X})\,\mathrm{vol}^n \tag{4.28}$$

If $\mathbf{Y}_1, \ldots, \mathbf{Y}_n$ are fields invariant under the flow generated by $\mathbf{X}$ then from (4.17)

$$\mathcal{L}_{\mathbf{X}}(\mathrm{vol})^n(\mathbf{Y}_1, \ldots, \mathbf{Y}_n) = \frac{d}{dt}\,\mathrm{vol}^n(\mathbf{Y}_1, \ldots, \mathbf{Y}_n)_{t=0}$$

and so div $\mathbf{X}$ *measures the logarithmic rate of change of volumes* along the flow. In local coordinates $\mathrm{vol}^n = \rho dx^1 \wedge \ldots \wedge dx^n$, $\rho(x) > 0$, and by Cartan's formula

$$\mathcal{L}_{\mathbf{X}}(\mathrm{vol})^n = d\{i_{\mathbf{X}}\,\mathrm{vol}^n\} = d\sum_r (-1)^{r-1}\rho dx^1 \wedge \ldots i_{\mathbf{X}} dx^r \wedge \ldots \wedge dx^n$$

$$= d\sum_r (-1)^{r-1}(\rho X^r) dx^1 \wedge \ldots \widehat{dx^r} \wedge \ldots \wedge dx^n$$

$$= \sum_r (-1)^{r-1}\left\{\frac{\partial}{\partial x^s}(\rho X^r) dx^s\right\} \wedge dx^1 \wedge \ldots \widehat{dx^r} \wedge \ldots \wedge dx^n$$

$$= \sum_r \left\{\frac{\partial}{\partial x^r}(\rho X^r)\right\} dx^1 \wedge \ldots \wedge dx^r \wedge \ldots \wedge dx^n$$

and thus

$$\mathrm{div}\,\mathbf{X} = \frac{1}{\rho}\sum_r \frac{\partial}{\partial x^r}(\rho X^r) \tag{4.29}$$

generalizing (2.88) of $\mathbb{R}^3$.

Note also that to the vector $\mathbf{X}$ and the volume form $\mathrm{vol}^n$ we may associate the $(n-1)$ form

$$\beta^{n-1} = i_{\mathbf{X}}\,\mathrm{vol}^n \tag{4.30}$$

and then Cartan's formula gives

$$d\beta^{n-1} = (\mathrm{div}\,\mathbf{X})\,\mathrm{vol}^n \tag{4.31}$$

generalizing (2.87) of $\mathbb{R}^3$.

We now use the Lie derivative formalism to complete our discussion of classical vector analysis in $\mathbb{R}^3$. Consider, for example, the vector identity for $\mathrm{curl}(\mathbf{A}\times\mathbf{B})$.

$$\mathrm{curl}(\mathbf{A}\times\mathbf{B}) \Leftrightarrow d i_{\mathbf{B}}\alpha^2 = \mathcal{L}_{\mathbf{B}}\alpha^2 - i_{\mathbf{B}} d\alpha^2$$

$$= \mathcal{L}_{\mathbf{B}}\alpha^2 - i_{\mathbf{B}}\,\mathrm{div}\,\mathbf{A}\,\mathrm{vol}^3 = \mathcal{L}_{\mathbf{B}}\alpha^2 - \mathrm{div}\,\mathbf{A}\; i_{\mathbf{B}}\,\mathrm{vol}^3$$

Now use (4.24).

$$\mathcal{L}_{\mathbf{B}}\alpha^2 = \mathcal{L}_{\mathbf{B}} i_{\mathbf{A}}\,\mathrm{vol}^3 = i_{\mathbf{A}}\mathcal{L}_{\mathbf{B}}\,\mathrm{vol}^3 + i_{[\mathbf{B},\mathbf{A}]}\,\mathrm{vol}^3$$

$$= \mathrm{div}\,\mathbf{B}\; i_{\mathbf{A}}\,\mathrm{vol}^3 + i_{[\mathbf{B},\mathbf{A}]}\,\mathrm{vol}^3$$

$$\Leftrightarrow (\mathrm{div}\,\mathbf{B})\mathbf{A} + [\mathbf{B}, \mathbf{A}]$$

Thus

$$\mathrm{curl}(\mathbf{A}\times\mathbf{B}) = (\mathrm{div}\,\mathbf{B})\mathbf{A} + [\mathbf{B}, \mathbf{A}] - (\mathrm{div}\,\mathbf{A})\mathbf{B} \tag{4.32}$$

In vector analysis books the term $\mathcal{L}_{\mathbf{B}}\mathbf{A} = [\mathbf{B}, \mathbf{A}]$ is written differently. We can write, in cartesian coordinates,

$$[\mathbf{B}, \mathbf{A}]^i = B^j \left( \frac{\partial A^i}{\partial x^j} \right) - A^j \left( \frac{\partial B^i}{\partial x^j} \right) = (D_{\mathbf{B}}\mathbf{A})^i - (D_{\mathbf{A}}\mathbf{B})^i$$

where $(D_{\mathbf{B}}\mathbf{A})^i := \mathbf{B} \cdot \operatorname{grad} A^i$. Thus they write the term $[\mathbf{B}, \mathbf{A}]$ as $\mathbf{B} \cdot \operatorname{grad} \mathbf{A} - \mathbf{A} \cdot \operatorname{grad} \mathbf{B}$ as if it made sense to talk about the gradient of a vector! This makes sense only in cartesian coordinates.

## Problems

**4.2(1)** Show that if $\alpha^1 = \sum_i a_i dx^i$ is a 1-form then

$$\mathcal{L}_X \alpha^1 = \sum_i \left\{ X^j \left( \frac{\partial a_i}{\partial x^j} \right) + a_j \left( \frac{\partial X^j}{\partial x^i} \right) \right\} dx^i$$

which should be compared with (4.6).

**4.2(2)** Show that if $\theta$ is a derivation and $A$ an antiderivation then

$$\theta \circ A - A \circ \theta$$

is an antiderivation. If $A$ and $B$ are antiderivations then

$$A \circ B + B \circ A$$

is a derivation.

**4.2(3)** Prove (4.24).

**4.2(4)** Prove (4.25) by expressing both sides in coordinates and using (2.58) and (2.35).

## 4.3. Differentiation of Integrals

How does one compute the rate of change of an integral when the domain of integration is also changing?

### 4.3a. The Autonomous (Time-Independent) Case

Let $\alpha^p$ be a $p$-form and $V^p$ an oriented compact submanifold (perhaps with boundary $\partial V$) of a manifold $M^n$. We consider a "variation" of $V^p$ arising as follows. We suppose that there is a flow $\phi_t : M^n \to M^n$, that is, a 1-parameter "group" of diffeomorphisms $\phi_t$, defined in a neighborhood of $V^p$ for small times $t$, and we define the submanifold $V^p(t) := \phi_t V^p$.

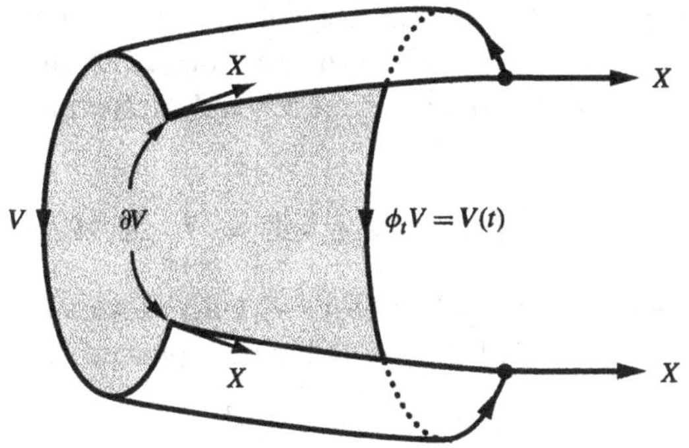


Figure 4.5

Let $\mathbf{X}_x = d\phi_t(x)/dt]_{t=0}$ be the resulting velocity field. We are interested in the time variation of the integral

$$I(t) = \int_{V(t)} \alpha^p = \int_V \phi_t^* \alpha$$

(see (3.17)). Differentiating

$$\begin{aligned} I'(t) &= \lim_{h\to 0} \frac{[I(t+h) - I(t)]}{h} \\ &= \lim_{h\to 0} \frac{\left[\int_V \phi_{t+h}^* \alpha - \int_V \phi_t^* \alpha\right]}{h} \\ &= \lim_{h\to 0} \left[\int_V \frac{\phi_t^*\{\phi_h^* \alpha - \alpha\}}{h}\right] \\ &= \lim_{h\to 0} \left[\int_{V(t)} \frac{\{\phi_h^* \alpha - \alpha\}}{h}\right] \\ &= \int_{V(t)} \lim_{h\to 0} \frac{\{\phi_h^* \alpha - \alpha\}}{h} \end{aligned}$$

Thus

$$\frac{d}{dt}\int_{V(t)} \alpha^p = \int_{V(t)} \mathcal{L}_{\mathbf{X}} \alpha^p \tag{4.33}$$

a remarkably simple and powerful formula! From Cartan's formula

$$\begin{aligned} \frac{d}{dt}\int_{V(t)} \alpha^p &= \int_{V(t)} i_{\mathbf{X}} d\alpha^p + d i_{\mathbf{X}} \alpha^p \\ &= \int_{V(t)} i_{\mathbf{X}} d\alpha^p + \oint_{\partial V(t)} i_{\mathbf{X}} \alpha^p \end{aligned} \tag{4.34}$$

When $\alpha$ is the volume form and $V^n$ is a compact region in $M^n$ we have

$$\begin{aligned} \frac{d}{dt}\int_{V(t)} \mathrm{vol}^n &= \int_{V(t)} d i_{\mathbf{X}}\, \mathrm{vol}^n = \int_{V(t)} \mathrm{div}\mathbf{X}\, \mathrm{vol}^n \\ &= \int_{\partial V} i_{\mathbf{X}}\, \mathrm{vol}^n \end{aligned} \tag{4.35}$$

a form of the **divergence theorem**. Let the volume form come from a Riemannian metric. Then, as in the derivation of (3.15) in the 2-dimensional case, letting **N** be the outward pointing normal to the boundary of $V^n$ and $\mathbf{X}_t$ the projection of **X** into the tangent space to $\partial V$

$$\int_{\partial V} i_{\mathbf{X}} \operatorname{vol}^n = \int_{\partial V} i_{\langle \mathbf{X},\mathbf{N}\rangle \mathbf{N}+\mathbf{X}_t} \operatorname{vol}^n = \int_{\partial V} \langle \mathbf{X},\mathbf{N}\rangle i_{\mathbf{N}} \operatorname{vol}^n$$

On $\partial V$, the form $i_{\mathbf{N}} \operatorname{vol}^n$, when applied to $n-1$ tangent vectors to $\partial V$, reads off the $(n-1)$-dimensional "volume" of the parallelopiped spanned, that is,

$$\operatorname{vol}^{n-1}_{\partial V} := i_{\mathbf{N}} \operatorname{vol}^n \tag{4.36}$$

is the **area** form for the boundary. We then have the usual form of the divergence theorem

$$\int_V \operatorname{div} \mathbf{X} \operatorname{vol}^n = \int_{\partial V} \langle \mathbf{X},\mathbf{N}\rangle \operatorname{vol}^{n-1}_{\partial V} \tag{4.37}$$

We emphasize that *the divergence theorem*, being a theorem about pseudo-$n$-forms, *holds whether $M^n$ is orientable or not.*

### 4.3b. Time-Dependent Fields

Consider a nonautonomous flow of water in $\mathbb{R}^3$, that is, a flow where the velocity field $\mathbf{v}(t,\mathbf{x}) = d\mathbf{x}/dt$ depends on time. We define a map $\phi_t : \mathbb{R}^3 \to \mathbb{R}^3$ as follows. If we observe a molecule at $\mathbf{x}$ when $t=0$, we let $\phi_t\mathbf{x}$ be the position of this same molecule $t$ seconds after 0. Consider $\phi_s[\phi_t\mathbf{x}]$. If we put $\mathbf{y} = \phi_t\mathbf{x}$ then $\phi_s\mathbf{y}$ is the point where the flow would take $\mathbf{y}$ $s$ seconds after time 0. This is usually not the same point as $\phi_{t+s}\mathbf{x}$ since the flow is time-dependent. A time-dependent flow of water is not a flow in the sense of 1.4a since it does not satisfy the 1-parameter group property. *A time-dependent vector field on a manifold $M^n$ does not generate a flow*!

Consider for example the contractions of $\mathbb{R}$ defined by $x \mapsto x(t) = \phi_t x := (1-t)x$, each of which is a diffeomorphism if $t \neq 1$. This does not define a flow, because it does not have the group property. The velocity vector at $x(t)$ and time $t$ are determined from

$$\frac{dx(t)}{dt} = -x = -\frac{x(t)}{(1-t)} \tag{4.38}$$

Thus $v(t,y) = -y/(1-t)$ is a *time-dependent* velocity field.

Suppose then that $v = v(t,x)$ is a time-dependent vector field on $M^n$. We apply a simple classical trick; *any tensor field A(t,x) on $M^n$ that is time-dependent should be considered as a tensor field on the product manifold $\mathbb{R} \times M^n$, where t is the coordinate for* $\mathbb{R}$. $\mathbb{R} \times M^n$ has local coordinates $(t = x^0, x^1, \ldots, x^n)$. A time-dependent vector field on $M^n$ is now an ordinary vector field $\mathbf{v} = \mathbf{v}(t,x)$ on $\mathbb{R} \times M^n$ since $t$ is now a coordinate on $\mathbb{R} \times M$. By solving the system of ordinary differential equations

$$\begin{aligned} &\frac{dx^i}{ds} = v^i(t,x), \qquad x^i(s=0) = x^i_0, \quad i = 1,\ldots,n \\ &\frac{dt}{ds} = 1, \qquad t(s=0) = t_0 \end{aligned} \tag{4.39}$$

we get a flow $\phi_s : \mathbb{R}\times M^n \to \mathbb{R}\times M^n$. If $\mathbf{v}(t,x)$ is the velocity field of a time-dependent flow of fluid in $M^n$, then the integral curves $s \mapsto \phi_s(t_0, x_0)$ on $\mathbb{R} \times M^n$ project down to yield the time-dependent "flow" on $M^n$; $\phi_s(t_0, x_0)$ is the position of the molecule at time $s + t_0$ that had been located at the point $x_0$ at time $t_0$.

In our example (4.38) we need to solve the s-*independent* system

$$\frac{dx}{ds} = -x/(1-t) \qquad x(s=0) = x_0 \tag{4.38$'$}$$

$$\frac{dt}{ds} = 1 \qquad t(s=0) = t_0$$

The solution is

$$x(s) = \left[\frac{(1-t_0-s)}{(1-t_0)}\right] x_0 \tag{4.38$''$}$$

$$t(s) = t_0 + s$$

and one verifies that $\phi(s) : \mathbb{R}^2 \to \mathbb{R}^2$ given by

$$\phi_s(t, x) = (t(s), x(s))$$

is indeed a flow. To see the path in $\mathbb{R}$ of a point that starts at $x_0$ at time 0, we merely put $t_0 = 0$, getting $x(s) = (1-s)x_0$, and forget the $t$ equation.

We now return to the general discussion. Note that the curves $s \mapsto \phi_s(t_0, x_0)$ of (4.39) are integral curves of the $s$-independent vector field

$$\mathbf{X} = \mathbf{v} + \frac{\partial}{\partial t}$$

> To discuss a time-dependent vector field $\mathbf{v}$ on $M^n$ we introduce the vector field $\mathbf{X} = \mathbf{v} + \partial/\partial t$ on $\mathbb{R} \times M^n$ and look at the flow on $\mathbb{R} \times M^n$ generated by this field. The path in $M^n$ traced out by a point that starts at $t = 0$ at $x_0$ consists of the projection into $M^n$ of the solution curve on $\mathbb{R} \times M^n$ starting at $(0, x_0)$.

We now recall an important **space–time** notation introduced in Section 3.5a. First note that in *any* manifold the operation of exterior differentiation

$$d(b_{\underset{\rightarrow}{I}} dx^I) = \left(\frac{\partial b_{\underset{\rightarrow}{I}}}{\partial x^j}\right) dx^j \wedge dx^I$$

can be written symbolically as $d = dx^j \wedge \partial/\partial x^j$; the operator $\partial/\partial x^j$ acts only on the coefficients. In a space–time $\mathbb{R} \times M^n$ with local coordinates $(t = x^0, \ldots, x^n)$ we have, for any form on $\mathbb{R} \times M^n$ (which may contain terms involving $dt$)

$$d\, b_{\underset{\rightarrow}{I}} dx^I = dt \wedge \left(\frac{\partial b_{\underset{\rightarrow}{I}}}{\partial t}\right) dx^I + dx^j \wedge \left(\frac{\partial b_{\underset{\rightarrow}{I}}}{\partial x^j}\right) dx^I$$

which we write symbolically as

$$d = dt \wedge \frac{\partial}{\partial t} + \mathbf{d} \tag{4.40}$$

where $\mathbf{d}$ is the **spatial** exterior derivative. We shall also write

$$X = \mathbf{v} + \frac{\partial}{\partial t} \tag{4.41}$$

using a boldfaced $\mathbf{v}$ to remind us that $\mathbf{v}$ is a spatial vector.

### 4.3c. Differentiating Integrals

Let $\phi_t : M^n \to M^n$ be a 1-parameter family of diffeomorphisms of $M$; we do **not** assume that they form a flow (i.e., they might not have the group property), but we do assume that $\phi_0$ is the identity and that $(t, \mathbf{x}) \to \phi_t\mathbf{x}$ is smooth as a function of $(t, \mathbf{x})$ on $\mathbb{R} \times M$. (In our previous example, $\phi_t\mathbf{x} = (1-t)\mathbf{x}$.)

Let $\alpha_t^p(\mathbf{x}) = \alpha^p(t, \mathbf{x})$ be a 1-parameter family of forms on $M$ and let $V^p$ be a $p$-dimensional submanifold of $M$. We wish to consider the $t$ derivative of $\int_{V(t)} \alpha$ where $V(t) = \phi_t V$.

$d\phi_t\mathbf{x}/dt$ is some $t$-dependent vector function $\mathbf{w}(t, \mathbf{x}) = \mathbf{w}(t, \phi_t^{-1}\phi_t\mathbf{x}) =: \mathbf{v}(t, \phi_t\mathbf{x})$ on $M$. This yields a time-dependent velocity field $d\mathbf{y}/dt = \mathbf{v}(t, \mathbf{y})$ on $M$. We consider this as a field on $\mathbb{R} \times M$ and we let $\alpha(t, \mathbf{x})$ be considered as a $p$-form on $\mathbb{R} \times M$ (with no $dt$ term).

Solving $d\mathbf{x}/ds = \mathbf{v}(t, \mathbf{x}), dt/ds = 1$ on $\mathbb{R} \times M$ (i.e., finding the integral curves of $X = \mathbf{v} + \partial/\partial t$) yields a flow $\Phi_s$ on $\mathbb{R} \times M$ and the curves $\phi_s(\mathbf{x})$ on $M$ are simply the projections of the curves $\Phi_s(0, \mathbf{x})$ on $\mathbb{R} \times M$. The 1-parameter family of submanifolds $V^p(s)$ of $M$ is the projection of the 1-parameter family $\Phi_s(0, V^p)$ of submanifolds of $\mathbb{R} \times M$.

**Theorem (4.42):** *Let $\phi_t : M^n \to M^n$ be a 1-parameter family of diffeomorphisms of $M$; we do not assume that they form a flow. Let $\alpha_t^p(\mathbf{x}) = \alpha^p(t, \mathbf{x})$ be a 1-parameter family of forms on $M$, let $V^p$ be a $p$-dimensional submanifold of $M$, and put $V(t) := \phi_t V$. Then*

$$\frac{d}{dt}\int_{V(t)} \alpha^p = \int_{V(t)} \frac{\partial \alpha}{\partial t} + i_{\mathbf{v}}\mathbf{d}\alpha + \mathbf{d}i_{\mathbf{v}}\alpha$$

*where $\mathbf{v}(t, \phi_t\mathbf{x}) = d\phi_t\mathbf{x}/dt$ is the $t$-dependent velocity field on $M$.*

PROOF: We again form $\mathbb{R} \times M^n$. $\alpha^p$ is now a $p$-form on $\mathbb{R} \times M^n$. $V^p(t)$ is now the projection of the submanifold $W(t) := \Phi_t(0, V)$ of $\mathbb{R} \times M^n$ that lies in the "spatial section" $\{t\} \times M^n$. Then $dt = 0$ when restricted to $W(t)$. The flow $\Phi_t$ on $\mathbb{R} \times M$ is generated by $X = \mathbf{v} + \partial/\partial t$. We then have, from (4.33),

$$\frac{d}{dt}\int_{V(t)} \alpha^p = \frac{d}{dt}\int_{W(t)} \alpha^p = \int_{W(t)} \mathcal{L}_X\alpha^p = \int_{W(t)} \mathcal{L}_{\mathbf{v}+\partial/\partial t}\alpha^p \tag{4.43}$$

We now write out (4.43) in the case at hand. Using (4.22) and $d = dt \wedge \partial/\partial t + \mathbf{d}$

$$\frac{d}{dt}\int_{W(t)} \alpha^p = \int_{W(t)} \mathcal{L}_{\mathbf{v}+\partial/\partial t}\alpha^p = \int_{W(t)} \mathcal{L}_{\mathbf{v}}\alpha^p + \frac{\partial \alpha}{\partial t}$$

$$= \int_{W(t)} \frac{\partial \alpha}{\partial t} + i_{\mathbf{v}}\mathbf{d}\alpha + \mathbf{d}i_{\mathbf{v}}\alpha$$

(since $\mathbf{v}$ does not involve $\partial/\partial t$ and $dt = 0$ on $W(t)$)

$$= \int_{V(t)} \frac{\partial \alpha}{\partial t} + i_{\mathbf{v}}\mathbf{d}\alpha + \mathbf{d}i_{\mathbf{v}}\alpha \quad \square$$

(Note that $\mathcal{L}_{\mathbf{v}}\alpha$ is the Lie derivative of $\alpha$ with respect to the vector field $\mathbf{v}$ "frozen" at time $t$, that is, we look at both $\alpha$ and $\mathbf{v}$ as fields fixed forever at time $t$!)

**Corollary (4.44):**

$$\frac{\partial}{\partial t}\phi_t^*\alpha = \phi_t^*\left\{\frac{\partial\alpha}{\partial t} + i_{\mathbf{v}}\mathbf{d}\alpha + \mathbf{d}i_{\mathbf{v}}\alpha\right\}$$
$$= \phi_t^*\left\{\frac{\partial\alpha}{\partial t} + \mathcal{L}_{\mathbf{v}}\alpha\right\}$$

This follows from $d/dt\int_V \phi_t^*\alpha^p = \int_V \phi_t^*\{\partial\alpha/\partial t + i_{\mathbf{v}}\mathbf{d}\alpha + \mathbf{d}i_{\mathbf{v}}\alpha\}$ with $V$ arbitrary.

## Problems

Let **A** and **B** be time-dependent vector fields on $\mathbb{R}^3$ and let $\rho(t, \mathbf{x})$ be a function. Show that (4.43) yields the following classical expressions for the time derivatives of line, surface, and volume integrals over moving domains.

**4.3(1)** $d/dt\int_C \mathbf{A}\cdot\mathbf{dx} = \int_C[\partial\mathbf{A}/\partial t - \mathbf{v}\times \text{curl}\,\mathbf{A} + \text{grad}(\mathbf{v}\cdot\mathbf{A})]\cdot d\mathbf{x}$

**4.3(2)** $d/dt\iint_S \mathbf{B}\cdot d\mathbf{S} = \iint_S[\partial\mathbf{B}/\partial t + (\text{div}\,\mathbf{B})\mathbf{v} - \text{curl}(\mathbf{v}\times\mathbf{B})]\cdot d\mathbf{S}$

**4.3(3)** $d/dt\iiint_U \rho\,\text{vol}^3 = \iiint_U[\partial\rho/\partial t + \text{div}(\rho\mathbf{v})]\,\text{vol}^3$

**4.3(4)** Show Faraday's law says $d/dt\iint_S \mathbf{B}\cdot d\mathbf{S} = -\oint_{\partial S}[\mathbf{E} + \mathbf{v}\times\mathbf{B}]\cdot d\mathbf{x}$ for a moving surface. $\mathbf{E} + \mathbf{v}\times\mathbf{B}$ is the **electromotive force.**

## Additional Problems on Fluid Flow

Consider a fluid flow in $\mathbb{R}^3$ with density $\rho(t, \mathbf{x})$ and velocity vector $\mathbf{v}(t, \mathbf{x})$. Problem 4.3(3) says conservation of mass is equivalent to

$$\frac{\partial\rho}{\partial t} + \text{div}(\rho\mathbf{v}) = 0$$

or

$$\mathcal{L}_X(\rho\,\text{vol}^3) = 0$$

These two expressions are equivalent since $i_X(\rho\beta^p) = i_{\rho X}\beta^p$.

In this section we shall use cartesian coordinates, but we shall still make an attempt to use the correct "variance" of the tensors involved.

Consider the linear momentum of a small region $U$. If $\nu$ is the velocity *covector*, $\nu = v_i dx^i$, the density of momentum is $\rho\nu$. *In* $\mathbb{R}^3$ *with cartesian coordinates* we attribute physical significance to the individual components of the momentum **P** of the moving region

$$P_i = \int_U v_i\rho\,\text{vol}^3$$

Since $\mathcal{L}_X(\rho\,\text{vol}^3) = 0$, we get ($v_i$ being a *function*)

$$\frac{dP_i}{dt} = \int_U \mathcal{L}_X(v_i\rho\,\text{vol}^3) = \int_U X(v_i)\rho\,\text{vol}^3 = \int_U\left[\mathbf{v} + \frac{\partial}{\partial t}\right](v_i)\rho\,\text{vol}^3$$
$$= \int_U\left[\frac{\partial v_i}{\partial t} + v^j\left(\frac{\partial v_i}{\partial x^j}\right)\right]\rho\,\text{vol}^3$$

$d\mathbf{P}/dt$ must equal the total force acting on $U$. (Newton's second law applies to particle mechanics. The generalization to continuum mechanics is due to Euler; see [T,T, footnote, p. 531].) Under the assumption of a "perfect" fluid, this consists of a body force (e.g., gravity) with mass density $\mathbf{f}$, and the pressure forces arising from the part of the fluid outside $U$. This latter is a vector integral $\mathbf{w} = -\int_{\partial U} p\mathbf{N}dS$. *Vector integrals make no sense on general manifolds* (how could we add two vectors located at different points?) but they can be defined in cartesian coordinates componentwise, that is, by putting $w^i = -\int_{\partial U} pN^i dS$. If the surface has local coordinates $u, v$, then, as in (3.14), $dS = \sqrt{g}du \wedge dv = \| \mathbf{n} \| du \wedge dv$. Thus $N^i dS = n^i du \wedge dv$. For example, from Problem 3.1(2) we have that $N^1 dS = \partial(y,z)/\partial(u,v) du \wedge dv = dy \wedge dz$. Thus in cartesian coordinates we may consider the symbolic **vector 2-form** $d\mathbf{S}$ with "components"

$$d\mathbf{S} = \mathbf{N}dS = (dy \wedge dz \qquad dz \wedge dx \qquad dx \wedge dy)^T$$

and then we could write $-\int_{\partial U} p\mathbf{N}dS = -\int_{\partial U} p\,d\mathbf{S}$. The first component of $\int_{\partial U} p\,d\mathbf{S}$ is

$$\int_{\partial U} p\,dy \wedge dz = \int_U dp \wedge dy \wedge dz = \int_U p_x dx \wedge dy \wedge dz$$

and likewise for the other components. Thus

$$\int_{\partial U} p\,d\mathbf{S} = \int_U \text{grad}\, p \text{vol}^3 \tag{4.45}$$

We conclude from Euler's version of the second law, applied to the arbitrarily small $U$

$$\frac{\partial v_i}{\partial t} + v^j\left(\frac{\partial v_i}{\partial x^j}\right) = -\left(\frac{1}{\rho}\right)\frac{\partial p}{\partial x^i} + f_i \tag{4.46}$$

where $f$ is the force density (per unit *mass)*. These are **Euler's equations**.

**4.3(5)** Assume that the body force density is derivable from a potential $\mathbf{f} = \text{grad}\,\phi$. Assume that the pressure is functionally related to the density, $p = p(\rho)$. (This is an "equation of state.") Then let $G(\rho)$ be a specific antiderivative of $dp/\rho$; we write this *symbolically* as $G(\rho) = \int dp/\rho = \int \rho^{-1}(dp/d\rho)d\rho$. Then $\partial G/\partial x^i = G'(\rho)\partial\rho/\partial x^i = \rho^{-1}(dp/d\rho)\partial\rho/\partial x^i = \rho^{-1}\partial p/\partial x^i$.

**(i)** Show that Euler's equations can then be written

$$\frac{\partial \nu}{\partial t} + \mathcal{L}_{\mathbf{v}}(\nu) = \mathbf{d}\left\{\frac{1}{2}\|\mathbf{v}\|^2 + \phi - \int \frac{dp}{\rho}\right\}$$

or (4.47)

$$\mathcal{L}_{\mathbf{v}+\partial/\partial t}(\nu) = \mathbf{d}\left\{\frac{1}{2}\|\mathbf{v}\|^2 + \phi - \int \frac{dp}{\rho}\right\}$$

where now $\mathcal{L}_{\mathbf{v}}(\nu)$ *is the Lie derivative of the 1-form* $\nu$ (we are no longer taking the Lie derivative of a function). Note that (4.47) *makes sense in any Riemannian manifold,* unlike (4.46) where $v^j(\partial v_i/\partial x^j)$ are not the components of a covector.

**(ii)** Conclude with Lord Kelvin that if $C(t)$ is a *closed* curve that follows the motion of the fluid, then the **circulation** $\oint_{C(t)} \nu$ *is constant in time.*

A *time-dependent* form $\alpha^p$ on $M^n$ is said to be **invariant under the flow** of the time-dependent vector field $\mathbf{v}$ provided

$$\mathcal{L}_{\mathbf{v}+\partial/\partial t}(\alpha) = \frac{\partial \alpha}{\partial t} + \mathcal{L}_{\mathbf{v}}\alpha = \frac{\partial \alpha}{\partial t} + i_{\mathbf{v}}\mathbf{d}\alpha + \mathbf{d}i_{\mathbf{v}}\alpha = 0$$

**(iii)** The **vorticity** 2-form for a flow in $\mathbb{R}^3$ is defined by

$$\omega^2 := \mathbf{d}\nu$$

Show (using $\mathbf{d} \circ \partial/\partial t = \partial/\partial t \circ \mathbf{d}$) that for a perfect fluid with $p = p(\rho)$ that the *vorticity form* $\omega^2$ *is invariant under the flow* (Helmholtz).

**(iv)** **Warning:** The vorticity *vector* $\omega = \text{curl}\,\mathbf{v}$, defined as usual by $\omega^2 = i_\omega \text{vol}^3$, is *not* usually invariant since the flow need not conserve the volume form. The mass form, $\rho\,\text{vol}^3$, however, *is* conserved. From $\omega = i(\omega/\rho)\rho\,\text{vol}^3$ we see that the vector $\omega/\rho$ should be invariant; that is, $\mathcal{L}_{\mathbf{v}+\partial/\partial t}(\omega/\rho) = 0$. *Show that this follows from* (4.24). Note that the *direction* of $\omega$ is invariant under the flow; physicists say that the "lines of $\omega$ " are "frozen" into the fluid.

**(v)** Let $V^3(t)$ be a compact region moving with the fluid. Assume that *at* $t = 0$ the vorticity 2-form $\omega^2$ vanishes when restricted to the boundary $\partial V^3(0)$; that is, $i^*\omega^2 = 0$, where $i$ is the inclusion of $\partial V$ in $\mathbb{R}^3$. (This does *not* say that $\omega^2$ itself vanishes, rather only that $\omega(\mathbf{u}, \mathbf{w}) = 0$ for $\mathbf{u}, \mathbf{w}$ tangent to $\partial V^3(0)$.) Show that the **helicity** integral

$$\int_{V(t)} \mathbf{v} \bullet \omega dx \wedge dy \wedge dz$$

is constant in time.

**4.3(6)** *Magnetohydrodynamics.* Define a **perfectly conducting fluid** as one with vanishing "electromotive intensity" $\mathcal{E}^1 - i_{\mathbf{v}}\mathcal{B}^2 = 0$ (otherwise there would be an infinite current flow).

**(i)** Show that $\mathcal{B}^2$ *is invariant under the flow*, $\mathcal{L}_{\mathbf{v}+\partial/\partial t}\mathcal{B}^2 = 0$ (and thus the lines of $\mathbf{B}$ are frozen into the fluid).

We are concerned with the case when the charge density $\sigma$ vanishes. Then the Lorentz force density (per unit *volume*) on the fluid is $-i_{\mathbf{J}}\mathcal{B}^2$ and so the external force density (per unit *mass*) is $f = -i_{\mathbf{J}}\mathcal{B}^2/\rho$. This is *not* derivable from a potential, and so Euler's equations become

$$\frac{\partial \nu}{\partial t} + \mathcal{L}_{\mathbf{v}}(\nu) = \mathbf{d}\left\{\frac{\|\mathbf{v}\|^2}{2} - \int \frac{dp}{\rho}\right\} - \frac{i_{\mathbf{J}}\mathcal{B}^2}{\rho}$$

**(ii)** Consider then a blob $U$ of perfectly conducting fluid with (moving) boundary $\partial U$ (e.g., the interface between the fluid and vacuum). Frequently one takes as boundary condition that $\mathcal{B}^2$ restricted to the boundary vanishes (i.e., $B_n = 0$). Show then that

$$\frac{d}{dt}\int_U \nu \wedge \mathcal{B} = 0$$

This result is due to Woltjer. See and compare with Moffat's treatment in [Mo].

---

## 4.4. A Problem Set on Hamiltonian Mechanics

Why phase space?

In Section 10.2 we shall talk about Lagrangian (i.e., tangent bundle) mechanics from first principles. In the present section we shall simply assume Lagrange's equations, and proceed to the Hamiltonian formulation in phase space. The following problems involve much of the machinery of forms and Lie derivatives that we have developed, and should be worked by the readers even if Hamiltonian mechanics is not their primary interest.

Let $M^n$ be the *configuration space* of a mechanical system; $M$ has local coordinates $q^1, \ldots, q^n$. The *phase space* is the cotangent bundle $T^*M$ with local coordinates $q^1, \ldots, q^n, p_1, \ldots, p_n$. Introduce the notation

$$x^i = q^i, \qquad x^{n+i} = p_i, \qquad i = 1, \ldots, n$$

On $T^*M$ we have the Poincaré 1-form (see 2.3d)

$$\lambda = p_i dq^i$$

and the resulting Poincaré 2-form

$$\omega^2 := d\lambda = dp_i \wedge dq^i$$

**Warning:** Many books call this form $-\omega^2$!

**Definition:** A 2-form $\omega^2$ on an even dimensional manifold $M^{2n}$ is called **symplectic** (and then $M$ is called a **symplectic manifold**) provided it satisfies

(i) $d\omega = 0$

(ii) $\omega$ is nondegenerate that is, the linear transformation associating to a vector $\mathbf{X}$ the 1-form $i_{\mathbf{X}}\omega^2$ is nonsingular. In local coordinates $x$, since $[i_{\mathbf{X}}\omega]_j = X^i \omega_{ij}$, this merely says det $(\omega_{ij}) \neq 0$.

As we shall see, *every cotangent bundle is a symplectic manifold.*

If $M^2$ is an orientable Riemannian *surface*, then an area 2-form $\text{vol}^2 = \omega^2$ is a symplectic form! The plane $\mathbb{R}^2 = \mathbb{R} \times \mathbb{R}$ and the cylinder $S^1 \times \mathbb{R}$ are the cotangent bundles, respectively, of the line $\mathbb{R}$ and the circle $S^1$. Closed (compact) orientable *surfaces* are symplectic but are never cotangent bundles since the vector space fibers of a cotangent bundle are never compact.

(Note that we demand that $\omega$ be a true form, not a pseudoform. On an orientable manifold, a pseudoform defines a true form by using a coordinate cover with positive Jacobians in each overlap.)

**Warning:** A symplectic form $\omega^2$ allows us to associate to each contravariant vector $\mathbf{X}$ a covariant vector $i_{\mathbf{X}}\omega$ with components $X^i \omega_{ij}$, and in this sense is similar to a Riemannian metric. *This similarity is very misleading* since the matrix $\omega$ is skew symmetric rather than symmetric. The remark $(i_{\mathbf{X}}\omega)(\mathbf{X}) = \omega(\mathbf{X}, \mathbf{X}) = 0$ shows in fact that in any Riemannian metric that one imposes on a symplectic manifold, *the contravariant version of $i_{\mathbf{X}}\omega$ is orthogonal to* $\mathbf{X}$!

**4.4(1)** Show that the Poincaré 2-form is symplectic. (You need only show that the 1-forms $i_{\partial/\partial x^i}\omega$ are linearly independent.)

**4.4(2)** Show that $\omega^n := \omega \wedge \ldots \wedge \omega = \pm n! dq^1 \wedge \ldots \wedge dq^n \wedge dp_1 \wedge \ldots dp_n$

Since $\omega$ is a well-defined 2-form on any cotangent bundle, this $2n$-form is actually independent of the local coordinates $q$ used on $M^n$. We call $\omega^n$ the Liouville or symplectic volume form for the phase space.

**4.4(3)** Clearly $\omega^n$ never vanishes. Show why this implies that $T^*M$ *is always orientable*, whether or not $M$ itself is orientable.

Since phase space is orientable we *need* not distinguish between forms and pseudoforms.

## 4.4a. Time-Independent Hamiltonians

Let $L = L(q, \dot{q})$ be a time-independent **Lagrangian**, a function on the tangent bundle. We have a map (see 2.3c) $P : TM \to T^*M$ given by $q^i = q^i$ and

$$p_i = \frac{\partial L}{\partial \dot{q}^i}$$

For our purposes we shall insist that this map is a diffeomorphism. Locally this means the following. Since for the pull-back

$$P^* dp_i = \left(\frac{\partial^2 L}{\partial \dot{q}^j \partial \dot{q}^i}\right) d\dot{q}^j + \left(\frac{\partial^2 L}{\partial q^j \partial \dot{q}^i}\right) dq^j$$

we have, from (2.51),

$$\begin{aligned} &P^*(dq^1 \wedge \ldots \wedge dq^n \wedge dp_1 \wedge \ldots \wedge dp_n) \\ &= \det\left(\frac{\partial^2 L}{\partial \dot{q}^j \partial \dot{q}^i}\right) dq^1 \wedge \ldots \wedge dq^n \wedge d\dot{q}^1 \wedge \ldots \wedge d\dot{q}^n \end{aligned}$$

Locally then, we have a diffeomorphism if the Lagrangian is "regular," that is, $\det(\partial^2 L / \partial \dot{q}^j \partial \dot{q}^i) \neq 0$.

**Lagrange's** equations, $\partial L/\partial q^i - d/dt(\partial L/\partial \dot{q}^i) = 0$ *in* $TM$, translate to **Hamilton's** equations in the phase space $T^*M$

$$\frac{dq^i}{dt} = \frac{\partial H}{\partial p_i} \qquad \frac{dp_i}{dt} = -\frac{\partial H}{\partial q^i} \tag{4.48}$$

where the **Hamiltonian** function is defined by

$$H(q, p) := p_i \dot{q}^i - L(q, \dot{q}) \tag{4.49}$$

It is assumed in this expression that $\dot{q}$ is expressed in terms of $q$ and $p$ by means of the inverse $T^*M \to TM$. For a proof one proceeds as follows, with an obvious notation. $dH = H_q dq + H_p dp$. But from (4.49) $dH = p d\dot{q} + \dot{q} dp - L_q dq - p d\dot{q}$. From Lagrange's equations, $L_q = dp/dt$. Comparing the two expressions for $dH$ yields Hamilton's equations. (The same proof works also when $L$ and $H$ are time dependent.) □

Let $\mathbf{X}$ be a time-independent vector field on $T^*M$,

$$\mathbf{X} = X^i \frac{\partial}{\partial q^i} + X^{i+n} \frac{\partial}{\partial p_i}$$

**4.4(4)** Show that the integral curves of **X**, that is, the solutions to

$$\frac{dq^i}{dt} = X^i \quad \text{and} \quad \frac{dp_i}{dt} = X^{i+n}$$

satisfy Hamilton's equations if and only if the vector field **X** satisfies

$$i_{\mathbf{X}}\omega = -dH \tag{4.50}$$

We shall refer to (4.50) again as Hamilton's equations and **X** will be called a **Hamiltonian vector field**. The flow $\phi_t : T^*M \to T^*M$ generated by **X** will be called a **Hamiltonian flow**.

**4.4(5)** Show that if **X** is Hamiltonian then

$$\mathcal{L}_{\mathbf{X}}\omega = 0 = \mathcal{L}_{\mathbf{X}}\omega^n \tag{4.51}$$

The right-hand side shows that *volumes in phase space are invariant under a Hamiltonian flow*; this is *Liouville's* theorem.

Under this time-independent Hamiltonian flow, $H$ is a *constant of the motion*, that is,

$$\frac{dH}{dt} = \mathbf{X}(H) = i_{\mathbf{X}}dH = i_{\mathbf{X}}(i_{-\mathbf{X}}\omega) = 0$$

This is merely a fancy way of saying

$$\frac{dH}{dt} = \left(\frac{\partial H}{\partial q^i}\right)\frac{dq^i}{dt} + \left(\frac{\partial H}{\partial p_i}\right)\frac{dp_i}{dt} = 0$$

from (4.48). $H$ is also called the **total energy**.

Look now at the "level sets" of the function $H$ in $T^*M$

$$V_E^{2n-1} := \{x = (q, p) \in T^*M \mid H(q, p) = E\}$$

If $dH \neq 0$ on $V_E$, then we know that $V_E$ is a $(2n - 1)$ dimensional submanifold of $T^*M$; it is called the **hypersurface of constant energy** $E$. By Sard's theorem of 1.3d, we know that for almost all $E$, $E$ is a regular value. In the following we shall assume that $V_E$ is a hypersurface of constant energy with $dH \neq 0$.

Since $dH/dt = 0$ along the flow lines of **X**, we conclude that **X** *is tangent to* $V_E$.

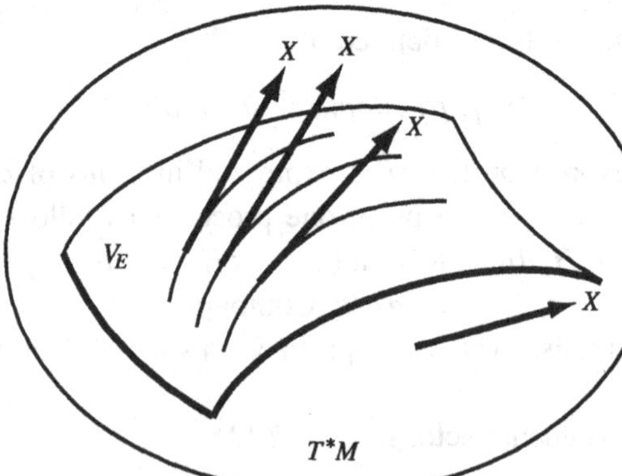


Figure 4.6

We know that $\mathcal{L}_{\mathbf{X}}\omega = d/dt[\phi_t^*\omega]_{t=0} = 0$. Then, for small $t$

$$\frac{d}{dt}[\phi_t^*\omega]_t = \lim_{h\to 0} h^{-1}[\phi_{t+h}^*\omega - \phi_t^*\omega]$$
$$= \phi_t^* \lim_{h\to 0} h^{-1}[\phi_h^*\omega - \omega]$$

that is,

$$\frac{d}{dt}[\phi_t^*\omega] = \phi_t^*\mathcal{L}_{\mathbf{X}}\omega$$

(and this is true for any form, any vector field). This also follows directly from Corollary (4.44). In our case then $\phi_t^*\omega_{x(t)} = \omega_{x(0)}$, and so

$$\phi_t^*\omega = \omega \tag{4.52}$$

holds for all small $t$ in any Hamiltonian flow.

**Definition:** A map $\phi : M \to M$ of a symplectic manifold is **canonical** if $\phi$ preserves $\omega$, that is, $\phi^*\omega = \omega$.

Thus

A Hamiltonian vector field $\mathbf{X}$ generates a local 1-parameter group of canonical transformations of phase space.

Since $\mathbf{X}$ is tangent to $V_E$, the integral curves of $\mathbf{X}$ that start on $V_E$ remain on $V_E$. Consequently

$$\phi_t : V_E \to V_E$$

We know that $\phi_t$ preserves Liouville volume on $T^*M$. We claim that *there is a* $(2n-1)$*-form* $\tau = \tau_V$ *on* $V_E$ *that is nonzero and is also invariant under* $\phi_t$! We see this as follows.

$dH \neq 0$ on $V_E$, and so $dH \neq 0$ in some $T^*M$ neighborhood of $x \in V_E$. We shall first construct a form $\sigma^{2n-1}$ in a neighborhood of $x$ so that

$$\omega^n = dH \wedge \sigma^{2n-1} \tag{4.53}$$

Since $dH \neq 0$, some $\partial H/\partial x^i \neq 0$. For simplicity we shall assume $\partial H/\partial q^1 \neq 0$. Introduce a local change of coordinates $y^1 = H$, $y^i = x^i$ for $i > 1$. Then

$$dH \wedge dq^2 \wedge \ldots \wedge dq^n \wedge dp_1 \wedge \ldots \wedge dp_n$$
$$= \left(\frac{\partial H}{\partial q^1}\right) dq^1 \wedge dq^2 \wedge \ldots \wedge dq^n \wedge dp_1 \wedge \ldots \wedge dp_n \neq 0$$

shows that this *is* an admissible change of coordinates. Put then

$$\sigma^{2n-1} = \left(\frac{\partial H}{\partial q^1}\right)^{-1} dq^2 \wedge \ldots \wedge dq^n \wedge dp_1 \wedge \ldots \wedge dp_n \tag{4.54}$$

Multipying by $\pm n!$ we shall get the desired form $\sigma$. *Since we are not concerned at all with this factor $\pm n!$ we shall simply omit all mention of it.*

The form $\sigma$ so constructed is a form on $T^*M$ defined near $x \in V_E$. Its construction was highly arbitrary. In an overlap of coordinate patches for $T^*M$ there is no hope for agreement. Problem **4.4(6)** shows, however, that this defect is not serious.

**4.4(6)** Let $i : V_E \to T^*M$ be the inclusion map. Let $\sigma^{2n-1}$ be any form satisfying (4.53). Show that the *restriction* (pull-back)

$$\tau^{2n-1} := i^*\sigma^{2n-1} \tag{4.55}$$

of $\sigma$ to $V_E$ is *independent of the choice of* $\sigma$. (Hint: Let $\sigma'$ be another choice. Show $i^*\sigma = i^*\sigma'$ by evaluating $dH \wedge (\sigma - \sigma')$ on a $2n$-tuple of vectors $(\mathbf{N}, \mathbf{T}_2, \dots, \mathbf{T}_{2n})$ where $\mathbf{N}$ is transverse to $V_E$ and the $\mathbf{T}$'s are tangent to $V_E$.)

To show that $\tau$ is invariant under the flow generated by $\mathbf{X}$ on $V_E$, we need only show that $\tau(\mathbf{T}_2, \dots, \mathbf{T}_{2n})$ is constant when the $\mathbf{T}$'s are tangent vectors to $V_E$ that are *invariant* under the flow. Let $\mathbf{N}$ be an invariant vector field that is transverse to $V_E$. Let $\mathbf{T}$ denote the $(2n-1)$-tuple $(\mathbf{T}_2, \dots, \mathbf{T}_{2n})$. Then $\omega(\mathbf{N}, \mathbf{T})$ is constant under the flow and so $(dH \wedge \sigma)(\mathbf{N}, \mathbf{T}) = dH(\mathbf{N})\sigma(\mathbf{T}) = dH(\mathbf{N})\tau(\mathbf{T})$ is constant. Since $H$ is invariant, $\mathcal{L}_{\mathbf{X}} H = \mathbf{X}(H) = 0, dH$ is also invariant. Thus $\tau(\mathbf{T}) =$ constant, as desired.

We now write down an expression for $\tau^{2n-1}$ that is found in books on statistical mechanics. In a coordinate patch $(q, p)$ of $T^*M$ near $x \in V_E$ we consider *any* Riemannian metric whose volume form is $\omega^n$ (modulo $\pm n!$). For example we can choose $ds^2 = \sum\{(dq^i)^2 + (dp_i)^2\}$; since $\sqrt{g} = 1$ we have

$$\mathrm{vol}^{2n} = dq^1 \wedge \dots \wedge dq^n \wedge dp_1 \wedge \dots dp_n$$

Of course these local *metrics* do not agree on overlaps, but from Problem 4.4(6) *our final result will be independent of such choices.* In any Riemannian metric, grad $H = \nabla H$ is normal to the level sets $H =$ constant, and so $\nabla H / \parallel \nabla H \parallel$ is a unit normal field to these submanifolds. Then the $(2n-1)$ forms $dS_V^{2n-1} = i_{\nabla H/\|\nabla H\|}\omega^n$ on $T^*M$ have the property that they restrict to the $(2n-1)$ area forms on each $H =$ constant. Whereas $dH$ is an invariant 1-form, the unit normal $\nabla H / \parallel \nabla H \parallel$ is not invariant since the metric $ds^2$ is not invariant (why should it be?). We claim, however, that the restriction $\tau^{2n-1}$ of

$$\sigma^{2n-1} := \parallel \nabla H \parallel^{-1} dS^{2n-1} = i_{\nabla_H/\|\nabla_H\|^2}\omega^n \tag{4.56}$$

to $V_E$ *is an invariant form for* $V_E$.

**4.4(7)** Show this. (Evaluate $dH \wedge \sigma$ on $(\nabla H / \parallel \nabla H \parallel, \mathbf{T})$, for $\mathbf{T}$ orthonormal and tangent to $V_E$.)

The expression (4.56) can be "understood" heuristically as follows.

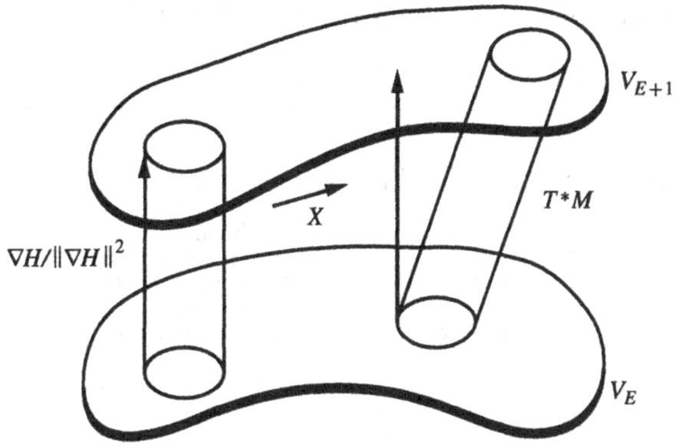


**Figure 4.7**

To flow from the level set $V_E$ to $V_{E+1}$ along the gradient lines of $H$, in 1 second, we solve the differential equations $dx/dt = \nabla H/ \parallel \nabla H \parallel^2$; see 2.1e. The right-hand side is a vector field of length $\parallel \nabla H \parallel^{-1}$. The region between these level sets is invariant under the Hamiltonian flow. A cylinder of gradient lines will have base area $\Delta S^{2n-1}$ and altitude $\parallel \nabla H \parallel^{-1}$. This will be sent by the Hamiltonian flow into an oblique cylinder of the same volume. Thus $\Delta S^{2n-1} \parallel \nabla H \parallel^{-1}$ is constant under the Hamiltonian flow, as required.

## 4.4b. Time-Dependent Hamiltonians and Hamilton's Principle

When $H = H(q, p, t)$ depends explicitly on time we consider $H$ as a function on the **extended phase space** $T^*M \times \mathbb{R}$. It is sometimes convenient to call the coordinates

$$q^i = x^i, \qquad p_i = x^{i+n}, \qquad t = x^{2n+1}$$

Hamilton's equations are still (4.48) but note now that

$$\frac{dH}{dt} = \left(\frac{\partial H}{\partial q^i}\right)\frac{dq^i}{dt} + \left(\frac{\partial H}{\partial p_i}\right)\frac{dp_i}{dt} + \frac{\partial H}{\partial t} = \frac{\partial H}{\partial t}$$

and $H$ is no longer a constant of the motion. Introduce new **Poincaré forms** on $T^*M \times \mathbb{R}$ (for interpretation see section 16.4b) by

$$\Lambda^1 = p_i dq^i - H dt \tag{4.57}$$

and

$$\Omega^2 = d\Lambda = dp_i \wedge dq^i - dH \wedge dt \tag{4.58}$$

where now $df = (\partial f/\partial q^i)dq^i + (\partial f/\partial p_i)dp_i + (\partial f/\partial t)dt$, and so on.

Consider a vector field on $T^*M \times \mathbb{R}$ of the type

$$X = X^i \frac{\partial}{\partial q^i} + X^{i+n}\frac{\partial}{\partial p_i} + \frac{\partial}{\partial t}$$

and thus along the integral curves of $X$ we have

$$X = \left(\frac{dq^i}{dt}\right)\frac{\partial}{\partial q^i} + \left(\frac{dp_i}{dt}\right)\frac{\partial}{\partial p_i} + \frac{\partial}{\partial t}$$

**4.4(8)** Show that Hamilton's equations together with $dH/dt = \partial H/\partial t$ are equivalent to

$$i_X \Omega = 0 \tag{4.59}$$

Such an $X$ will again be called a Hamiltonian vector field. It is

$$X = \left(\frac{\partial H}{\partial p_i}\right)\frac{\partial}{\partial q^i} - \left(\frac{\partial H}{\partial q^i}\right)\frac{\partial}{\partial p_i} + \frac{\partial}{\partial t} \tag{4.60}$$

Let $\phi_X : T^*M \times \mathbb{R} \to T^*M \times \mathbb{R}$ be the Hamiltonian flow generated by the field $X$ given by (4.60).

**4.4(9)** Show that

$$\mathfrak{L}_X \Omega = 0 \tag{4.61}$$

for $X$ Hamiltonian.

**4.4(10)** Let $C$ be a closed curve in $T^*M \times \mathbb{R}$. ($C$ need *not* be the boundary of any surface.)

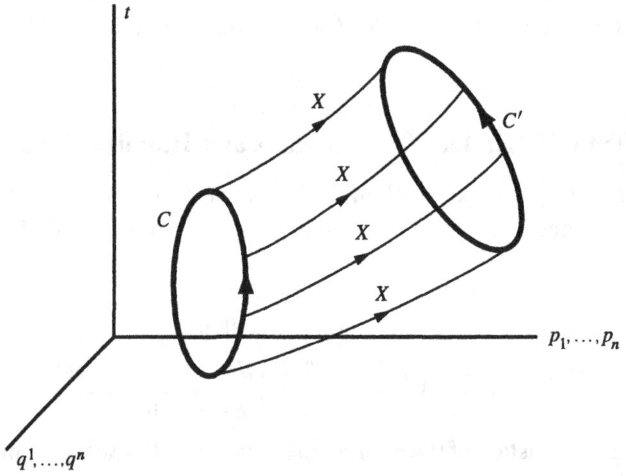


**Figure 4.8**

Let $C'$, as shown, be another closed curve that meets each orbit through $C$ once and only once (it *need not* be the push-forward of $C$). Show that

$$\oint_C p_i dq^i - H dt = \oint_{C'} p_i dq^i - H dt \tag{4.62}$$

(Hint: Look at the indicated surface with boundary swept out by the orbits through $C$)

**Definition:** Let $C$ be *any* oriented compact curve in $T^*M \times \mathbb{R}$. The **action** associated to $C$ is the line integral

$$S(C) = \int_C \Lambda = \int_C p_i dq^i - H dt \tag{4.63}$$

**Remark:** As all physics students know, and as we shall see in Section 10.2, Lagrange's equations result from Hamilton's principle, namely that the first variation of the "action" $\int_C L(q, \dot{q}, t)dt$ vanishes for the actual dynamical path $q = q(t)$ in configuration space. This integral should be thought of as being the integral of the Lagrangian function $L(q, \dot{q}, t)$ in $TM \times \mathbb{R}$ and where the curve $C$ in $TM \times \mathbb{R}$ is the **lift** of a curve $q = q(t)$ obtained by putting $\dot{q} = dq/dt$. Since we are restricting $\dot{q}$ to be $dq/dt$ in $TM \times \mathbb{R}$, $L(q, \dot{q}, t)dt$, though a 1-form on the lifted curve, is not to be considered a 1-form on $TM \times \mathbb{R}$. On the other hand, along this lifted curve we do have, from (4.49), $Ldt = (p\dot{q} - H)dt = pdq - Hdt$. This is the reason for calling the integral $\int pdq - Hdt$ the action integral in $T^*M \times \mathbb{R}$. We shall *not* restrict our curves in $T^*M \times \mathbb{R}$ to be lifted from $M$. Lagrange's equations are simply the Euler–Lagrange equations for $\int Ldt$, and we are now going to look at the result of putting the first variation of $\int pdq - Hdt$ equal to 0. It is not necessary to consider the Euler– Lagrange equations for this since $pdq - Hdt$ is a 1-form on $T^*M \times \mathbb{R}$ and *we already know how to differentiate integrals of forms* from (4.33). We proceed to the details.

Consider a curve $C_0 = C_0(u)$, $a \le u \le b$, in $T^*M \times \mathbb{R}$ parameterized by $u = t$ (in particular it is not a closed curve).

**Definition:** A **variation** of $C_0$ is a map $C$ of a rectangle in a $(u, \alpha)$ plane $\mathbb{R}^2$ into $T^*M \times \mathbb{R}$ such that $C(u, 0) = C_0(u)$.

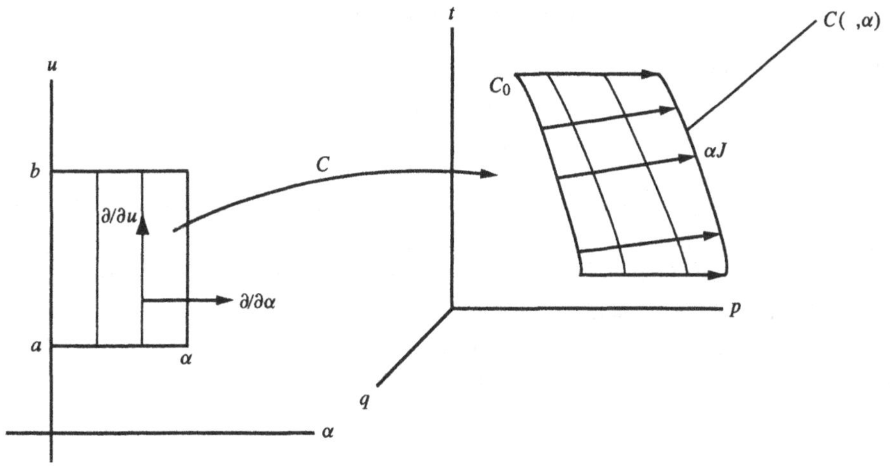


**Figure 4.9**

$u$ need not be $t$ when $\alpha \neq 0$. Denote the curves $u \to C(u, \alpha)$, $\alpha$ fixed, by $C_\alpha$. The vectors

$$C_*\left(\frac{\partial}{\partial u}\right) = \frac{\partial x(u, \alpha)}{\partial u}$$

are tangent to the varied curves and the vector field

$$C_*\left(\frac{\partial}{\partial \alpha}\right) = \frac{\partial x(u, \alpha)}{\partial \alpha}$$

at $\alpha = 0$ is called the **variation field**. We denote it by $J$.

We may compute the action along the varied curve $C_\alpha$; call it $S(\alpha)$. Suppose now *we restrict ourselves to variations that change neither $q$ nor time $t$ at the endpoints* (as indicated in our diagram). Thus $J$ has no $\partial/\partial q$ nor $\partial/\partial t$ component at $t = a$ and at $t = b$.

The **first variation of action** is by definition

$$S'(0) := \left[\frac{d}{d\alpha}\int_{C_\alpha} p_i dq^i - H dt\right]_{\alpha=0} = \int_{C_0} \mathcal{L}_{\partial x/\partial\alpha}\Lambda \tag{4.64}$$

**4.4(11)** Show that $S'(0) = \int_{C_0} i_J \Omega$.

**4.4(12)** Suppose that $S'(0) = 0$ for *all* such variation fields $J$. $C_0$ is parameterized by $t$. Show then that the tangent vector $T$ to $C_0$ must satisfy $i_T\Omega = 0$ and thus $C_0$ must be a solution to Hamilton's equations. This is **Hamilton's principle of stationary action** as formulated by Poincaré. (Hint: You may use the "fundamental lemma of the calculus of variations"; if $f$ is continuous and if $\int_a^b f(t)\alpha(t)dt = 0$ for all smooth functions $\alpha$ that vanish at $a$ and $b$, then $f(t) = 0$ for all $a \le t \le b$.) Classically one writes

$$\delta \int pdq - Hdt = 0$$

iff $C_0$ satisfies Hamilton's equations.

### 4.4c. Poisson brackets

Given a time-independent function $F$ on $T^*M$ we may associate a unique vector field $X_F$ by

$$dF = -i_{X_F}\omega$$

(when $F = H$ is the Hamiltonian, $X_F = X$ is the Hamiltonian vector field). This simply means that along the integral curves of $X_F$ we have $dq^i/dt = \partial F/\partial p_i$ and $dp_i/dt = -\partial F/\partial q^i$. Suppose that $G$, $X_G$ is another pair, $dG = -i_{X_G}\omega$. We define the **Poisson bracket** of the functions $F$ and $G$, written $(F, G)$, by taking the *derivative of $F$ as we move along the integral curves of $G$*, $(F, G) := X_G(F)$. In particular, the rate of change of a function $F$ along a Hamiltonian flow is

$$\frac{dF}{dt} = (F, H)$$

**4.4(13)** Show that $X_F$ generates canonical transformations, and

$$(F, G) = -\omega(X_F, X_G) = -(G, F)$$

and in coordinates

$$(F, G) = \sum_i \frac{\partial(F, G)}{\partial(q^i, p_i)}$$

**4.4(14)** Show, using Theorem (4.24), that $i_{[X_F, X_G]}\omega = d(F, G)$, and thus *the vector field associated to $(F, G)$ is $-[X_F, X_G]$*.

CHAPTER 5

# The Poincaré Lemma and Potentials

## 5.1. A More General Stokes's Theorem

We shall accept the following technical generalizations of results already proven.

Let $V^p$ be a compact oriented submanifold (perhaps with boundary) of $M^n$ and let $F : M^n \to W^m$ be a smooth map into a manifold $W^m$. The image $F(V)$ in $W$ need not be a submanifold. It might have self-intersections and all sorts of pathologies. Still, if $\beta^p$ is a form on $W$, it makes sense to talk of the integral of $\beta$ over $F(V)$ and in fact

$$\int_{F(V)} \beta^p = \int_V F^* \beta^p \tag{5.1}$$

which generalizes (3.17). In a sense, the right-hand side is the definition of the left-hand side. Then

$$\int_{F(V)} d\beta^{p-1} = \int_V F^* d\beta^{p-1} = \int_V dF^* \beta^{p-1}$$
$$= \int_{\partial V} F^* \beta^{p-1} = \int_{F(\partial V)} \beta^{p-1}$$

Then if we define $\partial F(V) = F(\partial V)$, we have the **generalized Stokes's theorem**

$$\int_{F(V)} d\beta^{p-1} = \int_{\partial F(V)} \beta^{p-1} \tag{5.2}$$

Actually one needs to integrate over manifolds with only "piecewise smooth" boundaries, such as a triangle, and also manifolds such as a solid cone. It is not easy to give a careful description of these objects. It is important that Stokes's theorem holds for very general objects, basically by approximating the object and its boundary by, say, manifolds with piecewise smooth boundaries ([A, M, R, box 7.2B]).

## 5.2. Closed Forms and Exact Forms

A form $\beta^p$ is **closed** if $d\beta = 0$. Thus

$$d\beta^0 = 0 \Leftrightarrow \beta^0 \quad \text{is a constant function}$$

$$d\beta^1 = 0 \Leftrightarrow (\partial_i b_j - \partial_j b_i) = 0 \quad \text{in } \mathbb{R}^3 \operatorname{curl} \mathbf{B} = 0$$

$$d\beta^2 = 0 \Leftrightarrow (\partial_i b_{jk} + \partial_j b_{ki} + \partial_k b_{ij}) = 0 \quad \text{in } \mathbb{R}^3 \operatorname{div} \mathbf{B} = 0$$

A form $\beta^p$ is **exact** if $\beta^p = d\alpha^{p-1}$, for some form $\alpha^{p-1}$.

The following observations are easy and important consequences of these definitions, $d^2 = 0$, and Stokes's theorem.

1. Every exact form is closed.
2. The product of two closed forms is closed.
3. The product of a closed form and an exact form is exact. (You are asked to prove this in Problem 5(1).)
4. The integral of an *exact* form over an orientable *closed* manifold (i.e., compact without boundary) is 0.
5. The integral of a *closed* form over the *boundary* of an oriented compact manifold is 0.

Although every exact form is closed, $\beta = d\alpha \Rightarrow d\beta = d^2\alpha = 0$, it is not true that every closed form is exact. A most important example is given by the 1-form

$$\beta^1 = (x^2 + y^2)^{-1}(xdy - ydx)$$

in $\mathbb{R}^2$. First note that this form is not defined in all of $\mathbb{R}^2$; certainly we must omit the origin. Thus the manifold in question is $\mathbb{R}^2 - 0$. One easily checks directly that $\beta^1$ is closed but it is easier to note that

$$\beta^1 = d \text{ “}\arctan\left(\frac{y}{x}\right)\text{”} = d\text{“}\theta\text{”}$$

This makes it seem as though $\beta$ is in fact exact, but this is not so; the 0-form “$\theta$” is not a *single-valued* function, and that is why we have introduced the quotation marks! It *is* single-valued if one introduces a “branch cut,” say the positive $x$ axis. Thus $\beta^1$ is exact on the portion $\mathbb{R}^2$–(positive $x$ axis). In particular $\beta$ is closed here. Clearly by choosing a different branch cut we can see that $d\beta^1 = 0$ on all of $\mathbb{R}^2 - 0$. But $\beta^1$ cannot be exact on all of $\mathbb{R}^2 - 0$, for if we consider the closed curve $C = x^2 + y^2 = 1$, oriented counterclockwise, then (dropping “ ”)

$$\oint_C \beta^1 = \oint_C d\theta = 2\pi$$

and then observation 4 shows that $\beta^1$ is not exact. Note that there is no contradiction with observation 5 since the circle $C$ is *not* the boundary of any *compact* surface in $\mathbb{R}^2 - 0$. It is true that $C = \partial$ (unit disc) in $\mathbb{R}^2$ but the unit disc has had its origin removed in $\mathbb{R}^2 - 0$. Thus the crucial point is that $C$ *is a closed curve in* $\mathbb{R}^2 - 0$ but *it is not the boundary of a compact surface in* $\mathbb{R}^2 - 0$!

Let us say that a manifold $M^n$ has **first Betti number** 0, written $b_1 = 0$, if, basically, every *closed* oriented piecewise smooth curve $C$ is the boundary of some compact oriented "surface"; that is, there is some piecewise smooth oriented surface (with boundary) $V^2$ and a map $F: V^2 \to M^n$ such that $\partial F(V) = C$. This concept, and its higher dimensional analogues (to be discussed more thoroughly in Chapter 13) was first introduced by Riemann. (The Italian mathematician Betti was a close friend of Riemann's.)

**Theorem (5.3):** *Let* $M^n$ *be a manifold with first Betti number* 0. *Then every closed 1-form* $\beta^1$ *on* $M^n$ *is exact.*

PROOF: The proof is essentially found in every calculus book in the case $M^n = \mathbb{R}^3$. We give a proof that uses our previously developed machinery for differentiating integrals.

We wish to exhibit a function $f$ such that $df = \beta^1$. Let $x \in M$ and let $y$ be a fixed point in $M$. Fix an oriented curve $C(y, x)$ that starts at $y$ and ends at $x$ and define

$$f(x) := \int_{C(y,x)} \beta^1$$

We note first that *$f$ is in fact independent of the curve chosen* to join $y$ to $x$, for if $C'(y, x)$ is another, then $C - C'$, that is, $C$ followed by $C'$ with orientation reversed, is a closed oriented curve. By hypothesis there is an oriented compact surface $F(V)$ such that $\partial F(V) = C$ and so

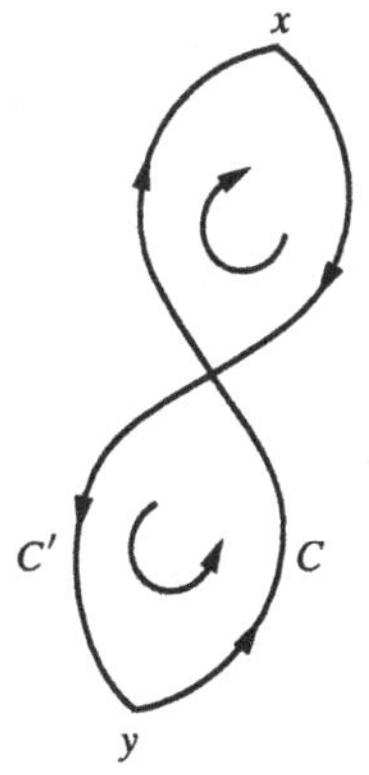


Figure 5.1

$$\int_C \beta - \int_{C'} \beta = \int_{C-C'} \beta = \oint_{\partial F(V)} \beta = \int_{F(V)} d\beta = 0$$

We can now compute $df$ at the variable point $x$. Let $\mathbf{v}_x$ be a vector at $x$. Take any vector field $\mathbf{v}$ that coincides with $\mathbf{v}_x$ at $x$, is defined in some neighborhood of the curve $C(y,x)$ and which vanishes at $y$. If $\phi_t$ is the flow generated by $\mathbf{v}$ then $\phi_t C(y,x)$ is a curve joining $y$ to $\phi_t x$, and we also have that $d\phi_t x/dt]_{t=0} = \mathbf{v}_x$. Then

$$df(\mathbf{v}) = \frac{d}{dt} f\{\phi_t x\}_{t=0} = \left[\frac{d}{dt}\int_{\phi_t C(y,x)} \beta\right]_{t=0}$$

$$= \int_{C(y,x)} \mathfrak{L}_{\mathbf{v}}\beta = \int_{C(y,x)} i_{\mathbf{v}} d\beta + d i_{\mathbf{v}}\beta = \int_{C(y,x)} d i_{\mathbf{v}}\beta$$

$$= i_{\mathbf{v}}\beta_x - i_{\mathbf{v}}\beta_y = i_{\mathbf{v}}\beta_x, \text{ since } \mathbf{v}_y = 0.$$

Thus $df(\mathbf{v}) = \beta(\mathbf{v})$, and so $df = \beta$. □

The following was the crucial ingredient of the proof.

**Corollary (5.4):** *In any manifold $M^n$, if $\beta^1$ is a 1-form whose integral over all closed curves vanishes, then $\beta^1$ is exact, $\beta^1 = df$.*

If a $p$-form $\beta^p$ is exact, $\beta^p = d\alpha^{p-1}$, we say that $\beta^p$ is derivable from the **potential** $\alpha^{p-1}$.

## 5.3. Complex Analysis

In the complex plane $M^2 = \mathbb{C}$, we introduce the complex coordinate $z = x + iy$. Then $dz = dx + idy$ is a complex valued 1-form with values 1 and $i$, respectively on $\partial/\partial x$ and $\partial/\partial y$. We may also consider the complex conjugate 1-form $d\bar{z} = dx - idy$, and then

$$dz \wedge d\bar{z} = -2i dx \wedge dy$$

Let $f(z,\bar{z}) = a(x,y) + ib(x,y)$ be a complex valued function on some open subset $U$ of $\mathbb{C}$. Then we can consider the 1-form

$$f(z,\bar{z})dz = (a+ib)(dx+idy) = (adx - bdy) + i(ady + bdx)$$

(This is *not* the most general 1-form since we have not included a term involving $d\bar{z}$.) If $C$, $z = z(t)$, is a curve, we may form the integral

$$\int_C f dz := \int_C (adx - bdy) + i\int_C (ady + bdx)$$

For exterior differential we get

$$d[fdz] = (da \wedge dx - db \wedge dy) + i(da \wedge dy + db \wedge dx)$$

$$= (-a_y - b_x) dx \wedge dy + i(a_x - b_y) dx \wedge dy$$

Thus

$f dz$ is closed iff $a$ and $b$ satisfy the Cauchy–Riemann equations, in $U$, that is, iff $f$ is **complex analytic** or **holomorphic**.

This can also be seen by the following formal calculation. By the chain rule we have the two differential operators

$$\frac{\partial}{\partial z} := \frac{1}{2}\left(\frac{\partial}{\partial x} - i\frac{\partial}{\partial y}\right)$$
$$\frac{\partial}{\partial \bar{z}} := \frac{1}{2}\left(\frac{\partial}{\partial x} + i\frac{\partial}{\partial y}\right)$$

Then $d[f dz] = (\partial f/\partial z)dz \wedge dz + (\partial f/\partial \bar{z})d\bar{z} \wedge dz = (\partial f/\partial \bar{z})d\bar{z} \wedge dz$, and so $f dz$ is closed iff $\partial f/\partial \bar{z} = 0$, that is, "$f$ does not depend on $\bar{z}$," and so $f$ is complex analytic.

$$\frac{\partial f}{\partial \bar{z}} = 0$$

is another form of the Cauchy–Riemann equations.

If $f dz$ is closed and we put $\alpha(z) = \int^z f dz$, the integral from a fixed point to $z$ along an arbitrary path, then $\alpha$ is the potential, $d\alpha = f dz$, provided it is single-valued, that is, provided the integral is independent of the path chosen. From (5.3 ) this will be the case provided $U$ has first Betti number 0. We shall see in Section 13.3 that asking $b_1 = 0$ for a manifold is a weaker condition than demanding that the manifold be simply connected. Simple connectivity is the usual condition imposed in complex analysis to ensure single-valuedness of the potential $\alpha$.

Note that to consider the behavior of $f$ at infinity we should consider $f$ as being defined on the Riemann sphere (see Section 1.2d) except perhaps at $\infty$ itself, that is, except at $w = 1/z = 0$. Since $z$ is a complex analytic function of $w$, $\partial z/\partial \bar{w} = 0$, and since $dz/dw \neq 0$ for our change of coordinates, we see from

$$\frac{\partial}{\partial \bar{w}} = \left(\frac{\partial z}{\partial \bar{w}}\right)\frac{\partial}{\partial z} + \left(\frac{\partial \bar{z}}{\partial \bar{w}}\right)\frac{\partial}{\partial \bar{z}}$$

that

$$\frac{\partial f}{\partial \bar{z}} = 0 \qquad \text{iff} \qquad \frac{\partial f}{\partial \bar{w}} = 0$$

This means that the notion of a function being complex analytic is well defined on the Riemann sphere, independent of which coordinate $z$ or $w$ is used.

In the complex plane $\mathbb{C}$, the residue of a *function* $f$ plays an important role in evaluating line integrals of $f$, but in the Riemann sphere it is the 1-form $f dz$ that is important, not its component $f$. For example, the function $f(z) = 1/z$ has residue 1 at the simple pole $z = 0$, and so $\oint_C dz/z = 2\pi i$ for any closed curve $C$ circling once $z = 0$ in the positive sense. But this curve also circles $z = \infty$ on the Riemann sphere, and the function $f = 1/z$ is described near $\infty$ by $f(z) = 1/z = w$ near $w = 0$. Thus the function $f = 1/z$ has a simple *zero* at $z = \infty$; its "residue" there is 0. One might then be mistakenly led to the contradiction that $\oint_C dz/z = 0$. The resolution lies with the 1-forms, not the functions:

$$\oint_C \left(\frac{1}{z}\right)dz = \oint_C w d\left(\frac{1}{w}\right) = \oint_C \left(-\frac{w}{w^2}\right)dw = -\oint_C \left(\frac{1}{w}\right)dw$$

which is again $2\pi i$ since $C$ circles $\infty$ in the negative sense. *We associate a residue to a 1-form, not a function!*

## 5.4. The Converse to the Poincaré Lemma

A closed 1-form $\beta^1$ on $M^n$ is exact if the first Betti number of $M^n$ vanishes, that is, if every closed oriented curve is the boundary of an oriented surface. On the 2 dimensional torus, neither closed curve $C$ nor $C'$ bounds a surface and thus we may

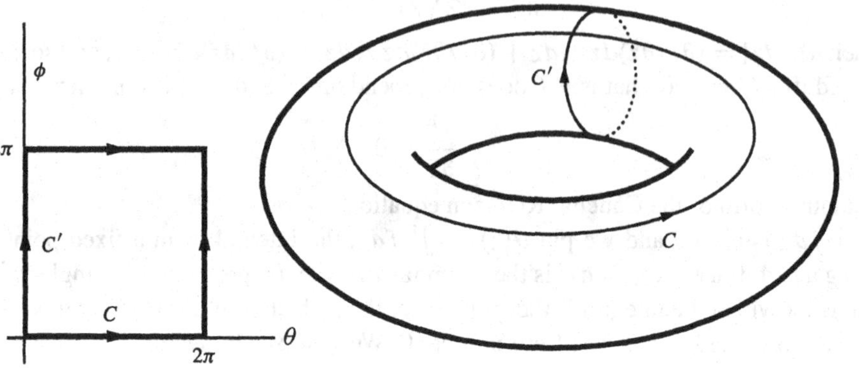


Figure 5.2

not expect that every closed 1-form is exact. In fact $d$"$\theta$" and $d$"$\phi$" are closed and $\oint d``\theta" = 2\pi = \oint d``\phi"$.

The fact that exact forms are closed, that is, $dd = 0$, is usually called *Poincaré's lemma*. It should be appreciated that Poincaré utilized this result before the *machinery* of exterior calculus had been developed! There is a partial converse to this result, namely, every closed form is *locally* exact. Precisely

**Theorem (5.5):** *If $d\beta^p = 0$, $p \geq 1$, in a neighborhood $U$ of $x \in M^n$, then there is some perhaps smaller neighborhood $U'$ of $x$ and a $(p-1)$ form $\alpha^{p-1}$ such that $\beta^p = d\alpha^{p-1}$ in $U'$.*

The following proof is *basically* a simple application of Cartan's formula for Lie derivatives. We give this proof because the same method is useful for other purposes. The reader might enjoy more an older proof, as is given, for example, in the book by Flanders [Fl].

PROOF: It is sufficient to prove this result in the case $M^n = \mathbb{R}^n$. This is because a sufficiently small neighborhood $U''$ of $x \in M^n$ is diffeomorphic to an open ball $V$ in $\mathbb{R}^n$ under a coordinate map $\phi : U'' \to V$. Since $\phi : U'' \to V$ is a diffeomorphism, $\phi^{-1}$ exists and $\beta^p = (\phi^{-1} \circ \phi)^* \beta^p = \phi^* \circ \phi^{-1*} \beta^p$. Then if $\beta$ is closed on $M$, $\phi^{-1*}\beta$ is closed on $V \subset \mathbb{R}^n$. If we have the converse of Poincaré on $V \subset \mathbb{R}^n$ then $\phi^{-1*}\beta = d\alpha$ shows $\beta = \phi^* d\alpha = d\phi^*\alpha$ as desired.

We may assume then that $\beta^p$ is a closed form on an open ball $U$ of $\mathbb{R}^n$. Consider (as in 4.3b) the deformation $\phi_t \mathbf{x} = (1-t)\mathbf{x}$; this time-dependent "flow" has $\phi_0 =$ the identity and $\phi_1$ is the map that sends *every* $\mathbf{x}$ to the origin. The velocity field is $\mathbf{v}(t, \mathbf{y}) = -\mathbf{y}/(1-t)$, for $t \neq 1$. First note that $\phi_0^*$ is the identity map and $\phi_1^*$ is the 0 map. Then considering $\beta = \beta(\mathbf{x})$ as a time-independent $p$-form on $\mathbb{R}^n$, we have

$$\beta(\mathbf{x}) = \phi_0^*\beta(\mathbf{x}) = \phi_0^*\beta(\phi_0\mathbf{x}) - \phi_1^*\beta(\phi_1\mathbf{x})$$
$$= \int_1^0 \frac{d}{ds}[\phi_s^*\beta(\phi_s\mathbf{x})]ds$$

To avoid subscripts upon subscripts upon..., let us introduce the following notation in this proof. We shall denote the vector $\mathbf{v}$ at $x$ by $\mathbf{v}(x)$ and we shall sometimes replace $\phi_t$ by $\phi(t)$. Also, for interior product we put $i_\mathbf{v} = i\{\mathbf{v}\}$. Then the previous expression for $\beta(\mathbf{x})$ becomes, using (4.44), $\mathbf{d}\beta = 0$ and $\partial\beta/\partial t = 0$

$$\int_1^0 \phi_s^*\mathbf{d}[i\{\mathbf{v}(\phi_s\mathbf{x})\}\beta(\phi_s\mathbf{x})]ds = \int_1^0 \mathbf{d}[\phi_s^* i\{\mathbf{v}(\phi_s\mathbf{x})\}\beta(\phi_s\mathbf{x})]ds$$

We should remark that this is not quite true. The vector field $\mathbf{v}(t, \mathbf{x})$ blows up at $t = 1$ (but note that $\phi_1^* = 0$). We should take the integral from $s = c$ to $s = 0$ and then let $c \to 1$. It will be apparent in our final formula (5.6) that the factor $(1 - t)^{-1}$ disappears. We proceed as if this difficulty were not present.

We may take the operator $\mathbf{d}$ outside the $s$ integral, yielding

$$\beta = \mathbf{d}\alpha^{p-1}, \qquad \alpha^{p-1} := \int_1^0 \phi_s^*[i\{\mathbf{v}(\phi_s\mathbf{x})\}\beta(\phi_s\mathbf{x})]ds \quad \square$$

Let us now write out the expression for $\alpha$ in detail. Put $\mathbf{y} = \phi_s(\mathbf{x}) = (1 - s)\mathbf{x}$. Then (in coordinates $y$ for $\mathbb{R}^n$)

$$i\{\mathbf{v}(\phi_s\mathbf{x})\}\beta(s, \mathbf{y}) = v^j(\mathbf{y})b_{j\underline{K}}(\mathbf{y})dy^K = -\frac{y^j}{(1-s)}b_{j\underline{K}}(\mathbf{y})dy^K$$

To take $\phi_s^*$ of this $(p - 1)$-form we must put everywhere $y^j = (1 - s)x^j$. We get $-x^j b_{j\underline{K}}((1 - s)\mathbf{x})dx^K(1 - s)^{p-1}$. Putting $\tau = (1 - s)$ gives

$$\alpha^{p-1} = \int_0^1 [\tau^{p-1}x^j b_{j\underline{K}}(\tau\mathbf{x})dx^K]d\tau \tag{5.6}$$

Note that the essential ingredient of the proof of the existence of a potential was the fact that at any point 0 of a manifold $M^n$ *there is a neighborhood of* 0 *that can be contracted to the point* 0; that is, there is a **deformation** $\mathbf{x} \mapsto \psi(t)\mathbf{x} = (1 - t)\mathbf{x}$ that collapses the neighborhood to the point 0 in 1 unit of time.

Note also that since *all* of $\mathbb{R}^n$ can be contracted to the origin, the result in $\mathbb{R}^n$ is *global*; if $d\beta^p = 0$ in all of $\mathbb{R}^n$ then $\beta^p$ is globally exact (if $p > 0$).

**Corollary (5.7):** *If div* $\mathbf{B} = 0$ *in* $\mathbb{R}^3$ *then* $\mathbf{B} = \operatorname{curl}\mathbf{A}$ *for some* $A$.

(See Problem 5.5(2) at this time.)

**Corollary (5.8):** *In* $M^n$, *a necessary and sufficient condition that one can solve locally the system of partial differential equations*

$$(\partial_i a_j - \partial_j a_i) = b_{ij} \qquad (\textit{with } b_{ji}(x) = -b_{ij}(x) \textit{ given})$$

*is that*

$$\partial_i b_{jk} + \partial_k b_{ij} + \partial_j b_{ki} = 0$$

## 5.5. Finding Potentials

In some simple situations one may exhibit potentials with very little effort. For example, consider the simplest case of the electric field due to a charge $q$ at the origin. In spherical coordinates $\mathbf{E} = (q/r^2)\partial/\partial r$ for $r > 0$. Using the euclidean metric in spherical coordinates in $\mathbb{R}^3 - 0$,

$$ds^2 = dr^2 + r^2(d\theta^2 + \sin^2\theta d\phi^2)$$

we see that $\mathcal{E} = (q/r^2)dr = \mathbf{d}(-q/r)$, for $r > 0$, exhibiting the *scalar* potential. The 2-form associated to $\mathbf{E}$ is the pseudoform

$$*\mathcal{E} = i_{\mathbf{E}}\,\mathrm{vol}^3$$

From Gauss's law $\mathbf{d}*\mathcal{E} = 4\pi\rho\,\mathrm{vol}^3$ we see that $*\mathcal{E}$ is closed for $r > 0$ since the charge density vanishes outside the origin. We compute directly a *vector* potential for $\mathbf{E}$ as follows. In spherical coordinates,

$$\mathrm{vol}^3 = r^2 \sin\theta dr \wedge d\theta \wedge d\phi$$

and so

$$*\mathcal{E} = i\left(\frac{q}{r^2}\frac{\partial}{\partial r}\right) r^2 \sin\theta dr \wedge d\theta \wedge d\phi = q \sin\theta d\theta \wedge d\phi$$

Thus, for example, $*\mathcal{E} = \mathbf{d}(-q\cos\theta d\phi)$ and $\mathcal{A}^1 = -q\cos\theta d\phi$ is a possible choice for potential. Note that spherical coordinates are badly behaved not only at the origin but at $\theta = 0$ and $\theta = \pi$ also, that is, along the entire $z$ axis. Hence $\mathcal{A}^1$ is a well-defined potential everywhere except the entire $z$ axis. Note however that we can also write $*\mathcal{E} = \mathbf{d}[q(1-\cos\theta)d\phi]$, and since $1 - \cos\theta = 0$ when $\theta = 0$, this expression

$$\mathcal{A}^1 = q(1 - \cos\theta)d\phi \tag{5.9}$$

is a well-defined potential everywhere except along the *negative* $z$ axis!

We certainly do *not* expect to find a potential $\mathcal{A}^1$ in the entire region $\mathbb{R}^3 - 0$, for if such an $\mathcal{A}^1$ existed we would have

$$\iint_V *\mathcal{E} = \iint_V \mathbf{d}\mathcal{A}^1 = \oint_{\partial V} \mathcal{A}^1 = 0$$

for any *closed* surface $V^2$ in $\mathbb{R}^3 - 0$. But if we choose $V^2$ to be the unit sphere about the origin we must have, by Gauss's law, that $\iint_V *\mathcal{E} = 4\pi q$! The **singularities** of $\mathcal{A}^1$ prevent us from applying Stokes's theorem to $V$.

We get the same result when we consider the magnetic field $\mathcal{B}^2$ due to a hypothetical magnetic *monopole* at the origin. This will be used when we discuss gauge fields in Section 16.4. The vector potential has a **Dirac string** of singularities along the negative $z$ axis.

## Problems

**5.5(1)** Prove that the product of a closed and an exact form is exact.

**5.5(2)** Write out what (5.6) says in terms of vectors, for $\beta^2$ in $\mathbb{R}^3$.

**5.5(3)** Consider the law of Ampere–Maxwell in the case of an infinitely long straight wire carrying a current $j$.

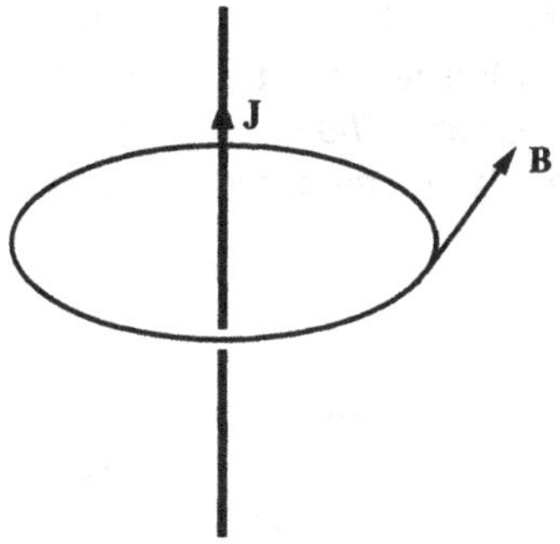


Figure 5.3

The steady state has $\partial{*\mathcal{E}}/\partial t = 0$ and we are reduced to Ampere's law $\oint *\mathcal{B} = 4\pi j$ for a curve as indicated, and $\mathbf{d}\mathcal{B}^2 = 0$. An immediate solution is suggested, $*\mathcal{B} = 2\,jd\phi$ . Introduce appropriate coordinates, show that $\mathbf{d}\mathcal{B}^2 = 0$, and exhibit directly the vector potential $\mathcal{A}^1$ in $\mathbb{R}^3$–wire. (You might wish to compare this with the usual treatments in textbooks.)

**5.5(4)** The unit 3-sphere $S^3 \subset \mathbb{R}^4$ can be parameterized by three angles $\alpha$, $\theta$, and $\phi$, where $\theta$ and $\phi$ are the usual spherical coordinates on the 2-sphere $S^2(\alpha)$ of radius $\sin\alpha$.

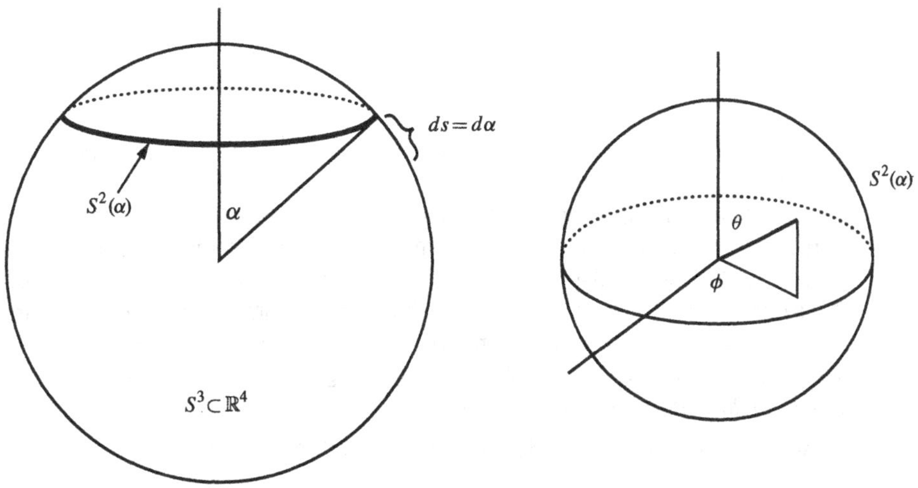


Figure 5.4

The Riemannian metric on $S^3$ is "clearly"

$$ds^2 = d\alpha^2 + \sin^2\alpha(d\theta^2 + \sin^2\theta d\phi^2)$$

Put a charge $q$ at the pole $N$ of $S^3$. **E** will certainly have the form $\mathbf{E} = E(\alpha)\partial/\partial\alpha$. Write down the resulting $*\mathcal{E} = i_{\mathbf{E}}\,\mathrm{vol}^3$. What form *must* the function $E = E(\alpha)$ have in order that $\mathbf{d}*\mathcal{E} = 0$ for $\alpha \neq 0, \pi$? Finish the determination of $*\mathcal{E}$ by computing $\int_{S(\alpha)} *\mathcal{E}$ (note that essentially no integration is needed if you know the area of the unit 2-sphere). Write down the electric covector $\mathcal{E}$ and verify $\mathbf{d}\mathcal{E} = 0$ and exhibit the scalar potential for $\mathcal{E}$, all for $\alpha \neq 0, \pi$. Put $\mathcal{B}^2 = 0$. *You have just verified Maxwell's equations* in the region outside the two poles. *Note that a "ghost" charge of $-q$ has appeared at the south pole!*

One could consider placing a charge $+q$ at the "north pole" of the projective space $\mathbb{R}P^3$.

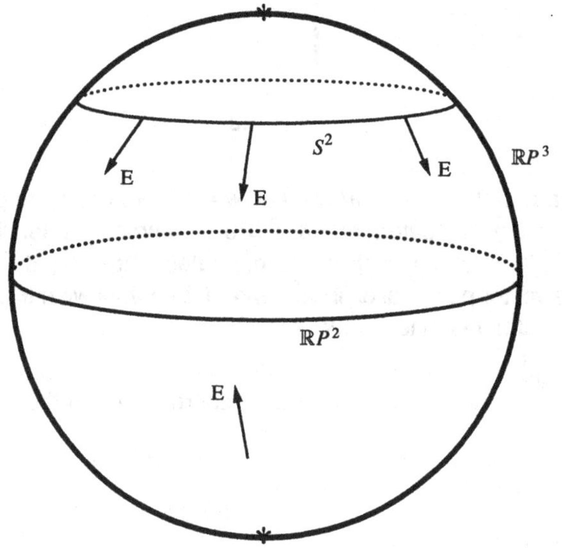


**Figure 5.5**

Since the "south pole" is now the same point, we have indicated the same charge there. The "equator" is really a projective plane $\mathbb{R}P^2$, since $\mathbb{R}P^3$ is $S^3$ with antipodal points identified. A 3-dimensional $\epsilon$-**neighborhood** of $\mathbb{R}P^2$, that is, points on $\mathbb{R}P^3$ that have distance $< \epsilon$ from $\mathbb{R}P^2$, has the indicated 2-sphere $S^2$ as boundary. (It *is* a 2-sphere since it is also the boundary of a 3-disc neighborhood of the north pole.) Gauss's theorem, applied to this neighborhood with boundary $S^2$, shows that there is a total charge of $-q$ inside $S^2$. Note that there is a **jump discontinuity** of **E** on $\mathbb{R}P^2$. This shows that a *ghost surface charge $-q$ must be distributed on the "equator" $\mathbb{R}P^2$!*

**5.5(5)** Show that in any *closed* manifold $M^3$, *the total charge vanishes!*

CHAPTER 6

# Holonomic and Nonholonomic Constraints

## 6.1. The Frobenius Integrability Condition

Can one always find a surface orthogonal to a family of curves in $\mathbb{R}^3$ ?

### 6.1a. Planes in $\mathbb{R}^3$

Given a smooth nonvanishing vector field in $\mathbb{R}^3$, by solving a system of ordinary differential equations one can always locally find a smooth family of integral curves, that is, nonintersecting curves that fill up a region and are always tangent to the vector field.

Given a smooth family of 2-planes $\Delta$ in $\mathbb{R}^3$, can one always find a smooth family of integral surfaces, that is, nonintersecting surfaces that fill up a region and are everywhere tangent to the planes? It is rather surprising that this is not always so! Suppose that one could find such integral surfaces.

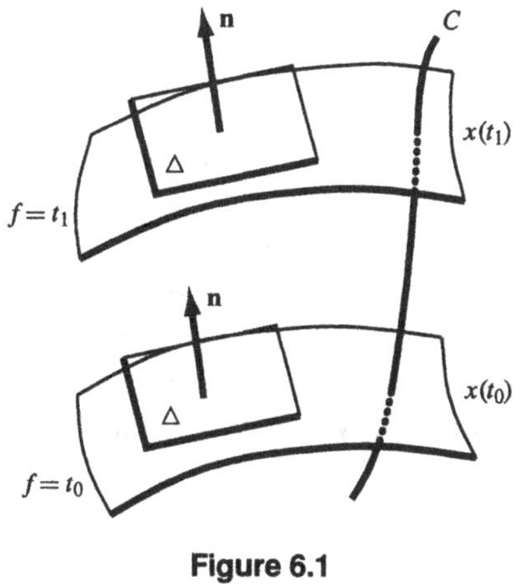


Figure 6.1

Let $C$, $\mathbf{x} = \mathbf{x}(t)$ be a parameterized curve that is transverse to the family of supposed integral surfaces (we can certainly find such a curve locally). Then locally we can define a function $f = f(\mathbf{x})$ whose level surfaces are surfaces of the family, namely, the level surface where $f = t_1$ consists of the supposed integral surface that is pierced by the transversal curve at parameter value $t = t_1$. But then $\nabla f$ must be along the given normal $\mathbf{n}$ to the planes, $\mathbf{n} = \lambda \nabla f$ for some function $\lambda$ (an "integrating factor"). In cartesian coordinates, the "normal" covector $\nu = n_i dx^i$ must satisfy $\nu = \lambda df$ and then $d\nu = d\lambda \wedge df = (d \log \lambda) \wedge \nu$, and we then recover **Euler's integrability condition**; if such integral surfaces exist, then

$$\nu \wedge d\nu = 0, \qquad \text{i.e., } \mathbf{n} \bullet \operatorname{curl} \mathbf{n} = 0$$

This condition, given entirely in terms of the field of normals, must be satisfied if integral surfaces are to exist.

Of course if $d\nu = 0$, $\nu = dg$ locally, and so $\mathbf{n}$ *is* normal to the surfaces $g =$ constant.

Consider the planes $\Delta$ normal to the vectors

$$\mathbf{n} = y\frac{\partial}{\partial x} - x\frac{\partial}{\partial y} + \frac{\partial}{\partial z} \sim (y, -x, 1)^T$$

Then $\nu = ydx - xdy + dz$ and so $\nu \wedge d\nu = -2dx \wedge dy \wedge dz \neq 0$; *the vectors* $\mathbf{n}$ *are not the normals to a family of surfaces*!

Classically, in cartesian coordinates, the planes $\Delta$ orthogonal to the vector $\mathbf{n}$ would be written

$$\nu = ydx - xdy + dz = 0$$

meaning not that the form $\nu$ is the form 0 but rather that at each point $(x_0, y_0, z_0)$ we are looking at *all* vectors $\mathbf{A} = (a^1, a^2, a^3)^T$ that satisfy

$$0 = \nu(\mathbf{A}) = i_{\mathbf{A}}\nu = y_0 a^1 - x_0 a^2 + a^3$$

clearly a 2-dimensional plane at $(x_0, y_0, z_0)$. The collection of all these planes at all points $\mathbf{x}$ in $\mathbb{R}^3$ is called the **distribution** associated to the 1-form $\nu$. (This is not to be confused with the generalized functions also called distributions.)

In general in $\mathbb{R}^3$ one would describe a family of planes by writing

$$\nu = P_1 dx^1 + P_2 dx^2 + P_3 dx^3 = 0 \tag{6.1}$$

where $P_1$, $P_2$, and $P_3$ are smooth functions. To "solve the total differential equation" (6.1) means to find surfaces $\mathbf{x} = \mathbf{x}(u^1, u^2)$ such that the pull-back of $\nu$ to these surfaces vanishes identically, that is, $P_i \partial x^i / \partial u^\alpha = 0$ for $\alpha = 1, 2$. We have seen that $\nu \wedge d\nu = 0$ is a necessary condition for this system of partial differential equations for $\mathbf{x} = \mathbf{x}(u^1, u^2)$ to possess a 1-parameter family of solutions. (We shall see shortly that this condition is also sufficient.) If we are given such a family of solutions, by taking a transversal curve $\mathbf{x} = \mathbf{x}(t)$ as earlier, this family of solutions can be described as the level sets $t =$ constant.

**Definition:** A $k$-dimensional **distribution** $\Delta_k$ on $M^n$ assigns in a smooth fashion to each $x \in M^n$ a $k$-dimensional subspace $\Delta_k(x)$ of the tangent space to $M^n$ at $x$. An $r$-dimensional **integral manifold** of $\Delta_k$ is an $r$-dimensional submanifold of $M^n$ that is everywhere tangent to the distribution. The distribution $\Delta_k$ is said to be

**(completely) integrable** if locally there are coordinates $x^1, \ldots, x^k, y^1, \ldots, y^{n-k}$ for $M^n$ such that the "coordinate **slices**" $y^1 = \text{constant}, \ldots, y^{n-k} = \text{constant}$ are $k$-dimensional integral manifolds of $\Delta_k$. Such a coordinate system $(x, y)$ will be called a **Frobenius chart** for $M$.

The fundamental question is clear. When is $\Delta_k$ completely integrable?

### 6.1b. Distributions and Vector Fields

Suppose that we are given a distribution $\Delta_k$ and a pair of vector fields $\mathbf{X}$ and $\mathbf{Y}$ on $M^n$ that are in the distribution $\mathbf{X} \in \Delta$ and $\mathbf{Y} \in \Delta$ at each point in an open set. Suppose now that the distribution is integrable. Then the two vector fields are always tangent to the integral manifolds. By the Corollary in 4.1c we conclude that the Lie bracket $[\mathbf{X}, \mathbf{Y}]$ is also in the distribution. We can describe this symbolically by saying that *if $\Delta_k$ is integrable* then

$$[\Delta, \Delta] \subset \Delta$$

It will turn out that this condition is also sufficient for showing integrability!

### 6.1c. Distributions and 1-Forms

Let $\theta^1$ be a 1-form that does not vanish at a point $x \in M^n$. The **annihilator** or **null space** of $\theta$ at $x$ is the $(n-1)$-dimensional subspace of $M^n_x$ defined by those vectors $\mathbf{X} \in M^n_x$ such that $\theta(\mathbf{X}) = 0$. Classically one writes $\theta = 0$ for this null space. (When discussing distributions it is common to call a 1-form $\theta$ a **Pfaffian**; $\theta = 0$ is then called a Pfaffian equation.) If $\theta_1, \ldots, \theta_r$ are $r = n - k$ linearly independent 1-forms at each point of an open subset of $M^n$, $\theta_1 \wedge \ldots \wedge \theta_r \neq 0$, then at each point the intersection of their null spaces forms an $n - r = k$ dimensional distribution $\Delta_k$. Thus

$$\mathbf{X} \in \Delta_k \quad \text{iff} \quad \theta_1(\mathbf{X}) = \ldots = \theta_r(\mathbf{X}) = 0$$

We may again write this distribution locally as $\theta_1 = 0, \ldots, \theta_r = 0$. We do not claim that every distribution can be *globally* defined by $r$ Pfaffians.

**Definition:** The distribution $\Delta$ is in **involution** if $[\Delta, \Delta] \subset \Delta$, that is, if the distribution is "closed under brackets."

We know that an integrable distribution is in involution.

If $\Delta_k$ is in involution, then for $\alpha = 1, \ldots, r$ we must have that for any pair of vector fields $\mathbf{X}, \mathbf{Y}$ that are in the distribution (see (4.25))

$$d\theta_\alpha(\mathbf{X}, \mathbf{Y}) = \mathbf{X}\{\theta_\alpha(\mathbf{Y})\} - \mathbf{Y}\{\theta_\alpha(\mathbf{X})\} - \theta_\alpha([\mathbf{X}, \mathbf{Y}]) = 0$$

We say then that if $\Delta$ is in involution, then "$d\theta_\alpha = 0$ when restricted to the distribution," that is, when we allow $d\theta_\alpha$ to be evaluated only on vectors of the distribution.

Conversely, suppose that $d\theta_\alpha = 0$ when restricted to $\Delta$, $\alpha = 1, \ldots, r$. Then $0 = d\theta_\alpha(\mathbf{X}, \mathbf{Y}) = \mathbf{X}(0) - \mathbf{Y}(0) - \theta_\alpha([\mathbf{X}, \mathbf{Y}])$ shows that $[\mathbf{X}, \mathbf{Y}] \in \Delta$, and so $[\Delta, \Delta] \subset \Delta$. We now give several rewordings of this result, all of which are important.

**Theorem (6.2):** *The following conditions are locally equivalent.*

*(i)* $\Delta$ *is in involution, that is,* $[\Delta, \Delta] \subset \Delta$.

*(ii)* $d\theta_\alpha$ *is the zero 2-form when restricted to* $\Delta$.

*(iii) There are 1-forms* $\lambda_{\alpha\beta}$ *such that* $d\theta_\alpha = \sum_\beta \lambda_{\alpha\beta} \wedge \theta_\beta$.

*(iv)* $d\theta_\alpha \wedge \Omega = 0$, *where* $\Omega = \theta_1 \wedge \ldots \wedge \theta_r$.

PROOF: We have already proved (i) $\Leftrightarrow$ (ii). (iii) $\Rightarrow$ (ii) since

$$\begin{aligned} d\theta_\alpha(\mathbf{X}, \mathbf{Y}) &= \sum_\beta \lambda_{\alpha\beta} \wedge \theta_\beta(\mathbf{X}, \mathbf{Y}) \\ &= \sum_\beta \lambda_{\alpha\beta}(\mathbf{X})\theta_\beta(\mathbf{Y}) - \sum_\beta \lambda_{\alpha\beta}(\mathbf{Y})\theta_\beta(\mathbf{X}) = 0 \end{aligned}$$

Conversely, suppose that all $d\theta_\alpha = 0$ when restricted to $\Delta$. Complete $\theta_1, \ldots, \theta_r$ locally to a basis for 1-forms by adjoining $\theta_{r+1}, \ldots, \theta_n$. Let $\mathbf{e}_1, \ldots, \mathbf{e}_n$ be the dual basis for vector fields. Then $\theta_\alpha(\mathbf{e}_i) = 0$ for $\alpha = 1, \ldots, r$ and $i = r+1, \ldots, n$ shows that $\mathbf{e}_{r+1}, \ldots, \mathbf{e}_n$ spans $\Delta$. Now expand $d\theta_\alpha$ in terms of the basis $\theta_1, \ldots, \theta_n$.

$$d\theta_\alpha = \sum_{1 \le \beta \le r} \lambda_{\alpha\beta} \wedge \theta_\beta + \sum_{r<i<j} \mu_\alpha^{ij} \theta_i \wedge \theta_j \tag{6.3}$$

for some coefficients $\lambda$ and $\mu$. Thus for $r < i < j$ we have $0 = d\theta_\alpha(\mathbf{e}_i, \mathbf{e}_j) = \mu_\alpha^{ij}$ and so $d\theta_\alpha = \sum_{1 \le \beta \le r} \lambda_{\alpha\beta} \wedge \theta_\beta$. This shows (ii) $\Rightarrow$ (iii) and so (ii) $\Leftrightarrow$ (iii).

It is immediate that (iii) $\Rightarrow$ (iv). Assume (iv). From (6.3)

$$0 = d\theta_\alpha \wedge \Omega = \sum_{r<i<j} \mu_\alpha^{ij} \theta_i \wedge \theta_j \wedge \Omega = \sum_{r<i<j} \mu_\alpha^{ij} \theta_i \wedge \theta_j \wedge \theta_1 \wedge \ldots \wedge \theta_r$$

But the $\theta$'s are independent; hence $\mu_\alpha^{ij} = 0$ for $r < i < j$. Thus (iv) $\Rightarrow$ (iii) and we are finished. □

In summary, we have seen that a distribution $\Delta_k$ can *locally* be described by either exhibiting $k$ linearly independent vector fields

$$\mathbf{X}_1, \ldots, \mathbf{X}_k$$

that span $\Delta_k$ at each point in a region, or by exhibiting $r = n - k$ linearly independent 1-forms

$$\theta_1, \ldots, \theta_r$$

whose common null space is $\Delta_k$. The system is *in involution* if either

$$[\Delta, \Delta] \subset \Delta$$

or $d\theta_\alpha = \sum_\beta \lambda_{\alpha\beta} \wedge \theta_\beta$ for some 1-forms $\lambda_{\alpha\beta}$. In this case we write

$$d\theta_\alpha = 0 \bmod \theta$$

meaning that $d\theta_\alpha$ becomes 0 when *all* of the $\theta_\alpha$ are put = 0.

We know that an integrable distribution is in involution. We now sketch a proof of the converse (usually attributed to Frobenius).

### 6.1d. The Frobenius Theorem

Let $\Delta_k$ be *any* smooth distribution of $k$-planes in $M^n$ and let (locally) $\{\mathbf{X}_A\}$, $A = 1, \ldots, k$ be smooth vector fields that span the distribution in some open set $U$ of $M^n$. Let $\phi_A$ be the local flow generated by the field $\mathbf{X}_A$. Given $x \in U$, we construct a $k$-dimensional submanifold of $M^n$ passing through $x$ as follows.

Let $D^k \subset \mathbb{R}^k$ be a small disc about the origin of $\mathbb{R}^k$ and let $t_1, \ldots, t_k$ be coordinates for $\mathbb{R}^k$ (for simplicity, we write indices on the $t$'s as subscripts). Define

$$\Phi : D^k \to M^n$$

by

$$\Phi(t) = \phi_k(t_k) \circ \phi_{k-1}(t_{k-1}) \circ \cdots \circ \phi_1(t_1)(x)$$

This is certainly defined if $t_1^2 + \ldots + t_k^2$ is small enough. We illustrate this for $k = 2$

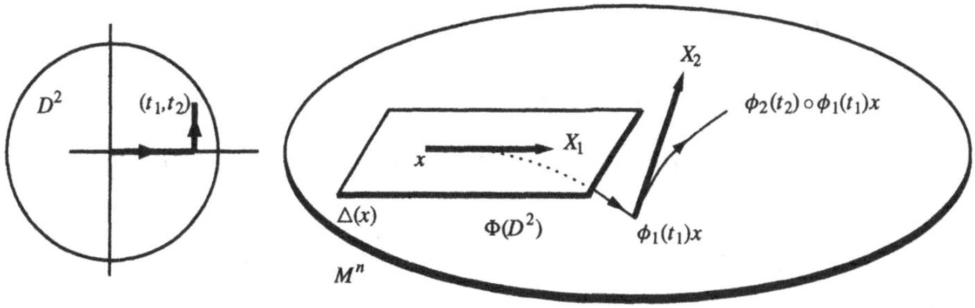


Figure 6.2

It should be clear (see Problem 6.1) that for the differential of $\Phi$ at $t = 0$, we have

$$\Phi_* : \mathbb{R}_0^k \to M_x^n$$

$$\Phi_*\left(\frac{\partial}{\partial t_A}\right) = \mathbf{X}_A \quad \text{at } x = \Phi(0) \tag{6.4}$$

and thus $\Phi_*\mathbb{R}_0^k = \Delta_k(x)$. Thus $\Phi(D^k)$ is tangent to $\Delta_k$ at the single point $x$.

**Definition:** A smooth map of manifolds $F : W^k \to M^n$ is an **immersion** and $F(W)$ is an **immersed submanifold** provided

$$F_* : W_w^k \to M_{F(w)}^n$$

is 1:1 (i.e., ker $F_* = 0$) at each $w \in W^k$.

In our case $\Phi_*$ is 1:1 at $0 \in \mathbb{R}^k$ and consequently 1:1 in some neighborhood of 0. Thus the map $\Phi : D^k \to M^n$ defines an immersed submanifold $\Phi(D^k)$ of $M^n$ provided $D^k$ is small enough.

**Frobenius Theorem: (6.5):** *If the distribution $\Delta_k$ is in involution*

$$[\Delta, \Delta] \subset \Delta$$

*then each such immersed disc $\Phi(D^k)$ is an integral manifold of $\Delta$ and this distribution is completely integrable.*

PROOF: In the following computation we shall denote the vector $\mathbf{X}$ at $x \in M^n$ by $\mathbf{X}(x)$ rather than $\mathbf{X}_x$. Since we are not using $\mathbf{X}$ as a differential operator there should be no confusion.

The essential point is to show that if $\Delta$ is in involution then $\Delta_k$ is tangent to $\Phi(D^k)$ at each point of this immersed disc. We already know, *without any assumption*, that $\Delta$ is tangent to the disc $\Phi(D)$ at $x = \Phi(0)$. From the definition of $\Phi$ (and again denoting $\phi_t$ by $\phi(t)$)

$$\Phi(t) = \phi_k(t_k) \circ \phi_{k-1}(t_{k-1}) \circ \cdots \circ \phi_1(t_1)(x)$$

we see that $\Phi_*$ takes the tangent vector $\partial/\partial t_A$ at $t$ into the vector

$$\begin{aligned} &\frac{\partial}{\partial h}[\phi_k(t_k) \circ \cdots \circ \phi_A(t_A + h) \circ \cdots \circ \phi_1(t_1)(x)]_{h=0} \\ &= \phi_k(t_k)_* \circ \cdots \circ \phi_A(t_A)_* \mathbf{X}_A \text{ (at the point } \phi_{k-1}(t_{k-1}) \circ \cdots \circ \phi_1(t_1)(x)) \end{aligned}$$

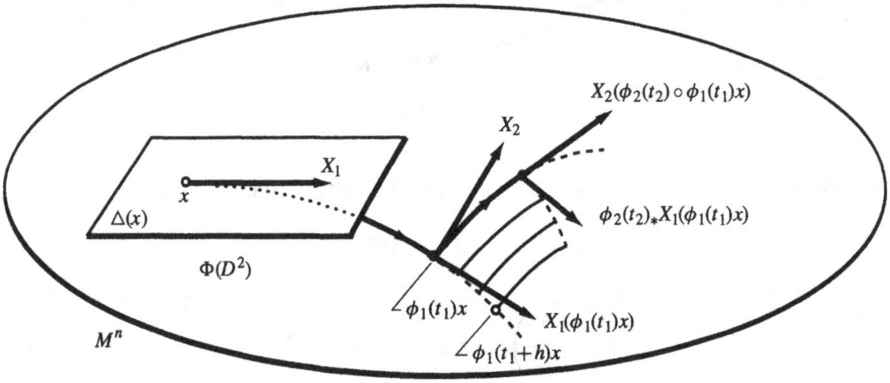


**Figure 6.3**

But this simply says that the tangent space to $\Phi(D^k)$ at $\Phi(t)$ has a basis given by

$$\begin{aligned} &\phi_k(t_k)_* \circ \cdots \circ \phi_2(t_2)_* \mathbf{X}_1(\phi_1(t_1)x) \\ &\phi_k(t_k)_* \circ \cdots \circ \phi_3(t_3)_* \mathbf{X}_2(\phi_2(t_2) \circ \phi_1(t_1)x) \\ &\cdots \\ &\mathbf{X}_k(\phi_k(t_k) \circ \cdots \circ \phi_1(t_1)x) \end{aligned}$$

*Thus we need only show that each flow $\phi_A(t)$ sends (via its differential) the distribution $\Delta_k$ into itself*! This will follow from $[\Delta, \Delta] \subset \Delta$ in the following manner.

Let $\mathbf{Y} \in \Delta(y)$. We must show that $[\phi_A(t)_* \mathbf{Y}] \in \Delta(\phi_A(t)y)$. Let $\Delta$ be defined by the Pfaffians $\theta_1 = 0, \ldots, \theta_r = 0$. We know that $\theta_\alpha(\mathbf{Y}) = 0$, $\alpha = 1, \ldots, r$. Let $\mathbf{Y}_t := \phi_A(t)_* \mathbf{Y}$ and put $\mathbf{X} := \mathbf{X}_A$. By construction, $\mathbf{Y}_t$ is invariant under the flow $\phi_A(t)$, and so

$$\mathcal{L}_{\mathbf{X}}(\mathbf{Y}_t) = 0 \qquad \text{along the orbit } \phi_A(t)y$$

Consider the real-valued functions

$$f_\alpha(t) = \theta_\alpha(\mathbf{Y}_t) = i_{\mathbf{Y}_t}\theta_\alpha, \qquad \alpha = 1, \ldots, r$$

Then, differentiating with respect to $t$

$$\begin{aligned} f'_\alpha(t) &= \mathbf{X}\{i_{\mathbf{Y}_t}\theta_\alpha\} = \mathcal{L}_{\mathbf{X}}\{i_{\mathbf{Y}_t}\theta_\alpha\}, \text{ which by (4.24)} \\ &= i_{\mathbf{Y}_t}\{i_{\mathbf{X}} d\theta_\alpha + d i_{\mathbf{X}}\theta_\alpha\} = i_{\mathbf{Y}_t} i_{\mathbf{X}} d\theta_\alpha \end{aligned}$$

since $i_{\mathbf{X}}\theta_\alpha = 0$. Since $\Delta$ is in involution, from part (iii) of (6.2) we have

$$\begin{aligned} f'_\alpha(t) &= i_{\mathbf{Y}_t} i_{\mathbf{X}}\left(\sum_\beta \lambda_{\alpha\beta} \wedge \theta_\beta\right) = i_{\mathbf{Y}_t}\left(\sum_\beta \lambda_{\alpha\beta}(\mathbf{X})\theta_\beta\right) \\ &= \sum_\beta \lambda_{\alpha\beta}(\mathbf{X})\theta_\beta(\mathbf{Y}_t) = \sum_\beta \lambda_{\alpha\beta}(\mathbf{X}) f_\beta(t) \end{aligned}$$

Thus the functions $f_\alpha$ satisfy the linear system

$$\begin{aligned} f'_\alpha(t) &= \sum_\beta \lambda_{\alpha\beta}(\mathbf{X}) f_\beta(t) \\ f_\alpha(0) &= \theta_\alpha(\mathbf{Y}) = 0 \end{aligned}$$

By the uniqueness theorem for such systems $f_\alpha(t) = 0$ and so $\theta_\alpha(\mathbf{Y}_t) = 0$. Thus $\mathbf{Y}_t \in \Delta$ for all $t$, as desired. Then $\Delta_k$ is tangent to $\Phi(D^k)$ at each point of this immersed disc.

To show complete integrability we must introduce coordinates for which our immersed discs are "slices" $y^1 = c^1, \ldots, y^{n-k} = c^{n-k}$. The procedure is very much like that followed in our introductory section (6.1a), where we introduced a coordinate $f = t$ by considering a curve transverse to the distribution. Here we must introduce a transverse $(n-k)$-dimensional manifold $W^{n-k}$ and we can let $y^1, \ldots, y^{n-k}$ be local coordinates on $W$. It can be shown, just as with integral curves of a smooth vector field, that the integral discs, through distinct points of W, will be disjoint if they are sufficiently small. This will be discussed more in Section 6.2. We shall not go into details. □

## Problems

**6.1(1)** Verify (6.4).

**6.1(2)** Show that a 1-dimensional distribution in $M^n$ is integrable. Why is this evident without using Frobenius?

## 6.2. Integrability and Constraints

Given a point on one curve of a family of curves, can one reach a nearby point on the *same* curve by a short path that is always perpendicular to the family?

### 6.2a. Foliations and Maximal Leaves

We know that if a distribution $\Delta_k$ on $M^n$ is in involution, $[\Delta, \Delta] \subset \Delta$, then the distribution is integrable; in the neighborhood of any point of $M$ one may introduce "Frobenius coordinates" $x^1, \ldots, x^k, y^1, \ldots, y^{n-k}$ for $M^n$ such that the "coordinate slices"

$$y^1 = \text{constant}, \ldots, y^{n-k} = \text{constant}$$

are $k$-dimensional integral manifolds of $\Delta_k$. The integral manifold through a given point $(x_0, y_0)$, of course, also exists outside the given coordinate system and might

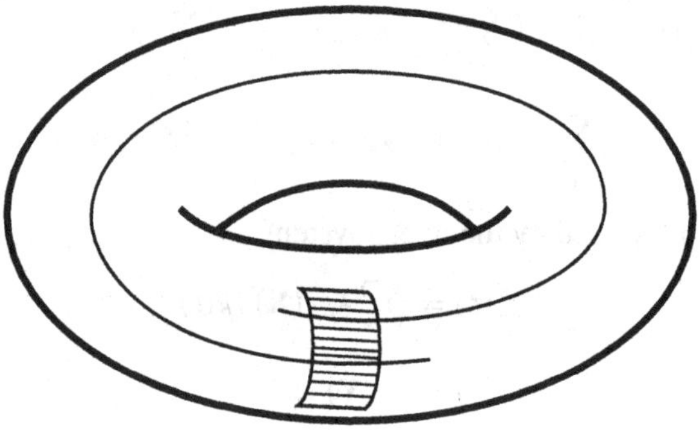

Figure 6.4

even return to the coordinate patch. If so, it will either reappear as the same slice or appear as a different one. For example, in the usual model of the torus $T^2$ as a rectangle in the plane (this time with sides of length 1) with periodic identifications, consider the

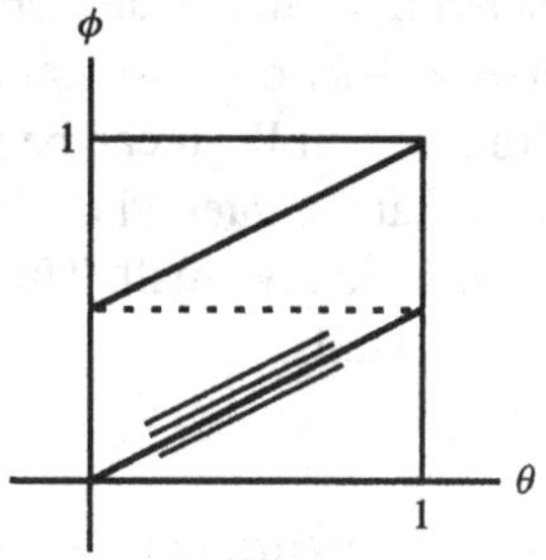


Figure 6.5

distribution $\Delta_1$ defined by $d\phi - kd\theta = 0$, where $k$ is a constant. The integral manifolds in this case are the straight lines in the rectangle with slope $k$. If $k = p/q$ is a rational number (we have illustrated the case $k = 1/2$) then the slice through $(0, 0)$ is a *closed* curve winding $q$ times around the torus in the $\theta$ direction and $p$ times around in the $\phi$ direction. On the other hand, if $k$ is irrational, then the integral curve leaving (0, 0) will never return to this point, but, it turns out, will lie dense on the torus. The integral curve will leave and reenter each Frobenius chart an infinite number of times, never returning to the same slice.

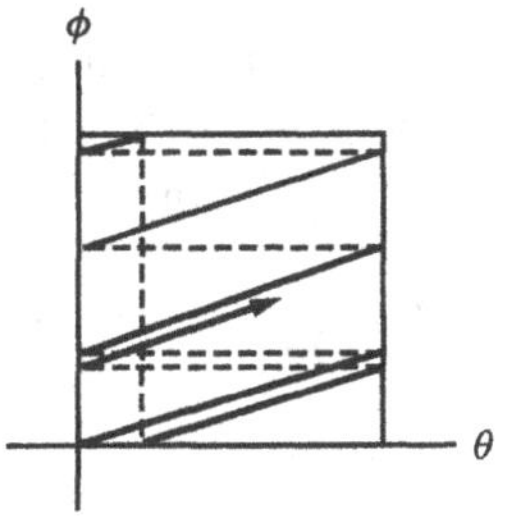


**Figure 6.6**

If a distribution $\Delta_k \subset M^n$ is integrable, then the integral manifolds define a **foliation** of $M^n$ and each *connected* integral manifold is called a **leaf** of the foliation. A leaf that is not properly contained in another leaf is called a **maximal leaf**. It seems clear from the preceding example with irrational slope that the maximal leaf through (0, 0) is not an *embedded* submanifold (see 1.3d); this is because the part of a maximal leaf that lies in a Frobenius chart consists of an infinite number of "parallel" line segments. There is no chance that we can describe all of these segments by a single equation $y = f(x)$. However, each "piece" of the leaf does look like a submanifold. The leaf through (0, 0) is the image of the real line under the map $F : \mathbb{R} \to T^2$ given by $\theta \to (\theta, k\theta)$; this is clearly an *immersion* since $F_*$ is 1:1 (see 6.1d).

We have just indicated one way in which an immersed submanifold can fail to be an embedded submanifold. There are two other commonly occurring instances.

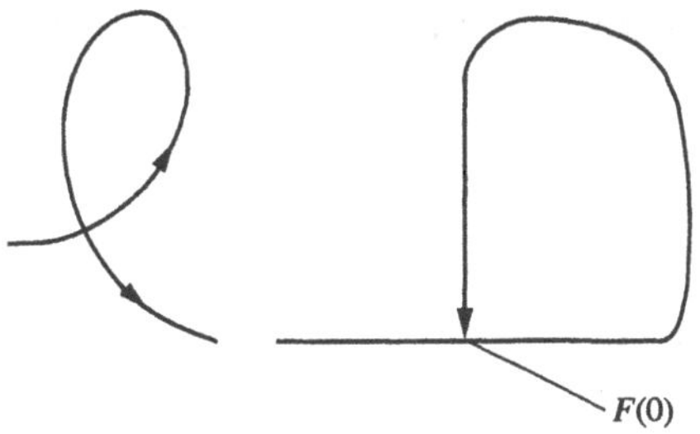


**Figure 6.7**

Both illustrated curves are immersions of the line $\mathbb{R}$ into the plane $\mathbb{R}^2$. In the first curve the map $F$ is not $1 : 1$ (even though $F_*$ *is* if the curve is parameterized so that the speed is never 0), whereas in the second curve, $F$ is $1 : 1$ but $F(0)$ is the limit of points $F(t)$ for $t \to \infty$. In neither case can one introduce local coordinates $x$, $y$ in $\mathbb{R}^2$ near the troublesome point so that the locus is defined by $y = y(x)$.

As we have seen in the case of $T^2$, a maximal leaf need not be an *embedded* submanifold. Chevalley, however, has proved the following.

**Theorem (6.6):** *A maximal leaf of a foliated manifold $M^n$ is a* $1:1$ *immersed submanifold; that is, there is a* $1:1$ *immersion* $F: V^k \to M^n$ *of some* $V^k$ *that realizes the given leaf globally.*

### 6.2b. Systems of Mayer–Lie

Classically the Frobenius theorem arose in the study of partial differential equations. An important system of such equations is the "system of Mayer–Lie"; we are to find functions $y^\beta = y^\beta(x)$, $\beta = 1, \ldots, r$, satisfying

$$\frac{\partial y^\beta}{\partial x^i} = b_i^\beta(x, y), \qquad i = 1, \ldots, k \tag{6.7}$$

with initial conditions

$$y^\beta(x_0) = y_0^\beta$$

where $b$ is a given matrix of functions. By equating mixed partial derivatives $\partial^2 y^\beta/\partial x^j \partial x^i = \partial^2 y^\beta/\partial x^i \partial x^j$ and using (6.7) we get the immediate **integrability conditions**

$$\left[\frac{\partial b_i^\beta}{\partial x^j} - \frac{\partial b_j^\beta}{\partial x^i}\right] = \sum_{\alpha=1}^{r}\left[\left(\frac{\partial b_j^\beta}{\partial y^\alpha}\right)b_i^\alpha - \left(\frac{\partial b_i^\beta}{\partial y^\alpha}\right)b_j^\alpha\right] \tag{6.8}$$

We wish to show that (6.8) is also a *sufficient* condition for a solution to exist.

Let $x^1, \ldots, x^k$ be coordinates in $\mathbb{R}^k$ and $y^1, \ldots, y^r$ be coordinates in $\mathbb{R}^r$. Then in $M^n = \mathbb{R}^k \times \mathbb{R}^r$ we consider the distribution $\Delta_k$ defined by the Pfaffians

$$\theta_\beta := dy^\beta - \sum_i b_i^\beta(x, y)dx^i = 0 \tag{6.9}$$

In Problem 6.2(1) you are asked to show that these 1-forms are independent.

The Frobenius integrability condition $d\theta_\beta = 0 \bmod \theta$ is simply the statement that $d\theta_\beta$ becomes 0 when all of the $\theta$'s are put equal to 0. In our case

$$d\theta_\beta = -d\sum b_i^\beta(x, y)dx^i = -\sum db_i^\beta \wedge dx^i$$

$$= -\sum_i\left[\sum_j\left(\frac{\partial b_i^\beta}{\partial x^j}\right)dx^j \wedge dx^i + \sum_\alpha\left(\frac{\partial b_i^\beta}{\partial y^\alpha}\right)dy^\alpha \wedge dx^i\right]$$

To put $\theta_\alpha = 0$ is to put $dy^\alpha = \sum_k b_k^\alpha dx^k$, and so, mod $\theta$,

$$d\theta_\beta = -\sum_{ij}\left(\frac{\partial b_i^\beta}{\partial x^j}\right)dx^j \wedge dx^i - \sum_{\alpha,i,j}\left(\frac{\partial b_i^\beta}{\partial y^\alpha}\right)b_j^\alpha dx^j \wedge dx^i$$

$$= -\sum_{ij}\left[\frac{\partial b_i^\beta}{\partial x^j} + \left(\frac{\partial b_i^\beta}{\partial y^\alpha}\right)b_j^\alpha\right]dx^j \wedge dx^i$$

and thus $d\theta_\beta = 0 \bmod \theta$ is simply the statement that the 2-form $d\theta_\beta$ above must be 0. This means that the coefficients of $dx^j \wedge dx^i$, *made skew symmetric in i and j*, must vanish. This gives exactly the naive integrability condition (6.8). Hence the distribution in $\mathbb{R}^k \times \mathbb{R}^r$ defined by (6.9) is completely integrable.

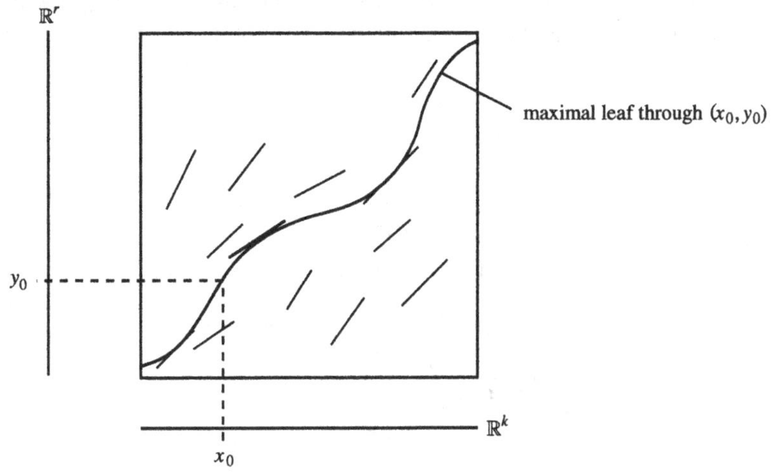


Figure 6.8

Let $V^k$ be the maximal leaf through $(x_0, y_0)$. One can easily see from (6.9) that the distribution is never "vertical": No nonzero vector of the form $a^\beta \partial/\partial y^\beta$ is ever in the distribution. It seems clear from the picture (and it is not difficult to prove) that this implies that the leaf through $(x_0, y_0)$ can be written in the form $y^\beta = y^\beta(x)$. For these functions we have that $\theta_\beta = 0$ when restricted to the leaf. Thus $dy^\beta = \sum b_i^\beta(x, y)dx^i$ and then $\partial y^\beta/\partial x^i = b_i^\beta(x, y)$ as desired.

## 6.2c. Holonomic and Nonholonomic Constraints

Consider a dynamical system with configuration space $M^n$ and local coordinates $q^1, \dots, q^n$. It may be that the configurations of the system may be constrained to lie on a submanifold of $M^n$. For example, a particle moving in $\mathbb{R}^3 = M^3$ may be constrained to move only on the unit sphere. In this case we have a single constraining equation $F(x, y, z) = x^2 + y^2 + z^2 = 1$. We may write this constraint in differential form $dF = 0 = xdx + ydy + zdz$. More generally we may impose constraints given by $r$ exact 1-forms, $dF_1 = 0, \dots, dF_r = 0$, constraining the configuration to lie on an $n - r$-dimensional submanifold $V^{n-r}$ of $M^n$, at least if $dF_1 \wedge \dots \wedge dF_r \neq 0$ on $V^{n-r}$. The constraints have reduced the number of "degrees of freedom" from $n$ to $n - r$. Still more generally, we may consider constraints given by $r$ independent Pfaffians that need not be exact

$$\theta_1 = 0, \dots, \theta_r = 0 \tag{6.10}$$

**Definition:** The constraints (6.10) are said to be **holonomic** or **integrable** if the distribution is integrable; otherwise they are nonholonomic or nonintegrable.

Of course, if the constraints are holonomic, then by the Frobenius theorem we may introduce local coordinates $x$, $y$ so that the system is constrained to the submanifolds $y^1 = \text{const.}, \ldots, y^r = \text{const.}$, and then the constraints can be equivalently written as $dy^1 = 0, \ldots, dy^r = 0$. Nonholonomic constraints are more puzzling. Consider the classic example of a vertical unit disc rolling on a horizontal plane "without slipping."

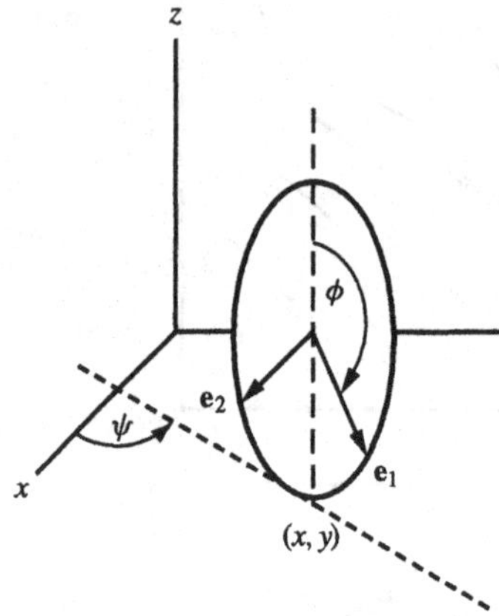


**Figure 6.9**

To describe the configuration of the disc completely we engrave an orthonormal pair of vectors $\mathbf{e}_1$, $\mathbf{e}_2$ in the disc and consider the endpoint of $\mathbf{e}_1$ as a distinguished point on the disc. The configuration is then completely described by

$$(x, y, \psi, \phi)$$

where $(x, y)$ are the coordinates of the center of the disc, $\phi$ is the angle that $\mathbf{e}_1$ makes with the vertical (positive rotations go from $\mathbf{e}_1$ to $\mathbf{e}_2$), and $\psi$ is the angle that the plane of the disc makes with the $x$ axis. (The line of intersection of the disc and the $xy$ plane is *directed* such that an increase of the angle $\phi$ will roll the disc in the positive direction along this line.) It is then clear that the configuration space of the disc is

$$M^4 = \mathbb{R}^2 \times S^1 \times S^1 = \mathbb{R}^2 \times T^2$$

The condition that the disc roll without slipping is expressed by looking at the motion of the center of the disc. It is

$$\theta_1 := dx - \cos\psi \, d\phi = 0 \tag{6.11}$$

$$\theta_2 := dy - \sin\psi \, d\phi = 0$$

It would seem that the constraints would reduce the degrees of freedom by 2, but *in a certain sense* this is not so. We can see that the constraints are nonholonomic as follows: $d\theta_1 = \sin\psi \, d\psi \wedge d\phi$ yields

$$d\theta_1 \wedge (\theta_1 \wedge \theta_2) = \sin\psi \, d\psi \wedge d\phi \wedge dx \wedge dy \neq 0$$

By (6.2), part (iv), the distribution is not integrable. Recall that in the case of integrable constraints we have integral manifolds, the leaves $V^k$, on which the system must remain. If we move (from a configuration point $p$) a small distance in a direction that *violates* the constraints, that is, along a curve whose tangent vector is not annihilated by all of the constraint Pfaffians $\theta_1, \ldots, \theta_r$, then we automatically end at a point $q$ on a *different* leaf. There is no way that one can move from $p$ to $q$ while *obeying the constraints* and

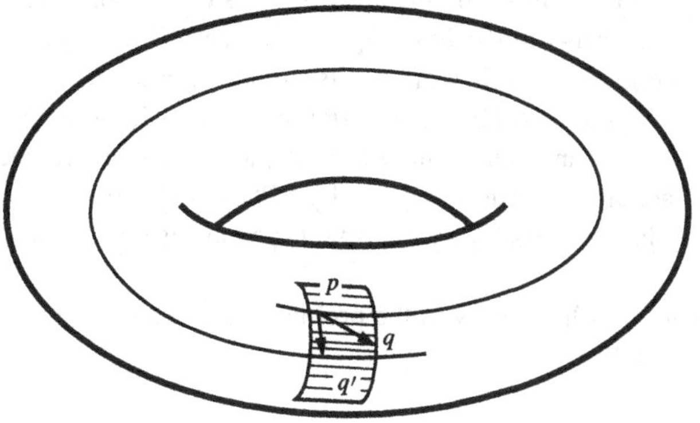


Figure 6.10

*remaining in the given Frobenius coordinate patch.* It is possible that an endpoint $q'$ lies on the same maximal leaf as $p$, but to go from $p$ to $q'$ while obeying the constraints requires a "long" path, that is, a path that leaves the coordinate patch. This is the meaning of the statement that in a holonomic system one has *locally* only $n - r$ degrees of freedom; we must stay on the $(n - r)$-dimensional leaf. It is also a fact that although a maximal leaf can return to an infinite number of different slices globally (as in $T^2$ with irrational slope) it cannot return to *every* slice in the coordinate patch. *Some points in the patch cannot be reached from p while obeying the constraints.*

This is *not* the case in our nonholonomic disc! Recall that the constraints demand rolling without sliding. Consider the disc in an initial state at the origin and lined up along the $x$ axis. Now violate the constraints by *sliding* the disc in the $y$ direction for an arbitrarily small distance. If the system were holonomic we could not *roll* the disc along a small path from the initial to the final configuration. But here we can!

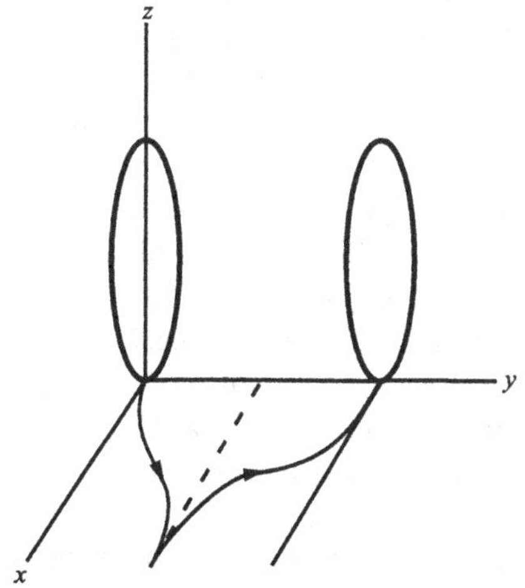


Figure 6.11

We have indicated a path in Fig. 6.11. You should convince yourself that you can obey the constraints and end up at a configuration that differs from the initial configuration by an increment in only one of the coordinates. We have illustrated the case when only $y$ has been changed. (A change in $\psi$ only is very easy since $dx = dy = d\phi = 0$ satisfies the constraints; this is simply revolving the disc about the vertical axis.) Thus, although the two constraints limit us "infinitesimally" to 2 degrees of freedom, we see that actually *all* neighboring states in a 4-dimensional region are "accessible" (by means of piecewise smooth curves) while obeying the constraints. In the general case of $r$ nonholonomic constraints in an $M^n$, there will be a set of states of dimension *greater* than $n - r$ that will be accessible from an initial state via short piecewise smooth paths obeying the constraints. The actual dimension is given by "Chow's theorem," to be discussed in Section 6.3g. We shall discuss a very important special case in thermodynamics in our next section.

For an application of holonomy to the problem of parking a car in a tight spot, see Nelson's book [N, p. 34]

## Problem

**6.2(1)** Show that the Pfaffians in (6.7) are linearly independent.

## 6.3. Heuristic Thermodynamics via Caratheodory

Can one go adiabatically from some state to any nearby state?

### 6.3a. Introduction

In this section we shall look at some elements of thermodynamics from the viewpoint of Frobenius's theorem and foliations. This was first done in 1909 by Caratheodory, who attempted (at the urging of Max Born) an axiomatic treatment of thermodynamics. His treatment had shortcomings; some were purely mathematical, stemming from the local nature of Frobenius's theorem. A careful axiomatic treatment of Caratheodory's approach has been given by J. B. Boyling [Boy]. My goal here is much more limited. I only wish to exhibit the geometrical setup that gives, in my view, the simplest *heuristic* picture for the construction of a *global entropy*, using the mathematical machinery that we have already developed. (My first introduction to the geometrical approach for a *local* entropy was from Bob Hermann; see his book [H].) I restrict myself to systems of a very simple type; I employ strong restrictions, which, however, are not uncommon in other treatments. I will use very specific constructions, for example, I will make use of familiar processes such as "stirring" and "heating at constant volume." We will accept Kelvin's version of the second law. This leads, through Caratheodory's mathematical characterization of a nonholonomic constraint, to the existence of the global entropy.

For supplementary reading I suggest chapter 22 of the book of Bamberg and Sternberg [B, S], but it should be remarked that their thermodynamic entropy is again only locally defined.

### 6.3b. The First Law of Thermodynamics

Consider, for example, a system of regions of fluids separated by "diathermous" membranes: membranes that allow only the passage of heat, not fluids. We assume the system to be connected.

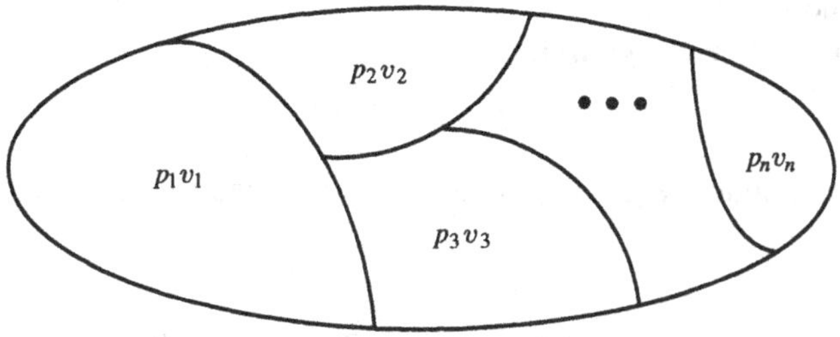


Figure 6.12

We assume that each state of the system is a *thermal equilibrium* state. Let $p_i, v_i$ be the (uniform) pressure and volume of the $i^{\text{th}}$ region. The "equations of state" (e.g., $p_i v_i = n_i RT_i$) at thermal equilibrium will allow us to eliminate all but one pressure, say $p_1$; thus a state, instead of being described by $p_1, v_1, \ldots, p_n, v_n$, can be described by the $(n+1)$-tuple $p_1, v_1, v_2, \ldots, v_n$. It is important to assume that there is a global **internal energy** function $U$ of the system that can be used instead of $p_1$. Our states then have $n+1$ coordinates

$$v_0 := U, v_1, v_2, \ldots, v_n$$

More generally, the *state space* is assumed to be an $n+1$-dimensional manifold $M^{n+1}$ with local coordinates of this type; $U$, however, is a *globally* defined energy function. In Section 6.3c we shall define the state space $M^{n+1}$ more carefully, but for the present we shall only be concerned with local behavior.

A path in $M^{n+1}$ represents a sequence of states, *each in equilibrium*. Physically, we are thus assuming very slow changes in time, that is, **quasi-static** transitions. We shall also need to consider non-quasi-static transitions, such as, "stirring." Such transitions start at some state $x$ and end at some state $y$, but since the intermediate states are not equilibrium states there is no path in $M^{n+1}$ joining $x$ to $y$ that represents the transition. These are "irreversible" processes. *Schematically*, we shall indicate such transitions by a *dashed line curve* joining $x$ to $y$.

On $M^{n+1}$ we assume the existence of a **work 1-form** $W^1$ describing the work done *by* the system during a quasi-static process.

$$W^1 = \sum_{i=1}^{n} p_i dv_i = \sum_{i=1}^{n} p_i(U, v_1, v_2, \ldots, v_n) dv_i$$

*Since we do not assume that $W^1$ is closed, the line integral of $W^1$ is in general dependent upon the path joining the endpoint states.*

We also assume the existence of a **heat 1-form**

$$Q^1 = \sum_{i=0}^{n} Q_i(U, v_1, v_2, \ldots, v_n) dv_i$$

(with again $v_0 = U$) representing heat added or removed from the system (quasi-statically). Again $Q^1$ is not assumed closed. We shall *assume* that $Q^1$ *never vanishes.* (In [$B$, $S$], $Q^1$ is derived, rather than postulated as here.)

We remark that in many books the 1-forms $Q^1$ and $W^1$ would be denoted by $đQ$ and $đW$, respectively. We shall never use this misleading and unnecessary notation; $Q^1$ and $W^1$ are in no sense exact.

The **first law of thermodynamics**

$$dU = Q^1 - W^1$$

associates a "mechanical equivalent energy" to heat and expresses **conservation of energy**.

### 6.3c. Some Elementary Changes of State

1. *Heating at constant volume*

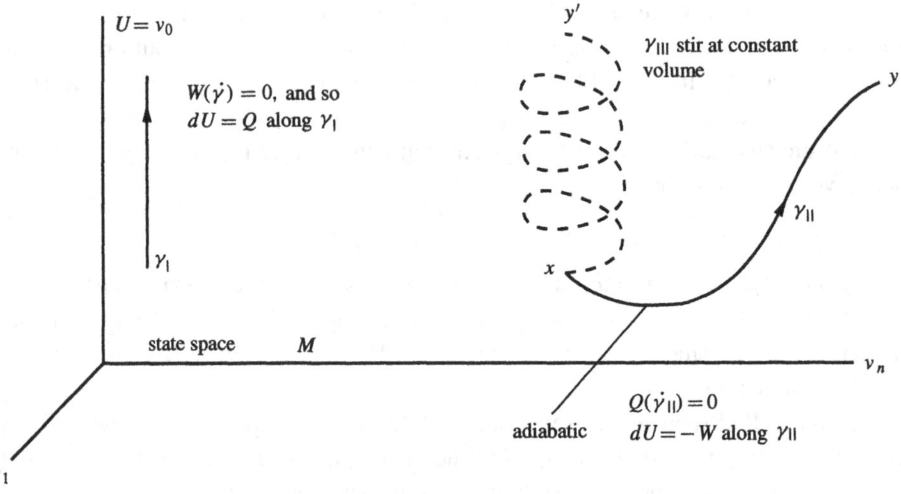


**Figure 6.13**

If $\gamma_I$ is a path representing heating at constant volume, then $dv_1 = 0, \dots, dv_n = 0$, and thus the work 1-form $W$ vanishes when evaluated on the tangent $\dot{\gamma}_I$. From conservation of energy $dU = Q$ along $\gamma_I$.

2. *Quasi-static adiabatic process.* Since no heat is added or removed in such a process we have $Q(\dot{\gamma}_{II}) = 0$ and so $dU = -W$.
3. *Stirring at constant volume.* This is an adiabatic process but since it is not quasi-static we cannot represent it by a curve in state space. We schematically indicate it by a dashed curve $\gamma_{III}$ joining the two end states $x$ and $y'$. $Q$ and $W$ make no sense for this process, but work *is* being done by (or on) the system, the amount of work being the difference of the internal energy $U(y') - U(x)$.

The preceding considerations suggest the following structure of the state space. We shall *assume* that there is a connected $n$-manifold, the **mechanical manifold** $V^n$, and a differentiable map $\pi$ of $M^{n+1}$ onto $V^n$ having the property that the differential $\pi_*$ is always onto. (Such a map is called a **submersion.**) Schematically

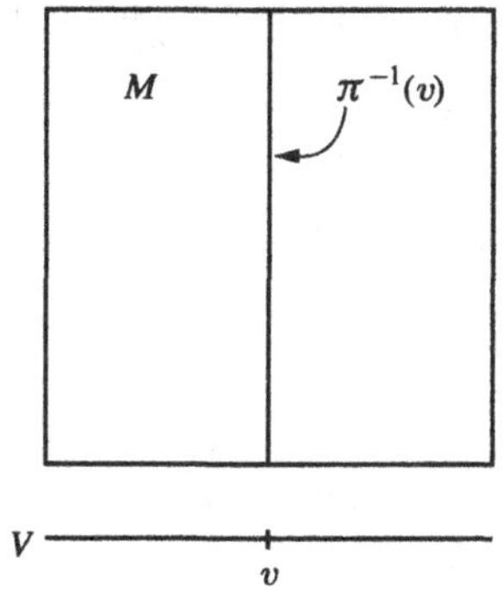


**Figure 6.14**

By the main theorem on submanifolds of Section 1.3d, if $v \in V^n$ then $\pi^{-1}(v)$ is a 1-dimensional embedded submanifold of $M^{n+1}$. We shall *assume* that each $\pi^{-1}(v)$ is *connected.* The manifold $V^n$ will be covered by a collection of *local* coordinate systems, typically denoted by $v^1, \ldots, v^n$. *$V^n$ takes the place of the volume coordinates used before.* The curves $\pi^{-1}(v)$ are the processes "heating and cooling at constant volume" employed previously. Since we have assumed that each such curve is connected, we are assuming that given any pair of states lying on $\pi^{-1}(v)$, one of them can be obtained from the other by "heating at constant volume." It is again assumed that the work 1-form $W^1$ on $M^{n+1}$ is 0 when restricted to $\pi^{-1}(v)$. On the other hand, the heat 1-form $Q^1$ is not 0 when restricted to these curves. The first law then requires that $dU = Q \neq 0$ for such processes. In particular it would be possible to parameterize each $\pi^{-1}(v)$ by internal energy $U$. Then $U, v^1, \ldots, v^n$ forms a local coordinate system for $M^{n+1}$ (with $U$ a global coordinate).

## 6.3d. The Second Law of Thermodynamics

A **cyclic** process is one that starts and ends at the same state. The *second law of thermodynamics, according to* Lord **Kelvin,** can be stated as follows.

> In no quasi-static cyclic process can a quantity of heat be converted *entirely* into its mechanical equivalent of work.

*The second law of thermodynamics, according to* **Caratheodory** (1909), says

> In every neighborhood of every state $x$ there are states $y$ that are not accessible from $x$ via quasi-static **adiabatic** paths, that is, paths along which $Q = 0$.

Caratheodory's assumption is weaker than Kelvin's:

**Theorem (6.12):** *Kelvin's version implies Caratheodory's.*

**PROOF:**

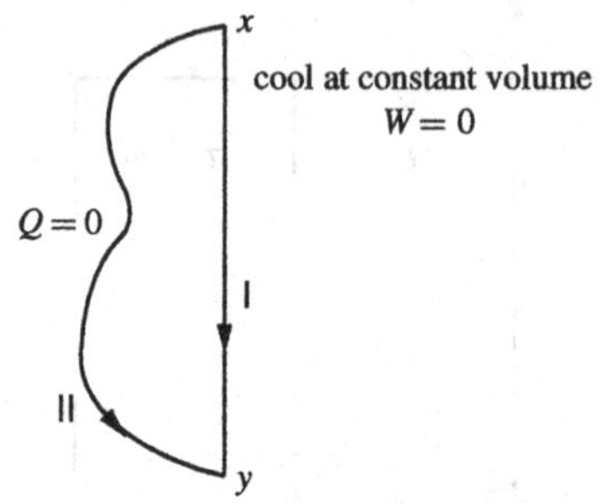


**Figure 6.15**

Given a state $x$, take a process of type $I$ by cooling at constant volume, $W = 0$, ending at a state $y$. We claim that there is no quasi-static *adiabatic* process II going from $x$ to $y$. Suppose that there were. We would then have

$$\int_{II} W = \int_{II} Q - dU = -\int_{II} dU = \int_{-II} dU = \int_{-I} dU = \int_{-I} Q$$

But this would say that the heat energy pumped into the system by going from $y$ to $x$ along $-I$, that is, by heating at constant volume, has been converted *entirely* into its mechanical equivalent of work $\int_{II} W$ by the hypothetical process $II$, contradicting Kelvin. □

Note in fact that no state on $I$ between $x$ and $y$ is quasi-statically adiabatically accessible from $x$.

An adiabatic quasi-static process is a curve characterized by the constraint $Q^1 = 0$. We know that if $Q = 0$ were a *holonomic* constraint then of course there would exist, in any neighborhood of a state $x$, other states $y$ not accessible from $x$ along such adiabatic paths, because the accessible points would all lie on the maximal leaf (integral manifold of codimension 1) through $x$. Does the existence of inaccessible points (i.e., the second law of thermodynamics) conversely imply that the distribution $Q = 0$ (the "adiabatic" distribution) must be integrable? Caratheodory showed that this is indeed the case by proving the following *purely mathematical* result.

**Caratheodory's Theorem (6.13):** *Let $\theta^1$ be a continuously differentiable non-vanishing 1-form on an $M^n$, and suppose that $\theta = 0$ is not integrable; thus at some $x_0 \in M^n$ we have*

$$\theta \wedge d\theta \neq 0$$

*Then there is a neighborhood $U$ of $x_0$ such that any $y \in U$ can be joined to $x_0$ by a piecewise smooth path that is always tangent to the distribution.*

**PROOF SKETCH:** An *indication* of why this should be is easily given. Since $\theta = 0$ is not integrable near $x_0$, we know that there is a pair of vector fields $\mathbf{X}$ and

**Y** defined near $x_0$, always tangent to the distribution $\theta = 0$ but such that [**X**, **Y**] is not in the distribution.

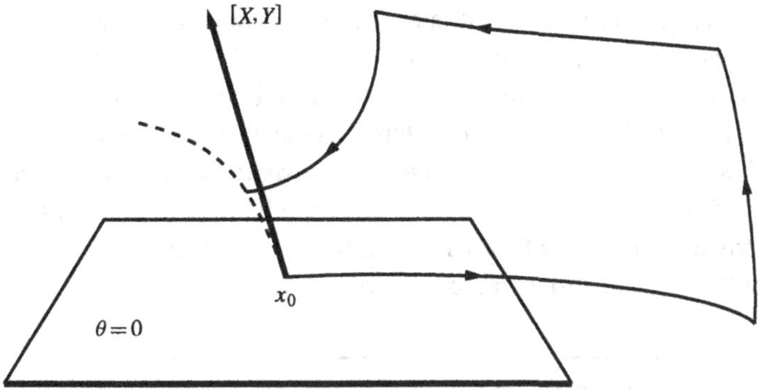


Figure 6.16

Let $\phi$ and $\psi$ be the flows generated by **X** and **Y** respectively. From 4.1c we know that the piecewise smooth integral curves

$$\psi(-\sqrt{t}) \circ \phi(-\sqrt{t}) \circ \psi(\sqrt{t}) \circ \phi(\sqrt{t}) x_0$$

have smooth segments tangent to the distribution $\theta = 0$, and have final endpoints lying on a curve whose tangent is [**X**, **Y**]. This direction is *transverse* to the distribution. *Thus, not only are points "along" $\theta = 0$ accessible from $x_0$, but a curve of points transverse to $\theta = 0$ is accessible also.* It is not difficult to show (using the machinery of the proof of the Frobenius theorem) that in fact *all* points in some neighborhood of $x_0$ are accessible (see [H]). □

We thus conclude from Caratheodory's mathematical theorem together with his version of the second law that

**Theorem (6.14):** *The adiabatic distribution $Q^1 = 0$ is integrable.*

Note that when the state space is 2-dimensional (with coordinates, say, $p_1$ and $v_1$) this is a tautology since *every 1-form in a 2-manifold defines an integrable distribution of curves.*

## 6.3e. Entropy

Since $Q^1 = 0$ is integrable, we know from 6.1a that there are *locally* defined functions $S$, called a **local entropy**, and $\lambda \neq 0$, on the state space $M^{n+1}$ such that $Q^1 = \lambda dS$. Since

$$\frac{Q}{\lambda} = dS$$

we say that $Q^1$ admits a local **integrating factor** $\lambda$ (since $dS$ is exact, $\int Q/\lambda$ is locally path-independent, that is, "integrable"). For thermodynamic purposes it is imperative that $\lambda$ and the entropy $S$ be *globally* defined, but the Frobenius theorem only yields local functions. If, for example, the foliation defined by $Q = 0$ has leaves that wind densely (as in a torus) then there is no way that a global function $S$ can exist, since such an $S$ must be constant on each maximal leaf. It is easy to see, however, that *Kelvin's second law of thermodynamics rules out the possibility of not only dense adiabatic leaves, but even leaves that "double back"*! For in the proof that "Kelvin implies Caratheodory," we saw that two states related by heating at constant volume cannot be joined by a quasi-static adiabatic. This says that *no* $\pi^{-1}(v)$ *can meet a maximal adiabatic leaf twice.*

It might be thought that the space $M^{n+1}$ and the adiabatic foliation must then be of a completely trivial nature. The following foliation of $\mathbb{R}^2$ by curves $Q^1 = 0$ gives some indication of the complications that could arise.

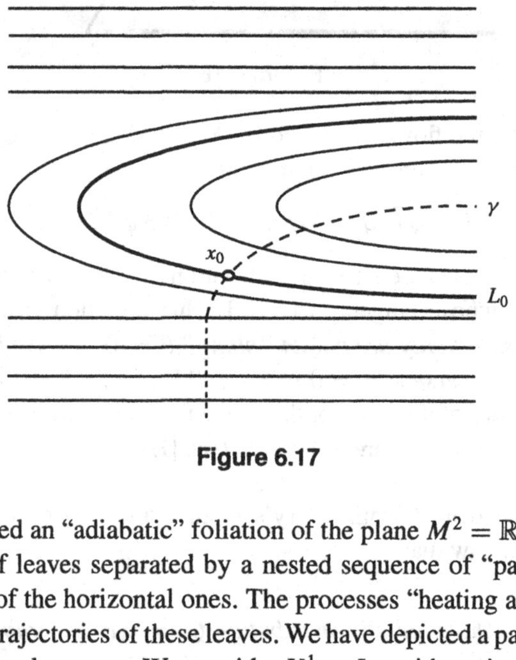


**Figure 6.17**

We have exhibited an "adiabatic" foliation of the plane $M^2 = \mathbb{R}^2$ consisting of two horizontal bands of leaves separated by a nested sequence of "paraboliclike" leaves asymptotic to two of the horizontal ones. The processes "heating at constant volume" are the orthogonal trajectories of these leaves. We have depicted a particular leaf $L_0$ and a particular transversal curve $\gamma$. We consider $V^1 = L_0$, with projection $\pi : M^2 \to V^1$ defined as follows: Move each point in the plane along the orthogonal trajectory through that point until you strike the leaf $L_0$. In particular, if we parameterize $L_0$ by a coordinate $v$ and if we let $v$ be constant on each orthogonal trajectory, then $v$ becomes a global "mechanical" coordinate on the state space $M^2$.

Return now to our quest for a global entropy. We attempt to construct a function $S$ such that $S$ is constant on each maximal adiabatic leaf $Q = 0$, as follows. As in 6.1a, we need a curve that is transverse to the leaves. Let $x_0$ be a given point in $M^{n+1}$, fixed once and for all, and let $\gamma = \gamma(U)$ be the curve $\pi^{-1}(\pi(x_0))$ obtained from $x_0$ by heating and cooling at constant volume, parameterized by internal energy $U$. Since $Q \neq 0$ along this curve (we are heating or cooling), it is transverse to the adiabatic leaves. *This is our transversal*! Let $L$ be a leaf that strikes $\gamma$ at the point $\gamma(U)$. We then define $S(x) = U$ for all $x$ in this leaf. This definition makes sense since we have already seen that the leaf $L$ cannot strike $\gamma$ a second time. *We have defined $S$ for all states that lie on adiabatic leaves that strike the basic transversal $\gamma$. If every maximal adiabatic leaf on $M^{n+1}$ met the basic transversal $\gamma$ then the function $S$ would be globally defined.* A general foliation will not have this property. For example, in our illustrated foliation of $\mathbb{R}^2$, we have exhibited the basic transversal $\gamma$ through $x_0$ and it is clear that this transversal does not meet any of the horizontal leaves at the top! Consequently, no state $y$ on one of these top leaves can be adiabatically deformed to have the same volume coordinate as $x_0$!

Sufficiently simple thermodynamical systems do not exhibit this behavior. Given two states $x_0$ and $y$, consisting of collections of contiguous bags of fluids, as in Fig. 6.12, we ought to be able to "massage" the bags in state $y$, quasi-statically and adiabatically, so that the final state $y'$ has the same volume coordinates as the state $x_0$. Thus the adiabatic leaf through $y$ would indeed strike the transversal through $x_0$ at the state $y'$.

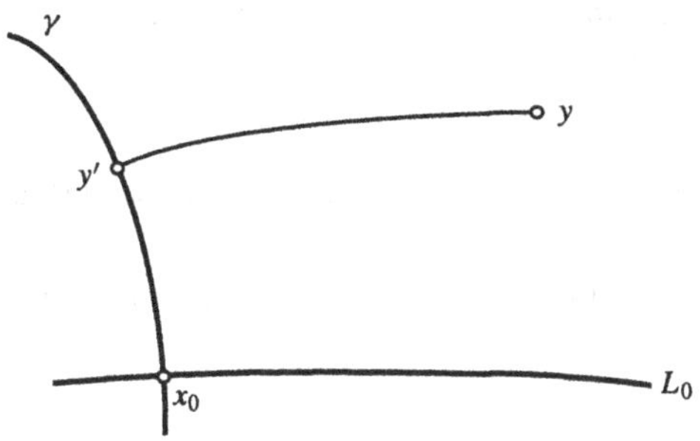


**Figure 6.18**

Furthermore, if, for instance, $U(y') \geq U(x_0)$, then by stirring at constant volume we could go adiabatically (but not quasi-statically) from $x_0$ to $y'$. If $U(y') \leq U(x_0)$ we could stir from $y'$ to $x_0$. This would say that given *any* pair of states $x$ and $y$, either $y$ is adiabatically accessible from $x$ or $x$ is adiabatically accessible from $y$, though not necessarily in quasi-static transitions.

Thus we shall *assume* that a basic transversal will strike every adiabatic leaf; we are then assured of the existence of a global entropy function $S$, which we assume smooth. By construction, then, $Q^1 = \lambda dS$ for some globally defined integrating factor $\lambda$. $\lambda \neq 0$ since $Q$ never vanishes. Since $S = U$ on $\gamma$ and $dU = Q$ along $\gamma$, we see $\lambda > 0$. As we shall see, $S$ is non-decreasing for each adiabatic process. $S$ is called an **empirical entropy**.

## 6.3f. Increasing Entropy

Experience shows that if we start at a state $y$ and "stir" the system adiabatically at constant volume (this cannot be done quasi-statically) we shall arrive at a state $x$ having the property that no *adiabatic* process (quasi-static or not) can return us to $y$; *we cannot "unstir" the system.*

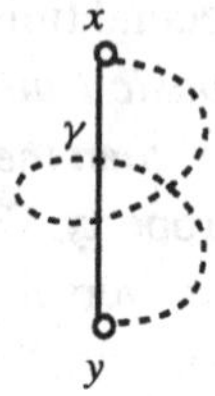


Figure 6.19

In Figure 6.19 we have stirred from $y$ to $x$. $U(x) > U(y)$. Note that $x$ can also be reached from $y$ by heating at constant volume.

We *assume* that if $x$ and $y$ are on $\pi^{-1}(v)$ and if $U(x) > U(y)$, then there is no adiabatic process, quasi-static or not, that will take us from $x$ to $y$.

**Theorem (6.15):** *If a state $y$ results from $x$ by any adiabatic process (quasi-static or not), then $S(y) \geq S(x)$.*

(Of course if the process is quasi-static then $dS = Q/\lambda = 0$ in the process.)

PROOF: Suppose that $S(x) > S(y)$ and that there is some adiabatic process $x \to y$ leading from $x$ to $y$.

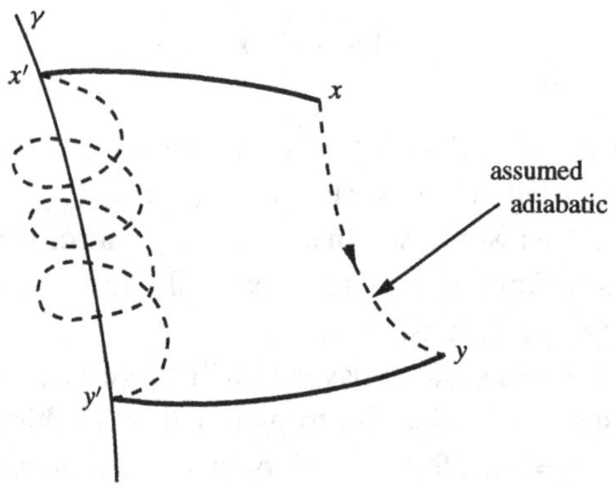


Figure 6.20

By deforming adiabatically we may move $x$ and $y$ quasi-statically to $x'$ and $y'$ on the basic transversal $\gamma$ through $x_0$. Then

$$S(x') = S(x) > S(y) = S(y')$$

But along the basic transversal $\gamma$ we have $S = U$, and so $U(x') > U(y')$. We could then stir adiabatically from $y'$ to $x'$. But then we could "unstir" by the adiabatic going from $x'$ to $x$ to $y$ to $y'$, a contradiction! Thus the adiabatic from $x$ to $y$ cannot exist. □

By assuming the existence of an **empirical temperature** and by combining simple systems into a single compound system (while introducing no "*adiabatic*" membranes) one can show that there is a specific *universal* choice for the integrating factor $\lambda$, called the **absolute temperature** $T$, *that depends only on the empirical temperature*. The resulting empirical entropy function $S$ is then *the* entropy

$$\frac{Q}{T} = dS$$

This is indicated in most books dealing with thermodynamics, for example, [B, S]. A careful mathematical treatment is given in Boyling's paper [Boy].

### 6.3g. Chow's Theorem on Accessibility

Let $\mathbf{Y}_\alpha, \alpha = 1, \ldots, n$, be vector fields on an $M^n$ that are linearly independent in the neighborhood of a point $P$. Then any point on $M$ sufficiently close to $P$ is accessible from $P$ by a sequence of broken integral curves of the fields $\mathbf{Y}_\alpha$; this was the significance of the computation (6.5), when coupled with the inverse function theorem.

In our sketch of Caratheodory's theorem (6.13) we have indicated a proof of the following: If vector fields $\mathbf{X}_1$ and $\mathbf{X}_2$ are tangent to a distribution $\Delta$ on an $M^n$, but $[\mathbf{X}_1, \mathbf{X}_2]$ is not, then by moving along a sequence of broken integral curves of $\mathbf{X}_1$ and $\mathbf{X}_2$ the endpoints trace out a curve tangent to $[\mathbf{X}_1, \mathbf{X}_2]$, which is transverse to $\Delta$. Thus

> points on integral curves of $[\mathbf{X}_1, \mathbf{X}_2]$ are accessible by broken integral curves of $\mathbf{X}_1$ and $\mathbf{X}_2$.

Let vector fields $\mathbf{X}_\alpha, \alpha = 1, \ldots, r$ span an $r$-dimensional distribution $\Delta$ on some neighborhood of $P$ on an $n$-manifold $M^n$. Suppose that $\Delta$ is not closed under brackets. *Adjoin* to the vector fields $\mathbf{X}_\alpha$ the vector fields $[\mathbf{X}_\alpha, \mathbf{X}_\beta]$ obtained from all the brackets. It may be that the new system of vector fields is still not closed under taking brackets; adjoin then all brackets of the new system, yielding a still larger system. *Suppose* that after a finite number of such adjoinings one is left with a distribution $D(\Delta)$ that has *constant* dimension $s \leq n$ and is closed under brackets, that is, is in involution. By Frobenius there is an immersed integral leaf $V^s$ of this distribution passing through $P$. From Caratheodory's theorem (6.13), points of this submanifold that are sufficiently close to $P$ are accessible from $P$ by broken integral curves of the *original* system $\mathbf{X}_\alpha$. Further, no points off the maximal leaf $V$ are accessible. This is the essential content of Chow's theorem. See [H] for more details.

PART TWO

# Geometry and Topology

CHAPTER 7

# $\mathbb{R}^3$ and Minkowski Space

## 7.1. Curvature and Special Relativity

What does the curvature of a world line signify in space–time?

### 7.1a. Curvature of a Space Curve in $\mathbb{R}^3$

We associate to a parameterized curve $C, \mathbf{x} = \mathbf{x}(t)$ in $\mathbb{R}^3$, its tangent vector $\dot{\mathbf{x}}(t) = (\dot{x}, \dot{y}, \dot{z})^T$. When $t$ is considered time, this tangent is the velocity vector $\mathbf{v}$, with speed $\| \mathbf{v} \| = v$. Introduce the arc length parameter $s$ by means of

$$\left(\frac{ds}{dt}\right)^2 = \| \dot{\mathbf{x}} \|^2 = v^2, \qquad s(t) = \int_0^t \| \dot{\mathbf{x}}(u) \| \, du$$

We then have the *unit* tangent vector $\mathbf{T} := d\mathbf{x}/ds = \dot{\mathbf{x}}\, dt/ds = \mathbf{v}/v$, that is, $\mathbf{v} = v\mathbf{T}$. For acceleration $\mathbf{a}$ we have

$$\mathbf{a} = \dot{\mathbf{v}} = \dot{v}\mathbf{T} + v\frac{\mathbf{d}T}{dt} = \dot{v}\mathbf{T} + v^2\frac{d\mathbf{T}}{ds}$$

Since $\mathbf{T}$ has constant length, $d\mathbf{T}/ds$ is orthogonal to $\mathbf{T}$ and so is normal to the curve $C$. If $d\mathbf{T}/ds \neq \mathbf{0}$, then its direction defines a unique unit normal to the curve called the **principal normal n**

$$\frac{d\mathbf{T}}{ds} = \kappa(s)\mathbf{n}(s) \tag{7.1}$$

where the function $\kappa(s) \geq 0$ is the **curvature** of $C$ at (parameter value) $s$. Then the acceleration

$$\mathbf{a} = \dot{v}\mathbf{T} + v^2\kappa(s)\mathbf{n} \tag{7.2}$$

lies in the **osculating plane**, the plane spanned by $\mathbf{T}$ and $\mathbf{n}$. To compute $\kappa$ in terms of the original parameter $t$ rather than $s$, note that

$$\begin{aligned} \mathbf{v} \times \mathbf{a} &= v\mathbf{T} \times (\dot{v}\mathbf{T} + v^2\kappa(s)\mathbf{n}) \\ &= v^3\kappa\mathbf{T} \times \mathbf{n} \end{aligned}$$

and so

$$\kappa = \frac{\| \mathbf{v} \times \mathbf{a} \|}{v^3}$$

See Problems 7.1(1) and (2).

We define the **curvature vector** by

$$\boldsymbol{\kappa} = \frac{d\mathbf{T}}{ds} = \kappa \mathbf{n}$$

We remark that when dealing with a *plane* curve, that is, a curve in $\mathbb{R}^2$, a slightly different definition that allows the curvature to be a signed quantity is usually used. If $\mathbf{T} = (\cos\alpha,\ \sin\alpha)^T$ is the unit tangent (where $\alpha$ is the angle from the $x$ axis to the tangent) then $\mathbf{T}^\perp = (-\sin\alpha,\ \cos\alpha)^T$ is the unit normal resulting from a counterclockwise rotation of the tangent. Then $d\mathbf{T}/ds = \tilde{\kappa}\mathbf{T}^\perp$ defines a signed curvature $\tilde{\kappa} = \pm\kappa$. But then

$$\frac{d\mathbf{T}}{ds} = \frac{d}{ds}(\cos\alpha,\ \sin\alpha)^T = (-\sin\alpha,\ \cos\alpha)^T \frac{d\alpha}{ds}$$

gives the familiar

$$\tilde{\kappa} = \frac{d\alpha}{ds}$$

It is shown in books on differential geometry that $\kappa$ and the osculating plane have the following geometric interpretations. To compute $\kappa(s)$ we consider the three nearby points $\mathbf{x}(s-\epsilon)$, $\mathbf{x}(s)$, and $\mathbf{x}(s+\epsilon)$ on $C$. If these points are not colinear (and generically they aren't) they determine a circle of some radius $\rho_\epsilon$ passing through $\mathbf{x}(s)$ and lying in some plane $P_\epsilon$. Under mild conditions, it is shown that $\lim_{\epsilon\to 0} P_\epsilon$ is the osculating plane and $\rho(s) = \lim_{\epsilon\to 0}\rho_\epsilon = 1/\kappa(s)$ is the **radius of curvature** of $C$ at $s$. (If $d\mathbf{T}/ds = \mathbf{0}$ at $s$, we say $\kappa(s) = 0$, $\rho = \infty$, and the osculating plane at $s$ is undefined.) Then (7.2) becomes

$$\mathbf{a} = \dot{v}\mathbf{T} + \left(\frac{v^2}{\rho}\right)\mathbf{n}$$

the classical expression for the tangential and normal components of the acceleration vector.

## 7.1b. Minkowski Space and Special Relativity

Minkowski space $M_0^4$ is $\mathbb{R}^4$ but endowed with the "pseudo-Riemannian" or "Lorentzian" metric or "arc length" (as discussed in Section 2.1d)

$$ds^2 = -c^2dt^2 + dx^2 + dy^2 + dz^2 \tag{7.3}$$

Here $c$ is the speed of light, and the coordinates $t = x^0, x = x^1, y = x^2, z = x^3$ for which $ds^2$ assumes the form (7.3) form an **inertial** coordinate system. (For physical motivation and further details see, for example, [Fr].) The metric tensor $g_{ij} = \langle \partial/\partial x^i, \partial/\partial x^j \rangle$ is then

$$(g_{ij}) = \mathrm{diag}(-c^2, 1, 1, 1) \tag{7.4}$$

**Warning:** Many books use the negative of this metric!

Let $x = (t, \mathbf{x})$ and let $d\mathbf{x} \bullet d\mathbf{x}$ be the usual dot product in $\mathbb{R}^3$. Then

$$ds^2 = -c^2 dt^2 + d\mathbf{x} \bullet d\mathbf{x}$$

Then a **4-vector**, that is, a tangent vector to $M_0^4$,

$$v = (v^0, \mathbf{v})^T = v^0 \partial/\partial t + v^\alpha \partial/\partial x^\alpha = v^0 \partial/\partial t + \mathbf{v}$$

is said to be

| | |
|---|---|
| **spacelike** | if $\langle v, v\rangle > 0$ |
| **timelike** | if $\langle v, v\rangle < 0$ |
| **lightlike** | if $\langle v, v\rangle = 0$ |

The path $x = x(t)$ of a mass point in $M_0^4$ is called its **world line**. Its tangent vector $dx/dt = (1, d\mathbf{x}/dt)^T = (1, \mathbf{v})^T$ is timelike since

$$\left\langle \frac{dx}{dt}, \frac{dx}{dt} \right\rangle = -c^2 + \mathbf{v} \bullet \mathbf{v} = -c^2 + v^2$$

and, as we shall see, $v < c$. Thus the tangent vector to the world line of a mass particle lies inside the **light cone** $\mathbf{x} \bullet \mathbf{x} = c^2 t^2$.

We shall call $\mathbf{v} = d\mathbf{x}/dt$ the **classical velocity vector**.

The analogue of the arc length parameter in $\mathbb{R}^3$ for the world line of a particle in $M_0^4$ is the **proper time** parameter $\tau$ defined by pulling back the tensor $-c^{-2} ds^2$ to the curve

$$\begin{aligned} d\tau^2 :&= -c^{-2} ds^2 = dt^2 - c^{-2} d\mathbf{x} \bullet d\mathbf{x} \\ &= \left(1 - \frac{v^2}{c^2}\right) dt^2 \end{aligned} \tag{7.5}$$

Define the **Lorentz factor** $\gamma$ by

$$\gamma := \left(1 - \frac{v^2}{c^2}\right)^{-1/2} = \frac{dt}{d\tau} \tag{7.6}$$

An analogue of the unit tangent in $\mathbb{R}^3$ is the **velocity 4-vector** $u$

$$u := \frac{dx}{d\tau} = \left(\frac{dt}{d\tau}, \frac{d\mathbf{x}}{d\tau}\right)^T = \gamma(1, \mathbf{v})^T \tag{7.7}$$

Note that

$$\langle u, u\rangle = \gamma^2(-c^2 + v^2) = -c^2 \tag{7.8}$$

We define, as usual, $\| A \|^2 := \langle A, A\rangle$ even though this may be negative! (When it is negative we shall never use its square root $\| A \|$.) A is said to be a **unit vector** if $\| A \|^2 = \pm c^2$; $u$ is a unit vector in the sense that one usually uses units in which the speed of light $c = 1$. The physical interpretation of the proper time parameter $\tau$ along

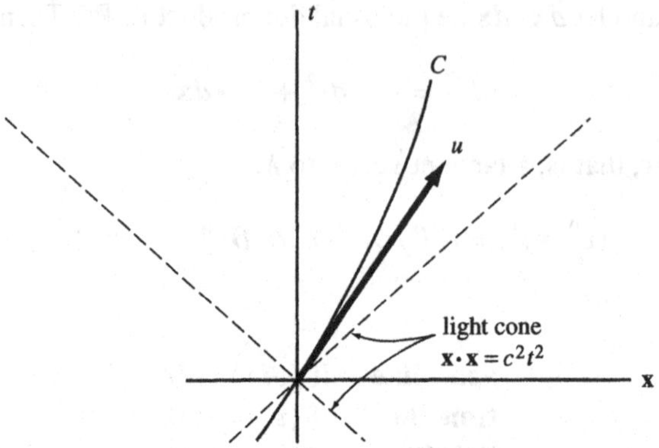


**Figure 7.1**

a world line $C$ is as follows (see [Fr, p. 18]):

$$\tau = \int \left(1 - \frac{v^2}{c^2}\right)^{1/2} dt$$

is the *time kept by an "atomic clock" moving with the particle along the world line C*. In particular, *coordinate time t* is the proper time kept by an atomic clock fixed at the spatial origin $\mathbf{x} = \mathbf{0}$ of the inertial coordinate system.

Associated with any particle is its **rest mass** $m_0$; this is an invariant (independent of coordinates, i.e., **observers**).

The **(linear) momentum** $P$ of the particle is the 4-vector

$$P := m_0 u = (m_0 \gamma, m\mathbf{v})^T \tag{7.9}$$

where

$$m := m_0 \gamma = m_0 \left(1 - \frac{v^2}{c^2}\right)^{-1/2}$$

is sometimes called the **relativistic mass;** $m$ is interpreted as the mass of the moving particle as viewed from the "fixed" inertial coordinate system. Note that $m \to \infty$ as $v \to c$, and, as we shall see in (7.15), an infinite classical force would be required to accelerate a mass to the speed of light. This is the justification for the assumption that $v < c$ for all massive particles.

Note that the momentum 4-vector has constant "length"

$$\| P \|^2 = \langle P, P \rangle = -m_0^2 c^2$$

If we define the **classical momentum** by $\mathbf{p} := m\mathbf{v}$ (with a *variable* mass!) then we can write $P = (m, \mathbf{p})^T$ and then $\langle P, P \rangle = -c^2 m^2 + p^2$, and so

$$m^2 c^2 = m_0^2 c^2 + p^2 \tag{7.10}$$

The analogue of the curvature vector $d\mathbf{T}/ds$ in $\mathbb{R}^3$ is the **curvature** or **acceleration** 4-vector

$$\frac{du}{d\tau}$$

The **Minkowski force** is the 4-vector defined by

$$f := \frac{dP}{d\tau} = \frac{d(m_0 u)}{d\tau} \tag{7.11}$$

Thus

$$f = \frac{d}{d\tau}(m, \mathbf{p})^T = \left(\frac{dm}{d\tau}, \gamma\frac{d\mathbf{p}}{dt}\right)^T = (f^0, \gamma \mathbf{f}_c)^T \tag{7.12}$$

where $\mathbf{f}_c := d\mathbf{p}/dt$ is the **classical force** in $\mathbb{R}^3$ and where $f^0$ is the $t = x^0$ component of $f$. Since $\langle P, P\rangle$ is a constant, $f = dP/d\tau$ must be orthogonal to $P$ (and thus to $u$) *in the Minkowski metric*

$$0 = \langle f, u\rangle = -c^2 f^0 \gamma + \gamma \mathbf{f}_c \bullet \gamma \mathbf{v}$$

that is,

$$f^0 = \left(\frac{\gamma}{c^2}\right)\mathbf{f}_c \bullet \mathbf{v} \tag{7.13}$$

The time component of the Minkowski 4-force is, except for a factor, the classical power (rate of doing work). Finally

$$f = \gamma(c^{-2}\mathbf{f}_c \bullet \mathbf{v}, \mathbf{f}_c)^T \tag{7.14}$$

Note that $f^0 = dm/d\tau = \gamma dm/dt$ shows that

$$\frac{dm}{dt} = c^{-2}\mathbf{f}_c \bullet \mathbf{v} \tag{7.15}$$

and so

$$d(c^2 m) = \mathbf{f}_c \bullet d\mathbf{x}$$

is the element of work done by the classical force. Classically this is the *energy* imparted to the particle. This leads us to associate to a mass $m$ an **energy** $E = mc^2$ and a **rest energy** $m_0c^2$. (7.10) becomes

$$E^2 = E_0^2 + c^2 p^2 \tag{7.16}$$

and we have

$$P = \left(\frac{E}{c^2}, \mathbf{p}\right)^T$$

Since $E/c^2$ appears as the time component of the momentum 4-vector, we see that special relativity unites the energy and classical momentum into a 4-vector, the momentum 4-vector.

The familiar startling effects of special relativity, such as length contraction and time dilation, are consequences of the geometry of Minkowski space. Their explanation rests on Einstein's simple analysis of the concept of time and simultaneity. This analysis was Einstein's monumental contribution to special relativity, and gave meaning to the ad hoc assumptions put forth previously by Lorentz, Poincaré, Larmor, and Fitzgerald; see [Fr].

### 7.1c. Hamiltonian Formulation

Consider a mass particle moving in $\mathbb{R}^3$ and suppose that the classical force is derivable from a time-independent potential $\mathbf{f}^c = -\nabla V$. From (7.15), $dm/dt = -c^{-2}\nabla V \cdot \mathbf{v} = -c^{-2}dV/dt$ along the world line, and consequently

$$H := mc^2 + V$$

is a constant of the motion and deserves the name **total energy**. In the phase space $\mathbb{R}^6$, $V$ is a function of $\mathbf{x} = q$ alone, and from (7.10) $mc^2 = (m_0^2c^4 + p^2c^2)^{1/2}$ is a function of $p$ alone. From (7.10) we have $2mc^2\partial m/\partial p_\alpha = 2p_\alpha$, showing that $\partial(mc^2)/\partial p_\alpha = p_\alpha/m = v_\alpha$, where $\alpha = 1, 2, 3$. Then

$$\frac{dx^\alpha}{dt} = v^\alpha = \frac{\partial(mc^2)}{\partial p_\alpha} = \frac{\partial(mc^2 + V)}{\partial p_\alpha} = \frac{\partial H}{\partial p_\alpha}$$

and

$$\frac{dp_\alpha}{dt} = f_\alpha^c = -\frac{\partial V}{\partial x^\alpha} = -\frac{\partial}{\partial x^\alpha}(mc^2 + V) = -\frac{\partial H}{\partial x^\alpha}$$

and thus we are able to put the equations of motion in Hamiltonian form provided we define the **Hamiltonian** $H$ to be the total energy.

## Problems

**7.1(1)** Compute the curvature of the helix $x = \cos\omega t$, $y = \sin\omega t$, $z = kt$, where $\omega$ and $k$ are constants.

**7.1(2)** Assume $\kappa \neq 0$; then $\mathbf{n}$ is well defined and we can define the **binormal** vector $\mathbf{B}$ to be the normal to the osculating plane, $\mathbf{B} = \mathbf{T} \times \mathbf{n}$. Show that $d\mathbf{B}/ds$ lies along $\mathbf{n}$, and hence the **torsion** $\tau$ is well defined by $d\mathbf{B}/ds = \tau(s)\mathbf{n}$. Then show that $d\mathbf{n}/ds = -\kappa(s)\mathbf{T} - \tau(s)\mathbf{B}$. (The equations for the arc length derivatives of $\mathbf{T}, \mathbf{n}$, and $\mathbf{B}$ constitute the **Serret–Frenet** formulas.)

**7.1(3)** Show that the action for a particle with $H = mc^2 + V$ is

$$\int p_\alpha dx^\alpha - Hdt = -m_0c^2 \int d\tau - \int Vdt$$

## 7.2. Electromagnetism in Minkowski Space

How can $\mathcal{E}^1$ and $\mathcal{B}^2$ be united to yield a 2-form in space-time?

### 7.2a. Minkowski's Electromagnetic Field Tensor

The Heaviside–Lorentz force law (3.36) becomes $\mathbf{f} = q[\mathbf{E} + (\mathbf{v}/c) \times \mathbf{B}]$ when we use units for which the speed of light $c$ is not necessarily 1. This spatial force vector can be completed to a Minkowski force 4-vector by using the prescription (7.14)

$$f = \gamma(c^{-2}\mathbf{f} \cdot \mathbf{v}, \mathbf{f})^T = \gamma q\left(c^{-2}\mathbf{E} \cdot \mathbf{v}, \mathbf{E} + \left(\frac{\mathbf{v}}{c}\right) \times \mathbf{B}\right)^T$$

The covariant expression for $f$, that is, the associated 1-form $f^1$, is, from (7.4),

$$f^1 = -\gamma q[i_{\mathbf{v}}\mathcal{E}^1]dt + \gamma q[\mathcal{E}^1 - i_{\mathbf{v}/c}(\mathcal{B}^2)]$$

Recall that the velocity 4-vector is $u = \gamma\partial/\partial t + \gamma\mathbf{v}$. In Problem 7.2(1) you are asked to show that $f^1$ can be written

$$f^1 = -qi_u F^2$$

where (7.17)

$$F^2 := \mathcal{E}^1 \wedge dt + c^{-1}\mathcal{B}^2$$

is the **electromagnetic field strength** 2-form.

The velocity 4-vector $u$ is intrinsic to the world line; since it is constructed using proper time $\tau$ rather than coordinate time $t$, all inertial coordinate systems will agree on the vector $u$ even though their local coordinate expressions for it will differ. The Lorentz force covector is intrinsic; this is a consequence of the *assumption* that $q[\mathbf{E}+(\mathbf{v}/c)\times\mathbf{B}]$ is an *accurate* discription of the *classical* force $\mathbf{f}_c$ acting on a charged particle even when moving at relativistic speeds! It follows then, from (7.17), that $F^2$ is intrinsic; that is, *$F^2$ is a covariant second-rank tensor!* This skew symmetric tensor was first introduced in 1907 by Minkowski.

*From this point on we shall revert to units in which the speed of light is unity*

$$c = 1$$

Written in full

$$F^2 = (E_1dx + E_2dy + E_3dz) \wedge dt + B_1dy \wedge dz + B_2dz \wedge dx + B_3dx \wedge dy \tag{7.18}$$

(Since the *spatial* part of the metric is euclidean we have $E_\alpha = E^\alpha$, etc.) If we write, as usual, $F^2 = \sum_{i<j} F_{ij}dx^i \wedge dx^j$, we see

$$(F_{ij}) = \begin{bmatrix} 0 & -E_1 & -E_2 & -E_3 \\ E_1 & 0 & B_3 & -B_2 \\ E_2 & -B_3 & 0 & B_1 \\ E_3 & B_2 & -B_1 & 0 \end{bmatrix}$$

The Lorentz force law (7.17) can then be written (from (2.76))

$$f_i = qF_{ij}u^j \tag{7.19}$$

Consider a second inertial coordinate system $t'$, $\mathbf{x}'$ (with identical orientation), representing an observer moving along the $x$ axis of the first observer with constant speed $v$. We assume that their spatial origins coincide when $t = t' = 0$. Elementary arguments (as in [Fr]) show that $y = y'$ and $z = z'$. We shall then only be concerned with the relations between $t$, $x$ and $t'$, $x'$. The basis vectors for the unprimed system are $\mathbf{e}_0 = (1, 0)^T$ and $\mathbf{e}_1 = (0, 1)^T$. The basis vector $\mathbf{e}_0'$ is of the form $(t, x)^T$ in the unprimed system; it must satisfy $-t^2+x^2 = -1$, and so it is of the form $(\cosh\alpha, \sinh\alpha)^T$. Likewise, to maintain Lorentz orthogonality, $\mathbf{e}_1'$ must be $(\sinh\alpha, \cosh\alpha)^T$. Thus, assuming a linear coordinate change, the coordinate systems are related by $t = t'\cosh\alpha + x'\sinh\alpha$ and $x =$

$t' \sinh\alpha + x' \cosh\alpha$. The spatial origin of the primed system, $x' = 0$, is moving so that $x = vt$. Thus $\tanh\alpha = v$. This allows us to express $\sinh\alpha$ and $\cosh\alpha$ in terms of $v$, yielding the usual expressions for the **Lorentz transformations** (with constant $v$ and $\gamma$)

$$t = \gamma(t' + vx') \qquad x = \gamma(x' + vt') \tag{7.20}$$
$$y = y' \qquad z = z'$$

One can check immediately that under such a coordinate change the volume form

$$\text{vol}^4 = dt \wedge dx \wedge dy \wedge dz = dt' \wedge dx' \wedge dy' \wedge dz'$$

is unchanged.

I wish to emphasize that Lorentz transformations in general are simply *the* changes of coordinates in $\mathbb{R}^4$ that leave the origin fixed and *preserve the form* $-t^2+x^2+y^2+z^2$.

If we make a Lorentz transformation (7.20), the local expression for the form $F^2$ in (7.18) will pull back to an expression $F^2 := \mathcal{E}'^1 \wedge dt' + \mathcal{B}'^2$. In Problem 7.2(2) you are asked to compute that

$$\begin{aligned} E_1' &= E_1 & B_1' &= B_1 \\ E_2' &= \gamma(E_2 - vB_3) & B_2' &= \gamma(B_2 + vE_3) \\ E_3' &= \gamma(E_3 + vB_2) & B_3' &= \gamma(B_3 - vE_2) \end{aligned} \tag{7.21}$$

showing, for example, that a pure electric field in a "fixed" system will yield both an electric and a magnetic field when viewed from a moving system. Since (see Problem 7.2(3))

$$F \wedge F = -2\mathbf{E} \cdot \mathbf{B}\, \text{vol}^4 \tag{7.22}$$

we see that $\mathbf{E} \cdot \mathbf{B}$ *is an invariant of such Lorentz transformations!* (If, however, we had allowed a change of orientation, then $\mathbf{E} \cdot \mathbf{B}$ would be replaced by its negative since $F \wedge F$ is a true 4-form and $\text{vol}^4$ is a pseudoform.)

### 7.2b. Maxwell's Equations

In Minkowski space we have (see (4.40))

$$d = \mathbf{d} + dt \wedge \frac{\partial}{\partial t}$$

Then, for $F^2 = \mathcal{E}^1(t, \mathbf{x}) \wedge dt + \mathcal{B}^2(t, \mathbf{x})$, we have

$$dF = \mathbf{d}\mathcal{E} \wedge dt + \mathbf{d}\mathcal{B} + dt \wedge \frac{\partial \mathcal{B}}{\partial t} = \left(\mathbf{d}\mathcal{E} + \frac{\partial \mathcal{B}}{\partial t}\right) \wedge dt + \mathbf{d}\mathcal{B} \tag{7.23}$$

and so

$$dF = 0 \Leftrightarrow \left\{ \begin{array}{c} \mathbf{d}\mathcal{E} = -\dfrac{\partial \mathcal{B}}{\partial t} \\ \text{and} \\ \mathbf{d}\mathcal{B} = 0 \end{array} \right\} \tag{7.24}$$

Thus $dF = 0$ *is equivalent to the first pair of Maxwell's equations.*

If there are no singularities in the field $F^2$, then, since Minkowski space is simply $\mathbb{R}^4$, the converse to the Poincaré lemma assures us that $F^2 = dA^1$ for some 1-form $A$. (Away from singularities, such an $A^1$ will exist locally.) Write

$$F^2 = dA^1 \tag{7.25}$$
$$A^1 = \phi dt + \mathcal{A}^1$$

where $\mathcal{A}^1 = A_\alpha(t, \mathbf{x})dx^\alpha$ and where Greek indices run from 1 to 3. Then $\mathcal{E}^1 \wedge dt + \mathcal{B}^2 = (\mathbf{d} + dt \wedge \partial/\partial t)(\phi dt + \mathcal{A}^1) = \mathbf{d}\phi \wedge dt + \mathbf{d}\mathcal{A}^1 + dt \wedge \partial\mathcal{A}^1/\partial t$ yields

$$\mathcal{E}^1 = \mathbf{d}\phi - \frac{\partial \mathcal{A}^1}{\partial t}$$

and (7.26)

$$\mathcal{B}^2 = \mathbf{d}\mathcal{A}^1$$

This yields the vector expressions $\mathbf{E} = \nabla\phi - \partial\mathbf{A}/\partial t$ and $\mathbf{B} = \operatorname{curl}\mathbf{A}$. $\phi$ is the scalar and $\mathbf{A}$ the vector potential. (*In most physics books $\nabla\phi$ occurs with a negative sign.*)

Consider a charged fluid (with charge density $\rho$) moving in $\mathbb{R}^3$ with local velocity vector $\mathbf{v}$. The current vector is $\mathbf{j} = \rho\mathbf{v}$; $\rho$ is the charge density as measured in the inertial system $x$. If $\rho_0 = \rho_0(t, \mathbf{x})$ is the *rest charge density*, that is, the density as measured by an observer moving instantaneously with the fluid, then

$$\rho = \rho_0\gamma$$

since the charge contained in a moving region must be independent of the observer and yet volumes are decreased by a factor $1/\gamma$ when viewed from a system in (relative) motion with speed $v$ (see [Fr], p. 112). Thus $\mathbf{j} = \rho_0\gamma\mathbf{v}$. Since $\rho_0$ is, by definition, independent of observer, we may construct an intrinsic 4-vector, the **current 4-vector**

$$J := \rho_0 u = (\rho_0\gamma, \rho_0\gamma\mathbf{v})^T = (\rho, \rho\mathbf{v})^T = (\rho, \mathbf{j})^T \tag{7.27}$$

We may then construct the associated **current 3-form**

$$\mathcal{S}^3 = i_J \operatorname{vol}^4 = i\left(\rho\frac{\partial}{\partial t} + \mathbf{j}\right) dt \wedge dx \wedge dy \wedge dz \tag{7.28}$$
$$= \rho dx \wedge dy \wedge dz - (j_1 dy \wedge dz + j_2 dz \wedge dx + j_3 dx \wedge dy) \wedge dt$$
$$\mathcal{S}^3 = \sigma^3 - \jmath^2 \wedge dt$$

In an important sense, $\mathcal{S}^3$ *is more basic than* $J$ (see Section (14.1c)).

We may now consider the second set of Maxwell equations. Define the pseudo-2-form $*F$ (where the star is *not* bold) as follows (the reason for this notation will be explained in Chapter 14):

$$*F^2 = -\boldsymbol{*}\mathcal{B} \wedge dt + \boldsymbol{*}\mathcal{E}$$

(see (3.41)). Then, as in (7.23)

$$d * F^2 = \mathbf{d}\boldsymbol{*}\mathcal{E} - \left(\mathbf{d}\boldsymbol{*}\mathcal{B} - \frac{\partial \boldsymbol{*}\mathcal{E}}{\partial t}\right) \wedge dt$$

Gauss's law and the law of Ampere–Maxwell then give

$$d * F^2 = 4\pi(\sigma^3 - j^2 \wedge dt) = 4\pi \mathcal{S}^3 \tag{7.29}$$

In particular

$$d\mathcal{S}^3 = 0 \tag{7.30}$$

and this is a reflection of **conservation of charge** (see [F, p. 111]).

We wish to make two final remarks.

1. Maxwell's equations are traditionally thought of as four independent axioms, but, remarkably, special relativity says that this is not so. Consider (7.23). Suppose, for instance, that every inertial observer notes that $\mathbf{d}\mathcal{B} = 0$. Then every inertial observer will see the 3-form $dF = (\mathbf{d}\mathcal{E} + \partial\mathcal{B}/\partial t) \wedge dt$, which is of the form $i_W \text{ vol}^4$, where the 4-vector $W$ *can have no time component*, $W^0 = 0$. But under a Lorentz transformation we will have $W'^0 = W^\alpha(\partial x'^0/\partial x^\alpha)$, and thus unless $W = 0$, *some* Lorentz transformation will yield a $W'^0 \neq 0$. Thus, *if every inertial observer sees* $\mathbf{d}\mathcal{B} = 0$, *then* $dF = 0$ *and* so *Faraday's law holds!* Likewise, *if Gauss's law is observed by every inertial observer, then so is Ampere–Maxwell.* This is comforting, since Gauss's law, for example, seems less sophisticated than Ampere–Maxwell.

2. We wish to emphasize the *Maxwell's equations* $dF = 0$ and $d * F = 4\pi\mathcal{S}$ *hold universally, in all materials.* Physicists and engineers usually introduce two *material dependent* fields, in our language a pseudo-1-form $\mathcal{H}^1$ and a pseudo-2-form $\mathcal{D}^2$, together with a material dependent current pseudo-3-form $\mathcal{C}^3$, and then write for Maxwell's equations $dF = 0$ and $d(-\mathcal{H} \wedge dt + \mathcal{D}) = 4\pi\mathcal{C}$. In the case of a "noninductive material," for example the vacuum, we have $\mathcal{H} = *\mathcal{B}$ and $\mathcal{D} = *\mathcal{E}$ and $\mathcal{C} = \mathcal{S}$, but in general the *macro*scopic fields $\mathcal{H}$ and $\mathcal{D}$ are related to the true *micro*scopic fields $\mathcal{B}$ and $\mathcal{E}$ by complicated "constitutive relations." We shall have no need for these new fields.

## Problems

**7.2(1)** Derive (7.17).

**7.2(2)** Derive (7.21).

**7.2(3)** Show (7.22) and show that $F^2 \wedge *F^2 = (|\mathbf{B}|^2 - |\mathbf{E}|^2)\text{ vol}^4$.

**7.2(4)** Show that (3.32) is equivalent to $d\mathcal{S}^3 = 0$.

**7.2(5)** All Lorentz transformations leave the 3 dimensional "unit hyperboloid" $t^2 - x^2 - y^2 - z^2 = 1$ of Minkowski space invariant. Show that

$$\frac{dx \wedge dy \wedge dz}{|t|}$$

is a volume form on this hyperboloid that is *invariant under Lorentz transformations.* (Hint: $H = t^2 - x^2 - y^2 - z^2$ is an invariant function. Use the *method* expressed by equation (4.53) of Hamiltonian mechanics.)

CHAPTER 8

# The Geometry of Surfaces in $\mathbb{R}^3$

The geometry or kinematics of this subject is a great contrast to that of the flexible line, and, in its merest elements, presents ideas not very easily apprehended, and subjects of investigation that have exercised, and perhaps overtasked, the powers of some of the greatest mathematicians.

Kelvin and Tait, *Elements of Natural Philosophy*

## 8.1. The First and Second Fundamental Forms

What is the length of a curve that leaves the north pole, ends at the south pole, and makes a constant angle with each meridian of longitude?

### 8.1a. The First Fundamental Form, or Metric Tensor

Let $M^2 \subset \mathbb{R}^3$ be a parameterized surface in space, $M^2 = F(U)$, where $U \subset \mathbb{R}^2$ and $F_*$ has rank 2. *Frequently we shall write* $u^1 = u$ *and* $u^2 = v$.

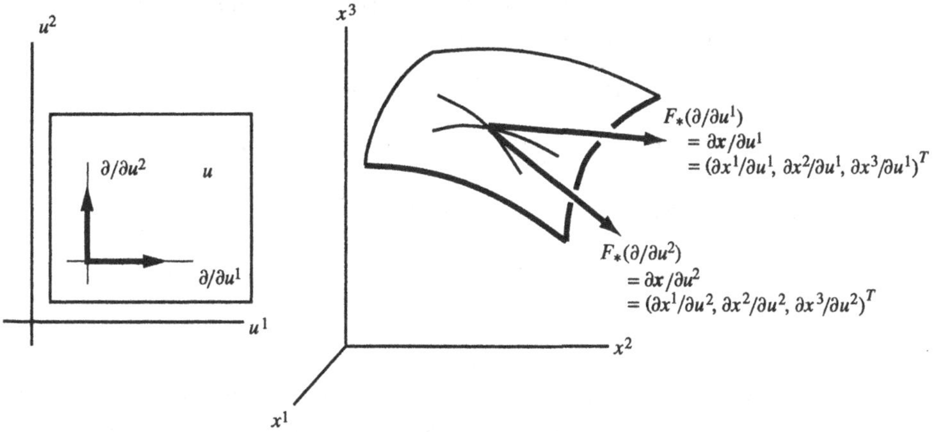


Figure 8.1

A curve $\mathbf{x} = \mathbf{x}(t)$ that lies on $M^2$ is the image of some curve $u^\alpha = u^\alpha(t)$ and so $\mathbf{x} = \mathbf{x}[u(t)]$. For velocity vector we have

$$\frac{d\mathbf{x}}{dt} = \left(\frac{\partial \mathbf{x}}{\partial u^\alpha}\right)\frac{du^\alpha}{dt} = \mathbf{x}_\alpha \frac{du^\alpha}{dt}$$

where

$$\mathbf{x}_\alpha := \frac{\partial \mathbf{x}}{\partial u^\alpha}, \qquad \alpha = 1, 2,$$

form a basis for the tangent space to $M^2$ at each point. A pair of tangent vectors has a euclidean scalar product

$$\langle \mathbf{A}, \mathbf{B} \rangle = \langle \mathbf{x}_\alpha A^\alpha, \mathbf{x}_\beta B^\beta \rangle = g_{\alpha\beta} A^\alpha B^\beta$$

where, as usual,

$$g_{\alpha\beta} = \langle \mathbf{x}_\alpha, \mathbf{x}_\beta \rangle = \sum_{i=1}^{3} \left( \frac{\partial x^i}{\partial u^\alpha} \right) \left( \frac{\partial x^i}{\partial u^\beta} \right) \tag{8.1}$$

We can then write, as in Section 2.7b,

$$ds^2 = \langle d\mathbf{x}, d\mathbf{x} \rangle = \langle \mathbf{x}_\alpha du^\alpha, \mathbf{x}_\beta du^\beta \rangle = g_{\alpha\beta} du^\alpha du^\beta \tag{8.2}$$

and this quadratic form associated to the metric tensor is called the **first fundamental form.** Note that we are, as usual, considering the coordinates $u^\alpha$ as functions on $M^2$, and $du^\alpha$ are 1-forms on $M$ with $du^\alpha(\mathbf{x}_\beta A^\beta) = A^\alpha$, and $ds^2$ is simply another name for the metric tensor $ds^2 = g_{\alpha\beta} du^\alpha \otimes du^\beta$ since

$$g_{\alpha\beta} du^\alpha \otimes du^\beta (\mathbf{A}, \mathbf{B}) = g_{\alpha\beta} A^\alpha B^\beta$$

The reason for this notation will become clear in a moment when we shall use a picture and ordinary arc length $ds$ to write down, with no computations, the metric tensor for the 2-sphere. But first, you must do it the hard way, from the definition (8.1).

The sphere of radius $a$ can be parameterized (except at the poles) by colatitude $\theta = u_1$ and the negative of the longitude, $\phi = u^2$. You are asked to show, in Problem 8.1(1), that for the sphere of radius $a$ we have

$$ds^2 = a^2(d\theta^2 + \sin^2\theta d\phi^2) \tag{8.3}$$

We define the length of a parameterized curve $u = u(t)$ on $M^2$ by

$$L = \int \| d\mathbf{x}/dt \| \, dt = \int \left[ g_{\alpha\beta}(u(t)) \left( \frac{du^\alpha}{dt} \right) \left( \frac{du^\beta}{dt} \right) \right]^{1/2} dt$$

The cosine of the angle between tangent vectors **A** and **B** is given by

$$\frac{\langle \mathbf{A}, \mathbf{B} \rangle}{\| \mathbf{A} \| \| \mathbf{B} \|} \tag{8.4}$$

and the angle between intersecting curves is the angle between their tangents. Thus the coordinate curves $v =$ constant and $u =$ constant are orthogonal iff $g_{uv} := g_{12} = 0$; in general they intersect at an angle

$$\cos^{-1} \frac{g_{uv}}{[g_{uu} g_{vv}]^{1/2}}$$

When the coordinate curves are orthogonal we interpret $ds^2 = g_{uu} du^2 + g_{vv} dv^2$ as an "infinitesimal" version of Pythagoras's rule. On the sphere of radius $a$, for example, we see immediately that (8.3) is the Pythagoras rule applied to the infinitesimal curved triangle.

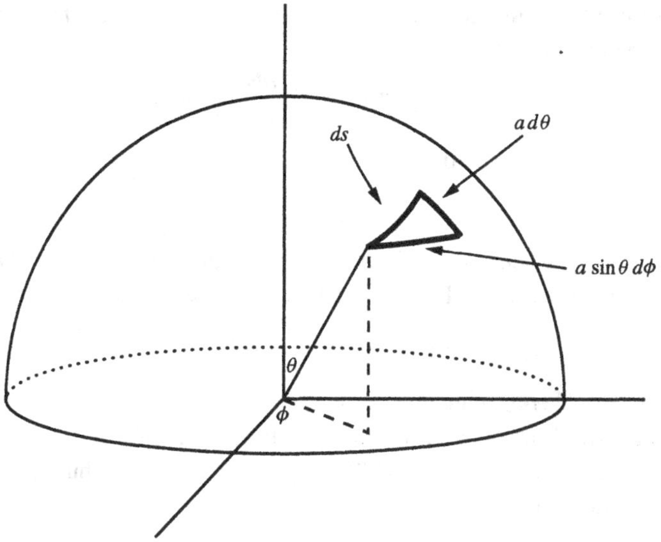


Figure 8.2

See Problem 8.1(2) at this time.

For element of area, from (2.72),

$$dS = \sqrt{g}du \wedge dv$$

See Problem 8.1(3).

Finally, we would like to make a remark on the classical notation $d\mathbf{x}$ appearing in (8.2). Classically $d\mathbf{x}$ is the "infinitesimal vector" with components $(dx, dy, dz)^T$, joining two infinitesimally distant points, and when we restrict the position vector $\mathbf{x}$ to end on the surface $M^2$ this vector $d\mathbf{x}$ is tangent to the surface. In *our* language, $d\mathbf{x}$ is a mixed tensor; in local coordinates for $M^2$,

$$d\mathbf{x} = \mathbf{x}_\alpha \otimes du^\alpha$$

(classically the tensor product sign is omitted). We shall think of this mixed tensor (linear transformation) as a **vector-valued 1-form,** that is, a 1-form whose value on any tangent vector $\mathbf{v}$ is a *vector*, rather than a scalar. For this particular vector valued 1-form, the value is again the vector $\mathbf{v}$,

$$d\mathbf{x}(\mathbf{v}) = (\mathbf{x}_\alpha \otimes du^\alpha)(\mathbf{v}) := \mathbf{x}_\alpha(du^\alpha(\mathbf{v})) = \mathbf{x}_\alpha v^\alpha = \mathbf{v}$$

## 8.1b. The Second Fundamental Form

Whenever we discuss the normal to a surface we shall assume that one of the two possible local normal fields has been chosen.

Let $\mathbf{N} = \mathbf{x}_u \times \mathbf{x}_v / \parallel \mathbf{x}_u \times \mathbf{x}_v \parallel$ be the unit normal to $M^2$ at a point $(u^1, u^2)$. Given any tangent vector $\mathbf{X} = \mathbf{x}_\alpha X^\alpha$ at $(u^1, u^2)$, let $u^\alpha = u^\alpha(t)$ be a curve on $M^2$ having $\mathbf{X}$ as tangent at $u^\alpha = 0$; $X^\alpha = du^\alpha/dt$. Then the derivative of $\mathbf{N}$ with respect to $\mathbf{X}$ is $d\mathbf{N}/dt = (\partial\mathbf{N}/\partial u^\alpha)(du^\alpha/dt) = \mathbf{N}_\alpha du^\alpha/dt = \mathbf{N}_\alpha X^\alpha$ (where again $\mathbf{N}_\alpha := \partial\mathbf{N}/\partial u^\alpha$)

and this vector is a **tangent** vector to $M^2$ since $\mathbf{N}$ is a *unit* vector. The assignment (the minus sign being traditional)

$$\mathbf{X} \mapsto -\mathbf{N}_\alpha X^\alpha = -X^\alpha \frac{\partial \mathbf{N}}{\partial u^\alpha} =: b(\mathbf{X})$$

defines then a linear transformation

$$b : M^2_{(u,v)} \to M^2_{(u,v)}$$

(Note that under $b$, $\mathbf{x}_\alpha$ is sent into $-\mathbf{N}_\alpha$ and that *if we reverse the choice of normal field, b will be sent into its negative.*) Let $(b^\alpha{}_\beta)$ be its matrix with respect to the basis $\{\mathbf{x}_\alpha\}$

$$b(\mathbf{x}_\beta) = \mathbf{x}_\alpha b^\alpha{}_\beta = -\mathbf{N}_\beta \tag{8.5}$$

These are called the **Weingarten** equations.

The bilinear form $B$ associated to the linear transformation $b$ is (as usual) defined by $B(\mathbf{X}, \mathbf{Y}) = \langle \mathbf{X}, b(\mathbf{Y})\rangle = \langle \mathbf{X}, -\mathbf{N}_\beta Y^\beta \rangle = -\langle \mathbf{x}_\gamma X^\gamma, \mathbf{N}_\beta Y^\beta\rangle$. Thus, as a tensor, $B$ is given by the **second fundamental form**

$$-\langle d\mathbf{x}, d\mathbf{N}\rangle = -\langle \mathbf{x}_\gamma, \mathbf{N}_\beta\rangle du^\gamma \otimes du^\beta$$

and the tensor product sign is usually omitted. Weingarten's equation can be written in terms of the vector-valued 1-form

$$d\mathbf{N} = \left[\frac{\partial \mathbf{N}}{\partial u^\beta}\right] \otimes du^\beta = -\mathbf{x}_\alpha b^\alpha{}_\beta \otimes du^\beta \tag{8.6}$$

Thus, along any curve $u = u(t)$ on the surface,

$$\frac{d\mathbf{N}}{dt} = -\mathbf{x}_\alpha b^\alpha{}_\beta \left(\frac{du^\beta}{dt}\right)$$

We may write for the second fundamental form, as in (2.39),

$$B = b_{\alpha\beta} du^\alpha du^\beta$$

where $b_{\alpha\beta} = g_{\alpha\gamma} b^\gamma{}_\beta$ is the covariant tensor associated to the linear transformation $b$. Then $b_{\alpha\beta} = B(\mathbf{x}_\alpha, \mathbf{x}_\beta) = \langle \mathbf{x}_\alpha, b(\mathbf{x}_\beta)\rangle = -\langle \mathbf{x}_\alpha, \mathbf{N}_\beta\rangle$, that is,

$$b_{\alpha\beta} = -\langle \mathbf{x}_\alpha, \mathbf{N}_\beta\rangle \tag{8.7}$$

This expression is inconvenient for computations since it involves the derivative of the *unit* vector $\mathbf{N}$ (which usually involves a complicated expression with square roots); we shall exhibit now a more useful formula. Put

$$\mathbf{x}_{\alpha\beta} := \frac{\partial^2 \mathbf{x}}{\partial u^\alpha \partial u^\beta}$$

Since $\mathbf{N}$ is a normal vector, $0 = \partial/\partial u^\beta \langle \mathbf{x}_\alpha, \mathbf{N}\rangle = \langle \mathbf{x}_{\alpha\beta}, \mathbf{N}\rangle + \langle \mathbf{x}_\alpha, \mathbf{N}_\beta\rangle = \langle \mathbf{x}_{\alpha\beta}, \mathbf{N}\rangle - b_{\alpha\beta}$, that is,

$$b_{\alpha\beta} = \langle \mathbf{x}_{\alpha\beta}, \mathbf{N}\rangle \tag{8.8}$$

which is *the* formula for computing $B$. In full, we have

$$b_{\alpha\beta} = \left(\frac{\partial^2 x}{\partial u^\alpha \partial u^\beta}, \frac{\partial^2 y}{\partial u^\alpha \partial u^\beta}, \frac{\partial^2 z}{\partial u^\alpha \partial u^\beta}\right)(N^1, N^2, N^3)^T$$

The linear transformation $b$ may then be computed from $b^\alpha{}_\beta = g^{\alpha\gamma} b_{\gamma\beta}$.

## Problems

**8.1(1)** Compute the metric for the sphere of radius $a$.

**8.1(2)** A "loxodrome" on a sphere of radius $a$ is a curve that makes a constant angle $\omega$ with each meridian of longitude. Usually it eventually winds around each pole. Compute the length of such a loxodrome by using $\theta$ as a parameter. (The tangent vector then has components $(1, d\phi/d\theta)$ and you may use (8.4) to determine $d\phi/d\theta$).)

**8.1(3)** Compute the area of the region on the Earth's surface bounded by latitudes 0° and 30° and longitude 0° and 45°.

**8.1(4)** Consider the surface $z = x^2 - 2y^2$ near the origin. Use $x = u^1, y = u^2$ for local coordinates. Compute the matrices $(g_{\alpha\beta})$ and $(b^\alpha{}_\beta)$ at $(0, 0)$. Save your computations for problem 8.2(2).

**8.1(5)** Let $M^2$ be a surface in $\mathbb{R}^3$ and let $x_0$ be a point on this surface. Choose new cartesian coordinates for $\mathbb{R}^3$ having $x_0$ as origin and such that the new $x^1, x^2$ plane is the tangent plane to $M$ at $x_0$. Use $x^1 = u^1$ and $x^2 = u^2$ as local coordinates near $x_0$. Show that $M$ near $x_0$ is described by the equations

$$x^3 = z(x^1, x^2) = (1/2) \sum_{\alpha,\beta=1,2} b_{\alpha\beta}(0)x^\alpha x^\beta$$
$$+ \text{ higher order in } x^1, x^2$$

exhibiting another geometric aspect of the second fundamental form.

## 8.2. Gaussian and Mean Curvatures

What do we mean by the curvature of a surface?

### 8.2a. Symmetry and Self-Adjointness

We recall from linear algebra that if $\mathcal{A}$ is a linear transformation in a vector space with scalar product, then the **adjoint** $\mathcal{A}^*$ of $\mathcal{A}$ is the linear transformation defined by $\langle \mathcal{A}\mathbf{X}, \mathbf{Y}\rangle = \langle \mathbf{X}, \mathcal{A}^*\mathbf{Y}\rangle$, and $\mathcal{A}$ is **self-adjoint** if $\mathcal{A} = \mathcal{A}^*$. In terms of the bilinear form $A$ associated to $\mathcal{A}$, $\mathcal{A}$ is self-adjoint provided

$$A(\mathbf{X}, \mathbf{Y}) = \langle \mathbf{X}, \mathcal{A}\mathbf{Y}\rangle = \langle \mathcal{A}\mathbf{X}, \mathbf{Y}\rangle = \langle \mathbf{Y}, \mathcal{A}\mathbf{X}\rangle = A(\mathbf{Y}, \mathbf{X})$$

that is, *a linear transformation $\mathcal{A}$ is self-adjoint iff the associated bilinear form $A$ is symmetric*. In components, $\mathcal{A}$ is self-adjoint iff $(A_{\alpha\beta})$ is symmetric, $A_{\alpha\beta} = A_{\beta\alpha}$. (You should convince yourself from the transformation laws for covariant and mixed tensors that such an equality is in fact independent of basis, whereas $A^\alpha{}_\beta = A^\beta{}_\alpha$ *might* hold in some basis but not another; it makes no sense to say that a mixed tensor is symmetric.)

From (8.8) we see that the second fundamental form $B$ is symmetric and thus *the linear transformation $b : M_u^2 \to M_u^2$ is self-adjoint!* As we shall now see, the special eigenvalue behavior of a self-adjoint transformation will have remarkable geometric consequences in the case of the linear transformation $b$.

### 8.2b. Principal Normal Curvatures

Let $\mathbf{x} = \mathbf{x}(s)$ define a curve $C$, parameterized by arc length, on the surface $M^2$ in $\mathbb{R}^3$. The unit tangent at $\mathbf{x}(0)$ is then $\mathbf{T} = d\mathbf{x}/ds = \mathbf{x}_\alpha du^\alpha/ds$. The curvature vector for $C$, *as a space curve*, at $\mathbf{x}(0)$ is

$$\boldsymbol{\kappa} = \kappa\mathbf{n} = \frac{d\mathbf{T}}{ds} = \mathbf{x}_{\alpha\beta}\left(\frac{du^\alpha}{ds}\right)\left(\frac{du^\beta}{ds}\right) + \mathbf{x}_\alpha \frac{d^2u^\alpha}{ds^2}$$

where $\mathbf{n}$ is the principal normal to $C$. The component of the curvature vector $\boldsymbol{\kappa} = \kappa\mathbf{n}$ in the direction of the unit surface normal $\mathbf{N}$ is then

$$\langle \kappa\mathbf{n}, \mathbf{N}\rangle = \langle \mathbf{x}_{\alpha\beta}, \mathbf{N}\rangle \left(\frac{du^\alpha}{ds}\right)\left(\frac{du^\beta}{ds}\right)$$

that is,

$$\langle \kappa\mathbf{n}, \mathbf{N}\rangle = b_{\alpha\beta}\left(\frac{du^\alpha}{ds}\right)\left(\frac{du^\beta}{ds}\right) = B(\mathbf{T}, \mathbf{T}) \tag{8.9}$$

There are, of course, an infinity of curves on $M^2$ that pass through $\mathbf{x}(0)$ with tangent $\mathbf{T}$, but (8.9) tells us that although these curves may have very different curvatures as space curves, *the component of the curvature vectors normal to the surface depends only on the tangent T and is the value of the second fundamental quadratic form B on T*!

In particular, let $\mathbf{T}$ be a unit tangent vector to $M$ at a point $p$.

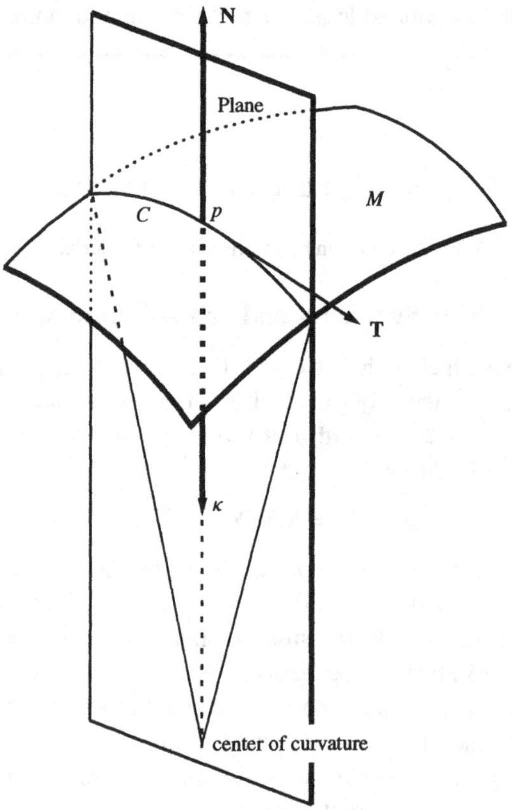


Figure 8.3

Let $P$ be the plane spanned by **T** and **N** at $p$. $P$ cuts out a curve $C$ on $M$, whose unit tangent is **T**. $C$ is a **normal section** of $M$ and of course it is a plane curve, lying as it does in $P$. Its curvature vector $\boldsymbol{\kappa} = \kappa \mathbf{n}$ (as a space curve) points from $p$ towards the center of curvature (at a distance $\kappa^{-1}$). Thus, for this normal section, from (8.9)

$$B(\mathbf{T}, \mathbf{T}) = \pm\kappa$$

where the + sign is used only if the curve $C$ is "curving" toward the chosen surface normal; for the indicated normal in our figure $B(\mathbf{T}, \mathbf{T}) = -\kappa$ is negative.

Now keep $p \in M$ fixed but rotate **T** in the tangent plane $M_p^2$; the curvatures $B(\mathbf{T}, \mathbf{T})$ will change in general. We define the **principal (normal) curvatures** of $M$ at $p$ by

$$\kappa_1(p) = \max B(\mathbf{T}, \mathbf{T}) \tag{8.10}$$
$$\kappa_2(p) = \min B(\mathbf{T}, \mathbf{T})$$

for unit $\mathbf{T} \in M_p^2$. The two directions $\mathbf{T}_\alpha, \alpha = 1, 2$, yielding these extrema are called the **principal directions** for $M$ at $p$. But $b$ is self-adjoint (i.e., $B$ is symmetric), and linear algebra (see Problem 8.2(1)) tells us the following:

**Theorem (8.11):** *$\kappa_1$ and $\kappa_2$ are the eigenvalues of $b$ and the corresponding principal directions $\mathbf{T}_\alpha$ are the eigenvectors*

$$b(\mathbf{T}_\alpha) = \kappa_\alpha \mathbf{T}_\alpha, \qquad \alpha = 1, 2$$

*If $\kappa_1 \neq \kappa_2$ then automatically the principal directions are orthogonal.*

(The orthogonality of the principal directions was known to Euler!)

Of course if $\kappa_1 = \kappa_2$ then all the normal curvatures at $p$ coincide; $p$ is then called an "umbilic" point. The usual round 2-sphere consists entirely of umbilic points.

### 8.2c. Gauss and Mean Curvatures: The Gauss Normal Map

We now define two measures of curvature of a surface $M^2$ at $p$.

$$\textbf{Gauss curvature} = K := \det b = \frac{\det(b_{\alpha\beta})}{\det(g_{\alpha\beta})} = \kappa_1\kappa_2$$

$$\textbf{Mean curvature} = H := \operatorname{tr} b = \sum b^\alpha{}_\alpha = \kappa_1 + \kappa_2$$

Note that since $b$ is sent into $-b$ under a change of normal, $H$ will be sent into its negative but $K$ is invariant under choice of normal!

**Warning**: Many authors define $H$ to be the true average $(\kappa_1 + \kappa_2)/2$.

Before discussing the significance of these quantities, we need some experience with computing them. See Problems 8.2(2), 8.2(3), and 8.2(4) at this time.

Note now the following. If $A : \mathbb{R}^n \to \mathbb{R}^n$ is a linear transformation and $\omega^n$ is any $n$-form, then

$$A^*\omega = \det(A)\omega \tag{8.12}$$

This follows from (2.65), or directly

$$\begin{aligned}\omega(A\mathbf{e}_1, \ldots, A\mathbf{e}_n) &= \omega(\mathbf{e}_i A^i{}_1, \ldots, \mathbf{e}_j A^j{}_n)\\ &= \omega(\mathbf{e}_i, \ldots, \mathbf{e}_j) A^i{}_1 \ldots A^j{}_n\\ &= \omega(\mathbf{e}_1, \ldots, \mathbf{e}_n)\epsilon_{i\ldots j} A^i{}_1 \ldots A^j{}_n = \omega(\mathbf{e}_1, \ldots, \mathbf{e}_n)\det A\end{aligned}$$

If $M^2 \subset \mathbb{R}^3$ is a surface with given normal field, we define the **Gauss (normal) map**

$$n : M^2 \to \text{unit sphere } S^2$$

by

$$n(p) = \mathbf{N}(p), \qquad \text{the unit normal to } M \text{ at } P$$

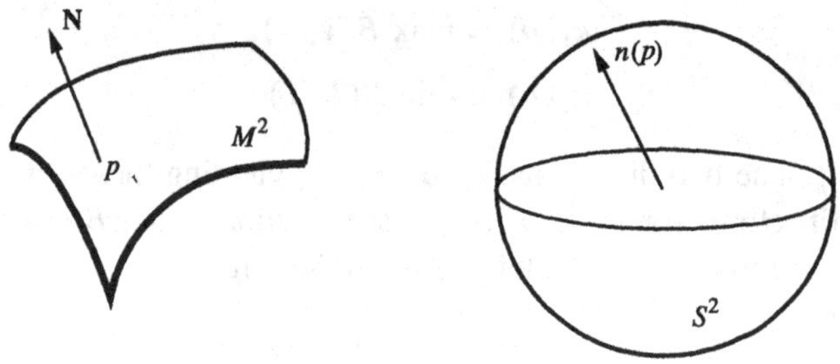


**Figure 8.4**

Define the positive orientation of $S^2$ by using the outward pointing normal. Let $\mathrm{vol}^2_M = i_\mathbf{N}\,\mathrm{vol}^3$ and $\omega^2 = \mathrm{vol}^2_S = i_\mathbf{n}\,\mathrm{vol}^3$ be the area forms for $M^2$ and $S^2$ respectively. Let $u$, $v$, be local coordinates for $M$. We wish to compute the pull-back of $\omega^2$ under the Gauss normal map. Note that the tangent plane to $M^2$ at $p$ is parallel to the tangent plane to $S^2$ at $n(p)$ and *we shall identify these two 2-dimensional vector spaces* by parallel translation in $\mathbb{R}^3$. (Note that under this identification, $\omega^2$ at $n(p)$ is the same as $\mathrm{vol}^2_M$ at $p$!) Thus, for example, $\partial\mathbf{x}/\partial u$ and $b(\partial\mathbf{x}/\partial u)$ may be identified with tangent vectors to $S^2$, and $b$ at $p$ can be considered as a linear transformation of the tangent plane to $S^2$ at $n(p)$. By the geometric meaning of the differential of the map $n : M^2 \to S^2$

$$n_*\left(\frac{\partial\mathbf{x}}{\partial u^\alpha}\right) = \frac{\partial}{\partial u^\alpha}(\mathbf{N}(u)) = \frac{\partial\mathbf{N}}{\partial u^\alpha} \tag{8.13}$$

and so, using (8.12),

$$\begin{aligned}(n^*\omega^2)\left(\frac{\partial\mathbf{x}}{\partial u}, \frac{\partial\mathbf{x}}{\partial v}\right) &= \omega^2\left(n_*\frac{\partial\mathbf{x}}{\partial u}, n_*\frac{\partial\mathbf{x}}{\partial v}\right)\\ &= \omega^2\left(\frac{\partial\mathbf{N}}{\partial u}, \frac{\partial\mathbf{N}}{\partial v}\right) = \omega^2\left(-b\left(\frac{\partial\mathbf{x}}{\partial u}\right), -b\left(\frac{\partial\mathbf{x}}{\partial v}\right)\right)\\ &= (\det b)\omega^2\left(\frac{\partial\mathbf{x}}{\partial u}, \frac{\partial\mathbf{x}}{\partial v}\right) = K\,\mathrm{vol}^2_M\left(\frac{\partial\mathbf{x}}{\partial u}, \frac{\partial\mathbf{x}}{\partial v}\right)\end{aligned}$$

Thus

$$n^*\,\mathrm{vol}^2_S = K\,\mathrm{vol}^2_M \tag{8.14}$$

This tells us that *the Gauss map is a local diffeomorphism in the neighborhood $U$ of any $p \in M^2$ at which $K(p) \neq 0$, and furthermore, if $U$ is positively oriented then $n(U)$ will be positively oriented on $S^2$* iff $K > 0$.

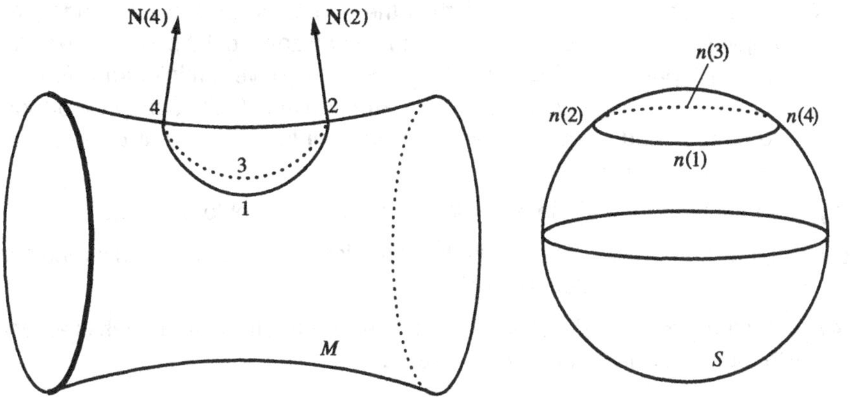


**Figure 8.5**

(8.14) exhibits the Gauss curvature as a "magnification factor" for areas under the normal map $n : M^2 \to S^2$, provided we consider area "signed" by the orientation.

$$\text{"signed" area of } n(U) \ := \int_{n(U)} \text{vol}_S^2 = \int_U n^* \text{vol}_S^2$$
$$= \int_U K \, \text{vol}_M^2$$

and thus

$$\lim_{U \to p} [\text{signed area of } n(U)/\text{area of } U] = K(p)$$

as the region $U$ shrinks down to the point $p$. *This was Gauss's original definition of* $K$. Note that $n$ reverses orientation iff the principal curvatures $\kappa_1$ and $\kappa_2$ at $p$ are of opposite sign, that is, iff $M^2$ is "saddle-shaped" at $p$.

## Problems

**8.2(1)** This problem gives a proof of the fundamental theorem on symmetric matrices. Let $b : \mathbb{R}^n \to \mathbb{R}^n$ be any self-adjoint linear transformation with symmetric bilinear form $B$. Let $S^{n-1}$ be the unit sphere in $\mathbb{R}^n$ and let $f : \mathbb{R}^n \to \mathbb{R}$ be the quadratic function $f(x) = B(x, x) = \langle x, bx \rangle$ but restricted to the unit sphere $S^{n-1}$. Since $S^{n-1}$ is compact (for this it is important that the metric on $\mathbb{R}^n$ is positive definite; we could not use a Minkowski metric where the "unit sphere" is in fact a hyperboloid), $f$ takes on its minimum value at some $e_1 \in S^{n-1}$. Let $x = x(t)$ be a curve on $S^{n-1}$ starting at $x(0) = e_1$. Let $\dot{x}$ denote the derivative with respect to $t$ at $t = 0$.

**(i)** Show that $\langle \dot{x}, be_1 \rangle = 0$. Since any tangent vector to $S^{n-1}$ at $e_1$ is of the form $\dot{x}$, this shows that $be_1$ is normal to $S^{n-1}$ at $e_1$, that is, $be_1 = \lambda_1 e_1$ for some real number $\lambda_1$. Thus $\lambda_1 = f(e_1)$. This argument shows in fact that *every critical point of f on $S^{n-1}$ is an eigenvector of b with a real eigenvalue and the eigenvalue is simply the value of f.*

Let $E_1$ be the subspace of $\mathbb{R}^n$ spanned by $e_1$ and let $E_1^\perp$ be the orthogonal subspace to $E_1$.

(ii) Show that $b : E_1^\perp \to E_1^\perp$ and thus the restriction of $b$ to $E_1^\perp$ is again a self-adjoint linear transformation (which we shall again call $b$). Then $f$ restricted to the unit sphere $S^{n-2} := S^{n-1} \cap E_1^\perp$ will again have a minimum value $\lambda_2 \geq \lambda_1$ attained at an eigenvector $e_2 \in E_1^\perp$. Proceed then to the subspace orthogonal to both $e_1$ and $e_2$, and so on. Induction will then show that $b$ has a basis of orthonormal eigenvectors.

**8.2(2)** Compute $K$ and $H$ at the origin for the surface in Problem 8.1(4).

**8.2(3)** What is the normal curvature for the direction $y = x$ at the origin for the surface $z = x^2 - 2y^2$ of Problem 8.1(4)?

**8.2(4)** Show that the normal curvature for a direction on an $M^2$ that makes an angle $\theta$ with the principal direction $\mathbf{T}_1$ is given by

$$\kappa(\theta) = \kappa_1 \cos^2\theta + \kappa_2 \sin^2\theta$$

**8.2(5)** For a surface $M^2$ given in "nonparametric form" $z = f(x, y)$ we can, of course, introduce $x = u$ and $y = v$ as coordinates. Show that

$$K = \frac{\det(f_{\alpha\beta})}{W^2}$$

and

$$H = W^{-3/2}[(1 + f_y^2)\, f_{xx} - 2\, f_x\, f_y\, f_{xy} + (1 + f_x^2)\, f_{yy}]$$

where $W := 1 + f_x^2 + f_y^2$

---

## 8.3. The Brouwer Degree of a Map: A Problem Set

Can you map a closed ball into itself so that *every* point is moved?

### 8.3a. The Brouwer Degree

In our previous section we discussed the Gauss normal map $n : M^2 \to S^2$. The situation of mapping a compact oriented manifold into another *of the same dimension* plays an important and recurring role in mathematics and its applications. We shall discuss the topological implications of this situation, first studied in detail by the Dutch mathematician L. E. J. Brouwer around the turn of the twentieth century.

Since our manifolds are oriented, we shall make no distinction between forms and pseudoforms.

Let $\phi : M^n \to V^n$ be a smooth map from one *closed oriented* manifold to another of the same dimension. Let $\omega^n$ be any $n$-form on $V$ subject to the single condition that it be normalized

$$\int_V \omega = 1$$

(Of course if $\int_V \omega \neq 0$ we may trivially normalize it.) The **(Brouwer) degree** of $\phi$ is defined by

$$\deg(\phi) = \int_M \phi^*\omega \tag{8.15}$$

Note that we may also write $\deg(\phi) = \int_{\phi(M)} \omega$; this tells us (in a sense to be clarified later) *how many times, algebraically, the image of M wraps around V.*

Our first task is to show that $\deg(\phi)$ is well defined, independent of the choice of the form $\omega$. We shall give only the barest sketch of this, relying on some "familiar" but nontrivial facts.

**Lemma (8.16):** *An n-form $\gamma^n$ on a closed oriented $V^n$ is exact iff its integral vanishes*

$$\int_V \gamma = 0$$

PROOF: Certainly if $\gamma = d\beta^{n-1}$, then, since $V$ has no boundary, $\int_V d\beta = 0$. Suppose then that $\int_V \gamma = 0$. We shall attempt to exhibit $\beta$. Introduce a Riemannian metric. We may assume that $\int_V \text{vol}^n = 1$. Write $\beta^{n-1} = i_{\mathbf{B}} \text{vol}^n$ for an as yet undetermined vector field **B**. If we write $\gamma = g \,\text{vol}^n$ in terms of a function $g$, we shall be done if we can solve $\text{div}\, \mathbf{B} = g$ for **B**. We shall determine **B** by writing $\mathbf{B} = \text{grad}\, f$ and then solving $\nabla^2 f = g$. It is a fact (see [W, p. 256]) that the Laplace operator on a compact manifold has a uniformly complete system of eigenfunctions; we have eigenfunctions $\{\alpha_k\}$, $\nabla^2 \alpha_k = -\lambda_k \alpha_k$, $0 = \lambda_0 < \lambda_1 \le \lambda_2 \le \dots$, where $\lambda_k \to \infty$, and any smooth $f$ can be expanded in terms of them, $f = \sum f_k \alpha_k$. This expansion converges pointwise, not just "in the mean." The only eigenfunction needed for the lowest eigenvalue $\lambda_0 = 0$ is the function $\alpha_0 = 1$, since $\int_V \| \text{grad}\, \alpha_0 \|^2 \text{vol} = \int_V \text{div}[\alpha_0 \,\text{grad}\, \alpha_0] \,\text{vol} - \int_V \alpha_0 \nabla^2 \alpha_0 \,\text{vol} = 0$ shows that $\alpha_0$ must be constant. The higher eigenvalues might have (finite) multiplicity greater than 1. We then expand $g = \sum g_k \alpha_k$. Then to solve $\nabla^2 f = g$ we need only solve for $f_k$ in the infinite system $-\lambda_k f_k = g_k$, $k = 0, 1, \dots$. This is trivial except for $k = 0$. Note, however, that the "Fourier coefficient" $g_0$ is the Hilbert space scalar product $(g, \alpha_0) = \int_V g \,\text{vol} = \int_V \gamma$, which by assumption vanishes. If we put $f_0 = 0$, then the desired $f$ has been exhibited. One can then show that the resulting $f$ is a solution to $\nabla^2 f = g$. □

We can now show that $\deg(\phi)$ is independent of the choice of $\omega^n$. This follows immediately on noting that if $\omega'$ is another choice, then, by the lemma, $\omega - \omega'$ is exact, so $\phi^*(\omega - \omega')$ is also exact and thus $\int_M \phi^*(\omega - \omega') = 0$.

The geometric significance of the degree is given by the following.

**Theorem (8.17):** *Let $y \in V$ be a regular value of $\phi : M^n \to V^n$; that is, $\phi_*$ at $\phi^{-1}(y)$ is onto. (Recall that Sard's theorem says that the regular values of $\phi$ are dense in V.) For each $x \in \phi^{-1}(y)$, $\phi_* : M_x \to V_y$ is also 1:1; that is, $\phi_*$ is an isomorphism. Put*

$$\text{sign}\, \phi(x) := \pm 1$$

*where the + sign is used iff $\phi_* : M_x \to V_y$ is orientation-preserving. Then*

$$\deg(\phi) = \sum_{x \in \phi^{-1}(y)} \text{sign}\, \phi(x) \tag{8.18}$$

**Corollary (8.19):** $\deg(\phi)$ *is an integer. From (8.15) we see that the sum in (8.18) is independent of the choice of the regular value y. Finally, since (8.15) shows that* $\deg(\phi)$ *varies continuously with* $\phi$*, and since it must be an integer,* $\deg(\phi)$ *remains constant under deformations of the map* $\phi$*.*

PROOF: First, we claim that since $y$ is regular *there are only a finite number of preimages* $x \in \phi^{-1}(y)$. We can see this as follows. It is known that compactness implies that every infinite sequence of points has a convergent subsequence. Thus if $\phi^{-1}(y)$ were infinite we could find a sequence $\{x_k\} \subset \phi^{-1}(y)$ that converges to some $x_\infty$. But then $\phi(x_\infty) = y$ and $x_\infty$ would be a regular point of $M$. Since $\phi_* : M_{x_\infty} \to V_y$ is 1:1, $\phi$ is (by the inverse function theorem) a diffeomorphism on some neighborhood $U_\infty$ of $x_\infty$. But since $x_k \to x_\infty$, $x_k \in U_\infty$ for all $k \geq$ some integer $R$. But then the two points $x_R$ and $x_\infty$ would both be sent to $y$ by $\phi$, contradicting $\phi$ is 1:1 on $U_\infty$.

For the rest of the proof it is good to have a simple example in view to keep track of the construction. We shall draw the case when $V^1 = S$ is the unit circle in the plane, and $M^1$ is a simple closed curve in the plane outside $S$ whose interior holds the origin. The map $\phi : M \to S$ moves each point of $M$ radially toward the origin until it strikes $S$. In this case the degree of $\phi$ is called the **winding number** of the curve $M$ about the origin.

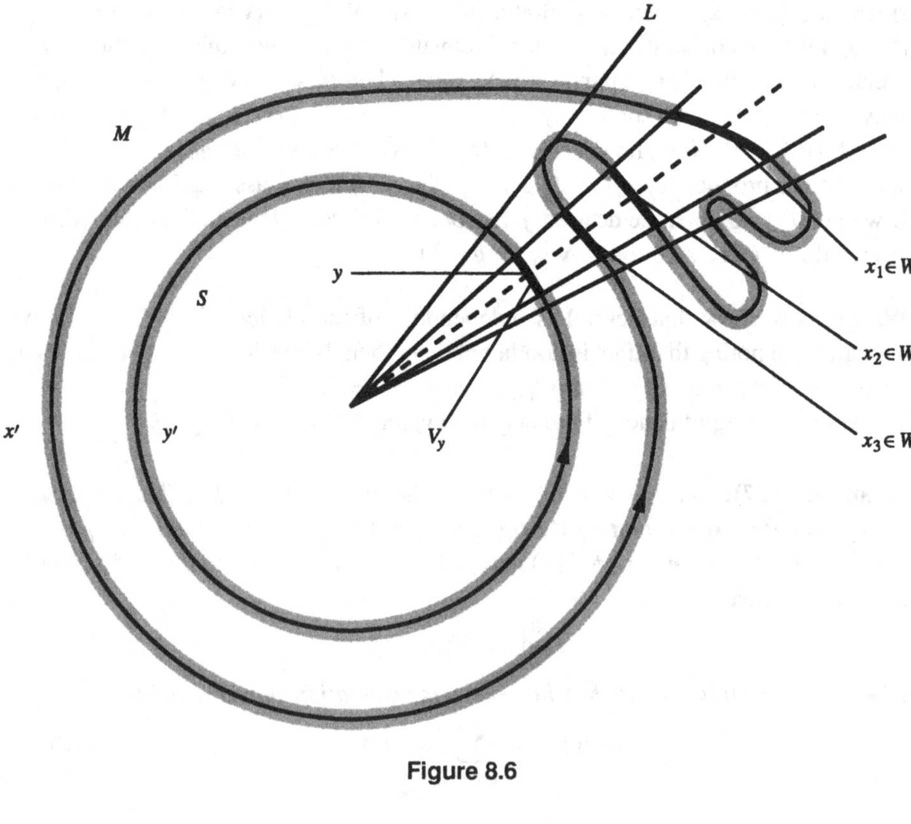


Figure 8.6

In our drawing we see that our indicated $y$ is a regular value since the radial line passing through $y$ is never tangent to $M$. (The line $L$, on the other hand, is tangent to $M$ at a critical point of $\phi$.) We have indicated the three inverse image points $x_i$ of $y$. Each is contained in a neighborhood $W_i$ that is projected diffeomorphically by $\phi$ onto a neighborhood $V_i$ of $y$ on $S$. These $W_i$ are indicated by thick segments on $M$. The complement of the union of these sets on $M$ is indicated by the fuzzy set, and the projection of this complement on $S$ is also made fuzzy. Note that only the neighborhood $W_2$ is such that its image has orientation opposite to that of $S$, and so (8.18) would yield $\deg(\phi) = 1 - 1 + 1 = 1$. This is also obvious from the choice $y'$ for regular value!

It is clear in our picture that the point $y$ has a neighborhood $V_y$ whose inverse image consists of a disjoint union of neighborhoods of the preimages $x_i$ of $y$, each being a diffeomorphic copy of $V_y$. This is the main fact that we shall need in the general case. The proof of this requires a topological argument, which we now present for those readers with a little background in topology.

Let $x_i, i = 1, \dots, N$ be the preimages of the regular $y \in M$ and let $W_i$ be disjoint neighborhoods of the $x_i$ that are sent diffeomorphically by $\phi$ onto neighborhoods $V_i$ of $y$. Let $V_y \subset (V_1 \cap V_2 \cap \dots \cap V_N)$ be a neighborhood so small that it does not meet the "fuzzy" set $\phi[M - (W_1 \cup W_2 \cup \dots \cup W_N)]$. (This is possible for the following reasons: $M - (W_1 \cup W_2 \cup \dots \cup W_N)$ is a closed subset of the compact $M$ and is hence itself compact. The continuous image of a compact set is compact, and hence closed in $V^n$. The point $y$ is in the complement $\mathcal{O}$ of this closed set, and $\mathcal{O}$ is indeed a neighborhood of $y$. Then define the neighborhood $V_y$ of $y$ by $V_y := \mathcal{O} \cap (V_1 \cap V_2 \cap \dots \cap V_N)$. *$V_y$ has the property that its inverse image under $\phi$ consists of disjoint neighborhoods $U_i := (\phi^{-1}\mathcal{O}) \cap W_i$ of $x_i$, each of which is diffeomorphic to $V_y$ under $\phi$.*

Now we shall take advantage of the fact that we may compute $\deg(\phi)$ by using *any* normalized form on $V^n$. Let $\omega^n$ be a normalized form on $V^n$ *whose support lies in $V_y$*, that is, $\omega = 0$ outside $V_y$ (e.g., we may use a "bump form" as in 3.2b) and let $y^1, \dots, y^n$ be local coordinates in $V_y$. Under the diffeomorphism $\phi$ restricted to each $U_i$, we may use the functions $y^\alpha$ as coordinates in $U_i$ (we are really using $y^\alpha \circ \phi$) and the map $\phi : U_i \to V_y$ *is then the identity map in these coordinates*! Note that $\phi(U_i)$ has the same orientation as $V_y$ iff $\operatorname{sign}\phi(x_i) = +1$. We then have, since $\phi^*\omega = 0$ outside the union of the $U_i$'s

$$\int_{V^n} \omega = \int_{V_y} \omega = 1$$

and

$$\deg(\phi) = \int_M \phi^*\omega = \sum_i \int_{U_i} \phi^*\omega = \sum_i \int_{\phi(U_i)} \omega = \sum_i \operatorname{sign}(x_i) \int_{V_y} \omega$$

as desired. □

**8.3(1)** The volume form on the unit sphere $S^n$ in $\mathbb{R}^{n+1}$ is $i_{\mathbf{r}} dx^1 \wedge \dots \wedge dx^{n+1} = \sum (-1)^{i-1} x^i dx^1 \wedge \dots \widehat{dx^i} \dots \wedge dx^{n+1}$. Show that the antipodal map $S^n \to S^n$ has degree $(-1)^{n+1}$.

### 8.3b. Complex Analytic (Holomorphic) Maps

Consider a map $f : \mathbb{C} \to \mathbb{C}$ given by analytic function $Z = f(z)$ in the complex plane. We consider $\mathbb{C}$ to be a *complex* 1-dimensional manifold; see Section 1.2d and Section 5.3. If we write $z = x + iy$ and $Z = u + iv$, then this map may be considered as a map $F : \mathbb{R}^2 \to \mathbb{R}^2$ given by $u = u(x, y)$ and $v = v(x, y)$, $u$ and $v$ satisfying the Cauchy–Riemann equations. The differential $f_*$ of the map $f$ at a point $z_1$ is a $1 \times 1$ matrix operating on complex 1-vectors, obtained as usual from $df(z(t))/dt = f'(z_1)dz/dt$, that is, at $z_1$

$$f_* = f'(z_1)$$

**8.3(2)** Let $f : \mathbb{C} \to \mathbb{C}$ be analytic. Show that the differential $f_* = f'(z_1) : \mathbb{C} \to \mathbb{C}$ as a complex $1 \times 1$ matrix is related to the real differential: $\mathbb{R}^2 \to \mathbb{R}^2$ by

$$\frac{\partial(u, v)}{\partial(x, y)} = | f'(z_1) |^2$$

and thus $f_*$ *is orientation-preserving if* $f'(z_1) \neq 0$.

Consider a *polynomial* map $P : \mathbb{C} \to \mathbb{C}$ of the complex plane to itself of the form $z = x + iy \to Z = u(x, y) + iv(x, y) = P(z) = z^n + a_{n-1}z^{n-1} + \cdots + a_0$. $\mathbb{C}$ is not compact and we therefore cannot discuss the Brouwer degree of this map. But $| z |^n \to \infty$ as $| z | \to \infty$ and since $P$ behaves like $z^n$ for $| z |$ large, we can see that $P$ extends to a continuous map (again called $P$) of the Riemann sphere (see Section 5c) into itself by putting $P(\infty) = \infty$. (Note, e.g., that $e^z$ does *not* extend to such a map; why?) We need to discuss the smoothness at $\infty$. Near $z = \infty$ we introduce the coordinate $w = 1/z$, and then our map can be expressed in the form

$$w \to W(w)$$

by

$$w = z^{-1} \to (z^n + a_{n-1}z^{n-1} + \cdots + a_0)^{-1}$$
$$= \frac{w^n}{(a_0 w^n + \cdots + a_{n-1}w + 1)} = W(w)$$

which is clearly smooth near $w = 0$. In fact $W$ is an analytic function of $w$ near $w = 0$. We may now discuss the Brouwer degree of this polynomial map of the Riemann sphere into itself.

**8.3(3)** Show that $z = \infty$ is neither a regular value nor a regular point of a polynomial $P$ if $n =$ degree of $P$ is $> 1$.

Deform the polynomial map by considering, for $0 \leq \epsilon \leq 1$, the smooth deformation $z \to z^n + \epsilon(a_{n-1}z^{n-1} + \cdots + a_0)$. In the $w$ patch this means $w \to w^n/[1 + \epsilon(a_0 w^n + \cdots + a_{n-1}w)]$. Note that this is smooth as a function of $w$ and $\epsilon$ near $w = 0$, and so we have defined a smooth deformation of the original polynomial map of the Riemann sphere.

**8.3(4)** Show that the Brouwer degree of the $n^{\text{th}}$-degree polynomial map of the Riemann sphere is the same as that of the map $z \to z^n$, $w \to w^n$. Then the value $Z = 1$ shows that this *degree is n*.

**8.3(5)** Show that if $F : M^n \to V^n$ has degree $\neq 0$, then $F$ is *onto*. Hence if $P$ is a nonconstant polynomial, then for some $z_1$, $P(z_1) = 0$. This is the **fundamental theorem of algebra.**

By factoring the polynomial by $(z - z_1)$, we see that $P$ has $n$ (not necessarily distinct) roots, and $P(z) = (z - z_1) \dots (z - z_n)$.

**8.3(6)** Use this to show that 0 is a regular value of $P$ iff $P$ has distinct roots.

## 8.3c. The Gauss Normal Map Revisited: The Gauss–Bonnet Theorem

From (8.14) we see that if $M^2$ is a closed submanifold of $\mathbb{R}^3$ then

$$\frac{1}{4\pi}\int_M K\,dA = \deg(n : M^2 \to S^2) \tag{8.20}$$

is the degree of the Gauss normal map and in particular is an *integer*! If we smoothly deform $M$, this integer must vary smoothly and thus it remains constant, even though $K$ itself will change! Recall, again from (8.14), that $u \in M$ is a regular point for the Gauss map provided $K(u) \neq 0$ and that $n$ preserves orientation iff $K(u) > 0$. This, together with (8.18), allows us to evaluate the left-hand side of (8.20), the so-called **total curvature** of $M$, merely by looking at a picture, as follows.

**8.3(7)** Show that $\int_M K\,dA = 4\pi(1 - g)$ for a surface of genus $g$, that is, the surface of a multidoughnut with $g$ holes

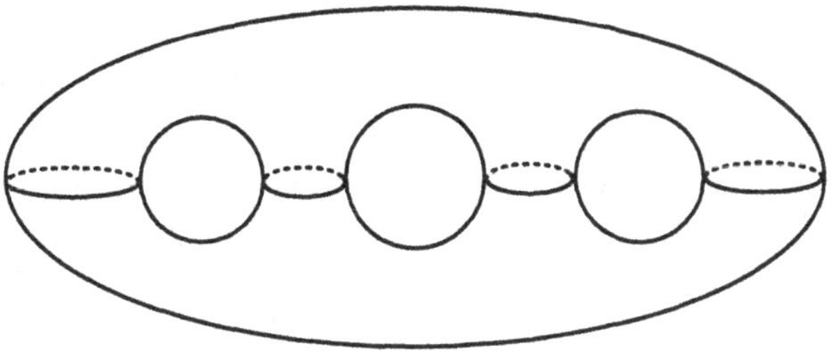

a surface of genus 3

**Figure 8.7**

This **Gauss–Bonnet theorem** is remarkable; *a deformation of the surface might change K pointwise and likewise the area form, yet the total curvature* $\int_M K\,dA$ *remains unchanged and is a measure of the genus of the surface!*

## 8.3d. The Kronecker Index of a Vector Field

Let $M^n$ be a closed submanifold of $\mathbb{R}^{n+1}$. It is a fact that $M^n$ is the boundary of a compact region $U$ of $\mathbb{R}^{n+1}$, $M^n = \partial U^{n+1}$. Then the orientation of $\mathbb{R}^{n+1}$ together with the outward-pointing normal defines an orientation of $M$. Let $\mathbf{v}$ be a *unit* vector field defined along $M$; it need not be tangent to $M$. It then defines a map $v : M^n \to S^n$ by $x \in M^n \mapsto \mathbf{v}(x) \in S^n$ (if $\mathbf{v}$ is always normal to $M$ then this is the Gauss map).

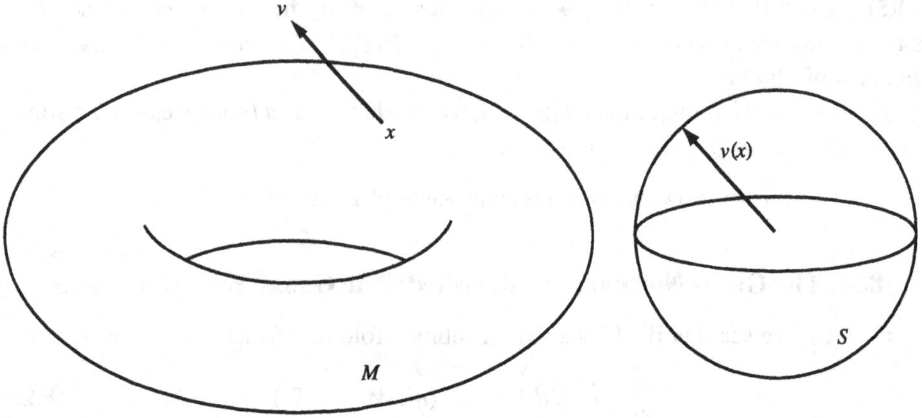


Figure 8.8

We define the **(Kronecker) index** of $\mathbf{v}$ on $M$ by

$$\text{index of } \mathbf{v} := \text{Brouwer degree of } v : M \to S$$

If $\mathbf{v}$ is any vector field on $M$ that never vanishes on $M$, we define the index of $\mathbf{v}$ to be the Kronecker index of $\mathbf{v}/\parallel \mathbf{v} \parallel$.

The following are four examples in the plane with $M^1$ itself the circle.

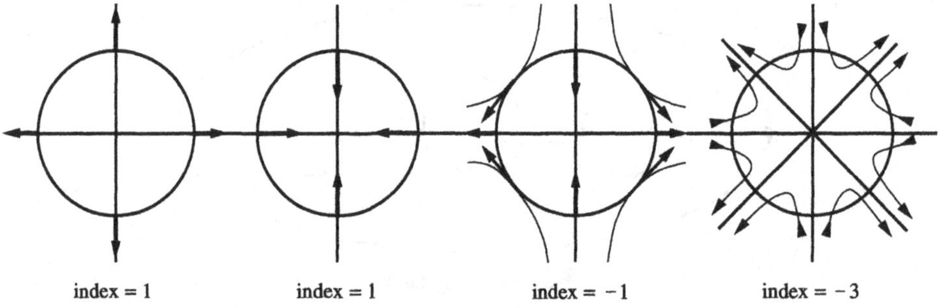


Figure 8.9

**8.3(8)** The vector fields on $S^n$ analogous to the first two depicted in the figure above are $\mathbf{v}(\mathbf{x}) = \mathbf{x}$ and $-\mathbf{x}$, respectively. Compute their Kronecker indices.

**8.3(9)** Use the integral definition of the Brouwer degree to show that *if* $\mathbf{v}$ *can be extended to be a nonvanishing vector field on all of the interior region* $U^{n+1}$, *then index* $(\mathbf{v}) = 0$. Thus none of the four fields illustrated can be extended to be nonvanishing on the disc.

**8.3(10)** Suppose that the unit vector field $\mathbf{v}$ on $M^n$ can be extended to be a smooth unit field on all of $U$ except for a finite number of points $\{P_\alpha\}$. Excise a small ball $B_\alpha$ centered at each $P_\alpha$ from $U$. Then $\mathbf{v}$ has an index on $M^n$ and also on each of the spheres

$\partial B_\alpha$ (with normal pointing out of $B_\alpha$). Show that the

$$\text{index of } \mathbf{v} \text{ on } M^n = \sum_\alpha (\text{index of } \mathbf{v} \text{ on } \partial B_\alpha)$$

We may then say that *the index of* $\mathbf{v}$ *on* $M$ *is equal to the sum of indices inside* $M$.

We have an immediate important fact.

**Theorem (8.21):** *If* $\mathbf{v}_t$ *is a smooth family of nonvanishing vector fields on* $M^n$ *with* $\mathbf{v}_0 = \mathbf{v}$ *and* $\mathbf{v}_1 = \mathbf{w}$, *then, since the index is an integer varying continuously with* $t$, *we have*

$$\text{index}(\mathbf{v}) = \text{index}(\mathbf{w})$$

**8.3(11)** Let $\mathbf{v}$ be a unit vector field on $M^n = S^n$ that never points to the center $O$. Show that

$$\text{index } (\mathbf{v}) = \text{index } (\mathbf{N}) = +1$$

In particular, *if the nonvanishing* $\mathbf{v}$ *is always tangent to* $M$, *then its index is* $+1$.

**8.3(12) The Brouwer fixed point theorem:** Show that *every smooth map* $\phi$ *of the closed* $(n+1)$ – ball $B^{n+1} = \{x \epsilon \mathbb{R}^{n+1} : \| x \| \leq 1\}$ *into itself has a fixed point.* (Hint: $B$ is a manifold with boundary $S^n$. Consider the vector field on $B$ given by $\mathbf{v}(x) =$ vector from $x$ to $\phi(x)$. On $S^n$, $\mathbf{v}$ never points toward the outer normal.)

Here is a simpler proof of the Brouwer fixed point theorem. If there is no fixed point, then the vector $\mathbf{v}$ from $x$ to $\phi(x)$ is never $\mathbf{0}$. We can then get a smooth map $r : B^{n+1} \to S^n$ by letting $r(x)$ be the point on $S^n$ where the directed line from $\phi(x)$ to $x$ strikes $S^n$. Note that $r$ is a **retraction**, that is, $r(x) = x$ for all $x$ on $S^n$. Let $\omega^n$ be any $n$-form on $S = S^n$ such that $\int_S \omega = 1$. $\omega$ is a form *on* $S$ and $d\omega = 0$; it need not be defined on $B^{n+1}$. Then $r^*\omega$ is an $n$-form on $B^{n+1}$ *that agrees with* $\omega$ *on* $S$. Note that $r(S) = S = \partial B^{n+1}$. Then

$$1 = \int_S \omega = \int_S r^*\omega = \int_{\partial B} r^*\omega = \int_B dr^*\omega$$

$$= \int_B r^* d\omega = \int_B r^* 0 = 0$$

This is a contradiction, as promised. □

Now let $u^1, \ldots, u^n$ be local coordinates for $M$. Just as in (8.13), since $\mathbf{v}(u)$ represents both the vector at $u$ and the position vector on $S^n$ at $\mathbf{v}(u)$, we have

$$v_*\left(\frac{\partial}{\partial u^\alpha}\right) = \frac{\partial \mathbf{v}}{\partial u^\alpha}$$

**8.3(13)** Show that

$$\text{index } (\mathbf{v}) = (A_n)^{-1} \int_M \text{vol}^{n+1}\left(\mathbf{v}, \frac{\partial \mathbf{v}}{\partial u^1}, \ldots, \frac{\partial \mathbf{v}}{\partial u^n}\right) du^1 \wedge \ldots \wedge du^n$$

where $\text{vol}^{n+1}$ is the volume form for $\mathbb{R}^{n+1}$, $A_n$ is the area of the unit sphere $S^n$, and we are using the traditional notation expressing the integral of an $n$-form $\alpha^n$ in terms of generic local coordinates, $\int_M \alpha^n = \int_M a_{1\ldots n}(u) du^1 \wedge \ldots \wedge du^n$. Note that

this expression for *index* ($\mathbf{v}$) *is in fact a general formula for computing the degree of any smooth map* $v : M^n \to S^n$ *of any compact oriented* $M^n$ *into* $S^n \subset \mathbb{R}^{n+1}$*!*

**8.3(14)** If $\mathbf{v}$ is nonvanishing but perhaps not unit, show that the integral on the right becomes

$$(A_n)^{-1} \int_M \| \mathbf{v}(u) \|^{-(n+1)} \operatorname{vol}^{n+1}\left(\mathbf{v}, \frac{\partial \mathbf{v}}{\partial u^1}, \ldots, \frac{\partial \mathbf{v}}{\partial u^n}\right) du^1 \wedge \ldots \wedge du^n$$

(This is not as *completely* trivial as it seems.) We then have

**Kronecker's Corollary (8.22):** *Let* $(n+1)$ *smooth functions* $f_1, \ldots, f_{n+1}$ *be defined on* $M^n$ *and its interior* $U^{n+1} \subset \mathbb{R}^{n+1}$ *with no common zeros on* $M^n$*. Let* $\det(f, df)$ *be the determinant of the* $(n+1) \times (n+1)$ *matrix whose* $j^{\text{th}}$ *row is* $(f_j, \partial f_j / \partial u^1, \ldots, \partial f_j / \partial u^n)$*. Then if*

$$\int_M (f_1^2 + \cdots + f_{n+1}^2)^{-(n+1)/2} \det(f, df) du^1 \wedge \cdots \wedge du^n \neq 0$$

*we may conclude that* $f_1 = 0, \ldots, f_{n+1} = 0$*, has a solution in* $U^{n+1}$.

### 8.3e. The Gauss Looping Integral

Let $C_\alpha : S^1 \to \mathbb{R}^3, \alpha = 1, 2$, be a pair of nonintersecting smooth closed curves in space, given by $\mathbf{r} = \mathbf{r}_1(\theta)$ and $\mathbf{r} = \mathbf{r}_2(\phi)$, respectively. Gauss wrote down an integral describing how the curves "link."

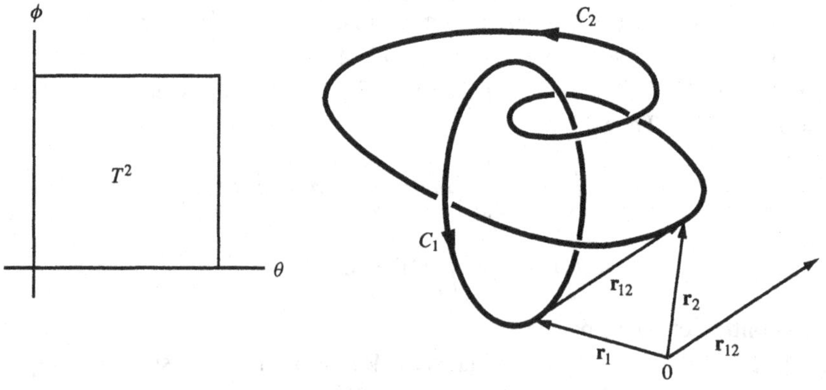


**Figure 8.10**

Consider the abstract torus $T^2 = S^1 \times S^1$ with coordinates $\theta, \phi$, and the map $L : T^2 \to S^2$ defined by

$$L(\theta, \phi) = \frac{\mathbf{r}_{12}(\theta, \phi)}{r_{12}(\theta, \phi)} := \frac{[\mathbf{r}_2(\phi) - \mathbf{r}_1(\theta)]}{\| \mathbf{r}_2(\phi) - \mathbf{r}_1(\theta) \|}$$

The Gauss **looping** or **linking** number of $C_1$ and $C_2$ is defined to be the integer

$$Lk(C_1, C_2) := \deg(L) : T^2 \to S^2$$

**8.3(15)** Show that the formula of Problem 8.3(14) translates to **Gauss's integral**

$$Lk(C_1, C_2) = (4\pi)^{-1} \int_{C_1} \left\{ \int_{C_2} r_{12}^{-3} (\mathbf{r}_{12} \times d\mathbf{r}_{12}) \right\} \bullet d\mathbf{r}_1$$

$$= (4\pi)^{-1} \int_0^{2\pi} \left[ \int_0^{2\pi} r_{12}^{-3} \left\{ \mathbf{r}_{12} \times \left( \frac{d\mathbf{r}_{12}}{d\phi} \right) \right\} d\phi \right] \cdot \left( \frac{d\mathbf{r}_1}{d\theta} \right) d\theta$$

where we choose the right-handed orientation for $\mathbb{R}^3$.

**8.3(16)** Now let $W^2$ be any orientable surface in $\mathbb{R}^3$ whose boundary is $C_1$. Choose the orientation of $W$ so that $\partial W^2 = C_1$. For the given orientation of $\mathbb{R}^3$ this picks out a preferred unit normal $\mathbf{N}$ to $W$.

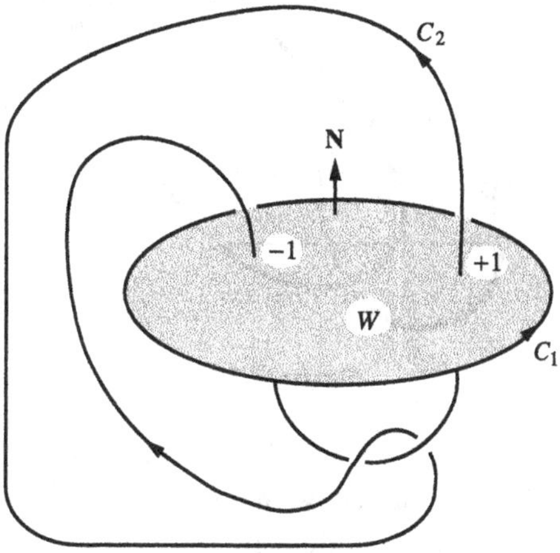


Figure 8.11

It is a fact that $C_2$ can always be moved slightly if necessary to ensure that it meets $W$ transversally. We may then consider the **intersection number** $W^2 \circ C_2$, defined to be the **signed** number of intersections of $C_2$ with $W^2$, an intersection carrying a $+$ sign only if $C_2$ is traversing $W^2$ in the same direction as $\mathbf{N}$. Then the linking number has the following interpretation.

**8.3(17)** Show that

$$Lk(C_1, C_2) = W^2 \circ C_2$$

Hint: A current of $I = 1$ in $C_2$ gives rise to a magnetic field at $\mathbf{r}_1$ given by the law of Biot–Savart

$$\mathbf{B}(\mathbf{r}_1) = \oint_{C_2} r_{12}^{-3} \mathbf{r}_{12} \times d\mathbf{r}_2$$

See Feynman's lectures [F, S, L, vol. II, pp. 14–10].

The intersection number $W^2 \circ C_2$ is a measure of how the curves link.

*It should be remarked that two wires can have linking number 0 and yet be physically inseparable*, as is indicated in our last illustration.

The preceding proof of 8.3(17) is very simple because of our acceptance of the Biot–Savart law; that is, we are assuming that the preceding integral for **B** indeed does satisfy Ampere's law! This law itself follows from Maxwell's equations, but the proof is not trivial. There are, for example complications arising from the familiar potential solutions of Poisson's equation since a wire is a limiting case of a volume distribution of current. A sketch of a purely mathematical proof, in terms of "solid angle," can be found in [C, J, p. 619 ff.] or in [Sp]. I prefer the following proof, which I learned from Michael Freedman; *it uses Theorem (8.17) directly* instead of Gauss's looping integral. For this we shall replace the intersection number by another measure of linking. We proceed as follows:

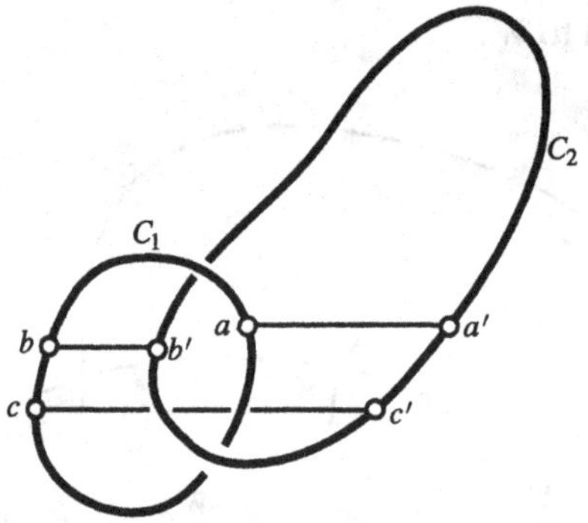


**Figure 8.12**

Two linking curves are shown. Move $C_1$ in a direction $aa'$ and keep moving it until it is far removed from $C_2$. We shall show that $\deg(L) : T^2 \to S^2$ is the (algebraic) number of times $C_1$ cuts through $C_2$ in this process.

First we must decide on a direction of motion. Pick any regular value of $L : T^2 \to S^2$. *This will be our direction*! We have drawn $(a, a')$ as a preimage on $T^2$ of this regular value; thus the segment from $a \in C_1$ to $a' \in C_2$ is in this regular direction. We have drawn the two other preimages $(b, b')$ and $(c, c')$. As we move $C_1$ in this given direction, in our picture, first $b$ will hit $b'$, then $a$ will hit $a'$, and finally $c$ will hit $c'$, and these will be the only meetings of these two curves in this example.

Look more closely at $a$ and $a'$. We have the two tangents $d\mathbf{r}_1/d\theta$ and $d\mathbf{r}_2/d\phi$ at $a = \mathbf{r}_1(\theta)$ and $a' = \mathbf{r}_2(\phi)$, respectively.

Again, the vector $aa'$ is $\mathbf{r}_{12}$. Since $\mathbf{r}_{12}/r_{12}$ is a regular value, it must be that $L_*(\partial/\partial\theta)$ and $L_*(\partial/\partial\phi)$ are linearly independent, and of course they are orthogonal to $\mathbf{r}_{12}$. Thus the vector $\mathbf{r}_{12}/r_{12} = [\mathbf{r}_2(\phi) - \mathbf{r}_1(\theta)]/r_{12}$ is a regular point of the map $L$ iff

$$\mathrm{vol}\left(\mathbf{r}_{12}, L_*\left(\frac{\partial}{\partial\theta}\right), L_*\left(\frac{\partial}{\partial\phi}\right)\right)$$

and hence

$$\mathrm{vol}\left(\mathbf{r}_{12}, -\frac{d\mathbf{r}_1}{d\theta}, \frac{d\mathbf{r}_2}{d\phi}\right)$$

are not 0, using, say, the right-hand orientation.

We shall say that $C_1$ cuts $C_2$ **positively** (resp. negatively) at $\mathbf{r}_2(\phi)$ if this "volume" is positive (resp. negative).

In our picture $(a' - a, -d\mathbf{r}_1/d\theta, d\mathbf{r}_2/d\phi)$ yields a positive cut. Similarly, $b'$ is again a positive cut and $c'$ is a negative cut.

*Thus the degree of the map L is precisely the number of times that the translated $C_1$ cuts $C_2$*, and we say that the curves are linked if the number of cuts is $\neq 0$. In our case the net number of cuts is $+1$.

## 8.4. Area, Mean Curvature, and Soap Bubbles

How can you determine the pressure inside an irregular bubble?

### 8.4a. The First Variation of Area

How does the area of a surface change as we move it in space? We consider this *very heuristically* at first. In the following picture we consider a very small curved rectangle on a positively curved surface whose sides, of length $l_1$ and $l_2$, are made up of lines of curvature; that is, they are in the two principal directions at the point $p$.

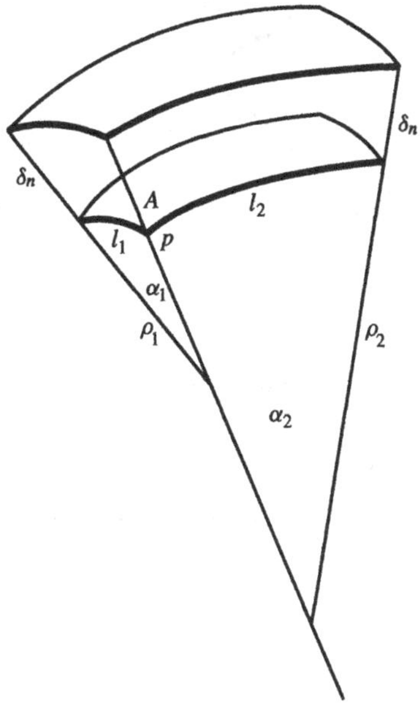


Figure 8.13

They are approximately arcs of circles of radius $\rho_1$ and $\rho_2$, the radii of principal curvatures. The area is approximately $A = l_1 l_2$. Move the whole rectangle in the normal direction a distance $\delta n$. The area changes approximately by $\delta A = \delta(l_1 l_2) = \delta l_1 l_2 + l_1 \delta l_2$. But $\delta l_1 \sim \alpha_1 \delta n = (l_1/\rho_1)\delta n$ and likewise for $\delta l_2$. Thus

$$\delta A \sim A(\rho_1^{-1} + \rho_2^{-1})\delta n = -AH\delta n$$

since the surface curves away from the normal. We now make a more careful study, for any surface, where the displacement need not be normal to the surface and can have a magnitude that varies on the surface. For this we simply consider a 1-parameter family of surfaces $M^2(t)$ in $\mathbb{R}^3$, a **variation** of an $M^2(0)$.

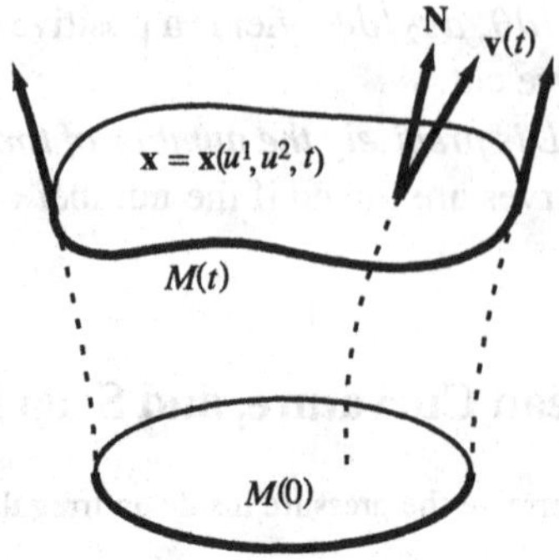


Figure 8.14

We assume that $M(0)$ is a compact manifold, perhaps with boundary. We wish to calculate how the area of $M(t)$ varies with $t$. There is a technical complication due to the fact that the surfaces $M(t)$ need not be disjoint. Schematically, reducing dimensions by 1

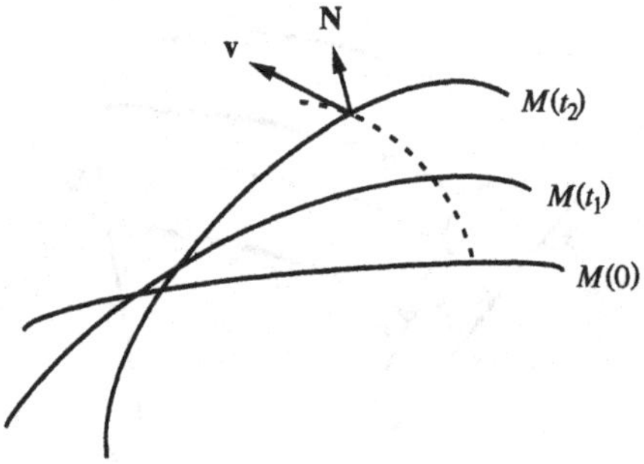


Figure 8.15

In this case the unit normals to the various $M(t)$ would not yield a well-defined vector field in $\mathbb{R}^3$, nor would the velocity ("variation") field $\partial \mathbf{x}/\partial t$. To prevent these complications we introduce an extra coordinate $t$ to the existing $\mathbb{R}^3$, as we did in 4.3b.

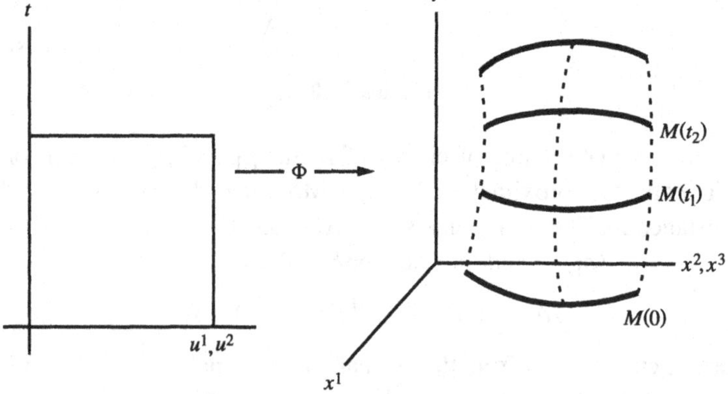


Figure 8.16

If $u^1, u^2$ are local coordinates on the base surface $M(0)$ and if we assign the same coordinates to corresponding points of $M(t)$, we then have a map $\Phi(u^1, u^2, t) = (\mathbf{x}(u^1, u^2, t), t)$ into $\mathbb{R}^4 = \mathbb{R}^3 \times \mathbb{R}$. There is then no trouble in extending the normals to define a vector field (again called **N**) in some neighborhood of the image of $\Phi$.

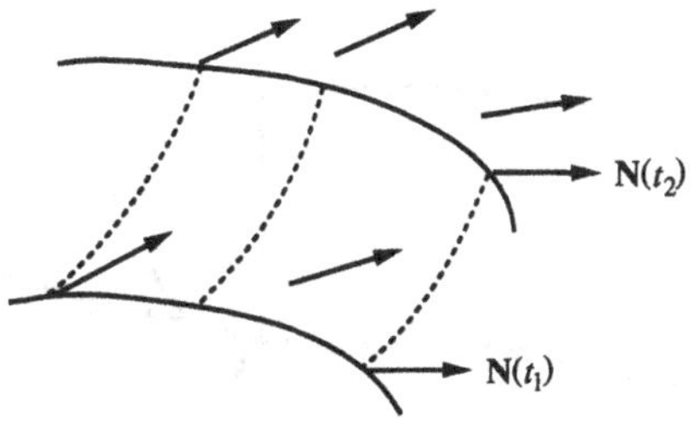


**Figure 8.17**

We may even keep the field **N** "horizontal," that is, with no $t$ component.

The same may be done with the velocity vectors $\mathbf{v} = \partial\mathbf{x}/\partial t$. Finally we may add $\partial/\partial t$ to this horizontal field to yield the space–time variation field $X = \mathbf{v} + \partial/\partial t$, as in (4.41).

We are now ready to compute the first variation of area. $\mathrm{vol}^3 = dx^1 \wedge dx^2 \wedge dx^3$ can be considered a 3-form in $\mathbb{R}^4$, and for area we have

$$A(t) = \int_{M(t)} i_{\mathbf{N}(t)}\, \mathrm{vol}^3$$

It would be possible to write down the Euler–Lagrange equations for this problem in the calculus of variations since $i_{\mathbf{N}(t)}\, \mathrm{vol}^3 = \sqrt{g} du^1 \wedge du^2$ has a "Lagrangian"

$$L\left(u^\alpha, x^j, \frac{\partial x^j}{\partial u^\alpha}\right) = \left\{\det \sum_j \left(\frac{\partial x^j}{\partial u^\alpha}\right)\left(\frac{\partial x^j}{\partial u^\beta}\right)\right\}^{1/2}$$

but it would be difficult to interpret geometrically the resulting expressions. We proceed instead directly, taking advantage of our machinery for differentiating integrals of forms in 4.3. From (4.43) we have

$$A'(t) = \int_{M(t)} \frac{\partial}{\partial t} i_{\mathbf{N}(t)}\, \mathrm{vol}^3 + \int_{M(t)} i_{\mathbf{v}} \mathbf{d} i_{\mathbf{N}(t)}\, \mathrm{vol}^3$$
$$+ \int_{\partial M(t)} i_{\mathbf{v}} i_{\mathbf{N}(t)}\, \mathrm{vol}^3$$

Look at each integral separately. First, since $\partial\mathbf{N}/\partial t$ is tangent to $M(t)$

$$\int_{M(t)} \frac{\partial}{\partial t} i_{\mathbf{N}(t)}\, \mathrm{vol}^3 = \int_{M(t)} i_{\partial\mathbf{N}/\partial t}\, \mathrm{vol}^3 = 0$$

Next, $\mathbf{d} i_{\mathbf{N}}(t)\, \mathrm{vol}^3 = \mathrm{div}\, \mathbf{N}\, \mathrm{vol}^3$, and the second integral becomes

$$\int_{M(t)} \langle \mathbf{v}, \mathbf{N} \rangle\, \mathrm{div}\, \mathbf{N}\, \mathrm{vol}^2$$

Finally, in the last integral, use arc length $s$ for parameter along $\partial M(t)$ and let $\mathbf{n}(s)$ be the unit vector field that is tangent to $M(t)$, normal to $\partial M(t)$, and points out of $M(t)$; thus in $\mathbb{R}^3$, $\mathbf{n}(s) = (d\mathbf{x}/ds) \times \mathbf{N}$.

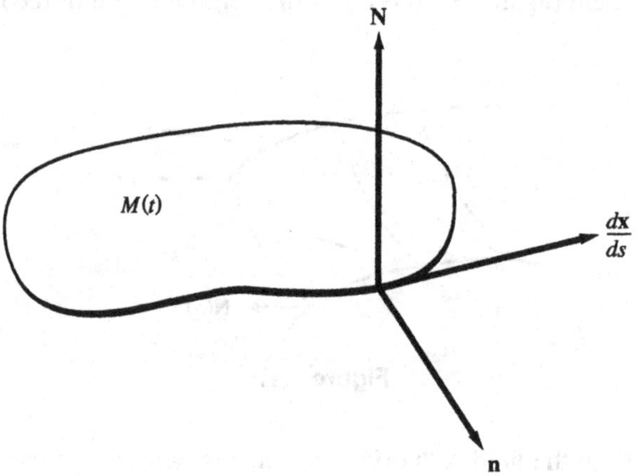


Figure 8.18

Then

$$\begin{aligned}\int_{\partial M(t)} i_{\mathbf{v}} i_{\mathbf{N}} \operatorname{vol}^3 &= \int_0^L i_{\mathbf{v}} i_{\mathbf{N}} \operatorname{vol}^3 \left(\frac{d\mathbf{x}}{ds}\right) ds \\ &= \int_0^L i_{\mathbf{N}} \operatorname{vol}^3 \left(\mathbf{v}, \frac{d\mathbf{x}}{ds}\right) ds \\ &= \int_0^L \operatorname{vol}^3 \left(\mathbf{N}, \mathbf{v}, \frac{d\mathbf{x}}{ds}\right) ds \\ &= \int_0^L \left\langle \left(\frac{d\mathbf{x}}{ds}\right) \times \mathbf{N}, \mathbf{v} \right\rangle ds \\ &= \int_0^L \langle \mathbf{n}, \mathbf{v} \rangle ds \\ &= \oint_{\partial M(t)} \langle \mathbf{n}, \mathbf{v} \rangle ds\end{aligned}$$

Thus

$$A'(t) = \int_{M(t)} \langle \mathbf{v}, \mathbf{N} \rangle \operatorname{div}\mathbf{N} \operatorname{vol}^2 + \oint_{\partial M(t)} \langle \mathbf{n}, \mathbf{v} \rangle ds \tag{8.23}$$

This formula confirms the rather obvious fact that there are two ways to increase the area of a surface with boundary. First, if the normals to the surface are diverging we should move the surface in the direction of the normals (note that this does not affect the boundary integral). Second, we may move the boundary outward at the boundary.

It is important for many purposes to realize that div $\mathbf{N}$ can be replaced essentially by the mean curvature of the surface.

$$\operatorname{div}\mathbf{N} = -H \tag{8.24}$$

**PROOF:** We shall give first a very useful expression for the divergence of any vector field in $\mathbb{R}^n$.

If $\mathbf{X}$ is a vector field and if $\mathbf{A}$ is a vector at a given point in $\mathbb{R}^n$, then the expression *in cartesian coordinates*

$$D_{\mathbf{A}}(\mathbf{X}) = \langle \mathbf{A}, D\mathbf{X} \rangle := A^j \left( \frac{\partial X^k}{\partial x^j} \right) \partial_k$$

is simply the derivative of $\mathbf{X}$ with respect to the vector $\mathbf{A}$. We claim that div $\mathbf{X}$ *is the trace of the linear transformation* $L_{\mathbf{X}} : \mathbb{R}^n \to \mathbb{R}^n$ defined by

$$L_{\mathbf{X}}(A) = D_{\mathbf{A}}(\mathbf{X}) \tag{8.25}$$

For, in our cartesian coordinates, $\operatorname{tr} L_{\mathbf{X}} = \sum_i \langle L_{\mathbf{X}}(\partial_i), \partial_i \rangle = \langle (\partial X^k / \partial x^i) \partial_k, \partial_i \rangle = \sum_i \partial X^i / \partial x^i = \operatorname{div}(\mathbf{X})$, as desired.

To compute div $\mathbf{N}$ we compute tr $(\mathbf{A} \to D_{\mathbf{A}}\mathbf{N})$, and linear algebra tells us that we may compute the trace of a linear transformation *using any basis*! We choose a basis adapted to the surface $M^2(t)$, namely $\mathbf{e}_1 = \partial \mathbf{x}/\partial u^1, \mathbf{e}_2 = \partial \mathbf{x}/\partial u^2$, and $\mathbf{e}_3 = \mathbf{N}$. Then from (8.5)

$$\mathbf{e}_1 \to \frac{\partial \mathbf{N}}{\partial u^1} = -\mathbf{e}_1 b^1{}_1 - \mathbf{e}_2 b^2{}_1$$

$$\mathbf{e}_2 \to \frac{\partial \mathbf{N}}{\partial u^2} = -\mathbf{e}_1 b^1{}_2 - \mathbf{e}_2 b^2{}_2$$

and we also have $D_{\mathbf{N}}\mathbf{N}$ is orthogonal to $\mathbf{N}$.

Thus

$$\operatorname{div} \mathbf{N} = -b^1{}_1 - b^2{}_2 = -H$$

as claimed. □

We then have **Gauss's formula for the first variation of area**

$$A'(t) = -\int_{M(t)} H \langle \mathbf{v}, \mathbf{N} \rangle \operatorname{vol}^2 + \oint_{\partial M(t)} \langle \mathbf{n}, \mathbf{v} \rangle ds \tag{8.26}$$

In the classical notation of the calculus of variations

$$\delta \mathbf{x} = \mathbf{v}(0) \qquad \delta x_N := \langle \delta \mathbf{x}, \mathbf{N} \rangle \qquad \delta x_n := \langle \delta \mathbf{x}, \mathbf{n} \rangle$$

$$\delta A := A'(0) = -\int_{M(0)} H \delta x_N dS + \oint_{\partial M(0)} \delta x_n ds \tag{8.27}$$

Note in particular that $A'(0)$ *depends only on* $\mathbf{v}(0)$, that is, the velocity vector at points of $M(0)$. In other words, given a surface $M(0)$ and a vector field $\mathbf{v}(0)$ defined along $M(0)$, extend $\mathbf{v}(0)$ in any smooth way you wish to be a vector field $\mathbf{v}$ in some neighborhood of $M(0)$. The flow generated by this vector field will define a variation $M(t)$ of $M(0)$, and the first variation of area, $A'(0)$, is given by Gauss's formula and is independent of the extension $\mathbf{v}$ chosen!

### 8.4b. Soap Bubbles and Minimal Surfaces

Consider a soap bubble blown on a pipe with perhaps irregular rim. (For the following physical considerations we shall use rather heuristic reasoning.)

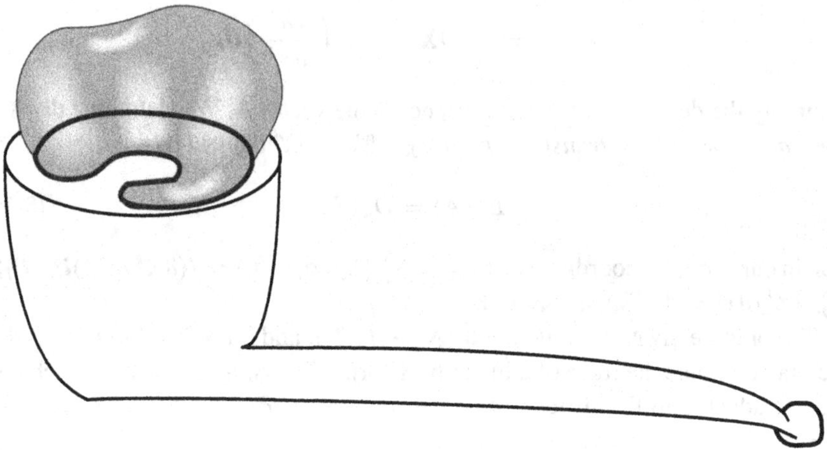

Figure 8.19

By blowing air in very slowly (*quasi-statically*, so that air inside has spatially constant pressure), the rate at which work is being done is given, in classical notation, by

$$\delta W = p\delta V$$

where $V$ is the volume of the bubble and $p$ is the difference in pressure, inside and out.

Consider a small piece of the soap film $M(0)$ as it sweeps out a small "cylinder" while being blown up for a short time.

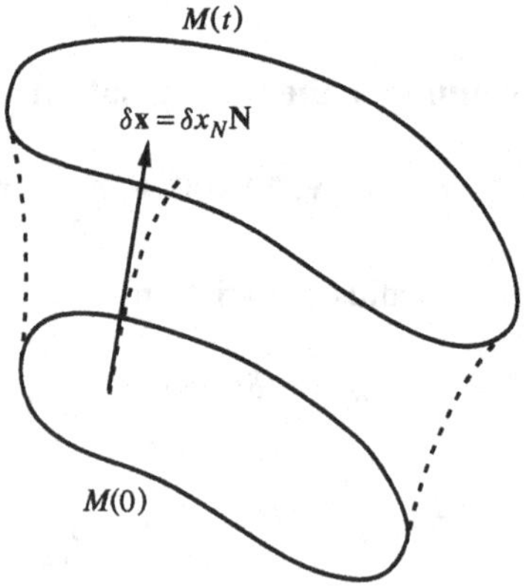


Figure 8.20

The pressure will force a normal displacement of the film of small amount $\delta \mathbf{x} = \delta x_N \mathbf{N}$. It is not hard to see that the small volume swept out will be approximately

$$\delta V = \int_{M(0)} \delta x_N dS$$

We then have

$$\delta W = p \int_{M(0)} \delta x_N dS$$

On the other hand, the work done against surface tension during the stretching of the film is approximately

$$\delta W = 2\sigma \delta A = -2\sigma \int_{M(0)} H \delta x_N dS$$

Here $\sigma$ is the coefficient of surface tension, the factor 2 arises since the film has an inside and an outside surface, and we have used Gauss's formula with $\delta x_n = 0$ since the displacements are normal to the surface. We conclude that $\int_{M(0)} (p + 2\sigma H) \delta x_N dS = 0$, and this must hold for each piece $M(0)$ of the bubble. Taking $M(0)$ to be an "infinitesimal" neighborhood of a point on the bubble, we conclude that $p + 2\sigma H = 0$ at each point of the bubble. We then have **Laplace's formula** for the pressure inside the bubble

$$p = -2\sigma H \tag{8.28}$$

(An air bubble in water has only one surface, in this case $p = -\sigma H$).

A soap bubble in equilibrium has spatially constant pressure inside (otherwise air would be in motion). Thus

A **soap bubble** in equilibrium describes a surface of **constant mean curvature** $H$.

For a spherical bubble of radius $R$, $H = \kappa_1 + \kappa_2 = -2/R$ if the outer normal is used. Then $p = 4\sigma/R$; the larger the bubble the smaller the pressure!

A soap **film** spanning a wire frame has the same pressure on both sides, and so $p = 0$. A soap film spanning a given curve $C$ describes a surface with *mean curvature* $H = 0$.

Any surface with mean curvature 0 is called a **minimal surface**. The name stems from the fact that a soap film spanning a curve tries to adjust itself so as to minimize its area. Mathematically we have the following.

**Theorem (8.29):** *Let $M^2$ be a compact surface in $\mathbb{R}^3$ with boundary curve $C = \partial M$. Then $M$ is a minimal surface, $H = 0$, if and only if the first variation of area vanishes $\delta A = 0$ for all variations of $M$ that leave the boundary $C$ fixed.*

This variational problem was first successfuly investigated by Lagrange. Experimental studies using soap films were carried out by the physicist Plateau.

The variational theorem is an immediate consequence of Gauss's formula. First note that the boundary integral vanishes since $\delta \mathbf{x} = 0$ on $C$. Next note that at a point of $M$ away from $C$, the variation $\delta x_N$ is quite arbitrary; this assures us that if the surface integral vanishes for all variations then we must have $H = 0$.

The preceding theorem assures us that a minimal surface yields a *critical point* for the area functional. To investigate the nature of the critical point (minimum, maximum, minimax, ...) one should look at the second variation $A''(0)$. One should also discuss whether a minimum is relative or absolute. It turns out that a sufficiently small piece of minimal surface yields an absolute minimum for area (keeping its boundary fixed). There are soap films that give a relative, though not absolute minimum for area. There are minimal surfaces that do not give even a relative minimum (i.e., they are "unstable," but such unstable surfaces cannot be realized by soap films). It would be better to call a surface with $H = 0$ a "stationary" surface, with no indication of minimality.

We conclude with two remarks. First, if $H = \kappa_1 + \kappa_2 = 0$ then $K = \kappa_1\kappa_2 \leq 0$, showing that a *minimal surface is always saddle-shaped.* Finally, a minimal surface of the form $z = f(x, y)$ satisfies, from Problem 8.2(5), the nonlinear partial differential equation

$$(1 + f_y^2) f_{xx} - 2 f_x f_y f_{xy} + (1 + f_x^2) f_{yy} = 0$$

the so-called minimal surface equation.

## Problem

**8.4(1)** Let $M^2$ be a minimal surface with boundary $\partial M = C$, and let $M$ be given in parametric form $\mathbf{x} = \mathbf{x}(u, v)$. Consider the variation ("dilation") of $M$ given by

$$\mathbf{x} = \mathbf{x}(u, v; t) = (1 + t)\mathbf{x}(u, v)$$

Note that this variation moves the boundary curve also.

**(i)** Show from $A = \int_M \| \mathbf{x}_u \times \mathbf{x}_v \| \, du dv$ that $A(t) = (1 + t)^2 A(0)$.

**(ii)** Show that 2 area $M^2 = \oint_C \mathrm{vol}^3(\mathbf{N}, \mathbf{x}, d\mathbf{x}/ds) ds = \oint_C \det(\mathbf{N}, \mathbf{x}, d\mathbf{x})$.

This formula is due to H. A. Schwarz and has the remarkable consequence that *the area of any minimal surface spanning C is completely determined by the normals to the surface at points of the boundary alone!*

## 8.5. Gauss's *Theorema Egregium*

Must every plane map of the Earth's surface have distortion?

### 8.5a. The Equations of Gauss and Codazzi

Let $M^2$ be a surface in $\mathbb{R}^3$ with local coordinates $u = u^1$ and $v = u^2$. Then the vectors $\mathbf{x}_\alpha = \partial\mathbf{x}/\partial u^\alpha$, for $\alpha = 1, 2$, give a basis for the tangent planes at each point of the coordinate patch. Of course $\mathbf{x}_{\alpha\beta} = \partial^2\mathbf{x}/\partial u^\beta \partial u^\alpha = \mathbf{x}_{\beta\alpha}$ need not be tangent to $M$. Decompose into tangential and normal parts

$$\mathbf{x}_{\alpha\beta} = \partial_\beta \partial_\alpha \mathbf{x} = \mathbf{x}_\gamma \Gamma^\gamma_{\beta\alpha} + \langle \mathbf{x}_{\alpha\beta}, \mathbf{N} \rangle \mathbf{N}$$

or

$$\mathbf{x}_{\alpha\beta} = \mathbf{x}_\gamma \Gamma^\gamma_{\beta\alpha} + b_{\alpha\beta}\mathbf{N} \tag{8.30}$$

where the coefficients $\Gamma^{\gamma}_{\alpha\beta} = \Gamma^{\gamma}_{\beta\alpha}$ are still to be determined. Now

$$\langle \mathbf{x}_{\alpha\beta}, \mathbf{x}_{\mu}\rangle = \langle \mathbf{x}_{\gamma}, \mathbf{x}_{\mu}\rangle \Gamma^{\gamma}_{\beta\alpha} = g_{\gamma\mu}\Gamma^{\gamma}_{\beta\alpha} =: \Gamma_{\beta\alpha,\mu}$$

Note that

$$\begin{aligned}\partial_\beta g_{\alpha\mu} &= \partial_\beta \langle \mathbf{x}_\alpha, \mathbf{x}_\mu\rangle = \langle \mathbf{x}_{\alpha\beta}, \mathbf{x}_\mu\rangle + \langle \mathbf{x}_\alpha, \mathbf{x}_{\mu\beta}\rangle \\ &= \Gamma_{\beta\alpha,\mu} + \Gamma_{\beta\mu,\alpha}\end{aligned} \tag{8.31}$$

We conclude $\partial g_{\alpha\mu}/\partial u^\beta + \partial g_{\beta\alpha}/\partial u^\mu - \partial g_{\mu\beta}/\partial u^\alpha = 2\Gamma_{\mu\beta,\alpha} = 2\Gamma^{\tau}_{\mu\beta}g_{\tau\alpha}$ and so

$$\Gamma^{\tau}_{\mu\beta} = \frac{1}{2}g^{\alpha\tau}\left(\frac{\partial g_{\alpha\mu}}{\partial u^\beta} + \frac{\partial g_{\beta\alpha}}{\partial u^\mu} - \frac{\partial g_{\mu\beta}}{\partial u^\alpha}\right) \tag{8.32}$$

the **Christoffel symbols** ("of the second kind").

Thus all the coefficients in **Gauss's surface equations** (8.30) have been evaluated in terms of the first and second fundamental forms $g$ and $b$. Gauss now took a further step by calculating the consequences of the identity $\mathbf{x}_{\alpha\beta\gamma} = \partial_\gamma\partial_\beta\partial_\alpha \mathbf{x} = \mathbf{x}_{\alpha\gamma\beta}$. In Problem 8.5(1) you are asked to show that

$$\mathbf{x}_{\alpha\beta\gamma} - \mathbf{x}_{\alpha\gamma\beta} = \mathbf{x}_\tau(R^{\tau}{}_{\alpha\gamma\beta} - U^{\tau}_{\alpha\beta\gamma}) + V_{\alpha\beta\gamma}\mathbf{N}$$

where (8.33)

$$R^{\tau}{}_{\alpha\gamma\beta} := \partial_\gamma \Gamma^{\tau}_{\beta\alpha} - \partial_\beta \Gamma^{\tau}_{\gamma\alpha} + \Gamma^{\tau}_{\gamma\mu}\Gamma^{\mu}_{\beta\alpha} - \Gamma^{\tau}_{\beta\mu}\Gamma^{\mu}_{\gamma\alpha}$$

is now called the **Riemann** or **Riemann–Christoffel curvature tensor**. $U$ and $V$ are given by

$$U^{\tau}{}_{\alpha\beta\gamma} = b^{\tau}{}_{\gamma}b_{\alpha\beta} - b^{\tau}{}_{\beta}b_{\alpha\gamma}$$

and

$$V_{\alpha\beta\gamma} = \Gamma^{\tau}_{\alpha\beta}b_{\tau\gamma} + \partial_\gamma b_{\alpha\beta} - \Gamma^{\tau}_{\alpha\gamma}b_{\tau\beta} - \partial_\beta b_{\alpha\gamma}$$

We then conclude that

$$R^{\tau}{}_{\alpha\gamma\beta} = b^{\tau}{}_{\gamma}b_{\alpha\beta} - b^{\tau}{}_{\beta}b_{\alpha\gamma}$$

and (8.34)

$$\partial_\gamma b_{\alpha\beta} - \Gamma^{\tau}_{\alpha\gamma}b_{\tau\beta} = \partial_\beta b_{\alpha\gamma} - \Gamma^{\tau}_{\alpha\beta}b_{\tau\gamma}$$

The first equations are called **Gauss's** equations and the second are called the equations of **Codazzi** and **Peterson**.

Only after Problem 8.5(1) will the reader fully appreciate that we have been using a very condensed notation that was not used at the time of Gauss. Gauss did not use indices. He wrote $ds^2 = Edu^2 + 2Fdudv + Gdv^2$ instead of $g_{\alpha\beta}du^\alpha du^\beta$, and $Ldu^2 + 2Mdudv + Ndv^2$ instead of $b_{\alpha\beta}du^\alpha du^\beta$, and so on.

The equations (8.34) are **integrability conditions**, that is, conditions that must be satisfied by $g_{\alpha\beta}(u, v)$ and $b_{\alpha\beta}(u, v)$ in order for these two matrices to be the first and second fundamental forms for a surface in $\mathbb{R}^3$. In fact, **Bonnet** showed that these conditions are also sufficient to ensure the local existence in $\mathbb{R}^3$ of a surface having a prescribed $g_{\alpha\beta}(u, v)$ and $b_{\alpha\beta}(u, v)$.

### 8.5b. The *Theorema Egregium*

Gauss's calculation of the first equation in (8.34) led him to one of the most important and surprising discoveries in all of mathematics. First , however, we need some background.

We are all familiar with geographical maps

$$\phi : S_a^2 \to \text{a portion of the plane } \mathbb{R}^2$$

where $S_a$ is a portion of the sphere of radius $a$. (We shall not be concerned here with the inaccuracies in approximating the Earth by a sphere.) Ideally one would hope for a map that preserves distances, up to a constant factor that for simplicity we shall take to be 1. The length of a curve $\mathbf{x} = \mathbf{x}(t)$ on the Earth's surface is

$$\int_0^1 \left\langle \frac{d\mathbf{x}}{dt}, \frac{d\mathbf{x}}{dt} \right\rangle^{1/2} dt$$

and its image in $\mathbb{R}^2$ has length

$$\int_0^1 \left\langle \phi_*\left(\frac{d\mathbf{x}}{dt}\right), \phi_*\left(\frac{d\mathbf{x}}{dt}\right) \right\rangle^{1/2} dt$$

We say that a local mapping $\phi : M^n \to V^n$ of Riemannian manifolds is a **local isometry** if $\phi_*$ preserves lengths of vectors

$$\langle \phi_*\mathbf{X}, \phi_*\mathbf{X} \rangle_V = \langle \mathbf{X}, \mathbf{X} \rangle_M$$

for all tangent vectors $\mathbf{X}$ to $M$. Note that $\phi_*$ then automatically preserves all scalar products, thanks to the identity

$$\langle \mathbf{X}, \mathbf{Y} \rangle = \frac{1}{2}\{\| \mathbf{X} + \mathbf{Y} \|^2 - \| \mathbf{X} \|^2 - \| \mathbf{Y} \|^2\}$$

If $\phi$ is a local isometry, then all lengths of curves, areas of regions, and angles between curves are preserved; in other words the map is **distortion-free**. Since $\phi_* M_p \to V_{\phi(p)}$ is then an **isomorphism** (i.e., 1–1 and onto), the inverse function theorem assures us that $\phi$ itself is a local diffeomorphism in the neighborhood of each point of $M$.

A familiar example is when a flat sheet of paper is rolled up into a cylinder or a cone; though the paper is "bent" there is basically no "stretching." Although the distances between points of the sheet are changed (considered as points in the ambient $\mathbb{R}^3$), the length of any curve on the flat sheet is the same as when it is rolled up; this is the meaning of bending without stretching!

If $\phi$ is a local isometry, one may transplant a local coordinate system $y$ near $\phi(p)$ back to a coordinate system $x$ near $p$ by

$$x^i(p) := y^i(\phi(p)) = y^i \circ \phi(p)$$

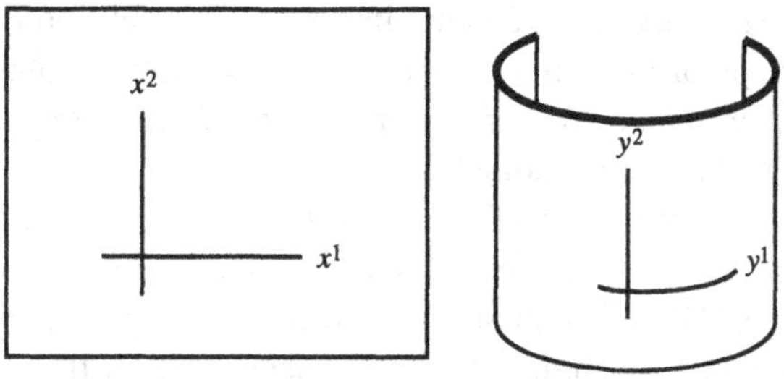


**Figure 8.21**

In terms of these associated coordinates, $\phi$ is given simply by $y^i = x^i$ and so

$$\phi_* \frac{\partial}{\partial x^i} = \frac{\partial}{\partial y^i}$$

Since $\phi$ is assumed to be a local isometry

$$g_{ij}^V(y) = \left\langle \frac{\partial}{\partial y^i}, \frac{\partial}{\partial y^j} \right\rangle_V = \left\langle \phi_* \frac{\partial}{\partial x^i}, \phi_* \frac{\partial}{\partial x^j} \right\rangle_V$$
$$= \left\langle \frac{\partial}{\partial x^i}, \frac{\partial}{\partial x^j} \right\rangle_M = g_{ij}^M(x)$$

that is, in the associated coordinates the metric tensors of $M$ and $V$ are identical at corresponding points. *But then the Christoffel symbols and the Riemann tensor*, which are *defined* in any Riemannian manifold using (8.32) and the second equation in (8.33), *are also identical at corresponding points* since they are constructed from the metric tensor alone!

Return now to our case of a surface $M^2$ in $\mathbb{R}^3$. Look carefully, with Gauss, at the first equation in (8.34). We have

$$R^{12}{}_{12} := g^{2\alpha} R^1{}_{\alpha 12} = g^{2\alpha}(b^1{}_1 b_{\alpha 2} - b^1{}_2 b_{\alpha 1})$$
$$= (b^1{}_1 b^2{}_2 - b^1{}_2 b^2{}_1) = \det b = K$$

But since $R^{12}{}_{12}$ is expressible entirely in terms of the metric tensor we have

**Gauss's *Theorema Egregium* (8.35):** *The Gauss curvature*

$$K = \kappa_1 \kappa_2 = R^{12}{}_{12}$$

*is an* ***isometry invariant****. In particular, if a surface is bent without stretching, then although the principal curvatures $\kappa_1$ and $\kappa_2$ may change, their product will not!*

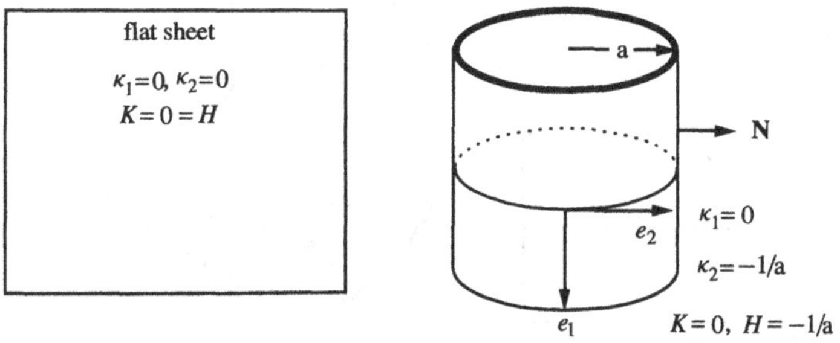


**Figure 8.22**

(Note that the *mean* curvature $H$ is not invariant!) We have an immediate familiar consequence for maps of the Earth. Since a sphere of radius $a$ has $K = 1/a^2 \neq 0$ we conclude that *every plane map of a portion of the Earth's surface must introduce distortions*, that is, cannot be an isometry.

Gauss's *theorema egregium* says that one measure of the curvature of a surface, $K$, can be expressed in terms of an object $R^{12}{}_{12}$ that is *completely determined by the metric tensor* of the surface. We call such an object **intrinsic**. In Equation (10.27) we shall exhibit geometric intrinsic formulas for $K$. (We shall see later that $R^{12}{}_{12}$ is essentially the only independent component of $R^{\alpha\beta}{}_{\gamma\delta}$.)

Riemann's generalization $R^{\alpha}{}_{\beta\gamma\tau}$ (the second equation in (8.33)) to $n$-dimensional manifolds defines, as we shall see again, an intrinsic measure of **curvature**. Curvature, in the space–time manifold of Einstein's general theory of relativity, as we shall see in Chapter 11, is a measure of the strength of the gravitational field.

Cartan generalized the notion of intrinsic curvature to general "vector bundles." In Yang–Mills's *gauge theories*, as we shall see, curvature becomes a measure of the "strength" of the gauge field.

This is just part of the legacy of Gauss's discovery.

## Problems

**8.5(1)** Using the surface equations (8.30) and the Weingarten equations (8.5), derive the Gauss and the Codazzi–Peterson equations (8.34).

**8.5(2)** Compute the curvature of the sphere with metric (8.3) the hard way: that is, show $R^{12}{}_{12} = 1/a^2$ directly from the second equation in (8.33). Later on we shall have much more efficient ways to compute.

## 8.6. Geodesics

How can we describe the "straightest" curves on a surface?

### 8.6a. The First Variation of Arc Length

Let $C$ be a curve on a surface $M^2$. We shall consider the first variation of arc length as we vary the curve. A variation $\mathbf{x}$ of $C$ is a map of a rectangle $R^2 = [0, L] \times (-1, +1)$ into $M$; $\mathbf{x} : R^2 \to M$

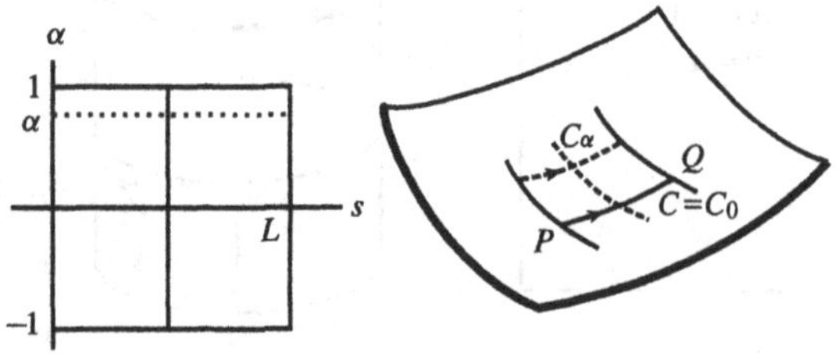


Figure 8.23

The map is described by $\mathbf{x} = \mathbf{x}(s, \alpha)$, where $\mathbf{x} = \mathbf{x}(s) = \mathbf{x}(s, 0)$ is the original curve $C = C_0$ *parameterized by arc length*, whose length is $L$. On the other hand, $s$ is *not* assumed to be arc length parameter for the curves $C_\alpha, \mathbf{x} = \mathbf{x}(s, \alpha)$, for fixed $\alpha \neq 0$, since such a parameterization would force all the $C_\alpha$ to have the same length $L$. The length of $C_\alpha$ is

$$L(\alpha) = \int_0^L \left\langle \frac{\partial \mathbf{x}(s,\alpha)}{\partial s}, \frac{\partial \mathbf{x}(s,\alpha)}{\partial s} \right\rangle^{1/2} ds$$

and so

$$\begin{aligned} L'(\alpha) &= \int_0^L \frac{\partial}{\partial \alpha} \left\langle \frac{\partial \mathbf{x}}{\partial s}, \frac{\partial \mathbf{x}}{\partial s} \right\rangle^{1/2} ds \\ &= \int_0^L \left\| \frac{\partial \mathbf{x}}{\partial s} \right\|^{-1} \left\langle \frac{\partial^2 \mathbf{x}}{\partial \alpha \partial s}, \frac{\partial \mathbf{x}}{\partial s} \right\rangle ds \end{aligned}$$

Since $s$ is arc length when $\alpha = 0$, we have $\| \partial \mathbf{x}(s, 0)/\partial s \| = 1$ and

$$\begin{aligned} L'(0) &= \int_0^L \left\langle \frac{\partial^2 \mathbf{x}}{\partial \alpha \partial s}, \frac{\partial \mathbf{x}}{\partial s} \right\rangle ds = \int_0^L \left\langle \frac{\partial^2 \mathbf{x}}{\partial s \partial \alpha}, \frac{\partial \mathbf{x}}{\partial s} \right\rangle ds \\ &= \int_0^L \frac{\partial}{\partial s} \left\langle \frac{\partial \mathbf{x}}{\partial \alpha}, \frac{\partial \mathbf{x}}{\partial s} \right\rangle ds - \int_0^L \left\langle \frac{\partial \mathbf{x}}{\partial \alpha}, \frac{\partial^2 \mathbf{x}}{\partial s^2} \right\rangle ds \end{aligned}$$

Thus we have the **first variation of arc length formula**

$$L'(0) = \langle \mathbf{J}, \mathbf{T} \rangle_Q - \langle \mathbf{J}, \mathbf{T} \rangle_P - \int_0^L \left\langle \mathbf{J}, \frac{\partial \mathbf{T}}{\partial s} \right\rangle ds \tag{8.36}$$

where $\mathbf{T} = \partial \mathbf{x}/\partial s(s, 0)$ is the unit tangent to $C = C_0$ and $\mathbf{J} = \partial \mathbf{x}/\partial \alpha(s, 0)$ is the **variation vector** along $C$.

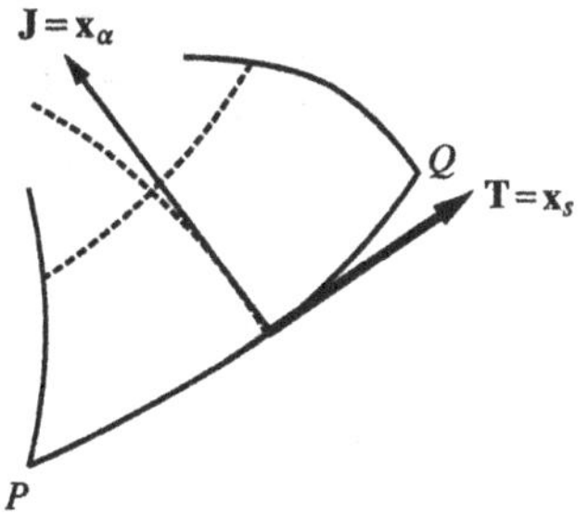


Figure 8.24

$C$ is said to be a **geodesic** if $L'(0) = 0$ for all variations that vanish at the endpoints $P$ and $Q$, that is, $\mathbf{x}(0, \alpha) = P$ and $\mathbf{x}(L, \alpha) = Q$ for all $\alpha$. For such variations $\mathbf{J} = 0$ at $P$ and $Q$ and the first variation vanishing yields

$$\int_0^L \left\langle \mathbf{J}, \frac{\partial \mathbf{T}}{\partial s} \right\rangle ds = 0$$

Both $\mathbf{T}$ and $\mathbf{J}$ are tangent vectors to the surface $M$, but of course $\partial \mathbf{T}/\partial s$ need not be. Since the variations allowed are very general (except at $P$ and $Q$)

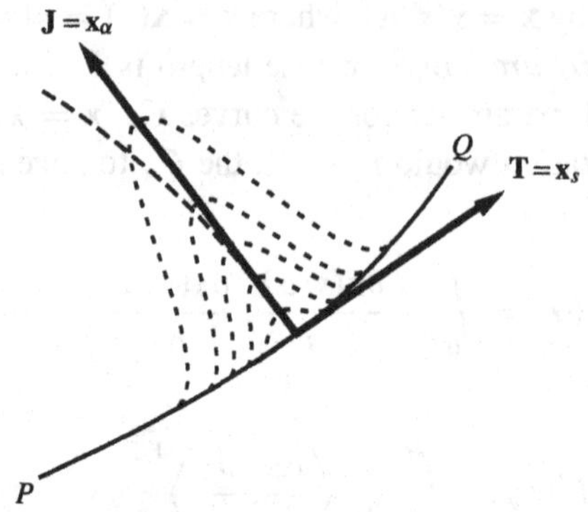


Figure 8.25

we conclude, by the fundamental lemma of the calculus of variations, that if $C$ is a geodesic then $\langle \mathbf{J}, \partial\mathbf{T}/\partial s\rangle = 0, 0 < s < L$, for every vector $\mathbf{J}$ that is *tangent to* $M$ along the geodesic $C$. Thus $\partial\mathbf{T}/\partial s$ must be normal to the surface $M^2$ along $C$. But $\partial\mathbf{T}/\partial s = \kappa\mathbf{n}$; we have derived **John Bernoulli's** characterization of geodesics of 1697:

**Theorem (8.37):** *$C$ on $M^2$ is a geodesic iff $C$, when considered as a space curve, has a principal normal* $\mathbf{n}$ *that is normal to $M$.*

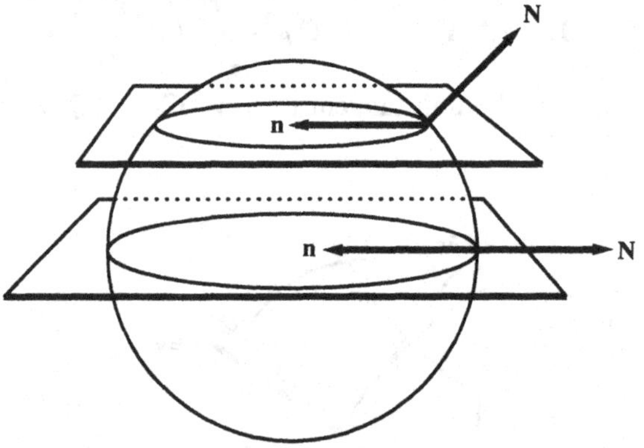


Figure 8.26

Thus if we cut out a circle on $S^2$ by slicing the sphere with a plane, the resulting circle will be a geodesic on $S^2$ iff it is a great circle.

### 8.6b. The Intrinsic Derivative and the Geodesic Equation

Let $\mathbf{X}$ be a vector field defined along a curve $C$ (parameterized by $t$) and tangent to $M^2$. $d\mathbf{X}/dt$ of course need not be tangent to $M$; we define a new derivative

$$\frac{\nabla\mathbf{X}}{dt} := \frac{d\mathbf{X}}{dt} - \left\langle \frac{d\mathbf{X}}{dt}, \mathbf{N}\right\rangle \mathbf{N} \tag{8.38}$$

Thus $\nabla\mathbf{X}/dt$ is the **tangential part** of $d\mathbf{X}/dt$, that is, the projection of $d\mathbf{X}/dt$ into the tangent space to $M^2$ at the given point. $\nabla\mathbf{X}/dt$ is called the **intrinsic derivative** (or sometimes the **covariant** derivative) of $\mathbf{X}$ along the curve $C$. This new type of derivative will be discussed in great detail shortly, but for the present we shall simply note that $\nabla\mathbf{T}/ds$ is the projection of the curvature vector $d\mathbf{T}/ds = \kappa\mathbf{n} = \boldsymbol{\kappa}$ of $C$, considered as a space curve, into the tangent plane. We shall denote this tangent vector by $\boldsymbol{\kappa}_g$ and call it the **geodesic curvature vector**; its magnitude $\kappa_g$ is called the geodesic curvature. Since $d\mathbf{T}/ds = \kappa\mathbf{n}$ is orthogonal to $\mathbf{T}$, so is $\boldsymbol{\kappa}_g$.

Geodesics are characterized by being curves $\mathbf{x} = \mathbf{x}(s)$ for which

$$\boldsymbol{\kappa}_g := \frac{\nabla\mathbf{T}}{ds} = 0 \tag{8.39}$$

A geodesic $C$ is then a curve for which *the derivative of the unit tangent has no component tangent to the surface.*

The first variation formula (8.36) then shows us that if $C$ is any curve, we may shorten it by moving the endpoints inward. If $C$ is not a geodesic in a neighborhood of some point $C(s)$, we may also shorten it by moving a small portion near $C(s)$ in the direction of its geodesic curvature vector $\boldsymbol{\kappa}_g$.

Finally, let us write out the geodesic equation $\nabla\mathbf{T}/ds = 0$ in local coordinates. For our curve $\mathbf{x} = \mathbf{x}(u(s))$

$$\begin{aligned}
\mathbf{T} = \frac{d\mathbf{x}}{ds} &= \left(\frac{\partial\mathbf{x}}{\partial u^\beta}\right)\left(\frac{du^\beta}{ds}\right) = \mathbf{x}_\beta\frac{du^\beta}{ds}\\
\frac{d^2\mathbf{x}}{ds^2} &= \mathbf{x}_{\beta\alpha}\left(\frac{du^\alpha}{ds}\right)\left(\frac{du^\beta}{ds}\right) + \mathbf{x}_\beta\frac{d^2u^\beta}{ds^2}\\
&= (\mathbf{x}_\gamma\Gamma^\gamma_{\alpha\beta} + b_{\alpha\beta}\mathbf{N})\left(\frac{du^\alpha}{ds}\right)\left(\frac{du^\beta}{ds}\right) + \mathbf{x}_\gamma\frac{d^2u^\gamma}{ds^2}
\end{aligned}$$

and so

$$\frac{\nabla\mathbf{T}}{ds} = \mathbf{x}_\gamma\left[\frac{d^2u^\gamma}{ds^2} + \Gamma^\gamma_{\alpha\beta}\left(\frac{du^\alpha}{ds}\right)\left(\frac{du^\beta}{ds}\right)\right] \tag{8.40}$$

Thus a curve $u = u(s)$ parameterized by arc length is a geodesic iff

$$\frac{d^2u^\gamma}{ds^2} + \Gamma^\gamma_{\alpha\beta}\left(\frac{du^\alpha}{ds}\right)\left(\frac{du^\beta}{ds}\right) = 0 \tag{8.41}$$

The fundamental theorem on differential equations tells us that this system, that is,

$$\begin{aligned}
\frac{du^\gamma}{ds} &= T^\gamma\\
\frac{dT^\gamma}{ds} &= -\Gamma^\gamma_{\alpha\beta}(u(s))T^\alpha T^\beta
\end{aligned}$$

has a unique solution $u^\gamma = u^\gamma(s)$ for given initial data $u^\gamma(0) = u^\gamma_0$ and $du^\gamma/ds(0) = T^\gamma_0$. Furthermore, as we shall see in the next section, $\mathbf{T}(s)$ automatically will have constant length, and thus $s$ will automatically be the arc length parameter if we start with a unit initial $\mathbf{T}$. *Thus there is a unique geodesic starting at each initial point with given initial unit tangent.* Since the system is nonlinear, *we may not insist that the solution exist for all parameter values s*!

A geodesic is a **critical point** for the length functional for curves joining two endpoints $P$ and $Q$. In Chapter 12 we shall discuss the nature of the critical point but we simply remark here that *if P and Q are sufficiently close then there is a unique geodesic joining them whose length is an absolute minimum.* A great circle on the 2-sphere that goes three-quarters of the way around the sphere is clearly a geodesic that does not yield an absolute minimum for the length of curves joining the endpoints; in fact, as we shall see in Chapter 12, it does not yield even a local minimum! A thorough analysis of geodesics is given in Milnor's book [M].

## 8.7. The Parallel Displacement of Levi-Civita

What should it mean to move a vector on a curved surface "parallel to itself" while it remains tangent to the surface?

Let $\mathbf{v}$ be a vector field in $\mathbb{R}^n$ defined along a curve $x = x(t)$. The derivative of this field is another vector field $d\mathbf{v}/dt$ along the curve, defined, as usual, by

$$\mathbf{v}'(t) = \frac{d\mathbf{v}(t)}{dt} = \lim_{h\to 0} \frac{[\mathbf{v}(t+h) - \mathbf{v}(t)]}{h}$$

We are clearly comparing a vector at one point, $x(t)$, with another vector at the second point $x(t+h)$. This is possible because $\mathbb{R}^n$, being an affine space, allows us to parallel translate a vector at a given point to any other point in $\mathbb{R}^n$. This process is not available to us in a general manifold $M^n$; the use of a local coordinate system to define parallelism (namely, keeping the components of a vector constant) would yield a definition strongly dependent on the coordinates used. This is intimately related to our discovery in Section 2.4e that the obvious notion of the derivative of a vector field $\partial v^j/\partial x^k$ using coordinates does not yield a tensor field.

If $M^n \subset \mathbb{R}^N$ is a submanifold of euclidean space, can we use the ambient space to define the notion of derivative of a vector field? Consider, for example, a surface in 3-space. Let $\mathbf{X}$ be a tangent vector to $M^2 \subset \mathbb{R}^3$ at a point $P$. Given a second point $Q$ on $M$, we may consider the vector $\mathbf{Y}$ at $Q$ obtained by parallel displacing $\mathbf{X}$ *in* $\mathbb{R}^3$ to the point $Q$. Of course $\mathbf{Y}$ in general will not be tangent to $M$ at $Q$; in fact, it may even be normal. If we used our previous definition to define the derivative $d\mathbf{X}/dt$ of a vector field along a curve, we would only recover the derivative in $\mathbb{R}^3$, yielding a vector field along the curve that is not tangent to the surface. Levi-Civita remedied this, yielding what we have called the *intrinsic derivative* $\nabla\mathbf{X}/dt$. If $\mathbf{X}$ is a vector field defined along a curve $C$ on $M^2 \subset \mathbb{R}^3$, $\mathbf{X}$ being tangent to $M$, we have defined $\nabla\mathbf{X}/dt$ to be the projection of $d\mathbf{X}/dt$ into the tangent plane to $M$. Writing

$$\mathbf{X} = X^\alpha(t)\mathbf{x}_\alpha(u(t))$$

we get

$$\frac{d\mathbf{X}}{dt} = \frac{dX^\alpha}{dt}\mathbf{x}_\alpha + X^\alpha \mathbf{x}_{\alpha\beta}\frac{du^\beta}{dt}$$

The Gauss surface equations (8.30) then yield

$$\frac{\nabla \mathbf{X}}{dt} = \left(\frac{\nabla X^\gamma}{dt}\right)\mathbf{x}_\gamma \tag{8.42}$$

where

$$\frac{\nabla X^\gamma}{dt} := \frac{dX^\gamma}{dt} + \left(\frac{du^\beta}{dt}\right)\Gamma^\gamma_{\beta\alpha}X^\alpha$$

is the $\gamma^{\text{th}}$ component of the intrinsic derivative of $\mathbf{X}$. (As such, it would be more reasonable to write $(\nabla\mathbf{X}/dt)^\gamma$, but we have used the traditional notation.)

Given the parameterized $C$, $u = u(t)$, and given an initial vector $\mathbf{X}_0$ tangent to $M^2$ at $u(0)$, there is a unique tangent vector field $\mathbf{X}(t)$ to $M$ along $C$ that satisfies the system of differential equations

$$\frac{\nabla X^\gamma}{dt} = 0 \tag{8.43}$$

with initial conditions $X^\gamma(0) = X^\gamma_0$. This solution exists for all parameter time $t$ since the system is linear. The unique solution $\mathbf{X}$ is called the **parallel translate** or **displacement** or **transport** of $\mathbf{X}_0$ along $C$, and (8.43) is called the equation of parallel translation.

Equation (8.41) then tells us that *the tangent vector to a geodesic parameterized by arc length is parallel displaced along the geodesic.*

Note that (8.43) merely tells us that $d\mathbf{X}/dt$ is always normal to the surface along the curve when $\mathbf{X}$ is parallel displaced.

The notion of intrinsic derivative is seen, from (8.42), to involve only the metric tensor, not the second fundamental form. This is the reason for the description "intrinsic." In particular, *the notions of intrinsic derivative and parallel displacement make sense on an abstract Riemannian surface*, even though the original motivation relied on a specific embedding $M^2 \subset \mathbb{R}^3$. Note also that the definition (8.43) *makes sense in a Riemannian manifold $M^n$ of any dimension*, since the definition of the Christoffel symbols (8.32) makes sense in any Riemannian manifold. It is not immediate, without looking at the transformation properties of the Christoffel symbols, that $\nabla X^\gamma/dt$, as given in (8.43), transforms as a contravariant vector, but this is indeed true. This discovery of Christoffel, in 1869, was the real beginning of tensor analysis. It wasn't until 1918 that Levi-Civita interpreted the intrinsic derivative in the case of an embedded surface as the tangential component of the usual derivative.

Since parallel displacement is intrinsic, *if $\phi : M^n \to V^n$ is an isometry and if $\mathbf{X}$ is parallel displaced along $C$ of $M$, then $\phi_*\mathbf{X}$ is parallel displaced along $\phi(C)$ in $V$.*

Furthermore, if $M^2 \subset \mathbb{R}^3$ and $W^2 \subset \mathbb{R}^3$ are two surfaces in space that are tangent along a common curve $C$, we see from (8.38) that *if $\mathbf{X}$ is parallel displaced along $C$ in $M$, then $\mathbf{X}$ is also parallel displaced along $C$ in $W$.*

For example, let $M^2 = S^2$ be the standard 2-sphere in $\mathbb{R}^3$ and let $C$ be a "small" circle of latitude. We wish to parallel displace a tangent vector $\mathbf{X}_0$ along $C$; we have chosen $\mathbf{X}_0$ to be pointing north.

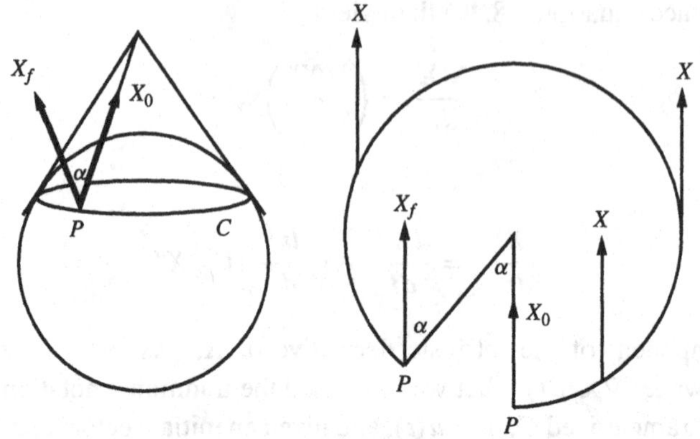


Figure 8.27

Let $V^2$ be the cone that is tangent to $S^2$ along $C$. Parallel translation along $C$ of $M$ is the same as parallel translation along $C$ considered as a curve on $V$. Any small portion of the cone that omits the vertex is isometric with a portion of the flat plane, as we see from cutting the cone along a generator and laying it out flat. This flattened version of the cone will have an "opening angle" $\alpha$ that is easily computed from the latitude of $C$. Parallel translation along $C$ is then the same as on the flattened cone. In the flattened cone one can introduce cartesian coordinates $x$, $y$, and in these coordinates the metric of the cone is $ds^2 = dx^2 + dy^2$. Clearly the Christoffel symbols for this flat metric all vanish and the equations of parallel translation are simply $dX^\gamma/dt = 0$; that is, *parallel translation in the flat plane is the usual parallelism of the euclidean plane.* We have indicated in our figure the parallel translation of $\mathbf{X}_0$ around the flattened cone, returning to $P$ with a final vector $\mathbf{X}_f$ that makes the opening angle $\alpha$ with the generator through $P$. When this flattened cone is then wrapped around the sphere again we see that *when $\mathbf{X}_0$ is parallel translated around the small circle of latitude C on the sphere, the vector $\mathbf{X}$ does not return to itself but rather to a vector $\mathbf{X}_f$ of the same length but rotated through the opening angle $\alpha$!*

We should note that if $C$ had been an equator of $S^2$, then the tangent cone would have been replaced by the tangent cylinder and then $\mathbf{X}_0$ would have then coincided with $\mathbf{X}_f$.

Since parallel displacement around a closed path does not necessarily return a vector to itself we conclude that, in general,

> *parallel displacement from a point P to a point Q will be dependent upon the choice of path joining P to Q!*

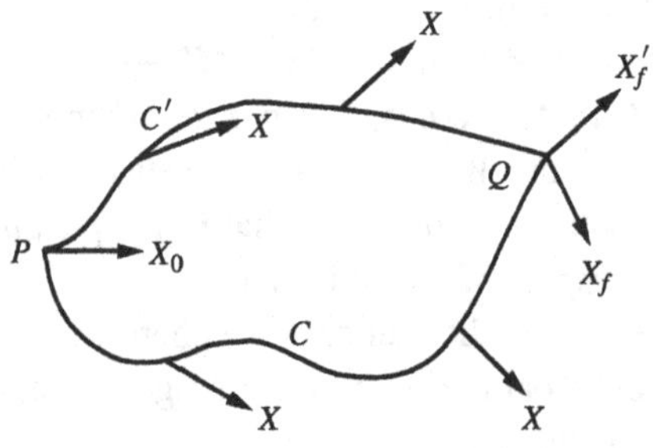


Figure 8.28

For this reason it makes no sense to ask whether a vector at $P$ is parallel to a vector at $Q$; *one can talk about parallelism only with respect to a specific path joining the two points.*

Finally, consider a pair of vectors $\mathbf{X}(t)$ and $\mathbf{Y}(t)$ defined along a curve $u = u(t)$ of a surface $M^2 \subset \mathbb{R}^3$ and tangent to $M$. Then, since $\nabla/dt$ is the tangential part of $d/dt$, we see

$$\frac{d}{dt}\langle \mathbf{X}(t), \mathbf{Y}(t)\rangle = \left\langle \frac{d\mathbf{X}}{dt}, \mathbf{Y}\right\rangle + \left\langle \mathbf{X}, \frac{d\mathbf{Y}}{dt}\right\rangle$$

yields

$$\frac{d}{dt}\langle \mathbf{X}(t), \mathbf{Y}(t)\rangle = \left\langle \frac{\nabla\mathbf{X}}{dt}, \mathbf{Y}\right\rangle + \left\langle \mathbf{X}, \frac{\nabla\mathbf{Y}}{dt}\right\rangle \tag{8.44}$$

(Although this important equation is in fact true in any Riemannian manifold, as we shall see, we have derived it only in the case of an embedded surface in $\mathbb{R}^3$.) In particular, if both $\mathbf{X}$ and $\mathbf{Y}$ are parallel displaced along $C$ we see that

$\langle \mathbf{X}(t), \mathbf{Y}(t)\rangle$ *is a constant under parallel displacement!*

If we let $\mathbf{Y} = \mathbf{T}$ be the unit tangent vector to a geodesic, we see that *a vector parallel displaced along a geodesic on a surface in $\mathbb{R}^3$ will make a constant angle with the geodesic.*

## Problems

**8.7(1)** The upper half plane $\{(x, y) : y > 0\}$ can be endowed with a particular abstract Riemannian metric, the **Poincaré metric**

$$ds^2 = y^{-2}\{dx^2 + dy^2\}$$

Parallel displace the initial vertical vector $X = \partial/\partial y$ at (0, 1) along the parameterized horizontal curve $C$; $x(t) = t$, $y(t) = 1$; that is, solve the differential equations (8.43).

**8.7(2)** (i) Let **w** be a unit vector, tangent to the surface, and defined along a curve $C$. Show that $\nabla\mathbf{w}/ds$ is orthogonal to **w**.

(ii) Let **v** be a vector that is parallel displaced along $C$ and let $\theta := \angle(\mathbf{v}, \mathbf{T})$ be the angle that $C$ makes with **v**. Recall that the geodesic curvature vector of $C$ is given by $\boldsymbol{\kappa}_g = \nabla\mathbf{T}/ds$, with length $\kappa_g$. Show that

$$\kappa_g = \left|\frac{d\theta}{ds}\right|$$

**B. Riemann**

CHAPTER 9

# Covariant Differentiation and Curvature

WE saw in Section 2.4 that the partial derivatives $\partial_j v^i$ of a vector field $\mathbf{v}$ do not form the components of a tensor. For a covariant vector field $\alpha^1$ we did show that we can construct a tensor by taking a combination of partial derivatives, $\partial_j a_k - \partial_k a_j$, the exterior derivative, but that $\partial_j a_k$ by themselves do not yield a tensor. Our goal in this chapter is to introduce an added structure to the notion of a manifold, a structure that will allow us to form a generalized derivative, a "covariant" derivative, taking vector fields into second-rank tensor fields.

## 9.1. Covariant Differentiation

### 9.1a. Covariant Derivative

Let us reformulate the concept of the intrinsic derivative of the last chapter.

Let $M^2$ be a *surface* in $\mathbb{R}^3$, and let $\mathbf{v}$ be a vector field that is tangent to $M$ and defined along a parameterized curve. Then the intrinsic derivative $\nabla \mathbf{v}/dt$ was defined to be the tangential part of the ordinary $\mathbb{R}^3$ derivative $d\mathbf{v}/dt$, and as such was again a tangent vector field to $M$ along the curve. We then define a **covariant derivative** as follows. Let $\mathbf{v}$ be a tangent vector *field* to $M$ defined now in some neighborhood of a point $p$, and let $\mathbf{X}$ be a tangent vector to $M$ at the single point $p$. Choose any curve on $M$ through $p$ whose tangent at $p$ is the vector $\mathbf{X}$, and define the covariant derivative $\nabla_{\mathbf{X}}\mathbf{v}$ at $p$ to be the intrinsic derivative $\nabla \mathbf{v}/dt$. In terms of coordinates we easily get

$$(\nabla_{\mathbf{X}}\mathbf{v})^\alpha = \left(\frac{\partial v^\alpha}{\partial u^\beta} + \Gamma^\alpha_{\beta\gamma} v^\gamma\right) X^\beta \tag{9.1}$$

which is clearly independent of the curve chosen to realize the given tangent vector $\mathbf{X}$ at $p$. The intrinsic derivative can then be expressed as the covariant dervative with respect to the tangent field $\mathbf{T} = d\mathbf{x}/dt$ to the curve

$$\frac{\nabla \mathbf{v}}{dt} = \nabla_{\mathbf{T}}\mathbf{v}$$

We have thus constructed the notion of a derivative of a tangent vector field $\mathbf{v}$ with respect to a vector $\mathbf{X}$ at $p$; the result is again a tangent vector at $p$. It is furthermore clear that if $\mathbf{X}$ is itself a tangent vector *field*, then $\nabla_{\mathbf{X}}\mathbf{v}$ is again a vector field. All this was possible because $M$ was a surface in $\mathbb{R}^3$, one already has a notion of derivative $d\mathbf{v}/dt$ in $\mathbb{R}^3$, and one also has the notion of orthogonal projection into the tangent space $M_p$ in $\mathbb{R}^3$.

A little reflection will show that we can again define $\nabla_{\mathbf{X}}\mathbf{v}$ when $M^n$ is any submanifold of any $\mathbb{R}^N$, using exactly the same procedure. In fact the coefficients $\Gamma$, *the Christoffel symbols, are defined exactly as before.*

Since the formulas for $\Gamma^{\alpha}_{\beta\gamma}$ make sense for any Riemannian manifold $M^n$, independent of whether or not it is embedded in some $\mathbb{R}^N$, it is reasonable to try to define the covariant derivative in a Riemannian $M^n$ again by the Formula (9.1), and indeed this does work. (In this case one would have to show, using the transformation properties of the metric tensor, that the components (9.1) do transform as the components of a vector, something that is geometrically immediate in the case of an embedded submanifold of $\mathbb{R}^N$.)

A covariant differentiation operation, defined fully in a moment, is also called a *connection.*

The connection in a Riemannian manifold in which the $\Gamma^{i}_{jk}$ are given by the Christoffel symbols is called the **Levi-Civita connection**, though Christoffel would be the natural name to associate with this connection.

It is important that we develop the concept of covariant derivative even when the manifold is not Riemannian. Later on we shall see that we shall need to differentiate objects that are much more general than tangent vector fields, and then the Christoffel symbols will be replaced by other quantities. For example, when discussing particle physics we shall have to differentiate wave functions, and we shall see that it is natural to define a covariant derivative in which the role of the Christoffel symbols is played by the electromagnetic vector potential $\mathbf{A}$! Part Three will be devoted to this concept of covariant differentiation in a "vector bundle," and the role of Christoffel symbols will be played frequently by certain physical fields, that is, by extra structures that are foreign to the unadorned notion of "manifold." For the present we shall only be dealing with quantities related to tangent vector fields. For this purpose, we generalize our preceding situation as follows. (The reader should verify that the indicated properties are indeed satisfied in the familiar case of a surface in $\mathbb{R}^3$ with the Levi-Civita derivative.)

**Definition:** Let $M^n$ be a manifold. An affine **connection** or **covariant differentiation** is an operator $\nabla$ that assigns to each pair consisting of a vector $\mathbf{X}$ at $p$ and a vector *field* $\mathbf{v}$ defined near $p$, a vector $\nabla_{\mathbf{X}}\mathbf{v}$ at $p$ that satisfies

$$\nabla_{\mathbf{X}}(a\mathbf{v}+b\mathbf{w}) = a\nabla_{\mathbf{X}}\mathbf{v} + b\nabla_{\mathbf{X}}\mathbf{w} \tag{9.2}$$

$$\nabla_{a\mathbf{X}+b\mathbf{Y}}\mathbf{v} = a\nabla_{\mathbf{X}}\mathbf{v} + b\nabla_{\mathbf{Y}}\mathbf{v}$$

and

$$\nabla_{\mathbf{X}}(f\mathbf{v}) = \mathbf{X}(f)\mathbf{v} + f\nabla_{\mathbf{X}}\mathbf{v} \qquad \text{("Leibniz rule")}$$

for all vector fields $\mathbf{v}$ and $\mathbf{w}$, functions $f$, and real numbers $a$ and $b$. We also demand that if $\mathbf{X}$ is a smooth vector field then $\nabla_{\mathbf{X}}\mathbf{v}$ is also a smooth vector field.

From the second equation we have that if $\mathbf{X} = \sum_i X^i \mathbf{e}_i$ then $\nabla_{\mathbf{X}} = \sum_i X^i \nabla_{\mathbf{e}_i}$. We shall write out what this says in terms of components. In our work up until now we have always used local coordinates $x$ to yield a basis $\partial/\partial x^i$ for the tangent vectors in a patch $U$. For *many* purposes, however, it is advantageous to use a more general basis. A **frame** of vector fields in a region $U$ consists of $n$ linearly independent smooth vector fields $\mathbf{e} =(\mathbf{e}_1, \dots, \mathbf{e}_n)$ in $U$. A special case is a **coordinate frame**, where $\mathbf{e}_i = \partial/\partial x^i$, for some coordinate system $x$ in $U$. First note that a frame $\mathbf{e}$ usually is *not* a coordinate frame, since $[\mathbf{e}_i, \mathbf{e}_j]$ is usually not 0 while $[\partial_i, \partial_j] = 0$. In fact we have

**Theorem (9.3):** *A frame* $\mathbf{e}$ *is locally a coordinate frame iff*

$$[\mathbf{e}_i, \mathbf{e}_j] = 0 \quad \textit{for all } i, j$$

PROOF: We need only show that this bracket condition implies the existence of functions $(x^i)$ such that $\mathbf{e}_i = \partial/\partial x^i$. Let $\sigma$ be the dual form basis. From (4.25)

$$d\sigma^i(\mathbf{e}_j, \mathbf{e}_k) = -\sigma^i([\mathbf{e}_j, \mathbf{e}_k]) \tag{9.4}$$

and so $d\sigma^i = 0$, for all $i$. Locally then each $\sigma^i$ is exact, $\sigma^i = dx^i$, for some functions $x^1, \dots, x^n$. Since $dx^1 \wedge \dots \wedge dx^n = \sigma^1 \wedge \dots \wedge \sigma^n \neq 0$, we see, from Corollary (1.16), that the $x$'s do form a local coordinate system. Since $\sigma = dx$ it follows that $\mathbf{e} = \partial/\partial x$. □

Let now $\mathbf{e} =(\mathbf{e}_1, \dots \mathbf{e}_n)$ be a frame of vector fields in a region $U$. We then have $\mathbf{X} = \mathbf{e}_j X^j$ and then from (9.2)

$$\nabla_{\mathbf{X}}(\mathbf{e}_k v^k) = X^j \mathbf{e}_i \omega^i_{jk} v^k + X^j \mathbf{e}_j(v^k)\mathbf{e}_k \tag{9.5}$$

where $\omega^i_{jk}$ is defined by

$$\nabla_{\mathbf{e}_j}\mathbf{e}_k = \mathbf{e}_i \omega^i_{jk} \tag{9.6}$$

In our surface case, when $\mathbf{e}_j = \partial_j$ was a coordinate frame, we had $\omega^i_{jk} = \Gamma^i_{jk}$.

**Warning:** As we shall see, it is not generally true that $\omega$ is symmetric in $j$ and $k$, $\omega^i_{jk} \neq \omega^i{}_{kj}$.

Since $\mathbf{X}(v^k) = dv^k(\mathbf{X})$, we may rewrite (9.5) as

$$\nabla_{\mathbf{X}}\mathbf{v} = \mathbf{e}_i\{dv^i(\mathbf{X}) + X^j \omega^i_{jk} v^k\}$$

The symbols $\omega^i_{jk}$ are called the **coefficients of the affine connection**, with respect to the frame $\mathbf{e}$. Using the dual basis $\sigma$ of 1-forms, we have $\nabla_{\mathbf{X}}\mathbf{v} = \mathbf{e}_i\{dv^i(\mathbf{X}) + \omega^i_{jk}\sigma^j(\mathbf{X})v^k\}$ or

$$\nabla_{\mathbf{X}}\mathbf{v} = \mathbf{e}_i\{dv^i + \omega^i_{jk}\sigma^j v^k\}(\mathbf{X}) \tag{9.7}$$

We wish to emphasize that *this makes sense in any frame* $\mathbf{e}$, and, as we shall see, for many purposes it will be important to employ frames that are not coordinate. For the present, however, it is an unnecessary complication. (For example, in a general frame, $df = f_{,j}\sigma^j$ for some coefficients $f_{,j}$ but $f_{,j}$ are not partial derivatives.)

*For the remainder of this Section 9.1 we shall restrict ourselves to the use of coordinate frames.*

When the frame $\mathbf{e}$ is a coordinate frame, $\mathbf{e}_i = \boldsymbol{\partial}_i = \partial/\partial x^i$, $\sigma^i = dx^i$,

$$\nabla_{\mathbf{X}}\mathbf{v} = \boldsymbol{\partial}_i \left\{ \frac{\partial v^i}{\partial x^j} + \omega^i_{jk} v^k \right\} dx^j(\mathbf{X})$$

that is,

$$(\nabla_{\mathbf{X}}\mathbf{v})^i = \left[ \frac{\partial v^i}{\partial x^j} + \omega^i_{jk} v^k \right] X^j \tag{9.8}$$

just as in (9.1). Since $\nabla_{\mathbf{X}}\mathbf{v}$ is assumed to be a vector, we conclude that

$$\nabla_j v^i = v^i_{/j} := \frac{\partial v^i}{\partial x^j} + \omega^i_{jk} v^k \tag{9.9}$$

form the components of a mixed tensor, the **covariant derivative** of the vector $\mathbf{v}$.

### 9.1b. Curvature of an Affine Connection

In the surface case, from (8.33) we see that curvature is at least related to the commutation of second covariant derivatives of vector fields. In Problem 9.1(1) you are asked to verify Equation (9.11).

**Theorem (9.10):** *Let $\mathbf{X}_p$, $\mathbf{Y}_p$, and $\mathbf{v}_p$ be vectors at a point $p$ of $M^n$ and let $\mathbf{X}$, $\mathbf{Y}$, and $\mathbf{v}$ be any extensions of these vectors to vector fields in some neighborhood $U$ of $p$. Form the vector field*

$$R(\mathbf{X}, \mathbf{Y})\mathbf{v} := \nabla_{\mathbf{X}}(\nabla_{\mathbf{Y}}\mathbf{v}) - \nabla_{\mathbf{Y}}(\nabla_{\mathbf{X}}\mathbf{v}) - \nabla_{[\mathbf{X},\mathbf{Y}]}\mathbf{v}$$

*in $U$. If we expand the vector fields in terms of a coordinate basis $\boldsymbol{\partial}$, then*

$$R(\mathbf{X}, \mathbf{Y})\mathbf{v} = \{R^i_{jkl} X^k Y^l v^j\}\boldsymbol{\partial}_i$$

*where, as in (8.33),* (9.11)

$$R^i_{jkl} := \partial_k \omega^i_{lj} - \partial_l \omega^i_{kj} + \omega^i_{kr}\omega^r_{lj} - \omega^i_{lr}\omega^r_{kj}$$

*Thus the value of the vector field $R(\mathbf{X}, \mathbf{Y})\mathbf{v}$ at $p$ is independent of the extensions of $\mathbf{X}$, $\mathbf{Y}$, and $\mathbf{v}$. From (9.11), the assignment*

$$\mathbf{v}_p \to R(\mathbf{X}_p, \mathbf{Y}_p)\mathbf{v}_p$$

*defines a linear transformation $R(\mathbf{X}, \mathbf{Y}) : M^n_p \to M^n_p$ called the* ***curvature transformation*** *for the pair $\mathbf{X}$, $\mathbf{Y}$; its matrix is given by $R(\mathbf{X}, \mathbf{Y})^i{}_j = R^i_{jkl} X^k Y^l$. Consequently, $R^i_{jkl}$ are the components of a mixed tensor of the fourth rank, the* ***Riemann tensor***. *We may write*

$$R(\mathbf{X}, \mathbf{Y}) = [\nabla_{\mathbf{X}}, \nabla_{\mathbf{Y}}] - \nabla_{[\mathbf{X},\mathbf{Y}]} \tag{9.12}$$

*where* $[\mathbf{X}, \mathbf{Y}]$ *is the Lie bracket of the extended vector fields and* $[\nabla_{\mathbf{X}}, \nabla_{\mathbf{Y}}] = \nabla_{\mathbf{X}}\nabla_{\mathbf{Y}} - \nabla_{\mathbf{Y}}\nabla_{\mathbf{X}}$ *is the commutator bracket of the covariant derivatives.*

(We have used the fact that since $R^i_{\ jkl}X^kY^l$ are the components of a second-rank mixed tensor for all $\mathbf{X}$ and $\mathbf{Y}$, it must be that $R^i_{\ jkl}$ are the components of a fourth-rank tensor. See Problem 9.1(2).)

From its definition it is clear that $R(\mathbf{X}, \mathbf{Y}) = -R(\mathbf{Y}, \mathbf{X})$, that is,

$$R^i_{\ jlk} = -R^i_{\ jkl} \tag{9.13}$$

## 9.1c. Torsion and Symmetry

Recall that the Lie bracket has components in a *coordinate frame* given by

$$[\mathbf{X}, \mathbf{Y}]^i = X^j\partial_j Y^i - Y^j\partial_j X^i = \mathbf{X}(Y^i) - \mathbf{Y}(X^i)$$

Compare this with the $i^{\text{th}}$ component of the difference of *covariant* derivatives. From (9.8)

$$(\nabla_{\mathbf{X}}\mathbf{Y} - \nabla_{\mathbf{Y}}\mathbf{X})^i = X^j\partial_j Y^i - Y^j\partial_j X^i + X^j(\omega^i_{\ jk} - \omega^i_{\ kj})Y^k$$

Now if $\mathbf{X}$ and $\mathbf{Y}$ are vector fields then so are $\nabla_{\mathbf{X}}\mathbf{Y} - \nabla_{\mathbf{Y}}\mathbf{X}$ and $[\mathbf{X}, \mathbf{Y}]$. We see that their difference, at a point $p$, is a vector,

$$\boldsymbol{\tau}(\mathbf{X}, \mathbf{Y})^i := X^j(\omega^i_{\ jk} - \omega^i_{\ kj})Y^k$$

that depends (bilinearly) *only on* $\mathbf{X}$ *and* $\mathbf{Y}$ *at* $p$. In other words, we have a well-defined "vector-valued 2-form" $\boldsymbol{\tau}$ the **torsion** form, defined by

$$\boldsymbol{\tau}(\mathbf{X}, \mathbf{Y}) := \nabla_{\mathbf{X}}\mathbf{Y} - \nabla_{\mathbf{Y}}\mathbf{X} - [\mathbf{X}, \mathbf{Y}] \tag{9.14}$$

(We started a discussion of vector-valued forms in Problem 4.3(5) and in Section 8.1a. We shall discuss this notion in more detail in Section 9.3a.) In terms of a *general* frame,

$$\boldsymbol{\tau} = \mathbf{e}_i \otimes \tau^i = \frac{1}{2}\mathbf{e}_i \otimes T^i_{\ jk}\sigma^j \wedge \sigma^k$$

where $T^i_{\ jk}$ are the components of a mixed tensor, the **torsion** tensor. In a *coordinate* frame, as we have seen,

$$T^i_{\ jk} := \omega^i_{\ jk} - \omega^i_{\ kj} \tag{9.15}$$

(This is rather surprising since, as we shall see, the $\omega^i_{\ jk}$ themselves do not form the components of a third-order tensor.)

We shall say that the connection is **torsion-free,** or **symmetric**, if the torsion tensor vanishes identically, $\boldsymbol{\tau} = 0$. In this case we have

$$\nabla_{\mathbf{X}}\mathbf{Y} - \nabla_{\mathbf{Y}}\mathbf{X} = [\mathbf{X}, \mathbf{Y}] \tag{9.16}$$

The reason for the description "symmetric" is as follows. From (9.15) we see that *in a coordinate frame,* $T^i_{\ jk} = 0$ means that the connection coefficients are symmetric in the two lower indices,

$$\omega^i_{\ jk} = \omega^i_{\ kj} \tag{9.17}$$

**Warning:** In a noncoordinate frame, (9.15) does *not* hold and consequently $\omega$ *need not be symmetric in the lower indices* when the torsion vanishes.

The Levi-Civita connection for a Riemannian manifold is symmetric because the Christoffel symbols satisfy $\Gamma^i_{jk} = \Gamma^i_{kj}$.

## Problems

**9.1(1)** Verify (9.11).

**9.1(2)** Show that if $A^i_{jkl}X^kY^l$ transforms as a mixed tensor $B^i_j$ *for all vectors* $\mathbf{X}$ *and* $\mathbf{Y}$, then $A^i_{jkl}$ transforms as a fourth-rank mixed tensor.

## 9.2. The Riemannian Connection

What distinguishes the Christoffel connection from the others?

In any manifold $M^n$ with an affine connection, that is, with a covariant differentiation operator $\nabla$, we can consider parallel displacement of a vector $\mathbf{Y}$ along a parameterized curve $x = x(t)$, defined again by

$$0 = \frac{\nabla \mathbf{Y}}{dt} = \partial_i Y^i_{/k}\left(\frac{dx^k}{dt}\right) = \partial_i\left\{\frac{\partial Y^i}{\partial x^k} + \omega^i_{kj}Y^j\right\}\left(\frac{dx^k}{dt}\right)$$

**Warning:** The connection coefficients $\omega^i_{jk}$ are usually denoted by $\Gamma^i_{jk}$. We, however, shall reserve this notation for the Christoffel symbols, that is, the Levi-Civita connection coefficients, *with respect to a coordinate frame.*

As we shall see later, there are an infinite number of distinct affine connections on any manifold. (In $\mathbb{R}^3$, e.g., one may choose functions $\omega^i_{jk}$ arbitrarily in the single coordinate patch.) If the manifold is *Riemannian*, however, there is one connection that is of special significance in that it relates parallel displacement with the Riemannian metric in an important way. In the case of a surface $M^2$ in $\mathbb{R}^3$, the Levi-Civita connection, first of all, was symmetric, and second, had the property that parallel displacement preserved scalar products of vectors (a consequence of Equation (8.44)).

**Theorem (9.18):** *On a Riemannian manifold there is a unique symmetric connection that satisfies*

$$\frac{d}{dt}\langle \mathbf{X}, \mathbf{Y}\rangle = \left\langle \frac{\nabla \mathbf{X}}{dt}, \mathbf{Y}\right\rangle + \left\langle \mathbf{X}, \frac{\nabla \mathbf{Y}}{dt}\right\rangle$$

*for any pair of vector fields defined along a parameterized curve, and this connection is the Riemannian connection; that is, in a coordinate frame,* $\omega^i_{jk} = \Gamma^i_{jk}$ *are the Christoffel symbols (8.32).*

PROOF: Consider the $k^{\text{th}}$ coordinate curve of a local coordinate system, parameterized by $x^k$, and let **X** and **Y** be two vector fields defined in a neighborhood of this curve. By hypothesis we have

$$\begin{aligned}\frac{\partial}{\partial x^k}(g_{ij}X^iY^j) &= g_{ij}X^i{}_{/k}Y^j + g_{ij}X^iY^j{}_{/k} \\ &= g_{ij}\left[\frac{\partial X^i}{\partial x^k} + \omega^i_{kl}X^l\right]Y^j + g_{ij}X^i\left[\frac{\partial Y^j}{\partial x^k} + \omega^j_{km}Y^m\right]\end{aligned}$$

Comparing this with the product rule expansion of $\partial/\partial x^k(g_{ij}X^iY^j)$ we see that $(\partial g_{ij}/\partial x^k)X^iY^j - g_{ij}\omega^i_{kl}X^lY^j - g_{ij}\omega^j_{km}X^iY^m = 0$. Changing dummy indices we get $[\partial g_{ij}/\partial x^k - g_{lj}\omega^l_{ki} - g_{il}\omega^l_{kj}]X^iY^j = 0$. Since this holds for all **X** and **Y** we conclude that

$$\frac{\partial g_{ij}}{\partial x^k} - g_{lj}\omega^l_{ki} - g_{il}\omega^l_{kj} = 0 \tag{9.19}$$

If we define $\omega_{kj,i} = g_{il}\omega^l_{kj}$ we then see that (9.19) is the same as Equation (8.31) in the surface case. If we now assume that $\omega^i_{kj}$ is symmetric in $k$ and $j$, as it is in the surface case, we are again led to (8.32); that is, the connection coefficients are indeed the Christoffel symbols. This shows that if a Riemannian connection exists, it is given by the Christoffel symbols.

We can then define a connection in each coordinate patch by putting $\omega^i_{jk}$ equal to the Christoffel symbol $\Gamma^i_{jk}$ for that patch. Our uniqueness result (that we have just proved) then shows that the local covariant derivatives in the patches agree in each overlap and thus we have a connection defined globally. □

The requirement $d/dt\langle \mathbf{X}, \mathbf{Y}\rangle = \langle \nabla\mathbf{X}/dt, \mathbf{Y}\rangle + \langle \mathbf{X}, \nabla\mathbf{Y}/dt\rangle$ easily implies the following. For two vector fields **X** and **Y**, and vector **T**, we may differentiate the function $\langle \mathbf{X}, \mathbf{Y}\rangle$ with respect to **T** and

$$\mathbf{T}\langle \mathbf{X}, \mathbf{Y}\rangle = \langle \nabla_{\mathbf{T}}\mathbf{X}, \mathbf{Y}\rangle + \langle \mathbf{X}, \nabla_{\mathbf{T}}\mathbf{Y}\rangle \tag{9.20}$$

The operation of covariant differentiation in a Riemannian manifold was introduced by Christoffel in 1869, following Riemann's paper of 1861 in which the curvature tensor was introduced. Levi-Civita, Hessenberg, and Weyl systematized the notion of manifold with an affine connection, independent of a Riemannian structure, in 1917 and 1918.

## 9.3. Cartan's Exterior Covariant Differential

How can we express connections and curvatures in terms of forms?

### 9.3a. Vector-Valued Forms

Cartan extended the notion of the exterior derivative of a $p$-form to that of the exterior "covariant" derivative of a "vector-valued $p$-form." This remarkable machinery is, as we shall see, ideally suited for computations involving the Riemann curvature tensor, and also seems to be the natural language for dealing with the gauge fields of present-day physics and the stress tensors of elasticity.

Let $A$ be a mixed tensor that is once contravariant and $p$ times covariant and that is skew symmetric in its covariant indices. Locally

$$A = \mathbf{e}_i \otimes \sum_{\underline{J}} A^i_{j_1 \ldots j_p} \sigma_1^j \wedge \ldots \wedge \sigma^{j_p}$$

Thus $A$ is of the form $A = \mathbf{e}_i \otimes \alpha^i$ where $\alpha^i$ is the $p$-form coefficient of $\mathbf{e}_i$. To $A$ we may then associate a **vector-valued $p$-form**, that is, a $p$-form (written $A$ or $\alpha$), whose values are vectors rather than scalars

$$\boldsymbol{\alpha}(\mathbf{v}_1, \ldots, \mathbf{v}_p) := \mathbf{e}_i \alpha^i(\mathbf{v}_1, \ldots, \mathbf{v}_p)$$

*We shall make no distinction between the tensor A and its associated vector-valued p-form $\boldsymbol{\alpha}$.*

Vector-valued forms occur frequently in classical vector analysis. In terms of cartesian coordinates, $d\mathbf{r} = (dx^1, dx^2, dx^3)^T$ is the vector-valued 1-form with values

$$d\mathbf{r}(\mathbf{v}) = (dx^1, dx^2, dx^3)^T(\mathbf{v}) = (dx^1(\mathbf{v}), dx^2(\mathbf{v}), dx^3(\mathbf{v}))^T$$
$$= (v^1, v^2, v^3)^T$$

that is, $d\mathbf{r}$ is the form that assigns to each vector the same vector! This comes from the mixed tensor (linear transformation) $I = \boldsymbol{\partial}_i \otimes dx^i$ whose matrix is the identity. Physicists think of $(dx^1, dx^2, dx^3)^T$ as a generic "infinitesimal" vector. The vector-valued 2-form (introduced in Problem 4.3(5))

$$d\mathbf{S} = (dy \wedge dz, dz \wedge dx, dx \wedge dy)^T$$

assigns to any pair of vectors the vector whose components are the signed areas of the parallelograms resulting from the projections of the vectors into the coordinate planes, that is, $d\mathbf{S}(\mathbf{A}, \mathbf{B}) = \mathbf{A} \times \mathbf{B}$.

A vector-valued 0-form is of course simply a vector.

## 9.3b. The Covariant Differential of a Vector Field

If $\mathbf{v}$ is a vector field in a manifold $M^n$ with affine connection, then we have seen that the coordinate patch expressions

$$\nabla_j v^i = v^i_{/j} := \frac{\partial v^i}{\partial x^j} + \omega^i_{jk} v^k$$

fit together to define a mixed tensor field, which we shall call the **covariant differential**, denoted by $\boldsymbol{\nabla}\mathbf{v}$

$$\boldsymbol{\nabla}\mathbf{v} = \boldsymbol{\partial}_i \otimes \nabla_j v^i dx^j = \boldsymbol{\partial}_i \otimes v^i_{/j} dx^j \tag{9.21}$$

This can be considered a vector-valued 1-form.

$$\boldsymbol{\nabla}\mathbf{v}(\mathbf{X}) = \boldsymbol{\partial}_i[(\nabla_j v^i dx^j)(\mathbf{X})] = \boldsymbol{\partial}_i[X^j \nabla_j v^i] \tag{9.22}$$

that is,

$$\boldsymbol{\nabla}\mathbf{v}(\mathbf{X}) := \boldsymbol{\nabla}_{\mathbf{X}}\mathbf{v}$$

In particular, if $\mathbf{e}$ is any frame of tangent vectors, we have, from (9.6), $\nabla \mathbf{e}_j(\mathbf{e}_i) = \mathbf{e}_k \omega^k_{ij}$. But $\mathbf{e}_k \otimes \omega^k_{rj}\sigma^r$ is a vector-valued 1-form that has the same value when applied to $\mathbf{e}_i$. We conclude that $\nabla \mathbf{e}_j = \mathbf{e}_k \otimes \omega^k_{rj}\sigma^r$. Finally, if we define the local matrix $\omega$ of **connection 1-forms** by

$$\omega^k{}_j := \omega^k_{rj}\sigma^r$$

we then have (9.23)

$$\nabla \mathbf{e}_j = \mathbf{e}_k \otimes \omega^k{}_j$$

Note that we may then write (9.7) in the form

$$\nabla_{\mathbf{X}}\mathbf{v} = \mathbf{e}_i\{dv^i + \omega^i{}_k v^k\}(\mathbf{X}) \tag{9.24}$$

and consequently

$$\nabla \mathbf{v} = \mathbf{e}_i \otimes \nabla v^i \tag{9.25}$$

where

$$\nabla v^i := dv^i + \omega^i{}_k v^k$$

It is immediate from (9.21) that if $f$ is a smooth function, then (recall that we occasionally prefer to write $\mathbf{v}f$ to the more usual $f\,\mathbf{v}$)

$$\nabla(\mathbf{v}f) = \mathbf{v} \otimes df + f\nabla \mathbf{v} \tag{9.26}$$

which we shall again refer to as the Leibniz rule.

### 9.3c. Cartan's Structural Equations

Let $\sigma$ be the basis of 1-forms dual to a given frame $\mathbf{e}$. Then $d\sigma^i$ can of course be written down with no mention of a connection, but if there is a connection we can write $d\sigma^i$ in the following manner. From (4.25) and (9.14)

$$\begin{aligned} d\sigma^i(\mathbf{e}_j, \mathbf{e}_k) &= \mathbf{e}_j\{\sigma^i(\mathbf{e}_k)\} - \mathbf{e}_k\{\sigma^i(\mathbf{e}_j)\} - \sigma^i([\mathbf{e}_j, \mathbf{e}_k]) \\ &= -\sigma^i([\mathbf{e}_j, \mathbf{e}_k]) = -\sigma^i\{\nabla_{\mathbf{e}_j}\mathbf{e}_k - \nabla_{\mathbf{e}_k}\mathbf{e}_j - \boldsymbol{\tau}(\mathbf{e}_j, \mathbf{e}_k)\} \\ &= -\sigma^i\{\mathbf{e}_r\omega^r_{jk} - \mathbf{e}_r\omega^r_{kj}\} + T^i_{jk} = -\{\omega^i_{jk} - \omega^i_{kj}\} + T^i_{jk} \end{aligned}$$

where $\boldsymbol{\tau} = 1/2\mathbf{e}_r \otimes T^r_{jk}\sigma^j \wedge \sigma^k$ is again the vector-valued torsion form. Then

$$d\sigma^i = \frac{1}{2}\sum_{j,k} d\sigma^i(\mathbf{e}_j, \mathbf{e}_k)\sigma^j \wedge \sigma^k = -(\omega^i_{jk}\sigma^j) \wedge \sigma^k + \frac{1}{2}T^i_{jk}\sigma^j \wedge \sigma^k$$

In terms of

$$\tau^i = \sum_{j<k} T^i_{jk}\sigma^j \wedge \sigma^k \tag{9.27}$$

we can write

$$d\sigma^i = -\omega^i{}_k \wedge \sigma^k + \tau^i \tag{9.28}$$

Equations (9.23) and (9.28) are **Cartan's structural equations**.

We shall abbreviate these as follows. Denote (as in (2.1)) the row matrix $(\mathbf{e}_1, \ldots, \mathbf{e}_n)$ by the matrix $\mathbf{e}$ and the column $(\sigma^1, \ldots, \sigma^n)^T$ by $\sigma$. The $n \times n$ matrix of connection 1-forms will be denoted by $\omega$

$$\omega = (\omega^i{}_j)$$

and the column vector of torsion 2-forms by $\tau$.

$$\tau = (\tau^1, \ldots, \tau^n)^T$$

Then we may write

$$\nabla \mathbf{e} = \mathbf{e} \otimes \omega$$

and (9.29)

$$d\sigma = -\omega \wedge \sigma + \tau$$

By $\omega \wedge \sigma$, for example, we mean the column matrix with 2-form entries $(\omega \wedge \sigma)^i = \sum_j \omega^i{}_j \wedge \sigma^j$, whereas $d\sigma$ is the column $(d\sigma^1, \ldots, d\sigma^n)^T$.

In our new notation, if $\mathbf{v}$ is a vector we may write $\mathbf{v} = \mathbf{e}\, v$ where $v$ is the column of components of $\mathbf{v}$, and then we may write (9.25) as

$$\nabla \mathbf{v} = \nabla(\mathbf{e}v) = \mathbf{e} \otimes \nabla v = \mathbf{e} \otimes (dv + \omega v) \tag{9.30}$$

### 9.3d. The Exterior Covariant Differential of a Vector-Valued Form

Let $\boldsymbol{\alpha}$ be a vector-valued $p$-form. Locally we have (in terms of a frame $\mathbf{e}$) $\boldsymbol{\alpha} = \mathbf{e}_i \otimes \alpha^i$, where each $\alpha^i = a^i{}_{\underrightarrow{J}}(x)\sigma^J$ is a locally defined $p$-form. We define its **exterior covariant differential,** the vector-valued $(p+1)$-form $\nabla\boldsymbol{\alpha}$, by demanding a Leibniz rule

$$\nabla\boldsymbol{\alpha} = \nabla(\mathbf{e}_i \otimes \alpha^i) = (\nabla\mathbf{e}_i) \otimes_\wedge \alpha^i + \mathbf{e}_i \otimes d\alpha^i$$

where the product $\otimes_\wedge$ is defined as follows:

$$(\nabla\mathbf{e}_i) \otimes_\wedge \alpha^i = (\mathbf{e}_k \otimes \omega^k{}_i) \otimes_\wedge \alpha^i := \mathbf{e}_k \otimes (\omega^k{}_i \wedge \alpha^i)$$

We drop this complicated notation and write $\otimes$ rather than $\otimes_\wedge$. Thus

$$\nabla\boldsymbol{\alpha} = \mathbf{e}_k \otimes (\omega^k{}_i \wedge \alpha^i) + \mathbf{e}_i \otimes d\alpha^i = \mathbf{e}_i \otimes (d\alpha^i + \omega^i{}_r \wedge \alpha^r)$$

In abbreviated notation with the column of $p$-forms $\alpha = (\alpha^1, \ldots \alpha^n)$ we may write

$$\nabla\boldsymbol{\alpha} = \mathbf{e} \otimes (d\alpha + \omega \wedge \alpha) \tag{9.31}$$

generalizing the vector field (i.e., vector-valued 0-form) case (9.30).

We have defined $\nabla\boldsymbol{\alpha}$ in terms of a local decomposition $\boldsymbol{\alpha} = \mathbf{e}_i \otimes \alpha^i$. It is not clear from this that $\nabla\boldsymbol{\alpha}$ is well defined, independent of the frame $\mathbf{e}$, but in fact we shall see later that this is indeed the case. We should remark that one can give a coordinate-free definition of $\nabla$ that is in the same spirit as the formula (4.27) for the exterior derivative of a scalar-valued exterior differential form

$$\nabla\alpha^p(\mathbf{Y}_0,\ldots,\mathbf{Y}_p) = \sum_r (-1)^r \nabla_{\mathbf{Y}_r}\{\alpha^p(\mathbf{Y}_0,\ldots,\widehat{\mathbf{Y}}_r,\ldots,\mathbf{Y}_p)\} + \sum_{r<s}(-1)^{r+s}\alpha^p([\mathbf{Y}_r,\mathbf{Y}_s],\ldots,\widehat{\mathbf{Y}}_r,\ldots,\widehat{\mathbf{Y}}_s,\ldots,\mathbf{Y}_p) \tag{9.32}$$

where we have again extended the vectors $\mathbf{Y}_r$ to be vector fields.

**Notation:** When dealing with vector-valued forms, we shall *usually* use Cartan's device of simply *omitting the tensor product* sign in equations such as (9.31); thus (9.31) will now be written

$$\nabla\alpha = \mathbf{e}(d\alpha + \omega\wedge\alpha) \tag{9.31$'$}$$

Furthermore, Cartan used the notation $d$ rather than $\nabla$; for example, Cartan would write his structure equation $\nabla\mathbf{e} = \mathbf{e}\otimes\omega$ as simply

$$d\mathbf{e} = \mathbf{e}\omega$$

$d\mathbf{e}$ would not be confused with an ordinary exterior derivative since it makes no *invariant* sense to take the exterior derivative of a vector field; one must use a covariant derivative. This notation is very convenient and is also used by many people, but *we shall not use it in this book.*

### 9.3e. The Curvature 2-Forms

$\nabla\mathbf{e} = \mathbf{e}\otimes\omega = \mathbf{e}\omega$ is a row matrix of local vector-valued 1-forms $\nabla\mathbf{e}_i$. We can then take the exterior covariant differential again

$$\nabla\nabla\mathbf{e} = \nabla(\mathbf{e}\omega) = (\nabla\mathbf{e})\omega + \mathbf{e}d\omega$$
$$= \mathbf{e}(\omega\wedge\omega + d\omega)$$

Thus if we define the local matrix $\theta$ of **curvature 2-forms** by

$$\theta := d\omega + \omega\wedge\omega \tag{9.33}$$

we have

$$\nabla\nabla\mathbf{e} = \mathbf{e}\otimes\theta = \mathbf{e}\theta$$

In full

$$\theta^i{}_j = d\omega^i{}_j + \omega^i{}_k\wedge\omega^k{}_j \tag{9.34}$$

Since the $\theta^i{}_j$ are 2-forms we may expand

$$\theta^i{}_j = \frac{1}{2}R^i_{jrs}\sigma^r\wedge\sigma^s \tag{9.35}$$

for some coefficients $R^i_{jrs}$. You are asked to show in Problem 9.3(1) that when $\mathbf{e} = \boldsymbol{\partial}$ is a coordinate frame, then the $R^i_{jrs}$ are given by Equation (8.33),

$$R^i_{jkl} := \partial_k\omega^i_{lj} - \partial_l\omega^i_{kj} + \omega^i_{kr}\omega^r_{lj} - \omega^i_{lr}\omega^r_{kj} \tag{9.36}$$

that is, *the* $R^i{}_{jrs}$ *are the components of the Riemann curvature tensor*! This of course is the reason for calling $\theta$ the matrix of curvature 2-forms.

Consider now a vector field $\mathbf{v} = \mathbf{e}v$. We have $\nabla \mathbf{v} = \mathbf{e}(dv + \omega v)$ and so from (9.30) we have

$$\nabla\nabla \mathbf{v} = \mathbf{e}[d(dv + \omega v) + \omega \wedge (dv + \omega v)]$$

Since $\omega$ is a matrix of 1-forms we then have

$$\nabla\nabla \mathbf{v} = \mathbf{e}[d\omega v - \omega \wedge dv + \omega \wedge dv + \omega \wedge \omega v]$$

that is,

$$\nabla\nabla \mathbf{v} = \mathbf{e} \otimes \theta v = \mathbf{e}\theta\, v \tag{9.37}$$

Note the remarkable fact that $\nabla\nabla\mathbf{v}$ depends linearly on $\mathbf{v}$ and *not at all on the derivatives of* $v$!

Some concluding remarks. Suppose that $M^n$ is a manifold that (like $\mathbb{R}^n$) can be covered by a single distinguished frame field $\mathbf{e}$. (Such a manifold is called **parallelizable.**) Define an affine connection by defining $\omega = 0$ for the distinguished frame $\mathbf{e}$, that is, $\nabla\mathbf{e} = 0$. Thus each of the vector fields $\mathbf{e}_i$ is covariant constant, or globally parallel. By construction the curvature of this connection vanishes, $\theta = 0$. $M^n$ is then said to admit a *distant parallelism.* Consider the 1-forms $\sigma$ dual to the frame $\mathbf{e}$. In general the forms $\sigma$ *will not all be closed.* Then $d\sigma = -\omega \wedge \sigma + \tau = \tau$ and the connection in general will have torsion. We thus see in this case of distant parallelism that torsion of the connection is a measure of misclosure of the orbits of the distinguished frame fields $\mathbf{e}$ (see Problem 4.1(3)).

Surveyors could introduce a frame of 3 orthonormal vectors in a small 3-dimensional neighborhood of a point on the irregular Earth's surface as follows: $\mathbf{e}_3$ is an upward pointing unit vector defined by a plumb line, $\mathbf{e}_1$ is a horizontal unit vector pointing magnetic north, and $\mathbf{e}_2 = \mathbf{e}_3 \times \mathbf{e}_1$ points "west." It is thus natural for surveyors to introduce (locally) a distinguished frame of vectors defining a distant parallelism with curvature 0, and this frame is not associated with any coordinate system; the torsion does not vanish! (For example, $\sigma^3 = \lambda(x)d\phi$ where $\phi$ is the gravitational potential.) When measuring, for instance, the difference in altitude of two nearby points they are essentially computing $\int_C \sigma^3$ along a curve joining the points. Note that if $C = \partial U$ is a closed curve, then $\oint_C \sigma^3 = \iint_U d\lambda \wedge d\phi = \iint_U \tau^3$ will not vanish in general; there is bound to be a natural misclosure in geodetic measurements! For more discussion of the use of Cartan's machinery in geodesy see Grossman's article [G].

## Problems

**9.3(1)** Verify (9.36).

**9.3(2)** $\mathbf{e} \otimes \sigma$ is a vector-valued 1-form that we have symbolically denoted by $d\mathbf{r}$. (In $\mathbb{R}^n$ it is the derivative of the vector-valued 0-form $\mathbf{r}$, but on a general manifold it isn't the *derivative* of anything.) Show that $\nabla d\mathbf{r} = \mathbf{e} \otimes \tau = \boldsymbol{\tau}$ is the vector-valued torsion 2-form. (Cartan would write $d^2 p = ddp = \tau$, where $p$ is the "position vector.")

---

## 9.4. Change of Basis and Gauge Transformations

What is a gauge transformation?

### 9.4a. Symmetric Connections Only

In the remainder of Part Two we shall be concerned almost exclusively with *symmetric* connections, $\tau = 0$. Cartan's equations then become

$$\nabla \mathbf{e} = \mathbf{e}\,\omega$$

and (9.38)

$$d\sigma = -\omega \wedge \sigma$$

### 9.4b. Change of Frame

We have defined the connection coefficients $\omega = (\omega^i{}_{jk})$ in terms of a given frame $\mathbf{e}$. If we demand that $\nabla$ have a basis-free significance, we shall have to require the $\omega$'s to have a special transformation property under a change of basis.

Let $\mathbf{e}' = \mathbf{e}P$ (i.e., $\mathbf{e}'_i = \mathbf{e}_j P^j{}_i$) be a change of basis, where $P = P(x)$ is a nonsingular $n \times n$ matrix function. Then for a vector $\mathbf{v}$ we have $\mathbf{v} = \mathbf{e}v = \mathbf{e}'v' = \mathbf{e}Pv'$. Thus

$$\mathbf{e}' = \mathbf{e}P \qquad (9.39)$$

$$v' = P^{-1}v$$

and since $\mathbf{e}\sigma = I = \mathbf{e}'\sigma' = \mathbf{e}P\sigma'$, we see that $\sigma = P\sigma'$

$$\sigma' = P^{-1}\sigma \qquad (9.40)$$

We demand that $\nabla$ be well defined, independent of basis. Thus $\nabla\mathbf{e} = \mathbf{e}\omega$ and $\nabla\mathbf{e}' = \mathbf{e}'\omega'$ must be compatible. Then $\nabla\mathbf{e}' = \nabla(\mathbf{e}P) = (\nabla\mathbf{e})P + \mathbf{e}dP = \mathbf{e}\omega P + \mathbf{e}dP$ must be the same as $\mathbf{e}'\omega' = \mathbf{e}P\omega'$. We must then have $\omega P + dP = P\omega'$, or

$$\omega' = P^{-1}\omega P + P^{-1}dP \qquad (9.41)$$

This is the *transformation rule for the matrix of connection 1-forms*. In terms of two *coordinate* frames, we have $dx'^i = (\partial x'^i/\partial x^j)dx^j$, and so $P$ is the inverse Jacobian matrix $P = \partial x/\partial x'$, and (9.41) states

$$\omega'^i{}_j = \left(\frac{\partial x'^i}{\partial x^r}\right)\omega^r{}_s\left(\frac{\partial x^s}{\partial x'^j}\right) + \left(\frac{\partial x'^i}{\partial x^r}\right)\left(\frac{\partial^2 x^r}{\partial x'^j \partial x'^s}\right)dx'^s$$

If we write, as usual, $\omega^i{}_j = \omega^i_{kj}dx^k$, then we could easily write out from this the transformation rule for the connection coefficients $\omega^i_{kj}$, found in all books on tensor analysis. We shall have no use for this expression. We do wish to point out that a linear transformation has a matrix that transforms as $A' = P^{-1}AP$, that is, as the first term in the right-hand side of (9.41). Thus $\omega$ does *not* transform as the matrix of a linear transformation and consequently $\omega^i_{kj}$ are *not* the components of a mixed tensor!

Look, on the other hand, at the matrix of curvature 2-forms $\theta$. $\theta' = d\omega' + \omega' \wedge \omega' = d(P^{-1}\omega P + P^{-1}dP) + (P^{-1}\omega P + P^{-1}dP) \wedge (P^{-1}\omega P + P^{-1}dP)$. From $P^{-1}P = I$ we see $dP^{-1}P + P^{-1}dP = 0$, or

$$dP^{-1} = -P^{-1}dPP^{-1} \tag{9.42}$$

You are asked in Problem 9.4(1) to put this in the expression for $\theta'$ and compare this with $\theta = d\omega + \omega \wedge \omega$, yielding finally

$$\theta' = P^{-1}\theta P \tag{9.43}$$

*Thus the matrix of curvature 2-forms transforms as the matrix of a linear transformation*! From (9.35) we can see from this that $R^i_{\ jrs}$ *are the components of a mixed tensor*, once contravariant and three times covariant.

This has the following consequence; *if* $\theta = 0$ *in some frame then* $\theta = 0$ *in every frame*! The same cannot be said of the connection forms $\omega$, as is evident from (9.41). See Problem 9.4(2).

Let us look at $\nabla$ applied to a vector field $\mathbf{v}$. We have seen in (9.30) that $\nabla\mathbf{v} = \mathbf{e}(dv + \omega v)$. One checks immediately from this that $\nabla(\mathbf{e}v)$ is indeed equal to $\nabla(\mathbf{e}'v')$. In terms of the column matrices involved we have, from (9.25), $\nabla\mathbf{v} = \mathbf{e}\nabla v = \mathbf{e}'\nabla' v'$, where $\nabla' v' = dv' + \omega' v'$. This says that $\nabla' v' = P^{-1}\nabla v$: that is, *the column* $\nabla v = dv + \omega v$ *transforms as the column of components of a (contravariant) vector.*

Let us introduce a more systematic notation. Let $\mathbf{e}_U$ and $\mathbf{e}_V$ be frames in open sets $U$ and $V$, respectively. We then have

$$\mathbf{e}_V = \mathbf{e}_U c_{UV} \tag{9.44}$$

in $U \cap V$, where $c_{UV}$ (formerly $P$), the **transition matrix function,**

$$c_{UV} : U \cap V \to Gl(n; \mathbb{R})$$

is a nonsingular matrix-valued function. Here $Gl(n; \mathbb{R})$ is the **general linear group**, the group of all nonsingular real $n \times n$ matrices. Of course $c_{VU} = c_{UV}^{-1}$. Then

$$\sigma_V = c_{VU}\sigma_U$$

If $\mathbf{v}$ is a vector field in $U \cap V$, then $\mathbf{v} = \mathbf{e}_U v_U = \mathbf{e}_V v_V$ says

$$v_V = c_{VU} v_U \tag{9.45}$$

is simply the transformation rule for the (column) components of a contravariant vector. The components $\omega$ transform as

$$\omega_V = c_{VU}\omega_U c_{UV} + c_{VU} dc_{UV} \tag{9.46}$$

and for curvature

$$\theta_V = c_{VU}\theta_U c_{UV} \tag{9.47}$$

To say that $\nabla\mathbf{v}$ is a vector-valued 1-form is to say the following: Put $(dv_U + \omega_U v_U) = \nabla_U v_U$, and so on. Then

$$v_V = c_{VU} v_U \quad \text{implies } \nabla_V v_V = c_{VU} \nabla_U v_U \tag{9.48}$$

In other words, $\nabla_V c_{VU} v_U = c_{VU} \nabla_U v_U$, or

$$\nabla_V \circ c_{VU} = c_{VU} \circ \nabla_U \tag{9.49}$$

We may then say that if $v$ transforms as a vector then so does $\nabla v$.

Finally a remark on physical terminology. A frame field $\mathbf{e}_U$ can be considered as giving a basis for the sections of the tangent bundle over the open set $U \subset M^n$; that is the meaning of the expansion $\mathbf{v}(x) = \mathbf{e}_U(x) v_U(x)$. Physics deals, as we shall see, with other "vector bundles." A frame of $n$ "vectors" in physics is sometimes called an $n$-bein. Thus a frame in Minkowski space is referred to as a 4-bein, or, in German, a *vier-bein*. A local change of basis, such as $\mathbf{e}_V = \mathbf{e}_U c_{UV}$, is called in physics a **gauge transformation**. A connection is an example of a **gauge field**, to be discussed at great length in Part Three. Equation (9.41) then tells how this particular gauge field transforms under a "change of gauge." Finally, (9.48) or (9.49) is said to exhibit *covariance* of the operation of covariant derivative.

## Problems

**9.4(1)** Prove (9.43).

**9.4(2)** Consider $\mathbb{R}^2$ with the standard metric $ds^2 = dx^2 + dy^2$. Thus $g_{ij} = \delta_{ij}$ in the coordinate frame $\mathbf{e} = (\partial/\partial x, \partial/\partial y)$. Thus $\omega = 0$ and $\theta = 0$. Now introduce polar coordinates $\mathbf{e}' = (\partial/\partial r, \partial/\partial\theta) = ([\partial x/\partial r]\partial/\partial x + [\partial y/\partial r]\partial/\partial y, \ldots)$. Write down the change of basis matrix $P$ and use $\omega' = P^{-1} dP$ to give

$$\omega' = \begin{bmatrix} 0 & -r d\theta \\ \dfrac{d\theta}{r} & \dfrac{dr}{r} \end{bmatrix}$$

Verify that $\theta' = 0$.

**9.4(3)** Let $\alpha = \mathbf{e}_U \alpha_U^p$ be the local expression, in terms of the frame $\mathbf{e}_U$, of a vector-valued $p$-form. If $\alpha$ is globally defined, we must have that $\alpha_V = c_{VU}\alpha_U$; that is, $\alpha$ transforms as the components of a vector. If we define, as in (9.30), $\nabla_U \alpha_U = d\alpha_U + \omega_U \wedge \alpha_U$, show that (9.49) holds again. This shows that $\nabla\alpha$, defined in (9.30), is well defined.

## 9.5. The Curvature Forms in a Riemannian Manifold

Why bother with noncoordinate frames?

### 9.5a. The Riemannian Connection

Note that *in a Riemannian manifold, one can take any frame and convert it to an orthonormal frame* by applying the Gram–Schmidt process. We shall see that many computations become much simpler if an orthonormal frame is employed. Let us look first at the connection forms.

Let us express the fundamental relation (9.20) in terms of a general frame **e**. We may write $d\langle \mathbf{e}_i, \mathbf{e}_j\rangle(\mathbf{e}_k) = \langle \nabla_{\mathbf{e}_k}\mathbf{e}_i, \mathbf{e}_j\rangle + \langle \mathbf{e}_i, \nabla_{\mathbf{e}_k}\mathbf{e}_j\rangle = \langle \mathbf{e}_r\omega^r_{ki}, \mathbf{e}_j\rangle + \langle \mathbf{e}_i, \mathbf{e}_r\omega^r_{kj}\rangle$, that is, $(dg_{ij})(\mathbf{e}_k) = g_{rj}\omega^r_{ki} + g_{ir}\omega^r_{kj}$. But $\omega^r_{ki} = \omega^r{}_i(\mathbf{e}_k)$ and $\omega^r_{kj} = \omega^r_j(\mathbf{e}_k)$. We conclude that $dg_{ij} = g_{rj}\omega^r{}_i + g_{ir}\omega^r{}_j$. If we **define**, as usual,

$$\omega_{ij} := g_{ir}\omega^r{}_j$$

then we have

$$dg_{ij} = \omega_{ij} + \omega_{ji} \tag{9.50}$$

as the basic relation for the compatibility of the connection with the Riemannian metric (i.e., parallel displacement preserves scalar products).

In particular, *if the frame is orthonormal,* $g_{ij} = \delta_{ij}$, *then the matrix of the connection 1-forms* (with both indices down) *is skew symmetric*

$$\omega_{ij} = -\omega_{ji} \tag{9.51}$$

for an orthonormal frame.

Look now at the curvature 2-forms in *any* frame. We define

$$\theta_{ij} := g_{ir}\theta^r{}_j \tag{9.52}$$

In an orthonormal frame of course we have $\omega^i{}_j = \omega_{ij}$, $\theta^i{}_j = \theta_{ij}$, and so forth. Thus in an orthonormal frame we have $\theta_{ij} = d\omega_{ij} + \omega_{ir}\wedge\omega_{rj} = -d\omega_{ji} - \omega_{jr}\wedge\omega_{ri} = -\theta_{ji}$. Hence in an orthonormal frame the $\theta$ matrix, with both indices down, is also skew symmetric. We claim that this is true in *any* frame! The matrix $(\theta_{ij})$ is, from (9.52), of the form $G\theta$, where $G$ is the matrix $(g_{ij})$. Under a change of basis $\theta$ transforms, from (9.43), as $\theta' = P^{-1}\theta P$, and the covariant tensor $(g_{ij})$ transforms as $G' = P^T G P$. Thus $G'\theta' = P^T G P P^{-1}\theta P = P^T(G\theta)P$. But this says that if $G\theta$ is skew symmetric in one frame (as it is in an orthonormal one) then *it is skew symmetric in every frame.*

$$\theta_{ij} = -\theta_{ji} \tag{9.53}$$

From (9.35) we see that for the purely covariant version of the Riemann curvature tensor

$$R_{ijrs} = -R_{jirs} \tag{9.54}$$

is skew symmetric not only in the second pair of indices, but also in the first!

**Theorem (9.55):** *Let* **e** *be an orthonomal frame field on a Riemannian manifold* $M^n$ *and let* $\sigma$ *be the dual frame field. Then the Riemannian (Levi-Civita) connection is given by the unique matrix* $\omega$ *of 1-forms that satisfies*

$$d\sigma = -\omega\wedge\sigma$$

*and*

$$\omega_{ij} = -\omega_{ji}$$

PROOF: Introduce local coordinates $x$ in the region covered by the frame. The Riemannian connection $\Gamma$ in these coordinates is given uniquely by the Christoffel symbols, $\Gamma^i_j = \Gamma^i_{kj}dx^k$. Under the change of frame to the frame $\mathbf{e}$, we get new unique connection forms $\omega$. Since the frame is orthonormal, $\omega$ is skew symmetric. Since the torsion vanishes, the second Cartan structural equation gives $d\sigma = -\omega \wedge \sigma$. This shows the existence of the matrix $\omega$. For the uniqueness of such $\omega$, see Problem 9.5(3). □

## 9.5b. Riemannian Surfaces $M^2$

Let $\mathbf{e}$ be an *orthonormal* frame over a portion of a 2- dimensional Riemannian manifold $M^2$. The matrix of Riemannian connection forms, $\omega = (\omega_{ij})$, is a skew symmetric 2 by 2 matrix of 1-forms. Thus $\omega_{12} = -\omega_{21}$ and $\omega_{11} = \omega_{22} = 0$; *$\omega$ is completely characterized by the single entry $\omega_{12}$*. The same is true of the matrix of curvature 2-forms $\theta = (\theta_{ij})$. Furthermore, $\theta_{12} = d\omega_{12} + \omega_{12} \wedge \omega_{22}$, that is,

$$\theta_{12} = d\omega_{12} \tag{9.56}$$

In particular, the curvature matrix of 2-forms is *exact*, $\theta = d\omega$, *in the entire region covered by the orthonormal frame.*

In Section 8.5 we discussed curvature, but always in the context of a coordinate system, that is, the frame was always a coordinate frame. We should note a simple fact about coordinates, in any dimension. If $x$ is a coordinate system with origin at $p$ and if $P$ is any nonsingular *constant* matrix, then $x' = Px$ defines a new coordinate system $x'$ for which $\boldsymbol{\partial}' = \boldsymbol{\partial}(\partial x/\partial x') = \boldsymbol{\partial} P^{-1}$. In particular, given any frame $\mathbf{e}$ at $p$, by an appropriate choice of $P$ we may find a new coordinate system $x'$ such that $\boldsymbol{\partial}' = \mathbf{e}$ at $p$; thus *if* $\mathbf{e}$ is a frame field in a region holding $p$, we may always find a coordinate system $x'$ whose coordinate frame *at the single point $p$* is $\mathbf{e}$!

Let $\mathbf{e}$ be an orthonormal frame at the point $p$ of $M^2$ (with dual frame $\sigma$). Let $x'$ be a coordinate system whose frame $\boldsymbol{\partial}'$ coincides with $\mathbf{e}$ at $p$. Since this coordinate system is orthonormal at $p$, we have, in the coordinate frame at $p$, $\theta'_{12} = \theta'^1{}_2 = \sum_{r<s} R'^1{}_{2rs} dx'^r \wedge dx'^s = R'^{12}{}_{12} dx'^1 \wedge dx'^2 = R'^{12}{}_{12}\,\mathrm{vol}^2 = K(p)\,\mathrm{vol}^2$, where $K = R'^{12}{}_{12}$ is the Gauss curvature of the Riemannian metric. But under the identity change of frame at $p$, $\boldsymbol{\partial} = \mathbf{e}$, we have $\theta_{12} = \theta'_{12}$. We thus have

$$\theta_{12} = d\omega_{12} = K\sigma^1 \wedge \sigma^2 = K\,\mathrm{vol}^2 \tag{9.57}$$

in any *orthonormal* frame.

This is a remarkable formula for it says that one can compute the Gauss curvature by simply computing the single 1-form entry $\omega_{12}$ in an orthonormal frame!

## 9.5c. An Example

Let us compute (using what we shall call **Cartan's method**) the Gauss curvature of a surface with a metric of the form

$$ds^2 = du^2 + G^2(u, v)dv^2 \tag{9.58}$$

This includes, for instance, the case of the sphere $ds^2 = a^2 d\theta^2 + a^2 \sin^2\theta d\phi^2$ computed in Problem 8.5(2). In fact, we shall see later that *on any surface we can introduce local coordinates in which the metric takes the form* (9.58).

The coordinate frame $\partial/\partial u$, $\partial/\partial v$ is orthogonal but not unit. For an orthonormal frame we would have $ds^2 = \sigma^1 \otimes \sigma^1 + \sigma^2 \otimes \sigma^2$, that is, $ds^2 = (\sigma^1)^2 + (\sigma^2)^2$. (These are not exterior products.) Clearly we should define

$$\sigma^1 = du \quad \text{and} \quad \sigma^2 = G(u, v)dv \tag{9.59}$$

(i.e., $\mathbf{e}_1 = \partial/\partial u$, $\mathbf{e}_2 = G^{-1}\partial/\partial v$). We wish to find the *unique* $\omega_{12} = -\omega_{21}$ satisfying (9.55). Put then $\omega_{12} = a(u, v)\sigma^1 + b(u, v)\sigma^2$ for as yet unknown functions $a$ and $b$. Then

$$d\sigma^1 = -\omega_{12} \wedge \sigma^2 = -(a\sigma^1 + b\sigma^2) \wedge \sigma^2 = -a\sigma^1 \wedge \sigma^2$$

But $d\sigma^1 = d(du) = 0$, and so $a = 0$ and $\omega_{12} = b\sigma^2$. Also

$$d\sigma^2 = -\omega_{21} \wedge \sigma^1 = \omega_{12} \wedge \sigma^1 = b\sigma^2 \wedge \sigma^1 = -b\sigma^1 \wedge \sigma^2$$

is to be compared with $d\sigma^2 = d(Gdv) = G_u du \wedge dv = (G_u/G)\sigma^1 \wedge \sigma^2$. Thus $b = -G_u/G$ and so

$$\omega_{12} = -\left(\frac{G_u}{G}\right)\sigma^2 = -G_u dv$$

$\theta_{12} = d\omega_{12} = -G_{uu} du \wedge dv = -(G_{uu}/G)\sigma^1 \wedge \sigma^2$. From (9.57) we see

$$K = -\frac{G_{uu}}{G} \quad \text{for metric } ds^2 = du^2 + G^2 dv^2 \tag{9.60}$$

The reader interested in elasticity might glance at this time at section **g** of the Appendix, where Cartan's methods are applied to Cauchy's equations of equilibrium.

## Problems

**9.5(1)** Use Cartan's method to compute the Gauss curvature of the Poincaré metric $ds^2 = y^{-2}(dx^2 + dy^2)$ in the upper half plane and check your result by first making a coordinate transformation and using formula (9.60) directly. Save your calculations for later use.

**9.5(2)** A curve in the plane, $y = f(x)$, with $f(x) > 0$, is revolved about the $x$ axis yielding a surface of revolution. Write down the metric of the surface in terms of $x$ and the angular parameter $\phi$ (using the pictorial infinitesimal version of Pythagoras's rule, as we illustrated for the 2-sphere in Section 8.1a). Compute the curvature.

**9.5(3)** To show *uniqueness* of the connection form matrix $\omega$ it is enough to show that the only solution to $\omega \wedge \sigma = 0$ and $\omega_{ij} = -\omega_{ji}$ is $\omega = 0$. Expand $\omega_{ij} = a_{ijk}\sigma^k$ where $a$ is *skew symmetric* in $(ij)$. But $0 = \omega_{ij} \wedge \sigma^j = a_{ijk}\sigma^k \wedge \sigma^j$ then shows that $a$ is *symmetric* in $(jk)$. Show that such a three-index symbol $a$ must vanish.

## 9.6. Parallel Displacement and Curvature on a Surface

When is parallel displacement independent of path?

We saw in Section 8.7 that parallel displacement of a vector between two points of a surface is path-dependent; that is, parallel displacement of a vector $\mathbf{v}_0$ around a closed curve results in a final vector $\mathbf{v}_f$ that might disagree with $\mathbf{v}_0$. This phenomenon is referred to as **holonomy** (and, as we shall see, is indeed related to the concept of holonomic and nonholonomic constraints studied in Chapter 6). We gave as an explicit example parallel displacement around a small circle on the 2-sphere. There is a remarkable result, in the case of surfaces, relating this holonomy $\mathbf{v}_f \neq \mathbf{v}_0$ with Gaussian curvature.

**Theorem (9.61):** *Let $U \subset M^2$ be a compact region in a Riemannian surface with piecewise smooth boundary $\partial U$. Assume that $U$ can be covered by a single orthonormal frame field $\mathbf{e}$ (e.g., $U$ may be contained in a coordinate patch). Let a unit vector $\mathbf{v}$ be parallel translated around $\partial U$, starting with an initial $\mathbf{v}_0$ and ending with $\mathbf{v}_f$. $\mathbf{e}$ defines an orientation in $U$. Then the angle $\Delta\alpha$ between $\mathbf{v}_0$ and $\mathbf{v}_f$ is given by*

$$\Delta\alpha = \iint_U K\,dS = \iint_U K\sigma^1 \wedge \sigma^2$$

PROOF:

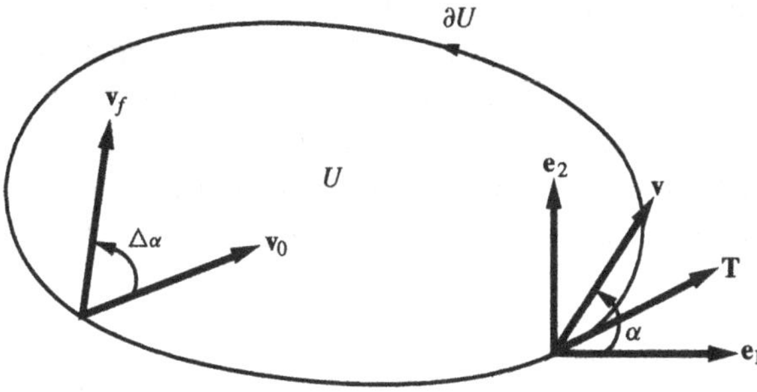


Figure 9.1

Parameterize $\partial U$, let $\mathbf{T}$ be the tangent, and let $\alpha = \measuredangle(\mathbf{e}_1, \mathbf{v})$. Although $\alpha$ (like $\mathbf{v}$) is not single-valued on $\partial U$, $d\alpha = (d\alpha/ds)ds$ is well defined and $\Delta\alpha = \measuredangle(\mathbf{v}_0, \mathbf{v}_f) = \oint_{\partial U} d\alpha$. Now $\mathbf{v} = \mathbf{e}_1 \cos\alpha + \mathbf{e}_2 \sin\alpha$ and so

$$\begin{aligned}\nabla\mathbf{v} &= \mathbf{e}(dv + \omega v) = \mathbf{e}_1(dv^1 + \omega_{12}v^2) + \mathbf{e}_2(dv^2 + \omega_{21}v^1)\\ &= \mathbf{e}_1(-\sin\alpha d\alpha + \omega_{12}\sin\alpha) + \mathbf{e}_2(\cos\alpha d\alpha + \omega_{21}\cos\alpha)\\ &= (-\mathbf{e}_1 \sin\alpha + \mathbf{e}_2\cos\alpha)(d\alpha - \omega_{12})\end{aligned}$$

To say that $\mathbf{v}$ is parallel displaced around $\partial U$ is to say $\nabla \mathbf{v}(\mathbf{T}) = 0$, that is, from the preceding,

$$d\alpha - \omega_{12} = 0 \quad \text{along } \partial U \tag{9.62}$$

(meaning that $d\alpha(\mathbf{T}) = \omega_{12}(\mathbf{T})$). Then

$$\Delta\alpha = \oint_{\partial U} d\alpha = \oint_{\partial U} \omega_{12} = \iint_U d\omega_{12}$$
$$= \iint_U \theta_{12} = \iint_U K\sigma^1 \wedge \sigma^2 \quad \square$$

Note that from (8.14) we have the following:

**Corollary (9.63):** *If $M^2 \subset \mathbb{R}^3$, then $\Delta\alpha =$ the signed area of the spherical image of $U$ under the Gauss normal map.*

A connection is said to be **flat** if the curvature $= 0$

$$\theta = 0, \quad \text{or } R(\mathbf{X}, \mathbf{Y}) = 0$$

for all vectors **X** and **Y**.

**Corollary (9.64):** *Parallel displacement on a Riemannian surface is locally independent of path iff $M^2$ is flat, that is, $K = 0$.*

By "locally" we mean that we must restrict our closed path to be the boundary of a compact region, $C = \partial U$, that is covered by an orthonormal frame. Consider, for example, the Möbius band obtained by bending and sewing a flat strip of paper. Although the usual picture of the band in $\mathbb{R}^3$ appears curved, this 2-manifold with boundary has $K = 0$ since $K$ is a bending invariant. If, however, one parallel translates the vector $\mathbf{e}_2$ along the midcircle of the band one ends up with $\mathbf{e}_2(1) = -\mathbf{e}_2(0)$.

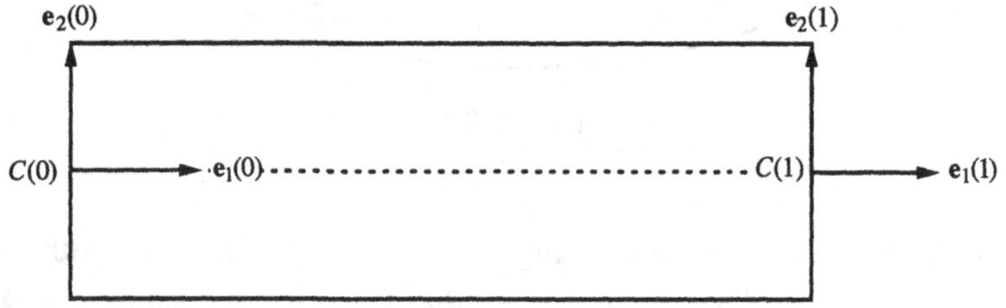


**Figure 9.2**

This does not contradict Theorem (9.61) since the midcircle $C$ does not bound any surface.

We remark further on the hypotheses of the theorem. It is crucial that there be an orthonormal frame that covers $U$, for we measure the variation of $\mathbf{v}$ by comparing $\mathbf{v}$

with $\mathbf{e}_1$ along $\partial U$. This requires $\mathbf{e}$ to be defined at least along $\partial U$. In order for $\omega_{12}$ to be defined inside $U$ we need, however, $\mathbf{e}$ to be defined in all of $U$. It turns out, however, that this is not a serious constraint, at least in the case of an orientable $U$, for the following reason. It can be shown that one can always find an orthonormal frame in any *noncompact* orientable 2-manifold. (It is *not* true that one can always cover it by a coordinate patch.) For example, given a closed orientable surface of genus $g$, if one removes a disc, *however small*, one can always cover the remaining surface with an orthonormal frame.

This has a remarkable consequence. Let $M^2$ be a compact oriented surface, and let $U$ be a small region on $M$, covered by an orthonormal frame $\mathbf{e}$, and with boundary an oriented curve $C = \partial U$. The complementary region $M - U$ is also a compact surface whose boundary is the oppositely oriented curve $-C$. As mentioned, $M - U$ can also be covered by an orthonormal frame $\mathbf{e}'$. Parallel displacement of a vector $\mathbf{v}$ around $C$ then gives an angular change $\Delta\alpha = \iint_U K dS$. But this vector is also being translated around $C = -\partial(M - U)$, and so $\Delta\alpha' = -\iint_{M-U} K dS$, where $\alpha' = \measuredangle(\mathbf{e}'_1, \mathbf{v})$. Thus

$$\iint_M K dS = \iint_U K dS + \iint_{M-U} K dS = \Delta(\alpha - \alpha')$$

But $d(\alpha - \alpha') = d\measuredangle(\mathbf{e}_1, \mathbf{v}) - d\measuredangle(\mathbf{e}'_1, \mathbf{v}) = d\measuredangle(\mathbf{e}_1, \mathbf{e}'_1)$, and so

$$\frac{1}{2\pi}\iint_M K dS = \text{total number of revolutions that } \mathbf{e}'_1 \text{ makes with respect to } \mathbf{e}_1 \text{ on going around } C. \tag{9.65}$$

In particular,

$$\frac{1}{2\pi}\iint_M K dS \quad \text{is an integer!} \tag{9.66}$$

Note that this "Gauss–Bonnet" theorem seems weaker than the Gauss normal map result (8.20), which says that $(1/4\pi)\iint_M K dS$ is an integer, but it should be appreciated that (9.66) holds for *any* (perhaps abstract) closed oriented Riemannian surface, whereas (8.20) holds only for surfaces embedded in $\mathbb{R}^3$. (We shall see in Section 12.2a that the real projective plane has a metric of curvature 1 that it inherits from the 2-sphere that covers it twice. The area of $\mathbb{R}P^2$ is half that of the sphere, that is, $2\pi$. Thus the integer in (9.66) is in this case 1. This tells us that $\mathbb{R}P^2$ cannot be embedded in $\mathbb{R}^3$ with this metric of curvature 1!) In Part Three we shall spend a great deal of time discussing this **topological quantization** rule and its generalizations and applications to physics. In particular we shall identify the integer involved in (9.66).

Finally, some remarks about flat manifolds. *Even a closed surface can be flat according to our definition*! The torus $T^2$ with the abstract Riemannian metric $ds^2 = d\theta^2 + d\phi^2$ clearly has curvature 0. This is certainly not the usual metric induced from an embedding in $\mathbb{R}^3$. In fact, we have the following:

**Theorem (9.67):** *The induced metric on any closed surface $M^2 \subset \mathbb{R}^3$ must have some point where $K > 0$.*

**PROOF:** We shall merely give a sketch. Let **x** be a point of $\mathbb{R}^3$ that is not on $M^2$. Since $M$ is compact, there is a point **y** on $M$ that is farthest from **x** (since every continuous function on a compact space achieves its maximum and minimum at points of the space). Then the 2-sphere centered at **x** and passing through **y** is tangent to $M$ at **y**

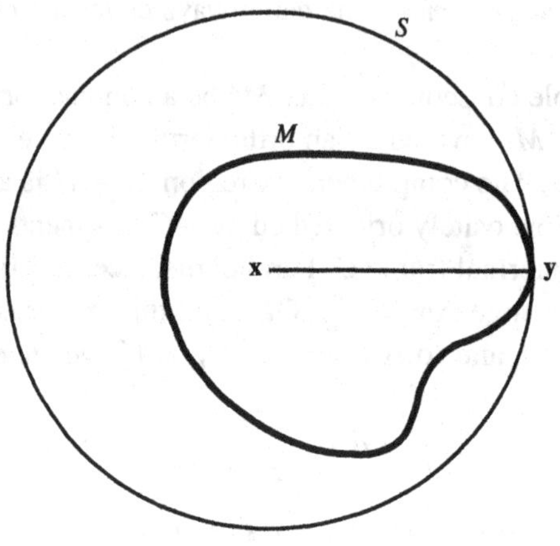


Figure 9.3

and $M$ lies entirely within the sphere. It should be geometrically clear that both principal curvatures of $M$ at **y** are of the same sign (since $M$ must be bending toward **x** at the farthest point) and of magnitudes greater than or equal to those of the 2-sphere. Thus, at **y** we have $K_M \geq \| \mathbf{y} - \mathbf{x} \|^{-2} > 0$. □

Although the flat metric on the torus is not that induced from an embedding in $\mathbb{R}^3$, it is remarkable that this metric *is* induced from the following embedding in $\mathbb{R}^4$, the so-called **Clifford embedding**:

$$x^1 = \cos\theta, \qquad x^2 = \sin\theta, \qquad x^3 = \cos\phi, \qquad x^4 = \sin\phi$$

for certainly then $ds^2 = \sum(dx^i)^2 = d\theta^2 + d\phi^2$. Note also that this torus is in fact a 2-dimensional submanifold of the 3-sphere $\sum(x^i)^2 = 2$ in $\mathbb{R}^4$.

## Problems

**9.6(1)** What is wrong with the following argument found in many books? A vector **v** is parallel displaced around a small closed curve $C = \partial U^2$ in an $n$-dimensional manifold $M^n$. Then $dv^i = -\omega^i{}_j v^j$ along $C$. Thus the total change in $v^i$ on going around $C$ is given by

$$\Delta v^i = \oint_C dv^i = -\oint \omega^i{}_j v^j$$

$$= -\iint_U d(\omega^i{}_j v^j)$$

$$= -\iint_U d(\omega^i{}_j)v^j - \omega^i{}_j \wedge dv^j$$

$$= -\iint_U [d\omega^i{}_k + \omega^i{}_j \wedge \omega^j{}_k]v^k$$

$$= -\iint_U \theta^i{}_k v^k$$

$$= -\iint_U \frac{1}{2} R^i_{krs} v^k dx^r \wedge dx^s$$

## 9.7. Riemann's Theorem and the Horizontal Distribution

When is $ds^2 = \sum (dx^j)^2$?

### 9.7a. Flat metrics

Linear algebra tells us that a *constant* quadratic form $Q = Q_{ij}dx^i dx^j$ in $\mathbb{R}^n$ can always be reduced to diagonal form $Q = \sum \lambda_i (dz^i)^2$ by an orthogonal change of coordinates, $z^i = P^i{}_j x^j$ (see Problem 8.2(1)). If $Q$ is positive definite, we can make a further (non-orthogonal linear) transformation $y^i = z^i \sqrt{\lambda_i}$ that will reduce $Q$ to a sum of squares $Q = \sum (dy^i)^2$. We may say that a *constant* Riemannian metric can always be reduced to the "flat" or "euclidean" form. Suppose now that we have a *variable* Riemannian metric $g_{ij}(x)dx^i dx^j$ in a coordinate patch of an $M^n$. By the previous arguments, we may always make a linear change of coordinates $y^i = P^i{}_j x^j$ so that the metric will take the form $\sum (dy^i)^2$ *at a single point*, say the origin. Is it possible that by making perhaps a non-linear change of coordinates $y = y(x)$ we can put the metric in the locally **euclidean** or **flat** form $\sum (dy^i)^2$ in the entire coordinate patch, or at least *in some neighborhood of the origin*?

It was for precisely such considerations that Riemann was led to introduce his curvature tensor; we know that if one could introduce such coordinates $y$, then $g_{ij} = \delta_{ij}$ in those coordinates, the Christoffel symbols would vanish and so the curvature tensor in the $y$ coordinates would vanish. Since the curvature tensor *is* a tensor, it would have to vanish in the $x$ system as well; in order that a Riemannian metric can be reduced to the locally euclidean form, the Riemann tensor must vanish. Riemann also noted that the converse is also true. We shall now discuss all these matters from a more geometrical viewpoint.

### 9.7b. The Horizontal Distribution of an Affine Connection

Parallel displacement of a vector $\mathbf{v}$ along a parameterized path $C$ in $M^n$ is described by the local system of differential equations

$$\frac{dv^i}{dt} + \omega^i_{jk}(x)v^k\left(\frac{dx^j}{dt}\right) = 0$$

The functions $(x(t), v(t))$ define a curve $C'$ in the tangent bundle $TM$ to $M$ that lies "over" the curve $C$ (recall that $(x, v)$ are local coordinates for $TM$).

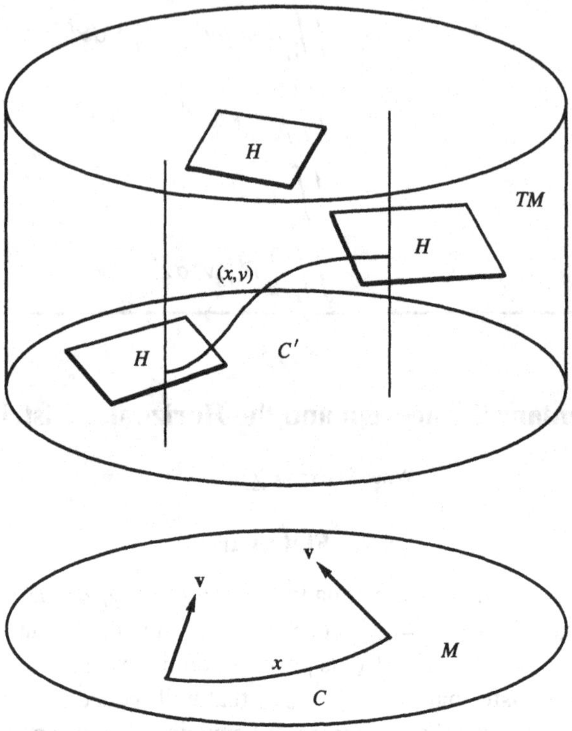


**Figure 9.4**

Since the projection map $\pi : TM \to M$ is of the form $(x, v) \to x$, that is, since we are allowing ourselves to use $x$ for coordinates in both $M$ and $TM$, the pull-back of the connection forms $\omega$ on $M$ to $TM$ is given by the same expressions as $\omega$ in $M$

$$\pi^*(\omega^i{}_k) = \pi^*(\omega^i_{jk} dx^j) = \omega^i_{jk} dx^j$$

For this reason we shall frequently omit the pull-back symbol $\pi^*$. Then parallel displacement tells us that the lifted curve $C'$ is that curve in $TM$ over $C$ having the property that the following 1-forms *in TM*

$$\mu^i := dv^i + \omega^i{}_k v^k$$

vanish when restricted to $C'$, $\mu^i[(dx^r/dt)\partial/\partial x^r + (dv^r/dt)\partial/\partial v^r)] = 0$. We write simply

$$\mu^i = dv^i + \omega^i{}_k v^k = 0 \tag{9.68}$$

as the *equations describing parallel displacement.*

The Pfaffian equations $\mu^i = 0, i = 1, \ldots, n$, define a distribution $H$ in $TM$. Since $\mu^1 \wedge \ldots \wedge \mu^n = dv^1 \wedge \ldots \wedge dv^n +$ terms involving the $dx^j$, we see that $\mu^1, \ldots, \mu^n$ are linearly independent, and thus the distribution is a distribution of $n$-planes in the $2n$-dimensional $TM$. Furthermore, it is clear that no nonzero "vertical" vector $a^j \partial/\partial v^j$, that is, a vector tangent to a fiber $\pi^{-1}(x)$, is never in this distribution. This implies that at every point the $n$-plane distribution is complementary to the vertical $n$-planes that are tangent to the fibers. There is usually no natural Riemannian metric in $TM$ and thus it makes no sense to talk of $H$ as being orthogonal to the fibers; still, we shall refer to $H$ as being the **horizontal distribution.**

We should remark that although $H$ has been defined using local coordinates and while we certainly cannot expect the individual forms $\mu^i$ to have intrinsic meaning, the distribution $H$ does have global meaning since it has been constructed using parallel displacement. Analytically, if $\mu'^i = dv'^i + \omega'^i{}_k v'^k$ are the forms in an overlapping patch, then, under the change of frame $\partial' = \partial P$ in $M$, we have $v' = P^{-1}v$ and then, from (9.41)

$$\mu' = dv' + \omega' v' = d(P^{-1}v) + (P^{-1}\omega P + P^{-1}dP)P^{-1}v = dP^{-1}v + P^{-1}dv$$
$$+ P^{-1}\omega v + P^{-1}dPP^{-1}v = P^{-1}(dv + \omega v) = P^{-1}\mu$$

Thus $\mu = 0$ iff $\mu' = 0$, and $H$ is well defined. Hence

**Theorem (9.69):** *A connection for $M$ yields a distribution of $n$-planes $H$ in $TM$ (the horizontal distribution) that is transverse to the fibers. A curve $C'$ in $TM$ represents parallel translation of a vector along a curve $C$ in $M$ iff $C'$ covers $C, \pi C' = C$, and $C'$ is tangent to the distribution $H$.*

To say that **v** returns to itself after being parallel translated around a closed curve $C$ in $M$ is to say that the "lift $C'$ of $C$ to $TM$ via **v**," that is, $x = x(t)$, $v = v(t)$, is itself a *closed* curve tangent to $H$.

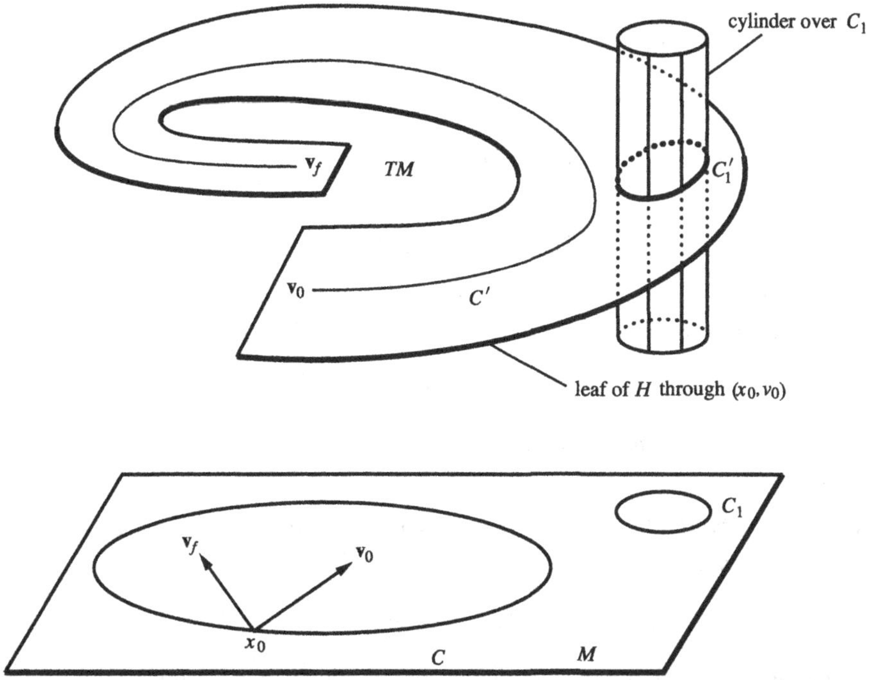


Figure 9.5

*If* the distribution $H$ is *integrable*, and if we choose a closed curve $C_1$ that is so small that its lift $C_1'$ lies in a Frobenius chart (see 6.1a), then $C_1'$ will also have to be closed since it will have to lie on a small portion of a leaf of the foliation; see the figure. This need not be the case if the curve $C$ is "long," as illustrated. On the other hand, if $H$ is not integrable, we do not expect a closed curve $C$ to have a closed lift $C'$.

When is the horizontal distribution $H$ integrable?

**Theorem (9.70):** *The horizontal distribution H is integrable (and consequently parallel displacement is locally independent of path), iff the curvature vanishes, that is, $M^n$ is flat.*

PROOF: $H$ is defined briefly as $\mu = dv + \omega v = 0$. Then

$$d\mu = d^2v + d\omega v - \omega \wedge dv = d\omega v - \omega \wedge (\mu - \omega v)$$
$$= (d\omega + \omega \wedge \omega)v - \omega \wedge \mu = \theta v \bmod \mu$$

where by "mod $\mu$" we mean the result of putting $\mu = 0$ (see 6.1c). Thus $d\mu = \theta v = 0 \bmod \mu$ if $\theta = 0$. Thus $H$ is integrable if the curvature vanishes. On the other hand, if $H$ is integrable, then, from Theorem (6.2),

$$0 = d\mu^i \wedge \mu^1 \wedge \ldots \wedge \mu^n$$
$$= (\theta^i{}_j v^j - \omega^i{}_j \wedge \mu^j) \wedge \mu^1 \wedge \ldots \wedge \mu^n$$
$$= \theta^i{}_j v^j \wedge \mu^1 \wedge \ldots \wedge \mu^n$$
$$= \theta^i{}_j v^j \wedge (dv^1 + \omega^1{}_k v^k) \wedge \ldots \wedge (dv^n + \omega^n{}_r v^r)$$
$$= \theta^i{}_j v^j \wedge dv^1 \wedge \ldots \wedge dv^n + \quad \text{terms where some } dv^j \text{ is missing}$$

Hence $\theta^i{}_j v^j = 0$ for $i = 1, \ldots, n$, and all $v$. Thus $\theta = 0$. □

## 9.7c. Riemann's Theorem

**Theorem (9.71):** *In a Riemannian manifold, one can introduce local coordinates y such that the metric assumes the euclidean or "flat" form*

$$ds^2 = (dy^1)^2 + \ldots (dy^n)^2$$

*iff the curvature vanishes, $\theta = 0$.*

PROOF: The "only if" part has already been discussed in 9.7a. Suppose now that the curvature vanishes. Then the horizontal distribution $H$

$$\mu^i = dv^i + \Gamma^i_k v^k = dv^i + \Gamma^i_{jk} v^k dx^j = 0$$

is integrable. (Here $\Gamma$ are the coefficients of the affine connection with respect to the coordinate frame $\partial/\partial x$, that is, the Christoffel symbols.) Since $H$ is transverse to the fibers $\pi^{-1}(x)$ of $TM$, this means (as in the system of Mayer–Lie of Section 6.2b) that we may locally solve the system of partial differential equations

$$\frac{\partial v^i}{\partial x^j} + \Gamma^i_{jk} v^k = 0 \tag{9.72}$$

$$v^i(x_0) = v^i_0 \quad \text{prescribed}$$

In particular, given $x_0$ and given $n$ linearly independent vectors $\mathbf{e}^0_1, \ldots, \mathbf{e}^0_n$ at $x_0$, we may find vector fields $\mathbf{e}_1, \ldots, \mathbf{e}_n$ coinciding with $\mathbf{e}^0$ at $x_0$ and each satisfying (9.72); that is, each is **covariant constant**

$$\frac{\nabla \mathbf{e}_r}{\partial x^j} := \nabla \mathbf{e}_r\left(\frac{\partial}{\partial x^j}\right) = 0 \tag{9.73}$$

for all $r$ and $j$. Thus if we let $\omega$ be the connection forms with respect to the new frame $\mathbf{e}$, we have $\nabla \mathbf{e} = \mathbf{e}\omega = 0$, and so $\omega = 0$.

Note that we have actually shown, so far, the following.

**Theorem (9.74):** *For any affine connection with curvature 0, one can find a local frame of covariant constant vector fields.*

Finally, consider the 1-forms $\sigma$ dual to the frame $\mathbf{e}$. If the connection is symmetric, as it is in the Riemannian case, we have $d\sigma^i = -\omega^i{}_j \wedge \sigma^j = 0$, and so each of the 1-forms $\sigma^i$ is closed and, by Poincaré, locally exact. Thus there are local functions $y^1, \ldots, y^n$ such that $\sigma^i = dy^i$. This means that $\mathbf{e}_i = \partial/\partial y^i$. In the Riemannian case, if the $\mathbf{e}^0$ had been chosen orthonormal at $x_0$, then the frame fields $\mathbf{e}$ would also be othonormal in the entire $y$ coordinate patch since $d\langle \mathbf{e}_i, \mathbf{e}_j\rangle = \langle \nabla \mathbf{e}_i, \mathbf{e}_j\rangle + \langle \mathbf{e}_i, \nabla \mathbf{e}_j\rangle = 0$. Since the coordinate frame $\partial/\partial y$ is orthonormal we have $ds^2 = (dy^1)^2 + \ldots (dy^n)^2$. □

A final remark. Let $M^2$ be the frustum of a cone that is tangent to a small circle $C$ on the round 2-sphere. The cone is flat, yet we have seen, when first discussing parallel displacement, that parallel displacement of a vector along $C$ does not return the vector to itself; there is no covariant constant vector field on the flat cone! This does not violate Riemann's theorem since that theorem only *locally* exhibits a flat frame.

# CHAPTER 10

# Geodesics

How rapidly do nearby geodesics separate?

## 10.1. Geodesics and Jacobi Fields

### 10.1a. Vector Fields Along a Surface in $M^n$

Let $\mathbf{x} : U \subset \mathbb{R}^2 \to M^n$ be a differentiable map of a rectangle in the plane into $M^n$. We call this map a (parameterized) surface even though we put *no* demands on the rank of the differential $\mathbf{x}_*$; that is, $\partial \mathbf{x}/\partial u^1$ and $\partial \mathbf{x}/\partial u^2$ may be dependent.

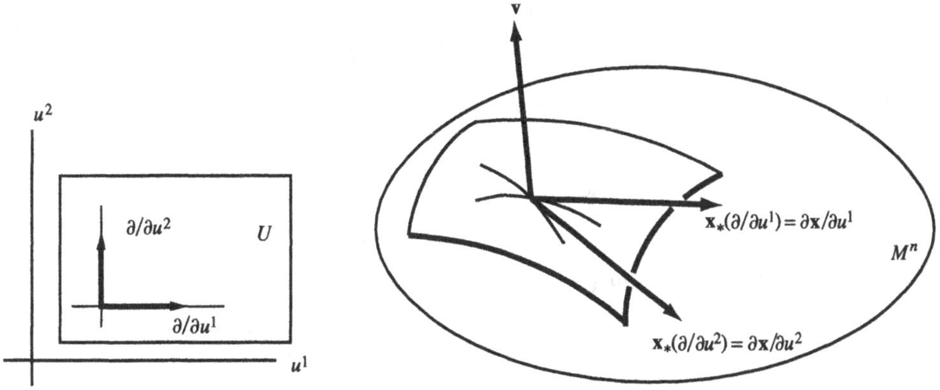


Figure 10.1

Let us again put $u^1 = u$ and $u^2 = v$.

A smooth map $\mathbf{v} : U \subset \mathbb{R}^2 \to M$ that assigns to each $(u, v)$ in the rectangle a tangent vector $\mathbf{v}(u, v)$ to $M$ at $\mathbf{x}(u, v)$ will be called a **vector field along the surface x**. In particular, $\partial \mathbf{x}/\partial u$ and $\partial \mathbf{x}/\partial v$ are both vector fields along $\mathbf{x}$ that happen to be "tangent" to the surface. Of course $[\partial/\partial u, \partial/\partial v] = 0$ in $U$, but we cannot talk of $[\partial \mathbf{x}/\partial u, \partial \mathbf{x}/\partial v]$ since the two entries in the bracket are not vector fields on $M^n$. If they were vector fields, we could consider their bracket, and if $M^n$ had a torsion-free connection this bracket could be expressed in terms of covariant derivatives (see (9.16)). Even when they are not vector fields we still have that, for example, $\partial \mathbf{x}/\partial v$ is defined along the orbit of $\partial\mathbf{x}/\partial u$, that is, the $u$-curves. The following is an important computational tool that replaces (9.16).

**Theorem (10.1):** *Let* $\mathbf{x}$ *be a surface in a manifold* $M^n$ *with a symmetric connection. Then we have, as vector fields along the surface,*

$$\frac{\nabla}{\partial u}\left(\frac{\partial \mathbf{x}}{\partial v}\right) = \frac{\nabla}{\partial v}\left(\frac{\partial \mathbf{x}}{\partial u}\right)$$

PROOF: Let $x^1, \ldots, x^n$ be local coordinates for $M$. Then, for example, $\partial\mathbf{x}/\partial v = (\partial x^i/\partial v)\boldsymbol{\partial}_i$, where $\boldsymbol{\partial}_i = \boldsymbol{\partial}/\boldsymbol{\partial} x^i$. If we fix $v$, then taking the covariant derivative of $\partial\mathbf{x}/\partial v$ along the $u$-curve gives, from Leibniz,

$$\frac{\nabla}{\partial u}\left(\left(\frac{\partial x^i}{\partial v}\right)\boldsymbol{\partial}_i\right) = \left(\frac{\partial^2 x^i}{\partial u \partial v}\right)\boldsymbol{\partial}_i + \left(\frac{\partial x^i}{\partial v}\right)\nabla_{\partial\mathbf{x}/\partial u}(\boldsymbol{\partial}_i)$$

Now $\partial\mathbf{x}/\partial u = (\partial x^j/\partial u)\boldsymbol{\partial}_j$ and using $\nabla_{\boldsymbol{\partial}_j}\boldsymbol{\partial}_i = \omega^k_{ji}\boldsymbol{\partial}_k$ yields $\nabla/\partial u((\partial x^i/\partial v)\boldsymbol{\partial}_i) = (\partial^2 x^i/\partial u\partial v)\boldsymbol{\partial}_i + (\partial x^i/\partial v)(\partial x^j/\partial u)\omega^k_{ji}\boldsymbol{\partial}_k$, which is symmetric in $u$ and $v$ since $\omega^k_{ji} = \omega^k_{ij}$. □

The next result is a replacement for Theorem (9.10).

**Theorem (10.2):** *If* $\mathbf{w}$ *is a vector field defined along the surface,*

$$\frac{\nabla}{\partial u}\left(\frac{\nabla \mathbf{w}}{\partial v}\right) - \frac{\nabla}{\partial v}\left(\frac{\nabla \mathbf{w}}{\partial u}\right) = R\left(\frac{\partial \mathbf{x}}{\partial u}, \frac{\partial \mathbf{x}}{\partial v}\right)\mathbf{w}$$

*where* $R(\partial\mathbf{x}/\partial u, \partial\mathbf{x}/\partial v)$ *is the curvature transformation defined in Theorem (9.10).*

PROOF: $\mathbf{w} = w^i(u, v)\boldsymbol{\partial}_i$. Then

$$\begin{aligned}\frac{\nabla}{\partial u}\left(\frac{\nabla \mathbf{w}}{\partial v}\right) &= \frac{\nabla}{\partial u}\left\{\left(\frac{\partial w^i}{\partial v}\right)\boldsymbol{\partial}_i + w^i\frac{\nabla\boldsymbol{\partial}_i}{\partial v}\right\} \\ &= \left(\frac{\partial^2 w^i}{\partial u\partial v}\right)\boldsymbol{\partial}_i + \left(\frac{\partial w^i}{\partial v}\right)\frac{\nabla\boldsymbol{\partial}_i}{\partial u} + \left(\frac{\partial w^i}{\partial u}\right)\frac{\nabla\boldsymbol{\partial}_i}{\partial v} \\ &\quad + w^i\frac{\nabla}{\partial u}\left(\frac{\nabla\boldsymbol{\partial}_i}{\partial v}\right)\end{aligned}$$

and so

$$\begin{aligned}&\frac{\nabla}{\partial u}\left(\frac{\nabla \mathbf{w}}{\partial v}\right) - \frac{\nabla}{\partial v}\left(\frac{\nabla \mathbf{w}}{\partial u}\right) \\ &\quad = w^i\left\{\frac{\nabla}{\partial u}\left(\frac{\nabla\boldsymbol{\partial}_i}{\partial v}\right) - \frac{\nabla}{\partial v}\left(\frac{\nabla\boldsymbol{\partial}_i}{\partial u}\right)\right\}\end{aligned} \tag{10.3}$$

But

$$\frac{\nabla \partial_i}{\partial v} = \nabla \partial_i \left\{ \left( \frac{\partial x^j}{\partial v} \right) \partial_j \right\} = \left( \frac{\partial x^j}{\partial v} \right) \nabla_{\partial_j} \partial_i .$$

Thus

$$\frac{\nabla}{\partial u} \left\{ \frac{\nabla}{\partial v} (\partial_i) \right\} = \left( \frac{\partial^2 x^j}{\partial u \partial v} \right) \nabla_{\partial_j} \partial_i + \left( \frac{\partial x^j}{\partial v} \right) \left( \frac{\partial x^k}{\partial u} \right) \nabla_{\partial_k} \nabla_{\partial_j} \partial_i$$

Then

$$\begin{aligned}
&\frac{\nabla}{\partial u} \left( \frac{\nabla \partial_i}{\partial v} \right) - \frac{\nabla}{\partial v} \left( \frac{\nabla (\partial_i)}{\partial u} \right) \\
&= \left\{ \left( \frac{\partial x^j}{\partial v} \right) \left( \frac{\partial x^k}{\partial u} \right) - \left( \frac{\partial x^j}{\partial u} \right) \left( \frac{\partial x^k}{\partial v} \right) \right\} \nabla_{\partial_k} \nabla_{\partial_j} \partial_i \\
&= \left( \frac{\partial x^k}{\partial u} \right) \left( \frac{\partial x^j}{\partial v} \right) \{ \nabla_{\partial_k} \nabla_{\partial_j} \partial_i - \nabla_{\partial_j} \nabla_{\partial_k} \partial_i \} \\
&= \left( \frac{\partial x^k}{\partial u} \right) \left( \frac{\partial x^j}{\partial v} \right) R(\partial_k, \partial_j)(\partial_i) \\
&= R \left\{ \left( \frac{\partial x^k}{\partial u} \right) \partial_k, \left( \frac{\partial x^j}{\partial v} \right) \partial_j \right\} (\partial_i) \\
&= R \left( \frac{\partial \mathbf{x}}{\partial u}, \frac{\partial \mathbf{x}}{\partial v} \right) (\partial_i)
\end{aligned}$$

Putting this in (10.3) yields (10.2). □

### 10.1b. Geodesics

We now return to the discussion of geodesics initiated in Section 8.6, but now we shall carry out the calculation *intrinsically* and in an $n$-dimensional Riemannian manifold $M^n$. Since our definition of covariant differentiation was tailored after the discussions in that section it should come as no great surprise that we can essentially mimic the calculations given there.

Let $C$ be a curve in the Riemannian $M^n$. To "vary" $C$ is to consider a surface $\mathbf{x} : [0, L] \times (-1, +1) \to M^n$ parameterized by $s$ and $\alpha$

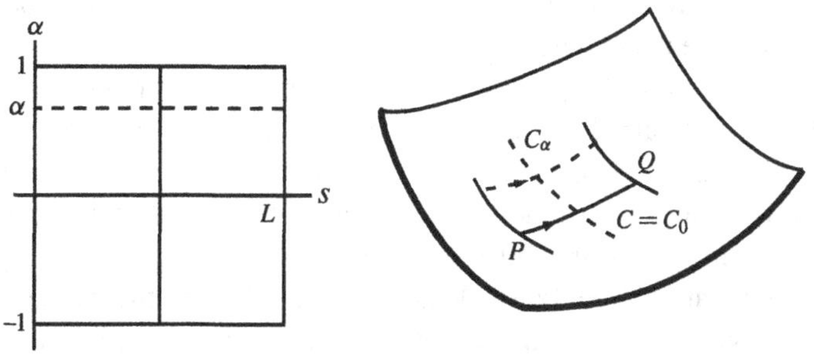


Figure 10.2

such that $\mathbf{x}(s, 0)$ describes the original curve $C$. The varied curve $C_\alpha$ is given by $s \mapsto \mathbf{x}(s, \alpha)$, where $s$ is arc length for $\alpha = 0$, that is, along the base curve, but not necessarily so when $\alpha \neq 0$. We proceed as in 8.6. The length of $C_\alpha$ is $L(\alpha) = \int_0^L \langle \partial\mathbf{x}/\partial s, \partial\mathbf{x}/\partial s\rangle^{1/2} ds$.

Since $M$ is Riemannian, we have

$$\frac{\partial}{\partial\alpha}\langle \mathbf{v}, \mathbf{w}\rangle = \left\langle \frac{\nabla \mathbf{v}}{\partial\alpha}, \mathbf{w}\right\rangle + \left\langle \mathbf{v}, \frac{\nabla \mathbf{w}}{\partial\alpha}\right\rangle$$

In the derivation of (8.36) we used $\partial^2\mathbf{x}/\partial\alpha\partial s = \partial^2\mathbf{x}/\partial s\partial\alpha$; this is now replaced by (10.1), that is, $\nabla/\partial\alpha(\partial\mathbf{x}/\partial s) = \nabla/\partial s(\partial\mathbf{x}/\partial\alpha)$. In Problem 10.1(1) you are asked to show that

$$L'(\alpha) = \int_0^L \left\langle \frac{\partial\mathbf{x}}{\partial s}, \frac{\partial\mathbf{x}}{\partial s}\right\rangle^{-1/2} \left\langle \frac{\nabla}{\partial s}\left(\frac{\partial\mathbf{x}}{\partial\alpha}\right), \frac{\partial\mathbf{x}}{\partial s}\right\rangle ds$$

and (10.4)

$$L'(0) = \langle \mathbf{J}, \mathbf{T}\rangle_Q - \langle \mathbf{J}, \mathbf{T}\rangle_P - \int_0^L \left\langle \mathbf{J}, \frac{\nabla\mathbf{T}}{\partial s}\right\rangle ds$$

Here $\mathbf{T} = \partial\mathbf{x}(s, 0)/\partial s$ is the unit tangent along $C$, $\mathbf{J} = \partial\mathbf{x}(s, 0)/\partial\alpha$ is the variation vector, and $P = \mathbf{x}(0, 0)$ and $Q = \mathbf{x}(L, 0)$ are the beginning and endpoints of $C$.

We now shall call *any* parameterized curve $C$, $x = x(t)$, a **geodesic** if

$$\frac{\nabla}{dt}\left(\frac{d\mathbf{x}}{dt}\right) = 0 \tag{10.5}$$

Note then that

$$\frac{d}{dt}\left\langle \frac{d\mathbf{x}}{dt}, \frac{d\mathbf{x}}{dt}\right\rangle = 2\left\langle \frac{d\mathbf{x}}{dt}, \frac{\nabla}{dt}\left(\frac{d\mathbf{x}}{dt}\right)\right\rangle = 0$$

This shows that $\| d\mathbf{x}/dt \| =$ constant, and so *the parameter t is, except for an additive constant, proportional to arc length.* We shall call such a parameter a **distinguished** or **affine** parameter.

A geodesic thus gives, from (10.4), the first variation of the arc length.

### 10.1c. Jacobi Fields

Let $C$ now be a geodesic, and let us vary $C$ by curves $C_\alpha$ where *each $C_\alpha$ is itself a geodesic, parameterized by a parameter s that is proportional to arc length.* The best example to keep in mind is probably the family of great circles on the round 2-sphere all passing through the north pole.

In talking about geodesic "separation" we are interested, as far as local coordinates $x$ go, in the behavior of a pair of points $x(s, \alpha)$ and $x(s, 0)$ as we increase $s$, that is, as we move along both geodesics at unit speed. The $n$-tuple $x^i(s, \alpha) - x^i(s, 0)$ has usually non-linear behavior as a function of $s$. Jacobi's equation, to be derived later, is the linear equation governing the linear approximation $\alpha\mathbf{J} = \alpha[\partial x(s, \alpha)/\partial\alpha]_{\alpha=0}$ to $[x(s, \alpha) - x(s, 0)]$.

Let us use the notation $\mathbf{T} = \partial\mathbf{x}(s, \alpha)/\partial s$ for the tangents to the geodesics along the curves and $\mathbf{J} = \partial\mathbf{x}(s, \alpha)/\partial\alpha$ for the variation vectors; although these usually are not vector fields on $M$, they are vector fields along the surface of variation. A differential equation satisfied by the variation vector field $\mathbf{J}(s, 0)$ can be obtained as follows.

Since each $C_\alpha$ is a geodesic we have $\nabla\mathbf{T}/\partial s = 0$ for all $\alpha$. Thus, from (10.2) and (10.1) we have

$$\begin{aligned} 0 &= \frac{\nabla}{\partial\alpha}\left(\frac{\nabla\mathbf{T}}{\partial s}\right) = \frac{\nabla}{\partial s}\left(\frac{\nabla T}{\partial\alpha}\right) + R(\mathbf{J},\mathbf{T})(\mathbf{T}) \\ &= \frac{\nabla}{\partial s}\left\{\frac{\nabla}{\partial\alpha}\left(\frac{\partial\mathbf{x}}{\partial s}\right)\right\} + R(\mathbf{J},\mathbf{T})(\mathbf{T}) \\ &= \frac{\nabla}{\partial s}\left\{\frac{\nabla}{\partial s}\left(\frac{\partial\mathbf{x}}{\partial\alpha}\right)\right\} + R(\mathbf{J},\mathbf{T})(\mathbf{T}) \\ &= \frac{\nabla}{\partial s}\left(\frac{\nabla J}{\partial s}\right) + R(\mathbf{J},\mathbf{T})(\mathbf{T}) \end{aligned}$$

or

$$\frac{\nabla^2\mathbf{J}}{\partial s^2} + R(\mathbf{J},\mathbf{T})(\mathbf{T}) = 0 \tag{10.6}$$

This is **Jacobi's equation of geodesic variation**. If we put $\alpha = 0$, it is a (complicated) second-order system of linear ordinary differential equations for $\mathbf{J}$ in terms of $s$. Any field $\mathbf{J}$ along a geodesic $C$ that satisfies (10.6) will be called a **Jacobi field** along $C$. It is not difficult to see that a Jacobi field always arises as the variation vector field resulting from varying the given geodesic by some 1-parameter family of geodesics. For such matters see [$M$].

In the case of a 2-dimensional surface $M^2$ this equation reduces to a simple form discovered by Jacobi. Let $C$ be a geodesic with unit tangent $\mathbf{T}$ and let $\mathbf{T}^\perp$ be a unit vector field along $C$ that is orthogonal to $\mathbf{T}$. $\mathbf{T}$ is parallel displaced along $C$ and, consequently, so is $\mathbf{T}^\perp$ (why?). Let $\mathbf{J}$ be a Jacobi field along $C$. We may expand

$$\mathbf{J}(s) = x(s)\mathbf{T} + y(s)\mathbf{T}^\perp$$

where $x$ and $y$ are the tangential and normal components of $\mathbf{J}$. Since $\nabla\mathbf{T}/ds = 0 = \nabla\mathbf{T}^\perp/ds$, Jacobi's equation becomes

$$\begin{aligned} \frac{\nabla^2\mathbf{J}}{ds^2} &= \frac{d^2x}{ds^2}\mathbf{T} + \frac{d^2y}{ds^2}\mathbf{T}^\perp = -R(x\mathbf{T} + y\mathbf{T}^\perp, \mathbf{T})\mathbf{T} \\ &= -R(y\mathbf{T}^\perp, \mathbf{T})\mathbf{T} = -yR(\mathbf{T}^\perp, \mathbf{T})\mathbf{T} \end{aligned}$$

Then

$$\frac{d^2y}{ds^2} = -y\langle R(\mathbf{T}^\perp, \mathbf{T})\mathbf{T}, \mathbf{T}^\perp\rangle$$

Let us express everything in terms of the orthonormal frame $\mathbf{e}_1 = \mathbf{T}$, $\mathbf{e}_2 = \mathbf{T}^\perp$ along $C$. Since $\langle R(\mathbf{X},\mathbf{Y})\mathbf{Z},\mathbf{W}\rangle = R^i_{jkl}X^kY^lZ^jW_i$ we see from (9.54) and (9.13) that

$$\langle R(\mathbf{e}_2,\mathbf{e}_1)\mathbf{e}_1,\mathbf{e}_2\rangle = R^2_{121} = R_{2121} = R_{1212} = K$$

Jacobi's equation becomes

$$d^2y/ds^2 + Ky = 0 \tag{10.7}$$

The function $y$ represents, roughly, how the "normal" separation of nearby geodesics is changing as we move along the geodesics. Consider, for example, the great circle $C$ of longitude zero on the 2-sphere defined by $\phi = 0$, starting at the north pole $\theta = 0$

and ending at the south pole $\theta = \pi$. We can vary $C$ by the meridians of longitude $\phi$ = constant; our parameter $\alpha = \phi$ in this case. Equation (10.7) in this unit sphere case becomes $d^2y/d\theta^2 + y = 0$ and since $y = 0$ at $\theta = 0$, the solution is $y = A\sin\theta$. We see just from this that the geodesics that were originally separating at the north pole tend to come together at the south pole. In fact, $\mathbf{J} = \partial\mathbf{x}/\partial\phi$, $\mathbf{T} = \partial\mathbf{x}/\partial\theta$, and $\mathbf{T}^{\perp}$ is $\partial\mathbf{x}/\partial\phi$ made unit. Then $y = \| \partial\mathbf{x}/\partial\phi \| = \sin\theta$.

In the $n$-dimensional case, $\mathbf{J}$ represents how the geodesics, in a 1-parameter family of geodesics, are separating. It is not true, however, even in 2 dimensions, that if $\mathbf{J}(s_0) = 0$ for some arc length value $s_0$, the geodesics have actually come together (as they did in the round $S^2$ case); it means only that the separation distance vanishes in the linear approximation at $s_0$.

From (10.7) it is clear that the sign of the Gauss curvature $K$ is crucial for understanding the behavior of nearby geodesics on a surface. If $K(u, v) > a^{-2} > 0$ is positive on $M^2$ then the Sturm theory of differential equations tells us that if $y(0) = 0$ then $y(s_0) = 0$ for some $s_0 < \pi a$, and thus a family of geodesics that start at the same point will meet again, in the linear approximation, before traveling a distance $\pi a$. On the other hand, if $K(u, v) \leq 0$, and if $y(0) = 0$, then $y(s)$ will never vanish again unless $y$ is identically 0. This does *not* mean that a pair of geodesics starting out from a point will not meet again; on the flat torus $ds^2 = d\theta^2 + d\phi^2$, the geodesic $\phi = 0$, and the geodesic $\theta = 0$ start at $(0, 0)$ and meet repeatedly at $(2\pi m, 2\pi n)$. It means only that a 1-parameter *family* will not come together. There are similar statements about the influence of the Riemannian curvature tensor on the "stability of geodesics" in $n$ dimensions. Arnold [A, p. 340 ff.] discusses the problem of long-range weather prediction using an infinite-dimensional version of Jacobi's equation.

### 10.1d. Energy

We have discussed geodesics in terms of yielding a critical point for the length functional $\int \| dx/dt \| \, dt$, that is, first variation zero; in classical language $\delta \int \| dx/dt \| \, dt = 0$. It is not difficult to see (in fact the computation is even simpler) that one also gets geodesics by varying the integrand $\| dx/dt \|^2$ instead

$$\delta \int \left\| \frac{dx}{dt} \right\|^2 dt = 0$$

(It should be noted that unlike the case of arc length, this integral depends on the parameter $t$ employed.) This new functional is called the **action** or **energy** for reasons that will become apparent in the next section. Some books (e.g., [M]) discuss energy rather than length, with final equations that are always rather similar to ours.

## Problems

**10.1(1)** Derive (10.4).

**10.1(2)** Consider the Poincaré upper half plane, $ds^2 = y^{-2}(dx^2 + dy^2)$. As in Problem 9.5(1) we have an orthonormal frame $\mathbf{e}_1 = y\partial/\partial x$, $\mathbf{e}_2 = y\partial/\partial y$. Show that the vertical lines are geodesics, $\nabla\mathbf{e}_2/ds = 0$, by using Cartan's equations

$\nabla \mathbf{e}_2/ds = \mathbf{e}_1\omega^1{}_2(\mathbf{e}_2)$. Then $\mathbf{J} = \partial/\partial x$ is a Jacobi field along the geodesic $x = 0$. Verify that Jacobi's equation (10.7) is indeed satisfied. Note that $\| \mathbf{J} \| \to \infty$ as $y \to 0$; that is, the vertical geodesics are separating as we approach the $x$ axis.

**10.1(3)** Show that a Jacobi field $\mathbf{J}$ that is orthogonal to its geodesic at two distinct parameter values $s = 0$ and $s = s_1 \neq 0$ (e.g., if $\mathbf{J}$ vanishes at $s = 0$ and $s = s_1$) must always be orthogonal to the geodesic. (Hint: Derive from (10.6) a second-order differential equation that is satisfied by $\langle \mathbf{J}, \mathbf{T} \rangle$.)

## 10.2. Variational Principles in Mechanics

Consider a double planar pendulum with arms of different lengths. Is there always a *periodic* motion where the top arm makes $p$ revolutions and the bottom makes $q$?

In Section 4.4 we discussed analytical dynamics in phase space, that is, the cotangent bundle $T^*M$ to the configuration space $M^n$. Our main purpose was to exhibit the usefulness of both exterior differential forms and the fact that Hamiltonian mechanics is, in a sense, the discussion of a particular vector field on $T^*M \times \mathbb{R}$ and its effect on the symplectic form $\omega^2$. Hamilton's variational principle in *phase space*, Problem 4.4(12), due, I believe, to Poincaré, was carried out using Lie derivatives to calculate the variations. In the present section we shall return to these considerations, but we shall emphasize more both the physical and geometric motivation and also the classical language of the variational calculations. We shall also include the relation between Hamilton's principle and the geodesics on the configuration space. We shall defer the *tensorial* properties of the variational calculus to Section 20.1.

We shall use a brief notation, omitting indices whenever possible; for example, we shall write $pdq$ rather than $p_i dq^i$.

### 10.2a. Hamilton's Principle in the Tangent Bundle

The configuration space of a dynamical system is an $n$-dimensional manifold $M^n$. Let $q^1, \dots, q^n$ be local coordinates in $M^n$.

The kinetic energy is frequently of the form $T = (1/2)g_{ij}(q)\dot{q}^i\dot{q}^j$, where $g_{ij}(q)$ is a positive definite matrix constructed out of a metric tensor for $M^n$ *and also the masses of the particles* of the system. For example, in the case of a particle moving in the plane with polar coordinates $q^1 = r$ and $q^2 = \theta$ we have $g_{\theta\theta} = mr^2$ since

$$T = \frac{m}{2}\left[\left(\frac{dr}{dt}\right)^2 + r^2\left(\frac{d\theta}{dt}\right)^2\right]$$

It is sometimes convenient to *use 2T to define a new Riemannian metric* for $M^n$, $ds^2 = g_{ij}(q)dq^i dq^j$. Thus $\langle \dot{q}, \dot{q} \rangle = 2T$. The momentum $p$ is the covariant version of the velocity, $p_i = g_{ij}\dot{q}^j$. The obvious expression of Newton's law of motion in the case when the forces are derived from a potential, $dp_i/dt = -\partial V/\partial q^i$, makes no sense since the right-hand side gives the components of the covector $-dV$, whereas the usual derivative of a covector (or a vector) along a curve has no intrinsic meaning. To remedy this we write the proposed "law" in contravariant form, $d\dot{q}^k/dt = -g^{ki}\partial V/\partial q^i = -(\text{grad } V)^k$, and then replace the ordinary derivative by an intrinsic or covariant derivative

$$\frac{\nabla \dot{q}}{dt} = -\text{grad } V \tag{10.8}$$

In coordinates

$$\frac{d\dot{q}^i}{dt} + \Gamma^i_{jk}\dot{q}^j\dot{q}^k = -g^{ik}\frac{\partial V}{\partial q^k}$$

It should not be surprising that Newton's law can be put in the form of a variational principle since the intrinsic derivative arose, in our treatment, when considering the variation of arc length.

Consider a variation $q = q(t, \alpha)$ of a parameterized curve $q = q(t)$ in $M$; we write

$$q(t, \alpha) = q(t) + \alpha\eta(t) \tag{10.9}$$

for some function $\eta$. Then $\partial q(t, \alpha)/\partial\alpha = \eta(t)$. Classically $[\partial q(t, \alpha)/\partial\alpha]_{\alpha=0} = \eta(t)$ is written $\delta q$, and is called a **virtual displacement**. Then the first derivative of the integral $\int_a^b V(q)dt$ is classically written

$$\begin{aligned}\delta\int_a^b V(q)dt &= \left(\frac{d}{d\alpha}\right)_{\alpha=0}\left[\int_a^b V(q)dt\right]\\ &= \int_a^b \left[\frac{\partial V(q)}{\partial q}\right]\left[\frac{\partial q}{\partial\alpha}\right]_{\alpha=0} dt\\ &= \int_a^b \left[\frac{\partial V(q)}{\partial q}\right]\eta dt = \int_a^b \left[\frac{\partial V(q)}{\partial q}\right]\delta q dt\end{aligned}$$

Consider now the variation of the kinetic energy $\int_a^b (1/2)\langle\dot{q}, \dot{q}\rangle dt$. The integrand is now a function $T$ of both $q$ (which appears in the metric tensor) and $\dot{q}$. We have computed the first variation of the more complicated $\int_a^b \langle\dot{q}, \dot{q}\rangle^{1/2} dt$ in (10.4). Essentially the same computation (but easier!) will give

$$\begin{aligned}\delta\int_a^b \frac{1}{2}\langle\dot{q}, \dot{q}\rangle dt &= \langle\delta q, \dot{q}\rangle(b) - \langle\delta q, \dot{q}\rangle(a)\\ &\quad - \int_a^b \left\langle\frac{\nabla\dot{q}}{dt}, \delta q\right\rangle dt\end{aligned} \tag{10.10}$$

We then see that Newton's law (10.8) is equivalent to the variational principle

$$\delta\int_a^b L dt := \delta\int_a^b (T - V)dt = 0 \tag{10.11}$$

provided

$$\delta q = 0 \quad \text{at } t = a \quad \text{and } t = b$$

We now accept as a generalization **Hamilton's principle** (10.11), for systems with a general Lagrangian $L = L(q, \dot{q}, t)$, at least in the case where all the forces are derived from a potential. We shall write down the associated Euler–Lagrange equations using classical notation.

$L$ is a function in the extended tangent bundle $TM \times \mathbb{R}$ of the configuration space $M$. Then a variation (10.9) of a curve $C$ in $M$ will yield a variation of the velocities. From (10.9), $\dot{q}(t,\alpha) := \partial q(t,\alpha)/\partial t = \dot{q}(t) + \alpha\dot{\eta}(t)$ and so

$$\delta\dot{q} = \dot{\eta} = (\delta q)^{\bullet} \tag{10.12}$$

Thus a curve $q(t)$ in $M$ yields a **lifted curve** $\{q(t), \dot{q}(t), t\}$ in $TM \times \mathbb{R}$ and we shall consider a variation of this lifted curve that arises, from (10.12), as the lift of the variation in $M$! We make no variation of the time parameter $t$. Then, in classical language (all integrations going from $t = a$ to $t = b$)

$$\begin{aligned}
\delta \int L(q, \dot{q}, t)dt &= \int \left\{ \left(\frac{\partial L}{\partial q}\right)\delta q + \left(\frac{\partial L}{\partial \dot{q}}\right)\delta\dot{q} \right\} dt \\
&= \int \left\{ \left(\frac{\partial L}{\partial q}\right)\delta q + \left(\frac{\partial L}{\partial \dot{q}}\right)\frac{\partial}{\partial t}(\delta q) \right\} dt \\
&= \int \left\{ \left(\frac{\partial L}{\partial q}\right)\delta q + \frac{\partial}{\partial t}\left[\left(\frac{\partial L}{\partial \dot{q}}\right)(\delta q)\right] \right\} dt - \int \left[\frac{\partial}{\partial t}\left(\frac{\partial L}{\partial \dot{q}}\right)\right]\delta q dt \\
&= \int \left\{ \left(\frac{\partial L}{\partial q}\right) - \frac{\partial}{\partial t}\left(\frac{\partial L}{\partial \dot{q}}\right) \right\}\delta q dt + \left[\left(\frac{\partial L}{\partial \dot{q}}\right)(\delta q)\right]\Bigg|_a^b
\end{aligned} \tag{10.13}$$

Since we assume that the variations vanish at the endpoints, $\delta q(a) = \delta q(b) = 0$, and since the variations $\delta q$ inside are arbitrary, we get Lagrange's equations

$$\frac{\partial L}{\partial q} - \frac{d}{dt}\left(\frac{\partial L}{\partial \dot{q}}\right) = 0 \tag{10.14}$$

Since the parameter $\alpha$ no longer appears (we are evaluating the derivative at $\alpha = 0$) we have written $d/dt$ rather than $\partial/\partial t$.

### 10.2b. Hamilton's Principle in Phase Space

(10.11), that is, Hamilton's principle in $TM$, was the starting point of our treatment of mechanics in Section 4.4a. It led, in Problem 4.4(12) to Poincaré's version of Hamilton's principle in phase space $T^*M$. In classical language,

$$\delta \int pdq - Hdt = 0 \tag{10.15}$$

They are equivalent (at least when the map $p : TM \to T^*M$ given by $p = \partial L/\partial\dot{q}$ is invertible) since Lagrange's equations and Hamilton's canonical equations (4.48) are equivalent. However, the differences in these two versions of Hamilton's principle should be kept in mind.

In the variational principle leading to Lagrange's equations earlier we considered a curve $q = q(t)$ in $M$, its unique lift to $TM \times \mathbb{R}$ (using $\dot{q} = dq/dt$), and variations in the velocity variables that arose from the time derivatives of the variation of the coordinates,

$\delta\dot{q} = d/dt(\delta q)$.Thus a variation of the configuration space curve led to a unique variation of the lifted curve in $TM$. *The variations of $q$ and $\dot{q}$ are not independent!*

In Poincaré's version we deal directly with an *arbitrary* curve $C, q = q(t), p = p(t)$, lying in $T^*M \times \mathbb{R}$, *that does not necessarily correspond to a lifted curve in $TM$*. Thus if we solve for $\dot{q}$ in terms of $q$ and $p = \partial L/\partial \dot{q}$, that is, when we look at the curve in $TM$ corresponding to $C$, *$\dot{q}$ is not necessarily $dq/dt$*! Furthermore, the *variations $\delta q$ and $\delta p$ are arbitrary*: We deal with variations that are not the lifts of variations of curves in $M$. Although we do again require that $\delta q = 0$ at the endpoints, *we make no such requirements on $\delta p$*. Not only this, in the phase space version *we may even vary the time parameter $t$, provided $\delta t = 0$ at the endpoints*. Hamilton's principle in $T^*M$ is simpler; for one thing, $pdq - Hdt$ is simply a 1-form in the space $T^*M \times \mathbb{R}$, and it is a simple matter to differentiate the integral of a form using the Lie derivative. This is the reason why the symplectic form $\omega^2$ is conserved under the canonical flow.

Let us reproduce the derivation of (4.48), but given now in classical notation. Instead of (10.12) one writes $\delta\, dq = d\, \delta q$, and so forth. Then

$$
\begin{aligned}
\delta \int pdq - Hdt &= \int \delta pdq + p\delta dq - \delta H dt - H\delta dt \\
&= \int \delta pdq + pd(\delta q) - \left(\frac{\partial H}{\partial q}\delta q + \frac{\partial H}{\partial p}\delta p + \frac{\partial H}{\partial t}\delta t\right)dt - Hd(\delta t) \\
&= \int \delta pdq + \{d(p\delta q) - dp\delta q\} \\
&\quad -\left(\frac{\partial H}{\partial q}\delta q + \frac{\partial H}{\partial p}\delta p + \frac{\partial H}{\partial t}\delta t\right)dt - \{d(H\delta t) - dH\delta t\} \\
&= \int \left[-dp - \left(\frac{\partial H}{\partial q}\right)dt\right]\delta q + \left[dq - \left(\frac{\partial H}{\partial p}\right)dt\right]\delta p \\
&\quad + \left[-\left(\frac{\partial H}{\partial t}\right)dt + dH\right]\delta t + \int d[p\delta q - H\delta t]
\end{aligned} \tag{10.16}
$$

Since $\delta q = 0 = \delta t$ at the endpoints, the last integral vanishes. Since $\delta q$, $\delta p$, and $\delta t$ are now otherwise arbitrary, we conclude that $\delta \int pdq - Hdt = 0$ is equivalent to Hamilton's equations.

### 10.2c. Jacobi's Principle of "Least" Action

The kinetic energy $T$, as a function on $TM$, yields a Riemannian metric on $M$

$$\left\langle \frac{dq}{dt}, \frac{dq}{dt} \right\rangle = \langle \dot{q}, \dot{q} \rangle = 2T$$

We have already defined $L = T - V$, and so, since $p$ is the covector associated to $\dot{q}$, $H = p\dot{q} - L = \langle \dot{q}, \dot{q} \rangle - (T - V) = T + V$ is the total energy. *Assume that $H = H(q, p)$ is independent of time*, $\partial H/\partial t = 0$. We know from Hamilton's equations that $H$ is a constant of the motion. Thus the trajectory $C$ of the dynamical system, that is, $q = q(t)$, $p = p(t)$ satisfying $dq/dt = \partial H/\partial p$ and $dp/dt = -\partial H/\partial q$ in $T^*M$, lies on the constant energy locus

$$V_E = \{(q, p, t) : H(q, p) = \text{constant } E\}$$

Furthermore, assume that $dH \neq 0$ on $V_E$ (by Sard's theorem this is generically so). Then this locus $V_E$ is a $2n$-dimensional submanifold of $T^*M \times \mathbb{R}$. *We shall assume that the given trajectory $C$ is such that $E - V$ is always positive along $C$*. Project the curve $C$ down into the configuration space $M$, obtaining the curve $C'$, which describes the spatial configurations traced out by the dynamical system. We shall now vary the curve $C$ in $T^*M \times \mathbb{R}$ as follows. In Figure 10.3 we illustrate the special case of the 1 dimensional harmonic oscillator with $H = p^2 + q^2$.

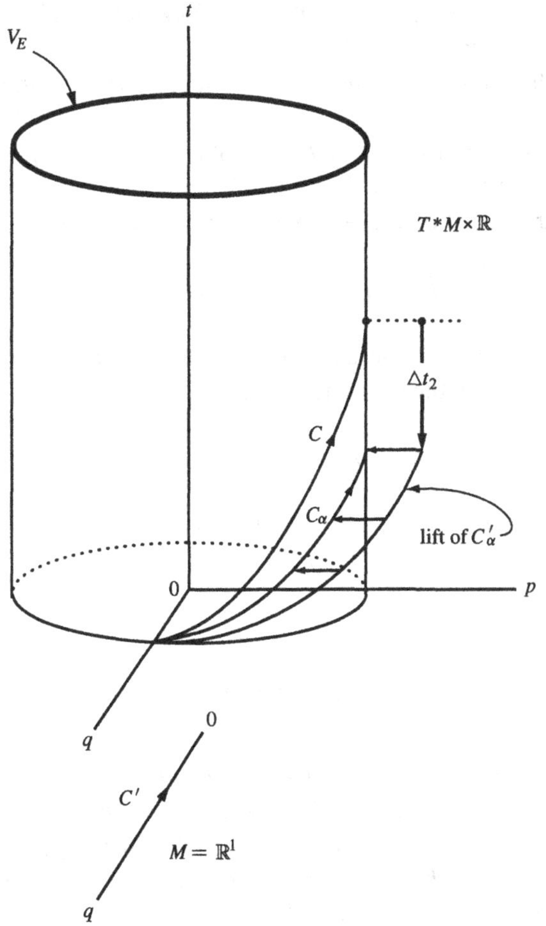


**Figure 10.3**

Let $C'_\alpha$ be a variation of the curve $C'$ always starting and ending at the same points as $C'$, that is, $\delta q = 0$ at the endpoints $q(t_1)$ and $q(t_2)$. *We are going to lift the varied curves $C'(\alpha)$ to yield a variation of $C$ that always lies on the hypersurface $H = E$*, by merely changing the speed at which we traverse $C'(\alpha)$ in $M$. We do this as follows. The curve $C'(\alpha)$ is some parameterized curve $q_\alpha = q(\tau)$. Consider the velocity $\dot{q} = dq/d\tau$ at the point $q(\tau)$. This determines a specific $p_\tau = \partial L/\partial \dot{q}$ in the momentum fiber over $q(\tau)$, that is, the vector space $\mathbb{R}^n$ of all covectors at $q(\tau)$, but this point in the fiber need not lie on $H = E$. The hypersurface $H = E$ intersects *this* fiber in the quadratic $(n-1)$-dimensional ellipsoid $T(p) = E - V(q(\tau))$ defined by the kinetic energy. We may assume that the constant $E - V(q(\tau))$ is positive, since this was true for the original curve $C$. Thus $p_\tau$ is a nonzero vector in the fiber $\mathbb{R}^n$ and so a *unique* positive multiple of it will end on the ellipsoid $T(p) = E - V(q(\tau))$. *This is the new momentum that we assign to the point $q(\tau)$ on $C'_\alpha$*; it is simply a positive multiple of the original $p_\tau$ on $C'_\alpha$. By doing this at each $q(\tau)$ on $C_\alpha$ we define a lift of $C'_\alpha$ that lies on $H = E$; that is, we have covered each $C'_\alpha$ by a curve $C_\alpha$ representing a motion with total energy $H = E$.

By construction, each $C_\alpha$ starts at the same $q$ and $t = t_1$ as does $C$ (with perhaps different $p$) and although all end at the same $q$ *they needn't all end at* $t = t_2$. The time $t = t_2(\alpha)$ is determined by the fact that the spatial locus $C'_\alpha$ is given together with the *speed* along this locus, since $H = E$.

Look now at Hamilton's principle in phase space and the variational calculation (10.16).

If all of the $C_\alpha$ ended at the same $t = t_2$, then Hamilton's principle would give $\delta \int_C p\,dq - H\,dt = 0$ since $C = C(0)$ is a Hamiltonian trajectory, but now we can expect the boundary term $\int d[p\delta q - H\delta t]$ to play a role. From (10.16)

$$\delta \int_C p\,dq - H\,dt = [p\delta q - H\delta t]_1^2$$

where 1 is the beginning point and 2 is the endpoint, all in $T^*M \times \mathbb{R}$. But $\delta q$ vanishes at both ends, and $\delta t = 0$ at the beginning, and so

$$\delta \int_C p\,dq - H\,dt = -E\delta t_2$$

rather than 0. On the other hand, since our varied curves all lie on $H = E$, we have directly

$$\delta \int_C p\,dq - H\,dt = \left(\delta \int_C p\,dq\right) - E\delta t_2$$

Comparing these expressions gives the following:

**Theorem (10.17):** *Consider all parameterized smooth curves $C'$ in configuration space $M$, $q = q(t)$, starting at $q_0$ and ending at $q_1$, each parameterized so that the total energy $H$ is a given constant $E$ along the path. Then $\int_{q_0}^{q_1} p\,dq$ is a functional of the path. A path $C'$ is the projection of a Hamiltonian trajectory in $T^*M \times \mathbb{R}$ (i.e., $C'$ is the trace in $M$ of a path of the dynamical system) iff*

$$\delta \int p\,dq = 0 \text{ at } C'$$

*for all variations having $H = E$ and $\delta q = 0$ at the given endpoints.*

This principle can be put in the following form. Along the curves $q = q(t)$ in $M$ parameterized by time, we have $pdq = p(dq/dt)dt = \| \dot{q} \|^2 dt = 2Tdt$, where $T$ is the kinetic energy. Thus, vaguely speaking, the trace of the dynamical system point in $q$-space is such that

$$\delta \int T dt = 0$$

among all curves with the same total energy $E$. (Note, however, that the $t$ interval of integration changes for curves in a variational family.) This is the **principle of least action of Maupertuis and Euler** (1744). Jacobi restated and proved the following version, using the language of geodesics.

If we have $H = E$ along the path, then $T = H - V(q) = E - V(q)$. Now

$$ds = \| \dot{q} \| dt = \sqrt{2}\sqrt{T}dt \tag{10.18}$$

is the element of arc length in $M$ given by the kinetic energy, and so $\sqrt{2}Tdt = \sqrt{2}\sqrt{T}\sqrt{T}dt = \sqrt{T}ds = [E - V(q)]^{1/2}ds$. We then have

**Theorem (10.19):** ***Jacobi's Principle of "Least" Action***

*The trace in M of a Hamiltonian trajectory of constant total energy E is a geodesic in M for the* ***Jacobi metric*** *given by* $d\rho := \sqrt{T}ds = [E - V(q)]^{1/2}ds$, *where ds is the standard metric given by the kinetic energy*

$$\delta \int d\rho = \delta \int [E - V(q)]^{1/2} ds = 0$$

Note that this metric is only defined on the part of $M$ where $E > V(q)$ (i.e., where the kinetic energy $T$ is $> 0$). If $V$ is bounded above on $M$, $V(q) < B$ for all points of $M$ (e.g., if $M$ is compact), then the metric makes sense for total energy $E > B$.

As we know, geodesics yield a vanishing first variation, but this need not be a *minimum* for the "action" $\int \| \dot{q} \|^2 dt$.

### 10.2d. Closed Geodesics and Periodic Motions

A geodesic $C$ on a manifold $M^n$ that starts at some point $p$ *might* return to that same point after traveling some arc length distance $L$. If it does, it will either cross itself transversally or come back tangent to itself at $p$. In the latter case the geodesic will simply retrace itself, returning to $p$ after traveling any distance that is an integer multiple of $L$. In such a case we shall call $C$ a **closed geodesic**. This is the familiar case of the infinity of great circles on the round 2-sphere.

If a 2-sphere is not perfectly round, but rather has many smooth bumps, it is not clear at all that there will be *any* closed geodesics, but, surprisingly, it can be proved that there are in fact at least three such closed geodesics! The proof is difficult.

Closed geodesics in mechanics are important for the following reason. The evolution of a dynamical system in time is described by a curve $q = q(t)$ being traced out in the configuration space $M$, and by Jacobi's principle, this curve is a geodesic in the

Jacobi metric $d\rho = [E - V(q)]^{1/2} ds$. Thus a closed geodesic in the configuration space corresponds to a **periodic motion** of the dynamical system. A familiar example is given by the case of a rigid body spinning freely about a principal axis of inertia.

Not all manifolds have closed geodesics.

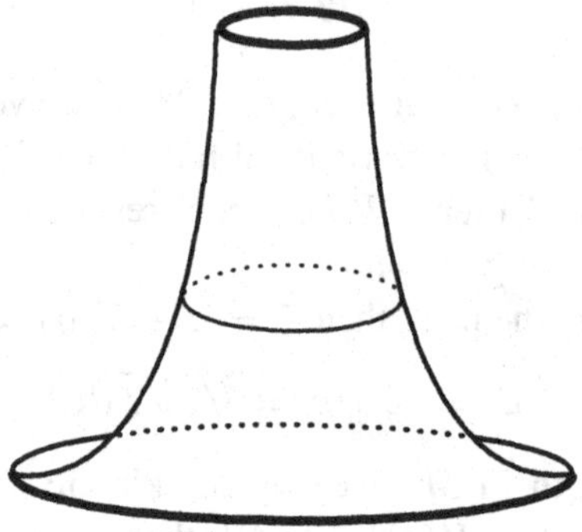

Figure 10.4

The infinite horn-shaped surface indicated has no closed geodesics. It is clear that the horizontal circles of latitude are not geodesics since the principal normal to such a curve is not normal to the surface. Furthermore, it is rather clear that *any* closed curve on this horn can be shortened by pushing it "north," and such a variation of the curve will have a negative first variation of arc length, showing that it could not be a geodesic. (One needs to be a little careful here; the equator on the round 2-sphere *is* a geodesic and it is shortened by pushing it north. The difference is that in this case the tangent planes at the equator are vertical and so the first variation of length is in fact 0; it is the *second* variation that is negative! We shall return to such matters in Chapter 12.)

One would hope that if a closed curve is not a geodesic, it could be shortened and deformed into one. A "small" circle of latitude on the northern hemisphere of the sphere, however, when shortened by pushing north, collapses down to the north pole. Somehow we need to start with a closed curve that cannot be "shrunk to a point," that is, perhaps we can succeed if we are on a manifold that is not simply connected (see Section 21.2a). But the circles of latitude on the horn-shaped surface in Figure 10.4 show that this is not enough; there is no "shortest" curve among those closed curves that circle the horn. We shall now "show" that if $M$ is a closed manifold (i.e., compact without boundary) that is not simply connected, then there is a closed geodesic. In fact a stronger result holds. We shall discuss many of these things more fully in Chapter 21.

We wish to say that two closed curves are "homotopic" if one can be smoothly moved through $M$ to the other. This can be said precisely as follows. Let $C_0$ and $C_1$ be two parameterized closed curves on $M^n$. Thus we have two maps $f_\alpha : S^1 \to M^n, \alpha = 0, 1$, of a circle into $M$. We say that these curves are **(freely) homotopic** provided these maps can be smoothly extended to a map $F : S^1 \times \mathbb{R} \to M$ of a cylinder $S^1 \times \mathbb{R}$ into $M$. Thus

$$F = F(\theta, t), \text{ with } F(\theta, 0) = f_0(\theta) \text{ and } F(\theta, 1) = f_1(\theta)$$

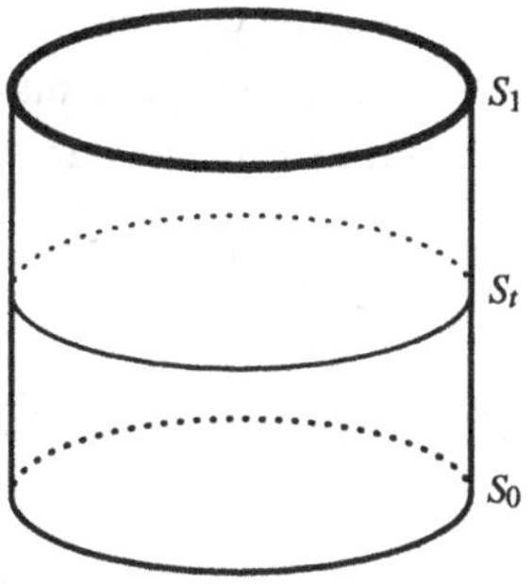


**Figure 10.5**

Thus $F$ interpolates between $f_0$ and $f_1$ by mapping the circle $S_t$ into $M$ by the map $f_t(\theta) = F(\theta, t)$.

Clearly the circles of latitude on the horn are homotopic.

Homotopy is an equivalence relation; if $C$ is homotopic to $C'$ (written $C \sim C'$) and $C' \sim C''$, then $C \sim C''$, and so on. Thus the collection of closed curves on $M$ is broken up into disjoint **homotopy classes** of curves. All curves $C$ that can be shrunk to a point (i.e., that are homotopic to the constant map that maps $S^1$ into a single point) form a homotopy class, the **trivial** class. If all closed curves are trivial the space $M$ is said to be **simply connected.**

On the 2-torus, with angular coordinates $\phi_1$ and $\phi_2$, the following can be shown. The

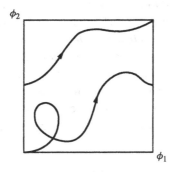

**Figure 10.6**

two basic curves $\phi_2 = 0$ and $\phi_1 = 0$ are nontrivial and are not homotopic. The closed curve indicated "wraps twice around in the $\phi_1$ sense and once in the $\phi_2$ sense"; we write that it is a curve of type (2, 1). Likewise we can consider curves of type $(p, q)$. All curves of type $(p, q)$ form a free homotopy class and this class is distinct from $(p', q')$ if $(p, q) \neq (p', q')$.

**Theorem (10.20):** *In each nontrivial free homotopy class of closed curves on a closed manifold $M^n$ there is at least one closed geodesic.*

The proof of this result is too long to be given here but the result itself should not be surprising; we should be able to select the shortest curve in any nontrivial free homotopy class; the *compactness* of $M$ is used here. If it were not a geodesic we could shorten it further. If this geodesic had a "corner," that is, if the tangents did not match up at the starting (and ending) point, we could deform it to a shorter curve by "rounding off the corner."

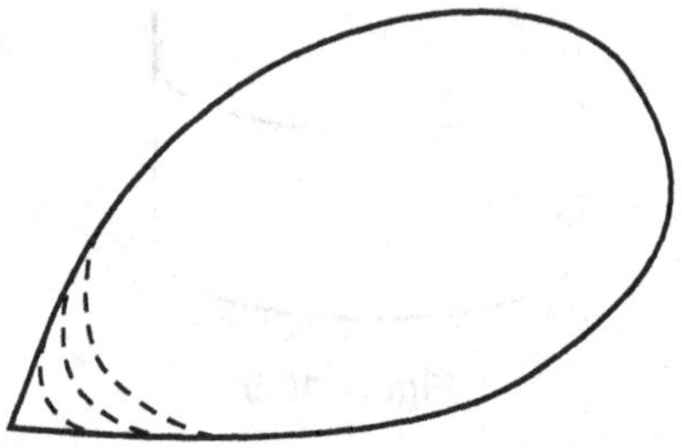

**Figure 10.7**

Finally we give a nontrivial application to dynamical systems ([A, p. 248].

Consider a planar double pendulum, as in Section 1.2b, but in an arbitrary potential field $V = V(\phi_1, \phi_2)$. The configuration space is a torus $T^2$. Let $B$ be the maximum of $V$ in the configuration space $T^2$. Then if the total energy $H = E$ is greater than $B$, the system will trace out a geodesic in the Jacobi metric for the torus. For any pair of integers $(p, q)$ there will be a closed geodesic of type $(p, q)$. Thus, given $p$ and $q$, if $E > B$ there is always a *periodic* motion of the double pendulum such that the upper pendulum makes $p$ revolutions while the lower makes $q$.

An application to rigid body motion will be given in Chapter 12.

Finally, we must remark that there is a far more general result than (10.20). Lyusternik and Fet have shown that there is a *closed geodesic on every closed manifold!* Thus *there is a periodic motion in every dynamical system having a closed configuration space, at least if the energy is high enough.* The proof, however, is far more difficult, and not nearly as transparent as (10.20). The proof involves the "higher homotopy groups"; we shall *briefly* discuss these groups in Chapter 22. For an excellent discussion of the closed geodesic problem, I recommend Bott's treatment in [Bo].

## 10.3. Geodesics, Spiders, and the Universe

Is our space flat?

### 10.3a. Gaussian Coordinates

Let $\gamma = \gamma(t)$ be a geodesic parameterized proportional to arc length; then $\| d\mathbf{x}/dt \|$ is a constant and $\nabla\dot{\mathbf{x}}/dt = 0$ along $\gamma$. There is a standard (but unusual) notation for this geodesic. Let $\mathbf{v}$ be the tangent vector to $\gamma$ at $p = \gamma(0)$; we then write

$$\gamma(t) = \exp_p(t\mathbf{v})$$

Then we have (10.21)

$$\frac{d}{dt}[\exp_p(t\mathbf{v})] = \frac{d\gamma}{dt}$$

is the tangent vector to $\gamma$ at the parameter value $t$.

*The point* $\exp_p(\mathbf{v})$ *is the point on the geodesic that starts at p, has tangent* $\mathbf{v}$ *at p, and is at arc length* $\|\mathbf{v}\|$ *from p.*

Of course if $t < 0$, we move in the direction of $-\mathbf{v}$. When $\mathbf{v}$ is a unit vector, $t$ is arc length along $\gamma$.

Since geodesics need not be defined for all $t$, $\exp_p(t\mathbf{v})$ may only make sense if $|t|$ is sufficiently small.

Given a point $p$ and a hypersurface $V^{n-1} \subset M^n$ passing through $p$, we may set up local coordinates for $M$ near $p$ as follows. Let $y^2, \ldots, y^n$ be local coordinates on $V$ with origin at $p$. Let $\mathbf{N}(y)$ be a field of unit normals to $V$ along $V$ near $p$. If from each $y \in V$ we construct the geodesic through $y$ with tangent $\mathbf{N}(y)$, and if we travel along this geodesic for distance $|r|$, we shall get, if $\epsilon$ is small enough, a map

$$(-\epsilon, \epsilon) \times V^{n-1} \to M^n$$

by

$$(r, y) \mapsto \exp_y(r\mathbf{N}(y))$$

and it can be shown ([M]) that this map is a diffeomorphism onto an open subset of $M^n$

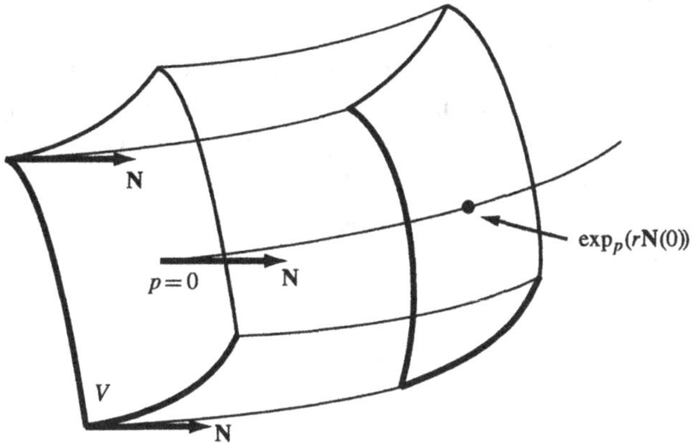


**Figure 10.8**

if $V^{n-1}$ and $\epsilon$ are small enough. This says, in particular, that any point $q$ of $M$ that is sufficiently close to $p$ will be on a *unique* geodesic of length $r < \epsilon$ that starts at some $y \in V$ and leaves orthogonally to $V$. If then $q = \exp_y(r\mathbf{N}(y))$, we shall assign to $q$ the **Gaussian coordinates** $(r, y^2, \ldots, y^n)$.

(As mentioned before, we recommend Milnor's book [M] for many of the topics in Riemannian geometry. We should mention, however, that Milnor uses an unusual notation. For example, Milnor writes

$$\mathbf{A} \vdash \mathbf{B}$$

instead of the usual covariant derivative $\nabla_{\mathbf{A}} \mathbf{B}$. Also *Milnor's curvature transformation* $R(\mathbf{X}, \mathbf{Y})$ *is the negative of ours.)*

We can then look at the hypersurface $V_r^{n-1}$ of all points $\exp_y(r\mathbf{N}(y))$ as $y$ runs through $V$ but with $r$ a small constant; this is the **parallel hypersurface** to $V$ at distance $r$.

***Gauss's Lemma* (10.22):** *The parallel hypersurface $V_r^{n-1}$ to $V^{n-1}$ is itself orthogonal to the geodesics leaving $V$ orthogonal to $V$.*

Put another way, this says:

**Corollary (10.23):** *The distribution $\Delta^{n-1}$ of hyperplanes that are orthogonal to the geodesics leaving $V^{n-1}$ orthogonally is completely integrable, at least near $V$.*

This is a local result; $\Delta^{n-1}$ isn't defined at points where distinct geodesics from $V^{n-1}$ meet (look at the geodesics leaving the equator $V^1 \subset S^2$).

PROOF OF GAUSS'S LEMMA: Let $\gamma_y$ be the geodesic leaving $V^{n-1}$ at the point $y$. It is orthogonal to $V$ at $y$ and we must show that it is also orthogonal to $V_r$ at the point $(r, y)$. Consider the 1-parameter variation of $\gamma$ given by the geodesics $s \mapsto \gamma_{y,\alpha}(s) := \exp_y(s\mathbf{N}(y^2 + \alpha, y^3, \ldots, y^n))$, for $0 \leq s \leq r$, emanating from the $y^2$ curve through $y$. The variation vector $\mathbf{J}$, in our Gaussian coordinate system, is simply $\partial/\partial y^2$. It is a Jacobi field along $\gamma$. By construction, all of these geodesics have length $r$. Thus the first variation of arc length is 0 for this variation. But Gauss's formula (10.4) gives $0 = L'(0) = \langle \mathbf{J}, \mathbf{T}\rangle(\gamma(r)) - \langle \mathbf{J}, \mathbf{T}\rangle(\gamma(0)) = \langle \mathbf{J}, \mathbf{T}\rangle(\gamma(r))$. Thus $\gamma$ is orthogonal to the coordinate vector $\partial/\partial y^2$ tangent to $V_r$ at $(r, y)$. The same procedure works for all $\partial/\partial y^i$. □

**Corollary (10.24):** *In Gaussian coordinates $r, y^2, \ldots, y^n$ for $M^n$ we have*

$$ds^2 = dr^2 + \sum_{\alpha,\beta=2}^{n} g_{\alpha\beta}(r, y)dy^\alpha dy^\beta$$

*since $\langle \partial/\partial r, \partial/\partial r\rangle = 1$ and $\langle \partial/\partial r, \partial/\partial y^\alpha\rangle = 0$.*

In particular, when $V^1$ is a curve on a surface $M^2$, the metric assumes the form

$$ds^2 = dr^2 + G^2(r, y)dy^2$$

promised in (9.58).

**Corollary (10.25):** *Geodesics locally minimize arc length for fixed endpoints that are sufficiently close.*

This follows since any sufficiently small geodesic arc can be embedded in a Gaussian coordinate system as an $r$ curve, where all $y$'s are constant. Then for any other curve lying in the Gaussian coordinate patch, joining the same endpoints, and parameterized by $r$

$$ds^2 = dr^2 + \sum_{\alpha,\beta=2}^{n} g_{\alpha\beta}(r, y)\frac{dy^\alpha}{dr}\frac{dy^\beta}{dr} \geq dr^2$$

since $(g_{\alpha\beta})$ is positive definite. The restriction that the curve be parametrized by $r$ can be removed; see [M].

### 10.3b. Normal Coordinates on a Surface

Let $p$ be a point on a Riemannian surface $M^2$. Let $\mathbf{e}, \mathbf{f}$ be an orthonormal frame at $p$. We claim that the map $(x, y) \in \mathbb{R}^2 \mapsto \Phi(x, y) = \exp_p(x\mathbf{e} + y\mathbf{f}) \in M$ is a diffeomorphism of some neighborhood of 0 in $\mathbb{R}^2$ onto a neighborhood of $p$ in $M^2$.

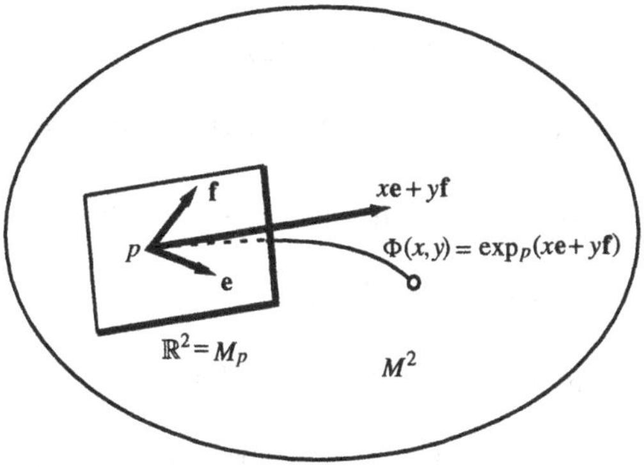


Figure 10.9

To see this we look at the differential $\Phi_*$ at 0. From (10.21)

$$\left.\frac{\partial}{\partial x}\right|_{(x,y)=0} \Phi(x, y) = \left.\frac{\partial}{\partial x}\right|_{x=0} \exp_0(x\mathbf{e}) = \mathbf{e}$$

Thus $\Phi_*(\partial/\partial x) = \mathbf{e}$ and likewise $\Phi_*(\partial/\partial y) = \mathbf{f}$, showing that $\Phi$ is a local diffeomorphism and thus that $x$ and $y$ can be used as local coordinates near $p$. These are (Riemannian) **normal coordinates,** with origin $p$. We can now introduce the analogue of polar coordinates near $p$ by putting $r^2 = x^2 + y^2$ and $x = r\cos\theta$, $y = r\sin\theta$. Thus if we keep $\theta$ constant and let $r \geq 0$ vary, we simply move along the geodesic $\exp_p[r(\cos\theta\mathbf{e} + \sin\theta\mathbf{f})]$, whereas if we keep $r$ constant, $\exp_p[r(\cos\theta\mathbf{e} + \sin\theta\mathbf{f})]$ traces out a closed curve of points whose distance along the radial geodesics is the constant $r$. We shall call this latter curve a **geodesic circle** of radius $r$, even though it itself is not a geodesic. We shall call $(r, \theta)$ **geodesic polar coordinates**. These are not good coordinates at the pole $r = 0$.

We can express the metric in terms of $(x, y)$ or $(r, \theta)$. In $(x, y)$ coordinates we have the form $ds^2 = g_{11}dx^2 + 2g_{12}dxdy + g_{22}dy^2$, whereas in $(r, \theta)$ we may write the metric in the form $ds^2 = g_{rr}dr^2 + 2g_{r\theta}drd\theta + G^2(r, \theta)d\theta^2$, for some function $G$. Now by keeping $\theta$ constant we move along a radial geodesic with arc length given by $r$, and thus $g_{rr} = 1$. By exactly the same reasoning as in Gauss's lemma this radial geodesic is orthogonal to the $\theta$ curves $r$ = constant; therefore $g_{r\theta} = 0$ and $ds^2 = dr^2 + G^2(r,\theta)d\theta^2$. By direct change of variables $x = r\cos\theta$ and $y = r\sin\theta$ in $ds^2 = g_{11}dx^2 + 2g_{12}dxdy + g_{22}dy^2$ we readily see that

$$G^2 = r^2[g_{11}\sin^2\theta - g_{12}\sin 2\theta + g_{22}\cos^2\theta]$$

where $g_{11} = 1 = g_{22}$ and $g_{12} = 0$ *at the origin*, since (**e**, **f**) is an orthonormal frame. Note then that $G^2(r,\theta)/r^2 \to 1$, uniformly in $\theta$, as $r \to 0$; in particular $G \to 0$ as $r \to 0$. Thus

$$\left.\frac{\partial G}{\partial r}\right]_0 = \lim \frac{G}{r} = 1$$

Also, $\partial^2 G/\partial r^2 = -KG$ follows from (9.60). We then have the Taylor expansion along a radial geodesic

$$G(r,\theta) = r - K(0)\frac{r^3}{3!} + \cdots \tag{10.26}$$

Thus the circumference $L(C_r)$ of the geodesic circle of radius $r$ is

$$L(C_r) = \int_0^{2\pi} \sqrt{g_{\theta\theta}}d\theta = 2\pi r - 2\pi K(0)\frac{r^3}{6} + \cdots$$

Likewise the area of the geodesic "disc" of radius $r$ is

$$A(B_r) = \iint \sqrt{g}drd\theta = \iint G(r,\theta)drd\theta = \pi r^2 - \frac{\pi}{12}K(0)r^4 + \cdots$$

These two expressions lead to the formulae, respectively, of Bertrand–Puiseux and of Diguet of 1848

$$\begin{aligned} K(0) &= \lim_{r\to 0}\left(\frac{3}{\pi r^3}\right)[2\pi r - L(C_r)] \\ &= \lim_{r\to 0}\left(\frac{12}{\pi r^4}\right)[\pi r^2 - A(B_r)] \end{aligned} \tag{10.27}$$

telling us that the Gauss curvature $K(p)$ is related to the deviation of the length and area of geodesic circles and discs from the expected euclidean values. See Problem 10.3(1).

There are analogous formulae in higher dimensions involving the curvature tensor.

### 10.3c. Spiders and the Universe

The expressions (10.27) give a striking confirmation of Gauss's *theorema egregium* since they exhibit $K$ as a quantity that can be computed in terms of measurements made intrinsically on the surface. There is no mention of a second fundamental form or of a bending of the surface in some enveloping space. A spider living on $M^2$ could mark off geodesic segments of length $r$ by laying down a given quantity of thread and experimenting to make sure that each of its segments is the shortest curve joining $p$ to its endpoint.

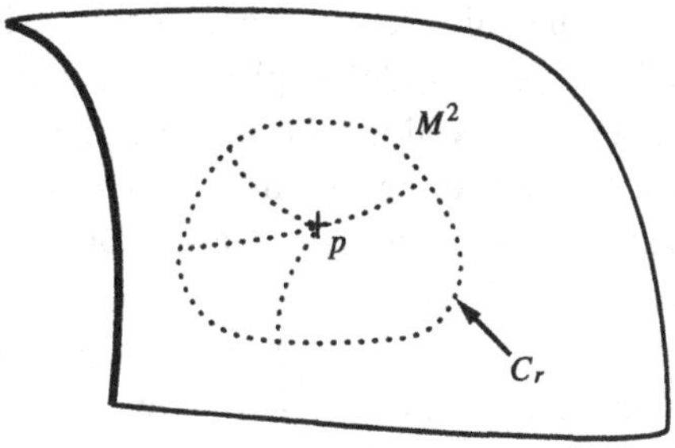


Figure 10.10

Then it could lay down a thread along the endpoints, forming a geodesic circle $C_r$ of radius $r$, and measure its length by the amount of thread used. Having already encountered the formula of Bertrand–Puiseux in its university studies, the spider could compute an approximation of $K$ at $p$, and all this without any awareness of an enveloping space!

What about us? We live in a 3-dimensional space, or a 4-dimensional space–time. To measure small spatial distances we can use light rays, reflected by mirrors, noting the time required on our atomic clocks (see Section 7.1b). A similar construction yields $ds^2$ for timelike intervals (see [Fr, p.10]). *Our world seems to be equipped with a "natural" metric.*

In ordinary affairs the metric seems flat; that is why euclidean geometry and the Pythagoras rule seemed so natural to the Greeks, but we mustn't forget that the sheet of paper on which we draw our figures occupies but a minute portion of the universe. (The Earth was thought flat at one time!) Is the curvature tensor of our space really zero? Can we compute it by some simple experiment as the spider can on an $M^2$ ? Gauss was the first to try to determine the curvature of our 3-space, using the following result of Gauss–Bonnet. Consider a triangle on an $M^2$ whose sides $C_1, C_2, C_3$, are geodesic arcs. Parallel

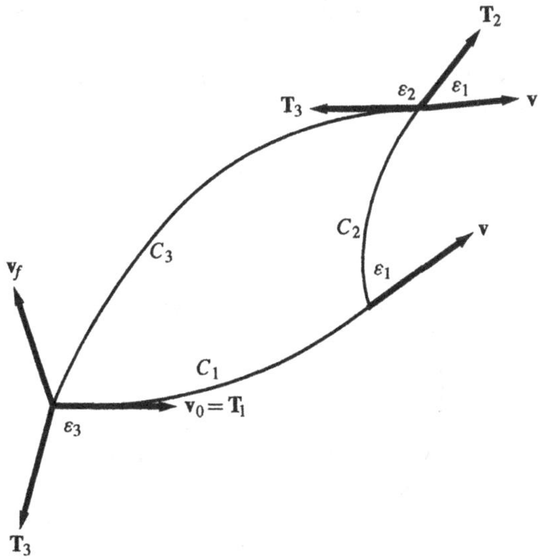


Figure 10.11

translate around this triangle the unit vector $\mathbf{v}$ that coincides with the unit tangent to $C_1$ at the first vertex. Since $\mathbf{T}_1$ is also parallel displaced, we have $\mathbf{v} = \mathbf{T}_1$ along all of $C_1$. Continue the parallel translation of $\mathbf{v}$ along the second arc; since this arc is a geodesic, we have that $\mathbf{v}$ will make a constant angle with this arc. This angle is $\epsilon_1$, the first exterior angle. Thus at the next vertex the angle from $\mathbf{v}$ to the new tangent $\mathbf{T}_3$ will be $\epsilon_1+\epsilon_2$. When we return to the first vertex we will have $\angle(\mathbf{v}_f, \mathbf{T}_1) = \epsilon_1+\epsilon_2+\epsilon_3$. Thus $2\pi - \angle(\mathbf{v}_0, \mathbf{v}_f) = \epsilon_1+\epsilon_2+\epsilon_3$ and so $\angle(\mathbf{v}_0, \mathbf{v}_f) = 2\pi-(\epsilon_1+\epsilon_2+\epsilon_3) =$ (the sum of the interior angles )$-\pi$. But from (9.61) we have that $\angle(\mathbf{v}_0, \mathbf{v}_f) = \iint K dS$ over the triangle. We conclude that

$$\iint K dS = \text{(the sum of the interior angles of the triangle with geodesic sides )} - \pi \tag{10.28}$$

This formula generalizes Lambert's formula of spherical geometry in the case when $M^2$ is a 2-sphere of radius $a$ and constant curvature $K = 1/a^2$. Of course the interior angle sum in a flat plane is exactly $\pi$ and (10.28) again exhibits curvature as indicating a breakdown of euclidean geometry.

Gauss considered a triangle whose vertices were three nearby peaks in Germany, the sides of the triangle being made up of the light ray paths used in the sightings. Presumably the sides, made up of light rays, would be geodesics in our 3-space. An interior angle sum differing from $\pi$ would have been an indication of a noneuclidean geometry, but no such difference was found that could not be attributed to experimental error.

Einstein was the first to describe the affine connection of the universe as a physical field, a *gauge field*, as it is called today. He related the curvature of space–time to a physical tensor involving matter, energy, and stresses and concluded that space–time is indeed curved. We turn to these matters in the next chapter.

## Problem

**10.3(1)** Use the first expression in (10.27) to compute the Gauss curvature of the round 2-sphere of radius $a$, at the north pole.

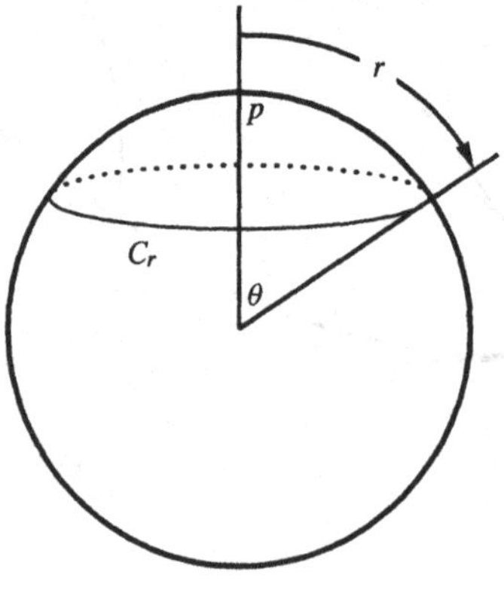


**Figure 10.12**

CHAPTER 11

# Relativity, Tensors, and Curvature

## 11.1. Heuristics of Einstein's Theory

What does $g_{00}$ have to do with gravitation?

### 11.1a. The Metric Potentials

Einstein's general theory of relativity is primarily a replacement for Newtonian gravitation and a generalization of special relativity. It cannot be "derived"; we can only speculate, with Einstein, by heuristic reasoning, how such a generalization might proceed. His path was very thorny, and we shall not hesitate to replace some of his reasoning, with hindsight, by more geometrical methods.

Einstein assumed that the actual space–time universe is some pseudo-Riemannian manifold $M^4$ and is thus a generalization of Minkowski space. In any local coordinates $x^0 = t, x^1, x^2, x^3$ the metric is of the form

$$ds^2 = g_{00}(t, \mathbf{x})dt^2 + 2g_{0\beta}(t, \mathbf{x})dt dx^\beta$$
$$+ g_{\alpha\beta}(t, \mathbf{x})dx^\alpha dx^\beta$$

where Greek indices run from 1 to 3, and $g_{00}$ must be negative. We may assume that we have chosen units in which the speed of light is unity *when time is measured by the local atomic clocks* (rather than the coordinate time $t$ of the local coordinate system). Thus an "orthonormal" frame has $\langle \mathbf{e}_0, \mathbf{e}_0 \rangle = -1$, $\langle \mathbf{e}_0, \mathbf{e}_\beta \rangle = 0$, and $\langle \mathbf{e}_\alpha, \mathbf{e}_\beta \rangle = \delta_{\alpha\beta}$.

**Warning**: Many other books use the negative of this metric instead.

To get started, Einstein considered the following situation. We imagine that we have massive objects, such as stars, that are responsible in some way for the preceding metric, and we also have a very small **test body**, a planet, that is so small that it doesn't appreciably affect the metric. We shall assume that the universe is **stationary** in the sense that it is possible to choose the local coordinates so that the metric coefficients do not depend on the coordinate time $t$, $g_{ij} = g_{ij}(\mathbf{x})$. In fact we shall assume more. A uniformly rotating sun might produce such a stationary metric; we shall assume that the metric has the further property that the mixed temporal–spatial terms vanish, $g_{0\beta} = 0$.

Such a metric

$$ds^2 = g_{00}(\mathbf{x})dt^2 + g_{\alpha\beta}(\mathbf{x})dx^\alpha dx^\beta \tag{11.1}$$

is called a **static** metric.

Along the world line of the test particle, the planet, we may introduce its **proper time** parameter $\tau$ by

$$d\tau^2 := -ds^2$$

As in Section 7.1b, it is assumed that proper time is the time kept by an atomic clock moving with the particle. Then

$$\left(\frac{d\tau}{dt}\right)^2 = -g_{00} - g_{\alpha\beta}\frac{dx^\alpha}{dt}\frac{dx^\beta}{dt}$$

We shall assume that the particle is moving very slowly compared to light; thus we put the spatial velocity vector equal to zero, $\mathbf{v} = d\mathbf{x}/dt \sim \mathbf{0}$, and consequently its unit velocity 4-vector is

$$u := \frac{dx}{d\tau} = \left(\frac{dt}{d\tau}\right)\left[1, \frac{d\mathbf{x}}{dt}\right]^T \sim \left(\frac{dt}{d\tau}\right)[1, \mathbf{0}]^T$$

or

$$u \sim (-g_{00})^{-1/2}[1, \mathbf{0}]^T$$

where, as is common, we allow ourselves to identify a vector with its components.

We shall also assume that the particle is moving in a very weak gravitational field so that $M^4$ is almost Minkowski space in the sense that

$$g_{00} \sim -1$$

We shall *not*, however, assume that the spatial derivatives of $g_{00}$ are necessarily small. Thus we are allowing for spatial inhomogeneities in the gravitational field.

The fact that all (test) bodies fall with the same acceleration near a massive body (**Galileo's law**) led Einstein to the conclusion that *gravitational force,* like centrifugal and Coriolis forces, is a *fictitious* force. A test body in free fall does not feel any force of gravity. It is only when the body is prevented from falling freely that the body feels a force. For example, a person standing on the Earth's solid surface does not feel the force of gravity, but rather the molecular forces exerted by the Earth as the Earth prevents the person from following its natural free fall toward the center of the planet.

Einstein *assumed then that a test body that is subject to no external forces (except the fictitious force of gravity) should have a world line that is a geodesic* in the space–time manifold $M^4$. Then, since $d\tau \sim dt$, the geodesic equation yields

$$\frac{d^2x^i}{dt^2} \sim \frac{d^2x^i}{d\tau^2} \sim -\Gamma^i_{jk}\frac{dx^j}{dt}\frac{dx^k}{dt} \sim -\Gamma^i_{00}$$

In particular, for $\alpha = 1, 2, 3$, we have

$$\begin{aligned}\frac{d^2x^\alpha}{dt^2} \sim -\Gamma^\alpha_{00} &= -\frac{1}{2}g^{\alpha j}\left(\frac{\partial g_{0j}}{\partial x^0} + \frac{\partial g_{0j}}{\partial x^0} - \frac{\partial g_{00}}{\partial x^j}\right)\\ &= \frac{1}{2}g^{\alpha j}\frac{\partial g_{00}}{\partial x^j}\\ &= \frac{1}{2}g^{\alpha\beta}\frac{\partial g_{00}}{\partial x^\beta}\end{aligned}$$

Thus

$$\frac{d^2x^\alpha}{dt^2} \sim \left[\text{grad}\left(\frac{g_{00}}{2}\right)\right]^\alpha$$

If now we let $\phi$ be the classical Newtonian gravitational potential, then we must compare the preceding with $d^2x^\alpha/dt^2 = [\text{grad}\,\phi]^\alpha$. (Note that physicists would write this in terms of $V = -\phi$.) This yields $g_{00}/2 \sim \phi$+constant. We have assumed $g_{00} \sim -1$; if we now assume that the gravitational potential $\phi \to 0$ "at infinity," we would conclude that

$$g_{00} \sim (2\phi - 1) \tag{11.2}$$

Thus Einstein concluded that $g_{00}$ *is closely related to the Newtonian gravitational potential*! But then what can we say of the other metric coefficients? Surely they must play a role although we have not yet exhibited this role. We then have the following comparisons:

1. *Newtonian* gravitation is governed by a single potential $\phi$. Newtonian gravitation is a **scalar** theory.
2. *Electromagnetism* is governed by a 4-vector potential $A$; see (7.25). Electromagnetism is a **vector** theory.
3. *Einstein's gravitation* is governed by the 10 "metric potentials" $(g_{ij})$. Gravitation is then a symmetric covariant second-rank **tensor** theory.

In (1), the potential $\phi$ satisfies a "field equation," namely Poisson's equation

$$\nabla^2\phi = -4\pi\kappa\rho \tag{11.3}$$

where $\rho$ is the density of matter and $\kappa$ is the gravitational constant.

In (2), $A$ can be chosen to satisfy a field equation of the form of a wave equation. If $\Box$ is the d'Alembertian, the Laplace operator in Minkowski space, we have

$$\Box A = 4\pi J$$

where $J$ is the current 1-form, the covariant version of the current 4-vector in (7.27). These matters will be discussed in more detail later.

What are the field equations satisfied by the $(g_{ij})$?

### 11.1b. Einstein's Field Equations

Consider now, instead of a single test particle, a "dust cloud" of particles having a density $\rho$. By dust we mean an idealized fluid in which the pressure vanishes identically. Lack of a pressure gradient ensures us that the individual molecules are falling freely under the influence of gravity. Each particle thus traces out a geodesic world line in $M^4$. We shall again restrict ourselves to static metrics (11.1).

First consider the Newtonian picture of this cloud in $\mathbb{R}^3$. Follow the "base" path $C_0$ of a particular particle and let $\delta\mathbf{x}_t$ be the variation vector, which classically joins the base particle at time $t$ to a neighboring particle at time $t$.

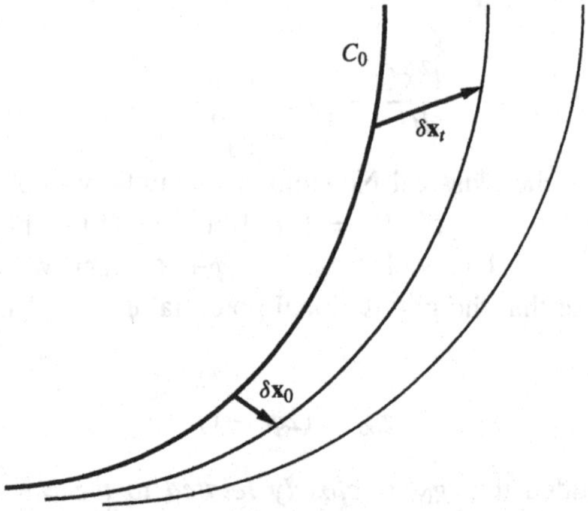


**Figure 11.1**

From 4.1b we know $(d/dt) \circ \delta = \delta \circ (d/dt)$, and so $(d^2/dt^2) \circ \delta = \delta \circ (d^2/dt^2)$. Thus from Newton's law (in cartesian coordinates)

$$\frac{d^2}{dt^2}(\delta x^\alpha) = \delta\left(\frac{d^2 x^\alpha}{dt^2}\right) = \delta\left(\frac{\partial \phi}{\partial x^\alpha}\right)$$

$$= \frac{\partial^2 \phi}{\partial x^\alpha \partial x^\beta}\delta x^\beta$$

$$\frac{d^2}{dt^2}(\delta x^\alpha) = \frac{\partial^2 \phi}{\partial x^\alpha \partial x^\beta}\delta x^\beta \qquad (11.4)$$

This is the equation of variation, a linear second-order equation for $\delta\mathbf{x}$ along $C_0$.

Now look at the same physical situation, but viewed in the 4-dimensional space–time $M^4$.

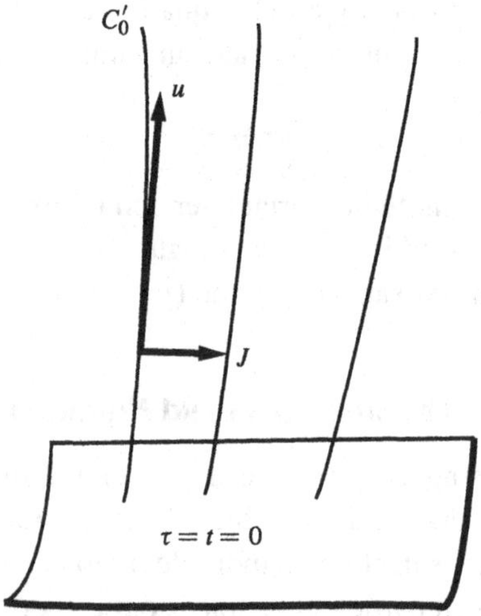


**Figure 11.2**

The particles now trace out world lines $C'$ in $M^4$ with unit 4-velocity $u$. The variation 4-vector $J_\tau$ "joins" the base particle at *proper* time $\tau$ to a nearby particle at the same proper time (we shall assume that all nearby particles have synchronized their atomic clocks at an initial $\tau = 0$ when $t = 0$). Since all of the world lines are geodesics, parameterized by "arc length," that is, proper time $\tau$, the variation vector $J$ is a *Jacobi field* and satisfies $\nabla^2 J/d\tau^2 = -R(J,T)T$. In a weak field and with small spatial velocities, we expect again that $\tau$ is approximately the coordinate time $t$, $\tau \sim t$. Then the Jacobi field $J$ will essentially have no time component, $J^0 \sim 0$, since it "connects" events at a common time $t$. By again looking at the Christoffel symbols with our smallness and static assumptions, we have (see Problem 11.1(1))

$$\frac{\nabla J^\alpha}{d\tau} \sim \frac{dJ^\alpha}{d\tau} \sim \frac{dJ^\alpha}{dt} \tag{11.5}$$

and Jacobi's equation becomes

$$\frac{d^2 J^\alpha}{dt^2} \sim -R^\alpha_{0\beta 0} J^\beta$$

If we now put $J^\beta = \delta x^\beta$ and compare this with (11.4) we get

$$-R^\alpha_{0\beta 0} \sim \frac{\partial^2 \phi}{\partial x^\beta \partial x^\alpha}$$

Consequently, since $R^j_{k00} = 0$,

$$\nabla^2 \phi = \sum_{1 \le \alpha \le 3} \frac{\partial^2 \phi}{\partial x^\alpha \partial x^\alpha} \sim -\sum_{1 \le \alpha \le 3} R^\alpha_{0\alpha 0} \sim -\sum_{0 \le i \le 3} R^i_{0i0}$$

In any $M^n$ with an affine connection we define the **Ricci tensor**, by contracting the full Riemann tensor

$$R_{jk} := R^i_{\ jik} \tag{11.6}$$

We shall show in Section 11.2 that $R_{jk} = R_{kj}$ *in the case of a (pseudo)-Riemannian manifold.* We then have

$$\nabla^2 \phi \sim -R_{00}$$

Poisson's equation yields

$$\nabla^2 \phi = -4\pi\kappa\rho \sim -R_{00} \tag{11.7}$$

for a slowly moving dust in a weak field. We see from this simple case that

*space–time $M^4$ must be curved in the presence of matter*!

As it stands, (11.7) "equates" the 00 component of a tensor, the Ricci tensor, with what is classically considered a scalar, a multiple of the density $\rho$. But in special relativity the density is not a scalar. Under a Lorentz transformation, mass $m_0$ gets transformed by the Lorentz factor, $m = m_0\gamma$ (see (7.9)). Also, 3-volumes transform as $\text{vol}^3 = \text{vol}_0^3/\gamma$ "since length in the direction of motion is contracted." Thus density transforms as $\rho = \rho_0\gamma^2$. This suggests that density is also merely one component of a second-rank tensor. Indeed just such a tensor, the *stress–energy–momentum tensor,* was introduced into special relativity. In classical physics there is the notion of the 3-dimensional symmetric "stress tensor" with components $S^{\alpha\beta}$ (see [Fr, chap. 6] for more details of the following). Consider the case of a perfect fluid; here $S^{\alpha\beta} = -p\delta^{\alpha\beta}$ where $p$ is the pressure. Let $\rho$ be the rest mass–energy density of the fluid and let $u$ be the velocity 4-vector of the fluid particles. Note that

$$\Pi^i{}_j := g^i{}_j + u^i u_j$$

*projects each 4-vector orthogonally into the 3-space orthogonal to u.* Then the **stress–energy–momentum tensor** for the fluid is defined by

$$T^{ij} := \rho u^i u^j + p(g^{ij} + u^i u^j) = (\rho + p)u^i u^j + pg^{ij} \tag{11.8}$$

In the case of a dust $p = 0$ and in the case of slowly moving particles the only nonvanishing component of $u$ is essentially $u^0 \sim 1$. Thus $T$ has essentially only one nonvanishing component $T^{00} \sim \rho$. Finally $T_{ij} = g_{ir}g_{js}T^{rs}$ also has one component $T_{00} = \rho$, since $g_{00} \sim -1$.

Equation (11.7) then can be stated as $R_{00} = 4\pi\kappa T_{00}$. Clearly this suggests a tensor equation, for all $i$, $j$

$$R_{ij} = 4\pi\kappa T_{ij}$$

These were the equations first proposed by Einstein in early November of 1915, for all types of matter undergoing any motion, although his path to these equations was far more tortuous than that indicated here. Furthermore, these equations are incorrect! In special relativity the tensor $T$ is known to have "divergence" 0, whereas the Ricci tensor does not usually have this property. These equations need to be amended in the same spirit as when Ampere's law was amended by the addition of Maxwell's displacement current in order to ensure conservation of charge. We shall discuss these matters in Section 11.2. Einstein arrived at the "correct" version at the end of that same November with **Einstein's equations**

$$R_{ij} - \frac{1}{2}g_{ij}R = 8\pi\kappa T_{ij} \tag{11.9}$$

In this equation we have introduced a second contraction of the Riemann tensor, the **(Ricci) scalar curvature**

$$R := g^{ij}R_{ij} = R^j{}_j \tag{11.10}$$

In order to handle the Einstein equations effectively we shall have to learn more about "tensor analysis," which was developed principally by Christoffel (covariant differentiation, the curvature tensor) and by Ricci. We turn to these matters in our next section.

### 11.1c. Remarks on Static Metrics

Some final comments.

1. Note that a light ray has a world line that, by definition, is always tangent to the light cone and so $ds^2 = 0$ along the world line. From (11.1) we conclude that $-g_{00} = g_{\alpha\beta}(dx^\alpha/dt^2)(dx^\beta/dt^2) = c^2$, the square of the speed of light when measured using

*coordinate* time $t$. Thus although the speed of light is *by definition* 1 when time is measured by using atomic clocks (i.e., proper time), its speed $c$ measured using coordinate time in a static universe varies and is given by

$$c = \sqrt{(-g_{00})} \sim \sqrt{(1-2\phi)} \sim (1-\phi)$$

*Thus the coordinate speed of light decreases as the gravitational potential increases.* Einstein realized this in 1912, three years before his field equations, and before he was aware of Riemannian geometry, and proposed then that $c$ be used as a replacement for the Newtonian potential!

2. Although the world line of a light ray is assumed to be a geodesic in *space–time, its spatial trace is not usually a geodesic in space!* We have just seen that $\sqrt{-g_{00}}$ is essentially the index of refraction. It can be shown (see [Fr]) that the spatial trace satisfies **Fermat's principle** of least time

$$\delta \int \frac{d\sigma}{\sqrt{-g_{00}}} = 0$$

where $d\sigma^2 = g_{\alpha\beta} dx^\alpha dx^\beta$ is the **metric of the spatial slice**. This is the "reason" for the observed curvature of light rays passing near the sun during a total eclipse.

3. We have given a crude heuristic "derivation" of (11.2), the relation between the metric coefficient $g_{00}$ and the classical Newtonian potential $\phi$. Note that in the "derivation" of $\nabla^2\phi \sim -R_{00}$ the Laplacian $\sum \partial^2\phi/\partial x^\alpha \partial x^\alpha$ that appears uses the flat metric rather than the correct Laplacian for the spatial metric

$$\nabla^2\phi = \frac{1}{\sqrt{h}} \frac{\partial}{\partial x^\alpha} \left( \sqrt{h} g^{\alpha\beta} \frac{\partial \phi}{\partial x^\beta} \right)$$

where we have put $h = \det(g_{\alpha\beta})$. In my book [Fr, p.22], I give a heuristic argument indicating that the classical potential $\phi$ is related to $g_{00}$ by

$$\phi \sim 1 - \sqrt{-g_{00}}$$

rather than (11.2). These two expressions are very close when $g_{00}$ is very near $-1$. The advantage of this new expression for $\phi$ is that it satisfies an *exact* equation in any static space–time, **Levi-Civita's equation**

$$\nabla^2 \sqrt{-g_{00}} = -R^0{}_0 \sqrt{-g_{00}}$$

where *the Laplacian is the correct one for the spatial metric.* $g_{00}$ itself, without the square root, does not satisfy any *simple* equation such as this. Poisson's equation then suggests an equation of the form $4\pi\kappa\rho^* = -R^0{}_0\sqrt{-g_{00}}$. In the case of a perfect fluid at rest, by using (11.8) and (11.9) it is shown [Fr, p. 32] that the "correct" density of mass–energy is, in this case

$$\rho^* = (\rho + 3p)\sqrt{-g_{00}}$$

4. Finally, in my book I give a heuristic "derivation" of Einstein's equations that automatically includes the term involving $R$. This is accomplished by looking at a spherical blob of water instead of a dust cloud. This more complicated situation works because it involves stresses, that is, pressure gradients, that were omitted in the dust cloud. The derivation also has the advantage that it does not use Einstein's assumption that free test particles have geodesic world lines; rather, this geodesic assumption comes out as a *consequence* of the equations.

Two other books that I recommend for reading in general relativity are [M, T, W] and [Wd].

## Problems

**11.1(1)** Verify (11.5).

**11.1(2)** Show that in the **Schwarzschild** *spatial* metric, with coordinates $r$, $\theta$, $\phi$, and constant $m$

$$g_{\alpha\beta}dx^\alpha dx^\beta = \left(1 - \frac{2m}{r}\right)^{-1} dr^2 + r^2(d\theta^2 + \sin^2\theta d\phi^2)$$

the function $U = (1 - 2m/r)^{1/2}$ satisfies Laplace's equation $\nabla^2 U = 0$.

# 11.2. Tensor Analysis

What is the divergence of the Ricci tensor?

## 11.2a. Covariant Differentiation of Tensors

In Equation (9.7) we have defined the covariant derivative $\nabla\mathbf{v}$ of a vector field $\mathbf{v}$; it is the mixed tensor with components in a coordinate frame given by

$$\nabla_j v^i = v^i_{/j} = \frac{\partial v^i}{\partial x^j} + \omega^i_{jk} v^k \tag{11.11}$$

(We must mention that many books use the notation $v^i_{;j}$ rather than $v^i_{/j}$.) We have also defined the *exterior* covariant differential of a (tangent) vector-valued $p$-form in Section 9.3d, taking such a form into a vector-valued-$(p+1)$-form. We are now going to define, *in a different way*, the covariant derivative of a general tensor of type $(p, q)$, that is, $p$ times contravariant and $q$ times covariant, the result being a tensor of type $(p, q+1)$. In the case of a vector-valued $p$-form (which is of type $(1, p)$) the result will be different from the exterior covariant differential in that it will *not* be skew symmetric in its covariant indices and so will not be a form.

The covariant derivative of a scalar field $f$ is defined to be the differential, $\nabla f = df$, with components $f_{/j} := \partial f/\partial x^j$.

We have already defined the covariant derivative $v^i_{/j}$ of a contravariant vector field. We define the covariant derivative $a_{i/j}$ of a covector field $\alpha$ so that the "Leibniz" rule holds; for the function $\alpha(\mathbf{v}) = a_i v^i$ we demand $\partial/\partial x^j(a_i v^i) = (a_i v^i)_{/j} = a_{i/j}v^i + a_i v^i_{/j}$. Using (11.11) we see that

$$\left(\frac{\partial a_i}{\partial x^j}\right)v^i + a_i\left(\frac{\partial v^i}{\partial x^j}\right) = a_{i/j}v^i + a_i\left(\frac{\partial v^i}{\partial x^j} + \omega^i_{jk}v^k\right)$$

and so

$$\nabla_j a_i = a_{i/j} := \frac{\partial a_i}{\partial x^j} - a_k \omega^k_{ji} \tag{11.12}$$

Note that $a_{i/j}$ is *not* skew in $i$, $j$.

Finally, we define the covariant derivative of a tensor of type $(p, q)$ by generalizing (11.11) and (11.12)

$$
\begin{aligned}
T^{i_1 \ldots i_p}{}_{j_1 \ldots j_q/k} := \frac{\partial}{\partial x^k} T^{i_1 \ldots i_p}{}_{j_1 \ldots j_q} \\
+ T^{r\, i_2 \ldots i_p}{}_{j_1 \ldots j_q} \omega^{i_1}_{kr} + T^{i_1 r \ldots i_p}{}_{j_1 \ldots j_q} \omega^{i_2}_{kr} + \cdots \\
- T^{i_1 \ldots i_p}{}_{r\, j_2 \ldots j_q} \omega^{r}_{kj_1} - T^{i_1 \ldots i_p}{}_{j_1 r \ldots j_q} \omega^{r}_{kj_2} - \cdots
\end{aligned}
\tag{11.13}
$$

Thus one repeatedly uses the rules (11.11) and (11.12) for each contravariant and each covariant index occurring in $T$.

One can show that this operation does indeed take a tensor field into another whose covariance has been increased by one. Furthermore it has the following two important properties.

1. Covariant differentiation obeys a *product rule*

$$(S^{\cdots}_{\cdots} T^{\cdots}_{\cdots})_{/k} = S^{\cdots}_{\cdots/k} T^{\cdots}_{\cdots} + S^{\cdots}_{\cdots} T^{\cdots}_{\cdots/k}$$

2. Covariant differentiation *commutes with contractions*. For example, the covariant derivative of the mixed tensor $T^i{}_j$ is

$$T^i{}_{j/k} = \frac{\partial T^i{}_j}{\partial x^k} + T^r{}_j \omega^i_{kr} - T^i{}_r \omega^r_{kj}$$

which is a third-rank tensor. Contract on $i$ and $j$ to get a covector

$$T^i{}_{i/k} = \frac{\partial T^i{}_i}{\partial x^k} + T^r{}_i \omega^i_{kr} - T^i{}_r \omega^r_{ki} = \frac{\partial T^i{}_i}{\partial x^k}$$

On the other hand, if we first contract on $i$ and $j$ in $T$, we get the scalar $T^i{}_i$, whose covariant derivative is $(T^i{}_i)_{/k} = \partial/\partial x^k (T^i{}_i)$ again.

**Warning**: As a result of the presence of the connection coefficients, the covariant derivative of a tensor with constant components in some coordinate system need not vanish.

See Problems 11.2(1), 11.2(2), and 11.2(3) at this time.

## 11.2b. Riemannian Connections and the Bianchi Identities

The principal property of the Riemannian connection is expressed by

$$\frac{\partial}{\partial x^k}(g_{ij} X^i Y^j) = g_{ij} X^i_{/k} Y^j + g_{ij} X^i Y^j_{/k}$$

and the left-hand side can now be written $(g_{ij} X^i Y^j)_{/k}$. On the other hand, we now know that this latter should be

$$(g_{ij} X^i Y^j)_{/k} = g_{ij/k} X^i Y^j + g_{ij} X^i_{/k} Y^j + g_{ij} X^i Y^j_{/k}$$

This says that the *metric tensor is covariant constant*!

$$g_{ij/k} = \frac{\partial g_{ij}}{\partial x^k} - g_{lj} \omega^l_{ki} - g_{il} \omega^l_{kj} = 0 \tag{11.14}$$

See Problem 11.2(4) at this time.

We define the **divergence** of a symmetric contravariant $p$-tensor field $T$ to be the symmetric $(p-1)$-tensor

$$(\text{Div}\, T)^{j_2 \dots j_p} = T^{i j_2 \dots j_p}{}_{/i} \tag{11.15}$$

We shall soon see that this agrees with div $\mathbf{v}$ when $T$ is a vector $\mathbf{v}$. See Problem 11.2(5) at this time.

We shall now derive two very important identities satisfied by the Riemann tensor. At first we shall not restrict ourselves to Riemannian or even symmetric connections. From Cartan's structural equations (9.28) we have

$$\begin{aligned} 0 = d(d\sigma) &= d(-\omega \wedge \sigma + \tau) = -d\omega \wedge \sigma + \omega \wedge d\sigma + d\tau \\ &= -d\omega \wedge \sigma + \omega \wedge (-\omega \wedge \sigma + \tau) + d\tau = -\theta \wedge \sigma + d\tau + \omega \wedge \tau \end{aligned}$$

or, using problem 9.4(3)

$$\nabla \tau = d\tau + \omega \wedge \tau = \theta \wedge \sigma \tag{11.16}$$

We are especially concerned with the case of a symmetric connection (i.e., $\tau = 0$). Then

$$\theta \wedge \sigma = 0 \tag{11.17}$$

But then $0 = \theta^i{}_j \wedge \sigma^j = 1/2 R^i_{jkr} \sigma^k \wedge \sigma^r \wedge \sigma^j$. This means that the coefficient of $\sigma^k \wedge \sigma^r \wedge \sigma^j$, made skew in $k$, $r$ and $j$, must vanish. Since $R^i_{jkr}$ is already skew in $k$ and $r$, this means

$$R^i_{jkr} + R^i_{rjk} + R^i_{krj} = 0 \tag{11.18}$$

Both (11.17) and (11.18) will be referred to as the **first Bianchi identities**, and we emphasize that *they require a symmetric connection.*

Recall that we have defined the Ricci tensor by $R_{jr} = R^i_{jir}$. From (11.18) we have $R_{jr} = -R^i_{rji}$, since $R^i_{irj} = g^{im} R_{mirj} = 0$ from skew symmetry of $R$ in $m, i$. But $R^i_{rji} = -R^i_{rij} = -R_{rj}$. We have thus shown that

$$R_{jr} = R_{rj} \tag{11.19}$$

*in a (pseudo-) Riemannian connection.*

For our second identity we again start out with a general connection. Then $d\theta = d(d\omega + \omega \wedge \omega) = d(\omega \wedge \omega) = d\omega \wedge \omega - \omega \wedge d\omega = (\theta - \omega \wedge \omega) \wedge \omega - \omega \wedge (\theta - \omega \wedge \omega) = \theta \wedge \omega - \omega \wedge \theta$, or

$$d\theta + \omega \wedge \theta - \theta \wedge \omega = 0 \tag{11.20}$$

which we call the **second Bianchi identity**, for *all* connections. Thus $d\theta^i{}_j + \omega^i{}_m \wedge \theta^m{}_j - \theta^i{}_m \wedge \omega^m{}_j = 0$. Writing this out in a coordinate frame we get

$$\left( \frac{\partial R^i_{jkr}}{\partial x^s} \right) dx^s \wedge dx^k \wedge dx^r + \omega^i_{pm} R^m_{juv} dx^p \wedge dx^u \wedge dx^v - R^i_{mab} \omega^m_{cj} dx^a \wedge dx^b \wedge dx^c = 0$$

Then

$$\left( \frac{\partial R^i_{jkr}}{\partial x^s} + R^m_{jkr} \omega^i_{sm} - R^i_{mkr} \omega^m_{sj} \right) dx^s \wedge dx^k \wedge dx^r = 0$$

But when the connection is symmetric,

$$(R^i_{jmr}\omega^m_{sk} + R^i_{jkm}\omega^m_{sr})dx^s \wedge dx^k \wedge dx^r = 0$$

Subtracting this from our previous expression gives

$$R^i_{jkr/s}dx^s \wedge dx^k \wedge dx^r = 0$$

Since $R^i_{jkr/s}$ is already skew in $k$ and $r$ we conclude

$$R^i_{jkr/s} + R^i_{jsk/r} + R^i_{jrs/k} = 0 \tag{11.21}$$

which we again call the **second Bianchi identity** for a *symmetric connection*. You are asked to show in Problem 11.2(6) that a consequence of (11.21) is

$$\frac{\partial R}{\partial x^s} = 2R^i{}_{s/i} \tag{11.22}$$

where $R$ is the scalar curvature (11.10).

Note that the mixed tensor version of Einstein's equation is $R^i{}_j - (1/2)\delta^i_j R = 8\pi\kappa T^i{}_j$. In special relativity the tensor $T$ has divergence 0 (see [Fr, p. 70]). Its divergence, from Einstein's equation, is given by $8\pi\kappa T^i{}_{j/i} = R^i{}_{j/i} - (1/2)\delta^i{}_{j/i}R - (1/2)\delta^i_j R_{/i} = R^i{}_{j/i} - (1/2)R_{/j} = 0$! Thus the mysterious $R$ term was included in Einstein's equation in order to ensure that Div $T = 0$ in general relativity also. See Problem 11.2(7) at this time.

**Warning**: In the case of a velocity field, the divergence theorem gives $\int_U \text{div}\,\mathbf{v}\,\text{vol} = \int_{\partial U}\langle \mathbf{v}, \mathbf{n}\rangle dS$. In particular, if div $\mathbf{v} = 0$ we have a *conservation theorem*: The rate of flow of volume into a region $U$ equals the rate leaving the region. *There is no analogue of this for the divergence of a tensor*! For example, $\int_U T^i{}_{j/i}$ vol makes no *intrinsic* sense; one cannot integrate a covector $T^i{}_{j/i}$ over a volume since one cannot add covectors based at different points. In spite of this, many books refer to Div $T = 0$ as a conservation law.

## 11.2c. Second Covariant Derivatives: The Ricci Identities

The covariant derivative of a vector field $\mathbf{Z}$ is a mixed tensor with components $Z^i_{/j}$. The covariant derivative of this mixed tensor is a tensor of third rank with components $Z^i_{/j/k}$, which is traditionally written

$$Z^i_{/jk} := Z^i_{/j/k}$$

We wish now to investigate $Z^i_{/jk} - Z^i_{/kj}$. Let $\mathbf{X}$ and $\mathbf{Y}$ be vectors at a point. Extend them to vector fields. We have $\mathbf{Z} = Z^i\boldsymbol{\partial}_i$, and so on. Then

$$\begin{aligned}\nabla_{\mathbf{X}}(\nabla_{\mathbf{Y}}\mathbf{Z}) &= \nabla_{\mathbf{X}}(Z^i_{/j}Y^j\boldsymbol{\partial}_i) = (Z^i_{/j}Y^j)_{/k}X^k\boldsymbol{\partial}_i \\ &= (Z^i_{/jk}Y^j + Z^i_{/j}Y^j_{/k})X^k\boldsymbol{\partial}_i\end{aligned}$$

Then, using symmetry of the connection,

$$[\mathbf{X}, \mathbf{Y}]^i = (\nabla_{\mathbf{X}}\mathbf{Y} - \nabla_{\mathbf{Y}}\mathbf{X})^i = X^j Y^i_{/j} - Y^j X^i_{/j}$$

we get

$$\begin{aligned}\nabla_{\mathbf{X}}(\nabla_{\mathbf{Y}}\mathbf{Z}) - \nabla_{\mathbf{Y}}(\nabla_{\mathbf{X}}\mathbf{Z}) &= [(Z^i_{/jk} - Z^i_{/kj})Y^j X^k + Z^i_{/j}(X^k Y^j_{/k} - Y^k X^j_{/k})]\partial_i \\ &= (Z^i_{/jk} - Z^i_{/kj})Y^j X^k \partial_i + Z^i_{/j}[X, Y]^j \partial_i \\ &= (Z^i_{/jk} - Z^i_{/kj})Y^j X^k \partial_i + \nabla_{[\mathbf{X},\mathbf{Y}]}\mathbf{Z}\end{aligned}$$

or $(Z^i_{/jk} - Z^i_{/kj})Y^j X^k \partial_i = R(\mathbf{X}, \mathbf{Y})\mathbf{Z} = (R(\mathbf{X}, \mathbf{Y})\mathbf{Z})^i \partial_i = R^i_{mkj} X^k Y^j Z^m \partial_i$. Thus

$$Z^i_{/jk} - Z^i_{/kj} = Z^m R^i_{mkj} \tag{11.23}$$

the **Ricci identities** for a *symmetric connection.* Mixed covariant derivatives do not commute. Note carefully the placement of the indices $j$ and $k$! This placement is more easily remembered if we write

$$\nabla_k \nabla_j Z^i - \nabla_j \nabla_k Z^i = Z^m R^i_{mkj}$$

In many books, the covariant derivative of a *tensor* is introduced before the notion of curvature, and then (11.23) is used to define the curvature tensor.

**Warning**: We may write

$$\nabla_{\partial_j}\mathbf{X} = X^i_{/j}\partial_i = (\nabla_j X^i)\partial_i \tag{11.24}$$

(Recall that $\nabla_{\partial_j}$ operates on *vectors* whereas $\nabla_j$ operates on the *components* of vectors.) It is easily seen, however, that in general

$$\nabla_{\partial_j}\nabla_{\partial_k}\mathbf{X} \neq (\nabla_j \nabla_k X^i)\partial_i = X^i_{/kj}\partial_i$$

It is true, however, that $\nabla_{\partial_j}\nabla_{\partial_k}\mathbf{X} - \nabla_{\partial_k}\nabla_{\partial_j}\mathbf{X} = (X^i_{/kj} - X^i_{/jk})\partial_i = X^m R^i_{mjk}\partial_i$. The second and third terms are equal by (11.23); the first and the third terms are equal by (10.2) when $u = x^j$ and $v = x^k$.

## Problems

**11.2(1)** Show that the *identity tensor* $\delta^i_j$ *is covariant constant,* $\delta^i_{j/k} = 0$.

**11.2(2)** Show directly from (9.19) that $g_{ij/k} = 0$.

**11.2(3)** Show that the **Codazzi equations** in (8.34) say that $b_{\alpha\beta/\gamma} = b_{\alpha\gamma/\beta}$.

**11.2(4)** Use $g^{ij}g_{jk} = \delta^i_k$ to show that $g^{ij}_{/r} = 0$.

**11.2(5)** Show that for a surface $M^2 \subset \mathbb{R}^3$ with mean curvature $H$, grad $H$ = Div $b$ where the second fundamental form $b$ is now considered as contravariant, $b^{ij}$.

**11.2(6)** Use (11.21) and contract several times to derive (11.22).

**11.2(7)** Let $T^{ij} := \rho u^i u^j + p(g^{ij} + u^i u^j)$ be the stress–energy–momentum tensor for a perfect fluid. Show that $T^{ij}_{/j} = 0$ yields the two sets of equations

$$\operatorname{div}(\rho u) = -p \operatorname{div} u$$

and

$$(\rho + p)\nabla_u u = -(\operatorname{grad} p)^{\perp}$$

where $^{\perp}$ denotes component orthogonal to $u$. The first equation replaces the flat space conservation of mass–energy, $\partial\rho/\partial t + \text{div}(\rho\mathbf{v}) = 0$; since div $u$ measures the change in 3-volume orthogonal to $u$ (see the 2-dimensional analogue in 8.23), $-p$ div $u$ gives the rate of work done by the pressure during expansion. The second equation is Newton's law, with mass density $\rho$ augmented by a small pressure term $p$ (really $p/c^2$). Thus $T^{ij}_{/j} = 0$ yields the **relativistic equations of motion.**

**11.2(8)** Show that in a symmetric connection, in the exterior derivative of a 1-form $d\alpha^1 = \sum_{j<k}(\partial_j a_k - \partial_k a_j)dx^j \wedge dx^k$, we may replace the partial derivatives by covariant derivatives

$$d\alpha = \sum_{j<k}(a_{k/j} - a_{j/k})dx^j \wedge dx^k$$

Show that if the connection is *symmetric*, then in the formula (2.55) *one may replace partial derivatives by covariant derivatives*

$$(d\alpha^P)_I = \sum_{j\underrightarrow{K}} \delta_I^{jK} a_{K/j} = \sum_{j\underrightarrow{K}} \delta_I^{jK} \nabla_j a_K$$

## 11.3. Hilbert's Action Principle

How does the scalar curvature $R$ vary with the metric.

### 11.3a. Geodesics in a Pseudo-Riemannian Manifold

Geodesics play an important role in relativity. We know that a geodesic in a *Riemannian* manifold is characterized by the property that there is a whole class of parameterizations $t$ such that $\nabla(d\mathbf{x}/dt)/dt = 0$ and all of these parameters are linear functions of the arc length parameter.

In general relativity we deal with a *pseudo*-Riemannian manifold. In our heuristics of relativity we needed to consider the world line of a "freely falling" moving body, and, since such bodies always travel at a speed less than that of light, the path is timelike (i.e., $d\tau^2 = -ds^2 > 0$). In terms of the proper time parameter $\tau$ we have, as the equation of the geodesic, $\nabla(d\mathbf{x}/d\tau)/d\tau = 0$. For a spacelike geodesic we may use $s$ instead of $\tau$ as parameter. A light ray, being the path of a photon, is the limiting case of a freely falling particle of vanishingly small mass; it is assumed that its world line is also a geodesic, called a **null geodesic** since $ds^2 = 0$. We may use neither $s$ nor $\tau$ for parameter. A parameter $\lambda$ for a null geodesic, for which $\nabla(d\mathbf{x}/d\lambda)/d\lambda = 0$, will be called, as before, a *distinguished or affine parameter* (see, e.g., [Fr, p. 92]).

### 11.3b. Normal Coordinates, the Divergence and Laplacian

Let $p$ be a point in a (pseudo-) Riemannian manifold $M^n$ and let $\mathbf{e}_1, \ldots, \mathbf{e}_n$ be an orthonormal frame at $p$. As in the 2-dimensional case considered in Section 10.3b, we may then introduce normal coordinates $y$ near $p$ by defining $\Phi : (\mathbb{R}^n = M_p^n) \to M^n$ by $\Phi(y) = \exp_p(\mathbf{e}_i y^i)$, for all sufficiently small $y$. The differential $\Phi_* : M_p \to M$ again has the property that $\Phi_*(\mathbf{e}_i) = \mathbf{e}_i$, and so the coordinate vectors $\partial/\partial y^i$ are orthonormal at the origin $p$; $(g_{ij}(0)) = \text{diag}(\pm 1, 1, \ldots, 1)$. The arguments to be given later hold in general, but we shall work in the case of a 4-dimensional space–time, as this is our immediate concern.

It should be clear that the geodesic that starts at $p$ with tangent vector $\mathbf{e}_i \lambda^i$ is given in these normal coordinates by the linear equations

$$y^i(t) = \lambda^i t, \qquad i = 0, \ldots, 3$$

It is also clear from the definition of the exponential map that $\| d\mathbf{y}/dt \|^2 = \lambda_1^2 + \lambda_2^2 + \lambda_3^2 - \lambda_0^2$ is a constant along each of the geodesics starting at $p$ and this constant vanishes only for the null geodesics tangent to the light cone, a submanifold of the vector space $M_p^4$ of codimension 1. By continuity, we conclude that $t$ is a distinguished parameter for each of the geodesics emanating from $p$. Since the preceding linear equations must satisfy the geodesic equations

$$\frac{d^2 y^i}{dt^2} = -\Gamma^i_{jk}(y(t)) \frac{dy^j}{dt} \frac{dy^k}{dt}$$

we must have $\Gamma^i_{jk}(\lambda^0 t, \ldots, \lambda^3 t)\lambda^j \lambda^k = 0$ for all $t$. In particular this holds at $p$, that is, $t = 0$, and for all $\lambda^i$. We conclude that

$$\Gamma^i_{jk}(p) = 0 \tag{11.25}$$

at the "pole" of the normal coordinate system. From (11.14) we have

$$\frac{\partial g_{ij}}{\partial y^k}(p) = 0 \tag{11.26}$$

*All first partial derivatives of the metric tensor vanish at the pole*!

As an application of the use of these coordinates, consider the divergence of a vector field $\mathbf{v}$. As in the Riemannian case

$$\text{div } \mathbf{v} = (|g|^{-1/2}) \frac{\partial}{\partial x^i}[|g|^{1/2} v^i]$$

At the pole of the normal coordinates we clearly have $\partial/\partial y^i |g|(p) = 0$ and thus at the pole we have $\text{div } \mathbf{v} = \partial v^i/\partial y^i$. Consider now the scalar $v^i_{/i}$. At the pole $v^i_{/i} = \partial v^i/\partial y^i + v^j \Gamma^i_{ij} = \partial v^i/\partial y^i$. But div $\mathbf{v}$ and $v^i_{/i}$ are well-defined scalars, independent of coordinates; we conclude that *in any coordinate system*

$$\text{div } \mathbf{v} = (|g|^{-1/2}) \frac{\partial}{\partial x^i}[|g|^{1/2} v^i] = v^i_{/i} \tag{11.27}$$

This in turn means that

$$\frac{\partial v^i}{\partial x^i} + v^i \frac{\partial}{\partial x^i} \log |g|^{1/2} = \frac{\partial v^i}{\partial x^i} + v^k \Gamma^i_{ik}$$

and so

$$\frac{\partial}{\partial x^k} \log |g|^{1/2} = \Gamma^i_{ik} \tag{11.28}$$

which is a frequently used formula.

We have already defined the gradient of a function $f$ to be the contravariant vector $(\text{grad} f)^i = g^{ij}\partial f/\partial x^j = g^{ij} f_{/j}$. The Laplacian of $f$ is then the scalar $\nabla^2 f = \text{div grad} f = (g^{ij} f_{/j})_{/i}$, or, since $g^{ij}{}_{/k} = 0$,

$$\nabla^2 f = g^{ij} f_{/ji} = g^{ij} f_{/ij} \tag{11.29}$$

Thus $\nabla^2 f = g^{ij}[\partial/\partial x^i(\partial f/\partial x^j) - \Gamma^k_{ij}(\partial f/\partial x^k)]$, or

$$\nabla^2 f = g^{ij}\left[\frac{\partial^2 f}{\partial x^i \partial x^j} - \Gamma^k_{ij}\frac{\partial f}{\partial x^k}\right] \tag{11.30}$$

As an example, consider a surface $M^2$ with local coordinates $u^1, u^2$, sitting in $\mathbb{R}^3$, with Cartesian coordinates $x^1, x^2, x^3$. For each $i$, $x^i$ is a *function* on $M$. In Problem 11.3(1) you are to show that

$$\nabla^2 x^i = HN^i \tag{11.31}$$

where $\nabla^2$ is the surface Laplacian, $H$ is the mean curvature, and $\mathbf{N}$ is the unit normal. In particular,

**Theorem (11.32):** *$M^2 \subset \mathbb{R}^3$ is a minimal surface iff each coordinate function $x^i$ is a surface harmonic function on $M$.*

### 11.3c. Hilbert's Variational Approach to General Relativity

Although the following approach will work in any dimension, we shall write out everything in the case of a 4-dimensional pseudo-Riemannian manifold $M^4$.

Let $R = g^{ik}R_{ik} = g^{ik}R^j{}_{ijk}$ be the scalar curvature. Since the determinant $g = \det g_{ij}$ is negative in a pseudo-Riemannian manifold, the volume form is

$$\sqrt{(-g)}d^4x := \sqrt{(-g)}dx^0 \wedge dx^1 \wedge dx^2 \wedge dx^3$$

We shall, with Hilbert, take the first variation of the functional

$$\int_M R\sqrt{(-g)}d^4x$$

for a 1-parameter family of metrics. For our purposes, it will be more convenient to vary the inverse of the metric

$$g^{ij}_\alpha = g^{ij}_0 + \alpha\eta^{ij} \tag{11.33}$$
$$\dot{g}^{ij} = \eta^{ij}$$

where the dot denotes differentiation with respect to $\alpha$ at $\alpha = 0$. We must compute

$$\left[\frac{d}{d\alpha}\int R\sqrt{(-g)}d^4x\right] = \int [R\sqrt{(-g)}]^\bullet d^4x \tag{11.34}$$

and where all integrals are over $M$. Now

$$[R\sqrt{(-g)}]^\bullet = [g^{ik}R_{ik}\sqrt{(-g)}]^\bullet$$
$$= [g^{ik}R_{ik}]^\bullet\sqrt{(-g)} + R[\sqrt{(-g)}]^\bullet \tag{11.35}$$

and

$$[g^{ik}R_{ik}]^{\bullet} = \dot{g}^{ik}R_{ik} + g^{ik}\dot{R}_{ik}$$

We then need $\dot{R}_{ik}$. From (9.11), omitting some indices,

$$R_{ik} = \left(\frac{\partial \Gamma^j_{ik}}{\partial x^j}\right) - \left(\frac{\partial \Gamma^j_{ij}}{\partial x^k}\right) + \Gamma\Gamma - \Gamma\Gamma$$

and so

$$\dot{R}_{ik} = \left(\frac{\partial \dot{\Gamma}^j_{ik}}{\partial x^j}\right) - \left(\frac{\partial \dot{\Gamma}^j_{ij}}{\partial x^k}\right) + \dot{\Gamma}\Gamma + \Gamma\dot{\Gamma} - \dot{\Gamma}\Gamma - \Gamma\dot{\Gamma}$$

We shall compute everything at the pole of a geodesic normal coordinate system for the base metric $g_0^{ik}$. Since $\Gamma = 0$ at the pole

$$\dot{R}_{ik} = \left(\frac{\partial \dot{\Gamma}^j_{ik}}{\partial x^j}\right) - \left(\frac{\partial \dot{\Gamma}^j_{ij}}{\partial x^k}\right) \tag{11.36}$$

at the pole. Although $(\Gamma^j_{ik})$ is not a tensor, we claim that $(\dot{\Gamma}^j_{ik})$ is a third-rank tensor. To see this we look at the transformation law (9.41) for a connection, $\omega'(\alpha) = P^{-1}\omega(\alpha)P + P^{-1}dP$. Differentiating and putting $\alpha = 0$ give $\dot{\omega}' = P^{-1}\dot{\omega}P$, and from this the tensorial nature of $\dot{\Gamma}$ follows. Thus at the pole we may write

$$\dot{R}_{ik} = \dot{\Gamma}^j_{ik/j} - \dot{\Gamma}^j_{ij/k} \tag{11.37}$$

and since this is a tensor equation it holds everywhere, in every coordinate system. In this equation, *all covariant derivatives are with respect to the base metric at* $\alpha = 0$. We may then write

$$g^{ik}\dot{R}_{ik} = (g^{ik}\dot{\Gamma}^j_{ik})_{/j} - (g^{ik}\dot{\Gamma}^j_{ij})_{/k} = \operatorname{div}\mathbf{W} \tag{11.38}$$

where $W^r := g^{ik}\dot{\Gamma}^r_{ik} - g^{ir}\dot{\Gamma}^j_{ij}$.

Look now at the second term in (11.35), $R[\sqrt{(-g)}]^{\bullet}$. To differentiate a determinant we use $\partial g/\partial g_{ik} = G^{ik}$ where $G^{ik}$ is the cofactor of the entry $g_{ik}$. This is clear upon expanding $g$ by the $k^{\text{th}}$ column. But the inverse matrix satisfies $g^{ik} = G^{ki}/g = G^{ik}/g$, and so

$$\frac{\partial g}{\partial g_{ik}} = g^{ik}g$$

Likewise $\partial g^{-1}/\partial g^{ik} = g_{ik}g^{-1}$, that is, $\partial g/\partial g^{ik} = -g_{ik}g$. Thus

$$\frac{\partial(-g)^{1/2}}{\partial g^{ik}} = -\frac{1}{2}g_{ik}(-g)^{1/2} \tag{11.39}$$

and so $\sqrt{(-g)}]^{\bullet} = (\partial(-g)^{1/2}/\partial g^{ik})(\partial g^{ik}/\partial\alpha) = -(1/2)g_{ik}\sqrt{(-g)}\dot{g}^{ik}$. Thus

$$\sqrt{(-g)}]^{\bullet} = -\frac{1}{2}g_{ik}\sqrt{(-g)}\dot{g}^{ik} \tag{11.40}$$

Finally

$$\delta \int R \operatorname{vol}^4 = \left[\frac{d}{d\alpha}\int R\sqrt{(-g)}d^4x\right]_{\alpha=0}$$
$$= \int \left[R_{ik} - \frac{1}{2}g_{ik}R\right]\dot{g}^{ik}\sqrt{(-g)}d^4x \tag{11.41}$$
$$+ \int \operatorname{div} \mathbf{W}\sqrt{(-g)}d^4x$$

By choosing a variation $\dot{g}^{ik}$ that vanishes outside some compact subregion of $M$ and applying the divergence theorem to a slightly larger region, we see that the last integral vanishes. Thus

$$\delta \int_M R \operatorname{vol}^4 = \left[\frac{d}{d\alpha}\int_M R\sqrt{(-g)}d^4x\right]_{\alpha=0}$$
$$= \int_M \left[R_{ik} - \frac{1}{2}g_{ik}R\right]\dot{g}^{ik}\sqrt{(-g)}d^4x \tag{11.42}$$

for all variations with compact support.

We define the (Hilbert) **action** for the gravitational field by

$$S_{\text{grav}} = \int_M L_{\text{grav}} d^4x := (8\pi\kappa)^{-1}\int_M R \operatorname{vol}^4 \tag{11.43}$$

(where $\kappa$ is again the gravitational constant), a nonlinear functional of the metric tensor. Let $S_{\text{nongrav}}$ be the action for the nongravitational fields that might be present, such as the electromagnetic fields; it is given by some Lagrange density

$$S_{\text{nongrav}} = \int_M L d^4x = \int_M \mathfrak{L} \operatorname{vol}^4$$

where $L = \mathfrak{L}\sqrt{(-g)}$. The **variational** or **functional derivative** $\delta S/\delta g^{ik}$ of a functional $S = \int_M L d^4x$ of the metric is defined through

$$\delta S = \delta \int_M L d^4x = \int_M \left(\frac{\delta L}{\delta g^{ik}}\right)\dot{g}^{ik} d^4x \tag{11.44}$$

where the variation is assumed to have compact support. In other words, putting $f_{,j} := \partial f/\partial x^j$, and so forth,

$$\frac{\delta L}{\delta g^{ik}} = \frac{\partial L}{\partial g^{ik}} - \left[\frac{\partial L}{\partial (g^{ik}_{,j})}\right]_{,j} + \left[\frac{\partial L}{\partial (g^{ik}_{,jr})}\right]_{,jr} - \cdots$$

is the usual Euler–Lagrange expression. Thus, from (11.41)

$$\frac{\delta L_{\text{grav}}}{\delta g^{ik}} = (8\pi\kappa)^{-1}\left[R_{ik} - \frac{1}{2}g_{ik}R\right]\sqrt{(-g)} \tag{11.45}$$

The **(stress)–energy–momentum** tensor of the gravitational field is defined to be 0 (since gravitation is a fictitious field); that of the nongravitational fields, $T_{ik}$, is defined by

$$T_{ik}\sqrt{(-g)} := -\frac{\delta L_{\text{nongrav}}}{\delta g^{ik}} \tag{11.46}$$

The total Lagrangian is $L = L_{\text{grav}} + L_{\text{nongrav}}$, and so

$$\frac{\delta L}{\delta g^{ik}} = \left\{(8\pi\kappa)^{-1}\left[R_{ik} - \frac{1}{2}g_{ik}R\right] - T_{ik}\right\}\sqrt{(-g)}$$

Then Einstein's equations (11.9) are equivalent to **Hilbert's action principle**

$$\delta\int_M [(8\pi\kappa)^{-1}R + \mathfrak{L}]\text{vol}^4 = 0 \tag{11.47}$$

It is natural to call $R\sqrt{(-g)}$ the **Lagrangian** of the gravitational field.

To understand the *geometric* meaning of Einstein's equations we must return to our study of second fundamental forms and curvature. We proceed to these matters in our next two sections.

## Problems

**11.3(1)** Use Gauss's surface equations to prove (11.31).

**11.3(2)** (i) Let **v** be a vector field in $\mathbb{R}^3$ defined along a surface $M^2$ in $\mathbb{R}^3$. If $x^1, x^2, x^3$, are *cartesian* coordinates for $\mathbb{R}^3$, we define the **vector integral** $\iint_M \mathbf{v}dS$ to be the vector **w** with components $w^i = \iint_M v^i dS$. Show that if $M^2$ is a *closed* surface with unit normal **N**, then

$$\iint_M H\mathbf{N}dS = 0$$

We considered the special integral $\iint \mathbf{N}dS = \iint d\mathbf{S}$ directly before Euler's equation (4.45). For a closed surface $M$ we have

$$\iint_M \mathbf{N}dS = 0$$

since, for example, $\iint_M N^1 dS = \iint_M dy \wedge dz = 0$. Thus, for any closed surface in $\mathbb{R}^3$, not only is the surface average of **N** zero, which is geometrically "clear," but also this average, when weighted by the mean curvature, also vanishes!

**11.3(3)** Let

$$L_{\text{em}} := -\frac{1}{8\pi}F_{ij}F^{ij}\sqrt{(-g)}$$

define the Lagrangian for the pure electromagnetic field, with associated action

$$S_{\text{em}} := -\frac{1}{8\pi}\int_M F_{ij}F_{rs}g^{ri}g^{sj}\sqrt{(-g)}d^4x$$

Show (recalling that $F_{ij}$ is independent of the metric) that the stress–energy–momentum tensor for the electromagnetic field is Minkowski's

$$T_{ij} = \frac{1}{4\pi}\left[F_{ik}F_j{}^k - \frac{1}{4}g_{ij}F_{rs}F^{rs}\right]$$

(Recall that locally $F^2 = dA^1$, where $A$ is the covector potential; we shall see later on that $A$ is usually globally defined. Thus $S_{\text{em}}$ can be expressed as a functional of $A$ and the metric. We shall see in Section 20.2c that $\delta S_{\text{em}}/\delta A = 0$ is simply a statement of Maxwell's equations in free space. Thus one obtains the *equations of motion* of the electromagnetic field, Maxwell's equations, by putting the first variation of the total action with respect to the potential equal to 0

$$\frac{\delta}{\delta A_j}[S_{\text{grav}} + S_{\text{em}}] = 0$$

Thus one varies the metric potentials in the total Lagrangian to obtain the gravitational (Einstein) field equations and one varies the electromagnetic potentials to obtain the electromagnetic (Maxwell) field equations!)

## 11.4. The Second Fundamental Form in the Riemannian Case

If you fold a sheet of paper once, why is the crease a straight line?

### 11.4a. The Induced Connection and the Second Fundamental Form

Let $V^r \subset M^n$ be a submanifold of a *Riemannian* manifold $M$. If we restrict the Riemannian metric of $M$, $\langle , \rangle$, to vectors tangent to $V$, we obtain a Riemannian metric for $V$, the **induced metric.**

Let $\nabla$ be the Riemannian connection for $M^n$ and let $V^{n-1}$ be an $(n-1)$-dimensional *hypersurface* of $M$. Define a new connection for $V$ as follows. Let $\mathbf{X}$ be tangent to $V$ at $p$ and let $\mathbf{Y}$ be a vector *field* tangent to $V$ near $p$. Let $x^1, \dots, x^n$ and $u^1, \dots, u^{n-1}$, be local coordinates for $M$ and $V$, respectively, near $p$. Then $\nabla_{\mathbf{X}}\mathbf{Y} = X^\alpha \nabla\mathbf{Y}/\partial u^\alpha$ makes

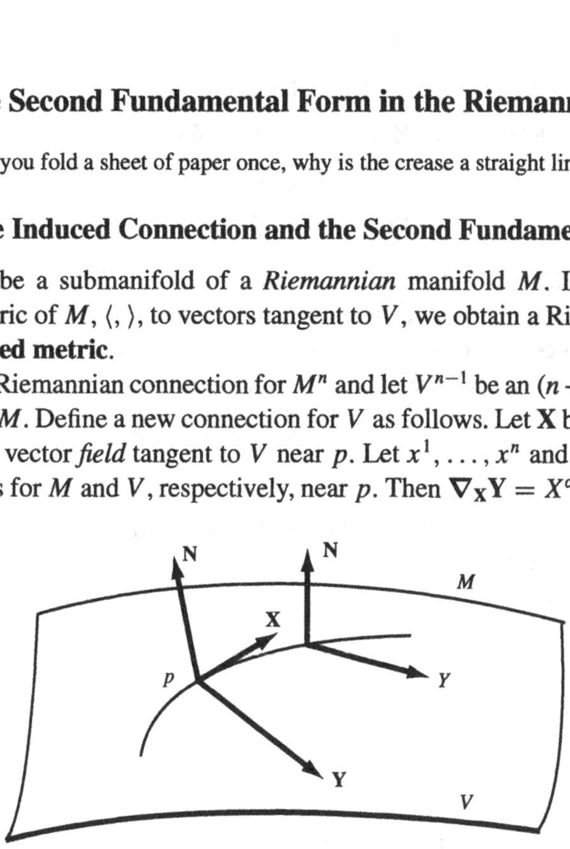


Figure 11.3

sense since $\mathbf{Y}$ is a vector field defined along $V$. Let $\mathbf{N}$ be a unit vector field along $V^{n-1}$ that is normal to $V$ and let $\mathbf{Z}$ be *any* vector field defined *along* $V$ (it needn't be tangent to $V$). Define, at $p$ in $V$

$$\begin{aligned} \nabla^V{}_{\mathbf{X}}\mathbf{Z} :&= \text{projection of } \nabla_{\mathbf{X}}\mathbf{Z} \text{ into the tangent space} V_p \\ &= \nabla_{\mathbf{X}}\mathbf{Z} - \langle \nabla_{\mathbf{X}}\mathbf{Z}, \mathbf{N}\rangle \mathbf{N}. \end{aligned} \tag{11.48}$$

In particular, to the vector fields $\mathbf{X}$ and $\mathbf{Y}$ *tangent* to $V$ we associate another tangent vector field $\nabla^V{}_{\mathbf{X}}\mathbf{Y}$. One checks immediately that (9.2) is satisfied by $\nabla^V$ and thus (11.48) defines a connection for $V^{n-1}$. We claim more: $\nabla^V$ is *the* Riemannian connection for the induced metric on $V^{n-1}$. You are asked to prove this in Problem 11.4(1). Notice that we have merely imitated Levi-Civita's construction in the case of a surface $V^2 \subset \mathbb{R}^3$.

What is the generalization of the second fundamental form? We proceed as in Section 8.1b. In the following **X**, **Y**, **Z**, are tangent to $V^{n-1}$ and **N** is a local unit normal to $V$. We define

$$b : V_p \to V_p$$

by

$$b(\mathbf{X}) = -\nabla_{\mathbf{X}}\mathbf{N} \tag{11.49}$$

Put

$$B(\mathbf{X}, \mathbf{Y}) := \langle \mathbf{X}, b(\mathbf{Y}) \rangle$$

Extending **X** and **Y** to be fields on $M$, we have

$$\begin{aligned} B(\mathbf{X}, \mathbf{Y}) &= \langle \mathbf{X}, b(\mathbf{Y}) \rangle = \langle \mathbf{X}, -\nabla_{\mathbf{Y}}\mathbf{N} \rangle \\ &= \mathbf{Y}(\langle \mathbf{X}, -\mathbf{N} \rangle) - \langle \nabla_{\mathbf{Y}}\mathbf{X}, -\mathbf{N} \rangle \\ &= \langle \nabla_{\mathbf{Y}}\mathbf{X}, \mathbf{N} \rangle = \langle \nabla_{\mathbf{X}}\mathbf{Y}, \mathbf{N} \rangle = B(\mathbf{Y}, \mathbf{X}) \end{aligned} \tag{11.50}$$

(why?). Then $b$ is again a *self-adjoint* linear transformation; $b$ has $(n-1)$ real eigenvalues $\kappa_1, \ldots, \kappa_{n-1}$ called **principal (normal) curvatures.** The eigen directions are called the **principal directions**, and they can always be chosen to be mutually orthogonal.

From (11.48) and (11.50) we have the **Gauss equations**

$$\nabla_{\mathbf{X}}\mathbf{Y} = \nabla^V_{\mathbf{X}}\mathbf{Y} + B(\mathbf{X}, \mathbf{Y})\mathbf{N} \tag{11.51}$$

generalizing the surface equations (8.30).

We shall say that $V^r \subset$ Riemannian $M^n$ is **geodesic at** $p$ provided every $M$-geodesic through $p$, tangent to $V$ at $p$, lies wholly in $V$. Thus all of the $V$-geodesics through $p$ are also $M$-geodesics!

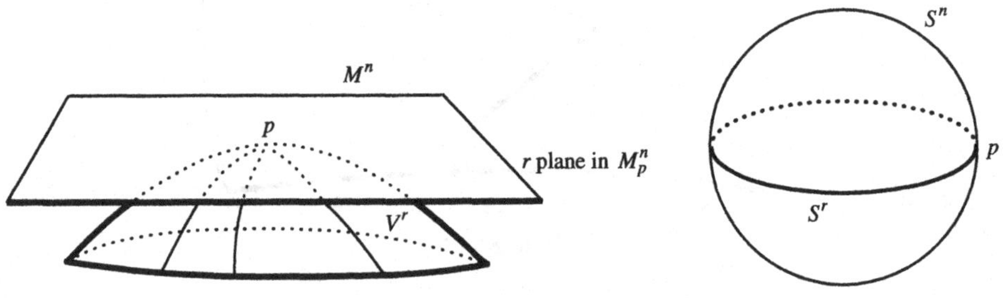


**Figure 11.4**

Then $V$ (if connected) is made up of geodesic segments of $M$ emanating from $p$, tangent to an $r$-plane in $M^n_p$. A plane in $\mathbb{R}^3$ and an equatorial $r$-sphere $S^r$ in $S^n$ are examples. Unlike in these examples, *it is not true in general that a $V$-geodesic starting at a point different from $p$ will still be an $M$-geodesic.*

If $V^{n-1}$ is geodesic at $p$, then at $p$

$$B(\mathbf{X}, \mathbf{X}) = \langle \nabla_{\mathbf{X}}\mathbf{X}, \mathbf{N} \rangle = 0 \tag{11.52}$$

since **X** can be extended to be the tangent to a geodesic of $V$ that is then also a geodesic of $M$. Thus *the second fundamental form $B$ of $V$ at $p$ is identically 0 if $V$ is geodesic at $p$.*

As in the case of a $V^2 \subset \mathbb{R}^3$, we define the **mean curvature** $H$ of $V^{n-1} \subset M^n$ by

$$H := \operatorname{tr} b = \kappa_1 + \cdots + \kappa_{n-1}$$

and this is again significant for considering variations of the $(n-1)$-volume of $V^{n-1}$. (In fact, you should be able to guess the generalization of Gauss's formula (8.26).) $V^{n-1} \subset M^n$ is said to be a **minimal submanifold** of $M$ if $H$ vanishes at all points of $V$. Note that if $V$ is geodesic at *every* point $p$ of $V$ (we then say that $V$ is **totally geodesic**) then $V$ is a minimal submanifold of $M$. Thus the equatorial $S^{n-1} \subset S^n$ is minimal in $S^n$. (Note, however, that $S^2$ does *not* have *minimum* area in $S^3$!)

The other invariants $\sum_{\alpha<\beta} \kappa_\alpha \kappa_\beta, \ldots, \kappa_1\kappa_2 \ldots \kappa_{n-1}$, are also useful, though not to the same extent as $K$ and $H$ for $V^2 \subset \mathbb{R}^3$. The last invariant of $b$, $\det b = \kappa_1\kappa_2 \ldots \kappa_{n-1}$, is *not* called the Gauss curvature. We shall talk more about some of these matters in our next section on relativity.

### 11.4b. The Equations of Gauss and Codazzi

$M^n$ has a connection $\nabla$ and curvature tensor $R$; $V^{n-1}$ has a connection $\nabla^V$ and curvature tensor $R^V$

$$R^V(\mathbf{X}, \mathbf{Y}) := [\nabla^V_{\mathbf{X}}, \nabla^V_{\mathbf{Y}}] - \nabla^V_{[\mathbf{X},\mathbf{Y}]}$$

How are their curvatures related? In other words, if $\mathbf{X}$, $\mathbf{Y}$, and $\mathbf{Z}$ are tangent to $V^{n-1}$, how are the vectors $R(\mathbf{X}, \mathbf{Y})\mathbf{Z}$ and $R^V(\mathbf{X}, \mathbf{Y})\mathbf{Z}$ related?

$R^V(\mathbf{X}, \mathbf{Y})\mathbf{Z}$ is certainly tangent to $V$ but there is no reason why $R(\mathbf{X}, \mathbf{Y})\mathbf{Z}$ should be. We can see their relation as follows.

Let $\partial_\alpha = \partial/\partial u^\alpha$, $\alpha = 1, \ldots, n-1$, be a local coordinate basis for $V^{n-1}$. Since these fields can be considered as vector fields defined along the submanifold $V$, we have, from (10.2)

$$[\nabla_{\partial_\alpha} \nabla_{\partial_\beta} - \nabla_{\partial_\beta} \nabla_{\partial_\alpha}]\partial_\gamma = R(\partial_\alpha, \partial_\beta)\partial_\gamma$$

On the other hand,

$$[\nabla^V_{\partial_\alpha} \nabla^V_{\partial_\beta} - \nabla^V_{\partial_\beta} \nabla^V_{\partial_\alpha}]\partial_\gamma = R^V(\partial_\alpha, \partial_\beta)\partial_\gamma$$

Now insert $\nabla_{\partial_\alpha}\partial_\gamma = \nabla^V_{\partial_\alpha}\partial_\gamma + \langle \nabla_{\partial_\alpha}\partial_\gamma, \mathbf{N}\rangle \mathbf{N}$, and take second derivatives, using $\nabla_{\partial_\beta}\mathbf{N} = -b(\partial_\beta)$. By a calculation entirely similar in spirit to Gauss's and yours in Problem 8.5(1) we get

$$\begin{aligned} &[\nabla_{\partial_\alpha} \nabla_{\partial_\beta} - \nabla_{\partial_\beta} \nabla_{\partial_\alpha}]\partial_\gamma \\ &= [\nabla^V_{\partial_\alpha} \nabla^V_{\partial_\beta} - \nabla^V_{\partial_\beta} \nabla^V_{\partial_\alpha}]\partial_\gamma + B(\partial_\alpha, \partial_\gamma)b(\partial_\beta) - B(\partial_\beta, \partial_\gamma)b(\partial_\alpha) \\ &+ \left\{ \frac{\partial}{\partial u^\alpha} B(\partial_\beta, \partial_\gamma) - B(\partial_\beta, \nabla^V_{\partial_\alpha}\partial_\gamma) - \frac{\partial}{\partial u^\beta} B(\partial_\alpha, \partial_\gamma) + B(\partial_\alpha, \nabla^V_{\partial_\beta}\partial_\gamma) \right\} \mathbf{N} \end{aligned} \tag{11.53}$$

The expression in the curly braces { } can be simplified. Our prescription (11.13) for taking the covariant derivative of a covariant tensor field can be shown to be equivalent to the following version of *Leibniz's rule*. For any $p$-times covariant tensor $T$, for vector

**X**, and for vector fields $\mathbf{Y}_1, \ldots, \mathbf{Y}_p$, then $T(\mathbf{Y}_1, \ldots, \mathbf{Y}_p)$ is a scalar field and we may differentiate it with respect to **X**. Then (11.13) says

$$\mathbf{X}T(\mathbf{Y}_1, \ldots, \mathbf{Y}_p) = (\nabla_X T)(\mathbf{Y}_1, \ldots, \mathbf{Y}_p) + \sum_r T(\mathbf{Y}_1, \ldots, \nabla_X \mathbf{Y}_r, \ldots, \mathbf{Y}_p) \tag{11.54}$$

(with a similar rule for any mixed tensor). Apply this to the manifold $V^{n-1}$ and the covariant tensor $B$ to get

$$\frac{\partial}{\partial u^\alpha} B(\partial_\beta, \partial_\gamma) = (\nabla^V_{\partial_\alpha} B)(\partial_\beta, \partial_\gamma) + B(\nabla^V_{\partial_\alpha} \partial_\beta, \partial_\gamma) + B(\partial_\beta, \nabla^V_{\partial_\alpha} \partial_\gamma)$$

Thus the expression in braces { } in (11.53) becomes, using (10.1),

$$(\nabla^V_{\partial_\alpha} B)(\partial_\beta, \partial_\gamma) - (\nabla^V_{\partial_\beta} B)(\partial_\alpha, \partial_\gamma) = B_{\beta\gamma//\alpha} - B_{\alpha\gamma//\beta} \tag{11.55}$$

where we use the double slash // for covariant differentiation using the connection $\nabla^V$. (This should be no surprise after Problem 11.2(3).) Then (11.53) can be written

$$\begin{aligned} R(\partial_\alpha, \partial_\beta)\partial_\gamma &= R^V(\partial_\alpha, \partial_\beta)\partial_\gamma \\ &\quad + B(\partial_\alpha, \partial_\gamma)b(\partial_\beta) - B(\partial_\beta, \partial_\gamma)b(\partial_\alpha) \\ &\quad + [B_{\beta\gamma//\alpha} - B_{\alpha\gamma//\beta}]\mathbf{N} \end{aligned} \tag{11.56}$$

Finally, we may multiply by $X^\alpha Y^\beta Z^\gamma$ and sum over $\alpha$, $\beta$, and $\gamma$ to get

$$\begin{aligned} R(\mathbf{X}, \mathbf{Y})\mathbf{Z} &= R^V(\mathbf{X}, \mathbf{Y})\mathbf{Z} + B(\mathbf{X}, \mathbf{Z})b(\mathbf{Y}) - B(\mathbf{Y}, \mathbf{Z})b(\mathbf{X}) \\ &\quad + [(\nabla^V_{\mathbf{X}} B)(\mathbf{Y}, \mathbf{Z}) - (\nabla^V_{\mathbf{Y}} B)(\mathbf{X}, \mathbf{Z})]\mathbf{N} \end{aligned} \tag{11.57}$$

which is a Riemannian generalization of (8.34).

On the right-hand side, only the last line is a vector normal to $V$. Since **X**, **Y**, and **Z** are tangent to $V$, we have two consequences. First

$$\langle R(\mathbf{X}, \mathbf{Y})\mathbf{Y}, \mathbf{X}\rangle = \langle R^V(\mathbf{X}, \mathbf{Y})\mathbf{Y}, \mathbf{X}\rangle + B(\mathbf{X}, \mathbf{Y})\langle b(\mathbf{Y}), \mathbf{X}\rangle - B(\mathbf{Y}, \mathbf{Y})\langle b(\mathbf{X}), \mathbf{X}\rangle$$

or

$$\langle R(\mathbf{X}, \mathbf{Y})\mathbf{Y}, \mathbf{X}\rangle = \langle R^V(\mathbf{X}, \mathbf{Y})\mathbf{Y}, \mathbf{X}\rangle + [B(\mathbf{X}, \mathbf{Y})]^2 - B(\mathbf{Y}, \mathbf{Y})B(\mathbf{X}, \mathbf{X}) \tag{11.58}$$

Now note that if we make a substitution, $\mathbf{X} \mapsto \mathbf{X}' = a\mathbf{X} + b\mathbf{Y}$ and $\mathbf{Y} \to \mathbf{Y}' = c\mathbf{X} + d\mathbf{Y}$, then it is easy to see that

$$\langle R(\mathbf{X}', \mathbf{Y}')\mathbf{Y}', \mathbf{X}'\rangle = (ad - bc)^2 \langle R(\mathbf{X}, \mathbf{Y})\mathbf{Y}, \mathbf{X}\rangle$$

On the other hand, if we let $\| \mathbf{X} \wedge \mathbf{Y} \|$ denote the area of the parallelogram spanned by **X** and **Y**

$$\begin{aligned} \| \mathbf{X} \wedge \mathbf{Y} \|^2 &= \| \mathbf{X} \|^2 \| \mathbf{Y} \|^2 \sin^2 \angle \mathbf{X}, \mathbf{Y} \\ &= \| \mathbf{X} \|^2 \| \mathbf{Y} \|^2 - \langle \mathbf{X}, \mathbf{Y}\rangle^2 \end{aligned}$$

then under the substitution we have $\| \mathbf{X}' \wedge \mathbf{Y}' \|^2 = (ad - bc)^2 \| \mathbf{X} \wedge \mathbf{Y} \|^2$. Consequently, if **X** and **Y** are independent and if we *let* $\mathbf{X} \wedge \mathbf{Y}$ *denote symbolically the 2-plane spanned by* **X** *and* **Y**, we then have that

$$K(\mathbf{X} \wedge \mathbf{Y}) := \langle R(\mathbf{X}, \mathbf{Y})\mathbf{Y}, \mathbf{X} \rangle \, \| \mathbf{X} \wedge \mathbf{Y} \|^{-2} \tag{11.59}$$

depends only on the plane $\mathbf{X} \wedge \mathbf{Y}$ and not the basis **X**, **Y** itself. This number, which is a function of 2-planes in the tangent spaces to $M^n$, is called the **(Riemannian) sectional curvature** for the plane $\mathbf{X} \wedge \mathbf{Y}$. By taking **X** and **Y** to be orthonormal, (11.58) can be written

$$K_M(\mathbf{X} \wedge \mathbf{Y}) = K_V(\mathbf{X} \wedge \mathbf{Y}) + [B(\mathbf{X}, \mathbf{Y})]^2 - B(\mathbf{Y}, \mathbf{Y})B(\mathbf{X}, \mathbf{X}) \tag{11.60}$$

which we shall call **Gauss's equation** for the hypersurface $V^{n-1} \subset M^n$.

Our second consequence of (11.57) is what we shall call the **Codazzi equation**

$$\langle R(\mathbf{X}, \mathbf{Y})\mathbf{Z}, \mathbf{N} \rangle = (\nabla^V_{\mathbf{X}} B)(\mathbf{Y}, \mathbf{Z}) - (\nabla^V_{\mathbf{Y}} B)(\mathbf{X}, \mathbf{Z}) \tag{11.61}$$

We now will show that these two equations reduce to the surface equations of the same name.

### 11.4c. The Interpretation of the Sectional Curvature

Suppose now that we consider a submanifold $V^r \subset M^n$ *that need not be of codimension 1*. We define, for any vector field **Z** defined along V and for any vector **X** tangent to $V$ at $p$

$$\nabla^V_{\mathbf{X}} \mathbf{Z} := \text{projection of } \nabla_{\mathbf{X}} \mathbf{Z} \text{ into the tangent space } V^r_p$$

The induced connection for $V^r$ is again defined at $p \in V$ by applying this formula in the case that $\mathbf{Z} = \mathbf{Y}$ is tangent to $V$.

The normal space to $V^r$ at $p$, $(V_p)^\perp \subset M_p$ now has dimension $n - r$; let $\mathbf{N}_A$, $A = 1, \ldots, n - r$, be normal vector fields along $V$ that are orthonormal. These will exist in some small $V$-neighborhood of $p$. Then

$$\nabla^V_{\mathbf{X}} \mathbf{Y} := \nabla_{\mathbf{X}} \mathbf{Y} - \sum_A \langle \nabla_{\mathbf{X}} \mathbf{Y}, \mathbf{N}_A \rangle \mathbf{N}_A \tag{11.62}$$

For *each* normal $\mathbf{N}_A$ we shall define a second fundamental linear transformation $b_A : V_p \to V_p$ by

$$b_A(\mathbf{X}) := -\nabla^V_{\mathbf{X}}(\mathbf{N}_A) \tag{11.63}$$

(Note that although $\nabla_{\mathbf{X}} \mathbf{N}_A$ is orthogonal to $\mathbf{N}_A$, we need $\nabla^V_{\mathbf{X}} \mathbf{N}_A$ in order to assure that it is tangent to $V$!) A calculation similar to that leading to (11.60) will now lead to **Gauss's equations**

$$K_M(\mathbf{X} \wedge \mathbf{Y}) = K_V(\mathbf{X} \wedge \mathbf{Y}) + \sum_A \{[B_A(\mathbf{X}, \mathbf{Y})]^2 - B_A(\mathbf{Y}, \mathbf{Y})B_A(\mathbf{X}, \mathbf{X})\} \tag{11.64}$$

Now let **X**, **Y** be any orthonormal pair of vectors tangent to $M^n$ at a point $p$. Consider the 2-dimensional *surface* $V^2 \subset M^n$ generated by all the geodesics of $M$ that are tangent to the 2-plane $\mathbf{X} \wedge \mathbf{Y}$ at $p$. This surface is geodesic at $p$, and just as in (11.52), all second fundamental forms must vanish at $p, b_A = 0$ at $p$. Thus, from (11.60) $K_M(\mathbf{X} \wedge \mathbf{Y}) = K_V(\mathbf{X} \wedge \mathbf{Y})$. Putting $\mathbf{X} = \mathbf{e}_1, \mathbf{Y} = \mathbf{e}_2$, we see

$$K_M(\mathbf{X} \wedge \mathbf{Y}) = K_V(\mathbf{X} \wedge \mathbf{Y})$$
$$= \langle R_V(\mathbf{e}_1, \mathbf{e}_2)\mathbf{e}_2, \mathbf{e}_1 \rangle = R^V_{1212} = K$$

is the Gaussian curvature of $V^2$ with its induced Riemannian metric. Thus

**Theorem (11.65):** *$K_M(\mathbf{X} \wedge \mathbf{Y})$ is the Gaussian curvature of the 2-dimensional geodesic disc $V^2$ generated by the geodesics of $M^n$ that are tangent to the plane* $\mathbf{X} \wedge \mathbf{Y}$.

In the special case when $M^3 = \mathbb{R}^3$ and $V^2$ is a surface in $\mathbb{R}^3$, $K_V(\mathbf{X}, \mathbf{Y})$ is simply the Gauss curvature $K_V = R^V{}_{1212}$ of $V^2$ and (11.60) says that $0 = K + (b_{12})^2 - b_{11}b_{22} = K - \det b$, since $\mathbf{X}$ and $\mathbf{Y}$ are orthonormal. This is Gauss's *theorema egregium*. For the Codazzi equations (11.61), in our $V^2 \subset \mathbb{R}^3$ case, $R = 0$ and the right-hand side say, from (11.55), $b_{\beta\gamma//\alpha} = b_{\alpha\gamma//\beta}$. From Problem 11.2(3) this is the usual Codazzi equation.

### 11.4d. Fixed Points of Isometries

Let $\Phi : M^n \to M^n$ be an isometry. The **fixed set**, that is, the set $F = \{x \in M | \Phi(x) = x\}$ of points left fixed by $\Phi$, can consist perhaps of several connected pieces or "components." Consider two points $x$ and $y$ in $F$ and consider the minimal geodesic $\gamma$ joining $x$ to $y$. We know from (10.25) that such a minimal geodesic will exist if $x$ and $y$ are sufficiently close, and furthermore this minimal geodesic is unique, again if $x$ and $y$ are sufficiently close. Since the length of $\Phi(\gamma)$ is the same as the length of $\gamma$, we see that $\Phi(\gamma)$ is again a minimal geodesic joining $x$ to $y$. By uniqueness $\Phi(\gamma) = \gamma$, that is, the entire minimal geodesic joining $x$ to $y$ lies in the fixed set $F$ provided that $x$ and $y$ are in $F$ and sufficiently close. In other words, if two fixed points of an isometry are sufficiently close, then the entire geodesic joining them is fixed. It is not difficult to see then (see [$K$]) that in fact

> the fixed set of an *isometry* consists of connected components, each of which is a *totally geodesic* submanifold.

As an example, the isometry of the unit sphere $x^2 + y^2 + z^2 = 1$ that sends $(x, y, z)$ to $(x, y, -z)$ has the equator as fixed set. The "same" isometry of $\mathbb{R}P^2$ has fixed set consisting of the "equator" and the "north pole."

## Problems

**11.4(1)** Let $\mathbf{X}, \mathbf{Y}, \mathbf{Z}$, be tangent vector fields to $V^{n-1}$. Extend them in any way you wish to be vector fields on $M^n$. Show that
(i) $\nabla^V_{\mathbf{X}}\mathbf{Y} - \nabla^V_{\mathbf{Y}}\mathbf{X}$ is the Lie bracket $[\mathbf{X}, \mathbf{Y}]$ on $V$ and thus the connection $\nabla^V$ is symmetric.

(ii) Show that

$$\mathbf{X}\langle\mathbf{Y},\mathbf{Z}\rangle = \langle\nabla^V_{\mathbf{X}}\mathbf{Y},\mathbf{Z}\rangle + \langle\mathbf{Y},\nabla^V_{\mathbf{X}}\mathbf{Z}\rangle$$

and hence $\nabla^V$ *is* the Levi-Civita connection for $V$.

**11.4(2)** If you fold a sheet of paper once, why is the crease a straight line?

---

## 11.5. The Geometry of Einstein's Equations

What does the second fundamental form have to do with the expansion of the universe?

### 11.5a. The Einstein Tensor in a (Pseudo-)Riemannian Space–Time

Let $\mathbf{e}_0, \ldots, \mathbf{e}_3$ be an "orthonormal" frame at a point of a pseudo-Riemannian $M^4$. The following relations can be found in [Fr, chap. 4]). There are sign differences from the Riemannian case (considered in every book on Riemannian geometry).

Recall that a null vector $\mathbf{X}$ has $\langle\mathbf{X},\mathbf{X}\rangle = 0$. For any *nonnull* vector $\mathbf{X}$ we define its **indicator** $\epsilon(\mathbf{X}) = \text{sign}\langle\mathbf{X},\mathbf{X}\rangle$. If $\mathbf{e}_i$ is a basis vector we shall write $\epsilon(i)$ rather than $\epsilon(\mathbf{e}_i)$; thus $\epsilon(0) = -1$.

The Ricci tensor in its covariant form defines a symmetric bilinear form

$$\text{Ric}(\mathbf{X},\mathbf{Y}) := R_{ij}X^iY^j$$

In particular (11.66)

$$R_{ij} = \text{Ric}(\mathbf{e}_i,\mathbf{e}_j)$$

The Ricci *quadratic* form can be expressed in terms of sectional curvatures

$$\text{Ric}(\mathbf{e}_i,\mathbf{e}_i) = \epsilon(i)\sum_{j\neq i} K(\mathbf{e}_i\wedge\mathbf{e}_j) \tag{11.67}$$

that is, the **Ricci curvature** for the unit vector $\mathbf{e}_i$ is (except for a sign) the sum of the sectional curvatures for the $(n-1)$-basis 2-planes that include $\mathbf{e}_i$. In particular, for a Riemannian *surface* $M^2$, $\text{Ric}(\mathbf{e}_1,\mathbf{e}_1) = K(\mathbf{e}_1\wedge\mathbf{e}_2) = K$ is simply the Gauss curvature.

The scalar curvature $R$ is also the sum of sectional curvatures

$$R = R^i{}_i = \sum_{i,j,\text{ with } i\neq j} K(\mathbf{e}_i\wedge\mathbf{e}_j) \tag{11.68}$$

In the case of a surface $R = K(\mathbf{e}_1\wedge\mathbf{e}_2) + K(\mathbf{e}_2\wedge\mathbf{e}_1) = 2K$.

The **Einstein tensor** is defined to be

$$G_{ij} := R_{ij} - \frac{1}{2}g_{ij}R \tag{11.69}$$

with associated quadratic form $G(\mathbf{X},\mathbf{X}) = R_{ij}X^iX^j - (1/2)\langle\mathbf{X},\mathbf{X}\rangle R$. One then has that the Einstein quadratic form is again a "sum" of sectional curvatures, $G(\mathbf{e}_i,\mathbf{e}_i) =$

$-\epsilon(i)\sum K(\mathbf{e}_i^{\perp})$, where $\mathbf{e}_i^{\perp}$ is a basis 2-plane that is orthogonal to $\mathbf{e}_i$. For example, for the timelike $\mathbf{e}_0$

$$G(\mathbf{e}_0, \mathbf{e}_0) = K(\mathbf{e}_1 \wedge \mathbf{e}_2) + K(\mathbf{e}_1 \wedge \mathbf{e}_3) + K(\mathbf{e}_2 \wedge \mathbf{e}_3) \tag{11.70}$$
$$8\pi\kappa T(\mathbf{e}_0, \mathbf{e}_0) = K(\mathbf{e}_1 \wedge \mathbf{e}_2) + K(\mathbf{e}_1 \wedge \mathbf{e}_3) + K(\mathbf{e}_2 \wedge \mathbf{e}_3)$$

The second equation follows from Einstein's equation (11.9).

In particular, if we are dealing with an electromagnetic field, the energy–momentum tensor (as given in Problem 11.3(3)) is

$$T_{ij} = \frac{1}{4\pi}\left[F_{ik}F_j{}^k - \frac{1}{4}g_{ij}F_{rs}F^{rs}\right] \tag{11.71}$$

Let us write out $T_{00} = T(\mathbf{e}_0, \mathbf{e}_0)$ in the case of Minkowski space. (We continue to use the convention that Greek indices run from 1 to 3 while the Roman run from 0 to 3; unfortunately this is counter to the notation in most physics books.) First, from Equation (7.18), note that $F_{0k}F_0{}^k = F_{0\alpha}F_0{}^\alpha = F_{0\alpha}g^{\alpha\beta}F_{0\beta} = E_\alpha E^\alpha = E^2$. Also $F_{rs}F^{rs} = 2(F_{0\beta}F^{0\beta} + \sum_{\alpha<\beta}F_{\alpha\beta}F^{\alpha\beta})$. But $F^{0\beta} = g^{\beta\alpha}F_{0\alpha}g^{00} = E^\beta$ and so $2F_{0\beta}F^{0\beta} = -2E_\beta E^\beta = -2E^2$. Since $F_{12} = B_3$, and so on, we have $\sum_{\alpha<\beta}F_{\alpha\beta}F^{\alpha\beta} = B_\alpha B^\alpha = B^2$, and so

$$F_{rs}F^{rs} = 2(B^2 - E^2) \tag{11.72}$$

Thus in Minkowski space, $T_{00} = (4\pi)^{-1}[E^2 + (1/4)2(B^2 - E^2)]$, or

$$T_{00} = \frac{1}{8\pi}(E^2 + B^2) \tag{11.73}$$

which is the classical energy density of the electromagnetic field (see Problem 11.5(1)). *In general*, $T_{00}$ is called the **energy density** of the nongravitational fields, as measured in the frame $\mathbf{e}$, and will be denoted by $\rho$

$$8\pi\kappa\rho = K(\mathbf{e}_1 \wedge \mathbf{e}_2) + K(\mathbf{e}_1 \wedge \mathbf{e}_3) + K(\mathbf{e}_2 \wedge \mathbf{e}_3) \tag{11.74}$$

Einstein's equation (11.69) says simply that the *indicated sum of sectional curvatures is a measure of the total nongravitational energy density*!

## 11.5b. The Relativistic Meaning of Gauss's Equation

In the space–time manifold $M^4$ we may introduce local coordinates $x^0 = t, x^1, x^2$, and $x^3$ in many ways. After such a selection has been made, the submanifolds $V^3(t)$ defined by putting $x^0 =$ the constant value $t$ are called the **spatial slices** of the coordinate system. These spatial slices are spatial in the sense that $\langle \mathbf{X}, \mathbf{X}\rangle > 0$ for each nonzero tangent vector to $V(t)$. On the other hand, the "unit" normal $\mathbf{N}$ to $V(t)$ will always be a timelike vector, $\langle \mathbf{N}, \mathbf{N}\rangle = -1$. Of course we could also consider other hypersurfaces, such as, those where $x^1 =$ constant and $\mathbf{N}$ is then spacelike, but our main concern here is with the spatial slices. The reader may refer to chapter 4 of [Fr] for further discussion.

Let $\mathbf{N} = \mathbf{e}_0$ be the unit normal to the *spatial slice* $V^3(t)$. Complete $\mathbf{N}$ to an orthonormal basis. We may consider the second fundamental form $b$ of $V(t)$, defined as in Section 11.4. We must now, however, be very careful with "signs." For example, if $\mathbf{e}$ is an "orthonormal" basis, then when we expand a vector in terms of this basis, $\mathbf{v} = \sum \mathbf{e}_i v^i$, we get $v^\alpha = \langle \mathbf{v}, \mathbf{e}_\alpha \rangle$ but $v^0 = -\langle \mathbf{v}, \mathbf{e}_0 \rangle$! Thus for our spatial slice $V(t)$ we have, rather than (11.48),

$$\nabla_{\mathbf{X}}\mathbf{Y} = \nabla^V_{\mathbf{X}}\mathbf{Y} - \langle \nabla_{\mathbf{X}}\mathbf{Y}, \mathbf{N}\rangle \mathbf{N} = \nabla^V_{\mathbf{X}}\mathbf{Y} - B(\mathbf{X}, \mathbf{Y})\mathbf{N} \tag{11.75}$$

This will then introduce minus signs into the Gauss equation (11.60)

$$K_M(\mathbf{X} \wedge \mathbf{Y}) = K_V(\mathbf{X} \wedge \mathbf{Y}) - [B(\mathbf{X}, \mathbf{Y})]^2 + B(\mathbf{Y}, \mathbf{Y})B(\mathbf{X}, \mathbf{X}) \tag{11.76}$$

We must now make a comment about self-adjoint linear transformations, for example, $b$, in the case of our *pseudo*-Riemannian metric $\langle\ ,\ \rangle$. When $M$ is pseudo-Riemannian, the proof in Problem 8.2(1) of the fundamental theorem on self-adjoint transformations $A : \mathbb{R}^n \to \mathbb{R}^n$ fails because the scalar product is not positive definite. The crucial point is that in this case the "unit sphere" $\langle \mathbf{x}, \mathbf{x} \rangle = 1$ is really a hyperboloid, and is thus not compact; there is, e.g., no assurance that the continuous function $f(\mathbf{x}) = \langle \mathbf{x}, A\mathbf{x} \rangle$ will attain its maximum at any point of this hyperboloid! *Thus a self-adjoint A need not have real eigenvalues*! For example, in Minkowski 2-space with metric diag$(-1, 1)$ the linear transformation with matrix

$$\begin{bmatrix} 0 & -1 \\ 1 & 0 \end{bmatrix}$$

is self-adjoint (since its covariant version is symmetric) with eigenvalues $\pm\, i$. We, however, are concerned here with the self-adjoint $b$ that maps the tangent space to $V(t)$ into itself. Since $V(t)$ is spacelike, $V(t)$ is a *Riemannian* submanifold of the pseudo-Riemannian space–time, and thus $b$ *will* have 3 real eigenvalues, and the corresponding eigenvectors, the principal directions, can be chosen orthonormal. By applying (11.76) to an orthonormal basis of eigenvectors $\mathbf{e}_1$, $\mathbf{e}_2$, and $\mathbf{e}_3$ of $b$, we get

$$K_M(\mathbf{e}_\alpha \wedge \mathbf{e}_\beta) = K_V(\mathbf{e}_\alpha \wedge \mathbf{e}_\beta) + \kappa_\alpha \kappa_\beta \tag{11.77}$$

Put this now into (11.74), where the sectional curvatures $K$ there are for $M^4$, that is, $K = K_M$. Einstein's equation becomes

$$\begin{aligned} 8\pi\kappa\rho &= K_V(\mathbf{e}_1 \wedge \mathbf{e}_2) + K_V(\mathbf{e}_1 \wedge \mathbf{e}_3) + K_V(\mathbf{e}_2 \wedge \mathbf{e}_3) \\ &\quad + (\kappa_1\kappa_2 + \kappa_1\kappa_3 + \kappa_2\kappa_3) \end{aligned}$$

or, from (11.68)

$$8\pi\kappa\rho = \frac{1}{2}R_V + (\kappa_1\kappa_2 + \kappa_1\kappa_3 + \kappa_2\kappa_3) \tag{11.78}$$

We shall think of this as *the* geometric version of Einstein's equation involving $T_{00}$. Let us put it in the proper perspective.

For a Riemannian surface $V^2 \subset \mathbb{R}^3$ we have $K = \kappa_1\kappa_2$, which we may now write as

$$0 = \frac{1}{2}R_V - \kappa_1\kappa_2$$

This is simply Gauss's *theorema egregium*, and, as we have just seen, is a consequence of the fact that the Einstein tensor $G$ of the flat $\mathbb{R}^3$ vanishes.

Consider a 3-dimensional submanifold $V^3$ of the flat *euclidean* 4-space $\mathbb{R}^4$. The statement that the Einstein tensor G of $\mathbb{R}^4$ vanishes can be written

$$0 = \frac{1}{2}R_V - (\kappa_1\kappa_2 + \kappa_1\kappa_3 + \kappa_2\kappa_3) \tag{11.79}$$

This is a 3-dimensional version of Gauss's *theorema egregium.*

If we consider instead a 3-dimensional *spacelike* submanifold $V^3 \subset M_0^4$ of *Minkowski* space, then there is only a simple sign change, yielding

$$0 = \frac{1}{2}R_V + (\kappa_1\kappa_2 + \kappa_1\kappa_3 + \kappa_2\kappa_3)$$

This is the *theorema egregium* for such a hypersurface of Minkowski space.

Consider now a 3-dimensional spatial section $V^3$ in the actual space–time manifold $M^4$ of our physical world. Einstein's equation (11.78) then says that

> *the combination (1/2) $R_V+(\kappa_1\kappa_2+\kappa_1\kappa_3+\kappa_2\kappa_3)$ is not 0, as it was in Minkowski space, but is rather a measure of the total nongravitational energy density of space–time!*

Note that $R_V$ is an *intrinsic* measure of curvature of the spatial section $V^3$, since it is constructed from the Riemann tensor of the Riemannian $V^3$. On the other hand, the $\kappa_\alpha$'s, being principal normal curvatures, measure how $V^3$ curves in the enveloping $M^4$; thus $(\kappa_1\kappa_2 + \kappa_1\kappa_3 + \kappa_2\kappa_3)$ is a measure of **extrinsic** curvature. As J. A. Wheeler put it, Einstein's equation (11.78) may be stated as follows:

> *The sum of the intrinsic and the extrinsic curvatures of a spatial section is a measure of the nongravitational energy density of space–time.*

Finally, I wish to elaborate on (11.78), putting it in the spirit of Gauss's *theorema egregium.* Let $p$ be a point of space–time and let **N** be a given unit timelike vector at $p$. Let $V^3$ be *any* spacelike hypersurface that is orthogonal to **N** at $p$; only its tangent plane at $p$ is prescribed. $V^3$ will have a scalar curvature $R_V$ at $p$ that depends strongly on the choice of $V^3$. $V^3$ will also have normal principal curvatures $\kappa_\alpha$ at $p$, and these again will depend on the choice of $V^3$. Gauss's generalized *theorema egregium* states that the combination $(1/2)R_V + (\kappa_1\kappa_2 + \kappa_1\kappa_3 + \kappa_2\kappa_3)$ does *not* depend on the choice of $V^3$, but is in fact equal to the value $G(\mathbf{N}, \mathbf{N}) = R_{ij}N^iN^j + (1/2)R$ of the Einstein quadratic form for $M^4$ evaluated on the given normal!

### 11.5c. The Second Fundamental Form of a Spatial Slice

Consider in space–time $M^4$ a coordinate system in which the metric assumes the form

$$ds^2 = g_{00}(t, \mathbf{x})dt^2 + h_{\alpha\beta}(t, \mathbf{x})dx^\alpha dx^\beta \tag{11.80}$$

Thus $g_{0\beta} = 0$ and $g_{\alpha\beta} = h_{\alpha\beta}$ *is the Riemannian metric induced on the slice* $V^3(t)$ defined by putting $t$ constant. (Such coordinates always exist; e.g., if we take an initial slice $V^3(0)$ and introduce Gaussian geodesic coordinates as in 10.3a, we can even make $g_{00} = -1$!) As we proceed along the $t$-lines we may contemplate $\partial/\partial t g_{\alpha\beta}(t, \mathbf{x})$.

$$\frac{\partial}{\partial t} g_{\alpha\beta} = \frac{\partial}{\partial t} \langle \boldsymbol{\partial}_\alpha, \boldsymbol{\partial}_\beta \rangle = \left\langle \frac{\nabla \boldsymbol{\partial}_\alpha}{\partial t}, \boldsymbol{\partial}_\beta \right\rangle + \left\langle \boldsymbol{\partial}_\alpha, \frac{\nabla \boldsymbol{\partial}_\beta}{\partial t} \right\rangle$$

$$= \left\langle \frac{\nabla \boldsymbol{\partial}_t}{\partial x^\alpha}, \boldsymbol{\partial}_\beta \right\rangle + \left\langle \boldsymbol{\partial}_\alpha, \frac{\nabla \boldsymbol{\partial}_t}{\partial x^\beta} \right\rangle$$

Put

$$\phi := (-g_{00})^{1/2}$$

then $\mathbf{N} = \phi^{-1} \boldsymbol{\partial}_t$ is the unit normal to each spatial slice $V^3(t)$.

$$\frac{\partial (h_{\alpha\beta})}{\partial t} = \left\langle \frac{\nabla (\phi \mathbf{N})}{\partial x^\alpha}, \boldsymbol{\partial}_\beta \right\rangle + \left\langle \boldsymbol{\partial}_\alpha, \frac{\nabla (\phi \mathbf{N})}{\partial x^\beta} \right\rangle$$

$$= \phi \left[ \left\langle \frac{\nabla \mathbf{N}}{\partial x^\alpha}, \boldsymbol{\partial}_\beta \right\rangle + \left\langle \boldsymbol{\partial}_\alpha, \frac{\nabla \mathbf{N}}{\partial x^\beta} \right\rangle \right] = -\phi [b_{\alpha\beta} + b_{\beta\alpha}]$$

Thus

$$\frac{\partial h_{\alpha\beta}}{\partial t} = -2 b_{\alpha\beta} \phi \tag{11.81}$$

In words,

> *$b_{\alpha\beta}$ is essentially the measure of the rate of change of the spatial metric $h_{\alpha\beta}$ as one moves along the normal to the slices, that is, in time!*

It should be clear from this that the second fundamental form will play a crucial role in discussing the *expansion of the universe* (see [Fr, chap. 12]).

Equation (11.81) is useful in the Riemannian $V^{n-1} \subset M^n$ case as well. See Problem 11.5(2).

## 11.5d. The Codazzi Equations

So far, in this section, we have discussed mainly the geometry of the Einstein equation $G_{00} = 8\pi\kappa T_{00}$, where $T_{00}$ is the (nongravitational) energy density. We now wish to discuss the geometry of $G_{0\beta} = 8\pi\kappa T_{0\beta}$.

Recall that we have already demonstrated certain symmetries of the covariant Riemann tensor; for example, $R_{ijkl}$ is skew in $(ij)$ and also in $(kl)$. The former is Equation (9.54). Using the Bianchi identity, you are asked in Problem 11.5(3) to show that there is also the symmetry

$$R_{ijkl} = R_{klij} \tag{11.82}$$

Back to relativity. Assume a metric of the form (11.80). The Codazzi equations are given in (11.61). If you write these out in coordinate form (as you are asked to in Problem 11.5(4)) you will get

$$(-g_{00})^{-1/2} R_{0\gamma\alpha\beta} = b_{\gamma\beta//\alpha} - b_{\gamma\alpha//\beta} \tag{11.83}$$

the double slash again denoting covariant derivatives in $V^3(t)$ (recall that $b$ is a tensor on $V^3(t)$, not $M^4$). Then

$$b^{\mu}{}_{\beta//\alpha} - b^{\mu}{}_{\alpha//\beta} = h^{\mu\gamma}(b_{\gamma\beta//\alpha} - b_{\gamma\alpha//\beta}) = h^{\mu\gamma}\phi^{-1}R_{0\gamma\alpha\beta}$$

where $(h^{\mu\gamma})$ is the inverse matrix to the 3-dimensional tensor $(h_{\alpha\beta})$. Since $(g^{ij})$ is a matrix of the form

$$\begin{bmatrix} -\phi^{-2} & 0 \\ 0 & \left(h^{\alpha\beta}\right) \end{bmatrix}$$

we may write

$$\phi^{-1}h^{\mu\gamma}R_{0\gamma\alpha\beta} = \phi^{-1}g^{\mu\gamma}R_{0\gamma\alpha\beta} = \phi^{-1}g^{\mu i}R_{0i\alpha\beta} = -\phi^{-1}R^{\mu}{}_{0\alpha\beta}$$

and so $R^{\mu}{}_{0\alpha\beta} = -\phi(b^{\mu}{}_{\beta//\alpha} - b^{\mu}{}_{\alpha//\beta})$. Then

$$R_{0\beta} = R^{i}{}_{0i\beta} = R^{\alpha}{}_{0\alpha\beta} = -\phi(b^{\alpha}{}_{\beta//\alpha} - b^{\alpha}{}_{\alpha//\beta})$$

Since $g_{0\beta} = 0$, Einstein's $G_{0\beta} = 8\pi\kappa T_{0\beta}$ gives

$$8\pi\kappa T_{0\beta} = \sqrt{(-g_{00})}(H\delta^{\alpha}_{\beta} - b^{\alpha}{}_{\beta})_{//\alpha} \tag{11.84}$$

which perhaps should be called the Einstein–Codazzi equation.

In the case of electromagnetism, in Minkowski space, $T_{0\beta} = (8\pi)^{-1}F_{0k}F_{\beta}{}^{k}$, and $F_{0k}F_{\beta}{}^{k} = F_{0\alpha}F_{\beta}{}^{\alpha} = -E_{\alpha}F_{\beta}{}^{\alpha} = -E^{\alpha}F_{\beta\alpha} = E^{\alpha}F_{\alpha\beta} = E^{\alpha}B_{\alpha\beta}$ = the $\beta^{\text{th}}$ component of $i_{\mathbf{E}}\mathcal{B}^2$, that is, $-\mathbf{E}\times\mathbf{B}$. By Problem 11.5(1), this is the negative of the momentum density of the field. *In general*, $-T_{0\beta}$ is defined to be the $\beta^{\text{th}}$ component of the **momentum density** of the nongravitational fields and the Einstein–Codazzi equation (11.84) relates this to the second fundamental form of the spatial slice.

### 11.5e. Some Remarks on the Schwarzschild Solution

We refer the reader to [Fr, chap. 5] for details of the following.

The Schwarzschild solution is a *static* solution of Einstein's equations corresponding to the gravitational field exterior to a single spherically symmetric static mass ball (e.g., the region outside the sun) in an otherwise empty universe. It is not hard to see that the metric for the entire universe must be of the form

$$ds^2 = g_{00}(r)dt^2 + g_{rr}(r)dr^2 + r^2(d\theta^2 + \sin^2\theta d\phi^2) \tag{11.85}$$

in spherical coordinates $r, \theta, \phi$ with the mass center at the origin. Note that *$dr$ does not measure radial distance* from the origin; the unknown $\sqrt{g_{rr}}dr$ does! On the other hand, $r^2(d\theta^2 + \sin^2\theta d\phi^2)$ is exactly the standard metric on the 2-sphere $S^2(r)$ of radius $r$ in $\mathbb{R}^3$ (i.e., the sphere of constant Gauss curvature $K = 1/r^2$). This sphere has area $4\pi r^2$. Thus *$r$ is a radial coordinate that is normalized* not so that it is distance from the origin but rather *so that the 2-sphere $r=a$ has area $4\pi a^2$*.

The metric coefficient $g_{rr}$ can be obtained as follows. From (11.78) we see that $R_V + 2(\kappa_1\kappa_2 + \kappa_1\kappa_3 + \kappa_2\kappa_3) = 16\pi\kappa\rho$, where $V^3$ is the spatial slice $t$ = constant and $R_V$ is the Ricci scalar curvature of $V$. But, from (11.81), the second fundamental form of a spatial slice vanishes in a static universe. We conclude that $R_V = 16\pi\kappa\rho$ and in particular $R_V = 0$ in the region outside the ball of matter.

We wish then to determine the metric coefficient $g_{rr}$ on the spherically symmetric $V^3$ with $R_V = 16\pi\kappa\rho(r)$. We may try to realize such a Riemannian $V^3$ as an embedded 3-manifold (again called $V^3$) in **euclidean** $\mathbb{R}^4 = \mathbb{R} \times \mathbb{R}^3$ with coordinates $w, r, \theta, \phi$, which respects spherical symmetry, that is, is invariant under the rotation group SO(3) acting on the space $\mathbb{R}^3$.

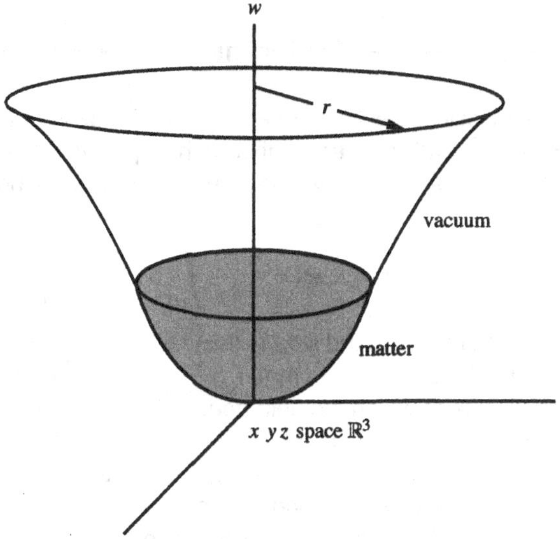


**Figure 11.5**

We assume a graph of the form $w = w(r, \theta, \phi) = w(r)$. Thus the slices $w$ = constant are simply 2-spheres, and the function $w$ of $r$ is to be determined so that $R_V = 16\pi\kappa\rho(r)$; since we are interested here in the region **exterior** to the ball, we shall not be concerned that $\rho$ is not known explicitly as a function of $r$.

For the entire $V^3$ sitting in $\mathbb{R}^4$, we may again apply Gauss's equation (11.79), where now the $\kappa$'s are the principal curvatures of $V^3 \subset \mathbb{R}^4$. It is easy to compute the normal curvatures for this 3-dimensional analogue of a surface of revolution, and in Chapter 5 of [Fr] it is shown that exterior to the ball, $w$ takes a parabolic form, yielding the **Flamm paraboloid**

$$w^2 = 8m(r - 2m) \quad \text{and} \quad g_{rr} = \left(1 - \frac{2m}{r}\right)^{-1}$$

where

$$m = \kappa \int_0^a 4\pi r^2 \rho(r) dr$$

is a measure of the "total mass" of the ball of coordinate "radius" a. Thus $V^3$ carries the spatial metric $(1 - 2m/r)^{-1} dr^2 + r^2(d\theta^2 + \sin^2\theta d\phi^2)$.

In Problem 11.1(2) it was shown that $U = (1 - 2m/r)^{1/2}$ is a solution to Laplace's equation in the spatial Schwarzschild metric, and, for large $r$, $U$ is of the form $U \sim$

$1 - m/r$. Thus $1 - (1 - 2m/r)^{1/2}$ is a good candidate for the "correct" gravitational potential in the exterior region. As in Section 11.1c, this suggests that $\sqrt{-g_{00}} = 1 - U = (1 - 2m/r)^{1/2}$ and so $g_{00} = -(1 - 2m/r)$. In [Fr] it is shown that this is in fact the solution demanded by the remaining Einstein equations. Thus in the external region we have the **Schwarzschild solution**

$$ds^2 = -\left(1 - \frac{2m}{r}\right)dt^2 + \left(1 - \frac{2m}{r}\right)^{-1} dr^2 + r^2(d\theta^2 + \sin^2\theta d\phi^2) \tag{11.86}$$

## Problems

**11.5(1)** Consider the classical electromagnetic field in $\mathbb{R}^3$, as in Section 3.5. Note that $\mathcal{E} \wedge * \mathcal{E} = \mathbf{E} \bullet \mathbf{E} vol^3$, and so differentiating with respect to time gives $\partial/\partial t(\mathcal{E} \wedge * \mathcal{E}) = 2\mathcal{E} \wedge \partial * \mathcal{E}/\partial t$. Likewise, we may compute $\partial/\partial t(\mathcal{B} \wedge * \mathcal{B})$. Show that for a fixed compact region $U$ of $\mathbb{R}^3$, we have

$$\frac{d}{dt}\frac{1}{8\pi}\int_U \mathcal{E} \wedge * \mathcal{E} + \mathcal{B} \wedge * \mathcal{B} = -\frac{1}{4\pi}\int_{\partial U} \mathcal{E} \wedge * \mathcal{B} - \int_U \mathcal{E} \wedge \mathfrak{j}^2 \tag{11.87}$$

The integrand on the left-hand side is $(8\pi)^{-1}(E^2 + B^2) \geq 0$ and is the classical *energy* density of the field. Note that $\int_U \mathcal{E}^1 \wedge \mathfrak{j}^2 = \int_U \mathbf{E} \bullet \mathbf{J} vol^3 \sim \int_U \mathbf{E} \bullet \rho \mathbf{v} vol^3$ represents the rate at which the field does work on the charges in the current. Then $(4\pi)^{-1} \int_{\partial U} \mathcal{E}^1 \wedge * \mathcal{B}^2 = \int_{\partial U} (4\pi)^{-1} \mathbf{E} \times \mathbf{B} \bullet d\mathbf{S}$ is interpreted as the flux of energy through $\partial U$. Relativistically, energy is the same as mass. But the flux of mass through a surface is given classically by the surface integral of the *momentum* density. (For example, in the case of a fluid with mass density $\rho$ we have $\int_{\partial U} \rho \mathbf{v} \bullet d\mathbf{S} = -d/dt \int_U \rho vol^3$.) Thus we may consider $(4\pi)^{-1}\mathbf{E} \times \mathbf{B}$, the **Poynting vector**, to be the *momentum* density of the field. Equation (11.87) is Poynting's theorem.

**11.5(2)** In the Riemannian case one puts $\phi = (g_{00})^{1/2}$, but (11.81) still holds. Show that

$$\frac{\partial}{\partial t}\sqrt{\det(h_{\alpha\beta})} = -\phi H \sqrt{\det(h_{\alpha\beta})}$$

(see (11.39)). Since $\sqrt{\det(g_{\alpha\beta})}dx^1 \wedge \ldots \wedge dx^{n-1}$ is the "area" form $dS^{n-1}$ for $V^{n-1}$ we may write for the first variation of area

$$\frac{d}{dt}\int_{V(t)} dS^{n-1} = -\int_{V(t)} \phi H dS^{n-1}$$

This is the $(n-1)$-dimensional version of (8.26), but where is the boundary term?

**11.5(3)** Prove (11.82).

**11.5(4)** Prove (11.83).

CHAPTER 12

# Curvature and Topology: Synge's Theorem

In Problem 8.3(7) it was shown that if $M^2$ is a closed surface in $\mathbb{R}^3$ then its curvature $K$ and its "genus" $g$ are related by

$$\frac{1}{2\pi}\int_M K\,dS = 2 - 2g \tag{12.1}$$

This is the Gauss–Bonnet theorem. In particular, when $M^2$ is a (perhaps) distorted torus (i.e., a surface of genus 1), then $(2\pi)^{-1}\int_M K\,dS = 0$. Thus it is not possible to embed the torus in $\mathbb{R}^3$ in such a way that its Gauss curvature is everywhere positive. This is not surprising; a few sketches of tori will "convince" one that there will always have to be saddle points somewhere. However, in Part Three, we shall see that (12.1) is true even for an *abstract* Riemannian metric (without any question of an embedding in $\mathbb{R}^3$). This is an example of a **global** or **topological** result, relating the purely "infinitesimal" notion of curvature to the topological notion of the genus of the surface.

In this brief chapter we will discuss a relation between curvature and the topological notion of simple connectivity, namely the theorem of J. L. Synge, one of the most beautiful results in global differential geometry of the twentieth century. In the process of proving Synge's theorem, we shall derive a formula, also due to Synge, for the second variation of arc length along a geodesic.

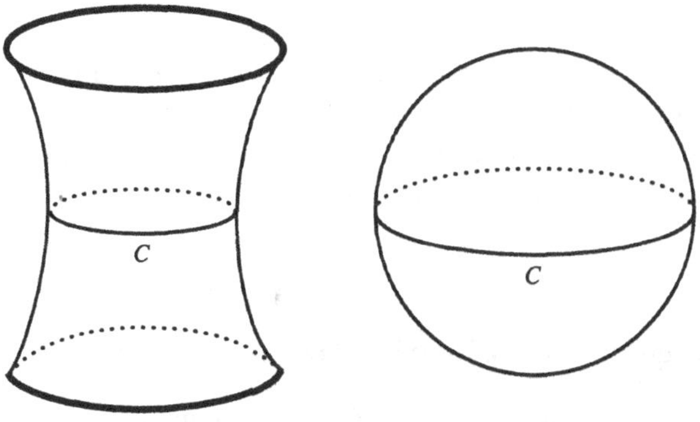


Figure 12.1

In the figure, we have drawn a closed geodesic $C$, first on a surface with negative curvature and then on a positively curved sphere. If we consider only variations of $C$ by smooth *closed* curves (where $P = Q$ and the tangents match up at $P$), it is clear from (10.4) that the first variation of arc length vanishes in both cases, the endpoint contributions cancel in the case of a closed geodesic. Still, we *can* shorten the equator $C$ on the sphere by pushing it north! We could say that in the "space $\Omega S^2$ of all smooth closed curves on $S^2$," the length functional $L$ has first derivative 0 at the point representing the equator $C$ but $C$ does not yield a relative minimum for $L$. We shall see, from Synge's formula, that in this case of positive curvature the *second variation is negative* for the variation pushing $C$ north, explaining why this geodesic is **unstable**. (A slippery rubber band stretched along the equator would contract if disturbed slightly.) It seems evident that the equator in the negatively curved surface is stable, yielding an (absolute) minimum for $L$, and this will also follow from Synge's formula.

## 12.1. Synge's Formula for Second Variation

What does curvature have to do with the stability of a geodesic?

### 12.1a. The Second Variation of Arc Length

We first introduce a notation that will simplify the appearance of our calculations. Consider, as in Section 10.1b, the variation of arc length. We have the tangent vector field $\mathbf{T} = \partial \mathbf{x}/\partial s$ and the variation field $\mathbf{J} = \partial \mathbf{x}/\partial \alpha$, both defined along the 2-dimensional variational surface. We shall write with some misgivings

$$\nabla_{\mathbf{T}}\mathbf{J} := \frac{\nabla \mathbf{J}}{\partial s} \quad \text{and} \quad \nabla_{\mathbf{J}}\mathbf{T} := \frac{\nabla \mathbf{T}}{\partial \alpha} \tag{12.2}$$

even though $\mathbf{T}$ and $\mathbf{J}$ are defined only along the variational surface. We shall also write, for instance, $\nabla_{\mathbf{T}}\mathbf{w}$ rather than $\nabla \mathbf{w}/\partial s$ when $\mathbf{w}$ is a field defined along the variational surface. Thus Lemmas (10.1) and (10.2) of Section 10.1 then take the form

$$\nabla_{\mathbf{T}}\mathbf{J} = \nabla_{\mathbf{J}}\mathbf{T}$$

and (12.3)

$$\nabla_{\mathbf{T}}\nabla_{\mathbf{J}}\mathbf{w} - \nabla_{\mathbf{J}}\nabla_{\mathbf{T}}\mathbf{w} = R(\mathbf{T}, \mathbf{J})\mathbf{w}$$

We now return to our consideration of arc length variation, started in Section 10.1. We suppose now that the base curve $C_0$, given by $\alpha = 0$, is a geodesic of length $L$. Recall that the parameter $s$ need be arc length only when $\alpha = 0$.

We shall only be concerned with the case in which the *first variation vanishes*, $L'(0) = 0$. From (10.4) we see that this requires, in this case of a geodesic base curve, that

$$\langle \mathbf{J}, \mathbf{T} \rangle_Q = \langle \mathbf{J}, \mathbf{T} \rangle_P$$

From the first equation of (10.4) we have, in our new notation

$$L'(\alpha) = \int_0^L \langle \mathbf{T}, \mathbf{T} \rangle^{-1/2} \langle \nabla_{\mathbf{T}}\mathbf{J}, \mathbf{T} \rangle ds$$

Then

$$L''(\alpha) = \int_0^L [-\langle \mathbf{T}, \mathbf{T} \rangle^{-3/2} \langle \nabla_{\mathbf{T}} \mathbf{J}, \mathbf{T} \rangle^2 + \langle \mathbf{T}, \mathbf{T} \rangle^{-1/2} \{ \langle \nabla_{\mathbf{J}} \nabla_{\mathbf{T}} \mathbf{J}, \mathbf{T} \rangle + \langle \nabla_{\mathbf{T}} \mathbf{J}, \nabla_{\mathbf{J}} \mathbf{T} \rangle \}] ds$$

Since $\| \mathbf{T} \| = 1$ when $\alpha = 0$, and since $\nabla_{\mathbf{T}} \mathbf{J} = \nabla_{\mathbf{J}} \mathbf{T}$

$$L''(0) = \int_0^L [-\langle \nabla_{\mathbf{T}} \mathbf{J}, \mathbf{T} \rangle^2 + \{ \langle \nabla_{\mathbf{J}} \nabla_{\mathbf{T}} \mathbf{J}, \mathbf{T} \rangle + \langle \nabla_{\mathbf{T}} \mathbf{J}, \nabla_{\mathbf{T}} \mathbf{J} \rangle \}] ds$$

Note that

$$\langle \nabla_{\mathbf{T}} \mathbf{J}, \nabla_{\mathbf{T}} \mathbf{J} \rangle - \langle \nabla_{\mathbf{T}} \mathbf{J}, \mathbf{T} \rangle^2 = \| \nabla_{\mathbf{T}} \mathbf{J} \|^2 - \| \nabla_{\mathbf{T}} \mathbf{J} \|^2 \cos^2 \theta$$

where $\theta$ is the angle between $\nabla_{\mathbf{T}} \mathbf{J}$ and $\mathbf{T}$. But this is simply the square of the area $\| (\nabla_{\mathbf{T}} \mathbf{J}) \wedge \mathbf{T} \|^2$ of the parallelogram spanned by these two vectors. Thus

$$L''(0) = \int_0^L \{ \langle \nabla_{\mathbf{J}} \nabla_{\mathbf{T}} \mathbf{J}, \mathbf{T} \rangle + \| (\nabla_{\mathbf{T}} \mathbf{J}) \wedge \mathbf{T} \|^2 \} ds \tag{12.4}$$

Look now at the first integrand

$$\langle \nabla_{\mathbf{J}} \nabla_{\mathbf{T}} \mathbf{J}, \mathbf{T} \rangle = \langle \nabla_{\mathbf{T}} \nabla_{\mathbf{J}} \mathbf{J}, \mathbf{T} \rangle + \langle R(\mathbf{J}, \mathbf{T}) \mathbf{J}, \mathbf{T} \rangle$$

But $\langle \nabla_{\mathbf{T}} \nabla_{\mathbf{J}} \mathbf{J}, \mathbf{T} \rangle = \partial / \partial s \langle \nabla_{\mathbf{J}} \mathbf{J}, \mathbf{T} \rangle - \langle \nabla_{\mathbf{J}} \mathbf{J}, \nabla_{\mathbf{T}} \mathbf{T} \rangle$, and so

$$\int_0^L \langle \nabla_{\mathbf{T}} \nabla_{\mathbf{J}} \mathbf{J}, \mathbf{T} \rangle ds = \langle \nabla_{\mathbf{J}} \mathbf{J}, \mathbf{T} \rangle_0^L$$

Equation (12.4) then becomes

$$L''(0) = \langle \nabla_{\mathbf{J}} \mathbf{J}, \mathbf{T} \rangle_0^L + \int_0^L \{ \langle R(\mathbf{J}, \mathbf{T}) \mathbf{J}, \mathbf{T} \rangle + \| (\nabla_{\mathbf{T}} \mathbf{J}) \wedge \mathbf{T} \|^2 \} ds$$

The statement, (9.54), that the covariant Riemann tensor is skew in the first two indices translates to the statement

$$\langle R(\mathbf{J}, \mathbf{T}) \mathbf{J}, \mathbf{T} \rangle = -\langle R(\mathbf{J}, \mathbf{T}) \mathbf{T}, \mathbf{J} \rangle$$

as one easily sees by expressing this in terms of components. Thus we finally have our principal formula, dating from the year 1925.

**Synge's Formula (12.5):** *For a variation of a geodesic in which the first variation vanishes,* $\langle \mathbf{J}, \mathbf{T} \rangle_Q = \langle \mathbf{J}, \mathbf{T} \rangle_P$, *we have*

$$L''(0) = \langle \nabla_{\mathbf{J}} \mathbf{J}, \mathbf{T} \rangle_0^L + \int_0^L \{ \| (\nabla_{\mathbf{T}} \mathbf{J}) \wedge \mathbf{T} \|^2 - \langle R(\mathbf{J}, \mathbf{T}) \mathbf{T}, \mathbf{J} \rangle \} ds$$

Note also that when the variation is *orthogonal to the geodesic*, that is, when $\langle \mathbf{J}, \mathbf{T} \rangle = 0$, then $\langle \nabla_{\mathbf{T}} \mathbf{J}, \mathbf{T} \rangle = \mathbf{T} \langle \mathbf{J}, \mathbf{T} \rangle - \langle \mathbf{J}, \nabla_{\mathbf{T}} \mathbf{T} \rangle = 0$, and $\| (\nabla_{\mathbf{T}} \mathbf{J}) \wedge \mathbf{T} \|^2$ becomes simply $\| \nabla_{\mathbf{T}} \mathbf{J} \|^2$.

**Corollary (12.6):** *For an othogonal variation of a geodesic we have*

$$L''(0) = \langle \nabla_{\mathbf{J}}\mathbf{J}, \mathbf{T}\rangle_0^L + \int_0^L \{\| \nabla_{\mathbf{T}}\mathbf{J} \|^2 - \langle R(\mathbf{J}, \mathbf{T})\mathbf{T}, \mathbf{J}\rangle\} ds$$

Recall (from (11.59)) that in a *Riemannian* manifold $M^n$, $\| \mathbf{A} \| \geq 0$ for all $\mathbf{A}$ and $M$ has negative sectional curvature if $\langle R(\mathbf{J}, \mathbf{T})\mathbf{T}, \mathbf{J}\rangle$ is negative whenever $\mathbf{T}$ and $\mathbf{J}$ are linearly independent. Consider a geodesic $C$ in such a space joining distinct points $P$ and $Q$. To see whether $C$ locally minimizes arc length between $P$ and $Q$ we consider a variation $\mathbf{J}$ that vanishes at $P$ and $Q$. Thus the endpoint contribution vanishes in Synge's formula. If $\mathbf{J}$ and $\mathbf{T}$ are not everywhere dependent along $C$ the integral will be positive. If $\mathbf{J} = f(s)\mathbf{T}$ along $C$, then the variation associated to $\mathbf{J}$ does not change the curve $C$ at all. From (12.5).

**Corollary (12.7):** *In a negatively curved Riemannian $M^n$, a nontrivial variation of a geodesic $C$ joining distinct points $P$ and $Q$ yields $L''(0) > 0$ and so $C$ is stable, that is, locally minimizes arc length.*

In the case of a *closed* geodesic, $\mathbf{J}$ need not vanish at $P = Q$, but both $\mathbf{T}$ and $\mathbf{J}$ match up at $P = Q$, and so the first variation still vanishes. Furthermore, $\langle \nabla_{\mathbf{J}}\mathbf{J}, \mathbf{T}\rangle_0^L = 0$. We conclude that $L''(0) \geq 0$, and $= 0$ only if $\mathbf{J}$ is a multiple of $\mathbf{T}$ along $C$; this would simply move the geodesic into itself.

**Corollary (12.8):** *In a negatively curved Riemannian $M^n$, each closed geodesic is stable.*

## 12.1b. Jacobi Fields

We shall reconsider the case of *distinct endpoints* when the variation field $\mathbf{J}$ vanishes at the endpoints and is orthogonal to $\mathbf{T}$. Then, as we have seen,

$$\langle \nabla_{\mathbf{T}}\mathbf{J}, \mathbf{T}\rangle = \mathbf{T}\langle \mathbf{J}, \mathbf{T}\rangle - \langle \mathbf{J}, \nabla_{\mathbf{T}}\mathbf{T}\rangle = \frac{\partial\langle \mathbf{J}, \mathbf{T}\rangle}{\partial s} = 0$$

and so

$$\int_0^L \| (\nabla_{\mathbf{T}}\mathbf{J}) \wedge \mathbf{T} \|^2 ds = \int_0^L \{\langle \nabla_{\mathbf{T}}\mathbf{J}, \nabla_{\mathbf{T}}\mathbf{J}\rangle - \langle \nabla_{\mathbf{T}}\mathbf{J}, \mathbf{T}\rangle^2\} ds$$
$$= \int_0^L \langle \nabla_{\mathbf{T}}\mathbf{J}, \nabla_{\mathbf{T}}\mathbf{J}\rangle ds$$

Synge's formula then reads

$$L''(0) = \int_0^L \{\langle \nabla_{\mathbf{T}}\mathbf{J}, \nabla_{\mathbf{T}}\mathbf{J}\rangle - \langle R(\mathbf{J}, \mathbf{T})\mathbf{T}, \mathbf{J}\rangle\} ds \tag{12.9}$$

But $\langle \nabla_{\mathbf{T}}\mathbf{J}, \nabla_{\mathbf{T}}\mathbf{J}\rangle = \mathbf{T}\langle \mathbf{J}, \nabla_{\mathbf{T}}\mathbf{J}\rangle - \langle \mathbf{J}, \nabla_{\mathbf{T}}\nabla_{\mathbf{T}}\mathbf{J}\rangle$ and the first term integrates to 0 since $\mathbf{J}$ vanishes at the endpoints. We then have

$$L''(0) = -\int_0^L \{\langle \nabla_{\mathbf{T}}\nabla_{\mathbf{T}}\mathbf{J} + R(\mathbf{J}, \mathbf{T})\mathbf{T}, \mathbf{J}\rangle\} ds \tag{12.10}$$

*for variations that vanish at the endpoints and are orthogonal to the geodesic.*

Note that if $\mathbf{J}$ is a *Jacobi field*, then $L''(0) = 0$. Thus, from Problem 10.1(3), *if we vary the geodesic $C$ by a 1-parameter family of geodesics passing through $P$ and $Q$, both the first and the second variations vanish*!

This has the following consequence. (Our treatment will be *very* brief; for a more careful treatment see, e.g., [Do, p. 423].)

First note that given *any* vector field $\mathbf{X} = \mathbf{X}(s)$ defined along a curve $C$, we can define a variation of $C$ having variation vector given by $\mathbf{X}$. There are many ways of forming such variations. For $x(s, \alpha)$ we may merely put $x(s, \alpha) = \exp_{x(s)} \alpha\mathbf{X}(s)$; that is, $x(s, \alpha)$ is the point on the geodesic starting at $x(s)$ on $C$ in the direction of $\mathbf{X}(s)$, and at distance $\| \alpha\mathbf{X}(s) \|$ from $x(s)$.

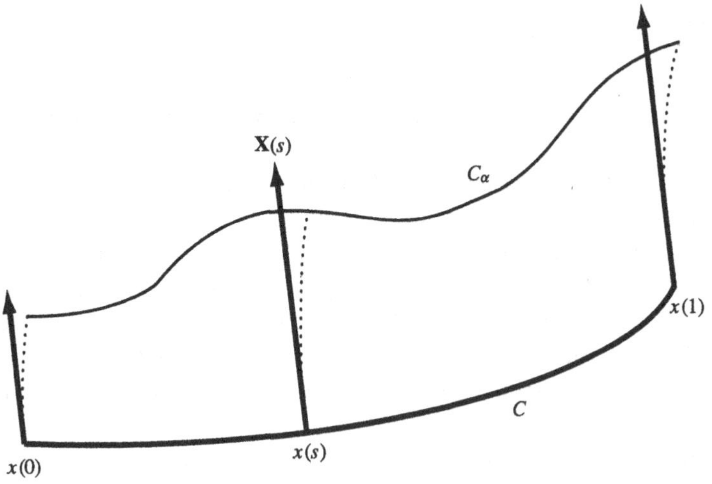

Figure 12.2

Suppose that there is a nontrivial *Jacobi* field $\mathbf{J}$ along the geodesic $C$ that vanishes at $P$ and at some point $P'$ between $P$ and $Q$; *we do not assume that $J$ vanishes at $Q$*. We call $P'$ a **conjugate point** to $P$ along the geodesic $C$.

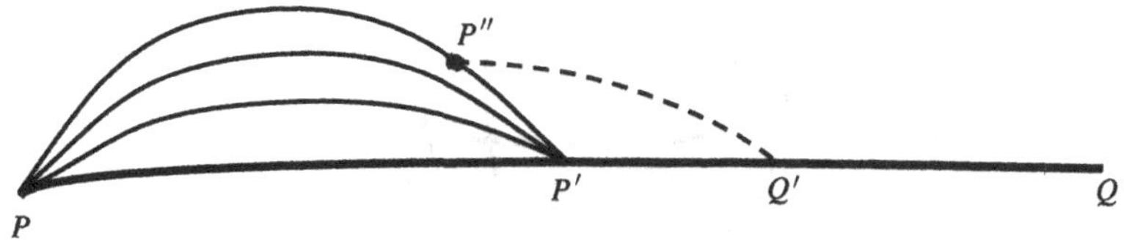

Figure 12.3

Using **J** we may construct a variation of the *portion* $PP'$ of $C$ as before. Note that *different variations having the same variation vector $J$ at $\alpha = 0$ will yield the same second variation formula* (12.10)! The varied curves $C_\alpha$ pass through $P'$ and, by (12.10), have the same length, to second order, as the arc $PP'$ of the base curve $C$.

The varied curves $C_\alpha$ meet $C = C(0)$ transversally if $\alpha$ is small enough; we see this as follows. We have already mentioned that a Jacobi field can be realized by varying $C$ by geodesics $C_\alpha$. If a geodesic $C_\alpha$ were tangent to the geodesic $C$ at some point $P'$, then $C_\alpha$ would coincide with $C$ and so $J \equiv 0$. Thus $C_\alpha$ is transversal to $C$ at $P'$.

Let then $P''$ be a point on $C_\alpha$, and $Q'$ a point on $C$, that are so close that there is a unique *minimal* geodesic $P''Q'$ joining them. Then the geodesic arc $P''Q'$ is strictly shorter than the broken arc $P''P'Q'$. This says then that the curve of broken arcs $PP''Q'Q$ is shorter than the original geodesic $PP'Q'Q = PQ$. The broken $PP''Q'Q$ can then be smoothed off to yield a smooth curve that is again shorter than $C$. We have "shown" that

**Theorem (12.11):** *If a geodesic arc $C$ contains a point $P'$ conjugate to the beginning point $P$ in its interior, then $C$ is not a minimizing geodesic; that is, $C$ is not stable.*

*Thus a geodesic cannot be minimizing after passing a point conjugate to the initial point!*

In fact Marston Morse has shown the following (see [$M$]). Let us say that the point $P'$ conjugate to $P$ has (Morse) **index** $\lambda$ iff there are exactly $\lambda$ linearly independent Jacobi fields along $C$ that vanish at both $P$ and $P'$. (This makes sense since the Jacobi equation is *linear* in **J**.) Suppose that $P'_1, \ldots, P'_r$ are exactly the conjugate points to $P$ that are between $P$ and $Q$, and that $P'_i$ has index $\lambda(i)$. We define the (Morse) **index of** $C$ to be $\sum_i \lambda(i)$, the sum of the indices of all conjugate points $P'$ interior to $PQ$. Then in a certain well-defined sense, *there are essentially $\sum_i \lambda(i)$ independent variations of $C$ that strictly decrease the arc length of $C$.*

For example, consider on the $n$ sphere the geodesic (great circle) $C$ that starts at the north pole $P$, passes through a point $Q$ on the equator, goes all the way around to $P$ again, and continues on to the point $Q$.

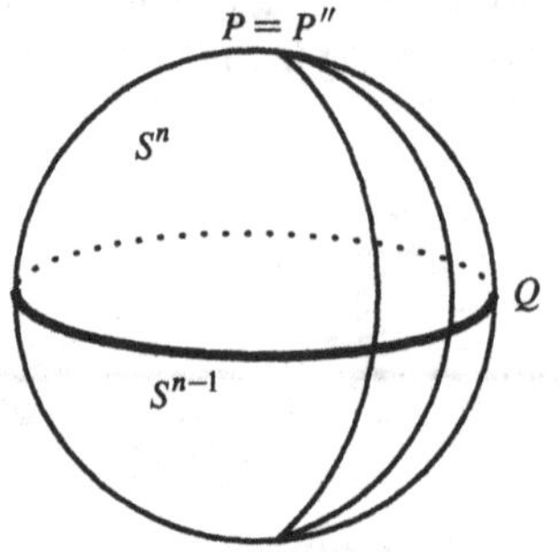


Figure 12.4

The first conjugate point to $P$ is the south pole $P'$; the next and last is $P'' = P$ itself (at arc length $2\pi$). For the arc $PP'$ there is an $(n-1)$-dimensional family of great circles (parameterized by the equator $S^{n-1}$); these yield an $(n-1)$-dimensional space of Jacobi fields vanishing at $P'$, and thus the index of the conjugate point $P'$ is $n-1$. These geodesics also yield variations of the segment $PP'P''$, and so $P''$ is conjugate to $P$ with index $n-1$. Thus the Morse index of the geodesic $PQP'P''Q$ is $2(n-1)$; *there are basically* $2n-2$ *independent variations of* $C$ *that decrease the length of* $C$.

## Problem

**12.1(1)** Use (11.82) and $[R(\mathbf{J},\mathbf{T})\mathbf{T}]^a = T^b R^a{}_{bcd} J^c T^d$ to show that the Jacobi linear transformation

$$\mathbf{J} \mapsto R(\mathbf{J},\mathbf{T})\mathbf{T}$$

is self-adjoint.

## 12.2. Curvature and Simple Connectivity

How is positive curvature related to simple connectivity?

### 12.2a. Synge's Theorem

**Theorem (12.12):** *Let* $M^{2n}$ *be an even-dimensional, orientable manifold with positive sectional curvatures,* $K(\mathbf{X}\wedge\mathbf{Y}) > 0$. *Then any closed geodesic is unstable, that is, can be shortened by a variation.*

For example, the equatorial great circle on the round 2-sphere can be shortened by pushing it north.

**Proof of Synge's theorem:** Let $C, x = x(s)$, be a closed geodesic. We first claim that we can find a unit vector field $\mathbf{J}$ along $C$ that is normal to $C$ and parallel displaced along $C$. This is proved as follows. Since parallel translation around $C$ will send the geodesic tangent $\mathbf{T}$ into itself, parallel translation around $C$ will also take the $(2n-1)$-dimensional plane of vectors normal to $\mathbf{T}$ into itself. Let $\mathbf{T}^\perp$ be the normal plane at $x(0)$. Parallel translation around $C$ will give a map $P : \mathbf{T}^\perp \to \mathbf{T}^\perp$. This map is linear since the differential equations of parallel translation are linear. We know that this map is an isometry; thus $P$ is given by an orthogonal matrix, $P^T = P^{-1}$. $P$ cannot reverse the orientation of $\mathbf{T}^\perp$, for if it did, since $\mathbf{T}$ is sent into itself, parallel translation would have reversed the orientation of the $2n$-dimensional tangent space to $M$ at $x(0)$, contradicting the assumption that $M$ is orientable. Thus $\det P = +1$. But the eigenvalues of $P$ either are real or occur in complex conjugate pairs, and since there are $2n-1$ of them, we conclude that there are an odd number of real eigenvalues. But each of these must be $\pm 1$, and yet $\det P =$ (the product of all the eigenvalues) $= 1$. Thus there is at least one eigenvalue $\lambda = +1$. But this means that some normal vector $\mathbf{J}$ must be sent into itself under the parallel translation; $\mathbf{J}(s)$ is a normal parallel displaced vector along $C$!

We may then construct a variation of $C$ by again considering the geodesics tangent to the vectors $\mathbf{J}$, that is,

$$x(s, \alpha) := \exp_{x(s)}\{\alpha \mathbf{J}(x(s))\}$$

By construction $(\partial x/\partial \alpha)(s, 0) = \mathbf{J}(x(s))$; that is, this variation has $\mathbf{J}$ as its variation vector. Look at Synge's formula (12.6). The boundary term vanishes since we have a closed curve. Further, $\nabla_{\mathbf{T}}\mathbf{J} = 0$ since $\mathbf{J}$ is parallel displaced. Thus

$$L''(0) = -\int_0^L K(\mathbf{T} \wedge \mathbf{J}) ds \tag{12.13}$$

since $\mathbf{T}$ and $\mathbf{J}$ are orthonormal. We conclude that $L''(0) < 0$. Since $L'(0) = 0$ for the geodesic we conclude that such a variation would decrease the length of the curve for small $\alpha$. $\square$

There *are* spaces with positive sectional curvatures. The usual paraboloid in $\mathbb{R}^3$ has positive curvature, and any deformation of it, if sufficiently small, will also. Likewise for the unit sphere (which is compact). The unit sphere $S^n \subset \mathbb{R}^{n+1}$ has sectional curvatures all unity. To see this we use the Gauss equation (11.60) applied to $M = \mathbb{R}^{n+1}$ and $V = S^n$. Since $K_M = 0$ for $M$ euclidean we have $K_V(\mathbf{X}, \mathbf{Y}) = B(\mathbf{Y}, \mathbf{Y})B(\mathbf{X}, \mathbf{X}) - \{B(\mathbf{X}, \mathbf{Y})\}^2$. For two orthogonal principal directions $\mathbf{X} = \mathbf{e}_1$ and $\mathbf{Y} = \mathbf{e}_2$ we would have $K_V(\mathbf{e}_1, \mathbf{e}_2) = \kappa_1\kappa_2$. But by symmetry, all principal curvatures for the round unit sphere must coincide, $\kappa_i = -1$ (using the outward-pointing normal). Thus all sectional curvatures for the unit $n$-sphere are $+1$.

For another example, consider the real projective $n$-space $\mathbb{R}P^n$. This is the space resulting from the unit $n$-sphere when antipodal pairs are identified. Any tangent vector $\mathbf{X}$ to $\mathbb{R}P^n$ corresponds to a pair of tangent vectors, $\mathbf{Y}$ and $-\mathbf{Y}$, to $S^n$ at antipodal points. These vectors have the same length, and thus there is no ambiguity in *defining* $\| \mathbf{X} \|$ to be the length in the Riemannian $S^n$ of either of the tangent vectors $\pm\mathbf{Y}$ "covering" $\mathbf{X}$. This defines a Riemannian metric for $\mathbb{R}P^n$. It should be clear that the 2:1 projection (identification) map $\pi : S^n \to \mathbb{R}P^n$ is then a local isometry, and thus the Riemann tensors of the two spaces agree at corresponding points, if we use local coordinates in $S^n$ that result from pulling back local coordinates in $\mathbb{R}P^n$ (see Section 8.5b). *Thus $\mathbb{R}P^n$ carries a Riemannian metric with sectional curvatures $K = 1$ again!*

We have mentioned in Section 10.2d that if a compact manifold is not simply connected, then among a free homotopy class of closed curves that cannot be shrunk to a point, there will be a shortest curve and it will be a closed geodesic. Thus we have Synge's theorem of 1936.

**Corollary (12.14):** *A compact, orientable, even-dimensional manifold with positive sectional curvatures is simply connected.*

### 12.2b. Orientability Revisited

The example $\mathbb{R}P^n$ is especially interesting with regard to Synge's corollary because *$\mathbb{R}P^n$ is not simply connected*! This can be seen as follows. An arc $C'$ on $S^n$ going from the north to the south pole projects down to yield a *closed* curve $C$ on $\mathbb{R}P^n$ since the north and south poles project to the same point (call it $N$) on $\mathbb{R}P^n$. *We claim that $C$ cannot be deformed to a point on $\mathbb{R}P^n$.* Let $C$ be parameterized, $x = x(t)$, with $x(0) = x(1) = N$. It should be clear that any deformation of $C$ can be "covered" by a deformation of $C'$ on $S^n$, using the identification. Under a deformation of $C$, the point $N$ might move to another point $N_\alpha$, and then the covered curve $C'_\alpha$ would start at one of the two points on $S^n$ covering $N_\alpha$ and end at its antipodal point $-N_\alpha$. If C could be deformed to a point, then eventually we would have to cover this single *point* curve $C_1$ at $N_1$ by a whole arc on $S^n$ going from a point over $N_1$ to its antipode. This is impossible since $N_1$ is covered only by two points on $S^n$. □

The fact that $\mathbb{R}P^3$ is not simply connected has the following application to mechanics ([A, p. 248]).

**Theorem (12.15):** *A rigid body in $\mathbb{R}^3$, fixed at one point and subject to any potential field, has a periodic motion for any sufficiently large total energy $E$.*

PROOF: A rigid body in $\mathbb{R}^3$ has the rotation group $SO(3)$ as configuration space (see Section 1.1d). For sufficiently large total energy $H = E$, the Jacobi metric (10.19) defines a Riemannian metric on all of $SO(3)$ in which the geodesics represent the motions of the system. But $SO(3)$ is topologically $\mathbb{R}P^3$ (see 1.2b, Example vii). Since $\mathbb{R}P^3$ is not simply connected, there exists a closed geodesic, and this corresponds to a periodic motion of the body. □

Does the fact that $\mathbb{R}P^{2n}$ is not simply connected contradict Synge's Corollary 1? $\mathbb{R}P^{2n}$ is compact, even-dimensional, and has positive sectional curvatures. *Thus even-dimensional projective spaces cannot be orientable*! This reaffirms the result of Problem 2.8(1).

Synge's method has another striking consequence for orientability. First note that *if $M^n$ is not orientable then there is some closed curve $C$ that cannot be deformed to a point (in particular $M$ is not simply connected!) and such that orientation is reversed on transporting an orientation around $C$.* To see this, suppose that $M$ is not orientable. Then it must be that transporting an orientation around *some* closed curve must lead to a reversal of orientation; otherwise it would be possible to transport an orientation uniquely from a given point to every other point, implying that $M$ was orientable. Let now orientation be reversed upon traversing a closed curve $C$. If we deform $C$ slightly to a curve $C_\alpha$, then, by continuity, orientation must be reversed also on traversing $C_\alpha$. Thus orientation would be reversed for every closed curve that is freely homotopic to $C$ (see Section 10.2d). But if we could deform $C$ to a point curve $C_1$, where orientation cannot be reversed, we would have a contradiction. □

Thus if $M^n$ is not orientable, there is, from Section 10.2d, a closed geodesic $C$ having the property that orientation is reversed upon traversing $C$ and $C$ is the shortest curve in its free homotopy class. In Problem 12.1(1) you are asked to prove the following:

**Corollary (12.16):** *If $M^{2n+1}$ is a compact, odd-dimensional manifold with positive sectional curvatures, then $M$ is orientable.*

This shows that the odd-dimensional projective spaces are orientable.

## Problem

**12.2(1)** Use Synge's *method* to prove Corollary (12.16).

CHAPTER 13

# Betti Numbers and De Rham's Theorem

When can we be certain that a closed form will be exact?

THE lack of simple connectivity is but one measure of topological complexity for a space. In this chapter we shall deal with others, the Betti numbers, and their relations with the potentials for closed exterior forms initiated in Chapter 5. This subject is a part of the discipline called algebraic topology.

## 13.1. Singular Chains and Their Boundaries

What does Stokes's theorem say for a Möbius band?

### 13.1a. Singular Chains

The **standard** (euclidean) $p$-**simplex** in $\mathbb{R}^p$ is the convex set $\Delta_p \subset \mathbb{R}^p$ generated by the $p+1$ points

$$P_0 = (0, \dots, 0),\ P_1 = (1, 0, \dots, 0), \dots,\ P_p = (0, \dots, 0, 1)$$

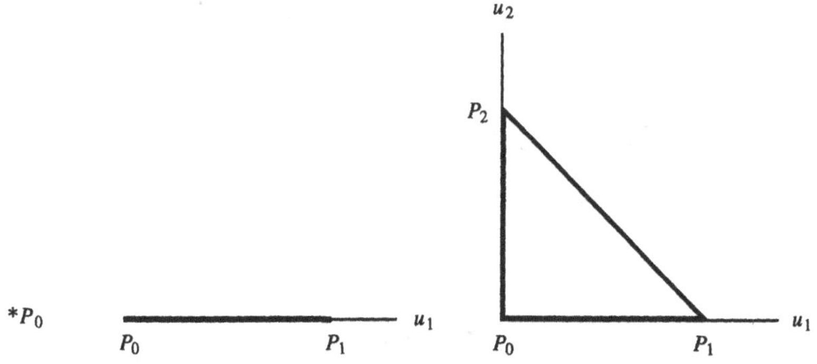

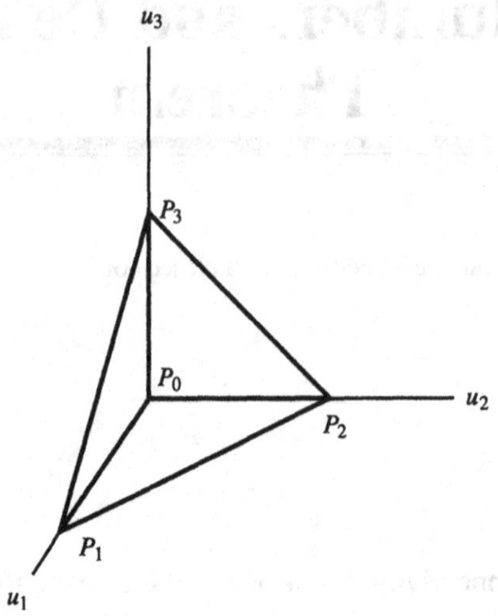


**Figure 13.1**

We shall write

$$\Delta_p = (P_0, P_1, \ldots, P_p)$$

A **singular** $p$-simplex in an $n$-manifold $M^n$ is a differentiable map

$$\sigma_p : \Delta_p \to M^n$$

of a standard $p$-simplex into $M$.

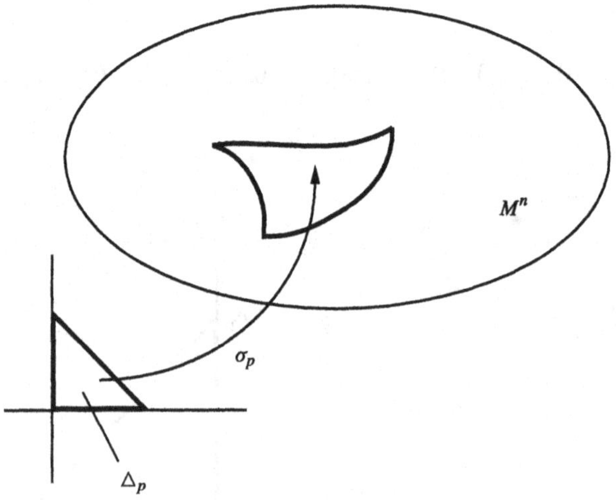


**Figure 13.2**

Note that a singular simplex is a special case of a parameterized subset discussed in Section 3.1b. *This is the natural object over which one integrates p-forms of M* via the pull-back

$$\int_{\sigma(\Delta)} \alpha^p := \int_{\Delta} \sigma^* \alpha^p$$

We emphasize that we put no restriction on the rank of the map $\sigma_p$; for example, the image of $\Delta_p$, which we shall also denote by $\sigma_p$, may be a single point of $M$.

Note that the $k^{\text{th}}$ **face** of $\Delta_p$

$$\Delta_{p-1}^{(k)} := (P_0, \ldots, \widehat{P_k}, \ldots, P_p)$$

that is, the face opposite the vertex $P_k$, is not a standard euclidean simplex, sitting as it does in $\mathbb{R}^p$ instead of $\mathbb{R}^{p-1}$. We shall rather consider it as a *singular* simplex in $\mathbb{R}^p$. In order to do this we must exhibit a specific map

$$f_k : \Delta_{p-1} \to \Delta_p$$

of $\Delta_{p-1}$ into $\mathbb{R}^p$, having the face as image. We do this in the following fashion. $f_k$ is the ***unique*** affine map (i.e., a linear map followed by a translation of origin) of $\mathbb{R}^{p-1}$ into $\mathbb{R}^p$ that sends $P_0 \to P_0, \ldots, P_{k-1} \to P_{k-1}, P_k \to P_{k+1}, \ldots, P_{p-1} \to P_p$.

If $\sigma : \Delta_p \to M^n$ is a singular simplex of $M$ and if $\phi : M^n \to V^r$ is a differentiable map, then the composition $\phi \circ \sigma : \Delta_p \to V^r$ defines a singular simplex of $V$. In particular $\sigma \circ f_k : \Delta_{p-1} \to M^n$ defines a singular $(p-1)$-simplex of $M$, the $k^{\text{th}}$ face of the singular $p$-simplex $\sigma$.

We define the **boundary** $\partial \Delta_p$ of the standard $p$-simplex, for $p > 0$, to be the *formal* sum of singular simplexes

$$\begin{aligned} \partial \Delta_p = \partial(P_0, P_1, \ldots, P_p) &:= \sum_k (-1)^k (P_0, \ldots, \widehat{P_k}, \ldots, P_p) \\ &= \sum_k (-1)^k \Delta_{p-1}^{(k)} \end{aligned} \tag{13.1}$$

whereas for the 0-simplex we put $\partial \Delta_0 = 0$. For example, $\partial(P_0, P_1, P_2) = (P_1, P_2) - (P_0, P_2) + (P_0, P_1)$.

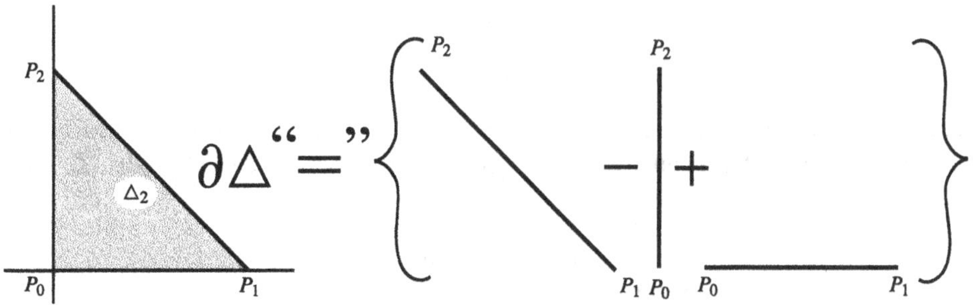


**Figure 13.3**

$\Delta_2 = (P_0, P_1, P_2)$ is an **ordered** simplex; that is, it is ordered by the given ordering of its vertices. From this ordering we may extract an *orientation*; the orientation of $\Delta_2$

is defined to be that of the vectors $\mathbf{e}_1 = P_1 - P_0$ and $\mathbf{e}_2 = P_2 - P_0$. Likewise, each of its faces is ordered by its vertices and has then an orientation. We think of the minus sign in front of $(P_0, P_2)$ as effectively *reversing the orientation* of this simplex. Symbolically,

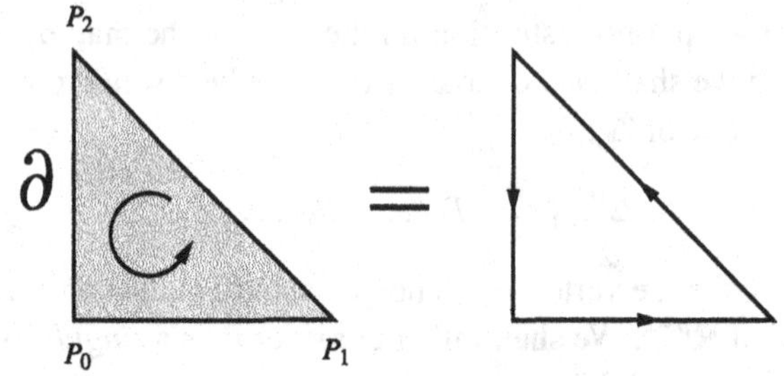


**Figure 13.4**

In this way the boundary of $\Delta_2$ corresponds to the boundary as defined in Section 3.3a, and, in fact, Stokes's theorem for a 1-form $\alpha^1$ in the plane says, for this $\Delta = \Delta_2$,

$$\int_{\partial\Delta} \alpha^1 = \iint_{\Delta} d\alpha^1$$

A similar result holds for $\Delta_3$. $\Delta_3 = (P_0, P_1, P_2, P_3)$ is an ordered simplex with orientation given by the three vectors $P_1 - P_0$, $P_2 - P_0$, and $P_3 - P_0$. As drawn, this is the right-hand orientation. $\partial\Delta_3$ has among its terms the "roof" $(P_1, P_2, P_3)$ and it occurs with a coefficient $+1$. The orientation of the face $+(P_1, P_2, P_3)$ is determined by the two vectors $P_2 - P_1$, and $P_3 - P_1$, which is the same orientation as would be assigned in Section 3.3a.

$\partial\Delta_p$, as a formal sum of simplexes with coefficients $\pm 1$, is not itself a simplex. It is an example of a new type of object, an **integer** $(p-1)$**-chain**. For topological purposes it is necessary, and no more difficult, to allow much more general coefficients than merely $\pm 1$ or integers. Let $G$ be any abelian, that is, commutative, group. The main groups of interest to us are

$G = \mathbb{Z}$, the group of integers

$G = \mathbb{R}$, the *additive* group of real numbers

$G = \mathbb{Z}_2 = \mathbb{Z}/2\mathbb{Z}$, the group of integers mod 2

The notation $\mathbb{Z}_2 = \mathbb{Z}/2\mathbb{Z}$ means that in the group $\mathbb{Z}$ of integers we shall identify any two integers that differ by an even integer, that is, an element of the subgroup $2\mathbb{Z}$. Thus $\mathbb{Z}_2$ consists of merely two elements

$$\mathbb{Z}_2 = \{\tilde{0}, \tilde{1}\} \quad \text{where} \quad \begin{array}{l} \tilde{0} \text{ is the equivalence class of } 0, \pm 2, \pm 4, \ldots \\ \tilde{1} \text{ is the equivalence class of } \pm 1, \pm 3, \ldots \end{array}$$

with addition defined by $\tilde{0}+\tilde{0} = \tilde{0}$, $\tilde{0}+\tilde{1} = \tilde{1}$, $\tilde{1}+\tilde{1} = \tilde{0}$. This of course is inspired by the fact that even + even = even, even + odd = odd, and odd + odd = even. We usually write $\mathbb{Z}_2 = \{0, 1\}$ and omit the tildes˜. Likewise, one can consider the group $\mathbb{Z}_p = \mathbb{Z}/p\mathbb{Z}$, the group of integers modulo the integer $p$, where two integers are identified if their difference is a multiple of $p$. This group has $p$ elements, written $0, 1, \dots, p-1$.

We define a (singular) $p$-chain on $M^n$, with **coefficients in the abelian group** $G$, to be a *finite* formal sum

$$c_p = g_1\,\sigma_p{}^1 + g_2\,\sigma_p{}^2 + \cdots + g_r\,\sigma_p{}^r \tag{13.2}$$

of singular simplexes $\sigma_p{}^s : \Delta_p \to M$, each with coefficient $g_s \in G$. This formal definition means the following. A $p$-chain is a function $c_p$ defined on *all* singular $p$-simplexes, with values in the group $G$, having the property that its value is $0 \in G$ for all but perhaps a finite number of simplexes. In (13.2) we have exhibited explicitly all of the simplexes for which $c_p$ is (possibly) nonzero and

$$c_p(\sigma_p{}^s) = g_s$$

We add two $p$-chains by simply adding the functions, that is,

$$(c_p + c'_p)(\sigma_p) := c_p(\sigma_p) + c'_p(\sigma_p)$$

The addition on the right-hand side takes place in the group $G$. In terms of the formal sums we simply add them, where of course we may combine coefficients for any simplex that is common to both formal sums. Thus the collection of all singular $p$-chains of $M^n$ with coefficients in $G$ themselves form an abelian group, the (**singular**) $p$-**chain group** of $M$ with coefficients in $G$, written $C_p(M^n; G)$.

A chain with integer coefficients will be called simply an *integer chain*.

The standard simplex $\Delta_p$ may be considered an element of $C_p(\mathbb{R}^p; \mathbb{Z})$; this $p$-chain has the value 1 on $\Delta_p$ and the value 0 on every other singular $p$-simplex. Then $\partial\Delta_p = \sum_k (-1)^k \Delta^{(k)}_{p-1}$ is to be considered an element of $C_{p-1}(\mathbb{R}^p; \mathbb{Z})$.

A **homomorphism** of an abelian group $G$ into an abelian group $H$ is a map $f : G \to H$ that commutes with addition (i.e., $f(g+g') = f(g) + f(g')$). On the left-hand side we are using addition in $G$; on the right-hand side the addition is in $H$. For example, $f : \mathbb{Z} \to \mathbb{R}$ defined by $f(n) = n\sqrt{2}$ is a homomorphism. $F : \mathbb{Z} \to \mathbb{Z}_2$, defined by $F(n) = \tilde{0}$ if $n$ is even and $\tilde{1}$ if $n$ is odd, describes a homomorphism. The reader should check that the only homomorphism of $\mathbb{Z}_2$ into $\mathbb{Z}$ is the **trivial** homomorphism that sends the entire group into $0 \in \mathbb{Z}$.

Let $F : M^n \to V^r$. We have already seen that if $\sigma$ is a singular simplex of $M$ then $F \circ \sigma$ is a singular simplex of $V$. We *extend* $F$ to be a homomorphism $F_* : C_p(M; G) \to C_p(V; G)$, the **induced chain homomorphism**, by putting

$$F_*(g_1\,\sigma_p{}^1 + \cdots + g_r\,\sigma_p{}^r) := g_1\,(F \circ \sigma_p{}^1) + \cdots + g_r\,(F \circ \sigma_p{}^r)$$

For a composition $F : M^n \to V^r$ and $E : V^r \to W^t$ we have

$$(E \circ F)_* = E_* \circ F_* \tag{13.3}$$

If $\sigma : \Delta_p \to M$ is a singular $p$-simplex, let its boundary $\partial\sigma$ be the *integer* $(p-1)$-chain defined as follows. Recall that $\partial\Delta_p$ is the integer $(p-1)$-chain $\partial\Delta_p = \sum_k (-1)^k\, \Delta^{(k)}_{p-1}$ on $\Delta_p$. We then define

$$\partial\sigma := \sigma_*(\partial\Delta) = \sum_k (-1)^k \sigma_*(\Delta^{(k)}{}_{p-1}) \tag{13.4}$$

Roughly speaking, *the boundary of the image of* $\Delta$ *is the image of the boundary* of $\Delta$! Finally, we define the boundary of any singular $p$-chain with coefficients in $G$ by

$$\partial \sum_r g_r \sigma^r_p := \sum_r g_r \partial \sigma^r_p \tag{13.5}$$

By construction we then have the **boundary homomorphism**

$$\partial : C_p(M; G) \to C_{p-1}(M; G) \tag{13.6}$$

If $F : M^n \to V^r$ and if $c_p = \sum g_r \sigma^r_p$ is a chain on $M$, then for the induced chain $F_* c$ on $V$ we have $\partial(F_* c) = \partial \sum g_r F_* \sigma^r = \sum g_r \partial(F_* \sigma^r) = \sum g_r (F \circ \sigma^r)_*(\partial\Delta) = \sum g_r F_*[\sigma^r_*(\partial\Delta)] = F_*[\sum g_r \sigma^r_*(\partial\Delta)] = F_* \partial c_p$. Thus

$$\partial \circ F_* = F_* \circ \partial \tag{13.7}$$

(Again we may say that the boundary of an image is the image of the boundary.) We then have a **commutative diagram**

$$\begin{array}{ccc} & F_* & \\ C_p(M;G) & \to & C_p(V;G) \\ \partial \downarrow & & \partial \downarrow \\ C_{p-1}(M;G) & \to & C_{p-1}(V;G) \\ & F_* & \end{array}$$

meaning that for each $c \in C_p(M; G)$ we have $F_* \partial c_p = \partial F_* (c_p)$.

Suppose we take the boundary of a boundary. For example, $\partial\partial(P_0, P_1, P_2) = \partial\{(P_1, P_2) - (P_0, P_2) + (P_0, P_1)\} = P_2 - P_1 - (P_2 - P_0) + P_1 - P_0 = 0$. This crucial property of the boundary holds in general.

**Theorem (13.8):**

$$\partial^2 = \partial \circ \partial = 0$$

PROOF: Consider first a standard simplex $\Delta_p$. From (13.1)

$$\begin{aligned} \partial\partial\Delta_p &= \sum_k (-1)^k \partial(P_0, \ldots, \widehat{P_k}, \ldots, P_p) \\ &= \sum_k (-1)^k \sum_{j<k} (-1)^j (P_0, \ldots, \widehat{P_j}, \ldots, \widehat{P_k}, \ldots, P_p) \\ &\quad + \sum_k (-1)^k \sum_{j>k} (-1)^{j-1} (P_0, \ldots, \widehat{P_k}, \ldots, \widehat{P_j}, \ldots, P_p) \\ &= 0 \quad \text{(cancellation in pairs)} \end{aligned}$$

But then, for a singular simplex, $\partial(\partial\sigma) = \partial(\sigma_*(\partial\Delta))$, which, from (13.7), is $\sigma_* \partial(\partial\Delta) = \sigma_*(0) = 0$. □

### 13.1b. Some 2-Dimensional Examples

1. The *cylinder* Cyl is the familiar rectangular band with the two vertical edges brought together by bending and then sewn together. We wish to exhibit a specific integer 2-chain

on Cyl. On the right we have the rectangular band and we have labeled six vertices. The

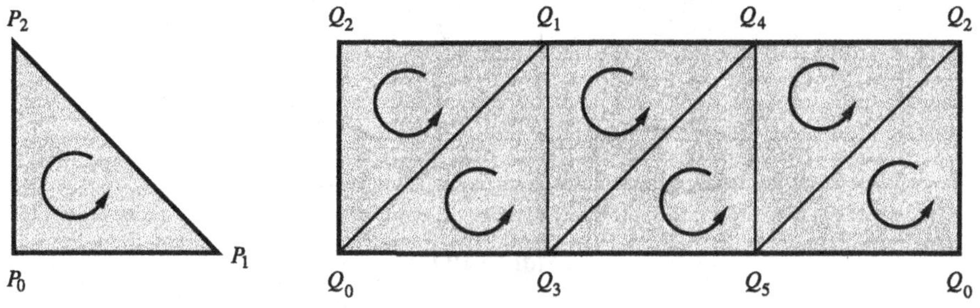


**Figure 13.5**

labels on the two vertical edges are the same, since the band is to be bent and the two edges are to be sewn, resulting in Cyl. On the band we have indicated six singular 2-simplexes. We shall always write a singular simplex with vertices in increasing order. For example, $(Q_1, Q_3, Q_4)$ is the singular simplex arising from the affine map of the plane into itself that assigns $(P_0, P_1, P_2) \to (Q_1, Q_3, Q_4)$. After the band is bent and sewn we shall then have a singular 2-simplex on Cyl that we shall again call $(Q_1, Q_3, Q_4)$. We have thus broken Cyl up into 2-simplexes, and *we have used enough simplexes* so that any 1- or 2-simplex is *uniquely* determined by its vertices.

We wish to write down a 2-chain where each simplex carries the orientation indicated in the figure. Since we always write a simplex with increasing order to its vertices, we put

$$\begin{aligned} c_2 = {} & (Q_0, Q_1, Q_2) - (Q_0, Q_1, Q_3) + (Q_1, Q_3, Q_4) - (Q_3, Q_4, Q_5) \\ & + (Q_2, Q_4, Q_5) + (Q_0, Q_2, Q_5) \end{aligned}$$

Then

$$\partial c_2 = (Q_0, Q_3) + (Q_3, Q_5) - (Q_0, Q_5) + (Q_2, Q_4) - (Q_1, Q_4) + (Q_1, Q_2)$$

We write this as $\partial c_2 = B + C$, where $B = (Q_0, Q_3) + (Q_3, Q_5) - (Q_0, Q_5)$ and $C = (Q_2, Q_4) - (Q_1, Q_4) + (Q_1, Q_2)$. $B$ and $C$ are two copies of a circle, with opposite orientations; $B$ is the bottom edge and $C$ the top. Denote the seam $(Q_0, Q_2)$ by $A$, and omitting all other simplexes, we get the following symbolic figure.

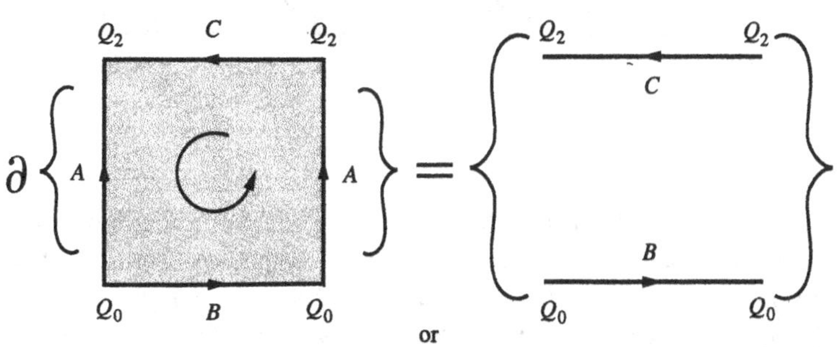


or

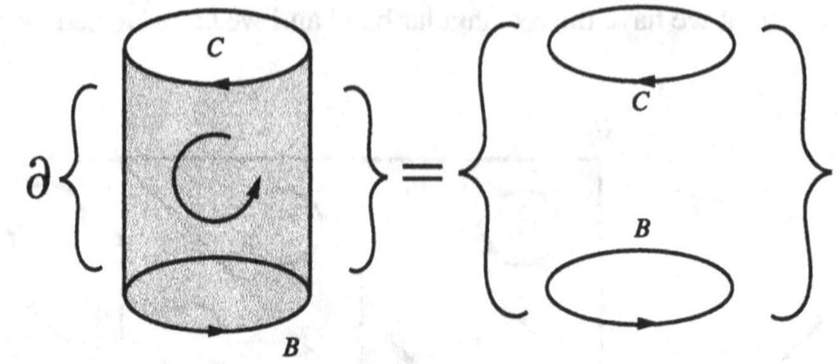


Figure 13.6

Note that in the lower figure the result is the same as would be obtained if we think of the cylinder as an oriented compact manifold with boundary, the boundary being then oriented as in Section 3.3a.

In the upper figure we have a rectangle with four sides. By denoting both vertical sides by the same curve $A$ we are implying that these two sides are to be identified by identifying points at the same horizontal level. The bottom curve $B$ and the top $C$, bearing different names, are not to be identified. As drawn, the bottom $B$, the top $C$, and the right-hand side $A$ have the correct orientation as induced from the given orientation of the rectangle, but the left-hand $A$ carries the opposite orientation. Symbolically, if we think of the 2-chain $c_2$ as defining the oriented manifold Cyl, we see from the figure that

$$\partial \, \text{Cyl} = B + A + C - A = B + C$$

the same result as our calculation of $\partial c_2$ given before with all of the simplexes. From the rectangular picture we see immediately that all of the "interior" 1-simplexes, such as $(Q_3, Q_4)$, *must* cancel in pairs when computing $\partial C_2$.

2. The *Möbius band* Mö. We can again consider a 2-chain $c_2$

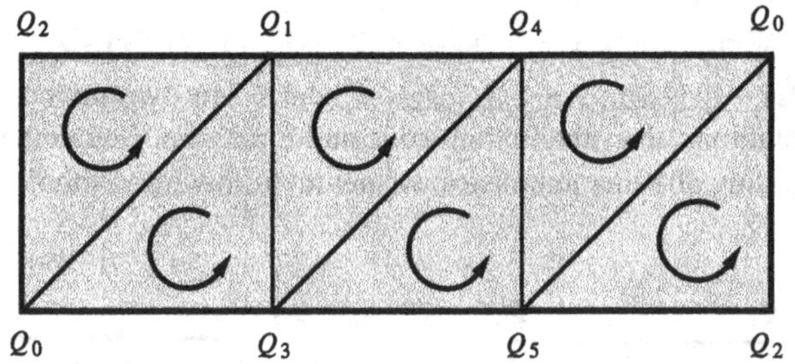


Figure 13.7

Note that the only difference is the right-hand edge, corresponding to the half twist given to this edge before sewing to the left hand edge; see Section 1.2b (viii). This $c_2$ is the same as in the cylinder except that the last term is replaced by its negative $-(Q_0, Q_2, Q_5)$. We can compute $\partial c_2$ just as before, but let us rather use the symbolic rectangle with identifications.

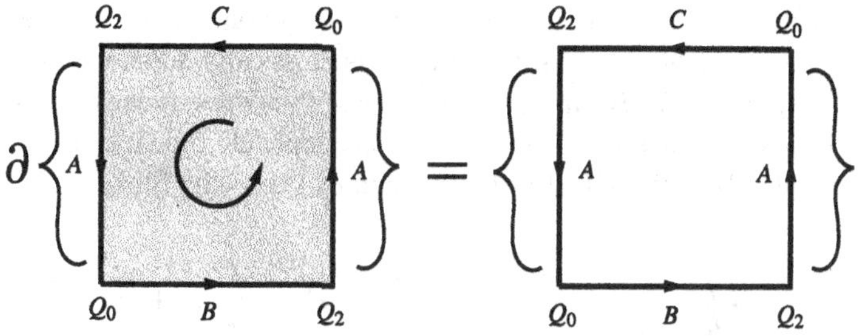


**Figure 13.8**

The boundary of the oriented rectangle is now

$$\partial \text{ Mö} = B + A + C + A = B + C + 2A$$

This is surely an unexpected result! If we think of the Möbius band as an integer 2-

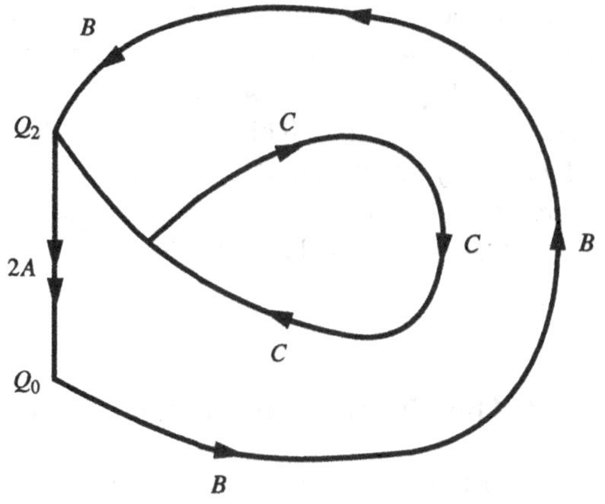


**Figure 13.9**

chain, as we did for the cylinder, then the "boundary," in the sense of algebraic topology, does *not* coincide with its "edge", that is, its boundary in the sense of "manifold with boundary." As a chain, one part of its boundary consists of the true edge, $B + C$, but note that although the point set $B + C$ is topologically a single closed curve it *changes its orientation* halfway around. It is even more disturbing that the rest of the boundary consists of an arc $A$ going from $Q_2$ to $Q_0$, traversed *twice*, and located along the seam of the band, not its edge!

The reason for this strange behavior is the fact that the Möbius band is *not orientable*. It is true that we have oriented each simplex, just as we did for the cylinder, but for the cylinder the simplexes were oriented **coherently**, meaning that adjacent simplexes, having as they do the same orientation, induce opposite orientations on the 1-simplex edge that is common to both. This is the reason that $\partial c_2$ on the cylinder has no 1-simplex in the interior; only the edge simplexes can appear in $\partial c_2$. On the Möbius band, however, the oriented simplexes $(Q_0, Q_1, Q_2)$ and $-(Q_0, Q_2, Q_5)$ induce the *same* orientation to their common $(Q_0, Q_2) = -A$ since these two 2-simplexes have opposite orientations! This is a reflection of the fact that the Möbius band is not orientable. We shall discuss this a bit more in our next section.

We have defined the integral of a 2-form over a compact oriented surface $M^2$ in Chapter 3, but we mentioned that the integral is classically defined by breaking up the manifold into pieces. *This is what is accomplished by construction of the 2-chain $c_2$!* Let $\alpha^1$ be a 1-form on the cylinder, oriented as in Example (1). The integral of $d\alpha$ over Cyl can be computed by writing Cyl as the 2-chain $c_2$. Applying Stokes's theorem to each simplex will give

$$\iint_{\mathrm{Cyl}} d\alpha^1 = \int_{\partial\mathrm{Cyl}} \alpha^1 = \int_{B+C} \alpha^1 = \int_B \alpha^1 + \int_C \alpha^1$$

just as expected. However, for the Möbius band, written as $c_2$,

$$\iint_{\mathrm{Mo}} d\alpha^1 = \int_{\partial\,\mathrm{Mo}} \alpha^1 = \int_{B+C+2A} \alpha^1 = \int_B \alpha^1 + \int_C \alpha^1 + 2\int_A \alpha^1$$

This formula, although correct, is of no value. The integral down the seam is not intrinsic since the position of the seam is a matter of choice. The edge integral is also of no value since we arbitrarily decide to change the direction of the path at some point. It should not surprise us that Stokes's theorem in this case is of no intrinsic value since the Möbius band is not orientable, and we have not defined the integral of a true 2-form over a nonorientable manifold in Chapter 3. If, however, $\alpha^1$ were a *pseudoform*, then when computing the integral of $d\alpha^1$ over the Möbius $c_2$, Stokes's theorem, as mentioned in Section 3.4d, would yield only an integral of $\alpha^1$ over the edge $B + C$. The fact that $B$ and $C$ carry different orientations is not harmful since the $\alpha$ that is integrated over $B$ will be the negative of the $\alpha$ that is integrated over $C$; this is clear from the two simplexes $(Q_0, Q_1, Q_2)$ and $-(Q_0, Q_2, Q_5)$.

## 13.2. The Singular Homology Groups

What are "cycles" and "Betti numbers"?

### 13.2a. Coefficient Fields

In the last section we have defined the singular $p$-chain groups $C_p\,(M^n; G)$ of $M$ with coefficients in the abelian group $G$, and also the boundary homomorphism

$$\partial : C_p(M; G) \to C_{p-1}\,(M; G)$$

Given a map $F : M^n \to V^r$ we have an induced homomorphism

$$F_* : C_p(M; G) \to C_p(V; G)$$

and the boundary homomorphism $\partial$ is "natural" with respect to such maps, meaning that

$$\partial \circ F_* = F_* \circ \partial$$

We also have $\partial^2 = 0$. *Notice the similarity with differential forms*, as $\partial$ takes the place of the exterior derivative $d$! We will look at this similarity in more detail later.

Many readers are probably more at home with vector spaces and linear transformations than with groups and homomorphisms. It will be comforting to know then that in many cases the chain groups are vector spaces, and not just abelian groups.

An abelian group $G$ is a **field**, if, roughly speaking, $G$ has not only an additive structure but an abelian multiplicative one also, with multiplicative identity element 1, and this multiplicative structure is such that each $g \neq 0$ in $G$ has a multiplicative inverse $g^{-1}$ such that $gg^{-1} = 1$. We further demand that multiplication is distributive with respect to addition. The most familiar example is the field $\mathbb{R}$ of real numbers. The integers $\mathbb{Z}$ do *not* form a field, even though there is a multiplication, since for example, $2 \in \mathbb{Z}$ does not have an *integer* multiplicative inverse. On the other hand, $\mathbb{Z}_2$ is a field if we define multiplication by $\tilde{0} \cdot \tilde{0} = \tilde{0}, \tilde{0} \cdot \tilde{1} = \tilde{0}$, and $\tilde{1} \cdot \tilde{1} = \tilde{1}$. In fact $\mathbb{Z}_p$ is a field whenever $p$ is a prime number. In $\mathbb{Z}_5$, the multiplicative inverse of 3 is 2.

When the coefficient group $G$ is a field, $G = K$, the chain groups $C_p(M^n; K)$ become vector spaces over this field upon defining, for each "scalar" $r \in K$ and chain $c_p = (\sum g_i \sigma^i{}_p) \in C_p(M^n; K)$

$$rc_p = \sum (rg_i)\sigma^i{}_p$$

The vector space of $p$-chains is infinite-dimensional since no finite nontrivial linear combination of distinct singular simplexes is ever the trivial $p$-chain 0.

From (13.5) we see that when $G = K$ is a field,

$$\partial : C_p(M; K) \to C_{p-1}\,(M; K)$$

is a *linear transformation.*

Finally, a notational simplification. When we are dealing with a specific space $M^n$ and also a specific coefficient group $G$, we shall frequently omit $M$ and $G$ in the notation for the chain groups and other groups to be derived from them. We then write, for example, $\partial : C_p \to C_{p-1}$.

### 13.2b. Finite Simplicial Complexes

At this point we should mention that there is a related notion of **simplicial complex** with its associated **simplicial** (rather than *singular*) **chains.** We shall not give definitions, but rather consider the example of the Möbius band. We have indicated a "triangulation" of the band into six singular 2-simplexes in Example (2) of the last section. Each of these simplexes is a *homeomorphic* copy of the standard simplex, unlike the general singular simplex. Suppose now that instead of looking at *all* singular simplexes on Mö we only allow these six 2-simplexes and allow only 1-simplexes that are edges of these 2-simplexes, and only the six 0-simplexes (i.e., vertices) that are indicated. We insist that *all* chains must be combinations of only these simplexes; these form the "simplicial" chain groups $\bar{C}_p$. Then $\bar{C}_0$(Mö; G) is a group with the six generators $Q_0, \ldots, Q_5$; $\bar{C}_1$ has twelve generators $(Q_0, Q_1), (Q_0, Q_2), \ldots, (Q_4, Q_5)$; and $\bar{C}_2$ has the six given triangles as generators. If we have a field $K$ for coefficients, then these chain groups become vector spaces of dimension 6, 12, and 6, respectively, and the simplexes indicated become basis elements. In terms of these bases we may construct the *matrix* for the boundary linear transformations $\partial : \bar{C}_p \to \bar{C}_{p-1}$. For example $\partial(Q_0, Q_1) = Q_1 - Q_0$ tells us that the 6 by 12 matrix for $\partial : \bar{C}_1(\text{Mö}; \mathbb{R}) \to \bar{C}_0\,(\text{Mö}; \mathbb{R})$ has first column $(-1, 1, 0, 0, 0, 0)^T$. The simplicial chain groups are of course much smaller than the singular ones, but in a sense to be described later, they already contain the essentials, as far as "homology" is concerned, in the case of compact manifolds.

### 13.2c. Cycles, Boundaries, Homology and Betti Numbers

Return to the general case of singular chains with a coefficient group $G$. We are going to make a number of definitions that might seem abstract. *In Section 13.3 we shall consider many examples.*

We define a (singular) $p$-**cycle** to be a $p$-chain $z_p$ whose boundary is 0. The collection of all $p$-cycles,

$$Z_p(M; G) := \{z_p \in C_p | \partial z_p = 0\} \qquad (13.9)$$
$$= \ker \partial : C_p \to C_{p-1}$$

that is, the **kernel** $\partial^{-1}(0)$ of the homomorphism $\partial$, is a subgroup of the chain group $C_p$ (called naturally the $p$-**cycle group**). When $G = K$ is a field, $Z_p$ is a vector subspace of $C_p$, the **kernel** or **nullspace** of $\partial$, and in the case of a finite simplicial complex this nullspace can be computed using Gauss elimination and linear algebra.

We define a **p-boundary** $\beta_p$ to be a $p$-chain that is the boundary of some $(p+1)$-chain. The collection of all such chains

$$B_p(M; G) := \{\beta_p \in C_p | \beta_p = \partial c_{p+1}, \quad \text{for some } c_{p+1} \in C_{p+1}\} \qquad (13.10)$$
$$= \operatorname{Im} \partial : C_{p+1} \to C_p$$

the **image** or **range** of $\partial$, is a subgroup (the $p$-**boundary group**) of $C_p$. Furthermore, $\partial\beta = \partial\partial c = 0$ shows us that $B_p \subset Z_p$ is a subgroup of the cycle group.

Consider a *real* $p$-chain $c_p$ on $M^n$, that is, an element of $C_p(M; \mathbb{R})$. Then $c_p = \sum b_i\, {\sigma_p}^{(i)}$, where $b_i$ are real numbers. If $\alpha^p$ is a $p$-form on $M$, it is natural to define

$$\int_{c_p} \alpha^p := \sum b_i \int_{\sigma^{(i)}} \alpha^p \qquad (13.11)$$

Then

$$\int_{c_p} d\alpha^{p-1} = \sum b_i \int_{\sigma^{(i)}} d\alpha^{p-1} = \sum b_i \int_{\partial\sigma^{(i)}} \alpha^{p-1} = \int_{\partial c_p} \alpha^{p-1} \qquad (13.12)$$

We shall mainly be concerned with integrating *closed* forms, $d\alpha^p = 0$, over $p$-*cycles* $z_p$. Then if $z_p$ and $z'_p$ differ by a boundary, $z - z' = \partial c_{p+1}$, we have

$$\int_z \alpha^p - \int_{z'} \alpha^p = \int_{z-z'} \alpha^p = \int_{\partial c} \alpha^p = \int_c d\alpha^p = 0 \qquad (13.13)$$

Thus, as far as *closed* forms go, *boundaries contribute nothing to integrals. When integrating closed forms, we may identify two cycles if they differ by a boundary.* This identification turns out to be important also for cycles with general coefficients, not just real ones. We proceed as follows.

If $G$ is an abelian group and $H$ is a subgroup, let us say that two elements $g$ and $g'$ of $G$ are **equivalent** if they differ by some element of $H$,

$$g' \sim g \quad \text{iff } g' - g = h \in H$$

Sometimes we will say $g' = g \bmod H$. The set of equivalence classes is denoted by $G/H$, and read $G$ **mod** $H$. If $g \in G$ we denote the equivalence class of $g$ in $G/H$ by $[g]$ or sometimes $g + H$. Such an equivalence class is called a **coset**. Any equivalence class $[\ ] \in G/H$ is the equivalence class of some $g \in G$, $[\ ] = [g]$; this $g$ is called a **representative** of the class but of course $[g] = [g+h]$ for all $h \in H$. Two equivalence classes can be added by simply putting $[g+g'] := [g]+[g']$. In this way we make $G/H$ itself into an abelian group, called the **quotient group**. This is exactly the procedure we followed when constructing the group $\mathbb{Z}_2 = \mathbb{Z}/2\mathbb{Z}$ of integers mod 2.

We always have a map $\pi : G \to G/H$ that assigns to each $g$ its equivalence class $[g] = g + H$. $\pi$ is, by construction, a homomorphism.

When $G$ is a vector space $E$, and $H$ is a subspace $F$, then $E/F$ is again a vector space. If $E$ is an inner product space, then $E/F$ can be identified with the orthogonal complement $F^\perp$ of $F$ and $\pi$ can be identified with the orthogonal projection into the

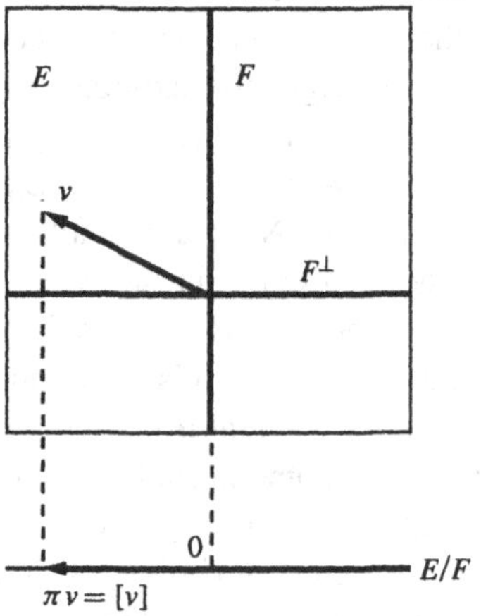


**Figure 13.10**

subspace $F^\perp$. If $E$ does not carry a specific inner product, then there is no natural way to identify $E/F$ with a subspace of $E$; *any* subspace of $E$ that is transverse to $F$ can serve as a model, but $E/F$ is clearly more basic than these nonunique subspaces.

Return now to our singular cycles. We say that two cycles $z_p$ and $z'_p$ in $Z_p(M; G)$ are equivalent or **homologous** if they differ by a boundary, that is, an element of the subgroup $B_p(M; G)$ of $Z_p(M; G)$. In the case of the cycles $Z_p$ and the subgroup $B_p$, the quotient group is called the $p^{\text{th}}$ **homology group,** written $H_p(M; G)$

$$H_p(M; G) := \frac{Z_p(M; G)}{B_p(M; G)} \tag{13.14}$$

When $G = K$ is a field, $Z_p$, $B_p$, and $H_p$ become vector spaces. We have seen that $Z$ and $B$ are infinite-dimensional, but *in many cases $H_p$ is finite-dimensional*! It can be shown, for example, that this is the case if $M^n$ is a *compact* manifold. Before discussing this, we mention a purely algebraic fact that will be very useful.

**Theorem (13.15):** *If $\phi : G_1 \to G_2$ is a homomorphism of abelian groups and if $\phi$ sends the subgroup $H_1$ of $G_1$ into the subgroup $H_2$ of $G_2$, then $\phi$ induces a* *homomorphism of the quotient groups*

$$\phi_* : \frac{G_1}{H_1} \rightarrow \frac{G_2}{H_2}$$

**PROOF:** The composition of the homomorphisms $\phi : G_1 \rightarrow G_2$ followed by $\pi : G_2 \longrightarrow G_2/H_2$ is a homomorphism $\pi \circ \phi : G_1 \rightarrow G_2/H_2$. Under this homomorphism $(g+h_1) \rightarrow \phi(g)+\phi(h_1)+H_2 = \phi(g)+H_2$, since $\phi(h_1) \in H_2$. Thus $\pi \circ \phi$ sends elements that are equivalent in $G_1$ (mod $H_1$) into elements that are equivalent in $G_2$ (mod $H_2$) and so we then have a homomorphism of $G_1/H_1$ into $G_2/H_2$; this is the desired $\phi_*$. □

We then have the following topological situation. If $M^n$ is a compact manifold, there is a **triangulation** of $M$ by a finite number of $n$-simplexes each of which is diffeomorphic to the standard $n$-simplex. This means that $M$ is a union of such $n$-simplexes and any pair of such simplexes either are disjoint or meet in a common $r$-subsimplex (vertex, edge, ...) of each. (We exhibited a triangulation explicitly for the Möbius band in Section 13.1b). These simplexes can be used to form a finite simplicial complex, for any coefficient group $G$, just as we did for the Möbius band. Since $\bar{C}_p$, $\bar{Z}_p$, and $\bar{B}_p$ are then finitely generated groups, so is $\bar{H}_p := \bar{Z}_p/\bar{B}_p$. Now any simplicial cycle can be considered a singular cycle (i.e., we have a homomorphism from $\bar{Z}_p$ to $Z_p$) and this homomorphism sends $\bar{B}_p$ to $B_p$. Thus we have an induced homomorphism of the simplicial homology class $\bar{H}_p$ to the singular homology class $H_p$. It is then a nontrivial fact that for compact manifolds $H_p = \bar{H}_p$; that is, *the $p^{th}$ singular homology group is isomorphic to the $p^{th}$ simplicial homology group*! (A homomorphism is an **isomorphism** if it is 1 : 1 and onto.) In particular the singular homology groups are also finitely generated (even though the singular cycles clearly aren't) and if $G$ is a field $K$, $H_p$ is finite-dimensional.

When $G$ is the field of real numbers, $G = \mathbb{R}$, the dimension of the vector space $H_p$ is called the $p^{\text{th}}$ **Betti number**, written $b_p = b_p(M)$

$$b_p(M) := \dim\ H_p(M; \mathbb{R}) \tag{13.16}$$

In words, $b_p$ *is the maximal number of p-cycles on M, no real linear combination of which is ever a boundary* (except for the trivial combination with all coefficients 0).

Let $F : M^n \rightarrow V^r$ be a map. Since, from (13.7), $F_*$ commutes with the boundary $\partial$, we know that $F_*$ takes cycles into cycles and boundaries into boundaries. Thus $F_*$ sends homology classes into homology classes, and we have an induced homomorphism

$$F_* : H_p(M; G) \rightarrow H_p(V; G) \tag{13.17}$$

Finally, we can see the importance of the homology groups. Suppose that $F : M^n \rightarrow V^n$ is a homeomorphism, then we have not only (13.17) but the homomorphism $F_*^{-1} : H_p(V; G) \rightarrow H_p(M, G)$ induced by the inverse map, and it is easily seen that these two homomorphisms are inverses. Thus $F_*$ is an isomorphism; *homeomorphic manifolds have isomorphic homology groups.* We say that the homology groups are **topological invariants**. Thus if we have two manifolds $M$ and $V$, and if *any* of their homology groups differ, for some coefficients $G$, then these spaces cannot be homeomorphic! Unfortunately, the converse is not true in general; that is, nonhomeomorphic manifolds can have the same homology groups.

## 13.3. Homology Groups of Familiar Manifolds

Is projective 3-space diffeomorphic to the 3-torus?

### 13.3a. Some Computational Tools

Any point $p \in M^n$ can be considered as a 0-chain. By the definition of the boundary operator $\partial p = 0$, and each point is a 0-cycle.

A smooth map $C : [0, 1] \to M$ is a curve in $M$; the image is compact since the image of a compact space (e.g., the unit interval) under a continuous map is again compact (see Section 1.2a). $C$ of course can be considered a singular 1-simplex, and we have $\partial C = C(1) - C(0)$. If $g \in G$, the coefficient group, then $\partial(gC)$ is the 0-chain $gC(1) - gC(0)$.

Suppose that $C : [a, b] \to M$ is a piecewise smooth curve. We may then break up the interval $[a, b]$ into subintervals on each of which the map is smooth. By reparameterizing the curve on each subinterval, we may consider the mappings of the subintervals as defining singular simplexes. We may then associate (nonuniquely) to our original curve a singular 1-chain, associating the coefficient $+1$ to each of the 1-simplexes. The boundary of this chain is clearly $C(b) - C(a)$, the intermediate vertices cancelling in pairs.

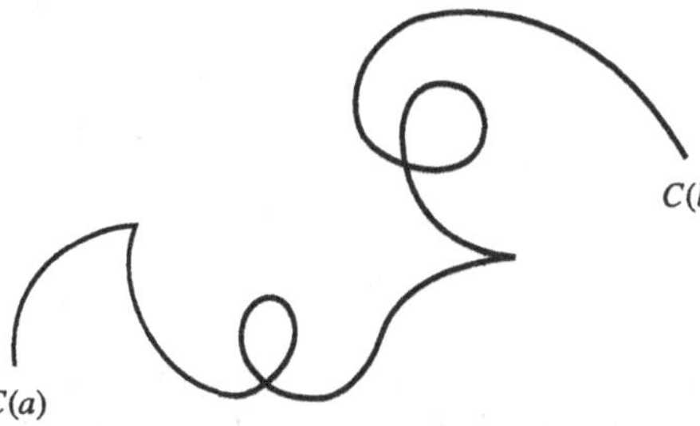


**Figure 13.11**

A manifold $M^n$ is said to be (**path-**)**connected** if any two points $p$ and $q$ can be joined by a piecewise smooth curve $C : [0, 1] \to M$; thus $C(0) = p$ and $C(1) = q$. This curve then generates a 1-chain, as in Figure 13.11. But then $\partial C = q - p$. Likewise $\partial(gC) = gq - gp$, where $gC$ is the 1-chain that associates $g \in G$ to each of the 1-simplexes. This shows that any two 0-simplexes with the same coefficient, in a *connected* manifold, are homologous. Also, since a 1-chain is merely a combination $C = \sum g_i C_i$, $\partial C = \sum\{g_i q_i - g_i p_i\}$, we see that no multiple $gp$ of a single point is a boundary, if $g \neq 0$. Thus any particular point $p$ of a connected space defines a 0-cycle that is not a boundary, and any 0-chain is homologous to a multiple $gp$ of $p$. We then have

$$H_0(M^n; G) = Gp \quad \text{for } M \text{ connected} \tag{13.18}$$

meaning that this group is the set $\{gp | g \in G\}$. For example, $H_0(M^n; \mathbb{Z})$ is the set $\{0, \pm p, \pm 2p, \pm 3p, \ldots\}$ and $H_0(M^n; \mathbb{R})$ is the 1-dimensional vector space consisting of all real multiples of the "vector" $p$. This vector space is isomorphic to the vector space $\mathbb{R}$, and we usually write $H_0(M^n; \mathbb{R}) = \mathbb{R}$. In particular, *a connected space has*

$0^{th}$ *Betti number* $b_0 = 1$. If $M$ is not connected, but consists of $k$ connected pieces, then $H_0(M^n; \mathbb{R}) = \mathbb{R}p_1 + \mathbb{R}p_2 + \cdots + \mathbb{R}p_k$, where $p_i$ is a point in the $i^{\text{th}}$ piece. In this case $b_0(M) = k$.

(We should mention that in topology there is the notion of a **connected** space; $M$ is connected if it cannot be written as the union of a pair of disjoint open sets. This is a weaker notion than pathwise connected, but for manifolds the two definitions agree.)

Next, consider a $p$-dimensional compact *oriented* manifold $V^p$ without boundary. By triangulating $V^p$ one can show that $V^p$ *always defines an integer p-cycle*, which we shall denote by $[V_p]$. For example, consider the 2-torus $T^2$.

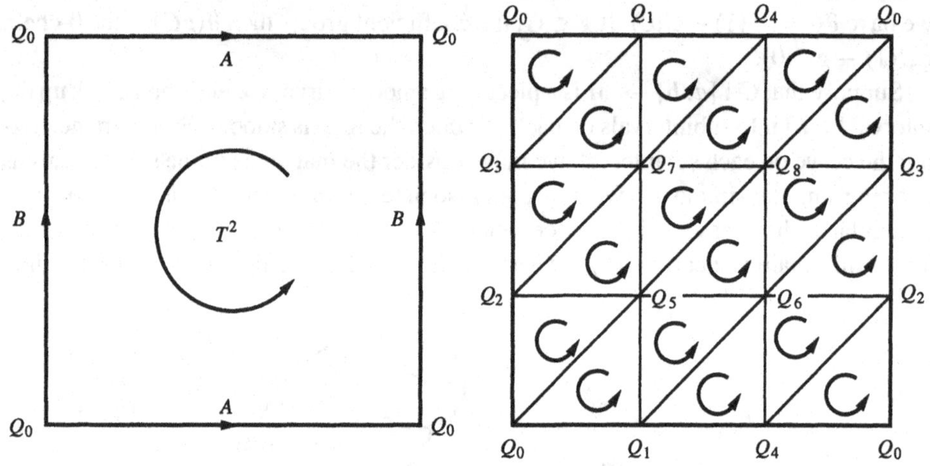


**Figure 13.12**

If we associate the integer $+1$ to each of the eighteen indicated oriented 2-simplexes, we get a chain $[T^2]$, for example, $[T^2](Q_5, Q_7, Q_8) = -1$.The boundary of this chain is clearly 0

$$\partial[T^2] = A + B - A - B = 0$$

and this same procedure will work for any compact orientable manifold.

On the other hand, consider a nonorientable closed manifold, the **Klein bottle** $K^2$. This surface cannot be embedded in $\mathbb{R}^3$ but we can exhibit an *immersion* with self-intersections. This is the surface obtained from a cylinder when the two boundary edges are sewn together after one of the edges is pushed through the cylinder. Abstractly, in

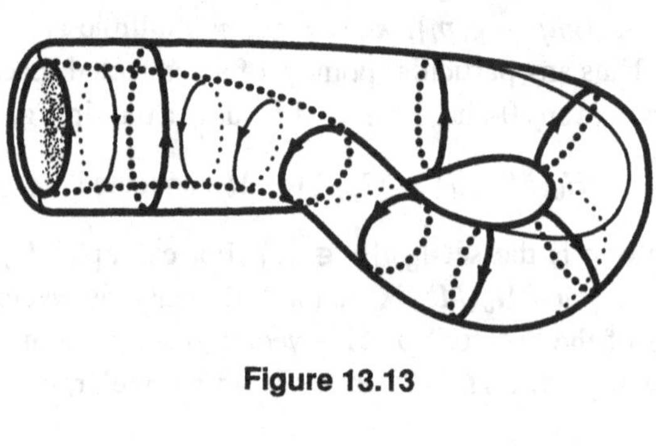

**Figure 13.13**

terms of a rectangle with identifications, we have the following diagram; note especially the directions of the arrows on the circle $B$.

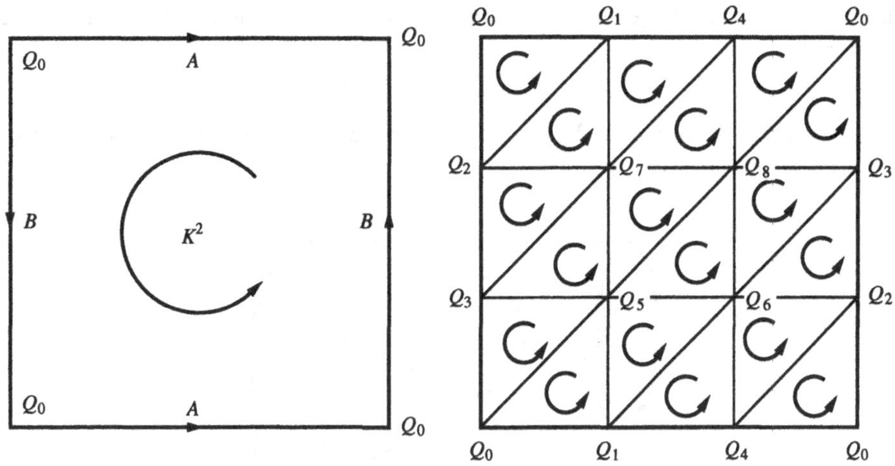


**Figure 13.14**

Orient each of the triangles as indicated and give to each oriented triangle the coefficient $+1$, yielding a singular 2-chain $[K^2]$. Now, however, we have

$$\partial[K^2] = A + B - A + B = 2B \neq 0$$

$[K^2]$ *is not a cycle*, even though the manifold has no boundary, that is, edge. This is a reflection of the fact that we have not been able to orient the triangles *coherently*; the Klein bottle is not orientable.

Note another surprising fact; the 1-cycle $B = (Q_0, Q_2) + (Q_2, Q_3) - (Q_0, Q_3)$ is not a boundary (using $\mathbb{Z}$ coefficients) but $2B$ *is*, $2B = \partial[K^2]$! Note also that if we had used real coefficients then $B$ itself would be a boundary since then $B = \partial(1/2)[K^2]$, where this latter chain assigns coefficient 1/2 to each oriented 2-simplex. Furthermore, if we had used $\mathbb{Z}_2$ coefficients, then $[K^2]$ would be a cycle, since $2B = 0 \bmod 2$. All these facts give some indication of the role played by the coefficient group $G$.

The following theorem in algebraic topology, reflecting the preceding considerations, can be proved.

**Theorem (13.19):** *Every closed oriented submanifold $V^p \subset M^n$ defines a p-cycle $g[V^p]$ in $H_p(M^n; G)$ by associating the same coefficient g to each oriented p-triangle in a suitable triangulation of $V^p$.*

*Thus a p-cycle is a generalization of the notion of a closed oriented submanifold.*

René Thom has proved a deep converse to (13.19) in the case of real coefficients.

**Thom's Theorem (13.20):** *Every real p-cycle in $M^n$ is homologous to a finite formal sum $\sum r_i V_i^p$ of closed oriented submanifolds with real coefficients.*

Thus, when looking for real cycles, we need only look at submanifolds.

Our next computational tool is concerned with deformations. In Section 10.2d we discussed deforming closed curves in a manifold. In a similar fashion we can deform submanifolds and more generally $p$-chains. We shall not go into any details, but merely mention the

**Deformation Theorem (13.21):** *If a cycle $z_p$ is deformed into a cycle $z'_p$, then $z'_p$ is homologous to $z_p$, $z'_p \sim z_p$.*

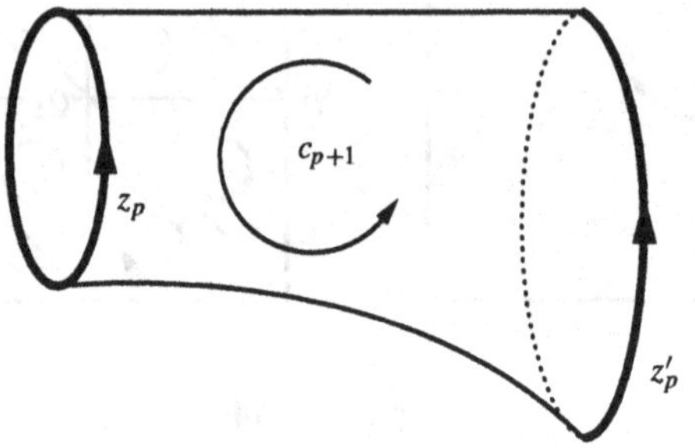


Figure 13.15

This follows from the fact that in the process of deforming $z_p$ into $z'_p$ one sweeps out a "deformation chain" $c_{p+1}$ such that $\partial c_{p+1} = z'_p - z_p$.

Our final tool is the following. For a closed $n$-manifold $M^n$, we know from Section 13.2c that the singular homology groups are isomorphic to the simplicial ones. But in the simplicial complex for $M^n$ there are no simplexes of dimension greater than $n$. Thus,

$$H_p(M^n; G) = 0 \quad \text{for } p > n \tag{13.22}$$

### 13.3b. Familiar Examples

1. $S^n$, the *$n$-sphere*, $n > 0$. $H_0(S^n; G) = G$ since $S^n$ is connected for $n > 0$. Since $S^n$ is a 2-sided hypersurface of $\mathbb{R}^{n+1}$ it is orientable, and since it is closed we have $H_n(S^n; G) = G$. If $z_p$ is a $p$-cycle, $0 < p < n$, it is homologous to a simplicial cycle in some triangulation of $S^n$. (The usual triangulation of the sphere results from inscribing an $(n+1)$-dimensional tetrahedron and projecting the faces outward from the origin until they meet the sphere.) In any case, we may then consider a $z_p$ that does not meet *some* point $q \in S^n$. We may then deform $z_p$ by pushing all of $S^n - q$ to the antipode of $q$, a single point. $z_p$ is then homologous to a $p$-cycle supported on the simplicial complex consisting of one point. But a point has nontrivial homology only in dimension 0. Thus $z_p \sim 0$ and

$$H_0(S^n; G) = G = H_n(S^n; G) \tag{13.23}$$
$$H_p(S^n; G) = 0, \quad \text{for } p \neq 0, n$$

The nonvanishing Betti numbers are $b_0 = 1 = b_n$.

2. $T^2$, the 2-*torus*. $H_0 = H_2 = G$.

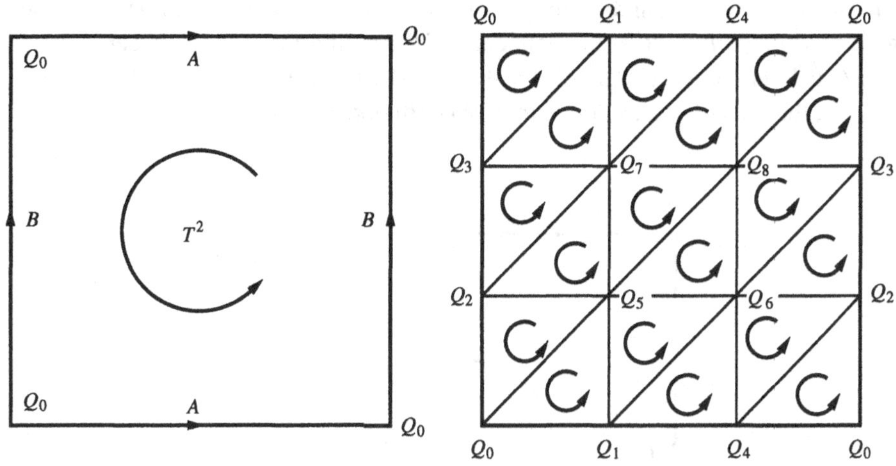


**Figure 13.16**

Orient each 2-simplex as indicated, as we did in Section 13.3a. $\partial[T^2] = A + B - A - B = 0$, confirming that we have an orientable closed surface. Any 1-cycle can be pushed out to the edge. It is clear that if we have a simplicial 1-cycle on the edge that has coefficient $g$ on, say, the simplex $(Q_1, Q_4)$, then this cycle will also have to have coefficient $g$ on $(Q_0, Q_1)$ and $-g$ on $(Q_0, Q_4)$, since otherwise it would have a boundary. Thus a 1-cycle on the edge will have the coefficient $g$ on the entire 1-cycle $A$. Likewise it will have a coefficient $g'$ on the entire 1-cycle $B$. It seems evident from the picture, and can indeed be shown, that no nontrivial combination of $A$ and $B$ can bound. (For example, in Figure 13.16 we may introduce the angular coordinate $\theta$ going around in the $A$ direction. Then $\oint_A$ "$d\theta$" $\neq 0$ shows that $A$ does not bound as a real 1-cycle.) We conclude that

$$H_0(T^2; G) = G = H_2(T^2; G) \tag{13.24}$$
$$H_1(T^2; G) = GA + GB$$

In particular, $H_1(T^2; \mathbb{R}) = \mathbb{R}A + \mathbb{R}B$ is 2-dimensional, $b_0 = b_2 = 1,\ b_1 = 2$.

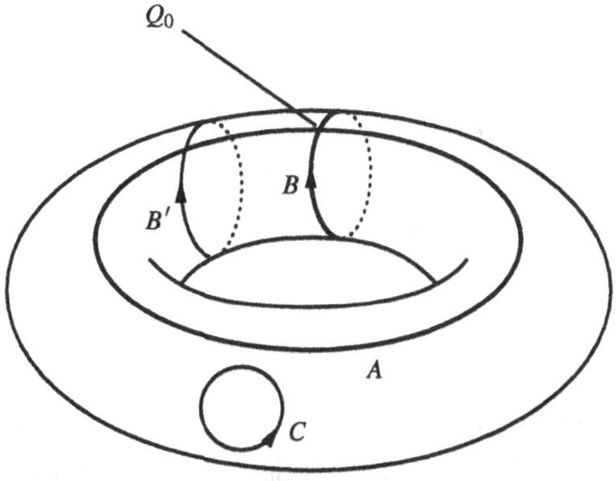


**Figure 13.17**

In the figure we have indicated the basic 1-cycles $A$ and $B$. The cycle $B'$ is homologous to $B$ since $B - B'$ is the boundary of the cylindrical band between them. The cycle $C$ is homologous to 0 since it is the boundary of the small disc.

3. $K^2$, the *Klein bottle*. Look at integer coefficients.

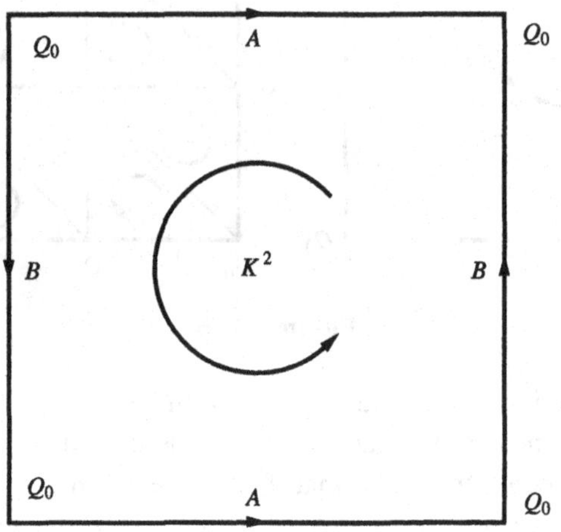


**Figure 13.18**

$H_0 = \mathbb{Z}$ but $H_2 = 0$ since $\partial[K^2] = A + B - A + B = 2B \neq 0$, the Klein bottle is a closed manifold but is not orientable. Again any 1-cycle can be pushed out to the edge, $z_1 \sim rA + sB$, $r$ and $s$ integers. Neither $A$ nor $B$ bound, but we do have the relation $2B \sim 0$. $A$ satisfies no nontrivial relation. Thus $A$ generates a group $\mathbb{Z}A$ and $B$ generates a group with the relation $2B = 0$; this is the group $\mathbb{Z}_2$. Hence

$$H_0(K^2; \mathbb{Z}) = \mathbb{Z}, \quad H_2(K^2; \mathbb{Z}) = 0 \tag{13.25}$$

$$H_1(K^2; \mathbb{Z}) = \mathbb{Z}A + \mathbb{Z}_2 B$$

If we used $\mathbb{R}$ coefficients we would get

$$H_0(K^2; \mathbb{R}) = \mathbb{R}, \quad H_2(K^2; \mathbb{R}) = 0 \tag{13.26}$$

$$H_1(K^2; \mathbb{R}) = \mathbb{R}A$$

since now $B = \partial(1/2)[K^2]$ bounds. Thus $b_0 = 1$, $b_1 = 1$, and $b_2 = 0$.

4. $\mathbb{R}P^2$, the *real projective plane*. The model is the 2-disc with antipodal identifications on the boundary circle. The upper and lower semicircles are two copies of the same

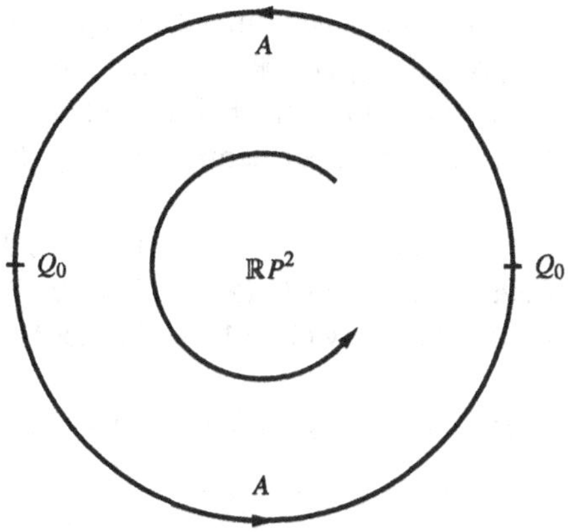


Figure 13.19

closed curve $A$. One should triangulate $\mathbb{R}P^2$ but we shall not bother to indicate the triangles. Orient all triangles as indicated. Clearly $H_0(\mathbb{R}P^2; \mathbb{Z}) = \mathbb{Z}$. Since $\partial[\mathbb{R}P^2] = 2A$, we see that the real projective plane is not orientable and $H_2(\mathbb{R}P^2; \mathbb{Z}) = 0$. $A$ is a 1-cycle and $2A \sim 0$.

$$H_0(\mathbb{R}P^2; \mathbb{Z}) = \mathbb{Z}, \qquad H_2(\mathbb{R}P^2; \mathbb{Z}) = 0 \tag{13.27}$$

$$H_1(\mathbb{R}P^2; \mathbb{Z}) = \mathbb{Z}_2 A$$

With real coefficients

$$H_0(\mathbb{R}P^2; \mathbb{R}) = \mathbb{R}, \qquad H_2(\mathbb{R}P^2; \mathbb{R}) = 0 \tag{13.28}$$

$$H_1(\mathbb{R}P^2; \mathbb{R}) = 0$$

and $b_0 = 1,\ b_1 = 0$, and $b_2 = 0$. $\mathbb{R}P^2$ has the same Betti numbers as a point!

5. $\mathbb{R}P^3$, *real projective* 3-*space*. The model is the solid ball with antipodal indentifications on the boundary 2-sphere. Note that this makes the boundary 2-sphere into a projective plane!

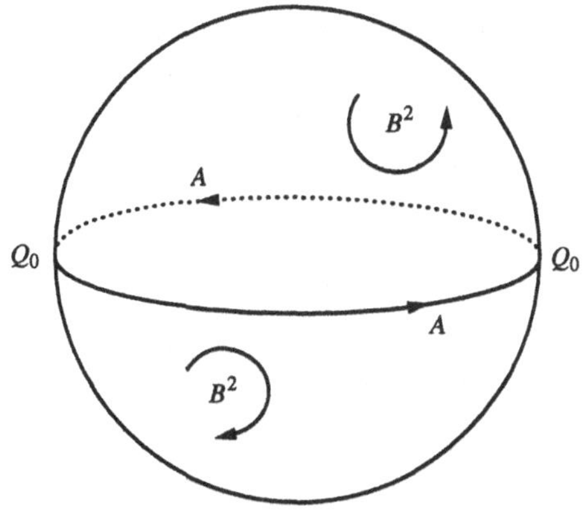


Figure 13.20

Orient the solid ball using the right-hand rule. The upper and lower hemispheres $B^2$ are two copies of the *same* projective plane $\mathbb{R}P^2$. Orient the identified hemispheres $B^2$ as indicated. Note that the orientation of the ball (together with the outward normal) induces the given orientation in the upper hemisphere but the opposite in the lower. Orient the equator $A$ as indicated.

$H_0(\mathbb{R}P^3; \mathbb{R}) = \mathbb{R}$. $A$ is a 1-cycle, but $\partial B^2 = 2A$ and so $H_1(\mathbb{R}P^3; \mathbb{R}) = 0$. $B^2$ is not a cycle since $\partial B^2 = 2A \neq 0$, and so $H_2(\mathbb{R}P^3; \mathbb{R}) = 0$. $\partial[\mathbb{R}P^3] = B^2 - B^2 = 0$. Hence $\mathbb{R}P^3$ is *orientable* (see Corollary (12.14)) and $H_3(\mathbb{R}P^3; \mathbb{R}) = \mathbb{R}$.

$$H_0(\mathbb{R}P^3; \mathbb{R}) = \mathbb{R} = H_3(\mathbb{R}P^3; \mathbb{R}) \qquad (13.29)$$

all others are 0

$\mathbb{R}P^3$ has the same Betti numbers as $S^3$! See Problem 13.3(1) at this time.

6. $T^3$, the 3-*torus*. The model is the solid cube with opposite faces identified.

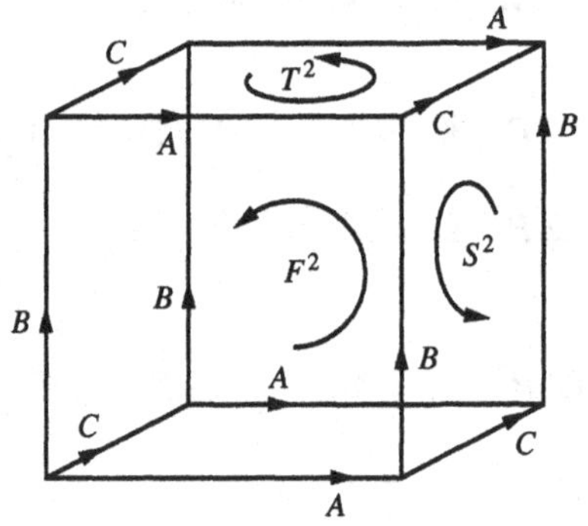


**Figure 13.21**

Note that the front, right side, and top faces (which are the same as the back, left side, and bottom faces) $F^2$, $S^2$, and $T^2$ become 2-toruses after the identification. Orient the cube by the right-hand rule. This induces the given orientation as indicated for the drawn faces but the opposite for their unlabeled copies. Orient the three edges $A$, $B$, and $C$ as indicated. $A$, $B$, and $C$ are 1-cycles. $F^2$, $S^2$, and $T^2$ have 0 boundaries just as in the case of the 2-torus. $\partial[T^3] = F^2 + S^2 + T^2 - F^2 - S^2 - T^2 = 0$ and so $T^3$ is orientable. We have

$$H_0(T^3; \mathbb{Z}) = \mathbb{Z} = H_3(T^3; \mathbb{Z})$$

$$H_1(T^3; \mathbb{Z}) = \mathbb{Z}A + \mathbb{Z}B + \mathbb{Z}C \qquad (13.24')$$

$$H_2(T^3; \mathbb{Z}) = \mathbb{Z}F^2 + \mathbb{Z}S^2 + \mathbb{Z}T^2$$

Using real coefficients we would get $b_0 = 1,\ b_1 = 3 = b_2,\ b_3 = 1$.

## Problems

**13.3(1)** Compute the homology groups of $\mathbb{R}P^3$ with $\mathbb{Z}$ coefficients.

**13.3(2)** A certain closed surface $M^2$ has as model an octagon with the indicated identifications on the boundary. Note carefully the directions of the arrows.

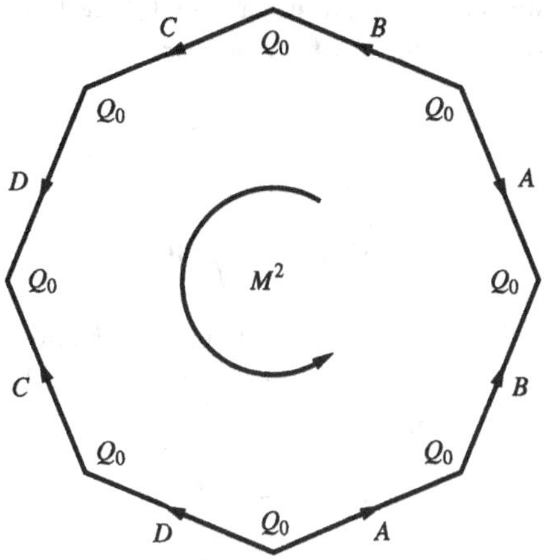


**Figure 13.22**

Write down $H_i(M^2; G)$ for $G = \mathbb{R}$ and $G = \mathbb{Z}$. What are the Betti numbers? Is the surface orientable?

## 13.4. De Rham's Theorem

When is a closed form exact?

### 13.4a. The Statement of de Rham's Theorem

In this section we shall only be concerned with homology with *real coefficients* $\mathbb{R}$ for a manifold $M^n$. The singular chains $C_p$, cycles $Z_p$, and homology groups $H_p$ then form real vector spaces.

We also have the real vector spaces of exterior differential forms on $M^n$.

$$A^p := \text{all (smooth) } p\text{-forms on } M$$
$$F^p := \text{the subspace of all closed } p\text{-forms}$$
$$E^p := \text{the subspace of all exact } p\text{-forms}$$

We have the linear transformation $\partial : C_p \to C_{p-1}$, with kernel $Z_p$ and image $B_{p-1}$ yielding $H_p = Z_p/B_p$. We also have the linear transformation $d : A^p \to A^{p+1}$ with kernel $F^p$ and image $E^{p+1} \subset F^{p+1}$, from which we may form the quotient

$$\mathcal{R}^p := \frac{F^p}{E^p} = \text{(closed } p\text{-forms)/(exact } p\text{-forms)} \qquad (13.30)$$

the **de Rham** vector space. $\mathcal{R}^p$ is thus the collection of equivalence classes of closed $p$-forms; two closed $p$-forms are identified iff they differ by an exact $p$-form. De Rham's theorem (1931) relates these two quotient spaces as follows.

Integration allows us to associate to each $p$-form $\beta^p$ on $M$ a *linear functional* $I\beta^p$ on the chains $C_p$ by $I\beta^p(c) = \int_c \beta^p$. We shall, however, only be interested in this linear functional when $\beta$ is *closed*, $d\beta^p = 0$, and when the chain $c = z$ is a *cycle*, $\partial z = 0$. We thus think of integration as giving a linear transformation from the vector space of closed forms $F^p$ to the dual space $Z_p^*$ of the vector space of cycles

$$I : F^p \to Z_p^*$$

by

$$(I\beta^p)(z) := \int_z \beta^p \tag{13.31}$$

Note that $\int_{z+\partial c} \beta^p = \int_z \beta^p$, since $\beta$ is closed. Thus $I\beta^p$ can be considered as a linear functional on the equivalence class of $z$ mod the vector subspace $B_p$. Thus (13.31) really gives a linear functional on $H_p$

$$I : F^p \to H_p^*$$

Furthermore, the linear functional $I\beta^p$ is the same linear functional as $I(\beta^p + d\alpha^{p-1})$, since the integral of an exact form over a cycle vanishes. In other words, (13.31) *really* defines a linear transformation from $F^p/E^p$ to $H_p^*$, that is, from the de Rham vector space to the dual space of $H_p$. This latter dual space is commonly called the $p^{\text{th}}$ **cohomology** vector space, written $H^p$

$$H^p(M; \mathbb{R}) := H_p(M; \mathbb{R})^* \tag{13.32}$$

Thus

$$I : \mathcal{R}^p \to H^p(M; \mathbb{R}) \tag{13.33}$$

In words, given a de Rham class $b \in \mathcal{R}^p$, we may pick as representative a closed form $\beta^p$. Given a homology class $\mathfrak{z} \in H_p$, we may pick as representative a $p$-cycle $z_p$. Then $I(b)(\mathfrak{z}) := \int_z \beta^p$, and this answer is independent of the choices made. Poincaré conjectured, and in 1931 de Rham proved

**de Rham's Theorem (13.34):** $I : \mathcal{R}^p \to H^p(M; \mathbb{R})$ *is an isomorphism. First, I is onto; this means that any linear functional on homology classes is of the form $I\beta^p$ for some closed p-form $\beta$. In particular, if $H_p$ is finite-dimensional, as it is when $M^n$ is compact, and if*

$$z_p^{(1)}, \ldots, z_p^{(b)} \qquad b = \text{the } p^{\text{th}} \text{ Betti number}$$

*is a p-cycle basis of $H_p$, and if $\pi_1, \ldots, \pi_b$ are arbitrary real numbers, then there is a closed form $\beta^p$ such that*

$$\int_{z_p^{(i)}} \beta^p = \pi_i \,, i = 1, \ldots, b \tag{13.35}$$

*Second, I is 1:1; this means that if $I(\beta^p)(z_p) = \int_z \beta^p = 0$ for all cycles $z_p$, then $\beta^p$ is exact,*

$$\beta^p = d\alpha^{p-1}$$

*for some form $\alpha^{p-1}$.*

The number $\pi_i$ in (13.35) is called the **period** of the form $\beta$ on the cycle $z_p^{(i)}$. Thus

*a closed p-form is exact iff all of its periods on p-cycles vanish.*

A finite-dimensional vector space has the same dimension as its dual space. Thus

**Corollary (13.36):** *If $M^n$ is compact, then dim $\mathcal{R}^p = b_p$, the $p^{\text{th}}$ Betti number. Thus $b_p$ is also the maximal number of closed p-forms on $M^n$, no linear combination of which is exact.*

The proof of de Rham's theorem is too long and difficult to be given here. Instead, we shall illustrate it with two examples. For a proof, see for example, [Wa].

### 13.4b. Two Examples

1. $T^2$, the 2-*torus*. $T^2$ is the rectangle with identifications on the boundary.

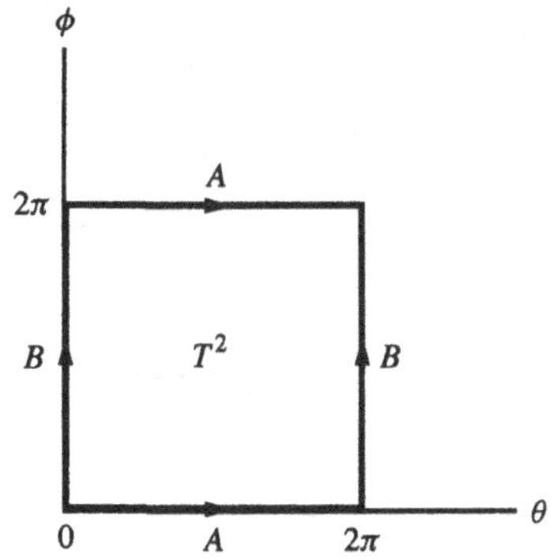


**Figure 13.23**

$\mathcal{R}^0$ consists of closed 0-forms, that is, constant functions, with basis $f = 1$.

$\mathcal{R}^1$ consists of closed 1-forms. Certainly $d\theta$ and $d\phi$ are closed 1-forms and these are not really exact since $\theta$ and $\phi$ are not globally defined functions, being multiple-valued. Since $H_1(T^2; \mathbb{R}) = \mathbb{R}A + \mathbb{R}B$, $A$ and $B$ give a basis for the 1-dimensional real homology. But then $\int_A d\theta/2\pi = 1$, $\int_B d\theta/2\pi = 0$, $\int_A d\phi/2\pi = 0$, and $\int_B d\phi/2\pi = 1$, show that *$d\theta/2\pi$ and $d\phi/2\pi$ form the basis in $\mathcal{R}^1 = H^1 = H_1{}^*$ that is dual to the basis $A, B$!*

$\mathcal{R}^2$ consists of closed 2-forms, but of course all 2-forms on $T^2$ are closed. $d\theta \wedge d\phi$ is closed and has period $\int\int_{[T]} d\theta \wedge d\phi = (2\pi)^2$. (Thus, in particular, it is not exact.) Since $H_2(T^2; \mathbb{R}) = \mathbb{R}\,[T^2]$, we see that $d\theta \wedge d\phi/4\pi^2$ is the basis of $\mathcal{R}^2$ dual to $[T^2]$.

This was all too easy because $\theta$ and $\phi$ are almost global coordinates on $T^2$.

2. The *surface of genus* 2.

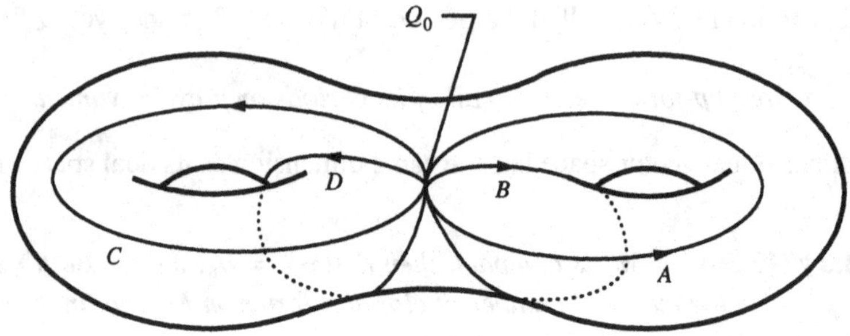


**Figure 13.24**

$\mathcal{R}^0$ has generator the constant function $f = 1$. $\mathcal{R}^2$ has generator any 2-form on $M^2$ whose integral over $[M^2]$ is different from 0, for example, the area 2-form in any Riemannian metric. We need then only consider $\mathcal{R}^1$.

This surface can be considered as an *octagon with identifications on the edges*. This can be seen as follows.

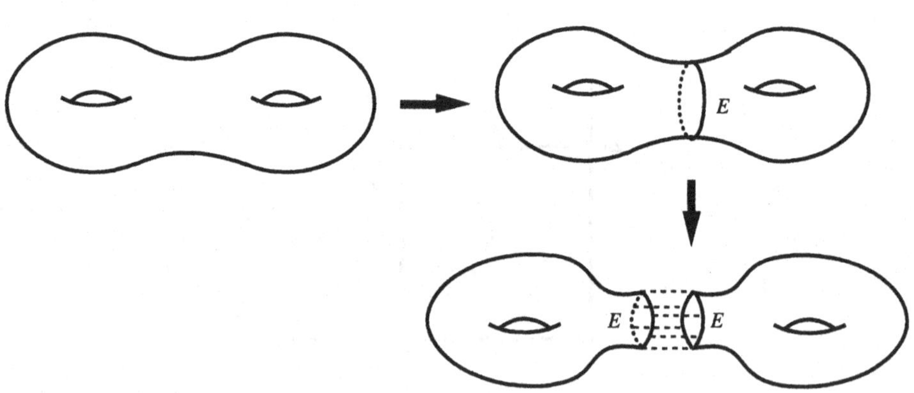


**Figure 13.25**

In the first step we merely narrow the neck. In the second step we cut the surface in two along the neck; the result is a left and a right torus, each with a disc removed, the disc in each case having the original neck circle $E$ as the edge. Of course these two curves must be identified.

We now represent each punctured torus as a rectangle with identifications and with a disc removed. All vertices are the same $Q_0$.

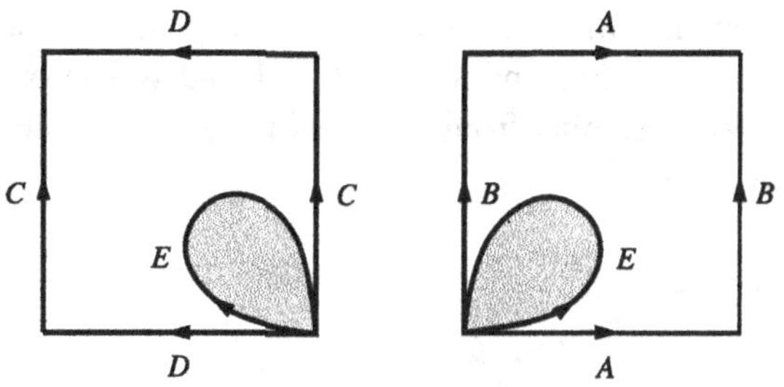


Figure 13.26

We now open up the punctured rectangles

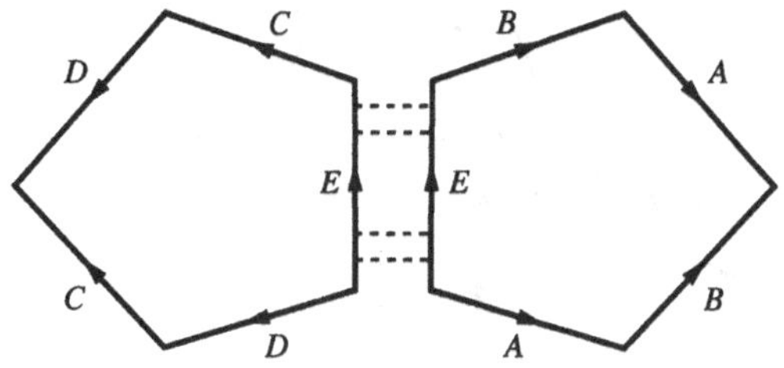


Figure 13.27

where again all vertices are the same $Q_0$. Finally we may sew the two together along the seam $E$, which now disappears

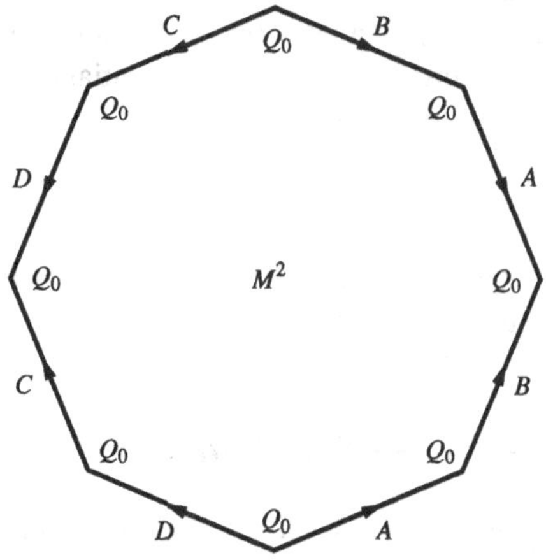


Figure 13.28

leaving the desired octagon with sides identified in pairs. (Note that this is *not* quite the surface that appeared in Problem 13.3(2) because of the identifications on the sides $B$.)

From this diagram the first homology is clearly

$$H_1(M^2; \mathbb{R}) = \mathbb{R}A + \mathbb{R}B + \mathbb{R}C + \mathbb{R}D, \; b_1 = 4$$

We now wish to exhibit the dual basis in $\mathcal{R}^1$. Suppose, for instance, we wish to construct a closed 1-form whose period on $A$ is 1 and whose other periods vanish. Take a thin band on $M^2$ stretching from the interval $pq$ on $A$ to the same points on the identified other copy of $A$.

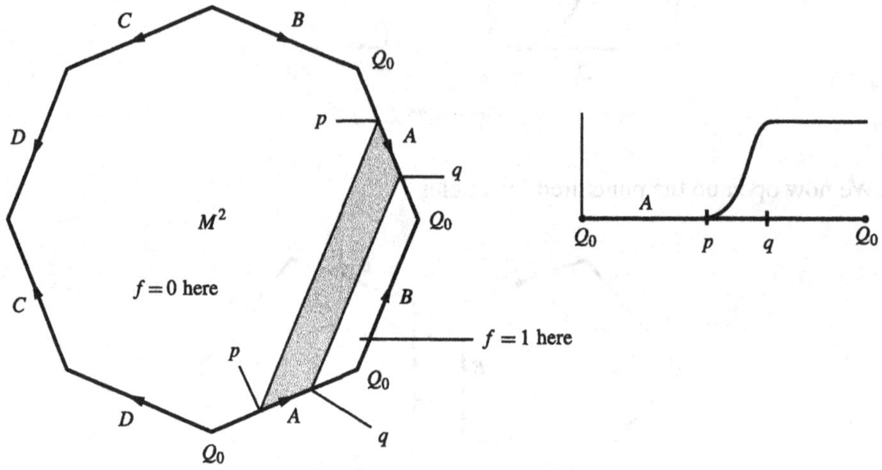


**Figure 13.29**

Define a "function" $f$ on $M^2$ as follows. Let $f = 0$ to the "left" of the band, let $f = 1$ to the "right" of the band, and let $f$ rise smoothly in the band to interpolate. This is not really a function on $M^2$ since, for example, the side $B$ is in both the left and the right regions. It *does* define a *multiple-valued* function; we could have $f$ starting with the value 0 to the left, and $f$ increases by 1 every time one crosses the band from left to right. Although $f$ is multiple-valued, its differential $df$ is a *well-defined* 1-form on all of $M^2$! By construction we have

$$\int_A df = 1, \quad \text{and} \quad \int_{B \text{ or } C \text{ or } D} df = 0$$

We have then exhibited the dual 1-form to the class $A$. Using other bands we can construct the remaining dual basis forms for $\mathcal{R}^1$.

## Problem

**13.4(1) (i)** Show that every map $F : S^2 \to T^2$ of a sphere into a torus has degree 0. Hint: Use "$d\theta$" $\wedge$ "$d\phi$" on $T^2$ and show pull-back is exact.

**(ii)** Put conditions on a closed $M^n$ to ensure that deg $F : M^n \to T^n$ must vanish.

CHAPTER 14

# Harmonic Forms

## 14.1. The Hodge Operators

What are Maxwell's equations in a curved space–time?

### 14.1a. The * Operator

On a (pseudo-)Riemannian manifold $M^n$ we introduce a *pointwise* scalar product between $p$-forms by

$$\langle \alpha^p, \beta^p \rangle := \alpha_{\underrightarrow{I}} \beta^I \tag{14.1}$$

where, as usual, $I = (i_1, \ldots, i_p)$, and $\rightarrow$ denotes that in the implied sum we have $i_1 < i_2 < \ldots < i_p$. It is not difficult to check that if $\mathbf{e} = \mathbf{e}_1, \ldots, \mathbf{e}_n$ is an orthonormal basis for tangent vectors at a point, then $\sigma^1, \ldots, \sigma^n$ is an orthonormal basis of 1-forms and also that

$$\sigma^{\underrightarrow{I}} = \sigma^{i_1} \wedge \sigma^{i_2} \wedge \ldots \wedge \sigma^{i_p}$$

yields an orthonormal basis for $p$-forms at the point for $i_1 < \cdots < i_p$.

We now introduce a **global** or **Hilbert space** scalar product by

$$(\alpha^p, \beta^p) := \int_M \langle \alpha^p, \beta^p \rangle \mathrm{vol}^n \tag{14.2}$$

whenever this makes sense; this will be the case when $M$ is compact, or, more generally, when $\alpha$ or $\beta$ has compact support.

We should remark at this time that the space of smooth $p$-forms on a Riemannian $M$ that satisfy $(\alpha^p, \alpha^p) < \infty$ form only a *pre*-Hilbert space since it is not complete; a limit of square integrable smooth forms need not even be continuous! To get a Hilbert space we must "complete" this space. We shall not be concerned here with such matters, and we shall continue to use the inaccurate description "Hilbert space." We shall even go a step further and use this denomination even in the pseudo-Riemannian case, when (,) is not even positive definite!

If $\alpha^1$ is a 1-form, we may look at its contravariant version $\mathbf{A}$, and to this vector we may associate the pseudo $(n-1)$-form $i_{\mathbf{A}}\,\mathrm{vol}^n$. In this way we associate to each 1-form a pseudo $(n-1)$-form. We are now going to generalize this procedure, associating to each $p$-form $\alpha^p$ a pseudo $(n-p)$-form $*\alpha$, the **(Hodge-) dual** of $\alpha$, as follows. If $\alpha^p = \alpha_{\underrightarrow{I}} dx^I$ then

$$*\alpha^p := \alpha^*_{\underrightarrow{J}} dx^J$$

where (14.3)

$$\alpha^*_{\underrightarrow{J}} := \sqrt{|g|}\alpha^K \epsilon_{\underrightarrow{K}\,\underrightarrow{J}}$$

If $f$ is a function we have

$$*(f\alpha^p) = f * \alpha^p \tag{14.4}$$

Written out in full

$$\alpha^*_{j_1 \ldots j_{n-p}} = \sqrt{|g|} \sum_{k_1 < \ldots < k_p} \alpha^{k_1 \ldots k_p} \epsilon_{k_1 \ldots k_p j_1 \ldots j_{n-p}}$$

and where the upper indices $K$ in $\alpha^K$ indicate that all of the covariant indices in $\alpha$ have been raised by the metric tensor,

$$\alpha^{k \ldots r} = g^{ks} \ldots g^{rt} \alpha_{s \ldots t}$$

For an important special case, the 0-form that is the constant function $f = 1$ has

$$*1 = \sqrt{|g|}\epsilon_{12\ldots n} dx^1 \wedge \ldots \wedge dx^n = \mathrm{vol}^n \tag{14.5}$$

Note that for a given $j_1 < j_2 < \ldots < j_{n-p}$, there is at most *one* nonvanishing term in the sum on the right side of (14.3), namely when $k_1 < \ldots < k_p$ is the complementary multiindex to $j_1 \ldots j_{n-p}$.

We then have

$$\alpha^p \wedge *\beta^p = (\alpha \wedge *\beta)_{12\ldots n} dx^1 \wedge \ldots \wedge dx^n$$

and

$$\begin{aligned}(\alpha \wedge *\beta)_{12\ldots n} &= \delta^{JK}_{12\ldots n} \alpha_{\underrightarrow{J}} (*\beta)_{\underrightarrow{K}} = \epsilon^{JK} \alpha_{\underrightarrow{J}} \sqrt{|g|} \beta^L \epsilon_{\underrightarrow{L}\,\underrightarrow{K}} \\ &= \sqrt{|g|} \alpha_{\underrightarrow{J}} \beta^J = \sqrt{|g|} \langle \alpha^p, \beta^p \rangle\end{aligned}$$

and so

$$\alpha^p \wedge *\beta^p = \langle \alpha^p, \beta^p \rangle \, \mathrm{vol}^n \tag{14.6}$$

This shows that indeed * takes forms into pseudoforms and conversely.

We have claimed that * generalizes the map $\alpha^1 \rightarrow i_{\mathbf{A}}\,\mathrm{vol}^n$. To see this, $i_{\mathbf{A}}\,\mathrm{vol}^n = i_{\mathbf{A}} \sqrt{|g|} \epsilon_{\underrightarrow{I}} dx^I = \sqrt{|g|} A^j \epsilon_{j\underrightarrow{K}} dx^{k_2} \wedge \ldots \wedge dx^{k_n}$

$$i_{\mathbf{A}}\,\mathrm{vol}^n = *\alpha^1 \tag{14.7}$$

Equation (14.3) is frequently awkward to apply; many times it is more convenient to use directly (14.6) together with the following. Let $\mathbf{e} = (\mathbf{e}_1, \ldots, \mathbf{e}_n)$ be an orthonormal frame of vectors (allowing $\| \mathbf{e}_1 \|^2 = -1$ in the case of a pseudo-Riemannian manifold); as we have mentioned, $\sigma^I$, for $I = (i_1, \dots, i_p)$, are then also orthonormal and $\sigma^1 \wedge \dots \wedge \sigma^n = \pm \text{vol}^n$. Thus, from (14.6),

$$*\sigma^I = \pm\sigma^J \tag{14.8}$$

where $J = (j_1, \dots, j_{n-p})$ is complementary to $I = (i_1, \dots, i_p)$.

Look, for example, at the electromagnetic field in a perhaps curved space–time manifold $M^4$. This will be discussed in more detail in Section 14.1c. We shall see there that the field is again described in local coordinates $(t, x)$ by the 2-form

$$F^2 = \mathcal{E}^1 \wedge dt + \mathcal{B}^2$$

where

$$\mathcal{E}^1 = E_1 dx^1 + E_2 dx^2 + E_3 dx^3$$

and

$$\mathcal{B}^2 = B_{23} dx^2 \wedge dx^3 + B_{31} dx^3 \wedge dx^1 + B_{12} dx^1 \wedge dx^2$$

Then, using the space–time metric, $*F^2$ will again be a 2-form, and so will be of the form

$$*F = *(\mathcal{E}^1 \wedge dt) + *\mathcal{B}^2 = [\star\mathcal{E}^1] - [\star\mathcal{B}^2 \wedge dt]$$

for some spatial 1-form $\star\mathcal{B}^2$ and some spatial 2-form $\star\mathcal{E}^1$. Let us find these forms in the special case of *Minkowski space*, without using (14.3).

$*$ takes $p$-forms into pseudo $(4 - p)$-forms. $\mathcal{B}^2 = B_1 dx^2 \wedge dx^3 + B_2 dx^3 \wedge dx^1 + B_3 dx^1 \wedge dx^2$, is a 2-form in Minkowski space–time. Since the coordinates are orthonormal and $\sqrt{|g|} = 1$, we can probably avoid the use of (14.3). $*(dx^2 \wedge dx^3)$ has the property that $(dx^2 \wedge dx^3) \wedge *(dx^2 \wedge dx^3) = \| dx^2 \wedge dx^3 \|^2 \, dt \wedge dx^1 \wedge dx^2 \wedge dx^3$. Since the $dx^\alpha$ are orthonormal and $\| dx^\alpha \|^2 = +1$ for $\alpha = 1, 2, 3$, we see that $\| dx^2 \wedge dx^3 \|^2 = \| dx^2 \|^2 \| dx^3 \|^2 = +1$, and so $*(dx^2 \wedge dx^3) = dt \wedge dx^1$. Likewise for the other two terms. We then have, from Equation (3.41),

$$*\mathcal{B}^2 = -(B_1 dx^1 + B_2 dx^2 + B_3 dx^3) \wedge dt = -(\star\mathcal{B}^2) \wedge dt$$

Note that $\star\mathcal{B}^2$ *is simply the star operator in* $\mathbb{R}^3$ (which takes $p$-forms to $(3 - p)$-forms) *applied to the 2-form* $\mathcal{B}^2$. In our older notation it is simply $\langle \, , \mathbf{B} \rangle$, as in Equation (3.41)!

Look now at the term $\mathcal{E}^1 \wedge dt = (E_1 dx^1 + E_2 dx^2 + E_3 dx^3) \wedge dt$. For example

$$*(dx^1 \wedge dt) = - \| dx^1 \wedge dt \|^2 \, dx^2 \wedge dx^3 = dx^2 \wedge dx^3$$

since $\| dt \|^2 = -1$. Thus $*(\mathcal{E}^1 \wedge dt) = E_1 dx^2 \wedge dx^3 + E_2 dx^3 \wedge dx^1 + E_3 dx^1 \wedge dx^2$, that is,

$$*(\mathcal{E}^1 \wedge dt) = \star\mathcal{E}^1$$

where $\star\mathcal{E}^1 = i_{\mathbf{E}} \, \text{vol}^3$ *results from applying the star operator of* $\mathbb{R}^3$ to $\mathcal{E}^1$.

*This explains our use of the notation* $*F^2$ *in Section 7.2b and the use of* $\star$ *in Section 3.5c*. This concludes our electromagnetic excursion for the moment.

In Problem 14.1(1) you are asked to show that

$$
\begin{aligned}
*(*\alpha^p) &= (-1)^{p(n-p)}\alpha \quad \text{if } M^n \text{ is Riemannian} \\
&\quad - (-1)^{p(n-p)}\alpha \quad \text{if } M^n \text{is pseudo-Riemannian}
\end{aligned} \tag{14.9}
$$

It is sufficient to verify these for terms of the form $\sigma^I$ and to assume these are orthonormal.

Finally, note the following. If **A** is a vector and $\alpha$ is its associated 1-form, then $*\alpha$ is a pseudo-$(n-1)$-form, and if $V^{n-1} \subset M^n$ is a transversally oriented hypersurface, then

$$
\int_V *\alpha^1 = \int_V i_{\mathbf{A}} \operatorname{vol}^n = \int_V \langle \mathbf{A}, \mathbf{N} \rangle dS^{n-1} \tag{14.10}
$$

In particular

$$
\int_V *df = \int_V \langle \nabla f, \mathbf{N} \rangle dS^{n-1}
$$

for any function $f$, and this last integral is the "surface" integral of the **normal derivative** $df/d\mathbf{N}$ over the hypersurface.

### 14.1b. The Codifferential Operator $\delta = d^*$

Exterior differentiation $d : \bigwedge^p M^n \to \bigwedge^{p+1} M^n$ sends $p$-forms to $(p+1)$-forms; in this section we shall exhibit an operator that decreases the degree of a form by one, and, in the case of a compact manifold, serves as the pre-Hilbert space *adjoint* of $d$. We thus want an operator

$$
d^* : \overset{p}{\bigwedge} \to \overset{p-1}{\bigwedge}
$$

such that (14.11)

$$
(d\alpha^{p-1}, \beta^p) = (\alpha^{p-1}, d^*\beta^p)
$$

Now $(d\alpha^{p-1}, \beta^p) = \int_M d\alpha^{p-1} \wedge *\beta^p$. Consider first the *Riemannian case*; we may then use the first equation in (14.9). Note then

$$
\begin{aligned}
d(\alpha \wedge *\beta) &= d\alpha \wedge *\beta + (-1)^{p-1}\alpha \wedge d * \beta \\
&= d\alpha \wedge *\beta + (-1)^{p-1}(-1)^{(n-p+1)(p-1)}\alpha \wedge * * d * \beta \\
&= d\alpha \wedge *\beta + (-1)^{n(p+1)}\alpha \wedge * * d * \beta
\end{aligned}
$$

and so

$$
d\alpha^{p-1} \wedge *\beta^p = (-1)^{n(p+1)+1}\alpha \wedge *(*d * \beta) + d(\alpha \wedge *\beta)
$$

with a similar result for the non-Riemannian case. We then define *whether M is compact or not and whether or not M has a boundary*

$$
\begin{aligned}
d^*\beta^p &:= (-1)^{n(p+1)+1} * d * \beta^p \quad \text{Riemannian} \\
&\quad\;\; (-1)^{n(p+1)} * d * \beta^p \quad \text{pseudo-Riemannian}
\end{aligned} \tag{14.12}
$$

and then

$$(d\alpha^{p-1}, \beta^p) - (\alpha^{p-1}, d^*\beta^p) = \int_M d(\alpha^{p-1} \wedge *\beta^p) \tag{14.13}$$

at least if $\alpha$ or $\beta$ has compact support. If $M^n$ is a *closed* manifold, then $d^*$, as defined in (14.12), *is the pre-Hilbert space adjoint of d.*

If $M$ is a compact manifold *with boundary* $\partial M$, let $i : \partial M \to M$ be inclusion. Then

$$(d\alpha^{p-1}, \beta^p) - (\alpha^{p-1}, d^*\beta^p) = \int_{\partial M} \alpha^{p-1} \wedge *\beta^p \tag{14.14}$$

and then $d^*$ is again the adjoint of $d$ if we restrict ourselves to one of two types of forms: those forms $\gamma$ that are 0 when restricted to the boundary, that is, $i^*\gamma = 0$, or those forms $\gamma$ whose dual $*\gamma$ is 0 when restricted to the boundary, $i^* * \gamma = 0$.

The operator $d^*$ is called the **codifferential** operator. The traditional notation for $d^*$ is $\delta$

$$\delta := d^*$$

but *we shall avoid this notation* since the symbol $d^*$ is more informative and *we prefer to reserve* $\delta$ *for the variational symbol.*

We shall need a coordinate expression for the $(p-1)$-form $d^*\beta^p$.

**Theorem (14.15):** $(d^*\beta^p)_K = -\beta^j{}_{K/j}$

We shall call the negative of the right-hand side the **Divergence** (with a capital $D$) of the form $\beta$

$$(\text{Div}\beta^p)_K := \beta^j{}_{K/j}$$

although sometimes it will look more like a curl! Note that this is the same definition as we gave for the Divergence of a *symmetric* tensor in Equation (11.15)! We only define the Divergence of a tensor that is either symmetric or skew symmetric.

PROOF: To show that two $(p-1)$-forms $\gamma$ and $\rho$ are identical we need only show given any small closed coordinate ball $B$ (disjoint from $\partial M$ if $M$ has a boundary) then for *all* $(p-1)$-forms $\alpha$ whose support lies in the *interior* of the ball, $\int_B \langle\alpha, \gamma\rangle *1 = \int_B \langle\alpha, \rho\rangle *1$, for if the volume integral of $\alpha_{\underrightarrow{I}}(\gamma^I - \rho^I)$ vanishes for all smooth $\alpha$ and for each small ball, then $\gamma - \rho = 0$. We shall verify (14.15) by showing that

$$\int_B \langle\alpha^{p-1}, d^*\beta^p\rangle * 1 = \int_B \langle\alpha^{p-1}, -\text{Div}\beta^p\rangle * 1$$

We may consider the new manifold-with-boundary $B$ instead of $M$. For this manifold the preceding integrals are inner products, and we must show, since $\alpha$ vanishes on the boundary of the ball,

$$(\alpha^{p-1}, d^*\beta^p) = (\alpha^{p-1}, -\text{Div}\beta^p)$$

Using Problem 11.2(1)

$$(\alpha^{p-1}, d^*\beta^p) = (d\alpha^{p-1}, \beta^p) = \int_B \langle d\alpha, \beta\rangle * 1 = \int_B (d\alpha)_{\underline{I}}\beta^I * 1$$

$$= \int_B \delta_{\underline{I}}^{jK}\alpha_{\underline{K}/j}\beta^I * 1 = \int_B (\delta_{\underline{I}}^{jK}\alpha_{\underline{K}}\beta^I)_{/j} * 1 - \int_B \alpha_{\underline{K}}\delta_{\underline{I}}^{jK}\beta^I_{/j} * 1$$

But

$$\delta_{\underline{I}}^{jK}\beta^I = \beta^{jK} \quad \text{(why?)}$$

and so

$$(\alpha^{p-1}, d^*\beta^p) = \int_B (\alpha_{\underline{K}}\beta^{jK})_{/j} * 1 - \int_B \alpha_{\underline{K}}\beta^{jK}{}_{/j} * 1$$

In the first integral, $C^j := \alpha_{\underline{K}}\beta^{jK} = [(p-1)!]^{-1}\alpha_K\beta^{jK}$ are the components of a contravariant vector $\mathbf{C}$, and then the integrand is the divergence of this vector. But $\int_B \operatorname{div}\mathbf{C} * 1 = \int_{\partial B}\langle \mathbf{C}, \mathbf{N}\rangle dS = 0$, since $\mathbf{C}$ vanishes on $\partial B$. Thus

$$(\alpha^{p-1}, d^*\beta^p) = -\int_B \alpha_{\underline{K}}\beta^{jK}{}_{/j} * 1 = \int_B \langle \alpha, -\mathrm{Div}\beta\rangle * 1$$

as desired. □

### 14.1c. Maxwell's Equations in Curved Space–Time $M^4$

We shall assume that the electromagnetic field is again described by an electromagnetic 2-form $F^2$. In *any* local coordinates ($t = x^0, \mathbf{x}$) we may decompose $F^2$ into a part that contains $dt$ and a part without $dt$; thus $F^2$ *defines* an electric 1-form $\mathcal{E}^1$ and a magnetic 2-form $\mathcal{B}^2$ through

$$F^2 = \mathcal{E}^1 \wedge dt + \mathcal{B}^2$$

but of course this decomposition depends on the coordinates used. We *postulate* that for any *bounding* 2-cycle $z^2 = \partial U^3$ in space–time $M^4$ we have

$$\int_{\partial U} F^2 = 0 \tag{14.16}$$

*If* $F$ is continuously differentiable, we conclude that $\int_U dF = 0$. Since $U$ can be chosen to be an arbitrarily small hypersurface with arbitrarily chosen normal, we see that we must then have

$$dF^2 = 0$$

This is the first set of Maxwell equations. If we write, as usual, $d = \mathbf{d} + dt \wedge \partial/\partial t$, $dF = 0$ yields the usual Maxwell equations (3.39) and (3.40), together with their primed differential versions. Note that the operator $d$ is independent of the metric of space–time.

We *postulate* that there is a current pseudo-3-form, with associated decomposition

$$\mathcal{S}^3 = \sigma^3 - j^2 \wedge dt$$

Since the notion of the charge contained in a region is independent of the metric, $\mathcal{S}^3$ is assumed given *independent of the metric*. Of course, $\mathcal{S}^3$ can be written in the form

$$\mathcal{S}^3 = i_J \text{ vol}^4$$

but the current 4-vector $J$ *will depend on the metric*! It is for this reason that $\mathcal{S}^3$ is more basic than $J$.

We then *postulate* that for any 3-cycle $Z^3$, bounding or not, we have

$$\int_Z \mathcal{S}^3 = 0 \tag{14.17}$$

If one applies this to the boundary of a solid space–time cylinder $Z = \partial\{V^3 \times [0, T]\}$ one sees that this is **conservation of charge** (this is Problem 14.1(4)).

We now *postulate* that

$$\int_{\partial U} *F = 4\pi \int_U \mathcal{S}^3 \tag{14.18}$$

for all 3-chains $U$. Note that this is compatible with (14.17). This is the second set of Maxwell equations. When $\mathcal{S}$ is smooth we see from the same argument as used after (14.16) that $\mathcal{S}^3$ is closed, $d\mathcal{S}^3 = 0$. Since the periods of $\mathcal{S}^3$ vanish, we conclude from de Rham that $\mathcal{S}^3$ is in fact exact, and postulate (14.18) says essentially that $*F^2$ is a "potential" for $\mathcal{S}^3$!

$$d * F^2 = 4\pi \mathcal{S}^3 \tag{14.19}$$

Since $*F$ is a pseudo-2 form we may define pseudoforms $\boldsymbol{*}\mathcal{E}^1$ and $\boldsymbol{*}\mathcal{B}^2$ by

$$*F = -(\boldsymbol{*}\mathcal{B}^2) \wedge dt + \boldsymbol{*}\mathcal{E}^1 \tag{14.20}$$

*It is no longer true that* $\boldsymbol{*}\mathcal{E}^1$ *and* $\boldsymbol{*}\mathcal{B}^2$ *are the Hodge duals (using the 3-space metric* $g_{\alpha\beta}$ *of the spatial section* $t = constant$*), of the forms* $\mathcal{E}^1$ *and* $\mathcal{B}^2$! If, for example, $g^{0\beta} \neq 0$, $\boldsymbol{*}\mathcal{B}^2$ *may involve* $\mathcal{E}$ *as well as* $\mathcal{B}$!

In the smooth case the second set of Maxwell's equations (14.19) are exactly as in Minkowski space, that is, (3.42′) and (3.43′). *Maxwell's equations in curved space are exactly as in flat space, once we accept* $*F$ *as defining the fields* $\boldsymbol{*}\mathcal{B}^2$ *and* $\boldsymbol{*}\mathcal{E}^1$.

### 14.1d. The Hilbert Lagrangian

The Hilbert action for Einstein's theory is essentially $\int_M R * 1$. Although the curvature matrix $\theta$ is a matrix of 2-forms, we haven't expressed either the Ricci tensor (which is symmetric) or the scalar curvature in terms of forms. Still it is possible to write the action in terms of forms; although the expression is awkward, it does occur in physics papers and the reader should be aware of it. We shall be very brief.

$\theta^a{}_b = R^a{}_{b(r<s)} dx^r \wedge dx^s$ is a matrix of 2-forms. Then $*\theta^a{}_b$ is defined to be the matrix obtained by taking the $*$ of each of the 2-forms, that is, $*$ *does not affect the indices* $^a{}_b$. Then

$$*\theta^a{}_b = |g|^{1/2} R^a{}_b{}^{cd} \epsilon_{(c<d)(r<s)} dx^r \wedge dx^s$$

and

$$\begin{aligned} dx^a \wedge dx^b \wedge *\theta_{ab} &= R_{ab}{}^{cd}|g|^{1/2}\epsilon_{(c<d)(r<s)}dx^a \wedge dx^b \wedge dx^r \wedge dx^s \\ &= R_{ab}{}^{cd}|g|^{1/2}\epsilon^{abrs}\epsilon_{(c<d)(r<s)}dx^0 \wedge dx^1 \wedge dx^2 \wedge dx^3 \\ &= R_{ab}{}^{cd}\epsilon^{abrs}\epsilon_{(c<d)(r<s)} * 1 = 2R_{a<b}{}^{cd}\epsilon^{abrs}\epsilon_{(c<d)(r<s)} * 1 \\ &= 2R_{a<b}{}^{ab} * 1 = R_{ab}{}^{ab} * 1 = R * 1 \end{aligned}$$

Thus

$$R * 1 = dx^a \wedge dx^b \wedge *\theta_{ab}$$

## Problems

**14.1(1)** Verify (14.9).

**14.1(2)** Show that for any $p$-form $\beta^p$

$$(\text{Div}\beta^p)^K = \beta^{jK}{}_{/j} = |g|^{-1/2}\partial/\partial x^j(|g|^{1/2}\beta^{jK})$$

**14.1(3)** Note that if $f$ and $g$ are functions then $\nabla^2 f = -d^*df$ and if $M$ is compact $(f, \nabla^2 g) = \int_M f\nabla^2 g * 1$. Apply Equation (14.14) in the case when $M^n$ is a compact manifold with boundary to obtain **Green's theorem**

$$\int_M (f\nabla^2 g - g\nabla^2 f) * 1 = \int_{\partial M} f * dg - g * df$$

**14.1(4)** Show that (14.17) does imply conservation of charge.

## 14.2. Harmonic Forms

Among all closed forms with a given set of periods, which one has the smallest global norm?

### 14.2a. The Laplace Operator on Forms

In $\mathbb{R}^n$ with cartesian coordinates, the Laplacian of a function $f$ is the familiar $\nabla^2 f = \sum(\partial^2 f/\partial x^i \partial x^i)$. We have given two equivalent invariant expressions for $\nabla^2$ on a Riemannian manifold in Equations (2.89) and (11.29).

The Laplacian of a $p$-form field is a more complicated matter. Consider a vector field **A**. In $\mathbb{R}^n$ with *cartesian* coordinates, one could define $\nabla^2\mathbf{A}$ to be the vector field whose components $(\nabla^2\mathbf{A})^i = \sum_j(\partial^2 A^i/\partial x^j \partial x^j)$ are simply the Laplacians of the components of **A**, considered as functions. In $\mathbb{R}^3$ this can be expressed in the usual form found in physics books,

$$\nabla^2\mathbf{A} = \text{grad div }\mathbf{A} - \text{curl curl }\mathbf{A} \tag{14.21}$$

We can write this expression in intrinsic form if we consider the covector $\alpha^1$ associated to **A**, instead of **A** itself. Note first that from Equation (14.15)

$$d^*\alpha^1 = -\text{div }\mathbf{A}$$

and so the covariant version of the first term in (14.21) is $-dd^*\alpha$. Furthermore, $d\alpha^1$ is the 2-form version of curl **A**. For any 2-form $\beta^2 = i(\mathbf{B})\,\mathrm{vol}$ we have, from (14.12), $d^*\beta^2 = (-1)^{(3)(3)+1} * d * \beta^2 = *d * \beta^2$. $*\beta^2$ is the 1-form version of **B** and so $d * \beta^2$ is the 2-form version of curl **B** and $*d * \beta^2$ is the 1-form version of curl **B**. Thus $-d^*d\alpha^1$ is the 1-form version of $-$curl curl **A**. Finally then (14.21) has as covariant version

$$\nabla^2\alpha^1 = -(dd^* + d^*d)\alpha^1$$

We shall define the Laplace operator $\Delta$ on $p$-form by the *negative* of the preceding, that is,

$$\Delta : \bigwedge^p \to \bigwedge^p \quad \text{by } \Delta := dd^* + d^*d \tag{14.22}$$

Occasionally we shall write $\nabla^2 := -\Delta$.

Note that from $d^2 = 0$ and $** = \pm 1$, we have

$$d^*d^* = \pm(*d*)(*d*) = 0 \quad \text{and so}$$

$$\Delta = (d + d^*)^2 \tag{14.23}$$

In Problem 14.2(1) you are asked to show the following in $\mathbb{R}^3$, using *brief* explanations as we did in deriving part 6 in the following

$\Delta$ in $\mathbb{R}^3$

1. $d^* f^0 = 0$.
2. $d^*\alpha^1 = -\operatorname{div}\mathbf{A}$.
3. $d^*\beta^2 = d^* i_{\mathbf{B}}\,\mathrm{vol}^3 = *i_{\mathrm{curl}\mathbf{B}}\,\mathrm{vol}^3$ is the 1-form version of curl **B**.
4. $d^*\gamma^3 = d^*(*g^0) = - * dg$ is the 2-form version of $-\operatorname{grad} g$.
5. $\Delta f^0 = -\nabla^2 f^0$.
6. $\Delta\alpha^1$ is the 1-form version of curl curl **A** $-$ grad div **A**.
7. $\Delta\beta^2 =$ is the 2-form version of curl curl **B** $-$ grad div **B**.
8. $\Delta(*f^0) = - * (\nabla^2 f)$.

## 14.2b. The Laplacian of a 1-Form

Let $\alpha^1 = a_i dx^i$ be a 1-form on a Riemannian $M^n$. We shall compute a coordinate expression for $\Delta\alpha = (dd^* + d^*d)\alpha$. First

$$d\alpha = \sum_{i<j}(\partial_i a_j - \partial_j a_i)dx^i \wedge dx^j = \sum_{i<j}(a_{j/i} - a_{i/j})dx^i \wedge dx^j$$

$$=: \sum_{i<j} c_{ij}dx^i \wedge dx^j$$

$$(d^*c)_j = -c^i{}_{j/i} = -a_j{}^{/i}{}_{/i} + a^i{}_{/ji}$$

where we have put

$$a_j{}^{/i} = g^{ik}a_{j/k}$$

Thus

$$(d^*d\alpha)_j = -a_j{}^{/i}{}_{/i} + a^i_{/ji}$$

Also

$$d^*\alpha = -a^r_{/r}$$

and so $d(d^*\alpha) = -a^i_{/ij}dx^j$, that is, $(dd^*\alpha)_j = -a^i_{/ij}$. Thus

$$(\Delta\alpha)_j = -a_j{}^{/i}{}_{/i} + a^i_{/ji} - a^i_{/ij}$$

By Ricci's identity (11.23)

$$(\Delta\alpha)_j = -a_j{}^{/i}{}_{/i} + a^k R^i_{kij} = -a_j{}^{/i}{}_{/i} + a^k R_{kj} \tag{14.24}$$

We conclude

$$\Delta\alpha = (-a_j{}^{/i}{}_{/i}dx^j) + (a_k R^k{}_j dx^j) \tag{14.25}$$

The first term in (14.24) looks, at first glance, as if we are taking the negative of the usual Laplacian of the component *function* $a_j$, but this is not so since $a_{j/i} = \partial_i a_j - a_k \Gamma^k_{ij}$, and this connection coefficient would not occur in the covariant derivative of a function. The first term in (14.25) is sometimes called a "rough" Laplacian, written $\tilde{\nabla}\tilde{\nabla}\alpha$. It differs from *the* Laplacian $\Delta\alpha$ (defined first by Kodaira and independently by Bidal and de Rham) by the second term in (14.25), *which does not involve any derivatives of* $\alpha$!

$$(\Delta\alpha)_j = -(\tilde{\nabla}\tilde{\nabla}\alpha)_j + a_k R^k{}_j \tag{14.26}$$

(14.25) and (14.26) are called **Weizenböck** formulae.

### 14.2c. Harmonic Forms on Closed Manifolds

Let $M^n$ be a compact *Riemannian* (rather than pseudo-Riemannian) manifold. Then the global inner product (,) is positive definite, for

$$(\alpha^p, \beta^p) = \int_M \alpha \wedge *\beta = \int_M \langle \alpha, \beta \rangle * 1$$

and at the pole of a geodesic coordinate system $\langle \alpha, \alpha \rangle = \sum (a_L)^2$. Thus $(\alpha, \alpha) \geq 0$, and vanishes only if $\alpha$ vanishes identically.

We say that a form $\alpha^p$ is **harmonic** if $\Delta\alpha = 0$. For a function (i.e., 0-form) this reduces to the usual notion.

Let $M^n$ be a *closed* manifold. If we again denote the formal adjoint of an operator $A$ on forms by $A^*$, then since $\Delta = (d+d^*)(d+d^*)$, we see that $\Delta$ is formally *self-adjoint*, $\Delta^* = \Delta$. Furthermore,

$$(\Delta\alpha^p, \alpha^p) = (dd^*\alpha + d^*d\alpha, \alpha) = (d^*\alpha, d^*\alpha) + (d\alpha, d\alpha) = \| d\alpha \|^2 + \| d^*\alpha \|^2$$

which is $\geq 0$ in our Riemannian case. Thus

$$\Delta\alpha = 0 \quad \text{iff } d\alpha = 0 \quad \text{and} \quad d^*\alpha = 0 \tag{14.27}$$

Harmonic forms on a *closed* manifold are both *closed* and *coclosed*!

This is far different from the situation in $\mathbb{R}^n$. For example, a closed 0-form is simply a *constant* function, yet harmonic functions in $\mathbb{R}^n$ need not be constant; the real part of *any* complex analytic function in the plane is harmonic!

The Laplace operator $\Delta : \bigwedge^p \to \bigwedge^p$ is an *elliptic* operator on a Riemannian manifold (for the notion of ellipticity and for the proof of Hodge's theorem later, see [Wa, chap. 6]); the main ingredient is that the metric tensor is positive definite. In *Minkowski space*, however, the Laplacian of a function becomes the **d'Alembertian**

$$\Delta f = \frac{\partial^2 f}{\partial t^2} - \nabla^2 f$$

where $\nabla^2$ is the spatial Laplacian; $\Delta$ in this case is the **wave-operator** and is *hyperbolic*. Difficult results in *elliptic* operator theory are needed for the following fundamental result:

**Hodge's Theorem (14.28):** *Let $M^n$ be a closed Riemannian manifold. Then the vector space of harmonic p-forms*

$$\mathcal{H}^p = \left\{ h \in \bigwedge^p | dh = 0 = d^* h \right\}$$

*is finite-dimensional, and **Poisson's equation***

$$\Delta \alpha^p = \rho^p$$

*has a solution $\alpha$ iff $\rho$ is orthogonal to $\mathcal{H}^p$*

$$(\rho^p, h^p) = 0 \quad \textit{for all } h^p \in \mathcal{H}^p$$

The finite dimensionality of $\mathcal{H}^p$ is a deep result on elliptic operators on closed manifolds. On the other hand, it is easy to see the necessity of the condition on $\rho$ in order that there be a solution to Poisson's equation; if $\Delta\alpha = \rho$, then for $h \in \mathcal{H}^p$,

$$(\rho, h) = (\Delta\alpha, h) = (\alpha, \Delta^* h) = (\alpha, \Delta h) = 0$$

The deep part is showing the sufficiency of this condition. Note also that in the case $p = 0$, that is, when we are dealing with functions, the harmonic function $h$ is then a constant, and the condition on $\rho$ is simply that

$$\int_M \rho \operatorname{vol}^n = (\rho, 1) = 0$$

that is, $\rho$ must have mean value 0 on $M$. This is of course necessary since

$$\int_M \Delta\alpha^0 \operatorname{vol}^n = -\int_M \operatorname{div}(\operatorname{grad} \alpha^0) \operatorname{vol}^n = 0$$

by the divergence theorem.

Suppose now that $\beta^p$ is an arbitrary $p$-form on the closed $M^n$. Let $h_1, h_2, \ldots, h_r$ be an orthonormal basis for the harmonic forms $\mathcal{H}^p$. Then

$$\beta - \sum_j (\beta, h_j) h_j =: \beta - h$$

is orthogonal to $\mathcal{H}^p$ and so, by Hodge's theorem, we can solve

$$\Delta\alpha^p = \beta^p - h^p$$

for $\alpha^p$. In other words, for any $\beta^p$ on $M^n$ we can write

$$\beta^p = d(d^*\alpha^p) + d^*(d\alpha^p) + h^p \tag{14.29}$$

Thus, *any p-form $\beta$ on the closed $M^n$ can be written as the sum of an exact form $d(d^*\alpha)$ plus a coexact form $d^*(d\alpha)$ plus a harmonic form.* Hence

$$\bigwedge^p = d\bigwedge^{p-1} + d^*\bigwedge^{p+1} + \mathcal{H}^p \tag{14.30}$$

Note further that the three subspaces are *mutually orthogonal*

$$(d\gamma, d^*\mu) = (d\gamma, h) = (d^*\mu, h) = 0$$

(14.30) is called the **Hodge decomposition** of $\bigwedge^p$.

Note that the decomposition (14.30) is *unique*. If we write $\beta = d\gamma + d^*\mu + h = d\gamma' + d^*\mu' + h'$, then orthogonality gives $d\gamma - d\gamma' = 0$, $d^*\mu - d^*\mu' = 0$, and $h - h' = 0$. Note also that we are *not* saying, for example, that $\gamma$ is unique, for clearly we can add to $\gamma^{p-1}$ any closed $(p-1)$-form; we are only saying that $d\gamma$ is a unique summand.

At first glance it might appear that (14.30) is a triviality, for we can see immediately that $\mathcal{H}^p$ *is* the orthogonal complement in $\bigwedge^p$ to the direct sum of the exact and coexact forms; if for some $p$-form $h$, $(d\gamma, h) = 0$ and $(d^*\mu, h) = 0$ for all $\gamma$ and $\mu$, then indeed $d^*h = 0 = dh$ and so $h$ is harmonic and thus $[d\bigwedge^{p-1} + d^*\bigwedge^{p+1}]^\perp = \mathcal{H}^p$. However, $\bigwedge^p$ is an infinite-dimensional space, and in infinite dimensions it is *not* necessarily true that if $A$ is a subspace then $A + A^\perp$ is the entire space! It *is* true that if $A$ is a *closed* subspace of a *Hilbert space*, then $A + A^\perp$ is the entire space. Thus to get the decomposition (14.30) one might first complete the pre-Hilbert space $\bigwedge^p$ to a Hilbert space, say the square integrable forms on $M^n$; we would have to consider forms that are not even continuous, and for such forms $d$ is not defined! In any case $[d\bigwedge^{p-1} + d^*\bigwedge^{p+1}]$ would not be a closed subspace. All these difficulties can be overcome by invoking elliptic operator theory, and we refer the reader again to [Wa] for this difficult material.

In the case of a closed 3-manifold we have $\beta^1 = d\phi^0 + d^*\mu^2 + h^1$, that is,

$$\mathbf{B} = \operatorname{grad}\phi + \operatorname{curl}\mathbf{M} + \mathbf{H}$$

that is, a smooth vector field can be written as the sum of a gradient, a curl, and a vector field that has both vanishing curl and divergence. Thus it is true that any vector field **B** can be written as the sum of a vector field with vanishing curl and a vector field with vanishing divergence. This version is also true in the noncompact $\mathbb{R}^3$, at least when the growth of **B** at infinity is controlled; this is the classical **Helmholtz decomposition**, which is so useful in vector analysis.

### 14.2d. Harmonic Forms and de Rham's Theorem

We now have the following picture illustrating the orthogonal Hodge decomposition on a closed manifold.

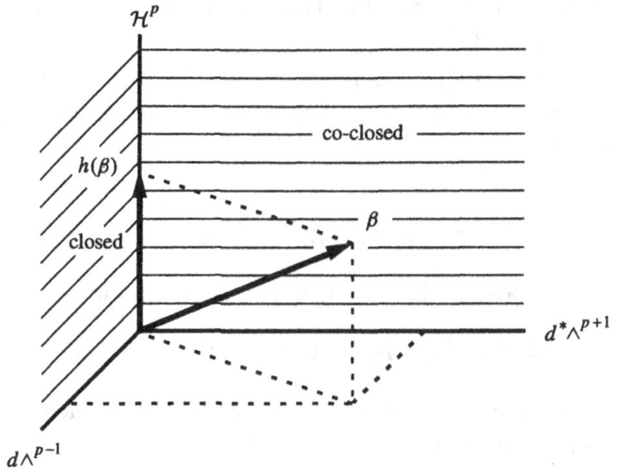


**Figure 14.1**

Any $p$-form $\beta$ may be written in the form $\beta^p = d\alpha^{p-1} + d^*\gamma^{p+1} + h^p$ where $h$ is harmonic. In particular, since the decomposition is orthogonal,

**Corollary (14.31):** *If $\beta^p$ is closed, $d\beta^p = 0$, on a closed manifold $M^n$, then*

$$\beta^p = d\alpha^{p-1} + h^p$$

*where $h^p$ is harmonic.*

Now $\beta$ and $\beta - d\alpha$ are in the same de Rham class. Thus

**Corollary (14.32):** *In each de Rham class $[\beta]$ there is a unique harmonic representative $h(\beta)$. Thus there exists a unique harmonic p-form with $b_p$ prescribed periods on a homology basis for the real p-cycles on $M^n$.*

Riemann was aware of this in the case of a closed surface. A "proof" goes along the following lines. Assume that one has a *closed* $p$-form $\beta^p$ on a closed manifold $M^n$. (Closed 1-forms on an $M^2$ with prescribed periods are easy to construct, as we did in Section 13.4b.) The 1-parameter family of forms $\beta^p(\epsilon) := \beta^p + \epsilon d\alpha^{p-1}$ are closed, with the same periods, for all $(p-1)$-forms $\alpha$. This yields a variation of $\beta$ with $\delta\beta = d\alpha$. Suppose that $\beta$ is the closed form with the prescribed periods *whose norm is a minimum. Dirichlet's principle* presumed that such a minimum norm element had to exist. Look then at the first variation as we vary $\alpha$

$$0 = \delta(\beta, \beta) = 2(\delta\beta, \beta) = 2(d\alpha, \beta) = 2(\alpha, d^*\beta)$$

Since this holds for all $\alpha$ we conclude that $\beta$ is not only closed, it is coclosed, $d^*\beta = 0$, and thus harmonic!

It was pointed out by Weierstrass that Dirichlet's principle was not always reliable, and thus the indicated proof is defective.

Note that the (difficult) Hodge decomposition justifies the norm claim since $\| \beta \|^2 = \| d\alpha \|^2 + \| h \|^2$ shows that in the de Rham class $[\beta]$, *the harmonic representative h has the smallest norm!*

### 14.2e. Bochner's Theorem

Let us say that $M^n$ has **positive Ricci curvature** if the Ricci tensor is positive definite,

$$\text{Ric}(\mathbf{X}, \mathbf{X}) = R_{ik} X^i X^k > 0 \quad \text{for all } \mathbf{X} \neq 0$$

This is a weaker condition than positive (sectional) curvature since this quadratic form represents a sum of sectional curvatures (see (11.67)).

**Bochner's Theorem (14.33):** *If the closed Riemannian $M^n$ has positive Ricci curvature, then a harmonic 1-form must vanish identically, and thus M has first Betti number $b_1 = 0$.*

PROOF: Let us compute, with Bochner, the Laplacian of the square of the pointwise length $\langle h, h \rangle = h_i h^i$ of any harmonic 1-form $h$. First,

$$[\text{grad}\langle h, h \rangle]_j = 2 h_{i/j} h^i$$

and so

$$\nabla^2 \frac{1}{2} \langle h, h \rangle = (h_i{}^{/j} h^i)_{/j} = h_i{}^{/j}{}_{/j} h^i + h_i{}^{/j} h^i{}_{/j}$$

$$= h_i{}^{/j}{}_{/j} h^i + h_{i/j} h^{i/j}$$

By (14.25) we have, since $\Delta h = 0$, $h_i{}^{/j}{}_{/j} = h^k R_{ki}$, and thus

$$\nabla^2 \frac{1}{2} \langle h, h \rangle = \text{Ric}(h, h) + h_{i/j} h^{i/j} \geq \text{Ric}(h, h) \geq 0$$

But then $0 = \int_M \nabla^2 (1/2) \langle h, h \rangle * 1 \geq \int_M \text{Ric}(h, h) * 1$ shows, since Ric is positive definite, that $h = 0$. □

Bochner's theorem should be compared to Synge's corollary (12.14). Before doing so, we need a general observation about closed curves.

A closed (oriented) curve $C$ on $M^n$ represents an element of the first homology group $H_1(M; G)$ for any coefficient group $G$. If $C$ is contractible to a point, then in the process of shrinking, the curve will sweep out a surface, of which it is the boundary. In other words, if a closed curve can be contracted to a point then this curve bounds, that is, trivial as a 1-cycle. (In particular, if $M$ is simply connected, then $H_1(M; G) = 0$.) The converse is not true.

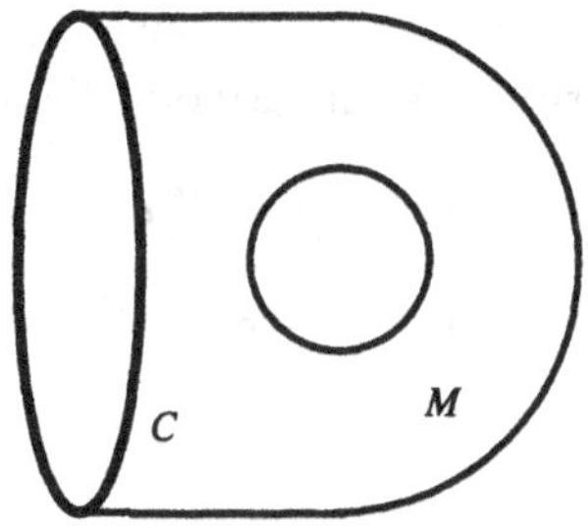


Figure 14.2

The edge $C$ of the punctured torus $M$ is clearly the boundary of the surface $M$, and so is homologically trivial, but it seems rather clear (and can be proved) that $C$ cannot be shrunk to a point because of the presence of the "hole."

As far as Betti numbers are concerned then, Bochner's theorem is stronger than Synge's corollary since positive Ricci curvature is weaker than positive sectional curvature, and also we do not require even dimensionality nor orientability, but it should be kept in mind that simple connectivity is a stronger notion than $b_1 = 0$.

## Problems

**14.2(1)** Derive all those equations (1) through (8) that have not been discussed previously.

**14.2(2)** Show that $\Delta$ commutes with $d$, $d^*$, and $*$.

**14.2(3)** Show that if $M^n$ is *closed and orientable* then $b_p = b_{n-p}$. This is a special case of **Poincaré duality**. Why do we need orientability? Illustrate with $b_0$ for the 2-torus and the Klein bottle.

## 14.3. Boundary Values, Relative Homology, and Morse Theory

What does topology have to do with the existence and uniqueness of physical fields?

The prime example of a manifold with boundary is the case of a bounded region in $\mathbb{R}^3$ with smooth boundary. If a fluid fills such a domain, with smooth walls forming the boundary, then the velocity vector field $\mathbf{v}$ is tangent to the boundary. If the flow is incompressible, then the velocity field has divergence 0. If further the flow is irrotational, then the velocity has curl 0 and the resulting velocity 1-form field $\nu$ is harmonic. We are interested in the existence of such fields and we shall find that with some type of prescribed *topological* restriction the solution becomes unique.

Note that in a compact manifold with boundary, Equation (14.14) shows that the operators $d$ and $d^*$ are not necessarily adjoints, and it is no longer true that $\Delta\alpha = 0$

iff $d\alpha = 0 = d * \alpha$. Furthermore, $\Delta$ *is no longer self-adjoint*. For physical problems involving forms we shall reserve the term harmonic **field** for forms that satisfy

$$d\alpha = 0 = d^*\alpha$$

Thus a harmonic 0-field is *constant*, whereas a harmonic function, that is 0-form, of course, need not be.

### 14.3a. Tangential and Normal Differential Forms

Let $M^n$ be a compact Riemannian manifold with boundary.

> A form $\alpha^p$ on $M$ is said to be **normal** to $\partial M$, or simply normal, provided the restriction $i^*\alpha$ of $\alpha$ to the boundary vanishes, $i^*\alpha = 0$

where $i : \partial M \to M$ is the inclusion map. Recall that this simply means that $\alpha(\mathbf{v}, \dots, \mathbf{w}) = 0$ when $\mathbf{v}, \dots, \mathbf{w}$ are all tangent to $\partial M$. If we suppose that $\partial M$ is locally defined in the coordinate system $x^1, \dots, x^n$ by putting $x^n = 0$, then

$$\alpha^p \text{ is normal} \quad \text{iff } \alpha^p = dx^n \wedge \gamma^{p-1}$$

for some form $\gamma$.

For example, a 1-form $\alpha^1$ is normal provided $\alpha^1 = a_n(x)dx^n$ (no sum!) at points of $\partial M$. If $\mathbf{T}$ is tangent to $\partial M$, then $0 = \alpha(\mathbf{T}) = \langle \mathbf{a}, \mathbf{T} \rangle$ shows that

$$\alpha^1 \text{ is normal} \quad \text{iff } \mathbf{a} \text{ is } \textit{normal} \text{ to } \partial M$$

where $\mathbf{a}$ is the contravariant version of $\alpha^1$. If, however, $\beta^{n-1}$ is an $(n-1)$-form, $\beta^{n-1} = i_{\mathbf{B}} \operatorname{vol}^n$, then $\beta$ is normal provided $\beta(\mathbf{T}_2, \dots, \mathbf{T}_n) = \operatorname{vol}^n(\mathbf{B}, \mathbf{T}_2, \dots, \mathbf{T}_n) = 0$ for tangent $\mathbf{T}_i$; and so

$$\beta^{n-1} = i_{\mathbf{B}} \operatorname{vol}^n \text{ is normal} \quad \text{iff } \mathbf{B} \text{ is } \textit{tangent} \text{ to } \partial M$$

A form $\alpha^p$ is said to be **tangent** to $\partial M$, or simply tangent, provided $*\alpha$ is normal, $i^* * \alpha = 0$.

Thus

$$\alpha^1 \text{ is tangent} \quad \text{iff } \mathbf{a} \text{ is } \textit{tangent} \text{ to } \partial M$$

while

$$\beta^{n-1} = i_{\mathbf{B}} \operatorname{vol}^n \text{ is tangent} \quad \text{iff } \mathbf{B} \text{ is } \textit{normal} \text{ to } \partial M$$

Note that from the remark following (14.14), $d^*$ *is the adjoint of* $d$ if we restrict ourselves either to tangential or to normal forms!

In the following we shall quote, without proofs, the versions of Hodge's theorem that have immediate applications to physical problems. My principal guide for the applications has been the mimeographed NYU notes [B, F, G] by A. Blank, K. Friedrichs, and H. Grad of 1957. For the (difficult) mathematics of harmonic forms on manifolds with boundary, the reader may consult [D, S] and [Fk].

## 14.3b. Hodge's Theorem for Tangential Forms

**Theorem (14.34):** *Let $M^n$ be a compact manifold with boundary. Let $z_1, \ldots, z_{b_p}$ be a basis for the p*th *homology vector space $H_p(M; \mathbb{R})$. Then there exists a unique tangent harmonic p-form field $\alpha^p$*

$$d\alpha^p = d^*\alpha^p = 0$$

*with prescribed periods $\int_z \alpha^p$ on the given basis.*

In other words, Hodge's original theorem holds for *tangential* harmonic fields in the case of a manifold with boundary!

**Example 1:** Let **v** be the velocity field for a steady incompressible, irrotational fluid flow inside a closed surface $V^2$ of genus $g$. As we have seen, $\nu^1$ is harmonic, $d\nu = d^*\nu = 0$, and $\nu$ is tangent to $\partial M$.

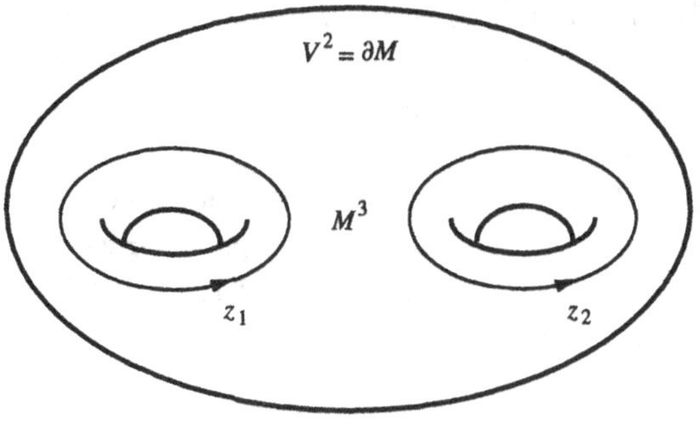


Figure 14.3

We shall illustrate the case for genus 2. $M^3$ is the solid "pretzel," and $\partial M$ is the surface of genus 2. It should be rather clear that a homology basis for $H_1(M; \mathbb{R})$ is given by the two indicated 1-cycles circling the "holes." The period of $\nu^1$ on a 1-cycle $z$, $\int_z \nu^1$, is called the **circulation** of **v** around $z$. Thus Hodge's theorem yields the following corollary, known to W. Thomson (Lord Kelvin).

**Corollary (14.35):** *There exists a unique incompressible irrotational flow inside a surface of genus g with prescribed circulations around the g holes.*

In particular, if all circulations vanish, then the fluid must be at rest! This is the only possibility in the case of a spherical surface since the solid ball has first Betti number 0.

**Example 2:** Let $M^3$ be the region *inside* a closed conducting surface $V_0$ and *outside* closed conducting surfaces $V_1$ and $V_2$.

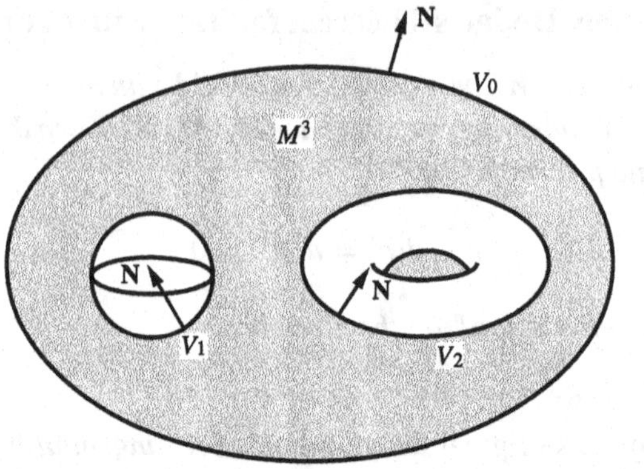


Figure 14.4

We have drawn the case when $V_0$ is a large ellipsoid, $V_1$ is an interior 2-sphere, and $V_2$ is an interior 2-torus. Consider an *electrostatic* problem in which there are no charges inside $M^3$; of course there may be charges interior to $V_1$ and/or $V_2$ or exterior to $V_0$. Then the electric field inside $M^3$ satisfies $\mathbf{d}{*}\mathcal{E} = 4\pi\sigma^3 = 0$ and $\mathbf{d}^*{*}\mathcal{E} = {*}\mathbf{d}**\mathcal{E} = {*}\mathbf{d}\mathcal{E}^1 = *[-\partial\mathcal{B}^2/\partial t] = 0$, and so $*\mathcal{E}$ is a harmonic 2-form in $M^3$. Since a tangential component of **E** would give rise to currents, that is, moving charges, in a conductor, the natural boundary condition for electrostatics is that **E** be normal to conducting surfaces. Thus $*\mathcal{E}$ *is a tangent harmonic 2-form field in* $M^3$.

Note that $\partial M^3 = V_0 + V_1 + V_2$, and thus a plausible (and correct) basis for $H_2(M^3; \mathbb{R})$ is, for example, $V_1$ and $V_2$. Thus there exists a unique electric field in $M^3$ with prescribed periods $\int\int *\mathcal{E}$ over $V_1$ and $V_2$. But the integral of $*\mathcal{E}$ over $V_i$ is $4\pi Q_i$, where $Q_i$ is the total charge inside $V_i$.

**Corollary (14.36):** *There exists a unique static electric field* **E** *in* $M^3$ *with preassigned charges in the cavities* $V_1$ *and* $V_2$. *The field is thus independent not only of charges outside* $V_0$ *("shielding"), but also of the exact placement of the charges in* $V_1$ *and in* $V_2$.

We should mention that Theorem (14.34) is a special case of a more general result. First recall that to say that $\alpha^p$ is "tangent" is to say that the restriction $i^*(*\,\alpha)$ of $*\alpha$ to the boundary vanishes. More generally, we could ask for a harmonic field $\alpha^p$ that has prescribed periods and such that $i^*(*\,\alpha)$ is a *prescribed* form $\gamma^{n-p}$ on $\partial M$. The special case $\gamma = 0$ would make $\alpha$ a tangent form. *We must put some restrictions on the form* $\gamma$ for the following reason. On $\partial M$ we have $d\gamma = di^* * \alpha = i^* d * \alpha = 0$, since $\alpha$ is coclosed. Hence $\gamma$ is *closed.* Furthermore, $\gamma$ is only defined on $\partial M$, but suppose that $z_{n-p}$ is a cycle on $\partial M$ *that bounds in* $M$, that is, $i_* z = \partial c$, for some $(n-p+1)$-chain $c$ on $M$. Then since the integral of $\gamma$ over $z$ is the same as the integral of $*\,\alpha$ over $z$, this integral must vanish, $*\,\alpha$ being closed on $M$. The following notion is due to $A$. Tucker.

**Definition (14.37):** An **admissible boundary value form** $\gamma^r$ on $\partial M$ is a *closed* form on $\partial M$ whose integral vanishes on every cycle $z_r$ on $\partial M$ *that bounds on* $M$.

The generalization of (14.34) is as follows. (For more along these lines see [D, S].)

**Theorem (14.38):** *There exists a unique harmonic field $\alpha^p$ on $M$ with prescribed periods and whose dual $*\,\alpha$ restricts on $\partial M$ to a prescribed admissible boundary value form $\gamma^{n-p}$.*

The uniqueness of $\alpha$ is simple (and was known to Lord Kelvin in the case $p = 1$).

PROOF OF UNIQUENESS: Let $\alpha^p$ be a solution and suppose $\beta^p$ is another with the same periods and whose dual $*\,\beta$ has the same boundary values $i^* * \beta = \gamma$. Then $\mu := \alpha - \beta$ is a *tangent* harmonic field with 0 periods. Since $d\mu^p = 0$, $\mu^p = d\nu^{p-1}$ for some $\nu$ (this is elementary if $p = 1$; otherwise it requires de Rham's theorem). We wish to show that $d\nu = 0$. But

$$(d\nu, d\nu) = \int_M d\nu \wedge *d\nu = \int_{\partial M} (\nu \wedge *d\nu) \pm \int_M \nu \wedge d * d\nu$$

Since $\mu = d\nu$ is tangent, $*\,d\nu$ is normal and the boundary integral vanishes. Also $d * d\nu = d * \mu = 0$ since $\mu$ is harmonic. □

## 14.3c. Relative Homology Groups

The topological "cycles" that we have been involved with so far are called **absolute** cycles. Given a compact manifold $M^n$ perhaps with boundary we can define a

**relative** $p$-**cycle** (mod $\partial M$)

to be a *p-chain* on *M whose boundary*, if there is one, *lies on* $\partial M$. Of course every (absolute) cycle is also a relative cycle.

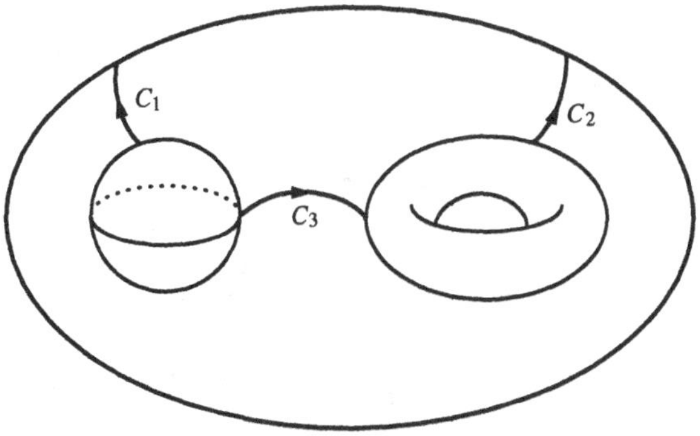


Figure 14.5

In Figure 14.5 the curves $C_1, C_2$, and $C_3$ are all relative 1-cycles (mod $\partial M = V_0 + V_1 + V_2$). We shall systematically disregard any chain that lies on $\partial M$. That is why we may think of a relative cycle as a cycle; we may disregard its boundary since it lies on $\partial M$.

We shall say that two relative $p$-cycles $c$ and $c'$ are **homologous** (mod $\partial M$) provided they differ by a true boundary plus, perhaps, a $p$-chain that lies wholly on $\partial M$; in other words, a **relative boundary** *is an absolute boundary plus any chain on* $\partial M$

$$c'_p \sim c_p \text{ if } c'_p - c_p = \partial w_{p+1} + v_p, \quad \text{where } v_p \subset \partial M$$

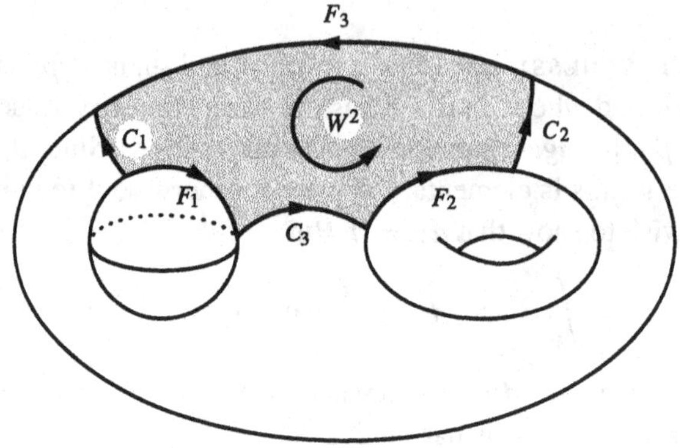


Figure 14.6

In Figure 14.6 we have drawn three more curves $F_1$, $F_2$, and $F_3$, all lying on $\partial M$, and also an oriented 2-chain $W^2$. Clearly $\partial W = -C_1 + F_1 + C_3 + F_2 + C_2 + F_3$. But the $F$ curves all lie on $\partial M$, and so we may say

$$\partial W = -C_1 + C_3 + C_2 \quad (\text{mod } \partial M)$$

We could then say that $C_3$ is homologous to $C_1 - C_2$ (mod $\partial M$)

$$C_3 \sim C_1 - C_2 \quad (\text{mod } \partial M)$$

Thus only $C_1$ and $C_2$ are *independent* relative cycles. (Of course we could have used $C_1$ and $C_3$, say.) Are there any more?

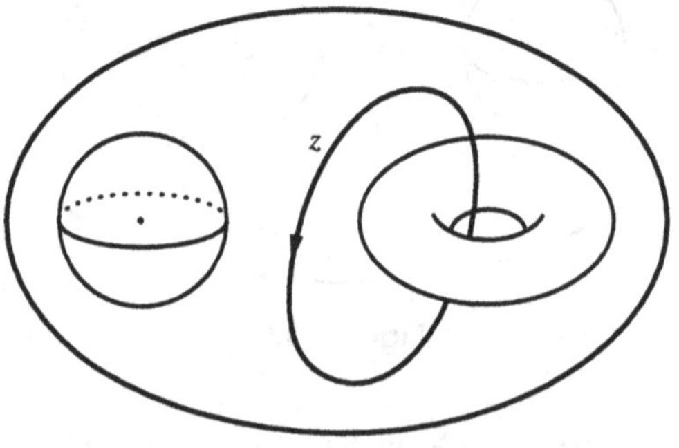


Figure 14.7

Consider the absolute 1-cycle $z$ that threads through the toroidal hole. It is an absolute cycle of $M$ that does not bound in $M$. However, as a *relative* 1-cycle it is trivial, that is, it bounds, since it is easily deformed in $M$ to lie on the torus $V_2 \subset \partial M$.

It can, in fact, be shown that $C_1$ and $C_2$ form a basis for the **relative homology group**, $H_1(M, \partial M; \mathbb{R})$, defined to be the relative cycles modulo the relative boundaries

$$H_1(M, \partial M; \mathbb{R}) = \mathbb{R}C_1 + \mathbb{R}C_2$$

### 14.3d. Hodge's Theorem for Normal Forms

**Theorem (14.39):** *Let $M^n$ be a compact manifold with boundary. Let $c_1, \dots, c_r$ be a basis for the relative p-cycles of M (mod $\partial M$)*

$$H_p(M, \partial M; \mathbb{R}) = \mathbb{R}c_1 + \cdots + \mathbb{R}c_r$$

*Then there exists a unique normal harmonic p-form $\alpha^p$ with prescribed periods*

$$\int_{c_i} \alpha^p$$

Note that if $c' \sim c(\text{mod}\, \partial M)$, that is, if $c' - c = \partial w^{p+1} + u^p$, where $u$ lies on $\partial M$, then if $\alpha^p$ is closed and normal

$$\int_{c'} \alpha - \int_c \alpha = \int_u \alpha = 0$$

since $\alpha^p = 0$ when $\alpha$ is restricted to $\partial M$! *Thus the indicated periods do not change when a $c_i$ is replaced by a homologous $c_i'$.*

**Example 2′:** In Example 2 earlier, consider the electric field 1-form $\mathcal{E}^1$ for the electrostatic field. It is a harmonic *normal* 1-form on $M^3$. Thus we may prescribe the line integrals $\int_{C_1} \mathcal{E}^1$ and $\int_{C_2} \mathcal{E}^1$. This means that instead of the charges in $V_1$ and $V_2$, the electric field in $M^3$ is uniquely determined equivalently by prescribing the *electrostatic potential differences* between the "inside" and the "outside" conductors!

**Example 1′:** In Example 1, we may consider the velocity vector $\mathbf{v}$ as defining a 2-form $\beta^2 = i_{\mathbf{v}}\, \text{vol}^3$. This is then a *normal* harmonic 2-form on $M^3$.

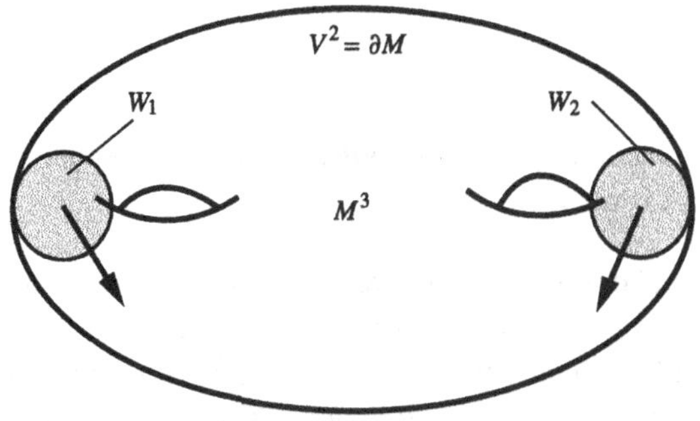


Figure 14.8

It should be "clear" that a basis for $H_2(M, \partial M; \mathbb{R})$ is given, say, by the two discs $w_1$ and $w_2$. Thus the harmonic normal $\beta^2$ is determined by *prescribing the fluid fluxes* $\int_{w_i} \beta^2 = \int_{w_i} \mathbf{v} \bullet d\mathbf{S}$, $i = 1, 2$, rather than the two circulations.

### 14.3e. Morse's Theory of Critical Points

We give here another application of relative homology groups. We shall not need these results for later portions of this book and so *this section can be omitted*, but this subject forms one of the most outstanding mathematical contributions in the twentieth century. We shall be very brief, referring the reader to Milnor's book [M] and Bott's expository paper [Bo] for more details and applications.

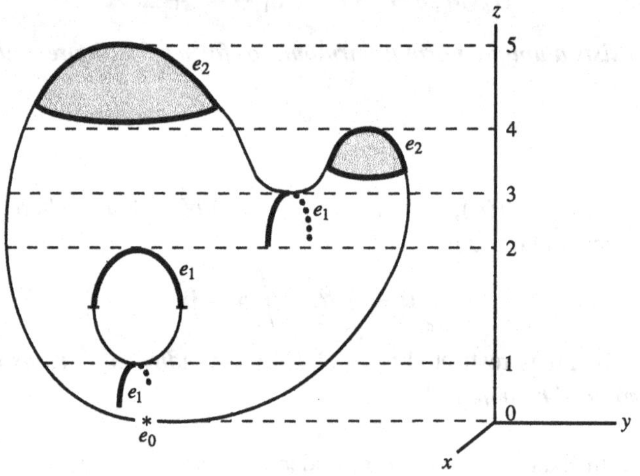


Figure 14.9

We have indicated here the height function $f = z$ on a bumpy torus. The critical points are at levels 0 (minimum); 1, 2, and 3 (saddles); and 4 and 5 (maxima). For any manifold $M^n$ with smooth real-valued function $f$, let us put

$$M_a := \{x \in M | f(x) \le a\}$$
$$M_a^- := \{x \in M | f(x) < a\}$$

We define

> a **value** $a$ of the function $f$ as **homotopically critical** if some relative homology group $H_i(M_a, M_a^-)$ is nonzero.

(For simplicity we shall use the real numbers $\mathbb{R}$ for coefficient group, but any coefficient field can be used.) We *claim* that the homotopically critical values in our example are exactly the critical values in the sense of Section 1.3d. Thus *in this example the critical values are precisely the levels at which new relative cycles appear* as we move "up" the manifold from the minimum to the maximum.

In our torus example, the relative maximum at level $z = 4$ has $H_2(M_4, M_4^-) = \mathbb{R}$ and we have exhibited a 2-disc $e_2$ at the critical point that is a generator for this homology group. We shall "prove" this later, but it should be *plausible* since any effort to slide this disc entirely into the lower region $M_4^-$ will require the boundary of the disc (which lies below $z = 4$) at some time to pass through the critical point; that is, the boundary will have to leave the lower region at some time. It should be clear that any 1-disc or 0-disc (point) on the level $M_4$ can be pushed away from the critical point into the lower region, so that $H_i(M_4, M_4^-) = 0$ for $i \neq 2$.

At a *noncritical* level $b$, say $z = 2.5$, it is "clear" that any chain on $M_b$ can be pushed into $M_b^-$ by a deformation along the negative gradient lines, similar to the Morse deformation of Section 2.1e. Thus $H_i(M_{2.5}, M_{2.5}^-) = 0$. In fact, if the regions $M_a$ are all assumed compact, and if there are no critical points $x_0$ with $d + \epsilon \geq f(x_0) \geq c - \epsilon$, then it can be shown that a modified Morse deformation (which does not move points $x$ with $f(x) \leq c - \epsilon$) can deform $M_d$ diffeomorphically into $M_c$.

At the level $z = 3$, it is again "clear" that the part of any chain away from the saddle point can be pushed down by following the negative gradient lines, but the critical point itself remains fixed. There is no continuous way to push the entire indicated 1-disc $e_1$ below level $z = 3$; $H_1(M_3, M_3^-) = \mathbb{R}$ with generator $e_1$ and the value 3 is again homotopically critical.

We have also indicated the remaining disc generators at the other homotopically critical levels. At the minimum we have a 0-disc (point) since $M_0^-$ is empty. We have verified our claim.

Note that the height function on the 1-dimensional manifold pictured

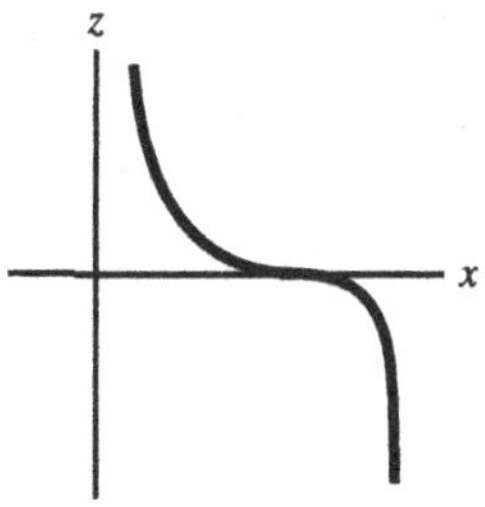


**Figure 14.10**

has a critical point at $z = 0$, an inflection point, but it is clear that this does not yield a *homotopically* critical value since any chain on $z \leq 0$ can be slid below $z = 0$. In a sense this critical point is **inessential** since a slight change in the function, say by tilting the $z$ axis very slightly (in the "correct" direction), will remove the critical point. In our toral example *all the ordinary critical values are homotopically critical, and vice versa*, and in fact as we shall see, this is true *whenever the critical points are* **nondegenerate** in the sense of having nonsingular Hessian matrices of second partial derivatives $H_{ij} = (\partial^2 f / \partial x^i \partial x^j)$,

$$\det(H_{ij}) \neq 0.$$

In this nondegenerate case we can, with Morse, write down the dimension of the nontrivial relative cycle at the critical point as follows. Since the Hessian is nonsingular, there is a maximal *subspace* of the tangent space at the critical point on which $H$ is negative definite, $H_{ij}v^i v^j \leq 0$. In terms of a Riemannian metric we are looking at the sum of the eigenspaces corresponding to the negative eigenvalues of $H^i{}_j = g^{ik}H_{kj}$. The dimension of the resulting subspace is called the (Morse-) **index** of the critical point, $\lambda :=$ number of negative eigenvalues (counted with multiplicity) and represents crudely the *dimension of the space of directions, at the critical point, in which the function is decreasing*. Then the relative cycle is the $\lambda$-cell $e_\lambda$ starting out tangent to the subspace. (We shall indicate in our next paragraph why $e_\lambda$ does not bound as a relative cycle.) For example, for the critical point at level 4, we can introduce new local coordinates $x, y$ (*with origin at the critical point*) on the torus such that $f = z = 4 - x^2 - y^2 +$ higher order, and so the Hessian is negative definite on the entire tangent space to $T^2$ at the critical point, the index is $\lambda = 2$, and the disc $x^2 + y^2 < \epsilon^2$ is the required generator for $\epsilon$ sufficiently small. For the critical point at level 3, in local coordinates $f = z = 3 - x^2 + y^2 +$ higher order, the Hessian has the new $x$ axis for negative eigenspace, the index is $\lambda = 1$, and $x^2 < \epsilon^2$ is the generating 1-disc.

Let us indicate why, for example, the relative 1-cycle $e_1$ at level $f = 1$ is *not trivial*. First note that near the critical point $f = 1 - x^2 + y^2 +$ higher order. The **Morse lemma** [M, p. 6] states that near a nondegenerate critical point, one may always introduce coordinates so that $f$ becomes exactly this form with the higher order terms removed; thus in new coordinates, which we shall again call $x, y, f$ is *exactly*

$$f(x, y) = 1 - x^2 + y^2$$

Look then at $g(x, y) := f(x, y) - 1 = -x^2 + y^2$. We are interested in relative cycles of the region $f \leq 1 \bmod f < 1$. Away from the critical point $x = 0 = y$ any chain on $f \leq 1$ can be pushed down into $f < 0$, and so discarded. We are then only interested in $f \leq 0$ *near the critical point*. In terms of the new coordinates we may deal with relative cycles on $g = y^2 - x^2 \leq 0$, that is, the shaded region in Figure 14.11.

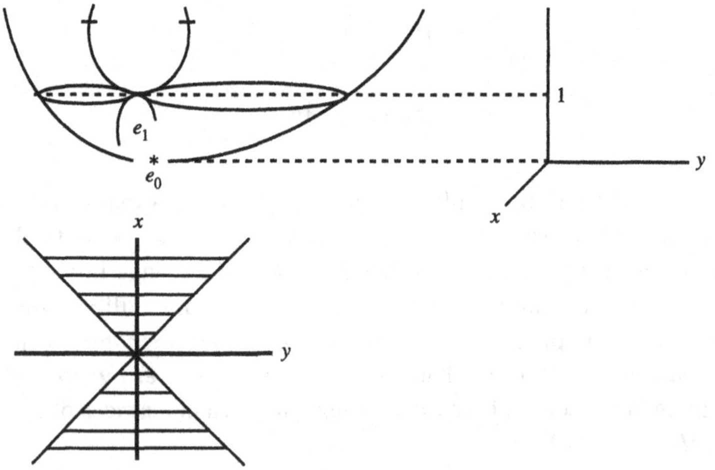


Figure 14.11

Now any chain on this shaded region $y^2 \le x^2$ can be deformed to lie on the $x$ axis, with no point $(x, y)$ with $g < 0$ ever leaving $g < 0$. Thus we are reduced to chains on the segment of the $x$ axis with $|x| \le 1$ modulo $x \neq 0$. But $x \neq 0$ on this segment can be pushed into the boundary $x = \pm 1$. Thus we are interested in the relative cycles of the segment $|x| \le 1$ modulo the boundary points. In the case of a critical point of index $\lambda$ we are interested in the relative homology of a closed $\lambda$-disc $B^\lambda$ modulo its boundary $(\lambda - 1)$-sphere $S^{\lambda-1}$. This rather clearly (as we shall see in Problem 22.3(3)) has only one nontrivial generator, $B^\lambda$, $H_\lambda(B^\lambda, S^{\lambda-1}) = \mathbb{R}B_\lambda$. In our toral case the only nontrivial generator of relative homology at the level $f = 1$ is the indicated 1-cell $e_1$, as claimed.

The fact that the nondegenerate critical points are homotopically critical, and so have topological significance, allowed Morse to give relations between the number of critical points on $M$ and the Betti numbers of $M$. Briefly we can proceed as follows. Introduce the $\lambda^{\text{th}}$ **Morse type number**

$$m_\lambda := \text{number of critical points of index } \lambda$$

For bookkeeping purposes only we form the formal polynomial in a variable $t$ with the type numbers as coefficients, the **Morse polynomial**

$$\mathfrak{M}(t) := \sum_{\lambda=0}^{n} m_\lambda t^\lambda$$

We also have the Betti numbers $b_\lambda = \dim.\ H_\lambda(M; \mathbb{R})$ and the formal **Poincaré polynomial**

$$P(t) := \sum_{\lambda=0}^{n} b_\lambda t^\lambda$$

**Morse's Theorem (14.40):** *Let $M^n$ be a closed manifold and $f : M \to \mathbb{R}$ a smooth function with only nondegenerate critical points. Then the Morse polynomial dominates the Poincaré polynomial; there is a polynomial $Q(t)$ with nonnegative coefficients and*

$$\mathfrak{M}(t) - P(t) = (1 + t)Q(t)$$

*In particular we have the "weak"* ***Morse inequalities***

$$m_\lambda \ge b_\lambda$$

*and equality*

$$\sum_{\lambda=0}^{n} (-1)^\lambda m_\lambda = \sum_{\lambda=0}^{n} (-1)^\lambda b_\lambda$$

In particular, *the total number of critical points on $M$ is bounded below by the sum of all the Betti numbers.*

In our toral example $b_0 = 1 = b_2$ and $b_1 = 2$, while $m_0 = 1, m_1 = 3$, and $m_2 = 2$ and

$$\mathfrak{M}(t) - P(t) = (1 + 3t + 2t^2) - (1 + 2t + t^2) = t + t^2 = (1 + t)t$$

By writing out $Q(t) = \sum_{\lambda=0}^{n-1} q_\lambda t^\lambda$ with $q_\lambda \geq 0$ it is not hard to see that we can successively derive the

**Strong Morse inequalities** (14.41)

$$\begin{aligned} m_0 &\geq b_0 \\ m_1 - m_0 &\geq b_1 - b_0 \\ &\cdots \\ m_n - m_{n-1} + \cdots \pm m_0 &= b_n - b_{n-1} + \cdots \pm b_0 \end{aligned}$$

**Proof sketch of (14.40):** For simplicity we assume that there is only one critical point at each critical level (this is generically so). At any level $f = a$ (critical or not) we shall consider the space $M_a$, the Morse polynomial $\mathfrak{M}(M_a; t)$, for *this* space, and the Poincaré polynomial, again just for this space, and we shall observe how these polynomials change, $\Delta P$, and so forth, as we pass through a critical point. It is clear, since topology changes only when passing through a critical point, that $\Delta\mathfrak{M}$ and $\Delta P$ are nonzero only when passing through a critical point.

Let $\mathfrak{M}(t)$ and $P(t)$ have the value 0 on the empty set, that is, below the absolute minimum At the absolute minimum we have a point and its index is 0. Thus on passing from the empty set to the set consisting only of the minimum point we have $\Delta\mathfrak{M} = 1$ and $\Delta P = 1$. (We shall keep our toral example in mind.) As we continue to higher values of $f$ we see the following. Consider passing though a critical point of index $\lambda$ at $f = a$, with its associated relative cycle, a disc $e_\lambda$ of dimension $\lambda$. There are two possibilities:

1. The boundary of this disc is a $(\lambda - 1)$-cycle (sphere) that *bounds* in $M_a^-$. (In the toral example the boundary of the 1-cell $e_1$ from the saddle at $f = 1$ is a pair of points that clearly bounds a 1-chain on $M_1^-$.) Let then $\partial e_\lambda = \partial c_\lambda$ where $c$ lies on $M_a^-$. Then $e_\lambda - c_\lambda$ is an *absolute* cycle on $M_a$. *It cannot bound in* $M_a$; if it did, $e_\lambda - c_\lambda = \partial c_{\lambda+1}$ would yield that $e_\lambda = \partial c_{\lambda+1} + c_\lambda$ and so $e_\lambda$ would be a trivial relative cycle, a contradiction. Thus in this case we have $\Delta\mathfrak{M} = t^\lambda$ and also $\Delta P = t^\lambda$ and so $\Delta(\mathfrak{M} - P) = 0$.
2. The boundary of the disc is a $(\lambda - 1)$-cycle (sphere) $S$ on $M_a^-$ that does *not* bound in $M_a^-$. But this says that $S$ is a nontrivial $(\lambda - 1)$-cycle on $M_a^-$ that bounds in $M_a$. Thus in this case $\Delta\mathfrak{M} = t^\lambda$ and $\Delta P = -t^{\lambda-1}$, and so $\Delta(\mathfrak{M} - P) = t^\lambda + t^{\lambda-1} = (1 + t)t^{\lambda-1}$.

These two cases show that on crossing a critical point of index $\lambda$, $\mathfrak{M} - P$ changes either by 0 or by $(1 + t)t^{\lambda-1}$. Since $\mathfrak{M}$ and $P$ start out equal on the empty set we have demonstrated (14.40). □

Note that in case (1) we can say that the relative cycle $e_\lambda$ on $M_a$ mod $M_a^-$ is **completable** to an absolute cycle on $M_a$. In this case we have shown that $\Delta(\mathfrak{M} - P) = 0$. Thus

**Corollary (14.41):** *If all the relative cycles from all the critical points are completable, then the Morse inequalities are equalities,* $m_\lambda = b_\lambda$.

In our toral example the 2-cell at level $f = 4$ is the only relative cycle that is not completable. This is reflected in $m_2 = 2 > b_2 = 1$ and $m_1 = 3 > b_1 = 2$.

If some critical points are degenerate, the Morse inequalities need not hold. In Problem 14.3(4) you will study a smooth function on the 2-torus $T^2$ that has only 3 critical points. (Of course there are *always* a max and a min on any compact space.)

A final comment. For a *continuous,* nondifferentiable function on a closed manifold $M$ we still have the notions of the absolute maximum and minimum values, but we cannot talk about minimaxes since we don't have partial derivatives at our disposal. Note, however, that we may **define** a homotopically critical value as earlier. We can also define a homotopically critical **point** to be a point $y$, at level $f(y) = a$, such that some homology group $H_i(M_a^- \cup \{y\}, M_a^-) \neq 0$.

## Problems

**14.3(1)** Consider a conducting surface of genus $g$ bounding a region $M^3$

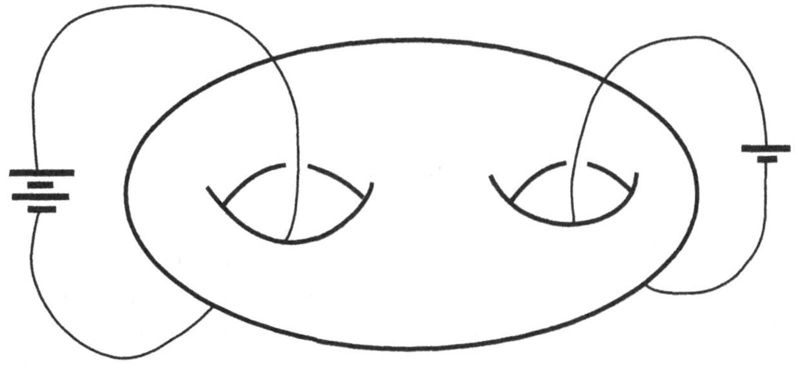

**Figure 14.12**

Let there be constant current loops in the exterior region, some of which thread through the holes. *Assume* that the appropriate boundary condition is that the normal component of **B** must vanish on the surface. Show that *there is a unique static magnetic field inside* $M^3$*, determined completely by the currents in the loops that thread the holes.*

**14.3(2)** Show that $d$ sends normal forms into normal forms and that $d^*$ sends tangent forms to tangent forms.

**14.3(3)** Let $\bigwedge_{nor}$ be the normal forms, and let $\bigwedge_{tan}$ be the tangent forms. It can be shown that the global orthogonal decomposition that replaces the Hodge decomposition (14.30) is

$$\overset{p}{\bigwedge} = d\left(\overset{p-1}{\underset{nor}{\bigwedge}}\right) + d^*\left(\overset{p+1}{\underset{tan}{\bigwedge}}\right) + \text{harmonic } p\text{-fields}$$

Show that these subspaces are indeed orthogonal.

**14.3(4)** We have drawn in Figure 14.13 a few level curves of a smooth function $f$ on the torus $T^2$ having a max at $f = 2$, a min at $f = -2$, and a single other critical point (at the four identified corners) at $f = 0$. It is clear that the corner point is critical since the level curves comprising $f = 0$ intersect there (and so grad $f$ must vanish there).

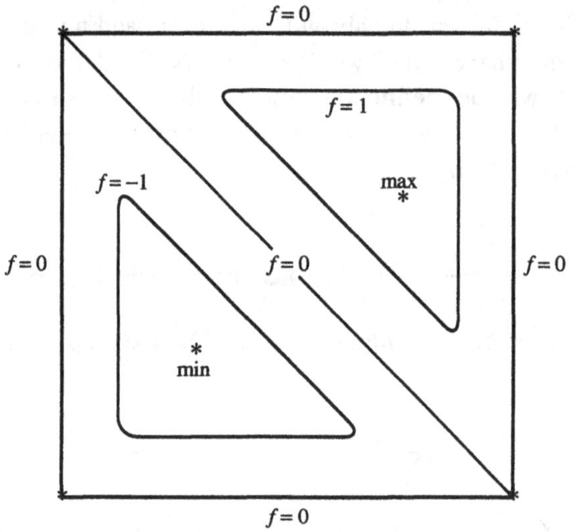


**Figure 14.13**

Continue this picture periodically in the plane so that the corner point is at the center. Show that from this center there are *three* directions for which the function *decreases* as one leaves the critical point, each pair being separated by a direction in which the function is *increasing*. This shows that the critical point is *degenerate*. This critical point is of the type of a monkey saddle; see [M, p. 8]. Find *two independent relative 1-cycles* in $H_1(T^2_0; T^2_{0-})$ *emanating from this critcal point.* (In a sense, then, this critical point counts as 2 critical points of index 1 each.)

**14.3(5)** Prove Morse's **lacunary** principle: if in the nondegenerate case we have $m_{\lambda-1} = 0 = m_{\lambda+1}$ for some $\lambda$, then $m_\lambda = b_\lambda$. (Hint: Write out the polynomial equation $\mathfrak{M}(t) - P(t) = (1 + t)\,Q(t)$ explicitly.)

PART THREE

# Lie Groups, Bundles, and Chern Forms

## CHAPTER 15

# Lie Groups

## 15.1. Lie Groups, Invariant Vector Fields and Forms

Is the unitary group $SU(n)$ connected?

### 15.1a. Lie Groups

Let $M(n \times n)$ be the set of all $n \times n$ real matrices. As in Section 1.1d, we shall associate to the matrix $x$ the point in $n^2$-dimensional euclidean space whose coordinates are $x_{11}, x_{12}, \ldots, x_{nn}$. Topologically then, $M(n \times n)$ is simply euclidean $n^2$ space! The **general linear group** $\mathrm{Gl}(n, \mathbb{R})$ is the group of all real $n \times n$ matrices $x = (x_{ij})$ with determinant $\det x \neq 0$. Since $\det x$ is an $n^{\text{th}}$-degree polynomial in the coordinates, it is a smooth function on $M(n \times n)$. Since the real numbers differing from 0 form an open set in $\mathbb{R}$, and since the inverse image of an open set under a continuous map is open, $\mathrm{Gl}(n\ , R)$ is an open subset of $M(n \times n)$. (This says that if $\det x \neq 0$ then $\det y \neq 0$ if $y$ is sufficiently near $x$.) Topologically $\mathrm{Gl}(n, \mathbb{R})$ is an open subset of euclidean space, and as such is an $n^2$-dimensional *manifold.* It is clear from $(xy)_{ij} = \sum x_{ik} y_{kj}$ that the product matrix has coordinates that are smooth functions of the coordinates of $x$ and $y$. From the formula for the inverse

$$x^{-1} = \frac{X}{\det x}$$

where $X_{ij}$ is the signed cofactor of $x_{ji}$, and the fact that $\det x \neq 0$, we see that the coordinates of $x^{-1}$ are also smooth functions of those of $x$. This leads us to the concept of a Lie group.

A **Lie group** is a differentiable manifold $G$ endowed with a "product," that is, a map

$$G \times G \to G \qquad (g, h) \to gh$$

making $G$ into a group. We demand that this map, as well as the "inversion map"

$$G \to G \qquad g \to g^{-1}$$

be differentiable.

In the following examples, the reader should verify that the given manifolds are indeed groups. For example, $\text{Gl}(n, \mathbb{R})$ is a group because, first, $\det x \neq 0$ and $\det y \neq 0$ implies $\det(xy) = (\det x)(\det y) \neq 0$, and second, $\det x^{-1} = (\det x)^{-1} \neq 0$.

**Examples:**

1. $G = \mathbb{R}$, the *additive* group of real numbers. The product here is addition of real numbers. This group is commutative, or "abelian."
2. $G = \mathbb{R}^+$, the *multiplicative* group of *positive* real numbers. This is again abelian.
3. $G = \text{Gl}(\text{n}, \mathbb{R})$, the general linear group of all $n$ by $n$ real matrices $g$ with $\det g \neq 0$. Similarly, we have the nonsingular complex matrices $\text{Gl}(n, \mathbb{C})$. By writing $z_{jk} = x_{jk} + iy_{jk}$, we see that $\text{Gl}(n, \mathbb{C})$ is a $2n^2$-dimensional open submanifold of $\mathbb{C}^{n^2} = \mathbb{R}^{2n^2}$. The notation $\text{Gl}(n)$ refers to either of the cases $\mathbb{R}$ or $\mathbb{C}$. $\text{Gl}(n)$ is not abelian for $n > 1$.
4. $G = \text{Sl}(n, \mathbb{R})$, the **special linear group** is the subgroup of $\text{Gl}(n, \mathbb{R})$ of matrices $x$ with $\det x = 1$. From Problem 1.1(3), we know that it is a submanifold of dimension $(n^2 - 1)$. For *any* matrix group, the adjective **special** means that $\det x = 1$.
5. $G = O(n)$, the **orthogonal group** of all real $n \times n$ matrices $x$ with $xx^T = I$. (Thus $\det x = \pm 1$.) $O(n)$ is clearly a subgroup of $\text{Gl}(n, \mathbb{R})$. In Section 1.1 we saw that it is also a submanifold of dimension $n(n-1)/2$. We also saw there that $O(n)$ is not connected, consisting of the *subgroup* $SO(n)$, the **rotation group**, where $\det x = +1$, and the disjoint sub*manifold* where $\det x = -1$. We shall show in Example (8) that, in fact, these two subsets are each connected. $G = SO(2)$, the rotation group of the plane, is especially easy to visualize. We are dealing with the matrices

$$R_2(\theta) = \begin{bmatrix} \cos\theta & -\sin\theta \\ \sin\theta & \cos\theta \end{bmatrix} \tag{15.1}$$

   and as such, $SO(2)$ is a curve parameterized by $\theta$ in $\mathbb{R}^4$, defined by $x_{11} = \cos\theta$, $x_{12} = -\sin\theta, x_{21} = \sin\theta$, and $x_{22} = \cos\theta$. As a manifold this curve is diffeomorphic to the circle $S^1$ in the plane defined by $x_1 = \cos\theta$ and $x_2 = \sin\theta$, and this is the way we usually think of $SO(2)$; to a rotation of the plane through an angle $\theta$ we associate the point on the unit circle $S^1$ at angle $\theta$. Sometimes we think of $SO(2)$ as the points $\exp(i\theta) = e^{i\theta}$ of the complex plane. To compose two rotations $e^{i\theta}$ and $e^{i\phi}$ we simply multiply $e^{i\theta}e^{i\phi} = e^{i(\theta+\phi)}$, that is, we add their angles. $SO(2)$ is abelian, whereas $SO(n)$, for $n > 2$, is not.
6. $G = U(n)$, the **unitary group**, consisting of complex $n \times n$ matrices $z = (z_{jk})$ with $z^\dagger := \bar{z}^T = z^{-1}$. The overbar denotes complex conjugation; the dagger denotes **(hermitian) adjoint**. The same type of argument that was used for $O(n)$ in Section 1 will show that $U(n)$ is a submanifold of complex $n^2$ space or real $2n^2$ space, and is thus a Lie group. We easily see that $\det z$ has absolute value 1. Note that $U(1)$ is the group of complex numbers $z = e^{i\theta}$ of absolute value 1, and thus $U(1)$ is isomorphic with $S^1$, that is, $SO(2)$. $U(1)$ is the only abelian unitary group.
7. $G = SU(n)$ is the **special unitary group**; $\det z = +1$.
8. $G = T^n$ is the abelian group of diagonal matrices of the form

$$z = \text{diag}[\exp(i\theta_1), \ldots, \exp(i\theta_n)] \tag{15.2}$$

This group is topologically $S^1 \times \cdots \times S^1$, the topological product of $n$ copies of the circle, and as such is an **$n$-torus**. Since the circle is connected (each point can be joined to the identity by a curve), it follows easily that $T^n$ is connected. From this we may see that the far more complicated group $U(n)$ is also connected! Before doing so, we note the following.

As a manifold, a Lie group is very special for the following reason. A Lie group always has two families of diffeomorphisms, the left and right **translations**. For $g \in G$, these translations are defined by

$$L_g : G \to G \qquad L_g(h) = gh$$

and (15.3)

$$R_g : G \to G \qquad R_g(h) = hg$$

It is clear that the mapping inverse to $L_g$ is simply $L_{g^{-1}}$.

**Theorem (15.4):** *$U(n)$ is connected.*

PROOF: Note that $T^n$ is clearly a subgroup (and consequently a subset) of $U(n)$. The familiar "principal axes theorem" of linear algebra states that any $g \in U(n)$ can be diagonalized by a unitary matrix. (**Proof**: Each such $g$ has eigenvalues of absolute value 1. Let $\mathbf{e}_1$ be an eigenvector with eigenvalue $\exp(i\theta_1)$. Let $\mathbf{e}_1^{\perp}$ be the orthogonal subspace to $\mathbf{e}_1$ in the Hermitian metric $\langle \mathbf{v}, \mathbf{w} \rangle = \sum v_k \overline{w}_k$. Since $g$ is an isometry, $g$ sends $\mathbf{e}_1^{\perp}$ into itself and so $g$ has an eigenvector $\mathbf{e}_2$ in this subspace with eigenvalue $\exp(i\theta_2)$. Continue with this process. In the eigenvector basis $\mathbf{e}_1, \mathbf{e}_2, \ldots, \mathbf{e}_n$, the linear transformation $g$ has matrix $z = \text{diag}(\exp(i\theta_1), \ldots, \exp(i\theta_n))$, as desired.) This means that given $g \in U(n)$, there exists an $h \in U(n)$ such that $h^{-1}gh = z = \text{diag}(exp(i\theta_1), \ldots,$ $\exp(i\theta_n))$. Then $g = hzh^{-1}$. (This says that $g \in (hT^nh^{-1})$, i.e., $g$ lies on the diffeomorphic copy of $T^n$ that results from left translating $T^n$ by $h$ and then right translating by $h^{-1}$.) Thus $g$ can be joined to the identity by a curve, by putting $\theta_j(t) = (1-t)\theta_j$. $U(n)$ is connected. □

The subgroup $T^n$ of $U(n)$ given by (15.2) is called a **maximal torus** of $U(n)$. Any **conjugate** $hT^nh^{-1}$ of this maximal torus is also called a maximal torus.

By the same type of reasoning we may deal with the rotation group.

**Theorem (15.5):** *$O(n)$ consists of two connected "components" and $SO(n)$ is the component holding the identity.*

PROOF: Consider first the case $SO(2n)$. The principal axes theorem states that any $g \in SO(2n)$ is "conjugate" to a block "diagonal" matrix with $2 \times 2$ rotation matrices down the diagonal

$$g = \text{diag}[R_2(\theta_1), \ldots, R_2(\theta_n)] \tag{15.6}$$

where $R_2(\theta_k)$ is as in (15.1). This simply says that after a suitable orthogonal change of basis in $\mathbb{R}^{2n}$, the rotation takes on the form of rotations in $n$ orthogonal 2-dimensional planes. In the case of $SO(2n+1)$ one adds a final diagonal entry of $+1$. We can arrive at this **canonical form** as follows. The possible real eigenvalues of $g \in SO(2n)$ are $\pm 1$, whereas the complex eigenvalues appear in complex conjugate pairs. If $+1$ is an eigenvalue then it must be a double eigenvalue since $\det g = 1$. The eigenspace for this double eigenvalue is a 2-plane $E_1$ on which $g$ takes the form $R_2(0)$. Likewise, if $-1$ is an eigenvalue, it also must be a double root and we get a 2-plane $E_{-1}$ on which $g$ takes the form $R_2(\pi)$. In both cases $g$ leaves invariant the orthogonal complementary $(2n-2)$-space. By continuing in this complementary subspace we either exhaust the entire $2n$-space or have left a remaining $2k$-space $\mathbb{R}^{2k}$ on which $g$ has *only complex eigenvalues*. Let $S^{2k-1}$ be the unit sphere in this subspace. The function $f(x) := \langle gx, x\rangle$ takes on its minimum at some point $x_0$ of the sphere. Now $gx_0$ does not lie along $x_0$ since $g$ has no real eigenvalue in $\mathbb{R}^{2k}$. *We claim that the plane spanned by $x_0$ and $gx_0$ is sent into itself by $g$*. By definition, $g$ sends $x_0$ into this plane; where does it send $gx_0$? Let $x(t)$ be a curve on $S^{2k-1}$ starting at $x_0$ and put $v = x'(0)$. Then $0 = f'(0) = \langle gv, x_0\rangle + \langle gx_0, v\rangle = \langle v, (g^T + g)x_0\rangle = \langle v, (g^{-1} + g)x_0\rangle$ for all tangent $v$. Thus $(g^{-1} + g)x_0 = \lambda x_0$ and so $g^2 x_0 = \lambda g x_0 - x_0$. Thus $g$ sends $g(x_0)$ into the plane spanned by $x_0$ and $gx_0$, as desired. But it is immediate that $g$ takes the form $R_2(\theta)$ on any invariant 2-plane. We may then continue with the complement of this plane in $\mathbb{R}^{2k}$.

Finally, in the case $SO(2n+1)$, any $g$ has $+1$ as an eigenvalue, with a corresponding eigenvector. We proceed with the complementary $\mathbb{R}^{2n}$ as earlier.

We continue with the proof of Theorem (15.5). The collection of all rotations of the form (15.6) (with a $+1$ included in the odd-dimensional case) forms again an $n$-dimensional torus $S^1 \times \cdots \times S^1$, a maximal torus $T^n$ of the rotation group. One then proceeds as in the $U(n)$ case to show that $SO(n)$ is connected.

$O(n)$ consists of the rotations $SO(n)$ and the **improper** orthogonal matrices $O^-(n)$ where the determinant is $-1$. But if we let $h = \mathrm{diag}(-1, 1, \ldots, 1) \in O^-$, then left translation $L_h$ by $h$ is a diffeomorphism of $O(n)$ that interchanges $SO(n)$ and $O^-(n)$, showing that these two subsets are diffeomorphic. □

Our final example, although not as intrinsically important as the preceding one will play an important role in our treatment because it will be possible to perfor explicit calculations. It is a nonabelian, noncompact, 2-dimensional Lie group.

**9.** $G = A(1)$, **the affine group of the line**, consists of those real $2 \times 2$ matrices

$$\begin{bmatrix} x & y \\ 0 & 1 \end{bmatrix}$$

with $x > 0$. The manifold for $A(1)$ can be considered as the "right half plane," thos $(x, y) \in \mathbb{R}^2$, with $x > 0$.

A **matrix group** is a subgroup of $\mathrm{Gl}(n)$ that is also a submanifold of Gl(n All of our previous examples are groups of matrices. Although there are importa Lie groups that cannot be realized as matrix groups, for our calculations *we sha occasionally pretend that our group is indeed a matrix group*, since the constructio and proofs are easier to visualize.

### 15.1b. Invariant Vector Fields and Forms

Lie groups are special as manifolds for the following reason. Given a tangent vector $\mathbf{X}_e$ to $G$ at the identity $e$, we may left or right translate $\mathbf{X}_e$ to each point of $G$, by means of the differentials

$$\mathbf{X}_g := L_{g*}\mathbf{X}_e$$

resp. (15.7)

$$\mathbf{X}_g := R_{g*}\mathbf{X}_e$$

yielding two nonvanishing vector fields on all of $G$! In fact, if we take a basis $\mathbf{X}_1, \ldots, \mathbf{X}_n$ for $G_e$ (the tangent space to $G$ at $e$), then we can left or right translate this basis to give $n$ linearly independent vector fields, such as,

$$L_{g*}\mathbf{X}_1, \ldots, L_{g*}\mathbf{X}_n \tag{15.8}$$

on all of $G$! In particular, *every Lie group is an orientable manifold*! Consider for instance, a closed orientable surface $M^2$ of genus $g$. We shall see in Section 16.2 that of these surfaces only the torus (genus 1) can support even a single *nonvanishing* tangent vector field. In fact $T^2$ supports two vector fields $\partial/\partial\theta$, $\partial/\partial\phi$, and the torus is indeed the commutative group $S^1 \times S^1$ with multiplication

$$(\theta_1, \phi_1)(\theta_2, \phi_2) \to (\theta_1 + \theta_2, \phi_1 + \phi_2)$$

Topologically, the only *compact* Lie group of dimension 2 is the torus. (The Klein bottle is nonorientable and admits a nonvanishing vector field, but not two independent ones!)

We shall say that a vector field $\mathbf{X}$ on $G$ is **left (right) invariant** if it is invariant under all left (right) translations, that is,

$$L_{g*}\mathbf{X}_h = \mathbf{X}_{gh}$$

resp. (15.9)

$$R_{g*}\mathbf{X}_h = \mathbf{X}_{hg}$$

You should convince yourself that if $\mathbf{X}_e$ is given, then (15.7) exhibits the unique left (resp. right) invariant field generated by $\mathbf{X}_e$.

Similarly, for example, an exterior $p$-form $\alpha$ on $G$ is **left invariant** if

$$L_g^*\alpha_{gh} = \alpha_h \tag{15.10}$$

and to get a left invariant form on all of $G$ one translates a form at $e$ over the entire group by

$$\alpha_g := L_{g^{-1}}{}^*\alpha_e \tag{15.11}$$

In the case of a matrix group, $L_{g*}\mathbf{X}_h$ is especially simple. Let $t \mapsto h(t)$ be a curve of matrices in $G$ with $h(0) = h$ and $h'(0) = \mathbf{X}_h$. Since $G \subset \mathrm{Gl}\,(n)$, this curve is simply a matrix $h$ whose entries $h_{jk}(t)$ are smooth functions of the parameter $t$. $h(t)$ describes a curve in $n^2$-dimensional euclidean space (real or complex). *Then* $\mathbf{X}_h$, *the tangent to this curve, is simply the matrix whose entries are the derivatives* at $t = 0$, $h'_{jk}(0)$. There is no reason to believe that this new matrix $h'$ associated to the point (matrix) $h$ will belong to the group $G$ (this will be illustrated in the case $A(1)$ later). Then for the constant matrix $g$, the curve $t \mapsto gh(t)$ will have for tangent vector at $t = 0$ the matrix

$$L_{g*}\mathbf{X}_h = gh'(0) = g\mathbf{X}_h$$

that is simply the *matrix product* of $g$ and $\mathbf{X}_h$.

**Example:** $G = A(1)$, (Example (9)). We may consider $A(1)$ either as a submanifold of $\mathbb{R}^4$ or as the right half plane, since the entries 0 and 1 at the bottom contribute nothing to our knowledge of the matrix. Since

$$\begin{bmatrix} x & y \\ 0 & 1 \end{bmatrix} \begin{bmatrix} x' & y' \\ 0 & 1 \end{bmatrix} = \begin{bmatrix} xx' & xy' + y \\ 0 & 1 \end{bmatrix}$$

we see that the right half plane is endowed with a rather unusual multiplication given in the top row of this matrix equation.

We shall identify

$$\begin{bmatrix} x & y \\ 0 & 1 \end{bmatrix} \in A(1) \quad \text{with } (x, y) \in R^2$$

and for tangent vectors we identify

$$\begin{bmatrix} \frac{dx}{dt} & \frac{dy}{dt} \\ 0 & 0 \end{bmatrix} \quad \text{with} \left( \frac{dx}{dt} \quad \frac{dy}{dt} \right)^T$$

which is the tangent vector $(dx/dt)\partial/\partial x + (dy/dt)\partial/\partial y$. Now let us *left* translate the vectors

$$\frac{\partial}{\partial x} \quad \text{and} \quad \frac{\partial}{\partial y}$$

at the identity $e$ to the point $(x, y)$. For $\partial/\partial x$ we consider the curve $h(t)$ given by

$$\begin{bmatrix} 1+t & 0 \\ 0 & 1 \end{bmatrix}$$

whose tangent at $e$ is $\partial/\partial x$. Then, letting $g$ be the matrix

$$\begin{bmatrix} x & y \\ 0 & 1 \end{bmatrix}$$

we have

$$L_{g*}\frac{\partial}{\partial x} = \frac{d}{dt}(gh(t))\}_{t=0} = \begin{bmatrix} x & 0 \\ 0 & 0 \end{bmatrix}$$

and this is indeed the left translate of $\partial/\partial x$ at the identity to the point $(x, y)$

$$\begin{bmatrix} x & y \\ 0 & 1 \end{bmatrix} \begin{bmatrix} 1 & 0 \\ 0 & 0 \end{bmatrix}$$

(Note that this matrix is *not* in $A(1)$; it is a tangent vector to $A(1)$). Thus the left translate of $\partial/\partial x$ to $(x, y)$ is

$$\mathbf{X} = x\frac{\partial}{\partial x}$$

To construct the left translate of $\partial/\partial y$ at $(1, 0)$ to the point $(x, y)$ we form

$$\begin{bmatrix} x & y \\ 0 & 1 \end{bmatrix} \begin{bmatrix} 0 & 1 \\ 0 & 0 \end{bmatrix} = \begin{bmatrix} 0 & x \\ 0 & 0 \end{bmatrix}$$

The result is the vector $x\partial/\partial y$. Thus a basis for the left invariant vector fields on $A(1)$ is given by the pair

$$\mathbf{X}_1 = x\frac{\partial}{\partial x} \qquad \mathbf{X}_2 = x\frac{\partial}{\partial y} \tag{15.12}$$

Next note that in any Lie group, if $\mathbf{X}_1, \ldots, \mathbf{X}_n$ is a basis for the left invariant vector fields and if $\sigma^1, \ldots, \sigma^n$ is the dual basis of 1-forms, then *this dual basis is automatically left invariant,* since

$$L_g^* \sigma_g(\mathbf{X}_e) = \sigma_g\{L_{g*}\mathbf{X}_e\} = \sigma_g(\mathbf{X}_g) = \sigma_e(\mathbf{X}_e)$$

shows that $L_g^*\sigma_g = \sigma_e$. The same argument shows that *if $\alpha^p$ is any p-form whose values on any p-tuple of left invariant vector fields are constant on G, then $\alpha$ is left invariant.*

Thus the basis of left invariant 1-forms dual to (15.12) is given by

$$\sigma^1 = \frac{dx}{x} \qquad \sigma^2 = \frac{dy}{x} \tag{15.13}$$

If $\alpha$ and $\beta$ are invariant under left translations then so are $d\alpha$ and $\alpha \wedge \beta$. Thus in $A(1)$

$$\sigma^1 \wedge \sigma^2 = \frac{dx \wedge dy}{x^2} \tag{15.14}$$

is a left invariant area form or left **Haar measure**; for any compact region $U \subset A(1)$, and for any $g \in A(1)$

$$\int\int_{gU} \frac{dx \wedge dy}{x^2} = \int\int_{U} \frac{dx \wedge dy}{x^2}$$

where $gU := L_g U$ is the left translate of the region $U$. This would not hold if the factor $x^{-2}$ were omitted.

## Problems

**15.1(1)** For the group $A(1)$, find the *right* invariant vector fields coinciding with $\partial/\partial x$ and $\partial/\partial y$ at $e$, find the dual right invariant 1-forms, and write down the right Haar measure.

**15.1(2)** $\mathbb{R}^4$ can be identified with the space of all real $2 \times 2$ matrices, identifying $x = (x^1, x^2, x^3, x^4)$ with the matrix (again called $x$)

$$\begin{bmatrix} x^1 & x^2 \\ x^3 & x^4 \end{bmatrix}$$

Sl(2, $\mathbb{R}$) can be considered as the submanifold $M^3$ of $\mathbb{R}^4$ defined by $\det(x) = 1$. Sl(2, $\mathbb{R}$)) acts *linearly* on $\mathbb{R}^4$, $g : \mathbb{R}^4 \to \mathbb{R}^4$, by $g(x) = gx$ (matrix multiplication).

(i) Compute the $4 \times 4$ matrix differential $g_*$ of $g$ and show that $\det g_* = 1$. This shows that the action of $G$ on $\mathbb{R}^4$ preserves the euclidean volume form.

(ii) $H(x) := \det(x)$ is of course a function on $\mathbb{R}^4$ that is invariant under the action of $G$. Use Equation (4.56) to write down a left invariant volume 3-form for all of $Sl(2, \mathbb{R})$.

## 15.2. One Parameter Subgroups

Does $e^{\theta J} = (\cos\theta)I + (\sin\theta)J$ look familiar?

A **homomorphism** of groups is a function

$$f : G \to H$$

that preserves products

$$f(g_1 g_2) = f(g_1) f(g_2)$$

In Section 13.1, we defined the special case of a homomorphism when the groups were abelian, and when the group "multiplication" was "addition."

As an example, the usual exponential function $f(t) = e^t$ defines (since $e^{s+t} = e^s e^t$) a homomorphism

$$\exp : \mathbb{R} \to \mathbb{R}^+$$

of the *additive* group of the reals to the *multiplicative group* of positive real numbers. Note that exp is also a differentiable map, and in this case it is 1:1 ( the homomorphism is **injective**), and also onto (**surjective**). We then say that exp is an **isomorphism** of Lie groups. exp is a diffeomorphism with inverse log: $\mathbb{R}^+ \to \mathbb{R}$.

A **1-parameter subgroup** of $G$ is by definition a differentiable homomorphism (in particular, a path)

$$g : \mathbb{R} \to G \qquad t \to g(t) \in G$$

of the additive group of the reals into the group $G$. Thus

$$g(s+t) = g(s)g(t) = g(t)g(s) \tag{15.15}$$

Consider now a 1-parameter subgroup of a matrix group $G$.

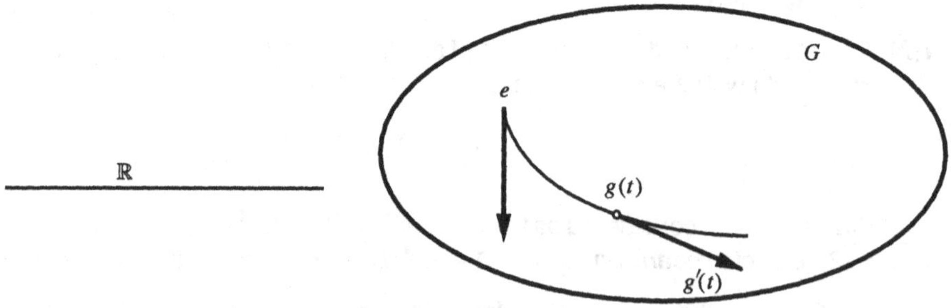


Figure 15.1

As matrices $g(t+s) = g(t)g(s)$, that is, $g_{ij}(t+s) = \sum_k g_{ik}(t)g_{kj}(s)$. Differentiate both sides with respect to $s$ and put $s = 0$,

$$g'(t) = g(t)\, g'(0) \tag{15.16}$$

Since $g'(0)$ is a constant matrix, the solution to this is

$$g(t) = g(0)\ \exp\{tg'(0)\}$$

where

$$\exp(S) = e^S := I + S + \frac{S^2}{2!} + \frac{S^3}{3!} + \cdots \tag{15.17}$$

It can be shown that this infinite series converges for all matrices $S$. Since $g(0) = e$ for any homomorphism $g : \mathbb{R} \to G$, we conclude that

$$g(t) = \exp\{tg'(0)\} \tag{15.18}$$

*is the most general form for a 1-parameter subgroup of a matrix group $G$.*

Equation (15.16) tells us how to proceed even if $G$ is not a matrix group, for it really says

$$g'(t) = L_{g(t)*}g'(0) \tag{15.19}$$

that is, the tangent vector $\mathbf{X}$ to the 1-parameter subgroup is *left translated* along the subgroup. Thus, given a tangent vector $\mathbf{X}_e$ at $e$ in $G$,

> the 1-parameter subgroup of $G$ whose tangent at $e$ is $\mathbf{X}_e$ is the integral curve through $e$ of the vector field $\mathbf{X}$ on $G$ resulting from left translation of $X_e$ over all of $G$.

The vector $\mathbf{X}_e$ is called the **infinitesimal generator** of the 1-parameter subgroup.

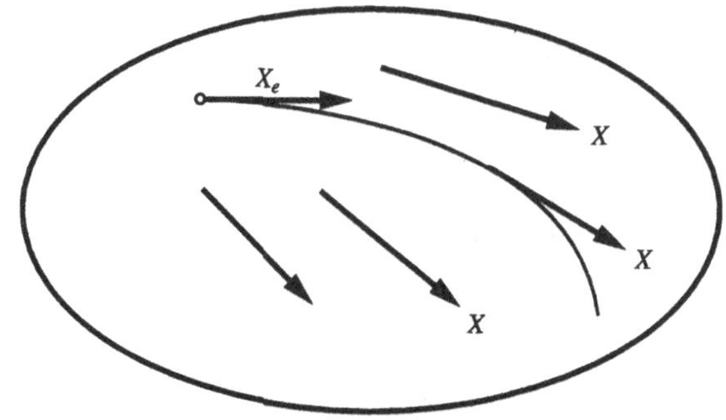


**Figure 15.2**

For *any* Lie group $G$ we shall denote the 1-parameter subgroup whose generator at $e$ is $\mathbf{X}_e$, by

$$g(t) := e^{t\mathbf{X}_e} = \exp\ t\mathbf{X}_e$$

just as we do in the case of a matrix group.

For example, in $A(1)$, to find the 1-parameter subgroup having tangent vector $(a\ b)^T$ at the identity, we left translate this vector over $A(1)$.

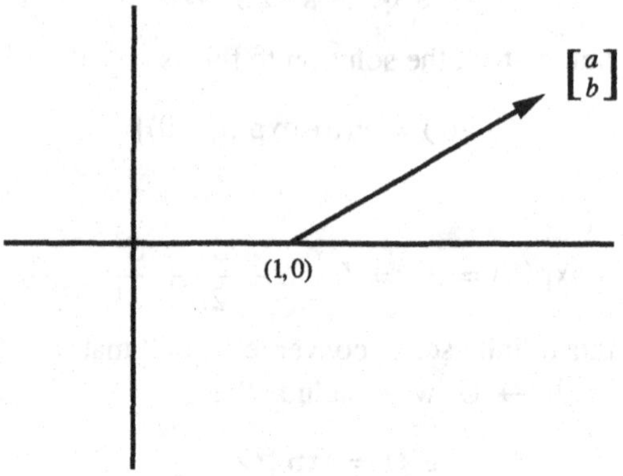


**Figure 15.3**

The left translate of $(a\partial/\partial x + b\partial/\partial y)$ to the point $(x, y)$ is, from (15.12), $(ax\partial/\partial x + bx\partial/\partial y)$. Then we need to solve

$$\frac{dx}{dt} = ax \qquad x(0) = 1 \tag{15.20}$$

$$\frac{dy}{dt} = bx \qquad y(0) = 0$$

The solutions are clearly straight lines $dy/dx = b/a$, but to see the parameterization we must solve (15.20) to get

$$x(t) = e^{at} \qquad y(t) = \frac{be^{at}}{a} - \frac{b}{a}$$

(which never reaches the $y$ axis).

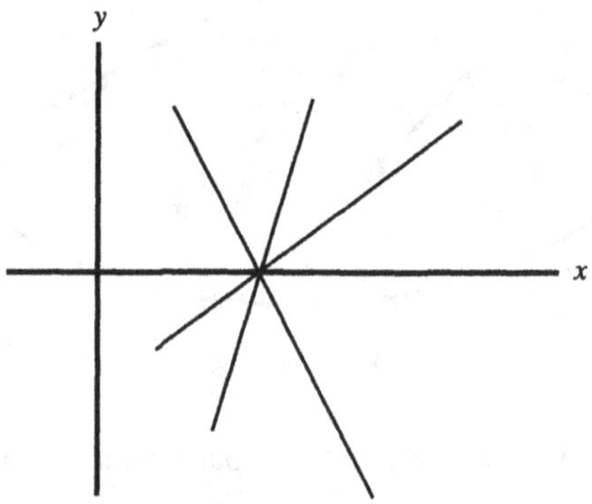


**Figure 15.4**

In Problem 15.2(2) you are asked to get this from the power series.

## Problems

**15.2(1)** We shall see in the next section that

$$J = \begin{bmatrix} 0 & -1 \\ 1 & 0 \end{bmatrix}$$

can be considered a tangent vector at the identity of the group Gl(2, $\mathbb{R}$). Use $J^2 = -I$, $J^3 = -J$, $J^4 = I$, to show

$$e^{\theta J} = (\cos\theta) I + (\sin\theta) J \quad \textit{i.e.},$$

$$\exp\begin{bmatrix} 0 & -\theta \\ \theta & 0 \end{bmatrix} = \begin{bmatrix} \cos\theta & -\sin\theta \\ \sin\theta & \cos\theta \end{bmatrix}$$

for all real $\theta$. This 1-parameter subgroup of Gl(2, $\mathbb{R}$) *is the entire subgroup of rotations of the plane, SO(2)!*

**Warning**: It makes no more sense to say $\exp \mathbf{S} = I + \mathbf{S}$ for $\mathbf{S}$ small than it does to say $e^x = 1 + x$ when $x$ is a small number. For example, $I + \theta J$ is *never* in $SO(2)$ for any $\theta \neq 0$.

**15.2(2)** Compute

$$\exp\, t \begin{bmatrix} a & b \\ 0 & 0 \end{bmatrix}$$

directly from the power series.

**15.2(3)** Consider the differential equation

$$x'(t) = \frac{dx(t)}{dt} = A(t)x(t)$$

$$x(0) = x_0$$

where $A(t)$ is an $n \times n$ matrix function of $t$ and $x(t)$ is a column matrix. It is known that if $A$ is actually a *constant* matrix, then the solution is $x(t) = \exp(tA)x_0$; this easily follows *formally* (i.e., disregarding questions of differentiating infinite series term by term, etc.) from the power series expansion of $\exp(tA)$. In the case of a $1 \times 1$ matrix *function* $A(t)$ the solution is of course

$$x(t) = \exp(\int_0^t A(\tau)d\tau)x_0$$

We claim that this same formula holds in the $n \times n$ case *provided* that the matrix A(t) *commutes with its indefinite integral* $B(t) := \int_0^t A(\tau)d\tau$ for all $t$. Verify this formally by looking at

$$x(t) := \exp[B(t)]x_0 = \left[ I + B(t) + \frac{1}{2!}\{B(t)B(t)\} + \cdots \right] x_0$$

and using $B'(t) = A(t)$.

## 15.3. The Lie Algebra of a Lie Group

What is the third Betti number of the eight-dimensional Sl(3, $\mathbb{R}$)?

### 15.3a. The Lie Algebra

Let $G$ be a Lie group. The tangent vector space $G_e$ at the identity $e$ plays an important role; we shall denote it by the script $\mathfrak{g}$

$$\mathfrak{g} := G_e$$

and call it (for reasons soon to be discussed) the **Lie algebra** of $G$.

Let $\mathbf{X}_R$, $R = 1, \ldots, N$, be a basis for $\mathfrak{g}$; $\mathbf{X}_R$ will also denote the left translation of this field to all of $G$. Since any left invariant vector field is determined by its value at $e$, the most general left invariant vector field is then of the form

$$\mathbf{X} = \sum v^R \mathbf{X}_R$$

where the $v^R$ are *constants*.

Let $\sigma^R$, $R = 1, \ldots, N$ be the dual basis of left invariant 1-forms on $G$; they are determined by their values on vectors from $\mathfrak{g}$. The most general left invariant $r$-form on $G$ is of the form

$$\alpha^r = \sum_{I} a_{i_1 \ldots i_r} \sigma^{i_1} \wedge \ldots \wedge \sigma^{i_r}$$

It is again determined by its values on $r$-tuples from $\mathfrak{g}$. It is constant when evaluated on left invariant vector fields and $a_I$ are constants.

Recall the notion of Lie derivative or Lie bracket of two vector fields on a manifold $M$; see Equation (4.4).

**Theorem (15.21):** *The Lie bracket* $[\mathbf{X}, \mathbf{Y}]$ *of two left invariant vector fields is again left invariant.*

PROOF: A vector field $\mathbf{X}$ is left invariant iff $\sigma(\mathbf{X})$ is constant on $G$ whenever $\sigma$ is a left invariant 1-form. If $\sigma$ is a left invariant 1-form then

$$d\sigma(\mathbf{X}, \mathbf{Y}) = -\sigma([\mathbf{X}, \mathbf{Y}]) \tag{15.22}$$

Since $d\sigma$ is left invariant, the left-hand side is constant. □

We may then write

$$[\mathbf{X}_R, \mathbf{X}_S] = \mathbf{X}_T C^T_{RS} \qquad C^T_{RS} = -C^T_{SR} \tag{15.23}$$

for some **structure constants** $C^T_{RS}$ (dependent on the basis $\{\mathbf{X}_R\}$).

In Problem 15.3(1) you are asked to prove the following.

**Theorem (15.24):** *The **Maurer–Cartan** equations*

$$d\sigma^U = -\sum_{R<S} C^U_{RS}\sigma^R \wedge \sigma^S$$

$$= -\frac{1}{2}\sum_{R,S} C^U_{RS}\sigma^R \wedge \sigma^S$$

*hold, and* $d^2\sigma^U = 0$ *yields the **Jacobi identity***

$$C^U_{RS}C^R_{LM} + C^U_{RM}C^R_{SL} + C^U_{RL}C^R_{MS} = 0$$

This Jacobi identity for left invariant 1-forms is also a consequence of a general Jacobi identity for vector fields on any manifold $M^n$. If $\mathbf{X}$, $\mathbf{Y}$, and $\mathbf{Z}$ are any three vector fields on a manifold, then as differential operators on functions $f$, $[\mathbf{X}, \mathbf{Y}](f) = \mathbf{X}(\mathbf{Y}(f)) - \mathbf{Y}(\mathbf{X}(f))$, and so on. Then the following Jacobi identity is immediate.

$$[[\mathbf{X}, \mathbf{Y}], \mathbf{Z}] + [[\mathbf{Z}, \mathbf{X}], \mathbf{Y}] + [[\mathbf{Y}, \mathbf{Z}], \mathbf{X}] = 0 \qquad (15.25)$$

and in the case of a Lie group this gives (15.24) via (15.23).

We now make the vector space $\mathfrak{g} = G_e$ into a "Lie *algebra*" by defining a product

$$\mathfrak{g} \times \mathfrak{g} \to \mathfrak{g}$$

as follows. Let $\mathbf{X} \in \mathfrak{g}$, $\mathbf{Y} \in \mathfrak{g}$. Extend them to be *left* invariant vector fields $\mathbf{X}'$, $\mathbf{Y}'$ on all of $G$, and then define the product of $\mathbf{X}$ and $\mathbf{Y}$ to be the Lie bracket

$$[\mathbf{X}, \mathbf{Y}] := [\mathbf{X}', \mathbf{Y}']_e$$

This product satisfies the relation $[\mathbf{X}, \mathbf{Y}] = -[\mathbf{Y}, \mathbf{X}]$ and the Jacobi identity (15.25).

We shall see later on that there are three vectors $\mathbf{X}$, $\mathbf{Y}$, $\mathbf{Z}$ in the Lie algebra of $SO(3)$ that satisfy $[\mathbf{X}, \mathbf{Y}] = \mathbf{Z}$ and $[\mathbf{X}, \mathbf{Z}] = -\mathbf{Y}$. Then $[\mathbf{X}, [\mathbf{X}, \mathbf{Y}]] = -\mathbf{Y}$, while $[[\mathbf{X}, \mathbf{X}], \mathbf{Y}] = 0$, and thus *the Lie algebra product is not associative*!

We shall consistently identify the Lie algebra $\mathfrak{g}$ with the $N(= \dim G)$ dimensional vector space of *left* invariant fields on $G$.

Classically the Lie algebra $\mathfrak{g}$ was known as the "infinitesimal group" of $G$, for classically a vector was thought of roughly as going from a point to an "infinitesimally nearby" point. $\mathfrak{g}$ then consisted of group elements infinitesimally near the identity! We shall *not* use this picture.

## 15.3b. The Exponential Map

**Theorem (15.26):** *For any matrix $A$, det $e^A = e^{trA}$.*

PROOF: Consider the matrix $A$ as a linear transformation of complex $n$-space $\mathbb{C}^n$. If $\lambda$ is an eigenvalue of $A$, $Av = \lambda v$, then from the power series for $e^A$ we see that $e^A v = e^\lambda v$. Thus $e^A$ has eigenvalues $\exp(\lambda_1), \ldots, \exp(\lambda_n)$, where $\lambda_1, \ldots, \lambda_n$ are the eigenvalues of $A$. Then, since the determinant is the product of the eigenvalues

$$\det \exp A = \prod \exp \lambda_i = \exp \sum \lambda_i = \exp\, trA \quad \square$$

**Theorem (15.27):** *The map* $\exp : \mathfrak{g} \to G$ *sending* $A \mapsto e^A$ *is a diffeomorphism of some neighborhood of* $0 \in \mathfrak{g}$ *onto a neighborhood of* $e \in G$.

PROOF: We shall give two proofs. For a matrix group, look at the differential of the exponential map applied to a vector $\mathbf{X} \in \mathfrak{g}$.

$$\exp_*(\mathbf{X}) = \frac{d}{dt}(\exp t\mathbf{X})_{t=0} = \frac{d}{dt}\left(I + t\mathbf{X} + \frac{1}{2}t^2\mathbf{X}^2 + \cdots\right)_{t=0} = \mathbf{X}$$

Thus $\exp_* : \mathfrak{g} \to \mathfrak{g}$ is the identity and exp is a local diffeomorphism by the inverse function theorem.

If $G$ is not a matrix group we would proceed as follows. Given $\mathbf{X}$ at $e$, $e^{t\mathbf{X}} = \exp(t\mathbf{X})$ is a curve through $e$ whose tangent vector at $t = 0$ is the vector $\mathbf{X}$ (recall that $\mathbf{e}^{t\mathbf{X}}$ is the integral curve through $e$ of the left invariant vector field $\mathbf{X}$). Thus again $\exp_*(\mathbf{X}) = \mathbf{X}$, and we proceed as previously. □

**Remark**: In a general Lie group, the 1-parameter subgroup $\exp(t\mathbf{X})$ is the integral curve of a vector field on $G$, and thus it would seem that this need only be defined for $t$ small. In this case of a left invariant vector field on a group, it can be shown that the curve exists for all $t$, just as it does in the matrix case.

## 15.3c. Examples of Lie Algebras

1. $G = \mathrm{Gl}(n, \mathbb{R})$. Let $M(n \times n)$ be the vector space of *all* real $n \times n$ matrices; $M(n \times n) \approx n^2$ dimensional Euclidean space. For $A \in M(n \times n)$

$$\det e^A = e^{\mathrm{tr} A} > 0$$

and therefore

$$\exp : M(n \times n) \to \mathrm{Gl}(n, \mathbb{R})$$

Since $\dim M(n \times n) = n^2 = \dim \mathrm{Gl}(n, \mathbb{R})$, we see that the Lie algebra of $\mathrm{Gl}(n, \mathbb{R})$ is

$$\mathfrak{gl}(n, \mathbb{R}) = M(n \times n)$$

We shall now use the fact that if $G$ is a matrix group, that is, a subgroup of $\mathrm{Gl}(n)$, then its Lie algebra $\mathfrak{g}$, being the tangent space to the submanifold $G$ of $\mathrm{Gl}(n, \mathbb{R})$, *is the largest subspace of* $M(n \times n)$ *such that* $\exp : \mathfrak{g} \to G$.

2. $G = SO(n)$. First we need two elementary facts about the exponential of a matrix. Since $e^A e^{-A} = (I + A + A^2/2! + \cdots)(I - A + A^2/2! - \cdots) = I$ we conclude

$$(e^A)^{-1} = e^{-A}$$

Next, from the power series it is evident that for transposes,

$$(\exp A)^T = \exp(A^T)$$

It is clear then that if $A$ is *skew symmetric*, $A^T = -A$, then

$$(\exp A)^{-1} = (\exp A)^T$$

and so exp $A \in O(n)$. Also, since det $e^A = e^{trA} = 1$ for a skew $A$, we see $e^A \in SO(n)$. Thus the skew symmetric matrices exponentiate to $SO(n)$ and the Lie algebra of $SO(n)$ is a vector subspace $\mathfrak{so}(n)$ of $\mathfrak{gl}(n)$ that contains the subspace of skew symmetric matrices.

Conversely, suppose that for some matrix $A \in \mathfrak{so}(n)$, that $e^A \in SO(n)$. Thus

$$\exp(A) = \exp(-A^T)$$

Since exp is a local diffeomorphism it is 1 : 1 in a neighborhood of $0 \in \mathfrak{gl}(n)$. Thus if $e^A$ is close enough to the identity then

$$A = -A^T$$

that is, $A$ is skew symmetric. Thus $\mathfrak{so}(n)$,

*the Lie algebra of $SO(n)$, is precisely the vector space of skew symmetric $n \times n$ matrices.*

One can also see this by looking at the tangent vector to a curve $g(t)$ in $SO(n)$ that starts at $e$. Since $gg^T = e$, we have $g'(0) + g'(0)^T = 0$, showing that $g'(0)$ is skew symmetric.

3. $G = U(n)$, the group of unitary matrices, $u^{-1} = u^\dagger$, where $\dagger$ is the hermitian adjoint, that is, the transpose complex conjugate. Then note that if $A$ is skew hermitian, $A^\dagger = -A$, then $e^A \in U(n)$ from the same reasoning. We conclude that

   *$\mathfrak{u}(n)$ is the vector space of skew hermitian matrices.*

4. $G = SU(n)$, the special unitary group of unitary matrices with det $u = 1$. Since a skew hermitian matrix $A$ has purely imaginary diagonal terms we conclude that det $e^A = e^{trA}$ has absolute value 1. However if $A$ also has trace 0 we see that $e^A$ will lie in $SU(n)$.

   *$\mathfrak{su}(n)$is the space of skew hermitian matrices with trace* 0

5. $G = Sl(n, \mathbb{R})$, the real matrices $g$ with det $g = 1$

   *$\mathfrak{sl}(n, \mathbb{R})$ is the space of all real matrices with trace* 0

### 15.3d. Do the 1-Parameter Subgroups Cover $G$?

Given $g \in G$, is there always an $A \in \mathfrak{g}$ such that $e^A = g$? In other words,

is the map exp : $\mathfrak{g} \to G$ onto?

It can be shown that this is indeed the case when $G$ is *connected* and *compact*. (It is clear that a 1-parameter subgroup must lie in the connected piece of $G$ that contains the identity.) Sl(2, $\mathbb{R}$) is *not* compact. For $g \in$ Sl (2, $\mathbb{R}$)

$$g = \begin{bmatrix} x & y \\ z & w \end{bmatrix} \qquad xw - yz = 1$$

that is, the coordinates $x, y, z, w$ satisfy the preceding simple quadratic equation. This locus is not compact since, for example, $x$ can take on arbitrarily large values. You are asked, in Problem 15.3(2), to show that any $g$ in Sl(2, $\mathbb{R}$) with trace$< -2$ is never of the form $e^A$ for any $A$ with trace 0, that is, for any $A \in \mathfrak{sl}(2, \mathbb{R})$.

This result is somewhat surprising since we shall now show that $Sl(2, R)$ *is* connected!

$$g = \begin{bmatrix} x & y \\ z & w \end{bmatrix}$$

in Sl(2, R) can be pictured as a pair of column vectors $(x\ z)^T$ and $(y\ w)^T$ in $\mathbb{R}^2$ spanning a parallelogram of area 1. Deform the *lengths* of both so that the first becomes a unit vector, keeping the area 1. This deforms Sl(2, $\mathbb{R}$) into itself. Next, "Gram–Schmidt" the second so that the columns are orthonormal. This can be done continuously; instead of forming $\mathbf{v} - \langle \mathbf{v}, \mathbf{e} \rangle \mathbf{e}$ one can form $\mathbf{v} - t\langle \mathbf{v}, \mathbf{e} \rangle \mathbf{e}$. The resulting matrix is then in the subgroup $SO(2)$ of Sl(2, $\mathbb{R}$); that is, it represents a rotation of the plane. We have shown that

> *we may continuously deform the 3-dimensional group Sl(2, R) into the 1-dimensional subgroup of rotations of the plane, all the while keeping the submanifold $SO(2)$ pointwise fixed!*

This last group, described by an angle $\theta$, is topologically a circle $S^1$, which is connected. This shows that Sl(2, R) is connected. □

In fact we have proved much more. Suppose that $V^k$ is a submanifold of $M^n$. (In the preceding SO(2) $= V^1 \subset M^3 =$ Sl (2, $\mathbb{R}$).) Suppose further that $V$ is a **deformation retract** of $M$; that is, there is a continuous 1-parameter family of maps $r_t : M \to M$ having the properties that

1. $r_0$ is the identity,
2. $r_1$ maps all of $M$ into $V$
   and
3. each $r_t$ is the identity on V.

Then, considering homology with any coefficient group, we have the homomorphism $r_{1*} : H_p(M; G) \to H_p(V; G)$, since $r$ will send cycles into cycles, and so on; see (13.17). If $z_p$ is a cycle on $M$ and if $r_1(z_p)$ bounds in $V$, then $z_p$ bounds in $M$ since under the deformation, $z_p$ is homologous to $r_t(z_p)$; see the deformation lemma (13.21). Thus $r_{1*}$ is 1 : 1. Furthermore, any cycle $z'_p$ of $V$ is in the image of $r_{1*}$ since $z'_p = r_1(z'_p)$. Thus $r_{1*}$ is also onto, and hence

**Theorem (15.28):** *If $V \subset M$ is a deformation retract, then $V$ and $M$ have isomorphic homology groups*

$$H_p(M; G) \approx H_p(V; G)$$

Since $SO(2)$ is topologically a circle $S^1$, we have

**Corollary (15.29):** $H_0(Sl(2, \mathbb{R}), \mathbb{Z}) \approx \mathbb{Z} \approx H_1(Sl(2, \mathbb{R}), \mathbb{Z})$ *and all other homology groups vanish.*

## Problems

**15.3(1)** Prove (15.24).

**15.3(2)** Let $A$ be real, $2 \times 2$, *with trace* 0. The Cayley–Hamilton theorem for a $2 \times 2$ matrix says that $A$ satisfies its own characteristic equation

$$A^2 - (\mathrm{tr}\, A)A + (\det A)I = 0$$

hence

$$A^2 = -\rho I \qquad \rho := \det A$$

(The proof of the Cayley–Hamilton theorem for a $2 \times 2$ matrix can be done by direct calculation. One can also verify it in the case of a diagonal matrix, which is trivial, and then invoke the fact that the matrices that can be diagonalized are "dense" in the set of all matrices, since matrices generically have distinct eigenvalues.) Show that

$$e^A = \begin{cases} (\cos\sqrt{\rho})I + (\sqrt{\rho})^{-1}(\sin\sqrt{\rho})A & \text{if } \rho > 0 \\ (\cosh\sqrt{|\rho|})I + (\sqrt{|\rho|})^{-1}(\sinh\sqrt{|\rho|})A & \text{if } \rho < 0 \end{cases}$$

and, of course, $e^A = I + A$ if $\rho = 0$. Conclude then that

$$tre^A \geq -2$$

Thus, in particular

$$g = \begin{bmatrix} -2 & 0 \\ 0 & -\frac{1}{2} \end{bmatrix}$$

is never of the form $e^A$ for $A \in \mathfrak{sl}(2, \mathbb{R})$. In particular, *this g does not lie on any 1-parameter subgroup of* Sl(2, R).

**15.3(3)** (i) Does Sl $(n, \mathbb{R})$ have an interesting deformation retract? Is Sl $(n, \mathbb{R})$ connected?

(ii) What are the integer homology groups of the 8-dimensional manifold Sl $(3, \mathbb{R})$?

(iii) What can we say about Gl $(n, \mathbb{R})$? Is it connected?

# 15.4. Subgroups and Subalgebras

How can one find subgroups of $G$ by looking at $\mathfrak{g}$?

## 15.4a. Left Invariant Fields Generate Right Translations

Let $\mathbf{X}$ be a left invariant vector field on the Lie group $G$. If $\mathbf{X}_e$ is the value of $\mathbf{X}$ at $e$, then

$$\exp(t\mathbf{X}_e)$$

is the 1-parameter subgroup generated by $\mathbf{X}_e$. We know that this curve is the integral curve of the field $\mathbf{X}$ that starts at the identity $e$. Since $\mathbf{X}$ is left invariant, the integral curve that starts at the generic point $g \in G$ must be the curve $g(t) := L_g \exp(t\mathbf{X}_e) = g \exp(t\mathbf{X}_e)$.

On the other hand, $\mathbf{X}$, as a vector field on a manifold $G$, generates a *flow* $\phi_t : G \to G$ (at least if $t$ is small enough), whose velocity field is again $\mathbf{X}$. Thus it must be that $\phi_t(g) = g \exp(t\mathbf{X}_e)$. Hence

**Theorem (15.30):** *The flow generated by the left invariant field X is the 1-parameter group of right translations*

$$\phi_t(g) = g \exp(t\mathbf{X}_e)$$

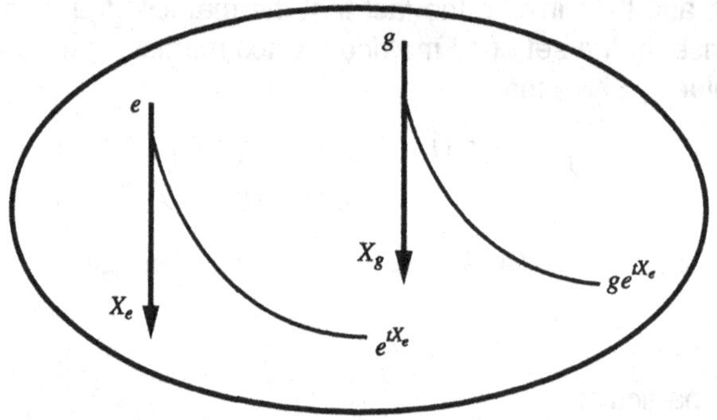


**Figure 15.5**

Since a *right* invariant vector field $\mathbf{Y}$ is then automatically invariant under the flow generated by a left invariant field $\mathbf{X}$, we conclude that their bracket vanishes

$$[\mathbf{X}^{\text{left}}, \mathbf{Y}^{\text{right}}] = 0 \tag{15.31}$$

Of course, by the same reasoning, right invariant fields generate left translations.

## 15.4b. Commutators of Matrices

Recall that the Lie algebra $\mathfrak{g}$, as a vector space, is simply the tangent space to $G$ at $e$, but as an algebra it is identified with the *left* invariant vector fields on $G$. (Of course this is merely a convention; we could have used right invariant fields just as well.) If $\mathbf{X} \in \mathfrak{g}$ and $\mathbf{Y} \in \mathfrak{g}$, then their Lie bracket

$$[\mathbf{X}, \mathbf{Y}] = \mathcal{L}_{\mathbf{X}}\mathbf{Y} \in \mathfrak{g}$$

is given by the Lie derivative, or, as first-order differential operators

$$[\mathbf{X}, \mathbf{Y}](f) = \mathbf{X}(\mathbf{Y}f) - \mathbf{Y}(\mathbf{X}f)$$

associated with the left invariant fields $\mathbf{X}$ and $\mathbf{Y}$.

If $G$ is a *matrix* group, each $\mathbf{X} \in \mathfrak{g}$ is itself a matrix (not in $G$ but rather in the tangent space to $G$ at $e$). For example, we have seen that if $G = SO(n)$ then $\mathbf{X}$ is a skew symmetric matrix. We claim then that $[\mathbf{X}, \mathbf{Y}]$ *is merely the commutator product of the matrices*

$$[\mathbf{X}, \mathbf{Y}] = XY - YX \tag{15.32}$$

To see this we use Theorem (4.12). We have, at $e = I \in \mathrm{Gl}(n, \mathbb{R})$

$$[\mathbf{X}, \mathbf{Y}] = \lim_{t \to 0} \frac{\{\phi^Y_{-t} \circ \phi^X_{-t} \circ \phi^Y_t \circ \phi^X_t(I) - (I)\}}{t^2}$$

where $\phi^{\mathbf{X}}_t$ refers to the flow generated by **X**, and so on. Since **X** and **Y** are left invariant, their flows are *right* translations,

$$\phi^{\mathbf{X}}_t(g) = g \ \exp(t\mathbf{X})$$

Thus

$$[\mathbf{X}, \mathbf{Y}] = \lim_{t \to 0} \frac{\{\exp(t\mathbf{X}) \exp(t\mathbf{Y}) \exp(-t\mathbf{X}) \exp(-t\mathbf{Y}) - I\}}{t^2} \tag{15.33}$$

In Problem 15.4(1) you are asked to show that this indeed does reduce to the commutator of the matrices.

This shows, for example, that if $X$ and $Y$ are skew symmetric matrices then so is $XY - YX$.

## 15.4c. Right Invariant Fields

All that we have said about left invariant fields can be redone for right invariant ones. Right invariant fields ("right fields" for short) generate left translations. We have defined the Lie algebra $\mathfrak{g}$ to be essentially the vector space of left fields, and then

$$[\mathbf{X}_i, \mathbf{X}_j] = \mathbf{X}_k C^k_{ij}$$

What would this become if we had used right fields instead?

Let $\{\mathbf{X}_j(e)\}$ be a basis for $G_e$ and extend them to left fields $\{\mathbf{X}_j(g)\}$ on all of $G$,

$$\mathbf{X}_i(g) = L_{g*}\mathbf{X}_i(e)$$

Let $\{\mathbf{Y}_i(e)\}$ coincide with the **X**'s at $e$ and extend them to *right* fields on $G$,

$$\mathbf{Y}_i(g) = R_{g*}\mathbf{Y}_i(e) = R_{g*}\mathbf{X}_i(e)$$

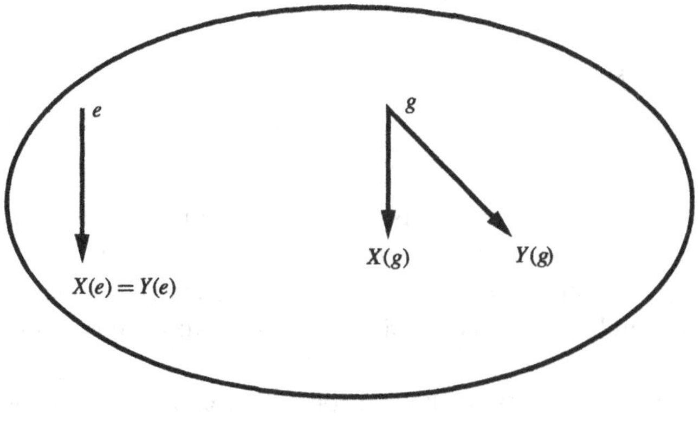


Figure 15.6

We are interested in the "right" structure constants

$$[\mathbf{Y}_i, \mathbf{Y}_j] = \mathbf{Y}_k D^k_{ij}$$

We calculate these for a matrix group, though the result holds in general.

The flow generated by $\mathbf{Y}_i$ consists of left translations. Repeating the steps going into Problem 15.4(1), but using right fields $\mathbf{Y}$, we see

$$[\mathbf{Y}_1, \mathbf{Y}_2]_{\text{right}} = Y_2 Y_1 - Y_1 Y_2$$

as matrices. We conclude (since $\mathbf{Y} = \mathbf{X}$ at $e$)

$$[\mathbf{Y}_i, \mathbf{Y}_j] = -\mathbf{Y}_k C^k_{ij}$$

and the *right structure constants are merely the negatives of the left*!

By "Lie algebra" we shall always mean the algebra of *left* invariant fields.

### 15.4d. Subgroups and Subalgebras

We are interested in subgroups of a Lie group. (We have already discussed 1-parameter subgroups.) For example SO($n$) is a subgroup

$$\mathrm{SO}(n) \subset \mathrm{Gl}\,(n, \mathbb{R})$$

of the general linear group and it is an embedded submanifold (we showed this in Section 1.1d). For a subgroup $H \subset G$ to qualify as a **Lie subgroup** we shall demand that $H$, if not embedded, is at least an *immersed* submanifold. The 2-torus, consisting of points

$$(e^{i\theta}, e^{i\phi}) \in S^1 \times S^1$$

is a 2-dimensional abelian group

$$(e^{i\theta}, e^{i\phi}) \circ (e^{i\alpha}, e^{i\beta}) = (e^{i(\theta+\alpha)}, e^{i(\phi+\beta)})$$

with a 1-parameter subgroup

$$H = (e^{irt}, e^{ist})$$

where $r$ and $s$ are real numbers. As discussed in Section 6.2a, if $s/r$ is irrational this curve winds *densely* on the torus; thus $H$ in this case is an immersed, not embedded submanifold. This is not a *closed* subset of the torus since its closure (obtained by adjoining its accumulation points) would be the entire torus, but it still qualifies as a Lie subgroup.

The tangent space $\mathfrak{gl}(n, \mathbb{R})$ to Gl($n, \mathbb{R}$) consists of all $n \times n$ matrices, whereas the tangent space $\mathfrak{so}(n)$ consists of skew symmetric $n \times n$ matrices.

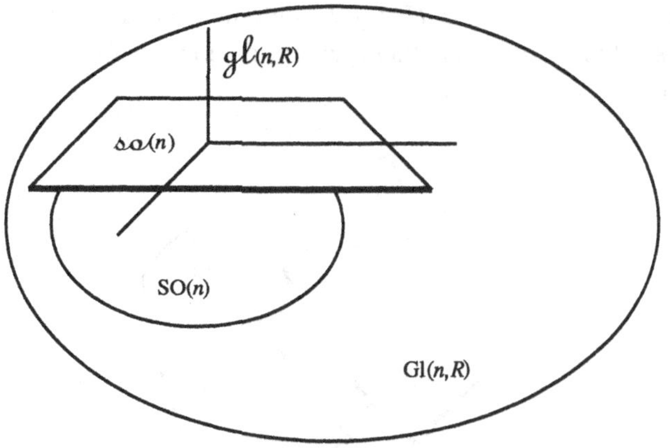


**Figure 15.7**

Let **X** and **Y** be skew symmetric matrices. Left translate them over all of $Gl(n, \mathbb{R})$. Since the resulting vector fields are tangent at $g \in SO(n)$ to $SO(n)$, so is their bracket [**X**, **Y**]. In particular $[\mathbf{X}, \mathbf{Y}]_e \in \mathfrak{so}(n)$. This says that $\mathfrak{so}(n)$ is not only a vector subspace of $\mathfrak{gl}(n, \mathbb{R})$, it is a **subalgebra**.

In general, if $H$ is a subgroup of $G$, then the Lie algebra $\mathfrak{h}$ of $H$ is a subalgebra of $\mathfrak{g}$. The converse of this is also true and of immense importance.

**Theorem (15.34):** *Let $G$ be a Lie group with Lie algebra $\mathfrak{g}$. Let $\mathfrak{h} \subset \mathfrak{g}$ be a vector subspace of $\mathfrak{g}$ that is also a subalgebra*

$$[\mathfrak{h}, \mathfrak{h}] \subset \mathfrak{h}.$$

*Then there is a subgroup $H \subset G$ whose Lie algebra is the given $\mathfrak{h} \subset \mathfrak{g}$.*

**Example:** For any $n \times n$ real matrices $X$ and $Y$ their commutator $XY - YX$ has trace 0. Thus the traceless $n \times n$ matrices form a subalgebra of $\mathfrak{gl}(n, \mathbb{R})$ and there is a corresponding subgroup; it is, of course, $Sl(n, \mathbb{R})$.

**PROOF:** Given the vector subspace $\mathfrak{h} \subset \mathfrak{g}$, left translate $\mathfrak{h}$ over all of $G$, yielding a distribution $\Delta$. Let $\mathbf{X}_1, \ldots, \mathbf{X}_r$ be left invariant fields spanning $\Delta$ everywhere. Since $\mathfrak{h}$ is a *subalgebra*

$$[\mathbf{X}_i, \mathbf{X}_j] \in \Delta$$

Thus $\Delta$ is in involution and is then completely integrable by the theorem of Frobenius. From Chevalley's theorem (6.6), we can construct the "maximal leaf" of this foliation passing through the identity; that is, there is a manifold $V^r$ and a 1 : 1 immersion $F : V^r \to G$ such that $H := F(V)$ is always tangent to the distribution $\Delta$ and passes through $e \in G$. We claim that $H$ is a subgroup of $G$; that is, *$H$ is closed under the $G$ operations of multiplication and taking inverse.*

Let $h_1$ and $h_2$ be in the leaf $H$. By the definition of $\Delta$, left translation of $H$ by $h_1$ must send the leaf into another (perhaps distinct) leaf $h_1H$ of the foliation, $h_1h_2 \in h_1H$.

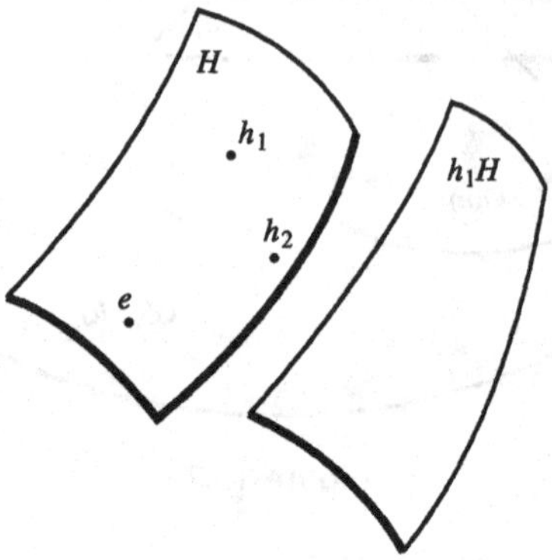


Figure 15.8

However $h_1e = h_1$ shows that $h_1$ is in both leaves $H$ and $h_1H$ and since $H$ is maximal it must be that $H = h_1H$. In particular $h_1h_2 \in H$, as desired. A similar argument (Problem 15.4(2)) shows $H$ is closed under taking inverses. □

## Problems

**15.4(1)** Use (15.33) and (15.17) to show $[X, Y] = XY - YX$ as matrices. (You needn't justify (legitimate!) manipulations with infinite series.)

**15.4(2)** Show that $H$ is closed under taking inverses.

**15.4(3)** Show that the skew hermitian $n \times n$ matrices ($A^\dagger = -A$) with trace 0 form a subalgebra of $\mathfrak{gl}(n, \mathbb{C})$. Identify the subgroup. Is there a group whose Lie algebra consists exactly of the hermitian matrices?

## CHAPTER 16

# Vector Bundles in Geometry and Physics

On the Earth's surface, the number of peaks minus the number of passes plus the number of pits is generically 2.

## 16.1. Vector Bundles

What is a "twisted product"?

### 16.1a. Motivation by Two Examples

1. *Vector fields on M.* A section of the tangent bundle $TM^n$ to $M^n$ is simply a vector field $\mathbf{w}$ on $M$. *Locally,* that is, in a coordinate patch $(U; u^1, \ldots, u^n)$, $\mathbf{w}$ is given by its component functions $w_U^1(u), \ldots, w_U^n(u)$ with respect to the coordinate basis $\boldsymbol{\partial}/\boldsymbol{\partial u}$, but of course these functions are defined only on $U$, not all of $M$. In another patch $V$, the same field is described by another $n$-tuple $w_V^1(v), \ldots, w_V^n(v)$. At a point $p$ in the overlap $U \cap V$ these two $n$-tuples are related by

$$w_V^i(p) = [c_{VU}(p)]^i_j w_U^j(p)$$

where $c_{VU} = \partial v/\partial u$ is the Jacobian matrix. Thus a section of $TM$ serves as a generalization of the ordinary notion of an $n$-tuple of functions $F = (f^1, \ldots, f^n) : M^n \to \mathbb{R}^n$ defined on an $n$-manifold, where now we assign a different $n$-tuple of functions in each patch, but we insist on a recipe telling us when two $n$-tuples are describing the same "vector" at a point common to two patches. The bundle $TM$ is, in a sense, the home in which all the sections live.

Not all $n$-tuples are to be considered as tangent vectors, for there are other bundles "over" $M$. The cotangent bundle $T^*M$ uses a different recipe; its $c_{VU}$ is $[\partial u/\partial v]^T$.

2. *The normal bundle to the midcircle of the Möbius band.*

Consider the Möbius band Mö$^2$ (a 2-manifold whose boundary is a single closed curve) and the midcircle submanifold $M^1 = S^1$. We are interested in the collection of all tangent vectors to Mö *along* $S^1$ that are normal to $S^1$. We shall call this collection the **normal bundle** $N(S^1)$ to $S^1$ in Mö. Clearly we have a map

$$\pi : N(S^1) \to S^1$$

that sends each normal vector to the point in $S^1$ where it is based.

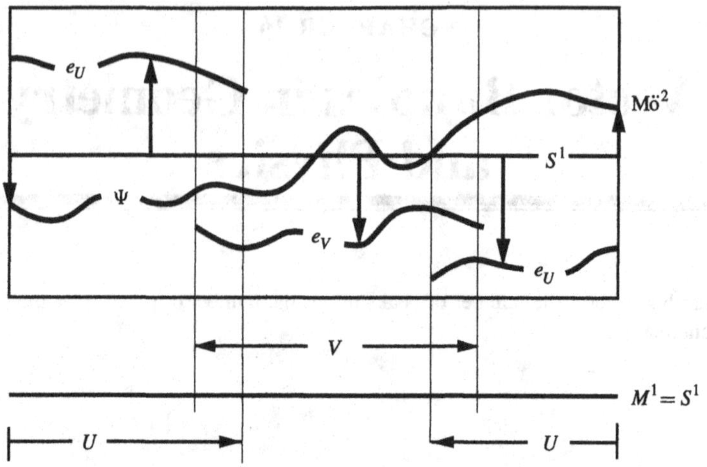


**Figure 16.1**

It should be clear that we cannot find a continuous normal vector field to $S^1$ that is *everywhere nonzero*, since if it points down at the left endpoint it must point up at the right endpoint because of our identifications. We have illustrated this with the normal field $\mathbf{\Psi}$. If we wished to describe this field by a "component" $\Psi$ we might proceed as follows.

Select (arbitrarily) smooth nonvanishing normal vector fields $\mathbf{e}_U$ and $\mathbf{e}_V$ over patches $U$ and $V$ of $S^1$, $U$ and $V$ being chosen so that their union is all of $M$. Then at any point $p \in U \cap V$ we have

$$\mathbf{e}_V(p) = \mathbf{e}_U(p) c_{UV}(p)$$

where $c_{UV}$ is a smooth *nonvanishing* $1 \times 1$ matrix defined in $U \cap V$. Note that this is the same notation that we used when talking about the tangent bundle; see equation (9.44). Also note that we may describe the nonvanishing of the "matrix" $c_{UV}$ as saying that

$$c_{UV} : U \cap V \to \mathrm{Gl}(1, \mathbb{R})$$

Let $\mathbf{\Psi}$ be a smooth normal field to $S^1$. Then in $U$ we have $\mathbf{\Psi}(p) = \mathbf{e}_U(p)\psi_U(p)$ and in $V$, $\mathbf{\Psi}(p) = \mathbf{e}_V(p)\psi_V(p)$, for smooth functions $\psi_U$ and $\psi_V$. In the overlap

$$\mathbf{\Psi}(p) = \mathbf{e}_U(p)\psi_U(p) = \mathbf{e}_V(p)\psi_V(p)$$

and so

$$\psi_V(p) = c_{VU}(p)\psi_U(p)$$

where $c_{VU} = c_{UV}^{-1}$. Thus a normal vector field to $S^1 \subset \text{Mö}^2$ is described not by a single "component" function $\psi$ on $S^1$, but by a component function $\psi_U$ in $U$ and by a component function $\psi_V$ in $V$, both related by the transition matrix $c_{VU}$.

Note that the local fields $\mathbf{e}_U$ and $\mathbf{e}_V$ allow us to say that the normal bundle $N(S^1)$ is **locally a product**, in the following sense. The part of the bundle consisting of normal vectors based in the patch $U$ is diffeomorphic to $U \times \mathbb{R}$ under the map $\Phi_U : U \times \mathbb{R} \to$

$N(S^1)$ defined by $\Phi_U(p, \psi) = \mathbf{e}_U(p)\psi$. Similarly $\Phi_V : V \times \mathbb{R} \to N(S^1)$ makes the part of $N(S^1)$ based in $V$ into a product.

Although $N(S^1)$ is *locally* a product, it is *globally twisted*, for the entire $N(S^1)$ is not itself a product $S^1 \times \mathbb{R}$. There is no continuous way to assign a unique normal vector to a pair $(p, \psi)$ for $\psi$ a fixed real number, as $p$ ranges over all of $S^1$. $N(S^1)$ is thus a **twisted product** of $S^1$ and $\mathbb{R}$.

If we were to consider the vectors normal to a curve $M^1$ in a Riemannian manifold $W^n$, we would have to find $(n-1)$ local normal fields $\mathbf{e}_1^U, \ldots, \mathbf{e}_{n-1}^U$ in each patch $U$ of $M^1$, and a normal field $\boldsymbol{\Psi}$ would then be described by an $(n-1)$-tuple of components $\psi_U^1, \ldots, \psi_U^{n-1}$. We shall consider this in Section 16.1d.

To generalize the notion of a $K$-tuple of functions on $M^n$ we introduce the general notion of a vector bundle over $M$.

### 16.1b. Vector Bundles

A (real or complex) rank $K$ **vector bundle** $E$ over a **base** manifold $M^n$ consists of a manifold $E$ (**the bundle space**) and a differentiable map, **projection**

$$\pi : E \to M$$

such that $E$ is a **local product space** in the following sense.

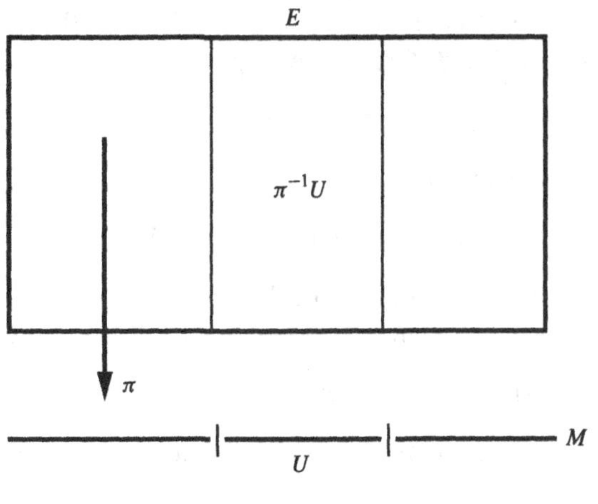


Figure 16.2

There is a covering of $M^n$ by open sets $\{U, V, \ldots\}$. There is a $K$-dimensional *vector space* (the **fiber**) $\mathbb{R}^K$ or $\mathbb{C}^K$ (and for definiteness we shall assume it to be $\mathbb{R}^K$) *equipped with its standard basis*

$$e_1, \ldots, e_K$$

We demand that for each open set $U$ in the covering

$$U \times \mathbb{R}^K \text{ is diffeomorphic to the “part of the bundle over } U\text{,” } \pi^{-1}U$$

that is, there are diffeomorphisms

$$\Phi_U : U \times \mathbb{R}^K \to \pi^{-1}U$$
$$\Phi_U(p, y) \in \pi^{-1}(p) \tag{16.1}$$

(In the case of the tangent bundle $E = TM^n$, if $\mathbf{e}^U$ is a frame in $U$ then $\Phi_U(p, y) = \sum_{1 \le i \le K} \mathbf{e}_i^U y^i$.) A point $s \in \pi^{-1}(U)$ then is represented, via $\Phi_U^{-1}$, by a point $p$ in $U$ and a $K$-tuple of real numbers $y$, the latter being the **fiber coordinates** of $s$. For $\pi(s) \in (U \cap V)$, we demand that the two sets of fiber coordinates be related by a nonsingular *linear* transformation

$$c_{VU}(p) : U \cap V \to \mathrm{Gl}(K)$$

that depends differentiably on $p$

$$y_V = c_{VU}(p) y_U$$

that is, (16.2)

$$y_V^i = c_{VU}(p)^i{}_j y_U^j$$

Note that each **fiber over $p$**, $\pi^{-1}(p)$, is a $K$-dimensional vector space but it is *not* identified with $\mathbb{R}^K$ *until* the patch $U$ holding $p$ is specified; only then can we use $\Phi_U^{-1}$ to make the identification. (*In the tangent bundle we can not read off the components of a vector until we have picked out a specific frame.*) Note also that in the name "rank $K$ vector bundle," $K$ refers to the dimension of the *fiber*, not the bundle space $E$.

A **(cross) section** of $E$ is a differentiable map

$$s : M \to E$$

such that $s(p)$ lies over $p$, that is,

$$\pi \circ s = \text{identity} : M \to M$$

Locally, over $U$, one describes a section $s$ by giving its vector components $y_U(p)$, subject to the requirement (16.2) in an overlap. In a triple overlap we have

$$y_W(p) = c_{WV}(p) y_V(p) = c_{WV}(p) c_{VU}(p) y_U(p)$$

and so $c_{WU} = c_{WV} c_{VU}$. Thus the **transition functions** $\{c_{VU}\}$ satisfy

$$c_{VU}(p) = c_{UV}(p)^{-1}$$

and (16.3)

$$c_{WV}(p) c_{VU}(p) c_{UW}(p) = I$$

Conversely, let $M$ be a manifold with a covering $\{U, \ldots\}$, and suppose that we are given matrix-valued functions $c_{VU}$ in each overlap

$$c_{VU} : U \cap V \to \mathrm{Gl}(K)$$

that satisfy (16.3). Then we may construct a vector bundle over $M$ whose transition functions are these $c_{VU}$ as follows. Take the *disjoint* collection of manifolds

$$\{U \times \mathbb{R}^K, V \times \mathbb{R}^K, \ldots\}$$

one for each patch. These are to be considered disjoint even though the patches can overlap. Now we make identifications:

$$(p, y_U) \in (U \times \mathbb{R}^K) \text{ is to be identified with}$$

$$(p', y_V) \in (V \times \mathbb{R}^K) \quad \text{iff } p' = p \quad \text{and} \quad y_V = c_{VU}(p) y_U$$

It can be shown that the resulting identification space $E$ is indeed a $K$-dimensional vector bundle over $M$ with $\{c_{VU}\}$ as transition matrices. This is the procedure we used for construction of the tangent bundle; from $X_V^i(x) = (\partial x_V^i / \partial x_U^j) X_U^j$ we see, from (16.2), that

$$c_{VU}(x) = \frac{\partial x_V}{\partial x_U} \tag{16.4}$$

Tangent bundle $T$

On the other hand, for the cotangent bundle, $a_i^V = a_j^U \partial x_U^j / \partial x_V^i = \sum_j [(\partial x_U / \partial x_V)^T]_{ij} a_j^U$ shows that

$$c_{VU}(x) = \left( \frac{\partial x_U}{\partial x_V} \right)^T = \left[ \left( \frac{\partial x_V}{\partial x_U} \right)^{-1} \right]^T \tag{16.5}$$

Cotangent bundle $T^*M$

Two bundles whose transition matrices are inverse transposes are said to be **dual** vector bundles.

If $E$ and $E'$ are vector bundles over the same base manifold $M$, then the **tensor product bundle** $E \otimes E'$ is defined to be the vector bundle with transition matrices $c_{VU} \otimes c'_{VU}$. This means the following. A point in $\pi^{-1}(U)$ has vector components $y_U = (y_U^1, \ldots, y_U^K)$, a point in $\pi'^{-1}(U)$ has vector components $z_U = (z_U^1, \ldots, z_U^L)$, and a point in the tensor product bundle has the $KL$ vector components

$$(y_U \otimes z_U)^{i\alpha} := y_U^i z_U^\alpha \tag{16.6}$$

and by definition

$$(c_{VU} \otimes c'_{VU})(y_U \otimes z_U) := (c_{VU} y_U) \otimes (c'_{VU} z_U)$$

For example, *the mixed tensors, once contravariant and once covariant (i.e., the linear transformations), form the vector bundle $TM \otimes T^*M$.*

### 16.1c. Local Trivializations

A bundle space $E$ is locally a product manifold. The diffeomorphisms $\Phi_U : U \times \mathbb{R}^K \to \pi^{-1}(U)$ that exhibit the local product structure allow one immediately to exhibit $K$ sections $\mathbf{e}_\alpha(p) := \Phi_U(p, \mathscr{e}_\alpha)$ over $U$, where again $\mathscr{e}_1 = (1, 0, \ldots, 0)^T$, and so on, and these sections are *linearly independent* in the sense that at each $p \in U$ the vectors $\mathbf{e}_\alpha(p)$ in the vector space $\pi^{-1}(p)$ (the fiber over $p$) are independent. The $\mathbf{e}_\alpha$ form a **frame** of sections.

Note that one frequently proceeds in the reverse direction. For example, we made the collection of vectors normal to the midcircle of the Möbius band into a rank 1 vector bundle by first picking out distinguished "sections"; this then defined the maps $\Phi$. In general, suppose that we have two manifolds $E$ and $M^n$ and a map $\pi : E \to M$ of $E$ onto $M$. Suppose that each $\pi^{-1}(p)$ is a vector space $\approx \mathbb{R}^K$. Suppose further that there is a covering $\{U, V, \ldots\}$ of $M$ and there are smooth maps $\mathbf{e}_\alpha^U : U \to E, \alpha = 1, \ldots, K$ such that $\pi \circ \mathbf{e}_\alpha^U$ is the identity map on $U$ and the $\mathbf{e}_\alpha(p)$ are independent for each $p \in U$. Define then $\Phi_U : U \times \mathbb{R}^K \to \pi^{-1}(U)$ by $\Phi_U(p, e_\alpha y^\alpha) = \mathbf{e}_\alpha(p) y^\alpha$. By construction, each $\Phi_U$ is a diffeomorphism that is linear on the "fiber" $\mathbb{R}^K$ for $p$ fixed. Then in an overlap $U \cap V$ we may define $c_{VU}(x) : \mathbb{R}^K \to \mathbb{R}^K$ by the linear map

$$y_V = c_{VU}(p) y_U := \Phi_V^{-1} \circ \Phi_U(p, y_U)$$

(16.2) is then automatically satisfied and we have made $E$ into a vector bundle over $M^n$ and the $\mathbf{e}_\alpha^U$ yield a frame of sections over $U$. □

We shall frequently denote a point of $M$ by $x$, rather than $p$; we are *not* implying that $x$ is a local coordinate, *though that will often be the case*. The most general cross section over $U$ is then of the form

$$\boldsymbol{\Psi} = \mathbf{e}_\alpha^U \psi_U^\alpha(x)$$

where the $\psi_U^\alpha(x)$ are component *functions*. We abbreviate this with matrix notation

$$\boldsymbol{\Psi}(x) = \mathbf{e}^U(x) \psi_U(x)$$

$$\mathbf{e}^U(x) = (\mathbf{e}_1^U(x), \ldots, \mathbf{e}_K^U(x))$$

$$\psi_U(x) = \begin{bmatrix} \psi_U^1(x) \\ \vdots \\ \psi_U^K(x) \end{bmatrix}$$

If $\boldsymbol{\Psi}$ is a cross section over $U \cap V$, then in $U \cap V$ we have

$$\boldsymbol{\Psi}(x) = \mathbf{e}^U(x) \psi_U(x) = \mathbf{e}^V(x) \psi_V(x) \tag{16.7}$$

$$\psi_V(x) = c_{VU}(x) \psi_U(x)$$

If we can find a frame $\mathbf{e}$ of sections over *all* of $M$, we say that the bundle is a **product bundle**, or is **trivial**. In this case

$$\Phi(x; \psi^1, \ldots, \psi^K) = \sum_{\alpha=1}^{k} \mathbf{e}_\alpha(x) \psi^\alpha$$

yields a *diffeomorphism*

$$\Phi : M \times \mathbb{R}^K \to E$$

making $E$ *globally* a product manifold. In particular,

*a 1-dimensional vector bundle (a **line** bundle) with a single nonvanishing global section is a trivial bundle.*

In a nontrivial bundle, the maps $\Phi_U : U \times \mathbb{R}^K \to \pi^{-1}(U)$ make the *portion* of the bundle over $U$ into a trivial bundle; each $\Phi_U$ is thus called a **local trivialization.**

We shall see in Section 16.2 that the tangent bundle to the 2-sphere $TS^2$ does not even possess a single nonvanishing section and so *$TS^2$ is not trivial.* On the other hand, the tangent bundle $TG$ to a Lie group has a frame of global sections given by left translating a basis of $\mathfrak{g}$ over all of $G$; thus *the tangent bundle to a Lie group is trivial*! (**Remark:** If the *tangent* bundle to a manifold $M$ is trivial, we say that $M$ is **parallelizable**.)

Note that every vector bundle $E$ has a global section, the zero section, defined locally in each $U$ by $\psi^1(x) = 0, \ldots, \psi^K(x) = 0$. In Problem 16.1(1) you are asked to give the 1-line proof.

### 16.1d. The Normal Bundle to a Submanifold

Consider a Riemannian manifold $V^{n+K}$ and a submanifold $M^n \subset V$. We define the **normal bundle** $N(M)$ to $M$ in $V$ to consist of those tangent vectors to $V$ that are based on $M$ and are orthogonal to $M$.

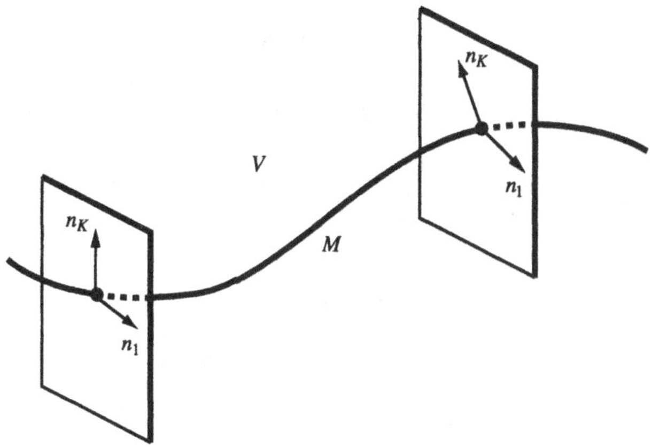


Figure 16.3

(In the figure, $M$ is drawn as a curve.) It should be "clear" that if $U \subset M$ is small enough one can find $K$ smooth fields $\mathbf{n}_1^U, \ldots, \mathbf{n}_K^U$ of *normal* vectors to $M$ that are linearly independent at each point of $U$. Then, if

$$\pi : E = N(M^n) \to M$$

denotes the normal bundle

$$\Phi : U \times \mathbb{R}^K \to \pi^{-1}(U)$$

is again defined by

$$\Phi(x : \lambda^1, \ldots, \lambda^K) = \sum_{\alpha=1,\ldots,K} \mathbf{n}_\alpha^U(x)\lambda^\alpha$$

The $\lambda$'s are the components of a normal vector in the patch $U$. In a patch $V$ we would have a new frame $\{\mathbf{n}_\alpha^V\}$ and in an overlap $U \cap V$ the frames would be related by a $K \times K$ matrix function $\mathbf{n}^V = \mathbf{n}^U c_{UV}$, and a normal vector would have two sets of components $\lambda_U$ and $\lambda_V$ related by $\lambda_V = c_{VU}\lambda_U$, where $c_{VU}(x) = c_{UV}^{-1}(x) \in Gl(K, \mathbb{R})$. If we had chosen the frames $\mathbf{n}^U$ and $\mathbf{n}^V$ to be orthonormal, then $c_{UV}(x) \in O(K)$.

For example, the normal line bundle to the 2-sphere $M^2 = S^2 \subset \mathbb{R}^3 = V^3$ is trivial, $N(S^2) = S^2 \times \mathbb{R}$, since we have a global nonvanishing section given by the outward-pointing unit normal.

As we have seen, the normal bundle to the central circle $M = S^1$ of the Möbius band $V^2$ is not trivial.

The normal bundle $N(S^1)$ to the indicated circle $S^1 \subset \mathbb{R}P^2$ is clearly itself an infinite

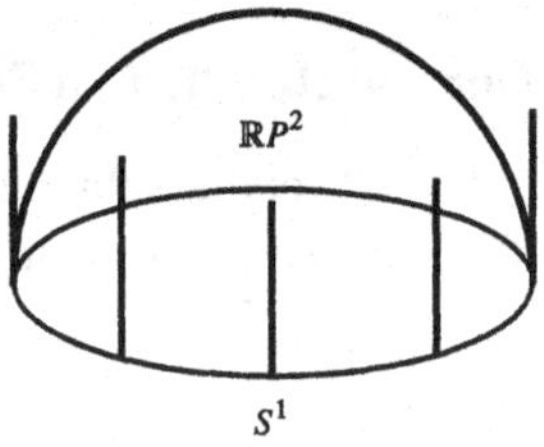


**Figure 16.4**

Möbius band (the lengths of the vectors are not bounded). For this $S^1 \subset \mathbb{R}P^2$, $N(S^1)$ is not trivial.

If we use as model of $\mathbb{R}P^2$ the disc with antipodal points identified, this $S^1$ can be deformed into a diameter. $N(S^1)$ is not trivial.

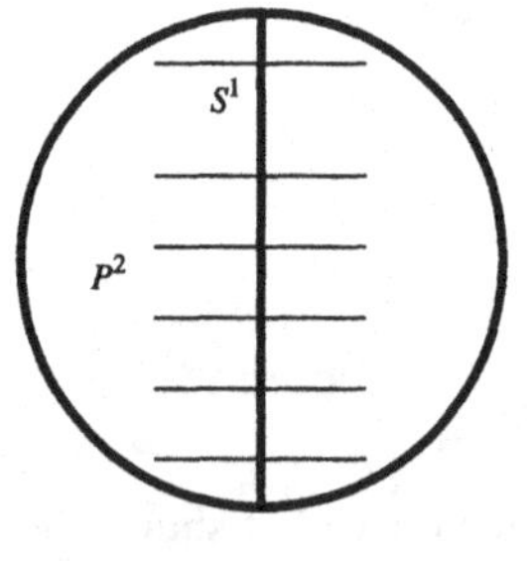


**Figure 16.5**

Let $C$, $\mathbf{x} = \mathbf{x}(t)$, $0 \le t \le 1$, be a *closed* curve in $\mathbb{R}^3$. Its normal bundle is a rank-2 vector bundle over $C$. Pick an orthonormal frame $\mathbf{n} = \mathbf{n}(0)$ of two normal vectors $\mathbf{n}_\alpha$ at $p = \mathbf{x}(0)$. Transport this frame continuously around all of $C$, always remaining orthonormal and orthogonal to $C$, arriving at $p = \mathbf{x}(1)$ but with perhaps a different frame $\mathbf{n}(1)$ from the original. Since $\mathbb{R}^3$ is orientable, and since the tangent $\mathbf{T}$ has returned to itself, it must be that $\mathbf{n}(0)$ and $\mathbf{n}(1)$ define the same orientation in the normal plane at $p$. This means that $\mathbf{n}(0)$ and $\mathbf{n}(1)$ are related by an $SO(2)$ matrix $g$, $\mathbf{n}(0) = \mathbf{n}(1)g$. We are now going to redefine the normal framing along the last $\epsilon$ seconds of the curve so that the framings match up at $t = 0$ and $t = 1$. Since $SO(2)$ is connected, we can find a curve of $2 \times 2$ matrices $g = g(s)$, $1 - \epsilon \le s \le 1$, in $SO(2)$, such that

$g(1-\epsilon) = I$ and $g(1) = g$. Now redefine the normal frame on the last part of $C$ by putting $\mathbf{m}(s) = \mathbf{n}(s)g(s)$, yielding a framing with agreement at $t = 0$ and $t = 1$. (By choosing the curve $g(s)$ to have $s$-derivative 0 at $s = 1 - \epsilon$ and at $s = 1$ we can even make the framing smooth.) *The normal bundle to a closed curve in* $\mathbb{R}^3$ *is trivial*!

## Problems

**16.1(1)** Show that the zero section is indeed always a section.

**16.1(2)** $\mathbb{R}P^3$ is the solid ball with boundary points identified antipodally. Is the normal bundle to the circle $S^1 \subset \mathbb{R}P^3$ trivial?

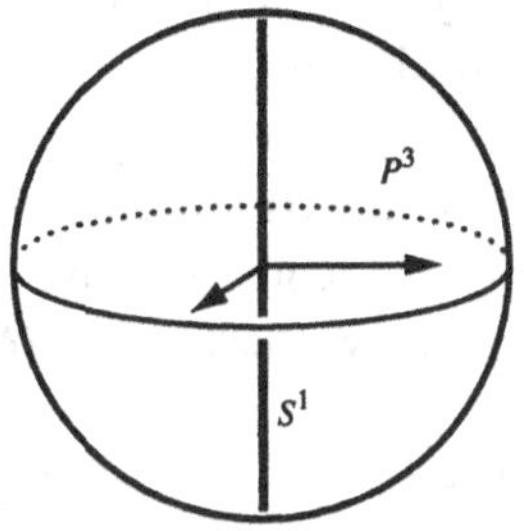


**Figure 16.6**

**16.1(3)** Is the normal bundle to $\mathbb{R}P^2$ in $\mathbb{R}P^3$ trivial?

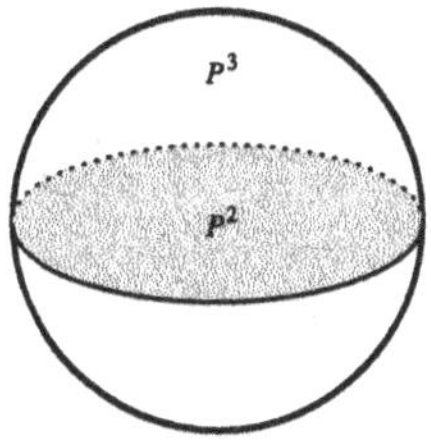


**Figure 16.7**

**16.1(4)** Is the normal bundle to a closed curve in an $M^n$ trivial? (Consider the cases $M$ orientable and $M$ not orientable.)

## 16.2. Poincaré's Theorem and the Euler Characteristic

Can you comb the hair on a sphere so that the directions vary smoothly and such that no hair sticks straight out radially?

Before discussing further properties of general vector bundles, we shall acquaint ourselves with the most important result on the sections of the tangent bundle to a surface.

For further discussion the reader may consult Arnold's book on differential equations [A2, chap. 5].

### 16.2a. Poincaré's Theorem

Let $M^2$ be a closed (compact without boundary) surface and let $\mathbf{v}$ be a *tangent* vector field to $M$ having at most a finite number of points $p$ where the vector field vanishes, $\mathbf{v}(p) = 0$. *Generically* this is so for the following reasons. The vanishing of a vector field requires, locally, the simultaneous vanishing of two functions $v^1$ and $v^2$ of the two coordinate variables $x$ and $y$, and generically these two zero sets intersect in isolated points. Compactness (as in the proof of Theorem (8.17)) then demands that there be only a finite number of zeros.

Let $p$ be a zero for $\mathbf{v}$. We may assume that $p$ is the origin of a local coordinate system $x, y$. Let $S$ be a small coordinate circle, $x^2 + y^2 = \epsilon^2$, where by "small" we mean that $p$ is the only zero inside $S$. Introduce a *Riemannian metric* in the coordinate patch. For example you *may* wish to use $ds^2 = dx^2 + dy^2$. We may *orient* the patch by demanding that $x, y$ be a positively oriented system. We may then consider the angle that $\mathbf{v}$ makes with the first coordinate vector $\partial_x = \partial/\partial x$ at each point $(x, y)$ on $S$

$$\theta(x, y) = \measuredangle(\partial_x, \mathbf{v}) := \cos^{-1}\left\{\frac{\langle \partial_x, \mathbf{v}\rangle}{\| \partial_x \| \| \mathbf{v} \|}\right\}$$

We then have the following situation. Let $S_0$ be a unit circle in an (abstract) $\mathbb{R}^2$; $S_0$ is parameterized by an angle $\phi$. We then have a map $S \to S_0$ defined by $\phi(x, y) =$ the preceding angle $\theta(x, y)$. This map has a Brouwer degree, called, as in Section 8.3d, the (Kronecker) index of $\mathbf{v}$ at the zero $p$, written $j_{\mathbf{v}}(p) = j(p)$. Of course it simply represents the *number of times that* $\mathbf{v}$ *rotates as the base of* $\mathbf{v}$ *moves around the circle* $S^1$. As such

$$j(p) = \frac{1}{2\pi}\int_S d\theta(x, y) \tag{16.8}$$

In Section 8.3d we have illustrated the indices of four vector fields at the origin of $M^2 = \mathbb{R}^2$.

We have made several rather arbitrary choices in the previous procedure, a Riemannian metric, a coordinate system, and a closed curve $S$ in the patch enclosing the zero. But the index varies continuously with the choices, and since it is an integer, it is in fact independent of the choices.

In particular we may replace the circle $S$ by a piecewise smooth triangle enclosing the zero.

Note that we may compute the index even when the field $\mathbf{v}$ does not vanish inside the curve $S$, but the index will then be 0; see Problem 8.3(9).

Finally, note that we may also consider a vector field that is smooth in a region except for an isolated "singular" point $p$; for example, the electric field grad$(1/r)$ of a charge in $\mathbb{R}^3$ is smooth everywhere except at the charge. By the same procedure as at a zero, we may again define the index $j_{\mathbf{v}}(p)$ of the vector field at the singularity.

By a **singularity** of a vector field $\mathbf{v}$ we shall mean any point at which $\mathbf{v}$ is not smooth *or* at which $\mathbf{v} = \mathbf{0}$.

A zero of a smooth vector field is not a singularity in the ordinary sense. In our present situation it is called a singularity because the *direction* field defined by the vector is undefined at a zero.

**Poincaré's Theorem (16.9):** *Let* $\mathbf{v}$ *be a vector field with perhaps a finite number of singularities on a closed surface* $M^2$. *Then the sum of the indices of* $\mathbf{v}$ *at the singular points*

$$\sum_p j_{\mathbf{v}}(p) = \chi_{\mathbf{v}}(M)$$

*is in fact independent of the vector field and is a topological invariant* $\chi$.

For reasons discussed in the next section, $\chi$ will be called the Euler characteristic.

Before looking at the proof, let us look at some examples on the 2-sphere. The vector field $\partial/\partial\theta$ tangent to the lines of longitude on the 2-sphere has a singularity at the north and south poles. At the north pole the field looks like the "source" in Section 8.3d of index 1 while the south pole is a "sink," also of index 1. Thus $\chi(S^2) = 2$ in this case. We can also consider the vector field $\partial/\partial\phi$ tangent to the parallels of latitude, again with singularities at the poles. The indices are easily seen again to be both $+1$, verifying the theorem. Poincaré's theorem implies the following, which we have mentioned many times in the past:

**Corollary (16.10):** *Every vector field on* $S^2$ *has a singularity. Thus every smooth section of the tangent bundle of the 2-sphere must be zero somewhere, and hence this bundle is not a product bundle.*

This has been paraphrased as "You can't comb the hair on a 2-sphere."

In our two fields $\partial/\partial\theta$ and $\partial/\partial\phi$ on $S^2$, both fields had two singularities. We shall now exhibit a field on $S^2$ with a single singularity (zero) with, of course, index $+2$.

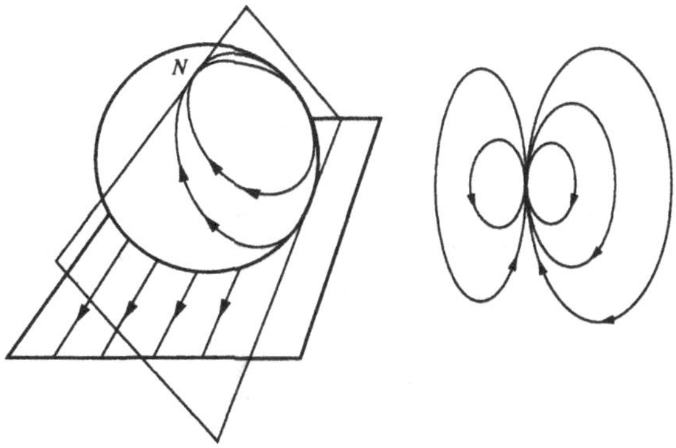


Figure 16.8

This field is obtained from a parallel field $\partial/\partial u$ in the $u$, $v$ plane by stereographically projecting (from the north pole) the field onto the tangent sphere. We have drawn the integral curves rather than the vector field itself. At the right of the figure we have shown a view from the top, and one easily sees that the index at the north pole is indeed +2. We can investigate this analytically as follows.

Consider the sphere as the Riemann sphere, as in Section 1.2d. In the complex $w$ plane $\mathbb{C}$ tangent to the sphere at the south pole, we have the velocity field of the flow $dw/dt = 1$, that is, $du/dt = 1$ and $dv/dt = 0$. When we stereographically project this flow onto the Riemann sphere we get the parallel-like flow near the south pole $w = 0$. Near the north pole $z = 0$ we get

$$\frac{dz}{dt} = \left(\frac{dz}{dw}\right)\left(\frac{dw}{dt}\right) = -\left(\frac{1}{w^2}\right) = -z^2$$

As we go around the path $z = e^{i\theta}$ about $z = 0$, the vector $-z^2 = -e^{2i\theta}$ makes 2 circuits, yielding the desired index 2.

**PROOF OF POINCARÉ'S THEOREM:** The following proof is due to Heinz Hopf, who also proved the higher-dimensional version. We shall discuss this in Section 16.2c.

We shall first prove the theorem in the case when $M^2$ is *orientable*; in the following section we shall then discuss briefly the nonorientable case.

Choose any Riemannian metric for all of $M^2$ (see Section 3.2d).

Let $\mathbf{v}$ and $\mathbf{w}$ be two vector fields on $M$, each having a finite number of singularities. Some singularities of $\mathbf{v}$ may coincide with those of $\mathbf{w}$.

We know that $M$ can be triangulated (see Section 13.2c). By choosing the triangles to be very small (e.g., by subdividing them) and by moving them around slightly, we may insist that (i) each triangle lies completely in some coordinate patch $(x_\alpha, y_\alpha)$; (ii) the singularities of $\mathbf{v}$ and $\mathbf{w}$ lie in the interiors of triangles, not on edges or vertices; and (iii) there is at most one singularity of $\mathbf{v}$ and at most one singularity of $\mathbf{w}$ in the interior of any triangle. Then if $\Delta$ is a triangle lying in a patch $(x_\alpha, y_\alpha)$, we have the Kronecker index integers

$$j_{\mathbf{v}}(\Delta) := \frac{1}{2\pi}\oint_{\partial\Delta} d\theta_{\mathbf{v}}$$

and

$$j_{\mathbf{w}}(\Delta) := \frac{1}{2\pi}\oint_{\partial\Delta} d\theta_{\mathbf{w}}$$

where $\theta_{\mathbf{v}}(x_\alpha, y_\alpha) = \measuredangle(\partial/\partial x_\alpha, \mathbf{v})$ and $\theta_{\mathbf{w}}(x_\alpha, y_\alpha)$ are computed with the chosen Riemannian metric. Note that if $\Delta$ lies in two patches, both coordinate systems will yield, as we know, the same indices. Then

$$\chi_{\mathbf{v}} := \sum_{\Delta} j_{\mathbf{v}}(\Delta) := \frac{1}{2\pi}\sum_{\Delta}\oint_{\partial\Delta} d\theta_{\mathbf{v}}$$

and

$$\chi_{\mathbf{w}} := \sum_{\Delta} j_{\mathbf{w}}(\Delta) := \frac{1}{2\pi}\sum_{\Delta}\oint_{\partial\Delta} d\theta_{\mathbf{w}}$$

are the sums of the indices for the two vector fields, since, for example, if $\mathbf{v}$ has no singularity in $\Delta$ then $j_{\mathbf{v}}(\Delta) = 0$. Thus their difference is

$$\chi_{\mathbf{v}} - \chi_{\mathbf{w}} = \frac{1}{2\pi} \sum_{\Delta} \oint_{\partial\Delta} \{d\theta_{\mathbf{v}} - d\theta_{\mathbf{w}}\}$$

Now $\theta_{\mathbf{v}}(x_\alpha, y_\alpha)$ and $\theta_{\mathbf{w}}(x_\alpha, y_\alpha)$ depend strongly on the coordinate patch used.

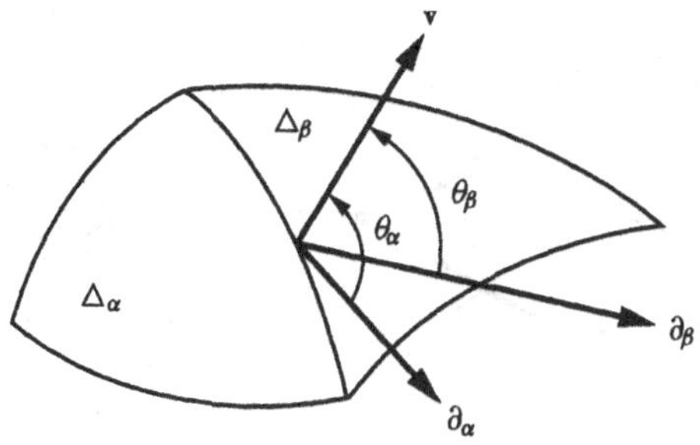


**Figure 16.9**

For example, if $\Delta_\alpha$ and $\Delta_\beta$ are adjacent triangles in patches $(x_\alpha, y_\alpha)$ and $(x_\beta, y_\beta)$, then the angle that $\mathbf{v}$ makes with the first coordinate vector $\partial/\partial x_\alpha$ is different from the angle it makes with the first coordinate vector $\partial/\partial x_\beta$. However,

$$\theta_{\mathbf{v}}(x_\alpha, y_\alpha) - \theta_{\mathbf{w}}(x_\alpha, y_\alpha) = \measuredangle(\mathbf{w}, \mathbf{v})$$

is the *same* as the difference constructed in the $\beta$ patch, since the preceding difference is merely the angle from $\mathbf{w}$ to $\mathbf{v}$, which is determined by the Riemannian metric, independent of patch! Taking the differential of both sides

$$d\theta_{\mathbf{v}} - d\theta_{\mathbf{w}} = d\measuredangle(\mathbf{w}, \mathbf{v})$$

is a well-defined 1-form on each edge of each $\partial\Delta$, *independent of the patch used.* Then

$$\chi_{\mathbf{v}} - \chi_{\mathbf{w}} = \frac{1}{2\pi} \sum_{\Delta} \oint_{\partial\Delta} d\measuredangle(\mathbf{w}, \mathbf{v})$$

Since $M$ is assumed orientable, we may assume that the coordinate patches have positive overlap Jacobians, and thus adjacent triangles $\Delta_\alpha$ and $\Delta_\beta$ will have the same orientation. But then

$$\sum_{\Delta} \oint_{\partial\Delta} d\measuredangle(\mathbf{w}, \mathbf{v}) = 0$$

because each common edge will be traversed twice in opposite directions. Thus $\chi_{\mathbf{v}} = \chi_{\mathbf{w}}$, as desired, and their common value will be called $\chi(M)$.

Note that if $F : M^2 \to V^2$ is a diffeomorphism, then $F_*$ will take the vector field $\mathbf{v}$ on $M$ into a vector field $F_*\mathbf{v}$ on $V$, and it is easy to see that the index of $\mathbf{v}$ at $p$ is the same as the index of $F_*\mathbf{v}$ at $F(P)$. Hence $\chi(M) = \chi(V)$ is a diffeomorphism invariant. □

We shall now see how this integer is related to the topology of $M^2$.

### 16.2b. The Stiefel Vector Field and Euler's Theorem

We now know that we may evaluate $\chi(M^2)$ on any closed orientable surface by looking at *any* vector field with a finite number of singularities and summing the indices. Stiefel constructed the following vector field on any $M^2$.

Take again a triangulation of $M^2$. Imagine that $M^2$ is the sea level surface of a planet; we shall now construct a mountain range on the planet.

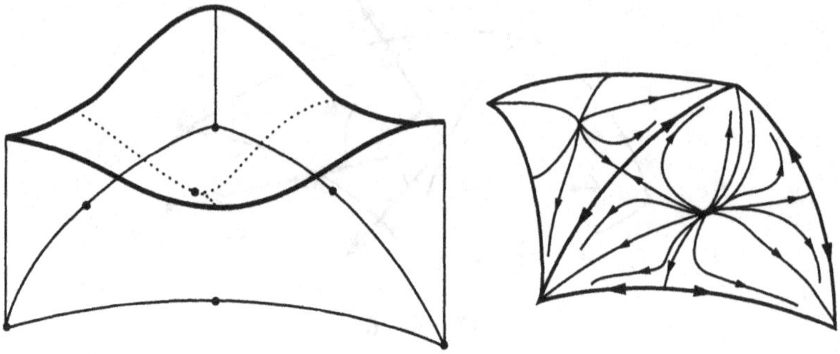

Figure 16.10

Put a mountain peak of height 2 at each vertex, a pit at the "midpoint" of each triangle at sea level 0, and a mountain pass of height 1 at the midpoint of each edge. The height of the land above sea level then defines a function on $M^2$, and if we are careful there will be a maximum at each vertex, a minimum at each face midpoint, and a minimax (saddle) at each edge midpoint. In the right-hand of the figure we have drawn the gradient lines for this function. The gradient vector has a zero at each peak, pass, and pit, and the indices there are $+1$, $-1$, and $+1$, respectively. Thus for this vector field

$$\chi = \text{no. peaks} - \text{no. passes} + \text{no. pits}$$

and we have proved

**Euler's Theorem (16.11):** *For all triangulations of the closed $M^2$ we have that the **Euler characteristic***

$$\chi(M^2) := \textit{no. vertices} - \textit{no. edges} + \textit{no. faces}$$

*is independent of the triangulation.*

From the triangulation of the 2-torus in Section 13.3a we see that $\chi(T^2) = 0$. Thus it would not contradict Poincaré's theorem if there were a field on the torus with no singularities, and of course there is, $\mathbf{v} = \partial/\partial\theta$.

We conclude with three brief remarks.

Consider the projective plane $\mathbb{R}P^2$. It is nonorientable, but it is "covered" twice by the orientable 2-sphere, since $\mathbb{R}P^2$ is $S^2$ with antipodal points identified. (We shall discuss coverings more in Section 21.2.) Thus we have a $2:1$ map $\pi : S^2 \to \mathbb{R}P^2$ that locally is a diffeomorphism.

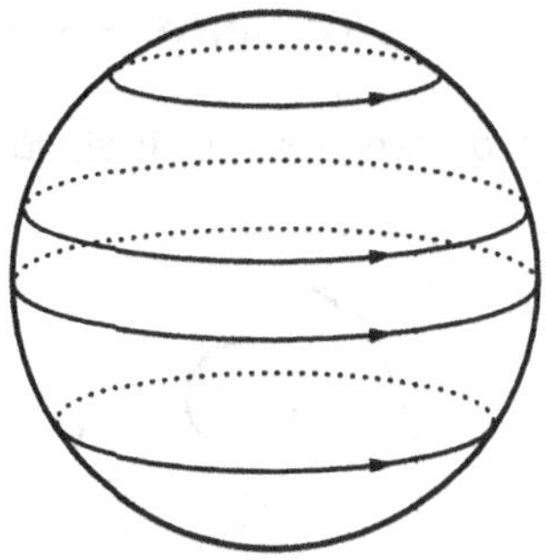

**Figure 16.11**

Consider any vector field $\mathbf{v}$ on $\mathbb{R}P^2$. There is a unique vector field $\mathbf{w}$ on $S^2$ such that $\pi_* \mathbf{w} = \mathbf{v}$. In the figure, $\mathbb{R}P^2$ is the *upper hemisphere* with antipodal identifications on the *equator*, and $\mathbf{v}$ is the vector field on $\mathbb{R}P^2$ that rotates around the "north pole" (there is no south pole on $\mathbb{R}P^2$). The field $\mathbf{w}$ rotates around both poles on $S^2$. The singularity on $\mathbb{R}P^2$ at the pole has index $+1$ and it is covered by two singularities on $S^2$, each with the same index $+1$. Thus $\sum j_{\mathbf{v}} = 1$ and $\sum j_{\mathbf{w}} = 2$. On the other hand, it is evident that if we take a triangulation of $\mathbb{R}P^2$ where each triangle is small, in the sense that each triangle will be covered by two disjoint triangles on $S^2$, then the Euler characteristics, computed via vertices, edges, and faces, as in (16.11), will satisfy $2 = \chi(S^2) = 2\chi(\mathbb{R}P^2)$. Thus Poincaré's theorem holds on the nonorientable $\mathbb{R}P^2$ also and $\chi(\mathbb{R}P^2) = 1$. This illustrates a general fact (discussed in Section 21.2d):

Each nonorientable manifold $M^n$ has a "2-sheeted" orientable covering manifold whose Euler characteristic is $2\chi(M^n)$.

This allows us to prove Poincaré's theorem for nonorientable surfaces as well.

Second, Hopf has proved the $n$-dimensional version of Poincaré's theorem. To a vector field $\mathbf{v}$ on an $M^n$ with an isolated singularity $p$, we may again assign an index $j(p)$ by taking a small $(n-1)$-sphere and considering again the Kronecker index of $\mathbf{v}$ on this $S^{n-1}$. We may look at a triangulation of $M^n$ and define the Euler characteristic

$$\chi(M^n) = (\text{no. 0-simplexes}) - (\text{no. 1-simplexes}) + (\text{no. 2-simplexes})$$
$$- \cdots + (-1)^n\,(\text{no. } n\text{-simplexes})$$

and we again have

**Hopf's Theorem (16.12):** *For any closed $M^n$ and any vector field $\mathbf{v}$ on $M^n$ with isolated singularities, we have $\sum j_{\mathbf{v}}(p) = \chi(M^n)$.*

The proof is considerably more difficult (see [G, P] or [M2])

Finally, a necessary condition for there to exist a vector field on $M^n$ without any singularities is clearly $\chi(M^n) = 0$. Hopf has also shown that this is sufficient; if $\chi(M^n) = 0$ then there is some $\mathbf{v}$ on $M^n$ with no singularities. One may again consult [M2].

## Problems

**16.2(1)** Let $M_g^2$ be a surface of genus $g$. Let it stand on a table and let $h$ be the

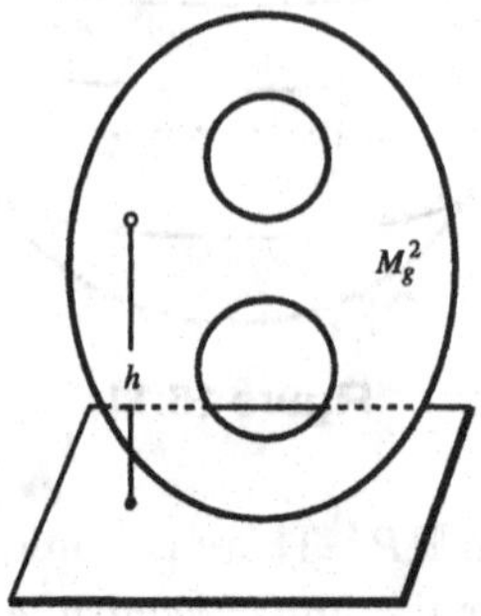


Figure 16.12

function on $M_g^2$ measuring the height above the table. By looking at the vector field grad $h$ on $M_g^2$, show that

$$\chi(M_g^2) = 2 - 2g$$

**16.2(2)** Consider a function with only nondegenerate critical points (in the sense of Morse, Section 14.3e) on a surface $M^2$. Its gradient vector at a critical point has Kronecker index 1, −1, or 1 if it is, respectively, a minimum, saddle, or maximum (see Figure 8.9). Show that the Poincaré–Stiefel pits − passes + peaks theorem, together with Problem 16.2(1), yields Morse's equality in Theorem 14.40.

## 16.3. Connections in a Vector Bundle

How can the tangent bundle to an orientable surface be considered a complex line bundle?

### 16.3a. Connection in a Vector Bundle

Let $\pi : E \to M^n$ be a rank-$K$ vector bundle (real or complex). We shall introduce the concept of a connection for such a bundle by imitating the procedure used in Section 9.3 for the tangent bundle.

A section $\Psi$ of $E$ assigns to each trivializing patch $U \subset M^n$ (i.e., patch over which $E$ is trivial) components $\psi_U$ such that in an overlap

$$\psi_V = c_{VU} \psi_U$$

A vector-valued $p$-form, in Section 9.3, associated to each $p$-tuple of *tangent vectors* to $M$ another element of the same *tangent bundle TM* over the same point. An $E$-valued $p$-form will associate to each $p$-tuple of *tangent vectors* $\mathbf{v}_1, \ldots, \mathbf{v}_p$ to $M$ at $x \in M$ an element of the *bundle* $E$ over $x$.

An **E-valued *p*-form** $\Psi$ assigns to each trivializing patch $U \subset M^n$ a $K$-tuple of ordinary exterior $p$-form $\psi_U$, such that in an overlap we have

$$\psi_V = c_{VU}\psi_U \tag{16.13}$$

For example, if $\alpha^p$ is a globally defined $p$-form on $M^n$, and if $\mathbf{\Psi}$ is a global section of $E$, then $\Psi := \alpha^p \otimes \mathbf{\Psi}$ defines a $p$-form section of $E$ by

$$\psi_U(\mathbf{v}_1, \ldots, \mathbf{v}_p) = \alpha^p(\mathbf{v}_1, \ldots, \mathbf{v}_p)\mathbf{\Psi}$$

A **connection** $\boldsymbol{\nabla}$ for $E$ is an operator taking sections $\mathbf{\Psi}$ of $E$ into $E$-valued 1-forms $\boldsymbol{\nabla}\mathbf{\Psi}$ such that the Leibniz rule holds; if $f$ is a function, then

$$\boldsymbol{\nabla}(\mathbf{\Psi} f) = (\boldsymbol{\nabla}\mathbf{\Psi})f + \mathbf{\Psi} \otimes df \tag{16.14}$$

Let $\mathbf{e} = (\mathbf{e}_1, \ldots, \mathbf{e}_K)$ be a frame of sections of $E$ over the trivializing patch $U$. Then $\boldsymbol{\nabla}\mathbf{e}_\alpha$ is an $E$-valued 1-form, and thus is of the form

$$\boldsymbol{\nabla}\mathbf{e} = \mathbf{e} \otimes \omega$$

or (16.15)

$$\boldsymbol{\nabla}\mathbf{e}_\alpha = \mathbf{e}_\beta \otimes \omega^\beta{}_\alpha$$

where

$$\omega = (\omega^\alpha{}_\beta) = \left(\sum_{i=1}^{n} \omega_i{}^\alpha{}_\beta(x)dx^i\right)$$

is some $K \times K$ matrix of 1-forms on $U$. (We shall try to use consistently Greek letters $\alpha$, $\beta$, and so on, or Roman capitals for fiber indices $1, \ldots, K$ and Roman lowercase $i, j \ldots$ for $M^n$ indices $1, \ldots, n$.) We shall also *frequently omit the tensor product sign.* Note that the **connection coefficients** $\omega^\alpha_{i\beta}$ have a *mixture of fiber and manifold indices*. Here we are assuming that $(x^i)$ are local coordinates for $U \subset M$. For a section $\mathbf{\Psi} = \mathbf{e}\psi$, we have by Leibniz

$$\boldsymbol{\nabla}(\mathbf{\Psi}) = \boldsymbol{\nabla}(\mathbf{e}_\alpha\psi^\alpha) = \boldsymbol{\nabla}(\mathbf{e}\psi) = (\boldsymbol{\nabla}\mathbf{e})\psi + \mathbf{e}(d\psi)$$

that is,

$$\boldsymbol{\nabla}\mathbf{\Psi} = \boldsymbol{\nabla}(\mathbf{e}\psi) = \mathbf{e} \otimes \nabla\psi$$

where

$$\nabla\psi = d\psi + \omega\psi \tag{16.16}$$

In full,

$$\nabla\psi^\alpha = d\psi^\alpha + \omega^\alpha{}_\beta\psi^\beta$$

The boldfaced $\boldsymbol{\nabla}$ operates on *sections*, whereas $\nabla = \nabla_U$ operates on the *components* of sections over the patch $U$.

Suppose now that $\boldsymbol{\Psi}$ is a section over $U \cup V$. In order that $\boldsymbol{\nabla}$ be well defined, we require in $U \cap V$ what physicists call **covariance**, that is,

$$\psi_V = c_{VU}\psi_U \Rightarrow \nabla_V\psi_V = c_{VU}\nabla_U\psi_U \tag{16.17}$$

where $c_{VU}$ is the $K \times K$ matrix $c_{VU} = (c^{\beta}_{VU\alpha})$ in $\mathrm{Gl}(K)$. As in Section (9.4b) this requires

$$\omega_V = c_{UV}^{-1}\omega_U c_{UV} + c_{UV}^{-1} dc_{UV} \tag{16.18}$$

Note that in our conventions

$$\begin{aligned}\psi_V &= c_{VU}\psi_U \\ \mathbf{e}_V &= \mathbf{e}_U c_{VU}^{-1} = \mathbf{e}_U c_{UV}\end{aligned} \tag{16.19}$$

As usual, we define the **covariant derivative** $\boldsymbol{\nabla}_X\boldsymbol{\Psi}$ of the section $\boldsymbol{\Psi}$ of $E$ with respect to the tangent vector $\mathbf{X}$ on $M^n$ by

$$\begin{aligned}\boldsymbol{\nabla}_X(\boldsymbol{\Psi}) := (\boldsymbol{\nabla}\boldsymbol{\Psi})(\mathbf{X}) &= (\mathbf{e} \otimes \nabla\psi)(\mathbf{X}) \\ &= \mathbf{e}[\nabla\psi(\mathbf{X})]\end{aligned} \tag{16.20}$$

Thus

$$\begin{aligned}\boldsymbol{\nabla}_X\boldsymbol{\Psi} &= \mathbf{e}[d\psi + \omega\psi](\mathbf{X}) \\ &= \mathbf{e}[\mathbf{X}(\psi) + \omega(\mathbf{X})\psi]\end{aligned} \tag{16.21}$$

where

$$\omega^{\alpha}{}_{\beta}(\mathbf{X}) = \omega^{\alpha}_{i\beta} dx^i(\mathbf{X}) = \omega^{\alpha}_{i\beta} X^i$$

Then

$$\boldsymbol{\nabla}_X\boldsymbol{\Psi} = \mathbf{e}_{\alpha}\left(X^i \frac{\partial\psi^{\alpha}}{\partial x^i} + X^i\omega^{\alpha}_{i\beta}\psi^{\beta}\right)$$

We then write

$$\begin{aligned}\boldsymbol{\nabla}_X\boldsymbol{\Psi} &= \mathbf{e}\nabla_{\mathbf{X}}\psi \\ \boldsymbol{\nabla}_X\psi &= X^i\nabla_i\psi\end{aligned}$$

where (16.22)

$$\nabla_i\psi^{\alpha} = \frac{\partial\psi^{\alpha}}{\partial x^i} + \omega^{\alpha}_{i\beta}\psi^{\beta}$$

We have defined the covariant differential on sections of $E$ (in a sense, on *0-forms* whose values are in $E$). As in Section 9.3d, we now let $\boldsymbol{\nabla}$ send $E$-valued $p$ forms into $E$-valued $(p+1)$-forms by defining the **exterior covariant differential** (again denoted by $\boldsymbol{\nabla}$)

$$\boldsymbol{\nabla}(\boldsymbol{\Psi} \otimes \alpha^p) = \boldsymbol{\nabla}\boldsymbol{\Psi} \wedge \alpha^p + \boldsymbol{\Psi} \otimes d\alpha^p \tag{16.23}$$

where, as in 9.3d, we write $\wedge$ rather than $\otimes\wedge$.

**Curvature** is introduced as before

$$\nabla^2(\mathbf{e}) = \nabla(\mathbf{e} \otimes \omega) = \mathbf{e} \otimes \theta$$

where

$$\theta = d\omega + \omega \wedge \omega \tag{16.24}$$

$$\theta^\alpha{}_\beta = d\omega^\alpha{}_\beta + \omega^\alpha{}_\gamma \wedge \omega^\gamma{}_\beta = \frac{1}{2} R^\alpha{}_{\beta ij} dx^i \wedge dx^j$$

Note the *mixture of indices* in the curvature tensor.

There is no notion of *torsion* in a connection for a general vector bundle.

As a simple example, consider the normal 2-plane bundle to a curve $M^1$, $\mathbf{x} = \mathbf{x}(t)$ in $\mathbb{R}^3$. If $\boldsymbol{\nu} = \boldsymbol{\nu}(t)$ is normal to $M$ along $M$, we wish $\nabla\boldsymbol{\nu} = (\nabla\boldsymbol{\nu}/dt)dt$ to be a normal vector valued 1-form on $M$. Let $\mathbf{d}$ be the usual differential operator for $\mathbb{R}^3$; it is the covariant differential for the tangent bundle for $\mathbb{R}^3$ with the usual euclidean flat metric. We should then put

$$\nabla\boldsymbol{\nu} := \mathbf{d}\boldsymbol{\nu} - \langle \mathbf{d}\boldsymbol{\nu}, \mathbf{T}\rangle \mathbf{T} \tag{16.25}$$

where $\mathbf{T}$ is the unit tangent to $M$. For a local description, let $\mathbf{n}_1$ and $\mathbf{n}_2$ be two normal vector fields along $M$ that are orthonormal. Then the prescription (16.25) translates to

$$\nabla\boldsymbol{\nu} = \langle \mathbf{d}\boldsymbol{\nu}, \mathbf{n}_1\rangle \mathbf{n}_1 + \langle \mathbf{d}\boldsymbol{\nu}, \mathbf{n}_2\rangle \mathbf{n}_2$$

In particular, since $\mathbf{dn}_1$ is orthogonal to $\mathbf{n}_1$, $\nabla\mathbf{n}_1 = \mathbf{n}_\alpha \omega^\alpha{}_1 = \langle \mathbf{dn}_1, \mathbf{n}_2\rangle \mathbf{n}_2$, shows that $\omega^1{}_1 = 0$ and $\omega^2{}_1 = \langle \mathbf{dn}_1, \mathbf{n}_2\rangle = \langle \mathbf{dn}_1/dt, \mathbf{n}_2\rangle dt$. When $t = s$ is arc length along the curve $M$, and when $\mathbf{n}_1$ is chosen to be the *principal normal* $\mathbf{n}$ to the curve, then, as in Problem 7.1(2), $\mathbf{n}_2 = \mathbf{T} \times \mathbf{n}_1$ is the *binormal* $\mathbf{B}$, and $-\langle \mathbf{dn}_1/ds, \mathbf{n}_2\rangle$ is the *torsion* $\tau$ of the space curve. Thus $\nabla\mathbf{n} = -\mathbf{B}\tau(s)ds$ and $\nabla\mathbf{B} = \mathbf{n}\tau(s)ds$.

## 16.3b. Complex Vector Spaces

Quantum mechanics deals almost exclusively with complex wave functions and $K$ component wave functions, in other words, with sections of **complex** vector bundles. (We shall consider quantum mechanics in Section 16.4.)

Consider the complex plane $\mathbb{C}$ with coordinate $z = x + iy$. $\mathbb{C}$ is a 1-dimensional vector space because we allow complex scalars, but $\mathbb{C}$ can also be thought of as a real 2-dimensional vector space $\mathbb{R}^2$,

$$z = x + iy \Leftrightarrow \begin{bmatrix} x \\ y \end{bmatrix}$$

and addition of complex numbers corresponds to vector addition in $\mathbb{R}^2$. The interesting thing about $\mathbb{C}$ is that it has a fascinating product

$$z_1 z_2 = (x_1 + iy_1)(x_2 + iy_2) = (x_1x_2 - y_1y_2) + i(x_1y_2 + x_2y_1)$$

Of course, this can be expressed entirely in real terms

$$\begin{bmatrix} x_1 \\ y_1 \end{bmatrix} \circ \begin{bmatrix} x_2 \\ y_2 \end{bmatrix} = \begin{bmatrix} x_1x_2 - y_1y_2 \\ x_1y_2 + x_2y_1 \end{bmatrix}$$

In particular, multiplication in $\mathbb{C}$ by the unit $i$ translates in real terms to a linear transformation

$$J : \mathbb{R}^2 \to \mathbb{R}^2$$

whose matrix is

$$J = \begin{bmatrix} 0 & -1 \\ 1 & 0 \end{bmatrix}$$

with, naturally, $J^2 = -I$. Similarly, complex $K$ space, $\mathbb{C}^K$, the vector space (with complex scalars) of complex $K$-tuples

$$z = (z_1, \ldots, z_K)^T = (x_1 + iy_1, \ldots, x_K + iy_K)^T$$

can be considered as $\mathbb{R}^{2K}$ under the identification

$$z \Leftrightarrow (x_1, y_1, x_2, y_2, \ldots, x_K, y_K)^T = \mathbf{x}$$

and then multiplication by $i$ in $\mathbb{C}^K$, $z \mapsto iz$, is translated into a linear transformation $J : \mathbb{R}^{2K} \to \mathbb{R}^{2K}$ with matrix

$$J = \begin{bmatrix} 0 & -1 & & & & & \\ 1 & 0 & & & 0 & & \\ & & 0 & -1 & & & \\ & & 1 & 0 & & & \\ & 0 & & & \ddots & & \\ & & & & & 0 & -1 \\ & & & & & 1 & 0 \end{bmatrix} \tag{16.26}$$

again with $J^2 = -I$.

Note that $J : \mathbb{R}^{2K} \to \mathbb{R}^{2K}$ is an isometry with respect to the usual metric

$$\langle \mathbf{x}, \mathbf{x}' \rangle = \langle J\mathbf{x}, J\mathbf{x}' \rangle$$

since it merely rotates each coordinate plane $x_\alpha, y_\alpha$ through 90 degrees.

Now let $F^{2k}$ be any *real* even-dimensional vector space with an inner product $\langle , \rangle$ and let

$$J : F \to F$$

be any linear *isometry* of $F$ (orthogonal transformation) that is also an **anti-involution**, that is,

$$J^2 = -I$$

Clearly the eigenvalues of $J$ are $\pm i$, and so det $J = 1$. Thus $J \in SO(2k)$ and assumes the form (15.6) in suitable orthonormal coordinates $(x_1, y_1, x_2, y_2, \ldots, x_k, y_k)$. But $J$ is *skew symmetric*,

$$\langle J\mathbf{x}, \mathbf{x}' \rangle = \langle J^2\mathbf{x}, J\mathbf{x}' \rangle = \langle \mathbf{x}, -J\mathbf{x}' \rangle$$

Equation 15.6 tells us that in these coordinates $J$ has matrix (16.26), since each $\theta_k$ must be $\pi/2$. Then one can introduce *complex* coordinates in $F$ by putting $z_\alpha = x_\alpha + iy_\alpha$, and $J : F \to F$ then corresponds to multiplication by $i$.

In particular, $\mathbb{R}^2$ with $J$ as earlier can be considered a complex 1-dimensional vector space $\mathbb{C}^1 = \mathbb{C}$, which can be called a **complex line.**

### 16.3c. The Structure Group of a Bundle

In a vector bundle each $c_{UV}(x) \in \mathrm{Gl}(n)$. We have seen that for a Riemannian manifold $M^n$, we may choose $c_{UV}(x) \in O(n)$ by using orthonormal frames. In a general bundle, it may be possible to choose the $c_{UV}(x)$ such that they all lie in a specific Lie group $G$

$$c_{UV} : U \cap V \to G$$

We then say that $G$ is the **structure group** of the bundle.

Let $M^2$ be an *oriented* Riemannian *surface*. We can cover $M$ by patches $U, V, \ldots$ each of which supports a positively oriented orthonormal frame $\{\mathbf{e}_U\}, \{\mathbf{e}_V\}, \ldots$ of tangent vectors. Suppressing the patch index,

$$\mathbf{e}_U = (\mathbf{e}_1, \mathbf{e}_2)$$

is a positively oriented orthonormal frame in $U$. It is then clear that each transition matrix for $E = TM$ is a rotation matrix

$$c_{UV}(x) = \begin{bmatrix} \cos\alpha(x) & -\sin\alpha(x) \\ \sin\alpha(x) & \cos\alpha(x) \end{bmatrix} \in SO(2)$$

We may say that the orthonormal frames have allowed us to **reduce** the structure group from $\mathrm{Gl}(2, \mathbb{R})$ to $SO(2)$.

### 16.3d. Complex Line Bundles

Define $J$ acting on the tangent planes of an *oriented surface*, $J : M^2_p \to M^2_p$, simply to be rotation through a right angle in the positive sense; thus

$$J\mathbf{e}_1 = \mathbf{e}_2 \tag{16.27}$$

$$J\mathbf{e}_2 = -\mathbf{e}_1$$

and of course $J^2 = -I$. It is clear that $J$ is *globally* defined; in an overlap $U \cap V$ the action of $J$ using the frame $\mathbf{e}_V$ coincides with the action of $J$ using $\mathbf{e}_U$. *Thus $J$ allows us to consider each fiber in $TM^2$ as a complex line!* The real vector

$$\mathbf{e}_1 \in M^2_p \approx \mathbb{R}^2$$

can be considered as a complex basis vector

$$\mathbf{e} := \mathbf{e}_1 \in M^2_p \approx \mathbb{C}^1$$

of the complex line $M^2_p$. Then

$$i\mathbf{e} = J\mathbf{e}_1 = \mathbf{e}_2$$

In terms of these bases $\mathbf{e}^U = \mathbf{e}_1^U$, $\mathbf{e}^V = \mathbf{e}_1^V, \ldots$, the previous $SO(2)$ transition matrices, $c_{UV}(p)$, become simply the complex numbers

$$c_{UV}(p) = e^{i\alpha(p)}$$

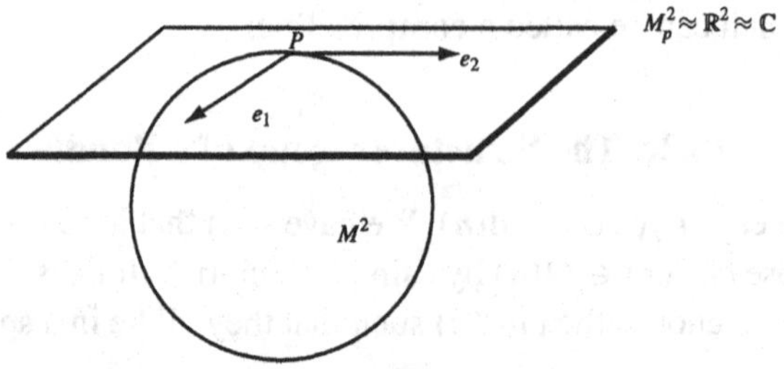


**Figure 16.13**

*The tangent bundle to an oriented Riemannian surface can be considered as a complex line bundle! The structure group of this bundle is now $U(1)$, the unitary group in 1 variable!*

The Riemannian connection for $M^2$ is a connection for the real 2-dimensional tangent bundle. In terms of the *orthonormal* frames $\mathbf{e}^U, \mathbf{e}^V, \ldots$, we have

$$\nabla \mathbf{e}_i = \mathbf{e}_j \otimes \omega^j{}_i = \mathbf{e}_j \otimes \omega_{ji}$$

and we also know $\omega_{ij} = -\omega_{ji}$; thus

$$\nabla \mathbf{e}_1 = \mathbf{e}_2 \otimes \omega_{21} \tag{16.28}$$

A connection matrix for a complex *line* bundle would be a $1 \times 1$ matrix, that is, a single 1-form, which we shall denote by $\omega^c$ ($^c$ for complex). We should then have, in our line bundle version of $TM^2$ (where $\mathbf{e} = \mathbf{e}_1$)

$$\nabla^c \mathbf{e} = \mathbf{e} \otimes \omega^c$$

and since $\mathbf{e}_2 = i\mathbf{e}_1$ we can rewrite (16.28) as

$$\nabla^c \mathbf{e} = \nabla \mathbf{e}_1 = i\mathbf{e}_1 \otimes \omega_{21} = \mathbf{e}_1 \otimes i\omega_{21}$$

or

$$\nabla^c \mathbf{e} = \mathbf{e} \otimes \omega^c \tag{16.29}$$

$$\omega^c := i\omega_{21} = -i\omega_{12}$$

Does this mean that $\omega^c = -i\omega_{12}$ defines a connection for this complex line bundle version of $TM^2$? For this to be true we certainly must have that $\nabla^c$ commutes with multiplication by complex constants

$$\nabla^c(i\psi) = i\nabla^c\psi$$

for any cross section $\psi$ (i.e., any vector field on $M^2$). For example

$$\nabla^c(i\mathbf{e}) = \nabla \mathbf{e}_2 = \mathbf{e}_1 \otimes \omega_{12} = i\mathbf{e}_1 \otimes (-i\omega_{12})$$

$$= i(\mathbf{e} \otimes \omega^c) = i\nabla^c \mathbf{e}$$

as desired. This connection will be discussed further in Problem 18.2(2).

What is the curvature for this complex line bundle connection? It is the *single* 2-form

$$\theta^c = d\omega^c + \omega^c \wedge \omega^c = d\omega^c$$
$$= d(-i\omega_{12}) = -id\omega_{12} = -i\theta_{12}$$

or

$$\theta^c = -i\theta_{12} = -iK\sigma^1 \wedge \sigma^2 \tag{16.30}$$

where again $K$ is the Gauss–Riemann curvature $R^{12}{}_{12}$ of $M^2$.

## Problem

**16.3(1)** If $\nabla$ and $\nabla'$ are connections for bundles $E$ and $E'$ respectively over $M$ then a connection for $E \otimes E'$ can be given by

$$\nabla''_{\mathbf{X}}(\Phi \otimes \Psi) = (\nabla_{\mathbf{X}}\Phi) \otimes \Psi + \Phi \otimes (\nabla'_{\mathbf{X}}\Psi)$$

for local sections $\Phi = e_a\phi^a$ and $\Psi = e'_R\Psi^R$. Show that for $\Lambda = e_a \otimes e'_R\lambda^{aR}$

$$\nabla''_j(\lambda^{aR}) = \partial_j(\lambda^{aR}) + \omega^a{}_{jb}\lambda^{bR} + \omega'^R{}_{jS}\lambda^{aS}$$

# 16.4. The Electromagnetic Connection

What does the electromagnetic field have to do with parallel displacement of a wave function?

## 16.4a. Lagrange's Equations without Electromagnetism

In Section 10.2a we showed that Lagrange's equations for a massive particle, $dp/dt = \partial L/\partial q$, with $p = \partial L/\partial \dot{q}$, follow from Newton's equations $\nabla\dot{q}/dt = -\operatorname{grad} V$. Although both sides of Newton's equations are contravariant vectors along the extremal $q = q(t)$, it is not true that both sides of Lagrange's equations are covectors along the extremal, since $dp/dt$ is an ordinary derivative (rather than a covariant derivative) and also

$$\frac{\partial L}{\partial q^k} = \frac{1}{2}\left\{\frac{\partial g_{ij}(q)}{\partial q^k}\right\}\dot{q}^i\dot{q}^j - \frac{\partial V}{\partial q^k} \tag{16.31}$$

is not a covector field because of the first term. To remedy this we may consider the covariant derivative of the momentum covector

$$\frac{\nabla p_i}{dt} = \frac{\nabla\{g_{ij}\dot{q}^j\}}{dt} = g_{ij}\left\{\frac{\nabla\dot{q}^j}{dt}\right\} = -g_{ij}g^{jk}\frac{\partial V}{\partial q^k}$$

that is,

$$\frac{\nabla p}{dt} = -\frac{\partial V}{\partial q} \tag{16.32}$$

This is a geometric version of Lagrange's equations; the left side differs from $dp/dt$ in that a covariant derivative is used; the right side uses the potential function $V$ rather than Lagrangian $L$.

Let us verify that (16.32) really reproduces Lagrange's equations, by computing what $dp_i/dt - p_j\Gamma^j_{ki}\dot{q}^k = -\partial V/\partial q^i$, that is, $dp_i/dt = p_j\Gamma^j_{ki}\dot{q}^k - \partial V/\partial q^i$, says.

$$\begin{aligned}\frac{dp_i}{dt} &= g_{jr}\dot{q}^r\frac{1}{2}g^{js}\left\{\frac{\partial g_{sk}}{\partial q^i}+\frac{\partial g_{si}}{\partial q^k}-\frac{\partial g_{ki}}{\partial q^s}\right\}\dot{q}^k - \frac{\partial V}{\partial q^i}\\ &= \frac{1}{2}\dot{q}^s\dot{q}^k\left\{\frac{\partial g_{sk}}{\partial q^i}+\frac{\partial g_{si}}{\partial q^k}-\frac{\partial g_{ki}}{\partial q^s}\right\} - \frac{\partial V}{\partial q^i}\\ &= \frac{1}{2}\dot{q}^s\dot{q}^k\left\{\frac{\partial g_{sk}}{\partial q^i}\right\} - \frac{\partial V}{\partial q^i} = \frac{\partial L}{\partial q^i}\end{aligned}$$

from (16.31). Combining this with

$$p_i = g_{ij}\dot{q}^j = \frac{\partial}{\partial \dot{q}^i}\left\{\frac{1}{2}g_{rs}\dot{q}^r\dot{q}^s\right\} = \frac{\partial T}{\partial \dot{q}^i} = \frac{\partial L}{\partial \dot{q}^i}$$

then yields Lagrange's equations, as promised. It is important that $\partial V/\partial \dot{q} = 0$.

## 16.4b. The Modified Lagrangian and Hamiltonian

Consider a charged particle moving in an $M^3$ with no external electromagnetic field present. Let $L = L(x, \dot{x}) = T - V$ be the Lagrangian. The particle then obeys Lagrange's equations $dp/dt = \partial L/\partial x$, where $p := \partial L/\partial \dot{x}$ is the **kinematical** momentum, that is, the covariant version of the velocity.

Suppose now that an electromagnetic field is present also. The particle then suffers not only the original force $-\partial V/\partial x$ but also an additional Lorentz force whose contravariant version is $e(\mathbf{E}+\mathbf{v}\times\mathbf{B})$. This additional force is not the gradient of a potential and so we cannot get the complete Lagrangian equations of motion merely by adding a new potential term $V'$ (although we could if the magnetic field were not present). It turns out, though, that we can write the equations in Lagrangian and Hamiltonian forms if we make a more sophisticated change. For this purpose we shall consider a massive charged particle, moving perhaps relativistically in $\mathbb{R}^3$.

We shall first make some *heuristic* remarks (inspired by comments of Weyl [Wy, pp. 52, 99]) concerning the notion of the Lagrangian in particle mechanics and the changes when an electromagnetic field is present. Unlike the total energy $T + V$, which is frequently a constant of the motion, the Lagrangian $T - V$ seemingly was introduced merely to make Lagrange's equations take a simpler mathematical form. Although introduced long ago, I feel that its physical significance could not be appreciated before the introduction of special relativity.

Introduce units for which the speed of light is unity, $c = 1$. Special relativity associates to the world line of a massive particle its energy-momentum 4-vector $P = (E, p)^T$. $E = m \sim m_0 + (1/2)m_0 v^2 + \dots$, which, except for a constant $m_0$, reduces to the classical kinetic energy $T$ for low speeds. If the classical force is derivable from a classical backround potential $V$, $f_c = -\nabla V$, special relativity suggests that we should augment the energy $E$ by $V$, yielding a "total energy" $H := E + V \sim m_0 + (T + V)$, as in section 7.1c. We may then form, as a first attempt, the "total energy momentum 4-vector" $(H, p)^T$. Put $H_0 := H - m_0 \sim T + V$. The 1-form associated to $(H, p)^T$, i.e., the total energy momentum covector is then

$$p_\alpha dx^\alpha - Hdt = p_\alpha dx^\alpha - H_0 dt - m_0 dt$$

which is the extended Poincaré 1-form or action 1-form (4.57) augmented by a term $-m_0 dt$ *which does not alter the equations of motion.* Along the world line, this 1-form is

$$\left[p_\alpha\left(\frac{dx^\alpha}{dt}\right) - H_0\right]dt - m_0 dt = Ldt - m_0 dt$$

**The Lagrangian action integrand** $Ldt$ **is,** except for a disposable exact differential, **the total energy-momentum 1-form,** in the sense of special relativity, along the world line! This, I believe, explains the significance of the Lagrangian in the principle of least action!

There is a disquieting feature of the above argument; we took a 4-covector $p_\alpha dx^\alpha - Edt$ and added to its time component $-E$ a scalar $-V$. This violates "Lorentz covariance"; one cannot add a scalar to one component of a covector. (This does not mean that the above procedure is invalid; it makes perfectly good sense if we agree to use only those Lorentz transformations that do not involve time, for example the usual changes of spatial coordinates traditionally used in non-relativistic mechanics.)

The situation is much more satisfactory when the backround field is the electromagnetic field, with covector potential $A = \phi dt + A_\alpha dx^\alpha$ or vector potential $(-\phi, \mathbf{A})^T$. In this case $-V = e\phi$ can be added to $-E$ provided we add $e\mathbf{A}$ to $\mathbf{p}$, since it makes Lorentz sense to add two 4-vectors together! The resulting covector, the total energy-momentum 1-form is simply

$$(p_\alpha + eA_\alpha)dx^\alpha - (E - e\phi)dt - m_0 dt$$

with Lagrangian

$$Ldt = \left[(p_\alpha + eA_\alpha)\left(\frac{dx^\alpha}{dt}\right) - (E - e\phi)\right]dt - m_0 dt$$

This also suggests that if one has a classical dynamical system, with Hamiltonian $H$ and no electromagnetism, then to get the Hamiltonian equations when electromagnetism is introduced one simply defines a new Hamiltonian by $H^* := H - e\phi$ and new momenta by $p^*{}_\alpha := p_\alpha + eA_\alpha$. But then

$$p_\alpha^* dx^\alpha - H^* dt = p_\alpha dx^\alpha - Hdt + e(A_\alpha dx^\alpha + \phi dt)$$

and the extended Poincaré 2-form should be redefined to be

$$\Omega^* := d(p_\alpha^* dx^\alpha - H^* dt) = \Omega + eF \tag{16.33}$$

(It can then be shown that Hamilton's equations are now $i_X \Omega^* = 0$, where $X = (dx/dt)\partial/\partial x + (dp/dt)\partial/\partial p + \partial/\partial t$ uses the original $p$ rather than the augmented $p^*$.)

We are now finished with our heuristic discussion and we proceed with our formal verification of these hopes.

**Theorem (16.34):** *Let $H = H(q, p, t)$ be the Hamiltonian for a charged particle when no electromagnetic field is present. Let an electromagnetic field be introduced. Define a new* **canonical** *momentum variable $p^*$ in $T^*M \times \mathbb{R}$ by*

$$p^*{}_\alpha := p_\alpha + eA_\alpha(t, q)$$

*and a new Hamiltonian*

$$H^*(q, p^*, t) := H(q, p, t) - e\phi(t, q) = H(q, p^* - eA, t) - e\phi(t, q)$$

*Then the particle of charge e satisfies new Hamiltonian equations*

$$\frac{dq}{dt} = \frac{\partial H^*}{\partial p^*} \quad \text{and} \quad \frac{dp^*}{dt} = -\frac{\partial H^*}{\partial q} \quad \text{and} \quad \frac{dH^*}{dt} = \frac{\partial H^*}{\partial t}$$

PROOF: Compare the solutions of the original system

$$\frac{dq}{dt} = \frac{\partial H}{\partial p} \quad \text{and} \quad \frac{dp}{dt} = -\frac{\partial H}{\partial q}$$

and the new system

$$\frac{dq}{dt} = \frac{\partial H^*}{\partial p^*} \quad \text{and} \quad \frac{dp^*}{dt} = -\frac{\partial H^*}{\partial q}$$

At a point $(q, p, t) = (q, p^* - eA, t)$ we have

$$\frac{\partial H^*}{\partial p^*} = \frac{\partial H(q, p^* - eA)}{\partial p^*} = \frac{\partial H(q, p)}{\partial p} = \frac{dq}{dt}$$

and so the velocities $dq/dt$ are identical in both systems.

Denote $-\partial H/\partial q$ by $f$, the force in the original system. Then

$$\frac{dp_\alpha^*}{dt} + \frac{\partial H^*}{\partial q^\alpha} = \left(\frac{dp_\alpha}{dt} + e\frac{dA_\alpha}{dt}\right) + \frac{\partial H}{\partial q^\alpha} + \frac{\partial H}{\partial p_\beta}\left[\frac{\partial(-eA_\beta)}{\partial q^\alpha}\right] - e\frac{\partial \phi}{\partial q^\alpha}$$

$$= \left(\frac{dp_\alpha}{dt} + e\frac{dA_\alpha}{dt}\right) - f_\alpha - e\left(\frac{dq^\beta}{dt}\right)\left(\frac{\partial A_\beta}{\partial q^\alpha}\right) - e\frac{\partial \phi}{\partial q^\alpha}$$

But

$$\left(\frac{dq^\beta}{dt}\right)\left(\frac{\partial A_\beta}{\partial q^\alpha}\right) = \left(\frac{dq^\beta}{dt}\right)(\partial_\alpha A_\beta) = \left(\frac{dq^\beta}{dt}\right)[(\partial_\alpha A_\beta - \partial_\beta A_\alpha) + \partial_\beta A_\alpha]$$

$$= \left(\frac{dq^\beta}{dt}\right)[F_{\alpha\beta} + \partial_\beta A_\alpha] = (\mathbf{v} \times \mathbf{B})_\alpha + (\partial_\beta A_\alpha)\left(\frac{dq^\beta}{dt}\right)$$

$$= (\mathbf{v} \times \mathbf{B})_\alpha + \frac{dA_\alpha}{dt} - \frac{\partial A_\alpha}{\partial t}$$

Thus

$$\frac{dp_\alpha^*}{dt} + \frac{\partial H^*}{\partial q^\alpha} = \frac{dp_\alpha}{dt} - f_\alpha - e\left[(\mathbf{v} \times \mathbf{B})_\alpha - \frac{\partial A_\alpha}{\partial t} + \frac{\partial \phi}{\partial q^\alpha}\right]$$

$$= \frac{dp_\alpha}{dt} - f_\alpha - e[(\mathbf{v} \times \mathbf{B}) + \mathbf{E}]_\alpha$$

Hence $dp^*/dt = -\partial H^*/\partial q$ is equivalent to the original system augmented by the Lorentz force, as desired. □

The Lagrangian $L$ and the Hamiltonian $H$ are related by $L(q, \dot{q}) = p\dot{q} - H(q, p)$. Along a lifted curve $\dot{q} = dq/dt$ we then have

$$L(q, \dot{q})dt = pdq - H(q, p)dt$$

In terms of our new Hamiltonian, we should define

$$\begin{aligned} L^*(q, \dot{q})dt &:= p^*dq - H^*(q, p^*)dt \\ &= (p_\alpha + eA_\alpha)dq^\alpha - [H(q, p) - e\phi]dt \\ &= [p_\alpha dq^\alpha - H(q, p)dt] + e[\phi dt + A_\alpha dq^\alpha] \end{aligned} \tag{16.35}$$

**Corollary (16.36):** *A particle in an electromagnetic field satisfies Lagrange's equations* $\partial L^*/\partial q - d/dt(\partial L^*/\partial \dot{q}) = 0$ *with new Lagrangian*

$$L^*(q, \dot{q}) = L(q, \dot{q}) + e[\phi + A_\alpha \dot{q}^\alpha]$$

*that is,*

$$L^*dt = Ldt + e[\phi dt + \mathfrak{A}] = Ldt + eA^1$$

## 16.4c. Schrödinger's Equation in an Electromagnetic Field

In the present section we shall *remove the mass term from the metric;* that is, the kinetic energy of a particle is the familiar

$$T = \frac{1}{2}m\langle \dot{q}, \dot{q}\rangle = \frac{p^2}{2m}$$

Consider a charged particle, of mass $m$, moving in a potential field $V$ in $\mathbb{R}^3$, with no external electromagnetic field present. If we neglect spin, the electron is commonly represented in quantum mechanics by a wave function

$$\psi(x) = \psi(\mathbf{x}, t)$$

a complex-valued time-dependent function on $\mathbb{R}^3$.

**Schrödinger's equation** states that the wave functions evolve in time according to

$$i\hbar\frac{\partial \psi}{\partial t} = H\psi \tag{16.37}$$

where the **Hamiltonian operator** $H$ is defined as follows.

The Hamiltonian of a particle in classical mechanics is given by

$$H(\mathbf{x}, \mathbf{p}) = \frac{p^2}{2m} + V(\mathbf{x}) \tag{16.38}$$

where $\mathbf{p}$ is the canonical momentum. Schrödinger then postulates that *in Cartesian coordinates* the **canonical** momenta $p_\alpha$ are represented by the differential operators

$$p_\alpha = -i\hbar\frac{\partial}{\partial x^\alpha} \tag{16.39}$$

The potential $V$ is simply the multiplicative operator $\psi \mapsto V(x)\psi$, and (16.38) becomes, in Cartesian coordinates in $\mathbb{R}^3$,

$$i\hbar\frac{\partial\psi}{\partial t} = -\left(\frac{\hbar^2}{2m}\right)\sum_\alpha \frac{\partial^2\psi}{(\partial x^\alpha)^2} + V\psi \tag{16.40}$$

If the particle has charge $e$ and there is an additional external electromagnetic field present, then (16.34) says that (16.38) should be replaced by

$$H(\mathbf{x}, \mathbf{p}^*) = \frac{(p^*{}_\alpha - eA_\alpha)^2}{2m} + V(\mathbf{x}) - e\phi$$

and the canonical momenta $p^*{}_\alpha$ should be replaced, when $x$ are Cartesian coordinates, by $p^*{}_\alpha = -i\hbar\partial/\partial x^\alpha$. Schrödinger's equation becomes

$$i\hbar\frac{\partial\psi}{\partial t} = \frac{1}{2m}\sum_\alpha\left[-i\hbar\frac{\partial}{\partial x^\alpha} - eA_\alpha\right]^2\psi + V\psi - e\phi\psi \tag{16.41}$$

If we write this in the form

$$i\hbar\left[\frac{\partial}{\partial t} - \left(\frac{ie}{\hbar}\right)\phi\right]\psi = -\left(\frac{\hbar^2}{2m}\right)\sum_\alpha\left[\frac{\partial}{\partial x^\alpha} - \left(\frac{ie}{\hbar}\right)A_\alpha\right]^2\psi + V\psi$$

we may then write

$$i\hbar\nabla_0\psi = -\left(\frac{\hbar^2}{2m}\right)\sum_\alpha \nabla_\alpha\nabla_\alpha\psi + V\psi$$

where

$$\nabla_0 := \frac{\partial}{\partial t} - \left(\frac{ie}{\hbar}\right)\phi$$

and (16.42)

$$\nabla_\alpha := \frac{\partial}{\partial x^\alpha} - \left(\frac{ie}{\hbar}\right)A_\alpha$$

We may write, instead of the last two definitions,

$$\nabla_j := \frac{\partial}{\partial x^j} - \left(\frac{ie}{\hbar}\right)A_j \tag{16.43}$$

We then have the following situation. We originally thought of $\psi$ as being a complex function on $\mathbb{R}^4$, that is, a section of the *trivial* complex line bundle over $\mathbb{R}^4$. Schrödinger's equation involves the vector potential $A^1 = A_j dx^j = \phi dt + A_\alpha dx^\alpha$. The vector potential is not uniquely determined; we may, if we wish, use a different choice $A^1_U$ in each of several patches $U$ in $\mathbb{R}^4$. If we do so, then in each patch $U$ we shall have a different Schrödinger equation, satisfied by a local solution $\psi_U$. *This is precisely the situation we met when we considered sections of a complex line bundle over* $\mathbb{R}^4$! Equation (16.43) then takes on the appearance of a *covariant derivative*

$$\nabla_j\psi := \frac{\partial\psi}{\partial x^j} + \omega_j\psi$$

where (16.44)

$$\omega_j := -\left(\frac{ie}{\hbar}\right)A_j$$

If $\omega$ is to be a connection, *what is the bundle*? Consider two choices $A_U$ and $A_V$ in overlapping patches of $\mathbb{R}^4$. In these patches we have

$$\omega_U = -\left(\frac{ie}{\hbar}\right)A_U, \qquad \omega_V = -\left(\frac{ie}{\hbar}\right)A_V$$

Since the electromagnetic field 2-form, $F = dA$, is well defined, it must be that $A_V - A_U$ is a closed 1-form on $U \cap V$, and, *if this intersection is simply connected*, $A_V - A_U$ is exact,

$$A_V = A_U + df_{UV} \quad \text{in } U \cap V$$

where $f_{UV}$ is a real single-valued function on $U \cap V$. Then

$$\omega_V = \omega_U - \left(\frac{ie}{\hbar}\right)df_{UV}$$

But a connection in a bundle transforms by (16.18), and when the bundle is a complex line bundle, the $c_{UV}$ are $1 \times 1$ complex matrices and the transformation rule becomes

$$\omega_V = \omega_U + c_{UV}^{-1}dc_{UV} \tag{16.45}$$

Thus we may choose $\log c_{UV}(x) = -(ie/\hbar)f_{UV}$, that is,

$$c_{UV}(x) = \exp\left\{-\left(\frac{ie}{\hbar}\right)f_{UV}\right\} \tag{16.46}$$

*If* $c_{UV}c_{VW}c_{WU} = 1$ *is satisfied* then (16.46) defines a line bundle whose cross sections will be our local wave functions.

*A wave function is then not a single complex-valued function $\psi$ but rather a collection $\psi_U, \psi_V, \ldots$ of functions such that in an overlap $U \cap V$*

$$\psi_V(x) = c_{VU}(x)\psi_U(x) = \exp\left\{\left(\frac{ie}{\hbar}\right)f_{UV}\right\}\psi_U(x) \tag{16.47}$$

This brings us back to the starting point of gauge theories in quantum mechanics, namely

**Weyl's principle of gauge invariance (16.48):** *If $\psi$ satisfies Schrödinger's equation (16.41), which involves the potential A, then*

$$\exp\left[\left(\frac{ie}{\hbar}\right)f(x)\right]\psi$$

*satisfies Schrödinger's equation when A has been replaced by*

$$A + df$$

To see this let $U = V$ but let us choose $A_V = A_U + df$. Then Weyl's principle simply says that if

$$i\hbar\nabla_0^V\psi_V = \left(-\frac{\hbar^2}{2m}\right)\sum_\alpha \nabla_\alpha^V\nabla_\alpha^V\psi_V + V\psi_V$$

then (16.17), that is, $c_{VU}\nabla^U = \nabla^V c_{VU}$, shows that *this same equation holds* (with $\nabla^V \mapsto \nabla^U$) *when* $\psi_V$ *is replaced by* $\psi_U$! Note that without the notion of a connection the verification of this would be messy; Schrödinger's equation, when written out without covariant derivatives, involves

$$\nabla_\mu \nabla_\mu \psi = \frac{\partial^2 \psi}{\partial x^{\mu 2}} - \left(\frac{ie}{\hbar}\right)\left[A_\mu \frac{\partial \psi}{\partial x^\mu} + \frac{\partial}{\partial x^\mu}(A_\mu \psi)\right] - \left(\frac{e^2}{\hbar^2}\right) A_\mu A_\mu \psi.$$

(16.17) is the crucial simplification.

It should be clear that Weyl's principle is *not restricted to Schrödinger's equation*; "covariance," that is, (16.17), is the essential ingredient.

Note that the transition functions (16.46) for our bundle are complex numbers of absolute value 1; the structure group of the given line bundle is the group $U(1)$. This implies that $|\psi_V|^2 = |\psi_U|^2$ and consequently *the probability interpretation of* $|\psi|^2$ *in quantum mechanics can be maintained.*

The *curvature* of the connection from (16.44) is essentially the electromagnetic field 2-form.

$$\theta = d\omega = -\left(\frac{ie}{\hbar}\right) dA^1 = -\left(\frac{ie}{\hbar}\right) F^2 \tag{16.49}$$

$$= -\left(\frac{ie}{\hbar}\right)[\mathcal{E} \wedge dt + \mathcal{B}]$$

Finally, we shall make some remarks about Schrödinger's equation in curvilinear coordinates. Consider a Riemannian manifold $M$, the most important case being $\mathbb{R}^3$ with a curvilinear coordinate system. Let $E$ be a complex vector bundle with connection $\omega$, for example the wave function bundle with $\omega = -ieA/\hbar$. We suppress the bundle indices on $\omega$ and on $\psi$. For covariant derivative we have $\psi_{/j} = \partial_j \psi + \omega_j \psi$. This represents a cross section of the bundle $E \otimes T^*M$, that is, the bundle of covariant vectors on $M$ whose values are in $E$. As we have seen in Problem 16.3(1), the covariant derivative $\psi_{/jk} = (\psi_{/j})_{/k}$ of this tensor will involve not only the connection $\omega$ for $E$ but also the Riemannian connection $\Gamma$

$$\begin{aligned}
\psi_{/jk} = (\psi_{/j})_{/k} &= \partial_k \psi_{/j} + \omega_k \psi_{/j} - \Gamma^r_{kj} \psi_{/r} \\
&= \partial_k(\partial_j \psi + \omega_j \psi) + \omega_k(\partial_j \psi + \omega_j \psi) - \Gamma^r_{kj}(\partial_r \psi + \omega_r \psi) \\
&= [\partial_k \partial_j \psi - \Gamma^r_{kj} \partial_r \psi] + \{\partial_k(\omega_j \psi) + \omega_k(\partial_j \psi + \omega_j \psi) - \Gamma^r_{kj} \omega_r \psi\}
\end{aligned}$$

which is now a covariant second rank tensor on $M$ with values in $E$. Then the "Laplacian"

$$\nabla^2 \psi := g^{jk} \psi_{/jk}$$

is again simply *a section of* $E$. In slightly more detail

$$\nabla^2 \psi = g^{jk} \psi_{/jk} = g^{jk}[\partial_k \partial_j \psi - \Gamma^r_{kj} \partial_r \psi] + g^{jk}\{\,\}$$

The term involving the square brackets $g^{jk}[\,]$ is simply the Laplacian of the "function" $\psi$, using the Riemannian connection

$$\frac{1}{\sqrt{g}} \frac{\partial}{\partial x^j}\left[\sqrt{g} g^{jk}\left(\frac{\partial \psi}{\partial x^k}\right)\right]$$

(see equation (11.30)). A candidate then for Schrödinger's equation for a charged particle in an electromagnetic field on $M$ would be

$$i\hbar\left(\frac{\partial}{\partial t} - \frac{ie\phi}{\hbar}\right)\psi = -\left(\frac{\hbar^2}{2m}\right)g^{jk}\psi_{/jk} + V\psi$$

**Summary.** When no electromagnetic field is present, the Hamiltonian is of the form.

$$\frac{1}{2m}\sum_{\alpha} g^{\alpha\beta} p_\alpha p_\beta + V$$

in curvilinear coordinates or on a Riemannian manifold $M$. We replace $p_\alpha$ by the Riemannian covariant derivative $-i\hbar\nabla^M_\alpha$. Schrödinger's equation becomes

$$i\hbar\frac{\partial\psi}{\partial t} = \left(-\frac{\hbar^2}{2m}\right)\frac{1}{\sqrt{g}}\frac{\partial}{\partial x^\alpha}\left[\sqrt{g}g^{\alpha\beta}\left(\frac{\partial\psi}{\partial x^\beta}\right)\right] + V\psi$$

*The only effect of introducing an electromagnetic field now is to replace the trivial wave function bundle by the bundle E with connection* $\omega = -ieA/\hbar$, and we must use the *full* covariant derivative (using both $\Gamma$ and $\omega$)

$$i\hbar\psi_{/0} = \left(-\frac{\hbar^2}{2m}\right)g^{\alpha\beta}\psi_{/\alpha\beta} + V\psi$$

In this procedure there is no need to first introduce the new canonical momenta $p^*$ in the classical system augmented by the electromagnetic field!

## 16.4d. Global Potentials

In most problems involving electromagnetics the vector potential 1-form $A^1$ is globally defined. We can see this as follows. Consider first a smooth electromagnetic field $F^2$ in all of Minkowski space, $M_0^4$. Since $M_0^4 = \mathbb{R}^4$ has second Betti number $b_2 = 0$, de Rham's theorem assures us that there is a potential 1-form $A^1$ for the closed 2-form $F^2$, $F^2 = dA^1$. Usually, however, there are singularities of $F^2$, located, for example, at the moving point charges. We cannot apply de Rham's theorem to singular forms; thus in order to use de Rham's theorem we must first remove the singularities of $F^2$ from $M_0^4$, leaving an open subset $U$ of $M_0^4$. Now, however, there is no reason to assume that $b_2(U) = 0$; for example, a fixed charge at the origin of $\mathbb{R}^3$ yields an entire $t$ axis of singularities in $M_0^4$, and the 2-sphere in $\mathbb{R}^3$ surrounding the origin is a 2-cycle of $U = M_0^4-$ (the $t$ axis) that does not bound. In spite of the fact that $U$ may have nonbounding 2-cycles, we still have the following:

**Theorem (16.50):** *Consider a region $U$ of a general relativistic space–time $M^4$ that has a global time coordinate, that is, $U$ is of the form $V^3 \times \mathbb{R}$ with $V^3$ a spacelike hypersurface and $t$ a global coordinate for $\mathbb{R}$. Suppose that the magnetic field $\mathcal{B}^2$ vanishes at time $t = 0$. Then $F^2$ has a globally defined potential $A^1, dA^1 = F^2$, on all of $U$.*

**PROOF:** $F^2$ is closed, $dF = 0$. By de Rham's theorem, we need only show that the integral of $F^2$ over each 2-cycle $z$ of $U$ vanishes. But $z$ can be deformed, by the deformation $\phi_\alpha(\mathbf{x}, t) = (\mathbf{x}, (1-\alpha)t)$, into a homologous *spatial* cycle $z'$ that lies in the hypersurface where $t = 0$. Since $F^2 = \mathcal{E}^1 \wedge dt + \mathcal{B}^2$ restricts to 0 on the deformed cycle, $\int_z F^2 = \int_{z'} F^2 = 0$. □

For a simpler discussion in $\mathbb{R}^3$, see Problem 16.4(1).

In the standard cosmological models, the Friedmann universes, there is a global time coordinate (see [F, chap. 12]). Thus the only way $F^2$ can avoid having a global potential today in these models is for there to have existed, *since the time of the big bang,* a nonbounding 2-cycle, and a magnetic field with nonzero flux through this cycle. (Some of the Friedmann models do have $b_2 \neq 0$; for example, there are models where the spatial sections $V^3$ are flat 3-dimensional tori $T^3$, and others with $V^3$ closed manifolds with negative curvature and $b_2 \neq 0$.)

### 16.4e. The Dirac Monopole

If there *is* a global potential $A^1$, there is then no *necessity* for introducing a bundle whose sections will serve as local wave functions, since one global patch $U$ will suffice. It may very well be though that when considering other fields, for example the Yang–Mills fields, to be discussed later, we shall not be so fortunate, and in that case we shall be forced to introduce bundles and connections, as we have had to do in the case of gravitation in general relativity. There is, however, a much simpler situation that requires bundles, namely the Dirac magnetic monopole (which, however, has never been shown to exist).

Consider then an electron moving in $\mathbb{R}^3 - \{O\}$ in the field of a magnetic monopole of strength $q$ fixed at the origin. The $\mathbf{B}$ field for this monopole is $\mathbf{B} = (q/r^2)\partial/\partial r$, that is (see equation (5.9)),

$$\mathcal{B}^2 = i_{\mathbf{B}} \operatorname{vol}^3 = \mathbf{d}[q(1-\cos\theta)d\phi]$$

Thus

$$\mathcal{A}_U^1 = q(1-\cos\theta)d\phi$$

in the region $U = \mathbb{R}^3 - \{\text{negative } z \text{ axis}\}$. We shall need also to consider points on the negative $z$ axis (except for the origin). In the region $V = \mathbb{R}^3 - \{\text{positive } z \text{ axis}\}$ we can use $\theta' = \pi - \theta$ and $\phi' = -\phi$ as coordinates and get

$$\mathcal{A}_V^1 = -q(1+\cos\theta)d\phi$$

Maxwell's equations hold everywhere on $\mathbb{R}^3 - \{0\}$.

Since $\mathcal{A}_U^1$ does not agree with $\mathcal{A}_V^1$ in $U \cap V = \mathbb{R}^3 - \{z \text{ axis}\}$, we shall be *forced* to introduce the electromagnetic bundle and connection of Section 16.4c. In Problem 16.4(2) you are asked to show that the transition function for this **monopole bundle** is

$$c_{VU} = \exp\left(-\frac{2ieq\phi}{\hbar}\right) \tag{16.51}$$

Note that this is not single-valued unless *Dirac's quantization condition*

$$\frac{2eq}{\hbar} \text{ must be an integer} \tag{16.52}$$

is satisfied. If this condition is not satisfied we shall have failed in our attempt to construct a bundle. Since there are only two patches $U$ and $V$, Equation (16.3) is automatically satisfied. Thus if (16.52) holds, the monopole bundle will exist.

That $c_{VU}$ is not in general single-valued is a reflection of the fact that in this case $U \cap V = \mathbb{R}^3 - \{\text{entire } z \text{ axis}\}$ is certainly not simply connected (more to the point, its first Betti number does not vanish). It is true that by using more sets (whose intersections are simply connected) to cover $\mathbb{R}^3 - \{0\}$, we could find transition functions that would be single-valued without requiring (16.52), but it would turn out that *it would not be possible to satisfy the crucial equation* $c_{UV}c_{VW}c_{WU} = 1$. In fact, we shall prove in Section 17.4, from a general Gauss–Bonnet theorem, that for *any* complex line bundle over $\mathbb{R}^3$-origin, the curvature *must* satisfy

$$\frac{i}{2\pi}\int_{S^2} \theta^2 = \text{integer} \tag{16.53}$$

The unit sphere $S^2$ is a generator for the second homology group $H_2(\mathbb{R}^3 - \{0\}, \mathbb{Z})$. (Note that we have already proved

$$\frac{i}{2\pi}\int_{M^2} \theta^2 = \text{integer}$$

in the geometrical case when the complex line bundle is the tangent bundle to the oriented closed surface $M^2$; see (9.66) and (16.30)!) For the monopole bundle, from (16.49),

$$\theta = -\frac{ie}{\hbar}\mathcal{B} = -\frac{ie}{\hbar}i_{\mathbf{B}}\,\text{vol}^3 = -\frac{ieq}{r^2\hbar}i_{\partial/\partial r}\,\text{vol}^3$$

and thus the integral in (16.53)) becomes

$$-\left(\frac{i}{2\pi}\right)\left(\frac{ieq}{\hbar}\right)(4\pi) = \frac{2eq}{\hbar}$$

Thus, as noted first, I believe, by Sniatycki [Sy]

> *if Dirac's condition (16.52) is not satisfied, there will be no complex line bundle whose sections can serve as wave "functions" for the electron in the field of a magnetic monopole.*

(For a description of the monopole bundle, see Section 17.4c.) This yields a **quantization condition**, relating the charge on a monopole to that of the electrons.

More generally, it will be shown in Chapter 17 that the flux of $e\mathcal{B}/2\pi\hbar$ through any closed oriented surface, for any magnetic field, must be an integer.

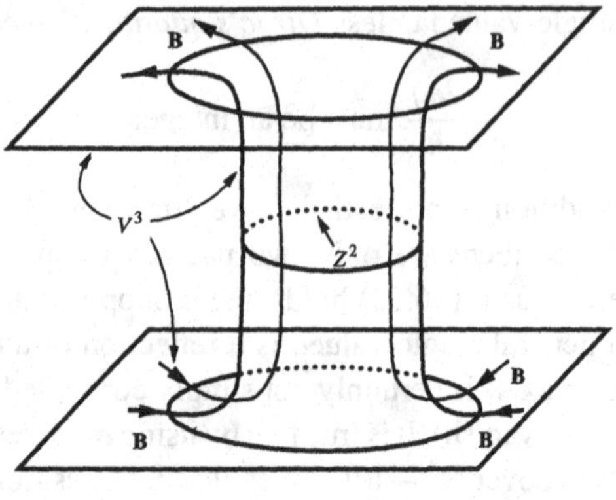


Figure 16.14

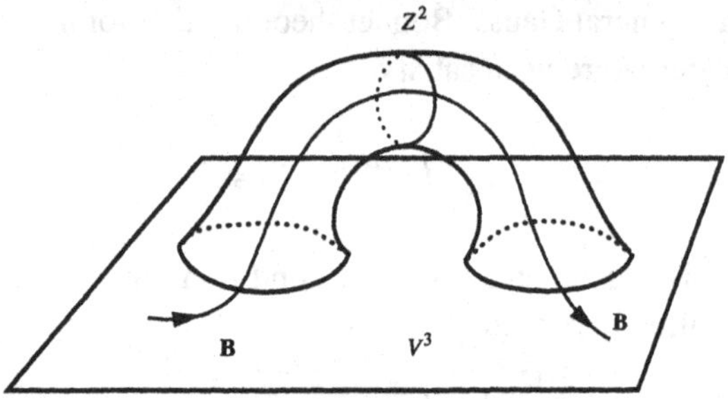


Figure 16.15

In Figure 16.14 we have indicated a $V^3$ that consists of two separated horizontal sheets (two "separate" universes) that are joined by a **wormhole** cylinder $S^2 \times [0, 1]$; we have indicated one of the spherical sections $Z^2$ going around the "throat" of the wormhole. A magnetic field goes from the bottom sheet through the lower "mouth," threads through the throat, and comes out of the top mouth. In this example $\int_Z \mathcal{B}^2 \neq 0$. Figure 16.15 is similar except that the wormhole joins two distant portions of the "same" universe, and again **B** has a nonzero flux through the throat. In both cases there is no global **A** and the flux of **B** through the throat must be quantized in terms of $e$.

Finally, we wish to emphasize one point. If there *is* a monopole, then from (16.51) we see that $\psi_V = \exp(-2ieq\phi/\hbar)\psi_U \neq \psi_U$. Thus the electron wave function $\psi$ cannot be defined (and single-valued) everywhere and it must, rather, be considered as a *section* of the monopole bundle with at least two patches. Presumably, then, we should expect that other types of fields that interact with elementary particles might demand that wave functions be replaced by sections of bundles, just as we do not expect that every manifold should be covered by a single coordinate patch.

### 16.4f. The Aharonov–Bohm Effect

At first sight the electromagnetic connection $\omega = -ieA^1/\hbar$ seems nonphysical since classically the vector potential $A^1$, changing with each choice of gauge, was regarded only as a mathematical tool for describing the physical electromagnetic field $F^2 =$

$dA^1$. We have noted, however, a similar situation in general relativity; the Levi-Civita connection, a gauge field, can be thought of as merely a preliminary mathematical step on the way to its derivative, the Riemann curvature tensor, describing the strength of the gravitational field. However, this gravitational connection is a physical field in the sense that it, with no use of its derivatives, governs parallel displacement. We shall now see that the electromagnetic connection, that is, the vector potential $A^1$, although not a classical physical field, is a physical field in quantum mechanics, and this will be illustrated with the famous Aharonov–Bohm effect.

With a solenoid carrying a current $j$, the circulation of the magnetic field about a closed loop $C$ going around the coil is, by Ampere's law, $\oint_C * \mathcal{B} = 4\pi j$. When $j$ is very small, if the wire is tightly wound, the magnetic field inside the coil can be substantial while $*\mathcal{B}$ outside is very small. In the simplified version of an infinitely long, infinitely tightly wound solenoid, it is assumed that the magnetic field inside is constant and parallel to the axis of the coil, and the magnetic field outside the coil is vanishingly small.

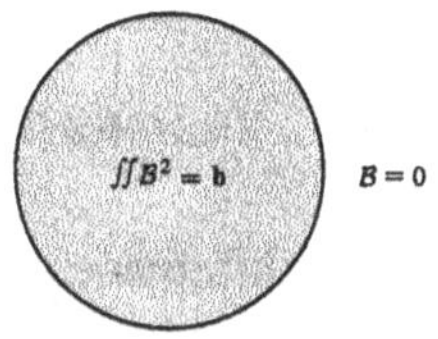


**Figure 16.16**

Let the magnetic flux inside the coil be $\iint \mathcal{B} = b$. Then $\mathcal{A}^1 = bd\theta/2\pi$ is a well-defined vector potential in the region *exterior* to the coil, designed to satisfy both $\oint_C \mathcal{A} = b$ and $d\mathcal{A} = 0$. See Problem 16.4(4) for the potential inside the coil.

It is possible to detect the effect of $\mathcal{A}$ on an electron constrained to the exterior region even though $\mathbf{B} = 0$ in this region; this is the Aharonov–Bohm effect. A brief explanation in terms of path integrals is as follows. (We assume here a slight familiarity with Feynman's method. For more details the reader is referred to Feynman's lectures [F, L, S, vols. II and III], Rabin's article [R ], and the excellent book [Fe] by Felsager. For insight into the path integral formalism (without mentioning integrals!) see Feynman's remarkable book [F].)

An electron is emitted from a source, passes through one of two slits in a screen, moves along a curve $\gamma$, and strikes a screen behind the solenoid at a point $y$. The "probability amplitude" for this process is proportional to the exponential of the classical action for the path $\gamma$

$$\Phi[\gamma] = \exp\left(\frac{i}{\hbar}\int_\gamma L dt\right) \tag{16.54}$$

The principal contribution to the amplitude for going from $x$ to $y$ is given by this expression when the path $\gamma$ is a classical path of "least" action. Since the electromagnetic field vanishes in this exterior region, the classical path will be a straight line from the slit in the screen. We exhibit the two classical paths $C$ and $C'$ (both must be taken into account since we don't know which slit the electron chooses).

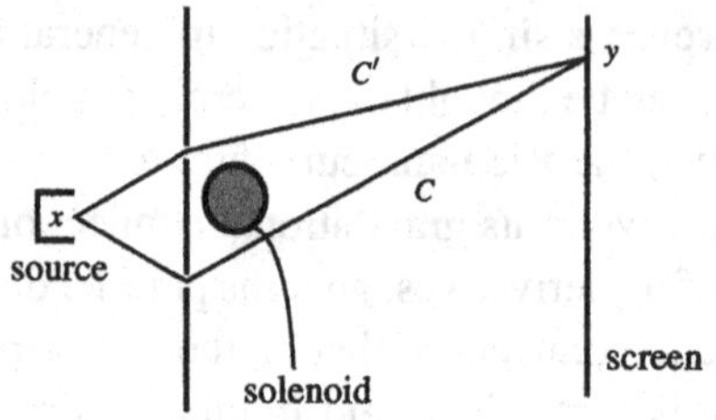


**Figure 16.17**

The **phase** of the complex number $\Phi$ is its angle or argument. The phase *difference*, due to paths $C$ and $C'$, is responsible for the interference pattern observed at the screen. Look at the cases when there is no current in the coil and when the current is flowing. In the first case the Lagrangian is $L_0$. After the current is turned on, there is a vector potential $\mathcal{A}$ present outside the solenoid. Corollary (16.36) then tells us to replace $L = L_0$ by $L = L_0 + e(\dot{x}^\alpha A_\alpha + \phi)$; that is, we replace $Ldt$ by $Ldt + e\mathbf{A} \cdot d\mathbf{x}$, since the scalar potential vanishes. In this new situation the phase difference becomes

$$\left(\frac{1}{\hbar}\right)\left(\oint_{C-C'} L_0 dt + e\mathbf{A} \cdot d\mathbf{x}\right)$$

and this differs from the original phase difference only by

$$\frac{e}{\hbar}\iint \mathcal{B} = \frac{eb}{\hbar}$$

Since this is *independent of* $y$ for any pair $C$, $C'$ of classically extremal paths, Aharonov and Bohm concluded that the original interference pattern will simply be shifted by a constant amount, in spite of the fact that the electron feels no magnetic force in the exterior region! This shift has actually been observed. The field $A$, and thus the connection $\omega$, *are* physical fields in quantum mechanics.

## Problems

**16.4(1)** Let $z^2$ be a closed surface in $\mathbb{R}^3$ that lies outside the singularities of the electromagnetic field. Show directly from Maxwell's equations that $\int_z \mathcal{B}^2$ is constant in time. This shows that if $\mathcal{B}^2$ vanished sometime in the past, then $\mathbf{B} = \text{curl}\,\mathbf{A}$ in the nonsingular set.

**16.4(2)** Derive (16.51).

**16.4(3)** In the monopole bundle with $2eq/\hbar$ an integer $\neq 0$, the function $\psi_U = 1$ is a cross section over $U = \mathbb{R}^3 -$ {negative $z$ axis}. Can $\psi_U$ be extended to be a cross section over all of $\mathbb{R}^3 - \{0\}$? (Look at the proposed $\psi_V$ at points of the negative $z$ axis.)

**16.4(4)** Assume a constant axial magnetic field $Bdx \wedge dy$ inside the coil whose axis is the $z$ axis. (Thus $B = b/\pi a^2$, where $a$ is the radius of the coil.) Of course $Bxdy$ is a covector potential, but to match up with our external potential use cylindrical coordinates $r, \theta, z$, and show that another choice of a covector potential 1-form is given by $A^1 = (B/2)r^2 d\theta$. What is the length $\| \mathbf{A} \|$ of this choice for **A**? What is the length of the exterior version for **A** used in the text? Why don't they match up *smoothly*?

**16.4(5)** Show the **gauge invariance** of Feynman's prescription (16.54) as follows: Let a particle in an electromagnetic field have probability amplitude $\psi_U(x, 0)$ of being at $x \in U$ at time 0. Then the probability amplitude that the particle will traverse a path $\gamma$ from $x$ to $y$, arriving at $y$ at time $t$, is, in Feynman's view

$$\psi_U(x,0)\exp\left[\frac{i}{\hbar}\int_\gamma L dt + e\mathbf{A}_U \bullet d\mathbf{x}\right] = \psi_U(x,0)\exp\left[\frac{i}{\hbar}\int_\gamma L dt\right]\exp\int_\gamma -\omega_U$$

There is a similar expression if we use a different gauge $\psi_V(x, 0) = c_{VU}(x)$ $\psi_U\,(x, 0)$. Show that these two gauges yield compatible results.

CHAPTER 17

# Fiber Bundles, Gauss–Bonnet, and Topological Quantization

A vector bundle is a family of vector spaces parameterized by points in the base space. How do we parameterize a family of manifolds, say Lie groups?

## 17.1. Fiber Bundles and Principal Bundles

### 17.1a. Fiber Bundles

The tangent bundle $TM^n$ to a Riemannian manifold is a vector bundle associated to $M$; it is locally of the form $U \times \mathbb{R}^n$. We have had occasion also to consider the set of *unit* vectors tangent to $M$; that is, we may consider, in each fiber $\pi^{-1}(p) \approx \mathbb{R}^n$ of $TM$ (a vector space with scalar product), the unit sphere $S^{n-1}(p) \subset \pi^{-1}(p)$. The collection of all these unit spheres $S^{n-1}(p)$, as $p$ ranges over $M$, forms a new manifold, called the unit tangent bundle $T_0M$ in Section 2.2b. We again have a projection $\pi : T_0M \to M$. The term bundle refers to the fact that the space is again locally a product in the following sense: $T_0M^n$ is the collection of all $(n-1)$spheres $S^{n-1}(p)$ in all of the tangent spaces to $M$, but there is no *natural* way to identify points in $S(p)$ with points in $S(q)$ for distinct points $p$ and $q$ in $M$. Choose an orthonormal frame $\mathbf{e}^U = (\mathbf{e}_1, \ldots, \mathbf{e}_n)$ in a patch $U$ of $M$ and take a fixed unit sphere $S$ in some euclidean space $\mathbb{R}^n$. We may then identify each tangent sphere $S(p)$, at $p \in U$, with the fixed sphere $S$, by identifying $\mathbf{v} = (\mathbf{e}_i v^i) \in S^{n-1}(p)$ with $s = (v^1, \ldots, v^n) \in S^{n-1} \subset \mathbb{R}^n$; thus $(p, \mathbf{v})$ is identified with $s$. We then have a diffeomorphism

$$\Phi_U : U \times S^{n-1} \to \pi^{-1}(U) \subset T_0M$$

exhibiting the local product structure. Of course if we go into another patch $V$, using a new frame $\mathbf{e}^V$, then we shall get a different identification. This space $T_0M$ is a "fiber bundle," but not a vector bundle, because the fiber $S^{n-1}$ is a manifold that is not a vector space. We may now define this new notion in general. $T_0M$ is atypical, since it is a subbundle of the vector bundle $TM$.

A **fiber bundle** consists of the following: There are a manifold $F^k$ (called the *fiber*), a manifold $E$ (the *bundle space*) and a manifold $M^n$ (the *base space*) together with a map

$$\pi : E \to M^n$$

of $E$ onto $M$. We demand that $E$ is locally a product space in the following sense: There is a covering of $M^n$ by open sets $U, V, \ldots$, such that $\pi^{-1}(U)$ is diffeomorphic to $U \times F$; there is a diffeomorphism

$$\Phi_U : U \times F \to \pi^{-1}(U) \tag{17.1}$$

with $\Phi_U(p, y_U) \in \pi^{-1}(p)$ for each $y_U \in F$. Then for each $p \in U$ the assignment $y \in F \to \Phi_U(p, y) \in \pi^{-1}(p)$ is a diffeomorphism; that is, *the fiber $\pi^{-1}(p)$ over $p$ is a diffeomorphic copy of the fiber $F$ of the bundle.* In an overlap $U \cap V$ a point $e \in \pi^{-1}(U \cap V)$ will have two representations

$$e = \Phi_U(p, y_U) = \Phi_V(p, y_V)$$

and we demand that

$$y_V = c_{VU}(p)[y_U] \tag{17.2}$$

where $c_{VU}(p) : F \to F$ is a *diffeomorphism* of the fiber. In the case of a vector bundle, $F = \mathbb{R}^K$ or $\mathbb{C}^K$, each $c_{VU}(p) : \mathbb{R}^K \to \mathbb{R}^K$ was a *linear* transformation, but now, of course, $F$ is a manifold and need not have a linear structure.

The set of all diffeomorphisms of a manifold $F$ clearly form a group in the sense of algebra. (It is not a Lie group; e.g., the diffeomorphisms of $\mathbb{R}^2$ form, in a sense, an *infinite*-dimensional manifold). If all the maps $c_{UV}(p)$ lie in a subgroup $G$ of the group of all diffeomorphisms of $F$ we say that $G$ is the **(structure) group** of the fiber bundle.

In the case of the unit tangent bundle $T_0M$ to a Riemannian manifold, by using orthonormal frames as we did earlier, each $c_{VU}(p) : S^{n-1} \to S^{n-1}$ is the restriction of an orthogonal transformation $\mathbb{R}^n \to \mathbb{R}^n$ to the unit sphere $S^{n-1} \subset \mathbb{R}^n$. Thus by employing orthonormal frames we reduce the structure group of the fiber from the group of all diffeomorphisms of $S^{n-1}$ to the subgroup $O(n)$ of orthogonal transformations of the sphere.

In the case of the normal real line bundle to the midcircle $M^1$ of the Möbius band (see Figure 16.1), we may choose *unit* sections $\mathbf{e}_U$ and $\mathbf{e}_V$ in $U$ and $V$, respectively. On the two pieces of the intersection $U \cap V$ we have in one case $\mathbf{e}_U = \mathbf{e}_V$ and in the other $\mathbf{e}_U = -\mathbf{e}_V$. *Thus the structure group of this normal bundle is the 2-element multiplicative group* $\{\pm 1\}$, which is easily seen to be another version of the additive group $\mathbb{Z}_2$.

Of course we still demand

$$c_{VU}(p) = c_{UV}(p)^{-1}$$

and (17.3)

$$c_{UW} \circ c_{WV} \circ c_{VU} = \text{identity on } U \cap V \cap W$$

As in the case of a vector bundle, a fiber bundle over $M$ can be constructed as soon as transition functions satisfying (17.3) are prescribed.

A (local cross) section is again a map $s : U \to E$ such that $\pi \circ s =$ identity. A **section** $s$ is simply a collection of maps $\{s_U : U \to F\}$ such that in $U \cap V$ we have $s_U(p) = c_{UV}(p)[s_V(p)]$.

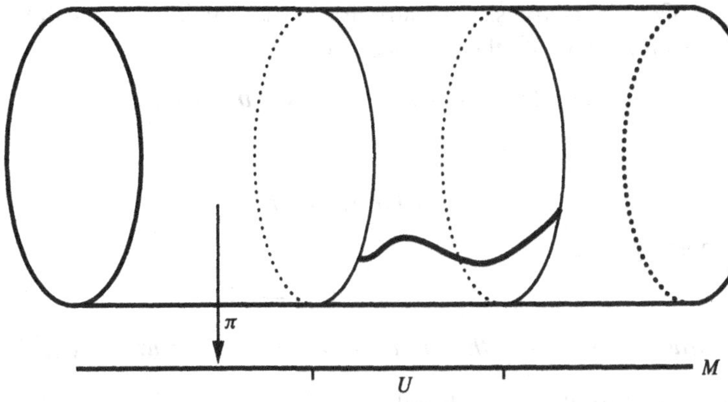


**Figure 17.1**

In Section 17.2 we shall see many examples of fiber bundles.

### 17.1b. Principal Bundles and Frame Bundles

Let $M^n$ be Riemannian and let *FM* be the collection of all orthonormal *frames* $\mathbf{f}_1, \ldots, \mathbf{f}_n$ of vectors at points of $M$. $\pi : FM \to M$ assigns to each frame $\mathbf{f} = (\mathbf{f}_1, \ldots, \mathbf{f}_n)$ the point $p$ of $M$ at which the frame is located. What is the fiber $\pi^{-1}(p)$ over $p$? Let $\mathbf{e}$ be a given frame at $p$; then the most general frame $\mathbf{f}$ at $p$ is of the form

$$\mathbf{f} = \mathbf{e}g \qquad \text{i.e.,} \qquad \mathbf{f}_\beta = \mathbf{e}_\alpha g^\alpha{}_\beta$$

where the matrix $[g^\alpha{}_\beta] = g \in G = O(n)$. Thus *after* a single frame $\mathbf{e}$ at $p$ has been chosen, the fiber $\pi^{-1}(p)$ of all orthonormal frames at $p$ can be identified with the structure group $G = O(n)$ of orthogonal $n \times n$ matrices. The *fiber* for the frame bundle *FM* is the Lie group $O(n)$. How do we exhibit the local trivialization (product structure)? Let $\mathbf{e}$ be an orthonormal frame field on an open set $U \subset M$; for example, we can apply the Gram–Schmidt process to a coordinate frame in a patch. Then a general orthonormal frame $\mathbf{f}$ on $U$ is uniquely of the form

$$\mathbf{f}(p) = \mathbf{e}_U(p) g_U(p) \tag{17.4}$$

Thus the frame $\mathbf{f}$ in $U$ is completely described by giving the point $p$ and the matrix $g_U$.

The local trivialization

$$\Phi_U : U \times G \to \pi^{-1}(U)$$

assigns to each $p \in U$ and each $g \in G$ the frame

$$\Phi_U(p, g) := \mathbf{e}_U(p) g$$

In an overlap, the same frame (17.4) will have another representation

$$\mathbf{f}(p) = \mathbf{e}_V(p) g_V(p)$$

where (17.5)

$$\mathbf{e}_V(p) = \mathbf{e}_U(p) c_{UV}(p)$$

$c_{UV}(p) \in G = O(n)$ is the transition matrix for the tangent bundle (recall that in $TM$, for a vector $\mathbf{y} = \mathbf{e}_U y_U = \mathbf{e}_V y_V$, then $y_V = c_{VU} y_U$). Then

$$\mathbf{f} = \mathbf{e}_U(p) g_U(p) = \mathbf{e}_V(p) g_V(p) = \mathbf{e}_U(p) c_{UV}(p) g_V(p)$$

gives

$$g_U(p) = c_{UV}(p) g_V(p) \tag{17.6}$$

Thus the diffeomorphism

$$c_{UV}(p) : [G = O(n)] \to G$$

is simply *left translation of $G$ by the (transition) orthogonal matrix $c_{UV}(p)$*!

> In general we shall say that a fiber bundle
>
> $$\{P, M, \pi, F, G\}$$
>
> is a **principal bundle** *if the fiber $F$ is the same as the group $G$*, and if the transition functions $c_{UV}(x)$ act on $F = G$ by *left translations.*

The frame bundle $FM$ is the principal bundle **associated** with the tangent (vector) bundle $TM$.

> By exactly the same procedure, given any vector bundle $E \to M$ with fiber $\mathbb{R}^k$, we can, by considering **frames** of $k$ linearly independent local cross sections, construct the associated principal bundle $P$ whose fiber is the structure group of the original vector bundle.

### 17.1c. Action of the Structure Group on a Principal Bundle

The frame bundle has a remarkable property that is not shared with the tangent vector bundle: *the structure group $G$ acts in a natural way as a group of transformations on $FM$*. Let $g \in G$ be a given matrix and let $\mathbf{f} = (\mathbf{f}_1, \ldots, \mathbf{f}_n)$ be a frame at $p$, that is; $\mathbf{f}$ is a point in $FM$. Then we can let $g$ send this point $\mathbf{f}$ into the new point $g(\mathbf{f}) := \mathbf{f}g$ by the usual

$$(\mathbf{f}g)_\beta = \mathbf{f}_\alpha g^\alpha{}_\beta \tag{17.7}$$

Note that this assignment is *intrinsic*: we have not used the local product structure! *There is*, however, *no natural action of $G$ on the tangent bundle itself.* For example, if $M^3$ is 3-dimensional and if $\mathbf{v}$ is a tangent vector at $p$ and if $g$ is a $3 \times 3$ matrix, what would you like $g(\mathbf{v})$ to be? We cannot assign a column to $\mathbf{v}$ without first assigning a basis for $M^3(p)$, and assigning a particular basis is very unnatural! It is because $FM$ *is* the space of bases that we succeeded in (17.7).

This works for any principal bundle, namely:

> **Theorem (17.8):** *The structure group $G$ of a principal bundle $P$ acts "from the right" on $P$*
>
> $$(\mathbf{f} \in P, g \in G) \to (\mathbf{f}g) \in P$$
>
> *without fixed points when $g \neq e$, and preserves fibers (i.e., $\pi(\mathbf{f}g) = \pi(\mathbf{f})$).*

**PROOF:** We first define the action locally. Let $\mathbf{f} \in P$ and let $\pi(\mathbf{f}) = p$ lie in some open $U$ over which $P$ is trivial

$$\Phi_U(U \times G) = \pi^{-1}U$$

Then we can write uniquely

$$\mathbf{f} = \Phi_U(p, f_U)$$

that is, $\mathbf{f}$ has the local "coordinate" $f_U \in G$. We define $\mathbf{f}g$ to be the point with local coordinate $f_U g$,

$$(\mathbf{f}g)_U = f_U g$$

This is in fact coordinate independent, for in an overlap $U \cap V$, $\mathbf{f}$ would have $f_V = c_{VU}(p) f_U$ and then

$$(\mathbf{f}g)_V = f_V g = c_{VU}(p) f_U g = c_{VU}(p)(\mathbf{f}g)_U \quad \square$$

We see in this proof that the essential point is that *left translations* in $G$ (say by $c_{VU}$) *commute with right translations* (say by $g$).

We can use the same *notation* in a principal bundle that we used in the frame bundle. Over $U$ we may consider the local section $\mathbf{e}_U$

$$\mathbf{e}_U(p) := \Phi_U(p, I)$$

where $I$ is the identity matrix in $G$. Then for any point $\mathbf{f} \in \pi^{-1}(p)$ we may write

$$\mathbf{f} = \Phi_U(p, f_U) = \Phi_U(p, I f_U) = \mathbf{e}_U(p) f_U \tag{17.9}$$

for a unique $f_U \in G$.

Each right action $\mathbf{f} \to \mathbf{f}g$ is a diffeomorphism : $P \to P$. Let the 1-parameter subgroup $e^{tA}$, $A \in \mathfrak{g}$, act. The resulting velocity vector field on $P$ is then

$$A^* \text{at } \mathbf{f} := \frac{d}{dt}[\mathbf{f}e^{tA}]_{t=0}$$

In terms of the local product structure, $\mathbf{f} = \mathbf{e}_U f_U$, and then

$$A^*(\mathbf{f}) = \frac{d}{dt}[\mathbf{e}_U f_U e^{tA}]_{t=0}$$

The action $\mathbf{f} \to \mathbf{f}e^{tA}$ on $P$ is completely described by the action in $G$

$$f_U \to f_U e^{tA}$$

whose velocity vector at $f_U \in G$ is

$$\frac{d}{dt}[f_U e^{tA}]_{t=0} = f_U A$$

the left translate of $A$ to $f_U$. The vector field $A^*$ on the principal bundle $P$ generated by $A \in \mathfrak{g}$ is said to be the **fundamental vector field** associated to $A$.

## 17.2. Coset Spaces

What do subgroups and cosets have to do with fiber bundles?

### 17.2a. Cosets

Let $G$ be a Lie group and $H \subset G$ a subgroup. The **left coset space** is the set of equivalence classes of elements of $G$

$$g \approx g' \qquad \text{iff} \qquad g' = gh$$

for some $h \in H$. Thus $g \approx g'$ iff $g^{-1}g' \in H$. (We discussed this in the case of abelian groups in Section 13.2c; we called it there the quotient space. In abelian groups one may write $g' \approx g$ iff $g' = g + h$.)

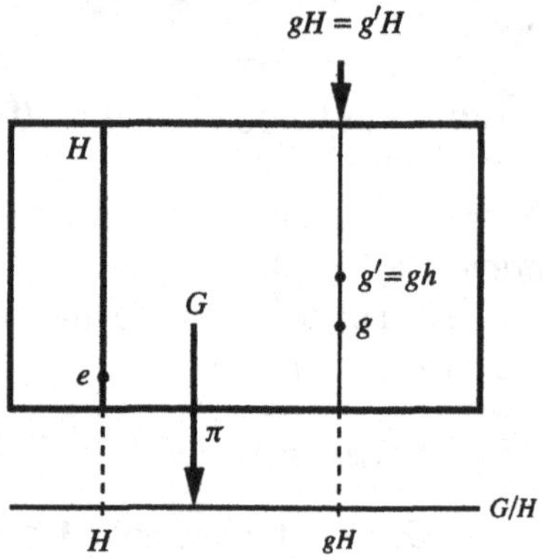


Figure 17.2

Thus we identify all elements of $G$ that lie on the same *left* translate

$$gH := \{gh | h \in H\}$$

of the subgroup $H$. We denote the equivalence class of the element $g \in G$ by $[g]$ or else $gH$. The map that sends $g$ into its equivalence class will be denoted by $\pi$.

Many familiar spaces are in fact coset spaces! Let us say that a group $G$ **acts (as a transformation group)** on a space $M$ provided there is a map

$$G \times M \to M$$
$$(g, x) \mapsto gx$$

such that

$$(gg')x = g(g'x) \quad \text{and } ex = x$$

If, furthermore, given any pair $x$, $y$, of points of $M$, there is at least one $g \in G$ that takes $x$ to $y$, $gx = y$, we say that $G$ acts **transitively** on $M$.

**Example**: $SO(3)$ acts transitively on the 2-sphere

$$SO(3) \times S^2 \to S^2$$

as the group of rotations.

**Fundamental Principle (17.10):** *Let $G$ act **transitively** on a set $M$. Let $x_0 \in M$ and let $H \subset G$ be the subgroup leaving $x_0$ **fixed**,*

$$H = \{g \in G | gx_0 = x_0\}$$

*$H$ is called the **stability**, or **isotropy**, or **little** subgroup of $x_0$. Then the points of $M$ are in $1:1$ correspondence with the left cosets $\{gH\}$ of $G$.*

The space of left cosets is again written $G/H$. Unlike the case when $G$ is abelian, $G/H$ is usually not itself a group.

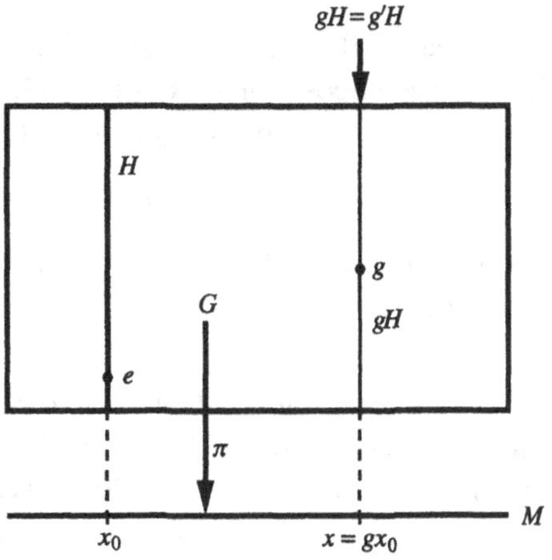


**Figure 17.3**

PROOF: Let $x_0$ be a point of $M$. Associate to $g \in G$ the point $x = gx_0$ where $g$ takes the distinguished point $x_0$. Since $ghx_0 = x$ also, for all $h \in H$, we see that under this assignment, the whole coset $gH$ is associated to this same $x$. We then have a correspondence $G/H \to M$. Conversely, to each $x \in M$ we may associate $\{g \in G : gx_0 = x\}$, which is easily seen to be an entire coset of $G$. □

**Example**: $SO(3)$ acts transitively on the 2-sphere $M = S^2$. Let $x_0$ be the north pole, $x_0 = (0, 0, 1)^T$. The little group of $x_0$ is clearly the 1-parameter subgroup of rotations about the $z$ axis.

$$H = SO(2) \subset SO(3) \qquad SO(2) = \begin{bmatrix} \cos\theta & -\sin\theta & 0 \\ \sin\theta & \cos\theta & 0 \\ 0 & 0 & 1 \end{bmatrix}$$

for all $\theta$. We conclude that

$$SO(3)/SO(2) \approx S^2$$

In our usual picture of $SO(3)$ as the ball with identifications, Figure 17.4, $SO(2)$ is the curve $C$. Note that all rotations through $\pi$ about axes in the $xy$ plane send $x_0$ to the south pole. Thus the coset of the rotation diag $(1, -1, -1)$ is the curve $C'$.

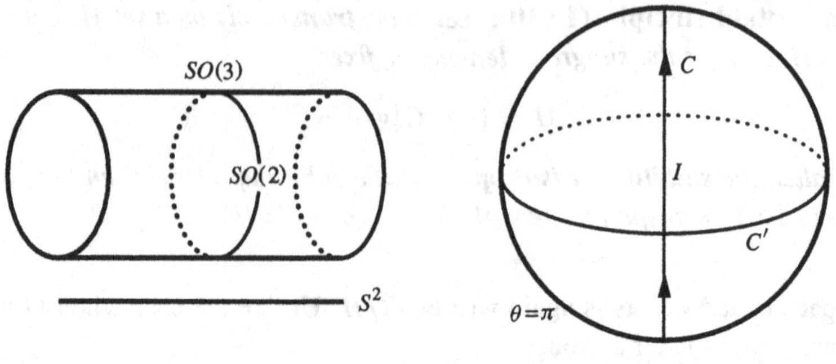


Figure 17.4

The coset $C'$ is *not* a subgroup; it does not contain the identity.

Note that in any (left) coset decomposition $\pi : G \to G/H$, the subgroup $H$ acts on $G$ *from the right* as a group of transformations of $G$ that sends each coset into itself

$$h \in H \text{ sends } g \in G \text{ into } gh; \qquad gH \mapsto gHh = gH$$

The following is a very important fact. We shall not prove this theorem here but we will make some comments about it.

**Theorem (17.11):** *Let G be a Lie group and let H be a closed subgroup (i.e., H contains its accumulation points). Then G/H can be made into a manifold of dimension dim G* − dim *H. Furthermore, G is a* ***principal bundle*** *with structure group H and base space M = G/H and π : G → G/H is the projection of the bundle space onto the base.*

A coset space $M = G/H$ of a Lie group is called a **homogeneous space.**

For example, $S^2$ is a homogeneous space, being the coset space $SO(3)/SO(2)$ of dimension 3 − 1.

**Remarks on the Proof of Theorem (17.11):** We indicate briefly why the cosets in a neighborhood of the coset $eH = H$ can be considered a manifold of dimension $\dim G - \dim H$.

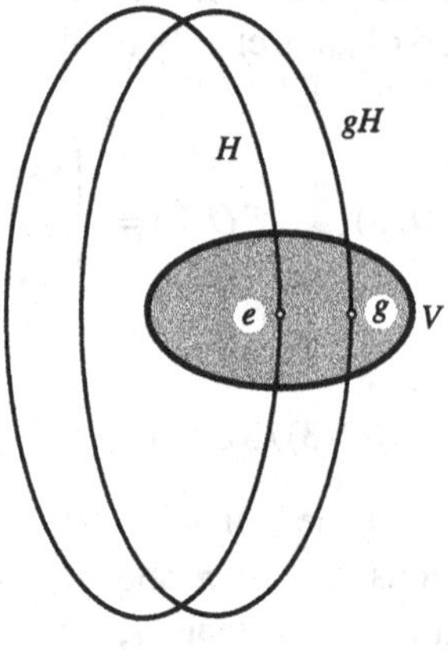


Figure 17.5

Let $V$ be an embedded submanifold of $G$, passing through $e$, transverse to $H$, and of dimension complementary to $H$ (a "normal disc"). An essential fact that can be proved is that if $V$ is sufficiently small, each coset $gH$ of $H$ will either miss $V$ or else strike $V$ in exactly one point. For this, it is important that the subgroup $H$ be *closed* in $G$; if, e.g., $H$ were a line winding densely on the torus $G = T^2$ of Section 15.4d, then surely if $H$ met the transversal $V$ once it would meet it an infinite number of times!). If a coset of $H$ meets $V$ we may say that this coset is near $H$. A coset near $H$ is of the form $gH$ for some unique $g \in V$. This shows that the points of $G/H$ "near $eH$" are in $1:1$ correspondence with the points of the "slice" $V$. Locally $G/H$ is a manifold of the same dimension as $V$, that is, of dimension (dim $G-$dim $H$). For details see, for example, Warner's book [Wa].

### 17.2b. Grassmann Manifolds

The real projective plane $\mathbb{R}P^2$ is the set of *unoriented* lines through the origin of $\mathbb{R}^3$ ($S^2$ is the set of oriented lines). The orthogonal group $O(3)$ acts transitively on the space of lines. Let $\ell_0$ be the $x$ axis line. The subgroup that sends this line into itself consists of orthogonal matrices that either leave the $x$ axis pointwise fixed or else reverses the $x$ axis. These orthogonal matrices automatically send the $yz$ plane into itself, that is, they act as $O(2)$ does on the $yz$ plane. Since $O(1)$ consists of the two numbers $\{-1, 1\}$, we can write the isotropy subgroup of $\ell_0$ as

$$\begin{bmatrix} 0(1) & 0 \\ 0 & 0(2) \end{bmatrix} = 0(1) \times 0(2) \subset 0(3)$$

Thus $\mathbb{R}P^2$ may be identified with the coset space

$$\mathbb{R}P^2 = \frac{O(3)}{O(1) \times O(2)}$$

The dimension of a cartesian product $M^r \times V^s$ of manifolds is $(r+s)$. Thus $\mathbb{R}P^2$ *is* a manifold of dimension $3 - (0+1)$.

The set of unoriented $k$-planes through the origin of $\mathbb{R}^n$ is called a **Grassmann manifold** and is frequently denoted by $\mathrm{Gr}(k, n)$. (**Beware**: there are different notations.) Thus $\mathrm{Gr}(1, 3) = \mathbb{R}P^2$.

## Problems

**17.2(1)** Exhibit $\mathrm{Gr}(k, n)$ as a coset space and compute its dimension.

**17.2(2)** $SO(3)$ acts transitively on $\mathbb{R}P^2$. Let $\ell_0$ be the unoriented $z$ axis. Show that we can write $\mathbb{R}P^2$ as the coset space $SO(3)/H$, where $H$ is the subgroup $C \cup C'$ consisting of the two curves $C$ and $C'$ considered in Figure 17.4.

**17.2(3)** We know that the collection of all frames of $n$ orthonormal vectors at the origin of $\mathbb{R}^n$ can be identified with the group $O(n)$. Show more generally that the *space of all orthonormal $k$-frames* $(\mathbf{f}_1, \dots, \mathbf{f}_k)$ at the origin of $\mathbb{R}^n$ forms a homogeneous space that can be written $O(n)/O(n-k)$. This space is called a ***Stiefel manifold***. What does this say about $S^{n-1}$?

## 17.3. Chern's Proof of the Gauss–Bonnet–Poincaré Theorem

What is an "Index Theorem" ?

### 17.3a. A Connection in the Frame Bundle of a Surface

Let $M^2$ be an oriented surface with a Riemannian metric. Its frame bundle is an $\mathbb{R}^2$ bundle with structure group $SO(2)$. Although we could proceed with this real 2-plane bundle, for our purposes it is more convenient to use instead the complex line bundle version of Section 16.3c. We shall, however omit the superscript $^c$ when discussing the connection and the curvature. There should be no confusion since $\omega$ and $\theta$ will carry no matrix indices, being $1 \times 1$ matrices of forms.

Let then $E = TM$ be the complex tangent line bundle to $M^2$. As in Section 16.3c, the structure group of this bundle is the unitary group $U(1)$, that is, the complex numbers $e^{i\alpha}$ of absolute value 1. A frame at a point $p$ is simply a *unit tangent vector* $\mathbf{e}$ at this point. Let *FM* be the frame bundle, with fiber and group the circle $G = U(1)$. (Note that in this simple case, *FM* is simply the unit tangent bundle to $M^2$!) For $g \in G$

$$g = e^{i\alpha} \tag{17.12}$$

Let $\mathbf{e}_U$ be a frame, that is, a unit vector field on $U$ (i.e., $\mathbf{e}_U$ is a section of *FM* over $U$). As in Section 16.3c (and omitting the tensor product sign and the superscript $^c$), the connection form $\omega$ is a *single* pure imaginary 1-form

$$\nabla \mathbf{e}_U = \mathbf{e}_U \otimes \omega_U = \mathbf{e}_U \omega_U$$

The fact that $\omega$ is pure imaginary, that is, skew hermitian, arose because we demanded that *parallel translation preserves lengths*. To see this, consider the section $\mathbf{e}_U$. Let $x = x(t)$ be a curve in $U$ starting at $x(0)$. Parallel translate $\mathbf{e}_U(x(0))$ along $x(t)$ yielding a unit vector field (frame) $\hat{\mathbf{e}}(t)$. Then $\hat{\mathbf{e}}(t) = \mathbf{e}_U(x(t))g(t)$ for some $g(t) \in U(1)$; that is, $g(t)$ is in the structure group. But then

$$\begin{aligned} 0 = \frac{\nabla \hat{\mathbf{e}}}{dt} &= \left(\frac{\nabla \mathbf{e}}{dt}\right) g + \mathbf{e}\frac{dg}{dt} \\ &= \mathbf{e}\omega\left(\frac{d\mathbf{x}}{dt}\right) g + \mathbf{e}\frac{dg}{dt} \end{aligned}$$

and so

$$\omega\left(\frac{d\mathbf{x}}{dt}\right) = -\left(\frac{dg}{dt}\right) g^{-1} \tag{17.13}$$

for this particular $g = g(t)$ defining parallel translation along $x = x(t)$. Thus the value of $\omega$ on the tangent vector is $-(dg/dt)g^{-1}$. But if $g(t) = e^{i\alpha(t)}$, then $-(dg/dt)g^{-1} = -i(d\alpha/dt)$ is pure imaginary, that is, in the Lie algebra to $U(1)$

$$\omega\left(\frac{d\mathbf{x}}{dt}\right) \in \mathfrak{g} = \mathfrak{u}(1) \tag{17.14}$$

It should not surprise us that $(dg/dt)g^{-1}$ is in $\mathfrak{g} = \mathfrak{u}(1)$ since $dg/dt$ is a tangent vector to $G$ at $g$ and $g^{-1}$ right translates it back into the Lie algebra. *The Riemannian condition* *on our $\mathbb{C}^1$ bundle connection demands that the connection 1-form $\omega$ takes its values in the Lie algebra of the structure group.*

In Section 16.3 we defined the general notion of a connection for a vector bundle $E$. The connection allowed us to differentiate *a section of E with respect to a tangent vector to the base space M*, that is, $\omega$ is a matrix of local 1-forms *on M*. Now we shall define a $1 \times 1$ matrix $\omega^*$ of global 1-forms *on the 3-dimensional principal bundle space FM*!

Let $\mathbf{e}_U$ be a section of $FM$ over $U$, that is, a frame on $U$, and let $\mathbf{f}$ be another section. Then

$$\mathbf{f}(x) = \mathbf{e}_U(x) g_U(x)$$

for some $g_U(x) \in U(1)$. The local "coordinates" for $\mathbf{f}$ are then $(x, \alpha)$, where $\alpha$ is the angular variable in (17.12) for $g = g_U$. The local coordinates for $\mathbf{e}_U$ are $(x, g = e)$, i.e., $\alpha = 0$.

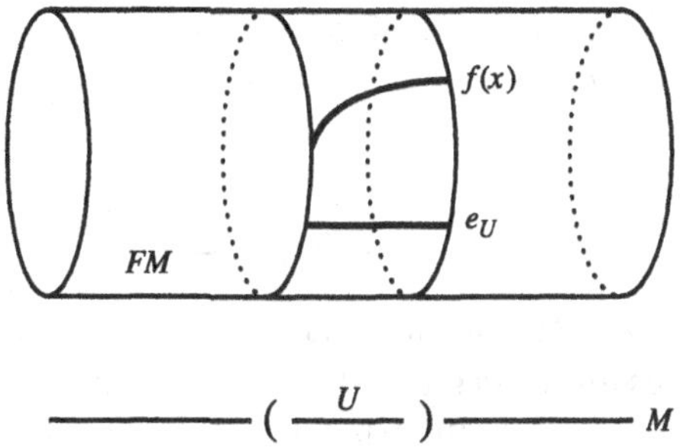


**Figure 17.6**

Then $\nabla \mathbf{f} = \nabla(\mathbf{e}_U g_U) = \mathbf{e}_U \omega_U g_U + \mathbf{e}_U dg_U = \mathbf{e}_U g_U g_U^{-1} \omega_U g_U + \mathbf{e}_U g_U g_U^{-1} dg_U$, or

$$\nabla \mathbf{f} = \mathbf{f} \otimes \{g_U^{-1} \omega_U g_U + g_U^{-1} dg_U\} \tag{17.15}$$

But $g_U^{-1} \omega_U g_U = \omega_U$ (it is crucial here that $U(1)$ is commutative) and $g^{-1} dg = i d\alpha$, and so

$$\nabla \mathbf{f} = \mathbf{f} \otimes \{\omega_U + i d\alpha\} \tag{17.16}$$

Note that $d\alpha$ can be considered a local 1-form on the frame bundle since $\alpha$ is a local coordinate in $\pi^{-1}U$. $\pi^*\omega$ is also, but we usually simply write $\omega$ for $\pi^*\omega$ since $x^1, x^2, \alpha$ are local coordinates for $\pi^{-1}U$

$$\omega = \Gamma_j dx^j = \pi^* \omega$$

for some functions $\Gamma_j$ on $U$. Thus we can **define** the local $1 \times 1$ matrix of 1-forms $\omega^*$ *on* $\pi^{-1}U$ by

$$\omega^*{}_U := \omega_U + i d\alpha \tag{17.17}$$

Since $\omega^*$ is again pure imaginary, this is now a $1 \times 1$ matrix of 1-forms on $\pi^{-1}U \subset FM$ that still takes its values in $\mathfrak{u}(1)$.

Now notice something remarkable. Since $\nabla \mathbf{f}$ has a geometric meaning independent of the frame used (in fact, using the real forms, putting $\omega + i d\alpha = -i\omega_{12} + i d\alpha =$

$-i\{\omega_{12} - d\alpha\} = 0$ defines parallel translation see (9.62)), the following should not be surprising.

**Theorem (17.18):** *On an overlap $\omega_U{}^* = \omega_V{}^*$, and thus the collection $\{\omega_U{}^*\}$ defines a $\mathfrak{g} = \mathfrak{u}(1)$ valued 1-form $\omega^*$ on all of the principal bundle FM. $\omega^*$ is called the* ***connection form on the frame bundle*** *FM.*

PROOF: Let $\mathbf{e}_V$ be a section over $V$. $\mathbf{e}_V = \mathbf{e}_U c_{UV}$ where $c_{UV} = e^{i\beta}$, for some $\beta$. Then a section $\mathbf{f}$ has two representations $\mathbf{f} = \mathbf{e}_U g_U = \mathbf{e}_V g_V$, where $g_U = e^{i\alpha}$ and $g_V = e^{i(\alpha-\beta)}$. Then at the point $\mathbf{f}$ of $FM$ $\omega^*{}_V = \omega_V + id(\alpha - \beta)$. But $\omega_V = c_{UV}^{-1}\omega_U c_{UV} + c_{UV}^{-1} dc_{UV} = \omega_U + e^{-i\beta}de^{i\beta} = \omega_U + id\beta$. Thus

$$\omega^*{}_V = \omega_U + id\alpha = \omega^*{}_U$$

Thus, although $\{\omega_U\}$ are only locally defined $\mathfrak{g}$-valued 1-forms, and $\{d\alpha_U\}$ are only locally defined 1-forms, the combinations $\{\omega_U^*\}$ match up to define a *global* $\mathfrak{g}$-valued 1-form $\omega^*$ on $FM$. □

But then

$$\theta^* := d\omega^* = \pi^* d\omega = \pi^*\theta \tag{17.19}$$

is also globally defined on *FM*; we shall call this the **curvature form on the frame bundle**. It is not new to us that $\theta^*$ is globally defined on *FM* since we already knew that $\theta = -i\theta_{12} = -iK\sigma^1 \wedge \sigma^2$ is globally defined on $M^2$. What *is* new and so important is Chern's observation;

**Theorem (17.20):** *The lift $\pi^*\theta$ of the curvature 2-form to $FM^2$ is* ***globally exact*** *on FM*

$$\theta^* = -\pi^* i\theta_{12} = -\pi^* iK\sigma^1 \wedge \sigma^2 = d\omega^*$$

We have seen that $\int_M \theta_{12}$ usually does not vanish, and thus $\theta$ itself on $M$ is usually *not* exact!

## 17.3b. The Gauss–Bonnet–Poincaré Theorem

**Theorem (17.21):** *Let $M^2$ be a closed Riemannian surface and let $\mathbf{v}$ be a vector field on M having a finite number of singularities at $p_1, \ldots, p_N$. Then*

$$\frac{1}{2\pi}\iint_M K\sigma^1 \wedge \sigma^2 = \chi(M^2) = \sum_\alpha j_{\mathbf{v}}(p_\alpha)$$

Note that since the left-hand side is independent of $\mathbf{v}$, so is the right-hand side. This is Poincaré's theorem (16.9). Since the right-hand side is independent of the abstract Riemannian metric used on $M^2$,

$$\iint_M K dA \textit{ must be independent of the metric.}$$

This is the **Gauss–Bonnet theorem**. In (8.20) we proved this for an *embedded* surface $M^2 \subset \mathbb{R}^3$.

The proof we shall give is due to S. S. Chern, who proved a far more general result. We shall talk about some of these generalizations later on. Chern's proof shows the equality of the integral with the index sum; we have already shown that the index sum is the Euler characteristic in (16.9).

PROOF: We shall prove the theorem when $M$ is orientable (and oriented); the nonorientable case can be handled by the standard trick of passing to the 2-sheeted orientable covering, discussed in Section 16.2b.

First remove small discs $\{D_a\}$ centered at the singularities. Then $\mathbf{f} = \mathbf{v}/\|\mathbf{v}\|$ is a unit vector field, that is, a frame on $M - \cup D_a$. We then have a section

$$\mathbf{f} : M^2 - \cup D_a \to FM^2$$

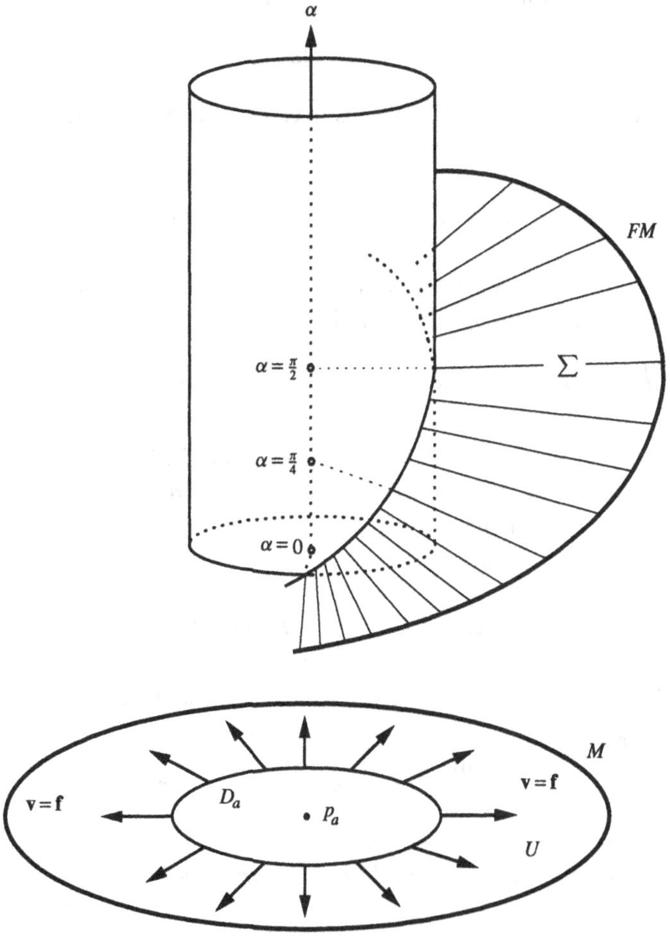


Figure 17.7

(**Remark**: The frame bundle on this $M^2$ is clearly the same as the unit tangent bundle; *for higher-dimensional generalizations it is important to keep the frame version in mind.*)

Let

$$\Sigma^2 = \mathbf{f}(M^2 - \cup D^2) \subset FM^2$$

be the image of the punctured $M$ under the section $\mathbf{f}$; it is a 2-dimensional submanifold of $FM$ diffeomorphic to $M - \cup D_a$, since $\pi \circ \mathbf{f}$ is the identity map. Then, since $\omega^*$ (not $\omega$) is *globally defined*

$$\begin{aligned} -i \iint_{M-\cup D_a} K dA &= -i \iint_{\pi\Sigma} K dA = - \iint_{\Sigma} \pi^*(iK dA) \\ &= \iint_{\Sigma} d\omega^* = \int_{\partial\Sigma} \omega^* = \int_{\partial\Sigma} \pi^*\omega - d\alpha \end{aligned} \tag{17.22}$$

Let the disc $D_a$ lie in the coordinate patch $U$ and let $S_a = \partial D_a$. Let $\mathbf{e}_U$ be a frame in the open $U$. We may express the $\omega_U$ of this frame in terms of the local coordinates $x^1, x^2$ (which are unrelated to the frame $\mathbf{e}_U$),

$$\omega = \gamma_i(x) dx^i$$

The part of the boundary of $\sum$ that lies over $D_a$ is over $S_a$; call this portion of $\partial \sum$ simply $\sigma_a$. Then in (17.22)

$$\oint_{\sigma_a} \pi^*\omega = \oint_{\pi\sigma_a} \omega = \oint_{S_a} \gamma_i(x) dx^i$$

and if we let the disc $D_a$ shrink down to the point $p_a$ this last integral will vanish in the limit. Thus as each $D_a$ shrinks to its $p_a$

$$\iint_M iK dA = \lim \iint_{M-\cup D_a} iK dA = - \lim \oint_{\partial\Sigma} i d\alpha \tag{17.23}$$

Consider again the part $\sigma_a$ of $\partial \sum$ that lies over $S_a$. In terms of the section $\mathbf{e}_U$ given by the frame, the section $\mathbf{f}$ is $\mathbf{f} = \mathbf{e}_U e^{i\alpha}$. Note that the part of $\partial \sum$ that lies over $S_a$ has orientation *opposite* to $\partial D_a$ (whose normal points out of $D_a$), $\partial \sum = \mathbf{f}(-D_a)$. Furthermore

$$\oint_{\mathbf{f}(\partial D_a)} d\alpha = \oint_{\partial D_a} d\measuredangle(\mathbf{e}_U, \mathbf{f}) \tag{17.24}$$

is simply $2\pi$ (index of $\mathbf{v}$ at $p_a$) $= 2\pi j(p_a)$. Then from (17.23)

$$\iint K dA = - \lim \sum_a \oint_{-\mathbf{f}(\partial D_a)} d\alpha = 2\pi \sum_a j_{\mathbf{v}}(p_a) \quad \square$$

**Corollary (17.25):** *If $M^2$ is a closed Riemannian manifold then*

$$\frac{1}{2\pi} \iint_M K dA \textit{ is an integer}$$

### 17.3c. Gauss–Bonnet as an Index Theorem

From Problem 16.2(1) we know that the Euler characteristic $\chi(M^2) = 2 - 2g$ is expressible in terms of the genus $g$ of the surface. In Section 13.4 we showed that a closed orientable surface of genus 2 has first Betti number $b_1 = 4$, and we have indicated the generators $A$, $B$, $C$, and $D$. The same type of picture shows that a closed orientable surface of genus $g$ has $b_1(M_g) = 2g$. If we recall that $b_0 = 1$ (since $M_g$ is connected), and that $b_2 = 1$ (since $M_g$ is closed and orientable), we see that the Euler characteristic (defined in (16.11)) can be written

$$\chi(M_g) = b_0 - b_1 + b_2 \tag{17.26}$$

in terms of homology! This, and its $n$-dimensional version, was proved by Poincaré. We shall discuss this further in Problem 22.3(2). Finally we may write the Gauss–Bonnet theorem (17.21) in the form

$$\frac{1}{2\pi}\iint_M K\,dA = b_0 - b_1 + b_2 \tag{17.27}$$

On the left-hand side we have a curvature, a local quantity involving derivatives of the metric tensor, quantities associated to the *tangent bundle* of $M$. Its integral (divided by $2\pi$) is simply a number. The right-hand side exhibits this number as an *integer* involving *dimensions of homology groups* of $M$. Recall from Hodge's theorem that $b_p(M)$ is also equal to the dimension of the space of harmonic $p$-forms, which is nothing other than the dimension of the kernel of the Laplace operator

$$b_p(M) = \dim\ker\Delta : \bigwedge^p(M) \to \bigwedge^p(M)$$

In physics, the kernel of an operator is called the space of **zero modes**. Thus, basically, an integral of the curvature of the tangent bundle of $M$ is related to the number of zero modes of differential operators constructed from this bundle. This is the first and most famous example of an **index theorem**. The **Atiyah–Singer index theorem** is a vast generalization of (17.27) replacing the tangent bundle by other bundles (we shall consider a few examples in our next section), the Gauss curvature by higher-dimensional curvature forms (some of which will be discussed in Chapter 22), and replacing the Laplacian by other *elliptic* differential operators associated with the bundle in question. The Atiyah–Singer theorem must be considered a high point of geometrical analysis of the twentieth century, but is *far* too complicated to be considered in this book. The reader may consult for instance, [Ro].

## 17.4. Line Bundles, Topological Quantization, and Berry Phase

How does a wave function change under an adiabatic transition?

### 17.4a. A Generalization of Gauss–Bonnet

Let $E$ be any complex *line* bundle over a manifold $M^n$ of any dimension. We suppose that the structure group $G$ is $U(1)$; $\psi_V = e^{i\alpha}\psi_U$. Let $\omega$ be a $U(1)$ connection; that is,

$\omega$ takes its values in $\mathfrak{g} = \mathfrak{u}(1)$. Thus $\omega(\mathbf{X})$ is *skew hermitian* (pure imaginary) for all tangent vectors $\mathbf{X}$ to $M^n$.

If $\mathbf{\Psi} = \mathbf{e}_U \psi_U$ and $\mathbf{\Phi} = \mathbf{e}_U \phi_U$ are sections, then,

$$\overline{\psi}_V \phi_V = e^{-i\alpha} \overline{\psi}_U e^{i\alpha} \phi_U = \overline{\psi}_U \phi_U$$

allows us to define a hermitian scalar product in each fiber by

$$\langle \mathbf{\Psi}, \mathbf{\Phi} \rangle := \overline{\psi}_U \phi_U$$

with associated norm $\| \mathbf{\Psi} \|^2 = |\psi_U|^2$. We then say that $E$ is a **hermitian** line bundle. If we put, as usual, $\nabla\psi = d\psi + \omega\psi$, then

$$\begin{aligned}\langle \nabla\mathbf{\Psi}, \mathbf{\Phi} \rangle + \langle \mathbf{\Psi}, \nabla\mathbf{\Phi} \rangle &= d\overline{\psi}\phi + \overline{\psi} d\phi \\ &\quad + \overline{\omega\psi}\phi + \overline{\psi}\omega\phi = d(\overline{\psi}\phi)\end{aligned}$$

since $\overline{\omega} = -\omega$. This is the analogue of the basic Riemannian condition that parallel translation preserves scalar products.

Let $\mathbf{e}_U$ be a "frame" over $U$, that is, a section of $E$ of norm 1. Then $\mathbf{\Psi} = \mathbf{e}_U \psi_U = \mathbf{e}_U e^{i\alpha}$ is the most general frame over $U$. Thus over $U$, the fiber coordinate in the *frame bundle* $FE$, that is, the principal bundle associated to $E$, is simply the angle $\alpha$. The frame bundle is a circle bundle over $M^n$, a bundle whose fibers are circles $S^1$. We may now proceed as we did in the case of the tangent bundle to the surface.

Let $V^2$ be a *closed oriented surface embedded in* $M^n$. The part of the bundle $FE$ over points of $V^2$ defines a bundle over $V^2$, which we shall again call $FE$; it is the same circle bundle but "restricted" to $V^2$. We wish to consider a smooth section $\mathbf{\Psi}$ of $FE$ over the closed surface $V^2$, but we know from the tangent bundle case that such a section might not exist over *all* of $V^2$. We might try to construct such a section by first taking a section $\mathbf{s} : V^2 \to E$ of the complex line bundle, and then putting $\mathbf{\Psi} = \mathbf{s} / \| \mathbf{s} \|$ at those $p \in V^2$ where $\mathbf{s} \neq 0$. A section $\mathbf{s}$ defines a 2-dimensional submanifold $\mathbf{s}(V^2)$ of the 4-dimensional manifold $E_V$, the part of the bundle $E$ over $V$. The 0-section $\mathbf{o}$ defines another 2-dimensional submanifold $\mathbf{o}(V^2)$ of $E_V$. *Generically*, a submanifold $V^r$ and a submanifold $W^s$ in an $N$-manifold, if they intersect, will intersect in a submanifold of dimension $(r + s - N)$, just as in the case of affine linear subspaces of a vector space. The section $\mathbf{s}$ and the section $\mathbf{o}$ are generically then going to intersect in a 0-dimensional set, that is, a finite set of points, which may be empty. *Thus*, just as in the case of the tangent bundle, *we expect to be able to find a nonvanishing section of E, and a resulting section of FE, over all of $V^2$ except perhaps over a finite set of points $p_1, \ldots, p_N$.* (The precise argument for such constructions will be taken up in Chapter 22.)

Let then $\mathbf{\Psi}$ be such a section. As in Section 17.3b we construct the connection form $\omega^* = \pi^*\omega + i d\alpha$, where $\alpha$ is the local fiber circle coordinate (recall that $\omega$ is now pure imaginary). Then, as in (17.24), we define the **index** of $\mathbf{\Psi} = \mathbf{e}\psi = \mathbf{e}e^{i\alpha}$ at the zero $p_k$ to be

$$j_{\mathbf{\Psi}}(p_k) := \frac{1}{2\pi} \oint_{\partial D} d\alpha$$

which is simply the *degree* of the map $\psi : \partial D_k \to S^1$. Then, just as in the proof of Theorem (17.21), we conclude that $(i/2\pi) \iint_V \theta^2 = \sum j_\Psi(p)$ is an integer! We have sketched a proof of the following theorem of Chern.

**Theorem (17.28):** *Let E be a hermitian line bundle, with (pure imaginary) connection $\omega^1$ and curvature $\theta^2$, over a manifold $M^n$. Let $V^2$ be any closed oriented surface embedded in $M^n$. Then*

$$\frac{i}{2\pi}\iint_V \theta^2$$

*is an integer and represents the sum of the indices of any section* **s** $: V^2 \to E$ *of the part of the line bundle over $V^2$; it is assumed that* **s** *has but a finite number of zeros on V. $i\theta/2\pi$ is the* **Chern form** *of E.*

This then proves *Dirac's quantization condition* (16.53).

Geometrically this integer represents (algebraically) the **number of times that the section** $s$ **intersects the 0-section o**, counted with multiplicity. By this we mean the following: Let $E$ be a rank-$n$ vector bundle over an $M^n$. We assume that $M$ is oriented and that the vector space fibers of $E$ can be oriented in a continuous fashion. (This will be the case if the structure group $G$ of the bundle is a connected group, such as $SO(n)$, or a unitary group. On the other hand, as discussed in Section 17.1a, the real line bundle given by the normal vectors to the midcircle $M^1$ of the Möbius band has structure group given by the 2-element group $Z_2$, which is not connected, yielding fibers that cannot be oriented continuously.) Let $x^1, \dots, x^n$ be positively oriented local coordinates in $M$, and $u^1, \dots, u^n$ be positively oriented fiber coordinates, near the intersection point $x = 0, u = 0$, of the sections **s** and **o**. **s** can be described by the $n$ functions $u = u(x)$. We say that the section **s** meets the **o** section **transversally** (or that **s** has a **nondegenerate zero**) if the Jacobian determinant $\partial(u)/\partial(x)$ is nonzero at $x = 0$. From $du^j/dt = (\partial u^j/\partial x^k)(dx^k/dt)$ we see that transversality simply means that the sections do not have any nontrivial tangent vector in common at the intersection. In this case we define the **local intersection number** to be $+1$ (resp. $-1$) provided the Jacobian is positive (resp. negative). The (total) **intersection number** is the sum of all the local intersection numbers at all intersections of the sections.

Consider, for example, a complex line bundle $E$ over the Riemann sphere. We may use $z = x + iy$ for local coordinates on $V^2 = S^2$ near $z = 0$ ($S^2$ is a complex manifold) and $\zeta = u + iv$ for fiber coordinates. The section **s** can be described by giving $u = u(x, y)$, $v = v(x, y)$, or more briefly $\zeta = \zeta(z, \bar{z})$, where we do *not* assume that $\zeta$ is holomorphic in $z$. If, however, $\zeta$ *is* a holomorphic function of $z$ (we would then say that **s** is a **holomorphic section**) then by the Cauchy–Riemann equations we have $\partial(u, v)/\partial(x, y) = |\zeta'(z)|^2 \geq 0$. Thus if a holomorphic section is not tangent to the 0-section, $\zeta'(0) \neq 0$, we conclude *the local intersection number is* $+1$.

Consider as a specific example the tangent bundle $E = TS^2$ of the Riemann sphere as a complex line bundle. Use $z$ as coordinate near 0 on $S^2$ and $w$ as coordinate near $\infty$. Let $\zeta$ be a fiber coordinate over the $z$ patch. On the Riemann sphere we have the vector field coming from $dz/dt = z^2$. It has (Kronecker) index $j = 2$ at $z = 0$ and 0 at $z = \infty$

(since $dw/dt = -1$ at $z = \infty$). How can we think of this in terms of intersections? The part of $TS^2$ over the $z$ patch is simply $\mathbb{C}^2$ with coordinates $(z, \zeta)$. We wish to see how the section $\zeta = \zeta(z) = z^2$ intersects the section $\zeta = 0$. Clearly these sections are tangent (i.e., nontransversal). By a slight deformation, however, we may replace this section by one with transverse intersections; for example consider the section defined by $\zeta = z^2 - a$ (i.e., $dz/dt = z^2 - a$), for some small $a \neq 0$. Near $z = \infty$ the field is $dw/dt = -1 + a/z^2$, and again has no zero; the zero at $z = 0$ has been replaced by two zeros at the square roots of $a$. In this holomorphic case, as we have seen, *both zeros have local intersection number* $+1$, yielding $+2$ as the total intersection number of the perturbed section with the 0-section. As we let $a \to 0$ the two zeros coalesce, and in this sense we say that the original section meets the 0-section with intersection number 2. This agrees with the Kronecker index $j$.

Note that this is very different from the usual *real* situation. In the real plane $\mathbb{R}^2$ the curve $y = x^2$ is tangent to the $x$ axis, but if we lift the curve slightly to $y = x^2 + a$ (for $a > 0$) there is then no intersection at all. On the other hand, if we drop the curve, $y = x^2 - a$, we get two intersections but the intersection at $x = \sqrt{a}$ is 1 whereas that at $x = -\sqrt{a}$ is $-1$, again yielding a *total intersection number* 0. For a good discussion of Kronecker indices and intersection numbers I recommend [G, P, chap. 3].

Return now to the general situation of Theorem (17.28). Note that when the second Betti number $b_2$ of $M^n$ is zero, for example, when each closed surface $V^2$ bounds, the integral condition in (17.28) is simply $\iint_{V=\partial B} \theta^2 = \iiint_B d\theta^2 = 0$. The integer in (17.28) can be nonzero only when $M^n$ has nontrivial homology in dimension 2; (17.28) is a **topological quantization condition.**

We may paraphrase (17.28) as follows. The curvature $\theta$ of a hermitian complex line bundle is a pure imaginary closed 2-form on the base space $M^n$ having the property that $i\theta/2\pi$ has integral periods on any basis of the integral second homology group $H_2(M^n; \mathbb{Z})$. We say that $i\theta/2\pi$ defines an **integral** cohomology class of $M$. There is a remarkable converse to this, whose proof is beyond the scope of this book.

**Theorem (17.29):** *Let $\beta^2$ be a real, closed 2-form defining an integral cohomology class on some manifold $M^n$. Then there exists a hermitian line bundle $E$ over $M$, and a $U(1)$ connection $\omega$ for this bundle, such that $-2\pi i\beta$ is the curvature form on $M$ for the bundle $E$.*

Thus each closed 2-form on $M$ with integral periods is essentially the curvature form for some hermitian line bundle over $M$. Furthermore, one can define a notion of "equivalent bundles" and then if $M$ is *simply connected*, the constructed line bundle $E$ is unique (up to equivalence)! The proof requires the introduction of more machinery (sheaf theory) and will not be given here.

## 17.4b. Berry Phase

We are going to be concerned with *complex* line bundles, but a *real* example will give us a good picture to start out with.

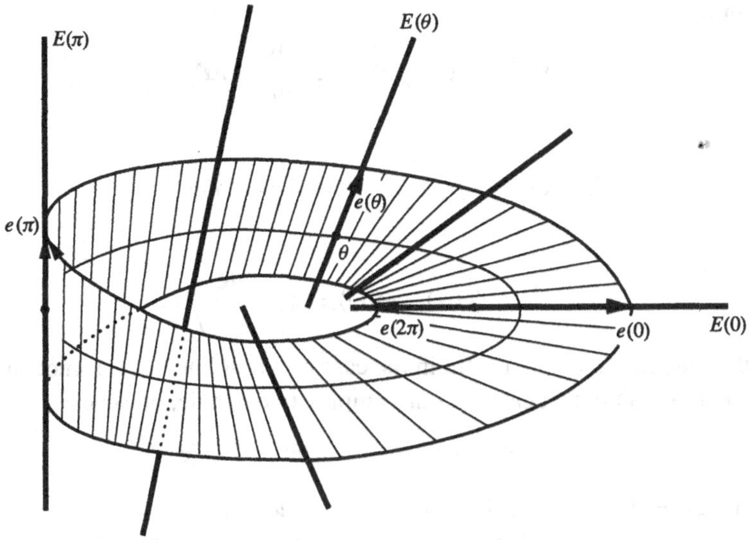


Figure 17.8

Consider an infinite Möbius band (i.e., each generating straight line is infinite) immersed in $\mathbb{R}^3$, with central circle $V^1$ given by $x^2 + y^2 = 1$, $z = 0$ and parameterized by $\theta$. The infinite real line of the Möbius band passing through $(\cos\theta, \sin\theta, 0)$ can be identified with a real 1-dimensional subspace of $\mathbb{R}^3$ by translating it to the origin, yielding a 1-parameter family of real 1-dimensional subspaces $E_\theta$ of $\mathbb{R}^3$. We can pick out smoothly a real unit vector $\mathbf{e}(\theta)$ in $E_\theta$ in some neighborhood of $\theta = 0$ (unique up to multiplication by $\pm 1$), but since $\mathbf{e}(2\pi)$ will be the negative of $\mathbf{e}(0)$, we see that we can't find $\mathbf{e}(\theta)$ smoothly for all $\theta$.

Look now at the following more general geometric situation. Consider a *complex* inner product space; in our main example it will be infinite-dimensional but for easy visualization we take $\mathbb{C}^n$ with the usual hermitian scalar product $\langle z, w\rangle = \sum \bar{z}_j w_j$. Suppose that for each point $p$ in a $K$-dimensional **parameter manifold** $V^K$, we may assign a complex 1-dimensional subspace ("line") $E_p$ of $\mathbb{C}^n$. We thus have a $K$-parameter family of complex lines. If $\alpha = (\alpha^1, \ldots, \alpha^K)$ is a local coordinate system for $V$, we may describe the family by $E_p = E_\alpha$. We assume that the lines $E_p$ vary smoothly with $p$, and so locally $E_\alpha$ depends smoothly on $\alpha$. Each $E_\alpha$ is simply a copy of the complex plane $\mathbb{C}$, and of course we can pick out a unit basis vector $\mathbf{e}(\alpha)$ in each $E_\alpha$, and $\mathbf{e}(\alpha)$ is unique up to multiplication by a complex number $e^{ir(\alpha)}$ of absolute value 1. Since $E_\alpha$ varies smoothly with $\alpha$, in some $\alpha$-neighborhood of, say $\alpha = 0$, we may pick the bases $\mathbf{e}(\alpha)$ to vary smoothly with $\alpha$. We may assume that the coordinate patches $\alpha$ are so small that $\mathbf{e}_\alpha$ is smooth in the entire patch. *The family $E_p$ forms a complex line bundle $E$ over the $K$-dimensional parameter space.*

A local section of this bundle is simply a complex vector field $\mathbf{v} = \mathbf{e}(\alpha)v(\alpha)$. We define a **covariant differentiation** by simply taking the *projection of the usual derivative in $\mathbb{C}^n$ along $E_\alpha$*. This is clearly intrinsic, independent of the basis $\mathbf{e}_\alpha$ chosen.

In terms of the basis

$$\nabla \mathbf{v} = \mathbf{e}\langle \mathbf{e}, d\mathbf{v}\rangle = \mathbf{e}\left\langle \mathbf{e}, \frac{\partial \mathbf{v}}{\partial \alpha^k}\right\rangle d\alpha^k$$

Thus

$$\nabla \mathbf{e}(\alpha) = \mathbf{e}(\alpha)\langle \mathbf{e}(\alpha), d\mathbf{e}(\alpha)\rangle = \mathbf{e}(\alpha)\omega^1$$

where (17.30)

$$\omega^1 := \langle \mathbf{e}(\alpha), d\mathbf{e}(\alpha)\rangle = \left\langle \mathbf{e}, \frac{\partial \mathbf{e}}{\partial \alpha^k}\right\rangle d\alpha^k$$

Note that this would *not* be useful in the case of a *real* line bundle since $\partial \mathbf{e}(\alpha)/\partial \alpha^j$ is orthogonal to $\mathbf{e}(\alpha)$ in that case. In our complex line bundle, however,

$$0 = d\langle \mathbf{e}, \mathbf{e}\rangle = \langle d\mathbf{e}, \mathbf{e}\rangle + \langle \mathbf{e}, d\mathbf{e}\rangle = 2\mathrm{Re}\langle \mathbf{e}, d\mathbf{e}\rangle$$

shows not that $\langle \mathbf{e}, d\mathbf{e}\rangle$ vanishes, but only that it is *pure imaginary*.

In a coordinate patch overlap we have $\mathbf{e}(\beta) = \mathbf{e}(\alpha)c_{\alpha\beta}$ for some function $c_{\alpha\beta}(p)$ of absolute value 1, and so our bundle *E is a hermitian bundle, with structure group U(1) and connection* $\omega$. See Problem 17.4(2) at this time.

The curvature of this connection is given (see Problem 17.4(3)) by

$$\theta^2 = d\omega = d\langle \mathbf{e}(\alpha), d\mathbf{e}(\alpha)\rangle = \langle d\mathbf{e}(\alpha), d\mathbf{e}(\alpha)\rangle$$

meaning (17.31)

$$\theta^2 = i\mathrm{Im}\left\langle \frac{\partial \mathbf{e}(\alpha)}{\partial \alpha^j}, \frac{\partial \mathbf{e}(\alpha)}{\partial \alpha^k}\right\rangle d\alpha^j \wedge d\alpha^k$$

*Then Theorem (17.28) gives topological quantization conditions in this purely geometric situation*!

In Section 17.4c we shall apply this machinery to Dirac's monopole bundle, but for the present we shall consider examples investigated by Berry in [B]. First we shall study a finite dimensional situation.

**Example:** Let $H = H(\alpha_1, \ldots, \alpha_K) = H(\alpha)$ be an $n \times n$ hermitian matrix that depends smoothly on $K$-parameters $\alpha_j$. (We may think of this as perturbing a given hermitian matrix $H(0)$.) $H(\alpha)$ operates on $\mathbb{C}^n$ and has $n$ real eigenvalues for each $\alpha$. We shall assume that the *lowest* eigenvalue $\lambda_1(\alpha) f$ or $H(\alpha)$ is nondegenerate for each $\alpha$; thus there is a unique complex 1-dimensional eigenspace $E_\alpha \subset \mathbb{C}^n$ picked out for each $\alpha$. We assume that the set of lowest eigenvalues $\{\lambda_1(\alpha)\}$ is separated from the higher eigenvalues. Note first that $\lambda_1$ *depends smoothly on* $\alpha$. To see this, observe that the characteristic polynomial $f(\lambda, \alpha) = \det[\lambda I - H(\alpha)]$ is a smooth function of both $\lambda$ and $\alpha$. Fix $\alpha = \alpha^0$, and let $\lambda_1$ be the unique lowest eigenvalue; thus $f(\lambda_1, \alpha^0) = 0$. Since $\lambda_1$ is a simple root, we have

$$f(\lambda, \alpha^0) = (\lambda - \lambda_1)(\lambda - \lambda_2)\ldots(\lambda - \lambda_n)$$

and $\lambda_1$ differs from $\lambda_j$ for $j \neq 1$. Hence $f = 0$ and $\partial f/\partial \lambda \neq 0$ at $\lambda = \lambda_1$ and $\alpha = \alpha^0$. From the implicit function theorem we conclude that $\lambda_1$ is a smooth function of $\alpha$ in some $\alpha$ neighborhood of $\alpha^0$.

It can be shown (see [Ka] for more details) that the 1-dimensional eigenspace $E_\alpha$ of the lowest eigenvalue $\lambda_1(\alpha)$ also depends smoothly on $\alpha$. A sketch is as follows. Since $H(\alpha)$ is hermitian we may write it in the form $H(\alpha) = \sum_j \lambda_j(\alpha)P_j(\alpha)$, where $P_j(\alpha)$ is the orthogonal projection onto the eigenspace for $\lambda_j(\alpha)$. Hence for any complex number $z$ we have $H(\alpha) - zI = \sum_j [\lambda_j(\alpha) - z]P_j(\alpha)$. Then for the **resolvent** $[H(\alpha) - zI]^{-1}$ we have

$$[H(\alpha) - zI]^{-1} = \sum_j [\lambda_j(\alpha) - z]^{-1} P_j(\alpha)$$

Thus if $C$ is a closed curve enclosing positively the set of lowest eigenvalues $\{\lambda_1(\alpha)\}$ but excluding the higher eigenvalues, we have, for each $\alpha$

$$\oint_C [H(\alpha) - zI]^{-1} dz = \sum_j \left\{ \oint_C dz/[\lambda_j(\alpha) - z] \right\} P_j(\alpha)$$
$$= -2\pi i P_1(\alpha)$$

exhibiting $P_1(\alpha)$ as a smooth function of $\alpha$. Thus the first eigenspace, $E_\alpha$, being the image of $\mathbb{C}^n$ under $P_1(\alpha)$, is smooth in $\alpha$. □

Again, a unit eigen*vector* $\mathbf{e}(\alpha)$ for $\lambda_1(\alpha)$ is determined only up to multiplication by a complex number of absolute value 1, as in our general situation.

Berry considered the following infinite-dimensional quantum situation. Let $\mathcal{H}$ be a *complex* "Hilbert space" of functions on $M^n$ with hermitian scalar product $\langle\phi|\psi\rangle = \langle\psi|\phi\rangle^-$, where $^-$denotes complex conjugation. Typically, in $M^n = \mathbb{R}^n$

$$\langle\phi|\psi\rangle = \int_M \overline{\psi}(x)\phi(x)dx$$

for a suitable class of functions.

States $\psi$ in quantum mechanics are *normalized*, $\langle\psi|\psi\rangle = 1$. The wave function for a state classically is determined up to multiplication by a constant factor $e^{i\lambda}$ of absolute value 1.

Berry considers a quantum analogue of our example in which a Hamiltonian operator $H = H(\alpha)$, acting on $\mathcal{H}$, depends smoothly on the points $\alpha$ in a $K$-dimensional parameter space $V^K$. Locally the point $\alpha$ is again described by coordinates $\alpha = (\alpha^1, \dots, \alpha^K)$. (For example, in the Aharonov–Bohm situation $\alpha = \alpha^1$ could be the flux $b$ through the solenoid, or we might have several solenoids with such varying fluxes.) The spectrum of $H(\alpha)$ is assumed to satisfy the requirements of our example. We again are led to a complex line bundle $E$ over space–time $M^4$, whose fibers are the complex 1-dimensional subspaces $E_\alpha \subset \mathcal{H}$ given by the eigenspaces of lowest energy of $H(\alpha)$.

Now let $C$ be a curve in parameter space, locally of the form $\alpha = \alpha(t)$, starting at $\alpha = 0$. Consider a solution $\psi(\mathbf{x}, t)$ of Schrödinger's equation $i\hbar d\psi/dt = H[\alpha(t)]\psi$ on $M$ that starts out at $t = 0$ with $\psi(\mathbf{x}, 0)$ a lowest energy eigenfunction of $H(0)$. The **adiabatic theorem** [Si] assures us that in the *limit* of $\alpha$ changing "infinitely slowly," the solution $\psi(\mathbf{x}, t)$ will remain an eigenfunction of lowest energy of $H(\alpha(t))$. If the curve $\alpha = \alpha(t)$ is a closed curve, the solution $\psi$ will then return to $\psi(\mathbf{x}, 0)$ except for a phase factor, and this phase factor was determined by Berry as follows.

Let $\phi_\alpha \subset E_\alpha \subset \mathcal{H}$ be a smooth choice of unit basis of $E_\alpha$ in the $\alpha$ patch; $\phi_\alpha$ satisfies $H(\alpha)\phi_\alpha = \lambda_\alpha \phi_\alpha$ and replaces the $\mathbf{e}(\alpha)$ of our previous example. From the adiabatic theorem, for very slowly changing $\alpha(t)$, $\psi(t)$ can be approximated by a multiple of the eigenstate $\phi_{\alpha(t)}$. Berry writes, *for this particular path $C$ in parameter space,*

$$\psi \sim \exp\left[\left(-\frac{i}{\hbar}\right)\int_0^t \lambda_\alpha(\tau)d\tau\right] \exp[i\gamma(\alpha(t))]\phi_{\alpha(t)} \tag{17.32}$$

(For a more careful treatment of the adiabatic limit see [Si].) The energy exponential is the usual **dynamical** phase factor (taking into account the fact that $\lambda$ is changing in time along the path) and the second (as yet unknown) exponential $\exp[i\gamma(\alpha(t))]$ is to account for the bases $\phi_\alpha$ having rather arbitrarily assigned phases. Inserting (17.32) into Schrödinger's equation yields **Berry's equation**

$$i\phi_\alpha \frac{d\gamma}{dt} + \left(\frac{\partial\phi}{\partial\alpha^j}\right)\left(\frac{d\alpha^j}{dt}\right) = 0$$

or

$$i\phi_\alpha d\gamma = -d\phi_\alpha$$

as $\mathcal{H}$-valued 1-forms *along* $C$ in parameter space, where $d = d\alpha^j \partial/\partial\alpha^j$. But then

$$\langle\phi|d\phi\rangle = -id\gamma$$

Barry Simon [Si] noticed that this can be written down in terms of connections. From (17.30) we have, along $C$,

$$d\gamma = i\omega \tag{17.33}$$

where $\omega$ is the connection in terms of the frame $\phi$. We shall call $\omega$ the **Simon connection** (avoiding the temptation to call it the Berry–Barry connection).

Thus if $C$ is a closed curve in a coordinate patch of parameter space, and *if $C$ bounds*, that is, $C = \partial S$ for a compact oriented surface $S$ in this patch, then the Berry phase factor for $C$ is given, from Simon's viewpoint, as

$$\gamma(C) = \int_C d\gamma = i\int_C \omega = i\iint_S \theta \tag{17.34}$$

$$= -\mathrm{Im} \iint_S \left\langle \frac{\partial\phi}{\partial\alpha^j} \middle| \frac{\partial\phi}{\partial\alpha^k} \right\rangle d\alpha^j \wedge d\alpha^k$$

Note in particular that $\gamma(\alpha)$ need not return to itself after completing a loop in parameter space, and likewise, neglecting the dynamical phase factor, for the wave function $\psi$. This was one of Berry's principal conclusions, and it should be mentioned that the final expression in (17.34) appears explicitly in Berry's paper but is not there related to curvature.

In Problem 17.4(4) you are asked to show that *$e^{i\gamma}\phi_\alpha$ is parallel displaced along $C$*. This gives geometric meaning to Berry's ansatz (17.32) and also to the adiabatic theorem.

For an application of the connection (17.30) and the quantization condition (17.28) to the "quantum Hall effect," see [Si].

### 17.4c. Monopoles and the Hopf Bundle

In Section 16.4 we discussed the Dirac monopole, which, for each integer $n = 2eq/\hbar$, requires a special hermitian complex line bundle, $H_n$, defined over $\mathbb{R}^3$ with the origin deleted. The unit sphere $S^2$ surrounds the monopole, and for our purposes it is sufficient to consider the part of the bundle that lies over $S^2$, which we shall again call $H_n$. The case $n = 0$ corresponds to the trivial bundle; the most important case is when $n = \pm 1$, that is, when $q = \pm 1/2e\hbar$. We shall look at the case $n = -1$. This complex line bundle $H_{-1}$ over $S^2$ is *not* the tangent bundle since the integral of $i\theta^2/2\pi$ over $S^2$ is $-1$, whereas for the tangent bundle the integral is the Euler characteristic 2. It is remarkable that Heinz Hopf investigated the appropriate bundles for purely geometric reasons (as we shall see in Section 22.4c) at about the same time as Dirac's work on monopoles!

Consider $S^2$ as being the Riemann sphere, that is the complex projective line $\mathbb{C}P^1$ of Section 1.2d and Problem 1.2(3). To a point $(z_0, z_1) \neq (0, 0)$ in $\mathbb{C}^2$ we associate the line $(\lambda z_0, \lambda z_1)$ of all complex multiples of this point. This line is described by the point in $\mathbb{C}P^1$ whose *homogeneous* coordinates are $[z_0, z_1]$. In the patch $U$ of $S^2$ where $z_0 \neq 0$ we introduce the complex coordinate $z = z_1/z_0$, whereas in $V$, where $z_1 \neq 0$, we use $w = z_0/z_1$.

The complex lines through the origin of $\mathbb{C}^2$ are parameterized by the points of $\mathbb{C}P^1$ and thus these lines form a complex line bundle over $S^2$, called the **Hopf bundle**. In a sense this bundle is "tautologous"; a point in $\mathbb{C}P^1$ represents a complex line in $\mathbb{C}^2$, and we may then associate to this point its complex line! Let us look at the local product structure.

When $z_0 \neq 0$, the line through $(z_0, z_1)$ has homogeneous coordinates $[z_0, z_1] = [1, z_1/z_0] = [1, z]$. To the point in $U \subset S^2$ with coordinate $z$, we may associate the vector $(1, z)^T$ in this line of $\mathbb{C}^2$. We call the resulting *unit* vector

$$\mathbf{e}_U(z) = \frac{(1, z)^T}{(1 + |z|^2)^{1/2}} \tag{17.35}$$

This defines a unit section of the part of the Hopf bundle over $U$ in $S^2$. Likewise, over $V$ we have $[z_0, z_1] = [z_0/z_1] = [w, 1]$ with section

$$\mathbf{e}_V(w) = \frac{(w, 1)^T}{(1 + |w|^2)^{1/2}} \tag{17.36}$$

Thus the transition functions $\mathbf{e}_V = \mathbf{e}_U c_{UV}$ are given through

$$\mathbf{e}_V(w) = \frac{(w, 1)^T}{(1 + |w|^2)^{1/2}} = \frac{w\left(1, w^{-1}\right)^T}{(1 + |w|^2)^{1/2}}$$
$$= \frac{z^{-1}(1, z)^T}{(1 + |z|^{-2})^{1/2}} = \frac{\mathbf{e}_U(z)|z|}{z}$$

Thus

$$c_{VU}(z) = \frac{z}{|z|} = e^{i\phi} \tag{17.37}$$

where $z = |z|e^{i\phi} = re^{i\phi}$ in terms of polar coordinates in $U$, that is, the upper plane in Figure 1.16. These transition functions are exactly those of the monopole bundle with $2eq/\hbar = -1$, as we see from Equation (16.51). *The monopole bundle $H_{-1}$ with $2eq/\hbar = -1$ is the tautologous Hopf bundle over* $\mathbb{C}P^1$.

$H_1$ will have transition functions $c_{VU} = e^{-i\phi}$. This is the *dual bundle* to $H_{-1}$. Consider now the tensor product bundle of $H_{-1}$ with itself, $H_{-1} \otimes H_{-1}$. This tensor product of two line bundles is again a line bundle; if $\zeta$ and $\eta$ are sections of $H_{-1}$, then $(\zeta\eta)_V = (c_{VU}\zeta_U)(c_{VU}\eta_U) = e^{2i\phi}(\zeta\eta)_U$ shows that $H_{-1} \otimes H_{-1} = H_{-2}$. In this way we can get all of the monopole bundles from tensor products of $H_1$ and $H_{-1}$, that is, from the Hopf bundle over $\mathbb{C}P^1$ and its dual.

We may now consider the Simon connection for the Hopf bundle. $\mathbb{C}^2$ carries the standard hermitian metric $\langle (a,b)^T, (c,d)^T \rangle = \bar{a}c + \bar{b}d$. Let us compute $\omega_U = \langle \mathbf{e}_U(z), d\mathbf{e}_U(z) \rangle$ in the patch $U$. Note that $U$ is simply a copy of the complex plane. Introduce polar coordinates $z = re^{i\phi}$. In Problem 17.4(5) you are asked to compute, from (17.35), that

$$\omega_U(z) = \frac{ir^2 d\phi}{(1+r^2)} \tag{17.38}$$

$$\theta_U = d\omega_U = \frac{2ir dr \wedge d\phi}{(1+r^2)^2} \tag{17.39}$$

and

$$\iint_S \frac{i\theta}{2\pi} = -1 \tag{17.40}$$

## Problems

**17.4(1)** Take as line bundle the tangent bundle to the Riemann sphere. Let $\phi_z$ (resp. $\phi_w$) be a fiber coordinate in the $z$ (resp. $w$) patch. Show that the transition function is $c_{wz} = -z^{-2}$. Since $|\phi_w|^2 \neq |\phi_z|^2$, we do *not* get a hermitian metric in the fibers by defining $\| \phi_z \|^2 = |\phi_z|^2$, and so on. It is true that $|w|^{-2}|\phi_w|^2 = |z|^{-2}|\phi_z|^2$ but these "metrics" blow up at the poles. Show that $(1+|z|^2)^{-2}|\phi_z|^2 = (1+|w|^2)^{-2}$ $|\phi_w|^2$. This expression then yields an Hermitian metric in the fibers.

**17.4(2)** Verify that $\omega$ in (17.30) does transform as a $U(1)$ connection.

**17.4(3)** Show (17.31).

**17.4(4)** Show that $e^{i\gamma(\alpha)}\phi_\alpha$ is parallel displaced along $C$.

**17.4(5)** Show (17.38), (17.39), and (17.40). The integral over $S^2 = \mathbb{C}P^1$ is the same as the integral over the entire $U$ plane since only the single point at infinity is missing.

CHAPTER 18

# Connections and Associated Bundles

IN this chapter we shall recast our previous machinery of connections, making more use of the fact that the connection and curvature forms take their values in the Lie algebra of the structure group. This will lead not only to a more systematic treatment of some topics that were previously handled in a rather ad hoc fashion, but also, in our following chapters, to generalizations of the Gauss–Bonnet–Poincaré theorem and to closer contact with the machinery used in physics.

## 18.1. Forms with Values in a Lie Algebra

What do we mean by $g^{-1}dg$?

### 18.1a. The Maurer–Cartan Form

If $E$ is a vector bundle over $M$, then the connection form $\omega = (\omega^R{}_S)$ and the curvature form $\theta = (\theta^R{}_S)$ are locally defined matrices of 1- and 2-forms, respectively. If the Lie group $G$ is the structure group of the bundle, that is, if each transition matrix $c_{UV} = (c_{UV}{}^R{}_S)$ is a matrix in $G$, then, as in (17.14), we usually require that $\omega$ and $\theta$ take their values in the Lie algebra $\mathfrak{g}$ of $G$; thus, e.g., $(\omega^R{}_S(\mathbf{X}))$ is a matrix in $\mathfrak{g}$ for each tangent vector $\mathbf{X}$ to $M$. For example, in a Riemannian $M$, by restricting the frames of the tangent bundle to be orthonormal, the Levi-Civita connection satisfies $\omega_{ij} = -\omega_{ji}$; that is, $\omega$ has its values in $\mathfrak{o}(n)$, the Lie algebra to $O(n)$. If we think of $\omega$ as being a form that takes its values in the fixed vector space $\mathfrak{g}$, rather than as a matrix of 1-forms, we shall have an equivalent picture that is in many ways more closely related to the terminology used in physics.

Let $M^n$ be a manifold and let $G$ be a Lie group with Lie algebra $\mathfrak{g}$. We shall consider locally defined exterior forms $\phi$ on $M$ taking values in the fixed vector space $\mathfrak{g}$.

First we define a $\mathfrak{g}$ valued 1-form on $G$ itself. Let $\{\mathbf{E}_R\}$ be a basis for $\mathfrak{g}$ and let $\{\mathbf{X}_R\}$ be the left invariant fields on $G$ obtained by left translating the $\mathbf{E}$'s. Let $\{\sigma^R\}$ be left invariant 1-forms on $G$ forming, at each $g \in G$, a basis dual to $\{\mathbf{X}_R\}$. Then

$$\Omega := \mathbf{E}_R \otimes \sigma^R \tag{18.1}$$

defined by

$$\Omega(\mathbf{Y}_g) = \mathbf{E}_R \sigma^R(\mathbf{Y}_g) = \mathbf{E}_R Y^R$$

takes a vector $\mathbf{Y} = \mathbf{X}_R Y^R$ at $g \in G$ and left translates it back to the identity. This is the **Maurer–Cartan 1-form** on $G$.

Classically this would be written differently. On any manifold, Cartan wrote $dp$ for the vector-valued 1-form at $p \in M$ that takes each vector $\mathbf{Y}$ at $p$ into itself. In coordinates it is the mixed tensor $\{\delta^i_j\}$

$$dp = \frac{\partial}{\partial x^i} \otimes dx^i = \frac{\partial}{\partial x^i} \otimes \delta^i_j dx^j$$

Then Cartan would write

$$\Omega = g^{-1} dg \tag{18.2}$$

Thus $dg$ takes $\mathbf{Y}$ at $g$ into $\mathbf{Y}$, and $g^{-1}$ left translates $\mathbf{Y}$ back to $e$. We should write

$$\Omega = (L_{g-1})_* \circ dg$$

For a matrix group, each $\mathbf{E}_r$ is simply a matrix of a certain type (e.g., skew symmetric for $G = O(n)$). By *construction* we have the following:

**Theorem (18.3):** *In any matrix group $G$, $\Omega = g^{-1}dg$ is a matrix with left invariant 1-form entries.*

For example, in $SO(2)$, for

$$g(\theta) = \begin{bmatrix} \cos\theta & -\sin\theta \\ \sin\theta & \cos\theta \end{bmatrix}$$

we have

$$g^{-1}dg = \begin{bmatrix} \cos\theta & \sin\theta \\ -\sin\theta & \cos\theta \end{bmatrix} \begin{bmatrix} -\sin\theta d\theta & -\cos\theta d\theta \\ \cos\theta d\theta & -\sin\theta d\theta \end{bmatrix}$$

or

$$g^{-1}dg = \begin{bmatrix} 0 & -d\theta \\ d\theta & 0 \end{bmatrix} = \begin{bmatrix} 0 & -1 \\ 1 & 0 \end{bmatrix} \otimes d\theta$$

and $d\theta$ *is* a rotation invariant 1-form on the circle $SO(2)$.

The usual "proof" that $g^{-1}dg$ is a matrix of left invariant 1-forms is as follows: Let $h$ be a given (fixed) group element. Then for variable $g$, $L_h^* \Omega_{hg} = (hg)^{-1} d(hg) = g^{-1}h^{-1}h dg = g^{-1}dg = \Omega_g$, as claimed.

Similarly $dgg^{-1}$ is a matrix of right invariant 1-forms. See Problem 18.1(1) at this time.

### 18.1b. $\mathfrak{g}$-Valued $p$-Forms on a Manifold

The most general $p$-form on $U \subset M$ with values in the Lie algebra of a Lie group $G$ is of the form

$$\phi = \mathbf{E}_R \otimes \phi^R$$

where each $\phi^R$ is an ordinary exterior $p$-form on $U$. Thus if $\mathbf{X}$ is a $p$ tuple of tangent vectors to $M$ at a point of $U$, then $\phi(\mathbf{X}) = \mathbf{E}_R \phi^R(\mathbf{X})$ is in $\mathfrak{g}$. (Note that $R$ refers to the $\mathbf{E}_R$ involved, not to the degree of $\phi$.) Since the $\mathbf{E}$'s do not vary (lying in the fixed vector space $\mathfrak{g}$), it is natural to **define**

$$d\phi = d(\mathbf{E}_R \otimes \phi^R) := \mathbf{E}_R \otimes d\phi^R \tag{18.4}$$

a $\mathfrak{g}$-valued $p+1$ form on $M$.

Multiplication of such forms is not so clear a process because, for example, in the case of $\mathfrak{g} = o(n)$, the product of two skew symmetric matrices is not necessarily skew symmetric. Instead we shall define the (Lie) **bracket** of forms, and we shall see that this includes a desirable product. We **define**

$$[\phi, \psi] = [\mathbf{E}_R \otimes \phi^R, \mathbf{E}_S \otimes \psi^S] := [\mathbf{E}_R, \mathbf{E}_S] \otimes \phi^R \wedge \psi^S \tag{18.5}$$

As an example, consider the Maurer–Cartan 1-form on $M = G$. From the Maurer–Cartan equations (15.24)

$$d\Omega = \mathbf{E}_R \otimes \left(-\frac{1}{2} C^R_{ST} \sigma^S \wedge \sigma^T\right)$$

while

$$\begin{aligned}[\Omega, \Omega] &= [\mathbf{E}_S \otimes \sigma^S, \mathbf{E}_T \otimes \sigma^T] = [\mathbf{E}_S, \mathbf{E}_T] \otimes \sigma^S \wedge \sigma^T \\ &= \mathbf{E}_R \otimes C^R_{ST} \sigma^S \wedge \sigma^T\end{aligned}$$

and so

$$d\Omega + \frac{1}{2}[\Omega, \Omega] = 0 \tag{18.6}$$

which will again be called the **Maurer–Cartan equation**.

**Remark:** (18.5) defines the bracket by means of a basis but it is not difficult to give an intrinsic definition. If $\mathbf{X}_I = \mathbf{X}_1, \ldots, \mathbf{X}_{p+q}$ are tangent vectors to $M^n$, we could have defined

$$\begin{aligned}[\phi, \psi](\mathbf{X}_I) &:= \delta_I^{JK} [\phi(\mathbf{X}_{\underrightarrow{J}}), \psi(\mathbf{X}_{\underrightarrow{K}})] \\ &= \delta_I^{JK} [\mathbf{E}_R \otimes \phi^R(\mathbf{X}_{\underrightarrow{J}}), \mathbf{E}_S \otimes \psi^S(\mathbf{X}_{\underrightarrow{K}})] \\ &= \delta_I^{JK} [\mathbf{E}_R, \mathbf{E}_S] \phi^R(\mathbf{X}_{\underrightarrow{J}}) \psi^S(\mathbf{X}_{\underrightarrow{K}}) = [\mathbf{E}_R, \mathbf{E}_S] \phi^R \wedge \psi^S(\mathbf{X}_I)\end{aligned} \tag{18.5'}$$

We have some immediate consequences of our definitions.

$$\begin{aligned}[\psi, \phi] &= [\mathbf{E}_S, \mathbf{E}_R] \otimes \psi^S \wedge \phi^R \\ &= -[\mathbf{E}_R, \mathbf{E}_S] \otimes \psi^S \wedge \phi^R\end{aligned}$$

and so

$$[\psi, \phi] = (-1)^{pq+1}[\phi, \psi] \tag{18.7}$$

Also

$$d[\phi, \psi] = [\mathbf{E}_R, \mathbf{E}_S] \otimes (d\phi^R \wedge \psi^S + (-1)^p \phi^R \wedge d\psi^S)$$

that is,

$$d[\phi, \psi] = [d\phi, \psi] + (-1)^p[\phi, d\psi] \tag{18.8}$$

Finally, we need to interpret the bracket in *the case of a matrix group*. For example, the Maurer–Cartan 1-form $\Omega$ for the affine group of the line, $G = A(1)$, is

$$\begin{bmatrix} \frac{dx}{x} & \frac{dy}{x} \\ 0 & 0 \end{bmatrix} = \begin{bmatrix} 1 & 0 \\ 0 & 0 \end{bmatrix} \otimes \frac{dx}{x} + \begin{bmatrix} 0 & 1 \\ 0 & 0 \end{bmatrix} \otimes \frac{dy}{x}$$
$$= \mathbf{E}_1 \otimes \frac{dx}{x} + \mathbf{E}_2 \otimes \frac{dy}{x}$$

*In general*, when $\{\mathbf{E}_R\}$ are matrices,

$$\begin{aligned} [\phi, \psi] &= [\mathbf{E}_R, \mathbf{E}_S]\, \phi^R \wedge \psi^S \\ &= (\mathbf{E}_R\mathbf{E}_S - \mathbf{E}_S\mathbf{E}_R) \otimes \phi^R \wedge \psi^S \\ &= (\mathbf{E}_R \otimes \phi^R) \wedge (\mathbf{E}_S \otimes \psi^S) - \mathbf{E}_S\mathbf{E}_R \otimes \phi^R \wedge \psi^S \end{aligned}$$

where in the first term of the last line we are simply multiplying the matrices but using the *exterior* product of the entries. (This is always what we did in the method of moving frames, e.g., when considering $\theta = d\omega + \omega \wedge \omega$.) For example, in $A(1)$

$$\Omega \wedge \Omega = g^{-1}dg \wedge g^{-1}dg = \begin{bmatrix} \frac{dx}{x} & \frac{dy}{x} \\ 0 & 0 \end{bmatrix} \wedge \begin{bmatrix} \frac{dx}{x} & \frac{dy}{x} \\ 0 & 0 \end{bmatrix}$$
$$= \begin{bmatrix} 0 & \left(\frac{dx \wedge dy}{x^2}\right) \\ 0 & 0 \end{bmatrix} = \mathbf{E}_2 \otimes \left(\frac{dx \wedge dy}{x^2}\right)$$

Continuing with our computation

$$[\phi, \psi] = (\mathbf{E}_R \otimes \phi^R) \wedge (\mathbf{E}_S \otimes \psi^S) - \mathbf{E}_S\mathbf{E}_R \otimes (-1)^{pq} \psi^S \wedge \phi^R$$

that is,

$$[\phi, \psi] = \phi \wedge \psi - (-1)^{pq} \psi \wedge \phi \tag{18.9}$$

as **matrices.**

For example, *if p is odd*

$$[\phi, \phi] = \phi \wedge \phi + \phi \wedge \phi = 2\phi \wedge \phi$$

as matrices.

Note that if either $\phi$ or $\psi$ is of even degree, then $[\phi, \psi]$, as a matrix, is the usual commutator, but using the wedge $\wedge$ as product. If both are *odd*, then $[\phi, \psi]$ is the **anticommutator**

$$[\phi, \psi] = \{\phi, \psi\} := \phi \wedge \psi + \psi \wedge \phi$$

Consider, for example, a Riemannian manifold with locally defined connection forms $\omega$. We may restrict ourselves to the use of orthonormal frames, in which case $\omega$ takes values in $\mathfrak{o}(n) = \mathfrak{so}(n)$. Thus when employing orthonormal frames, $\omega$ is a skew symmetric matrix of forms and of course $d\omega$ is also. But why should $\omega \wedge \omega$ be a skew symmetric matrix? It is because $\omega \wedge \omega$ is in fact the same as $1/2[\omega, \omega]$! This shows that *curvature* $\theta$ can be written

$$\theta = d\omega + \frac{1}{2}[\omega, \omega] \tag{18.10}$$

and of course is $\mathfrak{so}(n)$-valued. Likewise, the second Bianchi identity

$$d\theta + \omega \wedge \theta - \theta \wedge \omega = 0$$

again makes sense in the Lie algebra setting since it says

$$d\theta + [\omega, \theta] = d\theta + \omega \wedge \theta - (-1)^{1 \cdot 2}\theta \wedge \omega = 0 \tag{18.11}$$

### 18.1c. Connections in a Principal Bundle

In Section 17.3, a crucial role was played by the notion of a connection in the principal bundle of frames to a Riemannian surface. Now we develop this machinery for the case of the principal bundle of frames of sections of an arbitrary vector bundle.

Let $E$ be a real or complex rank-$K$ vector bundle over a manifold $M^n$, the structure group being a Lie group $G$. Thus the transition functions $c_{UV}(x)$ are *linear* transformations of $\mathbb{R}^K$ or $\mathbb{C}^K$ into itself.

To say that $G$ is the structure group means, effectively, that there is in each trivializing patch $U$ of $M$ a distinguished collection of frames of sections (e.g., orthonormal), which we may call $G$ frames, and such that any two such frames $\mathbf{e}_U$ and $\mathbf{e}_V$ in an overlap $U \cap V$ are related by $\mathbf{e}_V = \mathbf{e}_U g$ for $g \in G$. What do we mean then by a $G$ connection?

Let $\{\omega_U\}$ be a connection for this vector bundle. If $\mathbf{e}_U$ is a $G$ frame of $K$ sections

$$\nabla \mathbf{e}_U = \mathbf{e}_U \otimes \omega_U = \mathbf{e}_U \omega_U$$

where $\omega^\alpha_{U\beta} = \omega^\alpha_{i\beta}(x)dx^i$ is a matrix of 1-forms on $U$. Let $C$ be any curve in $U$ and *let* $\mathbf{f}$ *be a* $G$ *frame at the single point* $C(0)$. Parallel displace $\mathbf{f}$ along $C$.

> To say that $\omega$ is a **G connection** is to demand that the *parallel displaced* $\mathbf{f}$ *is a* $G$ *frame along all of* $C$

and this must be true for all curves $C$. Let $\mathbf{f}$ be such a parallel displaced $G$ frame.

If we now write, along $C$,

$$\mathbf{f}(t) = \mathbf{e}_U g(t)$$

we have, as in (17.15),

$$\frac{\nabla \mathbf{f}}{dt} = \mathbf{f} \otimes \left[ g^{-1}\omega\left(\frac{dx}{dt}\right)g + g^{-1}\frac{dg}{dt}\right] \tag{18.12}$$

Since the entire frame is parallel translated along a curve $x = x(t)$ in $U$, we must have

$$g^{-1}\omega\left(\frac{dx}{dt}\right)g = -g^{-1}\frac{dg}{dt} \tag{18.13}$$

where the term on the right is an element of $\mathfrak{g}$. Then the first term

$$g^{-1}\omega\left(\frac{dx}{dt}\right)g \tag{18.14}$$

is also in $\mathfrak{g}$, and since $L_{g-1*} \circ R_{g*}$ certainly sends $\mathfrak{g}$ into itself we have

$$\omega\left(\frac{dx}{dt}\right) \in \mathfrak{g}$$

Thus to demand that $\omega$ is a $G$ connection is to require that $\omega_U$ *is a* $\mathfrak{g}$*-valued* 1*-form* on $U$.

Of course the curvature is then also $\mathfrak{g}$-valued.

$$\theta_U = d\omega_U + \frac{1}{2}[\omega_U, \omega_U]$$

Under a change of frame

$$\mathbf{e}_V = \mathbf{e}_U c_{UV} \tag{18.15}$$

$$\omega_V = c_{UV}^{-1}\omega_U c_{UV} + c_{UV}^{-1}dc_{UV}$$

$$\theta_V = c_{UV}^{-1}\theta_U c_{UV} \tag{18.16}$$

The transformation rule for $\theta$ was exhibited in (9.41) for the tangent bundle; the proof here is the same.

Consider now the principal bundle $P$ of frames of sections of the vector bundle $E$. This fiber bundle has for fiber $F$ the structure group $G$ and the transition functions $c_{UV}$ are the same as for $E$; now, however, they operate on $G$ by left translation,

$$g \in G \text{ is sent to } c_{UV}(x)g$$

We now define the connection forms $\omega^*$ in $P$; these are $\mathfrak{g}$-valued 1-forms on $P$ rather than $M$. The local frame $\mathbf{e}_U$ of sections of $E$ can be thought of as a section of the bundle $P$ over $U$. For a point $\mathbf{f} \in P$ over the point $x \in U$ we can write

$$\mathbf{f} = \mathbf{e}_U(x)g_U(x) \tag{18.17}$$

for a unique $g_U \in G$. From (18.12) we are encouraged to define

$$\omega_U^*(x, g_U) := g_U^{-1}\pi^*\omega_U(x)g_U + g_U^{-1}dg_U \tag{18.18}$$

which is the nonabelian version of (17.17). We usually omit the $\pi^*$ coming from $\pi : P \to M$.

The local section $\mathbf{e}_U$ of $P$ over $U$ gives us a local product structure $U \times G$ for $\pi^{-1}U$; in fact (18.17) assigns to the point $\mathbf{f}$ in $P$ the local "coordinates" $x$ in $U$ and $g_U$ in $G$. A tangent vector at $(x, g_U)$ in $P$ is a velocity vector $(dx/dt, dg_U/dt)$ to some curve in $P$. $g_U^{-1}\pi^*\omega_U g_U$ applied to this velocity vector yields $g_U^{-1}\omega_U(dx/dt)g_U$, an element of $\mathfrak{g}$. $g_U^{-1}dg_U$ applied to this same velocity vector yields $g_U^{-1}dg_U(dg_U/dt) = g_U^{-1}dg_U/dt$, which is again an element of $\mathfrak{g}$. Thus $\omega_U^*$ is a local $\mathfrak{g}$-valued 1-form on $P$. Both terms in $\omega_U^*$ depend on the choice of section $e_U$.

**Theorem (18.19):** *In $\pi^{-1}(U \cap V)$ we have $\omega_U^* = \omega_V^*$ and thus $\{\omega_U^*\}$ defines a global $\mathfrak{g}$-valued 1-form $\omega^*$ on the principal bundle $P$.*

(In $P$ we may then consider the distribution of $n$-planes transversal to the fibers, defined by $\omega^* = 0$. This distribution is called the **horizontal distribution**, reminiscent of that appearing in the tangent bundle discussed in Section 9.7. Many books take this distribution as the starting point for their discussion of connections.)

PROOF: See Problem 18.1(2).

We then define the *global* $\mathfrak{g}$-valued curvature 2-form $\theta^*$ on $P$ by

$$\begin{aligned}\theta^* :&= d\omega^* + \frac{1}{2}[\omega^*, \omega^*] \\ &= d\omega^* + \omega^* \wedge \omega^*\end{aligned} \tag{18.20}$$

Note that unlike in the case of a tangent bundle of a surface (where the group $G$ was abelian) *we cannot expect $\theta^*$ to be globally exact or even closed!*

Of course we also have local curvature forms $\theta_U = d\omega_U + \omega_U \wedge \omega_U = d\omega + (1/2)[\omega, \omega]$ on $M$ from the vector bundle connection. As in (9.47) one can show

$$\begin{aligned}\theta^*(x, g_U) &= g_U^{-1}\pi^*\theta_U g_U \\ &= g_U^{-1}\theta_U g_U \quad \square\end{aligned} \tag{18.21}$$

## Problems

**18.1(1)** Exhibit the left invariant and the right invariant 1-forms on the affine group of the line (Example (i), Section 15.1) by means of $g^{-1}dg$ and $dgg^{-1}$.

**18.1(2)** Prove (18.19). (At a given $\mathbf{f} \in \pi^{-1}(U \cap V)$, $\mathbf{f} = \mathbf{e}_U g_U = \mathbf{e}_V g_V$, $\mathbf{e}_V = \mathbf{e}_U c_{UV}$, and so on. Use the transformation rule (18.16) for the vector bundle.)

## 18.2. Associated Bundles and Connections

What does it mean to take the covariant derivative of $\sqrt{g}$?

### 18.2a. Associated Bundles

Let $P$ be a **principal** bundle over $M^n$ with fiber = group = $G$, and with local transition matrices $c_{UV} : U \cap V \to G$. Let $\rho : G \to Gl(N)$ be some **representation** of the structural group $G$; thus each $\rho(g)$ is an $N \times N$ matrix operating on $\mathbb{C}^N$ and $\rho$ is a homomorphism

$$\rho(gg') = \rho(g)\rho(g') \quad \text{and} \quad \rho(g^{-1}) = [\rho(g)]^{-1}$$

(For example, we may represent $G = U(1)$ as a subgroup of Gl(2) by putting $\rho(e^{i\theta}) = \mathrm{diag}(e^{i\theta}, e^{3i\theta})$.)

We then define a new *vector* bundle $\pi : P_\rho \to M^n$ **associated to $P$ through the representation** $\rho$, with fiber $\mathbb{C}^N$, by making identifications in $U \times \mathbb{C}^N$ and $V \times \mathbb{C}^N$

$$(x, \psi_V) \sim (x, \psi_U) \quad \text{iff} \qquad \psi_V = \rho(c_{VU}(x))\psi_U$$

Thus we construct a new vector bundle by simply using the *new transition matrices* $\rho(c_{UV})$ rather than $c_{UV}$.

We frequently have the following situation. Let $E$ be a vector bundle $\pi : E \to M^n$ with transition functions

$$c_{UV} : U \cap V \to G \subset \mathrm{Gl}(K)$$

each $c_{UV}(x)$ being a $K \times K$ matrix. Thus in $\pi^{-1}(U \cap V)$ we are identifying (for $y_U$ and $y_V$ $K$-tuples of real or complex numbers)

$$(x, y_V) \approx (x, y_U) \quad \text{iff } y_V = c_{VU}(x) y_U$$

We may then form the *principal frame bundle* $P$ over $M$ by considering the $K$-tuples of local independent sections $\mathbf{e}^U_\alpha(p) := \Phi_U(p, e_\alpha)$, as in Section 16.1c, and the general frame over $U$ is of the form $\mathbf{f} = \mathbf{e}_U g_U$. Recall then that in an overlap we have $\mathbf{e}_V = \mathbf{e}_U c_{UV}$ and so $\mathbf{f} = \mathbf{e}_U g_U = \mathbf{e}_V g_V = \mathbf{e}_U c_{UV} g_V$, showing that $g_U = c_{UV} g_V$. Thus the principal frame bundle of $K$-tuples of sections of $E$ again has transition functions $c_{UV}$, which now act on $G$ by left translations. If now $\rho : G \to \mathrm{Gl}(N)$ is a representation of $G$ we may form the vector bundle $E_\rho := P_\rho$ associated to $P$ through $\rho$, which has transition functions $\rho(c_{UV})$, and we may say that $E$ and $E_\rho$ are also **associated** through the representation $\rho$.

**Example 1:** $E$ is the tangent bundle to $M^n$ and $\tau^* : \mathrm{Gl}(n) \to \mathrm{Gl}(n)$ is the representation $\tau^*(g) = g^* := (g^{-1})^T$. The old transition matrices are $c_{VU} = \partial x_V / \partial x_U$ and the associated functions are $\tau^*(c_{VU}) = c^*_{VU} = [\partial x_U / \partial x_V]^T$. Thus we are making the identification

$$a_i^V = \left[\frac{\partial x_U}{\partial x_V}\right]^T_{ij} a_j^U = a_j^U \left[\frac{\partial x_U^j}{\partial x_V^i}\right]$$

and $E_\tau$ is thus the *cotangent bundle!* In *general*, if $E$ is a vector bundle and $\tau^*$ is the representation $\tau^*(g) = (g^{-1})^T$, then the associated vector bundle is called the **dual** bundle to $E$.

**Example 2:** Let $E$ again be the tangent bundle to $M^n$. Let $G = \mathrm{Gl}(n)$ act on mixed second order tensors $\mathbb{R}^n \otimes \mathbb{R}^{n*}$ as follows. Let $\tau : \mathbb{R}^n \to \mathbb{R}^n$ be the standard representation $\tau(g)(v)^i = g^i{}_j v^j$ and let $\tau^* : \mathbb{R}^{n*} \to \mathbb{R}^{n*}$ be the dual representation $\tau^*(g)(a)_i = a_j (g^{-1})^j{}_i$ given in Example 1. Then $G$ acts on mixed tensors, say $\mathbf{v} \otimes \alpha$ in $\mathbb{R}^n \otimes \mathbb{R}^{n*}$, by the **tensor product representation** $\tau \otimes \tau^*$

$$(\tau \otimes \tau^*)(g)(\mathbf{v} \otimes \alpha) : = \tau(g)(\mathbf{v}) \otimes \tau^*(g)(\alpha)$$

$$= \partial_i (g^i{}_r v^r a_s g^{-1s}{}_j) \otimes dx^j$$

where $\mathbf{v} = \partial_r v^r$ and $\alpha = a_s dx^s$. The resulting bundle $E_{\tau\otimes\tau^*}$ is then the familiar *bundle of mixed second-rank tensors* on $M^n$.

In a similar manner, essentially

*all the tensor fields considered previously were sections of vector bundles that were associated to the tangent bundle through some tensor product representation of the structure group of the tangent bundle or its dual!*

### 18.2b. Connections in Associated Bundles

A connection in a vector bundle $E$ assigns to the patch $U$ of $M^n$ a $\mathfrak{g}$-valued 1-form $\omega_U$, that is, a matrix of 1-forms. This matrix acts on a section, given by the $K$-tuple $y$, yielding a $K$-tuple of 1-forms

$$(\omega y)^R = \omega^R{}_S y^S = \omega^R_{jS} y^S dx^j$$

We then have the covariant differential

$$\nabla_U y_U = dy_U + \omega_U y_U \tag{18.22}$$

and in each overlap

$$\nabla_V y_V = c_{VU} \nabla_U y_U$$

Suppose now that we have a representation $\rho : G \to \mathrm{Gl}(N)$ of the structure group of $E$. Since $\mathfrak{g}$ is the tangent space to the manifold $G$ at $e$ and $\mathfrak{gl}(N)$ is the tangent space to $\mathrm{Gl}(N)$ at $\rho(e) = I$, the differential $\rho_*$ yields a linear transformation $\rho_* : \mathfrak{g} \to \mathfrak{gl}(N)$. If $\mathbf{S} \in \mathfrak{g}$, the 1-parameter subgroup generated by $\mathbf{S}$ is $\exp(t\mathbf{S})$. Since $\rho$ is a *homomorphism*, the image curve $\rho[\exp(t\mathbf{S})]$ is a 1-parameter subgroup of $\mathrm{Gl}(N)$, and so is again of the form $\exp(t\mathbf{Y})$ for some $\mathbf{Y} \in \mathfrak{gl}(N)$. But the tangent to $\rho[\exp(t\mathbf{S})]$ at $I$ is, by the definition of the differential, simply $\rho_*(\mathbf{S})$, and so $\mathbf{Y} = \rho_*(\mathbf{S})$ and

$$\rho[\exp(t\mathbf{S})] = \exp[t\rho_*(\mathbf{S})] \tag{18.23}$$

For example, in the homomorphism $\rho : U(1) \to \mathrm{Gl}(2, \mathbb{C})$ given by $\rho(e^{i\theta}) = \mathrm{diag}(e^{i\theta}, e^{3i\theta})$, $i \in \mathfrak{u}(1)$ gets sent into the $2 \times 2$ matrix $\rho_*(i) = \mathrm{diag}(i, 3i)$.

In the homomorphism $g \to \rho(g) = \tau^*(g) = (g^{-1})^T$ of $G$ into itself, $\exp(t\mathbf{S})$ gets sent into $\exp(-t\mathbf{S}^T)$, and so $\rho_*(\mathbf{S}) = -\mathbf{S}^T$.

Let now $E_\rho$ be the bundle associated to $E$ through a representation $\rho : G \to \mathrm{Gl}(N)$. We define an associated connection for $E_\rho$ by using as connection form in $U$

$$\Omega_U := \rho_* \omega_U \tag{18.24}$$

which is defined as follows. Let $\mathbf{X}$ be a tangent vector to $M^n$. Then $\omega_U(\mathbf{X}) \in \mathfrak{g}$ and so $\rho_*[\omega_U(\mathbf{X})] \in \rho_*(\mathfrak{g}) = \mathfrak{gl}(N)$. Then we define

$$[\rho_* \omega_U](\mathbf{X}) := \rho_*[\omega_U(\mathbf{X})] \tag{18.25}$$

In particular

$$\Omega_j := \Omega(\partial_j) = (\rho_* \omega)_j = (\rho_* \omega)(\partial_j) = \rho_* \omega_j \tag{18.26}$$

**Theorem (18.27):** $\{\Omega_U\}$ *defines a connection for the bundle* $E_\rho$ *associated to* $E$ *via the representation* $\rho$.

Before looking at the proof we consider two examples. Let $\omega$ be the connection form for the tangent bundle $E = TM^n$.

**Example 1′:** We have seen in Example 1 that the *cotangent* bundle is associated with the representation $\tau^*(g) = (g^{-1})^T$. We have also seen that $\rho_*(\mathbf{S}) = -\mathbf{S}^T$ for all $\mathbf{S} \in \mathfrak{g}$. Hence $\Omega = -\omega^T$ is the connection form for the cotangent bundle, that is, $\Omega_j = -(\Gamma_j)^T$. Thus for covariant derivative we get

$$a_{i/j} = \partial_j a_i - a_R \Gamma^R_{ji}$$

which agrees with Equation (11.12).

**Example 2′:** As in Example 2, consider the vector bundle of mixed second-order tensors associated to the tangent bundle through the representation $\rho = \tau \otimes \tau^*$. For any 1-parameter subgroup $g = e^{t\mathbf{S}}$ of $G$ we have

$$\rho(\exp t\mathbf{S})(\mathbf{v} \otimes \alpha) = (\exp t\mathbf{S}\mathbf{v}) \otimes (\exp -t\mathbf{S}^T \alpha)$$

and thus

$$\rho_*(\mathbf{S})(\mathbf{v} \otimes \alpha) = \frac{d}{dt}[\rho(\exp t\mathbf{S})(\mathbf{v} \otimes \alpha)]_{t=0} = (\mathbf{S}\mathbf{v}) \otimes \alpha - \mathbf{v} \otimes (\mathbf{S}^T \alpha)$$

We may write

$$\rho_*\mathbf{S} = \mathbf{S} \otimes I - I \otimes \mathbf{S}^T$$

and then

$$\Omega_j = \omega_j \otimes I - I \otimes \omega_j^T \tag{18.28}$$

Thus

$$A^R_{S/j} = \partial_j A^R_S + \Gamma^R_{jK} A^K_S - \Gamma^K_{jS} A^R_K$$

which is the familiar rule (11.13) for the covariant derivative of a mixed tensor.

**Proof of Theorem (18.27):** Let us put

$$\rho_{UV} := \rho(c_{UV})$$

for the transition matrices of the new bundle. We must show

$$\Omega_V(\mathbf{X}) = \rho_{UV}^{-1} \Omega_U(\mathbf{X}) \rho_{UV} + \rho_{UV}^{-1} d\rho_{UV}(\mathbf{X})$$

Now

$$\Omega_V(\mathbf{X}) = \rho_*[\omega_V(\mathbf{X})] = \rho_*[c_{UV}^{-1} \omega_U(\mathbf{X}) c_{UV} + c_{UV}^{-1} dc_{UV}(\mathbf{X})]$$

Consider the two terms on the right-hand side. For brevity, let us write

$$\omega_U \text{ instead of } \omega_U(\mathbf{X})$$

Now $\omega_U \in \mathfrak{g}$ is the tangent vector at $e \in G$ to a 1-parameter subgroup $g(t) :=$ $\exp(t\omega_U)$. From the geometric meaning of the differential (and using the fact that we are at a fixed point $x \in U$)

$$\rho_*(c_{UV}^{-1}\omega_U c_{UV}) = \frac{d}{dt}[\rho(c_{UV}^{-1}g(t)c_{UV})]_{t=0}$$

which, since $\rho$ is a homomorphism,

$$= \frac{d}{dt}[\rho^{-1}(c_{UV})\rho\{g(t)\}\rho(c_{UV})]_{t=0}$$

$$= \rho^{-1}(c_{UV})\frac{d}{dt}[\rho\{g(t)\}]_{t=0}\rho(c_{UV})$$

$$= \rho^{-1}(c_{UV})\rho_*(\omega_U)\rho(c_{UV}) = \rho_{UV}^{-1}\Omega_U\rho_{UV}$$

Consider now the second term $\rho_*(c_{UV}^{-1}dc_{UV})$. Let $x = x(t)$ be a curve on $M$ having $\mathbf{X}$ as tangent vector at $t = 0$. We then have a curve in the Lie group $G$

$$c_{UV}^{-1}(x(0))c_{UV}(x(t))$$

that starts at the identity with tangent $c_{UV}^{-1}dc_{UV}(\dot{\mathbf{X}})$

$$\rho_*[c_{UV}^{-1}dc_{UV}(\mathbf{X})] = \frac{d}{dt}\rho[c_{UV}^{-1}(0)c_{UV}(x(t))]_{t=0}$$

$$= \rho_{UV}^{-1}(0)\frac{d}{dt}[\rho_{UV}(x(t))]_{t=0} = \rho_{UV}^{-1}d\rho_{UV}(\mathbf{X})$$

and we are finished. □

From Theorem (18.27) we have that the covariant differential for the associated bundle is then

$$\nabla\psi = d\psi + (\rho_*\omega)\psi \tag{18.29}$$

and automatically

$$\nabla_V\psi_V = \rho(c_{VU})\nabla_U\psi_U$$

For covariant derivative

$$\nabla_j\psi = \partial_j\psi + (\rho_*\omega_j)\psi \tag{18.30}$$

If we do not suppress the fiber indices

$$\nabla_j\psi^R = \partial_j\psi^R + (\rho_*\omega_j)^R{}_S\psi^S$$

### 18.2c. The Associated *Ad* Bundle

We may let $G$ *act as a group of linear transformations on its own Lie algebra* by

$$Ad(g)\mathbf{Y} := L_{g*} \circ R_{g^{-1}*}\mathbf{Y} = g\mathbf{Y}g^{-1} \tag{18.31}$$

for all $\mathbf{Y} \in \mathfrak{g}$. Thus

$$Ad : \mathrm{G} \to \mathrm{Gl}(\mathfrak{g})$$

and one checks immediately that this *is* a representation of $G$, called the **adjoint representation**. The subgroup $Ad(G) \subset Gl(\mathfrak{g})$ is called the **adjoint group** of $G$.

For example, when $G$ is abelian (e.g., the $n$-torus), $Ad(G)$ reduces to the single identity transformation; this follows immediately upon differentiating with respect to $t$ the relation $ge^{t\mathbf{Y}}g^{-1} = e^{t\mathbf{Y}}$. In Chapter 19 we shall see that $Ad(SU(2))$ is isomorphic to the group $SO(3)$.

Since $Ad : G \to Gl(\mathfrak{g})$, its differential at the identity takes $\mathfrak{g}$ into the tangent space at 0 to the vector space $\mathfrak{g}$, $gl(\mathfrak{g})$, that is, all linear transformations of $\mathfrak{g}$

$$Ad_* : \mathfrak{g} \to \text{linear transformations of } \mathfrak{g} \text{ into itself}$$

and we can compute this as follows.

Take the curve (1-parameter subgroup of $G$) $g(t) = e^{t\mathbf{X}}$ starting at the identity of $G$. This yields the 1-parameter group of linear transformations of $\mathfrak{g}$ given by

$$Ade^{t\mathbf{X}}(\mathbf{Y}) = e^{t\mathbf{X}}\mathbf{Y}e^{-t\mathbf{X}}$$

The tangent vector to this curve in $\mathfrak{g}$, at $t = 0$, is, when translated to 0,

$$Ad_*(\mathbf{X})(\mathbf{Y}) = \frac{d}{dt}[e^{t\mathbf{X}}\mathbf{Y}e^{-t\mathbf{X}}]_{t=0} = [\mathbf{X}, \mathbf{Y}]$$

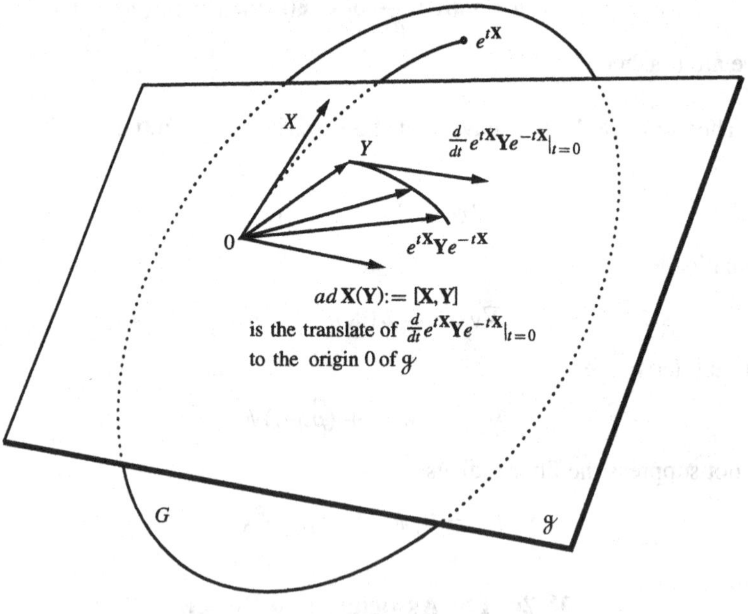


**Figure 18.1**

Let us write $ad(\mathbf{X})$ for the linear transformation $\mathfrak{g} \to \mathfrak{g}$ given by $\mathbf{Y} \to [\mathbf{X}, \mathbf{Y}]$. Thus $Ad_*(\mathbf{X}) = ad(\mathbf{X}) = [\mathbf{X}, \;]$

Recall that a 1-parameter group $h(t)$ has an infinitesimal generator $S$ such that $h(t) = e^{tS}$, and that $S = h'(0)$. Thus we have shown that for fixed $\mathbf{X}$, the 1-parameter group $Ade^{t\mathbf{X}}$ has infinitesimal generator $S = ad(\mathbf{X})$. In summary

$$\begin{aligned} Ade^{t\mathbf{X}} &= e^{tad\mathbf{X}} \\ ad(\mathbf{X}) &:= Ad_*(\mathbf{X}) \\ ad(\mathbf{X})\mathbf{Y} &:= [\mathbf{X}, \mathbf{Y}] \end{aligned} \tag{18.32}$$

We can then write

$$\begin{aligned} Ad(e^{t\mathbf{X}})\mathbf{Y} &= \left[ I + tad(\mathbf{X}) + \frac{t^2}{2!}ad(\mathbf{X})ad(\mathbf{X}) + \cdots \right]\mathbf{Y} \\ &= \mathbf{Y} + t[\mathbf{X}, \mathbf{Y}] + \frac{t^2}{2!}[\mathbf{X}, [\mathbf{X}, \mathbf{Y}]] + \cdots \end{aligned} \tag{18.33}$$

Returning to a vector bundle $E$ with structure group $G$ and connection $\omega$, we can now consider the bundle associated to $E$ through the adjoint representation $Ad : G \to \mathrm{Gl}(\mathfrak{g})$; $Ad(G)$ acts on the new fiber $\mathfrak{g}$. Then if $E$ had transition functions $c_{UV}$, the $Ad(G)$ bundle has transition matrices $Adc_{UV} : \mathfrak{g} \to \mathfrak{g}$

$$Adc_{UV}(x)\mathbf{Y} = c_{UV}(x)\mathbf{Y}c_{VU}(x)$$

and connection (18.34)

$$(Ad_*\omega_j)\mathbf{Y} = [\omega_j, \mathbf{Y}]$$

Then the covariant differential and derivatives are

$$\begin{aligned} \nabla y &= dy + [\omega, y] \\ \nabla_j y &= \partial_j y + [\omega_j, y] \end{aligned} \tag{18.35}$$

where $y = \{y_U\}$ is a section of the $Ad$ bundle $E_{Ad}$; that is, each $y_U(x) \in \mathfrak{g}$ and $y_V(x) = c_{VU}(x)y_U(x)c_{UV}(x)$

## Problems

**18.2(1)** The *co*tangent bundle $T^*M^n$ has transition functions $c_{VU} = (\partial x_U/\partial x_V)^T$ in $G = \mathrm{Gl}(n; \mathbb{R})$. $G$ acts on the 1-dimensional vector space $\mathbb{R}$ via the **determinant representation** det; $g \in G$ sends $r \in \mathbb{R}$ to $\det(g)r$. One may then consider the *real* line bundle, the **determinant** line bundle, associated to $T^*M$ via det.

**(i)** Show that any globally defined exterior $n$-form on $M^n$ can be used to define a cross section of this new bundle.

**(ii)** If $\omega$ is a connection form for the *tangent* bundle $TM^n$ (for example, the Levi-Civita connection for a Riemmanian $M$) show that $-\mathrm{tr}\omega$ is the associated connection for the determinant bundle and thus the covariant derivative of a section $\phi$ is given by

$$\nabla_j\phi = \phi_{/j} = \partial\phi/\partial x^j - \mathrm{tr}(\omega_j)\phi$$

**(iii)** If $M^n$ is Riemannian, the volume $n$-form $\text{vol}^n = \sqrt{g}dx$ is a pseudoform. The **volume** bundle is the line bundle with transition functions $c'_{VU} = |c_{VU}| = |\det(\partial x_U/\partial x_V)|$. $\{\sqrt{g_U}\}$ defines a global section of this bundle. Show that $-\text{tr}\omega$ is again a connection for this bundle and show that the section $\{\sqrt{g_U}\}$ defined by the volume form is *covariant constant!* This is the interpretation of Equation (11.28)!

**18.2(2)** The tangent bundle to an orientable surface has transition functions

$$c_{UV} = \begin{bmatrix} \cos\theta & -\sin\theta \\ \sin\theta & \cos\theta \end{bmatrix}$$

when orthonormal frames are employed. Consider the representation

$$\rho : SO(2) \to U(1)$$

defined by $\rho(c_{UV}) = e^{i\theta}$. This defines an associated bundle; *it is simply the tangent bundle considered as a complex line bundle.* If $(\omega_{jk})$ is the $\mathfrak{so}(2)$ matrix of connection forms, show that $i\omega_{21}$ is the connection for the associated line bundle. This agrees with (16.29).

---

## 18.3. r-Form Sections of a Vector Bundle: Curvature

Where do the curvature forms live?

### 18.3a. r-Form sections of E

In this section we generalize the notion of a (tangent) vector-valued $r$-form that played such an important role in Cartan's method in Section 9.3 and following.

An ***r*-form section of a vector bundle** $E$ over $M^n$ is by **definition** a collection of $r$-forms $\{\phi_U\}$, $\phi_U$ defined on the patch $U \subset M$ and having values in the fixed fiber $\mathbb{C}^K$ or $\mathbb{R}^K$ of $E$, such that in an overlap

$$\phi_V = c_{VU}\phi_U$$

that is, (18.36)

$$\phi_V(\mathbf{v}_1, \ldots, \mathbf{v}_r) = c_{VU}(x)\phi_U(\mathbf{v}_1, \ldots, \mathbf{v}_r)$$

for all tangent vectors $\mathbf{v}_1, \ldots, \mathbf{v}_r$ to $M^n$ at $x \in U \cap V$. (Thus, if $\mathbf{v}_1, \ldots, \mathbf{v}_r$ are sections of the tangent bundle $TM$, then $\{\phi_V(\mathbf{v}_1, \ldots, \mathbf{v}_r)\}$ defines a section of the bundle $E$!) *Each $\phi_U$ is simply a column of local r-forms*

$$\phi_U(x) = [\phi_U^R(x)] = [\phi_U^1(x), \ldots, \phi_U^K(x)]^T$$

$$\phi_U^R = \phi_{U\underline{I}}^R dx^I = \phi_{Ui_1<\cdots<i_r}^R(x)dx^{i_1} \wedge \ldots \wedge dx^{i_r}$$

We define the **exterior covariant differential** of $\phi$ (generalizing (9.29)), $\nabla\phi$, to be the collection of $r+1$ forms

$$(\nabla\phi_U)_B = \delta_B^{jI}\{\partial_j\phi_{\underline{I}} + \omega_j\phi_{\underline{I}}\}$$

that is, (18.37)

$$(\nabla\phi^R)_B = \delta_B^{jI}\{\partial_j\phi_{\underline{I}}^R + \omega_{jS}^R\phi_{\underline{I}}^S\}$$

(Recall that $\omega_j = \omega(\partial_j) \in \mathfrak{g}$ is a matrix). Note that this merely says

$$(\nabla\phi)^R = d\phi^R + \omega^R{}_S \wedge \phi^S \tag{18.38}$$

It follows, as usual, that $\nabla\phi$ is an $(r+1)$-form section of $E$, that is,

$$\nabla\phi_V = c_{VU}\nabla\phi_U$$

that is, we have *covariance*. Note that $\omega$ *is the connection for* $E$, not $TM$!

## 18.3b. Curvature and the *Ad* Bundle

We know that the local curvature forms $\theta_U = d\omega_U + \frac{1}{2}[\omega_U, \omega_U]$ of the vector bundle $E$ are $\mathfrak{g}$-valued 2-forms, that is, matrices of forms. These local $\mathfrak{g}$-valued forms, however, do not fit together to yield a global form; rather, they transform as $\theta_V = c_{VU}\theta_U c_{VU}^{-1}$

$$\theta_V = Ad(c_{VU})\theta_U \tag{18.39}$$

Thus

**Theorem (18.40):** *The collection of local curvature forms* $\{\theta_U\}$ *fit together to give a global 2-form section of the Ad(G) bundle!*

(To exhibit curvature as an $N$-tuple rather than a matrix, one introduces a basis $\{\mathbf{E}_R\}$ of the Lie algebra and writes $\theta = \sum \mathbf{E}_R\theta^R$.)

Consider the exterior covariant differential of $\theta$ in the $Ad$ bundle associated to $E$. Let $\underset{\to}{I} = (i_1 < i_2)$, $K = (k_1k_2k_3)$. Then $\theta_{\underset{\to}{I}} \in \mathfrak{g}$ and we have

$$(\nabla\theta)_K = \delta_K^{jI}\{\partial_j\theta_{\underset{\to}{I}} + Ad_*\omega_j(\theta_{\underset{\to}{I}})\} = \delta_K^{jI}\{\partial_j\theta_{\underset{\to}{I}} + [\omega_j, \theta_{\underset{\to}{I}}]\} = d\theta_K + [\omega, \theta]_K$$

from (18.5′). Thus

$$\nabla\theta = d\theta + [\omega, \theta] = d\theta + \omega\wedge\theta - (-1)^2\theta\wedge\omega$$

and since $\theta = d\omega + \frac{1}{2}[\omega, \omega] = d\omega + \omega\wedge\omega$, we have again

$$\nabla\theta = 0 \quad \text{(Bianchi identity)} \tag{18.41}$$

In general, for any $p$-form section of the $Ad(G)$ bundle

$$\nabla F^p = dF^p + [\omega, F^p] = dF^p + \omega\wedge F^p - (-1)^p F^p\wedge\omega \tag{18.42}$$

A $p$-form section of the $Ad$ bundle will be said to be a $p$-form **of type** $Ad(G)$.

Physicists traditionally do not deal with exterior forms and thus they are forced to exhibit the space–time tensor indices. On the other hand, they usually suppress the Lie algebra index. For a 1-form $F^1$ they would write

$$(\nabla F^1)_{jk} = \partial_j F_k - \partial_k F_j + [\omega_j, F_k] - [\omega_k, F_j] \tag{18.43}$$

and for a 2-form $(\nabla F^2)_{ijk} = \delta_{ijk}^{rs<t}(\partial_r F_{st} + [\omega_r, F_{st}])$, that is,

$$\begin{aligned}(\nabla F^2)_{ijk} &= \partial_i F_{jk} + \partial_k F_{ij} + \partial_j F_{ki} \\ &\quad + [\omega_i, F_{jk}] + [\omega_k, F_{ij}] + [\omega_j, F_{ki}]\end{aligned} \tag{18.44}$$

## Problems

**18.3(1)** Show that for any *p*-form of type $Ad(G)$

$$\nabla^2\psi = \nabla\nabla\psi = [\theta, \psi] \tag{18.45}$$

**18.3(2)** Let $\phi^p$ and $\psi^q$ be form *sections of an Ad(G) bundle,* associated to a vector bundle with transition matrices $c_{UV}$. Assume that $G$ is a matrix group (i.e., a subgroup of Gl($N$)). Since $G$ is a subgroup of Gl($N$), each $c_{UV}(x)$ is a matrix in Gl($N$) and we may think of $\phi$ and $\psi$ as form sections of the $Ad$Gl($N$) bundle. Think of them, as usual, as collections of locally defined matrices $\{\phi_U\}$, $\{\psi_U\}$ of forms. Then $\phi_U \wedge \psi_U$ is a local matrix of $(p+q)$-forms, and though its values need not be in $\mathfrak{g}$, they will be in $\mathfrak{gl}(N)$, which is simply the space of all $N \times N$ matrices. Show that $\phi \wedge \psi$ is a $(p+q)$-form section of the associated $Ad$Gl($N$) bundle and show then that (18.42) yields the Leibniz rule

$$\nabla(\phi \wedge \psi) = (\nabla\phi) \wedge \psi + (-1)^p \phi \wedge (\nabla\psi) \tag{18.46}$$

In particular, for any exterior power of the *curvature* form

$$\nabla(\theta \wedge \theta \wedge \ldots \wedge \theta) = 0$$

**18.3(3)** Show that if $\phi$ is a *p*-form section of an $AdG$ bundle then tr$\phi$ *is an ordinary exterior p-form on M.*

**18.3(4)** We have seen in Section 17.1c that given a constant $g \in G$ there is a *right* action of $g$ on the principal bundle; locally it was defined in "coordinates" by $g_U \to g_U g$, and then it was shown that this was compatible with the bundle structure. One cannot get a *left* action by this process; however, we can do the following. Consider G-valued *functions* $h_U : U \to G$ on each trivialization patch $U \subset M$. Let $h_U$ act on $\pi^{-1}(U)$ of the principal bundle $P$ by

$$g_U(x) \to h_U(x) g_U(x) \tag{18.47}$$

Show that these local actions fit together to give a global transformation of $P$ into itself provided

$$h_V = c_{VU} h_U c_{VU}^{-1} \tag{18.48}$$

Thus we have the following. Consider the *fiber* bundle associated to the principal bundle $P$, whose fiber is again $G$ but where $G$ acts on itself *not* by left translation, but by the **adjoint action,** adjoint$_g : G \to G$, of $G$ on $G$ (*not* on $\mathfrak{g}$)

$$\text{adjoint}_g(g') := g g' g^{-1} \tag{18.49}$$

This bundle is called the **gauge bundle**. *Thus the left action (18.48) is globally defined provided* $\{h_U\}$ *defines a cross section of the gauge bundle.* The left action is again called a **gauge transformation**, but we shall not discuss here the relation with the gauge transformations of Section 9.4b. We do *not* claim that any such section other than $h = e$ exists for a given bundle $P$.

CHAPTER 19

# The Dirac Equation

Spin is what makes the world go 'round.

## 19.1. The Groups $SO(3)$ and $SU(2)$

How does SU(2) act on its Lie algebra?

FOR physical and mathematical motivation for this section (which involves nonrelativistic quantum mechanics) we refer the reader to some remarks of Feynman and of Weyl. Specifically, Feynman [FF, pp. 8, 9], in his section entitled "Degeneracy," shows that a process involving a specific choice of direction in space requires that the process be described not by a single wave function $\psi$ but rather by a multicomponent column vector of wave functions $\Psi = (\psi_1, \ldots, \psi_N)^T$. He then indicates [pp. 9–12], roughly speaking, that since the *physics* cannot depend on the choice of cartesian coordinates $(x^1, x^2, x^3)$ of space, the $N$-tuples must transform under some representation $\rho : SO(3) \to U(N)$ of the rotation group $SO(3)$ of space. This is not quite accurate; since $e^{i\gamma}\Psi$ represents the same wave function (when $\gamma$ is a constant), $\rho$ is only a "ray" representation, $\rho(g)\rho(h) = e^{i\gamma(g,h)}\rho(gh)$ for a function $\gamma(g, h)$. Weyl [Wy, p. 183] shows that this can be made into a genuine representation, except that it is (perhaps) double-valued. We shall show in this section that there is a natural $2 : 1$ homomorphism $\pi$ of the special unitary group $SU(2)$ onto $SO(3)$, thus yielding a (perhaps double valued) representation of $SU(2)$ into $U(N)$. An argument of Weyl [pp. 183–4] indicates that a multiple-valued representation of a *simply-connected* group is actually single-valued. We have seen in Section 12.2 that $SO(3)$ is *not* simply connected, but we shall show in Section 19.1c that $SU(2)$ *is* simply connected, and thus the wave vectors $\Psi$ transform via a true representation of the "covering group" $SU(2)$ of $SO(3)$. The relationship with "spinors" will be discussed in Section 19.2.

A concrete physical example (the Stern–Gerlach experiment) is discussed by Feynman in [F, L, S, vol. III, chap. 6].

## 19.1a. The Rotation Group $SO(3)$ of $\mathbb{R}^3$

Rotations of $\mathbb{R}^3$ about the $z$ axis form a 1-parameter subgroup

$$R(\theta) = \begin{bmatrix} \cos\theta & -\sin\theta & 0 \\ \sin\theta & \cos\theta & 0 \\ 0 & 0 & 1 \end{bmatrix} = \exp\theta E_3$$

for some $E_3 \in \mathfrak{so}(3)$. Then

$$E_3 = \frac{d}{d\theta}\exp\theta E_3\Big]_{\theta=0} = \frac{d}{d\theta}R(\theta)\Big]_{\theta=0}$$
$$= \begin{bmatrix} 0 & -1 & 0 \\ 1 & 0 & 0 \\ 0 & 0 & 0 \end{bmatrix}$$

and likewise for $E_1$ and $E_2$. We use as a basis for $\mathfrak{so}(3)$

$$E_1 = \begin{bmatrix} 0 & 0 & 0 \\ 0 & 0 & -1 \\ 0 & 1 & 0 \end{bmatrix}, \qquad E_2 = \begin{bmatrix} 0 & 0 & 1 \\ 0 & 0 & 0 \\ -1 & 0 & 0 \end{bmatrix}, \qquad E_3 = \begin{bmatrix} 0 & -1 & 0 \\ 1 & 0 & 0 \\ 0 & 0 & 0 \end{bmatrix} \tag{19.1}$$

For Lie algebra, we compute

$$[E_1, E_2] = E_3 \qquad [E_2, E_3] = E_1 \qquad [E_3, E_1] = E_2$$

that is,

$$[E_i, E_j] = \sum_k \epsilon_{ijk} E_k \tag{19.2}$$

and then these are the structure constants of $SO(3)$

$$c_{ij}^k = \epsilon_{ijk}$$

Consider now a 1-parameter group of rotations with angular velocity $\boldsymbol{\omega}$, $\omega = d\theta/dt$.

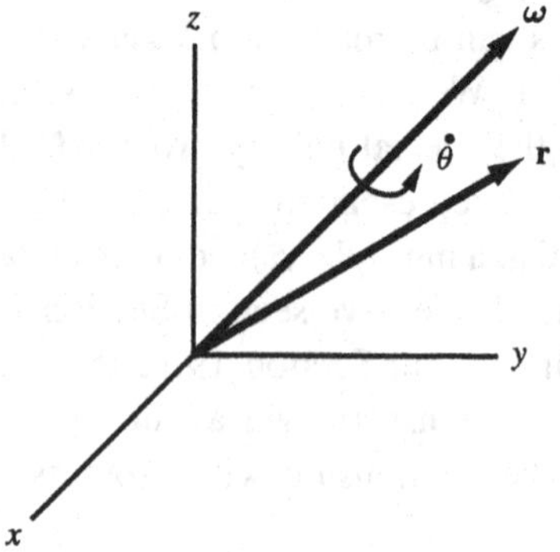


Figure 19.1

Then

$$\frac{d\mathbf{r}}{dt}\Big]_{t=0} = \boldsymbol{\omega}\times\mathbf{r}(0)$$

On the other hand, this 1-parameter subgroup is of the form

$$R(t) = e^{tS}$$

for some skew symmetric matrix $S$, and so $\mathbf{r}(t) = R(t)\mathbf{r}(0) = e^{tS}\mathbf{r}(0)$

$$\left.\frac{d\mathbf{r}}{dt}\right]_{t=0} = S\mathbf{r}(0)$$

and we conclude that

$$S(\mathbf{r}) = \boldsymbol{\omega}\times\mathbf{r}$$

Note in particular that the skew symmetric matrices $E_1$, $E_2$, and $E_3$ are simply the matrices of the linear transformations

$$E_j(\mathbf{r}) = \mathbf{e}_j \times (\mathbf{r})$$

where $\{\mathbf{e}_j\}$ is the standard basis of $\mathbb{R}^3$. Then we can write symbolically

$$R(t) = \exp(E_j\omega^j t) =: \exp(\mathbf{E}\bullet\boldsymbol{\omega}t) \tag{19.3}$$

in this case of constant angular velocity.

In terms of an *angle* of rotation, $\theta = td\theta/dt$, and the unit vector $\mathbf{n}$ along the axis $\boldsymbol{\omega}$,

$$R(\theta) = \exp(\theta\mathbf{E}\bullet\mathbf{n}) \tag{19.4}$$

represents a rotation through an angle $\theta$ about an axis with unit normal $\mathbf{n}$.

### 19.1b. $SU(2)$: The Lie algebra $\mathfrak{su}(2)$

$\mathfrak{su}(2) = \mathfrak{g}$ consists of skew hermitian matrices with trace 0. Then $i\mathfrak{g}$ is the vector space of **hermitian** matrices with trace 0, to be considered as a *real* 3-dimensional vector space (i.e., it is closed under multiplication by real numbers). A basis for $i\mathfrak{g}$ is given by the **Pauli matrices**

$$\sigma_1 = \begin{bmatrix} 0 & 1 \\ 1 & 0 \end{bmatrix}, \qquad \sigma_2 = \begin{bmatrix} 0 & -i \\ i & 0 \end{bmatrix}, \qquad \sigma_3 = \begin{bmatrix} 1 & 0 \\ 0 & -1 \end{bmatrix} \tag{19.5}$$

For example, $\exp(\theta\sigma_3/i) = \exp\operatorname{diag}(-i\theta, i\theta) = \operatorname{diag}(e^{-i\theta}, e^{i\theta})$ describes a complete 1-parameter subgroup of $SU(2)$ for $0 \le \theta \le 2\pi$. Note that the commutation relations are given by

$$[\sigma_j, \sigma_k] = 2i\epsilon_{jkl}\sigma_l \tag{19.6}$$

which is the *same* as for $SO(3)$ if one uses $\sigma_j/2i$ as new basis for $\mathfrak{g} = \mathfrak{su}(2)$. We shall soon see that $SU(2)$ is simply connected. Lie group theory states that there is then a homomorphism from $SU(2)$ onto $SO(3)$. (These groups are then locally "the same": The proof is an application of the Frobenius theorem.) We shall exhibit the classical homomorphism

$$Ad : SU(2) \to SO(3)$$

Thus we claim that

*the adjoint representation $Ad(g)\mathbf{Y} = g\mathbf{Y}g^{-1}$ of $SU(2)$ on its 3-dimensional Lie algebra $\mathfrak{su}(2)$ yields (see Theorem (19.12)) the standard representation of $SO(3)$ on $\mathbb{R}^3$.*

We start out by looking more carefully at the Lie algebra $su(2) = \mathfrak{g}$. $\mathfrak{g}$ and $i\mathfrak{g}$ are to be considered as 3-dimensional vector spaces over *real* coefficients; $i\mathfrak{g}$ has a basis given by the $\sigma$'s and $\{\sigma_\alpha/i\}$ give a basis for $\mathfrak{g}$. Define a map

$$_* : \mathbb{R}^3 \to i\mathfrak{g} \qquad \mathbf{x} \mapsto \mathbf{x}_* \tag{19.7}$$

$$\mathbf{x}_* = x \bullet \sigma = x^R \sigma_R = \begin{bmatrix} z & x - iy \\ x + iy & -z \end{bmatrix}$$

This linear transformation maps $\mathbb{R}^3$ *onto* the space of *traceless* hermitian matrices and has inverse given by

$$x = \frac{1}{2}\mathrm{tr}(\mathbf{x}_*\sigma_1) \qquad y = \frac{1}{2}\mathrm{tr}(\mathbf{x}_*\sigma_2) \qquad z = \frac{1}{2}\mathrm{tr}(\mathbf{x}_*\sigma_3) \tag{19.8}$$

Under the map$_*$

$$\mathbf{e}_1 = (1, 0, 0)^T \mapsto \sigma_1 \qquad \mathbf{e}_2 = (0, 1, 0)^T \mapsto \sigma_2 \qquad \mathbf{e}_3 = (0, 0, 1)^T \mapsto \sigma_3$$

*We shall use $_*$ to identify points $\mathbf{x}_*$ in $i\mathfrak{g}$ (i.e., hermitian traceless matrices) with points $\mathbf{x}$ of $\mathbb{R}^3$.*

From

$$\mathrm{tr}(\sigma_1\sigma_1) = \mathrm{tr}(\sigma_2\sigma_2) = \mathrm{tr}(\sigma_3\sigma_3) = 2$$

and

$$\mathrm{tr}(\sigma_j\sigma_k) = 0 \text{ if } j \neq k$$

we see that if we define a **real scalar product** in $i\mathfrak{g}$ by

$$\langle h, h' \rangle := \mathrm{tr}(hh') \tag{19.9}$$

then the Pauli matrices form an orthogonal basis (of lengths $\sqrt{2}$).

Recall that every Lie group $G$ acts on its Lie algebra $\mathfrak{g}$ by the adjoint action

$$Ad : G \to Gl(\mathfrak{g}) \qquad Ad(g)(\mathbf{X}) = g\mathbf{X}g^{-1}$$

for all $\mathbf{X} \in \mathfrak{g}$. Each $Ad(g)$ is a *linear* transformation.

In our case we consider instead the action of $SU(2)$ on the hermitian traceless matrices $i\mathfrak{g}$, and *we shall still call this Ad.* $Ad(u)$ is the linear transformation $Ad(u) : i\mathfrak{g} \to i\mathfrak{g}$

$$u \in SU(2) \quad \text{sends } \mathbf{x}_* \in i\mathfrak{g} \text{ into } u\mathbf{x}_*u^{-1} \tag{19.10}$$

For each $2 \times 2$ $u \in SU(2)$ we are associating a $3 \times 3$ matrix

$$Ad(u) : \mathbb{R}^3 \to \mathbb{R}^3$$

using the identification $_*$ of (19.7). Note that $Ad$ is a representation of $SU(2)$ by $3 \times 3$ matrices,

$$Ad(uu')(\mathbf{x}_*) = uu'\mathbf{x}_*(uu')^{-1} = Ad(u) \circ Ad(u')(\mathbf{x}_*)$$

Note further that

$$\langle Ad(u)\mathbf{x}_*, Ad(u)\mathbf{x}_*\rangle = \mathrm{tr}(u\mathbf{x}_*u^{-1}u\mathbf{x}_*u^{-1}) = tr(\mathbf{x}_*\mathbf{x}_*)$$
$$= \langle \mathbf{x}_*, \mathbf{x}_*\rangle$$

and so $Ad$ is a representation of $SU(2)$ by *orthogonal* $3 \times 3$ matrices. We claim that these matrices also have determinant $+1$. To see this (and more) we shall discuss the topology of $SU(2)$.

### 19.1c. $SU(2)$ is Topologically the 3-Sphere

The usual ("fundamental") representation of $SU(2)$ is by $2 \times 2$ complex unitary matrices with unit determinant

$$\begin{bmatrix} u_{11} & u_{12} \\ u_{21} & u_{22} \end{bmatrix}$$

We shall show that $SU(2)$ is topologically the 3-sphere $S^3$. $S^3$ can be pictured as the set of unit vectors in $\mathbb{C}^2 \approx \mathbb{R}^4$

$$S^3 = \{(z_1, z_2)^T : |z_1|^2 + |z_2|^2 = 1\}$$

Note that $SU(2) : S^3 \to S^3$; this is the meaning of being unitary. Note further that $SU(2)$ acts transitively on $S^3$, for $(1, 0)^T \in S^3$ can be sent into a generic point $(z_1, z_2)^T \in S^3$ by

$$u = \begin{bmatrix} z_1 & -\bar{z}_2 \\ z_2 & \bar{z}_1 \end{bmatrix} \in SU(2) \tag{19.11}$$

(In fact, the second column is the *unique* vector in $\mathbb{C}^2$ that is hermitian–orthogonal to $(z_1, z_2)^T$ and is such that $\det u = 1$.)

From (17.10) we know that topologically

$$S^3 \approx \frac{SU(2)}{H}$$

where $H$ is the stability subgroup of the point $(1, 0)^T$. But, as we see in (19.11), $H$ is simply the $2 \times 2$ identity matrix $I$. Thus

$$SU(2) \approx S^3$$

topologically. In fact we have seen that *the correspondence* $SU(2) \to S^3$ *is given simply by sending the matrix u into its first column*

$$u \mapsto (u_{11}, u_{21})^T$$

In particular $SU(2) = S^3$ is *connected.* Since $Ad(u)$ is an orthogonal matrix, $\det Ad(u)$ is $\pm 1$. Since it is continuous in $u$ and always $\pm 1$ on the connected $S^3$ we see that the determinant is $+1$. Thus $Ad(u) \in SO(3)$

$$Ad : SU(2) \to SO(3)$$

### 19.1d. *Ad* : *SU*(2) → *SO*(3) in More Detail

**Theorem (19.12):** *The representation $Ad : SU(2) \to SO(3)$ given in (19.10) is onto; that is, every rotation in $\mathbb{R}^3$ is of the form (19.10). Furthermore, this representation is 2:1; that is, for each rotation R there are exactly two matrices $\pm u \in SU(2)$ such that $Ad(\pm u) = R$.*

PROOF: Let $u(t)$ be a 1-parameter subgroup of $SU(2)$; it is of the form $u(t) = e^{th/i}$, where $h$ is a hermitian $2 \times 2$ matrix. This produces a 1-parameter subgroup of $SO(3)$ under our identification of $i\mathfrak{g}$ with $\mathbb{R}^3$ (i.e., $\mathbf{x}_* \sim \mathbf{x}$)

$$Adu(t)\mathbf{x} \sim Adu(t)\mathbf{x}_* = e^{-ith}\mathbf{x}_* e^{ith}$$

The velocity vector at $\mathbf{x} \in \mathbb{R}^3$ is given by

$$\begin{aligned}\frac{d}{dt}Adu(t)\mathbf{x}_*|_{t=0} &= \frac{d}{dt}e^{-ith}\mathbf{x}_* e^{ith}|_{t=0} = -i[h, \mathbf{x}_*]\\ &= -i[h^j\sigma_j, x^k\sigma_k] = -ih^j x^k[\sigma_j, \sigma_k]\\ &= 2h^j x^k \epsilon_{jkl}\sigma_l \sim 2(\mathbf{h} \times \mathbf{x})^l \sigma_l\end{aligned}$$

The angular velocity vector of the 1-parameter group $Adu(t)\mathbf{x}$ in $\mathbb{R}^3$ is then $\omega = 2\mathbf{h}$, and from (19.3)

$$Ad\exp\left(\frac{\sigma}{i}\bullet\mathbf{h}t\right)\mathbf{x}_* \sim R(t)x = \exp(\mathbf{E}\bullet 2\mathbf{h}t)\mathbf{x} \tag{19.13}$$

We have just verified that

$$Ad_*(\sigma_\alpha/2i) = E_\alpha \tag{19.14}$$

and this is not very surprising considering the remarks after (19.6).

For example, as we have seen, the vector $\mathbf{h} = \sigma(0, 0, 1)^T$, that is, $\mathbf{h} = \sigma_3$, generates the 1-parameter subgroup of $SU(2)$

$$\exp\theta\frac{\sigma_3}{i} = \begin{bmatrix} e^{-i\theta} & 0 \\ 0 & e^{i\theta} \end{bmatrix}$$

and this corresponds, under $Ad$, to the 1-parameter subgroup of rotations of $\mathbb{R}^3$ (see Problem (15.2(1)))

$$\exp 2\theta E_3 = \exp\begin{bmatrix} 0 & -2\theta & 0 \\ 2\theta & 0 & 0 \\ 0 & 0 & 0 \end{bmatrix} = \begin{bmatrix} \cos 2\theta & -\sin 2\theta & 0 \\ \sin 2\theta & \cos 2\theta & 0 \\ 0 & 0 & 1 \end{bmatrix}$$

Note that $\exp(\theta\sigma_3/i)$ describes a *simple closed curve* in $SU(2)$ for $0 \le \theta \le 2\pi$, and $\exp(2\theta E_3)$ yields *two full rotations* in this same $\theta$ range!

Since every rotation of $\mathbb{R}^3$ is a rotation about some axis, that is, is of the form $R = \exp(\mathbf{E}\bullet\omega\theta)$, we see from (19.13) that $Ad\exp(\sigma/2i\bullet\omega\theta) = R$, and $Ad$ is indeed onto.

It is immediate that if $Ad(u) = R$ then $Ad(-u) = R$ also, so that the $Ad$ representation is at least 2 : 1; that is, it is **not faithful**. It is an elementary result of group theory that

If $\phi : G \to G'$ is a homomorphism of $G$ *onto* $G'$, then $G'$ is isomorphic to the coset space $G/H$, where $H = \phi^{-1}(e')$ is the kernel.

This is basically our "fundamental principle" (17.10), for $G$ acts on $G'$ by $(g, g') \mapsto \phi(g)g'$ and the stability subgroup of $e' \in G'$ is the kernel $H = \phi^{-1}(e')$. In our case we need to know that the kernel of the $Ad$ homomorphism consists precisely of the two $2 \times 2$ matrices $\pm I$; we will then know, from (17.11), that $SU(2)$ is a fiber bundle over $SO(3)$ with fiber always consisting of exactly two points $\pm u$. This should not surprise us since topologically $SU(2)$ is $S^3$, $SO(3)$ is the projective space $\mathbb{R}P^3$, and $\mathbb{R}P^3$ results from $S^3$ by identifying pairs of antipodal points!

Look then for those special unitary $u$ such that $Ad(u)$ is the identity rotation in $\mathbb{R}^3$. We thus need $u\mathbf{x}_* u^{-1} = \mathbf{x}_*$, for all hermitian $\mathbf{x}_*$ with trace 0. In particular $\sigma_\alpha^{-1} u \sigma_\alpha = u$, for each Pauli matrix. Writing $u$ in the form (19.11) and putting $\alpha = 1$ will show that $z_1$ must be real. Putting $\alpha = 3$ will yield that $z_2 = 0$. Thus $u$ must be of the form $\pm I$, as desired. □

## 19.2. Hamilton, Clifford, and Dirac

Why is it that a full rotation is something whereas two full rotations is nothing?

### 19.2a. Spinors and Rotations of $\mathbb{R}^3$

We saw in the last section that there is a representation of $SU(2)$ as a group of rotations of $\mathbb{R}^3$

$$
\begin{aligned}
Ad : SU(2) &\to SO(3) \\
\exp\left(\frac{\sigma}{2i} \cdot \mathbf{A}\theta\right) &\mapsto \exp(\mathbf{E} \cdot \mathbf{A}\theta)
\end{aligned}
\tag{19.15}
$$

for any $\mathbf{A} = (A_1, A_2, A_3)$, and that the mapping $Ad$ is exactly $2:1$. Thus to a rotation of $\mathbb{R}^3$ about an axis given by a unit vector $\mathbf{A}$ through an angle $\theta$ radians one associates two $2 \times 2$ unitary matrices with determinant 1,

$$
\exp\left[\frac{\sigma}{2i} \cdot \mathbf{A}\theta\right] \quad \text{and} \quad \exp\left[\frac{\sigma}{2i} \cdot \mathbf{A}(\theta + 2\pi)\right]
$$

In other words, $SO(3)$ not only has the usual representation by $3 \times 3$ matrices, it *also has a double-valued representation by* $2 \times 2$ *matrices acting on* $\mathbb{C}^2$.

The complex vectors $(\psi_1, \psi_2)^T \in \mathbb{C}^2$ on which $SO(3)$ acts in this double-valued way are called (2-component) **spinors**. Mathematicians do not like double-valued *anythings*; they prefer to say that $SU(2)$ *furnishes* naturally *a spinor representation of the 2-fold cover of* $SO(3)$. When $SU(2)$ is thought of as the 2-fold cover of $SO(3)$, it is called the **spinor group** Spin (3).

The topological reason that $SO(3)$ can admit a nontrivial double-valued representation is that $SO(3)$ is not simply connected. The reasoning is very much like that used in complex function theory when showing that a region in the complex plane supporting a multiple-valued analytic function cannot be simply connected. The 1-parameter subgroup of $SO(3)$

$$\theta \mapsto \begin{bmatrix} \cos\theta & -\sin\theta & 0 \\ \sin\theta & \cos\theta & 0 \\ 0 & 0 & 1 \end{bmatrix}$$

for $0 \leq \theta \leq 2\pi$ is a closed curve $C$ in $SO(3) = \mathbb{R}P^3$.

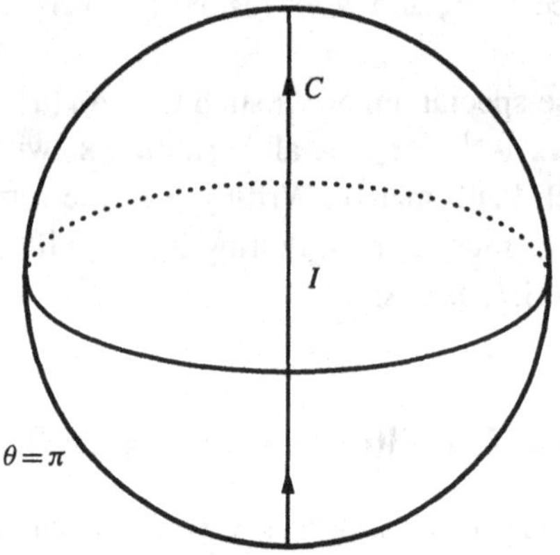


Figure 19.2

This curve can be deformed into the curve $A$ of Section 13.3b, Example (5), and *this curve cannot be shrunk to a point.* This subgroup is generated by $\mathbf{E} \cdot \mathbf{A} = E_3$. It is covered in the group $SU(2)$ by the portion of the 1-parameter subgroup generated by $\sigma/2i \cdot \mathbf{A} = \sigma_3/2i$

$$\theta \to \exp\frac{\sigma_3\theta}{2i} = \begin{bmatrix} e^{-i\theta/2} & 0 \\ 0 & e^{i\theta/2} \end{bmatrix} \tag{19.16}$$

for $0 \leq \theta \leq 2\pi$. This is *not* a closed curve in $SU(2)$ since it starts at $I$ and ends at $-I$. Of course, if we make 2 complete rotations in $\mathbb{R}^3$, this 1-parameter subgroup in $SU(2)$ that covers it will be a closed curve on $SU(2) = S^3$. This curve on $S^3$ can be shrunk to a point (why?), and by "projecting down" we can use this to shrink the curve representing 2 full rotations of $\mathbb{R}^3$ to a point.

> *In this way one can distinguish between a full rotation (which of course brings every point of $\mathbb{R}^3$ back to its original position) and two full rotations of $\mathbb{R}^3$ about an axis!*

This truly mysterious fact can be experienced on at least three different levels. We shall mention two manifestations here; after discussing the Dirac equation we shall discuss the significance for particle physics.

The two remarks to follow are related to the topological fact that the closed curve $A$ in $SO(3)$ has the property that it cannot be shrunk to a point, whereas any even multiple of it can be.

1. *Physiologically*. This is the old "waiter with a platter" trick; see Feynman's treatment in [F, W, p. 29].
2. *Mechanically*. (This interpretation was given by Weyl.) We are going to show that the closed curve in $SO(3)$ described by rotating a rigid body *twice* about an axis through a given point $O$ of the body can be deformed into the point curve representing no rotation at all. Consider a mathematical cone in space, vertex at $O$, with axis always the $z$-axis, and with (half) opening angle $\alpha$. Consider another mathematical cone, congruent to the first, but this time fixed in the body with vertex at $O$. Move the body so that the body cone rolls around the space cone.

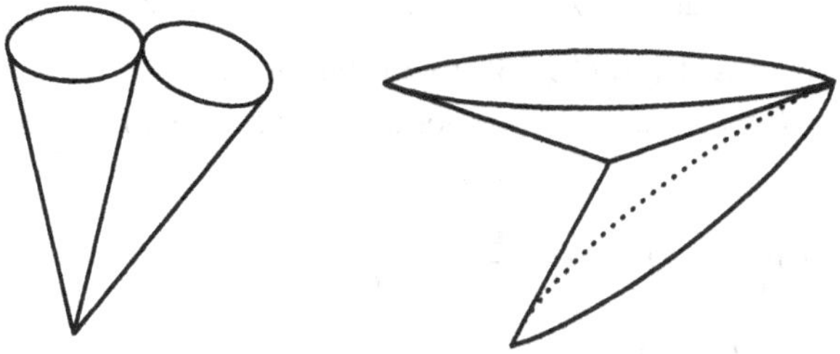

**Figure 19.3**

If the opening angle $\alpha$ is *very small*, then on looking down on the space cone one can see that when the body cone has come around to its original position, the body has made approximately *two* full revolutions about the $z$ axis, and as $\alpha$ tends to 0 the body rotation tends exactly to two revolutions. On the other hand, if we use an opening angle $\alpha$ that is almost $\pi/2$, then the cones are very flat and the body cone will be seen to wobble, with hardly any rotation at all, and in the limit as $\alpha \to \pi/2$ the body remains motionless! Thus, when using $\alpha$ as a deformation parameter, the curve representing a rotation through $4\pi$ radians about the $z$ axis ($\alpha = 0$) can be deformed into the point curve representing no rotation ($\alpha = \pi/2$).

See also the picture in Wald's book [Wd, p. 346].

For an application to rotating electrical machinery you may read about an invention of D. Adams in the article [Sto].

## 19.2b. Hamilton on Composing Two Rotations

The relation (19.15) is a powerful tool for investigating the product of two rotations. This is a consequence of the fact that that the Pauli matrices satisfy very simple product relations

$$\sigma_1\sigma_2 = i\sigma_3 \qquad \sigma_2\sigma_3 = i\sigma_1 \qquad \sigma_3\sigma_1 = i\sigma_2 \tag{19.17}$$
$$(\sigma_1)^2 = (\sigma_2)^2 = (\sigma_3)^2 = I$$

The infinitesimal generators $E_j$ of $SO(3)$ satisfy nothing like this; for example, $(E_1)^2 = \text{diag}(0, -1, -1)$. From (19.17) one gets not only the commutation relations (19.6) but also the *anticommutator* $2 \times 2$ matrices

$$\{\sigma_i, \sigma_j\} := \sigma_i\sigma_j + \sigma_j\sigma_i = 2\delta_{ij} I \tag{19.18}$$

In Problem 19.2(1) you are asked to use the commutation and anti-commutation formulas to show the following. For any pair of vectors **A**, **B** in $\mathbb{R}^3$

$$(\sigma \bullet \mathbf{A})(\sigma \bullet \mathbf{B}) = (\mathbf{A} \bullet \mathbf{B})I + i\sigma \bullet (\mathbf{A} \times \mathbf{B}) \tag{19.19}$$

and if **A** is a *unit* vector

$$(\sigma \bullet \mathbf{A})^2 = I$$

For unit **A**

$$\exp\left(\frac{\sigma}{2i} \bullet \mathbf{A}\theta\right) = \cos\left(\frac{\theta}{2}\right) I - i \sin\left(\frac{\theta}{2}\right) \sigma \bullet \mathbf{A} \tag{19.20}$$

corresponds, as we know from (19.15), to a rotation $R_1$ of $\mathbb{R}^3$ about the axis $\sum \mathbf{e}_j A^j$ through an angle of $\theta$ radians. Let **B** be another unit vector with corresponding rotation $R_2$. Show

$$R_1 R_2 = \exp\left(\frac{\sigma}{2i} \bullet \mathbf{A}\theta\right) \exp\left(\frac{\sigma}{2i} \bullet \mathbf{B}\phi\right) \tag{19.21}$$
$$= \left[\cos\frac{\theta}{2}\cos\frac{\phi}{2} - \left(\sin\frac{\theta}{2}\sin\frac{\phi}{2}\right)\mathbf{A} \bullet \mathbf{B}\right] I$$
$$- i\sigma \bullet \left[\sin\frac{\theta}{2}\cos\frac{\phi}{2}\mathbf{A} + \cos\frac{\theta}{2}\sin\frac{\phi}{2}\mathbf{B} + \sin\frac{\theta}{2}\sin\frac{\phi}{2}(\mathbf{A} \times \mathbf{B})\right]$$

*This expression* (via (19.20)) *exhibits explicitly the (cosine of) the rotational angle and then the axis for the rotation* $R_1 R_2$.

The expression (19.21) was known to Hamilton in terms of his *quaternions* rather than Pauli matrices. We shall discuss the relation between these "algebras" next. For more information and nice pictures see the chapter on spinors in [M, T, W].

Finally note that we have mentioned before that the exponential map $\exp : \mathfrak{g} \to G$ is *onto* in the case of a connected compact group such as $SU(2)$. Thus (19.20) shows that every $u \in SU(2)$ can be written in the form

$$u = aI + i\sigma \bullet \mathbf{C}$$

where $a^2 + \| \mathbf{C} \|^2 = 1$. This expression is unique since $aI$ is real and $i\sigma \bullet \mathbf{C}$ is skew hermitian.

### 19.2c. Clifford Algebras

Let us abstract some of the properties of the Pauli matrices that will be important for generalizations. We shall be *very* informal.

First note that $\sigma_1$, $\sigma_2$, and $\sigma_3$ span a 3-dimensional vector space $V^3$ under addition and under multiplication by *real* scalars; $V^3$ is the space of trace-free hermitian matrices. In this vector space there is a quadratic form $\langle , \rangle$ given by half that in (19.9), that is, $\langle h, h' \rangle = (1/2) \operatorname{tr} hh'$, and then

$$\langle \sigma_j, \sigma_k \rangle = g_{jk} = \delta_{jk}$$

Furthermore, there is a multiplication (in this case matrix multiplication) in $V^3$ but $V^3$ is *not* closed under this multiplication. For examples, $\sigma_1 \sigma_1 = I$ is not in $V^3$ and

$\sigma_1\sigma_2 = i\sigma_3$ is not in $V^3$ ($i$ is not real). Suppose that we now try to "close" this system. We adjoin the new matrix $e_4 = \sigma_1\sigma_2$ and all its real multiples. Continuing, we define (in no particular order)

$$e_1 = \sigma_1 \qquad e_2 = \sigma_2 \qquad e_3 = \sigma_3$$

$$e_4 = e_1e_2 \qquad e_5 = e_2e_3 \qquad e_6 = e_1e_3$$

$$e_7 = e_1e_2e_3 = iI \qquad e_8 = (e_1)^2 = I$$

From the anticommutation relations (19.18), we see, for example, that $e_1e_2 = -e_2e_1$, and so we needn't adjoin $e_2e_1$. Note also that $e_8 := e_1e_1 = I$ also follows directly from (19.18). We may now form the real 8-dimensional vector space with basis given by $e_1, \ldots, e_8$. From (19.18) alone we see that this new real vector space is closed under products (e.g., $e_7e_1 = e_1e_2e_3e_1 = -e_1e_2e_1e_3 = e_1e_1e_2e_3 = e_2e_3 = e_5$). Note further that in a monomial expression (such as $e_7e_1$) any repeated basis element of $V^3$ (such as $e_1$) can be eliminated by using (19.18), yielding an expression having two fewer basis elements.

The vector space (of $2 \times 2$ matrices) of real linear combinations of these $e$'s is 8-dimensional and forms, as can be verified, an associative **algebra**, that is, a vector space with a composition (called *product*) that is associative and is distributive with respect to addition. In fact, in this case, this 8-dimensional vector space is simply the algebra of *all* complex $2 \times 2$ matrices! This algebra is **generated** by the Pauli matrices, and will be called the **Pauli algebra**.

**Definition (19.22):** If $C_n$ is an associative algebra (over $\mathbb{R}$) with "unit" $I$, generated by an $n$-dimensional vector subspace $V^n$, if $\langle , \rangle$ is *any* real quadratic form on $V^n$, and if $V^n$ has a basis $e_1, \ldots, e_n$ satisfying

$$e_je_k + e_ke_j = 2g_{jk}I \tag{19.23}$$

where $g_{jk} := \langle e_j, e_k\rangle$, then $C_n = C(V^n)$ is called the **Clifford algebra generated by** $V^n$ with the quadratic form $\langle , \rangle$.

Note that we put *no* requirements on the quadratic form $\langle , \rangle$, but of course the resulting Clifford algebra will depend on the choice of $\langle , \rangle$. For example, consider a Clifford algebra generated by an $n$-dimensional vector space $V^n$ with *quadratic form* $\langle , \rangle$ *identically* 0. Then we have $e_je_k = -e_ke_j$, for all $j, k$, and of course $(e_j)^2 = 0$. The resulting Clifford algebra is simply the *exterior algebra* based on the vector space $V^n$!

In general, as a vector space, $C_n$ is generated by expressions of the form $e_ie_j \ldots e_k$. Each $(e_j)^2$ is a multiple $g_{jj}$ of the identity, and thus commutes with everything. Also, as we have seen, we needn't consider expressions containing a repeated basis vector $e_j$. From (19.18) we need only consider expressions $e_ie_j \ldots e_k$ that are ordered, $i < j < \ldots < k$. It is then obvious that *as a vector space* (i.e., neglecting the product structure), the Clifford algebra $C(V^n)$ is isomorphic to the exterior algebra $\bigwedge(V^n)$ and thus has dimension $2^n$.

For example, the Pauli algebra, as a vector space, is isomorphic to the exterior algebra on $\mathbb{R}^3$ with abstract basis given by $\sigma_1$, $\sigma_2$, and $\sigma_3$, but of course the exterior product is far different from the product of Pauli matrices, that is, the "Clifford" product.

In an exterior algebra, the real coefficients, that is, the scalars or 0-forms, span a 1-dimensional subspace. In a Clifford algebra, the scalar multiples of the "unit" I form a 1-dimensional subspace that can be identified with the coefficient field $\mathbb{R}$.

To form a Clifford algebra with generators $e_1, \ldots, e_n$ and quadratic form $\langle , \rangle$, we simply consider all "formal expressions," $e_i e_j \ldots e_k$ with $i < j < \ldots < k$, and impose the "relations" (19.18). It can be shown that the result is indeed a Clifford algebra. Let us look at some examples.

$C_0$ is the algebra over $\mathbb{R}$ with *no* other generators; thus there are no $e_j$'s. $C_0 = \mathbb{R}$ is simply the algebra of *real numbers*.

Let $V^1$ be a 1-dimensional vector space with basis $e_1$, and quadratic form $\langle e_1, e_1 \rangle = -1$. Form the 2-dimensional vector space with formal basis consisting of $e_1$ and a new vector "$e_1 e_1$" satisfying (19.23), $(e_1)^2 = (-1)I$. Thus we are adjoining to $V^1$ a 1-dimensional vector space to accomodate the scalars (i.e., all real multiples of $-1$). The basis element $e_1$ will be called $i$, the element $(e_1)^2$ will be identified with the real number $-1$, and the 2-dimensional vector space over $\mathbb{R}$ is simply the algebra of **complex numbers** $a + bi$, $C_1 = \mathbb{C}$.

Let $V^2$ be a real 2-dimensional vector space with basis $e_1, e_2$, and quadratic form $\langle e_j, e_k \rangle = -\delta_{jk}$. We write $e_1 = \mathbf{j}$, $e_2 = \mathbf{k}$. We adjoin a 1-dimensional vector space to accommodate the scalars $\mathbf{j}^2 = \mathbf{k}^2 = (-1)I$. We adjoin another 1-dimensional vector space to house the new element $\mathbf{i} := \mathbf{jk} = -\mathbf{kj}$ (from (19.18)). Then $\mathbf{ijk} = \mathbf{i}^2 = \mathbf{jkjk} = -\mathbf{jjkk} = -I$, which is not a new element. Thus we needn't adjoin anything else. $C_2$ is Hamilton's 4-dimensional algebra of **quaternions** $a + b\mathbf{i} + c\mathbf{j} + d\mathbf{k}$.

Let $V^3$ be a 3-dimensional vector space with basis $\sigma_1, \sigma_2, \sigma_3$ and this time with scalar product $\langle \sigma_j, \sigma_k \rangle = +\delta_{jk}$. We have discussed this case previously. Adjoining products of pairs $\sigma_j \sigma_k$ satisfying (19.18) yields a 1-dimensional space of scalars (e.g., $\sigma_1^2 = I$) and a 3-dimensional space spanned by $\sigma_1 \sigma_2 = -\sigma_2 \sigma_1$, and so forth. Another 1-dimensional vector space is adjoined to house $i := \sigma_1 \sigma_2 \sigma_3$. $C_3$ is the **Pauli algebra** (but note our choice of scalar product).

### 19.2d. The Dirac Program: The Square Root of the d'Alembertian

We wish to emphasize that *we are continuing to use our choice of metric in Minkowski space,*

$$ds^2 = -dt^2 + dx^2 + dy^2 + dz^2$$

that is,

$$(g_{jk}) = (\eta_{jk}) := \text{diag}(-1, +1, +1 + 1)$$

although most treatments of quantum mechanics use the negative of this form.

Schrödinger's equation (16.40) treats time and space differently and is thus not relativistic. The first relativistic wave equation was proposed by Schrödinger, but was abandoned by him. It was then reintroduced by Klein and Gordon and is now called the **Klein–Gordon** equation. For a particle of mass $m$ it is

$$\Box\psi := g^{jk}\partial_j \partial_k \psi = m^2 \psi \tag{19.24}$$

that is,

$$-\frac{\partial^2\psi}{\partial t^2} + \frac{\partial^2\psi}{\partial x^2} + \frac{\partial^2\psi}{\partial y^2} + \frac{\partial^2\psi}{\partial z^2} = m^2\psi$$

Dirac wanted to have an equation that was *first order* in $t$, as in the nonrelativistic Schrödinger equation. Special relativity would then demand that it be first order in the spatial variables $x$, $y$, and $z$. Thus, Dirac was led to construct a *first-order* differential operator

$$\partial\!\!\!/ = \gamma^j \partial_j$$

with some constant coefficients $\gamma^j$ such that

$$\Box\psi = \partial\!\!\!/(\partial\!\!\!/\psi)$$

that is, to construct a "square root" of the d'Alembertian. Then we could solve the Klein–Gordon equation by first solving **Dirac's equation** (using the physicist's convention of putting $\hbar = 1$)

$$\partial\!\!\!/\psi = \gamma^j \partial_j \psi = m\psi \tag{19.25}$$

Then

$$\Box\psi = \partial\!\!\!/(\partial\!\!\!/\psi) = m^2\psi$$

as desired.

We then need

$$\Box = \partial\!\!\!/\ \partial\!\!\!/ = (\gamma^j\partial_j)(\gamma^k\partial_k)$$
$$= \gamma^j\gamma^k\partial_j\partial_k = \frac{1}{2}(\gamma^j\gamma^k + \gamma^k\gamma^j)\partial_j\partial_k$$

requiring that we put

$$\gamma^j\gamma^k + \gamma^k\gamma^j = 2g^{jk} = 2\eta^{jk}$$

that is, the $\gamma$'s *cannot be scalars* $(\gamma^1\gamma^2 = -\gamma^2\gamma^1)$. *The* $\gamma$*'s appear to generate a Clifford algebra*! It is then clear from Dirac's equation (19.25) that the wave function $\psi$ cannot be a single-component complex function since the **Clifford numbers** $\gamma^j$ would then take the complex numbers $\partial_j\psi$ into a Clifford number $\gamma^j(\partial_j\psi)$ that could not be equated with the complex number $m\psi$. Somehow *the Clifford numbers must act on the wave functions in a less trivial fashion.*

For relativistic purposes, Dirac also wanted "covariance" under Lorentz transformations. We now turn to these matters.

## Problem

**19.2(1)** Derive (19.19, 20, and 21). Let $R_1$ be a rotation of $\pi/2$ about the $z$ axis, and let $R_2$ be a rotation of $\pi/2$ about the $y$ axis. Describe $R_1 R_2$.

## 19.3. The Dirac Algebra

What is the topology of the Lorentz group?

### 19.3a. The Lorentz Group

Our treatment of Dirac 4-component spinors to follow owes much to Bleecker's book, *Gauge Theory and Variational Principles* [Bl]. Our metric is, however, of *opposite sign.*

The **Lorentz group** is by definition the group of linear isometries of Minkowski space $M_0^4$

$$L = \{\text{real } 4 \times 4 \text{ matrices } B | \langle Bx, By\rangle = \langle x, y\rangle\}$$

with metric $(\eta_{jk}) = \text{diag}(-1, +1, +1, +1)$. In matrix notation,

$$\langle x, y\rangle = x^T \eta y$$

and then

$$x^T \eta y = (Bx)^T \eta B y = x^T B^T \eta \text{By}$$

requires

$$B^T \eta B = \eta \tag{19.26}$$

We see that $\det B = \pm 1$. Let $e_0, e_1, e_2, e_3$ be an orthonormal basis. Since $\langle Be_0, Be_0\rangle = -1$ and $Be_0 = [B^0{}_0, B^1{}_0, B^2{}_0, B^3{}_0]^T$, we see that $(B^0{}_0)^2 \geq 1$. Bleecker shows that $L$ breaks up into 4 connected components (pieces)

$$L_0 = L+\uparrow: \det B > 0 \quad \text{and} \quad B^0{}_0 \geq 1$$

$$L-\uparrow: \det B < 0 \quad \text{and} \quad B^0{}_0 \geq 1$$

$$L+\downarrow: \det B > 0 \quad \text{and} \quad B^0{}_0 \leq -1$$

$$L-\downarrow: \det B < 0 \quad \text{and} \quad B^0{}_0 \leq -1$$

where $L_0$ is the component holding the identity. This is clearly the component consisting of Lorentz transformations that preserve not only the orientation of Minkowski space ($\det B > 0$) but also the direction of time ($B^0{}_0 > 0$). Thus the orientation of 3-space is also preserved.

Consider the **Lie algebra** $\ell$ of $L$. Write $B = e^{tS}$. Then $(e^{tS})^T \eta e^{tS} = \eta$, and differentiating with respect to $t$ and putting $t = 0$ yield $S^T \eta + \eta S = 0$. Since $\eta^T = \eta$, this says $(\eta S)^T = -\eta S$. This merely says that *when we lower the upper index of S by means of the Lorentz metric, the resulting covariant second-rank tensor is skew symmetric*!

$$S_{jk} := \eta_{jl} S^l{}_k = -S_{kj}$$

Thus $\dim L = \dim SO(4) = 6$.

$SO(3)$ is covered twice by $SU(2)$. We shall now indicate why $L_0$ *is covered twice by* $Sl(2, \mathbb{C})$, the complex $2 \times 2$ matrices with determinant $+1$ (which, of course, is again 6 dimensional). Let

$$H(2, \mathbb{C}) := \{2 \times 2 \text{ matrices } A | A^\dagger := (\overline{A})^T = A\}$$

be the 4-dimensional vector space (over $\mathbb{R}$) of $2 \times 2$ **hermitian** matrices with *no requirement on the trace*. For a basis for $H(2, \mathbb{C})$ we augment the Pauli matrices by the unit matrix

$$\tau_0 = \sigma_0 := I \qquad \tau_\alpha := \sigma_\alpha, \qquad \alpha = 1, 2, 3 \tag{19.27}$$

Define now a new map $_*: M_0^4 \to H(2, \mathbb{C})$ by

$$x \in M \mapsto x_* := x^T \tau = x^j \tau_j = x^0 \tau_0 + \mathbf{x} \cdot \boldsymbol{\sigma} \tag{19.28}$$

$$x_* = \begin{bmatrix} x^0 + z & x - iy \\ x + iy & x^0 - z \end{bmatrix}$$

We can solve for $x$

$$x^j = \frac{1}{2} \operatorname{tr}(x_* \tau_j) \tag{19.29}$$

Easily

$$\det x_* = -\langle x, x \rangle \tag{19.30}$$

We shall also have need for *another* identification of $M_0^4$ with $H(2, \mathbb{C})$, namely

$$x^* := x^T \eta \tau = -x^0 \tau_0 + \mathbf{x} \cdot \boldsymbol{\sigma}$$

Then (19.31)

$$x^* = \begin{bmatrix} -x^0 + z & x - iy \\ x + iy & -x^0 - z \end{bmatrix}$$

and one computes

$$\det x^* = -\langle x, x \rangle \tag{19.32}$$
$$x_* x^* = x^* x_* = \langle x, x \rangle I$$

The two maps $_*$ and $^*$ allow us to think of Minkowski space as being simply $H(2, \mathbb{C})$ in two ways. By using $_*$ we have the following.

**Theorem (19.33):** *The assignment to* $A \in Sl(2, \mathbb{C})$ *of the linear map* $\Lambda$ *of Minkowski space*

$$\Lambda(A) : M_0^4 = H(2, \mathbb{C}) \to H(2, \mathbb{C})$$

$$\Lambda(A)(x)_* := A x_* A^\dagger = A x_* \overline{A}^T \tag{19.34}$$

*yields a* $2:1$ *homomorphism of* $Sl(2, \mathbb{C})$ *onto* $L_0$

$$\Lambda : Sl(2, \mathbb{C}) \to L_0$$

Note that $\Lambda$ is similar to $Ad : SU(2) \to SO(3)$; in fact when $A$ is in the subgroup $SU(2)$ of $Sl(2,\mathbb{C})$ we have $A^\dagger = \overline{A}^T = A^{-1}$. Before proceeding to the proof of (19.33) we shall investigate this similarity in more detail. For the notion of "deformation retract," see Section 15.3d.

**Theorem (19.35):** *$SU(2)$ is a deformation retract of $Sl(2,\mathbb{C})$ and $SO(3)$ is a deformation retract of $L_0$.*

**Proof sketch:** $A \in Sl(2,\mathbb{C})$ can be thought of as a pair of complex vectors $[a_{11}, a_{21}]^T$ and $[a_{12}, a_{22}]^T$ spanning an "area" $\det A = 1$. By the usual Gram–Schmidt-like process used in Section 15.3d in the case of $Sl(2,\mathbb{R})$ (but using a hermitian scalar product instead) we may deform $Sl(2,\mathbb{C})$ into its subgroup $SU(2)$, all the while keeping $SU(2)$ pointwise fixed. $SU(2)$ is thus a deformation retract of $Sl(2,\mathbb{C})$.

For the Lorentz group we proceed as follows, using familiar facts about Lorentz transformations. Consider the upper sheet $H^3$ of the "unit" hyperboloid in Minkowski space,

$$-x_0^2 + \mathbf{x} \cdot \mathbf{x} = -1$$

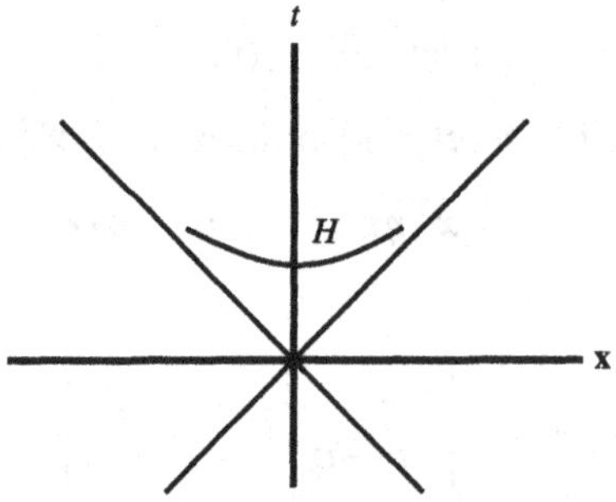


**Figure 19.4**

Each Lorentz transformation in $L_0$ takes $H$ into itself since Lorentz transformations preserve the Minkowski metric. By a suitable Lorentz transformation $\Lambda \in L_0$, we may take the unit vector $(1\ \mathbf{0})^T \in H$ along the $t$ axis into any other given vector $(t\ \mathbf{x})^T$ of $H$, since any timelike vector can be along the $t$ axis for some inertial observer. Thus $L_0$ acts transitively on $H$. The stability subgroup of $(1\ \mathbf{0})^T$ is immediately determined to be

$$\begin{bmatrix} 1 & 0 \\ 0 & SO(3) \end{bmatrix}$$

which we call $1 \times SO(3)$, or, more simply, $SO(3)$; this is simply the subgroup of all spatial rotations of $\mathbb{R}^3$ in Minkowski space. Thus $H$ is diffeomorphic to the coset space $L_0/SO(3)$. In other words (see Theorem (17.11)) *$L_0$ is a principal fiber bundle over the base space H, with fiber $SO(3)$.*

Note that the upper hyperboloidal sheet $H$ is diffeomorphic to $\mathbb{R}^3$ (under the projection $(t, \mathbf{x}) \to (0, \mathbf{x})$) and so is contractible to a point. We now invoke the following

**Theorem (19.36):** *If $E^{n+k}$ is a bundle over a base space $M^n$, if $M$ is contractible to a point $p \in M$, then $E$ has the fiber over $p$ as a deformation retract.*

In particular, $SO(3)$ is a deformation retract of $L_0$. We shall not prove (19.36) here; a detailed proof can be found in Steenrod's book [St]. The following picture, in the case at hand, makes it seem plausible.

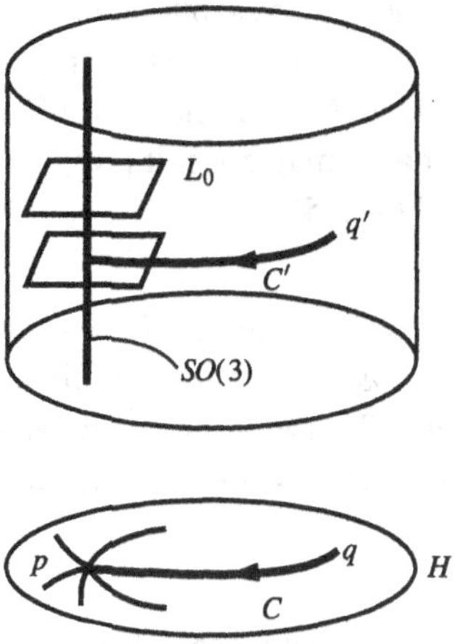


**Figure 19.5**

Take a Riemannian metric for $E$. Each fiber is a submanifold of $E$. Consider the "horizontal" distribution $\Delta$ of $(n-k)$-planes in $E$ that are orthogonal to the fibers. $M$ can be contracted to a point $p$. Let $q'$ be a point in $E$ and let $C$ be the curve swept out in $M$ as $q = \pi(q')$ is deformed to $p$. There is apparently then a unique curve $C'$ covering $C$, starting at $q'$, *tangent to* $\Delta$, and ending at some point in the fiber over $p$. (This is similar to the picture of parallel displacement described in Section 9.7b.) In this way, we deform $E$ into $\pi^{-1}(p)$.

What is wrong with this sketch? We simply note that in the general case, if the distribution $\Delta$ is not chosen with some care, that is, if the metric in $E$ misbehaves, then the curve $C'$ covering $C$ *may never reach the fiber* $\pi^{-1}(p)$.

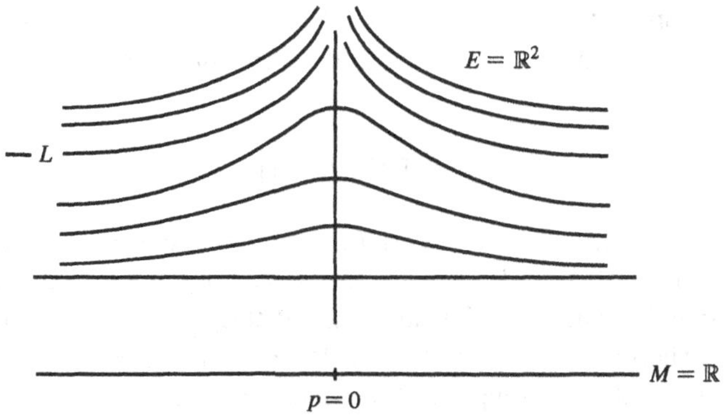


**Figure 19.6**

For example, in the usual projection $\pi : \mathbb{R}^2 \to \mathbb{R}$ given by $(x, y) \to x$, $E = \mathbb{R}^2$ is a bundle over $M = \mathbb{R}$. We have chosen a strange metric in $\mathbb{R}^2$ and have indicated the integral curves of the "horizontal" distribution, that is, the orthogonal trajectories to the vertical fibers.The integral curve labled $L$ is asymptotic to the $y$ axis, and all the integral curves above $L$ are also. The integral curves below $L$ are bell-shaped, with the highest point of the bells tending to infinity as the integral curve is chosen closer and closer to $L$. For $C$ we may take the interval $[-1, 0]$, ending at $p = 0$. This clearly can be covered by arcs of the bell-shaped curves, but on the leaf $L$ and above one will never reach the $y$ axis! Rather than use the subspaces orthogonal to the fibers, one should introduce a *connection* in the *fiber bundle* and then use parallel translation to cover curves in the base space. The reader may consult [No, chap. 2] for details.

This concludes our sketch of (19.35).

**Corollary (19.37):** *$Sl(2, \mathbb{C})$ is both connected and simply connected, since $SU(2)$ is. $L_0$ is connected and each closed curve in $L_0$ is homotopic to a curve in $SO(3)$ representing a multiple of a full rotation in $\mathbb{R}^3$ about some axis, say the $z$ axis. The even multiples are homotopic to a constant; the odd multiples are homotopic to a full rotation.*

**Proof of (19.33):** First note that since $\det A = 1$,

$$\begin{aligned}\langle \Lambda(A)x, \Lambda(A)x\rangle &= -\det\{\Lambda(A)x\}_* = -\det Ax_*A^\dagger \\ &= -\det A \det x_* \det A^\dagger = -\det x_* = \langle x, x\rangle\end{aligned}$$

and so $\Lambda(A)$ *is* a Lorentz transformation. $\det \Lambda(A) = \pm 1$ (since every Lorentz transformation preserves $\pm$ the volume form $dx^0 \wedge dx^1 \wedge dx^2 \wedge dx^3$). To show that the determinant is $+1$ we need only know that $Sl(2, \mathbb{C})$ is connected, and this was proved in (19.37). Since $A = I$ yields a Lorentz transformation $I$ with ${B^0}_0 = 1$, connectedness of $Sl(2, \mathbb{C})$ shows us that ${B^0}_0 \geq 1$ for all $A \in SL(2, \mathbb{C})$; that is, $\Lambda$ maps $Sl(2, \mathbb{C})$ into $L_0$.

It is immediate that $\Lambda$ is a homomorphism, as you are asked to show in Problem 19.3(1).

We must show that $\Lambda$ maps $Sl(2, \mathbb{C})$ *onto* $L_0$. First look at the differential of $\Lambda$ at the identity of $Sl(2, \mathbb{C})$. $S \in \mathfrak{sl}(2, \mathbb{C})$ means that $S$ is a complex $2 \times 2$ matrix with trace 0. Then $\Lambda_* S$ is the linear transformation of Minkowski space corresponding to

$$x_* \to \frac{d}{dt}[e^{tS} x_* (e^{tS})^\dagger]_{t=0}$$

But this is simply $Sx_* + x_*S^\dagger = Sx_* + (Sx_*)^\dagger$, that is, twice the hermitian part of $Sx_*$. We claim that $\Lambda_*$ is $1 : 1$ at $I$. Otherwise, for some $S \neq 0$, $Sx_*$ is *skew* hermitian for all hermitian $2 \times 2$ matrices $x_*$. Putting $x_* = I$ shows then that $S$ would be skew hermitian, (i.e., that $S \in \mathfrak{su}(2)$). Thus if $\Lambda_* S = 0$ then $S \in \mathfrak{su}(2)$. But $\Lambda$ restricted to the subgroup $SU(2)$ is a local diffeomorphism into $SO(3)$, as we have seen in Theorem (19.12). Thus $\Lambda_*$ is $1 : 1$ at $I$. It is not difficult to see that the group property, that is, the fact that $\Lambda$ is a homomorphism, would show that $\Lambda_*$ is 1:1 at all points of $Sl(2, \mathbb{C})$. Thus $\Lambda$ is a local diffeomorphism near each point of $Sl(2, \mathbb{C})$. We conclude that the image $U := \Lambda[Sl(2, \mathbb{C})]$ is an *open* subgroup of $L_0$ of the same dimension 6, since the image of an open set under a homeomophism is again open. But $L_0$ is then the *disjoint* union of the open cosets of $U$. It is plausible, and can be proved (see [S]), that a space in which any two points can be connected by an arc, say $L_0$, cannot be written as a disjoint union of two or more *open* subsets. It must be that there is only one coset, that is, $\Lambda[Sl(2, \mathbb{C})] = L_0$, and thus $\Lambda$ is onto.

We need only show then that $\Lambda$ is $2:1$. Ker $\Lambda$ consists of those $A \in Sl(2, \mathbb{C})$ such that $Ax_*A^\dagger = x_*$ for all hermitian $x_*$. Putting $x_* = I$ shows $A^\dagger = A^{-1}$, that is, $A \in SU(2)$. But we have already seen in Theorem (19.12) that $A = \pm I$. Thus

$$L_0 = \frac{Sl(2, \mathbb{C})}{\{\pm I\}}$$

and we are finished. □

As $SU(2) \to SO(3)$ yielded a double-valued spinor representation of the rotation group, so $\Lambda$ yields a double-valued **spinor** representation of the Lorentz group $L_0$. *It is simply the usual representation of Sl(2,* $\mathbb{C}$*) as* $2 \times 2$ *matrices*. This spinor representation of the Lorentz group $L_0$ will be denoted by

$$\mathfrak{D}\left(\frac{1}{2}, 0\right)$$

## 19.3b. The Dirac Algebra

We have seen in Section 19.2 that the Pauli matrices (without $\sigma_0$) generate a Clifford algebra $C_3$

$$\sigma_\alpha \sigma_\beta + \sigma_\beta \sigma_\alpha = 2\delta_{\alpha\beta} I$$

and that Dirac's program requires a $C_4$. There is a rather standard procedure leading from a $C_n$ to a $C_{n+1}$. We shall only be concerned with going from the Pauli algebra to $C_4$. There is a complication due to the Pauli algebra using the metric $\delta_{\alpha\beta}$ in $\mathbb{R}^3$ while relativity requires that we use the Lorentz metric $\eta_{jk}$ in $M_0^4$. We proceed, with Bleecker, as follows.

In the case of the Pauli algebra, the map $_*: \mathbb{R}^3 \to 2 \times 2$ matrices can be thought of as a map $\sigma : \mathbb{R}^3 \to \mathfrak{gl}(2, \mathbb{C})$

$$\sigma(\mathbf{x}) = \mathbf{x}_* = \boldsymbol{\sigma} \cdot \mathbf{x}$$

For example, $\sigma(1, 0, 0)^T = \sigma_1$, and so on.

We now define a map $\gamma : \mathbb{R}^4 \to \mathfrak{gl}(4, \mathbb{C})$ (i.e., all $4 \times 4$ complex matrices), by

$$\gamma(x) = \begin{bmatrix} 0 & x_* \\ x^* & 0 \end{bmatrix} \tag{19.38}$$

(The meaning of this will be discussed in the next section.)

In particular

$$\gamma_1 := \gamma(e_1) = \begin{bmatrix} 0 & \sigma_1 \\ \sigma_1 & 0 \end{bmatrix} = \begin{bmatrix} 00 & 01 \\ 00 & 10 \\ 01 & 00 \\ 10 & 00 \end{bmatrix} \tag{19.39}$$

$$\gamma_2 := \gamma(e_2) = \begin{bmatrix} 0 & \sigma_2 \\ \sigma_2 & 0 \end{bmatrix}$$

$$\gamma_3 := \gamma(e_3) = \begin{bmatrix} 0 & \sigma_3 \\ \sigma_3 & 0 \end{bmatrix}$$

$$\gamma_0 := \gamma(e_0) = \begin{bmatrix} 0 & I \\ -I & 0 \end{bmatrix}$$

the famous Dirac matrices. (This is one particular representation of the Dirac matrices. There are others in use.)

The matrices $\gamma$ generate a Clifford algebra. In fact we have

**Theorem (19.40):** *For all $x \in M_0^4$, $y \in M_0^4$, we have*

$$\gamma(x)\gamma(y) + \gamma(y)\gamma(x) = 2\langle x, y\rangle I$$

*where $\langle , \rangle$ is the Lorentz metric.*

PROOF: Both sides of (19.40) are bilinear symmetric functions of $x$ and $y$. For any such function $f$ we have

$$4f(x, y) = f(x + y, x + y) - f(x - y, x - y)$$

and it is thus sufficient to verify (19.40) when the arguments $x$ and $y$ are the same. But

$$\gamma(x)\gamma(x) = \begin{bmatrix} 0 & x_* \\ x^* & 0 \end{bmatrix} \begin{bmatrix} 0 & x_* \\ x^* & 0 \end{bmatrix} = \begin{bmatrix} x_*x^* & 0 \\ 0 & x^*x_* \end{bmatrix}$$

$$= \begin{bmatrix} \langle x, x\rangle I & 0 \\ 0 & \langle x, x\rangle I \end{bmatrix} = \langle x, x\rangle \begin{bmatrix} I & 0 \\ 0 & I \end{bmatrix}$$

as desired. □

## Problem

**19.3(1)** Show that $\Lambda : Sl(2, \mathbb{C}) \to L_0$ is a homomorphism.

## 19.4. The Dirac Operator $\partial\!\!\!/$ in Minkowski Space

What is a Dirac spinor?

**Warning:** Our choice of metric signature has always been $(- + ++)$ as this is most convenient for discussing the geometry of general relativity. Approximately half of the physics books use this convention also in general relativity. Most physics books, however, when discussing (special) relativistic *quantum mechanics*, use the metric with signature $(+ - --)$. In particular, their d'Alembertian is the negative of ours. This introduces the imaginary unit $i$ into many equations. For example they would write the Dirac equation (19.48) below as $i\gamma^j\partial_j\psi = m\psi$. There are so many different conventions in use for the Dirac matrices that we feel that this will not cause much more confusion than is already present in the literature. We are mainly concerned with the concepts involved in this subtle subject and feel that a change of signature at this time would only put an added burden on the reader.

### 19.4a. Dirac Spinors

In the last section we exhibited the Dirac matrices $\gamma$ generating a Clifford algebra $C_4$, the Dirac algebra. The space $\mathbb{C}^4$ on which these $\gamma$'s operate will be the space of values of our wave functions, that is, *a wave function $\psi$ will be a column of 4 complex functions.* The Dirac algebra will allow us to construct a square root of the d'Alembertian, $\partial\!\!\!/ = \sum \gamma_j\partial_j$. There is a serious problem remaining; we have constructed $\gamma_j$ by using a specific frame in Minkowski space. We shall *choose* $\gamma_j$ to be the *same matrix in each frame* $\boldsymbol{\partial}$ *because there is no preferred frame in* $M_0^4$. Since $\gamma_j$ is the same matrix in each frame $\boldsymbol{\partial}$ and since $\partial_j$ is frame-*dependent*, it is clear that $\partial\!\!\!/ = \sum \gamma_j\partial_j$ would represent a different *operator* in each frame! In order to avoid this the *"functions" $\psi$ on which $\partial\!\!\!/$ operate must themselves be made to be frame-dependent*! Let us see how the $\psi$'s are to transform.

We have defined the matrix $\gamma(X)$ for each *4-tuple* $X$ by

$$\gamma(X) = \begin{bmatrix} 0 & X_* \\ X^* & 0 \end{bmatrix}$$

and by definition of $\gamma_j$

$$\gamma(X) = X^j\gamma_j$$

Consider a Lorentz transformation $\Lambda$ of $M_0^4$

$$X' = \left(\frac{\partial x'}{\partial x}\right)X = \Lambda X$$

The Lorentz transformation $\Lambda$ will correspond, under

$$\Lambda : Sl(2,\mathbb{C}) \to L_0$$

to two matrices $\pm A \in Sl(2,\mathbb{C})$; *pick one of them.* By (19.34)

$$[\Lambda(A)(X)]_* = AX_*A^\dagger$$

**Lemma (19.41):** *The* $2 \times 2$ *matrix associated to* $\Lambda(A)(X)$ *under* $^*$ *is*

$$\Lambda(A)(X)^* = A^{\dagger-1}X^*A^{-1}$$

PROOF: Recall from (19.32) that $X_*X^* = X^*X_* = \langle X, X\rangle I$, and $\det X_* = -\langle X, X\rangle$. Thus if $X$ is not lightlike

$$X^* = \langle X, X\rangle X_*^{-1}$$

But if we prove (19.41) when $X$ is not lightlike, it will follow for all $X$ by continuity and the fact that any vector in the light cone is the limit of spacelike vectors. Assume then that $\langle X, X\rangle \neq 0$. Then

$$\begin{aligned}\Lambda(A)(X)^* &= \langle\Lambda(A)X, \Lambda(A)X\rangle[\Lambda(A)(X)_*]^{-1}\\ &= \langle X, X\rangle[AX_*A^\dagger]^{-1} = \langle X, X\rangle A^{\dagger-1}X_*^{-1}A^{-1}\\ &= A^{\dagger-1}X^*A^{-1} \quad \square\end{aligned}$$

**Theorem (19.42):** *Let* $\rho : Sl(2, \mathbb{C}) \to Gl(4, \mathbb{C})$ *be the representation of* $Sl(2, \mathbb{C})$ *by* $4 \times 4$ *complex matrices defined by*

$$\rho(A) = \begin{bmatrix} A & 0 \\ 0 & A^{\dagger-1} \end{bmatrix}$$

*Then the Dirac matrices satisfy*

$$\gamma(\Lambda(A)X) = \rho(A)\gamma(X)\rho(A)^{-1} \tag{19.43}$$

(**Note:** $X$ and $\Lambda(A)X$ are the components of the *same* vector $\mathbf{X}$ in the two Lorentz coordinate systems $\mathbf{e}$ and $\mathbf{e}' = \mathbf{e}\Lambda^{-1}$.)

PROOF:

$$\begin{aligned}\gamma(\Lambda(A)X) &= \begin{bmatrix} 0 & \Lambda(A)X_* \\ \Lambda(A)X^* & 0 \end{bmatrix}\\ &= \begin{bmatrix} 0 & Ax_*A^\dagger \\ A^{\dagger-1}X^*A^{-1} & 0 \end{bmatrix} = \begin{bmatrix} A & 0 \\ 0 & A^{\dagger-1} \end{bmatrix}\begin{bmatrix} 0 & X_* \\ X^* & 0 \end{bmatrix}\begin{bmatrix} A^{-1} & 0 \\ 0 & A^\dagger \end{bmatrix}\\ &= \rho(A)\gamma(X)\rho(A^{-1}) = \rho(A)\gamma(X)\rho(A)^{-1} \quad \square\end{aligned}$$

How do we interpret this result? If $\mathbf{X}$ is a tangent vector to $M_0^4$ we may define the matrix $\gamma(\mathbf{X}) = \gamma(X)$ by expressing $\mathbf{X}$ as a 4-tuple. *This depends on the Lorentzian frame* $\mathbf{e}$ *in which* $\mathbf{X} = \mathbf{e}X$ *is expressed.* If, however, for each Lorentz transformation $\Lambda$ of $M_0^4$ we also make a change of frame in $V = \mathbb{C}^4$ given by the change of basis matrix $\rho(A)$, then we see from (19.43) that $\gamma(X) = \gamma(\mathbf{X})$ *is then a well-defined linear transformation* $\gamma(X) : V \to V$ *that is independent of the Lorentz frame.* This follows since the matrix B of a linear transformation changes under a change of frame $\rho$ precisely by $B \mapsto \rho B\rho^{-1}$.

Equation (19.43) is written in physics books as follows. Let $\Lambda^i{}_j$ be the entries of the matrix $\Lambda(A)$. Then by our definitions

$$\gamma(\Lambda(A)X) = \Lambda^i{}_j X^j \gamma_i$$

and

$$\rho(A)\gamma(X)\rho(A)^{-1} = \rho(A)X^j\gamma_j\rho(A)^{-1}$$

yield, from (19.43)

$$\Lambda^i{}_j\gamma_i = \rho(A)\gamma_j\rho(A)^{-1} \tag{19.44}$$

As mentioned before, the usual representation of $Sl(2, \mathbb{C})$ by $2 \times 2$ matrices $A$ is called the spinor representation when thought of as a two-valued representation of $L_0$ and it is denoted then by $\mathfrak{D}(1/2, 0)$. The representation using $A^{\dagger-1}$ instead of $A$ is called the **cospinor** representation and is denoted by $\mathfrak{D}(0, 1/2)$. Two component spinors $\psi_L$, transforming under $A$, are also called **left-handed**, whereas two component cospinors $\psi_R$ transforming under $A^{\dagger-1}$, are called **right-handed**.

In order for $\gamma(\mathbf{X})$ to be a well-defined linear transformation

$$\psi \in V \mapsto \gamma(\mathbf{X})\psi \in V$$

$\psi = (\psi_L, \psi_R)^T$ must be a **4-component spinor** or **Dirac spinor**; that is, it must transform via the representation $\rho$ in (19.42)

$$\psi \mapsto \rho(A)\psi$$

for each Lorentz transformation $\Lambda$ of $M_0^4$. In summary

**Corollary (19.45):** *A Lorentz transformation $\Lambda : M_0^4 \to M_0^4$ must always be accompanied by a change of basis $\rho(A) : \mathbb{C}^4 \to \mathbb{C}^4$ (as given in (19.42)) in spinor space. Only then will $\gamma$ $(X)$ act on Dirac spinors.*

The representation $\rho$ of $Sl(2, \mathbb{C})$ is written $\mathfrak{D}(1/2, 1/2)$ and is the direct sum of $\mathfrak{D}(1/2, 0)$ and $\mathfrak{D}(0, 1/2)$.

## 19.4b. The Dirac Operator

Consider $M_0^4$ with a given Lorentzian coordinate system $x$. A "wave function" $\psi$ will be a Dirac spinor, that is, a function on $M_0^4$ taking its values in $\mathbb{C}^4$ and transforming as in (19.45). In terms of a (two-component) left-handed spinor $\psi_L$ and a right-handed spinor $\psi_R$

$$\psi = (\psi^1, \psi^2, \psi^3, \psi^4)^T = (\psi_L^1, \psi_L^2, \psi_R^1, \psi_R^2)^T$$
$$= (\psi_L, \psi_R)^T$$

As usual we shall write

$$\gamma^j := g^{jk}\gamma_k$$

One verifies easily that these new $\gamma$'s also satisfy the Clifford relations

$$\gamma^j\gamma^k + \gamma^k\gamma^j = 2g^{jk}I = 2\eta^{jk}I$$

We define the **Dirac operator** $\partial\!\!\!/$ sending wave functions into wave functions by

$$\partial\!\!\!/\psi := \gamma^j \frac{\partial \psi}{\partial x^j} \tag{19.46}$$
$$\partial\!\!\!/ = \gamma^j \partial_j$$

where, as in (19.38),

$$\gamma_k := \gamma(e_k) = \begin{bmatrix} 0 & \sigma_k \\ \pm\sigma_k & 0 \end{bmatrix}$$

This defines $\partial\!\!\!/$ in terms of the Lorentzian coordinates $x$. What happens if we consider the same definition using a system $x' = \Lambda x = (\partial x'/\partial x)x$? Then

$$\psi' = \rho(A)\psi$$

where $A$ is a *constant* matrix. We then have

$$\begin{aligned}
\partial\!\!\!/'\psi' &= \gamma_j g'^{jk}\frac{\partial \psi'}{\partial x'^k} = \gamma_j g'^{jk}\rho(A)\frac{\partial \psi}{\partial x'^k} \\
&= \gamma_j g'^{jk}\rho(A)\left(\frac{\partial x^i}{\partial x'^k}\right)\frac{\partial \psi}{\partial x^i} \\
&= \gamma_j \frac{\partial x'^j}{\partial x^r}\rho(A) g^{ri}\frac{\partial \psi}{\partial x^i}
\end{aligned}$$

which, from (19.44), yields

$$\partial\!\!\!/'\psi' = \rho(A)\gamma_r g^{ri}\frac{\partial \psi}{\partial x^i}$$

Then

$$\psi' = \rho(A)\psi \Rightarrow \quad \partial\!\!\!/'\psi' = \rho(A)\,\partial\!\!\!/\psi \tag{19.47}$$

shows that

*the Dirac operator $\partial\!\!\!/$ is a well-defined first-order differential operator on 4-component spinors of type $\mathfrak{D}(1/2, 1/2)$ in Minkowski space !*

From (19.39) and $(g^{jk}) = (\eta^{jk}) = \text{diag}[-1, +1, +1, +1]$

$$\gamma^0 = \begin{bmatrix} 0 & -I \\ +I & 0 \end{bmatrix} \qquad \gamma^\alpha = \begin{bmatrix} 0 & \sigma_\alpha \\ \sigma_\alpha & 0 \end{bmatrix}$$

Finally

$$\begin{aligned}
\partial\!\!\!/ &= \begin{bmatrix} 0 & -I\partial_0 + \sigma_1\partial_1 + \sigma_2\partial_2 + \sigma_3\partial_3 \\ I\partial_0 + \sigma_1\partial_1 + \sigma_2\partial_2 + \sigma_3\partial_3 & 0 \end{bmatrix} \\
&= \begin{bmatrix} 0 & -I\partial_0 + \boldsymbol{\sigma}\cdot\boldsymbol{\partial} \\ I\partial_0 + \boldsymbol{\sigma}\cdot\boldsymbol{\partial} & 0 \end{bmatrix}
\end{aligned}$$

Thus the Dirac equations (19.25) become the *coupled* system

$$\partial\!\!\!/\psi = m\psi \tag{19.48}$$

or

$$(-\partial_t + \boldsymbol{\sigma} \cdot \boldsymbol{\partial})\psi_R = m\psi_L$$
$$(\partial_t + \boldsymbol{\sigma} \cdot \boldsymbol{\partial})\psi_L = m\psi_R$$

Note that for a **massless** particle these equations decouple and we can get by with a single equation for a 2-component spinor $\psi_L$ of type $\mathfrak{D}(1/2, 0)$, $(\partial_t + \boldsymbol{\sigma} \cdot \boldsymbol{\partial})\psi_L = 0$. This is **Weyl's equation**, which was found later to be an equation applicable to the *neutrino*.

## 19.5. The Dirac Operator in Curved Space–Time

Does it make sense to say that a body, on returning from a long trip through the wormholes of space, has made an "odd number of full rotations"?

### 19.5a. The Spinor Bundle

Consider now a pseudo-Riemannian 4-manifold $M^4$ rather than Minkowski space. We suppose that there are patches $\{U, V, \ldots\}$ on $M^4$ and orthonormal frame ("vierbein") fields $e_U, e_V, \ldots$ on each. Thus

$$\langle \mathbf{e}_j^U, \mathbf{e}_k^U \rangle = \eta_{jk}$$

and in an overlap we shall assume

$$\mathbf{e}_V(x) = \mathbf{e}_U(x) c_{UV}(x)$$

where $c_{UV} : U \cap V \to L_0$. (Recall that this is only one of the four components of the full Lorentz group; we are assuming that $M^4$ is both space- and time-"orientable"). We shall need to construct some analogue of the space of 4 component spinors. In our discussion in $M_0^4$ of the Dirac spinors, we associated with a Lorentz transformation $\Lambda$ the matrix $A$, one of the two $2 \times 2$ matrices of $Sl(2, \mathbb{C})$ covering $\Lambda$. There was no problem in doing this since we were dealing with a single constant matrix $\Lambda$. Now, however, we shall have to choose for each $\Lambda(x) = c_{UV}(x)$ a matrix $A(x) = c'_{UV}(x)$ in $Sl(2, \mathbb{C})$ from among the two $\pm A(x)$ covering it, and we shall have to do this in a continuous fashion. The transition functions $c_{UV}(x)$ for the tangent bundle certainly satisfy the requirement (16.3), but it is not at all clear that the $c'_{UV}(x)$ can be chosen consistently to satisfy it because of the ambiguity $\pm A$.

*If* this can be done, then we say that we have "lifted" the structure group of the tangent bundle of $M^4$ from the Lorentz group to the group $Sl(2, \mathbb{C})$ and that $M^4$ has a **spin structure**.

This would have the following consequence.

Let $M^4$ be a pseudo-Riemannian manifold that is both space- and time-orientable; we may then assume that the tangent bundle has structure group $L_0$. Let $\mathbf{e}$ and $\mathbf{f}$ be frames at a given point $p$. Then there is a unique $\Lambda \in L_0$ such that $\mathbf{f} = \mathbf{e}\Lambda$. If $\mathbf{f}(t)$ is a 1-parameter family of frames at $p$ such that $\mathbf{f}(0) = \mathbf{f}(1) = \mathbf{e}$, then $\mathbf{f}(t) = \mathbf{f}(0)\Lambda(t)$ yields a closed curve $t \mapsto \Lambda(t)$ in $L_0$ starting and ending at $I$. $Sl(2, \mathbb{C})$ is a 2-fold cover of $L_0$, and thus this curve is covered by a unique curve $t \to A(t)$ in $Sl(2, \mathbb{C})$ starting at $I$. (Visualize this by analogy to $SU(2) = S^3$, the 2-fold cover of $SO(3) = \mathbb{R}P^3$, as in section 19.2a.) We know, from Corollary (19.37), that $SL(2, \mathbb{C})$, like $SU(2)$, is simply connected, whereas $L_0$, like $SO(3)$, has the property that the closed curve $t \to \Lambda(t)$ is homotopic either to a full rotation about some axis, say the $z$ axis, or to a constant map. The covering curve $t \to A(t)$ detects the difference; $A(1) = I$ if $t \to \Lambda(t)$ describes an even number of full rotations, whereas $A(1) = -I$ if $t \to \Lambda(t)$ describes an odd number of full rotations. All this is for a 1-parameter family of frames $\mathbf{e}(t)$ *at a given point* $x$. No spin structure is required.

Suppose now that $p$ is in a patch $U$ covered by a Lorentzian frame field $\mathbf{e}_U$. (This patch need *not* be a coordinate frame.) Take the frame $\mathbf{f}(p) = \mathbf{e}_U(p)$ at $p$ and transport it arbitrarily but continuously around some closed curve $C = C(t)$, $0 \le t \le 1$, lying in $U$, again returning to the same frame $\mathbf{f}(p)$. We can compare $\mathbf{f}(C(t))$ with $\mathbf{f}(p) = \mathbf{f}(C(0))$ as follows. Identify all frames $\mathbf{e}_U$ at points of $U$ with the single frame $\mathbf{e}_U$ at $p$. Then by comparing $\mathbf{f}(C(t))$ with $\mathbf{e}_U$ at $C(t)$, $\mathbf{f}(C(t)) = \mathbf{e}_U(C(t))\Lambda(t)$, we again trace out a closed curve $t \mapsto \Lambda(t)$ in $L_0$. The resulting curve in $L_0$ can again be uniquely covered by a curve in $Sl(2, \mathbb{C})$ starting at $I$. In this way we may be tempted to say that if $A(1) = -I$ then the frame has made an odd number of rotations, whereas if $A(1) = I$ it has made an even number of rotations. *Unfortunately this result might depend on the choice of frames* $\mathbf{e}_U$ *in* $U$! To see this, consider a *spatial* example, rather than space-time, replacing $L_0$ by $SO(3)$ and $Sl(2, \mathbb{C})$ by $SU(2)$. Let $M^3$ be the 3-torus $T^3$, with angular coordinates $x, y, z$. Let $U = T^3$ and let $\mathbf{e}_U$ be the frame $\partial/\partial x$, $\partial/\partial y$, and $\partial/\partial z$. Then with the preceding identification, the frame $\mathbf{f} = \mathbf{e}_U$ along the closed $z$-curve $(0, 0, z)$ would make no rotation at all. We may consider a new frame field $\mathbf{e}_V$ on $V = T^3$ defined by $\mathbf{e}_V = \mathbf{e}_U c_{UV}$, where

$$c_{UV}(z) = \begin{bmatrix} \cos z & -\sin z & 0 \\ \sin z & \cos z & 0 \\ 0 & 0 & 1 \end{bmatrix}$$

This frame coincides with $\mathbf{e}_U$ on $z = 0$ but rotates once about it as one moves along the $z$ circuit. Clearly the frame $\mathbf{f} = \mathbf{e}_U$ along the $z$ circuit now makes one complete rotation with respect to the $\mathbf{e}_V$ frame, that is, by identifying frames in $T$ by means of the $\mathbf{e}_V$ frames. We see that the contradiction arises because the $\mathbf{e}_U$ and $\mathbf{e}_V$ frames *are* related by $SO(3)$ transformations $c_{UV}$; they are *not* related by $SU(2)$ transformations. We cannot decide whether $\mathbf{e}_U$ and $\mathbf{e}_V$ at $(0, 0, 0) = (0, 0, 2\pi)$ are related by the identity $I$ in $SU(2)$ or by $-I$ in $SU(2)$! The same problem would arise in space–time. We also see that this problem in the patch would not arise if we restricted ourselves to frames in $T$ that *can* be related by SL(2,$\mathbb{C}$) transformations, that is, by frames that "do not make full rotations about each other."

If $M^4$ has a *spin structure*, that is, if $Sl(2, \mathbb{C})$ is the structure group, and if we transport a frame $\mathbf{f}$ around *any* closed path $C$ in $M^4$, returning to the same Lorentzian frame, then we *can* decide whether the frame has made an even or an odd number of complete rotations! For we may consider the $Sl(2, \mathbb{C})$ *frame bundle* to $M$, that is, the frame bundle but using the structure group $Sl(2, \mathbb{C})$. The curve $C$ in $M$ is then covered by a unique curve in this frame bundle, defined by **f**. Upon returning to the starting point of $C$, the lifted curve will return either to its starting point, corresponding to an even number of rotations, or to a point in the frame bundle related to the initial point by $-I \in Sl(2, \mathbb{C})$, corresponding to an odd number of rotations.

In our spatial toral illustration $T^3$ just considered, $T^3$ is covered by a single frame field $\mathbf{e}_U$, and this frame field does define a spin structure. $T^3$ can also be covered by the single frame field $\mathbf{e}_V$ and so this also defines a spin structure on $T^3$, *but it is a different spin structure*! On the other hand, $T^3$ *does not admit any spin structure that includes both frame fields* $\mathbf{e}_U$ *and* $\mathbf{e}_V$, as we have seen; we cannot lift $c_{UV}(z)$ uniquely to $SU(2)$ for all $0 \le z \le 2\pi$.

This has the following remarkable physical manifestation: We assume that our space–time $M^4$ carries a spin structure (for if $M$ does *not* admit a spin structure we will not be able to consider the Dirac equation). For example, we may assume that space–time is simply Minkowski space $M_0$. As we have seen in Corollary (19.45), the electron wave "functions," 4-component Dirac spinors $\psi$ defined over $M_0^4$, will be, in fact, cross sections of a bundle over $M$ associated to the tangent bundle through the representation $\rho$ of Theorem (19.42). Thus the structure group of the wave function bundle is $Sl(2, \mathbb{C})$, rather than $L_0$. These spinors will then have the property that a complete rotation of $\mathbb{R}^3$ will send a spinor $\psi$ not into itself but rather to its negative $-\psi$. Aharonov and Susskind [A, S] have devised a hypothetical experiment illustrating this. Two cubical devices can theoretically be constructed so that when they are brought together and aligned at a common face, a current will flow from one to the other, and if the cubes are then separated slightly and one of the cubes is rotated through $2\pi$ about their common axis and then brought back in contact as before, *current will again flow but in the reverse direction*! Even in the case of a general space–time $M^4$ *with spin structure*, the cubes can be separated, one of the cubes can be transported along any closed curve, and upon return the direction of the current flow will tell us of the number (modulo 2) of "rotations" made by the traveling cube!

The "obstruction" to having a spin structure can be measured by the cohomology groups of $M$, but we shall only remark that a spin structure exists if for example, $H_2(M; \mathbb{Z}_2)$, the second homology group with $\mathbb{Z}_2$ coefficients (see Section 13.2), vanishes. Obstruction theory will be discussed more in Chapter 22.

If $M$ *does* have a spin structure, then we may replace the Lorentz structure group by $Sl(2, \mathbb{C})$; the fiber for the tangent bundle of $M^4$ is still $\mathbb{R}^4$. (Recall that $Sl(2, \mathbb{C})$ acts on $\mathbb{R}^4$ as follows. To a 4-tuple $x$ one associates a $2 \times 2$ hermitian matrix $x_* = x^0\sigma_0 + \mathbf{x} \cdot \boldsymbol{\sigma}$. Then $A \in Sl(2, \mathbb{C})$ acts on $x$ by sending $x_*$ to $Ax_*A^{-1}$ and then reading off the 4-tuple that corresponds to this hermitian matrix.) If $c_{UV}$ are the Lorentzian transition functions for the tangent bundle, we shall let $c'_{UV}$ be the $Sl(2, \mathbb{C})$ transition functions. We then construct the *new* 4-component

**Dirac spinor bundle** $\mathbb{S} = \mathbb{S}M$

whose fiber is $\mathbb{C}^4$ and whose transition functions

$$\rho_{UV} : U \cap V \to Gl(4, \mathbb{C})$$

are given by

$$\rho_{UV}(x) = \rho(c'_{UV}(x)) = \begin{bmatrix} c'_{UV}(x) & 0 \\ 0 & c'_{UV}(x)^{\dagger -1} \end{bmatrix} \tag{19.49}$$

(See the discussion following (16.3) for the construction of this bundle.) This Dirac spinor bundle $\mathbb{S}$ is simply the *vector bundle associated to the* $Sl(2, \mathbb{C})$ *tangent bundle via the representation* $\rho$ of Theorem (19.42)!

This spinor bundle is the bundle whose sections $\psi$ will serve as *wave "functions"* on $M^4$.

*From this point on we shall assume that M does admit a spin structure and that one has been chosen.*

The Dirac operator construction $\partial\!\!\!/ = \gamma^j \partial_i$ in $M_0^4$ will not work in our curved $M^4$; in our proof that $\partial\!\!\!/'\psi' = \rho(A)\ \partial\!\!\!/\psi$ for $M_0$ we used the fact that the matrices $A \in Sl(2, \mathbb{C})$ were constant (global Lorentz transformations were used since $M_0$ is covered by global coordinate systems). We shall now have to replace $\partial_j \psi = \partial\psi/\partial x^j$ by some sort of *covariant* derivative. The Riemannian connection on $M^4$ won't work because $TM$ and $\mathbb{S}M$ are different bundles. What we need is a connection in this bundle $\mathbb{S}M$ that is associated to $TM$ through the *double-valued* representation $\rho$ of $L_0$.

### 19.5b. The Spin Connection in $\mathbb{S}$M

Let $M^4$ be a pseudo-Riemannian manifold with a Lorentzian connection. Thus for any tangent vector $\mathbf{X}$ to $M^4$, $\omega_U(\mathbf{X}) \in \mathfrak{l}_4$, the Lie algebra of the Lorentz group. We are assuming that $M^4$ has a spin structure (we may then consider $Sl(2, \mathbb{C})$ as the structure group of the tangent bundle) and we want a connection for the associated spin bundle $\mathbb{S}M$ of wave functions given by the Dirac spinor representation

$$\rho : Sl(2, \mathbb{C}) \to Gl(4, \mathbb{C})$$

First we need to construct a connection for the tangent bundle whose structure group is $Sl(2, \mathbb{C})$ rather than $L_0$. Let $\omega$ be the connection form for the Lorentzian tangent bundle; this is simply the Levi-Civita or Christoffel connection. Since $\Lambda : Sl(2, \mathbb{C}) \to L_0$ is a $2 : 1$ cover, to $\omega_U(\mathbf{X}) \in \mathfrak{l}_4$ there is a unique $\omega'(\mathbf{X}) \in \mathfrak{sl}(2, \mathbb{C})$ such that $\Lambda_* \omega'_U(\mathbf{X}) = \omega_U(\mathbf{X})$ (there are two "vectors" "above" $\omega_U(\mathbf{X})$ but only one of them starts at $I \in Sl(2, \mathbb{C})$). It is not difficult to see that the $\mathfrak{sl}(2, \mathbb{C})$-valued local 1-forms $\omega'_U$ so defined form the connection forms for the tangent bundle to $M^4$ whose structure group is $Sl(2, \mathbb{C})$. One only needs to show that

$$\Lambda_*[\omega'_V(\mathbf{X})] = \Lambda_*[A^{-1}\omega'_U(\mathbf{X})A + A^{-1}dA(\mathbf{X})]$$

since $\Lambda_*$ is $1 : 1$. The proof is very similar to that in Theorem (18.27). $\omega'$ *will be exhibited explicitly* in (19.53).

We now have a connection for the tangent bundle $TM^4$ with structure group $Sl(2, \mathbb{C})$ and we have a representation $\rho$ of $Sl(2, \mathbb{C})$ given by $4 \times 4$ matrices. The Dirac 4-component spinor bundle is associated to the tangent bundle through the representation $\rho$. The prescription for constructing the associated connection in $\mathbb{S}M$ is given by

(18.24). We need to find an $\Omega$ such that

$$\rho_* \omega' = \Omega \tag{19.50}$$

which is short for $\rho_* \omega'(\mathbf{X}) = \Omega(\mathbf{X})$, where $\mathbf{X}$ is tangent to $M^4$ at $x$. We shall exhibit $\Omega$ by an explicit calculation.

First we need to calculate $\Lambda_* : \mathfrak{sl}(2, \mathbb{C}) \to \mathfrak{l}_4$, identifying the Lie algebra of $Sl(2, \mathbb{C})$ with that of the Lorentz group. $\mathfrak{sl}(2, \mathbb{C})$ consists of all $2 \times 2$ complex matrices $z$ with trace 0. By writing

$$z = \frac{(z + z^\dagger)}{2} + \frac{(z - z^\dagger)}{2}$$

as a sum of a hermitian plus an antihermitian matrix, both with trace 0, we see that a basis for $\mathfrak{sl}(2, \mathbb{C})$ can be taken to be the $\sigma_\alpha$'s divided by $i$ and the $\sigma_\alpha$'s, $\alpha = 1, 2, 3$. Since $i\sigma_1 = \sigma_2\sigma_3$, and so on, and $\sigma_\alpha = \sigma_0\sigma_\alpha$, where $\sigma_0 = \tau_0 = I$, we prefer to write this basis as

$$-\sigma_2\sigma_3, -\sigma_3\sigma_1, -\sigma_1\sigma_2, \sigma_0\sigma_1, \sigma_0\sigma_2, \sigma_0\sigma_3 \tag{19.51}$$

Note that *the first three give the standard basis for the $SU(2)$ subgroup of $Sl(2\mathbb{C})$.*

The identity component of the Lorentz group is generated by rotations and "boosts." The infinitesimal rotations have a basis given by the matrices $E_\alpha$ of (19.1), where $\alpha$ runs from 1 to 3, but augmented by zeros in the $0^{\text{th}}$ row and $0^{\text{th}}$ column. *For our purposes it is preferable to introduce a minus sign in the $E_\alpha$'s.* The resulting $4 \times 4$ matrix obtained from $-E_1$ will be called $E_{23}$, $-E_2$ will yield $E_{31}$, and $-E_3$ will yield $E_{12}$.

A **boost** in the 01 plane is given by the $2 \times 2$ matrix

$$\begin{bmatrix} 0 & 1 \\ 1 & 0 \end{bmatrix}$$

augmented by 0 elsewhere. We then have as basis for $\mathfrak{l}_4$ the matrices $E_{23}$, $E_{31}$, $E_{12}$, $E_{01}$, $E_{02}$, $E_{03}$, where

$$E_{23} = \begin{bmatrix} 0 & 0 & 0 & 0 \\ 0 & 0 & 0 & 0 \\ 0 & 0 & 0 & 1 \\ 0 & 0 & -1 & 0 \end{bmatrix}, \ldots, \qquad E_{03} = \begin{bmatrix} 0 & 0 & 0 & 1 \\ 0 & 0 & 0 & 0 \\ 0 & 0 & 0 & 0 \\ 1 & 0 & 0 & 0 \end{bmatrix}$$

Each $E_{\alpha\beta}$ is a skew symmetric matrix and we shall define $E_{\beta\alpha} := -E_{\alpha\beta}$. The $E_{0\beta}$ are symmetric matrices and we define $E_{\beta 0} := E_{0\beta}$.

The homomorphism $\Lambda : Sl(2\mathbb{C}) \to L_0$ is given by $[\Lambda(A)x]_* = Ax_*A^\dagger$, and so if $h \in \mathfrak{sl}(2, \mathbb{C})$,

$$[\Lambda_*(h)x]_* = \frac{d}{dt}[\exp(\text{th})x_* \exp(\text{th}^\dagger)]_{t=0}$$

We have essentially done this calculation for $h = \sigma_\alpha / i$ in (19.13). We have, under $\Lambda_*$,

$$\sigma_2\sigma_3 \to 2E_{23} \qquad \sigma_3\sigma_1 \to 2E_{31} \qquad \sigma_1\sigma_2 \to 2E_{12} \tag{19.52}$$

Let us now calculate where $\Lambda_*$ sends $\sigma_0\sigma_\alpha = \sigma_\alpha$. Since $\sigma_\alpha^\dagger = \sigma_\alpha$ we get now an *anticommutator*

$$\begin{aligned}\frac{d}{dt}[\exp(t\sigma_\alpha)x_* \exp(t\sigma_\alpha)]_{t=0} &= \{\sigma_\alpha, x_*\}\\ &= \{\sigma_\alpha, \sigma_0 x^0 + \sigma_\beta x^\beta\} = 2\sigma_\alpha x^0 + \{\sigma_\alpha, \sigma_\beta\}x^\beta\\ &= 2\sigma_\alpha x^0 + 2\delta_{\alpha\beta}\sigma_0 x^\beta = 2\sigma_\alpha x^0 + 2\sigma_0 x^\alpha\end{aligned}$$

For example, if $\alpha = 1$, $\Lambda_*\sigma_1$ is the infinitesimal Lorentz transformation that sends $(x^0, x^1, x^2, x^3, )^t$ to $(2x^1, 2x^0, 0, 0)^T$, and so $\Lambda_*\sigma_0\sigma_1 = 2E_{01}$.

$$\sigma_0\sigma_\beta \to 2E_{0\beta} \tag{19.53}$$

(19.52) and (19.53) describe $\Lambda_*$ completely.

Let $\omega = (\omega^i{}_j)$ be the Levi-Civita connection for the pseudo-Riemannian $M^4$, using an orthonormal frame **e**. Using $\omega^1{}_0 = \omega_{10} = -\omega_{01} = \omega^0{}_1$, and so on, we have

$$\omega = \begin{bmatrix} 0 & \omega^0{}_1 & \omega^0{}_2 & \omega^0{}_3 \\ \omega^0{}_1 & 0 & \omega^1{}_2 & \omega^1{}_3 \\ \omega^0{}_2 & -\omega^1{}_2 & 0 & \omega^2{}_3 \\ \omega^0{}_3 & -\omega^1{}_3 & -\omega^2{}_3 & 0 \end{bmatrix}$$

In terms of the matrices $E$ we have

$$\omega = \sum_{i<j} E_{ij}\omega^i{}_j$$

Now use $\omega^0{}_\beta = \omega^{0\beta}$, (19.52) and (19.53) to get

$$\omega = \Lambda_*\omega' \qquad \omega' = \frac{1}{2}\sum_{i<j}\sigma_i\sigma_j\omega^{ij} \tag{19.54}$$

and *this exhibits the* $Sl(2, \mathbb{C})$ *connection form* $\omega'$, whose values are trace-free $2 \times 2$ hermitian matrices.

Now we must compute $\Omega = \rho_*\omega'$.

From

$$\rho(A) = \begin{bmatrix} A & 0 \\ 0 & A^{\dagger -1}\end{bmatrix}$$

we see

$$\rho_*(h) = \begin{bmatrix} h & 0 \\ 0 & -h^\dagger\end{bmatrix}$$

for all $h \in \mathfrak{sl}(2, \mathbb{C})$. Then

$$\begin{aligned}\Omega &= \frac{1}{2}\sum_{i<j}\omega^{ij}\begin{bmatrix}\sigma_i\sigma_j & 0 \\ 0 & -\sigma_j\sigma_i\end{bmatrix}\\ &= \frac{1}{2}\sum_\beta \omega^{0\beta}\begin{bmatrix}\sigma_\beta & 0 \\ 0 & -\sigma_\beta\end{bmatrix} + \frac{1}{2}\sum_{\alpha<\beta}\omega^{\alpha\beta}\begin{bmatrix}\sigma_\alpha\sigma_\beta & 0 \\ 0 & \sigma_\alpha\sigma_\beta\end{bmatrix}\end{aligned}$$

But

$$\gamma_0\gamma_\beta = \begin{bmatrix} \sigma_\beta & 0 \\ 0 & -\sigma_\beta \end{bmatrix} \quad \text{and} \quad \gamma_\alpha\gamma_\beta = \begin{bmatrix} \sigma_\alpha\sigma_\beta & 0 \\ 0 & \sigma_\alpha\sigma_\beta \end{bmatrix}$$

then shows that $\Omega = (1/2)\sum_\beta \gamma_0\gamma_\beta\omega^{0\beta} + (1/2)\sum_{\alpha<\beta} \gamma_\alpha\gamma_\beta\omega^{\alpha\beta}$.

Thus the **spin connection** in the spinor bundle is given by

$$\begin{aligned} \Omega &= \frac{1}{4}\omega^{jk}\gamma_j\gamma_k = \frac{1}{4}\omega_{jk}\gamma^j\gamma^k \\ &= \frac{1}{8}\omega_{jk}[\gamma^j, \gamma^k] \end{aligned} \tag{19.55}$$

recalling that $\omega_{ij} = -\omega_{ji}$. The *covariant derivative* in the spinor bundle is then

$$\frac{\nabla\psi}{dt} = \frac{d\psi}{dt} + \frac{1}{4}\omega_{jk}\left(\frac{dx}{dt}\right)\gamma^j\gamma^k\psi \tag{19.56}$$

and the curved **Dirac operator** applied to $\psi$ is

$$\gamma^i\left[e_i(\psi) + \frac{1}{4}\omega^j_{ik}\gamma_j\gamma^k\psi\right] = \partial\!\!\!/\psi + \frac{1}{4}\omega^j_{ik}\gamma^i\gamma_j\gamma^k\psi \tag{19.57}$$

In the presence of an electromagnetic field with covariant 4-vector potential $A$ and $A_j = A(e_j)$, then as in (16.43) the flat Dirac operator $\partial\!\!\!/$ would be replaced by

$$\partial\!\!\!/ - \left(\frac{ie}{\hbar}\right)\gamma^j A_j$$

## CHAPTER 20

# Yang–Mills Fields

## 20.1. Noether's Theorem for Internal Symmetries

How do symmetries yield conservation laws?

In Section 10.2 we discussed Hamilton's variational principle for a dynamical system consisting of a finite number of particles. We shall now consider variational problems associated with a continuum or "field." We are frequently concerned with a multiple integral variational problem roughly of the form

$$\delta \int_M L_0(x, \phi, \phi_x) dx^0 \wedge dx^1 \wedge \ldots \wedge dx^n = 0$$

where both the field $\phi$ and the domain of integration $M$ might be varied; that is, we consider variations $\delta\phi$ and $\delta x$. In physics, one calls a variation $\delta x$ of the domain an **external** variation, whereas field variations are called **internal**. We have considered external variations when dealing with arc length (geodesics) and with area (minimal surfaces); in both cases we dealt with the variations directly, rather than writing down the Euler–Lagrange equations. In this section we shall investigate the *tensor* nature of *internal* variations in more detail and also the effect of such variations that leave the Lagrangian invariant.

$\phi$ will usually be an $N$-tuple $\phi^a(t, \mathbf{x}) = \phi^a(x)$ of functions, that is, the local representation of a section of some vector bundle $E$. In the case of a Dirac electron, we have seen that $E$ is the bundle of complex 4-component Dirac spinors over a perhaps curved space–time. If $E$ is not a trivial bundle (or if we insist on using curvilinear coordinates) we shall have to deal with the fact that the derivatives $\partial\phi^a/\partial x^j$ do not form a tensor.

### 20.1a. The Tensorial Nature of Lagrange's Equations

Let $M^{n+1}$ be a (pseudo-) Riemannian manifold and let $E$ be a vector bundle over $M$; for definiteness we shall let the fiber be $\mathbb{R}^N$. A section of this bundle over $U \subset M$ is described by $N$ real-valued functions $\{\phi_U^a\}$, where $\phi_V = c_{VU}\phi_U$ and $c_{VU}(x)$ is an $N \times N$ transition matrix function, $c^a_{VUb}$. A "Lagrangian" is a single "function" $L_0(x, \phi, \phi_x)$ of

$x$, and section $\phi$, and (for our present purposes) its first derivative matrix $\phi_x := \partial\phi^a/\partial x^j$. (We have given the local description of the section and its first derivatives in a patch $U$. Higher derivatives may occur, as they did in Hilbert's approach to relativity in Section 11.3. In that case the bundle was the vector bundle of covariant symmetric second-rank Lorentzian tensors; that is, the sections $\phi$ were pseudo-Riemannian metric tensors $g_{ij}$ on $M^4$, and the Lagrangian $L_0$ was $R|g|^{1/2}$, involving second derivatives of the metrics.)

We are concerned with the **action** integral

$$S = \int_M L_0(x, \phi, \phi_x)dx$$

where $dx = dx^0 \wedge dx^1 \wedge dx^2 \wedge \ldots \wedge dx^n$. For this to be independent of coordinates, we shall assume that for each given $\phi$, $L_0 dx$ is a pseudo-$(n+1)$-form on $M$. In terms of the volume form $\sqrt{g}dx$ (for simplicity we omit the absolute value sign on $g$) we write $L_0 dx = \mathcal{L}_0\sqrt{g}dx$, and so

$$S = \int_M \mathcal{L}_0(x, \phi, \phi_x)\sqrt{g}dx$$

$\mathcal{L}_0$ is a true function or scalar, classically called the Lagrangian **density**. For the gravitational field, Hilbert's $\mathcal{L}_0$ is the scalar curvature $R$.

We shall vary the *section* $\phi$. *We shall assume that the metric of M and any connections used in E are not varied.* We are interested in the first variation of the action, and we shall use the same classical notation as we used in Section 10.2, but we shall emphasize here the tensorial nature of this process.

First note that $\mathcal{L}$ is to be a scalar constructed out of first partial derivatives $\partial_j\phi^a = \partial\phi^a/\partial x^j$ of the section $\phi$. The collections of *partial derivatives* $\partial_j\phi^a$ do *not* form a tensorial object (for example, in the case when $E = TM$) and consequently it is not clear how one is to construct a scalar $\mathcal{L}_0$! Frequently, however, there will be a connection in the *bundle* E and then we can construct instead the covariant derivatives

$$\nabla_j\phi^a = \phi^a_{/j} = \frac{\nabla\phi^a}{\partial x^j}$$

These *do* fit together to form a 1-form section of $E$ (as described in Section 18.3), that is, a section of the bundle $E \otimes T^*M$, a generalized tensor. (Note that $^a$ is an $E$ index, not a $TM$ index; thus it makes no sense to ask whether $^a$ is a "contravariant" index since contravariant in our sense refers to the tangent bundle only!) There is then hope for constructing a scalar out of $\phi^a_{/j}$. For example, suppose that the structure group of the bundle $E$ is $SO(N)$ and that the connection $\omega$ has its values in $\mathfrak{so}(N)$; that is, $\omega$ is skew symmetric. Then if $g_{ij}$ is the metric tensor for $M$ we may form $\sum_a \phi^a_{/j}\phi^a_{/k}g^{jk}$, and it is not difficult to see that this is indeed a scalar. This scalar could be written $\| \nabla\phi \|^2$ and might be called the square of the "gradient" of the section $\phi$.

Thus we shall assume that $E$ has a connection for the given structure group $G$, and that from $\mathcal{L}_0$ we may form a new Lagrangian $\mathcal{L}$ constructed using covariant derivatives,

$$\mathcal{L} = \mathcal{L}(x, \phi, \nabla\phi) = \mathcal{L}(x, \phi, \phi_{/x}) = \mathcal{L}_0(x, \phi, \phi_x)$$

rather than partial derivatives. (This will *not* always be the case. In Hilbert's variational approach to relativity, the fields $\phi$ are the components of the metric tensor. $\mathcal{L} = R$, the scalar curvature, is expressible in terms of partial derivatives of the metric tensor but not the *covariant* derivatives of the metric tensor, which are all identically 0!)

From $\phi^a_{/j} = \partial_j \phi^a + \omega^a_{jb}\phi^b$ and $\delta\partial_j = \partial_j\delta$ (as in Equation (10.12)) and from the fact that the connection $\omega$ is assumed unaffected by a variation $\delta\phi$ of $\phi$, we see immediately that

$$\delta(\phi^a_{/j}) = (\delta\phi^a)_{/j}$$

Then

$$\delta \int_M \mathfrak{L}(x, \phi, \phi_{/j})\sqrt{g}dx = \int_M \left[\left(\frac{\partial \mathfrak{L}}{\partial \phi^a}\right)\delta\phi^a + \left(\frac{\partial \mathfrak{L}}{\partial \phi^a_{/j}}\right)\delta(\phi^a_{/j})\right]\sqrt{g}dx \tag{20.1}$$

Now in an overlap $U \cap V$ we have

$$\frac{\partial \mathfrak{L}}{\partial \phi^a_V} = \left(\frac{\partial \mathfrak{L}}{\partial \phi^b_U}\right)\left(\frac{\partial \phi^b_U}{\partial \phi^a_V}\right)$$

But $\phi^b_U = c^b_{UVc}\phi^c_V$ shows that $\partial\phi^b_U/\partial\phi^c_V = c^b_{UVc}$, and so

$$\frac{\partial \mathfrak{L}}{\partial \phi^a_V} = \left(\frac{\partial \mathfrak{L}}{\partial \phi^b_U}\right)c^b_{UVa} \tag{20.2}$$

Hence if $\phi$ is a section of the bundle $E$, then $\{\partial\mathfrak{L}/\partial\phi^a\}$ *defines a section of the dual bundle* $E^*$. But $\delta\phi$ is a section of $E$ (being basically a difference of sections) and so the contraction

$$\left(\frac{\partial \mathfrak{L}}{\partial \phi^a}\right)\delta\phi^a$$

occurring in the first integrand of (20.1) is a scalar. Since $\delta \int_M \mathfrak{L}(x, \phi, \phi_{/j})\sqrt{g}dx$ is a scalar by hypothesis, it must be that the contraction

$$\left(\frac{\partial \mathfrak{L}}{\partial \phi^a_{/j}}\right)\delta(\phi^a_{/j})$$

must also be a scalar. Since $\delta(\phi^a_{/j})$ is a section of $E \otimes T^*M$, it must be that

$$\left(\frac{\partial \mathfrak{L}}{\partial \phi^a_{/j}}\right) \text{ defines a section of } E^* \otimes TM \tag{20.3}$$

Our usual rules of tensor analysis apply in this situation. For example, we have the connection $\omega$ for $E$. Then $-\omega^T$ defines the connection for $E^*$ (see Example 1′ following Theorem (18.27)). We have the standard Riemannian connections $\Gamma$ and $-\Gamma^T$ for $TM$ and $T^*M$. Thus, as discussed in Problem 16.3(1), we have a connection in any tensor product of the bundles $TM, T^*M, E, E^*, \ldots$. For example, $(\partial\mathfrak{L}/\partial\phi^a_{/j})\delta\phi^b$ defines a section of $E^* \otimes TM \otimes E$; for simplicity let us call it $A^{bj}_a$. It is of the form $A^{bj}_a = B^j_a C^b$. Thus $a$ is an $E^*$ index, $b$ in an $E$ index, and $j$ is a $TM$ index. Its covariant derivative, again written $A^{bj}_{a/k}$, is a section of $(E^* \otimes TM \otimes E) \otimes T^*M$ and would be given by

$$A^{bj}_{a/k} = \partial_k A^{bj}_a - A^{bj}_c \omega^c_{ka} + \omega^b_{kc} A^{cj}_a + \Gamma^j_{ki} A^{bi}_a$$

We may invoke the Leibniz rule $(B^j_a C^b)_{/k} = B^j_{a/k} C^b + B^j_a C^b_{/k}$, where the covariant derivative of $B$ involves both $\omega$ and $\Gamma$, whereas that of $C$ involves only $\omega$. Covariant differentiation commutes with contractions, and so on.

We now proceed with our calculation of the first variation. From (20.1) and $\delta(\phi^a_{/j}) = (\delta\phi^a)_{/j}$ we have

$$\delta \int_M \mathcal{L}(x, \phi, \phi_{/j})\sqrt{g}dx = \int_M \left[\left(\frac{\partial \mathcal{L}}{\partial \phi^a}\right)\delta\phi^a + \left(\frac{\partial \mathcal{L}}{\partial \phi^a_{/j}}\right)(\delta\phi^a)_{/j}\right]\sqrt{g}dx \tag{20.4}$$

$$= \int_M \left[\left(\frac{\partial \mathcal{L}}{\partial \phi^a}\right) - \left(\frac{\partial \mathcal{L}}{\partial \phi^a_{/j}}\right)_{/j}\right]\delta\phi^a\sqrt{g}dx$$

$$+ \int_M \left[\left(\frac{\partial \mathcal{L}}{\partial \phi^a_{/j}}\right)\delta\phi^a\right]_{/j}\sqrt{g}dx$$

Now $(\partial\mathcal{L}/\partial\phi^a_{/j})\delta\phi^b$ is a section of $E^*\otimes TM\otimes E$, and contraction yields that $\partial\mathcal{L}/\partial\phi^a_{/j})\delta\phi^a$ is a section of $TM$, that is, is an *ordinary contravariant vector field* $X^j$ on $M$ and we may then write

$$\left[\left(\frac{\partial \mathcal{L}}{\partial \phi^a_{/j}}\right)\delta\phi^a\right]_{/j} = \operatorname{div}\left[\left(\frac{\partial \mathcal{L}}{\partial \phi^a_{/j}}\right)\delta\phi^a\right] \tag{20.5}$$

If $M$ is compact with boundary, we have

$$\delta \int_M \mathcal{L}(x, \phi, \phi_{/j})\sqrt{g}dx = \int_M \left[\left(\frac{\partial \mathcal{L}}{\partial \phi^a}\right) - \left(\frac{\partial \mathcal{L}}{\partial \phi^a_{/j}}\right)_{/j}\right](\delta\phi^a)\sqrt{g}dx \tag{20.6}$$

$$+ \int_{\partial M} \left[\left(\frac{\partial \mathcal{L}}{\partial \phi^a_{/j}}\right)\delta\phi^a\right]N_j dS$$

where $\mathbf{N}$ is the unit normal to the boundary and $dS = i_{\mathbf{N}}\sqrt{g}dx$ is the $n$-dimensional area form. Thus if the first variation vanishes for all variations vanishing on $\partial M$, we have the **(Euler–) Lagrange** equations

$$\frac{\delta\mathcal{L}}{\delta\phi^a} := \left(\frac{\partial \mathcal{L}}{\partial \phi^a}\right) - \left(\frac{\partial \mathcal{L}}{\partial \phi^a_{/j}}\right)_{/j} = 0 \tag{20.7}$$

where the left-hand side, called the **functional** or **variational derivative**, *defines a section of* $E^*$. It is convenient to define

$$\operatorname{div}\left(\frac{\partial \mathcal{L}}{\partial \nabla\phi}\right) := \left(\frac{\partial \mathcal{L}}{\partial \phi^a_{/j}}\right)_{/j}$$

which is not a scalar but rather a section of $E^*$. Without components, we may write (20.7) in the form

$$\frac{\delta\mathcal{L}}{\delta\phi} = \frac{\partial\mathcal{L}}{\partial\phi} - \operatorname{div}\left(\frac{\partial \mathcal{L}}{\partial \nabla\phi}\right)$$

## 20.1b. Boundary Conditions

Suppose that $\phi$ satisfies Lagrange's equations. Then we see immediately from (20.6) that if we demand that $\delta\phi = 0$ on $\partial M$, then $\delta S = 0$. The condition

$$\delta\phi = 0 \quad \text{on } \partial M$$

is called an **essential** or **imposed** boundary condition; we simply *prescribe the value of* $\phi$ *on the boundary*. We see, however, that the boundary conditions

$$\left(\frac{\partial \mathfrak{L}}{\partial \phi^a_{/j}}\right) N_j = 0, \quad a = 1, \ldots, N$$

will also yield $\delta S = 0$ when $\phi$ satisfies Lagrange's equations. These are called the **natural** boundary conditions. See Problem 20.1(1) at this time.

## 20.1c. Noether's Theorem for Internal Symmetries

Suppose now that we have a 1-parameter group of **symmetries** of the Lagrangian, that is, we suppose that $\mathfrak{L}$ is invariant under a 1-parameter group of *fiber motions* $\phi \mapsto \phi(\alpha)$. We shall mainly be interested in the case when there is a 1-parameter subgroup $g(\alpha) = e^{\alpha \mathbf{E}}$ of the structure group, $\mathbf{E} \in \mathfrak{g}$, and $\mathfrak{L}$ is invariant under $\phi \mapsto g(\alpha)\phi$. (In the case of the Dirac electron, we shall see that the Lagrangian is invariant under the $U(1)$ action $\psi \mapsto e^{i\alpha}\psi$ on spinors $\psi$.) In this case $g$ is a matrix function $g^a{}_b(\alpha)$ of $\alpha$. In a given local patch $U$ of $M$, the section $\phi$ is represented by a column $\phi^a$ and then the symmetry would be of the form $\phi^a(\alpha) = g^a{}_b(\alpha)\phi^b = (e^{\alpha \mathbf{E}})^a{}_b\phi^b$. Then $\delta\phi^a = (\partial\phi^a/\partial\alpha)_{\alpha=0} = E^a{}_b\phi^b$. The symmetry assumption yields $\delta S = 0$. Thus if $\phi$ is a **critical** section, that is, *if* $\phi$ *satisfies Lagrange's equations*, then for any compact submanifold $M'$ of $M$ with boundary $\partial M'$ we have, from (20.4) and (20.5),

$$\int_{M'} \left[\left(\frac{\partial \mathfrak{L}}{\partial \phi^a_{/j}}\right)\delta\phi^a\right]_{/j} \sqrt{g}dx = \int_{M'} \operatorname{div}\left[\left(\frac{\partial \mathfrak{L}}{\partial \phi^a_{/j}}\right)\delta\phi^a\right]\sqrt{g}dx = 0$$

Since $M'$ is arbitrary, we conclude

**Noether's Theorem for Internal Symmetries (20.8):** *If* $\phi$ *satisfies Lagrange's equations and if* $\delta\phi$ *is a variation by symmetries of the Lagrangian, then*

$$\operatorname{div}\left[\left(\frac{\partial \mathfrak{L}}{\partial \phi^a_{/j}}\right)\delta\phi^a\right] = 0$$

**Corollary (20.9):** *For the 1-parameter group* $e^{\alpha\mathbf{E}}$ *of symmetries we have*

$$\operatorname{div}\left[\left(\frac{\partial \mathfrak{L}}{\partial \phi^a_{/j}}\right)E^a_b\phi^b\right] = \left[\left(\frac{\partial \mathfrak{L}}{\partial \phi^a_{/j}}\right)E^a_b\phi^b\right]_{/j}$$

*and thus the vector field* $\boldsymbol{J}$

$$J^j := \left(\frac{\partial \mathfrak{L}}{\partial \phi^a_{/j}}\right)E^a_b\phi^b$$

*has divergence 0.*

We shall mention an application of this to the Dirac equation in Section 20.2.

### 20.1d. Noether's Principle

The *principle* behind Noether's theorem is of more applicability and importance than the specific formula given in (20.8). All internal first-variation problems lead to an expression of the form

$$\delta \int_U L dx = \int_U \left[\frac{\delta L}{\delta \phi}\right] \delta\phi dx + \int_U G(x, \phi, \delta\phi) dx$$

(for all compact regions $U \subset M$) where the form of the functional derivative $\delta L/\delta\phi$ depends on the number of derivatives $\phi_{/j}, \phi_{/jk}, \ldots,$ appearing in L. A solution to the variational problem satisfies the Euler–Lagrange equations $\delta L/\delta\phi = 0$. If then we have a variation that leaves $\int_U L dx$ invariant, that is, is a group of internal *symmetries*, then we must have $G(x, \phi, \delta\phi) = 0$ for the solution $\phi$. This identity can be called Noether's theorem, and is frequently of the form $\operatorname{div} \mathbf{J} = 0$ for some vector field $\mathbf{J}$.

We shall illustrate this with the familiar cases of geodesics and minimal surfaces. A geodesic $M^1$ in $W^n$ is a solution to the variational problem

$$\delta \int_M \left\langle \frac{d\mathbf{x}}{dt}, \frac{d\mathbf{x}}{dt} \right\rangle^{1/2} dt = 0$$

for variations $\delta\mathbf{x}$ vanishing at any pair of prescribed endpoints $p$ and $q$ of $M$. (This does not fit into the scheme of (20.8); e.g., $M$ is the image in the $n$-dimensional $W^n$ of the unit $t$-interval; furthermore, $x$ takes the place of the field $\phi$, but the $x$'s are local coordinates in the manifold $W$, which is not a vector bundle.) $x$ satisfies the Euler equations $\nabla\mathbf{T}/ds = 0$. Consider a vector field $\mathbf{J}$ on $W$ that generates a 1-parameter group of *isometries* (e.g., the rotations of the round 2-sphere $W^2$). Such a field is called a **Killing** field, after the mathematician Killing, and its flow clearly leaves the Lagrangian $[g_{jk}(dx^j/dt)(dx^k/dt)]^{1/2}$ invariant. However, this "variation" $\delta\mathbf{x} = \mathbf{J}$ does not vanish at the endpoints. The first variation formula (10.4) has "boundary" terms, and yields, since $\nabla\mathbf{T}/ds = 0$, the result $\langle\delta\mathbf{x}, \mathbf{T}\rangle(p) = \langle\delta\mathbf{x}, \mathbf{T}\rangle(q)$. Since this holds for all $p, q$ on $M$, we have

$$\langle\delta\mathbf{x}, \mathbf{T}\rangle \text{ is constant along the solution } M^1 \tag{20.10}$$

and we can make this look more like (20.9) by saying

$$d\langle\delta\mathbf{x}, \mathbf{T}\rangle = 0$$

where $d$ is the differential for the 1-manifold $M$. Thus a *Killing field* $\delta\mathbf{x}$ *has constant scalar product with the unit tangent to any geodesic.* See Problems 20.1 (2 and 3) for some applications of this result.

Consider now the generalization of a minimal surface in $\mathbb{R}^3$. $M^r$ is a **minimal submanifold** of the Riemannian $W^n$ provided

$$\delta \int_U \mathrm{vol}^r = 0$$

for each compact region $U$ of $M$ and each variation $\delta\mathbf{x}$ that vanishes on $\partial U$. We considered the case when $M^2$ is a surface in $\mathbb{R}^3$ in Section 8.4, where we derived the first variation formula of Gauss. We accept the higher-dimensional version of this in the form

$$\delta \int_U \mathrm{vol}^r = -\int_U \langle \mathbf{H}, \delta \mathbf{x}_N \rangle\, \mathrm{vol}^r + \int_{\partial U} \delta x_T \mathrm{vol}^{r-1}_{\partial U} \tag{20.11}$$

where **H** is a type of **mean curvature vector** that is *normal* to $M$, $\delta x_T$ is the component of the variation vector $\delta\mathbf{x}$ along the unit outward-pointing normal $\mathbf{n}$ to $\partial U$ that is *tangent* to $U$, that is, $\delta x_T = \langle \delta \mathbf{x}, \mathbf{n} \rangle$. (For a derivation of (20.11) see, e.g., [L].) The mean curvature **H** is more complicated than in the case of a surface in $\mathbb{R}^3$ since the normal space to $M$ is of dimension $n - r$ rather than 1, but we shall not be concerned with it at this time. The boundary term, however, should be completely evident. The formula then says that a minimal submanifold $M$ must have mean curvature $\mathbf{H} = 0$. For a minimal $M$ and a general variation we have $\delta \int_U \mathrm{vol}^r = \int_{\partial U} \delta x_T \mathrm{vol}^{r-1}_{\partial U}$. Since $\mathrm{vol}^{r-1}_{\partial U} = i_{\mathbf{n}} \mathrm{vol}^r$ we can write this as

$$\delta \int_U \mathrm{vol}^r = \int_{\partial U} \langle \delta \mathbf{x}, \mathbf{n} \rangle i_{\mathbf{n}} \mathrm{vol}^r = \int_{\partial U} \langle \delta \mathbf{x}_T, \mathbf{n} \rangle i_{\mathbf{n}} \mathrm{vol}^r$$
$$= \int_{\partial U} i(\delta \mathbf{x}_T) \mathrm{vol}^r$$

where $\delta \mathbf{x}_T$ is projection of $\delta \mathbf{x}$ tangent to $M$. We now apply Noether's principle; if $\delta \mathbf{x}$ is a *Killing vector field* on $W^n$, that is, the generator of isometries, *then the tangential part of* $\delta \mathbf{x}$ *is a vector field on M whose M-divergence is* 0

$$d_M i(\delta \mathbf{x}_{\mathbf{T}}) \mathrm{vol}^r = 0 \tag{20.12}$$

In the next sections we shall give some physical applications.

## Problems

**20.1(1)** Let $\rho$ be a given function on a compact Riemannian manifold with boundary. Consider the variational problem for a scalar function $\phi$

$$\delta \int_M [g^{jk} \phi_{/j} \phi_{/k} + 2\rho\phi] \sqrt{g}\, dx = 0$$

Find the Euler–Lagrange equations and the essential and natural boundary conditions. These should all be expressed in familiar, classical language.

**20.1(2)** The flow generated by a *Killing field* $X$ is a 1-parameter group $\phi_t$ of isometries. Thus if $Y$ and $Z$ are fields that are *invariant* under the flow, $\langle Y, Z \rangle = g_{ij} Y^i Z^j$ is independent of $t$ along an orbit of $X$.

**(i)** Show that in the Riemannian connection, *Jacobi's equation* of variation (4.10) can be written

$$\frac{\nabla Y}{dt} = \nabla_Y X$$

**(ii)** Show then that $d\langle Y, Z \rangle / dt = 0$ translates into $(X_{i/j} + X_{j/i}) Y^i Z^j = 0$, and, since $Y$ and $Z$ can be chosen arbitrarily at a given point,

$$X_{i/j} + X_{j/i} = 0$$

These are **Killing's equations**, satisfied by every Killing vector field.

**(iii)** Use these equations to show directly that (20.10) holds.

**(iv)** Let $p$ be a point at which $\| X \|^2 = \langle X, X \rangle$ achieves its maximum $\neq 0$. Thus $T\langle X, X \rangle = 2\langle \nabla_T X, X \rangle = 0$ for every vector $T$ at $p$. Let $T$ be the unit tangent to a geodesic through $p$ with arc length parameter $s$. Show that $d^2\langle X, X \rangle / ds^2 = 2\langle -R(X, T)T, X \rangle + 2\langle \nabla_T X, \nabla_T X \rangle$. By considering $(n-1)$ such unit tangents $T_\alpha$, which, together with $X$, are orthonormal at $p$, show that

$$\sum_\alpha \frac{d^2\langle X, X \rangle}{ds^2} = -2R_{ij}X^iX^j + 2\sum_\alpha \langle \nabla_{T_\alpha} X, \nabla_{T_\alpha} X \rangle$$

We conclude **Nomizu's** theorem:

> If $M^n$ has negative definite Ricci curvature then no Killing field $X \neq 0$ can achieve its maximum length at any point of $M$. In particular, we have another theorem of **Bochner**: *A compact M with negative Ricci curvature has no nontrivial Killing vector field.*

For example, the Killing field $\partial/\partial x$ on the Poincaré upper half plane (see Problem 10.1(2)) has a length that tends to infinity as we approach the $x$ axis.

**20.1(3)** Let the curve $y = y(x)$ in the $xy$ plane of $\mathbb{R}^3$ be revolved about the $x$ axis, yielding a *surface of revolution* $M^2$. We may use $x$ and the cylindrical angle $\theta$ (the polar angle in the $yz$ plane) as coordinates for $M$.

**(i)** Write down (using the picture) $ds^2$ for this surface. Clearly $\mathbf{J} = \partial/\partial\theta$ is a Killing vector field on $M^2$, since it generates the rotations about the $x$ axis. Consider a geodesic $C, \theta = \theta(x)$ on $M^2$ and let $\alpha(x)$ be the angle that this geodesic makes with the lines of latitude, that is, the $\theta$ curves.

**(ii)** Derive **Clairaut's** relation

$$y \cos\alpha = \text{constant along } C$$

Consider an infinite horn-shaped surface of revolution given by $y = f(x), -\infty < x < +\infty$, where $f$ is increasing, $f'(x) > 0$ for $-\infty < x < +\infty$, and $f(x) \to 0$ as $x \to -\infty$.

**(iii)** Show that a geodesic that crosses the latitude circle at $x = 0$ and is not orthogonal to this circle will lie in the region $x \geq -a^2$, for some $a$. What is the best value for $a^2$? What happens in the region $x > 0$?

**20.1(4)** *Geodesics in the Poincaré upper half plane.* The Poincaré metric in $M^2 = \{(x, y) | y > 0\}$ is $ds'^2 = y^{-2}\{dx^2 + dy^2\}$. Since the metric coefficients $g'_{\alpha\beta}$ are independent of $x$, $\partial/\partial x$ is a Killing vector field. Since $dy^2/y^2 \leq (dx^2 + dy^2)/y^2$, the vertical lines $x =$ constant are clearly minimizing geodesics. We are interested in the other geodesics.

Let $\mathbf{T}$ be the unit tangent to a geodesic and let $\alpha$ be the angle that $\mathbf{T}$ makes with $\partial/\partial x$, all in the Poincaré metric.

**(i)** Show that $y^{-1}\cos\alpha =$ constant $k$ along the geodesic.

**(ii)** Show directly from the metric that a horizontal line cannot be locally minimizing, and hence is not a geodesic.

**(iii)** Show that if two Riemannian metrics $ds$ and $ds'$ on a space are **conformally related**, meaning $ds'^2 = \lambda^2 ds^2$ for some smooth function $\lambda$, then angles measured with $ds$ coincide with angles measured with $ds'$.

Since the Poincaré metric is conformally related to the euclidean metric $ds^2 = dx^2 + dy^2$, we see that the angle $\alpha$ in part (i) is the same as the euclidean angle. Use now the *euclidean* metric $ds_0$. But in the euclidean metric $dy/ds_0 = \sin\alpha$ along a curve.

**(iv)** Conclude that $d\alpha/ds_0 = -k$, and thus the geodesic has constant euclidean curvature, and is thus an arc of a *circle* (of perhaps infinite radius). Show that if the geodesic is not a vertical line, then $k \neq 0$, and so it is not straight. Then at the highest point $y_0, k = 1/y_0$. Show that the euclidean circle strikes the $x$ axis orthogonally. Thus *the geodesics of the Poincaré metric are euclidean circles (or vertical lines) that meet the x axis orthogonally.*

---

## 20.2. Weyl's Gauge Invariance Revisited

What can global symmetries tell us about background fields?

We remind the reader that our formulas will differ sometimes by factors of $i$ from those of most books since we are using the metric signature $(-+++)$.

We shall also use the physicist's convention of frequently putting

$$\hbar = 1$$

Our remarks about quantization, especially "second quantization," will be extremely brief and sketchy.

### 20.2a. The Dirac Lagrangian

We shall exhibit a Lagrangian whose Euler equations are the Dirac equations for a free electron (i.e., an electron not interacting with any other field) in Minkowski space $M_0^4$.

First we shall need to construct scalars out of 4-component spinors $\psi = (\psi_1, \psi_2, \psi_3, \psi_4)^T$. Recall that $\psi^\dagger = \overline{\psi}^T$ is the hermitian conjugate row matrix. Then $\psi^\dagger\phi$ is a hermitian bilinear form that is invariant under *unitary* transformations of $\mathbb{C}^4$, but, as we shall see, it is not invariant under the Dirac representation $\rho(A) : \mathbb{C}^4 \to \mathbb{C}^4$ of (19.42)

$$\rho(A) = \begin{bmatrix} A & 0 \\ 0 & A^{\dagger-1} \end{bmatrix}$$

that accompanies each Lorentz transformation $\Lambda$ of $M_0^4$. We remedy this as follows.

Recall the Dirac matrices (with our choice of signature)

$$\gamma^0 = \begin{bmatrix} 0 & -I \\ +I & 0 \end{bmatrix} \qquad \gamma^\alpha = \begin{bmatrix} 0 & \sigma_\alpha \\ \sigma_\alpha & 0 \end{bmatrix}$$

It is clear that $\gamma^\alpha$ is hermitian whereas $\gamma^0$ is skew hermitian, and thus $i\gamma^0$ is hermitian. We now define the **Dirac conjugate spinor** (or **adjoint** spinor) to $\psi$ by

$$\widetilde{\psi} := \psi^\dagger i\gamma^0 \tag{20.13}$$

(The factor $i$ appears because of our choice of signature.) Since $i\gamma^0$ is a hermitian matrix, the bilinear form

$$\widetilde{\psi}\phi = \psi^\dagger i\gamma^0\phi \tag{20.14}$$

is again hermitian. This form, however, is *not definite* because of the switching of components resulting from $\gamma^0$. We claim that

**Theorem (20.15):** *The form $\widetilde{\psi}\phi$ is invariant under the Dirac representation $\rho$. Thus it is a scalar under Lorentz transformations.*

PROOF: One sees immediately that

$$\rho(A)^\dagger\gamma^0\rho(A) = \gamma^0 \tag{20.16}$$

and so, abbreviating $\rho(A)$ to $\rho$, we have $(\rho\psi)^\dagger i\gamma^0(\rho\phi) = \psi^\dagger\rho^\dagger i\gamma^0\rho\phi = \psi^\dagger i\gamma^0\phi$, as desired. □

Since $\rho^\dagger\rho \neq I$, it is clear that $\psi^\dagger\phi$ is not Lorentz invariant.
Since $\partial\!\!\!/\psi$ is a Dirac spinor if $\psi$ is (this is the content of (19.47)), we conclude

**Corollary (20.17):** *$\widetilde{\psi}\partial\!\!\!/\psi$ and $\widetilde{\psi}\psi$ are Lorentzian scalars.*

For an electron of mass $m$ we may try to form a Lagrangian by $\widetilde{\psi}\partial\!\!\!/\psi - m\widetilde{\psi}\psi$. As we shall see, the first term needs to be made more symmetrical in $\psi$ and $\widetilde{\psi}$. The **Dirac Lagrangian** is defined by

$$\mathfrak{L}_e = \frac{1}{2}[\widetilde{\psi}\gamma^j\partial_j\psi - (\partial_j\widetilde{\psi})\gamma^j\psi] - m\widetilde{\psi}\psi \tag{20.18}$$

where $\partial_j\widetilde{\psi}$ is really $(\partial_j\psi)^\sim = (\partial_j\psi)^\dagger i\gamma^0$. We claim that the Euler equations for the Dirac action

$$\int_M \mathfrak{L}_e dx$$

yield the Dirac equations (19.48). First note that $\psi$ consists of four complex fields $\psi^j$ in $M_0^4$. Since these are complex, we may write them in terms of their real and imaginary parts, yielding eight real fields to be varied independently. It is simpler (and equivalent) to allow the eight complex fields $\psi$ and $\widetilde{\psi}$ to be varied independently. These eight fields comprise the section $\phi = (\phi^a)$ appearing in (20.4). In Problem 20.2(1) you are asked to show that the Euler equations for the Dirac action yield the Dirac equations for $\psi$

$$\not{\partial} = m\psi$$

and the conjugate (20.19)

$$(\partial_j \widetilde{\psi})\gamma^j = -m\widetilde{\psi}$$

It is clear from (20.17) that the Dirac Lagrangian is invariant under the 1-parameter group of "gauge transformations"

$$\psi \mapsto e^{i\alpha}\psi, \quad \text{and} \quad \widetilde{\psi} \mapsto e^{-i\alpha}\widetilde{\psi} \tag{20.20}$$

where $\alpha$ is any real *constant*. Under this variation $\delta\psi = i\psi$ and $\delta\widetilde{\psi} = -i\widetilde{\psi}$. Noether's theorem (20.8) then shows that the 4-vector $J$ defined by

$$J^k := -ie\widetilde{\psi}\gamma^k\psi \tag{20.21}$$

has *vanishing divergence* in Minkowski space (the electron charge $-e$ is put in for future needs) *provided* that $\psi$ is a solution to the Dirac equation. Thus for the spatial slice $V^3(t)$ we have

$$\begin{aligned} \frac{d}{dt}\int_{V(t)} J^0 dx \wedge dy \wedge dz &= \int_{V(t)} \frac{\partial J^0}{\partial t}\, dx \wedge dy \wedge dz \\ &= -\int_{V(t)} \partial_\alpha J^\alpha dx \wedge dy \wedge dz \end{aligned}$$

If we assume that the wave function $\psi$ vanishes sufficiently rapidly at spatial infinity, the last integral vanishes by the divergence theorem and we have that

$$\int_{V(t)} e\psi^\dagger (i\gamma^0)^2 \psi \mathrm{vol}^3 = \int_{V(t)} e\psi^\dagger \psi \mathrm{vol}^3$$

is constant in time. As we shall see in Section 20.2c, if we think of $\psi$ as a *classical* (unquantized) field, this integral is interpreted as the *electric charge*, $e\psi^\dagger\psi$ is the charge density, and then $J^k$ is interpreted as the electric current vector.

### 20.2b. Weyl's Gauge Invariance Revisited

A guiding principle of Einstein's theories of relativity is that the laws of physics should be expressed in a form that is independent of any particular coordinate system used. Let us first look at a simple example in Newtonian gravitation to see how coordinate changes can be used to infer information about interactions.

Consider a "small" laboratory in free fall in our space, distant from any sizable bodies. With respect to a small cartesian coordinate system attached to the laboratory, a small test particle in *free fall* satisfies Newton's equations $d^2\mathbf{x}/dt^2 = 0$. With respect to a second cartesian system that is moving *uniformly* with respect to the first, that is, $\mathbf{x}' = \mathbf{x} - \mathbf{k}t$, where $\mathbf{k}$ is a constant, we again have the same Newtonian law $d^2\mathbf{x}'/dt^2 = 0$. We may say that uniform translation is a symmetry of our system. Newton of course realized this. He maintained that there are distinguished coordinate systems in our universe, those that are at rest with respect to "absolute space" and those that are moving uniformly with respect to it, and his laws hold in any such system. If however we allow $\mathbf{k}$ to vary in time, for example $\mathbf{k} = (1/2)\mathbf{g}_0 t$, then $\mathbf{x}' = \mathbf{x} - (1/2)\mathbf{g}_0 t^2$, and Newton's equations become, in the new coordinate system, $d^2\mathbf{x}'/dt^2 = -\mathbf{g}_0$. This additional term is simply telling us that our new coordinate system is accelerating with respect to Newton's absolute space. Bishop Berkeley and, later, Ernst Mach rejected the notion of absolute space; they would say that the new coordinate system was accelerating with respect to the bulk of matter in the universe, the distant matter in the universe, or as they would say, the "fixed stars," and this is the interpretation preferred today. The additional term, $-\mathbf{g}_0$ in this case, is informing us of the *existence of the gravitational influence of the distant matter*, even if we had been unaware of the notion of gravitation! Even when the gravitational *force* vanishes, as it does for all intents and purposes in our free-fall laboratory located at a great distance from matter, the distant matter still informs the laboratory, through gravitation, of which coordinate systems are to be considered as (approximately) inertial. I believe that if space were devoid of even this distant matter, Newton's laws would make no sense, since there would then be no intrinsic notion of an accelerating frame or that of an inertial frame. There would be no notion of the "mass" of a test particle, since mass is measured via accelerations. *Newton's laws of motion are an indication of some "background field," gravitation, that is interacting with the test particle*, and presumably these laws need amending when this background field is taken into account, particularly when the "strength" of the field does not vanish. We have learned of this background field through the fact that Newton's laws do *not* remain invariant under *non-uniform* changes of coordinates.

Newtonian mechanics takes place not in matter-free space but rather space with a "uniform" distribution of distant matter.

> *Similarly, the Minkowski space of special relativity is not general relativity with no matter present, but rather an approximation in general relativity of a region in curved space far from a uniform distribution of distant matter.*

Consider the Dirac electron in Minkowski space $M_0^4$. A *free* electron is postulated to satisfy the Dirac equation (20.19), derivable from the Lagrangian (20.18). The Dirac equation may be thought of as a replacement for Newton's law. Both (20.19) and (20.18) are invariant under (global) Lorentz transformations of $M_0^4$, but not under more general space–time coordinate changes. To allow for the general coordinate changes we proceed as we did in Section 19.5; we *change the Dirac equation* by replacing the Dirac operator by introducing the Riemannian connection for true space–time and replacing partial derivatives by covariant derivatives, yielding the new Dirac operator (19.57)

$$\partial\!\!\!/\psi + \frac{1}{4}\Gamma_i{}^j{}_k \gamma^i \gamma_j \gamma^k \psi$$

The second term, involving $\Gamma$ and $\psi$, is an **interaction** term, telling us how the gravitational field interacts with the electron field.

### 20.2c. The Electromagnetic Lagrangian

Physicists, following Weyl in 1929, have carried this principle a step further. For simplicity we shall neglect the very small gravitational interaction, that is, we shall put

$\Gamma = 0$, thus returning to the original Dirac equation (20.19). Instead of considering a change of (space–time) coordinates $x$, we shall look at a change of the *field* (fiber) coordinate $\psi$, that is, a *gauge transformation*. Although quantum mechanics assigns a physical, measurable meaning to each absolute value $|\psi^a|$, the argument or **phase** of $\psi^a = |\psi^a| \exp(i\theta_a)$, that is, $\theta_a$, has no such meaning; *one cannot measure the phase of a wave function or spinor*. Both Dirac's equation and his Lagrangian are invariant under the *global* gauge transformation

$$\psi \mapsto e^{i\alpha}\psi = (e^{i\alpha}\psi^1, e^{i\alpha}\psi^2, e^{i\alpha}\psi^3, e^{i\alpha}\psi^4)^T$$

the term *global* meaning (in physics terminology) that $\alpha$ is a *constant*. This invariance is crucial since a global change of phase of all of the wave functions in quantum mechanics must leave the physics unchanged. A global gauge transformation is a symmetry of the Dirac equation.

Since the phase of $\psi$ is not measurable, we *should* be able to have invariance under a *local* gauge transformation, where $\alpha = \alpha(x)$ varies with the space–time point $x$! Clearly the Dirac equation and Lagrangian are *not* invariant under such a substitution because of the appearance of terms involving $d\alpha$. It must be that *there is some background field that is interacting with the electron*. This background field will manifest itself through the appearance of a connection. Since each component $\psi^a$ of $\psi$ is undergoing the same phase transformation, we shall forget the 4-component nature of $\psi$ and simply write $\psi \mapsto e^{i\alpha(x)}\psi$. If we think of this as a *change of frame in a complex line bundle* with transition functions $g^{-1} = c_{VU}(x) = e^{i\alpha(x)}$, then we need a connection in this line bundle that transforms as $\omega \mapsto g^{-1}\omega g + g^{-1}dg = \omega + g^{-1}dg = \omega - i d\alpha$. If we define the real field, that is, 1-form, $A$, by $\omega = -iA$, then $A' = A + d\alpha(x)$. Thus our unknown background field $A$ transforms in the same way as the vector potential in electromagnetism, *suggesting* (with hindsight) *that we identify the background field with electromagnetism*! (Of course we could have written $\omega = -ikA$ for any real constant $k$. Comparison with classical mechanics, as in Section 16.4, leads to the choice $k = e/\hbar = e$.) The new Dirac operator is then

$$\partial\!\!\!/_A := \gamma^j(\partial_j + \omega_j) = \partial\!\!\!/ - ie\gamma^j A_j \tag{20.22}$$

If we now replace $\partial\!\!\!/$ by $\partial\!\!\!/_A$ in the Lagrangian (20.18) we get a new Lagrangian, which now contains terms involving the field A.

$$\begin{aligned}\mathfrak{L}_e &= \frac{1}{2}[\widetilde{\psi}\gamma^j(\partial_j - ieA_j)\psi - (\partial_j + ieA_j)\widetilde{\psi}\gamma^j\psi] - m\widetilde{\psi}\psi \\ &= \frac{1}{2}[\widetilde{\psi}\gamma^j\partial_j\psi - (\partial_j\widetilde{\psi})\gamma^j\psi] - m\widetilde{\psi}\psi - (ie)A_j\widetilde{\psi}\gamma^j\psi\end{aligned} \tag{20.23}$$

since $(\partial_j - ieA_j)^\dagger = \partial_j + ieA_j$. Note that the last term is, from (20.21),

$$\omega_j\widetilde{\psi}\gamma^j\psi = -ieA_j\widetilde{\psi}\gamma^j\psi = A_jJ^j$$

Quantum mechanics then dictates that the A field is also to be considered as an *independent* field in its own right; that is, we are also to allow variations of the new Lagrangian involving variations of $A$. To get nontrivial field equations for $A$ we need to have "kinetic" terms, terms involving first derivatives of $A$ with respect to $t$. To maintain

Lorentz invariance we shall need all first derivatives $\partial_j A_k$ in the Lagrangian. These partial derivatives do not yield a gauge covariant quantity; one cannot form a gauge invariant scalar for the Lagrangian simply by taking $\sum(\partial_j A_k)^2$. Geometry tells us that the **curvature** $\theta^2 = d\omega + \omega \wedge \omega = d\omega$, with components $-ie(\partial_j A_k - \partial_k A_j) =: -ieF_{jk}$, is the correct tensor to use, rather than $\partial_j A_k$. We then add some multiple of the square of this *electromagnetic field strength* $F^2$ to the Lagrangian. Our choice of $-(1/16\pi)$ for this multiple will be vindicated shortly. *This is our final Lagrangian.*

$$\mathfrak{L} = \mathfrak{L}_e + \mathfrak{L}_{em} := \frac{1}{2}[\widetilde{\psi}\gamma^j \partial_j \psi - (\partial_j \widetilde{\psi})\gamma^j \psi] \qquad (20.24)$$
$$- m\widetilde{\psi}\psi + A_j J^j - \frac{1}{16\pi} F_{jk}F^{jk}$$

Look now at the variational equations involving $\delta A$. Note first that for variations $\delta A$ vanishing outside a small region

$$\delta \int \frac{1}{16\pi} F_{jk}F^{jk}\mathrm{vol}^4 = \delta \int \frac{1}{8\pi} F \wedge *F$$
$$= \delta \frac{1}{8\pi}(F, F) = \delta \frac{1}{8\pi}(dA, dA) = \frac{(\delta dA, dA)}{4\pi}$$
$$= \frac{(d\delta A, F)}{4\pi} = \frac{(\delta A, d^* F)}{4\pi}$$

Also

$$\delta \int A_j J^j \mathrm{vol}^4 = \int \delta A_j J^i \mathrm{vol}^4 = (\delta A, *\mathcal{S}^3)$$

where $\mathcal{S}^3 := i_J \mathrm{vol}^4$. We conclude then that $d^* F = 4\pi * \mathcal{S}^3$. But $d^* F = *d*F$, from (14.12), and so we have $d*F = 4\pi\mathcal{S}$. Since $dF = 0$, we conclude that *variation of the A field yields Maxwell's equations provided that we identify J of (20.21) with the electric current density*. Charge conservation $d\mathcal{S} = 0$ follows. In summary,

> *the Dirac Lagrangian (20.18) admits the global symmetry group (20.20). If we insist that the Lagrangian should admit local symmetries, when α is not constant, then Weyl's procedure leads to the introduction of the "electromagnetic field" A; Maxwell's equations (and charge conservation) then follow!*

### 20.2d. Quantization of the *A* Field: Photons

We have now a Lagrangian involving the two fields $\psi$ and $A$. Quantum mechanics then requires that these fields be *quantized*; that is, these fields in some sense are to be represented by operators and one performs "second quantization" (see, e.g., [Su, chap. 7]). The quanta of these fields, which automatically appear, are interpreted as **particles** associated with the fields. Very *roughly* we have the following. The $\psi$ field yields again the electron. The $\psi^\dagger$ field also yields a particle, the **positron**, which had been predicted earlier by Dirac just on the basis of his new equation. The "gauge field" $A$ yields another "new" particle, the **photon**. *Physicists then say that the electromagnetic* *force between electrons is "explained" by the exchange of these new "gauge particles," the photons, between electrons.*

We should also remark that in the process of quantization the current (20.21) gets replaced by a new operator; in particular, the density becomes the electron–positron charge density, rather than simply the electron charge density.

In a few sentences, the guiding principle, the **gauge principle,** for studying the force between particles can be stated as follows. If a proposed Lagrangian of some matter field $\psi$ is invariant under global (but not local) gauge transformations, alter the Lagrangian by replacing partial derivatives by covariant derivatives (introducing a new **gauge field**, a connection $\omega$, or potential $A$, whose transformation rule is compatible with the gauge transformations); the Lagrangian then has local gauge invariance. Then add to the resulting Lagrangian a new term proportional to the square of the "length" of the curvature $d\omega + 1/2[\omega, \omega]$ of the gauge field (to be more fully explained in the next section) so that gauge invariance is not destroyed. Variations with respect to $\psi$ yield the field equations for $\psi$ and variations with respect to $\omega$ yield the field equations for the gauge field. Then when one quantizes the gauge field, the quanta of this field are identified as particles, and the force between the particles of the original matter field $\psi$ is explained by the exchange of these gauge particles.

This principle was first applied by Yang and Mills, and we turn to this now.

## Problems

**20.2(1)** Derive (20.19) as Euler equations for the Dirac action.

**20.2(2)** Show from (20.3) that $J^k = \widetilde{\psi} i\gamma^k \psi$ is a contravariant 4-vector field. Prove this also by looking at the transformation properties of $J_k = \widetilde{\psi} i\gamma_k \psi$, using (20.16) and (19.44).

**20.2(3)** Show that the term $\int A_i J^i \text{vol}^4$ is gauge-invariant if $J$ has compact support.

## 20.3. The Yang–Mills Nucleon

How did the groups $SU(2)$ and $SU(3)$ appear in particle physics?

### 20.3a. The Heisenberg Nucleon

Heisenberg postulated that the proton $p$ and the neutron $n$ behave identically with respect to the "strong" interactions between nuclei. These forces are much stronger than electromagnetic effects on the charged proton. Suppose then, with Heisenberg, we neglect completely all electromagnetic properties. He then considered $p$ and $n$ as being two states of the same particle, the "nucleon," represented by two 4-component spinor functions, again denoted by $p$ and $n$. We shall not here be concerned with the spinor components, but shall write schematically

$$\psi = (p, n)^T$$

where $p$ and $n$ are now complex-valued functions of space–time. Thus a nucleon that is, in the estimation of some observer, definitely a proton at a given point would have $n = 0$ there; a neutron would have $p = 0$. Heisenberg felt that an observer is free to make a *global* linear change in the components $(p, n)^T$, keeping $|p|^2 + |n|^2$ invariant; for example, the nucleon could be called a proton at any given space–time point. In essence, then, Heisenberg demanded that the (unknown) strong force Lagrangian for the nucleon must be invariant under the generalized gauge transformation

$$\psi \mapsto u\psi = \begin{bmatrix} u_{11} & u_{12} \\ u_{21} & u_{22} \end{bmatrix} \begin{bmatrix} p \\ n \end{bmatrix}$$

where, since $|p|^2 + |n|^2$ is to be unchanged, $u \in U(2)$ is a (constant) unitary matrix. Since

$$(p, n)^T \quad \text{and} \quad (e^{ia}p, e^{ia}n)^T$$

represent the same nucleonic mixture we may eliminate this special phase transformation by restricting $u$ to have determinant 1; the symmetry group of the strong Lagrangian then consists of constant matrices $u \in SU(2)$ and the nucleon admits $SU(2)$ as a *global* gauge group. (As I learned from Meinhard Mayer, Heisenberg actually thought not in terms of $SU(2)$ but rather the spin "representation" of $SO(3)$!)

### 20.3b. The Yang–Mills Nucleon

Yukawa, in 1935, introduced the idea that one should explain the strong nuclear force between nucleons by assuming that the force arises from the exchange of certain particles, **mesons**, unobserved at that time, just as the force between electrons results from the exchange of photons. Yang and Mills in 1954 suggested that we can arrive at exchange mesons by assuming that the correct Lagrangian for the nucleon will admit $SU(2)$ as a *local* symmetry group, rather than the global one of Heisenberg. Weyl's principle will then require a gauge field, that is, a connection.

Recall that when we studied (in Section 16.4e) an electron moving in the background field of a magnetic monopole, the vector potential was not globally defined and had to be defined in patches of $M_0^4$. The nuclear field, analogous to the electromagnetic field, is completely unknown. There is a good chance that any "potential" for this field will again only be defined in patches, and likewise for the $\psi$ field. Thus the nucleon field should be considered not as a $\mathbb{C}^2$ function on space–time but rather as a section of a $\mathbb{C}^2$ vector bundle, whose structure group is $SU(2)$. Of course the bundle might be trivial, but it is no more work to consider the general case. Gauge transformations are simply changes of frames in the fibers of the bundle. In this new unknown bundle the Yang–Mills covariant derivative will be locally of the form

$$\nabla_j = \frac{\partial}{\partial x^j} + \omega_j$$

where $\omega_j = \omega(\partial/\partial x^j)$ and $\omega = (\omega^a{}_b) = dx^j \omega_j{}^a{}_b$ is an $\mathfrak{su}(2)$-valued connection 1-form. $\mathfrak{su}(2)$ consists of skew hermitian matrices with trace 0 and so has a basis consisting of imaginary multiples of the Pauli matrices $\{i\sigma_a\}$, $a = 1, 2, 3$

$$\sigma_1 = \begin{bmatrix} 0 & 1 \\ 1 & 0 \end{bmatrix}, \qquad \sigma_2 = \begin{bmatrix} 0 & -i \\ i & 0 \end{bmatrix}, \qquad \sigma_3 = \begin{bmatrix} 1 & 0 \\ 0 & -1 \end{bmatrix}$$

Thus each $\omega_j$ is of the form

$$\omega_j = -iq\sigma_a A^a_j = -iq\boldsymbol{\sigma} \cdot \mathbf{A}_j \tag{20.25}$$
$$= -iq\{\sigma_1 A^1_j + \sigma_2 A^2_j + \sigma_3 A^3_j\}$$

where we have completely suppressed the matrix indices.

We have thus been forced to introduce *three* new covariant vector fields $\mathbf{A}^1$, $\mathbf{A}^2$, $\mathbf{A}^3$, the **Yang–Mills** fields, to mediate the force between nucleons. The strength of the force is reflected in the **coupling constant** $q$, replacing the charge in the case of electromagnetism. Our covariant derivative is

$$\nabla_j \psi = \frac{\partial \psi}{\partial x^j} - iq\boldsymbol{\sigma} \cdot \mathbf{A}_j \psi \tag{20.26}$$

where again $\psi = (p, n)^T$

One then must introduce "kinetic terms" in the Lagrangian involving derivatives of the **A** fields, that is, of the connection $\omega$. The natural candidate for "derivative" of $\omega$ is of course the curvature

$$\theta = d\omega + \frac{1}{2}[\omega, \omega] = d\omega + \omega \wedge \omega$$

Then

$$\theta_{jk} = \theta(\boldsymbol{\partial}_j, \boldsymbol{\partial}_k) = d\omega(\boldsymbol{\partial}_j, \boldsymbol{\partial}_k) + \omega \wedge \omega(\boldsymbol{\partial}_j, \boldsymbol{\partial}_k)$$
$$= \partial_j \omega_k - \partial_k \omega_j + \omega_j \omega_k - \omega_k \omega_j$$
$$\theta_{jk} = \partial_j \omega_k - \partial_k \omega_j + [\omega_j, \omega_k] \tag{20.27}$$

(**Caution:** Each $\omega_j$ is an ordinary matrix, *not* a matrix of 1-forms!) Introducing the *matrices*

$$A_j := \boldsymbol{\sigma} \cdot \mathbf{A}_j$$

we get

$$\theta_{jk} = -iqF_{jk}$$

where (20.28)

$$F_{jk} := \partial_j A_k - \partial_k A_j - iq[A_j, A_k]$$

is again a trace–free hermitian $2 \times 2$ matrix, the **field strength** of the Yang–Mills field.

We must remark that Yang and Mills were unaware, at the time, of the notion of curvature of a vector bundle; the bracket term in (20.28) was added because they knew that *some* term was needed to give a nonabelian version of electromagnetism! For an interview with Yang on the history, see [Z].

In our former notation

$$\theta^a{}_b = \frac{1}{2} R^a{}_{bjk} dx^j \wedge dx^k$$

and so $\theta_{jk}$ is the skew hermitian matrix with ${}^\alpha{}_\beta$ entry $R^\alpha{}_{\beta jk}$. We wish to construct a *scalar* from $\theta$. The analogue of the Ricci tensor $R^\alpha{}_{\beta\alpha k}$ makes no sense (why?), and so the scalar curvature analogue doesn't exist. An obvious scalar can be constructed quadratically from the Riemann tensor, namely $R^\alpha{}_{\beta jk} R^\beta{}_\alpha{}^{jk}$ (the indices ${}^{jk}$ here have been raised by the Minkowski metric tensor), which is essentially the trace of the matrix $\sum_{jk} \theta_{jk}\theta^{jk}$. One then adds to the Lagrangian a kinetic term proportional to this trace

$$\text{tr}\,(F_{jk}F^{jk})$$

We shall discuss this more thoroughly in the next section.

After second quantization the fields $\mathbf{A}^1, \mathbf{A}^2, \mathbf{A}^3$ yield three particles, the exchange particles that mediate the nuclear force.

This model of nuclear forces is now obsolete. The currently accepted version holds that the nucleons are not fundamental; each is made up of **quarks.** Each "flavored" quark $\psi$ appears in three different **color** states $\psi = (R, B, G)^T$, analogous to the two nucleon states $(p, n)^T$. The gauge group is then the 8-dimensional $SU(3)$. Its Lie algebra of traceless skew-hermitian $3 \times 3$ matrices has a basis given by $\{i\lambda_b\}$, where $\lambda_b$ are the hermitian **Gell-Mann** matrices; see [Su, p. 245]. The connection is of the form $\omega_j = -ig_s\lambda_a A^a_j$, where there are now 8 covariant vector fields $\mathbf{A}^1, \ldots, \mathbf{A}^8$, and the "charge" $g_s$ is called the strong coupling constant. There are then 8 gauge fields, the **gluons,** that yield the forces between quarks.

### 20.3c. A Remark on Terminology

We have related the connection matrices $\omega$ to the gauge potentials $A$ by

$$\omega = -iqA$$

$q$ is called a generalized **charge.** Now it follows from the transformation rule for a connection that if $\omega$ is a connection for a bundle $E$ then a multiple $a\omega$ of $\omega$ is again a connection for $E$ only if $a = 1$

$$a\omega' = g^{-1}a\omega g + ag^{-1}dg$$

Thus if $\omega$ is a connection, $A = (i/q)\omega$ is *not* a connection, and it transforms in a slightly different way

$$A' = g^{-1}Ag + \left(\frac{i}{q}\right)g^{-1}dg$$

In spite of this, physicists almost always refer to $A$ as the connection, and $F = (i/q)\theta$ as the curvature.

## Problem

**20.3(1)** Show that if $\omega$ and $\omega'$ are connections for $E$ then their convex combination

$$(1-a)\omega + a\omega'$$

is also a connection for $E$ for each real $a$.

## 20.4. Compact Groups and Yang–Mills Action

What if the group is a compact group other than $SU(N)$?

### 20.4a. The Unitary Group Is Compact

**Theorem (20.29):** *The group $U(n)$ is compact*

PROOF: Consider $U(n)$ as the subset of complex $n^2$ space satisfying $uu^\dagger = u\overline{u}^T = I$, that is,

$$\sum_j u_{ij}\overline{u}_{kj} = \delta_{ik}$$

In particular

$$\sum_{k,j} |u_{kj}|^2 = \sum_{k,j} u_{kj}\overline{u}_{kj} = \sum_k \delta_{kk} = n$$

Thus $U(n)$ consists of points that lie on the sphere $\| u \| = \sqrt{n}$ and is therefore a *bounded* subset of complex $n^2$ space. It is also clear that the limit of a sequence of unitary matrices is again unitary, and so $U(n)$ is a *closed*, bounded (i.e., compact) set (see Section 1.2a). □

### 20.4b. Averaging over a Compact Group

We have seen that the left and right invariant 1-forms on the affine group of the line, $A(1)$, do not always coincide. This is to be expected in general. Let $\{\sigma^j\}$ and $\{\tau^j\}$ be bases for the left invariant and right invariant 1-forms on $G$ that coincide at $e$. The corresponding *Haar measures*

$$\sigma^1 \wedge \ldots \wedge \sigma^N \quad \text{and} \quad \tau^1 \wedge \ldots \wedge \tau^N$$

will in general be different, as they are in $A(1)$. This cannot happen in a compact group.

**Theorem (20.30):** *In a compact Lie group, the left and right Haar measures coincide (the Haar measure is* ***bi-invariant****).*

PROOF: Let $\omega = \sigma^1 \wedge \ldots \wedge \sigma^N$ be the left invariant volume form and let $\mathbf{e}$ be an orthonormal basis of left invariant vector fields; in particular $\omega(\mathbf{e}) = 1$. To say that $\omega$ is not right invariant is to say that for some right translate, $\omega(\mathbf{e}g^{-1}) := \omega(R_{g^{-1}*}\mathbf{e}) = c \neq 1$. But then $\omega(g\mathbf{e}g^{-1}) = c$. By replacing $g$ by $g^{-1}$ if necessary we may assume $c > 1$. Thus under this adjoint action $Ad(g)$, the orthonormal $\mathbf{e}$ at the identity is sent into a frame at the identity with volume $c > 1$. Under $Ad(g^n)$, the frame $\mathbf{e}$ is sent into a frame with volume $c^n \to \infty$, as $n \to \infty$. This means that the continuous function $F : G \to \mathbb{R}$ defined by $F(g) = \omega(g\mathbf{e}g^{-1})$ is not bounded on $G$. But a continuous real-valued function on a compact space is bounded, a contradiction. □

Given a compact group $G$ with bi-invariant volume form $\omega$, the integral of a continuous function $f : G \to \mathbb{C}$ is usually written

$$\int_G f\omega = \int_G f(g)\omega_g$$

where $\omega_g$ is the volume form at $g$. (This is similar to the notation $\int f(x)dx$.) When $\omega$ has been normalized so that the total volume of $G$ is 1, $\int_G \omega = 1$, then $\int_G f\omega$ is simply the **average** of $f$ on $G$ and plays a central role in many aspects of Lie theory.

**Theorem (20.31):** *For any continuous function $f$ and for all $g$ in the compact group $G$ we have*

$$\int_G f(hg)\omega_h = \int_G f(gh)\omega_h = \int_G f(h)\omega_h$$

PROOF: Consider first $\int_G f(hg)\omega_h$. Right translation $R_g : G \to G$ sends $h \mapsto hg$. Since $\omega$ is right invariant

$$\omega_h = R_g^*\omega_{hg} = (R_g^*\omega)_h$$

that is,

$$R_g^*\omega = \omega$$

Also $f(hg) = f \circ R_g(h) = (R_g^* f)(h)$, and so the function $F$ defined by $F(h) = f(hg)$ is simply $F = R_g^* f$. Hence

$$\int_G f(hg)\omega_h = \int_G F\omega = \int_G (R_g^* f) \wedge R_g^*\omega$$
$$= \int_G R_g^*(f \wedge \omega) = \int_{R_g G} f\omega = \int_G f(h)\omega_h$$

since $R_g G = G$. The proof for $\int_G f(gh)\omega_h$ is similar since $\omega$ is left invariant as well. □

In many books this proof is written as follows: The statement that $\omega$ is right invariant is written

$$\omega_{hg} = \omega_h$$

Then

$$\int_G f(hg)\omega_h = \int_G f(hg)\omega_{hg} = \int_G f(h)\omega_h \tag{20.32}$$

replacing the dummy variable $hg$ by the dummy $h$.

### 20.4c. Compact Matrix Groups Are Subgroups of Unitary Groups

Let $G$ be a *compact* group of $n \times n$ matrices. We can consider the matrices as linear transformations of $\mathbb{C}^n$ (think of them as being complex matrices). Let $(\,,)$ be *any* hermitian scalar product in $\mathbb{C}^n$ (e.g., $(\mathbf{z}, \mathbf{w}) = (\sum \bar{z}_j w_j)$). The matrices will *not*, in general, preserve this scalar product (i.e., the matrices will not be unitary with respect to this metric). We claim, however, that the *averaged scalar product will be invariant.*

For given $X, Y$ in $\mathbb{C}^n$, we define the new scalar product

$$\langle X, Y\rangle := \int_G (hX, hY)\omega_h \tag{20.33}$$

This is of the form $\langle X, Y\rangle = \int_G f(h)\omega_h$. Then, from (20.31), for $g \in G$

$$\langle gX, gY\rangle = \int_G (hgX, hgY)\omega_h = \int_G f(hg)\omega_h = \langle X, Y\rangle$$

as desired. *Thus the compact matrix group acts by unitary transformations with respect to this new scalar product.* After choosing a new basis for $\mathbb{C}^n$ that is orthonormal in this metric, the matrices will be unitary in the usual sense. In this sense we may consider any given compact matrix group as a subgroup of the unitary group, (More accurately, it is *similar* to such a subgroup.)

## 20.4d. *Ad* Invariant Scalar Products in the Lie Algebra of a Compact Group

Let $G = U(n)$, the group of unitary matrices, $g^\dagger := \overline{g}^T = g^{-1}$. Then $\mathfrak{g} = \mathfrak{u}(n)$ is the space of skew hermitian matrices, $X^\dagger = \overline{X}^T = -X$.

*We shall always consider Lie algebras as real vector spaces.*

Define a real scalar product $\langle , \rangle$ in the vector space $\mathfrak{u}(n)$ by

$$\langle X, Y\rangle := -\mathrm{tr}XY = -X_{ij}Y_{ji} \tag{20.34}$$

(This agrees with that used for SU(2) in (19.9).) In Problem 20.4(1) you are asked to show that this form on $\mathfrak{g} = \mathfrak{u}(n)$ is real, symmetric, and positive definite.

Note that this scalar product in $\mathfrak{u}(n)$ is *invariant under the adjoint action* of $G = U(n)$ on $\mathfrak{g}$; for $u \in U(n)$

$$\langle uXu^{-1}, uYu^{-1}\rangle = -\mathrm{tr}\, uXYu^{-1} = -\mathrm{tr}XY = \langle X, Y\rangle$$

Now let $G$ be any *compact* $n \times n$ matrix group. As we have seen, $G$ may be considered a subgroup of $U(n)$, and then, as we have seen in Section 15.4d, $\mathfrak{g}$ is a *subalgebra* of $\mathfrak{u}(n)$. Then for $X, Y$ in $\mathfrak{g}$ we will have that $\langle X, Y\rangle = -trXY$ *is a real scalar product in* $\mathfrak{g}$ *that is invariant under the adjoint action of* $G$ *on* $\mathfrak{g}$! For $X, Y, Z$ in $\mathfrak{g}$

$$\langle e^{tX}Ye^{-tX}, e^{tX}Ze^{-tX}\rangle = \langle Y, Z\rangle$$

Differentiating and putting $t = 0$ gives, from (18.32),

$$\langle [X, Y], Z\rangle + \langle Y, [X, Z]\rangle = 0$$

that is, (20.35)

$$ad(X) : \mathfrak{g} \to \mathfrak{g} \quad \text{is skew adjoint}$$

and note that *this holds for any group whose Lie algebra is endowed with a scalar product invariant under the adjoint action!*

### 20.4e. The Yang–Mills Action

Let $\pi : E \to M^n$ be a vector bundle with *compact* structure group $G \subset U(N)$. We are mainly concerned with the case $M^n = M^4$ = space–time. In the original Yang–Mills model, $G = SU(2)$.

If $\omega$ is a connection in $E$ then

$$\omega = -iqA \tag{20.36}$$

expresses $\omega$ in terms of the "gauge field" or "potential" $A$ and a "coupling constant" or generalized "charge" $q$. Since $G \subset U(N)$, $\omega_j = \omega(\partial_j)$ is skew hermitian and $A_j$ is hermitian. For curvature

$$\theta = d\omega + \frac{1}{2}[\omega, \omega] = -iqF \tag{20.37}$$

$$F_{jk} = \partial_j A_k - \partial_k A_j - iq[A_j, A_k]$$

and $F$ is the **field strength**. It also is hermitian.

In our computations we shall use $\omega$ and $\theta$; when we are done we may convert to $A$ and $F$! Our constants might differ from those used in physics.

We define the **Yang–Mills** (briefly, Y–M) **action functional** by

$$S[\omega] := \frac{1}{4}\int_M -\text{tr}\,(\theta_{jk}\theta^{jk})\text{vol}^n \tag{20.38}$$

Note that for each $j, k$, $\theta_{jk} = (R^\alpha{}_{\beta jk})$ is a skew-hermitian *matrix*, that is, $\theta_{jk} \in \mathfrak{g}$, and $-\text{tr}\,(\theta_{jk}\theta^{jk})$ is the scalar product in $\mathfrak{g}$ of these matrices. The indices in $\theta^{jk}$ have been raised by $g^{jk}$, the pseudo-Riemannian metric in $M^n$. We wish to write this action using the curvature *forms*, rather than matrices. The curvature forms are

$$\theta_U = (\theta^\alpha{}_\beta) = \frac{1}{2}R^\alpha{}_{\beta jk}dx^j_U \wedge dx^k_U$$

Each matrix $\theta_U$ is a matrix of locally defined 2-forms $\theta^\alpha{}_\beta$. Each of these 2-forms $\theta^\alpha{}_\beta$ has a Hodge dual $(n-2)$-form $*\theta^\alpha{}_\beta$ from the pseudo-Riemannian metric on $M^n$, and we know from (14.6) that

$$R^\alpha{}_{\beta jk}R^\nu{}_\eta{}^{jk}\,\text{vol}^n = 2!\theta^\alpha{}_\beta \wedge *\theta^\nu{}_\eta$$

We can then write the action as

$$S[\omega] = -\frac{1}{2}\int_M \theta^\alpha{}_\beta \wedge *\theta^\beta{}_\alpha = -\frac{1}{2}\int_M \text{tr}\,\theta \wedge *\theta \tag{20.39}$$

$$= \frac{1}{2}(\theta, \theta)$$

where we have defined a Hilbert space scalar product (,) on $\mathfrak{g} \subset \mathfrak{u}(N)$-valued *p-forms* by

$$(\theta^p, \phi^p) := -\int_M \text{tr}\,\theta \wedge *\phi \tag{20.40}$$

This makes sense whenever $\theta$ and $\phi$ are $p$-form sections of an $Ad(U(N))$ bundle since $\text{tr}\,[c\theta c^{-1} \wedge c{*}\phi c^{-1}] = \text{tr}\,[\theta \wedge *\phi]$.

How does $S$ depend on the connection $\omega$? Take a 1-parameter family of connections $\omega = \omega(\epsilon)$ with "velocity" $\delta\omega := \omega'(0)$. For first variation (*keeping the metric on M fixed*)

$$\begin{aligned}\delta S[\omega] : &= \frac{d}{d\epsilon}\{S[\omega(\epsilon)]\}_{\epsilon=0} = \frac{1}{2}\delta(\theta,\theta) = (\delta\theta,\theta)\\ &= \left(\delta\left\{d\omega + \frac{1}{2}[\omega,\omega]\right\},\theta\right)\\ &= \left(d\delta\omega + \frac{1}{2}[\delta\omega,\omega] + \frac{1}{2}[\omega,\delta\omega],\theta\right)\\ S'[\omega] &= (d\delta\omega + [\omega,\delta\omega],\theta)\end{aligned} \tag{20.41}$$

since $\omega$ and $\delta\omega$ are 1-forms; see (18.7). Now if $\omega_1$ and $\omega_2$ are connections their difference $\Delta\omega$ is a $\mathfrak{g}$-valued 1-form that transforms as

$$\Delta\omega_V = c_{VU}\Delta\omega_U c_{VU}^{-1}$$

and is thus a 1-form section of the *Ad* bundle associated to the $G$ bundle $E$. Likewise, $\delta\omega$ is a 1-form section and of course the curvature $\theta$ is a 2-form section of this same bundle. But then

$$d\delta\omega + [\omega,\delta\omega] = \nabla\delta\omega$$

is the covariant differential of $\delta\omega$, see (18.42). We then have, from (20.41),

$$\begin{aligned}\delta\theta &= \nabla(\delta\omega)\\ S'[\omega] &= (\nabla\delta\omega,\theta) = (\delta\omega,\nabla^*\theta)\end{aligned} \tag{20.42}$$

where $\nabla^*$ is the Hilbert space adjoint to $\nabla$.

As usual we demand that $S'[\omega] = 0$ for all variations $\delta\omega$ of $\omega$. This gives

$$\nabla^*\theta = 0 \quad \text{(Yang–Mills)}$$

with, of course (20.43)

$$\nabla\theta = 0 \quad \text{(Bianchi)}$$

the latter holding for *any* connection.

These equations clearly generalize Maxwell's equations in the case when the current $J$ vanishes.The coordinate expressions for these appear in Section 20.5.

## Problem

**20.4(1)** Show that (20.34) is real, symmetric, and positive definite.

## 20.5. The Yang–Mills Equation

How do the Yang–Mills equations compare with Maxwell's?

### 20.5a. The Exterior Covariant Divergence $\nabla^*$

We have seen in (20.42) that the Y–M curvature $\theta = -iqF$, a $\mathfrak{g}$-valued 2-form, must satisfy $\nabla^*\theta = 0$, where $\nabla^*$ is the Hilbert space adjoint of the covariant exterior differential $\nabla$ for the *Ad* bundle. We shall compute a coordinate expression for this analogous to the formula (14.15) for scalar-valued forms. $\nabla^*$ satisfies

$$\int \text{tr}\,\langle d\delta\omega + [\omega, \delta\omega], \theta\rangle \text{vol} = \int \text{tr}\,\langle \delta\omega, \nabla^*\theta\rangle \text{vol} \tag{20.44}$$

for all 1-form sections $\delta\omega$ and all 2-form sections $\theta$ of the *Ad* bundle. Here $\langle , \rangle$ is the pseudo-Riemannian (pointwise) scalar product. We can also write this as

$$\int \text{tr}\,\{d\delta\omega + [\omega, \delta\omega]\} \wedge *\theta = \int \text{tr}\,\delta\omega \wedge *\nabla^*\theta \tag{20.45}$$

All the forms involved take their values in the fixed vector space $\mathfrak{g}$ and both $d$ and $*$ commute with taking traces ($*$ only affects the manifold indices $i, j, \ldots$, not the fiber indices $\alpha, \beta, \ldots$). Consider the left-hand side of (20.45). The first term is

$$\int \text{tr}\,\{d\delta\omega \wedge *\theta\} = \int d\delta\omega^\alpha{}_\beta \wedge *\theta^\beta{}_\alpha = (d\delta\omega^\alpha{}_\beta, \theta^\beta{}_\alpha) \tag{20.46}$$

$$= (\delta\omega^\alpha{}_\beta, d^*\theta^\beta{}_\alpha) = \int \text{tr}\,\delta\omega_k\{d^*\theta\}^k \text{vol}$$

assuming as usual that the boundary integral involving $\delta\omega$ vanishes. The second term on the left-hand side of (20.45) can be computed using

$$\begin{aligned}[\omega, \delta\omega]_{jk} &= \{\omega \wedge \delta\omega + \delta\omega \wedge \omega\}(\boldsymbol{\partial}_j, \boldsymbol{\partial}_k) \\ &= \omega_j\delta\omega_k - \omega_k\delta\omega_j + \delta\omega_j\omega_k - \delta\omega_k\omega_j\end{aligned}$$

for then $[\omega, \delta\omega]_{jk}\theta^{jk} = 2[\omega_j, \delta\omega_k]\theta^{jk}$ (since $j$ and $k$ are form indices, $\theta^{jk} = -\theta^{kj}$). Then

$$\begin{aligned}\int \text{tr}\,[\omega, \delta\omega] \wedge *\theta &= \frac{1}{2}\int \text{tr}\,[\omega, \delta\omega]_{jk}\theta^{jk}\text{vol} \\ &= \int \text{tr}\,[\omega_j, \delta\omega_k]\theta^{jk}\text{vol} = -\int \langle[\omega_j, \delta\omega_k], \theta^{jk}\rangle\text{vol}\end{aligned}$$

where $\langle , \rangle$ is the scalar product in $\mathfrak{g}$. From (20.35) we can write this as

$$= \int \langle\delta\omega_k, [\omega_j, \theta^{jk}]\rangle\text{vol} = \int -\text{tr}\,\delta\omega_k[\omega_j, \theta^{jk}]\text{vol}$$

Combining this with (20.46) gives

$$\int \text{tr}\,\delta\omega_k\{(d^*\theta)^k - [\omega_j, \theta^{jk}]\}\text{vol} = \int \text{tr}\,\delta\omega \wedge *\nabla^*\theta$$

But from (14.15) $(d^*\theta)^k = -\theta^{jk}{}_{/j}$, where this covariant derivative is with respect to the pseudo-Riemannian connection on $M$, *not* the bundle connection. $\theta$ *is to be considered as a second rank tensor on M with extra indices from* $\mathfrak{g}$ *that are not considered in this covariant derivative!* Finally we have the coordinate expression of the Y–M equation $\nabla^*\theta = 0$

$$(\nabla^*\theta)^k = -\{\theta^{jk}{}_{/j} + [\omega_j, \theta^{jk}]\} = 0 \tag{20.47}$$

where, we emphasize, all indices are manifold indices; $\omega_j$ and $\theta^{jk}$ are matrices whose indices have been suppressed

$$\omega_j = (\omega^\alpha_{j\,\beta}) \qquad \theta^{jk} = (R^\alpha{}_\beta{}^{jk})$$

We remark, though we shall have no use for it, that the expression (20.47) can be written as the negative of a tensorial type of divergence. The ${}^{\alpha k}_{\beta}$ component of (20.47) can be obtained from $(\theta^{\alpha}{}_{\beta})^{jk} = R^{\alpha}{}_{\beta}{}^{jk}$. Thus $\theta^{jk}{}_{|j} + [\omega_j, \theta^{jk}]$ becomes

$$\begin{aligned}\theta^{jk}{}_{|j} + \omega_j\theta^{jk} - \theta^{jk}\omega_j &= \theta^{jk}{}_{|j} + \omega^{\alpha}_{j\gamma} R^{\gamma}{}_{\beta}{}^{jk} - R^{\alpha}{}_{\gamma}{}^{jk}\omega^{\gamma}_{j\beta} \\ &= (\partial_j R^{\alpha}{}_{\beta}{}^{jk} + \Gamma^{j}_{jr} R^{\alpha}{}_{\beta}{}^{rk} + \Gamma^{k}_{jr} R^{\alpha}{}_{\beta}{}^{jr}) \\ &\quad + \omega^{\alpha}{}_{j\gamma} R^{\gamma}{}_{\beta}{}^{jk} - R^{\alpha}{}_{\gamma}{}^{jk}\omega^{\gamma}_{j\beta}\end{aligned}$$

Note that we could then write (20.47) as

$$(\nabla^*\theta)^{\alpha}{}_{\beta}{}^{k} = -R^{\alpha}{}_{\beta}{}^{jk}{}_{//j} = 0 \tag{20.48}$$

where we are considering $R^{\alpha jk}_{\beta}$ as the components of a tensor of type $E \otimes E^* \otimes TM \otimes TM$, and $_{//}$ denotes the covariant derivative of such a tensor, using $\omega$ for the bundle part and $\Gamma$ for the tangent bundle part.

### 20.5b. The Yang–Mills Analogy with Electromagnetism

If we now put $\omega = -iqA$ and $\theta = -iqF$, then we have seen in (20.37)

$$F_{jk} = \partial_j A_k - \partial_k A_j - iq[A_j, A_k]$$

generalizes the situation in electromagnetism, where the action is (when no sources are present) essentially $\int F_{jk}F^{jk}\mathrm{vol}^4$. The Y–M action is, except for a constant,

$$\begin{aligned} S[A] &\sim \int \mathrm{tr}\ F_{jk}F^{jk}\ \mathrm{vol}^n \qquad (20.49)\\ &= \int \mathrm{tr}\ (\partial_j A_k - \partial_k A_j - iq[A_j, A_k])(\partial^j A^k - \partial^k A^j - iq[A^j, A^k])\mathrm{vol}^n\end{aligned}$$

Whereas the electromagnetic action is quadratic in the fields $A$, the Y–M action also contains cubic and quartic terms. The Y–M equation $\nabla^*\theta = 0$ and the Bianchi equation $\nabla\theta = 0$ are, from (18.44) and (20.47),

$$F^{jk}{}_{|j} - iq[A_j, F^{jk}] = 0$$

and (20.50)

$$\partial_i F_{jk} + \partial_k F_{ij} + \partial_j F_{ki} - iq\{[A_i, F_{jk}] + [A_k, F_{ij}] + [A_j, F_{ki}]\} = 0$$

It is instructive to compare these with Maxwell's equations in $M_0^4$ with metric $\{-1, 1, 1, 1\}$. We shall write the Y–M fields for $G = SU(n)$ as follows. We give the usual electromagnetic names to the components of $F$

$$F^{0i} = E^i \qquad i = 1, 2, 3 \qquad F^{12} = B^3, \ldots$$

even though $\mathbf{E}$ and $\mathbf{B}$ are now 3-vectors with hermitian $n \times n$ matrix components. Look, for example, at Y–M for $k = 0$. We have

$$F^{j0}{}_{|j} - iq[A_j, F^{j0}] = 0$$

that is,

$$\operatorname{div} \mathbf{E} = iq(\mathbf{A} \bullet \mathbf{E} - \mathbf{E} \bullet \mathbf{A}) \tag{20.51}$$

This is the analogue of Gauss's equation. We see that even though we started out without external sources,

$$iq(\mathbf{A} \bullet \mathbf{E} - \mathbf{E} \bullet \mathbf{A})$$

plays the role of a "charge density." Thus the Y–M field **E** and the potential **A** combine to act as a *source* for the Y–M field! The *nonabelian* nature of the structure group $SU(n)$, that is, $[\mathbf{A}, \mathbf{E}] \neq 0$, allows this to happen!

Look again at Y–M, this time considering only a spatial index $k = \beta = 1, 2$ or 3:

$$F^{0\beta}{}_{|0} + F^{\alpha\beta}{}_{|\alpha} - iq[A_0, F^{0\beta}] - iq[A_\alpha, F^{\alpha\beta}] = 0$$

that is,

$$\operatorname{curl} \mathbf{B} = \frac{\partial \mathbf{E}}{\partial t} - iq(A_0\mathbf{E} - \mathbf{E}A_0) + iq(\mathbf{A} \times \mathbf{B} + \mathbf{B} \times \mathbf{A}) \tag{20.52}$$

replacing Ampere–Maxwell. Note that there are two extra contributions to a "current" other than the displacement current.

The Y–M equations thus yield generalizations of the laws of Gauss and of Ampere–Maxwell, without external sources.

Similarly, in Problem 20.5(1) you are asked to derive the analogues of the laws of Faraday and of the absence of magnetic monopoles from the Bianchi identity

$$\operatorname{curl} \mathbf{E} + \frac{\partial \mathbf{B}}{\partial t} = iq\{(\mathbf{A} \times \mathbf{E} + \mathbf{E} \times \mathbf{A}) + (A_0\mathbf{B} - \mathbf{B}A_0)\} \tag{20.53}$$

and

$$\operatorname{div} \mathbf{B} = iq(\mathbf{A} \bullet \mathbf{B} - \mathbf{B} \bullet \mathbf{A}) \tag{20.54}$$

Note that "magnetic charge density" can exist in a nonabelian Y–M field!

### 20.5c. Further Remarks on the Yang–Mills Equations

It is clear that if $\phi$ is a $p$-form section of any $Ad\,(G)$ bundle, then tr $\phi$ is an ordinary $p$-form on $M$ since tr $(c_{VU}\phi c_{VU}^{-1}) = \operatorname{tr} \phi$.

Note that if $G = SU(N)$ then for any $p$-form section $\phi$ of the $Ad\,(G)$ bundle (for example the curvature 2-form) we must have tr $\phi = 0$. However, if $\psi$ is another form section, then $\phi \wedge \psi$ does not take its values in $\mathfrak{g}$. Although tr $(\phi \wedge \psi)$ is again a form on $M$ it need not be 0. Furthermore, there are times when one uses groups other than $SU(N)$.

**Theorem (20.55):** *Let $\phi$ be a p-form section of an Ad (G) bundle. Then*

$$d\operatorname{tr} \phi = \operatorname{tr} \nabla\phi$$

PROOF:

$$\nabla\phi = d\phi + [\omega, \phi] = d\phi + \omega \wedge \phi - (-1)^p \phi \wedge \omega$$

and so

$$\begin{aligned}\operatorname{tr}\nabla\phi &= \operatorname{tr} d\phi + \omega^\alpha{}_\beta \wedge \phi^\beta{}_\alpha - (-1)^p \phi^\beta{}_\alpha \wedge \omega^\alpha{}_\beta \\ &= \operatorname{tr} d\phi \quad \square\end{aligned}$$

The following is clearly the analogue of (14.12). Recall that we are using the Hilbert space scalar product (20.40).

**Theorem (20.56):** *For any form section of an Ad(G) bundle*

$$\nabla^*\phi = \pm{*}\nabla{*}\phi$$

PROOF: Let $\gamma$ be a $(p-1)$-form section of the Ad bundle with small support. Then, from (18.46)

$$\begin{aligned}(\nabla\gamma, \phi) &= -\int \operatorname{tr}(\nabla\gamma \wedge {*}\phi) = -\int \operatorname{tr}\nabla(\gamma \wedge {*}\phi) \pm \int \operatorname{tr}(\gamma \wedge \nabla{*}\phi) \\ &= -\int d\operatorname{tr}(\gamma \wedge {*}\phi) \pm \int \operatorname{tr}(\gamma \wedge \nabla{*}\phi)\end{aligned}$$

Since $\gamma$ has small support, the first integral vanishes by Stokes's theorem. We conclude

$$(\nabla\gamma, \phi) = \pm \int \operatorname{tr}(\gamma \wedge {*}{*}\nabla{*}\phi = \pm(\gamma, {*}\nabla{*}\phi) \quad \square$$

The actual sign is given as in (14.12).

**Definition (20.57):** *A* **Yang–Mills field** *A* is one that satisfies

$$\nabla^* F = 0$$

**Definition (20.58):** Any field strength $F_{jk} = \partial_j A_k - \partial_k A_j - iq[A_j, A_k]$ that satisfies

$$F = \begin{cases} {*}F \text{ is called } \textbf{self-dual} \\ -{*}F \text{ is called } \textbf{anti-self-dual} \end{cases}$$

Since any field strength satisfies the Bianchi equation $\nabla F = 0$, we see that $\nabla{*}F = 0$ if $F$ is self- or anti-self-dual. *A self- or anti-self-dual field strength is automatically the field strength of a Yang–Mills field!*

## Problems

**20.5(1)** Supply the details of the electromagnetic analogues (20.53) and (20.54) for the Bianchi equations.

**20.5(2)** The electromagnetic analogues can also be derived using exterior forms. Fill in the details in the following.

Decompose $A$ into temporal and spatial parts $A = \phi dt + \mathbf{A}^1$. Here $\phi = A_0$ is a $\mathfrak{g}$-valued function and $\mathbf{A}^1$ is a $\mathfrak{g}$-valued 1-form. As usual we write $d = \mathbf{d} + dt \wedge \partial/\partial t$. Then $F^2 = (i/q)\theta^2 = dA - iqA \wedge A$ yields, after writing $F^2 = \mathbf{E}^1 \wedge dt + \mathbf{B}^2$ with $\mathfrak{g}$-valued forms $\mathbf{E}^1$ and $\mathbf{B}^2$, the "electric" and "magnetic" parts of the field strength.

$$\mathbf{E}^1 = \mathbf{d}\phi - \frac{\partial \mathbf{A}^1}{\partial t} - iq[\mathbf{A}^1, \phi]$$

$$\mathbf{B}^2 = \mathbf{d}\mathbf{A}^1 + \mathbf{A}^1 \wedge \mathbf{A}^1 = \mathbf{d}\mathbf{A}^1 + \frac{1}{2}[\mathbf{A}^1, \mathbf{A}^1]$$

Then the Bianchi equations $\nabla F^2 = dF^2 + [\omega^1, F^2] = 0$ will yield

$$\mathbf{d}\mathbf{E}^1 + \frac{\partial \mathbf{B}^2}{\partial t} = iq\{[\mathbf{A}^1, \mathbf{E}^1] + [\phi, \mathbf{B}^2]\}$$

$$\mathbf{d}\mathbf{B}^2 = iq[\mathbf{A}^1, \mathbf{B}^2]$$

For the Yang–Mills equation $\nabla^* F = \pm{*}\nabla{*}F = 0$, we put ${*}F^2 = -\mathbf{*B}^2 \wedge dt + \mathbf{*E}^1$ for $\mathfrak{g}$-valued forms $\mathbf{*B}^2$ and $\mathbf{*E}^1$; the bold $\mathbf{*}$ is the spatial Hodge operator. Then

$$0 = \nabla{*}F^2 = d(-\mathbf{*B}^2 \wedge dt + \mathbf{*E}^1) + [-iq\mathbf{A}^1, (-\mathbf{*B}^2 \wedge dt + \mathbf{*E}^1)]$$

yields

$$\mathbf{d*E}^1 = iq[\mathbf{A}^1, \mathbf{*E}^1]$$

and

$$\mathbf{d*B}^2 = \frac{\partial \mathbf{*E}^1}{\partial t} + iq\{[\mathbf{A}^1, \mathbf{*B}^2] - [\phi, \mathbf{*E}^1]\}$$

**20.5(3)** Let $M^4$ be compact and suppose that the support of $\delta\omega$ does not meet the boundary (if any) of $M$. Use $\delta\theta = \nabla(\delta\omega)$ and Theorem (20.56) to show that

$$\delta \int_M \operatorname{tr}(\theta \wedge \theta) = \pm\delta(\theta, *\theta) = 0$$

Thus if $\int_M \operatorname{tr}(\theta \wedge \theta)$ is added to a given action integral, the action will be altered *but the variational equations will be unchanged!* We shall study the 4-form $\operatorname{tr}(\theta \wedge \theta)$ extensively in our remaining chapters.

---

## 20.6. Yang–Mills Instantons

How can the Brouwer degree distinguish between two Yang–Mills vacua?

### 20.6a. Instantons

Consider a quantum particle interacting with a Yang–Mills field in Minkowski space. This particle is described by a "wave funtion" $\psi$, a cross section of a complex $\mathbb{C}^N$

vector bundle $E$ over Minkowski space $M = M_0^4$. We assume that the structural group is $SU(n)$; thus $G = SU(n)$ acts on $\mathbb{C}^N$ via some representation. For our purposes it is sufficient to consider the standard representation on $\mathbb{C}^n$. The bundle has a Y–M connection $\omega = -iqA$ and a curvature $\theta = -iqF$, where $A$ and $F$ are hermitian matrix valued local forms on $M_0^4$. In $U \subset M$ we have a frame of sections

$$\Psi_U = \mathbf{e}_U = (\mathbf{e}_1^U, \ldots, \mathbf{e}_n^U)$$

and $\omega_U$ and $\theta_U$. In an overlap $\mathbf{e}_V = \mathbf{e}_U c_{UV}$, $c_{UV}(x) \in SU(n)$.

In this section we shall be concerned with the background Y–M field, rather than with the particle. The action for this Y–M field alone is essentially

$$\int_M -\text{tr}\, F_{jk} F^{jk} * 1 \sim \int_M (\| \mathbf{E} \|^2 - \| \mathbf{B} \|^2) * 1$$

where we have given the electromagnetic analogue on the right (Problem 7.2(3)).

For certain purposes it is useful in physics to replace the Minkowski metric of space–time by the 4-dimensional *euclidean* metric $+dt^2 + d\mathbf{x} \cdot d\mathbf{x}$. This will not be discussed here. (See e.g., [C, chap. 7]. This chapter of Coleman's book will also overlap with some of the topological material that we shall discuss later.) The action is then called the **euclidean** action. We shall be concerned with Y–M fields having finite euclidean action

$$\int_M (\| \mathbf{E} \|^2 + \| \mathbf{B} \|^2) * 1 < \infty$$

(Note that the euclidean version of the electromagnetic Lagrangian is the energy density of the electromagnetic field.) Such fields are called **instantons** since they "vanish" as $|t| \to \infty$. An example of an instanton is given in [I, Z, sec. 12-1-3].

For simplicity, to avoid the limiting values of boundary integrals, we assume that the field strength $E^2 + B^2$ not only dies off at infinity but has support lying inside some 3- sphere $S^3$ centered at the origin of $\mathbb{R}^4$.

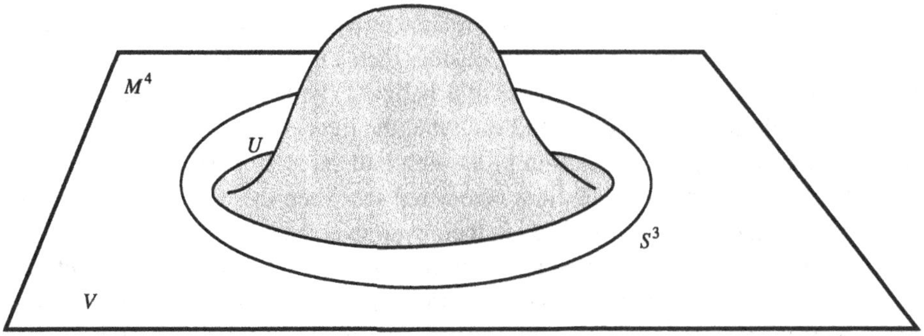


**Figure 20.1**

(This does not make sense in electromagnetism in $M_0^4$ since an electromagnetic field in free space would radiate out to infinity and would be present for all $t$.)

Let $U$ be a coordinate patch holding this $S^3$ and its interior and let $V$ be a coordinate patch holding $S^3$, extending to $\infty$, and such that $F = 0$ in $V$. We assume that $V$ is the exterior to some sphere inside $S^3$.

In the "exterior region" $V$ we have $\theta = 0$. We claim that we can make a change of frame over all of $V$ (in the wave "function" vector bundle $E$, *not* in Minkowski space) so that in the new frame $\omega_V = 0$!

This should not be a complete surprise; it is a *global* version of Riemann's theorem (9.70) on curvature 0, but for an arbitrary vector bundle. To see it, let $\omega'$ be the original connection form for $V$. We wish to find a $g : V \to SU(n)$ so that

$$\omega_V := g^{-1}\omega' g + g^{-1}dg = 0$$

that is,

$$dg + \omega' g = 0 \tag{20.59}$$

Can we solve this 1-form system for $g = g^r_s(x)$? Using the symbol $\approx$ to signify mod $(dg + \omega' g)$ as arises in the Frobenius theorem

$$\begin{aligned} d(dg + \omega' g) &= d\omega' g - \omega' \wedge dg \\ &\approx d\omega' g - \omega' \wedge (-\omega' g) \approx (d\omega' + \omega' \wedge \omega')g \\ &\approx \theta' g = 0, \quad \text{in } V \end{aligned}$$

By Frobenius we may *locally* solve (20.59) *uniquely* for $g$, subject to any initial $g_0 = g(p)$ at $p \in V$.

Suppose that we have two solutions, $g$ and $h$, in two overlapping patches. Then $dg = -\omega' g$ and $dh = -\omega' h$, and so

$$\begin{aligned} d(g^{-1}h) &= -g^{-1}dgg^{-1}h + g^{-1}dh \\ &= g^{-1}\omega' gg^{-1}h - g^{-1}\omega' h = 0 \end{aligned}$$

Thus two overlapping solutions are always related by a *constant* matrix $k \in SU(n)$, $h = gk$, at least if the overlap is connected! Consider then a path $C : [0, 1] \to V$ that starts at $p$. Cover this path by a finite number of 4-balls $B_\alpha$ (lying in $V$) each small enough to support a solution $g_\alpha$ to (20.59) and such that the intersections of consecutive balls are connected. Let $g_0$ be the solution in the first ball $B_0$ at $p$. Let $g_1$ be a solution in the next ball $B_1$. $B_1$ intersects $B_0$ in a connected set. Then there is a constant matrix $k_1 \in SU(n)$ such that $g_1(x) = g_0(x)k_1$ in their overlap and it is clear that $g_1' := g_1 k_1^{-1}$ is a new solution of (20.59) in $B_1$ that agrees with $g_0$ in their overlap. We have continued the solution into the second ball. Proceed to the third ball and so forth. In this way we continue the given solution in the initial ball to all points of $V$. Is this well defined? If $C$ is a *closed* curve that returns to $p$, the final solution could be a $g_0'$ that differs from $g_0$; this is the same situation as in analytic continuation of an analytic function in the complex plane! However, the region in $\mathbb{R}^4$ that is exterior to a ball is *simply connected*, and just as analytic continuation is unique in such a region (seen by shrinking the closed curve to a point), so it is in our situation. Thus a global solution $g : V \to SU(n)$ to (20.60) exists in all of the exterior region $V$ and

$$\omega_V = 0$$

when we use the new frame of sections $\mathbf{e}_V = \mathbf{e}'_V g$. □

Note that the *original* connection $\omega'$ is of the form

$$\omega' = -dgg^{-1} \tag{20.60}$$

and is said to be **pure gauge.**

Since $\omega_V = 0$ on $U \cap V$, and in particular on $S^3$,

$$\omega_U = c_{VU}{}^{-1}\omega_V c_{VU} + c_{VU}{}^{-1} dc_{VU} = c_{VU}{}^{-1} dc_{VU}$$

We again write this in a simplified form, $c_{VU} = g$,

$$\omega_U = g^{-1} dg \quad \text{on } S^3 \tag{20.61}$$

where $g : S^3 \to SU(n)$ are new matrices, *not* those of (20.60).

We then have the following situation: Look at the part of the wave function bundle that lies over the sphere $S^3$. Over $S^3$ we have two frame fields given, the "flat" frame $\mathbf{e}_V$ and the frame $\mathbf{e}_U$ over $U$. The flat frame consists of sections $\mathbf{e}^V{}_1, \ldots, \mathbf{e}^V_n$ each of which is covariant constant

$$\nabla \mathbf{e}^V_b = \mathbf{e}^V_a \omega^a_{Vb} = 0$$

that is, these sections are parallel displaced along $S^3$. We are comparing the $U$-frames $\mathbf{e}_U$ with these covariant constant frames along $S^3$,

$$\mathbf{e}_U(x) = \mathbf{e}_V(x) c_{VU}(x) = \mathbf{e}_V(x) g(x) \tag{20.62}$$

and, consequently the matrices g(x) define a mapping

$$g : S^3 \to SU(n) \tag{20.63}$$

This situation is similar to that encountered in Chern's proof of Poincaré's index theorem (17.21). Let us go back and reconsider Chern's proof in the light of our Y–M field with finite action.

## 20.6b. Chern's Proof Revisited

Consider, instead of a closed $M^2$ as in Section 17.3, a curved "wormhole" version $M^2$ of the plane, but such that the curvature vanishes in the region $V$ exterior to some circle $S^1$. The bundle we are considering is the tangent bundle $TM^2$ to the orientable surface

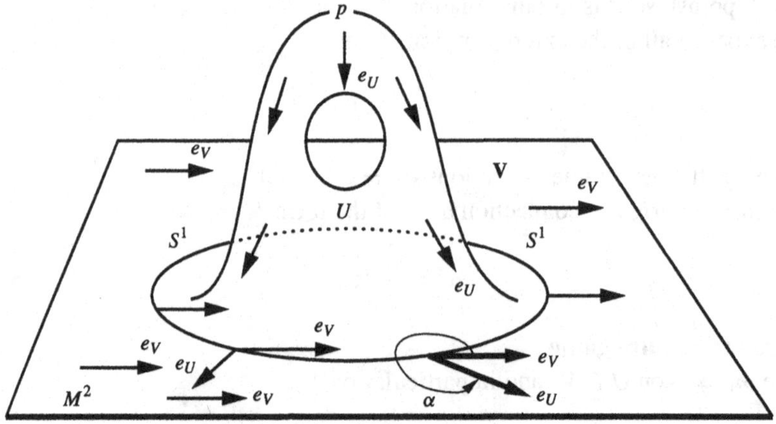


Figure 20.2

$M^2$, but considered as a complex $\mathbb{C}$ line bundle. By using "orthonormal" frames $\mathbf{e}_U$, $\mathbf{e}_V$, we may consider the structural group of this bundle to be $U(1)$. We have indicated the "flat" covariant constant frame $\mathbf{e}_V$ in the exterior region.

**Warning:** Unlike the case when $M^n$ has dimension $n \geq 3$, the region $V$ is *not* simply connected. One cannot always find a global flat frame in this region $V$. For example, $M^2$ is flat in the *conical* region in the following figure, but a parallel displaced vector will not return to itself after traversing $S^1$

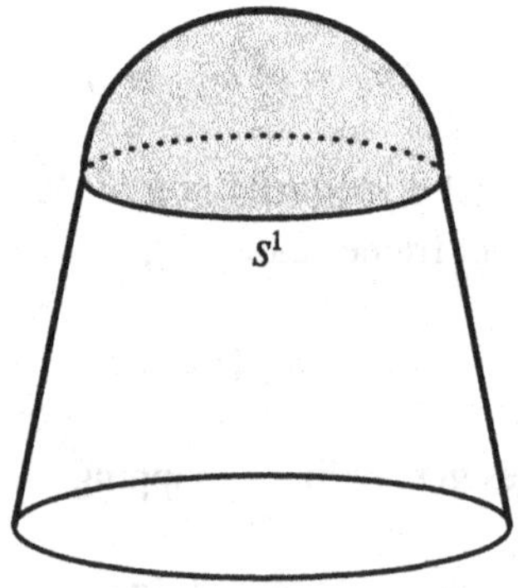


Figure 20.3

as we saw in Section 8.7. In fact this picture is the *geometric analogue of the Aharonov–Bohm effect*, discussed in Section 16.4f. Using the electromagnetic connection, the curvature inside the coil is constant, since the magnetic field $\mathbf{B}$ is constant there; this corresponds to the constantly curved spherical cap. Furthermore, the exterior to the coil corresponds to the flat conical region. Since $\omega = -iebd\theta/2\pi\hbar$ in the exterior region, the equation of parallel translation in the electron wave function bundle is $d\psi - ieb\psi d\theta/2\pi\hbar = 0$. Hence $\psi = \exp(ieb\theta/2\pi\hbar)$ is covariant constant *but is not single-valued* unless the flux $b$ takes on very special values!

(**Notational comment:** In the case of a section of a vector bundle with structure group $G$, parallel translation along a parameterized curve $x = x(t)$ is still defined by $d\psi + \omega\psi = 0$, that is,

$$\frac{d\psi}{dt} = -\omega\left(\frac{dx}{dt}\right)\psi \tag{20.64}$$

for the matrix-valued connection 1-form $\omega$. Since $\int \omega(dx/dt)dt$ also lies in $g$, we see from Problem 15.2(3) that *if the structure group G is commutative* then the solution to (20.64) is $\psi(t) = \exp[-\int_0^t \omega(dx/d\tau)d\tau]\psi(0)$. If $G$ is not commutative, there is no such formula, but physicists *write* the solution in the form

$$\psi(t) = P\exp\left[-\int \omega\left(\frac{dx}{dt}\right)dt\right]\psi(0)$$

The symbol $P$ indicates an operation called **path ordering**. It is important to realize that *this can simply be considered a notation for the operation that sends an initial* $\psi(0)$ *into the unique solution* $\psi(t)$ *of* (20.64). (We shall not use this notation.)

In our wormhole, Figure 20.2, we have chosen $V$ so that a global covariant constant frame $\mathbf{e}_V$ *does* exist, as it does in the Y–M example.

In the curved region $U$ we have indicated a cross section $\mathbf{e}_U$ that has singularities at the critical points of the height function; the top $p$ is one of them. (The field looks like the normalized velocity field for molasses oozing down from the top.)

For our complex line bundle version of the tangent bundle we have, as in 17.3a, the connection $\omega$ and curvature

$$\theta = d\omega = -iKdA \tag{20.65}$$

On the circle $S^1$ we have $\mathbf{e}_U = \mathbf{e}_V e^{i\alpha}$ and so

$$g(x) = e^{i\alpha} \tag{20.66}$$

and

$$\omega_U = g^{-1}dg = id\alpha$$

In the situation of Poincaré's theorem, Chern considered a closed surface. In our case

$$\int_M KdA = \int_U KdA$$

since $K$ vanishes outside $U$. In Chern's proof

$$\int_M KdA = 2\pi\sum_p j_p(\mathbf{e}_U)$$

whereas in our nonclosed $M^2$, using $KdA = d\omega_{12}$, Chern's proof would give

$$\begin{aligned}\frac{1}{2\pi}\int_U KdA &= \sum_p j_p(\mathbf{e}_U) + \frac{1}{2\pi}\oint_{S^1}\omega_{12}^U \\ &= \sum_p j_p(\mathbf{e}_U) - \frac{1}{2\pi}\oint_{S^1} d\alpha \end{aligned}\tag{20.67}$$

We may then write (for future reference)

$$\frac{i}{2\pi}\iint_M \theta = \sum_p j_p(\mathbf{e}_U) + \frac{i}{2\pi}\oint_{S^1}\omega$$

or (20.68)

$$\frac{i}{2\pi}\iint_M \theta = \sum_p j_p(\mathbf{e}_U) - \frac{1}{2\pi}\oint_{S^1} d\alpha$$

(20.68) tells us that we get the same result as in the closed $M^2$ case except for a *boundary term describing how many times the given cross section rotates around the flat section* $\mathbf{e}_V$*!*

$$\frac{1}{2\pi}\iint_M K\,dA = \sum_p j_p(\mathbf{e}_U) - \frac{1}{2\pi}\oint_{S^1} d\angle(\mathbf{e}_V, \mathbf{e}_U) \tag{20.69}$$

Note that this last "rotation number" is exactly the *degree of the map*

$$g : S^1 \to S^1 \quad \text{defined by } x \to g(x) = e^{i\alpha}$$

Now in our Y–M situation we have a similar map, at least in the case when $G = SU(2)$, for then (20.63) involves a map

$$g : S^3 \to SU(2) = S^3 \tag{20.70}$$

and this map indeed does have a degree, called the **winding number of the instanton.**

In our Y–M case we shall **assume** that the frame $\mathbf{e}_U$ in the wave function bundle has *no singularities inside* $S^3$.

We draw a surface analogue consisting of a flat cylinder $V$ with a hemispherical cap (a diffeomorphic copy of $\mathbb{R}^2$) $U$. In $V$ we put the flat vertically oriented "frame" $\mathbf{e}_V$, whereas in the cap $U$ we may put a singularity-free field $\mathbf{e}_U$, for example, as follows. In Section 16.2a we introduced a vector field on $S^2$ having a single singularity of index 2 at the north pole. The field $\mathbf{e}_U$ is simply the part of this field that lives on the southern hemisphere.

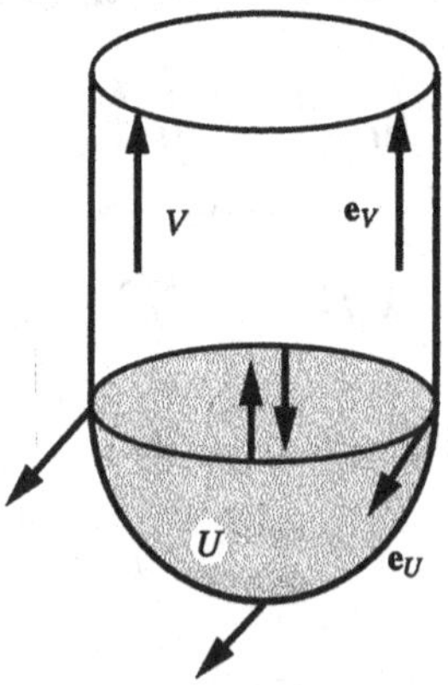


Figure 20.4

In Problem 20.6(1) you are asked to verify (20.69) in this case.

In the Chern situation, in (20.69), if there are no $\mathbf{e}_U$ singularities, we see that the degree of the boundary map is completely described by the integral of the curvature! *Can the corresponding Y–M degree in (20.70) be evaluated by looking at the curvature, that is, the field strength, of the Yang–Mills field?* The answer *yes* will be proven in Section 21.2; it was first given by Chern a decade before the paper of Yang and Mills.

## 20.6c. Instantons and the Vacuum

In Yang–Mills we may consider the **vacuum state** in which the *field strength* $F$ or $\theta$ *vanishes*. One must not conclude that nothing of interest can be associated to such a vacuum. In the geometric analogue we may consider a flat surface; the connection $\omega$ replaces the gauge field $A$ and the curvature $\theta = 0$ replaces field strength $F = 0$. In the example considered previously of the frustrum of a flat cone, tangent to a 2-sphere along a small circle $S^1$, we may delete the spherical cap completely. This corresponds to the exterior region in the Aharonov–Bohm effect. We have seen that parallel translation about $S^1$ does not return a vector to itself, in spite of the fact that the connection is flat. There is more information in the flat connection than is read off from the 0 curvature alone! Likewise there is more information in a gauge field $A$ for a vacuum than can be read from the vanishing field strength.

Before considering the Yang–Mills vacuum we shall look at another geometric analogue. In the following figure we have again drawn the 2-dimensional analogue, a flat surface, but instead of using the "flat" (covariant constant) frame (pointing, for example, constantly in the $t$ direction) we use a frame that is time-independent, is flat at spatial infinity, and rotates (in this case) once about the flat frame along each spatial slice.

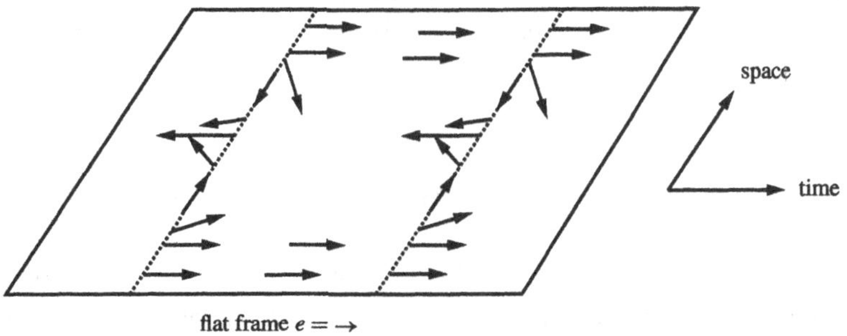


**Figure 20.5**

We have **gauge transformed** the flat frame **e** to a new one, **e**$g$, where $g : \mathbb{R} \to U(1) = S^1$ maps each spatial slice so that $g(-\infty) = g(\infty) = 1$. (The field, i.e., connection, is again "pure gauge," $\omega = g^{-1}dg$.) We assume, again for simplicity, that for each spatial section $t =$ constant we have $g(x) = 1$ for $|x| \geq a$ for some $a$. This vacuum solution in $\mathbb{R}^2$ is *not deformable, while remaining flat at spatial infinity*, into the identically flat frame vacuum for the following reason.

The function $g$ maps the spatial slice $\mathbb{R}$ into $S^1$. We may stereographically project $\mathbb{R}$ onto a circle $S^1$–(north pole) by projecting from the north pole. In this way we may consider g as being defined on $S^1$–(north pole). Since $g$ is identically 1 in some neighborhood of the pole we can extend $g$ to the entire circle $S^1$. (This can be thought of in the following way. By identifying all $x$ for $|x| \geq a$ with the point $x = a$ on the section $t$ = constant, this section becomes topologically a circle $S^1$. We have "compactified" $\mathbb{R}$ to a circle and since $g = 1$ for $|x| \geq$ a, $g$ extends to this compactification.) This gives, for each $t$, a map $g : S^1 \to U(1) = S^1$, which in this case has degree 1 by construction. If our vacuum solution were to be deformable to the flat vacuum solution, while keeping $|x| \geq a$ flat, then $1 = \deg g : S^1 \to U(1) = S^1$ would have to equal that of the flat vacuum case, which clearly has degree 0. This is a contradiction. We thus have two **inequivalent vacua**. Similarly, we could get a vacuum frame that winds $k$ times around the flat frame.

In the 4-dimensional Yang–Mills case (with $G = SU(2)$) there will likewise be an infinity of inequivalent vacua, each one characterized by the degree or "winding number" of the map $g : S^3 \to SU(2) = S^3$ arising from the spatial slice $\mathbb{R}^3$ "compactified" to $S^3$; this is discussed more in Problem 20.6(2). Physicists then interpret an instanton with winding number $k$, that is, degree $k$ given in (20.70), as representing a *non*vacuum field **tunneling** between a vacuum at $t = -\infty$ with winding number $n$, and a vacuum at $t = +\infty$ with winding number $n + k$ (see [C, L, sect. 16.2] or [I, Z sect. 12-1-3]). We discuss the *geometry* of this situation in Problem 20.6(2).

Further significance of the winding number of the instanton will be sketched in Chapter 21.

We have seen why $g : S^3 \to SU(2)$ has a degree. To understand why $g : S^3 \to SU(n), n \geq 2$, has an associated "degree," and to understand Chern's results when there are singularities, we need to delve more into topology, in particular the topology of Lie groups, "homotopy groups," and "characteristic classes." Homotopy groups arise also in other aspects of physics (see, e.g., [Mi]). We shall proceed with this program in the next chapter.

## Problems

**20.6(1)** Verify (20.69) in the case of our specific example of the cylinder with a cap.

**20.6(2)** Consider an instanton. Let $\mathbf{e}_U$ be the frame in the interior U; we shall *assume that* $\mathbf{e}_U$ *can be extended to be a nonsingular frame in all of* $\mathbb{R}^4$. Let $\mathbf{e}_V$ be a flat vacuum frame in the exterior $V$ of the instanton, and let, as in (20.70), $g : S^3 \to SU(2)$, mapping the surface of the instanton into the group, have degree $k$. Recall that $k$ is called in physics the *winding number of the instanton.*

(i) Show that if $\mathbf{e}_V$ can be extended to a frame on all of $\mathbb{R}^4$ then $k = 0$ (Hint: Generalize Problem 8.3(9).). Thus in general $\mathbf{e}_V$ cannot be extended.

Consider a 3-dimensional "can" $W^3$ surrounding the instanton, lying entirely in the vacuum region $V$, and with ends $D$ and $D^*$ at two spatial slices $t = \pm$"$\infty$". Let the side of the can be given by $\| \mathbf{x} \| = a$.

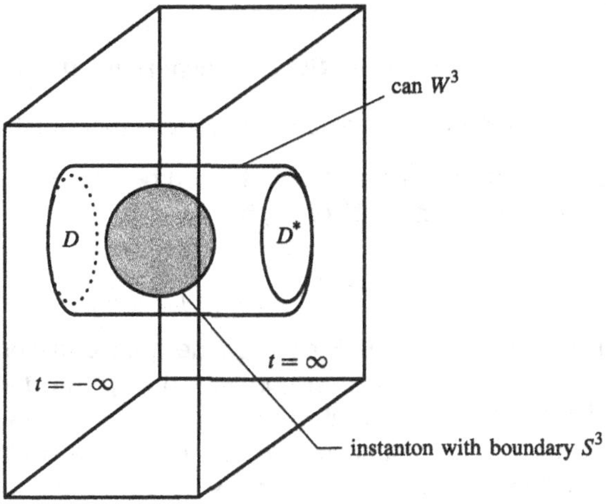


**Figure 20.6**

$g$ is defined on the can $W^3$ and, in fact, on the entire 4-dimensional region that is inside the can and outside $S^3$. *Assume that $g$ takes a constant value, say $g = e$, on an entire region $\| \mathbf{x} \| \geq (a - \epsilon)$ containing the sides of the can.* The can then can be smoothed off near the ends $D$ and $D^*$, yielding a smooth 3-dimensional manifold diffeomorphic to a 3-sphere and such that $g = e$ everywhere on this new can except on the portions of $D$ and $D^*$ where $\| \mathbf{x} \| < a - \epsilon$.

We shall now apply the theory of the Brouwer degree. $g$ maps the 3-disc $D$ into $SU(2) = S^3$ and maps $\partial D$ into a single point $g = e$. This means that if, in $D$, we identify all of $\partial D$ to a single point (the "point at $\infty$") then we can consider this new space as a 3-sphere, and we have a map $g$ of this 3-sphere into SU(2). This map has a Brouwer degree that can be evaluated by looking at inverse images of some regular value $u \in SU(2)$, $u \neq e$. Call this degree $\deg(-\infty) = n$. Similarly we can look at the disc $D^*$ and assign a degree $\deg(+\infty) = n + k$, for some integer $k$. In physics books these integers are called the *winding numbers* of the vacua at $t = -\infty$ and at $t = +\infty$, respectively. On the other hand, the entire can $W^3$ is a smooth version of a 3-sphere, and we have the degree of $g$ mapping this can into $SU(2)$. The 2-dimensional analogue is

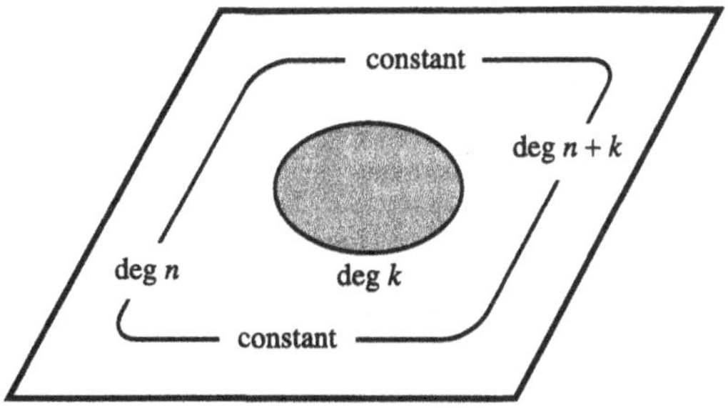


**Figure 20.7**

(ii) Show why

$$\deg(g : W^3 \to SU(2)) = \deg(+\infty) - \deg(-\infty)$$

(iii) Show why this degree $k$ is also the winding number of the instanton.

**20.6(3)** *The Winding Number of a Vacuum.* Let $\lambda \neq 0$ be any constant and $g : \mathbb{R}^3 \to SU(2)$ map a spatial section $\mathbb{R}^3$ of $\mathbb{R}^4$, by

$$g(\mathbf{x}) = \exp\left[\frac{i\pi \mathbf{x} \cdot \sigma}{(\| \mathbf{x} \|^2 + \lambda^2)^{1/2}}\right]$$

We can think of this $g$ as defining a gauge transformation of the classical vacuum (where $\omega = 0$) to a new one with $\omega = g^{-1}(\mathbf{x})dg(\mathbf{x})$, in the spatial section $\mathbb{R}^3$ defined by $t = +\infty$. We claim that this vacuum has winding number $= \pm 1$. To show this we first show that $g(\mathbf{x})$ tends to a *constant* $SU(2)$ group element limit (independent of $\mathbf{x}$) as $\| \mathbf{x} \| \to \infty$.

(i) What is this limit? (Hint: Use (19.20), which holds for *unit* $\mathbf{A}$.) *Now* we are allowed to compute the winding number using (8.18).

(ii) Show that only the origin $\mathbf{x} = \mathbf{0}$ is mapped by $g$ onto $I \in SU(2)$ and show that $\mathbf{0}$ is a regular point by using (19.20) applied to the line $\mathbf{x} = t\mathbf{A}$, where $\mathbf{A}$ is a unit vector. We have then shown that this vacuum has winding number $\pm 1$.

In [I,Z], sect. 12-1-3, an instanton solution that tunnels between a vacuum with winding number 0 and the vacuum of this problem is given.

CHAPTER 21

# Betti Numbers and Covering Spaces

## 21.1. Bi-invariant Forms on Compact Groups

Why is it that the 1-parameter subgroups of a compact Lie group are geodesics?

Samelson's article [Sam] is a beautiful exposition on the topology of Lie groups as it was known up to 1951.

### 21.1a. Bi-invariant $p$-Forms

Recall that a form or vector field on $G$ is said to be bi-invariant if it is both left and right invariant. For example, on the affine group $G = A(1)$ of the line, $dx/x$ is bi-invariant.

**Theorem (21.1):** *If $\alpha^p$ is a bi-invariant p-form, then $\alpha$ is closed,*

$$d\alpha = 0$$

PROOF: Let $\sigma^1, \ldots, \sigma^n$ and $\tau^1, \ldots, \tau^n$ be bases of the left and the right invariant 1-forms, respectively, and let $\sigma^j = \tau^j$ at the identity. Since the left and right structure constants are negatives of each other (see Section 15.4c), $d\sigma^i = -1/2C^i_{jk}\sigma^j \wedge \sigma^k$ and $d\tau^i = 1/2C^i_{jk}\tau^j \wedge \tau^k$. Let $\alpha^p$ be bi-invariant

$$\alpha^p = a_{\underset{\rightarrow}{I}}\sigma^I$$

where $a_{\underset{\rightarrow}{I}}$ are constants. Since $\alpha$ is also right invariant,

$$\alpha^p = a_{\underset{\rightarrow}{I}}\tau^I$$

Now compute $d\alpha$ at $e$ from both expressions. □

### 21.1b. The Cartan *p*-Forms

In Section 18.1a we have defined the Maurer–Cartan matrix of 1- forms

$$\Omega := g^{-1}dg$$

When $G$ is the affine group of the line, $G = A(1)$,

$$\Omega = \begin{bmatrix} \frac{dx}{x} & \frac{dy}{x} \\ 0 & 0 \end{bmatrix}$$

We can also consider exterior powers $\Omega^2 = \Omega \wedge \Omega, \Omega^3 = \ldots$. For example,

$$\begin{bmatrix} \frac{dx}{x} & \frac{dy}{x} \\ 0 & 0 \end{bmatrix} \wedge \begin{bmatrix} \frac{dx}{x} & \frac{dy}{x} \\ 0 & 0 \end{bmatrix} = \begin{bmatrix} 0 & \frac{dx \wedge dy}{x^2} \\ 0 & 0 \end{bmatrix}$$

which has the left invariant volume form for its only nontrivial entry.

We define the **Cartan *p*-forms** $\Omega_1, \Omega_2, \ldots, \Omega_{n=\dim G}$ by

$$\Omega_p := \mathrm{tr}\Omega^p = \mathrm{tr}\{g^{-1}dg \wedge g^{-1}dg \wedge \ldots \wedge g^{-1}dg\} \tag{21.2}$$

These are, of course, (scalar) *left* invariant *p*-forms on $G$. For $G = A(1)$, $\Omega_1 = dx/x$ and $\Omega_2 = 0$.

**Theorem (21.3):** *The Cartan p-forms are bi-invariant, and hence closed,* $d\Omega_p = 0$. *Furthermore,* $\Omega_{2p} = 0$.

PROOF: For constant $k \in G$,

$$\begin{aligned} &\mathrm{tr}\{(gk)^{-1}d(gk) \wedge (gk)^{-1}d(gk) \wedge \ldots\} \\ &= \mathrm{tr}\{k^{-1}(g^{-1}dg \wedge g^{-1}dg \wedge \ldots g^{-1}dg)k\} \\ &= \Omega_p \end{aligned}$$

and so they are also right invariant. Next note $\Omega_2 = \mathrm{tr}(\Omega \wedge \Omega) = \Omega_{ij} \wedge \Omega_{ji} = -\Omega_{ji} \wedge \Omega_{ij} = -\Omega_2$, and so $\Omega_2 = 0$. Similarly, $\Omega_{2p} = 0$, all $p$. □

The Cartan 3-form plays an especially important role. Since $\Omega(\mathbf{X}) = \mathbf{X}$, all $\mathbf{X} \in \mathfrak{g}$,

$$(\Omega \wedge \Omega)(\mathbf{X}, \mathbf{Y}) = \Omega(\mathbf{X})\Omega(\mathbf{Y}) - \Omega(\mathbf{Y})\Omega(\mathbf{X}) = [\mathbf{X}, \mathbf{Y}]$$

and thus

$$(\Omega \wedge \Omega) \wedge \Omega(\mathbf{X}, \mathbf{Y}, \mathbf{Z}) = [\mathbf{X}, \mathbf{Y}]\mathbf{Z} + [\mathbf{Z}, \mathbf{X}]\mathbf{Y} + [\mathbf{Y}, \mathbf{Z}]\mathbf{X} \tag{21.4}$$

Taking the trace of this and using $[\mathbf{X}, \mathbf{Y}]\mathbf{Z} = \mathbf{XYZ} - \mathbf{YXZ}$, and so on, give

$$\Omega_3(\mathbf{X}, \mathbf{Y}, \mathbf{Z}) = 3\mathrm{tr}([\mathbf{X}, \mathbf{Y}]\mathbf{Z}) \tag{21.5}$$

When $G$ is **compact** we can express this in terms of the *Ad* invariant scalar product (20.34) in $\mathfrak{g} \subset \mathfrak{u}(N)$

$$\Omega_3(\mathbf{X}, \mathbf{Y}, \mathbf{Z}) = -3\langle[\mathbf{X}, \mathbf{Y}], \mathbf{Z}\rangle \tag{21.6}$$

(21.4) brings up a point. Consider $G = SO(3)$, and let $\{\mathbf{E}_i\}$ be the basis (19.1). Then

$$\Omega \wedge \Omega \wedge \Omega(\mathbf{E}_1, \mathbf{E}_2, \mathbf{E}_3) = \mathbf{E}_1^2 + \mathbf{E}_2^2 + \mathbf{E}_3^2 \tag{21.7}$$

and this matrix is *not* in the Lie algebra $\mathfrak{so}(3)$! (Recall that in Section 18.1b we defined the *bracket* of $\mathfrak{g}$-valued forms to remedy this situation.) The matrix (21.7) is called a **Casimir element**.

### 21.1c. Bi-invariant Riemannian Metrics

Let $\langle , \rangle_e$ be a scalar product in $\mathfrak{g}$ that is *Ad invariant*; for example, when $G = U(n)$, $\langle X, Y\rangle_e = -\text{tr}XY$. Thus the Lie algebra of every compact group has such an invariant scalar product. Define then a Riemannian metric on the *group* $G$ by "left translation," that is,

$$\langle \mathbf{X}_g, \mathbf{Y}_g\rangle := \langle L_{g^{-1}*}\mathbf{X}_g, L_{g^{-1}*}\mathbf{Y}_g\rangle_e = \langle g^{-1}\mathbf{X}_g, g^{-1}\mathbf{X}_g\rangle_e$$

By construction, this metric is left invariant. We claim that it is also right invariant. For

$$\langle \mathbf{X}_e g^{-1}, \mathbf{Y}_e g^{-1}\rangle = \langle g\mathbf{X}_e g^{-1}, g\mathbf{Y}_e g^{-1}\rangle = \langle \mathbf{X}_e, \mathbf{Y}_e\rangle$$

by *Ad* invariance! We have shown

**Theorem (21.8):** *There is a bi-invariant Riemannian metric on every compact Lie group.*

The group $A(1)$ is not compact. $\sigma^1 = dx/x$ and $\sigma^2 = dy/x$ are left invariant. Hence

$$\sigma^1 \otimes \sigma^1 + \sigma^2 \otimes \sigma^2 = \frac{dx^2 + dy^2}{x^2}$$

is a *left* invariant Riemannian metric on $A(1)$. (Note that this is the Poincaré metric on the "right half plane"; see Problem 8.7(1).) This metric is *not* right invariant, and in fact there are no bi-invariant metrics on this group.

**Theorem (21.9):** *In any bi-invariant metric on a group, the geodesics are the 1-parameter subgroups and their translates.*

PROOF: Let $\mathbf{X}$ be a left invariant field on $G$. We shall show that each integral curve of $\mathbf{X}$ is a geodesic in a bi-invariant metric.

Since $\mathbf{X}$ generates *right* translations, $\mathbf{X}$ is a Killing field (see Section 20.1c). Let $C$ be a geodesic that is tangent to $\mathbf{X}$ at a point $g$. We need only show that $\mathbf{X}$ is everywhere tangent to $C$. By Noether's theorem, $\mathbf{X}$ and the unit tangent $\mathbf{T}$ to $C$ have a constant scalar product $\langle \mathbf{X}, \mathbf{T}\rangle$ along $C$. $\mathbf{T}$ has unit length and $\mathbf{X}$, being *left* invariant, also has constant length. Since $\mathbf{X}$ and $\mathbf{T}$ are tangent at $g$, it must be that $\mathbf{X}$ and $\mathbf{T}$ are tangent everywhere along $C$. □

Thus in a group with a bi-invariant Riemannian metric, a geodesic through $e$ is of the form $\exp(t\mathbf{X})$, where $\mathbf{X}$ is the tangent at $e$. *This was the* (very meager) *motivation for denoting geodesics in a Riemannian manifold by* $\exp(t\mathbf{X})$!

One says that a Riemannian manifold $M^n$ is **geodesically complete** if every geodesic segment $C(t) = \exp_{C(0)}(t\mathbf{X})$ can be extended for all parameter values $t$. The euclidean plane $\mathbb{R}^2$ is complete but the euclidean plane $\mathbb{R}^2 - 0$ with the origin deleted is not; the geodesic $\exp_{(-1,0)}(t\partial/\partial x)$ does not exist for $t = 1$ because of the hole at the origin. The Poincaré upper half plane is complete; even though there is an edge at $y = 0$, this edge is "at an infinite distance" from any point of the manifold.

It is a fact that if $M$ is *compact* then it is automatically geodesically complete. Furthermore

**Theorem of Hopf–Rinow (21.10):** *If $M^n$ is geodesically complete, then any pair of points can be joined by a geodesic of minimal length.*

For a proof of these two facts see Milnor's book [M].

In a compact group $G$ we may introduce a bi-invariant metric, and then the 1-parameter subgroups are geodesics. Thus

**Theorem (21.11):** *Every point in a compact connected Lie group $G$ lies on at least one 1-parameter subgroup.*

As we have seen in the case $G = Sl(2, \mathbb{R})$ in Problem 15.3(2), compactness is essential.

### 21.1d. Harmonic Forms in the Bi-invariant Metric

**Theorem (21.12):** *In a bi-invariant metric on a compact connected Lie group $G$, the bi-invariant forms coincide with the harmonic forms.*

The proof will be broken into several parts.

**Lemma:** *In a bi-invariant metric, the Hodge $*$ operator commutes with left and right translations*

$$* \circ L_g^* = L_g^* \circ * \quad \text{and} \quad * \circ R_g^* = R_g^* \circ *$$

PROOF: We wish to show that $L_g^* * \beta_{gh} = * L_g^* \beta_{gh}$ for every form $\beta$ at every point $gh$. Thus it suffices to show that for any form $\alpha$ at $h$ we have $\alpha_h \wedge L_g^* * \beta_{gh} = \alpha_h \wedge * L_g^* \beta_{gh}$. Define $\alpha_{gh}$ by $\alpha_h = L_g^* \alpha_{gh}$. Since the metric is bi-invariant, so is the volume form $\omega$. Recall that $(\alpha \wedge *\beta)_{gh} = \langle \alpha_{gh}, \beta_{gh} \rangle \omega_{gh}$. Then

$$\begin{aligned} \alpha_h \wedge L_g^* * \beta_{gh} &= L_g^* \alpha_{gh} \wedge L_g^* * \beta_{gh} = L_g^* (\alpha_{gh} \wedge * \beta_{gh}) \\ &= L_g^* (\langle \alpha_{gh}, \beta_{gh} \rangle \omega_{gh}) \end{aligned}$$

$$= \langle \alpha_{gh}, \beta_{gh} \rangle \omega_h \quad \text{(since } \langle \alpha_{gh}, \beta_{gh} \rangle \text{ is a number)}$$

$$= \langle L_g^* \alpha_{gh}, L_g^* \beta_{gh} \rangle \omega_h = \langle \alpha_h, L_g^* \beta_{gh} \rangle \omega_h = \alpha_h \wedge *L_g^* \beta_{gh}$$

as desired. Similarly for right translations. □

**Lemma:** *Bi-invariant forms are harmonic in the bi-invariant metric.*

PROOF: If $\beta$ is bi-invariant then $\beta$ is closed, $d\beta = 0$. From our previous lemma, $*\beta$ is also bi-invariant; for example, $L_g^* * \beta_{gh} = *\beta_g$ shows that $*\beta$ is left invariant. Then $d*\beta = 0$, showing that $\beta$ is harmonic. □

(21.12) will then be proved when we show

**Lemma:** Harmonic forms in the bi-invariant metric are bi-invariant if $G$ is connected.

PROOF: First note that a left (right) translate of a harmonic form is harmonic, since $d(L_g^* h) = L_g^* dh = 0$ and $d(*L_g^* h) = dL_g^* *h = L_g^* d*h = 0$, because $*h$ is also harmonic. We claim that if $G$ is connected then in fact $L_g^* h_{gk} = h_k$, and so on. To see this, we need only show that both $h$ and $L_g^* h$ have the same periods; see Corollary (14.27). Let $z$ be a cycle on $G$ and let $g(t)$ be a curve in $G$ joining $e = g(0)$ with $g = g(1)$. Then

$$\int_z L_g^* h = \int_{gz} h$$

But $\{g(t)z\}$, for $0 \le t \le 1$ defines a *deformation* of $z = g(0)z$ into $gz = g(1)z$; thus these cycles are *homologous*, $gz - z = \partial c$, by the deformation theorem (13.21), and since $h$ is closed

$$\int_{gz} h = \int_z h + \int_{\partial c} h = \int_z h$$

as desired. □

## 21.1e. Weyl and Cartan on the Betti Numbers of $G$

The **center** of a group $G$ is the subgroup of elements that commute with all elements of the group. For example, the center of $U(n)$ is the 1-parameter subgroup $e^{i\theta} I$, whereas the center of $SU(n)$ consists of the $n$ scalar matrices $\lambda I$, where $\lambda$ is an $n^{\text{th}}$ root of unity.

**Weyl's Theorem (21.13):** *Let $G$ be a compact connected group. Then the first Betti number vanishes, $b_1(G) = 0$, iff the center of $G$ does not contain any 1-parameter subgroup.*

(In particular, $b_1 = 0$ for $SU(n)$ but not for $U(n)$.)

PROOF: Suppose first that the center of $G$ contains a 1-parameter group $e^{t\mathbf{X}}$, where $\mathbf{X}\in\mathfrak{g}$. Then $e^{t\mathbf{X}}g = ge^{t\mathbf{X}}$ for all $g$ in $G$. Differentiate with respect to $t$ and put $t = 0$, yielding $\mathbf{X}g = g\mathbf{X}$. Then the left invariant vector field $\mathbf{X}_g = L_{g*}\mathbf{X} = g\mathbf{X}$ on $G$ is also right invariant, and thus bi-invariant. In terms of a bi-invariant Riemannian metric on $G$, the covariant version of $\mathbf{X}$, that is, the 1-form $\alpha$ defined by $\alpha(\mathbf{Y}) = \langle \mathbf{X}, \mathbf{Y}\rangle$, is bi-invariant and hence harmonic. By Hodge's theorem $b_1 \geq 1$.

Suppose $b_1 \neq 0$. In a bi-invariant metric, there is then a harmonic, hence bi-invariant 1-form $\alpha \neq 0$. Its contravariant version is then a bi-invariant vector field $\mathbf{X}$, that is, $g\mathbf{X}_e = \mathbf{X}_e g$. Thus for all real $t$, $gt\mathbf{X}_e g^{-1} = t\mathbf{X}_e$. Then $\exp(t\mathbf{X}_e) = \exp(gt\mathbf{X}_e g^{-1}) = g\exp(t\mathbf{X}_e)g^{-1}$. Thus $\exp(t\mathbf{X}_e)$ is in the center of $G$. □

Since the center of $SO(3)$ consists only of the identity, Weyl's theorem yields $b_1 = 0$ for $G = SO(3)$. Of course we knew this from (13.25) and the fact that $SO(3)$ is topologically $\mathbb{R}P^3$. Although the first Betti number vanishes, $SO(3)$ is not simply connected. We shall see in Section 21.4 that a strengthening of this version of Weyl's theorem will yield information about the contractibility of closed curves in groups.

The following plays an important role in gauge theories, as we shall see in Section 22.1.

**Cartan's Theorem (21.14):** *If $G$ is a compact nonabelian Lie group, then the Cartan 3-form*

$$\Omega_3 = \operatorname{tr} g^{-1}dg \wedge g^{-1}dg \wedge g^{-1}dg$$

*is a nontrivial harmonic form. In particular $b_3(G) \neq 0$.*

PROOF: $\Omega_3$ is bi-invariant, hence harmonic, and $\Omega_3(\mathbf{X}, \mathbf{Y}, \mathbf{Z}) = -3\langle[\mathbf{X}, \mathbf{Y}], \mathbf{Z}\rangle$. We need only show that it is not identically 0. But the only way $\langle[\mathbf{X}, \mathbf{Y}], \mathbf{Z}\rangle$ can be 0 for all $\mathbf{Z}$ is if $\mathbf{XY} - \mathbf{YX} = [\mathbf{X}, \mathbf{Y}] = 0$ for all $\mathbf{X}$ and $\mathbf{Y}$ in $\mathfrak{g}$. But then, since $\mathbf{X}$ and $\mathbf{Y}$ commute, the power series shows

$$e^{\mathbf{X}}e^{\mathbf{Y}} = e^{\mathbf{X}+\mathbf{Y}} = e^{\mathbf{Y}}e^{\mathbf{X}}$$

In a compact connected group each $g \in G$ is an exponential, and so $G$ is abelian. □

Finally note the component form of $\Omega_3$. Let $\mathbf{e}$ be *any* left invariant basis and let $\sigma$ be the dual basis. In Problem 21.1(1) you are asked to show that

$$(\Omega_3)_{ijk} = -3C_{kij} = -3C_{ijk}$$

and thus

$$\Omega_3 = -\frac{1}{2}C_{ijk}\sigma^i \wedge \sigma^j \wedge \sigma^k$$

where $C^l_{ij}$ are the structure constants and where $C_{kij} := g_{kl}C^l_{ij}$. *When we use the bi-invariant metric tensor to lower the top index of the structure constant symbol, the resulting coefficients* $C_{kij}$ *are skew symmetric in all indices*, not just $i$ and $j$! This need not hold when $G$ is not compact.

## Problem

**21.1(1)** Compute the preceding component form of $\Omega_3$.

## 21.2. The Fundamental Group and Covering Spaces

In what sense does the torus cover the Klein bottle?

### 21.2a. Poincaré's Fundamental Group $\pi_1(M)$

Let $\gamma$ be a closed curve on a *connected* space $M$ that begins and ends at a given **base point** $p_0$. Such a curve can either be considered as a map of a circle into $M$ (that passes through $p_0$) or as a map $\gamma : [0, 1] \to M$ with $\gamma(0) = p_0 = \gamma(1)$. The latter seems more convenient. Consider now *all* such maps *with the same base point*. We shall identify two such "loops" $\gamma_1 = \gamma_1(\theta)$ and $\gamma_2 = \gamma_2(\theta)$, saying they are **homotopic**,

$$\gamma_1 \sim \gamma_2$$

provided they are homotopic via a homotopy that *preserves the base point*; thus there is an $F : [0, 1] \times [0, 1] \to M$, $F = F(\theta, t)$, with $F(0, t) = p_0 = F(1, t)$ for all $0 \leq t \leq 1$, and $F(\theta, 0) = \gamma_1(\theta)$, $F(\theta, 1) = \gamma_2(\theta)$. $t$ is the deformation parameter. We talked about this notion in Section 10.2d. If $\gamma$ is homotopic to a constant, we say $\gamma$ is **trivial** and write $g \sim 1$.

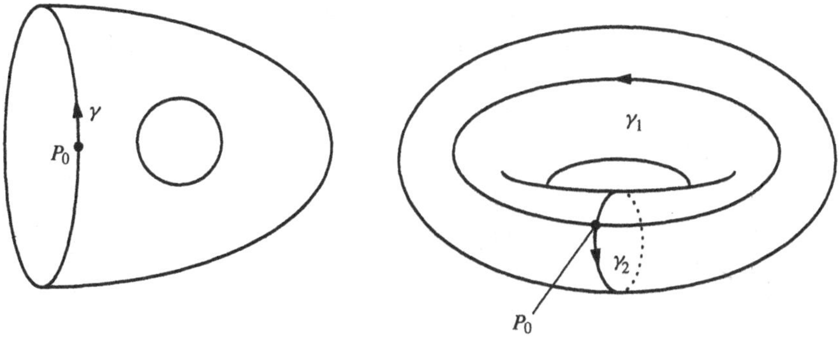


**Figure 21.1**

Note that in the left-hand figure, the loop $\gamma$ is *not* trivial as far as homotopy is concerned (try to contract it to the point $p_0$!) even though it *is* trivial in homology (it *is* the boundary of an orientable surface).

Given two loops $\gamma_1$ and $\gamma_2$ on $M$, by reparameterization (so that each loop is traversed with double speed) we may compose them to give a new loop, which is *traditionally written from left to right*

$$\gamma_1\gamma_2(\theta) := \gamma_1(2\theta), \quad \text{for } 0 \le \theta \le \frac{1}{2}$$
$$:= \gamma_2(2\theta - 1) \quad \text{for } \frac{1}{2} \le \theta \le 1$$

One can show that if $\gamma_1' \sim \gamma_1$ and if $\gamma_2' \sim \gamma_2$, then $\gamma_1'\gamma_2' \sim \gamma_1\gamma_2$. The homotopy classes of loops on $M$ form a group under "multiplication"

$$(\gamma_1, \gamma_2) \to \gamma_1\gamma_2$$

This is the **fundamental group** of $M$, written $\pi_1(M; p_0)$. It turns out that in a certain sense the resulting group is in fact independent of the base point, and one simply writes

$$\pi_1(M)$$

The identity 1 in this group is the homotopy class of the trivial loop (contractible to $p_0$). The inverse to a loop $\gamma$ is the same loop traversed in the opposite direction, $\gamma^{-1}(\theta) := \gamma(1 - \theta)$.

A space is **simply connected** if all loops are contractible to a point, that is, if the group $\pi_1(M)$ consists only of the identity.

Consider loops on the circle $M^1 = S^1$, and the resulting $\pi_1(S^1)$. These are homotopy classes of maps $\gamma : S^1 \to S^1$. We know that homotopic maps of the circle into itself have the same (Brouwer) degree; see Corollary (8.19). It can also be shown, though it is more difficult, that *maps of $S^1$ into itself having the same degree are homotopic.* Thus a loop $\gamma$ is characterized, as far as homotopy is concerned, by its degree (i.e., an integer). Since the map $\theta \to n\theta$ has degree $n$, we have

$$\pi_1(S^1) = \mathbb{Z}$$

It can be shown that the fundamental group of the 2-torus is generated by the familiar $A$ and $B$ of Figure 21.2. Briefly, any loop in the rectangle can be deformed (pushed) out to the edge. $\pi_1(T^2)$ is abelian because it is clear that the loop $A$ followed by $B$

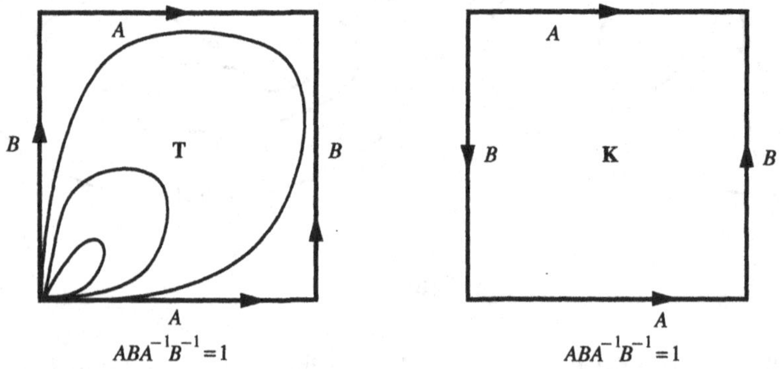


Figure 21.2

followed by $A^{-1}$ followed by $B^{-1}$, being a loop going around the edge of the rectangle, is contractible to $p_0$, that is, homotopic to the constant map; $ABA^{-1}B^{-1} = 1$ or $AB = BA$. *Thus $\pi_1(T^2)$ is the abelian group with generators $A$* and $B$.

For the Klein bottle $K$, on the other hand, we have, from Figure 21.2, $AB = B^{-1}A$. We say that $\pi_1(K)$ is the (nonabelian) group with 2 generators and the single relation $ABA^{-1}B = 1$.

The rotation group in the plane, $SO(2)$, is topologically $S^1$. $\pi_1\{SO(2)\} = \mathbb{Z}$. The rotation group in space, $SO(3)$, is topologically $\mathbb{R}P^3$

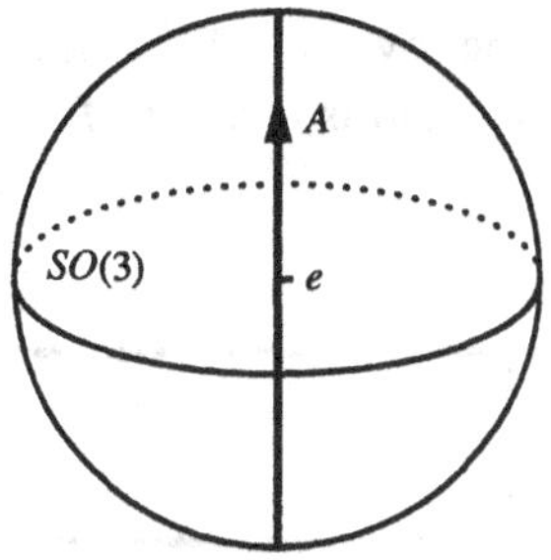


Figure 21.3

The 1-parameter subgroup $A$ of rotations about the $z$-axis is not contractible, $A \neq 1$, but $A^2 = AA = 1$; see Section 19.2a. Thus

$$\pi_1\{SO(3)\} = \mathbb{Z}_2 A \tag{21.15}$$

As we also have seen in Section 19.2a, that is why spinors can exist!

### 21.2b. The Concept of a Covering Space

We have discussed the notion of covering space informally several times in this book; now we shall need to be a bit more systematic.

> We shall say that a *connected* space $\overline{M}$ is a **covering** of the connected $M$, with covering or projection map $\pi : \overline{M} \to M$, if each $x \in M$ has a neighborhood $U$ such that the preimage $\pi^{-1}(U)$ consists of *disjoint* open subsets $\{U_\alpha\}$ of $\overline{M}$, each *diffeomorphic*, under $\pi : U_\alpha \to U$, with $U$.

We illustrate this in the case $\overline{M} = \mathbb{R}$; $M = S^1$ is the unit circle in the complex plane, and $\pi$ is the map $\pi(x) = \exp(2\pi i x)$

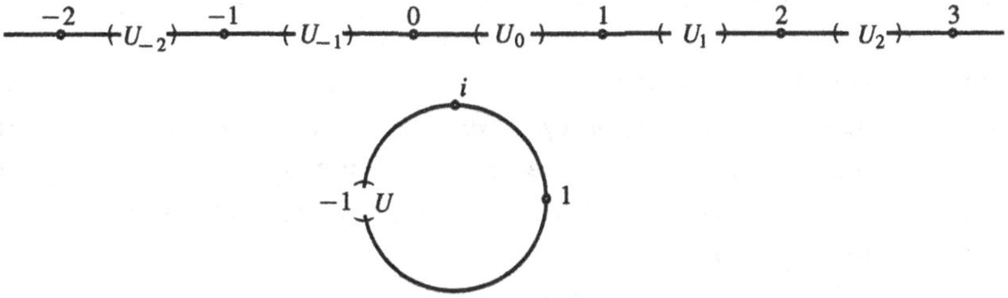


Figure 21.4

We have indicated a neighborhood $U$ of $i \in S^1$ and the preimages of $U$ in $\mathbb{R}$.

The notion of covering space can also be described in terms of fiber bundles as follows:

A *covering space* of a manifold $M$ is a *connected* space $\overline{M}$ that is a fiber bundle over $M$ with fiber $F$ a discrete set of points.

If $F$ has $k$ points we say that $\overline{M}$ is a $k$-fold or *$k$-sheeted* cover of $M$. Thus $\mathbb{R}$ is an infinite fold cover of $S^1$. The "fiber over $1 \in S^1$" is the infinite set of integers in $\mathbb{R}$.

The *edge* of a (finite) Möbius band is a circle $\overline{M} = S$ that is a 2-fold cover of the central *circle* $M$ of the band

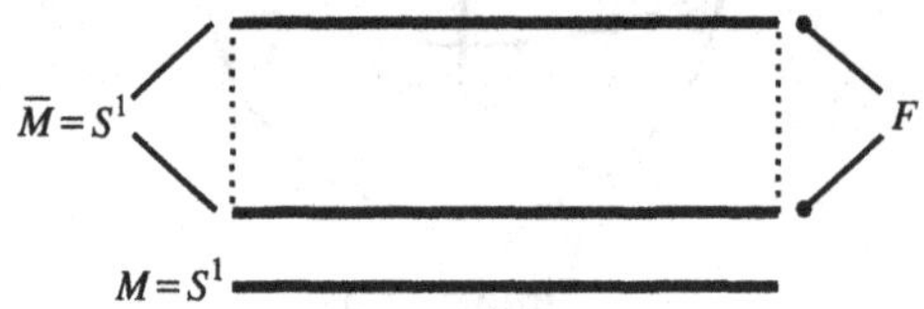


**Figure 21.5**

The $n$-sphere $S^n$ is a 2-fold cover of the projective $P^n(\mathbb{R})$. $SU(2)$ is a 2-fold covering space of $SO(3)$. $\mathbb{R}^n$ is an $\infty$-fold cover of the $n$-torus

$$T^n = S^1 \times \ldots \times S^1 \subset \mathbb{C}^n$$

$$\pi(x_1, \ldots, x_n) = (\exp[2\pi i x_1], \ldots, \exp[2\pi i x_n])$$

We shall now indicate how one can construct, in several ways, interesting covering spaces $\overline{M}$ for any manifold $M$ that is not simply connected. (It will turn out that a simply connected $M$ will have $M$ itself as its only covering.)

### 21.2c. The Universal Covering

Let $M^n$ be a connected manifold. The **universal covering manifold** $\overline{M}^n$ of $M^n$ is constructed as follows: Pick a base point $p_0$ in $M$. A point of this new space $\overline{M}$ is then defined to be an equivalence class of pairs $(p, \gamma)$, where $p$ is a point in $M$ and $\gamma : [0, 1] \to M$ is a path in $M$ starting at $p_0$ and ending at $p$, and where $(p, \gamma)$ is equivalent to $(p_1, \gamma_1)$ iff $p = p_1$ and the paths $\gamma$ and $\gamma_1$ are *homotopic*. This last requirement means simply that the closed path $\gamma\gamma_1^{-1}$ consisting of $\gamma$ followed by the reversal of $\gamma_1$ is deformable to the point $p_0$. We then automatically have a covering map $\pi : \overline{M} \to M$ defined by assigning to the pair $p, \gamma$ the endpoint $p = \gamma(1)$. To give a manifold structure to $\overline{M}$ we need to describe the local coordinate systems; we shall do this after the following simple example.

We illustrate all this with $M$ a 2-torus.

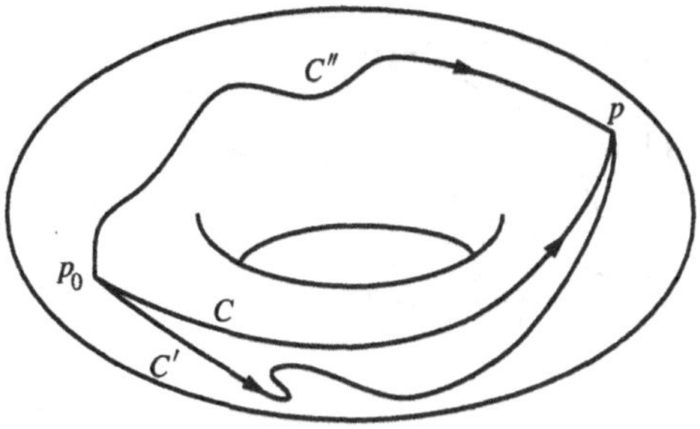


**Figure 21.6**

The curves $C$ and $C'$ are homotopic, but neither is homotopic to $C''$. Thus in our new space $\overline{M}$, the universal cover of $T^2$, the pair $p$, $C$ and $p$, $C'$ will define the same point $\overline{p}$ (to be described shortly) but $p$, $C''$ will be represented by a different point $\overline{p}'$.

In the general case, we need to describe the manifold structure of $\overline{M}$. We define a coordinate neighborhood of the pair $p, \gamma$ on $\overline{M}$ by first taking a *simply connected* coordinate neighborhood $U$ of $p$ on $M$. Then to a point $q$ in $U$ we assign a curve consisting of the given $\gamma$ followed by an arc $\gamma_{pq}$ in $U$ from $p$ to $q$. The homotopy class of $\gamma\gamma_{pq}$ is independent of the arc $\gamma_{pq}$ chosen since all arcs from $p$ to $q$ in $U$ are homotopic as a result of the simple connectivity of $U$. Then a "lifted" neighborhood $\overline{U}$ of $p, \gamma$ in $\overline{M}$, *by definition*, consists of the classes of all such curves $\gamma\gamma_{pq}$ for all $q$ in $U$. This is illustrated in the toral case that follows.

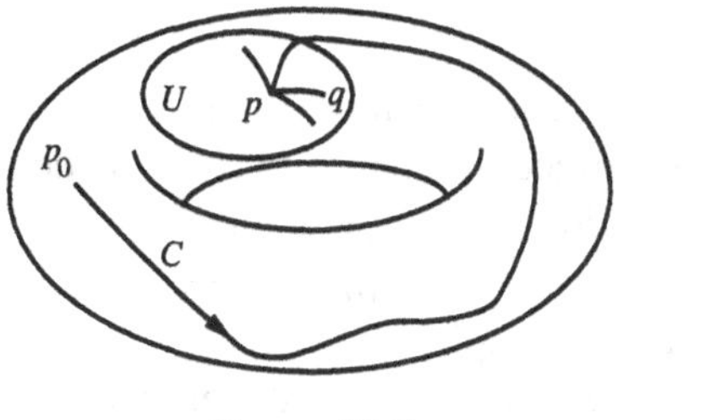


**Figure 21.7**

Since a pair $q, \gamma\gamma_{pq}$ is completely determined, *up to homotopy*, by the endpoint $q$, the points of $\overline{U}$ described are in 1 : 1 correspondence with the points $q$ in $U$. Since $U$ *is* a coordinate patch on $M$, we have succeeded in introducing local coordinates in the set $\overline{U}$; *the local coordinates of* $q, \gamma\gamma_{pq}$ *in* $\overline{M}$ *are simply the local coordinates in U of* $q$! We do this for all $p$ in $M$. By this construction, the map $\pi : \overline{M} \to M$ is such that *each* $\pi : \overline{U} \to U$ *is a diffeomorphism.*

Because $\pi : \overline{M} \to M$ is locally a diffeomorphism, any Riemannian metric in $M$ can be lifted by $\pi$ to yield a Riemannian metric in $\overline{M}$, since the local coordinates in $M$ yield the "same" local coordinates in $\overline{M}$. By this construction, $\pi$ is also a *local isometry*, and of course the curvatures coincide at $\overline{p}$ and $\pi(\overline{p})$.

Let us verify that *the universal cover of the torus* $T^2$ *is the plane* $\mathbb{R}^2$. To simplify our pictures, we shall consider new curves on the torus and illustrate with these.

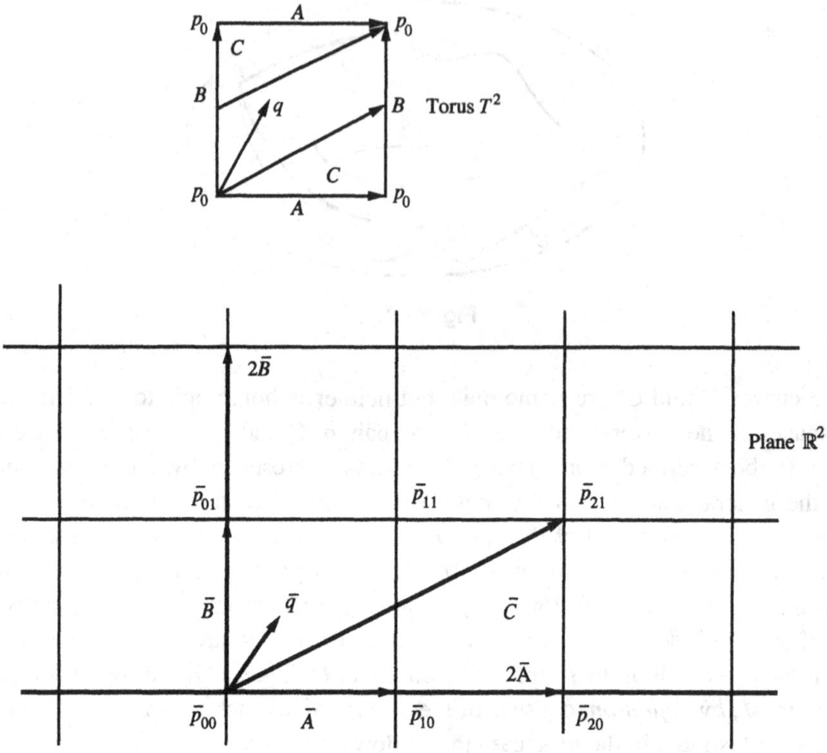


**Figure 21.8**

In the upper diagram we have drawn the torus in the usual way as a unit rectangle with opposite sides $A$ identified and opposite sides $B$ identified. We have drawn the closed curves $A$ and $B$ starting at the base point $p_0$.

In the lower $\mathbb{R}^2$ diagram the point $\overline{p}_{00}$ corresponds to the pair $p_0\gamma$ where $\gamma$ is the constant path whose locus is simply the point $p_0$. We know that a simply connected patch around $p_0$ in $T$ will be in $1:1$ correspondence with a patch around $\overline{p}_{00}$.

As we move along the curve $A$ from $p_0$ we also trace out a curve $\overline{A}$ starting out at $\overline{p}_{00}$. On the completion of $A$ in $T$ we return to the point $p_0$ again. Since, however, the curve $A$ in $T$ is not homotopic to the constant curve $p_0$, the pair $(p_0, \gamma = p_0)$ is not equivalent to the pair $p_0 A$. This means that the endpoint $\overline{p}_{10}$ of $\overline{A}$ *is not to be identified with its beginning point* $\overline{p}_{00}$! For the same reason, the vertical line through $\overline{p}_{10}$ is not to be identified with that through $\overline{p}_{00}$. Likewise, if one goes around $A$ twice on $T$, in $\overline{T}$ we end not at $\overline{p}_{00}$ nor at $\overline{p}_{10}$ but rather at a new point $\overline{p}_{20}$. The same procedure shows that on going around $B$ we trace out a curve $\overline{B}$ that ends at a new point $\overline{p}_{01}$, and so forth. We have also illustrated the case of a closed curve $C$ in $T$ that wraps twice around in the $A$ sense and once in the $B$ sense; its lift $\overline{C}$ in $\overline{T}$ ends at the point $\overline{p}_{21}$. We also know, by definition, that *any curve in* $\overline{T}$ *that starts at* $\overline{p}_{00}$ *and ends at* $\overline{p}_{21}$ *represents* (i.e., projects down via $\pi$ to) *a closed curve in* $T$ *that is homotopic to* $C$!

Thus although $T$ can be considered the plane with identifications $(x, y) \sim (x + n, y + m)$, the universal cover $\overline{T}$ is the plane without identifications, that is, $\overline{T} = \mathbb{R}^2$.

Note that when $M$ is simply connected, then by construction *its universal cover $\overline{M}$ coincides with $M$ itself*, since any pair of curves from $p_0$ to a point $p$ are homotopic.

## 21.2d. The Orientable Covering

This is the covering one obtains by using the same method as in the universal cover except that we now say that *a pair $p, \gamma$ is equivalent to a pair $p, \gamma_1$ iff when we transport an orientation from $p_0$ to $p$ along $\gamma_1$ we obtain the same orientation as along $\gamma$*, that is, if when we translate an orientation along the closed curve $\gamma\gamma_1^{-1}$ we return with the original orientation. As in the construction of the universal cover, it is important that we are dealing with homotopy classes; if a closed curve $C$ preserves orientation, and if $C'$ is homotopic to $C$, then $C'$ will also preserve orientation. If $M$ is orientable, then the covering obtained reduces to $M$ itself, but if $M$ is not orientable we obtain a new space $\overline{M}$. In any case $\overline{M}$ is called the orientable cover of $M$, for, as we shall see, this $\overline{M}$ is always orientable.

Consider, for instance, the Klein bottle, considered as a rectangle with the twisted identifications on the vertical sides

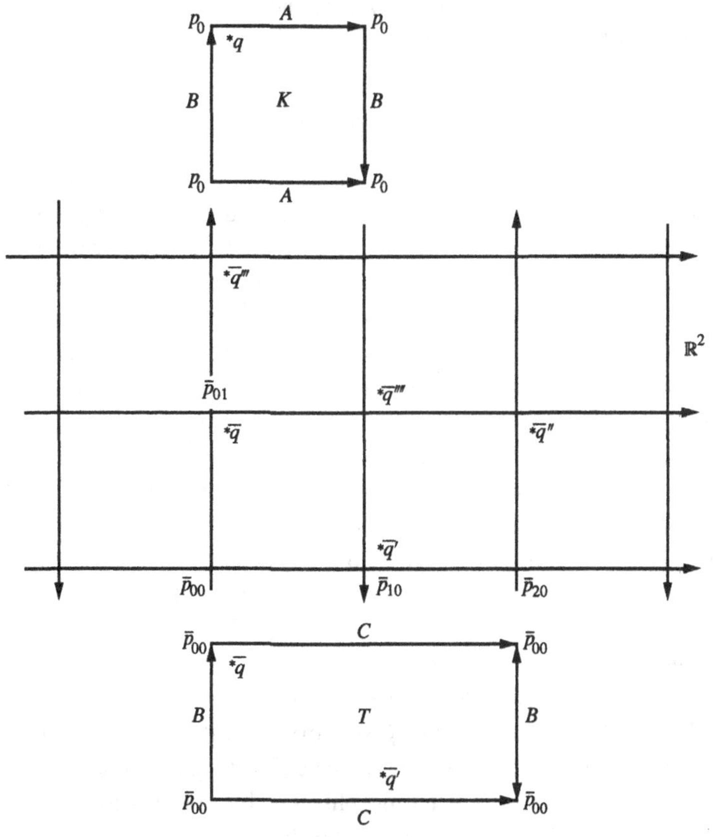


Figure 21.9

In the second diagram we have indicated how one can view the Klein bottle as the plane with twisted identifications; the point $q$ in $K$ corresponds to all of the points $\overline{q}, \overline{q}', \overline{q}'', \ldots$, in the plane.

As we move along the curve $A$ in $K$ starting at $p_0$, it is equivalent to moving along the segment $\overline{p}_{00}\overline{p}_{10}$ in $\mathbb{R}^2$. When we reach the point $p_0$ again in $K$ we note that we have traversed a closed path $A$ in $K$ along which the orientation has been reversed. This means that in our $\mathbb{R}^2$ picture of $K$, the point $\overline{p}_{10}$ is not to be identified with $\overline{p}_{00}$ in our model for this new covering space. If, however, we traverse the curve $A$ twice, the orientation is preserved; thus in the $\mathbb{R}^2$ picture the point $\overline{p}_{20}$ *is* to be identified with $\overline{p}_{00}$, but not to $\overline{p}_{10}$. On the other hand, $\overline{p}_{30}$, corresponding to traversing $A$ three times, is to be identified with $\overline{p}_{10}$, and so on.

On traversing $B$ the orientation is preserved; hence $\overline{p}_{01}$ is still to be identified with $\overline{p}_{00}$. It will then follow that in this new covering $\overline{K}$, horizontal lines are to be identified if they are separated by multiples of 1 unit, whereas vertical lines are to be identified *(without twisting)* if they are separated by multiples of 2 units. If we make such identifications in $\mathbb{R}^2$ we see that the resulting space is simply a torus $T$ of twice the area of $K$. *The two-sheeted-orientable cover of the Klein bottle is the torus*! We have drawn the torus in the last figure as a rectangle with the usual identifications on the boundary, and no other identifications, $\overline{q} \neq \overline{q}'$. $C$ is the closed curve that covers $A$ twice.

By the same arguments, it can be shown in general that the orientable cover of $M$ is either $M$ itself, if $M$ is orientable, or a 2-sheeted cover of $M$.

## 21.2e. Lifting Paths

Let $\pi : \overline{M} \to M$ be any covering of the manifold $M$. $\overline{M}$ and $M$ are locally diffeomorphic under the map $\pi$. The fiber over $p$, $\pi^{-1}(p)$, is a disconnected set of points. (It is useful to keep in mind the examples of the universal covering $\mathbb{R}^2$ over $T^2$, with fiber an infinite set of points, and the orientable cover $T^2$ over the Klein bottle $K^2$ with fiber a pair of points.) Let $\overline{p}$ be any point in this fiber. Let $C$ be a curve in $M$ starting at some $p$ and ending at some $q$. Since $M$ and $\overline{M}$ are locally diffeomorphic, there is a *unique* curve $\overline{C}$ that starts at $\overline{p}$ and covers $C$, $\pi(\overline{C}) = C$. Its endpoint $\overline{q}$ by construction is at some point in the fiber $\pi^{-1}(q)$. This defines the *lift* of $C$ to $\overline{M}$ that starts at $\overline{p}$.

If $C$ is closed, $q = p$, it may be that $\overline{C}$ is *not* closed; that is, it may be that $\overline{q} \neq \overline{p}$. This occurs in the universal covering iff the closed curve $C$ is not homotopic to the constant curve $p$; in the orientable cover it occurs when $C$ is a curve that reverses orientation. These follow essentially from the definitions of these covers. (In our definitions we based everything at a base point $p_0$, but it is not hard to see that we get similar behavior if we choose a new base point $p$.)

Consider now the case of the universal cover. Let $\overline{\gamma}$ be *any* closed curve *in* $\overline{M}$ that starts and ends at $\overline{p}$. It projects down to a closed curve $\gamma = \pi(\overline{\gamma})$ starting and ending at $p = \pi(\overline{p})$. Since the closed curve $\overline{\gamma}$ is a lift of $\gamma$, it must be that the curve $\gamma$ is homotopic to the constant map $p$ in $M$. As we deform $\gamma$ to the point $p$ we may cover this deformation, using the local diffeomorphism $\pi$, by a deformation of $\overline{\gamma}$ to the point $\overline{p}$. We have thus shown that $\overline{M}$ is simply connected.

Furthermore, by definition of the universal cover, the points of the fiber $\pi^{-1}(P_0)$ are in 1 : 1 correspondence with the distinct homotopy classes of closed curves in $M$ starting at $p_0$. Summarizing, we have shown

**Theorem (21.16):** *The universal cover $\overline{M}$ of $M$ is simply connected and the number of sheets in the covering is equal to the number of elements (the* **order***) of $\pi_1(M)$.*

If a manifold is not orientable, there is some closed curve that reverses orientation. By the same type of reasoning as in (21.16) we have the following explanation of the terminology that we have been using:

**Theorem (21.17):** *The orientable cover of $M$ is always orientable. The number of sheets is 1 if $M$ is orientable and 2 if $M$ is not orientable.*

## 21.2f. Subgroups of $\pi_1(M)$

The orientable cover of $M$ resulted from identifying two curves $\gamma$ and $\gamma_1$ from $p_0$ to $p$ iff the closed curve $\gamma\gamma_1^{-1}$ preserves orientation, that is, if the homotopy class of $\gamma\gamma_1^{-1}$ lies in the subgroup of $\pi_1(M)$ consisting of orientation preserving loops. Similarly, given *any subgroup* $G$ of $\pi_1(M)$, we may associate a covering space $M_G$ of $M$ as follows: We again consider pairs $p, \gamma$, and we identify $p, \gamma$ with $p, \gamma_1$ iff the homotopy class of the loop $\gamma\gamma_1^{-1}$ lies in the subgroup $G$. For example, when $G$ is the identity 1 of $\pi_1(M)$, the covering is the universal cover, whereas if $G$ is the subgroup of orientation-preserving loops the cover is the orientable cover.

## 21.2g. The Universal Covering Group

Let $\pi : \overline{G} \to G$ be the universal covering space of a Lie group $G$. We shall indicate why it is that $\overline{G}$ *itself is then a Lie group*! For example, $SU(2)$, being a simply connected cover of $SO(3)$, is the universal covering group of $SO(3)$. A simpler example is furnished by $\exp : \mathbb{R} \to S^1$ sending $\theta \in \mathbb{R}$ to $e^{i\theta}$. This is a homomorphism of the additive group of real numbers onto the multiplicative group of unit complex numbers. We have already seen in Section 21.2b that this makes $\mathbb{R}$ a covering manifold for $S^1$. Since $\mathbb{R}$ is simply connected, it is the universal covering group of $S^1$.

For identity in $\overline{G}$ we pick any point $\overline{e} \in \pi^{-1}(e)$ in the fiber over the identity $e$ of $G$. If $\overline{g}$ is any point in $\overline{G}$ we define $\overline{g}^{-1}$ as follows: $\overline{g}$ can be represented by a path $g(t)$ in $G$ joining the base point $e$ to the point $g(1) = g := \pi(\overline{g})$. Then the inverse path $g^{-1}(t)$ joins $e$ to $g^{-1}$. This path can be covered by a *unique* path in $\overline{G}$ that starts at $\overline{e}$. It ends at some point in $\pi^{-1}(g^{-1})$ and we define this point to be $\overline{g}^{-1}$.

Let $\overline{g}$ and $\overline{h}$ be points in $\overline{G}$; they can be represented by paths $C_g$ and $C_h$ joining $e$ to $g \in \pi(\overline{g})$ and to $h \in \pi(\overline{h})$. Consider the path $C_g$ followed by the left translate $gC_h$;

since $gC_h$ starts at $g$ and ends at $gh$, the composite path starts at $e$ and ends at $gh$. Its unique lift that starts at $\overline{e}$ ends in $\pi^{-1}(gh)$. This endpoint is defined to be $\overline{gh}$.

These basic constructions can be shown to yield the required universal covering group (see, e.g., [P, chapter viii]).

Note that since we may lift the Lie algebra $\mathfrak{g} = G(e)$ uniquely to $\overline{e}$, *the universal cover of G has the same Lie algebra as G.*

## 21.3. The Theorem of S. B. Myers: A Problem Set

A spiral curve in the plane can have curvature $\geq 1$ and infinite length. Can a surface in space have Gauss curvature $\geq 1$ and infinite area?

Let $M^n$ be a Riemannian manifold and consider a geodesic $C$ joining $p$ to $q$. Then the first variation of arc length vanishes, $L'(0) = 0$, for all variations whose variation vector $\mathbf{J} = \partial\mathbf{x}/\partial\alpha$ is *orthogonal* to $\mathbf{T}$. Consider the second variation in this case, as given by Synge's formula (12.6)

$$L''(0) = \langle \nabla_{\mathbf{J}}\mathbf{J}, \mathbf{T}\rangle_0^L + \int_0^L \{\| \nabla_{\mathbf{T}}\mathbf{J} \|^2 - \langle R(\mathbf{J}, \mathbf{T})\mathbf{T}, \mathbf{J}\rangle\} ds$$

We shall construct $(n-1)$ such variations as follows: Let $\mathbf{e}_2, \mathbf{e}_3, \ldots, \mathbf{e}_n$ be orthonormal vector fields that are parallel displaced along $C$ and orthogonal to $C$; this is possible since $\mathbf{T}$ is parallel displaced also. Define the $(n-1)$ variation vectors

$$\mathbf{J}_i(s) := f(s)\mathbf{e}_i(s)$$

where $f$ is a smooth function that *vanishes at the endpoints* $p$ and $q$. We may put $\mathbf{e}_1 := \mathbf{T}$ and use the $\mathbf{e}$'s as a basis along $C$.

**21.3(1)** Show that for $i = 2, \ldots, n$ we have for the $i^{\text{th}}$ variation vector

$$L_i''(0) = \int_0^L \{|f'(s)|^2 - |f(s)|^2 R^i{}_{1i1}\} ds$$

and

$$\sum_{i=2}^n L_i''(0) = \int_0^L \sum_{i=2}^n |f'(s)|^2 ds - \int_0^L |f(s)|^2 \text{Ric}(\mathbf{T}, \mathbf{T}) ds$$

Suppose now that the Ricci curvature is positive

$$\text{Ric}(\mathbf{T}, \mathbf{T}) \geq c > 0$$

and choose for variation function $f(s) = \sin(\pi s/L)$.

**21.3(2)** Show that

$$\sum_{i=2}^{n} L_i''(0) \le \frac{L}{2}\left[\frac{\pi^2(n-1)}{L^2} - c\right]$$

and conclude then that if the geodesic $C$ has length $L$ such that

$$L > \pi\left[\frac{(n-1)}{c}\right]^{1/2}$$

then $C$ can *not* be a length-minimizing geodesic from $p$ to $q$.

**21.3(3)** What does this say for the round $n$-sphere of radius a in $\mathbb{R}^{n+1}$?

Now suppose that $M$ is geodesically complete; the theorem of Hopf–Rinow (21.10) states that between *any* pair of points there is a minimizing geodesic. Let us say that a geodesically complete manifold has **diameter** $\Delta$ if any pair of points can be joined by a geodesic of length $\le \Delta$ and for some pair $p, q$ the minimizing geodesic has length exactly $\Delta$. We have proved

**Theorem of S. B. Myers (21.18):** *A geodesically complete manifold $M^n$ whose Ricci curvature satisfies*

$$Ric(\mathbf{T}, \mathbf{T}) \ge c > 0$$

*for all unit $\mathbf{T}$ has diameter $\le \pi[(n-1)/c]^{1/2}$.*

**Corollary (21.19):** *A geodesically complete $M^n$ with $Ric(\mathbf{T}, \mathbf{T}) \ge c > 0$ is a closed (compact) manifold. In particular its volume is finite.*

(In the case of 2 dimensions, $\mathrm{Ric}(\mathbf{T}, \mathbf{T}) = K$ is simply the Gauss curvature. The 2-dimensional version was proved by Bonnet in 1855.)

PROOF: For a given $p$ in $M$ the exponential map $\exp_p : M_p \to M$ is a smooth map of all of $\mathbb{R}^n$ into $M$, since $M$ is complete. By Myers's theorem the closed ball of radius $r > \pi[(n-1)/c]^{1/2}$ in $M(p)$ is mapped onto all of $M$. This closed ball is a compact subset of $\mathbb{R}^n$ and its image is again compact. □

**21.3(4)** The paraboloid of revolution $z = x^2 + y^2$ clearly has positive curvature (and can be computed from Problem 8.2(4)) and yet is not a closed surface. Reconcile this with (21.19).

Now let $M^n$ be geodesically complete with $\mathrm{Ric}(\mathbf{T}, \mathbf{T}) \ge c > 0$. It is thus compact. Let $\overline{M}$ be its universal cover. We use the local diffeomorphism $\pi : \overline{M} \to M$ to lift the metric to $\overline{M}$, and then, since $\pi$ is a local isometry, $\overline{M}$ has the same Ricci curvature. Every geodesic of $\overline{M}$ is clearly the lift of a geodesic from $M$, and so $\overline{M}$ is also geodesically complete. We conclude that $\overline{M}$ is also compact. We claim that this means that $\overline{M}$ is a *finite*-sheeted cover of $M$! Take a cover $\{U, V, \ldots\}$ of $M$ such that $U$ is the only set holding $p_0$ and $U$ is so small that it is diffeomorphic to each connected component of $\pi^{-1}(U)$. The inverse images of $U, V, \ldots$ form a cover of $\overline{M}$, where each connected component of $\pi^{-1}(U)$ is considered as a separate open set. It is clear that if $\overline{M}$ were infinite-sheeted then any subcovering of $\overline{M}$ would have to include the infinite collection in $\pi^{-1}(U)$. This contradicts the fact that $\overline{M}$ is compact. From (21.16) we have

**Myers's Corollary (21.20):** *If $M^n$ is complete with positive Ricci curvature bounded away from 0, then the universal cover of M is compact and $\pi_1(M)$ is a group of finite order.*

Thus given a closed curve $C$ in $M$, it may be that $C$ cannot be contracted to a point, but *some finite multiple* $kC$ of it *can be so contracted.* We have observed this before in the case $M = \mathbb{R}P^3$.

This should first be compared with Synge's theorem (12.12). It is stronger than Synge's theorem in that (i) $M$ needn't be compact, nor even-dimensional, nor orientable; and (ii) positive Ricci curvature $\text{Ric}(\mathbf{e}_1, \mathbf{e}_1)$, being a condition on a sum of sectional curvatures $\sum_{j>1} K(\mathbf{e}_1 \wedge \mathbf{e}_j)$, is a weaker condition than positive sectional curvature. On the other hand, Synge's conclusion is stronger, in that $\pi_1$, being finite, is a weaker conclusion than $\pi_1$ consisting of one element. Synge's theorem does not apply to $\mathbb{R}P^3$ whereas Myers's theorem does (and in fact the fundamental group here is the group with 2 elements $\mathbb{Z}_2$), but Myers's theorem tells us that even-dimensional spheres have a finite fundamental group whereas Synge tells us they are in fact simply connected.

There is a more interesting comparison with Bochner's theorem (14.33). Myers's theorem is in every way stronger. First, it doesn't require compactness; it derives it. Second, it concludes that some multiple $kC$ of a closed curve is contractible. Now in the process of contracting $kC$, $kC$ will sweep out a 2-dimensional deformation chain $c_2$ for which $\partial c_2 = kC$ see 13.3a(III), and so $C = \partial(k^{-1}c_2)$. This says that $C$ bounds as a real 1-cycle, and thus $b_1(M) = 0$. Thus Myers's theorem implies Bochner's. We have also seen in Section 21.2a that contractibility is a stronger condition than bounding, for a loop.

Although it is true that Myers's theorem is stronger than Bochner's, it has turned out that Bochner's *method*, using harmonic forms, has been generalized by Kodaira, yielding his so-called vanishing theorems, which play a very important role in complex manifold theory.

Finally, it should be mentioned that there are generalizations of Myers's theorem. Galloway [Ga] has relaxed the condition $\text{Ric}(\mathbf{T}, \mathbf{T}) \geq c > 0$ to the requirement that $\text{Ric}(\mathbf{T}, \mathbf{T}) \geq c + df/ds$ along the geodesic, where $f$ is a bounded function of arc length. $\text{Ric}(\mathbf{T}, \mathbf{T})$ need not be positive in this case in order to demonstrate compactness. Galloway uses this version of Myers's theorem to give conditions on a space–time that will ensure that the *spatial section of a space–time is a closed manifold*!

**21.3(5)** *Distance from a point to a closed hypersurface.* Let $V^{n-1}$ be hypersurface of the *geodesically complete* Riemannian $M^n$ and let $p$ be a point that does not lie on $V$. We may look at all the minimizing geodesics from $p$ to $q$, as $q$ ranges over $V$. The distance $L$ from $p$ to $V$ is defined to be the greatest lower bound of the lengths of these geodesics. Let $V$ be a *compact hypersurface without boundary*. Then it can be shown that this infimum is attained, that is, there is a point $q \in V$ such that the minimizing geodesic $C$ from $p$ to $q$ has length $L$. Parameterize $C$ by arc length $s$ with $p = C(0)$.

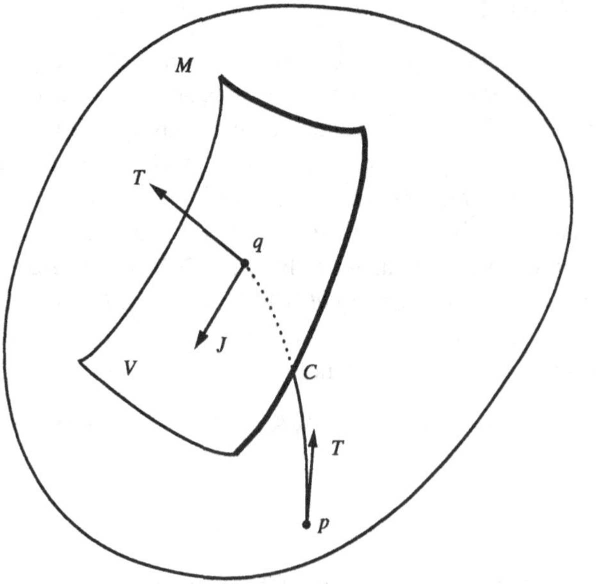


**Figure 21.10**

(i) Show from the first variation formula that $C$ strikes $V$ orthogonally. (This generalizes the result of Problem 1.3(3).)

(ii) Consider a variation vector field of the form $\mathbf{J}(s) = g(s)\mathbf{e}_2(s)$ where $\mathbf{e}_2$ is parallel displaced along $C$ and $g$ is a smooth function with $g(0) = 0$ and $g(L) = 1$. Then $L''(0)$ is of form $B(\mathbf{J}, \mathbf{J}) + \int_0^L \{|g'(s)|^2 - |g(s)|^2 R^2{}_{121}\} ds$, where $B(\mathbf{J}, \mathbf{J})$ is the **normal curvature** of $V$ at the point $q$ for direction $\mathbf{J}(L)$ and hypersurface normal $\mathbf{T}(L)$; see (11.50). By taking such variations based on $(n-1)$ parallel displaced orthonormal $\mathbf{e}_2, \ldots, \mathbf{e}_n$, all with the same $g$, and putting $g(s) = s/L$, show that

$$\sum_{i=2}^{n} L_i''(0) = H(q) + \frac{(n-1)}{L} - \left(\frac{1}{L^2}\right) \int_0^L s^2 \mathrm{Ric}(\mathbf{T}, \mathbf{T}) ds$$

where $H(q)$ is the *mean curvature* of $V$ at $q$ for normal direction $\mathbf{T}$.

(iii) Assume that $M$ has *positive Ricci curvature*, $\mathrm{Ric}(\mathbf{T}, \mathbf{T}) \geq 0$ (but we do *not* assume that it is bounded away from 0) and assume that $V$ is on the average *curving towards p* at the point $q$; that is, $h := H(q) < 0$. Show then that our *minimizing geodesic C must have length L at most* $(n-1)/h$.

In general relativity one deals with timelike geodesics that locally *maximize* proper time (because of the metric signature $-, +, +, +$). Our preceding argument is similar to analysis used there to prove the **Hawking singularity theorems,** but the pseudo-Riemannian geometry involved is really quite different from the Riemannian and forms a subject in its own right. For further discussion you may see, for example, [Wd, chaps. 8 and 9].

## 21.4. The Geometry of a Lie Group

What are the curvatures of a compact group with a bi-invariant metric?

### 21.4a. The Connection of a Bi-invariant Metric

Let $G$ be a Lie group endowed with a bi-invariant metric. (As we know from Theorem (21.8), such metrics exist on every compact group, and of course on any commutative group. The plane $G = \mathbb{R}^2$ can be considered the Lie group of translations of the plane itself; $(a, b) \in \mathbb{R}^2$ sends $(x, y)$ to $(x + a, y + b)$. This is an example of a noncompact Lie group with bi-invariant metric $dx^2 + dy^2$.)

To describe the Levi-Civita connection $\nabla_{\mathbf{X}}\mathbf{Y}$ we may expand the vector fields in terms of a left invariant basis. Thus we only need $\nabla_{\mathbf{X}}\mathbf{Y}$ in the case when $\mathbf{X}$ and $\mathbf{Y}$ are left invariant. From now on, *all vector fields* $\mathbf{X}, \mathbf{Y}, \mathbf{Z}, \dots$ *will be assumed left invariant.*

We know from Theorem (21.9) that the integral curves of a left invariant field are geodesics in the bi-invariant metric, hence $\nabla_{\mathbf{X}}\mathbf{X} = 0$. Likewise

$$0 = \nabla_{\mathbf{X}+\mathbf{Y}}(\mathbf{X} + \mathbf{Y}) = \nabla_{\mathbf{X}}\mathbf{Y} + \nabla_{\mathbf{Y}}\mathbf{X} = \nabla_{\mathbf{X}}\mathbf{Y} - \nabla_{\mathbf{Y}}\mathbf{X} + 2\nabla_{\mathbf{Y}}\mathbf{X} \tag{21.21}$$

that is,

$$2\nabla_{\mathbf{X}}\mathbf{Y} = [\mathbf{X}, \mathbf{Y}]$$

*exhibits the covariant derivative as a bracket* (but of course only for left invariant fields).

Look now at the curvature tensor

$$R(\mathbf{X}, \mathbf{Y})\mathbf{Z} = \nabla_{\mathbf{X}}\nabla_{\mathbf{Y}}\mathbf{Z} - \nabla_{\mathbf{Y}}\nabla_{\mathbf{X}}\mathbf{Z} - \nabla_{[\mathbf{X},\mathbf{Y}]}\mathbf{Z}$$

In Problem 21.4(1) you are asked to show that this reduces to

$$R(\mathbf{X}, \mathbf{Y})\mathbf{Z} = -\frac{1}{4}\,[[\mathbf{X}, \mathbf{Y}], \mathbf{Z}] \tag{21.22}$$

For sectional curvature, using (20.35),

$$\begin{aligned} -4\langle R(\mathbf{X}, \mathbf{Y})\mathbf{Y}, \mathbf{X}\rangle &= \langle [[\mathbf{X}, \mathbf{Y}], \mathbf{Y}], \mathbf{X}\rangle = -\langle \mathbf{Y}, [[\mathbf{X}, \mathbf{Y}], \mathbf{X}]\rangle \\ &= \langle \mathbf{Y}, [\mathbf{X}, [\mathbf{X}, \mathbf{Y}]]\rangle = -\langle [\mathbf{X}, \mathbf{Y}], [\mathbf{X}, \mathbf{Y}]\rangle \end{aligned}$$

or

$$K(\mathbf{X} \wedge \mathbf{Y}) = \frac{1}{4}\,\|\,[\mathbf{X}, \mathbf{Y}]\,\|^2 \tag{21.23}$$

Thus *the sectional curvature is always* $\geq 0$, *and vanishes iff the bracket of* $\mathbf{X}$ *and* $\mathbf{Y}$ *vanishes!*

For Ricci curvature, in terms of a basis of left invariant fields $\mathbf{e}_1, \dots, \mathbf{e}_n$

$$\mathrm{Ric}(\mathbf{e}_1, \mathbf{e}_1) = \sum_{j>1} K(\mathbf{e}_1 \wedge \mathbf{e}_j) = \frac{1}{4}\sum_j \|\,[\mathbf{e}_1, \mathbf{e}_j]\,\|^2$$

Thus $\mathrm{Ric}(\mathbf{X}, \mathbf{X}) \geq 0$ and $= 0$ iff $[\mathbf{X}, \mathbf{Y}] = 0$ for all $Y \in \mathfrak{g}$.

The **center** of the Lie *algebra* is by definition the set of all $\mathbf{X} \in \mathfrak{g}$ such that $[\mathbf{X}, \mathbf{Y}] = 0$ for all $\mathbf{Y} \in \mathfrak{g}$. Thus if the center of $\mathfrak{g}$ is trivial we have that the continuous function $\mathbf{X} \mapsto \text{Ric}(\mathbf{X}, \mathbf{X})$ is bounded away from 0 on the compact unit sphere in $\mathfrak{g}$ at the identity. But since the metric on $G$ is invariant under left translations, we then conclude that the Ricci curvature is positive and bounded away from 0 on all of $G$. From Myers's theorem we conclude

**Weyl's Theorem (21.24):** *Let $G$ be a Lie group with bi-invariant metric. Suppose that the center of $\mathfrak{g}$ is trivial. Then $G$ is compact and has a finite fundamental group $\pi_1(G)$.*

This improves (21.13) since it can be shown that if there is no 1-parameter subgroup in the center of $G$ then the center of $\mathfrak{g}$ is trivial; see Problem 21.4(2). Note also that the condition "the center of $\mathfrak{g}$ is trivial" is a purely algebraic one, unlike the condition for the center of the *group* appearing in Theorem (21.13).

## 21.4b. The Flat Connections

We have used the Levi-Civita connection for a bi-invariant Riemannian metric. When such metrics exist, this is by far the most important connection on the group. On *any* group we can consider the flat left invariant connection, defined as follows: Choose a basis $\mathbf{e}$ for the left invariant vector fields and define the connection forms $\omega$ to be 0, $\nabla\mathbf{e} = 0$. (There is no problem in doing this since $G$ is covered by this single frame field.) Thus we are forcing the left invariant fields to be covariant constant, and by construction the curvature vanishes, $d\omega + \omega \wedge \omega = 0$. This connection will have torsion; see Problem 21.4(3). Similarly we can construct the flat right invariant connection.

## Problems

**21.4(1)** Use the Jacobi identity to show (21.22).

**21.4(2)** Suppose that $\mathbf{X}$ is a nontrivial vector in the center of $\mathfrak{g}$; thus $ad\,\mathbf{X}(\mathbf{Y}) = 0$ for all $\mathbf{Y}$ in $\mathfrak{g}$. Fill in the following steps, using (18.32), showing that $e^{t\mathbf{X}}$ is in the center of $G$. First $e^{t\,ad\mathbf{X}}\mathbf{Y} = \mathbf{Y}$. Then $e^{t\mathbf{X}}\mathbf{Y}e^{-t\mathbf{X}} = \mathbf{Y}$. Thus $\exp(e^{t\mathbf{X}}\mathbf{Y}e^{-t\mathbf{X}}) = e^{\mathbf{Y}}$. Then $e^{t\mathbf{X}}$ is in the center of $G$.

**21.4(3)** Show that the torsion tensor of the flat left invariant connection is given by the structure constants $T^i_{jk} = -C^i_{jk}$.

CHAPTER 22

# Chern Forms and Homotopy Groups

How can we construct *closed* $p$-forms from the matrix $\theta^2$ of curvature forms?

## 22.1. Chern Forms and Winding Numbers

### 22.1a. The Yang–Mills "Winding Number"

Recall that in (20.62) and (20.63), we were comparing, on a distant 3-sphere $S^3 \subset \mathbb{R}^4$, the interior frame $\mathbf{e}_U$ with the covariant constant frame $\mathbf{e}_V$,

$$\mathbf{e}_U(x) = \mathbf{e}_V(x) g_{VU}(x)$$

$$g_{VU} : S^3 \to SU(n)$$

the gauge group being assumed $SU(n)$.

We saw in (21.14) that the Cartan 3-form on $SU(n)$

$$\Omega_3 = \operatorname{tr} g^{-1}dg \wedge g^{-1}dg \wedge g^{-1}dg$$

is a nontrivial harmonic form, and we now consider the real number obtained by pulling this form back via $g_{VU}$ and integrating over $S^3$

$$\int_{S^3} g_{VU}^*(\Omega^3) = \int_{g_{VU}(S^3) \subset SU(n)} \Omega_3 \tag{22.1}$$

We shall normalize the form $\Omega_3$; this will allow us to consider (22.1) as defining the degree of a map derived from $g_{VU}$.

Consider, for this purpose, the $SU(2)$ subgroup of $SU(n)$

$$SU(2) = SU(2) \times I_{n-2} := \begin{bmatrix} SU(2) & 0 \\ 0 & I_{n-2} \end{bmatrix} \subset SU(n)$$

The Cartan 3-form $\Omega_3$ of $SU(n)$ restricts to $\Omega_3$ for $SU(2)$, and we shall use as normalization constant

$$\int_{SU(2)} \Omega_3$$

which we proceed to compute.

$\Omega_3$ and the volume form $\mathrm{vol}^3$ on $SU(2)$, in the bi-invariant metric, are both bi-invariant 3-forms on the 3-dimensional manifold $SU(2)$; it is then clear that $\Omega_3$ is some *constant* multiple of $\mathrm{vol}^3$. From (19.9) we know that $i\sigma_1/\sqrt{2}, i\sigma_2/\sqrt{2}$, and $i\sigma_3/\sqrt{2}$ form an orthonormal basis for $\mathfrak{su}(2)$ with the scalar product $\langle X, Y\rangle = -\operatorname{tr} XY$ (recall that (19.9) defines the scalar product in $i\mathfrak{g}$, not $\mathfrak{g}$). Then, from (21.5) and (19.6)

$$\Omega_3(i\sigma_1, i\sigma_2, i\sigma_3) = 3\operatorname{tr}([i\sigma_1, i\sigma_2]i\sigma_3) = -3\operatorname{tr}(2i\sigma_3 i\sigma_3)$$
$$= 6\operatorname{tr}\sigma_3\sigma_3 = 12$$

Since the $i\sigma$'s/$\sqrt{2}$ are orthonormal, we have $\mathrm{vol}^3(i\sigma_1, i\sigma_2, i\sigma_3) = 2^{3/2}$. Thus we have shown

$$\Omega_3 = (2^{-3/2})12\,\mathrm{vol}^3 \tag{22.2}$$

What, now, is the volume of $SU(2)$ *in its bi-invariant metric*?

$SU(2)$ is the unit sphere $S^3$ in $\mathbb{C}^2 = \mathbb{R}^4$ where we assign to the $2 \times 2$ matrix $u$ its first column. The identity element $e$ of $SU(2)$ is the complex 2-vector $(1, 0)^T$ or the real 4-tuple $N = (1, 0, 0, 0)^T$. The *standard* metric on $S^3 \subset \mathbb{R}^4$ is invariant under the 6-dimensional rotation group $SO(4)$, and the stability group of the identity is the subgroup $1 \times SO(3)$. Thus $S^3 = SO(4)/SO(3)$. The standard metric is constructed first from a metric in the tangent space $S^3_N$ to $S^3$ at $N$ that is invariant under the stability group $SO(3)$ and then this metric is transported to all of $S^3$ by the action of $SO(4)$ on $SO(4)/SO(3)$. Since the stability group $SO(3)$ is *transitive* on the *directions* in $S^3_N$ at $N$, it should be clear that *this metric is completely determined once we know the length of a single nonzero vector $X$ in $S^3_N$*.

Of course $SU(2)$ acts transitively on itself $SU(2) = S^3$ by left translation. It also acts on its Lie algebra $S^3_e$ by the adjoint action (18.31), and we know that the bi-invariant metric on $SU(2)$ arises from taking the metric $\langle X, Y\rangle = -\operatorname{tr} XY$ at $e$ and left translating to the whole group. Now the adjoint action of $SU(2)$ on $S^3_e$ is a double cover of the rotation group $SO(3)$ (see Section 19.1d) and thus is transitive again on directions at $e$. We conclude then that the bi-invariant metric on $SU(2) = S^3$ is again determined by the length assigned to a single nonzero vector in $S^3_e = S^3_N$. The *bi-invariant metric on $SU(2)$ is simply a constant multiple of the standard metric* on $S^3$.

Consider the curve on $SU(2)$ given by $\operatorname{diag}(e^{i\theta}, e^{-i\theta})$; its tangent vector at $e$ is simply $i\sigma_3$ whose length in the bi-invariant metric is $\sqrt{2}$. The corresponding curve in $\mathbb{C}^2$ is $(e^{i\theta}, 0)^T$, which in $\mathbb{R}^4$ is $(\cos\theta, \sin\theta, 0, 0)^T$, whose tangent vector at $N$ is $(0\ 1\ 0\ 0)^T$ with length 1. Thus the bi-invariant metric is $\sqrt{2}$ times the standard metric on the unit sphere $S^3$. Since a great circle will then have bi-invariant length $2\pi\sqrt{2}$, we see that the bi-invariant metric is the same as the *standard metric on the sphere of radius* $\sqrt{2}$. (Note that this agrees with the sectional curvature result (21.23), $K(i\sigma_1 \wedge i\sigma_2) = (1/4) \parallel [i\sigma_1, i\sigma_2] \parallel^2 / \parallel i\sigma_1 \wedge i\sigma_2 \parallel^2 = (1/4) \parallel -2i\sigma_3 \parallel^2 / \parallel i\sigma_1 \wedge i\sigma_2 \parallel^2 = 1/2$.)

The volume of the *unit* 3-sphere is easily determined.

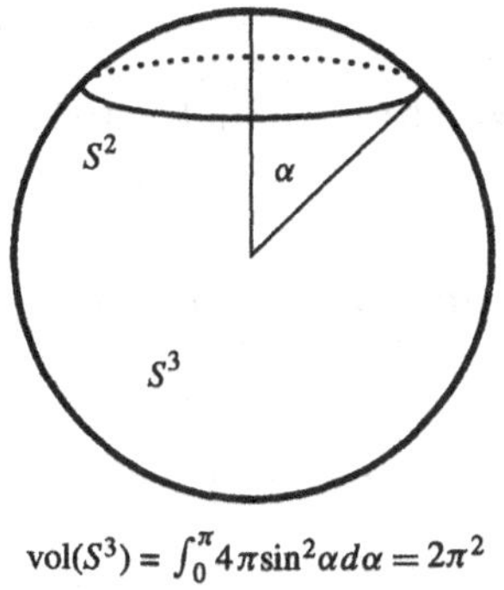


**Figure 22.1**

Thus our sphere of radius $\sqrt{2}$ has volume $(2^{3/2})2\pi^2$, and so

$$\int_{SU(2)} \Omega_3 = 24\pi^2$$

Finally we define the *winding number at infinity* of the instanton by

$$\frac{1}{24\pi^2}\int_{S^3} g_{VU}^* \Omega_3 = \frac{1}{24\pi^2}\int_{g_{VU}(S^3)} \Omega_3 \tag{22.3}$$
$$= \frac{1}{24\pi^2}\int_{g_{VU}(S^3)} \mathrm{tr}\, g^{-1}dg \wedge g^{-1}dg \wedge g^{-1}dg$$

This is the degree of the map $g_{VU}$ in the case when $G = SU(2)$. What it means in the case $SU(n)$ will be discussed later on in this chapter.

## 22.1b. Winding Number in Terms of Field Strength

Chern's expression (20.68) in the $U(1)$ case suggests the possibility of an expression for this winding number in terms of an integral of a 4-form involving curvature.

We shall *assume* that the Y–M potential $\omega_U$ is globally defined in $U$; that is, *$\omega_U$ has no singularities in $U$, $j(e_U) = 0$.*

Consider the following observation, holding for the curvature 2-form matrix for *any* vector bundle over *any* manifold:

$$\theta \wedge \theta = (d\omega + \omega \wedge \omega) \wedge (d\omega + \omega \wedge \omega)$$
$$= d\omega \wedge d\omega + d\omega \wedge \omega \wedge \omega + \omega \wedge \omega \wedge d\omega + \omega \wedge \omega \wedge \omega \wedge \omega$$

Use now

$$\mathrm{tr}(\omega \wedge \omega \wedge d\omega) = \mathrm{tr}(d\omega \wedge \omega \wedge \omega)$$

and, as in Theorem (21.3)

$$\mathrm{tr}(\omega \wedge \omega \wedge \omega \wedge \omega) = 0$$

Then

$$\operatorname{tr}\theta \wedge \theta = d\operatorname{tr}(\omega \wedge d\omega) + 2\operatorname{tr}(d\omega \wedge \omega \wedge \omega)$$

Also

$$d(\omega \wedge \omega \wedge \omega) = d\omega \wedge \omega \wedge \omega - \omega \wedge d\omega \wedge \omega + \omega \wedge \omega \wedge d\omega$$

and so

$$d\operatorname{tr}(\omega \wedge \omega \wedge \omega) = 3\operatorname{tr}(d\omega \wedge \omega \wedge \omega)$$

Thus we have shown

**Theorem (22.4):** *For any vector bundle over any $M^n$ we have*

$$\operatorname{tr}(\theta \wedge \theta) = d\operatorname{tr}\left\{\omega \wedge d\omega + \frac{2}{3}\omega \wedge \omega \wedge \omega\right\}$$

Thus $\operatorname{tr}\theta \wedge \theta$ is always *locally* the differential of a 3-form, the **Chern–Simons** 3-form. Of course $\omega$ is usually not globally defined.

Now back to our Y–M case considered in Section 20.6a. In that case $\theta$ vanishes on and outside the 3-sphere $S^3$, and so

$$\omega \wedge d\omega = \omega \wedge (\theta - \omega \wedge \omega) = -\omega \wedge \omega \wedge \omega$$

on and outside $S^3$. Then from (22.4)

$$\int_U \operatorname{tr}\theta \wedge \theta = \int_{\partial U = S^3} -\frac{1}{3}\operatorname{tr}\omega \wedge \omega \wedge \omega$$

But $\omega_U = g^{-1}dg$ on $S^3$; see (20.61). (22.3) then gives

**Theorem (22.5):** *The winding number of the instanton is given by*

$$\frac{1}{24\pi^2}\int_{S^3} \operatorname{tr}\omega_U \wedge \omega_U \wedge \omega_U = -\frac{1}{8\pi^2}\int_{\mathbb{R}^4} \operatorname{tr}\theta \wedge \theta$$

Note that $\operatorname{tr}\theta \wedge \theta$ is *not* the Lagrangian, which is basically $\operatorname{tr}\theta \wedge *\theta$

$$\begin{aligned} F \wedge F &= (F \wedge F)_{0123} dt \wedge dx \wedge dy \wedge dz \\ &= \sum_{i<j}\sum_{k<l} \epsilon^{ijkl} F_{ij} F_{kl} dt \wedge dx \wedge dy \wedge dz \end{aligned} \tag{22.6}$$

whereas

$$F \wedge *F = \sum_{j<k} F_{jk} F^{jk} dt \wedge dx \wedge dy \wedge dz$$

where the $F_{jk}$ are *matrices*. $\operatorname{tr}\theta \wedge \theta$ was introduced in Problem 20.5(3).

We have just shown that the winding number of an instanton is given, in terms of the Hilbert space scalar product (20.40), by $(8\pi^2)^{-1}(\theta, *\theta)$; this scalar product is defined since $\theta$ is assumed to have compact support. This is the degree of the map $g : S^3 \to SU(2)$ defined by the instanton. This degree is interesting for the following reason: The Y–M fields are critical points for the Y–M action functional. In particular, a connection $\omega$ yielding a (relative) minimum for $S$ will be a Y–M field. But by Schwarz's inequality (in the euclidean metric), the euclidean action $S$ on $M^4$ will satisfy

$$8\pi^2 \mid \deg(g) \mid = \mid (\theta, *\theta) \mid \leq \| \theta \| \| *\theta \| = \| \theta \|^2 = 2S$$

since $*$ is an isometry on forms. Thus *the degree yields a lower bound for the euclidean action!* Furthermore, we have equality iff $*\theta$ is proportional to $\theta$. Now $**\alpha = \alpha$, when $\alpha$ is a 2-form. It is easily seen that $*$ acting on our 2-forms in $M^4$ is self-adjoint in the scalar product (20.40). Thus $*$ has eigenvalues $\pm 1$ on the 2-forms and so $*\theta$ is proportional to $\theta$ only when $*\theta = \pm\theta$, that is, iff the connection is self-dual or anti-self-dual; see (20.58). In particular, the self-dual fields with degree $n$ and the anti-self-dual fields with degree $-n$ will both yield Y–M fields having minimum action among all fields of degree $\pm n$.

### 22.1c. The Chern Forms for a $U(n)$ Bundle

The topological significance of tr $\theta \wedge \theta$, generalizing Poincaré's theorem for closed surfaces, $\iint K dA = 2\pi\chi(M^2)$, was discovered by Chern and will be discussed later in this chapter. tr $\theta \wedge \theta$ is but one of a whole family of significant integrands, the Chern forms. We shall define these forms now and then proceed to the topological questions in our remaining sections.

Let $A$ be any $N \times N$ matrix of complex numbers operating on complex $N$-space $V = \mathbb{C}^N$. Consider the characteristic (eigenvalue) polynomial for $A$

$$\begin{aligned}\det(\lambda I - A) &= (\lambda - \lambda_1)(\lambda - \lambda_2)\ldots(\lambda - \lambda_N) \\ &= \lambda^N - (\lambda_1 + \cdots + \lambda_N)\lambda^{N-1} + \cdots \pm (\lambda_1\lambda_2\ldots\lambda_N)\end{aligned}$$

Putting $\lambda = -1$ yields

$$\begin{aligned}\det(I + A) &= \sum_{p=0}^{N} \left[\mathrm{tr} \bigwedge^{p} A\right] \\ &= 1 + (\mathrm{tr}\, A) + \left(\mathrm{tr} \bigwedge^{2} A\right) + \cdots + \left(\mathrm{tr} \bigwedge^{N} A\right)\end{aligned} \tag{22.7}$$

where

$$\begin{aligned}\mathrm{tr}\, A &:= \sum_i \lambda_i \\ \mathrm{tr} \bigwedge^{2} A &:= \sum_{i<j} \lambda_i \lambda_j \\ \mathrm{tr} \bigwedge^{3} A &:= \sum_{i<j<k} \lambda_i \lambda_j \lambda_k \\ \mathrm{tr} \bigwedge^{N} A &:= \lambda_1 \lambda_2 \ldots \lambda_N = \det A\end{aligned} \tag{22.8}$$

are the **elementary symmetric functions** of the eigenvalues of $A$. The reason for this notation is as follows: if $A : V \to V$ then we may let $A$ act on each of the exterior power spaces $\bigwedge^p V$ by the (linear) **exterior power operation** $\bigwedge^p A$

$$\left(\bigwedge^p A\right)(v_1 \wedge v_2 \wedge \ldots \wedge v_p) := Av_1 \wedge Av_2 \wedge \ldots \wedge Av_p$$

We then take the usual trace of $\bigwedge^p A$ on the space $\bigwedge^p V$. For example, $\bigwedge^N V$ is a 1-dimensional vector space and from (2.50)

$$\left(\bigwedge^N A\right)(v_1 \wedge v_2 \wedge \ldots \wedge v_N) = \det A(v_1 \wedge v_2 \wedge \ldots \wedge v_N)$$

and so $\operatorname{tr} \bigwedge^N A = \det A = \lambda_1 \ldots \lambda_N$.

Note that $(\lambda_1^k + \cdots + \lambda_N^k) = \operatorname{tr} A^k$ is simply the trace of the $k^{\text{th}}$ (ordinary matrix) power of the matrix. Thus, for example,

$$\operatorname{tr} \bigwedge^2 A = \sum_{r<s} \lambda_r \lambda_s = \frac{1}{2}\left[\left(\sum_j \lambda_j\right)^2 - \sum_j \lambda_j^2\right] \tag{22.9}$$

$$= \frac{1}{2}[(\operatorname{tr} A)^2 - \operatorname{tr} A^2]$$

In a similar manner it can be shown, using "Newton's identities," that each $\operatorname{tr} \bigwedge^k A$ can be expressed as a polynomial in $\operatorname{tr} A$, $\operatorname{tr}(A^2), \ldots, \operatorname{tr}(A^k)$. We shall return to this point in a moment.

Now let $E$ be a complex $\mathbb{C}^N$ bundle with structure group $U(N)$, base manifold $M^n$, and connection $\omega$.

Consider the result of *formally* substituting for $A$ in (22.7), the matrix of curvature 2-forms $\theta = \theta_U$ multiplied by $i/2\pi$

$$\det\left(I + \frac{i\theta}{2\pi}\right)$$

Thus we are looking at a matrix whose ${}^\alpha{}_\alpha$ entry is $1 + (i/2\pi)\theta^\alpha{}_\alpha$ and whose nondiagonal ${}^\alpha{}_\beta$ entry is $(i/2\pi)\theta^\alpha{}_\beta$ and where we expand out the determinant in the usual way with *products being replaced by* $\wedge$ *products*; since $\theta^j{}_k$ is a 2-form *there is no problem with ordering*. The result is a sum of forms of different degrees

$$\det\left(I + \frac{i\theta}{2\pi}\right) = 1 + \frac{i}{2\pi}\operatorname{tr}\theta + \cdots \tag{22.10}$$

$$:= 1 + c_1(E) + c_2(E) + \cdots + c_N(E)$$

where $c_r(E)$ is a $2r$-form on $U \subset M^n$, the $r^{\text{th}}$ **Chern form**.

The form $c_1$ is familiar

$$c_1 = \frac{i}{2\pi}\operatorname{tr}\theta = \frac{i}{2\pi}\theta^\alpha{}_\alpha \tag{22.11}$$

and in the case of a complex line bundle, $\theta^\alpha{}_\alpha$ is simply the 2-form $\theta$ appearing in Theorem (17.28). For the tangent complex line bundle to an oriented surface, $c_1(TM^2) = (1/2\pi)K\,dA$.

For $c_2$, from (22.9) we have

$$c_2 = -\frac{1}{8\pi^2}[\operatorname{tr}\theta \wedge \operatorname{tr}\theta - \operatorname{tr}(\theta \wedge \theta)] \tag{22.12}$$

Suppose that the bundle actually has the *special* unitary group $SU(N)$ for structure group, rather than $U(N)$. Since the Lie algebra then consists of *traceless* skew-hermitian matrices, $\operatorname{tr}\theta = 0$, and thus in this case

$$c_1(E) = 0$$

and furthermore

$$c_2(E) = \frac{1}{8\pi^2}\operatorname{tr}(\theta \wedge \theta)$$

*This is precisely the 4-form appearing in the winding number of an SU(2) instanton,* given in (22.5)!

In the general case, note that the matrices $\theta$ are only locally defined, and in an overlap $\theta_V = c_{VU}\theta_U c_{VU}^{-1}$. However

$$\det\left(I + \frac{i\theta_V}{2\pi}\right) = \det\left\{I + \frac{i}{2\pi}c_{VU}\theta_U c_{VU}^{-1}\right\}$$
$$= \det c_{VU}\left(I + \frac{i\theta_U}{2\pi}\right)c_{VU}^{-1} = \det\left(I + \frac{i\theta_U}{2\pi}\right)$$

shows that *each Chern form $c_r(E)$ is in fact a globally defined $2r$-form on all of $M^n$*!

In Problem 22.1(1) you are asked to show that each $c_r$ is a *real* form.

We can see that $c_1$ is a *closed* 2-form as follows: From $-2\pi i dc_1 = d\operatorname{tr}\theta = \operatorname{tr} d\theta$, and from Bianchi this is $\operatorname{tr}(\theta \wedge \omega - \omega \wedge \theta)$. But $\operatorname{tr}\omega \wedge \theta = \operatorname{tr}\theta \wedge \omega$ since $\theta$ is a 2-form. We conclude that $dc_1 = 0$, as claimed. It is even simpler to remark that locally $\theta = d\omega + \omega \wedge \omega$ and then

$$\operatorname{tr}\theta = d\operatorname{tr}\omega$$

since $\operatorname{tr}\omega \wedge \omega = -\operatorname{tr}\omega \wedge \omega = 0$. Thus $\operatorname{tr}\theta$ is locally exact, hence closed.

For an $SU(N)$ bundle, $c_2$ is locally the differential of the Chern–Simons 3-form given in Theorem (22.4), and so $c_2$ is a closed 4-form in this case. We can also see this directly for *any* $U(N)$ bundle, from the Bianchi identity. From (22.12)

$$-(8\pi^2)dc_2 = d[\operatorname{tr}\theta \wedge \operatorname{tr}\theta - \operatorname{tr}(\theta \wedge \theta)] = -d\operatorname{tr}(\theta \wedge \theta)$$

But, from (18.46) and (20.55), $d\operatorname{tr}(\theta \wedge \theta) = \operatorname{tr}\boldsymbol{\nabla}(\theta \wedge \theta) = 0$, since $\boldsymbol{\nabla}\theta = 0$.

As we have mentioned (but not proved), Newton's identities show that each $c_r$ is a polynomial in forms of the type $\operatorname{tr}(\theta \wedge \theta \wedge \ldots \wedge \theta)$; we have shown this for $c_1$ and $c_2$ and you are asked in Problem 22.1(2) to verify it for $c_3$. (For a derivation of the Newton identities, see [Ro, ex. 1, p. 132], but not before reading the remainder of this section.) Since $\boldsymbol{\nabla}$ of such a polynomial vanishes by Bianchi, we conclude that each Chern form is closed. We present a different proof of this important fact now.

**Theorem of Chern and Weil (22.13):** *Each $c_r$ is a closed $2r$-form and thus defines a real de Rham class. Furthermore, different connections for the $U(N)$ bundle will yield Chern forms that differ by an exact form and hence define the same de Rham cohomology class.*

PROOF: We sketch *briefly* a proof from Roe's book [Ro, p. 113].

We shall look at *formal* power series expansions. For example, the matrix $a = a(\theta) = I + q\theta$ considered previously, where $q = i/2\pi$, has a formal inverse. If we write $\theta^r := \theta \wedge \theta \wedge \ldots \wedge \theta$, $r$ factors

$$a^{-1} = (I + q\theta)^{-1} = \sum_r (-1)^r q^r \theta^r \tag{22.14}$$

This makes sense since it is only a finite series, $\theta \wedge \theta \wedge \ldots \wedge \theta$ vanishing when the number of factors exceeds half the dimension of the manifold $M$. Suppose now that we let the connection $\omega$ vary smoothly with a real parameter $t$, $\omega = \omega(t)$. Then both the curvature $\theta$ and the matrix $a$ vary with $t$. But for any nonsingular matrix $a(t)$ we have for the derivative of its determinant $| a(t) |$

$$\begin{aligned}\frac{d \mid a(t) \mid}{dt} &= \sum \left[\frac{\partial \mid a \mid}{\partial a_{jk}}\right] \frac{da_{jk}}{dt} \\ &= A^{jk} \dot{a}_{jk} = \mid a \mid (a^{-1})^{kj} \dot{a}_{jk} = \mid a \mid \operatorname{tr}[a^{-1} \dot{a}]\end{aligned}$$

where $A^{jk}$ is the signed cofactor of $a_{jk}$. Hence

$$\frac{d \log \mid a(t) \mid}{dt} = \operatorname{tr}[a^{-1} \dot{a}] \tag{22.15}$$

Thus, putting $\theta = d\omega + \omega \wedge \omega$, $\dot{\theta} = d\dot{\omega} + \omega \wedge \dot{\omega} + \dot{\omega} \wedge \omega$, $\dot{a} = q\dot{\theta}$

$$\frac{d \log \mid a(t) \mid}{dt} = \sum_r (-1)^r q^{r+1} \operatorname{tr}[\theta^r \wedge (d\dot{\omega} + \omega \wedge \dot{\omega} + \dot{\omega} \wedge \omega)]$$

One sees immediately by induction from Bianchi that

$$d\theta^r = \theta^r \wedge \omega - \omega \wedge \theta^r$$

for $r \geq 0$, with $\theta^0 = 1$.

Furthermore, $\operatorname{tr}[\theta^r \wedge \dot{\omega} \wedge \omega] = -\operatorname{tr}[\omega \wedge \theta^r \wedge \dot{\omega}]$, since $\theta^r \wedge \dot{\omega}$ is a form of odd degree. Hence

$$\frac{d \log \mid a(t) \mid}{dt} = \sum_r (-1)^r q^{r+1} \operatorname{tr}[\theta^r \wedge d\dot{\omega} + d\theta^r \wedge \dot{\omega}]$$

or

$$\frac{d \log \mid a(t) \mid}{dt} = d \sum_r (-1)^r q^{r+1} \operatorname{tr}[\theta^r \wedge \dot{\omega}] \tag{22.16}$$

exhibits $d \log | a(t) | /dt$ as the differential of a sum of forms (of various degrees). Note also that the forms on the right are indeed *globally defined forms on the base space M*, since both $\theta^r$ and $\dot{\omega}$ are forms of type Ad $G$; this was Problem 18.3(4).

As a first consequence of (22.16) note the following: If $\omega$ and $\omega'$ are two connections on $M$, then Problem 20.3(1) shows that their convex combination $\omega(t) = t\omega + (1-t)\omega'$ is again a connection. This gives a line in the affine space of all connections on $M$ that starts at $\omega'$ and ends at $\omega$. Now the flat connection $\omega = 0$, $\theta = 0$, is not necessarily a connection on $M$ for the given bundle (why?), but it *is* a connection on a *single coordinate patch U* of $M$. Then $\omega(t) = t\omega$ is a line of connections on $U$ joining any given connection $\omega = \omega(1)$ to the flat connection $\omega(0) = 0$. Since $a(0) = I$, we have, from (22.16),

$$\log \mid I + q\theta \mid = d \int_0^1 \left\{ \sum_r (-1)^r q^{r+1} \operatorname{tr}[\theta^r(t) \wedge \omega] \right\} dt$$

and so $\log | I + q\theta |$ is locally exact (being exact on $U$) hence *closed;* in fact it is of the form $\log | I + q\theta | = d\beta$ where

$$\beta := \int_0^1 \{[q \operatorname{tr}\omega - q^2 \operatorname{tr}\theta(t) \wedge \omega + \cdots]\}dt$$

But then

$$| I + q\theta | = \exp\log | I + q\theta | = \exp d\beta = 1 + d\beta + \frac{1}{2!} d\beta \wedge d\beta + \cdots$$

is again locally exact, except for the constant term, hence closed. We are finished with the first part of Theorem (22.13).

Consider now a pair of *global* connections $\omega$ and $\omega'$ on $M$ and the line $t\omega' + (1-t)\omega$ in the space of connections. From (22.16) we have $\log | a_{\omega'} | - \log | a_\omega | = d\gamma$ for a *globally defined* form $\gamma$ on $M$

$$\gamma = \int_0^1 \sum_r (-1)^r q^{r+1} \operatorname{tr}[\theta^r(t) \wedge (\omega' - \omega)]dt$$

Then

$$\begin{aligned} \frac{| a_{\omega'} |}{| a_\omega |} &= \exp\{\log | a_{\omega'} | - \log | a_\omega |\} = \exp d\gamma \\ &= 1 + d\gamma + \frac{1}{2!} d\gamma \wedge d\gamma + \cdots =: 1 + d\nu \end{aligned}$$

and so $| a_{\omega'} | - | a_\omega | = | a_\omega | \wedge d\nu$. But we have just seen that $| a_\omega | = \det(I + q\theta)$ is closed. Hence $| a_{\omega'} | - | a_\omega |$ is globally exact, proving the second part of the theorem. □

## Problems

**22.1(1)** Show directly from $\det(I + i\theta/2\pi)$ that each $c_r$ is a real form when the structure group is a subgroup of $U(N)$.

**22.1(2)** Express $c_3$ as a polynomial in $\operatorname{tr}\theta$, $\operatorname{tr}(\theta \wedge \theta)$, and $\operatorname{tr}(\theta \wedge \theta \wedge \theta)$.

## 22.2. Homotopies and Extensions

Is $SU(n)$ simply connected?

### 22.2a. Homotopy

In Section 10.2d we discussed when two closed curves in $M$ are homotopic. We now introduce the general concept of homotopic maps.

Let $f_0$ and $f_1$ be two maps of a space $W$ into $M^n$. We say that they are **homotopic** if there is a map $F : W \times I \to M$ of the "cylinder" $W \times [0, 1]$ into $M$ such that

$$F(w, 0) = f_0(w) \quad \text{and} \quad F(w, 1) = f_1(w)$$

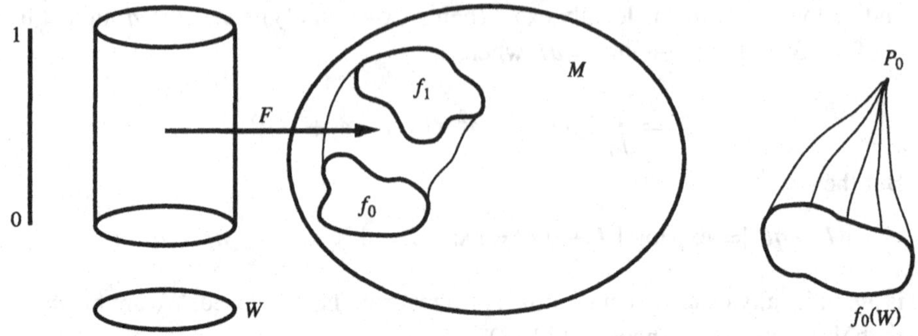


**Figure 22.2**

Each of the maps $f_t$, defined by $f_t(w) := F(w, t)$, is homotopic to the "original" map $f_0$. If $f_1$ maps all of $W$ into a single point $p_0$ we say that $f_0$ **is homotopic to the constant map** $p_0$.

We shall be especially concerned with the case when $W = S^k$ is the unit $k$-sphere, $k = 0, 1, 2, \ldots$, in $\mathbb{R}^{k+1}$, *even when* $k > n = \dim M$! $S^k$ is of course the boundary of the closed $(k+1)$-ball $D^{k+1}$ and the following simple observation will play a crucial role in our final section.

**Extension Theorem (22.17):** *$f : S^k \to M^n$ is homotopic to a constant map iff $f$ can be extended to a map of the ball*

$$f' : D^{k+1} \to M^n$$

PROOF: Suppose that $f' : D^{k+1} \to M$ extends $f : S^k \to M$; thus $f'(\mathbf{x}) = f(\mathbf{x})$ for $\| \mathbf{x} \| = 1$. Define $F : S^k \times I \to M$ by

$$F(\mathbf{x}, r) = f'\{(1-r)\mathbf{x}\}, \qquad \mathbf{x} \in S^k, \qquad 0 \le r \le 1$$

Then $F(\mathbf{x}, 0) = f'(\mathbf{x}) = f(\mathbf{x})$ and $F(\mathbf{x}, 1) = f'(\mathbf{0})$ shows that $f$ is homotopic to the constant map $f'(\mathbf{0})$.

Suppose, on the other hand, that $f (= f_0)$ is homotopic to the constant map $f_1(\mathbf{x}) = p_0 \in M$. Then we have a map $F : S^k \to M$ with $F(\mathbf{x}, 0) = f(\mathbf{x})$ and $F(\mathbf{x}, 1) = f_1(\mathbf{x}) = p_0$. Define an extension $f' : D^{k+1} \to M$ by $f'(r\mathbf{x}) = F(\mathbf{x}, 1-r)$ for $0 \le r \le 1$.

The extension theorem is important when discussing *defects*, see [Mi]. □

## 22.2b. Covering Homotopy

Let $\pi : E \to M^n$ be a vector bundle and let $f : W \to E$ be a map of a space $W$ into the bundle space $E$. Then we get a map $\underline{f} : W \to M^n$ into the base space by $\underline{f} := \pi \circ f$.

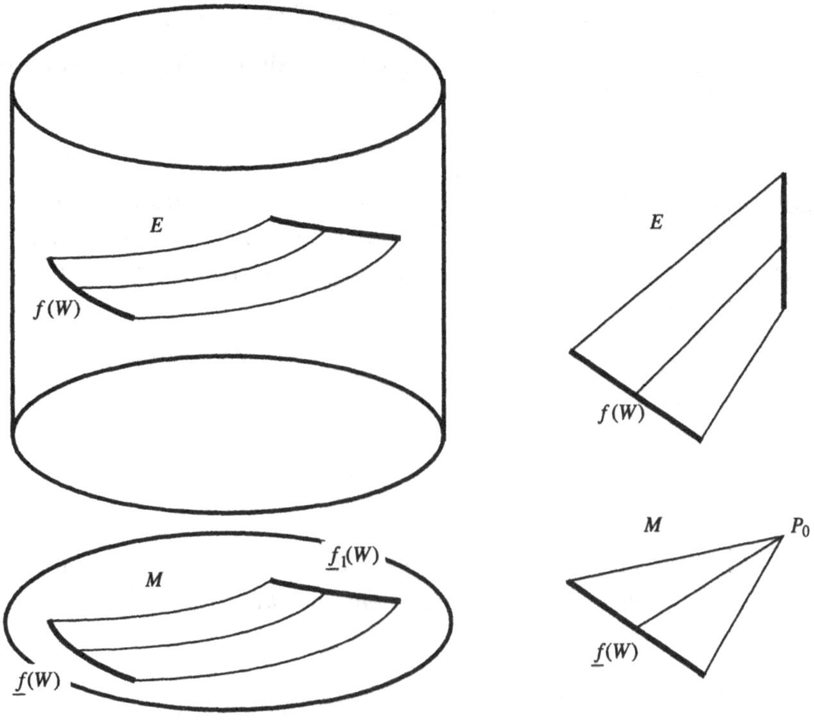


**Figure 22.3**

Suppose now that we have a homotopy $\underline{F}$ of $\underline{f}$ to a new map $\underline{f}_1 : W \to M$. We claim that *we can "cover" this homotopy by a homotopy of the original map* $f$; that is, there is a map $F : W \times I \to E$ such that $F(w, 0) = f(w)$ and $\pi F(w, t) = \underline{F}(w, t)$. A *sketch* goes as follows: Let the vector bundle $\pi : E \to M$ have a connection. Consider a fixed point $w \in W$ and look at the curve $\underline{C} : t \to \underline{F}(w, t)$ in $M$. There is a *unique lift* of this curve to a curve $C$ in $E$ starting at $f(w)$ that represents *parallel translation* along $\underline{C}$. In other words, we look at the unique curve in $E$ that starts at $f(w)$, lies over $\underline{C}$, and is tangent to the $n$-plane distribution defined locally by

$$d\psi^\alpha + \omega^\alpha{}_\beta \psi^\beta = 0$$

Note that if $\underline{f}$ is homotopic to the constant map $p_0$ (as in the second part of our figure) *it need not be that* $f$ *will be homotopic to a constant map*; the points $F(w, 1)$ of the lifted homotopy will lie on the fiber $\pi^{-1}(p_0)$ but will not necessarily reduce to a single point in the fiber.

What we have said for a vector bundle can also be shown to hold for a principal fiber bundle. The lifted curves are then tangent to the $n$-plane distribution

$$\omega^* = g^{-1}\omega g + g^{-1}dg = 0$$

It turns out that one can cover homotopies in *any* fiber bundle, without any use of a connection. In fact, one generalizes the notion of a fiber bundle to that of a **fiber space**; this is a space $P$ and a map $\pi : P \to M$ such that *homotopies can always be covered*, as defined earlier. Such spaces need *not* be local products.

### 22.2c. Some Topology of $SU(n)$

$SU(n)$ is represented by $n \times n$ matrices acting on $\mathbb{C}^n$. Since each $g \in SU(n)$ is unitary, $SU(n)$ sends the unit sphere $S^{2n-1} \subset \mathbb{C}^n$

$$S^{2n-1} = \{\mathbf{z} \in \mathbb{C}^n \mid |\, \mathbf{z}_1 \,|^2 + \cdots + |\, \mathbf{z}_n \,|^2 = 1\}$$

into itself. It is clear that $SU(n)$ acts transitively on $S^{2n-1}$, for the point $(1, 0, \ldots, 0)$ can be sent into the point $\mathbf{z} = (z_1, \ldots, z_n)$ simply by writing down some $g \in SU(n)$ having $\mathbf{z}^T$ as its first column. The isotropy subgroup for the point $(1, 0, \ldots, 0)$ is clearly the subgroup

$$\begin{bmatrix} 1 & 0 \\ 0 & SU(n-1) \end{bmatrix}$$

which we shall briefly denote simply by $SU(n-1)$.

$$S^{2n-1} = \frac{SU(n)}{SU(n-1)} \tag{22.18}$$

and in fact $SU(n)$ is a principal $SU(n-1)$ bundle over $S^{2n-1}$ (see Theorem (17.11)).

If $P$ is a fiber bundle over $M$ with fiber $F$ we shall write *symbolically*

$$F \to P \xrightarrow{\pi} M \tag{22.19}$$

and we shall frequently omit the projection map $\pi$. Thus we write

$$SU(n-1) \to SU(n) \to S^{2n-1} \tag{22.20}$$

**Theorem (22.21):** *If $F \to P \to M$ is a fiber bundle with connected $M$ and connected $F$, then $P$ is connected.*

PROOF: Let $p$ and $p_0$ be points in $P$. Project them down to points $\pi(p)$ and $\pi(p_0)$ in $M$. Since $M$ is connected there is a curve in $M$ joining these two points.

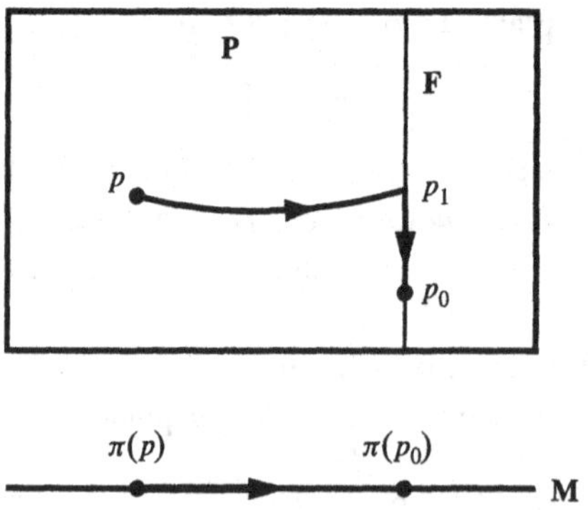


Figure 22.4

This curve can be considered a homotopy from the constant map of a point $w$ into $\pi(p)$, to the constant map of the point $w$ to $\pi(p_0)$. Cover this homotopy by a path from $p$ to the fiber through $p_0$, that is, from $p$ to some point $p_1$ in this fiber. Since this fiber is assumed connected we can find a curve in this fiber from $p_1$ to $p_0$. We have joined $p$ to $p_0$ by a succession of two paths in $P$. □

**Corollary (22.22):** *$SU(n)$ is connected.*

PROOF: $SU(1)$ is a single point. $SU(2)$ is a 3-sphere and is connected, as are all $k$-spheres for $k > 0$. From $SU(2) \to SU(3) \to S^5$ we see that $SU(3)$ is connected. Induction gives the corollary. □

See Problem 22.2(1) at this time.

Recall that we say that $M$ is simply connected provided every map of a circle into $M$ is homotopic to a constant map. During the homotopy, the closed curve gets "contracted" or "deformed" to the point.

**Theorem (22.23):** *Let $F \to P \to M$ be a fiber bundle whose fiber $F$ and base $M$ are simply connected. Then $P$ is simply connected.*

PROOF: Let $C$ be a closed curve in $P$. Project it down to a closed curve $\pi(C)$ in $M$. Since $M$ is simply connected, $\pi(C)$ can be contracted to a point $p_0$ in $M$. We may cover this homotopy by a deformation of $C$ into the fiber over $p_0$; that is, $C$ is deformed into a new closed curve lying in the fiber $\pi^{-1}(p_0)$. Since the fiber is simply connected, this new closed curve can be shrunk to a point in the fiber. Thus the composition of the two deformations deforms $C$ to a point, as desired. □

## Problems

**22.2(1)** Show that $SO(n)$ is connected.

**22.2(2)** We know that the cartesian product of connected manifolds is connected; this is the special case of (22.21) when $F \to M \times F \to M$ is simply a product bundle. In a product bundle we also have the converse (which is evident from a picture); if $M$ and $M \times F$ are connected, then $F$ is connected. That this need not be true when $M \times F$ is replaced by a twisted product, that is, a bundle $P$, may be seen as follows: Denote the principal frame bundle to a Riemannian 3-manifold $M^3$ by $O(3) \to FM \to M$. $O(3)$ is definitely not connected, being the disjoint union of $SO(3)$ and those $g \in O(3)$ with $\det g = -1$. In spite of this, show that if $M$ is connected and *not orientable*, then $FM$ is connected! In particular $FM$ in this case is not a product.

A simpler example $\mathbb{Z}_2 \to S^1 \to S^1$ is the 2-fold covering of a circle by itself. Show that this is realized in the case of the unit normal bundle $P$ to the central circle $S^1$ of the (infinite) Möbius band Mö.

**22.2(3)** Show that $SU(n)$ is simply connected.

## 22.3. The Higher Homotopy Groups $\pi_k(M)$

**Why is the alternating sum of Betti numbers equal to the Euler characteristic?**

### 22.3a. $\pi_k(M)$

We shall consider *continuous* maps $f : S^k \to M$ of a $k$-sphere into $M^n$. We shall always ask that some distinguished point on $S^k$, the "north pole," be sent into a distinguished **base point**, written $*$ in $M^n$. *We shall only consider* $k \geq 1$.

For technical reasons we consider $S^k$ to be the unit $k$-cube, $I^k = [0, 1] \times \cdots \times [0, 1]$, with the entire boundary $\dot{I}^k$ identified with a single point, the north pole.

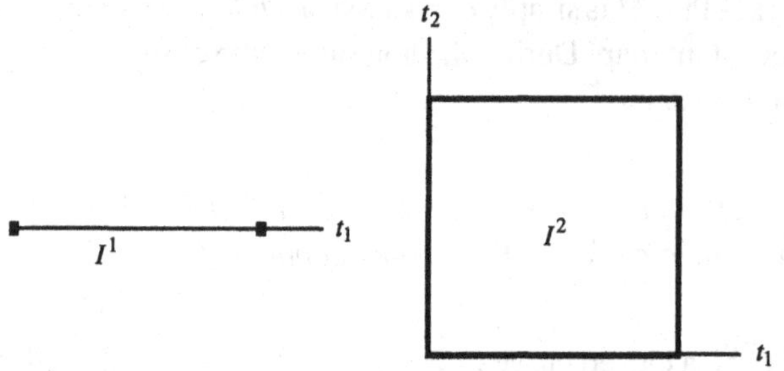


Figure 22.5

Then $f : S^k \to M$ is a map $f : I^k \to M$ such that $f(\dot{I}) = *$. In our diagrams the heavy portions are always mapped to $*$. To say that $f_0$ and $f_1$ are homotopic, $f_0 \sim f_1$, is to say that there is a map $F : I^k \times I \to M$ such that $F(y, 0) = f_0(y)$, $F(y, 1) = f_1(y)$, and $F(\text{north pole}, t) = *, 0 \leq t \leq 1$

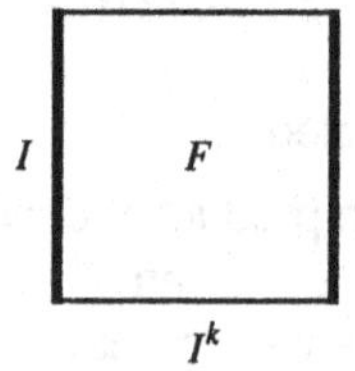


Figure 22.6

(Again the heavy portions are sent into the base point.)

We compose two maps $f : S^k \to M$ and $g : S^k \to M$ using the first coordinate, as we did for loops, but this time the result is written $f + g$:

$$
\begin{aligned}
(f + g)(t_1, \ldots, t_k) &= f(2t_1, t_2, \ldots t_k) & 0 \leq t_1 \leq \frac{1}{2} \\
&= g(2t_1 - 1, t_2, \ldots, t_k) & \frac{1}{2} \leq t_1 \leq 1
\end{aligned}
$$

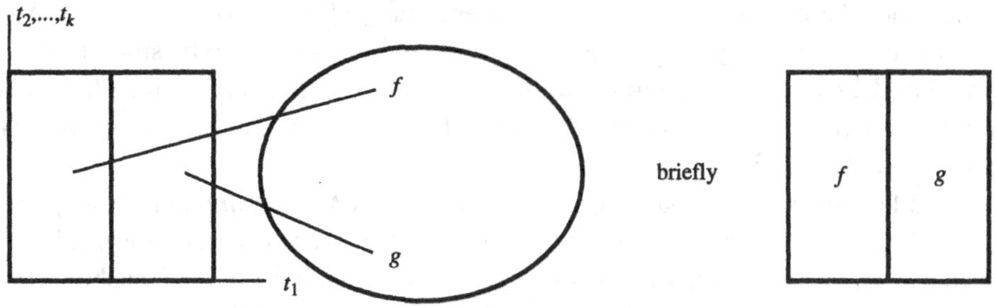


**Figure 22.7**

Again, two maps are to be identified if they are homotopic. The homotopy classes of such maps define **the k$^{\text{th}}$ homotopy group**

$$\pi_k(M,*) = \pi_k(M)$$

(It can be shown that if $f' \sim f$ and $g' \sim g$ then $f' + g' \sim f + g$.) The identity is represented by maps homotopic to the constant map $f = *$, and the inverse of the map $f(t_1, \ldots, t_n)$ is represented by $f(1 - t_1, t_2, \ldots, t_n)$. The composition is written additively since these classes of maps form a **commutative** group (*if* $k \geq 2$). The commutativity can be "seen" from the following sequence of homotopies where a

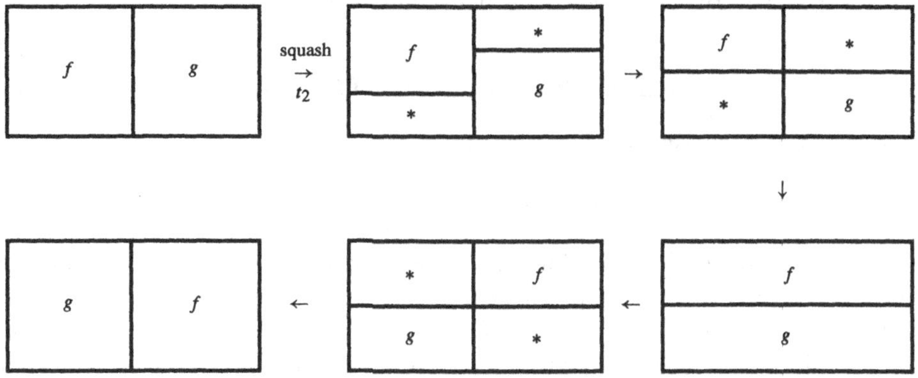


**Figure 22.8**

whole box labeled $*$ is to be sent into the base point. See [H,Y] for details. Note that this procedure will not work in the case $n = 1$; there is no room to maneuver. This is why the fundamental group $\pi_1$ can be nonabelian.

## 22.3b. Homotopy Groups of Spheres

$\pi_k(S^n)$ consists of homotopy classes of maps of a $k$-sphere into an $n$-sphere. We have already discussed $\pi_1(S^1) = \mathbb{Z}$ where the homotopy class is characterized by the Brouwer degree of the map. (We have shown that maps of different degrees are not homotopic, but we have not proved the converse.)

Consider the case $k < n$. It seems evident that $f(S^k)$ cannot cover all of $S^n$ if $k < n$ but this is actually *false* since we do not require our maps to be smooth! Peano constructed a curve, a *continuous* map of the interval [0, 1], whose image filled up an entire square $[0, 1] \times [0, 1]$; see [H,Y, p. 123]. This map cannot be smooth, as you will show in Problem 22.3(1).

It is a fact that a continuous map of a sphere into an $M^n$ is *homotopic* (via approximation) to a smooth one. Hence we may assume that $f(S^k)$ does not cover all of $S^n$ when $k < n$. Suppose then that the south pole of $S^n$ is not covered. By pushing away from the south pole we may push the entire image to the north pole; we have deformed the map into a constant map. Thus $\pi_k(S^n) = 0$ if $k < n$.

Consider the case $k = n$. We know that homotopic maps of an $n$-sphere into itself have the same degree. A theorem of Heinz Hopf says in fact that *maps of any connected, closed, orientable n-manifold $M^n$ into an n-sphere $S^n$ are homotopic if and only if they have the same degree* (the nontrivial proof can be found in [G, P]). Thus the homotopy classes of maps $S^n \to S^n$ are again characterized by an integer, the degree. Again, as for circles, one can construct a map of any integral degree. Thus we have, so far

$$\begin{aligned} \pi_k(S^n) &= 0 \quad \text{if } 0 < k < n \\ &= \mathbb{Z} \quad \text{if } k = n \end{aligned} \tag{22.24}$$

Hopf made the surprising discovery that there can be nontrivial maps of $S^k$ onto $S^n$ when $k > n > 1$! We shall discuss one in Section 22.4.

### 22.3c. Exact Sequences of Groups

A sequence of groups and homomorphisms

$$\cdots \to F \xrightarrow{f} G \xrightarrow{g} H \to \cdots$$

is said to be **exact at G** provided that the *kernel of g* (the subgroup of $G$ sent into the identity of $H$) coincides with the *image of f*, $f(F) \subset G$. In particular, we must have that the *composition $g \circ f : F \to H$ is the trivial homomorphism sending all of F into the identity element of H*. The (entire) sequence is **exact** if it is exact at each group. 0 will denote the group consisting of just the identity (if the groups are not abelian we usually use 1 instead of 0).

Some examples. If

$$0 \xrightarrow{f} H \xrightarrow{h} G$$

is exact at $H$ then $\ker h = \operatorname{im} f = 0$. Thus *$h$ is 1 : 1*. Since $h$ is 1 : 1, we may identify $H$ with its image $h(H)$; in other words *we may consider H to be a subgroup of G*. Ordinarily we do not label the homomorphism $0 \xrightarrow{f} H$; we would write simply $0 \to H \xrightarrow{h} G$.

If

$$H \xrightarrow{h} G \xrightarrow{g} 0$$

is exact at $G$ then im $h$ = ker $g$ = $G$, and so *$h$ is onto*. Again we would write $H \xrightarrow{h} G \to 0$.

If

$$0 \to H \xrightarrow{h} G \to 0$$

is exact, meaning exact at all the interior groups $H$, $G$, then $h$ is 1 : 1 and onto; that is, *$h$ is an isomorphism.*

Consider an exact sequence of three nontrivial *abelian* groups (a so-called **short** exact sequence)

$$0 \to F \xrightarrow{f} G \xrightarrow{g} H \to 0$$

Then ker $g$ is im $f$, which is considered the subgroup $F = f(F)$ of $G$, and $g$ maps $G$ onto all of $H$. Note that if $h = g(g_1)$ and $h = g(g_2)$, then $(g_1 - g_2) \in f(F) \approx F$. Thus $H$ may be considered as equivalence classes of elements of $G$, $g_2 \sim g_1$ iff $g_2 - g_1$ is in the subgroup $F$. In other words, *$H$ is the coset space $G/F$!* (See Sections 13.2c and 17.2a, but note that we are using *additive* notation for these abelian groups.)

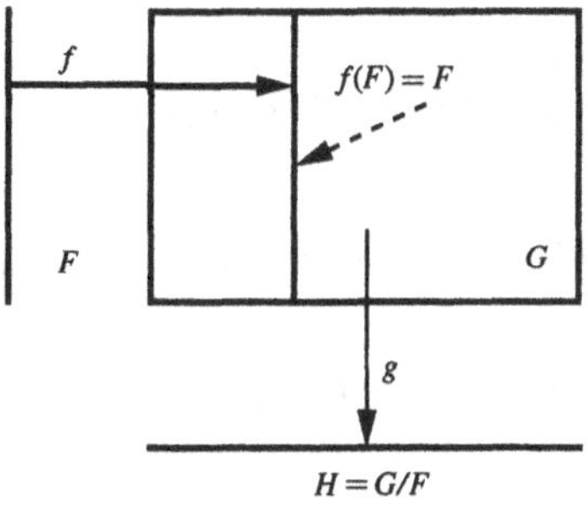


**Figure 22.9**

If the homomorphisms involved are understood, we frequently will omit them. For example, the exact sequence ($2\mathbb{Z}$ is the group of even integers)

$$0 \to 2\mathbb{Z} \to \mathbb{Z} \to \mathbb{Z}_2 \to 0$$

says that the even integers form a subgroup of the integers and $\mathbb{Z}_2 \approx \mathbb{Z}/2\mathbb{Z}$. The exact sequence

$$0 \to \mathbb{Z} \to \mathbb{R} \to S^1 \to 0$$

where the group of integers $\mathbb{Z}$ is considered as a subgroup of the additive reals, and where $\mathbb{R} \to S^1$ is the exponential homomorphism

$$r \in \mathbb{R} \mapsto \exp(2\pi i r)$$

onto the unit circle in the complex plane (a group under multiplication of complex numbers) exhibits the circle as a coset space

$$\mathbb{R}/\mathbb{Z} = S^1$$

In brief, a short exact sequence of abelian groups is always of the form

$$0 \to H \to G \to \frac{G}{H} \to 0 \tag{22.25}$$

where the first homomorphism is inclusion and the second is projection. (As we saw in Section 13.2c, $G/H$ is always a group when $G$ is abelian.)

We have two examples from homology theory, as in Section 13.2c

$$0 \to Z_k \to C_k \xrightarrow{\partial} B_{k-1} \to 0 \tag{22.26}$$
$$0 \to B_k \to Z_k \to H_k \to 0$$

See Problem 22.3(2).

### 22.3d. The Homotopy Sequence of a Bundle

For simplicity only, we shall consider a fiber bundle $F \to P \to M$ with *connected* fiber and base. If $F$ is *not* connected there is a change in only the last term of the following.

**Theorem (22.27):** *If the fiber $F$ is connected, we have the* ***exact sequence of homotopy groups***

$$\cdots \to \pi_k(F) \to \pi_k(P) \to \pi_k(M) \xrightarrow{\partial} \pi_{k-1}(F) \to \cdots$$

$$\cdots \xrightarrow{\partial} \pi_2(F) \to \pi_2(P) \to \pi_2(M) \xrightarrow{\partial} \pi_1(F) \to \pi_1(P) \to \pi_1(M) \to 1$$

The homomorphisms are defined as follows. Here we assume that the base point $x_0 = *_M$ of $M$ is the projection $\pi(*_P)$ of that of $P$, $F$ is realized via an inclusion $i : F \to P$ as the particular fiber that passes through $*_P$, and $*_F = *_P$.

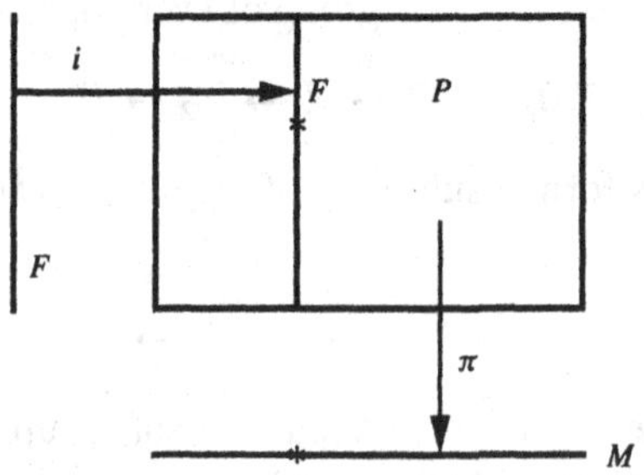


Figure 22.10

It should be clear that a continuous map $f : V \to M$ that sends base points into base points will **induce** a homomorphism $f_* : \pi_k(V) \to \pi_k(M)$, since a sphere that gets mapped into $V$ can then be sent into $M$ by $f$. This "explains" the homomorphisms

$$i_* : \pi_k(F) \to \pi_k(P)$$

and

$$\pi_* : \pi_k(P) \to \pi_k(M)$$

induced by the inclusion $i : F \to P$ and the projection $\pi : P \to M$. We must explain the remaining **boundary homomorphism** $\partial : \pi_k(M) \to \pi_{k-1}(F)$. We illustrate the case $k = 2$.

Consider $f : S^2 \to M$, defining an element of $\pi_2(M)$. This is a map of a square $I^2$ into $M$ such that the entire boundary $\dot{I}^2$ is mapped to a base point $x_0 \in M$.

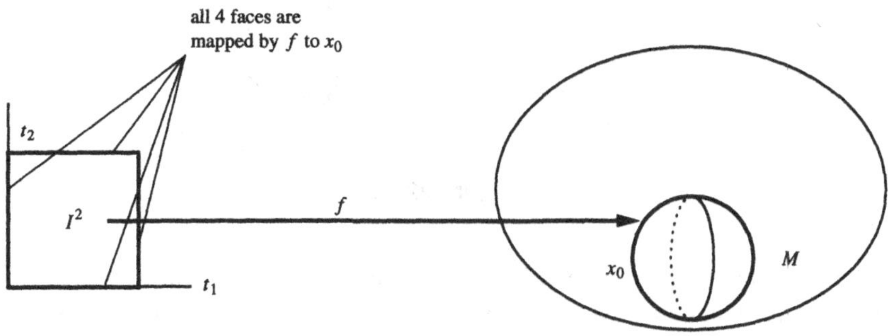


Figure 22.11

This map can be considered as a homotopy of the map given by restricting $f$ to the initial face $I$ defined by $t_2 = 0$.

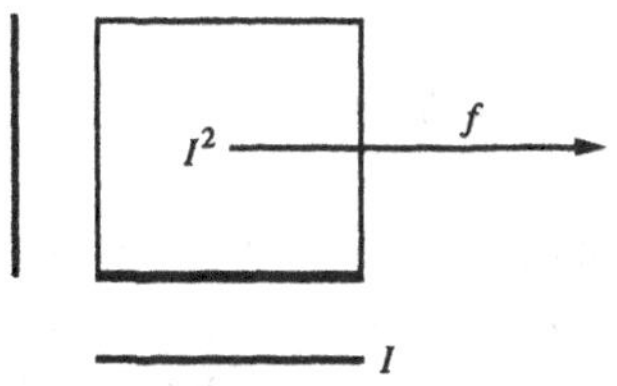


Figure 22.12

$f$ restricted to this face is of course the constant map $x_0$. The base point $*$ of $P$ lies over $x_0$. By the covering homotopy theorem, $f$ can be covered by a homotopy in $P$ of the constant map $I^1 \to *$.

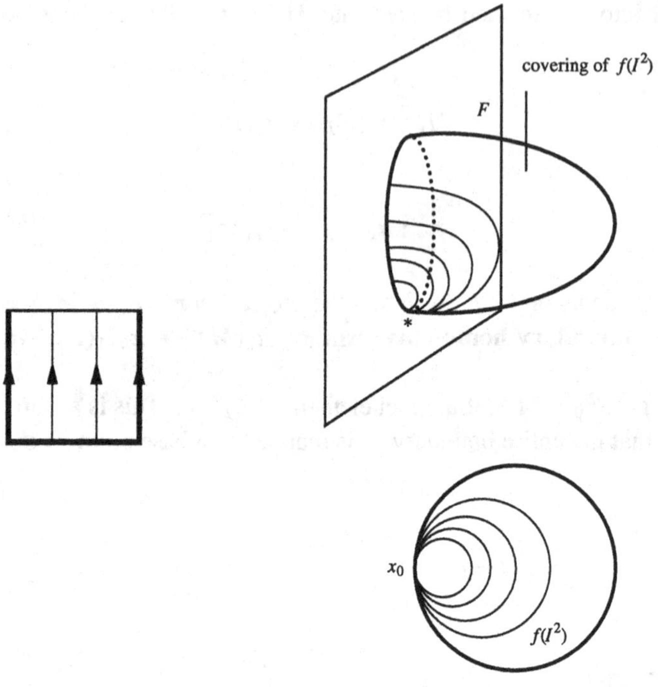


**Figure 22.13**

Under $f$, the two sides and the bottom of the square are mapped constantly to $x_0$ and the light vertical deformation curves are sent into closed curves on $f(I^2)$ since the top face is also sent to $x_0$. When these deformation curves are lifted into $P$ from $*$ they will become curves that start at $*$ and end at points of the fiber $\pi^{-1}(x_0) = F$ holding $*$, but they needn't be closed curves. Since the lines $t_1 = 0$ and $t_2 = 0$ are mapped to $x_0$, we see that these *endpoints* of the lifts of the deformation curves will form a *closed* curve in $F$, the image of some circle $S^1$ being mapped into $F$, that is, an element of $\pi_1(F)$. *This then is our assignment*

$$\partial : \pi_2(M) \to \pi_1(F)$$

Briefly speaking, *the lift of a k-sphere in M yields a k-disc in P whose boundary is a (k − 1)-sphere in F.*

We shall not prove exactness, though some parts are easy. For example, consider the portion $\pi_k(F) \to \pi_k(P) \to \pi_k(M)$. A $k$-sphere mapped into $F$ is of course also mapped into $P$. When this same sphere is projected down into $M$, the entire sphere is sent into a single point, and so is trivial. This shows that a sphere of $P$ in the image of $\pi_k(F) \to \pi_k(P)$ must always be in the kernel of $\pi_k(P) \to \pi_k(M)$. Conversely, if a sphere in $P$ is in the kernel of $\pi_k(P) \to \pi_k(M)$, then its image sphere in $M$ is contractible to the point $x_0$. By covering homotopy, the original sphere in $P$ can be deformed so as to lie entirely in the fiber $F$ over $x_0$; that is, it is in the image of $\pi_k(F) \to \pi_k(P)$. This shows that the homotopy sequence is indeed exact at the group $\pi_k(P)$. For proofs of exactness at the other groups (a few of which are easy) see [St].

Note that at the last stage, $\pi_1(P) \to \pi_1(M)$ *is onto because F has been assumed connected.* A circle on $M$ can be lifted to a curve in $P$ whose endpoints lie in $F$ and since $F$ is connected, these endpoints can be joined in $F$ to yield a closed curve in $P$ that projects down to the original circle.

### 22.3e. The Relation Between Homotopy and Homology Groups

The homology groups $H_k(M^n; \mathbb{Z})$ deal with cycles (think of *closed oriented* $k$-dimensional *submanifolds* of $M^n$); a cycle is homologous to 0 if it bounds a $(k+1)$-*chain.* The homotopy groups $\pi_k(M^n)$ deal with special cycles, namely $k$-*spheres* mapped into $M^n$. A $k$-sphere is homotopic to 0 if it can be shrunk to a point, that is, if the sphere bounds the image of a $(k+1)$-*disk*. This is the extension theorem (22.17). There are relations between these two groups. The following can be shown (but will not be used here).

Let $\pi_1$ be the fundamental group of a connected $M$. We know that $\pi_1$ is not always abelian. Let $[\pi_1, \pi_1]$ be the subgroup of $\pi_1$ generated by the commutators (elements of the form $aba^{-1}b^{-1}$). Then the quotient group $\pi_1/[\pi_1, \pi_1]$ turns out to be abelian and is isomorphic to the first homology group with integer coefficients

$$\frac{\pi_1}{[\pi_1, \pi_1]} \approx H_1(M^n; \mathbb{Z})$$

For the proof, see [G, H].

For the higher homotopy groups we have the **Hurewicz theorem** (Hurewicz was the inventor of these groups):

> Let $M$ be simply connected, $\pi_1 = 0$, and let $\pi_j(M)$, $j > 1$, be the first nonvanishing homotopy group. Then $H_j(M, \mathbb{Z})$ is the first nonvanishing homology group (for $j > 0$) and these two groups are isomorphic
>
> $$\pi_j(M^n) \approx H_j(M^n; \mathbb{Z})$$

The proof is difficult (see, e.g., [B, T]). As an example, we know that $S^n$ is simply connected for $n > 1$. Also, we know that $H_j(S^n; \mathbb{Z})$ is 0 for $1 \leq j < n$, and $H_n(S^n; \mathbb{Z}) = \mathbb{Z}$ (see (13.23)). Thus $\pi_j(S^n) = 0$, for $j < n$ and $\pi_n(S^n) = \mathbb{Z}$.

## Problems

**22.3(1)** Use Sard's theorem to show that if $f : V^k \to M^n$ is smooth and $k < n$, then $f(V)$ does not cover all of $M$.

**22.3(2)** Show that both sequences in (22.26) are exact. (Note that the first sequence is defined only for $k > 0$, but if we define $B_{-1} := 0$ the sequences make sense for all $k \geq 0$.)

Suppose we have a *compact manifold* and we consider the resulting finite *simplicial complex*, as in 13.2c. Suppose further that a *field* is used for coefficients. Then all the groups $C_k$, $Z_k$, $B_k$, $H_k$ are finite-dimensional vector spaces. Let $c_k$, $z_k$, $\beta_k$, and $b_k$ be their respective dimensions (recall that $b_k$ is the $k^{\text{th}}$

Betti number). For example $c_k$ is simply the number of $k$-simplexes in the complex. ($c_k$ is independent of the field used, but we know that $b_k$ depends on the field.) Note then that, for example, $b_k = \dim H_k = \dim(Z_k/B_k) = z_k - \beta_k$.

**(i)** Show that

$$c_k - b_k = \beta_k + \beta_{k-1}$$

for all $k \geq 0$. This is a *Morse*-type relation, as in Theorem (14.40), where now the Morse type number $m_k$ is replaced by $c_k$ and $q_k$ is replaced by $\beta_k$. We immediately have

$$c_k \geq b_k$$

that is, there are more $k$-simplexes than the $k^{\text{th}}$ Betti number. Furthermore, as in the Morse inequalities, we have for an $n$-dimensional *closed* manifold

$$\sum_{k=0}^{n}(-1)^k c_k = \sum_{k=0}^{n}(-1)^k b_k$$

This is **Poincaré's theorem**, expressing the Euler characteristic

$$\chi(M) = \sum_{k=0}^{n}(-1)^k c_k = (\text{no. vertices}) - (\text{no. edges}) + \cdots$$

*as the alternating sum of the Betti numbers.* A special case of this was noted in Problem 16.2(1).

**(ii)** What is the Euler characteristic of $S^n$, of $\mathbb{R}P^n$, of the Klein bottle?

**(iii)** Show that the Euler characteristic of a closed odd-dimensional orientable manifold vanishes (Hint: Problem 14.2(3)). Show that orientability is not really required by looking at the 2-sheeted orientable cover.

**22.3(3)** Let $A \subset M$ be a subspace of $M$. Recall (from Section 14.3) the relative homology groups $H_p(M; A)$ constructed from relative cycles $c_p$. A relative $p$-cycle $c_p$ is a *chain* on $M$ whose boundary, if any, lies on $A$. Two relative cycles $c$ and $c'$ are homologous if $c - c' = \partial m_{p+1} + a_p$, where $m$ is a chain on $M$ and $a$ is a chain on $A$. The **relative homology sequence for M mod A** is

$$\cdots \to H_{p+1}(M; A) \xrightarrow{\partial} H_p(A) \to H_p(M) \to H_p(M; A) \xrightarrow{\partial} H_{p-1}(A) \to \cdots$$

Here we are using the homomorphism induced by inclusion $A \to M$, the fact that any absolute cycle $z$ on $M$ is automatically a relative cycle, and the fact that the boundary of any relative cycle is a cycle of $A$ (which bounds on $M$ but not necessarily on A). We claim that the relative homology sequence is *exact*.

**(i)** Show that the composition of any two successive homomorphisms in the sequence is trivial.

**(ii)** Conclude the proof of exactness. (As an example, we show exactness at $H_p(M)$. From (i) we need only show that anything in the kernel of $H_p(M) \to H_p(M; A)$ must come from $H_p(A)$. But if the absolute cycle $z_p$ of $M$ is trivial as a relative cycle, we must have $z = \partial m_{p+1} + a_p$ or $z - a = \partial m$, which says that a is an absolute cycle on $A$ and the abolute cycle $z$ is homologous to it. Thus, as homology *classes* $a \to z$ and so $z$ is in the image of the homomorphism $H_p(A) \to H_p(M)$.) Simple pictures should be helpful.

(iii) By considering the sphere $S^{n-1} \subset B^n$ in the $n$-ball, and knowing the homology of $S$ and of $B$, show that

$$H_p(B^n, S^{n-1}) = \begin{matrix} 0 & \text{for } p < n \\ \mathbb{Z} & \text{for } p = n \end{matrix}$$

What is the generator of $H_n(B, S)$?

---

## 22.4. Some Computations of Homotopy Groups

How can one map a 3-sphere onto a 2-sphere in an essential way, that is, so that the map is not homotopic to a constant?

### 22.4a. Lifting Spheres from *M* into the Bundle *P*

In the definition of $\partial : \pi_k(M) \to \pi_{k-1}(F)$ in Theorem (22.27), we have explicitly shown the following (the sketch for $k = 2$ works for all $k \geq 1$; one now lifts the image of the $t_k$ lines instead of the $t_2$ lines).

**Sphere Lifting Theorem (22.28):** *Any map of a k-sphere into $M^n$ (with base point $x_0$) can be covered by a map of a k-disc into the bundlespace P, in which the boundary $(k-1)$-sphere is mapped into the fiber $F = \pi^{-1}(x_0)$.*

This has an important consequence for covering spaces. Recall that a covering space is simply a bundle over $M$ with a discrete fiber.

**Theorem (22.29):** *If $\pi : \overline{M} \to M$ is a covering space, then the homomorphism induced by projection*

$$\pi_* : \pi_k(\overline{M}, \overline{*}) \to \pi_k(M, *)$$

*is an isomorphism for $k \geq 2$. Furthermore, for $k = 1$*

$$\pi_* : \pi_1(\overline{M}, \overline{*}) \to \pi_1(M, *)$$

*is* $1 : 1$.

**PROOF:** We first show that $\pi_*$ is $1 : 1$. Let $f(I^k)$ be a map of a sphere into $\overline{M}$ that when projected down is homotopic to the constant map to $*$. This homotopy can be covered by a homotopy of $f$ into the fiber $\pi^{-1}(*)$. But if $k \geq 1$, the resulting map of a $k$-sphere into this fiber must be connected, and yet the fiber is discrete. It must be that the entire sphere is mapped to the single point $\overline{*}$. Thus if $\pi_* f$ is trivial, then $f$ itself is trivial, and $\pi_*$ is $1 : 1$ for all $k \geq 1$.

We now show that $\pi_*$ is onto for $k \geq 2$. Let $f(I^k)$ be a map of a $k$-sphere into $M$. This can be covered by a map $\overline{f}(I^k)$ of a $k$-disc into $\overline{M}$ whose boundary $(k-1)$-sphere lies in the discrete fiber $\pi^{-1}(*)$. If $k \geq 2$, this whole boundary must collapse to the point $*$. Thus $\overline{f}(I^k)$ is a map of a *k-sphere* into $\overline{M}$ that projects via $\pi$ to $f(I^k)$, and $\pi_*$ is onto. □

As simple corollaries we have

$$\begin{aligned} \pi_k(\mathbb{R}P^n) &= \pi_k(S^n) \\ \pi_k(T^n) &= \pi_k(\mathbb{R}^n) = 0 \\ \pi_k(\text{Klein bottle}) &= \pi_k(T^2) = 0 \end{aligned} \tag{22.30}$$

for all $k \geq 2$. In particular, *every map of a $k > 1$ sphere into a circle $T^1$ is contractible to a point!*

### 22.4b. *SU(n)* Again

In Corollary (22.22) and in Problem 22.2(3) we saw that $SU(n)$ is both connected and simply connected, $\pi_1 SU(n) = 0$. We now show that

$$\pi_2 SU(n) = 0$$

**PROOF:** From the fibering $SU(n-1) \to SU(n) \to S^{2n-1}$ we have the exact homotopy sequence

$$\cdots \to \pi_3 S^{2n-1} \to \pi_2 SU(n-1) \to \pi_2 SU(n) \to \pi_2 S^{2n-1} \to \cdots$$

For $n \geq 3$ this gives

$$0 \to \pi_2 SU(n-1) \to \pi_2 SU(n) \to 0$$

and so $\pi_2 SU(n) = \pi_2 SU(n-1) = \ldots = \pi_2 SU(2) = \pi_2 S^3 = 0$ □

In fact, E. **Cartan** has shown that *every map of a 2-sphere into any Lie group is contractible to a point*

$$\pi_2 G = 0 \quad \text{for every Lie group} \tag{22.31}$$

In Problem 22.4(1) you are asked to show that

$$\pi_3 SU(n) = \pi_3 SU(2) = \mathbb{Z} \quad \text{for } n \geq 2 \tag{22.32}$$

and thus *every map of a 3-sphere into SU(n), for $n \geq 3$, can be deformed to lie in an SU(2) subgroup*!

### 22.4c. The Hopf Map and Fibering

The starting point for Hurewicz's invention of the homotopy groups must have been related to Heinz Hopf's discovery of an *essential* map of $S^3$ onto $S^2$, that is, a map

$\pi : S^3 \to S^2$ that was not homotopic to a constant map. (We have seen in (22.30) that this cannot happen in the case of a 2-sphere mapped into a 1-sphere.) With our machinery we can easily exhibit this map. We know that when the group $SU(2)$ acts on its Lie algebra (or on the trace-free hermitian matrices) by the adjoint action, the resulting action covers the rotation group $SO(3)$ acting on $\mathbb{R}^3$. In particular, $SU(2)$ acts transitively on the spheres $S^2$ centered at the origin of its Lie algebra $\mathbb{R}^3$. The stability subgroup of the hermitian matrix $\sigma_3$ is immediately seen to be the subgroup

$$\begin{bmatrix} e^{i\theta} & 0 \\ 0 & e^{-i\theta} \end{bmatrix}$$

which is simply a circle group $S^1$. Thus we have the fibration

$$S^1 \to SU(2) \xrightarrow{\pi} S^2$$

From the homotopy sequence

$$0 = \pi_3 S^1 \to \pi_3 SU(2) \to \pi_3 S^2 \to \pi_2 S^1 = 0$$

we see that $\pi_3 S^2 = \pi_3 SU(2) = \pi_3 S^3 = \mathbb{Z}$, that is,

$$\pi_3 S^2 = \mathbb{Z}$$

and that the projection map

$$\pi : SU(2) \to S^2$$

the **Hopf** map, is essential.

We have shown that $S^3 = SU(2)$ is a fiber bundle over $S^2$ with (nonintersecting) circles as fibers. This is the **Hopf fibration**

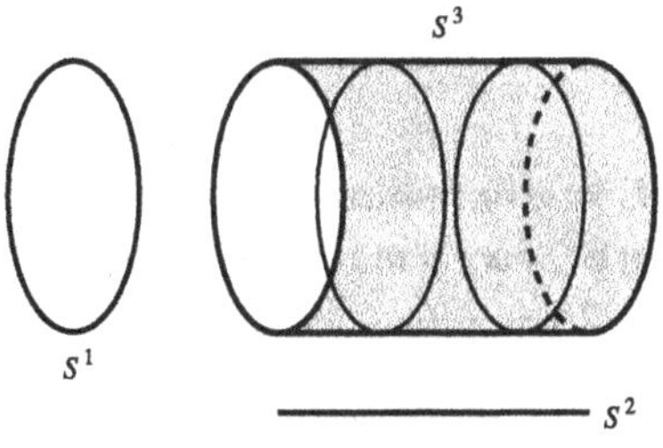


Figure 22.14

Here is another view of the Hopf map. Consider the unit 3-sphere $S^3$

$$| z_0 |^2 + | z_1 |^2 = 1$$

in $\mathbb{C}^2$. We then have a map $\pi : S^3 \to S^2$ defined by $(z_0, z_1) \to [z_0, z_1]$, where the latter pair denote the *homogeneous* coordinates of a point in $\mathbb{C}P^1$, that is, the Riemann sphere (see Section 17.4c). The inverse image of the point $[z_0, z_1]$ consists of those multiples $(\lambda z_0, \lambda z_1)$ *in* $S^3$, where $\lambda \in (\mathbb{C} - 0)$. Since $| z_0 |^2 + | z_1 |^2 = 1$, we see that $| \lambda |^2 = 1$, and so $\pi^{-1}[z_0, z_1]$ consists of all multiples $e^{i\theta}$ of $(z_0, z_1)$. This is a circle on $S^3$ passing through $(z_0, z_1)$. Thus $S^3$ can be considered as the subbundle of the Hopf complex line bundle (of Section 17.4c) consisting of unit vectors through the origin of $\mathbb{C}^2$, and then the Hopf map $\pi : S^3 \to S^2$ is simply the restriction of the projection map to this subbundle.

## Problems

**22.4(1)** Derive (22.32).

**22.4(2)** We know $\pi_1 SO(3) = \mathbb{Z}_2$. Use $SO(n)/SO(n-1) = S^{n-1}$ and induction to show that $\pi_1 SO(n) = \mathbb{Z}_2$ for $n \geq 3$.

**22.4(3)** We have stated the exact homotopy sequence for a fiber bundle in the case that the fiber is connected. When the fiber is not connected (as in the case of a covering) the only difference is that in the very last term, $\pi_* : \pi_1 P \to \pi_1 M$ need not be onto, so that we do *not* necessarily have that the sequence is exact at this last group $\pi_1 M$. Accept this fact and go on to show that Theorem (22.29) is an immediate consequence of this exact sequence.

## 22.5. Chern Forms as Obstructions

Given a closed orientable submanifold $V^4$ of $M^n$, why is $(1/8\pi^2) \int_V \mathrm{tr}(\theta \wedge \theta)$ always an integer?

### 22.5a. The Chern Forms $c_r$ for an $SU(n)$ Bundle Revisited

Let us rephrase some results that we have proved concerning Chern forms. First consider a $U(1)$ bundle.

**Theorem (22.33):** *Let E be a hermitian line bundle, with (pure imaginary) connection $\omega^1$ and curvature $\theta^2$, over a manifold $M^n$. Let $V^2$ be any closed oriented surface embedded in $M^n$. Then*

$$\frac{i}{2\pi} \int_V \theta^2 = \int_V c_1$$

*is an integer and represents the sum of the indices of any section* $\mathbf{s} : V^2 \to E$ *of the part of the line bundle over $V^2$; it is assumed that* $\mathbf{s}$ *has but a finite number of zeros on $V$.*

It is only when this integer vanishes that one can possibly find a nonvanishing section (that is, a frame over all of $V$).

Next, instantons are associated with $SU(2)$ bundles.

**Theorem (22.34):** *The winding number of the instanton is given by*

$$\frac{1}{24\pi^2}\int_{S^3} \operatorname{tr} \omega_U \wedge \omega_U \wedge \omega_U = -\int_{\mathbb{R}^4} c_2$$

This represents the "number of times the frame $\mathbf{e}_U$ on the boundary $S^3$ wraps around the frame $\mathbf{e}_V$ that is flat at infinity." It is only when this *integer* vanishes that the flat frame outside $S^3$ can be extended to the entire interior of $S^3$.

We have defined the Chern forms for a complex $U(n)$ bundle $E$ in Section 22.1c

$$\det\left(I + \frac{i\theta}{2\pi}\right) = 1 + c_1(E) + c_2(E) + \cdots \tag{22.35}$$

$$= 1 + \frac{i}{2\pi}\operatorname{tr}\theta - \frac{1}{8\pi^2}[(\operatorname{tr}\theta) \wedge (\operatorname{tr}\theta) - \operatorname{tr}(\theta \wedge \theta)] + \cdots$$

We have shown that each $c_r$ is closed, $dc_r = 0$, and thus defines a de Rham cohomology class, and that this cohomology class, with real coefficients, is independent of the connection used. The factor $i$ is introduced to make each of the forms real ($i\theta$ is hermitian). The factor $1/2\pi$ ensures that the "periods" of the Chern forms will be integers when evaluated on integral homology classes. We have already seen this in Theorem (17.24) for the case of $c_1$ for a complex line bundle over a surface and have verified a very special case of this for $c_2$ in Theorem (22.5). In this lecture we shall concentrate on the second Chern class $c_2$ but for a general $SU(k)$ bundle over a manifold.

### 22.5b. $c_2$ as an "Obstruction Cocycle"

Let $\mathbb{C}^k \to E \to M^n$ be complex vector bundle with connection. We shall be concerned with the case of most interest in physics, in which the structure group is the *special* unitary group $G = SU(k)$. We are going to evaluate

$$\int_{z_4} c_2$$

where $z_4$ is a 4-cycle on $M^n$ with integer coefficients. For simplicity we shall in fact assume that $z$ is represented by a closed oriented 4-dimensional submanifold of $M^n$.

Let us consider the problem of constructing a *frame* of $k$ linearly independent sections of the bundle $E$ *just over the cycle* $z$. Since $SU(k)$ is the structure group, *this is equivalent to constructing a section of the principal SU(k) bundle P associated to the part of E over* $z$. Each fiber is then a copy of $G = SU(k)$.

We shall attempt to find a *continuous* section, since it can be approximated then by a differentiable one.

Triangulate $z_4$ into simplexes $\Delta_4$, each of which is so small that the part of the bundle over it is trivial, $\pi^{-1}\Delta \approx \Delta \times G$. We *picture* sections as frames of vectors.

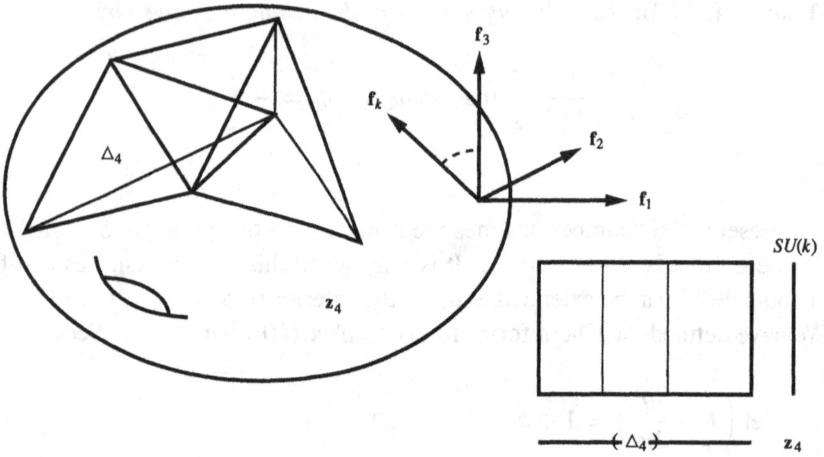


Figure 22.15

Now begin to construct a cross section. Over each 0-simplex (vertex) $\Delta_0$ we pick arbitrarily a point in $\pi^{-1}(\Delta_0)$. Thus *we have constructed a section of the bundle P over the* **"0-skeleton,"** that is, the union of all 0-simplexes.

Given $\Delta_0$, look at a 4-simplex $\Delta_4$ holding this vertex. The part of the bundle over $\Delta_4$ is trivial, $\pi^{-1}(\Delta_4) \approx \Delta_4 \times G$. To construct a *section over* $\Delta_4$ is simply to give a continuous map $f : \Delta_4 \to \Delta_4 \times G$ of the form $x \to (x, g(x))$ that extends the given $f$ over the 0-skeleton.

Let $\Delta_1$ be a 1-simplex of the triangulation. This is a map $\sigma$ of $I$ into $z_4$. Pick a $\Delta_4$ holding $\Delta_1$. $g$ is defined on the two vertices $P$ and $Q$ of $\Delta_1 = I$; that is, $g(P)$ and $g(Q)$ are two points in $G$.

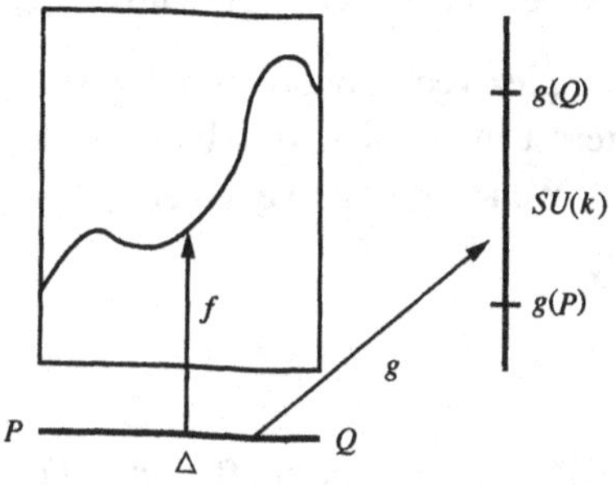


Figure 22.16

Since $G = SU(k)$ is *connected,* these two points can be joined by a curve $g : I \to G$. Then define $f : \Delta_1 \to \Delta_1 \times G$ by $f(t) = (\sigma(t), g(t))$. In this way *we have extended the cross section to each* $\Delta_1$ *and thus over the entire 1-skeleton.*

We now have the section $f$ defined on the *boundary* of each 2-simplex $\Delta_2$; can we extend to the entire $\Delta_2$?

Letting $\pi_G$ be the local projection of $\pi^{-1}\Delta_4 = \Delta_4 \times G$ onto $G$, we see that $\pi_G \circ f$ is a map of $\partial\Delta_2$, topologically a circle, into the group $SU(k)$. We know from the "extension theorem" (22.17) that this map can be extended to a map of the "disc" $\Delta_2$ if and only if it is homotopic to a constant map. But $SU(k)$ is *simply connected* (Problem 22.2(3)), and so any map of a circle into $G$ is homotopic to a constant map, and $\pi_G \circ f$ *can* be extended to a map $F : \Delta_2 \to G$. Define then a section over $\Delta_2$ by $f(x) = (x, F(x))$, an extension of $f$ over $\partial\Delta_2$. *We have extended f to the entire 2-skeleton of the 4-manifold z.*

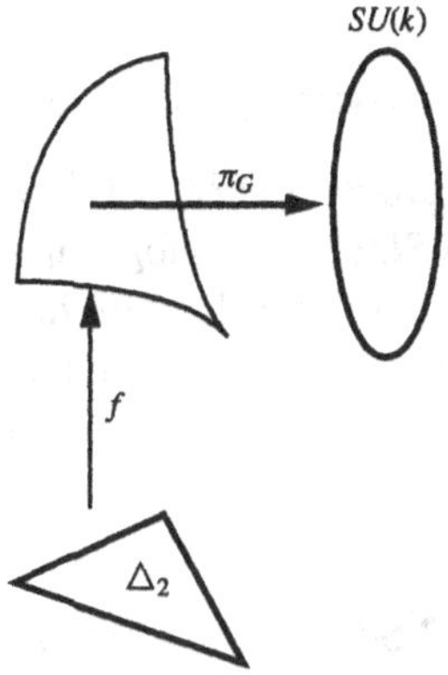


**Figure 22.17**

We have defined $f$ on the boundary of each 3-simplex. Since each $\Delta_3$ is topologically a 3-disc with boundary a topological 2-sphere $\Delta_2$, and $f_G = \pi_G \circ f$ is a map of $\partial\Delta_3$ into $G$, we know that this map can be extended to all of $\Delta_3$ if and only if $f_G : \partial\Delta_3 \to G$ is homotopic to a constant. But $\pi_2(SU(k)) = 0$, (22.31), and thus $f_G$ *is* homotopic to a constant. As before, *this allows us to extend the section* $f$ *to the entire 3-skeleton.*

$f$ is now defined on the 3-sphere boundary $\partial\Delta_4$ of each 4-simplex $\Delta_4$. But now $\pi_3(SU(k))) = \mathbb{Z}$, (22.32), and $f_G : \partial\Delta_4 \to SU(k)$ need not be homotopic to a constant. We have met with a possible **obstruction** to extending $f$ to the entire $\Delta_4$ in question! We "measure" this obstruction as follows: The homotopy class of $f_G : \partial\Delta_4 \to SU(k)$ is characterized, from (22.32), by an integer (call it $j(\Delta_4)$, and *we assign this integer to the 4-simplex* $\Delta_4$.

There is now a slight complication. Different $\Delta_4$'s in the 4-manifold $z_4$ will yield different trivializations of the bundle; that is, the $SU(k)$ coordinate in the frame bundle over $\Delta_4$ changes with the simplex $\Delta_4$. Consider for example, the case of $SU(2)$, which is topologically $S^3$. When we map $\partial\Delta_4$ into $SU(2)$ we shall be using different copies of $SU(2)$, that is, different 3-spheres over different simplexes. If we change the orientation of the 3-sphere, our integer $j$ will change sign. We shall assume that the fibers $SU(2)$ can be coherently "oriented." Similarly, we shall assume that the fibers $SU(k)$ can be coherently "oriented" so that the sign ambiguity in $\pi_3 SU(k)$ is not present. Steenrod called such a bundle **orientable**.

In this manner we assign to each 4-simplex in the triangulation of the 4-manifold $z$ a definite integer; thus we have a singular 4-chain on $z$ with integer coefficients, called the **obstruction cocycle**. The reader should note that we did not really use the fact that the fiber $F$ was $SU(k)$; *the only information that was used was that the fiber was connected and that* $\pi_j(F) = 0$ *for* $j = 1, 2,$ *and that* $\pi_3(F) = \mathbb{Z}$.

If each coefficient is 0 it is clear that one can extend the section to the interior of each $\Delta_4$, and in this case we have succeeded in finding a section on all of $z$! If, on the other hand, some of the integers are not 0, it still *may* be possible to start anew and succeed. *This will be the case if the sum* $\sum j(\Delta) = 0$, where the $\Delta$ are oriented so that $z_4 = \sum \Delta$. This can be shown using the cohomology theory of obstructions. *We wish, rather, to show how this sum can be expressed as an integral involving curvature.*

### 22.5c. The Meaning of the Integer $j(\Delta_4)$

The fact that $\pi_3(SU(k)) = \mathbb{Z}$, proved in (22.32), is a result of two things. First, each $SU(k), k \geq 3$, has the 3-sphere $SU(2)$ as a subgroup and then the homotopy sequence shows that this 3-sphere is a generator for the third homotopy group of $SU(k)$. In other words, every map of a 3-sphere into $SU(k)$ can be deformed so that its image lies on the $SU(2)$ subgroup! *But then a map* $f_G : (\partial \Delta_4 = S^3) \to S^3$ *has a degree, and this integer is* $j(\Delta_4)$.

### 22.5d. Chern's Integral

The partial cross section $f : \partial \Delta_4 \to P$ on the simplex $\Delta_4$ is defined only on its boundary, but we can immediately extend it to all of $\Delta_4$ with a small 4-ball $B_\epsilon$ about its barycenter $x_0$ removed; *we merely make the SU(k) coordinates* $f_G$ *constant along radial lines leading out to* $\partial \Delta_4$.

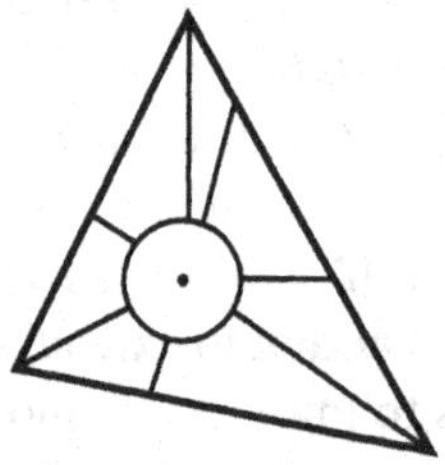

Figure 22.18

We shall now compute the integral of the second Chern form over $\Delta_4$; since $c_2$ is a smooth form and is independent of the section

$$\int_{\Delta_4} c_2 = \lim_{\epsilon \to 0} \int_{\Delta_4 - B_\epsilon} c_2$$

We will be brief since the procedure is similar to that in Section 17.3b.

Let $\sum_4 = f(\Delta_4 - B_\epsilon)$ be the "graph" of the local section. Then

$$\int_{\Delta_4 - B_\epsilon} c_2 = \int_{\pi \sum_4} c_2 = \int_{\sum_4} \pi^* c_2$$

Now

$$c_2 = -\frac{1}{8\pi^2}[\operatorname{tr}\theta \wedge \operatorname{tr}\theta - \operatorname{tr}(\theta \wedge \theta)]$$
$$= \frac{1}{8\pi^2} \operatorname{tr}(\theta \wedge \theta)$$

Let $\mathbf{e}_U$ be a frame on the open $U \subset M^n$ holding $\Delta_4$ for which $\pi^{-1}(U)$ is a product $U \times SU(k)$. $\theta = \theta_U$ is the local curvature 2-form for the vector bundle $E$. $c_2 = (1/8\pi^2)\operatorname{tr}\theta_U \wedge \theta_U$. Then at a frame $\mathbf{f} = \mathbf{e}_U g$ we have

$$\begin{aligned}\pi^* c_2 &= \frac{1}{8\pi^2}\operatorname{tr}\pi^*\theta_U \wedge \pi^*\theta_U \\ &= \frac{1}{8\pi^2}\operatorname{tr}[g^{-1}\pi^*\theta_U g \wedge g^{-1}\pi^*\theta_U g]\end{aligned}$$

and from (18.21)

$$\pi^* c_2 = \frac{1}{8\pi^2}\operatorname{tr}\theta^* \wedge \theta^*$$

where $\theta^*$ is the *globally defined* curvature form *on the frame bundle*, $\theta^* = d\omega^* + \omega^* \wedge \omega^*$, where again $\omega^*$ is globally defined. The same calculation that gave (22.4) shows

$$\pi^* c_2 = d\frac{1}{8\pi^2}\operatorname{tr}\left[\omega^* \wedge d\omega^* + \frac{2}{3}\omega^* \wedge \omega^* \wedge \omega^*\right] \tag{22.36}$$

Thus *the pull-back of $c_2$ to the frame bundle is the differential of a globally defined 3-form, the* Chern–Simons *form*. Thus, for the graph $\sum_4$ of our section $f$ over $z_4 - \cup B_\epsilon$,

$$\int_{\sum}\pi^* c_2 = \frac{1}{8\pi^2}\int_{\partial\sum}\operatorname{tr}\left[\omega^* \wedge d\omega^* + \frac{2}{3}\omega^* \wedge \omega^* \wedge \omega^*\right] \tag{22.37}$$

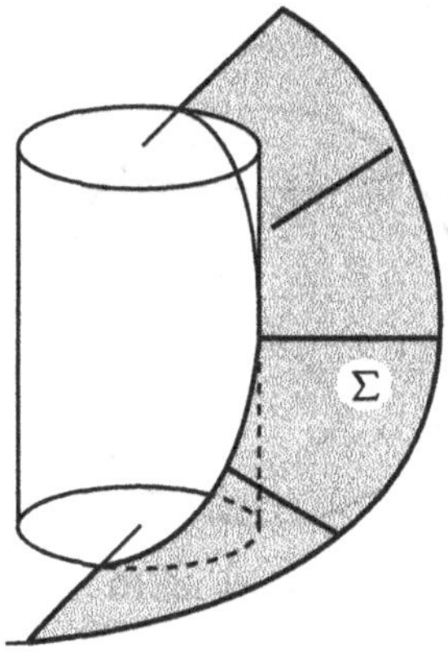


Figure 22.19

Recall that we have removed 4-balls from the 4-cycle $z_4$. The boundary of $\sum$ over the 4-cycle $z_4$ consists of the part of the section $f$ over the union of the boundary of the $\epsilon$-balls, but with orientation *opposite* to that of the balls (since $f$ is $1:1$, $\sum$ carries an orientation induced from that of $z$).

$$\int_{\sum}\pi^* c_2 = -\frac{1}{8\pi^2}\sum\int_{f(\partial B_\epsilon)}\operatorname{tr}\left[\omega^* \wedge d\omega^* + \frac{2}{3}\omega^* \wedge \omega^* \wedge \omega^*\right] \tag{22.38}$$

Now over $U$, for points of $\sum$

$$\omega^* = g^{-1}\pi^*\omega_U(x)g + g^{-1}dg \tag{22.39}$$

where the section $f$ is given by $f(x) = \mathbf{e}_U(x)g(x)$. The triple integral for that $B_\epsilon$ in $U$ will involve terms containing $g^{-1}dg$ and

$$\omega_U = (\omega^\alpha_{j\,\beta}(x)dx^j)$$

In the integral (22.38), gather together all those terms that do *not* involve any $dx$; one finds easily that the contribution of these terms is

$$\frac{1}{24\pi^2}\sum\int_{f(\partial B_\epsilon)} \operatorname{tr} g^{-1}dg \wedge g^{-1}dg \wedge g^{-1}dg \tag{22.40}$$

As in (22.3), we see that the integral in (22.38) over $f(\partial B_\epsilon)$ represents the number of times that the image $f(\partial B_\epsilon)$ of the $\epsilon$-sphere wraps around the $SU(2)$ subgroup of $SU(k)$! Furthermore, since the integrand, the Cartan 3-form $\Omega_3$, is closed, Stokes's theorem tells us that this also *represents the number of times that the image of* $\partial\Delta_4$ *wraps around* $SU(2)$ where $\Delta_4$ is the simplex holding the given singularity. But *this is precisely the index* $j(\Delta_4)$ *that occurred in the obstruction cocycle* in 22.5c. We have shown that

$$\int_{z-\cup B_\epsilon} c_2 = \sum j(\Delta_4) + \text{integrals involving } dx \tag{22.41}$$

Now we can let $\epsilon \to 0$. The left side tends to the integral of $c_2$ over the entire 4-cycle $z$.

We claim that the integrals involving $dx$ all tend to 0. Introduce coordinates $x^1$, $x^2$, $x^3$, $x^4$ with origin at the singularity in question.

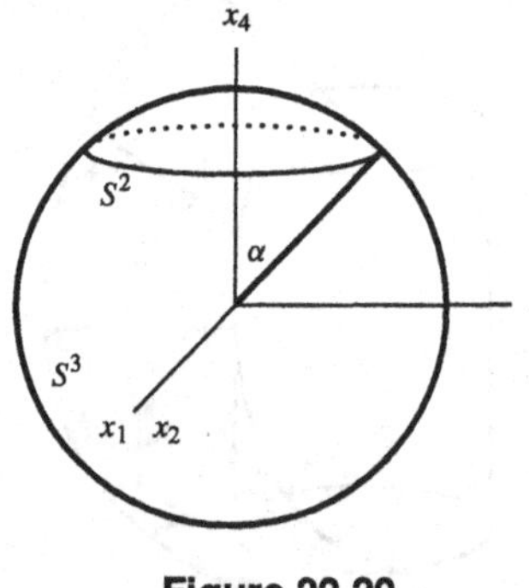


**Figure 22.20**

For $\partial B_\epsilon$ we choose the 3-sphere given by $\sum x_j^2 = \epsilon^2$. This can be parameterized by angles $\alpha, \theta$, and $\phi$; $x^1 = \epsilon \sin\alpha \sin\theta \cos\phi$, $x^2 = \epsilon \sin\alpha \sin\theta \sin\phi$, $x^3 = \epsilon \sin\alpha \cos\phi$, $x^4 = \epsilon \cos\alpha$. Each integral in (22.38), $\int_{f(\partial B)} A$, can be evaluated as the integral of a pull-back $\int_{\partial B} f^*A$. Let $\phi^1 = \alpha, \phi^2 = \theta, \phi^3 = \phi$. The pull-back of a term like $g^{-1}dg$ will be of the form $G_j(\phi)d\phi^j$ where the $G_j$ are differentiable and *independent of* $\epsilon > 0$ since we have extended the section to the interior of $\Delta_4$ keeping $g$ constant along radial lines $\phi =$ constant. Furthermore, since we have already taken care of the term involving $\Omega_3$, each integral will also involve $dx$ through $\omega = f^*\omega^*$, and $dx^i$ is of the form $(\partial x^i/\partial\phi^j)d\phi^j$, which will have a factor of $\epsilon$. Since the functions $\omega^\alpha_{j\,\beta}(x)$ are differentiable, we conclude that all the remaining integrals on the right-hand side of (22.39) vanish in the limit as $\epsilon \to 0$. We have proved the following special case of a theorem of **Chern.**

**Theorem (22.42):** *Let $P$ be a principal $SU(k)$ bundle over $M^n$. Then the integral*

$$\int_{z_4} c_2$$

*represents the following. There always exists a section $f : (z_4 - \cup p_\alpha) \to P$ over $z$ except, perhaps, for a finite number of points $\{p_\alpha\}$. About each $p_\alpha$ we construct a small 3-sphere $S^3_\alpha$ and map it into $SU(k)$ by means of the section $f$ followed by the local projection $\pi_G$ of the bundle into $SU(k)$. The image of $S_\alpha$ in $SU(k)$ can be deformed so as to lie on an $SU(2)$ subgroup. Let $j_\alpha(f)$ denote the number of times that the image covers $SU(2)$, that is,*

$$j_\alpha(f) := \text{Brouwer degree of } f \circ \pi_G : S^3_\alpha \to SU(2)$$

*Then*

$$\int_{z_4} c_2 = \sum j_\alpha(f)$$

*Thus $\sum j_\alpha(f)$ is independent of the section $f$! In particular, a section on all of $z$ exists only if $\int_z c_2 = 0$. It is also immediate that*

$$\int_{z_4} c_2$$

*is an integer for each integer cycle $z$! Furthermore, this integral*

$$\frac{1}{8\pi^2} \int_{z_4} \operatorname{tr} \theta \wedge \theta$$

*is independent of the $SU(k)$ connection used in the bundle!*

### 22.5e. Concluding Remarks

If our group $G = SU(k)$ had not been simply connected, for example, if it had been $U(k)$, then, in our construction, we would have met an obstruction to a section of the $k$-frame bundle already at the 2-skeleton. The problem then would have been to try to construct a section over a 2-cycle, rather than a 4-cycle. The measure of the obstruction then would be the integral of the *first* Chern form $c_1$ over the 2-cycle. It turns out that for a $U(k)$ bundle, the integral of $c_2$ over a 4-cycle measures the obstruction to constructing not a $k$-frame, but rather a $(k-1)$-frame section, that is, finding $(k-1)$ linearly independent sections of the original bundle. It is easy to see, however, that if the group is $SU(k)$, then a $(k-1)$-frame can then lead to a unique $k$-frame. For example, in $\mathbb{C}^2$, the most general unit vector orthogonal to $(1\ 0)^T$ is of the form $(0\ e^{i\theta})^T$ and is thus not unique, but if we demand that the pair $(1\ 0)^T$ and $(0\ e^{i\theta})^T$ have determinant $+1$ then $(0\ e^{i\theta})^T$ must reduce to the unique $(0\ 1)^T$. That is why we considered directly the search for a $k$-frame. The *general* situation is as follows.

**Chern's Theorem (22.43):** *Let $E$ be a complex vector bundle with structure group $U(k)$ and connection $\omega$ over $M^n$. Then each Chern form $c_r$ defines via de Rham an integral cohomology class, that is, the $2r$-form $c_r$ has integral periods* *on a basis of $H_{2r}(M; \mathbb{Z})$. This class is called a characteristic cohomology class and represents an obstruction to the construction of a cross section to a bundle associated to E, namely the bundle of $(k - r + 1)$ frames! Although the forms $c_r$ depend on the connection used in the bundle E, their periods do not.*

Note that we had considered orthonormal frames of $p$ vectors in $\mathbb{R}^n$ in Problem 17.2(3); the space of all such frames forms the (real) Stiefel manifold $O(n)/O(n-p)$. Similarly, the space of all orthonormal frames of $p$ *complex* vectors in $\mathbb{C}^n$ forms the **complex Stiefel manifold** $U(n)/U(n-p)$. For example, the 1-frames in $\mathbb{C}^n$ form the unit sphere $S^{2n-1}$, and it is easily seen that $S^{2n-1}$ is $U(n)/U(n-1)$, since $U(n)$ acts transitively on this sphere.

Besides Chern classes, there are other characteristic classes, the Stiefel–Whitney classes and Pontrjagin classes, which were defined before the Chern classes. We have dealt with the first and second Chern characteristic classes in terms of obstructions to constructing cross sections to $U(n)$ bundles. For many purposes modern treatments consider characteristic classes from a different, more axiomatic viewpoint. The interested reader might refer to [M, S] for such questions.

## Problems

**22.5(1)** Consider the real *unit* tangent bundle $T_0M^n$ to a compact orientable Riemannian $n$-manifold (see Section 2.2b). This fiber bundle has fiber $S^{n-1}$. Mimic our obstruction procedure to show that one can find a section on the $(n-1)$-skeleton of a triangulation, and then one can find a section on all of $M^n$ – except perhaps for a finite collection of points. Hopf's theorem (16.12) states that the index sum is the Euler characteristic.

**22.5(2)** The *unit* normal bundle to a closed surface $V^2$ embedded in a Riemannian $M^5$ is a 2-sphere bundle over $V^2$. Show that one can always find a section; that is, there is always a unit normal vector field to a $V^2$ in $M^5$. What about for a $V^2$ in an $M^4$?

APPENDIX A

# Forms in Continuum Mechanics

We shall assume the reader has read sections $\mathbb{O}$.p, $\mathbb{O}$.q, and $\mathbb{O}$.r of the Overview.

## A.a. The Equations of Motion of a Stressed Body

Let $x = (x^i)$ be a fixed **cartesian** coordinate system in $\mathbb{R}^3$ with coordinate basis vectors $\boldsymbol{\partial}_i$. Let $M^3(t)$ be a moving compact body acted on perhaps by surface forces on its boundary $\partial M$. We shall consider vector (contravariant or covariant) valued 2-forms, principally the Cauchy stress form

$$\mathfrak{t}^2 = \boldsymbol{\partial}_i \otimes \mathfrak{t}^i = \boldsymbol{\partial}_i \otimes t^{ij} i(\boldsymbol{\partial}_j)\mathrm{vol}^3$$

Consider a compact moving sub-body $\mathbf{B}(t)$ contained in the interior of $M(t)$. In cartesian coordinates the equations of motion of $\mathbf{B}(t)$ are obtained from equating the time rate of change of momentum of $\mathbf{B}(t)$ with the total "body" force (for example, gravity) acting on $\mathbf{B}(t)$ and the stress force acting on the boundary $\partial\mathbf{B}(t)$ arising from the stress force exerted on $\mathbf{B}(t)$ by the remainder of the body. Let

$$m^3 := \rho(x(t))\mathrm{vol}^3$$

be the mass 3-form. We assume conservation of mass (see 4.3c)

$$d/dt \int_{\mathbf{B}(t)} m^3 = \int_{\mathbf{B}(t)} \mathcal{L}_{\mathbf{v}+\partial/\partial t} m^3 = 0$$

Let $\mathbf{b}$ be the external force density (per unit mass); we have

$$\begin{aligned} d/dt \int_{\mathbf{B}(t)} v^i m^3 &= \int_{\mathbf{B}(t)} b^i m^3 + \int_{\partial\mathbf{B}(t)} \mathfrak{t}^i \\ &= \int_{\mathbf{B}(t)} b^i m^3 + \int_{\mathbf{B}(t)} d\mathfrak{t}^i \end{aligned} \tag{A.1}$$

Thus, using $\int_{\mathbf{B}(t)} \mathcal{L}_{\mathbf{v}+\partial/\partial t}(v^i m^3) = \int_{\mathbf{B}(t)} (\mathcal{L}_{\mathbf{v}+\partial/\partial t} v^i) m^3$, we have

$$\begin{aligned} &[\partial v^i/\partial t + v^j(\partial v^i/\partial x^j)]m^3 = b^i m^3 + d\mathfrak{t}^i \\ &d\mathfrak{t}^i = d[t^{i1} dx^2 \wedge dx^3 + t^{i2} dx^3 \wedge dx^1 + t^{i3} dx^1 \wedge dx^2] = [\partial t^{ij}/\partial x^j]\text{vol}^3 \end{aligned} \tag{A.2}$$

As mentioned in the derivation of (4.46), this *derivation* makes no sense in a general Riemannian manifold with curvilinear coordinates $u$, but the final formula above, Cauchy's equations of motion, can be rewritten so that they make sense in these situations, by replacing partial derivatives by covariant derivatives with respect to $u^j$

$$\left[\partial v^i/\partial t + v^j v^i{}_{/j}\right] m^3 = b^i m^3 + t^{ij}{}_{/j}\text{vol}^3 \tag{A.3}$$

Look now at equation (A.3) with any dual frames $\mathbf{e}$ and $\sigma$ for a Riemannian $M^n$. We are assuming that $\mathfrak{t} = \mathbf{e}_r \otimes \mathfrak{t}^r$ is an $(n-1)$-form section of the tangent bundle; thus from equation (9.31) we have

$$\nabla \mathfrak{t} = \nabla(\mathbf{e}_r \otimes \mathfrak{t}^r) = \mathbf{e}_r \otimes (d\mathfrak{t}^r + \omega^r{}_s \wedge \mathfrak{t}^s) = \mathbf{e}_r \otimes \nabla \mathfrak{t}^r$$

where $\omega$ is the connection form matrix for the frame $\mathbf{e}$ on $M^n$. If, temporarily, we use $r, s, \ldots$ for bundle indices and $i, j, \ldots$ for $M$ indices (both sets run in this case from 1 to $n$), we would write, as usual

$$\omega^r{}_s = \omega^r_{js} \sigma^j$$

Equation (A.3) can be written

$$[\partial v^r/\partial t + v^s v^r{}_{/s}] m^3 = b^r m^3 + \nabla \mathfrak{t}^r \tag{A.4}$$

which is our final form of Cauchy's equations. A specific computation involving equation (A.4) in spherical coordinates is given in section A.e.

## A.b. Stresses are Vector Valued $(n-1)$ *Pseudo*-Forms

In considering the stress force on a tiny hypersurface, the transverse orientation of the normal to the hypersurface must be given and, as we have seen in the opening paragraph of Section $\mathbb{C}$.p, the stress vector for a tiny hypersurface element is reversed if we change "sides" of the hypersurface, that is, if the transverse orientation is reversed. Thus the stress form is a *pseudo*-form. When no confusion can arise, we shall omit the statement that $\mathfrak{t}$ is a pseudo-form, rather than a true form.

As we have shown in ($\mathbb{C}$.44) and ($\mathbb{C}$.45), for most elastic bodies, the Cauchy stress tensor is symmetric, i.e., in cartesian coordinates $x^i$, *used for the form part as well as the vector part*, we have

$$dx^i \wedge \mathfrak{t}^j = dx^j \wedge \mathfrak{t}^i$$

and then

$$t^{ij} = t^{ji} \tag{A.5}$$

and since the stress tensor *is* a tensor, this last symmetry holds in *any* coordinates $x^i$.

## A.c. The Piola–Kirchhoff Stress Tensors $S$ and $P$

Consider a body $M^3$ in $\mathbb{R}^3$ and a diffeomorphism $M \to \Phi(M)$ of this body in $\mathbb{R}^3$. Let $\mathbf{B} \subset M$ be a compact portion of the original body with image $\Phi(\mathbf{B})$. We use *any* local coordinates $X^R$ for $\mathbf{B}$ and *any* local coordinates $x^r$ for $\Phi(\mathbf{B})$, and we write $\Phi$ in the form $x^r = x^r(X)$, using the notation of Section 2.7b.

It is traditional in engineering to use capital letters for the coordinates $X$ and the volume form VOL in the "reference" body $\mathbf{B}$ and lower case letters for the coordinates $x$ and vol in the "current" body $\Phi(\mathbf{B})$. For simplicity we shall initially use the same coordinates for the form and vector parts in the Cauchy stress form.

The Cauchy stress form $\mathfrak{t} = \boldsymbol{\partial}_i \otimes \mathfrak{t}^i = \boldsymbol{\partial}_i \otimes t^{ij} i(\boldsymbol{\partial}_j)\, \mathrm{vol}^3$ on $\Phi(\mathbf{B})$ is a vector valued 2-form on $\Phi(\mathbf{B})$. We define, as in (©.36), the **(second) Piola–Kirchhoff stress form** $\mathbb{S}$ on $\mathbf{B}$ by pushing the vector part $(\boldsymbol{\partial}_i)$ back to $\mathbf{B}$ via $(\Phi^{-1})_*$ (which is well defined since the diffeomorphism $\Phi$ is 1–1), and pulling the form part $\mathfrak{t}^i$ back to $\mathbf{B}$

$$\mathbb{S} = [(\Phi^{-1})_*(\boldsymbol{\partial}_i)] \otimes \Phi^* \mathfrak{t}^i = \boldsymbol{\partial}_C \Phi^*[(\partial X^C/\partial x^i)] \otimes \Phi^* \mathfrak{t}^i = \boldsymbol{\partial}_C \otimes \Phi^*[(\partial X^C/\partial x^i \mathfrak{t}^i]$$

(where $\Phi^*[(\partial X^C/\partial x^i]$ merely says express $\partial X^C/\partial x^i$ in terms of coordinates $X$), which is of the form

$$\mathbb{S} = \boldsymbol{\partial}_C \otimes S^{CA} i(\boldsymbol{\partial}_A)\mathrm{VOL} = \boldsymbol{\partial}_C \otimes \mathbb{S}^C \tag{A.6}$$

where

$$\mathbb{S}^C = \Phi^*[(\partial X^C/\partial x^i)\mathfrak{t}^i]$$

We know that the Cauchy stress tensor is symmetric, (©.44) and (©.45). What about $\mathbb{S}$? Note first that $\Phi^*(dx^a) = (\partial x^a/\partial X^A)dX^A$ and so $dX^A = \Phi^*[(\partial X^A/\partial x^a)dx^a]$. Then

$$\begin{aligned}
dX^A \wedge \mathbb{S}^B &= dX^A \wedge \Phi^*[(\partial X^B/\partial x^b)\mathfrak{t}^b] \\
&= \Phi^*[(\partial X^A/\partial x^a)dx^a] \wedge \Phi^*[(\partial X^B/\partial x^b)\mathfrak{t}^b] \\
&= \Phi^*[(\partial X^A/\partial x^a)(\partial X^B/\partial x^b)]\, \Phi^*(dx^a \wedge \mathfrak{t}^b) \\
&= \Phi^*[(\partial X^A/\partial x^a)(\partial X^B/\partial x^b)]\, \Phi^*(dx^b \wedge \mathfrak{t}^a) \\
&= \Phi^*[(\partial X^B/\partial x^b)dx^b] \wedge \Phi^*[(\partial X^A/\partial x^a)\mathfrak{t}^a] \\
&= dX^B \wedge \mathbb{S}^A
\end{aligned}$$

shows that $\mathbb{S}$ has the same symmetry as $\mathfrak{t}$, that is, the second Piola–Kirchhoff stress tensor is symmetric

$$\mathbb{S}^{AB} = \mathbb{S}^{BA} \tag{A.7}$$

Though we shall not require it, there is a **(first) Piola–Kirchhoff 2-form**

$$\mathcal{P} = \boldsymbol{\partial}_i \otimes \mathcal{P}^i = \boldsymbol{\partial}_i \otimes P^{iR} i(\boldsymbol{\partial}_R)\mathrm{VOL}^3 := \boldsymbol{\partial}_i \otimes \Phi^* \mathfrak{t}^i \tag{A.8}$$

i.e., we pull back the Cauchy stress 2-form to the reference body **B**, but we leave the vector value at $\Phi(\mathbf{B})$. Thus for vectors **V** and **W** at $X \in \mathbf{B}$,

$$\mathcal{P}(\mathbf{V}, \mathbf{W}) := \text{the vector } \mathfrak{t}(\Phi_* \mathbf{V}, \Phi_* \mathbf{W}) \text{ at } \Phi(X) \tag{A.9}$$

The tensor $(P^{iR})$ is called a "2-point tensor" since $i$ is an index on $\Phi(\mathbf{B})$ but $R$ is an index on **B**. For this reason one cannot talk about symmetry in the pair $(i, R)$.

## A.d. Strain Energy Rate

Consider a body $M^3$ in $\mathbb{R}^3$ with a given **cartesian** coordinate system $(X^A)$, and let $\mathbf{B} \subset M$ be our reference sub-body. Let $\phi_t : M \to \mathbb{R}^3$ be a be a 1-parameter family of diffeomorphisms of $M$ into $\mathbb{R}^3$ with $\phi_0$ equal to a map $\Phi$, and put $\mathbf{B}_t = \mathbf{B}(t) = \phi_t(\mathbf{B})$ and $M_t = \phi_t(M)$. Let $(x^a)$ be a cartesian coordinate system identical to $(X^A)$ to be used for the image points.

We suppose that there are no external body forces **b** such as gravity that act on $M_t$, but external surface forces on $\partial M_t$ will be transmitted to $\partial B_t$ via the Cauchy stresses in $\phi_t(M)$. We shall also assume that the deformations are so slow that we can neglect the velocities imparted to $M$ by the deformations. From (©.43) or (A.4) we have $d\mathfrak{t}^a = 0$ in $\mathbf{B}_t$.

An example to keep in mind is the very slow twisting of the cylinder presented in the Overview by surface forces on the ends $z = 0$ and $z = L$.

Since our coordinates are cartesian we can put indices up or down at our convenience.

The "power," i.e., the rate at which a force **F** does work in moving a particle with velocity **v** is $\mathbf{F} \cdot \mathbf{v}$. Then the rate at which the Cauchy stress forces do work on the boundary $\partial B_t$ at $t = 0$ is, putting $v^a = [dx^a/dt]_{t=0}$ and using $dt^a = 0$ and the symmetry (A.5)

$$\begin{aligned}
[dW/dt]_0 &= \int_{\partial\Phi(\mathbf{B})} v_a \mathfrak{t}^a = \int_{\Phi(\mathbf{B})} d(v_a \mathfrak{t}^a) = \int_{\Phi(\mathbf{B})} (dv_a \wedge \mathfrak{t}^a) \\
&= \int_{\Phi(\mathbf{B})} [(\partial v_a/\partial x^b)(dx^b \wedge \mathfrak{t}^a)] \\
&= \tfrac{1}{2} \int_{\Phi(\mathbf{B})} [(\partial v_a/\partial x^b) + (\partial v_b/\partial x^a)]\, dx^b \wedge \mathfrak{t}^a
\end{aligned} \tag{A.10}$$

The actual deformations $\phi_t$ of $M$ generically lead to a *time dependent* velocity field **v** but our power $[dW/dt]_0$ depends only on the velocity field **v** at time 0. Hence, for our calculation we may replace the deformations $\phi_t$ by the time flow (which we write as $\psi_t$) that results from flowing along the integral curves of the velocity field **v** frozen at $t = 0$. We again have (A.10) for this flow. Let us pause to interpret these partial derivatives in (A.10).

**Theorem (A.d):** *If* **v** *is a time independent vector field on a Riemannian $M^n$ then the Lie derivative of the metric tensor has components*

$$(\mathcal{L}_{\mathbf{v}} g)_{ab} = v_{a/b} + v_{b/a}$$

**PROOF:** Let $\psi_t$ be the flow generated by $\mathbf{v}$. Let $X$ and $Y$ be vector fields **invariant** under the flow $\psi$, i.e., $(\mathcal{L}_{\mathbf{v}}X)^a = v^r X^a{}_{/r} - X^r v^a{}_{/r} = 0$ and likewise for $Y$. Then,

$$
\begin{aligned}
(\mathcal{L}_{\mathbf{v}}g)(X,Y) &= (\mathcal{L}_{\mathbf{v}}g)_{ab}X^a Y^b \\
&= d/dt(g_{ab}X^a Y^b) \text{ (from (4.18)) and } g_{ab/r} = 0) \\
&= g_{ab}(X^a{}_{/r}v^r Y^b + X^a v^r Y^b{}_{/r}) \\
&= g_{ab}(X^r v^a{}_{/r} Y^b + X^a v^b{}_{/r} Y^r) = X^r v_{b/r} Y^b + X^a v_{a/r} Y^r \\
&= X^a (v_{b/a} + v_{a/b}) Y^b
\end{aligned}
$$

as desired.

Now in cartesian coordinates, $v_{b/a} = \partial v_b/\partial x^b$ and so (A.10) says, for our field $\mathbf{v}$ "frozen" at time $t=0$

$$
\begin{aligned}
[dW/dt]_0 &= \tfrac{1}{2}\int_{\Phi(\mathbf{B})} (\mathcal{L}_{\mathbf{v}}g)_{ab} dx^b \wedge \mathfrak{t}^a \\
&= \tfrac{1}{2}\int_{\Phi(\mathbf{B})} [(\partial/\partial t)_0(\psi_t^* g)]_{ab} dx^b \wedge \mathfrak{t}^a \\
&= \tfrac{1}{2}\int_{\mathbf{B}} \Phi^*\{[(\partial/\partial t)_0(\psi_t^* g)_{ab} dx^b] \wedge \mathfrak{t}^a\} \\
&= \tfrac{1}{2}\int_{\mathbf{B}} (\partial/\partial t)_0\{\Phi^*(\psi_t^* g)_{ab} dx^b\} \wedge \Phi^* t^a) \qquad \text{(A.11)}
\end{aligned}
$$

since $\Phi$ is time independent, which from (A.6) is

$$
\begin{aligned}
[dW/dt]_0 &= \tfrac{1}{2}\int_{\mathbf{B}} \{(\partial/\partial t)_0[(\psi_t \circ \Phi)^*(g)]_{ab}\}(\partial x^b/\partial X^B) dX^B \wedge (\partial x^a/\partial X^C)\mathfrak{S}^C \\
&= \tfrac{1}{2}\int_{\mathbf{B}} \{(\partial/\partial t)_0[(\psi_t \circ \Phi)^* g]_{CB}\} dX^B \wedge \mathfrak{S}^C \\
&= \tfrac{1}{2}\int_{\mathbf{B}} \{(\partial/\partial t)_0[(\psi_t \circ \Phi)^*(g) - G]_{CB}\} dX^B \wedge \mathfrak{S}^C
\end{aligned}
$$

since the metric $G_{CB}$ on the reference $\mathbf{B}$ is time independent. But $\mathfrak{S}^C = S^{CA} i(\partial_A)$ VOL and so

$$dX^B \wedge \mathfrak{S}^C = dX^B \wedge S^{CA} i(\partial_A)\text{VOL} = S^{CB}\text{VOL} \qquad \text{(A.12)}$$

Thus finally, for reference body $\mathbf{B}$, we have our main result

$$dW/dt = \int_{\mathbf{B}} S^{CB}[dE_{CB}/dt]\text{VOL} \qquad \text{(A.13)}$$

where $E$ is the Lagrange deformation tensor $E_{CB} = \frac{1}{2}[(\psi_t \circ \Phi)^*(g) - G]_{CB}$. □

If, during the deformation, no energy is dissipated, for example by heat flux, then this is the rate $dU/dt$ at $t=0$, at which energy $U$ is being stored in the body during this deformation by the surface forces on the boundary.

For the total amount of energy stored in a body during a deformation, we do *not* claim that the same amount of energy is stored during two deformations starting at the same initial state and ending at the same final state; we expect the result to depend on the specific **family** of deformations from the initial to the final state. When the result *is* independent of the path in the space of deformations, the material is called **hyperelastic**. This will be discussed in more detail in Appendix D.a.

## A.e. Some Typical Computations Using Forms by Hidenori Murakami

(1) *The equilibrium equations in spherical coordinates*. The metric is

$$ds^2 = dr^2 + r^2(d\theta^2 + \sin^2\theta\, d\phi^2) \tag{A.14}$$

with respect to the coordinate basis $[\partial/\partial r, \partial/\partial\theta, \partial/\partial\phi]$. To get an orthonormal basis it is immediately suggested that we define a new basis of 1-forms by

$$\sigma = \begin{bmatrix} \sigma^r \\ \sigma^\theta \\ \sigma^\phi \end{bmatrix} = \begin{bmatrix} dr \\ r\, d\theta \\ r\sin\theta\, d\phi \end{bmatrix} \tag{A.15}$$

with dual vector basis, which forms an orthonormal frame

$$\mathbf{e} = [\mathbf{e}_r\ \mathbf{e}_\theta\ \mathbf{e}_\phi] = [\partial/\partial r\ r^{-1}\partial/\partial\theta \quad (r\sin\theta)^{-1}\partial/\partial\phi] \tag{A.16}$$

The advantage of using an orthonormal frame and associated 1-form basis to express tensors is that their components give dimensionally correct physical components. A vector $\mathbf{v}$ has *tensor* components denoted by $[v^1\ v^2\ v^3]^T$ and *physical* components $[v^r\ v^\theta\ v^\phi]^T$:

$$\mathbf{v} = [\partial/\partial r\ \partial/\partial\theta\ \partial/\partial\phi] \begin{bmatrix} v^1 \\ v^2 \\ v^3 \end{bmatrix} = [\mathbf{e}_r\ \mathbf{e}_\theta\ \mathbf{e}_\phi] \begin{bmatrix} v^r \\ v^\theta \\ v^\phi \end{bmatrix}$$

Thus

$$v^r = v^1 \quad v^\theta = r v^2 \quad v^\phi = r\sin\theta\, v^3$$

The Cauchy vector valued stress 2-form in physical components is

$$\mathfrak{t}^2 = [\mathbf{e}_r\ \mathbf{e}_\theta\ \mathbf{e}_\phi] \begin{bmatrix} \mathfrak{t}^r \\ \mathfrak{t}^\theta \\ \mathfrak{t}^\phi \end{bmatrix} \tag{A.17}$$

where

$$\mathfrak{t}^r = t^{rr}\sigma^\theta \wedge \sigma^\phi + t^{r\theta}\sigma^\phi \wedge \sigma^r + t^{r\phi}\sigma^r \wedge \sigma^\theta$$
$$\mathfrak{t}^\theta = t^{\theta r}\sigma^\theta \wedge \sigma^\phi + t^{\theta\theta}\sigma^\phi \wedge \sigma^r + t^{\theta\phi}\sigma^r \wedge \sigma^\theta$$
$$\mathfrak{t}^\phi = t^{\phi r}\sigma^\theta \wedge \sigma^\phi + t^{\phi\theta}\sigma^\phi \wedge \sigma^r + t^{\phi\phi}\sigma^r \wedge \sigma^\theta$$

Similarly, the body force per unit mass of the deformed body is expressed in physical components:

$$\rho\, \mathbf{b}\, \mathrm{vol}^3 = [\mathbf{e}_r\ \mathbf{e}_\theta\ \mathbf{e}_\phi] \begin{bmatrix} \rho b^r \\ \rho b^\theta \\ \rho b^\phi \end{bmatrix} \sigma^r \wedge \sigma^\theta \wedge \sigma^\phi = \rho\{\mathbf{e}_r\, b^r + \mathbf{e}_\theta\, b^\theta + \mathbf{e}_\phi\, b^\phi\}\sigma^r \wedge \sigma^\theta \wedge \sigma^\phi$$

We shall need the matrix of connection 1-forms $\omega$ for our orthonormal basis. Letting $\partial/\partial\mathbf{x}$ be the cartesian basis, using $x = r\sin\theta\cos\phi$, $y = r\sin\theta\sin\phi$, $z = r\cos\theta$, we get

$$\begin{aligned}\mathbf{e}_r = \partial/\partial r &= (\partial x/\partial r)\partial/\partial x + (\partial y/\partial r)\partial/\partial y + (\partial z/\partial r)\,\partial/\partial z\\ &= \sin\theta\cos\phi\,\partial/\partial x + \sin\theta\sin\phi\,\partial/\partial y + \cos\theta\,\partial/\partial z\end{aligned}$$

with similar expressions for $\mathbf{e}_\theta$ and $\mathbf{e}_\phi$. We then have

$$\mathbf{e} = (\partial/\partial\mathbf{x})P \quad \text{or} \quad [\mathbf{e}_r\ \mathbf{e}_\theta\ \mathbf{e}_\phi] = [\partial/\partial\mathbf{x}\ \partial/\partial\mathbf{y}\ \partial/\partial\mathbf{z}]P$$

where $P$ is the orthogonal matrix

$$P = \begin{bmatrix} \sin\theta\cos\phi & \cos\theta\cos\phi & -\sin\phi \\ \sin\theta\sin\phi & \cos\theta\sin\phi & \cos\phi \\ \cos\theta & -\sin\theta & 0 \end{bmatrix}$$

The flat connection $\Gamma$ for the cartesian frame $\partial/\partial\mathbf{x}$ is $\Gamma = 0$. Under the change of frame $\mathbf{e} = \partial/\partial\mathbf{x}\ P$ we have the new connection matrix, as in (9.41)

$$\omega = P^{-1}\Gamma P + P^{-1}dP = P^T dP$$

yielding the skew symmetric matrix

$$\omega = \begin{bmatrix} 0 & -d\theta & -\sin\theta\,d\phi \\ d\theta & 0 & -\cos\theta\,d\phi \\ \sin\theta\,d\phi & \cos\theta\,d\phi & 0 \end{bmatrix} = \begin{bmatrix} 0 & -\sigma^\theta/r & -\sigma^\phi/r \\ \sigma^\theta/r & 0 & -\sigma^\phi\cot\theta/r \\ \sigma^\phi/r & \sigma^\phi\cot\theta/r & 0 \end{bmatrix} \tag{A.18}$$

An additional preparation is to compute $d\sigma$ of (A.15) and express the result with the unit 1-form basis using $d\sigma = -\omega\wedge\sigma$ or directly from (A.15)

$$d\sigma = \begin{bmatrix} d\sigma^r \\ d\sigma^\theta \\ d\sigma^\phi \end{bmatrix} \tag{A.19}$$

$$= \begin{bmatrix} d\,dr \\ dr\wedge d\theta \\ dr\wedge\sin\theta\,d\phi + d\theta\wedge r\cos\theta\,d\phi \end{bmatrix} = \frac{1}{r}\begin{bmatrix} 0 \\ \sigma^r\wedge\sigma^\phi \\ -\sigma^\phi\wedge\sigma^r + \cot\theta\,\sigma^\phi\wedge\sigma^\phi \end{bmatrix}$$

The equilibrium equation is expressed using Cartan's exterior covariant differential (9.31)

$$\nabla\mathfrak{t} + \rho\mathbf{b}\,\mathrm{vol}^3 = 0 \tag{A.20a}$$

In component form

$$\mathbf{e}_i(\nabla\mathfrak{t}^i + \rho b^i\mathrm{vol}^3) = \mathbf{e}_i(d\mathfrak{t}^i + \omega^i_j\wedge\mathfrak{t}^j + \rho b^i\mathrm{vol}^3) = 0 \tag{A.20b}$$

With respect to the orthonormal basis (A.16), the $\nabla\mathfrak{t}^2$ term of (A.20b) becomes

$$\begin{bmatrix} \nabla t^r \\ \nabla t^\theta \\ \nabla t^\phi \end{bmatrix} = \begin{bmatrix} dt^r \\ dt^\theta \\ dt^\phi \end{bmatrix} + \begin{bmatrix} 0 & -\sigma^\theta/r & -\sigma^\phi/r \\ \sigma^\theta/r & 0 & \sigma^\phi\cot\theta/r \\ \sigma^\phi/r & \sigma^\phi\cot\theta/r & 0 \end{bmatrix} \wedge \begin{bmatrix} t^r \\ t^\theta \\ t^\phi \end{bmatrix} \tag{A.21}$$

Look, for example, at the $r$-component using (A.17), (A.19), (A.15), and $\mathrm{vol}^3 = \sigma^r \wedge \sigma^\theta \wedge \sigma^\phi$

$$\begin{aligned}\nabla \mathfrak{t}^r &= d\mathfrak{t}^r - \sigma^\theta/r \wedge \mathfrak{t}^\theta - \sigma^\phi/r \wedge \mathfrak{t}^\phi \\ &= d(t^{rr}\sigma^\theta \wedge \sigma^\phi + t^{r\theta}\sigma^\phi \wedge \sigma^r + t^{r\phi}\sigma^r \wedge \sigma^\theta) - (t^{\theta\theta} + t^{\phi\phi})/r\, \sigma^r \wedge \sigma^\theta \wedge \sigma^\phi \\ &= dt^{rr} \wedge \sigma^\theta \wedge \sigma^\phi + t^{rr}(d\sigma^\theta \wedge \sigma^\phi - \sigma^\theta \wedge d\sigma^\phi) + \cdots - (t^{\theta\theta} + t^{\phi\phi})/r\ \mathrm{vol}^3 \\ &= \{(\partial t^{rr}/\partial r + 2t^{rr}/r) + (1/r)(\partial t^{r\theta}/\partial\theta + \cot\theta\, t^{r\theta}) + [1/(r\sin\theta)]\partial t^{r\phi}/\partial\phi \\ &\quad -(1/r)(t^{\theta\theta} + t^{\phi\phi})\}\ \mathrm{vol}^3\end{aligned}$$

The $\mathbf{e}_r$-component of the equilibrium equation is now obtained:

$$\begin{aligned}&(1/r^2)\partial(r^2 t^{rr})/\partial r + [1/(r\sin\theta)]\{\partial(t^{r\theta}\sin\theta)/\partial\theta + \partial t^{r\phi}/\partial\phi\} \\ &\quad - (1/r)(t^{\theta\theta} + t^{\phi\phi}) + \rho b^r = 0 \qquad \text{(A.22)}\end{aligned}$$

The $\mathbf{e}_\theta$- and $\mathbf{e}_\phi$-components are handled in the same way. The above procedure of using an orthonormal frame to write a physical equation, invented by Cartan, takes an order of magnitude less time for computations. Compared to classical tensor analysis, the following two tedious computations are eliminated: (i) computation of Christoffel symbols, and (ii) conversion of tensor components to physical components.

(2) *The rate of deformation tensor in spherical coordinates.* Consider the metric tensor

$$ds^2 = g_{ij}dx^i \otimes dx^j$$

in any Riemannian manifold. If $\mathbf{v} = (\boldsymbol{\partial}/\boldsymbol{\partial} x^i)v^i$ is a vector field then, from Theorem (A.d) and equation (4.16), the Lie derivative of the metric, measuring how the flow generated by $\mathbf{v}$ deforms figures, is given by

$$\mathcal{L}_{\mathbf{v}}(g_{ij}dx^i \otimes dx^j) = 2d_{ij}dx^i \otimes dx^j \qquad \text{(A.23a)}$$

where

$$2d_{ij} := v_{i/j} + v_{j/i} \qquad \text{(A.23b)}$$

defines the **rate of deformation** tensor, which plays an important role when discussing fluid or solid flow. We shall now compute this tensor in spherical coordinates, not by using covariant derivatives but rather by looking directly at the Lie derivative of the metric tensor

$$\mathcal{L}_{\mathbf{v}}(g_{ij}dx^i \otimes dx^j) = \partial/\partial\varepsilon\phi_\varepsilon^*(g_{ij}dx^i \otimes dx^j)|_{\epsilon=0}$$

where $\phi_\varepsilon$ is the flow generated by $\mathbf{v}$. We shall do this by using only the simplest properties of the Lie derivative. We have mainly discussed the Lie derivative of vector fields and *exterior* forms, where Cartan's formula (4.23) played an important role.

Equation (4.23) cannot be used here since we are dealing now with quadratic (symmetric) forms. However, we still have a product rule

$$\mathcal{L}_{\mathbf{v}}\alpha \otimes \beta = (\mathcal{L}_{\mathbf{v}}\alpha) \otimes \beta + \alpha \otimes \mathcal{L}_{\mathbf{v}}\beta$$

and the basic

$$\mathcal{L}_{\mathbf{v}}(f) = \mathbf{v}(f) = df(\mathbf{v})$$

for any function $f$. Also, using (4.23) for a 1-form basis

$$\mathcal{L}_{\mathbf{v}}\sigma^i = di_v\sigma^i + i_v d\sigma^i = dv^i + i_v d\sigma^i \tag{A.24}$$

The metric tensor for spherical coordinates using the unit 1-form basis (A.12) is

$$ds^2 = \sigma^r \otimes \sigma^r + \sigma^\theta \otimes \sigma^\theta + \sigma^\phi \otimes \sigma^\phi \tag{A.25}$$

The rate of deformation tensor $\mathbf{d}$ is symmetric and is expanded as follows:

$$\begin{aligned} 2\mathbf{d} &= \sigma^r \otimes [d_{rr}\sigma^r + d_{r\theta}\sigma^\theta + d_{r\phi}\sigma^\phi] + \sigma^\theta \otimes [d_{\theta r}\sigma^r + d_{\theta\theta}\sigma^\theta + d_{\theta\phi}\sigma^\phi] \\ &\quad + \sigma^\phi \otimes [d_{\phi r}\sigma^r + d_{\phi\theta}\sigma^\theta + d_{\phi\phi}\sigma^\phi] \end{aligned}$$

The above components are computed from the definition (A.23) by taking the Lie derivative of the metric tensor (A.25)

$$\begin{aligned} 2\mathbf{d} &= (\mathcal{L}_{\mathbf{v}}\sigma^r) \otimes \sigma^r + \sigma^r \otimes (\mathcal{L}_{\mathbf{v}}\sigma^r) + (\mathcal{L}_{\mathbf{v}}\sigma^\theta) \otimes \sigma^\theta + \sigma^\theta \otimes (\mathcal{L}_{\mathbf{v}}\sigma^\theta) \\ &\quad + (\mathcal{L}_{\mathbf{v}}\sigma^\phi) \otimes \sigma^\phi + \sigma^\phi \otimes (\mathcal{L}_{\mathbf{v}}\sigma^\phi) \end{aligned} \tag{A.26}$$

The Lie derivatives of the basis 1-forms are computed using (A.24) with (A.19) and (A.25)

$$\mathcal{L}_{\mathbf{v}}\sigma^r = dv^r + i_v d\sigma^r = (\partial v^r/\partial r)\sigma^r + (1/r)(\partial v^r/\partial\theta)\sigma^\theta + (1/r\sin\theta)(\partial v^r/\partial\phi)\sigma^\phi \tag{A.27a}$$

$$\begin{aligned} \mathcal{L}_{\mathbf{v}}\sigma^\theta &= dv^\theta + i_v d\sigma^\theta = (\partial v^\theta/\partial r)\sigma^r + (1/r)(\partial v^\theta/\partial\theta)\sigma^\theta + (1/r\sin\theta)(\partial v^\theta/\partial\phi)\sigma^\phi \\ &\quad + i_v((1/r)\sigma^r \wedge \sigma^\theta) \\ &= (\partial v^\theta/\partial r)\sigma^r + (1/r)(\partial v^\theta/\partial\theta)\sigma^\theta + (1/r\sin\theta)(\partial v^\theta/\partial\phi)\sigma^\phi \\ &\quad + (1/r)(v^r\sigma^\theta - v^\theta\sigma^r) \end{aligned} \tag{A.27b}$$

$$\begin{aligned} \mathcal{L}_{\mathbf{v}}\sigma^\phi &= dv^\phi + i_v d\sigma^\phi = (\partial v^\phi/\partial r)\sigma^r + (1/r)(\partial v^\phi/\partial\theta)\sigma^\theta + (1/r\sin\theta)(\partial v^\phi/\partial\phi)\sigma^\phi \\ &\quad + i_v(-(1/r)\sigma^\phi \wedge \sigma^r + (\cot\theta/r)\sigma^\theta \wedge \sigma^\phi) \\ &= (\partial v^\phi/\partial r)\sigma^r + (1/r)(\partial v^\phi/\partial\theta)\sigma^\theta + (1/r\sin\theta)(\partial v^\phi/\partial\phi)\sigma^\phi \\ &\quad - (1/r)v^\phi\sigma^r + (1/r)v^r\sigma^\phi + \cot\theta(v^\theta\sigma^\phi - v^\phi\sigma^\theta)(1/r) \end{aligned} \tag{A.27c}$$

By substituting (A.27) into (A.26) and collecting terms for each pair of basis 1-forms, the physical components of the rate of deformation tensor are obtained

$$\begin{aligned}
&d_{rr} = \partial v^r/\partial r \quad d_{\theta\theta} = (1/r)[(\partial v^\theta/\partial\theta) + v^r] \\
&d_{\phi\phi} = (1/r)[(1/\sin\theta)(\partial v^\phi/\partial\phi) + v^r + v^\theta \cot\theta] \\
&2d_{\theta r} = 2d_{r\theta} = (1/r)[(\partial v^r/\partial\theta) - v^\theta] + \partial v^\theta/\partial r \\
&2d_{\phi r} = 2d_{r\phi} = (1/r)[(\partial v^r/\partial\phi)(1/\sin\theta) - v^\phi] + (\partial v^\phi/\partial r) \qquad \text{(A.28)} \\
&2d_{\phi\theta} = 2d_{\theta\phi} = (1/r)[(\partial v^\theta/\partial\phi)(1/\sin\theta) + (\partial v^\phi/\partial\theta) - v^\phi \cot\theta]
\end{aligned}$$

It is to be noted here that if the velocity components are replaced by the displacement components, the formula (A.28) gives the relation between the infinitesimal strain tensor and the displacements.

(3) *The Lie derivative of the Cauchy stress 2-form, the Truesdell stress rate.* With respect to a coordinate frame, $\boldsymbol{\partial}_i = \boldsymbol{\partial}/\boldsymbol{\partial}x^i$, we define the Lie derivative of the vector valued

$$\mathfrak{t} = \boldsymbol{\partial}_i \otimes d\mathfrak{t}^i = \boldsymbol{\partial}_i \otimes (t^{ij} i_{\partial_j} \mathrm{vol}^3)$$

for a time-dependent vector field $\mathbf{v}$, by

$$\begin{aligned}
\mathcal{L}_{\partial_t+\mathbf{v}}(\partial_i \otimes t^{ij} i_{\partial_j}\mathrm{vol}^3) &= (\mathcal{L}_{\partial_t+\mathbf{v}}\partial_i) \otimes t^{ij} i_{\partial_j}\mathrm{vol}^3 + \partial_i \otimes (\mathcal{L}_{\partial_t+\mathbf{v}} t^{ij}) i_{\partial_j}\mathrm{vol}^3 \\
&\quad + \partial_i \otimes t^{ij}(\mathcal{L}_{\partial_t+\mathbf{v}} i_{\partial_j}\mathrm{vol}^3) \\
&= [\mathbf{v}, \partial_i] \otimes t^{ij} i_{\partial_j}\mathrm{vol}^3 + \partial_i \otimes (\mathcal{L}_{\partial_t+\mathbf{v}} t^{ij}) i_{\partial_j}\mathrm{vol}^3 \\
&\quad + \partial_i \otimes t^{ij}(\mathcal{L}_{\mathbf{v}} i_{\partial_j}\mathrm{vol}^3) \qquad \text{(A.29)}
\end{aligned}$$

Using (4.6) the bracket term becomes

$$[\mathbf{v}, \partial_i] \otimes t^{ij} i_{\partial_j}\mathrm{vol}^3 = -\boldsymbol{\partial}_m(\partial v^m/\partial x^i) \otimes t^{ij} i_{\partial_j}\mathrm{vol}^3 = -\boldsymbol{\partial}_i(\partial v^i/\partial x^k) t^{kj} i_{\partial_j}\mathrm{vol}^3 \qquad \text{(A.30)}$$

Also, the second term on the right of (A.29) becomes

$$\partial_i \otimes (\partial_t + \mathbf{v})(t^{ij}) i_{\partial_j}\mathrm{vol}^3 = \partial_i \otimes (\partial t^{ij}/\partial t + v^k \partial t^{ij}/\partial x^k) i_{\partial_j}\mathrm{vol}^3 \qquad \text{(A.31)}$$

Look now at the last term of (A.29). Using (4.24) and $[\mathbf{v}, \boldsymbol{\partial}_j] = -(\partial v^k/\partial x^j)\boldsymbol{\partial}_k$

$$\mathcal{L}_{\mathbf{v}} i_{\partial_j} = i_{[\mathbf{v},\partial_j]} + i_{\partial_j}\mathcal{L}_{\mathbf{v}} = -(\partial v^k/\partial x^j) i_{\partial_k} + i_{\partial_j}\mathcal{L}_v$$

And so

$$\begin{aligned}
\mathcal{L}_{\mathbf{v}} i_{\partial_j}\mathrm{vol}^3 &= [-(\partial v^k/\partial x^j) i_{\partial_k} + i_{\partial_j}\mathcal{L}_{\mathbf{v}}]\mathrm{vol}^3 \\
&= -(\partial v^k/\partial x^j) i_{\partial_k}\mathrm{vol}^3 + (\mathrm{div}\,\mathbf{v}) i_{\partial_j}\mathrm{vol}^3
\end{aligned}$$

and the last term in (A.29) becomes, using $t^{ij}(\partial v^k/\partial x^j) i_{\partial_k} = t^{im}(\partial v^j/\partial x^m) i_{\partial_j}$

$$\partial_i \otimes [t^{ij}(\mathrm{div}\,\mathbf{v}) - t^{im}(\partial v^j/\partial x^m)] i_{\partial_j}\mathrm{vol}^3 \qquad \text{(A.32)}$$

The stress rate (A.29) then becomes the **Truesdell** stress rate

$$\mathcal{L}_{\partial_t+\mathbf{v}}(\partial_i \otimes t^{ij} i_{\partial_j}\mathrm{vol}^3) = \partial_i \otimes \overset{\nabla}{t}{}^{ij} i_{\partial_j}\mathrm{vol}^3 \qquad \text{(A.33a)}$$

where

$$\overset{\nabla}{t}{}^{ij} := (\partial t^{ij}/\partial t) + v^k(\partial t^{ij}/\partial x^k) - (\partial v^i/\partial x^m)t^{mj} - t^{im}(\partial v^j/\partial x^m) + (\operatorname{div} \mathbf{v})t^{ij} \tag{A.33b}$$

In a similar manner, the stress rate of the covector valued Cauchy stress 2-form can be computed.

## A.f. Concluding Remarks

In 1923, Elie Cartan introduced a vector valued 3-form version of the stress tensor in Einstein's space–time $M^4$ and combined this with a vector valued version of Einstein's energy momentum tensor $\rho u^j u^k$. For these matters and much more, see the translations of Cartan's papers in the book [Ca], and especially A. Trautman's Foreword to that book.

L. Brillouin discussed the three index version of the stress tensor in $\mathbb{R}^3$ in his book [Br, p. 281 ff].

APPENDIX B

# Harmonic Chains and Kirchhoff's Circuit Laws

Chapter 14 deals with harmonic forms on a manifold. This involves analysis in infinite dimensional function spaces. In particular, the proof of Hodge's theorem (14.28) is far too difficult to be presented there, and only brief statements are given. By considering finite chain complexes, as was done in section 13.2b, one can prove a *finite dimensional analogue of Hodge's theorem* using only elementary linear algebra. In the process, we shall consider cohomology, which was only briefly mentioned in section 13.4a. In the finite dimensional version, the differential operator $d$ acting on differential forms is replaced by a "coboundary" operator $\delta$ acting on "cochains," and the geometry of $\delta$ is as appealing as that of the boundary operator $\partial$ acting on chains!

As an application we shall consider the Kirchhoff laws in direct current electric circuits, first considered from this viewpoint by Weyl in the 1920s. This geometric approach yields a unifying overview of some of the classical methods of Maxwell and Kirchhoff for dealing with circuits. Our present approach owes much to a paper of Eckmann [E], to Bott's remarks in the first part of his expository paper [Bo 2], and to the book of Bamberg and Sternberg [B, S], where many applications to circuits are considered.

We shall avoid generality, going simply and directly to the ideas of Hodge and Kirchhoff.

## B.a. Chain Complexes

A (real, finite) **chain complex** $C$ is a collection of real finite dimensional vector spaces $\{C_p\}$, $C_{-1} = 0$, and **boundary** linear transformations

$$\partial = \partial_p : C_p \rightarrow C_{p-1}$$

such that $\partial^2 = \partial_{p-1} \circ \partial_p = 0$. Chapter 13 is largely devoted to the (infinite dimensional) singular chain complex $C(M; R)$ on a manifold and the associated finite simplicial complex on a compact triangulated manifold. We shall illustrate most of the concepts with a chain complex on the 2-torus based *not* on simplexes (as in Fig. 13.16) but rather on another set of basic chains illustrated in Figure B.1. This chain complex is chosen not for its intrinsic value but rather to better illustrate the concepts.

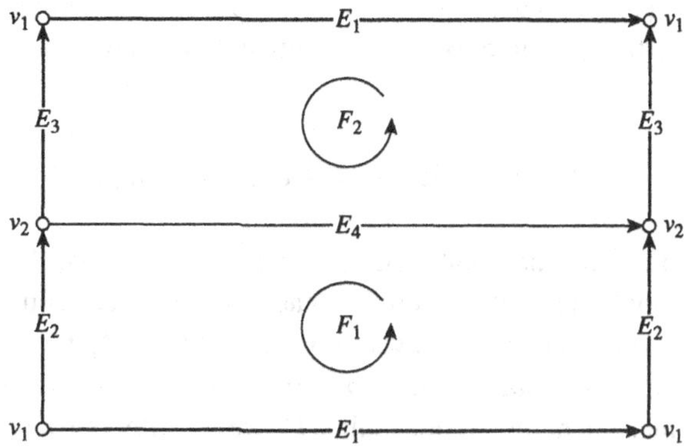


**Figure B.1**

The vector space $C_0$ is 2-dimensional with basis the vertices $v_1$ and $v_2$. $C_1$ is 4-dimensional with basis consisting of the two *circles* $E_1$ and $E_4$ and the two 1-simplexes $E_2$ and $E_3$, each carrying the indicated orientation. $C_2$ has as basis the two oriented *cylinders* $F_1$ and $F_2$. We call these eight basis elements **basic chains.**

A general 1-chain is a formal sum of the form $c = \sum a^i E_i$, where the $a^i$ are real numbers. This means that $c$ is a real valued function on the basis $\{E_i\}$ with values $c(E_i) = a^i$. Similarly for $C_0$ and $C_2$.

For boundary operators we are led to **define**

$$\partial = \partial_0(v_i) = 0 \quad i = 1, 2$$

$$\partial = \partial_1 E_1 = v_1 - v_1 = 0, \quad \partial_1 E_2 = v_2 - v_1, \quad \partial_1 E_3 = v_1 - v_2, \quad \partial_1 E_4 = v_2 - v_2 = 0$$

$$\partial = \partial_2 F_1 = E_1 + E_2 - E_4 - E_2 = E_1 - E_4, \quad \partial_2 F_2 = E_4 - E_1$$

and extend $\partial$ to the chain groups by linearity, $\partial \sum a^i E_i = \sum a^i \partial E_i$. Using the usual column representations for the bases, $E_3 = [0, 0, 1, 0]^T$, etc., we then have the matrices

$$\partial_0 = 0 \quad \partial_1 = \begin{bmatrix} 0 & -1 & 1 & 0 \\ 0 & 1 & -1 & 0 \end{bmatrix} \quad \partial_2 = \begin{bmatrix} 1 & -1 \\ 0 & 0 \\ 0 & 0 \\ -1 & 1 \end{bmatrix} \tag{B.1}$$

We may form the homology groups (vector spaces) of the chain complex. $H_p(C) := \ker(\partial_p)/\mathrm{Im}(\partial_{p+1})$, which are again cycles modulo boundaries. One sees easily that the bases of the homology vector spaces can be written

$$H_0 = \{v_1\} \quad H_1 = \{E_1, E_2 + E_3\} \quad H_2 = \{F_1 + F_2\}$$

yielding the same bases as (13.24) for the finite simplicial chains on the torus.

There is no reason to expect, however, that other decompositions of the torus will yield the same homology as the simplicial chains. For example, we could consider a new chain complex on the torus where $C_2$ has a single basic chain $T$, the torus itself, while $C_1 = 0$ and $C_0$ is the 1-dimensional space with basic 0-chain a single vertex $v$, and with all $\partial_p = 0$. The homology groups of this complex would be $H_0 = \{v\}$, $H_1 = 0$, and $H_2 = \{T\}$, which misses all the 1-dimensional homology of the torus. We have chosen our particular complex to better illustrate our next concept, the cochains.

## B.b. Cochains and Cohomology

A $p$-**cochain** $\alpha$ is a linear functional $\alpha : C_p \to \mathbb{R}$ on the $p$-chains. (In the case when $C_p$ is infinite-dimensional one does *not* require that $f$ vanish except on a finite number of basic chains!). The $p$-cochains form a vector space $C^p := C_p{}^*$, the dual space to $C_p$, of the same dimension. Thus *chains correspond to vectors while cochains correspond to covectors* or 1-forms. Cochains are not chains. However, after one has chosen a basis for $p$-chains (the basic chains), each chain is represented by a column $c = [c^1, \ldots c^N]^T$ and a cochain, with respect to the dual basis, may be represented by a row $\alpha = [a_1, \ldots . a_N]$. However, for our present purposes, some confusion will be avoided by *representing cochains also by columns*. Then the value of the cochain $\alpha$ on the chain $c$ is the matrix product $\alpha(c) = a^T c$. We may also think, in our finite dimensional case, of a chain as a function on cochains, using the same formula

$$c(\alpha) := \alpha(c) = a^T c \tag{B.2}$$

In our simple situation there will always be basic chains chosen so there is basically no difference between chains and cochains: both are linear functions of the basic chains, but just as we frequently want to distinguish between vectors and 1-forms, so we shall sometimes wish to distinguish between chains and cochains, especially in the case of Kirchhoff's laws.

We define a **coboundary** operator $\delta_p : C_p{}^* \to C_{p+1}{}^*$ to be the usual pull back of 1-forms under the boundary map $\partial_{p+1} : C_{p+1} \to C_p$. Ordinarily we would call this $\partial_{p+1}{}^*$, but as we shall soon see, * is traditionally used for the closely related "adjoint" operator.

$$\delta = \delta_p : C^p \to C^{p+1}$$

is defined by

$$\delta_p \alpha(c) := \alpha(\partial_{p+1} c) \quad \text{i.e.,} \quad (\delta\, a)_i = a_r \partial^r{}_i \tag{B.3}$$

or, briefly

$$\delta_p(\alpha) = (\partial_{p+1})^T a$$

for each $(p+1)$ chain $c$. As usual the matrix for $\delta_p$ is the transpose of the matrix for $\partial_{p+1}$, again operating on columns.

It is immediately apparent that

$$\delta^2 = \delta \circ \delta = 0 \tag{B.4}$$

If $\delta\alpha^p = 0$ we say that $\alpha$ is a $p$-**cocycle**, and if $\alpha = \delta\beta^{p-1}$ then $\alpha$ is a **coboundary**. It is clear that every coboundary is a cocycle.

In the case when $C_p$ is the infinite dimensional space of real singular chains on a manifold $\mathbf{M}^n$, then an exterior $p$-form $\alpha$ defines a linear functional by integration (called $I\alpha$ in our discussion of de Rham's theorem)

$$\alpha(c) = \int_c \alpha$$

and so defines a cochain. Then Stokes's theorem $d\alpha(c) = \alpha(\partial c)$ shows that *$d$ behaves as a coboundary operator*. A closed form defines a cocycle and an exact form a coboundary.

The analogue of the de Rham group, $\mathcal{R}^p$ = closed $p$-forms modulo exact $p$-forms, is called the $p^{\text{th}}$ (real) **cohomology** group for the chain complex

$$H^p = \ker \delta_p / \mathrm{Im}\, \delta_{p-1} \tag{B.5}$$

Consider the chain complex on $T^2$ pictured in Figure B.1. Consider the basic chains also as cochains; for example, $\mathcal{E}_1$ is the 1-cochain whose value on the chain $E_1$ is 1 and which vanishes on $E_2$, $E_3$ and $E_4$. Then $\delta\mathcal{E}_1(F_1) = \mathcal{E}_1(\partial F_1) = \mathcal{E}_1(E_1 + E_2 - E_4 - E_2) = 1$, while similarly $\delta\mathcal{E}_1(F_2) = -1$. Thus we can visualize $\delta\mathcal{E}_1$ as the 2-chain $F_1 - F_2$.

$$\delta\mathcal{E}_1 = F_1 - F_2$$

In words, to compute $\delta\mathcal{E}_1$ *as a chain*, we take the formal combination $\sum_r a^r F_r$ of exactly those basic 2-chains $\{F_r\}$ whose boundaries meet $E_1$, $a^r$ chosen so that $\partial(a^r F_r)$ contains $E_1$ with coefficient 1. Note that

$$\delta\mathcal{E}_2 = 0$$

since $F_1$ is the only basic 2-chain adjacent to $E_2$, but $\partial F_1 = E_1 - E_4$ does not contain $E_2$.

These remarks about $\delta\mathcal{E}_1$ and $\delta\mathcal{E}_2$ also follow immediately from the matrices in (B.1), putting $\delta_1 = \partial_2^T$.

Observe that $\delta\mathcal{E}_4 = F_2 - F_1$, and so

$$\delta(\mathcal{E}_1 + \mathcal{E}_4) = 0 \tag{B.6}$$

The 1-chain $E_1 + E_4$ is not only a cycle, it is a cocycle. We shall see in the next section that this implies that $E_1 + E_4$ *cannot bound*.

## B.c. Transpose and Adjoint

We shall continue to consider only *finite* dimensional chain complexes. We have identified chains and cochains by the choice of a basis (the "basic" chains). Another method we have used to identify vectors and covectors is to introduce a metric (scalar product). We continue to represent cochains by *column* matrices.

We may introduce an arbitrary (positive definite) scalar product $\langle\, ,\rangle$ in each of the chain spaces $C_p$. Given $\langle\, ,\rangle$ and given a choice of basic chains in $C_p$ we may then introduce, as usual, the "metric tensor" $g(p)_{ij} = \langle E_i, E_j \rangle$, yielding $\langle c, c' \rangle = c^i g_{ij} c'^j = c^T g c'$, and its inverse $g(p)^{-1}$ with entries $g(p)^{ij}$. This inverse yields a metric in the dual space of cochains, $\langle \alpha, \beta \rangle = a_i g^{ij} b_j = a^T g^{-1} b$. (The simplest case to keep in mind is when we choose basic chains and demand that they be declared orthonormal, i.e., when each matrix $g$ is the identity. This is what we effectively did in our previous section when considering the chain complex on the torus; $\langle E_j, E_k \rangle$ was the identity matrix.)

To the $p$-cochain with entries $(a_i)$ we may associate the $p$ chain with entries $(a^j)$, $a^j := g(p)^{jk} a_k$. Thus $g(p)^{-1} : C^p \to C_p$ "raises the index on a cochain" making it a chain, while $g(p) : C_p \to C^p$ "lowers the index on a chain" making it a cochain. We shall now deal mainly with cochains. *If a chain c appears in a scalar product we shall assume that we have converted c to a cochain.*

Let $A : V \to W$ be a linear map between vector spaces. The **transpose** $A^T$ is simply the pullback operator that operates on covectors in $W^*$.

$$A^T : W^* \to V^*$$

If we were writing covectors as row matrices, $A^T$ would be the same matrix as as $A$ but operating to the left on the rows, but since our covectors are columns we must now interchange the rows and columns of $A$, i.e., we write $w_R A^R{}_i = A^R{}_i w_R = (A^T)_i{}^R w_R$, and so

$$(A^T)_i{}^R := A^R{}_i$$

(Recall that in a matrix, the left-most index always designates the row.)

Suppose now that $V$ and $W$ are inner product vector spaces, with metrics $g_V = \{g(V)_{ij}\}$ and $g_W = \{g(W)_{RS}\}$ respectively. Then the **adjoint**

$$A^* : W \to V$$

of $A$ is classically defined by $\langle A(v), w \rangle_W = \langle v, A^*(w) \rangle_V$. $A^*$ is constructed as follows. To compute $A^*(w)$ we take the covector $g_W(w)$ corresponding to $w$, pull this back to $V^*$ via the transpose $A^T g_W(w)$, and then take the vector in $V$ corresponding to this covector, $g_V{}^{-1} A^T g_W(w)$. Thus $A^* = g_V{}^{-1} A^T g_W$. In components $(A^*)^j{}_R = g(V)^{jk} (A^T)_k{}^S g(W)_{SR} = g(V)^{jk} A^S{}_k g(W)_{SR}$. In summary

$$A^* = g_V{}^{-1} A^T g_W$$
$$A^{*\,j}{}_R = A_R{}^j := g(W)_{RS} A^S{}_k \, g(V)^{kj} \tag{B.7}$$

Note that in this formulation $A^*$ would reduce simply to the transpose of $A$ if bases in $V$ and $W$ were chosen to be orthonormal.

The coboundary operator and matrix have been defined in (B.3), $\delta_p = \partial_{p+1}{}^T$. The adjoint $\delta^*$ satisfies $\langle \delta(\alpha), \beta \rangle = \langle \alpha, \delta^*(\beta) \rangle$ . Then

$$\delta^* \circ \delta^* = 0$$

Consider $\delta_p : C^p \to C^{p+1}$. The metric in $C^p = C_p{}^*$ is the inverse $g(p)^{-1}$ of the metric $g(p)$ in $C_p$. Hence, from (B.7), $\delta^* = g(p) \delta^T g(p+1)^{-1} = g(p) \partial_{p+1} \, g(p+1)$. Since

$(\delta_p)^* : C^{p+1} \to C^p$, we prefer to call this operator $\delta^*{}_{p+1}$.

$$\delta^*{}_{p+1} := \delta_p{}^* = g(p)\partial_{p+1}g(p+1)^{-1} \tag{B.8}$$

Thus in any bases $\delta$ is $\partial^T$, and in orthonormal bases $\delta^* = \partial$.

## B.d. Laplacians and Harmonic Cochains

We now have two operators on cochains

$$\delta_p : C^p \to C^{p+1} \qquad \text{and} \qquad \delta^*{}_p : C^p \to C^{p-1}$$

If a cochain $\alpha$ satisfies $\delta^*\alpha = 0$ we shall, with abuse of language, call $\alpha$ a **cycle**. Similarly, if $\alpha = \delta^*\beta$, we say $\alpha$ is a **boundary**. We define the **laplacian** $\Delta : C^p \to C^p$ by

$$\Delta_p = \delta^*{}_{p+1}\delta_p + \delta_{p-1}\delta^*{}_p \tag{B.9}$$

or briefly

$$\Delta = \delta^*\delta + \delta\,\delta^*$$

Note that

$$\Delta = (\delta + \delta^*)^2 \text{ and } \Delta \text{ is self adjoint, } \Delta^* = \Delta.$$

A cochain $\alpha$ is called **harmonic** iff $\Delta\alpha = 0$. Certainly $\alpha$ is harmonic if $\delta^*\alpha = 0 = \delta\alpha$. Also, $\Delta\alpha = 0$ implies $0 = \langle(\delta^*\delta + \delta\,\delta^*)\alpha, \alpha\rangle = \langle\delta\alpha, \delta\alpha\rangle + \langle\delta^*\alpha, \delta^*\alpha\rangle$, and since a metric is positive definite we conclude that $\delta^*\alpha = 0 = \delta\alpha$.

*A cochain is harmonic if and only if it is a cycle and a cocycle.* (B.10)

Let $\mathcal{H}$ be the harmonic cochains. If $\gamma$ is orthogonal to all boundaries, $0 = \langle\gamma, \delta^*\alpha\rangle = \langle\delta\gamma, \alpha\rangle$, then $\gamma$ is a cocycle. Likewise, if $\gamma$ is orthogonal to all coboundaries, then $\gamma$ is a cycle. Thus if $\gamma$ is orthogonal to the subspace spanned by the sum of the boundaries and the coboundaries, then $\gamma$ is harmonic. Also, any harmonic cochain is clearly orthogonal to the boundaries and coboundaries. Thus the orthogonal complement of the subspace $\delta C^{p-1} \oplus \delta^* C^{p+1}$ is $\mathcal{H}^p$. A non-zero harmonic cochain is *never* a boundary nor a coboundary! For example, the cycle $E_1 + E_4$ of section B.b cannot be a boundary.

In our finite dimensional $C^p$, we then have the orthogonal ("Hodge") decomposition

$$C^p = \delta C^{p-1} \oplus \delta^* C^{p+1} \oplus \mathcal{H}^p$$

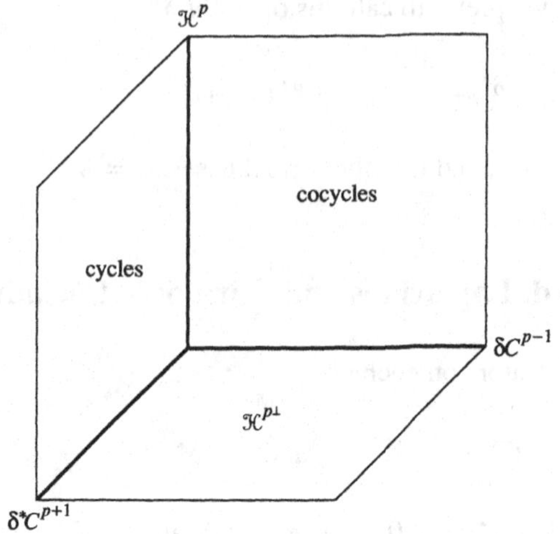


**Figure B.2**

Thus any cochain $\beta$ is of the form

$$\beta^p = \delta\alpha^{p-1} + \delta^*\gamma^{p+1} + h^p \tag{B.11}$$

The three cochains on the right are unique (though $\alpha$ and $\gamma$ need not be).

We can actually say more. The self-adjoint operator $\Delta = \delta^*\delta + \delta\,\delta^*$ has $\mathcal{H}$ as kernel and clearly sends all of $C^p$ into the subspace $\mathcal{H}^{p\perp} = \delta C^{p-1} \oplus \delta^* C^{p+1}$. Thus $\Delta : \mathcal{H}^{p\perp} \to \mathcal{H}^{p\perp}$ is $1:1$, and, since $\mathcal{H}^{p\perp}$ is finite dimensional, onto, and so $\Delta : C^p \to \mathcal{H}^{p\perp}$ is onto. Hence any element of $\mathcal{H}^{p\perp}$ is of the form $\Delta\alpha$ for some $\alpha$.

*Given any $\beta \in \mathcal{H}^\perp$ there is an $\alpha \in C$ such that $\Delta\alpha = \beta$*
*and $\alpha$ is unique up to the addition of a harmonic cochain.* (B.12)

"Poisson's equation" $\Delta\alpha = \beta$ has a solution iff $\beta \in \mathcal{H}^\perp$. Now let $\beta \in C^p$ be any $p$-cochain and let $H(\beta)$ be the orthogonal projection of $\beta$ into $\mathcal{H}$. Then $\beta - H(\beta)$ is in $\mathcal{H}^{p\perp}$ and

$$\beta - H(\beta) = \Delta\alpha = \delta\delta^*\alpha + \delta^*\delta\alpha \tag{B.13}$$

refines (B.11).

In particular, if $\beta$ is a cocycle, then, since the cycles are orthogonal to the coboundaries, we have the unique decomposition

$$\delta\beta = 0 \quad \Rightarrow \quad \beta = \delta\delta^*\alpha + H(\beta) \tag{B.14}$$

Thus,

*In the cohomology class of a cocycle $\beta$ there*
*is a unique harmonic representative. The* (B.15)
*dimension of $\mathcal{H}^p$ is* $\dim . H^p$.

There is a similar remark for cochains with $\delta^* z = 0$. Since we may always introduce a euclidean metric in the space of chains $C_p$, we can say

$$\delta z_p = 0 \quad \Rightarrow \quad z_p = \partial c_{p+1} + h_p \tag{B.16}$$

where $\partial h = 0 = \delta h$.

$$\begin{array}{l}\textit{In the homology class of a cycle } z \textit{ there} \\ \textit{is a unique harmonic representative } h, \textit{ i.e.,} \\ \textit{a chain that is both a cycle and a cocycle,} \\ \textit{and} \dim. H_p = \dim. \mathcal{H}^p = \dim. H^p.\end{array} \tag{B.17}$$

Three concluding remarks for this section. First, once we write down the matrices for $\partial$ and $\delta = \partial^T$, the harmonic chains, the nullspace of $\Delta$, can be exhibited simply by linear algebra, e.g., Gaussian elimination.

Second, it is clear from the orthogonal decomposition (B.16), that in the homology class of a cycle $z$, *the harmonic representative has the smallest norm*, $\|h\| \le \|z\|$. For our toral example, $E_1$ and $(E_1 + E_4)/2$ are in the same homology class, since $E_4 \approx E_1$ and $(E_1 + E_4)/2$ is harmonic from (B.6). While it seems perhaps unlikely that $E_1 + E_4$ is "smaller" than $2E_1$, recall that our basic chains are there declared orthonormal, and so $\|2E_1\| = 2$, while $\|E_1 + E_4\| 1 = \sqrt{2}$.

Finally, we write down the explicit expression for the laplacian of a 0-cochain $\phi^0$. This is especially simple since $\delta * \phi^0 = 0$. From (B.9) and (B.3) $\Delta\phi = \delta * \delta\phi = \delta_1 * \delta_0 \phi$, i.e.,

$$\Delta\phi = g(0)\partial_1 g(1)^{-1} \partial_1{}^T \phi \tag{B.18}$$

## B.e. Kirchhoff's Circuit Laws

Consider a very simple electric circuit problem. We have wire 1-simplexes forming a *connected* 1-dimensional chain complex with **nodes** (vertices) $\{v_j\}$ and **branches** (edges) $\{e_A\}$, each edge endowed with an orientation. The vertices and edges are the basic 0- and 1-chains. The circuit, at first, will be assumed purely **resistive**, i.e., each edge $e_A$ carries a resistance $R_A > 0$, but there are no coils or batteries or capacitors. We assume that there is an

**external** source of current $i(v_j) = i_j$ at each vertex $v_j$

which may be positive (coming in), negative (leaving), or zero. In Figure B.3 we have indicated the three non-zero external currents $i_2$, $i_4$, and $i_7$. The problem is to determine the current $I_A := I(e_A)$ in each edge after a steady state is achieved. Current is thus a real valued function of the oriented edges; it defines either a 1-chain or cochain, denoted by **I**.

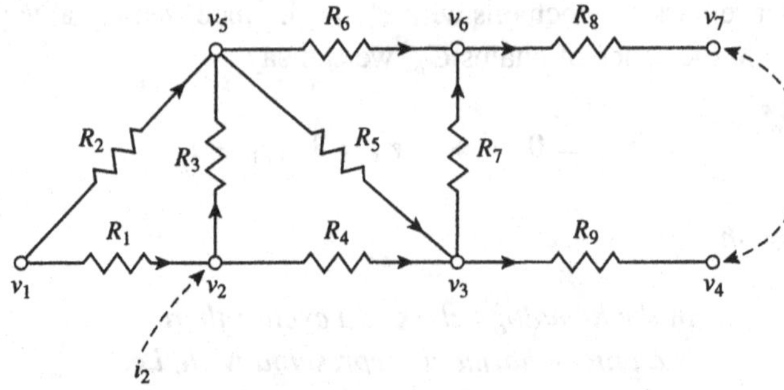


Figure B.3

In Figure B.3, $C_0$ has basis $\{v_1, \ldots, v_7\}$, $C_1$ has basis $\{e_1, \ldots e_9\}$.

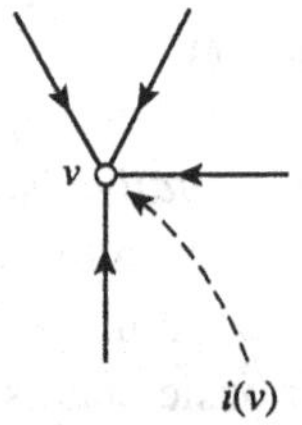


Figure B.4

**Kirchhoff's current law KCL** states that at any node $v$, the sum of all the currents flowing into $v$ from the wire edges and the external source must equal that leaving. But (see Figure B.4) the edges coming into $v$ form the coboundary of the vertex, and so $0 = \mathbf{I}(\delta v) + \mathbf{i}(v) = \partial\mathbf{I}(v) + \mathbf{i}(v)$. This suggests that the wire *currents* form a 1-*chain* (since we are taking a boundary) and

$$\partial\mathbf{I} = -\mathbf{i} \qquad \textbf{(KCL)}$$

The external currents $i$ form a 0-chain. We write $\mathbf{I}(e_A) = I^A$ and $i(v_j) = i^j$.

Kirchhoff's voltage law involves the electric field in each wire. Let

$$\mathcal{E}(e) = \int_e \mathcal{E}^1 = \int_e \mathbf{E}\cdot\mathbf{dx}$$

be the integral of the electric field over the basic 1-chain $e$. This is the **voltage drop** along branch $e$. Since we are dealing with steady state, i.e., static fields, we know that the electric field 1-form $\mathcal{E}^1$ is the differential of the electostatic potential $\phi$; see (7.26). Hence $\mathcal{E}(e) = \phi(\partial e) = \delta\phi(e)$. This suggests that we should consider *voltage* as a 1-*cochain*. We have then **Kirchhoff's voltage law**

$$\mathcal{E} = \delta\phi \qquad \textbf{(KVL)}$$

and the electrostatic potential at a vertex defines a 0-cochain $\phi$. Write $\mathcal{E}(e_A) = \mathcal{E}_A$ and $\phi(v_j) = \phi_j$. $\phi$ is defined only up to an additive constant.

Finally, **Ohm's law** says that the voltage drop across the resistor **R** is always **RI**. Since we are assuming at first that only resistances are present in each branch, we may say $\mathcal{E}_A = \mathbf{R}_A I^A$. (When batteries are present this will be amended; see (B.22). Since $\mathcal{E}$

is covariant and $\mathbf{I}$ is contravariant, *we interpret the resistances as determining a* **metric** *in* $C_1$, $\mathcal{E}_A = g(1)_{AB}\mathbf{I}^B$. Thus the metric tensor in the 1-chains is diagonal

$$g(1)_{AB} = \mathbf{R}_A\delta_{AB} \tag{B.19}$$
$$\mathcal{E} = g(1)\mathbf{I}$$

*We put the identity metric tensor in* $C_0$; thus the vertices $\{v_j\}$ are declared orthonormal and may be considered either as chains or as cochains.

Kirchhoff's laws then yield, for the electric potential 0-cochain $\phi$

$$\Delta\phi = \delta^*{}_1\delta_0\phi = \delta^*{}_1\mathcal{E}$$

From (B.9) we have

$$\Delta\phi = \delta^*{}_1\mathcal{E} = g(0)\partial_1 g(1)^{-1}\mathcal{E} \tag{B.20}$$

and from (B.19)

$$\Delta\phi = \partial\mathbf{I} = -\mathbf{i}$$

(In circuit theory, $\partial$ is called the **incidence** matrix and $\Delta$ the **admittance**.) If we can solve this Poisson equation for $\phi$, then we will know $\mathcal{E}$ in each $e_A$. Knowing this and the resistances, we get the current in each branch.

Is there always a solution? From (B.12) we know that a necessary and sufficient condition is that the 0-cochain $\mathbf{i}$ of external currents be a boundary, $\mathbf{i} = \delta^*{}_1$ (a 1-cochain $\beta$) $= \partial c$, where $c$ is the 1-chain version of $\beta$. Let $c = \sum c^A e_A$. Then $\partial c = \sum c^A \partial e_A = \sum c^A(v_A{}^+ - v_A{}^-)$, where $v_A{}^{\pm}$ are the vertices of $e_A$. Thus the sum of the coefficients of all the vertices in the boundary of a 1-chain vanishes. Conversely, in a chain complex that is connected (such as our circuit), meaning that any two vertices can be connected by a curve made up of edges, it is not hard to see that any collection of vertices with coefficients whose sum vanishes is indeed a boundary. We conclude

*There exists a solution to (B.20) iff the total external current entering the circuit equals the total external current leaving,* $\sum_k \mathbf{i}(v_k) = 0$ (B.21)

which is of course what is expected. The solution $\phi$ is unique up to an additive harmonic 0-cochain. We claim that a harmonic 0-cochain $f$ has the same value on each vertex in our connected circuit. For if $P$ and $Q$ are any vertices, let $c$ be a 1-chain with boundary $Q$–$P$. Then $f(Q)$–$f(P) = f(\partial c) = \delta f(c) = 0$, since $f$ is a cocycle. Hence, as to be expected, the potential $\phi$ is unique up to an additive constant.

Just to illustrate the computations, consider a pair of resistances in parallel, Figure B.5. We know that we need to have $i_2 = -i_1 := -i_0$. Put $v_1 = [1\,0]^T$, $v_2 = [0\,1]^T$, $e_1 = [1\,0]^T$, $e_2 = [0\,1]^T$, $\phi = [\phi_1\,\phi_2]^T$ and $i = [i_0 - i_0]^T$. The matrix $g(1)$ is the $2\times 2$ diagonal matrix with entries $R_1$ and $R_2$. We have $\partial e_1 = v_2 - v_1 = \partial e_2 = [-1\,1]^T$.

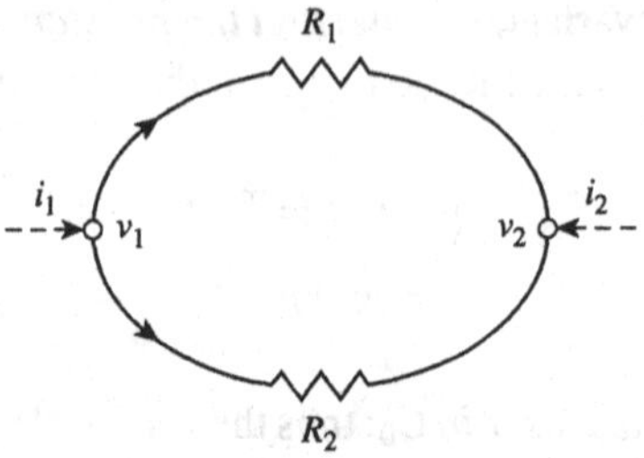


**Figure B.5**

Then $\Delta\phi = -i$ becomes, from (B.18),

$$\Delta\phi = \partial_1 g(1)^{-1}\partial_1{}^T\phi = -i$$

The laplacian matrix is

$$\begin{bmatrix} -1 & -1 \\ 1 & 1 \end{bmatrix} \begin{bmatrix} 1/R_1 & 0 \\ 0 & 1/R_2 \end{bmatrix} \begin{bmatrix} -1 & 1 \\ -1 & 1 \end{bmatrix} = (1/R_1 + 1/R_2) \begin{bmatrix} 1 & -1 \\ -1 & 1 \end{bmatrix}$$

Then $\Delta[\phi_1\, \phi_2]^T = [-i_0\, i_0]^T$ gives immediately

$$(1/R_1 + 1/R_2)(\phi_2 - \phi_1) = i_0$$

Since $\phi_2 - \phi_1$ is the voltage in both branches, this gives the familiar result that the equivalent resistance for the two resistances in parallel is $(1/R_1 + 1/R_2)^{-1}$.

Some words about *circuits with batteries but no external currents.* First a simplification of notation. Since only the 1-chains involve a non–standard metric (based on the resistances), we shall write $g$ rather than $g(1)$. Let $\mathbf{B}$ be the 1-cochain, with $B_A$ the voltage of the battery in edge $e_A$, $B_A$ being positive if the direction from the negative to the positive terminal yields the given orientation of $e_A$. Consider a closed loop formed by a battery of voltage $B$ and a resistor $R$ across the poles of the battery. By Ohm's law the integral of $\mathcal{E}^1$ over the resistor is $RI = B$. But the integral of $\mathcal{E}^1 = d\phi$ over the entire loop must vanish, and so the integral of $\mathcal{E}^1$ over the battery part of the loop must be $-B$. Thus when a battery is present in a branch $e_A$ we have, as expected, the voltage drop $\mathcal{E}_A = R_A I^A - B_A$. Kirchhoff's laws are then

$$\partial\mathbf{I} = 0 \quad \text{and} \quad \mathcal{E} = g\mathbf{I} - \mathbf{B} = \delta\phi \tag{B.22}$$

and then $\Delta\phi = \partial_1\, g^{-1}\mathcal{E} = \partial\mathbf{I} - \partial g^{-1}\mathbf{B} = -\partial[g^{-1}\mathbf{B}]$

$$\Delta\phi = -\partial[g^{-1}\mathbf{B}] \tag{B.23}$$

which always has a solution, since the boundaries are in $\mathcal{H}^\perp$.

Note also that

$$\mathcal{E}^T\mathbf{I} = \mathcal{E}(\mathbf{I}) = (\delta\phi)(\mathbf{I}) = \phi(\partial\mathbf{I}) = 0$$

which is **Tellegen's theorem**, saying that the total power loss $I^2R$ in the resistors is equal to the power $BI$ supplied by the batteries.

Further, we note the following. Look at (B.22), written as $\mathbf{B} = g\mathbf{I} - \delta\phi$. Since $\mathbf{I}$ is a 1-cycle, $\partial\mathbf{I} = 0$, its cochain version $g\mathbf{I}$ satisfies $\delta^*[g\mathbf{I}] = 0$. $\mathbf{B}$ is thus the sum of a cycle and a coboundary and the two summands $g\mathbf{I}$ and $\delta\phi$ are orthogonal. Thus, in Figure B.2, *the cochain version of* **I** *is the orthogonal projection* $\prod \mathbf{B}$ *of* **B** *into the subspace of cycles,* $I_A = \prod_A{}^C B_C$. Thus if we choose an orthonormal basis for the cycles, the "meshes," then given any battery cochain **B** we can easily project it orthogonally into the cycle space, and the resulting cochain is the current. (For the *chain I* we may write $I^A = \prod^{AC} B_C$.) This is a special case of **Weyl's method of orthogonal projection.**

The orthogonal projection operator $\prod$ is self-adjoint and depends only on the metric, i.e., the resistances in the given branches. In terms of the basic 1-chains $\{e_A\}$ we have $I^A = \prod^{AC} B_C$, where $\prod^{AC} = \prod^{CA}$. Consider a circuit where there is only one battery present, of voltage $V$, in branch $e_1$. Then the current present in branch $e_2$ is $I^2 = \prod^{21} B_1 = \prod^{21} V$. Remove this battery, put it in branch $e_2$, and look at the new current in branch $e_1$; $I'^1 = \prod^{12} V = I^2$! This surprising result is a special case of **Green's Reciprocity.**

Finally, we consider the modifications necessary when there are *constant current sources* $K_A$ present in parallel with the resistors in each branch $e_A$.

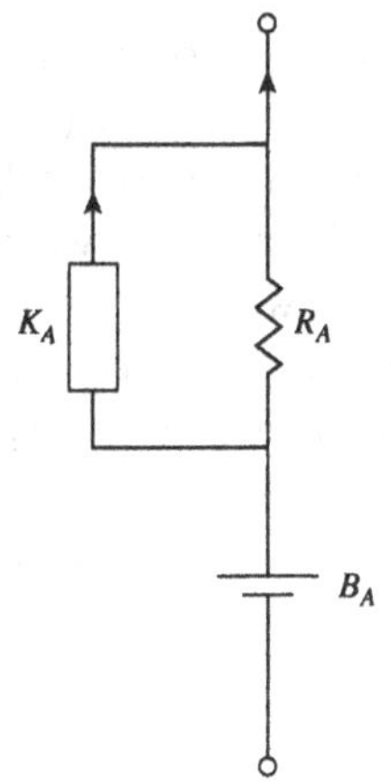


**Figure B.6**

We do *not* consider the current source $K_A$ as forming a new branch; $\{K_A\}$ forms rather a new 1-chain $\mathcal{K}$.

$$\mathcal{K}(e_A) = \mathcal{K}^A := K_A$$

If $I^A$ is the current in branch $e_A$, i.e., $I^A$ is the current entering $e_A$ at one node of $\partial e_A$ and leaving at the other node, then the current through the resistor $R_A$ is now $I^A - K_A$. The voltage drop along the resistor is then, by Ohm's law, $R_A(I^A - K_A)$, and thus $\mathcal{E}_A = R_A(I^A - K_A) - B_A$. Kirchhoff's laws become, since the total current entering a node is still 0,

$$\partial \mathrm{I} = 0 \qquad \text{and} \qquad \mathcal{E} = \delta\phi = g(\mathbf{I} - \mathcal{K}) - \mathbf{B} \tag{B.24}$$

Poisson's equation becomes

$$\Delta\phi = -\partial[g^{-1}\mathbf{B} + \mathcal{K}] \tag{B.25}$$

Orthogonal projection onto cycles now says

$$\mathrm{I}^A = \prod{}^{AC}(B_C + \mathcal{K}_C) \tag{B.26}$$

## APPENDIX C

# Symmetries, Quarks, and Meson Masses

At the end of Section 20.3b we spoke very briefly about "colored" quarks and the resulting Yang–Mills field with gauge group $SU(3)$. This was not, however, the first appearance of quarks. They appeared in the early 1960s in the form of "flavored" quarks, independently in the work of Gell–Mann and Zweig. Their introduction changed the whole course of particle physics, and we could not pass up the opportunity to present one of the most striking applications to meson physics, the relations among pion, kaon, and eta masses. This application involves only global symmetries, rather than the Yang–Mills feature of the colored quarks.

For expositions of particle physics for "the educated general reader" see, e.g., the little books ['t Hooft] and [Nam].

## C.a. Flavored Quarks

The description to follow will be brief and very sketchy; the main goal is to describe the almost magical physical interpretations physicists gave to the matrices that appear. My guide for much of this material is the book [L–S,K], with minor changes being made to harmonize more with the mathematical machinery developed earlier in the present book. As to mass formulas, while there are more refined, technical treatments (see, e.g., [We, Chap. 19]) applying (sometimes with adjustments required) to more mesons and to "baryons," the presentation given in Section C.f for the "$0^-$ meson octet" seems quite direct.

Flavored quarks generalize the notion of the Heisenberg nucleon of Section 20.3a The symmetry group there, $SU(2)$, is called **isotopic spin,** or briefly **isospin.** Isospin refers to the "internal" symmetry group $SU(2)$ and is not to be confused with the usual quantum mechanical spin [Su, Section 4.1], which refers to the space symmetry group $SO(3)$, but the terminology mimics that of ordinary spin. (Recall that $SU(2)$ is the twofold cover of $SO(3)$.) Thus since isospin for the nucleon has two states $p$ and $n$, we say that these nucleons have isotopic spin $I = 1/2$. In general (number of states) $= 2I + 1$. The diagonal normalized third Pauli matrix $I_3 = (1/2)\sigma_3$ is, except for a factor of $\sqrt{-1}$, an infinitesimal generator of $SU(2)$ and is called the **isotopic spin operator** $I_3$. $p$, being an eigenvector of $I_3$ with eigenvalue $1/2$, is said to be the nucleon state of isotopic spin $1/2$, while the neutron is the state of isotopic spin $-1/2$.

In the quark model the nucleon is no longer considered basic; it was proposed that nucleons and many other particles are composed of quarks. For our purposes, we need only consider particles at a given space–time point. (We shall not be considering kinematics nor quantum dynamics.) Associate with this point a complex three dimensional vector space **Q**, a copy of $\mathbb{C}^3$, with a *given* orthonormal basis and the usual hermitian metric $\langle z, w\rangle = \overline{z}^T w$. A **quark** is represented by a unit vector

$$(q^1\, q^2\, q^3)^T = (u\; d\; s)^T$$

in **Q**.

If $\mathbf{m} = (m_1, m_2, m_3)^T$ is any vector in **Q**, then it defines a (complex) linear functional $\mu$ on **Q** by $\mu(\mathbf{w}) = \langle \mathbf{m}, \mathbf{w}\rangle = \sum \overline{m}_j w_j$. (We may use subscripts throughout since our bases are orthonormal.) Thus the covariant version of the vector $\mathbf{m} = (m_1, m_2, m_3)^T$ is the covector given by the row matrix $\mu = (\overline{m}_1, \overline{m}_2, \overline{m}_3)$.

If $q = (u\; d\; s)^T$ is a quark, then its covector $q^* = (\overline{u}\, \overline{d}\, \overline{s})$ is assumed to describe the **antiquark** of $q$, written here as $q^*$ since its matrix is the hermitian adjoint of $q$. For formal "bookkeeping" purposes we will concentrate not on the individual quarks but on bases or frames of three quarks or antiquarks.

Let **u**, **d**, and **s** be the basis vectors of the given **Q**. These three quarks are called the **up, down**, and **strange flavored** quarks associated with this basis. A second basis related to this one by an $SU(3)$ change of basis will result in a new set of **u**, **d**, and **s** flavored quarks. These flavors are not to be confused with the colored quarks of Section 20.3b.

A quark **frame q** of orthonormal vectors in **Q**,

$$\mathbf{q} = [\mathbf{u}, \mathbf{d}, \mathbf{s}]$$

is written as in geometry (p. 250) as a formal row matrix (formal because the entries are quarks rather than numbers.)

Since the quarks **u**, **d**, and **s** are orthonormal, their three antiquarks $\mathbf{u}^*$, $\mathbf{d}^*$, and $\mathbf{s}^*$ form an orthonormal basis for the dual space $\mathbf{Q}^*$ and we can consider the formal **dual frame** of antiquarks,

$$\mathbf{q}^* = \begin{bmatrix} \mathbf{u}^* \\ \mathbf{d}^* \\ \mathbf{s}^* \end{bmatrix}$$

It was assumed that the part of the Lagrangian dealing with the strong force is invariant under an $SU(3)$ change of frame in **Q**. If, e.g., one observer believes the quark in question to be a down quark **d**, another could see it as an **s**. Thus, just as with the Heisenberg nucleon, **u**, **d**, and **s** are to be considered as three states of the same particle, the flavored quark. Invariance of the Lagrangian under the eight-dimensional group $SU(3)$ led to Gell–Mann's denomination of this theory as the "eight-fold way," using a phrase from Buddhist thought.

To view a nucleon as composed of quarks, quarks are assumed to have fractional electric charges

$$Q(\mathbf{u}) = 2/3, \qquad Q(\mathbf{d}) = Q(\mathbf{s}) = -1/3 \tag{C.1}$$

(Charge thus violates $SU(3)$ symmetry, but recall that $SU(3)$ symmetry is assumed only for the strong force, not the electromagnetic.) It turns out, e.g., that the proton $p$ is made up of three quarks, written $p = \mathbf{duu}$, whose total charge is $-1/3 + 2/3 + 2/3 = 1$. The neutron $n = \mathbf{ddu}$ has charge 0. The electric charge of an antiquark is always the negative of that of the quark. The antiproton $p^* = \mathbf{d^*u^*u^*}$ has charge $-1$.

## C.b. Interactions of Quarks and Antiquarks

A composite particle formed from a quark $\mathbf{q}$ and it its antiquark $\mathbf{q}^*$ is described by physicists by considering the tensor product $\mathbf{q}^* \otimes \mathbf{q}$ in $\mathbf{Q}^* \otimes \mathbf{Q}$.

Recall that if $\mathbf{e}$ is a basis for a vector space $\mathbf{Q}$ and if $\sigma$ is the dual basis for $\mathbf{Q}^*$, then for a vector $\mathbf{v} = \mathbf{e}_j v^j$ and covector $\boldsymbol{\alpha} = a_k \sigma^k$ we have $\boldsymbol{\alpha} \otimes \mathbf{v} = a_k \sigma^k \otimes \mathbf{e}_j v^j = a_k(\sigma^k \otimes \mathbf{e}_j) v^j$ and $\mathbf{Q}^* \otimes \mathbf{Q}$ thus has basis elements $\sigma^k \otimes \mathbf{e}_j$. Each basis element $\sigma^k \otimes \mathbf{e}_j$ defines a linear transformation sending $\mathbf{Q}$ into itself, $(\sigma^k \otimes \mathbf{e}_j)(\mathbf{v}) = \sigma^k(\mathbf{v})\mathbf{e}_j = v^k \mathbf{e}_j$, *but we shall largely ignore this aspect.* The formal matrix $\boldsymbol{\sigma} \otimes \mathbf{e}$ with entries $(\boldsymbol{\sigma} \otimes \mathbf{e})^k{}_j = \sigma^k \otimes \mathbf{e}_j$ forms a **frame** for $\mathbf{Q}^* \otimes \mathbf{Q}$.

*We shall be dealing entirely with the formal aspects of all these matrices.* $\mathbf{q}^*$ *is merely a formal column matrix,* $\mathbf{q}$ *is a row matrix,* $\mathbf{q}^* \otimes \mathbf{q}$ *is a* $3 \times 3$ *matrix, and SU(3) acts by* $g(\mathbf{q}^*) = g\mathbf{q}^*$, *and* $g(\mathbf{q}) = \mathbf{q}g^{-1}$. We are interested in antiquark–quark interactions forming composite particles. The appropriate frame is

$$\mathbf{q}^* \otimes \mathbf{q} = \begin{bmatrix} \mathbf{u}^* \\ \mathbf{d}^* \\ \mathbf{s}^* \end{bmatrix} \otimes [\mathbf{u}\ \ \mathbf{d}\ \ \mathbf{s}] = \begin{bmatrix} \mathbf{u^*u} & \mathbf{u^*d} & \mathbf{u^*s} \\ \mathbf{d^*u} & \mathbf{d^*d} & \mathbf{d^*s} \\ \mathbf{s^*u} & \mathbf{s^*d} & \mathbf{s^*s} \end{bmatrix}$$

In the $3 \times 3$ matrix on the right *we have omitted the tensor product sign* in each entry; e.g., $\mathbf{u^*u}$ is really $\mathbf{u}^* \otimes \mathbf{u}$: We are not interested in the fact that, e.g., the entries in the frame are themselves matrices Note also *that in the present case, the tensor product matrix is the same as the usual matrix product of the column matrix* $\mathbf{q}^*$ *and the row matrix* $\mathbf{q}$. This would not be the case for the product in the reverse order in which case the tensor product frame matrix would again be $3 \times 3$ while the matrix product would be a $1 \times 1$ matrix.

The three entries $\mathbf{u}$, $\mathbf{d}$, and $\mathbf{s}$ of the row matrix $\mathbf{q}$ are identified as the three states of the quark, while the entries in $\mathbf{q}^*$ are the states of the antiquark. *What particle or particles do the nine entries of the frame* $\mathbf{q}^* \otimes \mathbf{q}$ *represent?*

Any quark $q$ can be sent into any other quark $q'$ by some $g \in SU(3)$. This is why we consider the different flavors up, down, and strange as being different states of the *same* particle. The group $G = SU(3)$ acts on the tensor product frame by $\mathbf{q}^{*\prime} \otimes \mathbf{q}' = (g\mathbf{q}^*) \otimes (\mathbf{q}g^{-1}) = g(\mathbf{q}^* \otimes \mathbf{q})g^{-1}$, i.e., by the **adjoint** action $\mathrm{Ad}(G)$ as it does on a *linear transformation.*

If any two antiquark–quark frame matrices $A$ and $B$ were necessarily related by a $g \in SU(3)$, $B = gAg^{-1}$, then we could conclude that the nine entries in $\mathbf{q}^* \otimes \mathbf{q}$ are simply the nine states of a single particle. But this is not the case! Clearly the scalar matrices $C = \lambda I$, $C_{ij} = \lambda\delta_{ij}$, form a one-dimensional complex vector subspace of the nine-dimensional $\mathbb{C}^9$ (= space of complex $3 \times 3$ matrices) that is left **fixed** under the $G$ action, $gCg^{-1} = C$. We conclude that the states of at least two particles appear in the frame $\mathbf{q}^* \otimes \mathbf{q}$. Since Ad $(G)$ acts by isometries on $\mathbb{C}^9$, the orthogonal complement of the scalar matrices must also be **invariant**, i.e., sent into itself by Ad$(G)$. If $D$ is orthogonal to $I$, then $0 = \sum_{ij}\delta_{ij}D_{ij} = \text{tr}ID$, and so the orthogonal complement of the scalar matrices is the complex eight-dimensional subspace consisting of *trace-free* $3 \times 3$ matrices. (Clearly tr $A = 0$ iff tr $gAg^{-1} = 0$.) We say that the adjoint action or representation of $SU(3)$ on the space of $3 \times 3$ complex matrices is **reducible**, breaking up into its action on trace-free matrices and its trivial (i.e., identity) action on scalar matrices.

We should remark that if we had been looking, e.g., at antiquark–antiquark interactions, the frame $\sigma \otimes \sigma$ would again be a $3 \times 3$ matrix with $ij$ entry $\sigma^i \otimes \sigma^j$ and would transform under $G \in SU(3)$ to $G_{ri}\sigma^i \otimes G_{sj}\sigma^j = G_{ri}\sigma^i \otimes \sigma^j G^T_{js}$; i.e., $\sigma \otimes \sigma \to G\sigma \otimes \sigma G^{-T} = G\sigma \otimes \sigma \overline{G}^{-1}$, which *does not preserve traces* (because of the complex conjugation). Since $A \to GAG^T$ preserves *symmetry* and *antisymmetry*, this is the natural decomposition to use in this case.

We now decompose every $3 \times 3$ matrix $A$ into its trace-free and scalar parts,

$$A = [A - (1/3)\,\text{tr}A\, I] + (1/3)\,\text{tr}A\, I.$$

In particular, for the matrix $\mathbf{q}^* \otimes \mathbf{q}$ we have the scalar part

$$(1/3)\,\text{tr}\,\mathbf{q}^* \otimes \mathbf{q}\, I = (1/3)(\mathbf{u}^*\mathbf{u} + \mathbf{d}^*\mathbf{d} + \mathbf{s}^*\mathbf{s})I \tag{C.2}$$

and then the trace-free part becomes

$$\mathbf{X} := \mathbf{q}^* \otimes \mathbf{q} - (1/3)\,\text{tr}\,(\mathbf{q}^* \otimes \mathbf{q})I$$

$$= \tag{C.3}$$

$$\begin{bmatrix} \frac{1}{3}(2\mathbf{u}^*\mathbf{u} - \mathbf{d}^*\mathbf{d} - \mathbf{s}^*\mathbf{s}) & \mathbf{u}^*\mathbf{d} & \mathbf{u}^*\mathbf{s} \\ \mathbf{d}^*\mathbf{u} & \frac{1}{3}(-\mathbf{u}^*\mathbf{u} + 2\mathbf{d}^*\mathbf{d} - \mathbf{s}^*\mathbf{s}) & \mathbf{d}^*\mathbf{s} \\ \mathbf{s}^*\mathbf{u} & \mathbf{s}^*\mathbf{d} & \frac{1}{3}(-\mathbf{u}^*\mathbf{u} - \mathbf{d}^*\mathbf{d} + 2\mathbf{s}^*\mathbf{s}) \end{bmatrix}$$

Since the scalar matrix (C.2) never mixes with the matrix $\mathbf{X}$ under $SU(3)$ we can use it to define a new particle, the eta prime,

$$\eta' := (1/\sqrt{3})(\mathbf{u}^*\mathbf{u} + \mathbf{d}^*\mathbf{d} + \mathbf{s}^*\mathbf{s}) \tag{C.4}$$

Why does the factor $1/\sqrt{3}$ appear? The quark flavors $\mathbf{u}$, $\mathbf{d}$, and $\mathbf{s}$ are unit vectors in $\mathbf{Q}$, and likewise for the antiquarks in $\mathbf{Q}^*$. Thus $\mathbf{u}^*\mathbf{u}$, etc. are unit vectors in $\mathbf{Q}^* \otimes \mathbf{Q}$, and the three vectors in (C.4) are orthonormal. The factor $1/\sqrt{3}$ makes the $\eta'$ a unit vector. Since quarks and antiquarks have opposite charges, the $\eta'$ is a neutral particle. ['tHooft,

p. 46] interprets the sum in (C.4) as implying that the $\eta'$ is "*continuously changing from* **u*u** *to* **d*d** *to* **s*s**".

The nine entries of the matrix **X** of (C.3) can represent at most eight particles since the trace is 0. To understand the action of $G = SU(3)$ on **X** we notice the following. $G$ is acting by the adjoint action on the space of traceless matrices. Now $SU(3)$ acts by the adjoint action on its Lie algebra $\mathfrak{su}(3)$, which is the space of skew hermitian matrices of trace 0. This is a *real* eight-dimensional vector space (i.e., the scalars must be real numbers); if $B$ is skew hermitian then $(a+ib)B$ is the sum of a hermitian matrix $ibB$ and a skew hermitian matrix $aB$. Since every matrix $C$ is the sum of a hermitian plus a skew hermitian, $C = (1/2)(C+C^*)+(1/2)(C-C^*)$, we see that if we allow complex scalars in the real Lie algebra vector space $\mathfrak{su}(3)$, then this **complexified** vector space is just the space of *all* traceless $3 \times 3$ matrices, and the action of $SU(3)$ on this space is again the adjoint action. *Thus we may consider our particle matrix* **X** *as being in this complexification* $\mathfrak{su}(3)$. We shall now look at this in more detail.

## C.c. The Lie Algebra of $SU(3)$

Physicists prefer hermitian to skew hermitian matrices, since observables in quantum mechanics are represented by hermitian operators. Note also that our matrix **X** is formally hermitian. Gell–Mann chose for a basis of $\mathfrak{g} := \sqrt{-1}\,\mathfrak{su}(3)$, i.e., the traceless hermitian matrices

$$\lambda_1 = \begin{bmatrix} 0 & 1 & 0 \\ 1 & 0 & 0 \\ 0 & 0 & 0 \end{bmatrix} \qquad \lambda_2 = \begin{bmatrix} 0 & -i & 0 \\ i & 0 & 0 \\ 0 & 0 & 0 \end{bmatrix} \qquad \lambda_4 = \begin{bmatrix} 0 & 0 & 1 \\ 0 & 0 & 0 \\ 1 & 0 & 0 \end{bmatrix}$$

$$\lambda_5 = \begin{bmatrix} 0 & 0 & -i \\ 0 & 0 & 0 \\ i & 0 & 0 \end{bmatrix} \qquad \lambda_6 = \begin{bmatrix} 0 & 0 & 0 \\ 0 & 0 & 1 \\ 0 & 1 & 0 \end{bmatrix} \qquad \lambda_7 = \begin{bmatrix} 0 & 0 & 0 \\ 0 & 0 & -i \\ 0 & i & 0 \end{bmatrix}$$

$$\lambda_3 = \begin{bmatrix} 1 & 0 & 0 \\ 0 & -1 & 0 \\ 0 & 0 & 0 \end{bmatrix} \qquad \lambda_8 = \frac{1}{\sqrt{3}} \begin{bmatrix} 1 & 0 & 0 \\ 0 & 1 & 0 \\ 0 & 0 & -2 \end{bmatrix}$$

These matrices are orthonormal with the scalar product $\langle A, B\rangle := (1/2)\mathrm{tr}\, AB^* = (1/2)$ tr $AB$ in $\mathfrak{g}$. Note that $\lambda_k, k = 1, 2, 3$, are just the Pauli matrices with zeros added in the third rows and columns, and when exponentiated these $\{i\lambda_k\}$ generate the subgroup $SU(2) \subset SU(3)$ that leaves the third axis of $\mathbb{C}^3$ fixed.

Let us expand $\mathbf{X} = \sum_{1\le j\le 8} X^j\lambda_j$, with all $X^j$ real. The only $\lambda$ with entry in the (3,3) spot is $\lambda_8$, and thus $-(1/3)(\mathbf{u^*u} + \mathbf{d^*d} - 2\mathbf{s^*s}) = (X^8\lambda_8)_{33} = X^8(-2/\sqrt{3})$ and so $X^8 = (1/2\sqrt{3})(\mathbf{u^*u} + \mathbf{d^*d} - 2\mathbf{s^*s}) = \eta/\sqrt{2}$, where the particle $\eta$ is defined by the unit vector

$$\eta := (1/\sqrt{6})(\mathbf{u^*u} + \mathbf{d^*d} - 2\mathbf{s^*s}) \tag{C.5}$$

Then

$$X^8\,\lambda_8 = \begin{bmatrix} \dfrac{\eta}{\sqrt{6}} & 0 & 0 \\ 0 & \dfrac{\eta}{\sqrt{6}} & 0 \\ 0 & 0 & \dfrac{-2\eta}{\sqrt{6}} \end{bmatrix}$$

Then from (C.3) we get for $\mathbf{X}$

$$\begin{bmatrix} \frac{1}{2}(\mathbf{u^*u} - \mathbf{d^*d}) + \dfrac{\eta}{\sqrt{6}} & \mathbf{u^*d} & \mathbf{u^*s} \\ \mathbf{d^*u} & -\frac{1}{2}(\mathbf{u^*u} - \mathbf{d^*d}) + \dfrac{\eta}{\sqrt{6}} & \mathbf{d^*s} \\ \mathbf{s^*u} & \mathbf{s^*d} & -\dfrac{2\eta}{\sqrt{6}} \end{bmatrix}$$

Finally, we define three sets of particles (with explanation to follow):

$$\begin{aligned} &\{\pi^0 = (1/\sqrt{2})(\mathbf{u^*u} - \mathbf{d^*d}) \quad \pi^- = \mathbf{u^*d} \quad \pi^+ = \mathbf{d^*u}\} \\ &\{\mathbf{K}^- = \mathbf{u^*s} \quad \overline{\mathbf{K}}^0 = \mathbf{d^*s}\} \\ &\{\mathbf{K}^+ = \mathbf{s^*u} \quad \mathbf{K}^0 = \mathbf{s^*d}\} \end{aligned} \tag{C.6}$$

and then

$$\mathbf{X} = \begin{bmatrix} \dfrac{\pi^0}{\sqrt{2}} + \dfrac{\eta}{\sqrt{6}} & \pi^- & \mathbf{K}^- \\ \pi^+ & \dfrac{-\pi^0}{\sqrt{2}} + \dfrac{\eta}{\sqrt{6}} & \overline{\mathbf{K}^0} \\ \mathbf{K}^+ & \mathbf{K}^0 & \dfrac{-2\eta}{\sqrt{6}} \end{bmatrix} \tag{C.7}$$

## C.d. Pions, Kaons, and Etas

The seven particles listed in (C.6) and the eta in (C.5) have physical attributes that led to their identification in the particle world. First there is electric charge. For example $\pi^- = \mathbf{u^*d}$ has, from the quark charges (C.1) the charge $-2/3 - 1/3 = -1$. This is the reason for the minus sign attached to the $\pi$ symbol. Neutral charge is denoted by the exponent 0, as for example in $\pi^0$. This explains the exponents in (C.6). Note that, e.g., $\pi^-$ is the antiparticle of $\pi^+$ while $\pi^0$ is its own antiparticle. Physicists usually denote antiparticles by a complex conjugation overbar. $\overline{\mathbf{K}}^0$ is the antiparticle of $\mathbf{K}^0$ and is distinct from $\mathbf{K}^0$, as we shall soon see. These eight particles are among those called mesons, because of their masses being intermediate between those of electrons and protons.

The diagonal matrices diag$\{e^{i\theta}, e^{i\phi}, e^{-i(\theta+\phi)}\}$ form a two-dimensional, maximal *commutative*, connected subgroup of the eight-dimensional $SU(3)$, i.e., a maximal torus $T^2$. (The maximal torus of $U(n)$ was discussed in Theorem (15.4).) Note that the two generators $\lambda_3$ and $\lambda_8$ generate, by exponentiation, two 1-parameter subgroups of this torus. Thus $\lambda_3$ and $\lambda_8$ form an orthonormal, basis for the Lie algebra of $T^2$, the tangent space $\mathfrak{h}$ of $T^2$ at the identity. (The Lie algebra of the maximal torus of any Lie group $G$ is called the **Cartan subalgebra** $\mathfrak{h}$ of $\mathfrak{g}$.)

We now change slightly the normalization of four of the Gell–Mann matrices

$$I_k := (1/2)\lambda_k, \qquad k = 1, 2, 3$$

and (C.8)

$$Y := (1/\sqrt{3})\lambda_8 = \operatorname{diag}\{1/3, 1/3, -2/3\}$$

The $I$s generate the $SU(2)$ subgroup of $SU(3)$, call it $SU(2) \times 1$,

$$\begin{bmatrix} SU(2) & 0 \\ 0 & 1 \end{bmatrix}$$

called the **isospin** subgroup, and $Y$ is the generator of the 1-parameter subgroup of $SU(3)$ called **hypercharge**.

Since the $I$s and $Y$ are hermitian they represent "observables"; since further $I_3$ and $Y$ commute they are "compatible" [Su, p. 57], and so in a sense they can both be measured simultaneously.

The flavored quarks are eigenvectors of these operators:

$$I_3(\mathbf{u}) = I_3(1\,0\,0)^T = (1/2)\mathbf{u} \qquad I_3(\mathbf{d}) = (-1/2)\mathbf{d} \qquad I_3(\mathbf{s}) = 0$$

Likewise

$$Y(\mathbf{u}) = (1/3)\mathbf{u} \qquad Y(\mathbf{d}) = (1/3)\mathbf{d} \qquad Y(\mathbf{s}) = (-2/3)\mathbf{s}$$

Furthermore, if $q = (u\ d\ s)^T$ is a quark, then an infinitesimal generator $A$ of $SU(3)$, say $A = I_3$ or $A = Y$, is basically a differentiation operator, i.e.,

$$iA(q) := \frac{d}{dt} e^{iAt}(q)\Big|_{t=0} = \frac{d}{dt} e^{iAt} q\Big|_{t=0} = iAq$$

or briefly

$$A(q) = \frac{d}{dt}(e^{tA} q)\Big|_{t=0} = Aq$$

while if $q^* = (\overline{u}\ \overline{d}\ \overline{s})$ is an antiquark

$$A(q^*) = \frac{d}{dt}(q^* e^{-tA})\Big|_{t=0} = -q^* A$$

Thus $I_3(\mathbf{u}^*) = -(1\,0\,0)I_3 = (-1/2)\mathbf{u}^*$. In general, if the quark $q$ is an eigenvector of a Gell–Mann generator $\lambda$ then its antiquark $q^*$ is an eigenvector with oppositely signed eigenvalue.

Finally, since each generator is a differentiation

$$A(q \otimes q') = A(q) \otimes q' + q \otimes A(q')$$

Since **u**, **d**, **s**, **u***, **d***, and **s*** are eigenvectors of $I^3$ and $Y$, any composite particle built up from them will also be an eigenvector whose eigenvalues are the sums of the constituents. For example, $Y(\mathbf{K}^0) = Y(\mathbf{s}^*\mathbf{d}) = (2/3 + 1/3)\mathbf{K}^0 = \mathbf{K}^0$ while $Y(\overline{\mathbf{K}}^0) = -\overline{\mathbf{K}}^0$. This shows indeed that $\overline{\mathbf{K}}^0$ *and* $\mathbf{K}^0$ *are distinct particles.*

Isospin and hypercharge play a very important role in describing particles. The eigenvalues of $I_3$ and $Y$ (briefly $I_3$ and $Y$) are two numbers that one assigns to strongly interacting particles with the experimentally observed property that if several particles collide and become other particles, then the sum of the isotopic spins before collision is the same as after, and likewise for the hypercharge. These "conservation laws," together with Noether's conservation principle (20.9), suggest that both the isospin and the hypercharge groups might be symmetry groups of the strong force Lagrangian. This is the origin of the hope that $SU(3)$, which contains both as subgroups, might even be a large symmetry group, or at least an approximate one.

In Figure C.1, we exhibit graphically $I_3$ and $Y$ for each of the representations of $SU(3)$ that we have considered. The result will be called the **weight diagram** of the

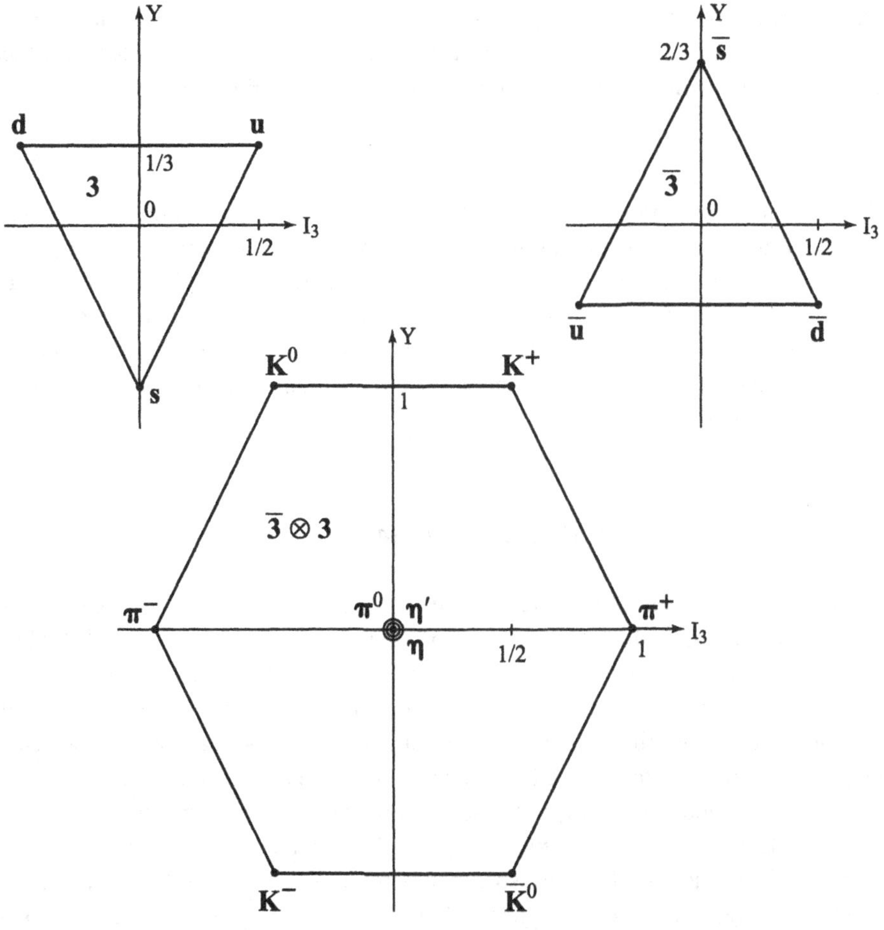


Figure C.1

representation. Since $I_3$ and $Y$ are a maximal set of (two) commuting operators, we shall have a two-dimensional graph for each representation. In Figure C.1, the representation **3** is the standard representation of $SU(3)$ on $\mathbb{C}^3$, i.e., on vectors $q = (u\, d\, s)^T$. (Physicists label the representations by their dimension, with or without an overbar.) We have drawn Cartesian axes labeled $I_3$ and $Y$ and have placed the particle **u** at the point with coordinates given by its eigenvalues $I_3(\mathbf{u}) = 1/2$ and $Y(\mathbf{u}) = 1/3$, etc.

The next representation is the representation labeled by physicists $\overline{\mathbf{3}}$; it is the representation on the dual space $\mathbb{C}^{3*}$, i.e., on antiquarks $q^* = (\overline{u}\, \overline{d}\, \overline{s})$. The eigenvalues here are the negatives of those in **3** and so the weight diagram is the reflection of that for **3** through the origin. We have also used the physicists' labels $\overline{\mathbf{u}}$ instead of $\mathbf{u}^*$, etc.

The final diagram is that for $\overline{\mathbf{3}} \otimes \mathbf{3}$. There are three particles $\pi^0$, $\eta$, and $\eta'$ at the origin, requiring a point surrounded by two circles. Note that this diagram is easily constructed graphically from the two previous ones because of the additivity of the eigenvalues. To construct it we take the whole diagram of **3**, translate it so that its origin is at a particle of $\overline{\mathbf{3}}$ (say $\overline{\mathbf{u}}$), erase that particle, and mark in the positions of the three particles of the translated **3**; then we repeat this operation at the two remaining particles of $\overline{\mathbf{3}}$. We have seen before that this representation is reducible, the particle $\eta'$ being fixed under all of $SU(3)$. If we remove this particle (the one-dimensional space **1** of scalar matrices) we get the weight diagram of the adjoint representation, denoted by **8**. It differs from that of $\overline{\mathbf{3}} \otimes \mathbf{3}$ only by having a point and one circle at the origin. Physicists say

$$\overline{\mathbf{3}} \otimes \mathbf{3} = \mathbf{8} + \mathbf{1} \tag{C.9}$$

There is (at least) one serious problem remaining. Our eight particles – the three pions, the four kaons, and the single eta – had been matched up by the physicists with the eight observed mesons with those names. While the observed particles in each category (e.g., the three pions ) have roughly the same mass, the masses of pions, kaons, and the eta differ widely. Since masses are coefficients that appear in the Lagrangian and the Lagrangian is assumed invariant under $SU(3)$, the assumption of $SU(3)$ invariance will have to be modified.

## C.e. A Reduced Symmetry Group

The mass of a pion is observed to be 140 MeV, the four kaons are at 495 MeV, and the eta has a mass of 550 MeV. (In comparison, the electron mass is about 1/2 MeV.) This suggests that the strange quark **s** might be considerably heavier than the up and the down quarks. On the other hand the equality of the three pion masses suggests that **u** and **d** have about the same mass. Individual quarks have never been seen; in fact there are reasons to believe that they will never be seen (quark "confinement"). It was then suggested that $SU(3)$ is too large to be the symmetry group for the strong interactions. Experimentally, however, isospin and hypercharge *are* conserved in strong interactions. This suggests that the isospin subgroup $SU(2) \times 1$ and the 1-parameter hypercharge subgroup $U(1) = \mathrm{diag}(e^{i\theta}, e^{i\theta}, e^{-2i\theta})$ of $SU(3)$ generate a more realistic symmetry group. Since $\lambda_8$ commutes with $\lambda_k$, for $k = 1, 2, 3$, it is clear that the three

$\lambda_k$s together with $\lambda_8$ form a Lie subalgebra of $\mathfrak{su}(3)$ and so, from (15.34), generate a four-dimensional subgroup, call it $SU(2) * U(1)$, of $SU(3)$. We shall identify this group, *but the identification will play no further role in our discussion since only the generators will be needed.* $SU(2) * U(1)$ consists of all products from the two subgroups, but since $SU(2) \times 1$ and $U(1)$ commute we need only consider the product of pairs $g \in SU(2) \times 1$ and $h = \text{diag}(e^{i\theta}, e^{i\theta}, e^{-2i\theta}) \in U(1)$. Let $\mathbf{a}$ be a $2 \times 2$ matrix in $SU(2)$. It is clear that to each of the products $(e^{i\theta}\mathbf{a}) \times e^{-2i\theta}$ we may associate the $U(2)$ matrix $e^{i\theta}\mathbf{a}$, and in fact this correspondence $SU(2) * U(1) \to U(2)$ is a homomophism onto all of $U(2)$. The kernel consists of those $\mathbf{a}$ and $\theta$ such that $e^{i\theta}\mathbf{a} =$ the $2 \times 2$ identity $I_2$; i.e., $\mathbf{a} = e^{-i\theta} I_2$. Since $\det \mathbf{a} = 1$ we have the two-element kernel with $\mathbf{a} = \pm I_2$, and so the group $SU(2) * U(1)$ can be considered as a two-sheeted covering group of $U(2)$.

Let now $G := SU(2) * U(1)$ be assumed to be the symmetry group for the strong interactions. It has generators $\lambda_k$, $k = 1, 2, 3$ (isospin), and $\lambda_8$ (hypercharge) and all operate again on the quarks $\mathbb{C}^3$, antiquarks $\mathbb{C}^{3*}$, and mesons $\mathbb{C}^{3*} \otimes \mathbb{C}^3$. Our basic meson frame is again $\mathbf{X}$ of (C.7).

$G$ can mix $\mathbf{u}$ and $\mathbf{d}$ but neither of these mixes with $\mathbf{s}$. Thus we may consider $\mathbf{u}$ and $\mathbf{d}$ as two states of the same particle, but $\mathbf{s}$ is assumed to be a different quark, with only one state.

A typical element $g$ of $SU(2) * U(1)$ is of the form

$$\begin{bmatrix} e^{i\theta} \begin{bmatrix} x & -\overline{w} \\ w & \overline{z} \end{bmatrix} & 0 \\ 0 & e^{-2i\theta} \end{bmatrix} \tag{C.10}$$

with $|z|^2 + |w|^2 = 1$ and the $2 \times 2$ submatrix in $SU(2)$. Consider the adjoint action of this matrix on $\mathbf{X}$. Since this operation is linear in $\mathbf{X}$ we may single out the particles in which we are interested. For the antikaons $K^-$ and $\overline{K}^0$ we may take for $\mathbf{X}$ the matrix

$$\begin{bmatrix} 0 & 0 & \mathbf{K}^- \\ 0 & 0 & \overline{\mathbf{K}^0} \\ 0 & 0 & 0 \end{bmatrix}$$

and we see easily that the adjoint action by (C.10) will produce mixtures of the $\mathbf{K}^-$ and the $\overline{\mathbf{K}}^0$. Similar results can be obtained for the $\mathbf{K}^+$ and the $\mathbf{K}^0$. Since the $(\mathbf{K}^-, \overline{\mathbf{K}}^0)$ do not mix with the $(\mathbf{K}^+, \mathbf{K}^0)$, we see that $(\mathbf{K}^+, \mathbf{K}^0)$ are to be considered as two states of a single particle and $(\mathbf{K}^-, \overline{\mathbf{K}}^0)$ are the two states of the antiparticle. (They are distinct particles, as we see from the weight diagram (Figure C.1) that their hypercharges are opposites.) Similarly, all three pions get mixed; they are three states of a single particle. Finally, the eta is completely unaffected by the adjoint action. We say that $(\mathbf{K}^+, \mathbf{K}^0)$ is a doublet, its antiparticle $(\mathbf{K}^-, \overline{\mathbf{K}}^0)$ is a doublet, the pion $(\pi^-, \pi^0, \pi^+)$ is a triplet, and $\eta$ is a singlet.

## C.f. Meson Masses

A **fermion** is a particle (e.g., an electron, proton, ...) whose wave function changes sign when an observer's coordinate system is rotated through a complete rotation (see p. 517), whereas a **boson** (e.g., a meson) has a wave function that returns to its original value under such a rotation. Particles composed of an odd number of fermions are again fermions but an even number will yield a boson. A neutron, made of three quarks, is a fermion. This leads us to think of quarks as fermions. A kaon, made of two quarks, is a boson.

Electrons and protons satisfy the Dirac equation, which can be "derived" from a Lagrangian (20.18). The coefficient of the squared wave function $|\Psi|^2$ is $m$, the mass of the fermion in question. Bosons are believed to satisfy something similar to the Klein–Gordon equation (19.24). To get this from a Lagrangian the coefficient of $|\Psi|^2$ must be the square of the mass, $m^2$. (Actually there is also a factor of $1/2$, but this will play no role in our discussion and so will be omitted.) We shall just accept the "rule" that *the coefficient of the squared term* $|\Psi|^2$ *in the Lagrangian involves* $m$ *for a fermion and* $m^2$ *for a boson.*

The classification of the particles that we have given followed from looking at *frames* of quarks, antiquarks, and mesons, i.e., $\mathbf{q}$, $\mathbf{q}^*$, and $\mathbf{X}$. A Lagrangian involves *components* (wave functions) rather than the basis elements (frames). For this reason we shall revert now to the component description of the meson matrix $\mathbf{X}$, which formally is simply the transpose,

$$X = \begin{bmatrix} \dfrac{\pi^0}{\sqrt{2}} + \dfrac{\eta}{\sqrt{6}} & \pi^+ & K^+ \\ \pi^- & \dfrac{-\pi^0}{\sqrt{2}} + \dfrac{\eta}{\sqrt{6}} & K^0 \\ K^- & \overline{K^0} & \dfrac{-2\eta}{\sqrt{6}} \end{bmatrix} \tag{C.11}$$

where the entries now are components, rather than basis elements. For example, $K^+ = u\,\overline{d}$.

We are interested in the masses of the mesons. We shall postulate a mass part $L_m$ of the total Lagrangian. In the original version, when $SU(3)$ was assumed, we could use a quadratic **Yukawa–Kemmer** type Lagrangian involving our meson matrix $X$, namely $L = \operatorname{tr} XX^*$, but as we shall soon see, this would result in all the mesons having the same mass. For the symmetry group $G = SU(2) * U(1)$ generated by isospin and hypercharge, we shall alter this by inserting an as yet to be determined $3 \times 3$ matrix $\mathfrak{M}$,

$$L_m = \operatorname{tr} X\,\mathfrak{M}\,X^* \tag{C.12}$$

To ensure that the mass coefficients are real we shall assume that $\mathfrak{M}$ is hermitian, for then $X\,\mathfrak{M}\,X^*$ will be hermitian and will have a real trace. Under a change of quark frame $\mathbf{q}$ used in $\mathbf{Q} = \mathbb{C}^3 = \mathbb{C}^2 \oplus \mathbb{C}^1$, $\mathfrak{M}$ is sent to $g\mathfrak{M}g^{-1}$, where $g \in G = SU(2) * U(1)$. Since there is no preferred frame, we insist that $\mathfrak{M}$ be unchanged under such a frame change, and so $\mathfrak{M} : \mathbb{C}^3 \to \mathbb{C}^3$ must commute with the $G$ action on $\mathbb{C}^3$. It is then not hard to see that $\mathfrak{M}$ *must* be of the form

$$\mathfrak{M} = \text{diag}(a, a, b) \tag{C.13}$$

where $a$ and $b$ are real numbers. In fact, we can apply elementary representation theory, in particular Schur's Corollary, as will be developed in Section D.c of Appendix D. An argument that is similar (but simpler) than that given there for the matrix $C$ in that section, applied to $\mathbb{C}^3 = \mathbb{C}^2 \oplus \mathbb{C}^1$ rather than $V = \mathbf{5} \oplus \mathbf{1}$, will show that $\mathfrak{M}$ must be of the form (C.13). The key point is that the action of G on $\mathbb{C}^3$ leaves both $\mathbb{C}^2$ and $\mathbb{C}^1$ invariant and this action is not further reducible. (I am indebted to Jeff Rabin for pointing out the uniqueness of this $\mathfrak{M}$.)

We then compute, using the fact that formally $X = X^*$,

$$\begin{aligned} L_m = \text{tr } X \mathfrak{M} X &= a(|\pi^0|^2 + \pi^-\pi^+ + \pi^+\pi^-) + (1/3)(a + 2b)|\eta|^2 \\ &+ (K^-K^+)b + (\overline{K}^0 K^0)b + (K^+K^-)a + (K^0\overline{K}^0)a \end{aligned}$$

Now the pion terms can be written

$$a(|\pi^0|^2 + |\pi^+|^2 + |\pi^-|^2)$$

and the kaon terms as

$$\begin{aligned} &(a + b)[(K^-K^+) + (\overline{K}^0 K^0)] \\ &= (1/2)(a + b)[|K^+|^2 + |K^-|^2 + |K^0|^2 + |\overline{K}^0|^2] \end{aligned}$$

We have chosen this arrangement since all the kaons must have the same mass since $\mathbf{K}^{\pm}$ are antiparticles, and so have the same mass, and since $(\mathbf{K}^+, \mathbf{K}^0)$ are the two states of a single particle (see the last paragraph of Section C.e.), and so have the same mass. Similar arguments follow for the pions. Then, since we are dealing with bosons,

$$\begin{aligned} m_\pi &:= \text{mass of any pion} = \sqrt{a} \\ m_\eta &:= \text{mass of eta} = \sqrt{[(a + 2b)/3]} \\ m_K &:= \text{mass of any kaon} = \sqrt{[(a + b)/2]} \end{aligned}$$

From these we see that

$$4m_K^2 = m_\pi^2 + 3m_\eta^2 \tag{C.14}$$

one of the famous **Gell–Mann/Okubo mass formulas.**

The *observed* masses of the pions and eta are $m_\pi \approx 140$ MeV and $m_\eta \approx 550$ MeV. Use these in (C. 14). Then (C. 14), i.e., *the assumption of symmetry group* $G = SU(2) * U(1)$ *together with the simple choice of G-invariant Lagrangian* (C.12), yields the prediction $m_K \approx 481$ MeV, which is less than 3% off from the observed 495 MeV.

APPENDIX D

# Representations and Hyperelastic Bodies

## D.a. Hyperelastic Bodies

In (A.13) we have shown that the rate at which energy is stored in a body during a given deformation from reference body $B(0)$, assuming no heat loss, is given by

$$\int_{B(0)} S^{AB}(dE_{AB}/dt)\ \mathrm{VOL} \tag{D.1}$$

where $S$ is the second Piola–Kirchhoff stress tensor, $E$ is the Lagrange deformation tensor (2.69), and the integral is over the fixed reference body $B(0)$. The entire integrand is not necessarily the time derivative of a function. The stress tensor $S$ is generally a complicated function of the deformation tensor $E$. In the linearized theory we assume generalized Hooke's coefficients $C$ and a relation of the form

$$S^{AB} = C^{AB\,JK} E_{JK} \tag{D.2}$$

where, since both $S$ and $E$ are *symmetric* tensors, $C$ is symmetric in $A$ and $B$ and also in $J$ and $K$. At each point there are thus 36 constants $C^{AB\,JK}$ involved. Let us now assume the **hyperelastic** condition

$$C^{AB\,JK} = C^{JK\,AB} \tag{D.3}$$

Then at each point of $B(0)$

$$S^{AB}(dE_{AB}/dt) = C^{AB\,JK} E_{JK}(dE_{AB}/dt) = d/dt\left(1/2\, C^{AB\,JK} E_{JK} E_{AB}\right)$$

and then (D.1) becomes

$$\int_{B(0)} S^{AB} dE_{AB}/dt\ \mathrm{VOL} = d/dt \int_{B(0)} U\ \mathrm{VOL}$$

where (D.4)

$$U = \tfrac{1}{2} C^{AB\,JK} E_{JK} E_{AB} = \tfrac{1}{2} S^{AB} E_{AB}$$

is the volume density of **strain energy**. As mentioned at the end of Section A.d, a body with such an energy function is called hyperelastic.

Note that in this linearized case we have

$$S^{AB} = \partial U / \partial E_{AB} \tag{D.5}$$

where the partial derivative is taken while keeping the coordinates $X$ of the reference body fixed.

We remark that in the general (nonlinear) case of a hyperelastic body we may in fact use (D.5) to *define* the stress tensor. That is, we can assume that there is some strain energy function $U(X, E)$ of the position $X$ and of the the Lagrange deformation tensor $E$ and then use (D.5) to define the second Piola–Kirchhoff stress tensor $S$.

*From now on we shall restrict ourselves to hyperelastic bodies in the linearized approximation with coefficients C satisfying (D.3) at each point.* Note that the number of independent $C^{AB\,JK}$ is now reduced from 36 to 21 components at each point of the body.

## D.b. Isotropic Bodies

In the following we shall be concerned only with $\mathbb{R}^3$ with an orthonormal basis. This allows us to forget distinctions of covariance and contravariance, though we shall frequently put indices in their "correct" place.

In the linear approximation, $S$ and $E$ are related as in (D.2). At each point we consider the real vector space $\mathbb{R}^6$ *of symmetric* $3 \times 3$ matrices. (D.2) says that $S \in \mathbb{R}^6$ is related to $E \in \mathbb{R}^6$ by a *linear* transformation $C : \mathbb{R}^6 \to \mathbb{R}^6$,

$$S = C(E) \tag{D.6}$$

Consider a given deformation tensor $E$ at a point. (For example, $E$ could result from a stretching along the $x$ axis and compressions along the $y$ and $z$ axes at the origin.) The result is a stress $S = C(E)$ at the point. Now consider the same physical deformation but oriented along different axes; call it $E'$. (In our example $E'$ could be stretching along the $y$ axis and compressions along the $x$ and $z$ axes, all with the same magnitudes as before.) The new stress is $S' = C(E')$. If we call the change of axes matrix $g \in SO(3)$, then the matrices $E$ and $E'$ are related by $E' = gEg^{-1}$, but we must not expect $S'$ to be $gSg^{-1}$; the material of the body might react, say, to compressions along the $x$ and $y$ axes in entirely different ways. If we do have $S' = gSg^{-1}$, in other words, if the (adjoint) action of $SO(3)$ on $3 \times 3$ symmetric matrices commutes with $C : \mathbb{R}^6 \to \mathbb{R}^6$,

$$gC(E)g^{-1} = C(gEg^{-1}) \tag{D.7}$$

and if this holds at each point of the body, we say that the body is elastically **isotropic**.

We now have the following situation for an isotropic hyperelastic body. The real $6 \times 6$ matrix $C : \mathbb{R}^6 \to \mathbb{R}^6$ has at most 21 independent entries and the matrix $C$ commutes with the adjoint action of $SO(3)$ on $\mathbb{R}^6$ (thought of as the space of symmetric $3 \times 3$ matrices). We shall sketch, in the remaining sections, how elementary representation theory shows that there are only two "Lamé" constants required to express $C$!

## D.c. Application of Schur's Lemma

Consider a representation $\mu$ of a compact group $G$ as a group of linear transformations of a *finite dimensional* vector space $V$ into itself; see Section 18.2a. Thus, for $g \in G$, $\mu(g) : V \to V$ and $\mu(gh) = \mu(g)\mu(h)$. When we are considering only one representation $\mu$ of $G$ on a vector space $V$, *we shall frequently call the representation* $V$ rather than $\mu$.

We have in mind for our application, the following:

**Example:** $G = SO(3)$, $V = \mathbb{R}^6$ is the real vector space of symmetric $3 \times 3$ matrices, and $\mu(g)$ acts on a matrix $E$ by the adjoint action, $\mu(g)(E) = gEg^{-1}$.

Since $G$ is compact, by averaging over the group (as in Section 20.4c), we may choose a scalar product in $V$ so that $\mu(g)$ acts on $V$ by unitary or orthogonal matrices, depending on whether $V$ is a complex or a real vector space.

The representation $\mu$ is **irreducible** if there is no nontrivial vector subspace $W$ that is invariant under all $\mu(g)$, i.e., $\mu(g) : W \to W$ for all $g \in G$.

If $\mu$ is reducible, then there is a nontrivial subspace $W \subset V$ that is invariant under $G$. In this case the orthogonal complement of $W$ is also invariant since $g$ acts by isometries. Then by choosing an orthonormal basis for $V$ such that the first $\dim(W)$ basis elements are in $W$ and the remaining are in the orthogonal complement of $W$, we see that each $\mu(g)$ is in block diagonal form. If $\mu$, when restricted to $W$, is reducible, we may break this reducible block into two smaller blocks. By continuing in this fashion we can reduce $V$ to a sum of mutually orthogonal invariant subspaces, each of which forms an *irreducible* representation of G.

In our example $V$ is the space of symmetric $3 \times 3$ matrices $E$. The deformation tensor $E_{JK}$ represents a covariant bilinear form and should transform as $\mu(g)(E) = gEg^T$, but since $g^T = g^{-1}$ for $g$ in $SO(3)$, we may think of $E$ as a linear transformation $:\mathbb{R}^3 \to \mathbb{R}^3$. (This is nothing more than saying $E_{JK} = E^J{}_K$ in an orthonormal basis). As a linear transformation, its trace tr $E$ will be invariant, and, just as we did for the meson matrix (C.3), we shall reduce the six-dimensional space of all symmetric $3 \times 3$ matrices into the sum of the trace-free symmetric matrices and its orthogonal complement of scalar matrices, which we could write in the same spirit as (C.9) as

$$V = \mathbf{5} \oplus \mathbf{1}$$
$$E = [E - (1/3)(\text{tr } E)I] + (1/3)(\text{tr } E)I \tag{D.8}$$

where we are now indicating the *real* dimensions. **1** is clearly irreducible, and we shall give a rather lengthy sketch showing that **5** is also.

**Schur's Lemma:** *Let* $(V, \mu)$ *and* $(W, \omega)$ *be two irreducible representations of G and let* $A : V \to W$ *be a linear transformation that commutes with the G actions on V and W in the sense that*

$$A[\mu(g)v] = \omega(g)A[v]$$

*Then either $A$ maps all of $V$ to $0 \in W$ or $A$ is 1–1 and onto. In this latter case we say that the representations $\mu$ and $\omega$ are* ***equivalent***.

PROOF: The commutativity of $A$ and the $G$ actions shows immediately that the subspaces $\ker(A) \subset V$ and $\mathrm{Im}(A) \subset W$ are invariant under the $G$ actions. Since $V$ is irreducible, $\ker(A)$ is either $V$, in which case $A(V) = 0$, or $\ker A = 0$, showing that $A$ is 1–1. In this last case, by irreducibility of $W$, we have $\mathrm{Im}(A) = W$. □

**Schur's Corollary:** *If $\mu$ is irreducible and if a linear transformation $C : V \to V$ commutes with each $\mu(g)$, and if $C$ has an eigenvector in $V$, then $C$ is a scalar matrix, $C = \lambda I$.*

Note that if $V$ is complex, $C$ will automatically have an eigenvector.

PROOF: Let $v$ be an eigenvector of $C$ with eigenvalue $\lambda$. Then $C - \lambda I : V \to V$ will also commute with the $G$ action on $V$. But $(C - \lambda I)v = 0$. By Schur's Lemma, $C - \lambda I = 0$. □

Return now to our elastic isotropic example. $V$ is the space of real symmetric $3 \times 3$ matrices. $C : V \to V$ is the linear map $S = C(E)$ in (D.6) relating stress to strain in the linear approximation. In terms of matrices, $S^{AB} = C^{AB\,JK} E_{JK}$. (D.7) says that $C$ commutes with the adjoint action of $G = SO(3)$ on $V$. Since $V$ is a real vector space, we must determine if $C$ has an eigenvector. But in the hyperelastic case, (D.3), i.e., $C^{AB\,JK} = C^{JK\,AB}$, says that $C$ is a *self-adjoint* (symmetric) matrix operating on $\mathbb{R}^6$,

$$\langle C(E), F\rangle = C^{AB\,JK} E_{JK} F_{AB} = E_{JK} C^{JK\,AB} F_{AB} = \langle E, C(F)\rangle$$

and so $C$ does have a real eigenvector. *Assume* for the present that $V = \mathbf{5} \oplus \mathbf{1}$ of (D.8) is a decomposition of $V$ into *irreducible* subspaces, i.e., that the real, trace-free, symmetric $3 \times 3$ matrices form an irreducible representation of the adjoint action of $SO(3)$. We shall prove this in our following sections. Note that isotropy (D.7) shows that the subspace $C(\mathbf{1})$ must be invariant under the $G$ action. Since $\mathbf{1}$ is $G$ invariant, the orthogonal projection $\prod C(\mathbf{1})$ of $C(\mathbf{1})$ into $\mathbf{5}$ must also be $G$ invariant. Since $\mathbf{5}$ is assumed irreducible, it must be that $\prod C(\mathbf{1}) = 0 \subset \mathbf{5}$, and so $C(\mathbf{1}) \subset \mathbf{1}$. Thus $C$ sends $\mathbf{1}$ into itself and, since $C$ is self-adjoint, $C : \mathbf{5} \to \mathbf{5}$. Then we may apply Schur's Corollary to the two cases, $C$ restricted to $\mathbf{5}$ and $C$ restricted to $\mathbf{1}$. In both cases $C$ is a scalar operator. $C$ restricted to $\mathbf{5}$ is multiplication by a real number $a$ and when restricted to $\mathbf{1}$ is multiplication by a real $b$. From (D.8) we may write

$$S = C(E) = a[E - (1/3)(\mathrm{tr}\, E)I] + (1/3)b(\mathrm{tr}\, E)I$$

but this is classically written in terms of the two **Lamé moduli** $\mu$ and $\lambda$ as

$$S_{AB} = 2\mu E_{AB} + \lambda(E^J_J)\delta_{AB} \tag{D.9}$$

which was essentially known already to Cauchy (see Truesdell [T, p. 306]).

## D.d. Frobenius–Schur Relations

Our only remaining task is to show that the trace-free, real, symmetric matrices **5** form an *irreducible* representation under the adjoint action of $SO(3)$. If our proof seems overly long it is because we are taking this opportunity to present very basic results about group representations. While our elasticity problem involves real representations, and real representations pose special problems (as in Schur's Corollary), we shall frequently use the notation of complex unitary representations (e.g., hermitian adjoint rather than transpose) but develop mainly those results that hold for real representations also, so that they can be applied to our problem.

For more about the Frobenius–Schur relations, see, e.g., the small book of Wu-Yi Hsiang [Hs] (but beware that his Theorem 2 on p. 6 has been labeled Theorem 1).

The principal tool is averaging over a compact group, as in Section 20.4c. If $(V, \mu)$ is a representation, then for each $g \in G$, $\mu(g)$ is a matrix and its average, with respect to a bi-invariant volume form $\omega$ *normalized so that the volume of G is* 1, is again a matrix $P : V \to V$,

$$P := \int_G \mu(g)\omega_g \tag{D.10}$$

meaning

$$P(v) = \int_G \mu(g)(v)\omega_g$$

for each vector $v \in V$. Clearly if $\mu(g)v = v$ for all $g$ then $P(v) = v$. Also

$$\mu(h)P(v) = \int_G \mu(hg)(v)\omega_g = \int_G \mu(g)(v)\omega_g = P(v) \tag{D.11}$$

shows that $P(v)$ is fixed under all $g$, and so $P : V \to V^G$, where $V^G$ is the subspace of all vectors fixed under all $\mu(g)$, the **fixed set** of the $G$ action. Finally, from (D.11) we see that

$$P^2(v) = P(P(v)) = \int_G \mu(h)P(v)\omega_h = \int_G P(v)\omega_h = P(v)$$

and so $P^2 = P$; i.e., $P$ is a *projection* of $V$ onto the fixed subspace $V^G$. Since this is a projection operator one sees immediately (by choosing a basis whose initial elements span $V^G$) that

$$\dim V^G = \operatorname{tr} P = \int_G \operatorname{tr} \mu(g)\omega_g \tag{D.12}$$

Let us look at some consequences of this formula. Let $U$ and $W$ be two vector spaces. Then $U \otimes W^*$ is the vector space of linear transformations of $W$ into $U$; $(u \otimes w^*)(z) = w^*(z)u$, for all $z \in W$. Suppose that $(U, \alpha)$ and $(W, \beta)$ are representations of $G$ on $U$ and $W$ respectively. Then the hermitian adjoint matrices $\beta^*(g) = \beta(g^*) = \beta(g^{-1})$ operate on $W^*$ by $\beta^*(g)(w^*) = \omega^*\beta(g^{-1})$. Thus $\alpha \otimes \beta^*$ is the representation sending the linear transformation $A = u \otimes w^*$ to the linear transformation $\alpha(g)u \otimes w^*\beta(g^{-1}) =$

$\alpha(g)A\beta(g^{-1})$. A linear transformation $A$ is fixed under this $G$ action iff $A$ commutes with the $G$ action,

$$(U \otimes W^*)^G = \text{those } A : W \to U \text{ such that } \alpha(g)A = A\beta(g) \tag{D.13}$$

For *any* representation $\mu$ the function $\chi_\mu : G \to \mathbb{C}$ defined by $\chi_\mu(\sigma) := \text{tr}\, \mu(\sigma)$ is traditionally called the **character** of the representation $\mu$.

We need one more simple fact. Given any two linear transformations $\alpha : U \to U$ and $\beta : W \to W$, then

$$\text{tr}\,(\alpha \otimes \beta) = \text{tr}\,(\alpha)\, \text{tr}\,(\beta)$$

since if $\{\mathbf{e}_j\}$ and $\{\mathbf{f}_a\}$ are bases for $U$ and $W$, then $\{\mathbf{e}_j \otimes \mathbf{f}_a\}$ is a basis for $V \otimes W$ and the coefficient of $\mathbf{e}_j \otimes \mathbf{f}_a$ in $\alpha \otimes \beta(\mathbf{e}_j \otimes \mathbf{f}_a)$ is $\alpha^j{}_j \beta^a{}_a$ (no sum).

Apply (D.12) in the case $V = U \otimes W^*$, and use the fact that $\beta(g^{-1})$ is the conjugate transpose of $\beta(g)$. We get

**Theorem (D.14):** *The dimension of the space of $A : W \to U$ that commute with the actions of $G$ is*

$$\int_G \chi_\alpha(g)\overline{\chi}_\beta(g)\omega_g$$

*In particular, if $(W, \beta)$ and $(U, \alpha)$ are irreducible and* inequivalent, *by Schur's Lemma this integral is 0.*

*On the other hand, if $U$ and $W$ are equivalent, there is at least one such map $A$ and so, in particular, for any representation $(V, \mu \neq 0)$, we have*

$$\int_G \chi_\mu(g)\overline{\chi}_\mu(g)\omega_g \geq 1$$

**Theorem (D.15):** *If $(V, \mu)$ is a representation and*

$$\int_G \chi_\mu(g)\overline{\chi}_\mu(g)\omega_g = 1$$

*then the representation is irreducible.*

PROOF: Suppose that $(V, \mu)$ is reducible. In Section D.c we showed that $V$ can be written as a direct sum of orthogonal, invariant, irreducible subspaces $V = \oplus V_\alpha$, and we can let $\mu_\alpha$ be the restriction of $\mu$ to $V_\alpha$. A simple example to keep in mind is a representation $\mu$ of $SO(2)$ (which as a manifold is the circle $S^1$ with angular coordinate $\theta$) acting on $V = \mathbb{R}^4$ by two $2 \times 2$ diagonal blocks, where $m$ and $n$ are nonnegative integers:

$$\mu(\theta) = \begin{bmatrix} \cos m\theta & -\sin m\theta & 0 & 0 \\ \sin m\theta & \cos m\theta & 0 & 0 \\ 0 & 0 & \cos n\theta & -\sin n\theta \\ 0 & 0 & \sin n\theta & \cos n\theta \end{bmatrix}$$

Call the $2 \times 2$ blocks $(V_1, \mu_1)$ and $(V_2, \mu_2)$. The two representations $\mu_1$ and $\mu_2$ are equivalent if and only if $m = n$. If $m = n$ we would write $V = 2V_1$ while if $m \neq n$ we would write $V = V_1 \oplus V_2$.

In the general case we can similarly write $V = \oplus\, m_j V_j$, where $V_j$ and $V_k$ are inequivalent if $j \neq k$. Then, from $\operatorname{tr} \mu = \sum_j m_j \operatorname{tr} \mu_j$ and (D.14) we have

$$\int_G \chi_\mu(g)\overline{\chi}_\mu(g)\omega_g = \sum m_j m_k \int_G \chi_j(g)\overline{\chi}_k(g)\omega_g = \sum m_j^2 \int_G \chi_j(g)\overline{\chi}_j(g)\omega_g$$

Thus if $\mu$ is reducible, i.e., $\sum m_j^2 \geq 2$, we would have that the integral is $\geq 2$. □

We remark that a *complex* irreducible representation will have

$$\int_G \chi(g)\overline{\chi}(g)\omega_g = 1$$

since by Schur's Corollary the matrices commuting with the $G$ action will be scalar and so have complex dimension 1 . On the other hand, the usual action of $SO(2)$ on $\mathbb{R}^2$ as in (15.0) is clearly a real irreducible representation that has for integral of $\operatorname{tr}^2$

$$\int_0^{2\pi} 4\cos^2\theta d\theta/2\pi = 2$$

corresponding to the fact that the two-dimensional subspace of real $2 \times 2$ matrices satisfying $x_{22} = x_{11}$ and $x_{21} = -x_{12}$ all commute with $SO(2)$.

## D.e. The Symmetric Traceless 3 × 3 Matrices Are Irreducible

(D.15) implies that we need only show

$$\int_{SO(3)} |\operatorname{tr} \operatorname{Ad} g|^2 \omega_g = 1 \tag{D.16}$$

where $SO(3)$ acts on $V = \mathbf{5}$, the space of traceless real symmetric matrices, by the adjoint action, $\operatorname{Ad}(g)A = gAg^{-1}$.

We have used before that $SO(3)$ can be realized as the real projective space $\mathbb{R}P^3$, pictured, e.g., as the solid ball of radius $\pi$ centered at the origin of $\mathbb{R}^3$ with antipodal points on the boundary sphere identified; see Example (vii) of Section 1.2b. The 1-parameter subgroups are the rays through the origin. This model is unsuitable for the integral (D.16) because in (D.16) the metric is the same as the metric on $\mathbb{R}P^3$, not $\mathbb{R}^3$. Since the unit sphere $S^3 \subset \mathbb{C}^2$ is the proper model for $SU(2)$ (see Chapter 19 and also p. 584), and since $SU(2)$ is the twofold cover of $SO(3)$, we shall use the "upper hemisphere" of $S^3$ as the model for $\mathbb{R}P^3$.

For example, the point $(e^{-i\beta}, 0) \in S^3 \subset \mathbb{C}^2$ represents both the matrix $u(\beta) \in SU(2)$ and the matrix $g(\beta) \in SO(3)$, where

$$u(\beta) = \begin{bmatrix} e^{-i\beta} & 0 \\ 0 & e^{i\beta} \end{bmatrix} \quad \text{and} \quad g(\beta) = \begin{bmatrix} \cos 2\beta & -\sin 2\beta & 0 \\ \sin 2\beta & \cos 2\beta & 0 \\ 0 & 0 & 1 \end{bmatrix}$$

This was shown in the example following the proof of Theorem (19.12). We then have the following picture (see Figure D.1) on the unit sphere $S^3 \in \mathbb{C}^2$ with Riemannian metric $ds^2 = d\alpha^2 + \sin^2\alpha(d\theta^2 + \sin^2\theta\, d\phi^2)$, where $\alpha$ is the colatitude, and the "north pole" is the identity matrix for both $SU(2)$ and $SO(3)$, and the "small sphere" $S^2(\alpha)$ at colatitude $\alpha$ has metric $\sin^2\alpha(d\theta^2 + \sin^2\theta\, d\phi^2)$ and area $4\pi \sin^2\alpha$. We will explain this diagram more in the following.

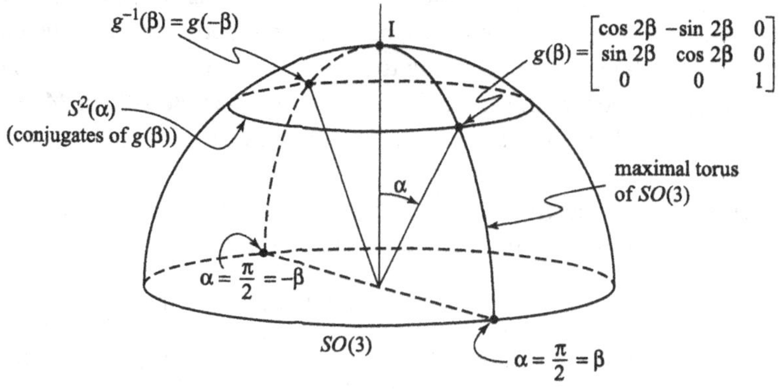


**Figure D.1**

The 1-parameter subgroup $u(\beta) = \text{diag}(e^{-i\beta}, e^{i\beta}) \subset SU(2), -\pi \leq \beta \leq \pi$, is a maximal torus of $SU(2)$ (see Theorem 15.4), and the image of this circle under

$$Ad : SU(2) \to SO(3)$$

(see Section 19.1b) covers twice the maximal torus of $SO(3)$ given by $g(\beta)$, for $-\pi/2 \leq \beta \leq \pi/2$. The parameter $\beta$ on this subgroup coincides with $\alpha$ for $\beta \geq 0$ and with $-\alpha$ for $\beta \leq 0$. ($\alpha$ is not a good coordinate at the identity.)

For any point $\sigma$ of a Lie group $G$ we can look at the **conjugates** of $\sigma$, i.e., the set of all group elements of the form $g\sigma g^{-1}$ as $g$ ranges over the group. This set $M_\sigma$ is thus the **orbit** of the point $\sigma$ under the adjoint action of $G$ on itself. The group elements that leave the point $\sigma$ fixed form the **centralizer** subgroup $C_\sigma$ of $\sigma$, those $g$ that commute with $\sigma$. Thus, from (17.10), the orbit points of $M_\sigma$ are in 1–1 correspondence with points of the quotient manifold $M_\sigma = G/C_\sigma$.

Consider Figure D.1 and the point $\sigma = g(\beta)$ on the maximal torus. Since $\text{Ad}_g : G \to G$ sending any $h$ to $ghg^{-1}$ is an isometry of the bi-invariant metric on $G$, and since $\text{Ad}_g$ leaves the identity $I$ fixed, $M_{g(\beta)}$ must lie on the sphere $S^2(\alpha)$ at constant distance from $I$. It is not difficult to see (see Section E.a of Appendix E) that $M_{g(\beta)}$ in fact coincides with this 2-sphere. This is not surprising; the centralizer of $g(\beta)$, $\beta \neq 0$ or $\pm\pi/2$, is exactly the maximal torus $T^1$, and $SO(3)/T^1 = SO(3)/SO(2) = S^2$.

If $\beta = 0$, we have the identity $I$ whose centralizer is all of $SO(3)$, and $SO(3)/SO(3)$ is the single point $I$.

If $\beta = \pi/2$, then the centralizer of diag$(-1, -1, 1)$ contains not only the maximal torus $T^1$ (on which it lies) but clearly also the elements diag$(1, -1, -1)$ and diag$(-1, 1, -1)$, which are rotations through 180° about the $x$ and the $y$ axes respectively. It is not hard to see, in fact, that all rotations through 180° about all axes in the $xy$ plane are in this centralizer. This curve of rotations is the curve $C'$ in Figure 17.4. The conjugate set of diag$(-1, -1, 1)$ is $[SO(3)/T \cup C']$. This is topologically $\mathbb{R}P^2$, because $SO(3)$ acts transitively on the space of lines through the origin of $\mathbb{R}^3$, and the subgroup leaving the $z$ axis invariant consists of all rotations about the $z$ axis, i.e., $T$, together with all rotations through 180° around all axes in the $xy$ plane (i.e., $C'$). In our Figure D.1 the conjugate set for $g(\pi/2) = \text{diag}(-1, -1, 1)$ is the equatorial 2-sphere with antipodal identifications, i.e., a projective plane! (The conjugacy orbits $M_\sigma = G/C_\sigma$ have very interesting topological properties in a general compact connected Lie group. For example, the Euler–Poincaré characteristic of $M_\sigma$ is equal to the number of times $M_\sigma$ intersects the maximal torus, as we easily noticed with $S^2$ and $RP^2$. See Theorem E.2 in Appendix E).

We return now to our integral (D. 16). Recall that each Ad$(\sigma)$ is a $5 \times 5$ matrix. Look at a general point $\sigma$ in $G = SO(3)$. The character $\chi$ has the property

$$\chi_\mu(g\sigma g^{-1}) = \text{tr}\, \mu(g\sigma g^{-1}) = \text{tr}\, [\mu(g)\mu(\sigma)\mu(g)^{-1}] = \text{tr}\, \mu(\sigma) = \chi_\mu(\sigma)$$

That is, $\chi$ *is constant on conjugacy orbits.* Thus our function $\chi_{\text{Ad}}(\sigma)$, the trace of the $5 \times 5$ matrix Ad$(\sigma)$, is constant on each of the 2-spheres $S^2(\alpha)$ of constant colatitude $\alpha$ In our volume integral, the two conjugacy sets at $\alpha = 0$ and $\alpha = \pi/2$ can be omitted. Note that these conjugacy sets to be omitted are precisely those passing through the only two points $g(0)$ and $g(\pi/2)$ of $T$ *whose centralizers are larger than T itself.* We can then evaluate our integral as follows, thanks to the fact that each remaining conjugacy sphere $M_{g(\beta)}$ meets $T$ orthogonally:

$$\int_{SO(3)} |\text{tr Ad}(g)|^2 \omega_g = 1/\pi^2 \int_0^{\pi/2} |\text{tr Ad}\, g(\beta)|^2 4\pi \sin^2(\beta) d\beta \qquad \text{(D.17)}$$

We integrate only from 0 to $\pi/2$ (i.e., only half of the maximal torus) to avoid counting the spheres $S^2(\beta)$ twice. The factor $\pi^{-2}$ is required since the Frobenius–Schur relations require that the volume of $G$ must be normalized to unity, and the total volume of our $SO(3)$ is

$$\int_0^{\pi/2} 4\pi \sin^2(\beta) d\beta = \pi^2$$

We now need to know the character function $\chi$ of Ad $g(\beta)$ along the maximal torus. A straightforward way is as follows. Write down a basis $\mathbf{E}_j$, $1 \le j \le 5$, of the real trace-free symmetric $3 \times 3$ matrices, starting say with $\mathbf{E}_1 = \text{diag}\,(1, -1, 0)$. For $g(\beta)$ on the maximal torus, compute $g(\beta)\mathbf{E}_j g(-\beta) = \sum \mathbf{E}_i a_{ij}(\beta)$, and take $\sum a_{jj}$. This calculation yields the result

$$\chi_{\text{Ad}}\, g(\beta) = 4\cos^2 2\beta + 2\cos 2\beta - 1$$

Finally our integral (D.17) becomes (with help, e.g., from Mathematica)

$$\frac{1}{\pi^2} \int_0^{\pi/2} |4\cos^2 2\beta + 2\cos 2\beta - 1|^2 4\pi \sin^2(\beta) d\beta = 1$$

showing indeed that the representation **5** of 3 × 3 real symmetric trace-free matrices is irreducible. □

One final remark should be noted. The character can be more easily computed by "general nonsense." Consider the following vector spaces of 3 × 3 real matrices:

$\mathbf{3} \otimes \mathbf{3} =$ all 3 × 3 matrices
$\mathbf{3} \circ \mathbf{3} =$ symmetric matrices
$\mathbf{3} \wedge \mathbf{3} =$ skew-symmetric matrices
$\mathbf{5} =$ trace-free symmetric matrices
$\mathbf{1} =$ scalar matrices

Then $\mathbf{3} \otimes \mathbf{3} = \mathbf{3} \circ \mathbf{3} \oplus \mathbf{3} \wedge \mathbf{3} = (\mathbf{5} \oplus \mathbf{1}) \oplus (\mathbf{3} \wedge \mathbf{3})$. But the Hodge star operator sends 2-forms to 1-forms, $* : \mathbf{3} \wedge \mathbf{3} \to \mathbf{3}$. In an orthonormal basis of $\mathbb{R}^3$, the star operator clearly commutes with the actions of $SO(3)$, which shows that $\mathbf{3} \wedge \mathbf{3}$ and $\mathbf{3}$ are equivalent representations, $\mathbf{3} \wedge \mathbf{3} = \mathbf{3}$. Taking traces of the representations, we get $(\text{tr}\, \mathbf{3})^2 = \text{tr}\, \mathbf{3} \otimes \mathbf{3} = \text{tr}\, \mathbf{5} + \text{tr}\, \mathbf{1} + \text{tr}\, \mathbf{3}$. Thus

$$\chi_5 = (\chi_3)^2 - \chi_3 - \chi_1 = [2\cos 2\beta + 1]^2 - [2\cos 2\beta + 1] - 1 = 4\cos^2 2\beta + 2\cos\beta - 1$$

which agrees with our previous calculation of $\chi_{\text{Ad}}\, g(\beta)$.

APPENDIX E

# Orbits and Morse–Bott Theory in Compact Lie Groups

*There once was a real classy Groupie*
*Who longed from the homotopyists to mut'nie*
*Bott appeared, it was Fate,*
*Made her period 8*
*By applying Morse Code to her Loopie.*

## E.a. The Topology of Conjugacy Orbits

We now wish to study in more detail the topology of conjugacy orbits in a compact Lie group $G$ with given maximal torus $T$. But first we present an example (more complicated than the $SO(3)$ case of Figure D.1) to keep in mind.

Let $G$ be the nine-dimensional unitary group $U(3)$. The subgroup of diagonal matrices $T = \{\text{diag}[\exp(i\theta_1), \exp(i\theta_2), \exp(i\theta_3)]\}$ is a three-dimensional maximal torus. Consider the diagonal matrix $\sigma = \text{diag}(-1, -1, 1)$. The subgroup $C_\sigma$ that commutes with $\sigma$, the centralizer of $\sigma$, is $U(2) \times U(1)$, which has dimension $4 + 1 = 5$. The conjugacy set of $\sigma$, $M_\sigma = \{u\sigma u^{-1}\}$ is, from (17.10), in 1:1 correspondence with the complex projective plane $\mathbb{C}P^2 = U(3)/U(2) \times U(1)$, the analogue of the real projective plane discussed in Section 17.2b. It has dimension $9 - 5 = 4$. This orbit $M_\sigma$ consists of unitary matrices with eigenvalues $-1, -1$, and $+1$. Thus $M_\sigma$ meets $T$ in the three points $\sigma$, $\text{diag}(-1, 1, -1)$ and $\text{diag}(1, -1, -1)$, the distinct permutations of the diagonal entries of $\sigma$, and, as we shall see in Theorem E.2, the Euler characteristic $\chi(\mathbb{C}P^2)$ is 3. The same argument would hold for any diagonal $\tau = \text{diag}(e^{i\theta}, e^{i\theta}, e^{i\phi})$ with exactly two distinct eigenvalues. $M_\tau$ would again be a complex projective plane. However, our example $\sigma$ is special in that $\sigma = \sigma^{-1}$, and so all of $M_\sigma$ is a component of the fixed set of the inversion isometry $i : G \to G, i(g) = g^{-1}$, and is thus a totally geodesic submanifold of $G$ (see Section 11.4d). On the other hand, if $\mu$ is a diagonal unitary with three distinct eigenvalues (such $\mu$ are dense on $T$), then the only $u$ commuting with $\mu$ will be diagonal, and so $C_\mu = T$, and $M_\mu = G/T = U(3)/T$, which has dimension $9 - 3 = 6$. The only matrices on $T$ that are conjugate to $\mu$ are the six distinct permutations of the diagonal elements of $\mu$, and we shall see that it must be that $\chi[U(3)/T] = 6$.

We now return to the general case of a compact lie group $G$ with maximal torus $T$ and some $\sigma \in T$. We know that $M_\sigma = \{g\sigma g^{-1}\}$ and we know that this set is in $1 : 1$ correspondence with the coset space $G/C_\sigma$, which is a manifold in its own right. Define a smooth map $F : G/C_\sigma \to G$ by $F(gC) := g\sigma g^{-1} \in M_\sigma \subset G$. First note that $F$ is

$1:1$, for if $g\sigma g^{-1} = F(gC) = F(hC) = h\sigma h^{-1}$ then $(h^{-1}g)\sigma = \sigma(h^{-1}g)$ says that $h^{-1}g \in C$, $g \in hC$, and so the coset $gC$ is the same coset as $hC$.

We now wish to show that this image $M_\sigma$ is an embedded submanifold. We show first that the differential $F_*$ maps no nonzero tangent vector to $G/C_\sigma$ at the single point $\sigma C$ into a zero tangent vector at the image point $\sigma$; i.e., that $F$ is an immersion at $\sigma C$. An example to keep in mind about failure of immersions is the map $f : \mathbb{R} \to \mathbb{R}^2$ given by $x(t) = t^2$, $y(t) = t^3$, which yields a curve with a cusp at the origin. This smooth map is not an immersion because $f_*\left(\frac{\partial}{\partial t}\right) = \left(\frac{dx}{dt}\right)\frac{\partial}{\partial x} + \left(\frac{dy}{dt}\right)\frac{\partial}{\partial y}$ vanishes at $t = 0$. This is the reason that a cusp can appear.

Since $G/C$ is made up of curves $t \to g(t)C$, a general tangent vector at $\sigma C$ is the velocity vector of a curve of the form $e^{tY}C$, where $Y$ is in the Lie algebra $\mathfrak{g}$ of $G$. The image of this curve under $F$ is $F(e^{tY}) = e^{tY}\sigma e^{-tY}$, whose velocity vector at $t = 0$ is $Y\sigma - \sigma Y$, and this, by the definition of the differential, is $F_*$ (velocity of $e^{tY}C$) . Suppose then that this $Y\sigma - \sigma Y = 0$. Then $Y = \sigma Y \sigma^{-1}$ and so $\exp(tY) = \exp(\sigma t Y \sigma^{-1}) = \sigma \exp(tY)\sigma^{-1}$ (from the power series). Thus the curve $\exp(tY)$ in $G$ lies in $C_\sigma$ and so the curve $e^{tY}C$ is a single point curve $\sigma C$ and has zero velocity at $t = 0$. Thus $F$ is an immersion at $\sigma C$. This implies that $F$ is an embedding of some $G/C$ neighborhood of $\sigma C$. But since each map $\mathrm{Ad}_g$ mapping $G \to G$ defined by $h \to ghg^{-1}$ is a diffeomorphism sending $M_\sigma$ onto itself, it is not hard to see that $F$ is locally an embedding near every point of $G/C$. Since $G$ is compact, the situation pictured in the second curve in Figure 6.7 cannot arise. It can be shown that $M_\sigma$ is a global embedded submanifold of $G$.

We now know, for $\sigma \in T$, that $M_\sigma$ is a submanifold of $G$ of dimension $\dim G/C_\sigma = \dim G - \dim C_\sigma \le \dim G - \dim T$, since $T \subset C_\sigma$. Thus $\dim T + \dim M_\sigma \le \dim G$.

We shall accept the fact that every conjugacy orbit $M_h$ must meet the maximal torus $T$. In the case of $U(n)$, with maximal torus

$$T = \{\mathrm{diag}(\exp(i\theta_1), \ldots, \exp(i\theta_m))\}$$

this is just the statement that every unitary matrix can be diagonalized, i.e., for every $u \in U(n)$ there is a $g \in U(n)$ such that $gug^{-1}$ is diagonal, i.e., in $T$. Thus each conjugacy orbit is of the form $M_\sigma$, where $\sigma \in T$.

Note also that our computation here has shown the following lemma.

**Lemma (E.1):** *The orthogonal complement to the tangent space to $C_\sigma$ at $e$ is mapped 1:1 and onto the tangent space to $M_\sigma$ at $\sigma$ under $Y \to \frac{d}{dt}e^{tY}\sigma e^{-tY}\big|_0 = Y\sigma - \sigma Y$.*

**Theorem (E.2):** *Each $M_\sigma$ meets $T$ orthogonally, and is even dimensional, and the Euler characteristic $\chi(M_\sigma)$ is the number of intersection points of $M_\sigma$ and $T$.*

PROOF: Let $\sigma \in T \subset C_a$. Let $Y$ be orthogonal to $C_\sigma$ at $e$. Then, in the bi-invariant metric in G, $Y\sigma$ and $\sigma Y$ are orthogonal to $C_\sigma$ at $\sigma$. We conclude from

Lemma E.1 that $M_\sigma$ is orthogonal to $C_\sigma$ at $\sigma$. A schematic picture is given in Figure E.1.

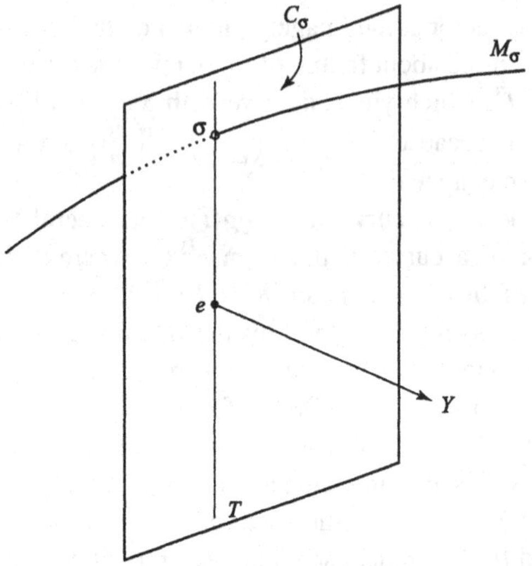


**Figure E.1**

We now compute the Euler characteristic of $M_\sigma$ by means of the Poincaré–Hopf theorem (16.12), using an argument that is a variation on ideas used by Weil and by Hopf and Samelson in the 1930s and 1940s . Let $W$ be a tangent vector to $T$ at the identity and consider the resulting 1-parameter group of isometries on $G$, $g \to g(t) = e^{tW} g e^{-tW}$. The velocity Killing field at any $g \in G$ is $w := Wg - gW$. Of course this field is tangent to each of the conjugacy orbits, in particular $M_\sigma$. Where are the zeros of this field $w$ on $G$ As computed previously, $w = 0$ at $g$ implies $e^{tW} g = g e^{tW}$, and so $g$ is in the centralizer of $e^{tW}$ for all $t$. Now we may choose the tangent vector $W$ to $T$ so that the 1-parameter group $e^{tW}$ lies dense on $T$ (see Section 6.2a). For this $W$, $g \in G$ is a zero if and only if $g$ is in the centralizer of the entire maximal torus $T$. It can be proved that *the centralizer of a maximal torus $T$ is exactly $T$ itself*, $C(T) = T$, see, e.g., [Hs, p. 45]. For this $W$ the zeros of the associated velocity field $w$ make up the entire maximal torus $T$. In particular, the zeros of the Killing field $w$ on $M_\sigma$ are the points where $M_\sigma$ meets $T$, these points being isolated since $M_\sigma$ meets $T$ orthogonally. What then is the Kronecker index of the field $w$ on $M_\sigma$ at such a meeting, say $\sigma$? Since the 1-parameter group $g(t)$ is a group of isometries leaving $\sigma$ fixed, the flow lines on $M_\sigma$ near $\sigma$ must be tangent to a small geodesic codimension 1-sphere $S$ on $M_\sigma$ centered at $\sigma$. Since $w$ is a nonvanishing tangent vector field to this sphere $S$, $S$ must have Euler characteristic 0, and so $S$ must be odd dimensional and $M_\sigma$ is even dimensional. By 8.3(11) the index of $w$ at $\sigma$ is $+1$ and the sum of the indices at the zeros of $w$ on $M_\sigma$ is exactly the number of intersection points of $M_\sigma$ and $T$. □

## E.b. Application of Bott's Extension of Morse Theory

We conclude with some remarks concerning how the topology of the orbits $M_\sigma$ is related to that of the entire group $G$. For this we shall use Bott's refinement of the presentation of Morse theory that was given in Section 14.3c.

For simplicity we restrict ourselves to the example $U(3)$ with which we started our discussion, but similar remarks hold for all the "classical groups," $U(n)$, $SO(n)$, $Sp(n)$, (but not $SU(n)$), with some modifications; see [Fr2]. The elements $g$ of order 2, $g^2 = I$, are exactly four orbits, $M_I = I$, $M_{-I} = -I$, and the two complex projective planes $M_\alpha$ and $M_\beta$, where $a = \text{diag}(1, 1, -1)$ and $\beta = \text{diag}(1, -1, -1)$. It is shown in [Fr2] that these points are also exactly the critical points of the function $f(g) = \text{Re tr}(g)$, the real part of the trace of the unitary matrix $g$, and we can call these the "critical orbits." For $M_\alpha$ and $M_\beta$, these are not isolated critical points but rather connected "nondegenerate critical *manifolds*" and one can apply Bott's extension of Morse theory (see, e.g, [Bo, Lecture 3]) to this situation.

Briefly, we require that the hessian matrix for $f$ be nondegenerate for directions orthogonal to the critical manifold. At a point $m$ of $M_\sigma$ we can look at the part of the tangent space to $G$ that is normal to $M_\sigma$ and note the number of independent normal directions from $m$ for which $f$ is decreasing, i.e., the dimension of the subspace on which the hessian form is negative definite. These directions span a subspace of the normal space to $M_\sigma$ at $m$ to be called the negative normal space. From nondegeneracy the dimension of these negative normal spaces will be constant along $M_\sigma$ and will be called the (Morse–Bott) **index** $\boldsymbol{\lambda(\sigma)}$ of the critical manifold $M_\sigma$. The collection of all of these subspaces at all $m \in M_\sigma$ form the **negative normal bundle** to $M_\sigma$. We ask that this bundle be **orientable**, meaning that the fibers can be oriented coherently as we range over the base space $M_\sigma$. (If they are not orientable, we may proceed but we may only use $\mathbb{Z}_2$ coefficients when talking about homology groups.) Look at the point $\alpha = \text{diag}(1, 1, -1)$ at which $f = 1$. Then the entire portion of the centralizer $C_\alpha$ given by $U(2) \times (-1) \subset U(2) \times U(1)$, except for $\alpha$ itself, lies in the region of $U(3)$ where $f < 1 = f(\alpha)$; $U(2) \times (-1)$ is "hanging down" from the critical point $\alpha$. Thus there are $\dim U(2) = 4$ independent directions at $\alpha$ along which $f$ decreases. Since $f$ is invariant under $\alpha \to g\alpha g^{-1}$, we see that this is true along all of $M_\sigma$, and so $\lambda(\alpha) = 4$. Similarly, the centralizer of $\beta$ is $U(1) \times U(2)$, the portion $U(1) \times \text{diag}(-1, -1)$ hangs down from $\beta$, and so the index in this case is $\lambda(\beta) = \dim U(1) = 1$. Of course $I$ is the isolated maximum point and $-I$ is the isolated minimum, and so $\lambda(I) = \dim U(3) = 9$ and $\lambda(-I) = 0$. Nondegeneracy can be proven for each of these critical manifolds.

We shall need some classical results about the topology of the complex projective plane $\mathbb{C}P^2$, but we shall be very brief. Since $\mathbb{C}P^2 = U(3)/U(2) \times U(1)$, it is a compact 4-manifold. Recall from Problem 1.2(3) that it is a complex manifold of complex dimension 2. To all points in $\mathbb{C}P^2$ with local *homogeneous* complex coordinates $[z_0, z_1, z_2]$, where $z_2 \neq 0$, we may assign the pair of genuine complex coordinates $(w_1 = z_0/z_2, w_2 = z_1/z_2)$. For example, these will be local coordinates near the point $[0, 0, 1]$ that represent the complex line along the $z_2$ "axis" (a copy of the ordinary $z_2$ plane). (Recall that $[z_0, z_1, z_2] \neq [0, 0, 0]$ represents the complex line through the origin of $\mathbb{C}^3$ that passes through the point $(z_0, z_1, z_2)$.) We use the schematic picture in Figure E.2.

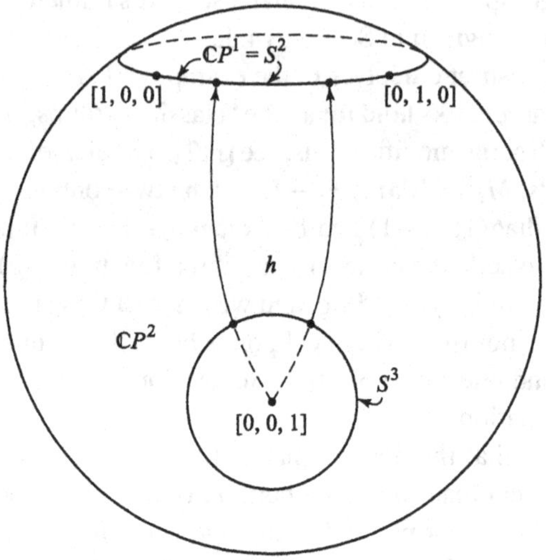


**Figure E.2**

The locus $z_2 = 0$ consists of all points $[z_0, z_1, 0]$ with $z_0, z_1$ not both 0, and thus is a complex projective line $\mathbb{C}P^1$ (i.e., a 2-sphere; see Problem 1.2(3)) with homogeneous coordinates $[z_0, z_1]$. We have aprojection map $h: \mathbb{C}P^2 - [0, 0, 1] \to \mathbb{C}P^1 = S^2$ defined by $[z_0, z_1, z_2] \to [z_0, z_1, 0]$. The locus $|w_1|^2 + |w_2|^2 = 1$ represents a 3-sphere $S^3$ in $\mathbb{C}P^2$ centered at $[0, 0, 1]$ and $h$, sending $S^3 \to \mathbb{C}P^1 = S^2$ by $(w_1, w_2) = [w_1, w_2, 1] \to [w_1, w_2]$, is simply the Hopf map of Section 22.4c. Note also that $h: \mathbb{C}P^2 - [0, 0, 1] \to \mathbb{C}P^1 = S^2$ is the endpoint of a deformation $h_t([z_0, z_1, z_2]) = [z_0, z_1, (1 - t)z_2]$ that deforms $\mathbb{C}P^2 - [0, 0, 1]$ onto $\mathbb{C}P^1$. Thus any cycle on $\mathbb{C}P^2 - [0, 0, 1]$ is homotopic to one on the subset $\mathbb{C}P^1$. In particular, any singular $j$-cycle on $\mathbb{C}P^2$, for $j < 4$, can clearly be pushed slightly to miss $[0, 0, 1]$ and can then be deformed into $\mathbb{C}P^1 = S^2$. But $S^2$ is simply connected. Thus, since $S^2$ has nontrivial homology only in dimensions 0 and 2, we have the following:

**Lemma (E.3):** *Any loop on $\mathbb{C}P^2$ is homotopic to a loop on $S^2$ and is thus deformable to a point; hence $\mathbb{C}P^2$ is simply connected and therefore orientable. $H_j(\mathbb{C}P^2, \mathbb{Z}) = 0$ for $j = 1$ and 3, and $H_2(\mathbb{C}P^2; \mathbb{Z}) = H_2(S^2, \mathbb{Z}) = \mathbb{Z}$. Since $\mathbb{C}P^2$ is a compact, orientable 4-manifold, $H_4(\mathbb{C}P^2) = \mathbb{Z}$. Since $\mathbb{C}P^2$ is simply connected, the negative normal bundles to $M_\alpha$ and $M_\beta$ are orientable.*

Note also that by pushing $S^3$ by means of the deformations $h_t$ we may move $S^3$ to $S^3_\varepsilon$ that lies at a small distance $\varepsilon$ (in some Riemannian metric) from $S^2$ (see Figure E.3). But the points of $\mathbb{C}P^2$ that are of distance $\leq \varepsilon$ from $\mathbb{C}P^1$, for $\varepsilon$ sufficiently small, form a normal 2-disc bundle, $N^4$ over $\mathbb{C}P^1$ in $\mathbb{C}P^2$, and the boundary $\partial N$ forms a normal

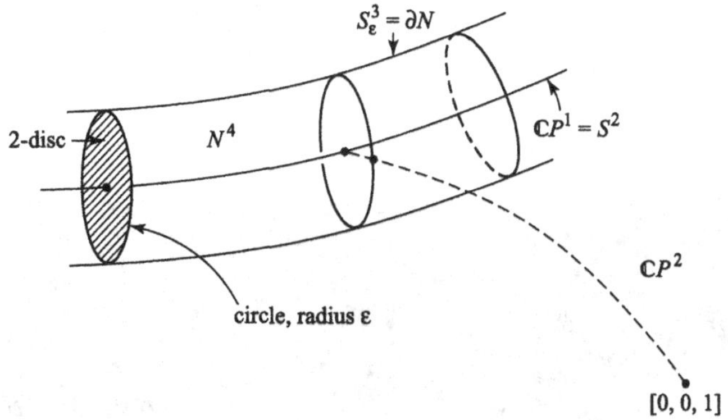


**Figure E.3**

circle bundle (circles of radius $\varepsilon$). *Thus $S^3_\varepsilon = \partial N$ is this normal circle bundle to $\mathbb{C}P^1$ and the deformation $h : \partial N \to \mathbb{C}P^1$ is a realization of the Hopf map $S^3_\varepsilon \to S^2$.*

Recall (see Section 14.3c) that the *Poincaré* polynomial of $U(3)$ is the polynomial in $t$ with coefficients the Betti numbers $b_j$ of $U(3)$, and so $P_{U(3)}(t) = \sum b_i t^i$. Bott's generalization of the *Morse* polynomial is constructed using the Poincaré polynomials and indices of the critical manifolds (where we write $P_\sigma$ for the Poincaré polynomial of $M_\sigma$)

$$\mathfrak{M}_B(t) = 1 + t^{\lambda(\beta)} P_\beta(t) + t^{\lambda(\alpha)} P_\alpha(t) + t^9$$

(Note that if each critical manifold reduces to an isolated critical point as in the original Morse case, then because $b_j$ (point) $= 0$ for $j > 0$, the coefficient of $t^\lambda$ is again simply the number of critical points of index $\lambda$.)

We have seen in Lemma E.3 that the Poincaré polynomial of $\mathbb{C}P^2$ is $1 + t^2 + t^4$, and so

$$\mathfrak{M}_B(t) = 1 + t(1 + t^2 + t^4) + t^4(1 + t^2 + t^4) + t^9 = (1 + t)(1 + t^3)(1 + t^5)$$

Bott's generalization of the Morse inequalities are then $\mathfrak{M}_B(t) \geq P_{U(3)}(t)$ but in [Fr2] it is shown that in $U(n)$, the "symplectic" groups $Sp(n)$, and $SO(n)$, these are in fact *equalities* (except that one must use $Z_2$ Betti numbers in the case of $SO(n)$, for $n > 3$, since, e.g., the negative normal bundles to real projective spaces are not always orientable). Thus the critical orbits $M_\sigma$ yield exactly the Betti numbers of the group, but with each $k$ cycle on $M_\sigma$ yielding a $[k + \lambda(\sigma)]$ cycle on $G$.

1)
If $\vec{v}$ is Killing
2)
If $\vec{v}$ Killing &
M Kähler then
M
3)
$H_{n+k}(\mathrm{Stein}^n) = 0$
vanishing
cycle
4) If G is a classical group and
$F = \{g \in G \mid g = g^{-1}\}$, then
U(3)
$CP^2, \lambda = 4$
$CP^1, \lambda = 1$
5) If $K_{M^4 \text{ Kähler}} > 0$, then $M^4 \sim CP^2$
6) If $V^{n-1}$ minimal in $M^n$, $\mathrm{Ric}(M) > 0$
then
$\pi_1(V) \longrightarrow \pi_1(M)$ is onto
7) If $-b \leq K_{M^3} \leq -c < 0$ and
if closed $V^2 \subset M^3$ has area
$A < \frac{4\pi}{3b-c}$, then A bounds in M
8)
$T^3$
max $\|\vec{E}\| \longrightarrow \infty$
as $t \longrightarrow \infty$ iff
does not bound
as a real 1-cycle.
9) A rigidly
handle

**"Final Exam"**

If students find things that are worthwhile in this book, it is
largely due to what I have learned from my own "teachers,"
among them Aldo Andreotti, Raoul Bott, S.S. Chern, Jim Eells,
G.C. Evans, Harley Flanders, Heinz Hopf, Charles Loewner, Hans Samelson,
Norman Steenrod and John Archibald Wheeler.

# References


[A, M, R] Abraham, R., Marsden, J., and Ratiu, T. *Manifolds, Tensor Analysis, and Applications*, Addison Wesley, 1983

[A, S] Aharonov, V., and Susskind, L. Observability of the sign change of spinors under $2\pi$ rotations. Phys. Rev. **158** (1967), 1237–38

[A] Arnold, V, I. *Mathematical Methods of Classical Mechanics*, Springer, 1978

[A2] ——. *Ordinary Differential Equations.* M.I.T. Press, 1978

[B, S] Bamberg, P. and Sternberg, S. *A Course in Mathematics for Students of Physics*, vol. 2, Cambridge, 1990

[B] Berry, M. Quantal phase factors accompanying adiabatic changes, Proc. R. Soc. Lond. A **392** (1984), pp. 45–57

[B, K, G] Blank, A., Friedrichs, K., and Grad, H. *Notes on Magneto-Hydrodynamics*, part V, N.Y.U. notes, (1957)

[Bl] Bleecker, David. *Gauge Theory and Variational Principles*, Addison Wesley, 1981

[B, T] Bott, R. and Tu, L. *Differential forms in Algebraic Topology*, Springer, 1982

[Bo] Bott, R. Lectures on Morse theory, old and new. Bul. Amer. Math. Soc. 7 (1982) pp. 331–358

[Bo2] Bott, R. On induced representations. Proc. Symp. in Pure Math. **48** (1988), pp. 1–13.

[Boy] Boyling, J.B. An axiomatic approach to classical thermodynamics, Proc. R. Soc., London, A **329** (1972), pp. 35–70

[Br] Brillouin, L. *Tensors in Mechanics and Elasticity*, Academic Press, 1964

[Ca] Cartan, E. *On Manifolds with an Affine Connection and the Theory of Relativity*, Bibliopolis, 1986

[C] Coleman, S. *Aspects of Symmetry*, Cambridge, 1985

[C, J] Courant, R. and John, F. *Introduction to Calculus and Analysis*, vol. 2, Wiley- Interscience, 1974

[Do] Do Carmo, M. *Differential Geometry of Curves and Surfaces*, Prentice-Hall, 1976

[D, F] Driver, B. and Frankel, T. On the growth of waves on manifolds, J. Math. Anal. Appl. **178** (1993), pp. 143–55

[D, S] Duff, G. and Spencer, D. Harmonic tensors on Riemannian manifolds with boundary, Ann. Math. **56** (1952), pp. 128–56

[E] Eckmann, B. Harmonische Funktionen und Randwertaufgaben in einem Komplex. Comm. Math. Helv. **17** (1945), pp. 240–55.

[Fe] Felsager, B. *Geometry, Particles and Fields*, Odense University Press, 1981
[F] Feynman, R. *Q.E.D.* Princeton, 1985
[FF] ——. *Theory of Fundamental Processes*, Benjamin, 1961
[F, L, S] Feynman, R., Leighton, R. and Sands, M. *The Feynman Lectures on Physics*, Vols I, II and III, Addison Wesley, 1964
[F, W] Feynman, R. and Weinberg, S. *Elementary Particles and the Laws of Physics*, Cambridge, 1987
[Fl] Flanders, H. *Differential Forms*, Academic Press, 1963
[Fr] Frankel, T. *Gravitational Curvature*, W.H. Freeman, 1979
[Fr 2] Frankel, T. Critical submanifolds of the classical groups and Stiefel manifolds, in *Differential and Combinatorial Topology*, Edited by Stewart Cairns, Princeton, 1965
[Fk] Friedrichs, K. Differential forms on Riemannian manifolds, Comm. Pure and Appl. Math. **8** (1955), pp. 551–90
[Ga] Galloway, G. A generalization of Myers' theorem and an application to relativistic cosmology, J. Diff. Geom. **14**, pp. 105–116, 1979
[G, F] Gelfand, I. and Fomin, S. *Calculus of Variations*, Prentice-Hall, 1963
[G, H] Greenberg, M. and Harper, J. *Algebraic Topology*, Benjamin, 1981
[G] Grossman, N. Holonomic measurables in geodesy. J. Geophysical Res. **79**. (1974), pp. 689–94
[G, P] Guillemin, V. and Pollack, A. *Differential Topology*, Prentice-Hall, 1974
[H] Hermann, R. *Differential Geometry and the Calculus of Variations*, 2nd ed., Mathematical Science Press, 1977
[H, O] Hehl, F.H. and Obukhov, Y.N. *Foundations of Classical Electrodynamics*, Birkhaüser, 2003
[H, Y] Hocking, J. and Young, G. *Topology*, Addison-Wesley, 1961
[Hs] Hsiang, W. Y. *Lectures on Lie Groups*, World Scientific, 2000
[I, Z] Itzykson, C. and Zuber, J-B. *Quantum Field Theory*, McGraw-Hill, 1980
[Ka] Kato, T. *Perturbation Theory for Linear Operators*, Springer, 1976
[K] Kobayashi, S. Fixed points of isometries, Nag. Math. J. **13** (1959), pp. 63–68
[K, N] Kobayashi, S. and Nomizu, K. *Foundations of Differential Geometry*, vols. 1 and 2. Wiley, New York, 1963
[L] Lawson, B. *Minimal Varieties in Real and Complex Geometries*. Presses de L'Université Montréal, 1974
[L-S, K] Levi Setti, R. and Lasinski, T. *Strongly Interacting Particles*,University of Chicago Press, 1973
[M, H] Marsden, J. and Hughes, T. *Mathematical Foundations of Elasticity*, Prentice-Hall, 1983
[Mi] Michel, L. Symmetry defects and broken symmetry. Rev. Mod. Phys. **51** (1980), pp. 617–51
[M] Milnor, J. *Morse Theory*, Princeton University Press, 1963
[M2] Milnor, J. *Topology from the Differentiable Viewpoint*, University Press of Virginia, 1965
[M, S] Milnor, J. and Stasheff, J. *Characteristic Classes*, Princeton, 1974
[M, T, W] Misner, C., Thorne, K., and Wheeler, J. *Gravitation*, Freeman, 1970
[Mo] Moffat, H. The degree of unknottedness of tangled vortex lines, J.Fluid Mech. **35** (1969), pp. 117–129
[Mu] Murnaghan, F. D. *Finite Deformation of an Elastic Solid*, Dover, 1951, republished 1967
[Nam] Nambu, Y. *Quarks*, World Scientific, 1985

[N, S] Nash, C. and Sen, S. *Topology and Geometry for Physicists*, Academic Press, 1983

[N] Nelson, E. *Tensor Analysis*, Princeton University Press, 1967

[No] Nomizu, K. *Lie Groups and Differential Geometry*, Math. Soc. Japan, 1956

[O] Osserman, R. *Poetry of the Universe*, Anchor Books, 1995

[R] Rabin, J. Introduction to quantum field theory for mathematicians, in *Geometry and Quantum Field Theory*, Edited by D. Fried and K. Uhlenbeck, Amer. Math Soc. 1995, pp. 183–269

[Ro] Roe, J. *Elliptic Operators, Topology and Asymptotic Methods*, Longman, 1988

[Sam] Samelson, H. *Topology of Lie Groups*, Bul. Amer. Math. Soc. **58** (1952), pp. 2–37

[Sam H] Samelson, H. Differential forms, the early days. Amer. Math. Monthly, **108** (2001) pp. 522–30

[S] Simmons, G. *Topology and Modern Analysis*, McGraw-Hill, 1963

[Si] Simon, B. Holonomy, the quantum adiabatic theorem, and Berry's phase. Phys, Rev. **51** (1983), pp. 2167–70

[Sp] Spivak, M. *A Comprehensive Introduction to Differential Geometry*, (5 volumes) Publish or Perish Press, 1979

[St] Steenrod, N. *Topology of Fiber Bundles*, Princeton University Press, 1951

[Sto] Stong, C.L. The amateur scientist, Scientific American **233** (December 1975), pp. 120–5

[Su] Sudbery, A. *Quantum Mechanics and the Particles of Nature*, Cambridge, 1986

[Sy] Sniatycki, J. Quantization of charge. J. Math. Phys. 15 (1974), pp. 619–20.

['t Hooft] 't Hooft, G. *In Search of the Ultimate Building Blocks*, Cambridge, 1997

[T] Truesdell, C. The influence of elasticity on analysis: the classic heritage. Bul. Amer. Math. Soc. **9** (1983), pp. 293–310

[T, T] Truesdell, C. and R. Toupin, R. The Classical Field Theories, *Handbuch der Physik*, III–I, 1960

[Wd] Wald, R. *General Relativity*, University of Chicago Press, 1984

[Wa] Warner, F. *Foundations of Differentiable Manifolds and Lie Groups*, Scott, Foresman, 1971

[We] Weinberg, S. *The Quantum Theory of Fields*, Vol II, Cambridge, 1996

[Wy] Weyl, H. *The Theory of Groups and Quantum Mechanics*, Dover, 1950

[W] Whittaker, E. *A History of the Theories of Aether and Electricity*, vol. 1, Harper, 1960

[Z] Zhang, D. Yang and contemporary mathematics. Math. Intelligencer **15** (1993), pp. 13–21

# Index